MW00571238

8th Edition

Calculus and Analytic Geometry

8th Edition

Calculus and Analytic Geometry Part II

George B. Thomas, Jr.
Massachusetts Institute of Technology

Ross L. Finney
U.S. Naval Postgraduate School

Addison-Wesley Publishing Company

Reading, Massachusetts Menlo Park, California New York
Don Mills, Ontario Wokingham, England Amsterdam Bonn
Sydney Singapore Tokyo Madrid San Juan Milan Paris

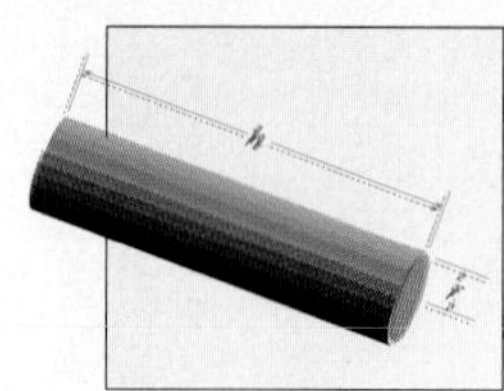

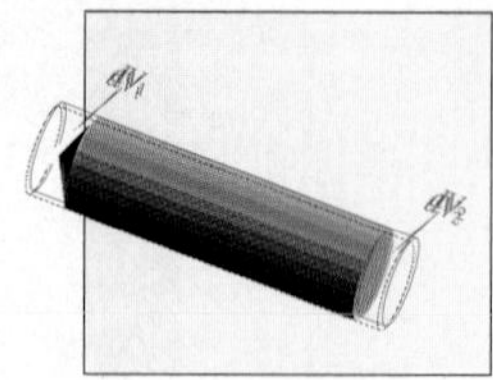

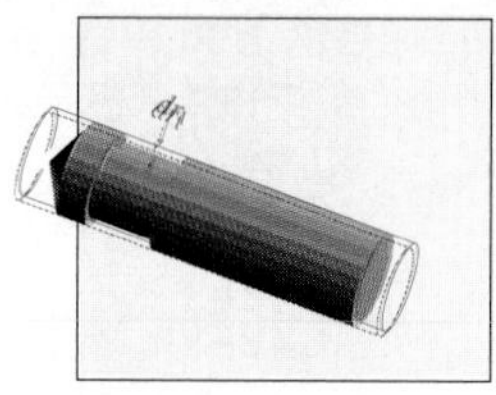

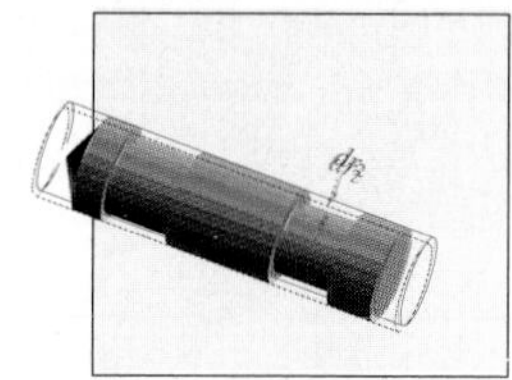

About the cover: Calculus is the mathematics of motion and change. We can use calculus to find out how rapidly the volume of a metal machine part changes as we cut a slot in it on a lathe. Or we can calculate the part's mass from a knowledge of its shape and density. Calculus can also tell us how precisely we must grind the metal to control its mass to one one-thousandth of a gram.

Sponsoring Editor:	Jerome Grant
Development Editor:	David M. Chelton
Managing Editor:	Karen Guardino
Production Supervisor:	Jack Casteel
Copy Editor:	Barbara Flanagan
Proofreader:	Joyce Grandy
Text Design:	Martha Podem
Art Consultant:	Joseph Vetere
Art Coordinator:	Connie Hulse
Electronic Illustration:	Tech-Graphics
Production Editorial Services:	Barbara Pendergast
Manufacturing Supervisor:	Roy Logan
Cover Design:	Marshall Henrichs

Library of Congress Cataloging-in-Publication Data

Thomas, George Brinton, 1914–
Calculus and analytic geometry / by George B. Thomas, Jr. and Ross L. Finney — 8th ed.
p. cm.
Includes index.
ISBN 0-201-52929-7 (set) – ISBN 0-201-53286-7 (pt. 1).
ISBN 0-201-53287-5 (pt. 2)
1. Calculus. 2. Geometry, Analytic. I. Finney, Ross L. II. Title
QA303, T42 1992 91–8848
515′, 15—cc20 CIP

Reprinted with corrections, April 1993.

3 4 5 6 7 8 9 10 DO 959493

Contents

Chapter 10 Vectors and Analytic Geometry in Space

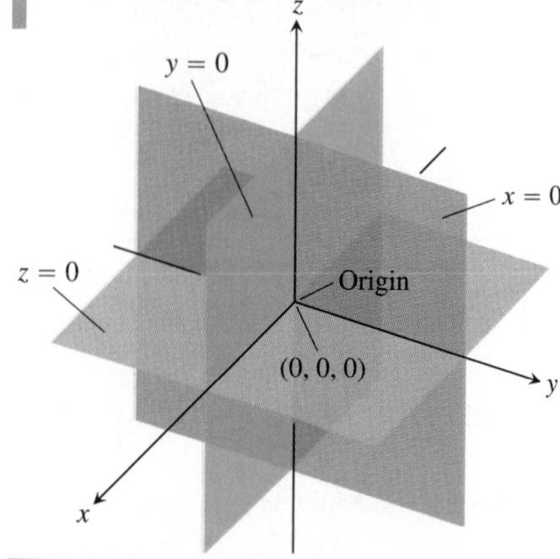

Chapter 11 Vector-Valued Functions and Motion in Space

Chapter 12 Functions of Two or More Variables and Their Derivatives

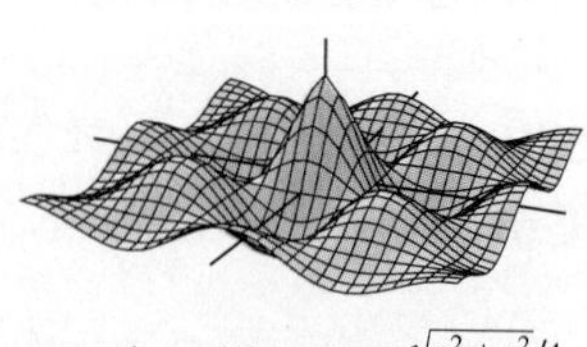
$z = (\cos x)(\cos y)\,e^{-\sqrt{x^2+y^2}/4}$

Chapter 13 Multiple Integrals

Chapter 14 Integration in Vector Fields

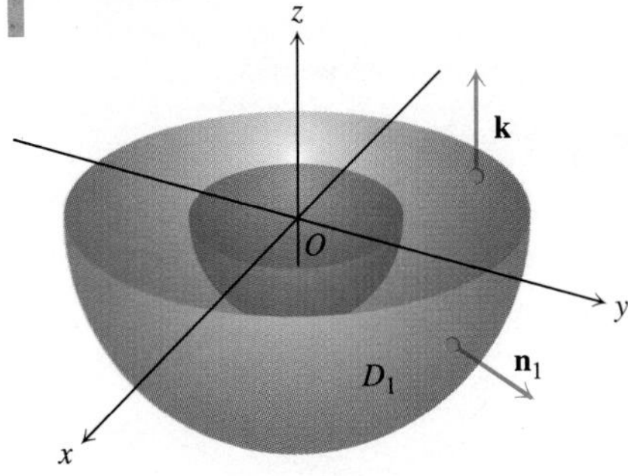

Chapter 15 Differential Equations

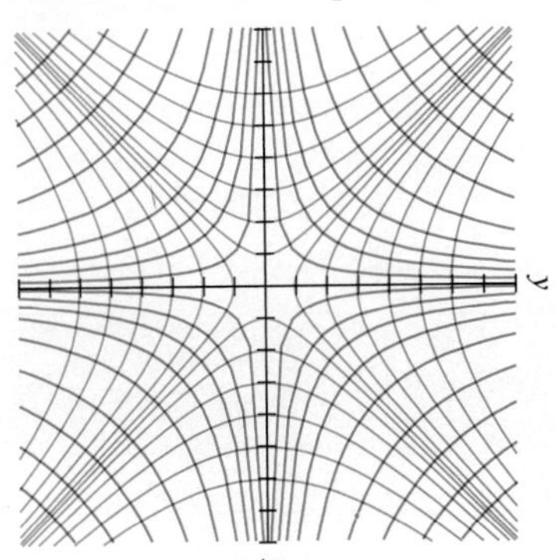

(Generated by Mathematica)

Appendices

Abridged Contents for Part I

8th Edition

Calculus and Analytic Geometry

Preface

This eighth edition of *Calculus and Analytic Geometry* is intended for the standard three-semester or four-quarter calculus sequence in the freshman and sophomore years of college. The prerequisites are the usual studies of algebra and trigonometry, which are reviewed at the beginning of Chapter 1 and in the Appendices at the end of the book.

Our goals in this edition have been to shorten the text without omitting significant topics; to continue our emphasis on visualization as an aid to understanding calculus; and to offer more options for integrating technology into the mainstream course. We believe the result is a more exciting and more motivating book, while being easier for instructors to adapt to their particular teaching situations. Throughout the revision, we have maintained the mathematical level of previous editions, presenting calculus as a practical tool students will use in their later academic and professional careers. Many applications in the worked examples and in the exercises have been updated, but continue to illustrate the core mathematics students need to learn.

Content Changes for the Eighth Edition

Chapter 1 has a shorter review of precalculus topics, allowing readers to get to the derivative a little sooner. Section 1.3 shows that hand-held calculators can do more than just compute numerical answers; they can also be used to explore patterns of mathematical expressions. The idea of target values in Section 1.4 helps set up the epsilon-delta definition of limit in Section 1.7. Circles and parabolas are discussed in Chapter 1 because they occur often in illustrations of calculus concepts in the first semester.

Chapter 2 begins with the formal definition of derivative and then moves quickly to show how derivatives are calculated easily with rules based on the limit theorems presented in Chapter 1. Section 2.7 on Newton's Method contains a brief (optional) discussion of the fascinating area of chaos theory, illustrating the surprising mathematics that can sometimes lie behind simple equations.

The Mean Value Theorem has been moved nearer the beginning of **Chapter 3** to give the theoretical justification before rather than after the applications of derivatives. The result is a shorter and tighter presentation of this material. At the same time, the chapter starts with related rates of change to give students earlier practice in applying derivatives to realistic situations.

Integration begins with the indefinite integral in **Chapter 4,** including a discussion of initial-value problems and mathematical modeling. The idea of using integration in important mathematical models is woven throughout **Chapter 5.** We have tried to emphasize the recurring pattern of formulating an appropriate Riemann sum, finding its limit, and comparing the limit with previously known results, so students will be better able to deal with kinds of problems they have not seen before.

Chapter 6 on transcendental functions includes material on hyperbolic functions, which has been reduced from the coverage in the previous edition. Section 6.6 on rates of growth has several examples on measuring the effectiveness of algorithms, with applications to computer science.

The presentation of techniques of integration in **Chapter 7** has been condensed without omitting any significant topics. Again, emphasis is on teaching general problem-solving methods applicable to situations not explicitly covered in the text, so that students will be prepared to solve problems in new contexts.

Sequences and series have been condensed into one chapter, **Chapter 8,** again without omitting major topics. (Several minor discussions have been deleted.) We have tried to emphasize the importance of Taylor and Maclaurin series from the start of the chapter, pointing to them as the motivating reason for studying some of the chapter's other topics.

The presentation of analytic geometry in **Chapter 9** has been combined with parametric equations, plane curves, and polar coordinates. While these topics are important in calculus, their coverage has been reduced to allow room for other topics now demanding place in the crowded mainstream calculus syllabus.

Chapters 10 and 11 on vectors and vector-valued functions remain largely unchanged from the previous edition. Drawing lessons to help students visualize and draw in three dimensions are incorporated into Chapter 11.

Functions of two or more variables are discussed in **Chapter 12,** in an organization meant to parallel the presentation of the single-variable material. Applications of partial differential equations to physics and engineering appear in several exercise sets.

Chapters 13 and 14 on multiple integrals and integration in vector fields are much the same as in the seventh edition. However, **Chapter 15** on differential equations has been cut back to serve more as an introduction to a topic that is usually taught as a separate course.

Features of the Text

Mathematical level: We have taken care to maintain the level of rigor of previous editions while striving for a more informal and accessible writing style. We try to explain things carefully without either belaboring the obvious or jumping ahead of the reader's understanding.

Art and design: Our introduction of four-color figures in the seventh edition was well received by our adopters. This edition is now completely four-color. As before, color is used pedagogically to highlight the most important parts of figures and help students visualize in three dimensions. We have increased the number of figures over the seventh edition to appeal to students' geometric intuition in explaining basic principles. Many complicated graphs and surfaces have been rendered in *Mathematica*® to ensure their accuracy. Overall, this edition has more figures and more exercises that require students to draw and interpret figures.

Applications: It has been a hallmark of this book through the years that we illustrate applications of calculus with real data based on situations students are likely to encounter later on. Most of our examples and exercises are directed toward science and engineering, such as calculating numerically the cross-section area of a solar-powered car (p. 306).

Worked examples: We have explicitly shown more of the steps involved in the worked examples, as well as the reasoning that leads from one step to the next. In addition, we have made a special point of correlating examples with the stated problem-solving strategies, so students see how these strategies work in practice.

Exercise sets: We have added new exercises in this edition at several levels of difficulty, from drill questions to challenging problems. Many of the new exercises require a calculator or computer grapher for their solution, and are so labeled. Our goal has not been to take a group of exercises and simply graph them, but to use the power of computer graphing to explore new ideas and new relations. We continue to include sets of **Review Questions** that ask students to think about the concepts presented without trying to calculate numerical answers. And we still end each chapter with a collection of **Miscellaneous Exercises** that extend the student's grasp of the material in new ways.

Integration of technology: Students do not need a calculator or computer grapher to use this book. However, for those who do have access to technological aids and would like to use them, we have included several technological features in this edition. We have already mentioned exercises for calculators and computer graphers. In addition, several optional computer programs, written in BASIC, have been added. Many sections of the text also include a list of programs from the **Calculus Explorer** that may be used as computational aids both for working the exercises in that section and for general exploration. Additional information about software for calculus is presented in the Supplements portion of this Preface.

Enhancements to learning: In the seventh edition we introduced a number of strategy boxes to summarize particular problem-solving methods, such as those used in solving related rates of change problems, optimization problems, and so on. We have expanded the number of these boxes to include more types of problems.

Several **Drawing Lessons** were also introduced in the seventh edition, aimed at helping students draw lines, planes, and curved surfaces in three dimensions. We have added lessons for drawing curves in two dimensions as well.

One further pedagogical feature of the seventh edition was the occasional use of flowcharts to summarize methods of attacking extended calculus problems, such as the problem of determining series convergence. This feature, too, has been expanded, to include flowcharts on such topics as l'Hôpital's rule and indefinite integration.

Supplements for the Instructor

Complete Solutions Manual: This two-volume supplement contains the worked-out solutions for all exercises in the text.

Complete Answer Book: Contains the answers to all exercises in the text.

OmniTest: Based on the learning objectives of the text, this computerized testing system allows the instructor to generate tests or quizzes easily. As an algorithm-driven system, OmniTest easily creates multiple versions of the same test by automatically inserting random numbers into model problems. While the numbers are random, they are constrained to produce reasonable answers. Questions may be selected in any combination of open-ended, multiple-choice, and true-false formats. Instructors may assign an instructor code and level of difficulty to each model problem, and may print tests with variable spacing. OmniTest is available for MacIntosh, and IBM PC/compatibles.

Printed Test Bank: Three versions of tests for each chapter in the text are included in multiple-choice, open-ended, and true-false formats. The Printed Test Bank also contains printed answer keys and student worksheets for each test.

Transparency Masters: Includes a selection of key definitions, theorems, proofs, formulas, tables, and figures. These may be copied onto transparency acetates for overhead projection.

Supplements for the Student

Student Study Guide: By Maurice Weir, Naval Postgraduate School. Organized to correspond with the text, this workbook in a semi-programmed format increases student proficiency.

Student Solutions Manual: By Thomas Cochran and Michael Schneider, of Belleville Area College. This manual is designed for the student and contains carefully worked-out solutions to all of the odd-numbered exercises in the text.

Options for Integrating Technology

Analyzer*: This program is a tool for exploring functions in calculus and many other disciplines. It can graph a function of a single variable and overlay graphs of other functions. It can differentiate, integrate, or iterate a function. It can find roots, maxima and minima, and inflection points, as well as vertical asymptotes. In addition, Analyzer* can compose functions, graph polar and parametric equations, make families of curves, and make animated sequences with changing parameters. It exploits the unique flexibility of the MacIntosh wherever possible, allowing input to be either numeric (from the keyboard) or graphic (with a mouse). Analyzer* runs on MacIntosh II, Plus, and SE computers.

The Calculus Explorer: Consisting of 27 programs ranging from functions to vector fields, this software enables the instructor and student to use the computer as an "electronic chalkboard." The Explorer is highly interactive and allows for manipulation of variables and equations to provide graphical visualization of mathematical relationships that are not intuitively obvious. The Explorer provides user-friendly operation through an easy-to-use menu-driven system, extensive on-line documentation, superior graphics capability, and fast operation. An accompanying manual includes sections covering each program, with appropriate examples and exercises. Available for IBM PC/compatibles.

InSight: A calculus demonstration software program that enhances understanding of calculus concepts graphically. The program consists of ten simulation scenarios. Each scenario presents an application and takes the user through the solution visually. The format is interactive. Available for IBM PC/compatibles.

The Student Edition of DERIVE®: A streamlined version of the professional program, the Student Edition of DERIVE is a powerful, yet exceptionally easy-to-use computer algebra system for numerical, symbolic, and graphical computation. Its menu-driven interface and on-line help make the software user-friendly and easy to learn. The accompanying manual is pedagogically oriented and introduces users to the capabilities of the program step by step. The Student Edition of DERIVE® frees students from performing long mathematical calculations by hand, thereby enhancing their learning experience by allowing them to spend more time in mathematical exploration. Available for IBM PC/compatibles.

The Student Edition of MathCAD®: A powerful free-form scratchpad, this Student Edition has all the problem-solving capabilities of the professional version of MathCAD®. Available for the IBM PC/compatibles.

Master Grapher and 3D Grapher: A powerful, interactive graphing utility for functions, polar equations, parametric equations and other functions in two and three variables. Prepared by Franklin Demana and Bert Waits of Ohio State University. Available for MacIntosh, Apple, and IBM PC/compatibles.

Mathematica® Laboratories for Calculus I Using Mathematica: By Margaret Höft, The University of Michigan–Dearborn. An inexpensive collection of *Mathematica* lab experiments consisting of material usually covered in the first term of the calculus sequence.

Math Explorations Series: Each manual provides problems and explorations in calculus. Intended for self-paced and laboratory settings, these books are an excellent complement to Thomas/Finney.

Exploring Calculus with a Graphing Calculator, Second Edition, by Charlene E. Beckmann and Ted Sundstrom of Grand Valley State University.

Exploring Calculus with Mathematica®, by James K. Finch and Millianne Lehmann of the University of San Francisco.

Exploring Calculus with Derive®, by David C. Arney of the United States Military Academy at West Point.

Exploring Calculus with the IBM® PC Version 2.0, by John B. Fraleigh and Lewis I. Pakula of the University of Rhode Island.

Acknowledgments

We would like to thank and acknowledge the contributions of our revision planning survey participants. Their comments and suggestions helped to shape our ideas for this Eighth Edition.

Linda Allen, Texas Tech University
William Arlinghaus, Lawrence Technological University
Lewis D. Blake, Duke University
Phyllis Boutilier, Michigan Technological University
Major James Boutner, U.S. Air Force Academy
Wayne Britt, Louisiana State University
David H. Carlson, San Diego State University
Misium Castroconde, University of California–Irvine
Robert Connelly, Cornell University
Robin Gottlieb, Harvard University
Hiroshi Gunji, University of Wisconsin–Madison
Leon M. Hale, University of Missouri–Rolla
Jennifer Johnson, University of Utah
Jeuel LaTorre, Clemson University
John Lawlor, University of Vermont
John C. Mainhuber, University of Maine
Francis J. Narcowick, Texas A & M University
Charles Okonkwo, Arizona State University
Cathryn Olsen, State University of New York–Buffalo
Sanford Segal, University of Rochester

Betty Travis, University of Texas–San Antonio
Constantine Tsatsos, New Jersey Institute of Technology
Paul Tseng, University of Washington
Jaak Vilms, Colorado State University
Bertram Walsh, Rutgers University
Dr. Richard Wheeden, Rutgers University

Many valuable contributions to this Eighth Edition were made by people who reviewed the manuscript as it developed through its various stages.

Martin Bartelt, Christopher Newport College
Therlene Boyett, Valencia Community College
Fred Brauer, University of Wisconsin–Madison
David Collingwood, University of Washington
John Erdman, Portland State University
David Furuto, Brigham Young University–Hawaii Campus
Stuart Goldenberg, California Polytechnic State University–San Luis Obispo
Ralph Grimaldi, Rose-Hulman Institute of Technology
Bernard Harris, Northern Illinois University
Arnold Insel, Illinois State University
David Johnson, Lehigh University
Cecilia Knoll, Florida Institute of Technology
Elton Lacey, Texas A & M University
James Lang, Valencia Community College
Melvin D. Lax, California State University–Long Beach
Millianne Lehmann, University of San Francisco
Stanley M. Luckawecki, Clemson University
Arthur Moore, Orange Coast College
Daniel Moran, Michigan State University
James Nicholson, Clemson University
James Osterburg, University of Cincinnati
F. J. Papp, University of Michigan–Dearborn
Thomas W. Rishel, Cornell University
J. Rod Smart, University of Wisconsin–Madison
Kirby C. Smith, Texas A & M University
Joseph Stephen, Northern Illinois University
Monty J. Strauss, Texas Tech University
Sally Thomas, Orange Coast College
Henry Zatzkis, New Jersey Institute of Technology

We would particularly like to express our gratitude to Richard A. Askey, University of Wisconsin–Madison, and Richard W. Hamming, U.S. Naval Postgraduate School, for their continuing and thoughtful advice, and to thank Curtis F. Gerald, California Polytechnic State University, Emeritus, for developing the computer programs in this edition.

We want to express our special appreciation for the generous advice and help given to us by Thomas Cochran and Michael Schneider, Department of Mathematics, Belleville Area College, as they developed the text's answer section and solutions manuals.

We owe a special thanks to Charles Slavin of the University of Maine at Orono and James Lang of Valencia Community College for proofreading the book in galley pages.

We would also like to thank Laura R. Finney for keyboarding the manuscript and for proofreading the book in manuscript, galleys, and pages.

We also wish to express our thanks to the many other contributors whose names we have not been able to mention.

Any errors that appear are the responsibility of the authors. We will appreciate having these brought to our attention.

State College, PA
Monterey, CA

G.B.T., Jr.
R.L.F.

Prologue: What Is Calculus?

Calculus is the mathematics of motion and change. Where there is motion or growth, where variable forces are at work producing acceleration, calculus is the right mathematics to apply. This was true in the beginnings of the subject, and it is true today.

Calculus was first created to meet the mathematical needs of the scientists of the seventeenth century. Differential calculus dealt with the problem of calculating rates of change. It enabled people to define slopes of curves, to calculate the velocities and accelerations of moving bodies, to find the firing angle that gave a cannon its greatest range, and to predict the times when planets would be closest together or farthest apart. Integral calculus dealt with the problem of determining a function from information about its rate of change. It enabled people to calculate the future location of a body from its present position and a knowledge of the forces acting on it, to find the areas of irregular regions in the plane, to measure the lengths of curves, and to locate the centers of mass of arbitrary solids.

Before the mathematical developments that culminated in the great discoveries of Sir Isaac Newton (1642–1727) and Baron Gottfried Wilhelm Leibniz (1646–1716), it took the astronomer Johannes Kepler (1571–1630) twenty years of concentration, record-keeping, and arithmetic to discover the three laws of planetary motion that now bear his name:

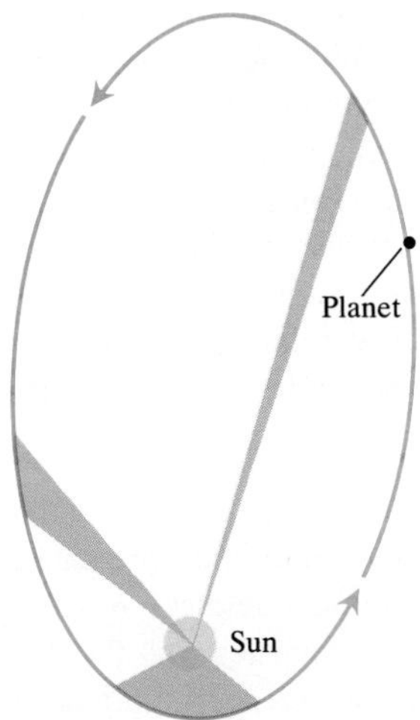

P.1 A planet moving about its sun. The shaded regions have equal areas. According to Kepler's second law, the planet takes the same amount of time to traverse the curved outer boundary of each region. The planet therefore moves faster near the sun than it does farther away.

1. Each planet travels in an ellipse that has one focus at the sun (Fig. P.1).
2. The radius vector from the sun to a planet sweeps out equal areas in equal intervals of time.
3. The squares of the periods of revolution of the planets about the sun are proportional to the cubes of their orbits' semimajor axes. If T is the length of a planet's year and a is the semimajor axis of its orbit, then the ratio T^2/a^3 has the same constant value for all planets in the solar system.

With calculus, deriving Kepler's laws from Newton's laws of motion is but an afternoon's work. Kepler described how the solar system worked—Newton and Leibniz, with their calculus, explained why.

Today, calculus and its extensions in mathematical analysis are far reaching indeed, and the physicists, mathematicians, and astronomers who first invented the subject would surely be amazed and delighted, as we hope you will be, to see what a profusion of problems it solves and what a wide range of fields now use it in the mathematical models that bring understanding about the universe and the world around us.

(a)

(b)

P.2 Calculus helped us predict that moons would travel in elliptical orbits about their planets; it also helped us to launch cameras and telescopes to observe the planets of our solar system. This photograph of Jupiter (a), taken on February 13, 1979, by the Voyager I space probe, shows two of its moons, Io (left) and Europa. Photograph (b) shows the Astro-1 group of telescopes on board the space shuttle Columbia in December 1990. One of the goals of this mission was to investigate the magnetic fields of Jupiter. We describe the effects of magnetic fields on moving electrical charges with calculus.

Economists use calculus to forecast global trends. Oceanographers use calculus to formulate theories about ocean currents and meteorologists use it to describe the flow of air in the upper atmosphere. Biologists use calculus to forecast population size and to describe the way predators like foxes interact with their prey. Medical researchers use calculus to design ultrasound and x-ray equipment for scanning the internal organs of the body. Space scientists use calculus to design rockets and explore distant planets. Psychologists use calculus to understand optical illusions in visual perception. Physicists use calculus to design inertial navigation systems and to study the nature of time and the universe. Hydraulic engineers use calculus to find safe closure patterns for valves in pipelines. Electrical engineers use it to design stroboscopic flash equipment and to solve the differential equations that describe current flow in computers. Sports equipment manufacturers use calculus to design tennis rackets and baseball bats. Stock market analysts use calculus to predict prices and assess interest rate risk. Physiologists use calculus to describe electrical impulses in neurons in the human nervous system. Drug companies use calculus to determine profitable inventory levels and timber companies use it to decide the most profitable time to harvest trees. The list is practically endless, for almost every professional field today uses calculus in some way.

"The calculus was the first achievement of modern mathematics," wrote John von Neumann (1903–1957), one of the great mathematicians of the present century, "and it is difficult to overestimate its importance. I think it defines more unequivocally than anything else the inception of modern mathematics; and the system of mathematical analysis, which is its logical development, still constitutes the greatest technical advance in exact thinking."*

**World of Mathematics*, Vol. 4 (New York: Simon and Schuster, 1960), "The Mathematician," by John von Neumann, pp. 2053-2063.

CHAPTER

8

INFINITE SERIES

Overview In this chapter, we study infinite polynomials called power series and, as a product of our study, develop one of the most remarkable formulas in all of mathematics. The formula, known as Taylor's formula, does two things for us. It shows how to calculate the value of an infinitely differentiable function like e^x at any point, just from its value and the values of its derivatives at the origin. And, as if that were not enough, the formula gives us polynomial approximations of differentiable functions of any order we want, along with their error formulas, all in a single equation (subject to the number of derivatives available).

Power series have many additional uses. They provide an efficient way to evaluate nonelementary integrals and they solve differential equations that give insight into heat flow, vibration, chemical diffusion, and signal transmission. What you will learn here sets the stage for the roles played by series of functions of all kinds in science and mathematics.

8.1 Limits of Sequences of Numbers

This section describes what it means for an infinite sequence of numbers to have a limit and shows how to find the limits of many of the sequences that arise in mathematics and applied fields.

Informally, a sequence is an ordered collection of things, but in this chapter the things will usually be numbers. We have seen sequences before, such as sequences $x_0, x_1, \ldots, x_n, \ldots$ of numerical approximations generated by Newton's method and the sequence $A_3, A_4, \ldots, A_n, \ldots$ of areas of n-sided regular polygons used to define the area of a circle. These sequences have limits, but many equally important sequences do not. The sequence $1, 2, 3, \ldots, n, \ldots$ of positive integers has no limit, nor does the sequence $2, 3, 5, 7, 11, 13, \ldots$ of prime numbers. We need to know when sequences do and do not have limits and how to find the limits when they exist. As with functions, we usually find out with theorems based on a formal definition. You will see a close parallel between what we do here and what we did with limits of functions in Chapter 1.

Definitions and Notation

Our starting point is a formal definition of sequence.

DEFINITION

An **infinite sequence** (or **sequence**) of numbers is a function whose domain is the set of integers greater than or equal to some integer n_0.

Usually n_0 is 1 and the domain is the set of all positive integers. But sometimes we want to start our sequences elsewhere. We might take $n_0 = 0$, for instance, when we begin Newton's method, or take $n_0 = 3$ when we work with n-sided polygons.

Sequences are defined the way other functions are, some typical rules being

$$a(n) = n - 1, \qquad a(n) = (-1)^{n+1}\frac{1}{n}, \qquad a(n) = \frac{n-1}{n} \tag{1}$$

(Example 1 and Fig. 8.1).

To indicate that the domains are sets of integers, we use a letter like n from the middle of the alphabet for the independent variable, instead of the x, y, z, and t used so widely in other contexts. The formulas in the defining rules, however, like those in (1), are often valid for domains much larger than the set of positive integers. This can be an advantage, as we shall see.

The number $a(n)$ is the ***n*th term** of the sequence, or the **term with index *n*.** For example, if $a(n) = (n - 1)/n$, then the terms are

First term	Second term	Third term		nth term	
$a(1) = 0$	$a(2) = \frac{1}{2}$,	$a(3) = \frac{2}{3}$,	...,	$a(n) = \frac{n-1}{n}$.	(2)

When we use the subscript notation a_n for $a(n)$, the sequence in (2) becomes

$$a_1 = 0, \qquad a_2 = \frac{1}{2}, \qquad a_3 = \frac{2}{3}, \qquad \ldots, \qquad a_n = \frac{n-1}{n}. \tag{3}$$

To describe sequences, we often write the first few terms as well as a formula for the nth term.

Example 1

We write	For the sequence whose defining rule is
$0, 1, 2, \ldots, n-1, \ldots$	$a_n = n - 1$
$1, \frac{1}{2}, \frac{1}{3}, \ldots, \frac{1}{n}, \ldots$	$a_n = \frac{1}{n}$
$1, -\frac{1}{2}, \frac{1}{3}, -\frac{1}{4}, \ldots, (-1)^{n+1}\frac{1}{n}, \ldots$	$a_n = (-1)^{n+1}\frac{1}{n}$
$0, \frac{1}{2}, \frac{2}{3}, \frac{3}{4}, \ldots, \frac{n-1}{n}, \ldots$	$a_n = \frac{n-1}{n}$
$0, -\frac{1}{2}, \frac{2}{3}, -\frac{3}{4}, \ldots, (-1)^{n+1}\left(\frac{n-1}{n}\right), \ldots$	$a_n = (-1)^{n+1}\left(\frac{n-1}{n}\right)$
$3, 3, 3, \ldots, 3, \ldots$	$a_n = 3$

The sequences of Example 1 are graphed here in two different ways: by plotting the numbers a_n on a horizontal axis and by plotting the points (n, a_n) in the coordinate plane.

The terms $a_n = n - 1$ eventually surpass every integer, so the sequence $\{a_n\}$ diverges, . . .

$a_n = n - 1$

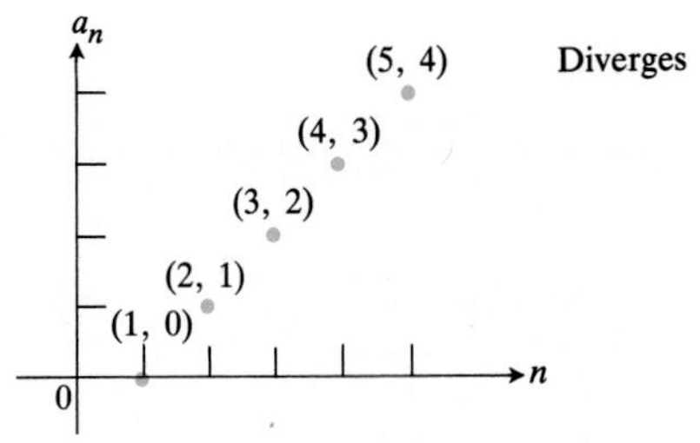

. . . but the terms $a_n = 1/n$ decrease steadily and get arbitrarily close to 0 as n increases, so the sequence $\{a_n\}$ converges to 0.

$a_n = \frac{1}{n}$

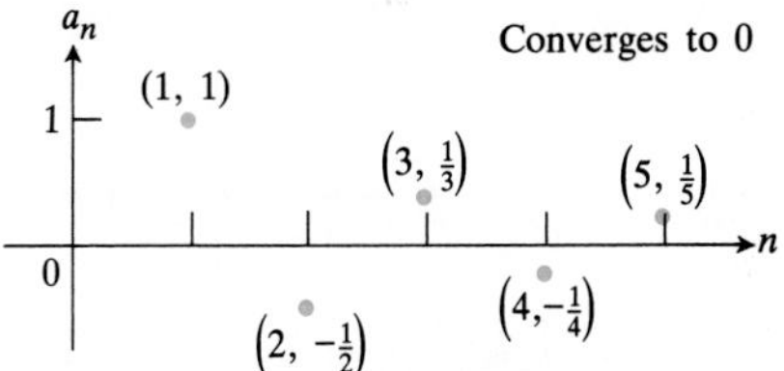

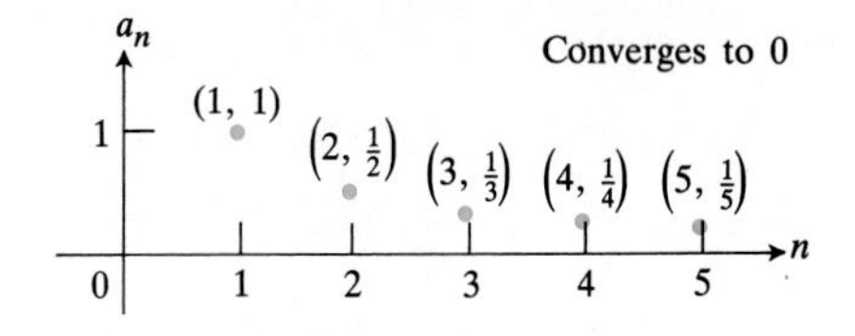

The terms $a_n = (-1)^{n+1}(1/n)$ alternate in sign but still converge to 0.

$a_n = (-1)^{n+1} \frac{1}{n}$

The terms $a_n = (n - 1)/n$ approach 1 steadily and get arbitrarily close as n increases, so the sequence $\{a_n\}$ converges to 1.

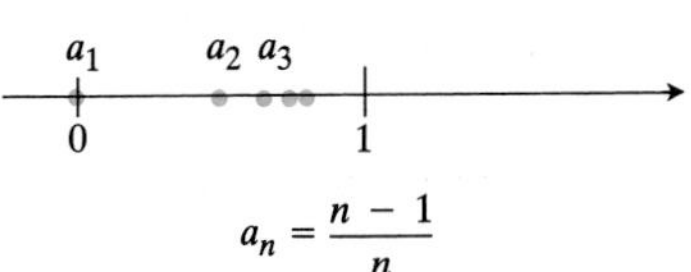

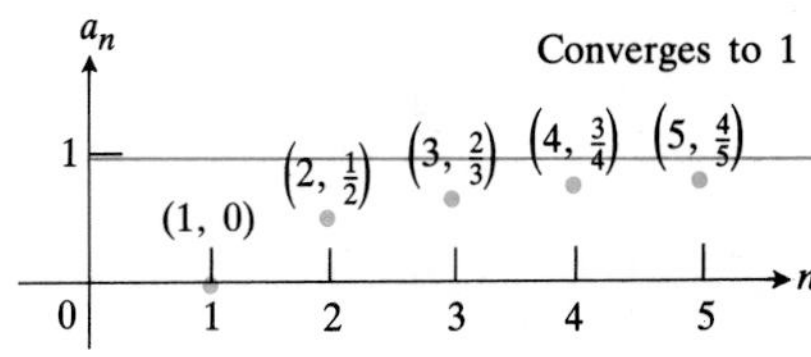

The terms $a_n = (-1)^{n+1}[(n - 1)/n]$ alternate in sign. The positive terms approach 1. But the negative terms approach -1 as n increases, so the sequence $\{a_n\}$ diverges.

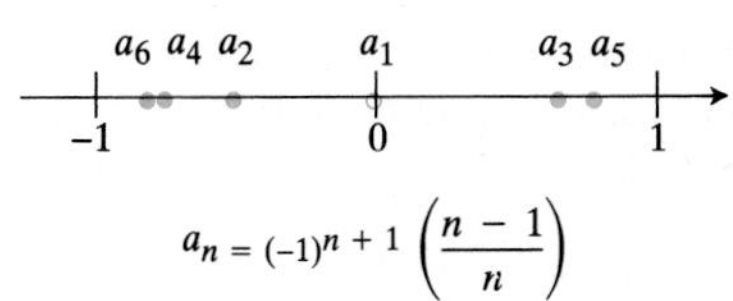

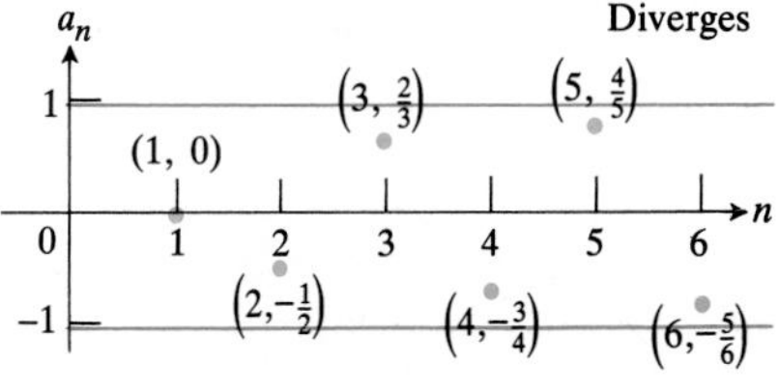

The terms in the sequence of constants $a_n = 3$ have the same value regardless of n; so we say the sequence $\{a_n\}$ converges to 3.

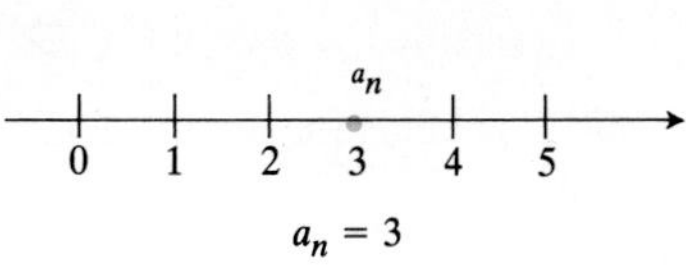

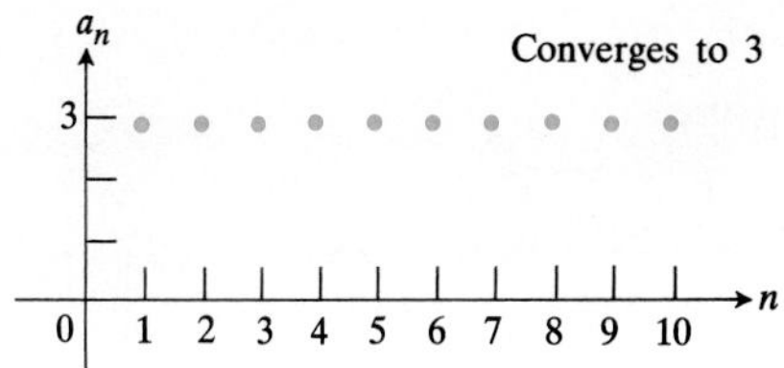

Notation We refer to the sequence whose nth term is a_n with the notation $\{a_n\}$ ("the sequence a sub n"). The second sequence in Example 1 is $\{1/n\}$ ("the sequence 1 over n"); the last sequence is $\{3\}$ ("the sequence 3").

Convergence and Divergence

As Fig. 8.1 shows, the sequences of Example 1 do not behave the same way. The sequences $\{1/n\}$, $\{(-1)^{n+1}(1/n)\}$, and $\{(n-1)/n\}$ each seem to approach a single limiting value as n increases, and $\{3\}$ is at a limiting value from the very first. On the other hand, terms of $\{(-1)^{n+1}(n-1)/n\}$ seem to accumulate near two different values, -1 and 1, while the terms of $\{n-1\}$ become increasingly large and do not accumulate anywhere.

To distinguish sequences that approach a unique limiting value L, as n increases, from those that do not, we say that the former sequences *converge*, according to the following definition.

DEFINITIONS

The sequence $\{a_n\}$ **converges** to the number L if to every positive number ϵ there corresponds an integer N such that for all n,

$$n > N \quad \Rightarrow \quad |a_n - L| < \epsilon. \tag{4}$$

If no such number and integer exist, we say that $\{a_n\}$ **diverges.**

If $\{a_n\}$ converges to L, we write $\lim_{n\to\infty} a_n = L$, or simply $a_n \to L$, and call L the **limit** of the sequence (Fig. 8.2).

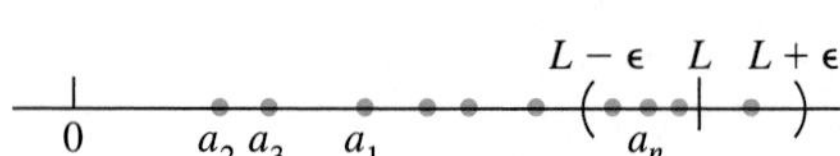

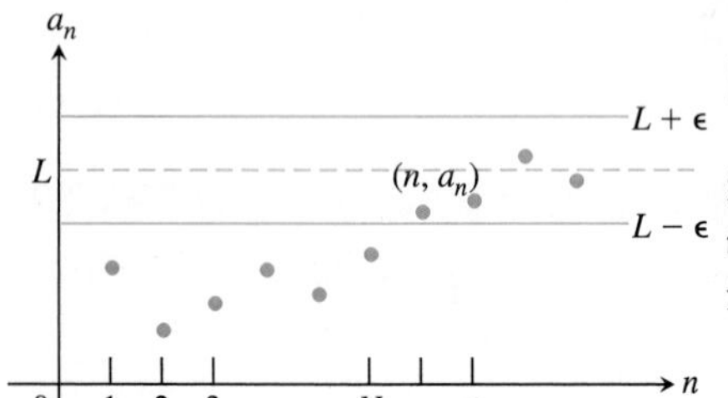

8.2 $a_n \to L$ if L is a horizontal asymptote of the sequence of points $\{(n, a_n)\}$. In this figure, all the a_n's after a_N lie within ϵ of L.

Example 2 Show that $\{1/n\}$ converges to 0.

Solution We set $a_n = 1/n$ and $L = 0$ in the definition of convergence. To show that $1/n \to 0$, we must show that for any $\epsilon > 0$, there exists an integer N such that for all n,

$$n > N \quad \Rightarrow \quad \left|\frac{1}{n} - 0\right| < \epsilon. \tag{5}$$

This implication will hold for all n for which

$$\frac{1}{n} < \epsilon \quad \text{or, equivalently,} \quad n > \frac{1}{\epsilon}.$$

Pick an integer N greater than $1/\epsilon$. Then any n greater than N will automatically be greater than $1/\epsilon$ and the implication in (5) will hold.

Example 3 Show that if k is any number, then $\{k\}$ converges to k.

Solution We set $a_n = k$ and $L = k$ in the definition of convergence. To show that $a_n \to k$, we must show that for any $\epsilon > 0$ there exists an integer N such that for all n,

$$n > N \quad \Rightarrow \quad |k - k| < \epsilon.$$

This implication holds for any integer N because $|k - k| = 0$ is less than every positive ϵ for all n.

$$a_n = (-1)^{n+1}\left(\frac{n-1}{n}\right)$$

Neither the ϵ-interval about 1 nor the ϵ-interval about -1 contains a complete tail of the sequence.

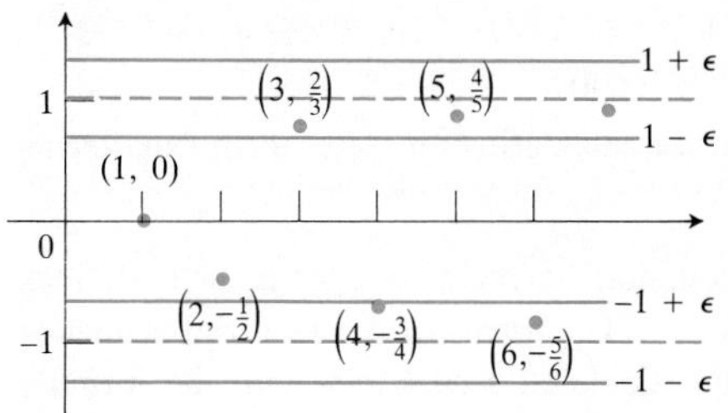

8.3 The sequence $\{(-1)^{n+1}[(n-1)/n]\}$ diverges.

Example 4 Show that $\{(-1)^{n+1}(n-1)/n\}$ diverges.

Solution Take a positive ϵ smaller than 1 so that the bands shown in Fig. 8.3 about the lines $y = 1$ and $y = -1$ do not overlap. Any $\epsilon < 1$ will do. Convergence to 1 would require every point of the graph beyond a certain index N to lie inside the upper band, but this will never happen. As soon as a point (n, a_n) lies within the upper band, every alternate point starting with $(n+1, a_{n+1})$ will lie within the lower band. Hence the sequence cannot converge to 1. Likewise, it cannot converge to -1. On the other hand, because the terms of the sequence get alternately closer to 1 and -1, they never accumulate near any other value. Therefore, the sequence diverges.

A **tail** of a sequence $\{a_n\}$ is the collection of all terms whose indices are greater than some integer N. In other words, a tail is one of the sets $\{a_n \mid n > N\}$. Another way to say that $a_n \to L$ is to say that every ϵ-interval about L contains a tail of the sequence. The convergence or divergence of a sequence has nothing to do with how a sequence starts out. It depends only on how the tails behave.

The behavior of $\{(-1)^{n+1}[(n-1)/n]\}$ is qualitatively different from that of $\{n-1\}$, which diverges because it outgrows every real number L. To describe the behavior of $\{n-1\}$ we write

$$\lim_{n\to\infty} (n-1) = \infty.$$

In speaking of infinity as a limit of a sequence $\{a_n\}$, we do not mean that the difference between a_n and infinity becomes small as n increases. We mean that a_n becomes numerically large as n increases.

Limits Are Unique

A sequence cannot converge to two different limits.

> If a sequence $\{a_n\}$ converges, then its limit is unique.

The argument goes like this: If $\{a_n\}$ converged to two different limits L_1 and L_2, we could take ϵ to be the positive number $|L_1 - L_2|/2$. Because $a_n \to L_1$, there would exist an integer N_1 such that for all n,

$$n > N_1 \quad \Rightarrow \quad |a_n - L_1| < \epsilon.$$

There would also exist an integer N_2 such that for all n,

$$n > N_2 \quad \Rightarrow \quad |a_n - L_2| < \epsilon.$$

For all n greater than both N_1 and N_2, we would then have

$$\begin{aligned}|L_1 - L_2| &= |L_1 - a_n + a_n - L_2| \le |L_1 - a_n| + |a_n - L_2| \\ &< \epsilon + \epsilon = 2(|L_1 - L_2|/2) \qquad \text{(Triangle inequality)} \\ &< |L_1 - L_2|.\end{aligned}$$

Factorial Notation

The notation $n!$ ("n factorial") means the product $1 \cdot 2 \cdot 3 \cdot \cdots \cdot n$ of the integers from 1 to n. Notice that $(n+1)! = (n+1) \cdot n!$. Thus, $4! = 1 \cdot 2 \cdot 3 \cdot 4 = 24$ and $5! = 1 \cdot 2 \cdot 3 \cdot 4 \cdot 5 = 5 \cdot 4! = 120$. We also define 0! to be 1. Factorials grow even faster than exponentials, as the following table suggests.

n	e^n(rounded)	$n!$
1	3	1
5	148	120
10	22,026	3,628,800
20	4.8×10^8	2.4×10^{18}

But this is absurd; no number is less than itself. Hence, if a sequence converges, its limit is unique.

Recursive Definitions

So far, we have calculated each a_n directly from the value of n. But sequences are often defined **recursively** by giving

1. the value of the first term (or first few terms);
2. a rule, called a **recursion formula,** for calculating any later term from terms that precede it.

The initial value $a_1 = 1$ and the recursion formula $a_{n+1} = a_n + 1$ defines the sequence of positive integers recursively. The terms of the factorial sequence $1, 2, 6, 24, \ldots, n!, \ldots$ are usually calculated recursively with the formula $(n+1)! = (n+1)n!$. The initial value $x_0 = 1$ and the recursion formula $x_{n+1} = x_n - [(\sin x_n - x_n^2)/(\cos x_n - 2x_n)]$ define a sequence that converges to the solution of the equation $\sin x = x^2$ (Section 2.7, Example 3). Recursion formulas arise regularly in computer programming, and we shall encounter them again when we preview numerical methods for solving differential equations (Section 15.9).

Subsequences

If the terms of one sequence occur in their given order among the terms of a second sequence, we call the first sequence a **subsequence** of the second.

Example 5 *Subsequences of the Sequence of Positive Integers*

a) The sequence $2, 4, 6, \ldots, 2n, \ldots$ of even integers

b) The sequence $1, 3, 5, \ldots, 2n+1, \ldots$ of odd integers

c) The sequence $2, 3, 5, 7, 11, \ldots$ of primes

Subsequences are important for two reasons. First, if a sequence $\{a_n\}$ converges to a limit L, then all subsequences also converge to L. If we know a sequence converges, it may be quicker for us to find or estimate its limit by choosing a rapidly convergent subsequence. The second reason is related to the first: If any subsequence of the original sequence diverges, or if two subsequences have different limits, then the original sequence diverges. The sequence $\{(-1)^{n-1}\}$ diverges because the subsequence $1, 1, 1, \ldots$ of odd-numbered terms converges to 1 while the subsequence $-1, -1, -1, \ldots$ of even-numbered terms converges to -1, which is a different limit.

Useful Theorems

The study of limits would be cumbersome if we had to answer every question about convergence by applying the definition directly. Fortunately, three theorems will make this largely unnecessary from now on. The first is the sequence version of Theorem 1 in Chapter 1.

THEOREM 1

The following rules hold if $\lim_{n\to\infty} a_n = A$ and $\lim_{n\to\infty} b_n = B$

1. *Sum Rule:* $\lim \{a_n + b_n\} = A + B$
2. *Difference Rule:* $\lim \{a_n - b_n\} = A - B$
3. *Product Rule:* $\lim \{a_n \cdot b_n\} = A \cdot B$
4. *Constant Multiple Rule:* $\lim \{k \cdot b_n\} = k \cdot B$ (Any number k)
5. *Quotient Rule:* $\lim \dfrac{a_n}{b_n} = \dfrac{A}{B}$ if $B \neq 0$

The limits are all taken as $n \to \infty$, and A and B are real numbers.

By combining Theorem 1 with Examples 2 and 3, we can proceed immediately to

$$\lim_{n\to\infty} \left(-\frac{1}{n}\right) = -1 \cdot \lim_{n\to\infty} \frac{1}{n} = -1 \cdot 0 = 0,$$

$$\lim_{n\to\infty} \left(\frac{n-1}{n}\right) = \lim_{n\to\infty} \left(1 - \frac{1}{n}\right) = \lim_{n\to\infty} 1 - \lim_{n\to\infty} \frac{1}{n} = 1 - 0 = 1,$$

$$\lim_{n\to\infty} \frac{5}{n^2} = 5 \cdot \lim_{n\to\infty} \frac{1}{n} \cdot \lim_{n\to\infty} \frac{1}{n} = 5 \cdot 0 \cdot 0 = 0,$$

$$\lim_{n\to\infty} \frac{4 - 7n^6}{n^6 + 3} = \lim_{n\to\infty} \frac{(4/n^6) - 7}{1 + (3/n^6)} = \frac{0 - 7}{1 + 0} = -7.$$

One general consequence of Theorem 1 is that every nonzero multiple of a divergent sequence $\{a_n\}$ diverges. For, suppose $\{ca_n\}$ were to converge for some number $c \neq 0$. Then $(1/c)\{ca_n\} = \{a_n\}$ would converge by the Constant Multiple Rule—but it does not.

The next theorem is the sequence version of the Sandwich Theorem.

THEOREM 2

The Sandwich Theorem for Sequences

If $a_n \leq b_n \leq c_n$ for all n beyond some index N, and if $\lim a_n = \lim c_n = L$, then $\lim b_n = L$ also.

An immediate consequence of Theorem 2 is that, if $|b_n| \leq c_n$ and $c_n \to 0$, then $b_n \to 0$ because $-c_n \leq b_n \leq c_n$. We use this fact in the next example.

Example 6 Since $1/n \to 0$, we know that

a) $\dfrac{\cos n}{n} \to 0$ because $0 \leq \left|\dfrac{\cos n}{n}\right| = \dfrac{|\cos n|}{n} \leq \dfrac{1}{n}$;

b) $\dfrac{1}{2^n} \to 0$ because $0 \leq \dfrac{1}{2^n} \leq \dfrac{1}{n}$;

c) $(-1)^n \dfrac{1}{n} \to 0$ because $0 \leq \left|(-1)^n \dfrac{1}{n}\right| \leq \dfrac{1}{n}$.

The application of Theorems 1 and 2 is broadened by a theorem stating that applying a continuous function to a convergent sequence produces a convergent sequence. We state the theorem without proof.

THEOREM 3

If $a_n \to L$ and if f is a function that is continuous at L and defined at all a_n, then $f(a_n) \to f(L)$.

Example 7 Show that $\sqrt{(n+1)/n} \to 1$.

Solution We know that $(n+1)/n \to 1$. Taking $f(x) = \sqrt{x}$ and $L = 1$ in Theorem 3 gives $\sqrt{(n+1)/n} \to \sqrt{1} = 1$.

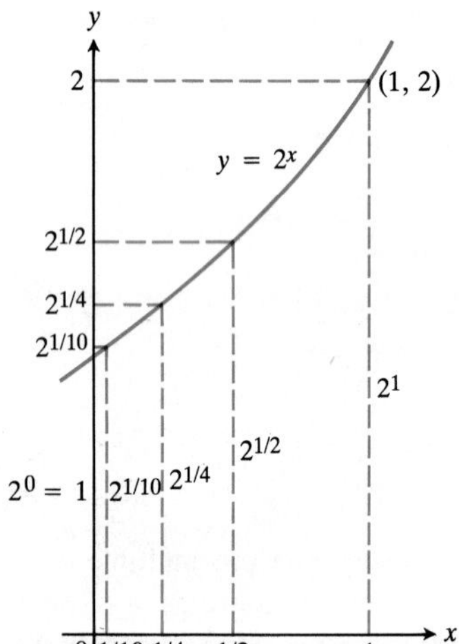

8.4 As $n \to \infty$, $x = 1/n \to 0$ and $y = 2^{1/n} \to 2^0 = 1$ (Example 8).

Example 8 Show that $2^{1/n} \to 1$ (Fig. 8.4).

Solution We know that $1/n \to 0$. Taking $f(x) = 2^x$ and $L = 0$ in Theorem 3 therefore gives $2^{1/n} \to 2^0 = 1$. Some sample values:

n	$2^{1/n}$
1	2
2	1.4142 13562
4	1.1892 07115
10	1.0717 73463
100	1.0069 55555
1000	1.0006 93387
10000	1.0000 69317

In Section 1.3, we saw that entering 2 in a calculator and pressing the square root key repeatedly produces a succession of numbers that approach 1. The result of Example 8 now tells us why. The successive square roots form a subsequence $2^{1/2}$, $2^{1/4}$, $2^{1/8}$, $2^{1/16}$, ... of the sequence $\{2^{1/n}\}$. The sequence converges to 1, so the subsequence does too.

The next theorem enables us to use l'Hôpital's rule to find the limits of some sequences. We state and prove the theorem first and then show how to apply it.

THEOREM 4

Suppose that $f(x)$ is a function defined for all $x \geq n_0$ and $\{a_n\}$ is a sequence such that $a_n = f(n)$ when $n \geq n_0$. If

$$\lim_{x\to\infty} f(x) = L, \qquad \text{then} \qquad \lim_{n\to\infty} a_n = L.$$

Proof Suppose that $\lim_{x\to\infty} f(x) = L$. Then for each positive number ϵ there is a number M such that for all x,

$$x > M \quad \Rightarrow \quad |f(x) - L| < \epsilon.$$

Let N be an integer greater than M and greater than or equal to n_0. Then

$$n > N \quad \Rightarrow \quad a_n = f(n) \quad \text{and} \quad |a_n - L| = |f(n) - L| < \epsilon.$$

Example 9 Show that $\lim_{n\to\infty}(\ln n)/n = 0$.

Solution The function $(\ln x)/x$ is defined for all $x \geq 1$ and agrees with the given sequence at positive integers. Therefore $\lim_{n\to\infty} (\ln n)/n$ will equal $\lim_{x\to\infty} (\ln x)/x$ if the latter exists. A single application of l'Hôpital's rule shows that

$$\lim_{x\to\infty} \frac{\ln x}{x} = \lim_{x\to\infty} \frac{1/x}{1} = \frac{0}{1} = 0.$$

We conclude that $\lim_{n\to\infty}(\ln n)/n = 0$.

When we use l'Hôpital's rule to find the limit of a sequence, we often treat n as a continuous real variable and differentiate directly with respect to n. This saves us from having to rewrite the formula for a_n as we did in Example 9.

Example 10 Find $\lim_{n\to\infty}(2^n/5n)$.

Solution By l'Hôpital's rule,

$$\lim_{n\to\infty} \frac{2^n}{5n} = \lim_{n\to\infty} \frac{2^n \cdot \ln 2}{5} = \infty.$$

Limits That Arise Frequently

The limits in Table 8.1 are useful and arise frequently. The first limit is from Example 9. The others are derived in Appendix 8.

TABLE 8.1

1. $\lim_{n\to\infty} \frac{\ln n}{n} = 0$	2. $\lim_{n\to\infty} \sqrt[n]{n} = 1$
3. $\lim_{n\to\infty} x^{1/n} = 1 \quad (x > 0)$	4. $\lim_{n\to\infty} x^n = 0 \quad (\lvert x\rvert < 1)$
5. $\lim_{n\to\infty}\left(1 + \frac{x}{n}\right)^n = e^x \quad$ (Any x)	6. $\lim_{n\to\infty} \frac{x^n}{n!} = 0 \quad$ (Any x)

In formulas (3)–(6), x remains fixed while $n \to \infty$.

Example 11 *Limits from Table 8.1*

1. $\dfrac{\ln(n^2)}{n} = \dfrac{2 \ln n}{n} \to 2 \cdot 0 = 0$ (Formula 1)
2. $\sqrt[n]{n^2} = n^{2/n} = (n^{1/n})^2 \to (1)^2 = 1$ (Formula 2)
3. $\sqrt[n]{3n} = 3^{1/n}(n^{1/n}) \to 1 \cdot 1 = 1$ (Formula 3 with $x = 3$, and Formula 2)
4. $\left(-\dfrac{1}{2}\right)^n \to 0$ $\left(\text{Formula 4 with } x = -\dfrac{1}{2}\right)$
5. $\left(\dfrac{n-2}{n}\right)^n = \left(1 - \dfrac{2}{n}\right)^n \to e^{-2}$ (Formula 5 with $x = -2$)
6. $\dfrac{100^n}{n!} \to 0$ (Formula 6 with $x = 100$)

Example 12 Does the sequence whose nth term is

$$a_n = \left(\frac{n+1}{n-1}\right)^n$$

converge? If so, find $\lim_{n\to\infty} a_n$.

Solution 1 *(L'Hôpital's Rule)* The limit leads to the indeterminate form 1^∞. To apply l'Hôpital's rule, we first change the form to $\infty \cdot 0$ by taking the natural logarithm of a_n (as in Section 6.4, Example 7):

$$\ln a_n = \ln\left(\frac{n+1}{n-1}\right)^n = n \ln\left(\frac{n+1}{n-1}\right).$$

Then,

$$\begin{aligned}
\lim_{n\to\infty} \ln a_n &= \lim_{n\to\infty} n \ln\left(\frac{n+1}{n-1}\right) && (\infty \cdot 0) \\
&= \lim_{n\to\infty} \frac{\ln\left(\frac{n+1}{n-1}\right)}{1/n} && \left(\frac{0}{0}\right) \\
&= \lim_{n\to\infty} \frac{-2/(n^2-1)}{-1/n^2} && \text{(l'Hôpital's rule)} \\
&= \lim_{n\to\infty} \frac{2n^2}{n^2-1} = 2.
\end{aligned}$$

Therefore,

$$a_n = e^{\ln a_n} \to e^2.$$

The limit exists and equals e^2.

Solution 2 (Less general, but works well in this case.) We have

$$\left(\frac{n+1}{n-1}\right)^n = \frac{\left(\frac{n+1}{n}\right)^n}{\left(\frac{n-1}{n}\right)^n} = \frac{\left(1+\frac{1}{n}\right)^n}{\left(1+\frac{-1}{n}\right)^n}.$$

Hence,

$$\begin{aligned}
\lim_{n\to\infty}\left(\frac{n+1}{n-1}\right)^n &= \frac{\lim_{n\to\infty}\left(1+\frac{1}{n}\right)^n}{\lim_{n\to\infty}\left(1+\frac{-1}{n}\right)^n} = \frac{e^1}{e^{-1}} && \left(\begin{array}{l}\text{Table 8.1,}\\ \text{Formula 5,}\\ \text{with } x = 1\\ \text{and with } x = -1\end{array}\right) \\
&= e^2
\end{aligned}$$

* Picard's Method for Finding Roots

The problem of finding the roots of the equation

$$f(x) = 0 \tag{6}$$

is equivalent to that of finding the solutions of the equation

$$g(x) = f(x) + x = x, \tag{7}$$

obtained by adding x to both sides of Eq. (6). Any value of x that satisfies Eq. (7) satisfies (6), and conversely. By this simple change we cast Eq. (6) into a form that may render it solvable on a computer by a powerful method called **Picard's method** (after the French mathematician Charles Émile Picard, 1856–1941).

If the domain of g contains the range of g, we can start with a point x_0 in the domain and apply g repeatedly to get

$$x_1 = g(x_0), \qquad x_2 = g(x_1), \qquad x_3 = g(x_2), \qquad \ldots . \tag{8}$$

Under simple restrictions that we shall describe shortly, the sequence generated by the recursion formula $x_{n+1} = g(x_n)$ will converge to a point x for which $g(x) = x$. This point solves the equation $f(x) = 0$ because

$$f(x) = g(x) - x = x - x = 0. \tag{9}$$

A point x for which $g(x) = x$ is a **fixed point** of g. We see in Eq. (9) that the fixed points of g are precisely the roots of f.

We begin with an example whose outcome we know so that we can see how the successive approximations work. You will find a BASIC program for Picard's method after the exercises at the end of this section.

Example 13 Solve the equation $(1/4)x + 3 = x$.

Solution By algebra we know that the solution is $x = 4$. To apply Picard's method, we take

$$g(x) = \frac{1}{4}x + 3,$$

choose a starting point, say $x_0 = 1$, and calculate the initial terms of the sequence $x_{n+1} = g(x_n)$. Table 8.2 lists the results. In ten steps, the solution of the original equation is found with an error of magnitude less than 3×10^{-6}. Figure 8.5 shows the geometry of the solution.

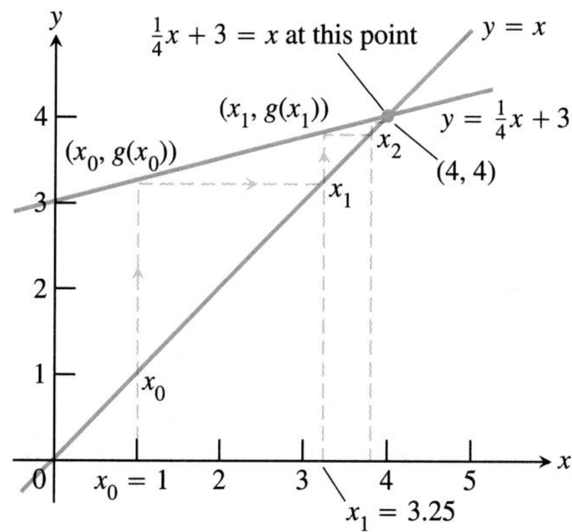

8.5 The geometric interpretation of the Picard solution of the equation $g(x) = (1/4)x + 3 = x$ in Example 13.

We start with $x_0 = 1$ and calculate the first y-value, $g(x_0)$. This becomes the second x-value, x_1. The second y-value, $g(x_1)$, becomes the third x-value, x_2, and so on. The process is shown here as a path (called the *iteration path*) that starts at $x_0 = 1$, moves up to $(x_0, g(x_0)) = (x_0, x_1)$, over to (x_1, x_1), up to $(x_1, g(x_1))$, and so on. The path converges to the point where the graph of g meets the line $y = x$, the point where $g(x) = x$.

TABLE 8.2
Successive iterates of $g(x) = (1/4)x + 3$, starting with $x_0 = 1$

x_n	$x_{n+1} = g(x_n) = (1/4)x_n + 3$
$x_0 = 1$	$x_1 = g(x_0) = (1/4)(1) + 3 = 3.25$
$x_1 = 3.25$	$x_2 = g(x_1) = (1/4)(3.25) + 3 = 3.8125$
$x_2 = 3.8125$	$x_3 = g(x_2) = 3.9531\ 25$
$x_3 = 3.9531\ 25$	$x_4 = 3.9882\ 8125$
$\vdots$	$x_5 = 3.9970\ 70313$
	$x_6 = 3.9992\ 67578$
	$x_7 = 3.9998\ 16895$
	$x_8 = 3.9999\ 54224$
	$x_9 = 3.9999\ 88556$
	$x_{10} = 3.9999\ 97139$
	$\vdots$

Example 14 Solve the equation $\cos x = x$.

Solution We take $g(x) = \cos x$, choose $x_0 = 1$ as a starting value, and use the recursion formula $x_{n+1} = g(x_n)$ to find

$$x_0 = 1, \qquad x_1 = \cos 1, \qquad x_2 = \cos(x_1), \ldots.$$

We can approximate the first 50 terms or so on a calculator in radian mode by entering 1 and pressing [cos] repeatedly. The display stops changing when $\cos x = x$ to the number of decimal places in the display.

Try it for yourself. As you press the key, notice that the successive approximations lie alternately above and below the fixed point

$$x = 0.7390\ 85133\ldots.$$

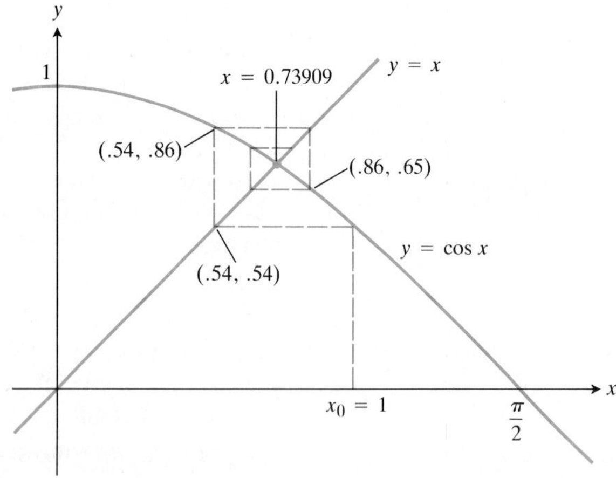

8.6 The solution of $\cos x = x$ by Picard's method starting at $x_0 = 1$ (Example 14).

Figure 8.6 shows that the values oscillate this way because the path of the procedure spirals around the fixed point.

Example 15 Picard's method will not solve the equation

$$g(x) = 4x - 12 = x.$$

As Fig. 8.7 shows, any choice of x_0 except $x_0 = 4$, the solution itself, generates a divergent sequence that moves away from the solution.

The difficulty in Example 15 can be traced to the fact that the slope of the line $y = 4x - 12$ exceeds 1, the slope of the line $y = x$. Conversely, the process worked in Example 13 because the slope of the line $y = (1/4)x + 3$ was numerically less than 1. A theorem from advanced calculus tells us that if $g'(x)$ is continuous on a closed interval I whose interior contains a solution of the equation $g(x) = x$, and if $|g'(x)| < 1$ on I, then any choice of x_0 in the interior of I will lead to the solution. (See Exercise 99 about what to do if $|g'(x)| > 1$.)

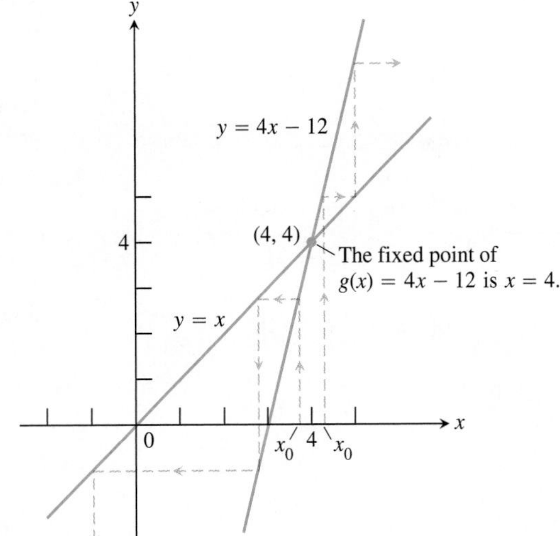

8.7 Applying the Picard method to $g(x) = 4x - 12$ will not find the fixed point unless x_0 is 4 itself (Example 15).

EXERCISES 8.1

Each of Exercises 1–4 gives a formula for the nth term a_n of a sequence $\{a_n\}$. Find the values of a_1, a_2, a_3, and a_4.

1. $a_n = \dfrac{1-n}{n^2}$

2. $a_n = \dfrac{1}{n!}$

3. $a_n = \dfrac{(-1)^{n+1}}{2n-1}$

4. $a_n = 2 + (-1)^n$

Each of Exercises 5–10 gives the first term or two of a sequence and a recursion formula for the remaining terms. Write out the first ten terms of each sequence.

5. $x_1 = 1, \quad x_{n+1} = x_n + (1/2^n)$

6. $x_1 = 1, \quad x_{n+1} = x_n/(n+1)$

7. $x_1 = 2, \quad x_{n+1} = (-1)^{n+1}x_n/2$

8. $x_1 = -2, \quad x_{n+1} = nx_n/(n+1)$

9. $x_1 = x_2 = 1, \quad x_{n+2} = x_{n+1} - x_n$

10. $x_1 = 2, \quad x_2 = -1, \quad x_{n+2} = x_{n+1}/x_n$

Which of the sequences $\{a_n\}$ in Exercises 11–70 converge and which diverge? Find the limit of each convergent sequence.

11. $a_n = 2 + (0.1)^n$

12. $a_n = \dfrac{n + (-1)^n}{n}$

13. $a_n = \dfrac{1-2n}{1+2n}$

14. $a_n = \dfrac{2n+1}{1-3\sqrt{n}}$

15. $a_n = \dfrac{1-5n^4}{n^4+8n^3}$

16. $a_n = \dfrac{n+3}{n^2+5n+6}$

17. $a_n = \dfrac{n^2-2n+1}{n-1}$

18. $a_n = \dfrac{1-n^3}{70-4n^2}$

19. $a_n = 1 + (-1)^n$

20. $a_n = (-1)^n\left(1 - \dfrac{1}{n}\right)$

21. $a_n = \left(\dfrac{n+1}{2n}\right)\left(1 - \dfrac{1}{n}\right)$

22. $a_n = \left(2 - \dfrac{1}{2^n}\right)\left(3 + \dfrac{1}{2^n}\right)$

23. $a_n = \dfrac{(-1)^{n+1}}{2n-1}$

24. $a_n = \left(-\dfrac{1}{2}\right)^n$

25. $a_n = \dfrac{\sin n}{n}$

26. $a_n = \dfrac{\sin^2 n}{2^n}$

27. $a_n = \sqrt{\dfrac{2n}{n+1}}$

28. $a_n = \sin\left(\dfrac{\pi}{2} + \dfrac{1}{n}\right)$

29. $a_n = \tan^{-1} n$

30. $a_n = \ln n - \ln(n+1)$

31. $a_n = \dfrac{n}{2^n}$

32. $a_n = \dfrac{3^n}{n^3}$

33. $a_n = \dfrac{\ln(n+1)}{\sqrt{n}}$

34. $a_n = \dfrac{\ln n}{\ln 2n}$

35. $a_n = 8^{1/n}$

36. $a_n = (0.03)^{1/n}$

37. $a_n = \left(1 + \dfrac{7}{n}\right)^n$

38. $a_n = \left(1 - \dfrac{1}{n}\right)^n$

39. $a_n = \dfrac{1}{(0.9)^n}$

40. $a_n = n\pi \cos n\pi$

41. $a_n = \sqrt[n]{10n}$

42. $a_n = \sqrt[n]{n^2}$

43. $a_n = \left(\dfrac{3}{n}\right)^{1/n}$

44. $a_n = (n+4)^{1/(n+4)}$

45. $a_n = \dfrac{\ln n}{n^{1/n}}$

46. $a_n = \sqrt[n]{4^n n}$

47. $a_n = \left(\dfrac{1}{3}\right)^n + \dfrac{1}{\sqrt{2^n}}$

48. $a_n = \sqrt[n]{3^{2n+1}}$

49. $a_n = \dfrac{n!}{n^n}$ (*Hint:* Compare the quotient with $1/n$.)

50. $a_n = \dfrac{(-4)^n}{n!}$

51. $a_n = \left(\dfrac{1}{n}\right)^{1/\ln n}$

52. $a_n = \dfrac{n!}{2^n \cdot 3^n}$

53. $a_n = \dfrac{n!}{10^{6n}}$

54. $a_n = \dfrac{3^n \cdot 6^n}{2^{-n} \cdot n!}$

55. $a_n = \tanh n$

56. $a_n = \sinh (\ln n)$

57. $a_n = \ln \left(1 + \dfrac{1}{n}\right)^n$

58. $a_n = \left(\dfrac{n}{n+1}\right)^n$

59. $a_n = \left(\dfrac{3n+1}{3n-1}\right)^n$

60. $a_n = \left(1 - \dfrac{1}{n^2}\right)^n$

61. $a_n = \sqrt[n]{\dfrac{x^n}{2n+1}}, \quad x > 0$

62. $a_n = \dfrac{\left(\frac{10}{11}\right)^n}{\left(\frac{9}{10}\right)^n + \left(\frac{11}{12}\right)^n}$

63. $a_n = \dfrac{1}{n}\displaystyle\int_1^n \frac{1}{x}\, dx$

64. $a_n = \displaystyle\int_1^n \frac{1}{x^p}\, dx, \quad p > 1$

65. $a_n = \sqrt[n]{n^2 + n}$

66. $a_n = \dfrac{(\ln n)^{200}}{n}$

67. $a_n = \dfrac{n^2}{2n-1} \sin \dfrac{1}{n}$

68. $a_n = n\left(1 - \cos \dfrac{1}{n}\right)$

69. $a_n = n - \sqrt{n^2 - n}$

70. $a_n = \dfrac{1}{\sqrt{n^2 - 1} - \sqrt{n^2 + n}}$

CALCULATOR In Exercises 71–74, experiment with a calculator to identify a value of N that will make the inequality hold for $n \geq N$.

71. $|\sqrt[n]{0.5} - 1| < 10^{-3}$

72. $|\sqrt[n]{n} - 1| < 10^{-3}$

73. $(0.9)^n < 10^{-3}$

74. $2^n/n! < 10^{-7}$

75. CALCULATOR *A recursive definition of $\pi/2$.* If you start with $x_1 = 1$ and define the subsequent terms of $\{x_n\}$ by the rule $x_n = x_{n-1} + \cos x_{n-1}$, you generate a sequence that converges rapidly to $\pi/2$. Try it. Figure 8.8 explains what is going on.

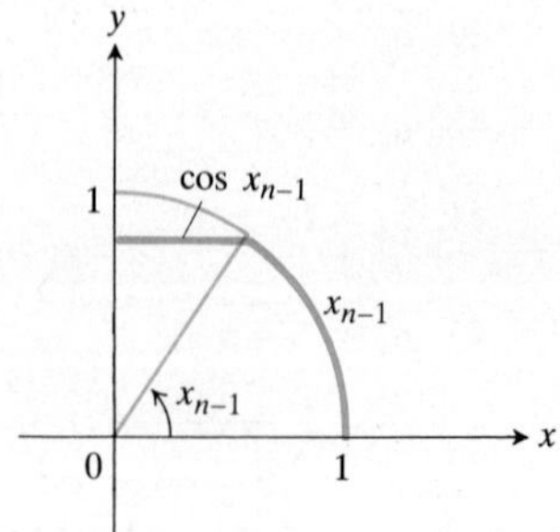

8.8 The length $\pi/2$ of the circular arc is approximated by $x_{n-1} + \cos x_{n-1}$ (Exercise 75).

76. The first term of a sequence is $x_1 = 1$. Each succeeding term is the sum of all those that come before it:

$$x_{n+1} = x_1 + x_2 + \cdots + x_n.$$

Write out enough early terms of the sequence to deduce a general formula for x_n that holds for $n \geq 2$.

77. A sequence of rational numbers is described as follows:

$$\frac{1}{1}, \frac{3}{2}, \frac{7}{5}, \frac{17}{12}, \ldots, \frac{a}{b}, \frac{a+2b}{a+b}, \ldots.$$

Here the numerators form one sequence, the denominators form a second sequence, and their ratios form a third sequence. Let x_n and y_n be, respectively, the numerator and the denominator of the nth fraction $r_n = x_n/y_n$.

a) Verify that $x_1^2 - 2y_1^2 = -1$, $x_2^2 - 2y_2^2 = +1$ and, more generally, that if $a^2 - 2b^2 = -1$ or $+1$, then

$$(a + 2b)^2 - 2(a + b)^2 = +1 \quad \text{or} \quad -1,$$

respectively.

b) The fractions $r_n = x_n/y_n$ approach a limit as n increases. What is that limit? (*Hint:* Use part (a) to show that $r_n^2 - 2 = \pm(1/y_n)^2$ and that y_n is not less than n.)

78. The **Fibonacci sequence** is defined recursively as follows: $x_1 = 1$, $x_2 = 1$, and $x_{n+2} = x_n + x_{n+1}$. It can also be described in another way that amounts to "zippering" two sequences together. (Although it may seem unnecessary, we found it easier to write a BASIC program doing it this way.) Let $a_n = x_{2n-1}$ and $b_n = x_{2n}$. Then the Fibonacci sequence can be written as

$$a_1, b_1, a_2, b_2, a_3, b_3, \ldots, a_k, b_k, a_{k+1}, b_{k+1}, \ldots.$$

The recursion formulas now become $a_1 = b_1 = 1$ and $a_{n+1} = a_n + b_n$, $b_{n+1} = a_{n+1} + b_n$. Verify that the following BASIC program will thus give the first 12 terms of the Fibonacci sequence, and fill in the rest of the table. (You do not need a computer to do the calculations.) The sequence is named for Leonardo Fibonacci (c. 1170–1240), who used the sequence to model successive generations of a rabbit population.

PROGRAM		n	a_n	b_n
10	a = 1	1	1	1
20	b = 1	2	2	
30	FOR n = 1 to 6	3		
40	PRINT n,a,b	4		
50	a = a + b	5		
60	b = a + b	6		
70	NEXT n			
80	END			

Note: The a on the right in line 60 comes from the a on the left in line 50.

79. *Sequences generated by Newton's method.* Newton's method, applied to a differentiable function $f(x)$, begins with a starting value x_0 and constructs from it a sequence of numbers $\{x_n\}$ that under favorable circumstances converges to a zero of f. The recursion formula for the sequence is

$$x_{n+1} = x_n - \frac{f(x_n)}{f'(x_n)}.$$

Do the following sequences converge? If so, to what value? In each case, begin by identifying the function f that generates the sequence.

a) $x_0 = 1, \quad x_{n+1} = x_n - \dfrac{x_n^2 - 2}{2x_n} = \dfrac{x_n}{2} + \dfrac{1}{x_n}$

b) $x_0 = 1, \quad x_{n+1} = x_n - \dfrac{\tan x_n - 1}{\sec^2 x_n}$

c) $x_0 = 1, \quad x_{n+1} = x_n - 1$

80. CALCULATOR Newton's method uses the formula $x_{n+1} = (x_n + a/x_n)/2$ to generate a sequence of approximations to the positive solution of the equation $x^2 - a = 0$, $a > 0$. Starting with $x_0 = 1$ and $a = 3$, calculate the successive terms of the sequence until you have approximated $\sqrt{3}$ as accurately as your calculator permits.

81. *Pythagorean triples.* A triple of positive integers a, b, and c is called a **Pythagorean triple** if $a^2 + b^2 = c^2$. Let a be an odd positive integer and let

$$b = \left\lfloor \frac{a^2}{2} \right\rfloor \quad \text{and} \quad c = \left\lceil \frac{a^2}{2} \right\rceil$$

be, respectively, the integer floor and ceiling for $a^2/2$.

a) Show that $a^2 + b^2 = c^2$ (Fig. 8.9). (*Hint:* Let $a = 2n + 1$ and express b and c in terms of n.)

b) By direct calculation, or by appealing to Fig. 8.9, find

$$\lim_{a\to\infty} \frac{\left\lfloor \dfrac{a^2}{2} \right\rfloor}{\left\lceil \dfrac{a^2}{2} \right\rceil}.$$

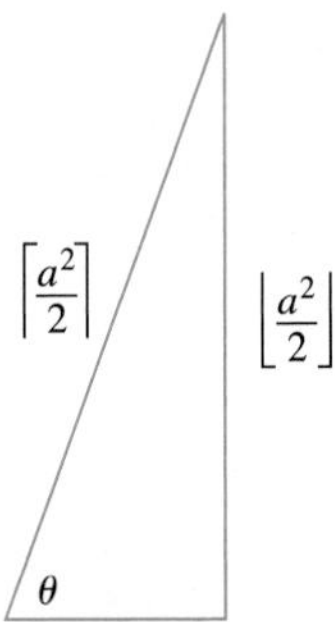

8.9 The right triangle for the Pythagorean triple in Exercise 81.

82. *The nth root of n!*

a) Show that $\lim_{n\to\infty} (2n\pi)^{1/(2n)} = 1$

and hence, using Stirling's approximation (Chapter 7, Miscellaneous Exercise 162a), that

$$\sqrt[n]{n!} \approx \frac{n}{e} \quad \text{for large values of } n.$$

b) CALCULATOR Test the approximation in (a) for $n = 40, 50, 60, \ldots$, as far as your calculator will allow.

83. Suppose that $f(x)$ is defined for all $0 \le x \le 1$, that f is differentiable, and that $f(0) = 0$. Define a sequence $\{a_n\}$ by the rule $a_n = nf(1/n)$. Show that $\lim_{n\to\infty} a_n = f'(0)$.

Use the result of Exercise 83 to find the limits of the sequences whose nth terms appear in Exercises 84–86.

84. $a_n = n \tan^{-1} \dfrac{1}{n}$ **85.** $a_n = n(e^{1/n} - 1)$

86. $a_n = n \ln \left(1 + \dfrac{2}{n}\right)$

87. a) Assuming that $\lim_{n\to\infty} (1/n^c) = 0$ if c is any positive constant, show that

$$\lim_{n\to\infty} \frac{\ln n}{n^c} = 0$$

if c is any positive constant.

b) Prove that $\lim_{n\to\infty} (1/n^c) = 0$ if c is any positive constant. (*Hint:* If $\epsilon = 0.001$ and $c = 0.04$, how large should N be to ensure that $|1/n^c - 0| < \epsilon$ if $n > N$?)

88. Prove Theorem 2.

89. Prove Theorem 3. (*Outline of proof:* Assume the hypotheses of the theorem and let ϵ be any positive number. For this ϵ there exists a $\delta > 0$ such that, for all x,

$$|x - L| < \delta \Rightarrow |f(x) - f(L)| < \epsilon.$$

For such a $\delta > 0$, there exists a positive integer N such that, for all n,

$$n > N \Rightarrow |a_n - L| < \delta.$$

What is the conclusion?)

90. *The zipper theorem.* Prove the "zipper theorem" for sequences: If $\{a_n\}$ and $\{b_n\}$ both converge to L, then the sequence

$$a_1, \quad b_1, \quad a_2, \quad b_2, \quad \ldots, \quad a_n, \quad b_n, \quad \ldots$$

converges to L.

✲ Picard's Method

CALCULATOR Use Picard's method to solve the equations in Exercises 91–96 to five decimal places.

91. $\sqrt{x} = x$ **92.** $x^2 = x$

93. $\cos x + x = 0$ **94.** $\cos x = x + 1$

95. $x - \sin x = 0.1$

96. $\sqrt{x} = 4 - \sqrt{1 + x}$ (*Hint:* Square both sides first.)

97. Solving the equation $\sqrt{x} = x$ by Picard's method finds the solution $x = 1$ but not the solution $x = 0$. Why? (*Hint:* Graph $y = x$ and $y = \sqrt{x}$ together.)

98. Solving the equation $x^2 = x$ by Picard's method finds the solution $x = 0$ but not the solution $x = 1$. Why? (*Hint:* Graph $y = x^2$ and $y = x$ together.)

99. *If $|g'(x)| > 1$, try g^{-1} instead.* There is more to be learned from Examples 13 and 15. Example 15 showed that we cannot apply Picard's method to find a fixed point of

$g(x) = 4x - 12$. But we can apply the method to find a fixed point of $g^{-1}(x) = (1/4)x + 3$ because the derivative of g^{-1} is 1/4, whose value is less than 1 in magnitude on any interval. In Example 13, we found the fixed point of g^{-1} to be $x = 4$. Now notice that 4 is also a fixed point of g, since

$$g(4) = 4(4) - 12 = 4.$$

In finding the fixed point of g^{-1}, we found the fixed point of g.

A function and its inverse always have the same fixed points. The graphs of the functions are symmetric about the line $y = x$ and therefore intersect the line at the same points.

We now see that the application of Picard's method is quite broad. For suppose g is one-to-one, with a continuous first derivative whose magnitude is greater than 1 on a closed interval I whose interior contains a fixed point of g. Then the derivative of g^{-1}, being the reciprocal of g', has magnitude less than 1 on I. Picard's method applied to g^{-1} on I will find the fixed point of g. As cases in point, find the fixed points of

a) $g(x) = 2x + 3$ b) $g(x) = 1 - 4x$

100. COMPUTER *The zeros of $x^2 - 2x - 3$.* Using a computer programmed for Picard's method (there is a program at the end of the exercises), try solving the equation $x^2 - 2x - 3 = 0$ in three different ways:

a) Write the equation as $x = \pm\sqrt{2x + 3}$ and take $g(x) = \pm\sqrt{2x + 3}$.

b) Write the equation as $x = (x^2 - 3)/2$ and take $g(x) = (x^2 - 3)/2$.

c) Write the equation as $x = 3/(x - 2)$ and take $g(x) = 3/(x - 2)$.

Explain what happens in each case.

EXPLORER PROGRAMS

Picard's Fixed Point Method	Generates graphs and iteration paths like those in Figs. 8.5, 8.6, and 8.7. In addition to being a powerful equation solver, the program enables you to toggle back and forth between the graphs of a function and its inverse.
Sequences and Series	Generates terms of one or two sequences and graphs them. You may define the sequences recursively or by giving formulas for their nth terms. Enables you to look for graphical and numerical evidence of convergence or divergence

✲ A COMPUTER PROGRAM FOR PICARD'S METHOD

The BASIC program listed below uses Picard's iteration method to solve the equation $g(x) = x$. You enter a formula for $g(x)$, a starting value for x, and a numerical tolerance (called TL) for how close $g(x)$ and x need to be before the iterations stop. The program also contains a limit (currently set at 100, but changeable) on how many iterations the computer should take before reporting nonconvergence.

Program Listing

Here is a listing of the program with $g(x) = \sqrt{2x + 3}$, to find the positive root of $x^2 - 2x - 3 = 0$.

	LINE	COMMENT
10	DEF FNG(X) = SQR(2 * X + 3)	Enters $g(x) = \sqrt{2x + 3}$.
20	LIMIT = 100	The program will count 100 steps before reporting nonconvergence.
30	INPUT "ENTER STARTING VALUE (X) "; X	Prompt for entering the starting value.
40	INPUT "ENTER VALUE FOR TERMINATION "; TL	Tells the computer when it has come close enough to stop; 10^{-6} might be good.
50	FOR ITER = 1 TO LIMIT	Iterations start at 1, and end at LIMIT if TL is not reached

(continued)

LINE		COMMENT
60	PRINT "AT ITER NO ";ITER; " X = "; X; " G(X) = "; FNG(X)	
70	IF ABS(X − FNG(X)) < TL THEN 120	
80	X = FNG(X)	Redefines x for the next step.
90	NEXT ITER	Closes the FOR NEXT loop.
100	PRINT "DID NOT CONVERGE IN "; LIMIT; " ITERATIONS"	
110	STOP	
120	PRINT "CONVERGED IN "; ITER; " ITERATIONS"	
130	END	

8.2 Infinite Series

In mathematics and science we often use infinite polynomials like

$$1 + x + x^2 + x^3 + \cdots + x^n + \cdots$$

to represent functions (the series above represents $1/(1 - x)$ for $|x| < 1$), evaluate nonelementary integrals, and solve differential equations. For any particular value of x, such a polynomial is calculated as an infinite sum of constants, a sum we call an *infinite series.* The goal of this and the next three sections is to learn to work with infinite series. Then, in Sections 8.6–8.8, we shall build on what we have learned to study infinite polynomials as infinite series of powers of x.

We begin by asking how to assign meaning to an expression like

$$1 + \frac{1}{2} + \frac{1}{4} + \frac{1}{8} + \frac{1}{16} + \cdots.$$

The way to do so is not to try to add all the terms at once (we cannot) but rather to add the terms one at a time from the beginning and look for a pattern in how these partial sums grow. When we do this, we find the following.

Partial sum		Value	
first:	$s_1 = 1$	1	
second:	$s_2 = 1 + \frac{1}{2}$	$2 - \frac{1}{2}$	
third:	$s_3 = 1 + \frac{1}{2} + \frac{1}{4}$	$2 - \frac{1}{4}$	
⋮			
nth:	$s_n = 1 + \frac{1}{2} + \frac{1}{4} + \cdots + \frac{1}{2^{n-1}}$	$2 - \frac{1}{2^{n-1}}$	(After some algebra)

Indeed there is a pattern. The partial sums form a sequence whose nth term is

$$s_n = 2 - \frac{1}{2^{n-1}},$$

and this sequence converges to 2. We say

"the sum of the series $1 + \frac{1}{2} + \frac{1}{4} + \cdots + \frac{1}{2^{n-1}} + \cdots$ is 2."

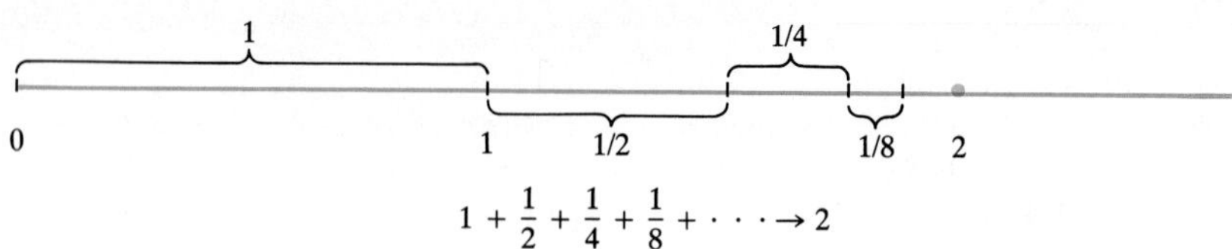

8.10 As the lengths 1, 1/2, 1/4, 1/8, . . . , are added one by one, the sum approaches 2.

Is the sum of any finite number of terms in the series 2? No. Can we actually add an infinite number of terms one by one? No. But we can still define their sum by defining it to be the limit of the sequence of partial sums as $n \to \infty$, in this case 2 (Fig. 8.10). Our knowledge of sequences and limits enables us to break away from the confines of finite sums.

DEFINITIONS

Given a sequence of numbers $\{a_n\}$, an expression of the form

$$a_1 + a_2 + a_3 + \cdots + a_n + \cdots \tag{1}$$

is an **infinite series.** The number a_n is the ***n*th term** of the series. The sequence $\{s_n\}$ defined by

$$\begin{aligned} s_1 &= a_1 \\ s_2 &= a_1 + a_2 \\ &\vdots \\ s_n &= a_1 + a_2 + \cdots + a_n = \sum_{k=1}^{n} a_k \end{aligned} \tag{2}$$

is the **sequence of partial sums** of the series, the number s_n being the ***n*th partial sum.** If the sequence of partial sums converges to a limit L, we say that the series **converges** and that its **sum** is L. In this case, we also write

$$a_1 + a_2 + \cdots + a_n + \cdots = \sum_{n=1}^{\infty} a_n = L. \tag{3}$$

If the sequence of partial sums of the series does not converge, we say that the series **diverges.**

When we begin to study a given series $a_1 + a_2 + \cdots + a_n + \cdots$, we might not know whether it converges or diverges. In either case, it is convenient to use sigma notation to write the series as

$$\sum_{n=1}^{\infty} a_n, \qquad \sum_{k=1}^{\infty} a_k, \qquad \text{or} \qquad \sum a_n.$$

The first of these is read "summation from n equals 1 to infinity of a_n"; the second as "summation from k equals 1 to infinity of a_k"; and the third as "summation a_n."

Geometric Series

Geometric series are series of the form

$$a + ar + ar^2 + \cdots + ar^{n-1} + \cdots = \sum_{n=1}^{\infty} ar^{n-1} \tag{4}$$

in which a and r are fixed real numbers and $a \neq 0$. The **ratio** r can be positive, as in

$$1 + \frac{1}{2} + \frac{1}{4} + \cdots + \frac{1}{2^{n-1}} + \cdots, \tag{5}$$

or negative, as in

$$1 - \frac{1}{3} + \frac{1}{9} - \cdots + (-1)^{n-1}\frac{1}{3^{n-1}} + \cdots. \tag{6}$$

If $r = 1$, the nth partial sum of the geometric series in (4) is

$$s_n = a + a(1) + a(1)^2 + \cdots + a(1)^{n-1} = na,$$

and the series diverges because $\lim_{n\to\infty} s_n = \pm\infty$, depending on the sign of a. If $r \neq 1$, we can determine the convergence or divergence of the series in the following way. We multiply the nth partial sum

$$s_n = a + ar + ar^2 + \cdots + ar^{n-1}$$

by r, obtaining

$$rs_n = ar + ar^2 + \cdots + ar^{n-1} + ar^n.$$

We then subtract rs_n from s_n. Most of the terms on the right cancel when we do this, leaving only

$$s_n - rs_n = a - ar^n \qquad \text{or} \qquad s_n(1 - r) = a(1 - r^n). \tag{7}$$

We solve for s_n, obtaining

$$s_n = \frac{a(1 - r^n)}{(1 - r)} \qquad (r \neq 1). \tag{8}$$

If $|r| < 1$, then $r^n \to 0$ as $n \to \infty$ (as we saw in Section 8.1), and $s_n \to a/(1 - r)$. In other words, the series converges to $a/(1 - r)$. If $|r| > 1$, then $|r^n| \to \infty$ and the series diverges.

If $|r| < 1$, the geometric series converges and

$$\sum_{n=1}^{\infty} ar^{n-1} = \frac{a}{1 - r}. \tag{9}$$

If $|r| \geq 1$, the series diverges.

Example 1 The geometric series with $a = 1/9$ and $r = 1/3$ is

$$\frac{1}{9} + \frac{1}{27} + \frac{1}{81} + \cdots = \frac{1}{9}\left(1 + \frac{1}{3} + \frac{1}{3^2} + \cdots\right) = \frac{1/9}{1 - (1/3)} = \frac{1}{6}.$$

Example 2 The geometric series with $a = 4$ and $r = -1/2$ is

$$4 - 2 + 1 - \frac{1}{2} + \frac{1}{4} - \cdots = 4\left(1 - \frac{1}{2} + \frac{1}{4} - \frac{1}{8} + \frac{1}{16} - \cdots\right)$$

$$= \frac{4}{1 + (1/2)} = \frac{8}{3}.$$

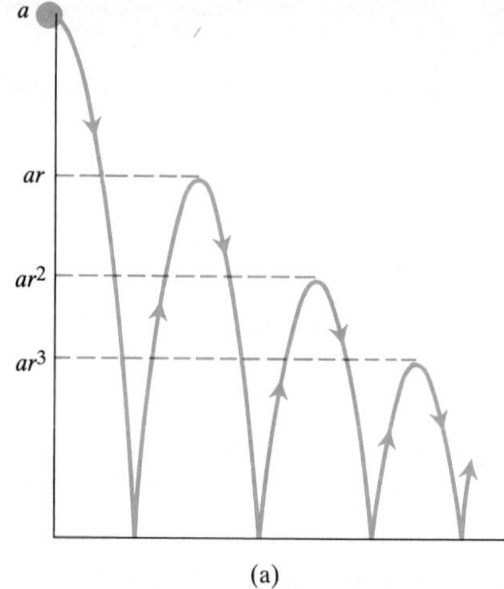

(a)

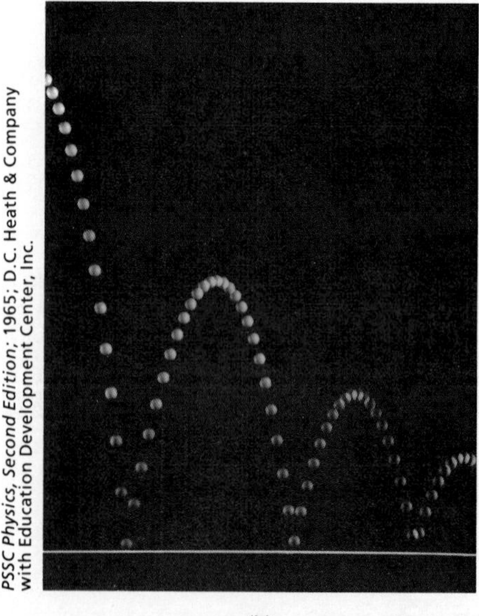

(b)

8.11 (a) Example 3 shows how to use a geometric series to calculate the total vertical distance traveled by a bouncing ball if the height of each rebound is reduced by the factor r. (b) A stroboscopic photo of a bouncing ball.

Example 3 You drop a ball from a meters above a flat surface. Each time the ball hits the surface after falling a distance h, it rebounds a distance rh, where r is positive but less than 1. Find the total distance the ball travels up and down (Fig. 8.11).

Solution The total distance is

$$s = a + \underbrace{2ar + 2ar^2 + 2ar^3 + \cdots}_{\text{This sum is } 2ar/(1-r).} = a + \frac{2ar}{1-r} = a\frac{1+r}{1-r}.$$

If $a = 6$ m and $r = 2/3$, for instance, the distance is

$$s = 6\frac{1 + (2/3)}{1 - (2/3)} = 6\left(\frac{5/3}{1/3}\right) = 30 \text{ m}.$$

Repeating Decimals

We use geometric series to explain why repeating decimals represent rational numbers.

Example 4 Express the repeating decimal 5.23 23 23 . . . as the ratio of two integers.

Solution

$$\begin{aligned} 5.23\ 23\ 23\ldots &= 5 + \frac{23}{100} + \frac{23}{(100)^2} + \frac{23}{(100)^3} + \cdots \\ &= 5 + \frac{23}{100}\underbrace{\left(1 + \frac{1}{100} + \left(\frac{1}{100}\right)^2 + \cdots\right)}_{1/(1-0.01)} \\ &= 5 + \frac{23}{100}\left(\frac{1}{0.99}\right) = 5 + \frac{23}{99} = \frac{518}{99}. \end{aligned}$$

Sums of Other Convergent Series

For geometric series we get a closed form for the nth partial sum and, from that, the formula $s = a/(1 - r)$. Unfortunately, such formulas are rare. What we usually have to do is test the series for convergence and then find some way to estimate the sums of the series that pass the test. Sometimes a convergent series represents a known function like $\sin x$ or $\ln x$. In other cases we might need a computer or calculator to estimate the sum.

The next example, however, is another of those rare cases in which we can find the series' sum from a formula for s_n.

Example 5 Determine whether $\sum_{n=1}^{\infty} [1/(n(n+1))]$ converges. If it does, find the sum.

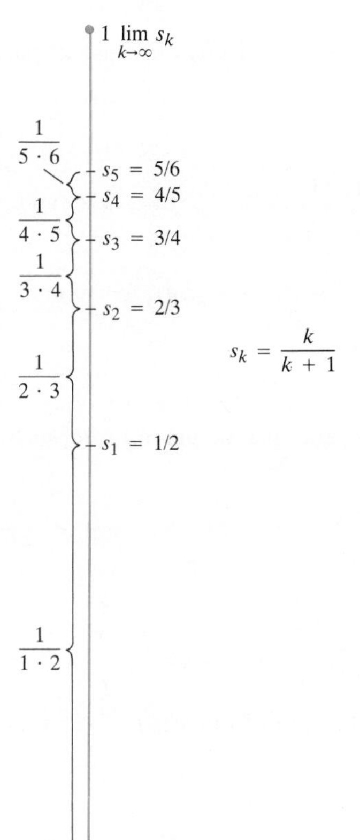

8.12 The sum of the first k terms of the series

$$\sum_{n=1}^{\infty} \frac{1}{n(n+1)}$$

in Example 5 is $k/(k+1)$, and the sum of the series is

$$\lim_{k\to\infty} \frac{k}{k+1} = 1.$$

Solution We look for a pattern in the sequence of partial sums that might lead us to a closed expression for s_k. The key to success here, as in the integration

$$\int \frac{dx}{x(x+1)} = \int \frac{dx}{x} - \int \frac{dx}{x+1},$$

is partial fractions. The observation that

$$\frac{1}{k(k+1)} = \frac{1}{k} - \frac{1}{k+1} \tag{10}$$

permits us to write the partial sum

$$\sum_{n=1}^{k} \frac{1}{n(n+1)} = \frac{1}{1\cdot 2} + \frac{1}{2\cdot 3} + \cdots + \frac{1}{k\cdot(k+1)}$$

as

$$s_k = \left(\frac{1}{1} - \frac{1}{2}\right) + \left(\frac{1}{2} - \frac{1}{3}\right) + \cdots + \left(\frac{1}{k} - \frac{1}{k+1}\right). \tag{11}$$

Removing parentheses and canceling the terms of opposite sign collapses the sum to

$$s_k = 1 - \frac{1}{k+1}. \tag{12}$$

We then see that $s_k \to 1$ as $k \to \infty$. The series converges, and its sum is 1 (Fig. 8.12).

$$\sum_{n=1}^{\infty} \frac{1}{n(n+1)} = 1.$$

Divergent Series

Geometric series with $|r| \geq 1$ are not the only series to diverge.

Example 6 The series

$$\sum_{n=1}^{\infty} n^2 = 1 + 4 + 9 + \cdots + n^2 + \cdots$$

diverges because the partial sums grow beyond every number L. After $n = 1$, the number $s_n = 1 + 4 + 9 + \cdots + n^2$ is greater than n^2.

Example 7 The series

$$\sum_{n=1}^{\infty} \frac{n+1}{n} = \frac{2}{1} + \frac{3}{2} + \frac{4}{3} + \cdots + \frac{n+1}{n} + \cdots$$

diverges because the partial sums eventually outgrow every preassigned number. Each term is greater than 1, so the sum of n terms is greater than n.

A series can diverge without its partial sums becoming large. The partial sums can oscillate between two extremes, for instance, as they do in the next example.

Example 8 The geometric series $\Sigma_{n=1}^{\infty}(-1)^{n+1}$ diverges because its partial sums alternate between 1 and 0:

$$s_1 = (-1)^2 = 1,$$
$$s_2 = (-1)^2 + (-1)^3 = 1 - 1 = 0,$$
$$s_3 = (-1)^2 + (-1)^3 + (-1)^4 = 1 - 1 + 1 = 1,$$

and so on.

The *n*th-Term Test for Divergence

We present a test for detecting the kind of divergence that occurs in Examples 6, 7, and 8.

> **The *n*th-Term Test for Divergence**
>
> If $\lim_{n\to\infty} a_n \neq 0$, or if $\lim_{n\to\infty} a_n$ fails to exist, then $\Sigma_{n=1}^{\infty} a_n$ diverges.

When we apply the *n*th-Term Test to the series in Examples 6, 7, and 8, we find that

$$\sum_{n=1}^{\infty} n^2 \quad \text{diverges because } n^2 \to \infty,$$

$$\sum_{n=1}^{\infty} \frac{n+1}{n} \quad \text{diverges because } \frac{n+1}{n} \to 1,$$

$$\sum_{n=1}^{\infty} (-1)^{n+1} \quad \text{diverges because } \lim_{n\to\infty} (-1)^{n+1} \text{ does not exist.}$$

The reason the *n*th-Term Test works is that $\lim_{n\to\infty} a_n$ must equal zero if $\Sigma\, a_n$ converges. To see why, let

$$s_n = a_1 + a_2 + \cdots + a_n$$

and suppose that $\Sigma\, a_n$ converges to S; that is, $s_n \to S$. When n is large, so is $n - 1$, and both s_n and s_{n-1} are close to S. Their difference, a_n, must then be close to zero. More formally, $a_n = s_n - s_{n-1} \to S - S = 0$.

Example 9 Determine whether each series converges or diverges. If it converges, find its sum.

a) $\displaystyle\sum_{n=1}^{\infty} \frac{n}{2n+5}$ b) $\displaystyle\sum_{n=1}^{\infty} \frac{5(-1)^n}{4^n}$

Solution

a) $\displaystyle\lim_{n\to\infty} \frac{n}{2n+5} = \frac{1}{2} \neq 0$. The series diverges by the *n*th-Term Test.

b) This is a geometric series with $a = -5/4$ and $r = -1/4$. It converges to

$$\frac{a}{1-r} = \frac{-5/4}{1+(1/4)} = -1.$$

A Necessary (but Not Sufficient) Condition for Convergence

We often state the nth-Term Test for divergence a shorter way.

> If $\Sigma_{n=1}^{\infty} a_n$ converges, then $a_n \to 0$.
>
> CAUTION This does *not* mean that Σa_n converges if $a_n \to 0$. A series Σa_n may diverge even though $a_n \to 0$. Thus, $\lim a_n = 0$ is a *necessary* but *not a sufficient* condition for the series Σa_n to converge.

Example 10 The series

$$1 + \underbrace{\frac{1}{2} + \frac{1}{2}}_{2 \text{ terms}} + \underbrace{\frac{1}{4} + \frac{1}{4} + \frac{1}{4} + \frac{1}{4}}_{4 \text{ terms}} + \cdots + \underbrace{\frac{1}{2^n} + \frac{1}{2^n} + \cdots + \frac{1}{2^n}}_{2^n \text{ terms}} + \cdots$$

diverges even though its terms form a sequence that converges to 0.

Whenever we have two convergent series, we can add them, subtract them, and multiply them by constants to make other convergent series. The next theorem gives the details.

THEOREM 5

> If $\Sigma a_n = A$ and $\Sigma b_n = B$, then
>
> 1. *Sum Rule:* $\sum (a_n + b_n) = A + B$
> 2. *Difference Rule:* $\sum (a_n - b_n) = A - B$
> 3. *Constant Multiple Rule:* $\sum ka_n = k \sum a_n = kA$ (Any number k)

As corollaries of Theorem 5 we have the following.

1. Every nonzero constant multiple of a divergent series diverges.
2. If Σa_n converges and Σb_n diverges, then $\Sigma(a_n + b_n)$ and $\Sigma(a_n - b_n)$ both diverge.

The proofs resemble the proofs of similar theorems discussed earlier, and we omit them.

Example 11

a) $$\sum_{n=1}^{\infty} \frac{4}{2^{n-1}} = 4\sum_{n=1}^{\infty} \frac{1}{2^{n-1}} = 4\frac{1}{1-(1/2)} = 8$$

b) $$\sum_{n=1}^{\infty} \frac{3^{n-1}-1}{6^{n-1}} = \sum_{n=1}^{\infty} \frac{1}{2^{n-1}} - \sum_{n=1}^{\infty} \frac{1}{6^{n-1}} = 2 - \frac{1}{1-(1/6)} = 2 - \frac{6}{5} = \frac{4}{5}$$

We can always add a finite number of terms to a series or delete a finite number of terms from a series without altering its convergence or divergence. If $\Sigma_{n=1}^{\infty} a_n$

converges, then $\Sigma_{n=k}^{\infty} a_n$ converges for any $k > 1$ and

$$\sum_{n=1}^{\infty} a_n = a_1 + a_2 + \cdots + a_{k-1} + \sum_{n=k}^{\infty} a_n. \tag{13}$$

Conversely, if $\Sigma_{n=k}^{\infty} a_n$ converges for any $k > 1$, then $\Sigma_{n=1}^{\infty} a_n$ converges. Thus,

$$\sum_{n=1}^{\infty} \frac{1}{5^n} = \frac{1}{5} + \frac{1}{25} + \frac{1}{125} + \sum_{n=4}^{\infty} \frac{1}{5^n} \tag{14}$$

and

$$\sum_{n=4}^{\infty} \frac{1}{5^n} = \left(\sum_{n=1}^{\infty} \frac{1}{5^n}\right) - \frac{1}{5} - \frac{1}{25} - \frac{1}{125}. \tag{15}$$

Reindexing

As long as we preserve the order of its terms, we can reindex any series without altering its convergence. To raise the starting value of the index h units, replace the n in the formula for a_n by $n - h$:

$$\sum_{n=1}^{\infty} a_n = \sum_{n=1+h}^{\infty} a_{n-h} = a_1 + a_2 + a_3 + \cdots.$$

To lower the starting value of the index h units, replace the n in the formula for a_n by $n + h$:

$$\sum_{n=1}^{\infty} a_n = \sum_{n=1-h}^{\infty} a_{n+h} = a_1 + a_2 + a_3 + \cdots.$$

It works like translation.

Example 12 We can write the geometric series that starts with

$$1 + \frac{1}{2} + \frac{1}{4} + \cdots$$

as

$$\sum_{n=0}^{\infty} \frac{1}{2^n}, \quad \sum_{n=5}^{\infty} \frac{1}{2^{n-5}}, \quad \text{or even} \quad \sum_{n=-4}^{\infty} \frac{1}{2^{n+4}}.$$

The partial sums remain the same no matter what indexing we choose.

We usually give preference to indexings that lead to simple expressions. We chose the indexing in Example 11(b) to begin with $n = 1$ to match the indexing in the geometric series formulas in Eqs. (4) and (9). Beginning with $n = 0$ instead would have produced the simpler formula

$$\sum_{n=0}^{\infty} \frac{3^n - 1}{6^n}.$$

EXERCISES 8.2

In Exercises 1–6, find a formula for the nth partial sum of each series and use it to find the series' sum if the series converges.

1. $2 + \frac{2}{3} + \frac{2}{9} + \frac{2}{27} + \cdots + \frac{2}{3^{n-1}} + \cdots$

2. $\frac{9}{100} + \frac{9}{100^2} + \frac{9}{100^3} + \cdots + \frac{9}{100^n} + \cdots$

3. $1 - \frac{1}{2} + \frac{1}{4} - \frac{1}{8} + \cdots + (-1)^{n-1}\frac{1}{2^{n-1}} + \cdots$

4. $1 - 2 + 4 - 8 + \cdots + (-1)^{n-1}2^{n-1} + \cdots$

5. $\frac{1}{2 \cdot 3} + \frac{1}{3 \cdot 4} + \frac{1}{4 \cdot 5} + \cdots + \frac{1}{(n+1)(n+2)} + \cdots$

6. $\frac{5}{1 \cdot 2} + \frac{5}{2 \cdot 3} + \frac{5}{3 \cdot 4} + \cdots + \frac{5}{n(n+1)} + \cdots$

In Exercises 7–14, write out the first few terms of each series to show how the series starts. Then find the sum of the series.

7. $\sum_{n=0}^{\infty} \frac{(-1)^n}{4^n}$

8. $\sum_{n=2}^{\infty} \frac{1}{4^n}$

9. $\sum_{n=1}^{\infty} \frac{7}{4^n}$

10. $\sum_{n=0}^{\infty} (-1)^n \frac{5}{4^n}$

11. $\sum_{n=0}^{\infty} \left(\frac{5}{2^n} + \frac{1}{3^n}\right)$

12. $\sum_{n=0}^{\infty} \left(\frac{5}{2^n} - \frac{1}{3^n}\right)$

13. $\sum_{n=0}^{\infty} \left(\frac{1}{2^n} + \frac{(-1)^n}{5^n}\right)$

14. $\sum_{n=0}^{\infty} \left(\frac{2^{n+1}}{5^n}\right)$

Use partial fractions to find the sum of each series in Exercises 15–18.

15. $\sum_{n=1}^{\infty} \frac{4}{(4n-3)(4n+1)}$

16. $\sum_{n=1}^{\infty} \frac{1}{(4n-3)(4n+1)}$

17. $\sum_{n=3}^{\infty} \frac{4}{(4n-3)(4n+1)}$

18. $\sum_{n=1}^{\infty} \frac{2n+1}{n^2(n+1)^2}$

Which series in Exercises 19–38 converge and which diverge? If a series converges, find its sum.

19. $\sum_{n=0}^{\infty} \left(\frac{1}{\sqrt{2}}\right)^n$

20. $\sum_{n=1}^{\infty} \ln \frac{1}{n}$

21. $\sum_{n=1}^{\infty} (-1)^{n+1} \frac{3}{2^n}$

22. $\sum_{n=1}^{\infty} (\sqrt{2})^n$

23. $\sum_{n=0}^{\infty} \cos n\pi$

24. $\sum_{n=0}^{\infty} \frac{\cos n\pi}{5^n}$

25. $\sum_{n=0}^{\infty} e^{-2n}$

26. $\sum_{n=1}^{\infty} \frac{n^2+1}{n}$

27. $\sum_{n=1}^{\infty} (-1)^{n+1} n$

28. $\sum_{n=1}^{\infty} \frac{2}{10^n}$

29. $\sum_{n=0}^{\infty} \frac{2^n - 1}{3^n}$

30. $\sum_{n=1}^{\infty} \left(1 - \frac{1}{n}\right)^n$

31. $\sum_{n=0}^{\infty} \frac{n!}{1000^n}$

32. $\sum_{n=0}^{\infty} \frac{1}{x^n}, \quad |x| > 1$

33. $\sum_{n=0}^{\infty} \left(\frac{e}{\pi}\right)^n$

34. $\sum_{n=0}^{\infty} \left(\frac{\ln(3/2)}{\ln 2}\right)^n$

35. $\sum_{n=1}^{\infty} \ln \left(\frac{n}{n+1}\right)$

36. $\sum_{n=1}^{\infty} \ln \left(\frac{n}{2n+1}\right)$

37. $\sum_{n=1}^{\infty} \left(\frac{1}{\sqrt{n}} - \frac{1}{\sqrt{n+1}}\right)$

38. $\sum_{n=0}^{\infty} \frac{e^{n\pi}}{\pi^{ne}}$

The series in Exercises 39–42 are geometric series. Find a and r in each case. Express the inequality $|r| < 1$ in terms of x and find the values of x for which the inequality holds and the series converges.

39. $\frac{1}{1+x} = \sum_{n=0}^{\infty} (-1)^n x^n$

40. $\frac{1}{1+x^2} = \sum_{n=0}^{\infty} (-1)^n x^{2n}$

41. $\frac{6}{3-x} = \sum_{n=0}^{\infty} 3\left(\frac{x-1}{2}\right)^n$

42. $\frac{2 + \sin x}{8 + 2\sin x} = \sum_{n=0}^{\infty} \frac{(-1)^n}{2}\left(\frac{1}{3 + \sin x}\right)^n$

In Exercises 43–48, find the values of x for which the given geometric series converges. Also, find the sum of the series (as a function of x) for those values of x.

43. $\sum_{n=0}^{\infty} 2^n x^n$

44. $\sum_{n=0}^{\infty} (-1)^n x^{2n}$

45. $\sum_{n=0}^{\infty} (-1)^n (x+1)^n$

46. $\sum_{n=0}^{\infty} \left(-\frac{1}{2}\right)^n (x-3)^n$

47. $\sum_{n=0}^{\infty} \sin^n x$

48. $\sum_{n=0}^{\infty} (\ln x)^n$

Express the numbers in Exercises 49–56 as ratios of integers.

49. $0.\overline{23} = 0.23\ 23\ 23\ldots$

50. $0.\overline{234} = 0.234\ 234\ 234\ldots$

51. $0.\overline{7} = 0.7777\ldots$

52. $0.\overline{d} = 0.dddd\ldots$, where d is a digit

53. $0.0\overline{6} = 0.06666\ldots$

54. $1.\overline{414} = 1.414\ 414\ 414\ldots$

55. $1.24\overline{123} = 1.24\ 123\ 123\ 123\ldots$

56. $3.\overline{142857} = 3.142857\ 142857\ldots$

57. A ball is dropped from a height of 4 m. Each time it strikes the pavement after falling from a height of h m it rebounds to a height of $0.75h$ m. Find the total distance the ball travels up and down.

58. *Continuation of Exercise 57.* Find the total number of seconds the ball in Exercise 57 is traveling. (*Hint:* The formula $s = 4.9t^2$ gives $t = \sqrt{s/4.9}$.)

59. The series in Exercise 5 can also be written as

$$\sum_{n=1}^{\infty} \frac{1}{(n+1)(n+2)} \quad \text{and} \quad \sum_{n=-1}^{\infty} \frac{1}{(n+3)(n+4)}.$$

Write it as a sum beginning with (a) $n = -2$, (b) $n = 0$, (c) $n = 5$.

60. The series in Exercise 6 can also be written as

$$\sum_{n=1}^{\infty} \frac{5}{n(n+1)} \quad \text{and} \quad \sum_{n=0}^{\infty} \frac{5}{(n+1)(n+2)}.$$

Write it as a sum beginning with (a) $n = -1$, (b) $n = 3$, (c) $n = 20$.

61. Find a closed-form expression for the nth partial sum of the series

$$\sum_{n=1}^{\infty} (-1)^{n+1}.$$

62. Find a closed-form expression for the nth partial sum of the series

$$\sum_{n=1}^{\infty} \ln\left(\frac{n}{n+1}\right).$$

(*Hint:* Write out the first few partial sums to see what is going on.)

63. Figure 8.13 shows the first five of an infinite sequence of squares. The outermost square has an area of 4 m^2. Each of the other squares is obtained by joining the midpoints of the sides of the square before it. Find the sum of the areas of all the squares.

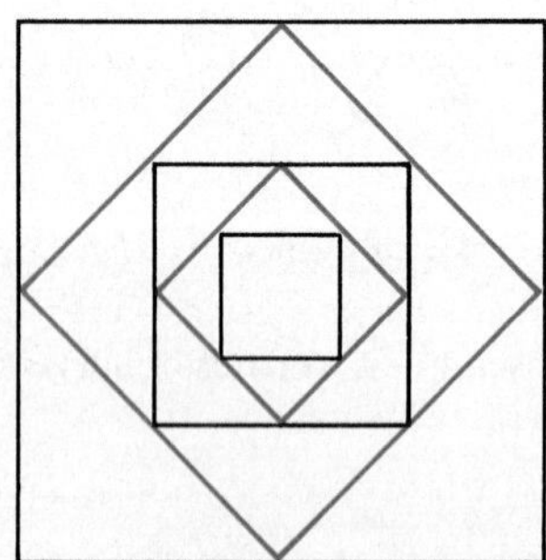

8.13 The first five squares in Exercise 63.

64. Figure 8.14 shows the first three rows of a sequence of rows of semicircles. There are 2^n semicircles in the nth row, each of radius $1/2^n$. Find the sum of the areas of all the semicircles.

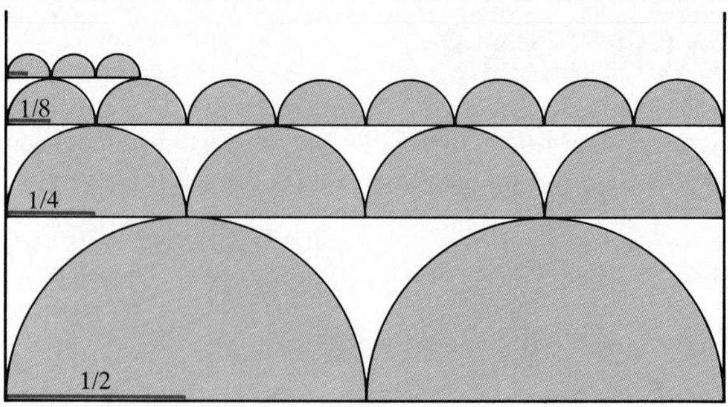

8.14 The semicircles in Exercise 64.

65. *Helga von Koch's snowflake curve.* Helga von Koch's snowflake (p. 144) is a curve of infinite length that encloses a region of finite area. To see why this is so, suppose the curve is generated by starting with an equilateral triangle whose sides have length 1.

a) Find the length L_n of the nth curve C_n and show that $\lim_{n\to\infty} L_n = \infty$.

b) Find the area A_n of the region enclosed by C_n and calculate $\lim_{n\to\infty} A_n$.

66. Make up an example of two divergent series whose term-by-term sum converges.

67. Show by example that $\Sigma\, (a_n/b_n)$ may diverge even though $\Sigma\, a_n$ and $\Sigma\, b_n$ converge and no b_n equals 0.

68. Show by example that $\Sigma\, (a_n/b_n)$ may converge to something other than A/B even when $A = \Sigma\, a_n$, $B = \Sigma\, b_n \neq 0$, and no b_n equals 0.

69. Find convergent geometric series $A = \Sigma\, a_n$ and $B = \Sigma\, b_n$ that illustrate the fact that $\Sigma\, a_n b_n$ may converge without being equal to AB.

70. Show that if $\Sigma\, a_n$ converges, and $a_n \neq 0$ for all n, then $\Sigma\, (1/a_n)$ diverges.

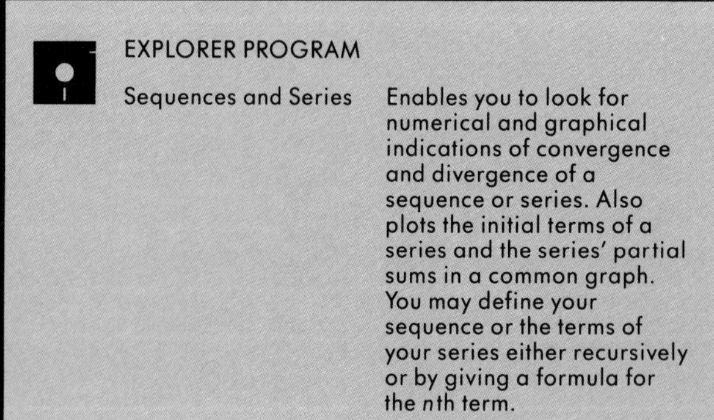

EXPLORER PROGRAM

Sequences and Series — Enables you to look for numerical and graphical indications of convergence and divergence of a sequence or series. Also plots the initial terms of a series and the series' partial sums in a common graph. You may define your sequence or the terms of your series either recursively or by giving a formula for the nth term.

8.3 Series Without Negative Terms: The Comparison and Integral Tests

Given a series Σa_n, we have two questions:

1. Does the series converge?
2. If it converges, what is its sum?

Much of the rest of this chapter is devoted to answering the first question. But as a practical matter, the second question is just as important; we come back to it later.

In this section and the next we study series that do not have negative terms. The reason for this restriction is that the partial sums of these series form nondecreasing sequences, and nondecreasing sequences *that are bounded from above* always converge. To show that a series of nonnegative terms converges, we need only show that there is some number beyond which the partial sums never go.

It may at first seem to be a drawback that this approach establishes the fact of convergence without producing the sum of the series in question. Surely it would be better to compute sums of series directly from nice formulas for their partial sums. But in most cases such formulas are not available, and in their absence we have to turn instead to the two-step procedure, as we said earlier, of first establishing convergence and then approximating the sum.

Nondecreasing Sequences

Suppose that Σa_n is an infinite series and that $a_n \geq 0$ for every n. Then, when we calculate the partial sums s_1, s_2, s_3, and so on, we see that each one is greater than or equal to its predecessor because $s_{n+1} = s_n + a_n$:

$$s_1 \leq s_2 \leq s_3 \leq \cdots \leq s_n \leq s_{n+1} \leq \cdots. \tag{1}$$

A sequence $\{s_n\}$ with the property that $s_n \leq s_{n+1}$ for every n is called a **nondecreasing sequence.**

There are two kinds of nondecreasing sequences—those that increase beyond any finite bound and those that don't. The former diverge to infinity, so we turn our attention to the other kind: those that do not grow beyond all bounds. Such a sequence is said to be **bounded from above,** and any number M such that $s_n \leq M$ for all n is called an **upper bound** of the sequence.

Example 1 If $s_n = n/(n+1)$, then 1 is an upper bound and so is any number greater than 1. No number smaller than 1 is an upper bound, so for this sequence 1 is the **least upper bound.**

A nondecreasing sequence that is bounded from above always has a least upper bound, but we shall not prove this fact. Instead, we shall prove that if L is the least upper bound, then the sequence converges to L. The following argument shows why L is the limit.

Suppose we plot the points $(1, s_1)$, $(2, s_2)$, $\ldots$, (n, s_n) in the xy-plane. If M is an upper bound of the sequence, all these points will lie on or below the line $y = M$ (Fig. 8.15). The line $y = L$ is the lowest such line. None of the points (n, s_n) lies above $y = L$, but some do lie above any lower line $y = L - \epsilon$, if ϵ is a positive number. The sequence converges to L because

a) $s_n \leq L$ for *all* values of n and

b) given any $\epsilon > 0$, there exists at least one integer N for which $s_N > L - \epsilon$.

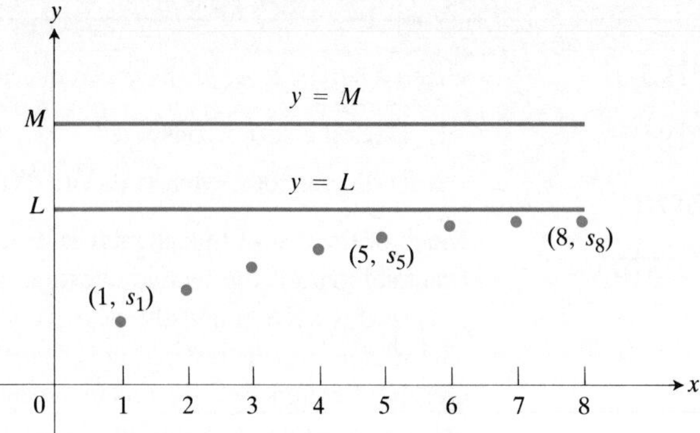

8.15 If the terms of a nondecreasing sequence have an upper bound M, they have a limit $L \leq M$.

The fact that $\{s_n\}$ is a nondecreasing sequence tells us further that

$$s_n \geq s_N > L - \epsilon \qquad \text{for all } n \geq N.$$

This means that *all* the numbers s_n beyond the Nth number lie within ϵ of L. This is precisely the condition for L to be the limit of the sequence s_n.

The facts for nondecreasing sequences are summarized in the following theorem. A similar result holds for nonincreasing sequences (Exercise 39).

THEOREM 6

The Nondecreasing Sequence Theorem

A nondecreasing sequence converges if and only if its terms are bounded from above. If all the terms are less than or equal to M, then the limit of the sequence is less than or equal to M as well.

Theorem 6 tells us that we can show that a series $\Sigma\, a_n$ of nonnegative terms converges if we can show that its partial sums are bounded from above. The question is how to find out in any particular instance whether the s_n's have an upper bound.

Sometimes we can show that the s_n's are bounded above by showing that each one is less than or equal to the corresponding partial sum of a series that is already known to converge. The next example shows how this can happen.

Example 2 The series

$$\sum_{n=0}^{\infty} \frac{1}{n!} = 1 + \frac{1}{1!} + \frac{1}{2!} + \frac{1}{3!} + \cdots \tag{2}$$

converges because its terms are all positive and less than or equal to the corresponding terms of

$$1 + \sum_{n=0}^{\infty} \frac{1}{2^n} = 1 + 1 + \frac{1}{2} + \frac{1}{2^2} + \cdots. \tag{3}$$

To see how this relationship leads to an upper bound for the partial sums of $\Sigma_{n=0}^{\infty}\,(1/(n!))$, let

$$s_n = 1 + \frac{1}{1!} + \frac{1}{2!} + \cdots + \frac{1}{n!}$$

and observe that, for each n,

$$s_n \leq 1 + 1 + \frac{1}{2} + \frac{1}{2^2} + \cdots + \frac{1}{2^n} < 1 + \sum_{n=0}^{\infty} \frac{1}{2^n} = 1 + \frac{1}{1 - (1/2)} = 3.$$

Thus the partial sums of $\Sigma_{n=0}^{\infty} (1/(n!))$ are all less than 3, so $\Sigma_{n=0}^{\infty} (1/(n!))$ converges.

The fact that 3 is an upper bound for the partial sums of $\Sigma_{n=0}^{\infty} (1/(n!))$ does not necessarily mean that the series converges to 3. As we shall see in Section 8.7, the series converges to e.

Nicole Oresme (1320–1382)

The argument we use to show the divergence of the harmonic series was devised by the French theologian, mathematician, physicist, and bishop Nicole Oresme (pronounced "or-*rem*"). Oresme was a vigorous opponent of astrology, a dynamic preacher, an adviser of princes, a friend of King Charles V, a popularizer of science, and a skillful translator of Latin into French. Oresme did not believe in Albert of Saxony's generally accepted model of free fall (Chapter 6, Miscellaneous Exercise 98) but preferred Aristotle's constant-acceleration model, the model that became popular among Oxford scholars in the 1330s and that Galileo eventually used three hundred years later.

Example 3 *The Harmonic Series.* The series

$$\sum_{1}^{\infty} \frac{1}{n} = 1 + \frac{1}{2} + \frac{1}{3} + \cdots + \frac{1}{n} + \cdots$$

is called the **harmonic series.** It diverges because there is no upper bound for its sequence of partial sums. To see why, imagine grouping the terms of the series in the following way:

$$1 + \frac{1}{2} + \underbrace{\left(\frac{1}{3} + \frac{1}{4}\right)}_{> \frac{2}{4} = \frac{1}{2}} + \underbrace{\left(\frac{1}{5} + \frac{1}{6} + \frac{1}{7} + \frac{1}{8}\right)}_{> \frac{4}{8} = \frac{1}{2}} + \underbrace{\left(\frac{1}{9} + \frac{1}{10} + \cdots + \frac{1}{16}\right)}_{> \frac{8}{16} = \frac{1}{2}} + \cdots.$$

The sum of the first two terms is 1.5. The sum of the next two terms is $1/3 + 1/4$, which is greater than $1/4 + 1/4 = 1/2$. The sum of the next four terms is $1/5 + 1/6 + 1/7 + 1/8$, which is greater than $1/8 + 1/8 + 1/8 + 1/8 = 1/2$. The sum of the next eight terms is $1/9 + 1/10 + 1/11 + 1/12 + 1/13 + 1/14 + 1/15 + 1/16$, which is greater than $8/16 = 1/2$. The sum of the next 16 terms is greater than $16/32 = 1/2$, and so on. In general, the sum of 2^n terms ending with $1/2^{n+1}$ is greater than $2^n/2^{n+1} = 1/2$. The sequence of partial sums is not bounded: If $n = 2^k$, the partial sum s_n is greater than $k/2$. The harmonic series diverges. (We shall see later that the nth partial sum is slightly greater than $\ln(n + 1)$.)

Notice that the nth-Term Test for divergence does not detect the divergence of the harmonic series. The nth term, $1/n$, goes to zero but the series still diverges.

Comparison Test for Convergence

We established the convergence of the series in Example 2 by comparing it with a series that was already known to converge. This kind of comparison is typical of a procedure called the Comparison Test for convergence of series of nonnegative terms.

Comparison Test for Series of Nonnegative Terms

Let $\Sigma\, a_n$ be a series with no negative terms.

a) **Test for convergence.** The series $\Sigma\, a_n$ converges if there is a convergent series $\Sigma\, c_n$ with $a_n \leq c_n$ for all $n > n_0$, for some positive integer n_0.

b) **Test for divergence.** The series $\Sigma\, a_n$ diverges if there is a divergent series of nonnegative terms $\Sigma\, d_n$ with $a_n \geq d_n$ for all $n > n_0$.

In part (a), the partial sums of the series $\Sigma\, a_n$ are bounded above by

$$M = a_1 + a_2 + \cdots + a_{n_0} + \sum_{n=n_0+1}^{\infty} c_n .$$

They therefore form a nondecreasing sequence with a limit L that is less than or equal to M.

In part (b), the partial sums for $\Sigma\, a_n$ are not bounded from above. If they were, the partial sums for $\Sigma\, d_n$ would be bounded by

$$M' = d_1 + d_2 + \cdots + d_{n_0} + \sum_{n=n_0+1}^{\infty} a_n$$

and $\Sigma\, d_n$ would have to converge instead of diverge.

To apply the Comparison Test to a series, we do not have to include the early terms of the series. We can start the test with any index N, provided we include all the terms of the series being tested from there on.

Example 4 We can establish the convergence of the series

$$5 + \frac{2}{3} + 1 + \frac{1}{7} + \frac{1}{2} + \frac{1}{3!} + \frac{1}{4!} + \cdots + \frac{1}{k!} + \cdots$$

by ignoring the first four terms and comparing the remainder with the convergent geometric series

$$\sum_{n=1}^{\infty} \frac{1}{2^n} = \frac{1}{2} + \frac{1}{4} + \frac{1}{8} + \cdots .$$

To apply the Comparison Test, we need to have on hand a list of series we already know about. Here is what we know so far:

Convergent series	Divergent series
Geometric series with $\lvert r \rvert < 1$	Geometric series with $\lvert r \rvert \geq 1$
Telescoping series like $\sum_{n=1}^{\infty} \frac{1}{n(n+1)}$	The harmonic series $\sum_{n=1}^{\infty} \frac{1}{n}$
The series $\sum_{n=0}^{\infty} \frac{1}{n!}$	Any series $\Sigma\, a_n$ with $\lim a_n \neq 0$

The next test, the Integral Test, will add some series to these lists.

The Integral Test

We introduce the Integral Test with a specific example, a series that is related to the harmonic series, but in which the nth term is $1/n^2$ instead of $1/n$.

Example 5 Does the series

$$\sum_{n=1}^{\infty} \frac{1}{n^2} = 1 + \frac{1}{4} + \frac{1}{9} + \frac{1}{16} + \cdots + \frac{1}{n^2} + \cdots \tag{4}$$

converge, or does it diverge?

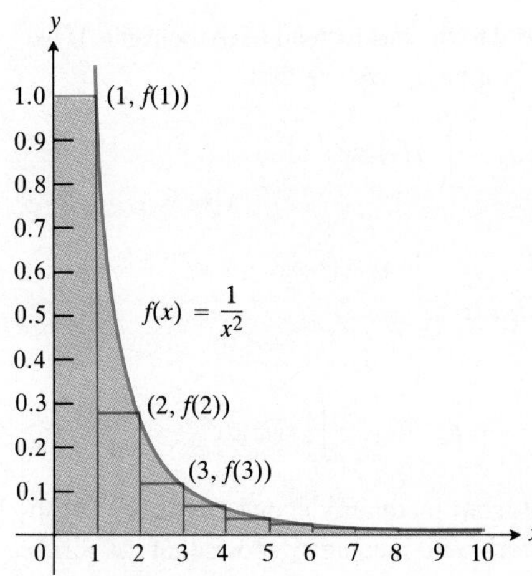

8.16 When $f(x) = 1/x^2$,

$$\begin{aligned} s_n &= f(1) + f(2) + \cdots + f(n) \\ &< f(1) + \int_1^n f(x)\,dx \\ &= 1 + \left(\frac{1}{1} - \frac{1}{n}\right) < 2. \end{aligned}$$

Solution When we studied improper integrals in Chapter 7, we learned that

$$\int_1^{\infty} \frac{1}{x}\,dx \text{ diverges} \qquad \text{and} \qquad \int_1^{\infty} \frac{1}{x^2}\,dx \text{ converges.}$$

If we can show that the sequence of partial sums of series (4) is bounded above, we can conclude that the series converges. Figure 8.16 suggests how we can find an upper bound for

$$s_n = \frac{1}{1^2} + \frac{1}{2^2} + \frac{1}{3^2} + \cdots + \frac{1}{n^2} = f(1) + f(2) + f(3) + \cdots + f(n),$$

where $f(x) = 1/x^2$.

The first term of the series is $f(1) = 1$, which we can interpret as the area of a rectangle of height $1/1^2$ and base the length of the interval $[0, 1]$ on the x-axis. The next rectangle, over the interval $[1, 2]$, has area $f(2) = 1/2^2$. That rectangle lies below the curve $y = 1/x^2$, so

$$f(2) = \frac{1}{2^2} < \int_1^2 \frac{1}{x^2}\,dx.$$

In the same way,

$$f(3) = \frac{1}{3^2} < \int_2^3 \frac{1}{x^2}\,dx, \ldots, f(n) = \frac{1}{n^2} < \int_{n-1}^{n} \frac{1}{x^2}\,dx. \tag{5}$$

Adding these inequalities, we get

$$f(2) + f(3) + \cdots + f(n) < \int_1^n (1/x^2)\,dx = -\frac{1}{x}\Big]_1^n = 1 - \frac{1}{n}.$$

To get s_n we must add $f(1) = 1$. When we do so, we have

$$s_n = f(1) + f(2) + f(3) + \cdots + f(n) < 2 - \frac{1}{n} < 2.$$

The sequence of partial sums $\{s_n\}$ is an increasing sequence that is bounded above, so it has a limit, and the given series converges. Its sum is known to be $\pi^2/6$, which is about 1.64493.

We now state and prove the Integral Test in more general terms.

Integral Test

Let $a_n = f(n)$ where $f(x)$ is a continuous, positive, decreasing function of x for all $x \geq 1$. Then the series $\Sigma\, a_n$ and the integral $\int_1^{\infty} f(x)\,dx$ both converge or both diverge.

Proof We start with the assumption that f is a decreasing function with $f(n) = a_n$ for every n. This leads us to observe that the rectangles in Fig. 8.17(a), which have areas $a_1, a_2, \ldots, a_n$, collectively enclose more area than that under the curve $y = f(x)$ from $x = 1$ to $x = n + 1$. That is,

$$\int_1^{n+1} f(x)\,dx \leq a_1 + a_2 + \cdots + a_n.$$

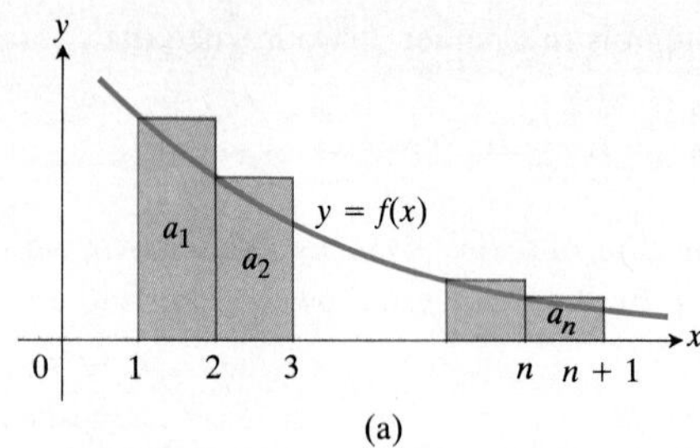

(a)

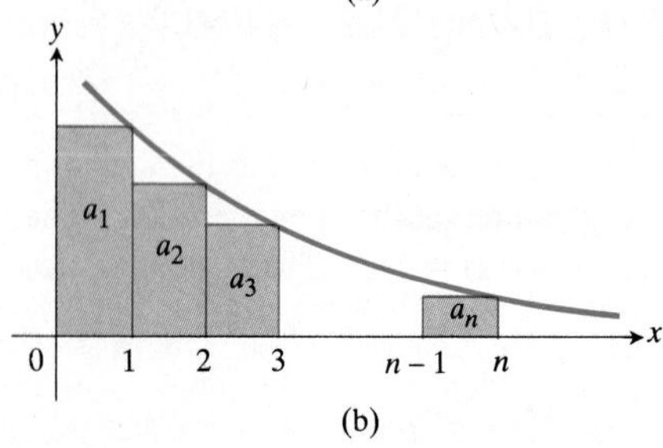

(b)

8.17 Subject to the conditions of the Integral Test, the series $\sum_{n=1}^{\infty} a_n$ and the integral $\int_1^\infty f(x)\,dx$ both converge or both diverge.

In Fig. 8.17(b) the rectangles have been faced to the left instead of to the right. If we momentarily disregard the first rectangle, of area a_1, we see that

$$a_2 + a_3 + \cdots + a_n \le \int_1^n f(x)\,dx.$$

If we include a_1, we have

$$a_1 + a_2 + \cdots + a_n \le a_1 + \int_1^n f(x)\,dx.$$

Combining these results gives

$$\int_1^{n+1} f(x)\,dx \le a_1 + a_2 + \cdots + a_n \le a_1 + \int_1^n f(x)\,dx. \tag{6}$$

If the integral $\int_1^\infty f(x)\,dx$ is finite, the right-hand inequality shows that $\Sigma\, a_n$ is also finite. But if $\int_1^\infty f(x)\,dx$ is infinite, then the left-hand inequality shows that the series is also infinite.

Hence the series and the integral are both finite or both infinite.

Example 6 *The p-series.* If p is a real constant, the series

$$\sum_{n=1}^{\infty} \frac{1}{n^p} = \frac{1}{1^p} + \frac{1}{2^p} + \frac{1}{3^p} + \cdots + \frac{1}{n^p} + \cdots \tag{7}$$

converges if $p > 1$ and diverges if $p \le 1$. To prove this, let $f(x) = 1/x^p$. Then, if $p > 1$, we have $-p + 1 < 0$ and

$$\int_1^\infty x^{-p}\,dx = \lim_{b\to\infty} \frac{x^{-p+1}}{-p+1}\bigg]_1^b = \frac{1}{1-p} \lim_{b\to\infty} (b^{-p+1} - 1) = \frac{1}{p-1},$$

which is finite. Hence the p-series converges if $p > 1$.

If $p = 1$, we have

$$1 + \frac{1}{2} + \frac{1}{3} + \cdots + \frac{1}{n} + \cdots,$$

which we already know diverges. Or, by the Integral Test,

$$\int_1^\infty x^{-1}dx = \lim_{b\to\infty} \ln x\bigg]_1^b = \infty,$$

and, since the integral diverges, the series diverges.

Finally, if $p < 1$, then the terms of the p-series are greater than the corresponding terms of the divergent harmonic series. Hence the p-series diverges, by the Comparison Test.

We have convergence for $p > 1$ but divergence for every other value of p.

The Limit Comparison Test

We now present a more powerful form of the Comparison Test, known as the Limit Comparison Test. It is particularly handy when we deal with series in which a_n is a rational function of n. The next example will show you what we mean.

Example 7 Do the following series converge, or diverge?

a) $\displaystyle\sum_{n=2}^{\infty} \frac{2n}{n^2 - n + 1}$ b) $\displaystyle\sum_{n=2}^{\infty} \frac{2n^3 + 100n^2 + 1000}{(1/8)n^6 - n + 2}$

Solution In determining convergence or divergence, only the tails count. And when n is very large, the highest powers of n in numerator and denominator are what count the most. So in (a), we reason this way:

$$a_n = \frac{2n}{n^2 - n + 1}$$

behaves approximately like $2n/n^2 = 2/n$, and, by comparing it with $\Sigma\ 1/n$, we guess that $\Sigma\ a_n$ diverges. In (b), we reason that a_n will behave approximately like $2n^3/(1/8)n^6 = 16/n^3$ and, by comparing it with $\Sigma\ 1/n^3$, a p-series with $p = 3$, we guess that the series converges.

To be more precise, in part (a) we take

$$a_n = \frac{2n}{n^2 - n + 1} \quad \text{and} \quad d_n = \frac{1}{n}$$

and look at the ratio

$$\frac{a_n}{d_n} = \frac{2n^2}{n^2 - n + 1} = \frac{2}{1 - \left(\frac{1}{n}\right) + \left(\frac{1}{n^2}\right)}.$$

Clearly, as $n \to \infty$, the limit is 2: $\lim\ (a_n/d_n) = 2$.

This means that, in particular, if we take $\epsilon = 1$ in the definition of limit, we know there is an integer N such that a_n/d_n is within 1 unit of this limit for all $n \geq N$:

$$2 - 1 \leq a_n/d_n \leq 2 + 1 \qquad \text{for} \quad n \geq N.$$

Thus $a_n \geq d_n$ for $n \geq N$. Therefore, by the Comparison Test, $\Sigma\ a_n$ diverges because $\Sigma\ d_n$ diverges.

In part (b), if we let $c_n = 1/n^3$, we can show that $\lim(a_n/c_n) = 16$.

Taking $\epsilon = 1$ in the definition of limit, we can conclude that there is an index N' such that a_n/c_n is between 15 and 17 when $n \geq N'$. Since $\Sigma\ c_n$ converges, so also does $\Sigma\ 17c_n$ and thus $\Sigma\ a_n$.

Our rather rough guesswork paved the way for successful choices of comparison series. We make all of this more precise in the following Limit Comparison Test.

Limit Comparison Test

a) **Test for convergence.** If $a_n \geq 0$ for $n \geq n_0$ and there is a convergent series $\Sigma\ c_n$ such that $c_n > 0$ and

$$\lim \frac{a_n}{c_n} < \infty, \tag{8}$$

then $\Sigma\ a_n$ converges.

b) **Test for divergence.** If $a_n \geq 0$ for $n \geq n_0$ and there is a divergent series $\Sigma\ d_n$ such that $d_n > 0$ and

$$\lim \frac{a_n}{d_n} > 0, \tag{9}$$

then $\Sigma\ a_n$ diverges.

A simpler version of the Limit Comparison Test combines parts (a) and (b) in the following way.

Simplified Limit Comparison Test

If the terms of the two series $\Sigma\, a_n$ and $\Sigma\, b_n$ are positive for $n \geq n_0$, and the limit of a_n/b_n is finite and positive, then both series converge or both diverge.

The Simplified Limit Comparison Test is the one we use most often.

Example 8 Which of the following series converge and which diverge?

a) $\frac{3}{4} + \frac{5}{9} + \frac{7}{16} + \frac{9}{25} + \cdots = \sum_{n=1}^{\infty} \frac{2n+1}{(n+1)^2}$

b) $\frac{101}{3} + \frac{102}{10} + \frac{103}{29} + \cdots = \sum_{n=1}^{\infty} \frac{100+n}{n^3+2}$

c) $\frac{1}{1} + \frac{1}{3} + \frac{1}{7} + \cdots = \sum_{n=1}^{\infty} \frac{1}{2^n - 1}$

Solution

a) Let $a_n = (2n+1)/(n^2+2n+1)$ and $d_n = 1/n$. Then

$$\sum d_n \text{ diverges} \qquad \text{and} \qquad \lim \frac{a_n}{d_n} = \lim \frac{2n^2+n}{n^2+2n+1} = 2,$$

so $\Sigma\, a_n$ diverges.

b) Let $a_n = (100+n)/(n^3+2)$. When n is large, this ought to compare with $n/n^3 = 1/n^2$, so we let $c_n = 1/n^2$ and apply the Limit Comparison Test:

$$\sum c_n \text{ converges} \qquad \text{and} \qquad \lim \frac{a_n}{c_n} = \lim \frac{n^3+100n^2}{n^3+2} = 1,$$

so $\Sigma\, a_n$ converges.

c) Let $a_n = 1/(2^n - 1)$ and $c_n = 1/2^n$. (We reason that $2^n - 1$ behaves somewhat like 2^n when n is large.) Then

$$\frac{a_n}{c_n} = \frac{2^n}{2^n - 1} = \frac{1}{1-(1/2)^n} \to 1 \qquad \text{as} \quad n \to \infty.$$

Because $\Sigma\, c_n$ converges, we conclude that $\Sigma\, a_n$ does also.

EXERCISES 8.3

Which series in Exercises 1–32 converge and which diverge? Give reasons for your answers.

1. $\sum_{n=1}^{\infty} \frac{1}{10^n}$ **2.** $\sum_{n=1}^{\infty} \frac{n}{n+2}$ **3.** $\sum_{n=1}^{\infty} \frac{\sin^2 n}{2^n}$ **4.** $\sum_{n=1}^{\infty} \frac{5}{n}$ **5.** $\sum_{n=1}^{\infty} \frac{1+\cos n}{n^2}$ **6.** $\sum_{n=1}^{\infty} -\frac{1}{8^n}$

7. $\sum_{n=2}^{\infty} \frac{\ln n}{n}$
8. $\sum_{n=1}^{\infty} \frac{1}{n\sqrt{n}}$
9. $\sum_{n=1}^{\infty} \frac{2^n}{3^n}$

10. $\sum_{n=0}^{\infty} \frac{-2}{n+1}$
11. $\sum_{n=1}^{\infty} \frac{1}{1+\ln n}$
12. $\sum_{n=1}^{\infty} \frac{1}{2n-1}$

13. $\sum_{n=1}^{\infty} \frac{2^n}{n+1}$
14. $\sum_{n=1}^{\infty} \left(\frac{n}{3n+1}\right)^n$
15. $\sum_{n=1}^{\infty} \frac{1}{\sqrt{n^3+2}}$

16. $\sum_{n=2}^{\infty} \frac{\sqrt{n}}{\ln n}$
17. $\sum_{n=1}^{\infty} \frac{n}{n^2+1}$
18. $\sum_{n=1}^{\infty} \frac{1}{n\sqrt[n]{n}}$

19. $\sum_{n=1}^{\infty} \left(1+\frac{1}{n}\right)^n$
20. $\sum_{n=1}^{\infty} \frac{\sqrt{n}}{n^2+1}$
21. $\sum_{n=1}^{\infty} \frac{1-n}{n \cdot 2^n}$

22. $\sum_{n=1}^{\infty} \frac{1}{(\ln 2)^n}$
23. $\sum_{n=1}^{\infty} \frac{1}{3^{n-1}+1}$

24. $\sum_{n=1}^{\infty} \frac{10n+1}{n(n+1)(n+2)}$

25. $\sum_{n=1}^{\infty} \frac{\tan^{-1} n}{n^{1.1}}$
26. $\sum_{n=2}^{\infty} n^2 e^{-n}$

27. $\sum_{n=1}^{\infty} \frac{1}{n(1+\ln^2 n)}$
28. $\sum_{n=3}^{\infty} \frac{1}{n(\ln n)\sqrt{(\ln n)^2-1}}$

29. $\sum_{n=1}^{\infty} \frac{1}{1+2+3+\cdots+n}$

30. $\sum_{n=1}^{\infty} \frac{n}{1+2^2+3^2+\cdots+n^2}$

31. $\sum_{n=1}^{\infty} \operatorname{sech} n$
32. $\sum_{n=1}^{\infty} \operatorname{sech}^2 n$

33. Determine whether the following series converges.

$$\sum_{n=1}^{\infty} \frac{1 \cdot 3 \cdot 5 \cdot \cdots \cdot (2n-1)}{2 \cdot 4 \cdot 6 \cdot \cdots \cdot (2n)}$$

You will see where this series comes from if you do Exercise 24 in Section 8.8.

34. For what value or values of a, if any, does the following series converge?

$$\sum_{n=1}^{\infty} \left(\frac{a}{n+2} - \frac{1}{n+4}\right)$$

35. CALCULATOR There is absolutely no empirical evidence for the divergence of the harmonic series even though we know it diverges. The partial sums, which satisfy the inequality

$$\ln(n+1) = \int_1^{n+1} \frac{1}{x}\,dx \le 1 + \frac{1}{2} + \cdots + \frac{1}{n}$$

$$\le 1 + \int_1^n \frac{1}{x}\,dx = 1 + \ln n$$

(Eq. 6), just grow too slowly. To see what we mean, suppose you had started with $s_1 = 1$ the day the universe was formed, thirteen billion years ago, and added a new term every *second*. About how large would s_n be today?

36. There are no values of x for which $\Sigma_{n=1}^{\infty}(1/nx)$ converges. Why?

37. Show that if $\Sigma_{n=1}^{\infty} a_n$ is a convergent series of nonnegative numbers, then $\Sigma_{n=1}^{\infty}(a_n/n)$ converges.

38. Show that if $\Sigma\, a_n$ and $\Sigma\, b_n$ are convergent series with $a_n \ge 0$ and $b_n \ge 0$, then $\Sigma\, a_n b_n$ converges. (*Hint:* From some integer on, $0 \le a_n$ and $b_n < 1$, so $a_n b_n \le a_n$.)

39. *Nonincreasing sequences.* A sequence of numbers $\{s_n\}$ in which $s_n \ge s_{n+1}$ for every n is called a **nonincreasing sequence.** A sequence $\{s_n\}$ is bounded from below if there is a finite constant M with $M \le s_n$ for every n. Such a number M is called a lower bound for the sequence. Deduce from Theorem 6 that a nonincreasing sequence that is bounded from below converges and that a nonincreasing sequence that is not bounded from below diverges.

40. *The Cauchy condensation test.* The Cauchy condensation test says: Let $\{a_n\}$ be a nonincreasing sequence ($a_n \ge a_{n+1}$ for all n) of positive terms that converges to 0. Then $\Sigma\, a_n$ converges if and only if $\Sigma\, 2^n a_{2^n}$ converges. For example, $\Sigma\,(1/n)$ diverges because $\Sigma\, 2^n \cdot (1/2^n) = \Sigma\, 1$ diverges. Show why the test works.

41. Use the Cauchy condensation test from Exercise 40 to show that

a) $\sum_{n=2}^{\infty} \frac{1}{n \ln n}$ diverges;

b) $\sum_{n=1}^{\infty} \frac{1}{n^p}$ converges if $p > 1$ and diverges if $p \le 1$.

42. *Logarithmic p-series.*

a) Show that

$$\int_2^{\infty} \frac{dx}{x(\ln x)^p} \qquad (p \text{ a positive constant})$$

converges if and only if $p > 1$.

Knowing this about the integral, what can you deduce about the convergence or divergence of the following series?

b) $\sum_{n=2}^{\infty} \frac{1}{n \ln n}$
c) $\sum_{n=2}^{\infty} \frac{1}{n(\ln n)^{1.01}}$

d) $\sum_{n=5}^{\infty} \frac{n^{1/2}}{(\ln n)^3}$
e) $\sum_{n=3}^{\infty} \frac{1}{n \ln(n^3)}$

f) $\sum_{n=2}^{\infty} \frac{1}{n(\ln n)^{(n+1)/2}}$

43. *Euler's constant.* Graphs like those in Fig. 8.17 suggest that as n increases there is little change in the difference between the sum

$$1 + \frac{1}{2} + \cdots + \frac{1}{n}$$

and the integral

$$\ln n = \int_1^n \frac{1}{x}\,dx.$$

To explore this idea, carry out the following steps.

a) By taking $f(x) = 1/x$ in inequality (6), show that

$$\ln(n+1) \leq 1 + \frac{1}{2} + \cdots + \frac{1}{n} \leq 1 + \ln n$$

or

$$0 < \ln(n+1) - \ln n \leq 1 + \frac{1}{2} + \cdots + \frac{1}{n} - \ln n \leq 1.$$

Thus, the sequence

$$a_n = 1 + \frac{1}{2} + \cdots + \frac{1}{n} - \ln n$$

is bounded from below and from above.

b) Show that

$$\frac{1}{n+1} < \int_n^{n+1} \frac{1}{x}\,dx = \ln(n+1) - \ln n,$$

and use this result to show that the sequence $\{a_n\}$ in part (a) is decreasing.

Since a decreasing sequence that is bounded from below converges (Exercise 39), the numbers a_n defined in (a) converge:

$$1 + \frac{1}{2} + \cdots + \frac{1}{n} - \ln n \to \gamma.$$

The number γ, whose value is 0.5772..., is called *Euler's constant.* In contrast to other special numbers like π and e, no other expression with a simple law of formulation has ever been found for γ.

44. *Prime numbers.* The prime numbers form a sequence $\{p_n\} = \{2, 3, 5, 7, 11, 13, 17, 19, \ldots\}$. It is known that $\lim_{n\to\infty} [(n \ln n)/p_n] = 1$. Using this fact, show that

$$\sum_{n=1}^{\infty} \frac{1}{p_n} = \frac{1}{2} + \frac{1}{3} + \frac{1}{5} + \frac{1}{7} + \frac{1}{11} + \cdots + \frac{1}{p_n} + \cdots$$

diverges. (See Exercise 42.)

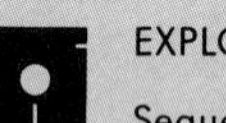

EXPLORER PROGRAM

Sequences and Series

Generates terms of one or two sequences and graphs them while you watch. You may define the sequences recursively or by giving formulas for their nth terms. Enables you to look for graphical and numerical evidence of convergence or divergence.

8.4 Series Without Negative Terms: The Ratio and Root Tests

Convergence tests that depend on comparing one series with another series or with an integral are called *extrinsic* tests. They are useful, but there are reasons to look for tests that do not require comparison. As a practical matter, we may not be able to find the series or function we need to make a comparison work. And, in principle, all the information about a given series should be contained in its own terms. We therefore turn our attention to *intrinsic* tests—those that depend only on the series at hand.

The Ratio Test

Our first intrinsic test, the Ratio Test, measures the rate of growth (or decline) of a series by examining the ratio a_{n+1}/a_n. For a geometric series, this rate of growth is a constant, and the series converges if and only if its ratio is less than 1 in absolute value. But even if the ratio is not constant, we may be able to find a geometric series for comparison, as in Example 1.

Example 1 Let $a_1 = 1$ and define a_{n+1} to be $a_{n+1} = \dfrac{n}{2n+1} a_n$.

Does the series $\Sigma\, a_n$ converge, or diverge?

Solution We begin by writing out a few terms of the series:

$$a_1 = 1, \qquad a_2 = \frac{1}{3} a_1 = \frac{1}{3}, \qquad a_3 = \frac{2}{5} a_2 = \frac{1 \cdot 2}{3 \cdot 5}, \qquad a_4 = \frac{3}{7} a_3 = \frac{1 \cdot 2 \cdot 3}{3 \cdot 5 \cdot 7}.$$

The series in Example 1 converges rapidly, as the following computer data suggest.

n	s_n
5	1.5607 5
10	1.5705 5
15	1.5707 89894
20	1.5707 96149
25	1.5707 96322
30	1.5707 96327
35	1.5707 96327

Each term is somewhat less than 1/2 the term before it, because $n/(2n+1)$ is less than 1/2. Therefore the terms of the given series are less than or equal to the terms of the geometric series

$$1 + \left(\frac{1}{2}\right) + \left(\frac{1}{4}\right) + \cdots + \left(\frac{1}{2}\right)^{n-1} + \cdots$$

that converges to 2. So our series also converges, and its sum is less than 2. Computer data show that the sum is approximately 1.5707 96327.

In proving the Ratio Test, we shall make a comparison with appropriate geometric series as in Example 1, but when we *apply* it we do not actually make a direct comparison.

The Ratio Test

Let $\Sigma\, a_n$ be a series with positive terms, and suppose that

$$\lim_{n\to\infty} \frac{a_{n+1}}{a_n} = \rho.$$

Then

a) the series *converges* if $\rho < 1$,

b) the series *diverges* if $\rho > 1$,

c) the series *may converge or it may diverge* if $\rho = 1$. (The test provides no information.)

Proof

a) $\rho < 1$. Let r be a number between ρ and 1. Then the number $\epsilon = r - \rho$ is positive. Since

$$\frac{a_{n+1}}{a_n} \to \rho,$$

a_{n+1}/a_n must lie within ϵ of ρ when n is large enough, say for all $n \geq N$. In particular,

$$\frac{a_{n+1}}{a_n} < \rho + \epsilon = r, \qquad \text{when } n > N.$$

That is,

$$\begin{aligned} a_{N+1} &< ra_N, \\ a_{N+2} &< ra_{N+1} < r^2 a_N, \\ a_{N+3} &< ra_{N+2} < r^3 a_N, \\ &\vdots \\ a_{N+m} &< ra_{N+m-1} < r^m a_N. \end{aligned}$$

These inequalities show that the terms of our series, after the Nth term, approach zero more rapidly than the terms in a geometric series with ratio

$r < 1$. More precisely, consider the series $\Sigma\, c_n$, where $c_n = a_n$ for $n = 1, 2, \ldots, N$ and $c_{N+1} = ra_N$, $c_{N+2} = r^2 a_N, \ldots, c_{N+m} = r^m a_N, \ldots$. Now $a_n \le c_n$ for all n, and

$$\sum_{n=1}^{\infty} c_n = a_1 + a_2 + \cdots + a_{N-1} + a_N + ra_N + r^2 a_N + \cdots$$
$$= a_1 + a_2 + \cdots + a_{N-1} + a_N(1 + r + r^2 + \cdots).$$

The geometric series $1 + r + r^2 + \cdots$ converges because $|r| < 1$, so $\Sigma\, c_n$ converges. Since $a_n \le c_n$, $\Sigma\, a_n$ also converges.

b) $\rho > 1$. From some index M on,

$$\frac{a_{n+1}}{a_n} > 1 \qquad \text{and} \qquad a_M < a_{M+1} < a_{M+2} < \cdots.$$

The terms of the series do not approach zero as n becomes infinite, and the series diverges by the nth-Term Test.

c) $\rho = 1$. The two series

$$\sum_{n=1}^{\infty} \frac{1}{n} \qquad \text{and} \qquad \sum_{n=1}^{\infty} \frac{1}{n^2}$$

show that some other test for convergence must be used when $\rho = 1$.

$$\text{For } \sum_{n=1}^{\infty} \frac{1}{n}: \qquad \frac{a_{n+1}}{a_n} = \frac{1/(n+1)}{1/n} = \frac{n}{n+1} \to 1.$$

$$\text{For } \sum_{n=1}^{\infty} \frac{1}{n^2}: \qquad \frac{a_{n+1}}{a_n} = \frac{1/(n+1)^2}{1/n^2} = \left(\frac{n}{n+1}\right)^2 \to 1^2 = 1.$$

In both cases $\rho = 1$, yet the first series diverges while the second converges (Section 8.3, Examples 3 and 6).

The Ratio Test is often effective when the terms of the series contain factorials of expressions involving n or expressions raised to the nth power or combinations of the two, as in the next example.

Example 2 Use the Ratio Test to investigate the convergence of the following series.

a) $\displaystyle\sum_{n=1}^{\infty} \frac{n!n!}{(2n)!}$ b) $\displaystyle\sum_{n=1}^{\infty} \frac{4^n n!n!}{(2n)!}$ c) $\displaystyle\sum_{n=0}^{\infty} \frac{2^n + 5}{3^n}$

Solution

a) If $a_n = n!n!/(2n)!$, then $a_{n+1} = (n+1)!(n+1)!/(2n+2)!$, and

$$\frac{a_{n+1}}{a_n} = \frac{(n+1)!(n+1)!(2n)!}{n!n!(2n+2)(2n+1)(2n)!}$$
$$= \frac{(n+1)(n+1)}{(2n+2)(2n+1)} = \frac{n+1}{4n+2} \to \frac{1}{4}.$$

The series converges because $\rho = 1/4$ is less than 1.

b) If $a_n = 4^n n!n!/(2n)!$, then

$$\frac{a_{n+1}}{a_n} = \frac{4^{n+1}(n+1)!(n+1)!}{(2n+2)(2n+1)(2n)!} \cdot \frac{(2n)!}{4^n n!n!}$$

$$= \frac{4(n+1)(n+1)}{(2n+2)(2n+1)} = \frac{2(n+1)}{2n+1} \to 1.$$

Because the limit is $\rho = 1$, we cannot decide on the basis of the Ratio Test alone whether the series converges or diverges. However, when we note that $a_{n+1}/a_n = (2n+2)/(2n+1)$, we conclude that a_{n+1} is always greater than a_n because $(2n+2)/(2n+1)$ is always greater than 1. Therefore, all terms are greater than or equal to $a_1 = 2$, and the nth term does not approach zero as $n \to \infty$. Hence, by the nth-Term Test, the series diverges.

c) For the series $\sum_{n=0}^{\infty} (2^n + 5)/3^n$,

$$\frac{a_{n+1}}{a_n} = \frac{(2^{n+1}+5)/3^{n+1}}{(2^n+5)/3^n} = \frac{1}{3} \cdot \frac{2^{n+1}+5}{2^n+5} = \frac{1}{3} \cdot \left(\frac{2+5 \cdot 2^{-n}}{1+5 \cdot 2^{-n}}\right) \to \frac{1}{3} \cdot \frac{2}{1} = \frac{2}{3}.$$

The series converges because $\rho = 2/3$ is less than 1.

This does *not* mean that 2/3 is the sum of the series. In fact,

$$\sum_{n=0}^{\infty} \frac{2^n+5}{3^n} = \sum_{n=0}^{\infty} \left(\frac{2}{3}\right)^n + \sum_{n=0}^{\infty} \frac{5}{3^n} = \frac{1}{1-(2/3)} + \frac{5}{1-(1/3)} = \frac{21}{2}.$$

The nth-Root Test

We return to the question "Does $\sum a_n$ converge?" When there is a simple formula for a_n, we can try one of the tests we already have. But consider the following example.

Example 3 Let $a_n = f(n)/2^n$, where

$$f(n) = \begin{cases} n & \text{if } n \text{ is a prime number,} \\ 1 & \text{otherwise.} \end{cases}$$

Does $\sum a_n$ converge?

Solution We write out several terms of the series:

$$\sum a_n = \frac{1}{2} + \frac{2}{4} + \frac{3}{8} + \frac{1}{16} + \frac{5}{32} + \frac{1}{64} + \frac{7}{128} + \cdots + \frac{f(n)}{2^n} + \cdots.$$

Clearly, this is not a geometric series. The nth term approaches zero as $n \to \infty$, so we do not know if the series diverges. The Integral Test does not look promising. The Ratio Test produces

$$\frac{a_{n+1}}{a_n} = \frac{1}{2}\frac{f(n+1)}{f(n)} = \begin{cases} \dfrac{1}{2} & \text{if neither } n \text{ nor } n+1 \text{ is a prime,} \\ \dfrac{1}{2n} & \text{if } n \text{ is a prime} \geq 3, \\ \dfrac{n+1}{2} & \text{if } n+1 \text{ is a prime} \geq 5. \end{cases}$$

The ratio sometimes is close to zero, sometimes is very large, and sometimes is 1/2. It has no limit because there are infinitely many primes. A test that will answer the question (affirmatively—yes, the series does converge) is the nth-Root Test. To apply it, we consider the following:

$$\sqrt[n]{a_n} = \frac{\sqrt[n]{f(n)}}{2} = \begin{cases} \dfrac{\sqrt[n]{n}}{2} & \text{if } n \text{ is a prime,} \\ \dfrac{1}{2} & \text{otherwise.} \end{cases}$$

Therefore,

$$\frac{1}{2} \le \sqrt[n]{a_n} \le \frac{\sqrt[n]{n}}{2}$$

and $\lim \sqrt[n]{a_n} = 1/2$ by the Sandwich Theorem. Because this limit is less than 1, the nth-Root Test tells us that the given series converges, as we shall now see.

The *n*th-Root Test

Let $\Sigma\, a_n$ be a series with $a_n \ge 0$ for $n \ge n_0$, and suppose that $\sqrt[n]{a_n} \to \rho$. Then

a) the series *converges* if $\rho < 1$,

b) the series *diverges* if $\rho > 1$,

c) the test is *not conclusive* if $\rho = 1$.

Proof

a) $\rho < 1$. Choose an $\epsilon > 0$ so small that $\rho + \epsilon < 1$ also. Since $\sqrt[n]{a_n} \to \rho$, the terms $\sqrt[n]{a_n}$ eventually get closer than ϵ to ρ. In other words, there exists an index $N \ge n_0$ such that

$$\sqrt[n]{a_n} < \rho + \epsilon \qquad \text{when } n \ge N.$$

Then it is also true that

$$a_n < (\rho + \epsilon)^n \qquad \text{for } n \ge N.$$

Now, $\Sigma_{n=N}^{\infty} (\rho + \epsilon)^n$, a geometric series with ratio $(\rho + \epsilon) < 1$, converges. By comparison, $\Sigma_{n=N}^{\infty} a_n$ converges, from which it follows that

$$\sum_{n=1}^{\infty} a_n = a_1 + \cdots + a_{N-1} + \sum_{n=N}^{\infty} a_n$$

converges.

b) $\rho > 1$. For all indices beyond some integer M, we have $\sqrt[n]{a_n} > 1$, so that $a_n > 1$ for $n > M$. The terms of the series do not converge to zero. The series diverges by the nth-Term Test.

c) $\rho = 1$. The series $\Sigma_{n=1}^{\infty} (1/n)$ and $\Sigma_{n=1}^{\infty} (1/n^2)$ show that the test is not conclusive when $\rho = 1$. The first series diverges and the second converges, but in both cases $\sqrt[n]{a_n} \to 1$.

Example 4 Which of the following series converges and which diverges?

a) $\displaystyle\sum_{n=1}^{\infty} \frac{n^2}{2^n}$ b) $\displaystyle\sum_{n=1}^{\infty} \frac{2^n}{n^2}$

Solution Series (a) converges because $\sqrt[n]{a_n} = \frac{\sqrt[n]{n^2}}{2} \to \frac{1}{2} < 1$. But series (b) diverges because $\sqrt[n]{b_n} = \frac{2}{\sqrt[n]{n^2}} \to 2 > 1$.

EXERCISES 8.4

Which series in Exercises 1–30 converge and which diverge? Give reasons for your answers.

1. $\sum_{n=1}^{\infty} \frac{n^2}{2^n}$
2. $\sum_{n=1}^{\infty} \frac{n!}{10^n}$
3. $\sum_{n=1}^{\infty} \frac{n^{10}}{10^n}$
4. $\sum_{n=1}^{\infty} n^2 e^{-n}$
5. $\sum_{n=1}^{\infty} \left(\frac{n-2}{n}\right)^n$
6. $\sum_{n=1}^{\infty} \frac{2+(-1)^n}{1.25^n}$
7. $\sum_{n=1}^{\infty} n!e^{-n}$
8. $\sum_{n=1}^{\infty} \frac{(-2)^n}{3^n}$
9. $\sum_{n=1}^{\infty} \left(1-\frac{3}{n}\right)^n$
10. $\sum_{n=1}^{\infty} \left(1-\frac{1}{n^2}\right)^n$
11. $\sum_{n=1}^{\infty} \frac{\ln n}{n^3}$
12. $\sum_{n=1}^{\infty} \left(\frac{1}{n}-\frac{1}{n^2}\right)$
13. $\sum_{n=1}^{\infty} \frac{\ln n}{n}$
14. $\sum_{n=1}^{\infty} \frac{n \ln n}{2^n}$
15. $\sum_{n=1}^{\infty} \frac{(n+1)(n+2)}{n!}$
16. $\sum_{n=1}^{\infty} e^{-n}(n^3)$
17. $\sum_{n=1}^{\infty} \frac{(n+3)!}{3!n!3^n}$
18. $\sum_{n=1}^{\infty} -\frac{n^2}{2^n}$
19. $\sum_{n=1}^{\infty} \frac{n!}{(2n+1)!}$
20. $\sum_{n=1}^{\infty} \frac{n!}{n^n}$
21. $\sum_{n=2}^{\infty} \frac{n}{(\ln n)^n}$
22. $\sum_{n=2}^{\infty} \frac{1}{(\ln n)^2}$
23. $\sum_{n=1}^{\infty} \frac{n! \ln n}{n(n+2)!}$
24. $\sum_{n=1}^{\infty} \frac{3^n}{n^3 2^n}$
25. $\sum_{n=1}^{\infty} \frac{(n!)^n}{(n^n)^2}$
26. $\sum_{n=1}^{\infty} \frac{(n!)^n}{n^{(n^2)}}$
27. $\sum_{n=1}^{\infty} \frac{n^n}{2^{(n^2)}}$
28. $\sum_{n=1}^{\infty} \frac{n^n}{(2^n)^2}$
29. $\sum_{n=1}^{\infty} \frac{1\cdot 3 \cdot \cdots \cdot (2n-1)}{4^n 2^n n!}$
30. $\sum_{n=1}^{\infty} \frac{1\cdot 3 \cdot \cdots \cdot (2n-1)}{[2\cdot 4 \cdot \cdots \cdot (2n)](3n+1)}$

Which of the series $\Sigma_{n=1}^{\infty} a_n$ defined by the formulas in Exercises 31–42 converge and which diverge? Give reasons for your answers.

31. $a_1 = 2, \quad a_{n+1} = \frac{1+\sin n}{n} a_n$
32. $a_1 = \frac{1}{3}, \quad a_{n+1} = \frac{3n-1}{2n+5} a_n$
33. $a_1 = 3, \quad a_{n+1} = \frac{n}{n+1} a_n$
34. $a_1 = 2, \quad a_{n+1} = \frac{2}{n} a_n$
35. $a_1 = -1, \quad a_{n+1} = \frac{1+\ln n}{n} a_n$
36. $a_1 = \frac{1}{2}, \quad a_{n+1} = \frac{n+\ln n}{n+10} a_n$
37. $a_n = \frac{2^n n! n!}{(2n)!}$
38. $a_n = \frac{(3n)!}{n!(n+1)!(n+2)!}$
39. $a_1 = 1, a_{n+1} = \frac{n(n+1)}{(n+2)(n+3)} a_n$
(*Hint:* Write out several terms, see which factors cancel, and then generalize.)
40. $a_1 = 1, \quad a_{n+1} = \frac{n}{(n-1)(n+1)} a_n$
41. $a_1 = a_2 = 1, \quad a_{n+1} = \frac{1}{1+a_n}$
42. $a_n = 1/3^n$ if n is odd, $a_n = n/3^n$ if n is even
43. Neither the Ratio nor the nth-Root Test helps with p-series, $p > 0$. Try each test on $\Sigma_{n=1}^{\infty} (1/n^p)$ to see what happens.

EXPLORER PROGRAM

Sequences and Series

Generates terms of one or two sequences and graphs them. You may define the sequences recursively or by giving formulas for their nth terms. Enables you to look for graphical and numerical evidence of convergence or divergence

8.5 Alternating Series and Absolute Convergence

A series in which the terms are alternately positive and negative is called an **alternating series.** Here are three examples:

$$1-\frac{1}{2}+\frac{1}{3}-\frac{1}{4}+\frac{1}{5}-\cdots+\frac{(-1)^{n+1}}{n}+\cdots \tag{1}$$

$$-2+1-\frac{1}{2}+\frac{1}{4}-\frac{1}{8}+\cdots+\frac{(-1)^n 4}{2^n}+\cdots \tag{2}$$

$$1-2+3-4+5-6+\cdots+(-1)^{n-1}n+\cdots \tag{3}$$

Series (1), called the **alternating harmonic series,** converges, as we shall see in a moment. Series (2), a geometric series with ratio $r=-1/2$, converges to $-2/[1+(1/2)]=-4/3$. Series (3) diverges because the nth term does not approach zero.

We prove the convergence of the alternating harmonic series by applying a general result known as the Alternating Series Theorem.

THEOREM 7

The Alternating Series Theorem (Leibniz's Theorem)

The series

$$\sum_{n=1}^{\infty}(-1)^{n+1}a_n = a_1 - a_2 + a_3 - a_4 + \cdots \tag{4}$$

converges if all three of the following conditions are satisfied:

1. The a_n's are all positive.
2. $a_n \geq a_{n+1}$ for all n.
3. $a_n \to 0$.

Proof If n is an even integer, say $n=2m$, then the sum of the first n terms is

$$\begin{aligned} s_{2m} &= (a_1-a_2)+(a_3-a_4)+\cdots+(a_{2m-1}-a_{2m}) \\ &= a_1-(a_2-a_3)-(a_4-a_5)-\cdots-(a_{2m-2}-a_{2m-1})-a_{2m}. \end{aligned} \tag{5}$$

The first equality shows that s_{2m} is the sum of m nonnegative terms, since each term in parentheses is positive or zero. Hence $s_{2m+2} \geq s_{2m}$, and the sequence $\{s_{2m}\}$ is nondecreasing. The second equality shows that $s_{2m} \leq a_1$. Since $\{s_{2m}\}$ is nondecreasing and bounded from above, it has a limit, say

$$\lim_{m\to\infty} s_{2m} = L. \tag{6}$$

If n is an odd integer, say $n=2m+1$, then the sum of the first n terms is

$$s_{2m+1} = s_{2m} + a_{2m+1}.$$

Since $a_n \to 0$,

$$\lim_{m\to\infty} a_{2m+1} = 0$$

and, as $m\to\infty$,

$$s_{2m+1} = s_{2m} + a_{2m+1} \to L + 0 = L. \tag{7}$$

When we combine the results of (6) and (7), we get

$$\lim_{n\to\infty} s_n = L.$$

∎

Example 1 The alternating harmonic series

$$\sum_{n=1}^{\infty} (-1)^{n+1}\frac{1}{n} = 1 - \frac{1}{2} + \frac{1}{3} - \frac{1}{4} + \cdots$$

satisfies the three requirements of Theorem 7; therefore it converges.

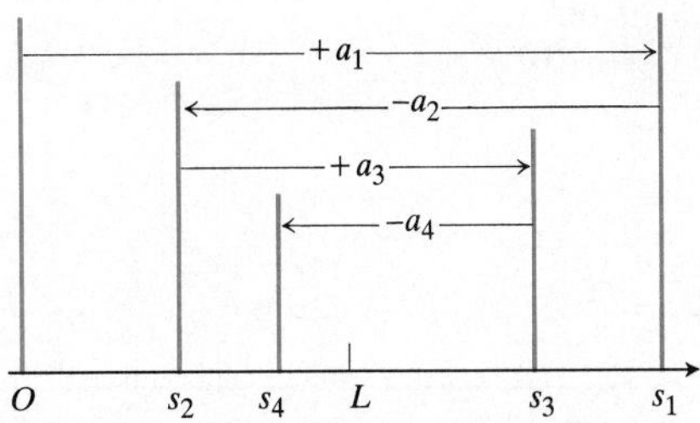

8.18 The partial sums of an alternating series that satisfies the hypotheses of Theorem 7 straddle their limit.

We can use a graphical interpretation of the partial sums to gain added insight into the way an alternating series converges to its limit L when the three conditions of Theorem 7 are satisfied. Starting from the origin of the x-axis (Fig. 8.18), we lay off the positive distance $s_1 = a_1$. To find the point corresponding to $s_2 = a_1 - a_2$, we back up a distance equal to a_2. Since $a_2 \le a_1$, we do not back up any farther than O at most. Next we go forward a distance a_3 and mark the point corresponding to $s_3 = a_1 - a_2 + a_3$. Since $a_3 \le a_2$, we go forward by an amount no greater than the previous backward step; that is, $s_3 \le s_1$. We continue in this seesaw fashion, backing up or going forward as the signs in the series demand. But each forward or backward step is shorter than (or at most the same size as) the preceding step, because $a_{n+1} \le a_n$. And since the nth term approaches zero as n increases, the size of step we take forward or backward gets smaller and smaller. We oscillate across the limit L, and the amplitude of oscillation approaches zero. The even-numbered partial sums $s_2, s_4, s_6, \ldots, s_{2m}$ increase toward L, while the odd-numbered sums $s_1, s_3, s_5, \ldots, s_{2m+1}$ decrease toward L. The limit L lies between any two successive sums s_n and s_{n+1} and hence differs from s_n by an amount less than a_{n+1}.

Because

$$|L - s_n| < a_{n+1} \qquad \text{for every } n, \tag{8}$$

we can make useful estimates of the sums of convergent alternating series.

THEOREM 8

The Alternating Series Estimation Theorem

If the alternating series $\sum_{n=1}^{\infty} (-1)^{n+1} a_n$ satisfies the three conditions of Theorem 7, then

$$s_n = a_1 - a_2 + \cdots + (-1)^{n+1} a_n$$

approximates the sum L of the series with an error whose absolute value is less than a_{n+1}, the numerical value of the first unused term. Furthermore, the remainder, $L - s_n$, has the same sign as the first unused term.

We leave the determination of the sign of the remainder for Exercise 53.

Example 2 We first try Theorem 8 on an alternating series whose sum we already know, namely the geometric series

$$\sum_{n=0}^{\infty} (-1)^n \frac{1}{2^n} = 1 - \frac{1}{2} + \frac{1}{4} - \frac{1}{8} + \frac{1}{16} - \frac{1}{32} + \frac{1}{64} - \frac{1}{128} \;\Big|\; + \frac{1}{256} - \cdots.$$

The theorem says that if we truncate the series after the eighth term, we throw away a total that is positive and less than 1/256. The sum of the first eight terms is 0.6640 625. The sum of the series is

$$\frac{1}{1-(-1/2)} = \frac{1}{3/2} = \frac{2}{3}.$$

The difference,

$$\frac{2}{3} - 0.6640\ 625 = 0.0026\ 04166\ 6\ldots,$$

is positive and less than

$$\frac{1}{256} = 0.0039\ 0625.$$

Absolute Convergence

DEFINITION

A series $\Sigma\, a_k$ **converges absolutely** (is **absolutely convergent**) if the corresponding series of absolute values, $\Sigma|a_k|$, converges.

The geometric series

$$1 - \frac{1}{2} + \frac{1}{4} - \frac{1}{8} + \cdots$$

converges absolutely because the corresponding series of absolute values

$$1 + \frac{1}{2} + \frac{1}{4} + \frac{1}{8} + \cdots$$

converges. But the alternating harmonic series, although it converges, does not converge absolutely. The corresponding series of absolute values is the (divergent) harmonic series.

DEFINITION

A series that converges but does not converge absolutely **converges conditionally.**

The alternating harmonic series converges conditionally.

Absolute convergence is important because, first, we have many good tests for convergence of series of positive terms. Second, if a series converges absolutely, then it converges. That is the thrust of the next theorem.

THEOREM 9

The Absolute Convergence Theorem

If $\Sigma_{n=1}^{\infty} |a_n|$ converges, then $\Sigma_{n=1}^{\infty} a_n$ converges.

Proof For each n,

$$-|a_n| \le a_n \le |a_n|, \quad \text{so} \quad 0 \le a_n + |a_n| \le 2|a_n|.$$

If $\Sigma_{n=1}^{\infty} |a_n|$ converges, then $\Sigma_{n=1}^{\infty} 2|a_n|$ converges and, by the Comparison Test, the nonnegative series $\Sigma_{n=1}^{\infty} (a_n + |a_n|)$ converges. The equality $a_n = (a_n + |a_n|)$

$-|a_n|$ now lets us express $\sum_{n=1}^{\infty} a_n$ as the difference of two convergent series:

$$\sum_{n=1}^{\infty} a_n = \sum_{n=1}^{\infty} (a_n + |a_n| - |a_n|) = \sum_{n=1}^{\infty} (a_n + |a_n|) - \sum_{n=1}^{\infty} |a_n|.$$

Therefore, $\sum_{n=1}^{\infty} a_n$ converges.

We can rephrase Theorem 9 to say that *every absolutely convergent series converges.* However, the converse statement is false. Many convergent series do not converge absolutely. The convergence of many series depends on the series' having infinitely many positive and negative terms arranged in a particular order.

Example 3 For $\sum_{n=1}^{\infty} (-1)^{n+1}\frac{1}{n^2} = 1 - \frac{1}{4} + \frac{1}{9} - \frac{1}{16} + \cdots$, the corresponding series of absolute values is

$$\sum_{n=1}^{\infty} \frac{1}{n^2} = 1 + \frac{1}{4} + \frac{1}{9} + \frac{1}{16} + \cdots.$$

The series converges because it is a p-series with $p = 2 > 1$. Therefore

$$\sum_{n=1}^{\infty} (-1)^{n+1}\frac{1}{n^2}$$

converges absolutely and

$$\sum_{n=1}^{\infty} (-1)^{n+1}\frac{1}{n^2}$$

converges.

Example 4 For $\sum_{n=1}^{\infty} \frac{\sin n}{n^2} = \frac{\sin 1}{1} + \frac{\sin 2}{4} + \frac{\sin 3}{9} + \cdots$, the corresponding series of absolute values is

$$\sum_{n=1}^{\infty} \left| \frac{\sin n}{n^2} \right| = \frac{|\sin 1|}{1} + \frac{|\sin 2|}{4} + \cdots,$$

which converges by comparison with $\sum_{n=1}^{\infty} (1/n^2)$ because $|\sin n| \le 1$ for every n. The original series converges absolutely; therefore it converges.

Alternating p-series

When p is a positive constant, the sequence $\{1/n^p\}$ is a decreasing sequence with limit zero. Therefore the alternating p-series

$$\sum_{n=1}^{\infty} \frac{(-1)^{n-1}}{n^p} = 1 - \frac{1}{2^p} + \frac{1}{3^p} - \frac{1}{4^p} + \cdots, \qquad p > 0$$

converges.

For $p > 1$, the series converges absolutely. For $0 < p \le 1$, the series converges conditionally.

Conditional convergence: $1 - \frac{1}{\sqrt{2}} + \frac{1}{\sqrt{3}} - \frac{1}{\sqrt{4}} + \cdots$

Absolute convergence: $1 - \frac{1}{2^{3/2}} + \frac{1}{3^{3/2}} - \frac{1}{4^{3/2}} + \cdots$

Rearranging Series

One other important fact about absolutely convergent series is the following theorem.

THEOREM 10

The Rearrangement Theorem for Absolutely Convergent Series

If $\Sigma_{n=1}^{\infty} a_n$ converges absolutely, and $b_1, b_2, \ldots, b_n, \ldots$ is any arrangement of the sequence $\{a_n\}$, then $\Sigma\, b_n$ converges and

$$\sum_{n=1}^{\infty} b_n = \sum_{n=1}^{\infty} a_n.$$

(For an outline of the proof, see Exercise 60.)

Example 5 As we saw in Example 3, the series

$$1 - \frac{1}{4} + \frac{1}{9} - \frac{1}{16} + \cdots + (-1)^{n-1}\frac{1}{n^2} + \cdots$$

converges absolutely. A possible rearrangement of the terms of the series might start with a positive term, then two negative terms, then three positive terms, then four negative terms, and so on: After k terms of one sign, take $k + 1$ terms of the opposite sign. The first ten terms of such a series look like this:

$$1 - \frac{1}{4} - \frac{1}{16} + \frac{1}{9} + \frac{1}{25} + \frac{1}{49} - \frac{1}{36} - \frac{1}{64} - \frac{1}{100} - \frac{1}{144} + \cdots.$$

The Rearrangement Theorem says that both series converge to the same value. In this example, if we had the second series to begin with, we would probably be glad to exchange it for the first, if we knew that we could. We can do even better: The sum of either series is also equal to

$$\sum_{n=1}^{\infty} \frac{1}{(2n-1)^2} - \sum_{n=1}^{\infty} \frac{1}{(2n)^2}.$$

(See Exercise 57.)

CAUTION ABOUT REARRANGEMENTS If we rearrange infinitely many terms of a conditionally convergent series, we can get results that are far different from the sum of the original series. The next example illustrates some of the things that can happen.

Example 6 Consider the alternating harmonic series

$$\frac{1}{1} - \frac{1}{2} + \frac{1}{3} - \frac{1}{4} + \frac{1}{5} - \frac{1}{6} + \frac{1}{7} - \frac{1}{8} + \frac{1}{9} - \frac{1}{10} + \frac{1}{11} - \cdots.$$

Here, the series of terms $\Sigma\,[1/(2n-1)]$ diverges to $+\infty$ and the series of terms $\Sigma(-1/2n)$ diverges to $-\infty$. No matter how far out in the sequence of odd-numbered terms we begin, we can always add enough positive terms to get an arbitrarily large sum. Similarly, with the negative terms, no matter how far out we start, we can add enough consecutive even-numbered terms to get a negative sum of arbitrarily large absolute value. If we wished to do so, we could start adding odd-numbered terms until we had a sum greater than $+3$, say, and then follow that with enough consecutive negative terms to make the new total less than -4. We could then add enough positive terms to make the total greater than $+5$ and follow with consecutive unused negative terms to make a new total less than -6, and so on. In this way, we could make the swings arbitrarily large in either direction.

Another possibility, with the same series, is to focus on a particular limit. Suppose we try to get sums that converge to 1. We start with the first term, 1/1, and then subtract 1/2. Next we add 1/3 and 1/5, which brings the total back to 1 or above. Then we add consecutive negative terms until the total is less than 1. We continue in this manner: When the sum is less than 1, add positive terms until the total is 1 or more; then subtract (add negative) terms until the total is again less than 1. This process can be continued indefinitely. Because both the odd-numbered terms and the even-numbered terms of the original series approach zero as $n \to \infty$, the amount by which our partial sums exceed 1 or fall below it approaches zero. So the new series converges to 1. The rearranged series starts like this:

$$\frac{1}{1} - \frac{1}{2} + \frac{1}{3} + \frac{1}{5} - \frac{1}{4} + \frac{1}{7} + \frac{1}{9} - \frac{1}{6} + \frac{1}{11} + \frac{1}{13} - \frac{1}{8} + \frac{1}{15} + \frac{1}{17} - \frac{1}{10}$$
$$+ \frac{1}{19} + \frac{1}{21} - \frac{1}{12} + \frac{1}{23} + \frac{1}{25} - \frac{1}{14} + \frac{1}{27} - \frac{1}{16} + \cdots.$$

The kind of behavior illustrated by this example is typical of what can happen with any conditionally convergent series. Moral: Add the terms of a conditionally convergent series in the order given.

FLOWCHART 8.1 Procedure for Determining Convergence

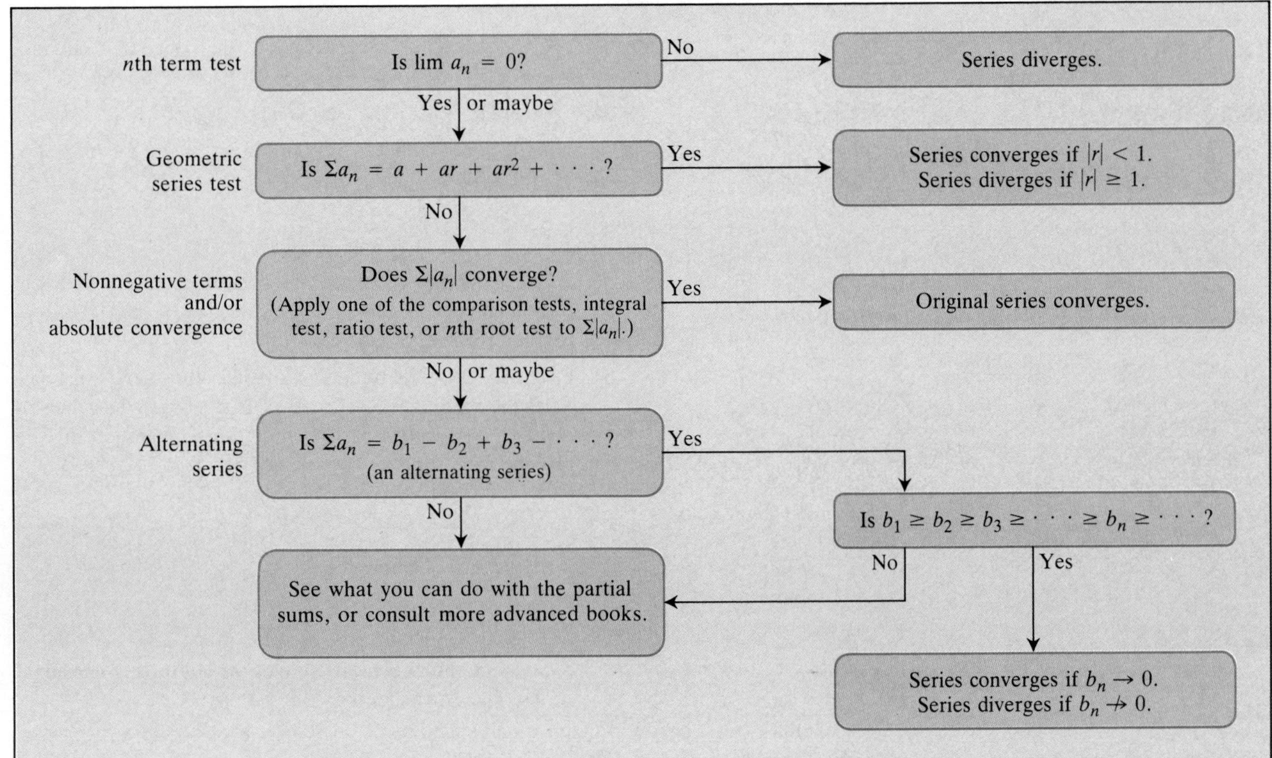

EXERCISES 8.5

Which of the alternating series in Exercises 1–12 converge and which diverge?

1. $\sum_{n=1}^{\infty} (-1)^{n+1} \frac{1}{n^{3/2}}$

2. $\sum_{n=2}^{\infty} (-1)^{n+1} \frac{1}{\ln n}$

3. $\sum_{n=1}^{\infty} (-1)^{n+1}$

4. $\sum_{n=1}^{\infty} (-1)^{n+1} \frac{10^n}{n^{10}}$

5. $\sum_{n=1}^{\infty} (-1)^{n+1} \frac{\sqrt{n}+1}{n+1}$

6. $\sum_{n=1}^{\infty} (-1)^{n+1} \frac{\ln n}{n}$

7. $\sum_{n=2}^{\infty} (-1)^{n} \log_n 2$

8. $\sum_{n=2}^{\infty} (-1)^{n+1} \frac{\ln n}{\ln n^2}$

9. $\sum_{n=1}^{\infty} (-1)^{n+1} \frac{3\sqrt{n+1}}{\sqrt{n}+1}$

10. $\sum_{n=1}^{\infty} (-1)^{n} \ln \left(1+\frac{1}{n}\right)$

11. $\sum_{n=1}^{\infty} (-1)^{n} \left(\sqrt{n+\sqrt{n}} - \sqrt{n}\right)$

12. $\sum_{n=1}^{\infty} (-1)^{n} \sqrt{n} \sin \frac{1}{n}$

Which of the series in Exercises 13–44 converge absolutely, which converge conditionally, and which diverge?

13. $\sum_{n=1}^{\infty} (-1)^{n+1} (0.1)^n$

14. $\sum_{n=1}^{\infty} (-1)^{n+1} \frac{1}{\sqrt{n}}$

15. $\sum_{n=1}^{\infty} (-1)^{n+1} \frac{n}{n^3+1}$

16. $\sum_{n=1}^{\infty} \frac{n!}{2^n}$

17. $\sum_{n=1}^{\infty} (-1)^{n} \frac{1}{n+3}$

18. $\sum_{n=1}^{\infty} (-1)^{n} \frac{\sin n}{n^2}$

19. $\sum_{n=1}^{\infty} (-1)^{n+1} \frac{3+n}{5+n}$

20. $\sum_{n=2}^{\infty} (-1)^{n} \frac{1}{\ln n^3}$

21. $\sum_{n=1}^{\infty} (-1)^{n+1} \frac{1+n}{n^2}$

22. $\sum_{n=1}^{\infty} \frac{(-2)^{n+1}}{n+5^n}$

23. $\sum_{n=1}^{\infty} n^2 (2/3)^n$

24. $\sum_{n=1}^{\infty} (-1)^{n+1} (\sqrt[n]{10})$

25. $\sum_{n=1}^{\infty} (-1)^{n} \frac{\tan^{-1} n}{n^2+1}$

26. $\sum_{n=2}^{\infty} (-1)^{n+1} \frac{1}{n \ln n}$

27. $\sum_{n=1}^{\infty} \left(\frac{1}{n} - \frac{1}{2n}\right)$

28. $\sum_{n=1}^{\infty} (-1)^{n+1} \frac{(0.1)^n}{n}$

29. $\sum_{n=1}^{\infty} (-1)^{n} \frac{n}{n+1}$

30. $\sum_{n=1}^{\infty} \frac{(-1)^n}{1+\sqrt{n}}$

31. $\sum_{n=1}^{\infty} \frac{-1}{n^2+2n+1}$

32. $\sum_{n=1}^{\infty} (5)^{-n}$

33. $\sum_{n=1}^{\infty} \frac{(-100)^n}{n!}$

34. $\sum_{n=2}^{\infty} (-1)^{n} \left(\frac{\ln n}{\ln n^2}\right)^n$

35. $\sum_{n=1}^{\infty} \frac{\cos n\pi}{n\sqrt{n}}$

36. $\sum_{n=1}^{\infty} \frac{\cos n\pi}{n}$

37. $\sum_{n=1}^{\infty} \frac{(-1)^n}{\sqrt{n}+\sqrt{n+1}}$

38. $\sum_{n=1}^{\infty} \frac{(-1)^{n+1}(n!)^2}{(2n)!}$

39. $\sum_{n=1}^{\infty} (-1)^{n} \frac{(2n)!}{2^n n!\, n}$

40. $\sum_{n=1}^{\infty} (-1)^{n} \frac{(n!)^2\, 3^n}{(2n+1)!}$

41. $\sum_{n=1}^{\infty} (-1)^{n} (\sqrt{n+1} - \sqrt{n})$

42. $\sum_{n=1}^{\infty} (-1)^{n} (\sqrt{n^2+n} - n)$

43. $\sum_{n=1}^{\infty} (-1)^{n} \operatorname{sech} n$

44. $\sum_{n=1}^{\infty} (-1)^{n} \operatorname{csch} n$

In Exercises 45–48, estimate the error involved in using the sum of the first four terms to approximate the sum of the entire series.

45. $\sum_{n=1}^{\infty} (-1)^{n+1} \frac{1}{n}$

46. $\sum_{n=1}^{\infty} (-1)^{n+1} \frac{1}{10^n}$

47. $\ln(1.01) = \sum_{n=1}^{\infty} (-1)^{n+1} \frac{(0.01)^n}{n}$

(In Section 8.8, you will see why the series adds to ln(1.01).)

48. $\frac{1}{1+t} = \sum_{n=0}^{\infty} (-1)^n t^n, \quad 0<t<1$

CALCULATOR Approximate the sums in Exercises 49 and 50 with an error of magnitude less than 5×10^{-6}.

49. $\sum_{n=0}^{\infty} (-1)^{n} \frac{1}{(2n)!}$ (As you will see in Section 8.7, the sum is cos 1, the cosine of 1 radian.)

50. $\sum_{n=0}^{\infty} (-1)^{n} \frac{1}{n!}$ (The sum is $\frac{1}{e}$.)

51. a) The series

$$\frac{1}{3} - \frac{1}{2} + \frac{1}{9} - \frac{1}{4} + \frac{1}{27} - \frac{1}{8} + \cdots + \frac{1}{3^n} - \frac{1}{2^n} + \cdots$$

does not meet one of the conditions of Theorem 7. Which one?

b) Find the sum of the series in (a).

52. CALCULATOR The limit L of an alternating series that satisfies the conditions of Theorem 7 lies between the values of any two consecutive partial sums. This suggests using the average

$$\frac{s_n + s_{n+1}}{2} = s_n + \frac{1}{2}(-1)^{n+2} a_{n+1}$$

to estimate L. Compute

$$s_{20} + \frac{1}{2} \cdot \frac{1}{21}$$

as an approximation to the sum of the alternating harmonic series. The exact sum is $\ln 2 = 0.6931\ldots$.

53. *The sign of the remainder of an alternating series that satis-*

fies the conditions of Theorem 7. Prove the assertion in Theorem 8 that whenever an alternating series satisfying the conditions of Theorem 7 is approximated with one of its partial sums, then the remainder (sum of the unused terms) has the same sign as the first unused term. (*Hint:* Group the remainder's terms in consecutive pairs.)

54. Show that the sum of the first $2n$ terms of the series

$$1-\frac{1}{2}+\frac{1}{2}-\frac{1}{3}+\frac{1}{3}-\frac{1}{4}+\frac{1}{4}-\frac{1}{5}+\frac{1}{5}-\frac{1}{6}+\cdots$$

is the same as the sum of the first n terms of the series

$$\frac{1}{1\cdot 2}+\frac{1}{2\cdot 3}+\frac{1}{3\cdot 4}+\frac{1}{4\cdot 5}+\frac{1}{5\cdot 6}+\cdots.$$

Do these series converge? What is the sum of the first $2n+1$ terms of the first series? If the series converge, what is their sum?

55. Show by example that $\sum_{n=1}^{\infty} a_n b_n$ may diverge even if $\sum_{n=1}^{\infty} a_n$ and $\sum_{n=1}^{\infty} b_n$ both converge.

56. Show that the alternating series

$$\sum_{n=2}^{\infty} \frac{(-1)^n}{\sqrt{n}+(-1)^n}$$

diverges. Which of the three conditions of Leibniz's theorem is not satisfied? (*Hint:* Show that

$$\frac{(-1)^n}{\sqrt{n}+(-1)^n}=\frac{(-1)^n}{\sqrt{n}}-\frac{1}{n+(-1)^n\sqrt{n}}$$

and write the original series as the difference of the two series.)

57. *Unzipping absolutely convergent series.*

a) Show that if $\sum_{n=1}^{\infty} |a_n|$ converges and

$$b_n=\begin{cases} a_n & \text{if } a_n \geq 0,\\ 0 & \text{if } a_n<0,\end{cases}$$

then $\sum_{n=1}^{\infty} b_n$ converges.

b) Use the results in (a) to show likewise that if $\sum_{n=1}^{\infty} |a_n|$ converges and

$$c_n=\begin{cases} 0 & \text{if } a_n \geq 0,\\ a_n & \text{if } a_n<0,\end{cases}$$

then $\sum_{n=1}^{\infty} c_n$ converges.

In other words, if a series converges absolutely, its positive terms form a convergent series, and so do its negative terms. Furthermore,

$$\sum_{n=1}^{\infty} a_n=\sum_{n=1}^{\infty} b_n+\sum_{n=1}^{\infty} c_n$$

because $b_n=(a_n+|a_n|)/2$ and $c_n=(a_n-|a_n|)/2$.

58. What is wrong here:

Multiply both sides of the alternating harmonic series

$$S=1-\frac{1}{2}+\frac{1}{3}-\frac{1}{4}+\frac{1}{5}-\frac{1}{6}+$$
$$\frac{1}{7}-\frac{1}{8}+\frac{1}{9}-\frac{1}{10}+\frac{1}{11}-\frac{1}{12}+\cdots$$

by 2 to get

$$2S=2-1+$$
$$\frac{2}{3}-\frac{1}{2}+\frac{2}{5}-\frac{1}{3}+\frac{2}{7}-\frac{1}{4}+\frac{2}{9}-\frac{1}{5}+\frac{2}{11}-\frac{1}{6}+\cdots.$$

Collect terms with the same denominator, as the arrows indicate, to arrive at

$$2S=1-\frac{1}{2}+\frac{1}{3}-\frac{1}{4}+\frac{1}{5}-\frac{1}{6}+\cdots.$$

The series on the right-hand side of this equation is the series we started with. Therefore, $2S=S$, and dividing by S gives $2=1$. (*Source: Riemann's Rearrangement Theorem* by Stewart Galanor, *Mathematics Teacher,* Vol. 80, No. 8, 1987, pp. 675–81.)

59. CALCULATOR In Example 6, suppose the goal is to arrange the terms to get a new series that converges to $-1/2$. Start the new arrangement with the first negative term, which is $-1/2$. Whenever you have a sum that is less than or equal to $-1/2$, start introducing positive terms, taken in order, until the new total is greater than $-1/2$. Then add negative terms until the total is less than or equal to $-1/2$ again. Continue this process until your partial sums have been above the target at least three times and finish at or below it. If s_n is the sum of the first n terms of your new series, plot the points (n, s_n) to illustrate how the sums are behaving.

60. *Outline of the proof of the Rearrangement Theorem (Theorem 10).* Let ϵ be a positive real number, let $L=\sum_{n=1}^{\infty} a_n$, and let $s_k=\sum_{n=1}^{k} a_n$. Show that for some index N_1 and for some index $N_2 \geq N_1$,

$$\sum_{n=N_1}^{\infty} |a_n|<\frac{\epsilon}{2} \quad \text{and} \quad |s_{N_2}-L|<\frac{\epsilon}{2}.$$

Since all the terms $a_1, a_2, \ldots, a_{N_2}$ appear somewhere in the sequence $\{b_n\}$, there is an index $N_3 \geq N_2$ such that if $n \geq N_3$, then $(\sum_{k=1}^{n} b_k)-s_{N_2}$ is at most a sum of terms a_m with $m \geq N_1$. Therefore, if $n \geq N_3$,

$$\left|\sum_{k=1}^{n} b_k-L\right| \leq \left|\sum_{k=1}^{n} b_k-s_{N_2}\right|+|s_{N_2}-L|$$
$$\leq \sum_{k=N_1}^{\infty} |a_k|+|s_{N_2}-L|<\epsilon.$$

EXPLORER PROGRAM

Sequences and Series	Generates terms of one or two sequences and graphs them. You may define the sequences recursively or by giving formulas for their nth terms. Enables you to look for graphical and numerical evidence of convergence or divergence.

8.6 Power Series

Now that we know how to test infinite series for convergence, we can study the infinite polynomials we mentioned at the beginning of Section 8.2. We call these infinite polynomials *power series* because they are defined as infinite series of powers of some variable, in our case x. Like polynomials, power series can be added, subtracted, multiplied, differentiated, and integrated to give new power series. They can be divided, too, but we shall not go into that. Almost any function with infinitely many derivatives can be represented by a power series, as long as the derivatives do not become too large. The series for e^x, $\sin x$, $\cos x$, $\ln(1 + x)$, $\tan^{-1}x$, and so on enable us to approximate the values of these functions as accurately as we please. We shall show you what these series are as the chapter continues.

Power Series and Convergence

We begin with a formal definition.

DEFINITIONS

A **power series** is a series of the form

$$\sum_{n=0}^{\infty} c_n x^n = c_0 + c_1 x + c_2 x^2 + \cdots + c_n x^n + \cdots \tag{1}$$

or

$$\sum_{n=0}^{\infty} c_n (x-a)^n = c_0 + c_1(x-a) + c_2(x-a)^2 + \cdots + c_n(x-a)^n + \cdots \tag{2}$$

in which the **center** a and the **coefficients** $c_0, c_1, c_2, \ldots, c_n, \ldots$ are constants.

Equation (1) is the special case obtained by taking $a = 0$ in Eq. (2).

Example 1 Taking all the coefficients to be 1 in Eq. (1) gives the geometric power series

$$\sum_{n=0}^{\infty} x^n = 1 + x + x^2 + \cdots + x^n + \cdots.$$

The series converges to $1/(1 - x)$ when $|x| < 1$. We express this fact by writing

$$\frac{1}{1-x} = 1 + x + x^2 + \cdots + x^n + \cdots \qquad \text{for} \quad -1 < x < 1. \tag{3}$$

Up to now we have used Eq. (3) as a formula to give us the sum of the series on the right. We now change the focus: We think of the partial sums of the series on the right as polynomials $P_n(x)$ that approximate the function on the left. For values of x near zero, we need take only a few terms of the series to get a good approximation. As we move toward $x = 1$ or -1, we must take more terms. Figure 8.19 shows the graphs of $y = 1/(1 - x)$, and the approximating polynomials $y = P_n(x)$ for $n = 0$, 1, and 2.

Example 2 The power series

$$1 - \frac{1}{2}(x-2) + \frac{1}{4}(x-2)^2 + \cdots + \left(-\frac{1}{2}\right)^n (x-2)^n + \cdots \tag{4}$$

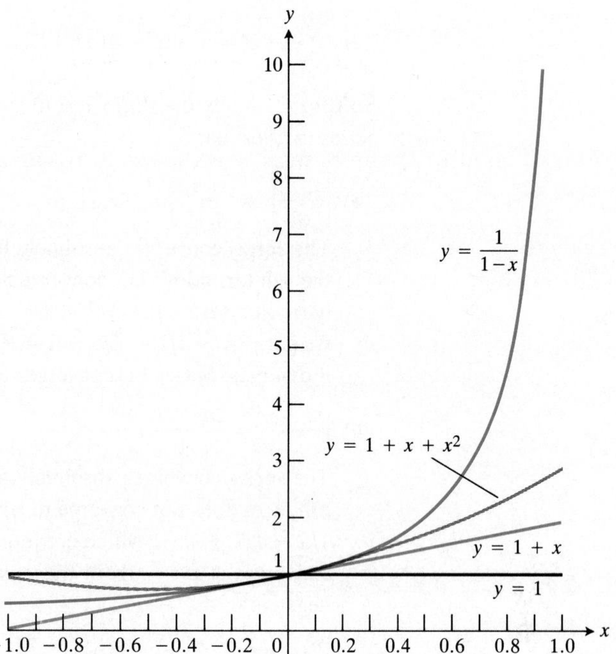

8.19 The graphs of $y = 1/(1 - x)$ and the approximating polynomials $P_0(x) = 1$, $P_1(x) = 1 + x$, and $P_2(x) = 1 + x + x^2$.

matches Eq. (2) with $a = 2$, $c_0 = 1$, $c_1 = -1/2$, $c_2 = 1/4, \ldots, c_n = (-1/2)^n$. This is a geometric series with first term 1 and ratio $r = -(x - 2)/2$. The series converges for $|(x - 2)/2| < 1$ or $0 < x < 4$. The sum is

$$\frac{1}{1-r} = \frac{1}{1 + \dfrac{x-2}{2}} = \frac{2}{x},$$

so we write

$$\frac{2}{x} = 1 - \frac{(x-2)}{2} + \frac{(x-2)^2}{4} - \cdots + \left(-\frac{1}{2}\right)^n (x-2)^n + \cdots, \qquad 0 < x < 4.$$

We use series (4) to generate polynomial approximations to $2/x$ (or $1/x$ if we divide by 2) for values of x near 2. We keep the powers of $(x - 2)$, instead of expanding them, and write the approximating polynomials as

$$P_0(x) = 1, \quad P_1(x) = 1 - \frac{1}{2}(x-2), \quad P_2(x) = 1 - \frac{1}{2}(x-2) + \frac{1}{4}(x-2)^2,$$

and so on. The higher powers of $(x - 2)$ decrease rapidly as n increases as long as $|x - 2|$ is small.

Example 3 For what values of x do the following series converge?

a) $\displaystyle\sum_{n=1}^{\infty} (-1)^{n-1} \frac{x^n}{n} = x - \frac{x^2}{2} + \frac{x^3}{3} - \cdots$

b) $\displaystyle\sum_{n=1}^{\infty} (-1)^{n-1} \frac{x^{2n-1}}{2n-1} = x - \frac{x^3}{3} + \frac{x^5}{5} - \cdots$

c) $\displaystyle\sum_{n=0}^{\infty} \frac{x^n}{n!} = 1 + x + \frac{x^2}{2!} + \frac{x^3}{3!} + \cdots$

d) $\sum_{n=0}^{\infty} n!\,x^n = 1 + x + 2!\,x^2 + 3!\,x^3 + \cdots$

Solution Apply the Ratio Test to the series $\Sigma|u_n|$, where u_n is the nth term of the series in question.

a) $\left|\frac{u_{n+1}}{u_n}\right| = \frac{n}{n+1}|x| \to |x|$

The series converges absolutely for $|x| < 1$. It diverges if $|x| > 1$ because the nth term does not converge to zero. At $x = 1$, we get the alternating harmonic series $1 - 1/2 + 1/3 - 1/4 + \cdots$, which converges. At $x = -1$ we get $-1 - 1/2 - 1/3 - 1/4 - \cdots$, the negative of the harmonic series; it diverges. Series (a) converges for $-1 < x \le 1$ and diverges elsewhere.

b) $\left|\frac{u_{n+1}}{u_n}\right| = \frac{2n-1}{2n+1}x^2 \to x^2$

The series converges absolutely for $x^2 < 1$. It diverges for $x^2 > 1$ because the nth term does not converge to zero. At $x = 1$ the series becomes $1 - 1/3 + 1/5 - 1/7 + \cdots$, which converges by the Alternating Series Theorem. It converges at $x = -1$ for the same reason. The value at $x = -1$ is the negative of the value at $x = 1$.

c) $\left|\frac{u_{n+1}}{u_n}\right| = \left|\frac{x^{n+1}}{(n+1)!}\cdot\frac{n!}{x^n}\right| = \frac{|x|}{n+1} \to 0$ for every x

The series converges absolutely for all x.

d) $\left|\frac{u_{n+1}}{u_n}\right| = \left|\frac{(n+1)!\,x^{n+1}}{n!\,x^n}\right| = (n+1)\,|x| \to \infty$ unless $x = 0$

The series diverges for all values of x except $x = 0$. The nth term does not approach zero unless $x = 0$.

Example 3 illustrates how we usually test a power series for convergence and the kinds of results we get.

How to Test a Power Series for Convergence

STEP 1: Use the Ratio Test (or nth-Root Test) to find the interval where the series converges absolutely. Ordinarily, this is an open interval

$$|x - a| < h \qquad \text{or} \qquad a - h < x < a + h.$$

In some instances, as in Example 3(c), the series converges for all values of x. These instances are not uncommon. In rare cases, the series may converge only at a single point, as in Example 3(d).

STEP 2: If the interval of absolute convergence is finite, test for convergence or divergence at each of the two endpoints, as in Example 2 and Examples 3(a) and (b). Neither the Ratio Test nor the nth-Root Test is useful at these points. Use a Comparison Test, the Integral Test, or the Alternating Series Theorem.

STEP 3: If the interval of absolute convergence is $a - h < x < a + h$, the series diverges (it does not even converge conditionally) for $|x - a| > h$, because for those values of x the nth term does not approach zero.

In the next section we shall see how series in powers of $(x - a)$ are generated by the values of a function f and its derivatives at $x = a$. Since $1/x$, $\sqrt{x}$, and $\ln x$ exist and have derivatives of all orders at $x = 1$, the power series we use to represent them can be in powers of $x - 1$. But they do not have power series representations in powers of x because they have no derivatives at $x = 0$.

To simplify the notation, the next theorem deals with the convergence of series of the form $\Sigma a_n x^n$. For series of the form $\Sigma a_n (x - a)^n$ we can replace $(x - a)$ by x' and apply the results to the series $\Sigma a_n (x')^n$.

THEOREM 11

The Convergence Theorem for Power Series

If

$$\sum_{n=0}^{\infty} a_n x^n = a_0 + a_1 x + a_2 x^2 + \cdots \tag{5}$$

converges for $x = c$ $(c \neq 0)$, then it converges absolutely for all $|x| < |c|$. If the series diverges for $x = d$, then it diverges for all $|x| > |d|$.

Proof Suppose the series

$$\sum_{n=0}^{\infty} a_n c^n \tag{6}$$

converges. Then

$$\lim_{n\to\infty} a_n c^n = 0.$$

Hence, there is an integer N such that $|a_n c^n| < 1$ for all $n \geq N$. That is,

$$|a_n| < \frac{1}{|c|^n} \qquad \text{for } n \geq N. \tag{7}$$

Now take any x such that $|x| < |c|$ and consider

$$|a_0| + |a_1 x| + \cdots + |a_{N-1} x^{N-1}| + |a_N x^N| + |a_{N+1} x^{N+1}| + \cdots.$$

There is only a finite number of terms prior to $|a_N x^N|$, and their sum is finite. Starting with $|a_N x^N|$ and beyond, the terms are less than

$$\left|\frac{x}{c}\right|^N + \left|\frac{x}{c}\right|^{N+1} + \left|\frac{x}{c}\right|^{N+2} + \cdots \tag{8}$$

because of (7). But the series in (8) is a geometric series with ratio $r = |x/c|$, which is less than 1 since $|x| < |c|$. Hence the series (8) converges, so the original series (6) converges absolutely. This proves the first half of the theorem.

The second half of the theorem follows from the first. If the series diverges at $x = d$ and converges at a value x_0 with $|x_0| > |d|$, we may take $c = x_0$ in the first half of the theorem and conclude that the series converges absolutely at d. But the series cannot converge absolutely and diverge at one and the same time. Hence, if it diverges at d, it diverges for all $|x| > |d|$. ∎

The Radius and Interval of Convergence

The examples we have looked at and the theorem we just proved lead to the conclusion that a power series always behaves in exactly one of the following three ways.

Possible Behavior of $\Sigma\, c_n\,(x - a)^n$

1. The series converges at $x = a$ and diverges elsewhere.
2. There is a positive number h such that the series diverges for $|x - a| > h$ but converges absolutely for $|x - a| < h$. The series may or may not converge at either of the endpoints $x = a - h$ and $x = a + h$.
3. The series converges absolutely for every x.

In case 2, the set of points at which the series converges is a finite interval, called the **interval of convergence.** We know from past examples that the interval may be open, half-open, or closed, depending on the particular series. But no matter which kind of interval it is, h is called the **radius of convergence** of the series, and $a + h$ is the least upper bound of the set of points at which the series converges. The convergence is absolute at every point in the interior of the interval. If $a = 0$, the interval is centered at the origin. If a power series converges absolutely for all values of x, we say that its radius of convergence is infinite. If it converges only at $x = a$, we say that the radius of convergence is zero.

Term-by-Term Differentiation of Power Series

A theorem from advanced calculus tells us that a power series can be differentiated term by term at each point in the interior of its interval of convergence.

THEOREM 12

The Term-by-Term Differentiation Theorem

If $\Sigma_{n=0}^{\infty} c_n(x - a)^n$ converges for $a - h < x < a + h$ for some $h > 0$, it defines a function f:

$$f(x) = \sum_{n=0}^{\infty} c_n(x - a)^n, \qquad a - h < x < a + h.$$

Such a function f has derivatives of all orders inside the interval of convergence. We can obtain the derivatives by differentiating the original series term by term:

$$f'(x) = \sum_{n=1}^{\infty} n\, c_n(x - a)^{n-1},$$

$$f''(x) = \sum_{n=2}^{\infty} n(n - 1)c_n(x - a)^{n-2},$$

and so on. Each of these derived series converges at every interior point of the interval of convergence of the original series.

Here is an example of how to apply term-by-term differentiation.

Example 4 Is there a more familiar name for the function f defined by the power series

$$f(x) = \sum_{n=0}^{\infty} \frac{(-1)^n x^{2n+1}}{2n + 1} = x - \frac{x^3}{3} + \frac{x^5}{5} - \cdots, \qquad -1 \le x \le 1?$$

A Word of Caution

Term-by-term differentiation might not work for other kinds of series. For example, the trigonometric series

$$\sum_{n=1}^{\infty} \frac{\sin(n!\,x)}{n^2}$$

converges for every x. But if we differentiate term by term we get the series

$$\sum_{n=1}^{\infty} \frac{n!\cos(n!\,x)}{n^2},$$

which diverges for all x.

Solution We differentiate the original series term by term and get

$$f'(x) = \sum_{n=0}^{\infty} \frac{(2n+1)(-1)^n x^{2n}}{2n+1} = \sum_{n=0}^{\infty} (-1)^n x^{2n} = 1 - x^2 + x^4 - x^6 + \cdots.$$

This is a geometric series with first term 1 and ratio $-x^2$, so

$$f'(x) = \frac{1}{1-(-x^2)} = \frac{1}{1+x^2}.$$

We can now integrate $f'(x) = 1/(1+x^2)$ to get

$$\int f'(x)\,dx = \int \frac{dx}{1+x^2} = \tan^{-1}x + C.$$

The series for $f(x)$ is zero when $x = 0$, so $C = 0$. Hence

$$f(x) = x - \frac{x^3}{3} + \frac{x^5}{5} - \frac{x^7}{7} + \cdots = \tan^{-1}x, \qquad -1 \le x \le 1. \tag{9}$$

The function is the restriction of $\tan^{-1}x$ to the interval $[-1, 1]$.

Notice that the original series in Example 4 converges at both endpoints of the original interval of convergence, but the differentiated series converges only inside the interval. This is all that the theorem guarantees.

Term-by-Term Integration of Power Series

Another theorem from advanced calculus states that a power series can be integrated term by term throughout its interval of convergence.

THEOREM 13

The Term-by-Term Integration Theorem

If $\sum_{n=0}^{\infty} c_n(x-a)^n$ converges when $a - h < x < a + h$ for some $h > 0$ and $f(x) = \sum_{n=0}^{\infty} c_n(x-a)^n$ for $a - h < x < a + h$, then the series $\sum_{n=0}^{\infty} c_n \frac{(x-a)^{n+1}}{n+1}$ converges for $a - h < x < a + h$ and

$$\int f(x)\,dx = \sum_{n=0}^{\infty} c_n \frac{(x-a)^{n+1}}{n+1} + C \qquad \text{for } a - h < x < a + h.$$

Example 5 The series

$$\frac{1}{1+t} = 1 - t + t^2 - t^3 \cdots$$

converges on the open interval $-1 < t < 1$. Therefore,

$$\ln(1+x) = \int_0^x \frac{1}{1+t}\,dt = t - \frac{t^2}{2} + \frac{t^3}{3} - \frac{t^4}{4} \cdots \bigg]_0^x$$

$$= x - \frac{x^2}{2} + \frac{x^3}{3} - \frac{x^4}{4} + \cdots, \qquad -1 < x < 1.$$

The latter series also converges at $x = 1$, but that was not guaranteed by the theorem.

Multiplication of Power Series

Still another theorem from advanced calculus states that power series can be multiplied term by term on any common interval of absolute convergence.

THEOREM 14

The Series Multiplication Theorem for Power Series

If $A(x) = \sum_{n=0}^{\infty} a_n x^n$ and $B(x) = \sum_{n=0}^{\infty} b_n x^n$ converge absolutely for $|x| < h$, and

$$c_n = a_0 b_n + a_1 b_{n-1} + a_2 b_{n-2} + \cdots + a_{n-1} b_1 + a_n b_0 = \sum_{k=0}^{n} a_k b_{n-k}, \quad (10)$$

then $\sum_{n=0}^{\infty} c_n x^n$ converges absolutely to $A(x)\,B(x)$ for $|x| < h$:

$$\left(\sum_{n=0}^{\infty} a_n x^n\right) \cdot \left(\sum_{n=0}^{\infty} b_n x^n\right) = \sum_{n=0}^{\infty} c_n x^n. \quad (11)$$

Example 6

Multiply

$$\sum_{n=0}^{\infty} x^n = 1 + x + x^2 + \cdots + x^n + \cdots = \frac{1}{1-x} \qquad \text{for } |x| < 1$$

by itself to get the power series for $1/(1-x)^2$ for $|x| < 1$.

Solution Let

$$A(x) = \sum_{n=0}^{\infty} a_n x^n = 1 + x + x^2 + \cdots + x^n + \cdots = 1/(1-x),$$

$$B(x) = \sum_{n=0}^{\infty} b_n x^n = 1 + x + x^2 + \cdots + x^n + \cdots = 1/(1-x),$$

and

$$c_n = \underbrace{a_0 b_n + a_1 b_{n-1} + \cdots + a_k b_{n-k} + \cdots + a_n b_0}_{n+1 \text{ terms, each with value } 1} = n + 1.$$

Then, by the Series Multiplication Theorem,

$$A(x) \cdot B(x) = \sum_{n=0}^{\infty} c_n x^n = \sum_{n=0}^{\infty} (n+1) x^n$$
$$= 1 + 2x + 3x^2 + 4x^3 + \cdots + (n+1) x^n + \cdots$$

is the series for $1/(1-x)^2$. The series all converge absolutely for $|x| < 1$.

EXERCISES 8.6

Each of Exercises 1–34 gives a formula for the nth term of a series. (a) Find the series' radius and interval of convergence. For what values of x does the series converge (b) absolutely; (c) conditionally?

1. x^n

2. $(x+5)^n$

3. $(-1)^n (x+1)^n$

4. $\dfrac{(x-2)^n}{n}$

5. $\dfrac{(x-2)^n}{10^n}$

6. $(2x)^n$

7. $\frac{nx^n}{n+2}$

8. $\frac{(-1)^n(x+2)^n}{n}$

9. $\frac{x^n}{n\sqrt{n}\,3^n}$

10. $\frac{(x-1)^n}{\sqrt{n}}$

11. $\frac{(-1)^n x^n}{n!}$

12. $\frac{3^n x^n}{n!}$

13. $\frac{x^{2n+1}}{n!}$

14. $\frac{(2x-3)^{2n+1}}{n!}$

15. $\frac{x^n}{\sqrt{n^2+3}}$

16. $\frac{(-1)^n x^n}{\sqrt{n^2+3}}$

17. $\frac{n x^n}{4^n(n^2+1)}$

18. $\frac{n(x-3)^n}{5^n}$

19. $\frac{\sqrt{n}\,x^n}{3^n}$

20. $\sqrt[n]{n}(x-1)^n$

21. $\left(1+\frac{1}{n}\right)^n x^n$

22. $(\ln n)x^n$

23. $n^n x^n$

24. $n!(x-4)^n$

25. $\frac{(-1)^{n+1}(x-2)^n}{n2^n}$

26. $(-2)^n(n+1)(x-1)^n$

27. $\frac{x^n}{n\ln n}$ (*Hint:* See Section 8.3, Exercise 41.)

28. $\frac{(x-1)^{2n+1}}{n^{3/2}}$

29. $\left(\frac{1}{|x|}\right)^n$

30. $\left(\frac{x^2+1}{3}\right)^n$

31. $(\cosh n)x^n$

32. $(\sinh n)x^n$

33. $(\tanh n)x^n$

34. $(\operatorname{sech} n)x^n$

In Exercises 35–38, find the series' interval of convergence and, within this interval, the sum of the series as a function of x.

35. $\sum_{n=0}^{\infty} \frac{(x-1)^{2n}}{2^n}$

36. $\sum_{n=1}^{\infty} (\ln x)^n$

37. $\sum_{n=0}^{\infty} \left(\frac{x^2-1}{2}\right)^n$

38. $\sum_{n=0}^{\infty} \left(\frac{\sqrt{x}}{2}-1\right)^n$

The geometric series in Exercises 39 and 40 have two intervals of convergence. Identify the intervals and find the sum of each series as a function of x.

39. $\sum_{n=0}^{\infty} \left(\frac{3}{2-x^2}\right)^n$

40. $\sum_{n=0}^{\infty} \left(\frac{1}{x-1}\right)^n$

41. For what values of x does the series

$$1-\frac{1}{2}(x-3)+\frac{1}{4}(x-3)^2+\cdots+\left(-\frac{1}{2}\right)^n(x-3)^n+\cdots$$

converge? What is its sum? What series do you get if you differentiate the given series term by term? For what values of x does the new series converge? What is its sum?

42. If you integrate the series in Exercise 41 term by term, what new series do you get? For what values of x does the new series converge and what is another name for its sum?

43. The series for $\tan x$,

$$\tan x = x+\frac{x^3}{3}+\frac{2x^5}{15}+\frac{17x^7}{315}+\frac{62x^9}{2835}+\cdots,$$

converges for $-\pi/2<x<\pi/2$.

a) Find the first five terms of the series for $\ln|\sec x|$. For what values of x should the series converge?

b) Find the first five terms of the series for $\sec^2 x$. For what values of x should this series converge?

c) Check your result in (b) by squaring the series given for $\sec x$ in Exercise 44.

44. The series for $\sec x$,

$$\sec x = 1+\frac{x^2}{2}+\frac{5}{24}x^4+\frac{61}{720}x^6+\frac{277}{8064}x^8+\cdots,$$

converges for $-\pi/2<x<\pi/2$.

a) Find the first five terms of a series for the function $\ln|\sec x+\tan x|$. For what values of x should the series converge?

b) Find the first four terms of a series for $\sec x\tan x$. For what values of x should the series converge?

c) Check your result in (b) by multiplying the series for $\sec x$ by the series given for $\tan x$ in Exercise 43.

45. *Equality of convergent power series.*

a) Show that if two power series $\Sigma_{n=0}^{\infty} a_n x^n$ and $\Sigma_{n=0}^{\infty} b_n x^n$ are convergent and equal for all values of x in an open interval $(-c, c)$, then $a_n = b_n$ for every n. (*Hint:* Let $f(x)=\Sigma_{n=0}^{\infty} a_n x^n = \Sigma_{n=0}^{\infty} b_n x^n$. Differentiate term by term to show that a_n and b_n both equal $f^{(n)}(0)/(n!)$.)

b) Show that if $\Sigma_{n=0}^{\infty} a_n x^n = 0$ for all x in an open interval $(-c, c)$, then $a_n = 0$ for every n.

46. *The sum of the series $\Sigma_{n=0}^{\infty}(n^2/2^n)$.* To find the sum of this series, express $1/(1-x)$ as a geometric series, differentiate both sides of the resulting equation with respect to x, multiply both sides of the result by x, differentiate again, multiply by x again, and set x equal to $1/2$. What do you get? (*Source:* David E. Dobbs' letter to the editor, *Illinois Mathematics Teacher,* Vol. 33, Issue 4, 1982, p. 27.)

47. *Convergence at endpoints.* Show by examples that the convergence of a power series at an endpoint of its interval of convergence may be either conditional or absolute.

8.7 Taylor Series and Maclaurin Series

This section shows how to find power series called Taylor series for functions that have infinitely many derivatives and shows how to control the errors involved in using the partial sums of these series as polynomial approximations of the functions they represent. Our work here continues our earlier work with linearizations and quadratic approximations. One of the important features of Taylor series is that they enable us to extend the domains of functions to include complex numbers. We shall go into that briefly at the end of the section.

DEFINITIONS

Let f be a function with derivatives of all orders throughout some interval containing a as an interior point. Then the **Taylor series generated by f at a** is

$$\sum_{k=0}^{\infty} \frac{f^{(k)}(a)}{k!}(x-a)^k = f(a) + f'(a)(x-a) + \frac{f''(a)}{2!}(x-a)^2 + \cdots + \frac{f^{(n)}(a)}{n!}(x-a)^n + \cdots. \quad (1)$$

The **Maclaurin series generated by f** is

$$\sum_{k=0}^{\infty} \frac{f^{(k)}(0)}{k!}x^k = f(0) + f'(0)x + \frac{f''(0)}{2!}x^2 + \cdots + \frac{f^{(n)}(0)}{n!}x^n + \cdots, \quad (2)$$

the Taylor series generated by f at $x = 0$.

Once we have found the Taylor series generated by a function f at a particular a, we can apply our usual tests to find where the series converges: usually in some interval $(a-h, a+h)$ or for all x. When the series does converge, we ask, "Does it converge to $f(x)$?"

Example 1 Find the Taylor series generated by $f(x) = 1/x$ at $a = 2$. Where, if anywhere, does the series converge to $1/x$?

Solution We need to compute $f(2)$, $f'(2)$, $f''(2)$, and so on. Taking derivatives we get

$$\begin{aligned}
f(x) &= x^{-1}, & f(2) &= 2^{-1} = \frac{1}{2},\\
f'(x) &= -x^{-2}, & f'(2) &= -\frac{1}{2^2},\\
f''(x) &= 2!\,x^{-3}, & \frac{f''(2)}{2!} &= 2^{-3} = \frac{1}{2^3},\\
f'''(x) &= -3!\,x^{-4}, & \frac{f'''(2)}{3!} &= -\frac{1}{2^4},\\
&\vdots\\
f^{(n)}(x) &= (-1)^n n!\,x^{-(n+1)}, & \frac{f^{(n)}(2)}{n!} &= \frac{(-1)^n}{2^{n+1}}.
\end{aligned}$$

The Taylor series is

$$f(2) + f'(2)(x - 2) + \frac{f''(2)}{2!}(x - 2)^2 + \frac{f'''(2)}{3!}(x - 2)^3 + \cdots + \frac{f^{(n)}(2)}{n!}(x - 2)^n + \cdots$$

$$= \frac{1}{2} - \frac{(x - 2)}{2^2} + \frac{(x - 2)^2}{2^3} - \cdots + (-1)^n \frac{(x - 2)^n}{2^{n+1}} + \cdots.$$

This is a geometric series with first term $1/2$ and ratio $r = -(x - 2)/2$. It converges absolutely for $|x - 2| < 2$ and its sum is

$$\frac{1/2}{1 + (x - 2)/2} = \frac{1}{2 + (x - 2)} = \frac{1}{x}.$$

In this example the Taylor series generated by $f(x) = 1/x$ at $a = 2$ converges to $1/x$ for $|x - 2| < 2$ or $0 < x < 4$.

Taylor Polynomials

The linearization and quadratic approximation of a twice-differentiable function f at a point a are the polynomials

$$P_1(x) = f(a) + f'(a)(x - a),$$

$$P_2(x) = f(a) + f'(a)(x - a) + \frac{f''(a)}{2}(x - a)^2.$$

If f has derivatives of higher order at a, then it has higher order polynomial approximations as well, one for each available derivative. These polynomials are called the Taylor polynomials of f.

DEFINITION

Let f be a function with derivatives of order k for $k = 1, 2, \ldots, N$ in some interval containing a as an interior point. Then for any integer n from 0 through N, the **Taylor polynomial** of order n generated by f at a is the polynomial

$$P_n(x) = f(a) + f'(a)(x - a) + \frac{f''(a)}{2!}(x - a)^2 + \cdots + \frac{f^{(k)}(a)}{k!}(x - a)^k + \cdots + \frac{f^{(n)}(a)}{n!}(x - a)^n. \tag{3}$$

We speak of a Taylor polynomial of *order n* rather than *degree n* because $f^{(n)}(a)$ may be zero. The first two Taylor polynomials of $\cos x$ at $x = 0$, for example, are $P_0(x) = 1$ and $P_1(x) = 1$. The first-order polynomial has degree zero, not one.

Special Property of Taylor Polynomials

What is so special about Taylor polynomials? We answer by looking for polynomials whose values at $x = a$ are equal to $f(a)$ and whose derivatives at $x = a$ are $f'(a)$, $f''(a)$, and so on. With that object in mind, we start with a polynomial of order n in powers of $(x - a)$ with undetermined coefficients $c_0, c_1, c_2, \ldots, c_n$:

$$P(x) = c_0 + c_1(x - a) + c_2(x - a)^2 + c_3(x - a)^3 + \cdots + c_n(x - a)^n. \tag{4}$$

Its derivatives of order 1, 2, . . . , n are

$$\begin{aligned}
P'(x) &= c_1 + 2c_2(x-a) + 3c_3(x-a)^2 + \cdots + nc_n(x-a)^{n-1} \\
P''(x) &= 2c_2 + (3)(2)c_3(x-a) + \cdots + n(n-1)c_n(x-a)^{n-2} \\
P'''(x) &= (3!)c_3 + \cdots + n(n-1)(n-2)c_n(x-a)^{n-3} \\
&\vdots \\
P^{(n)}(x) &= (n!)c_n
\end{aligned}$$

When we substitute $x = a$, the terms with $(x - a)$ become 0. We also want $P(a) = f(a)$, $P'(a) = f'(a)$, $P''(a) = f''(a)$, . . . , $P^{(n)}(a) = f^{(n)}(a)$. This leads to the equations

$$P(a) = f(a) = c_0, \qquad P'(a) = f'(a) = c_1,$$

$$P''(a) = f''(a) = (2!)c_2, \qquad P'''(a) = f'''(a) = (3!)c_3,$$

$$\vdots$$

$$P^{(n)}(a) = f^{(n)}(a) = (n!)c_n.$$

When we determine the coefficients c_k in this way, we get

$$c_0 = f(a), \quad c_1 = f'(a), \quad c_2 = \frac{f''(a)}{2!}, \quad c_3 = \frac{f'''(a)}{3!}, \quad \ldots, \quad c_n = \frac{f^{(n)}(a)}{n!}.$$

Substituting these values for $c_0, c_1, \ldots, c_n$ in Eq. (4) gives the Taylor polynomial $P_n(x)$ in Eq. (3). The special property of the Taylor polynomials is just this:

> The Taylor polynomial of order n and its first n derivatives have the same values that f and its first n derivatives have at $x = a$.

A function that has derivatives of all orders at $x = a$ generates a Taylor polynomial for every $n \geq 0$.

Example 2 Find the Taylor polynomials generated by $f(x) = e^x$ at $a = 0$.

Solution Expressed in terms of x, the given function and its derivatives are

$$f(x) = e^x, \qquad f'(x) = e^x, \qquad \ldots, \qquad f^{(n)}(x) = e^x,$$

so

$$f(0) = e^0 = 1, \qquad f'(0) = 1, \qquad \ldots, \qquad f^{(n)}(0) = 1,$$

and

$$P_n(x) = 1 + x + \frac{x^2}{2!} + \frac{x^3}{3!} + \cdots + \frac{x^n}{n!}.$$

See Fig. 8.20.

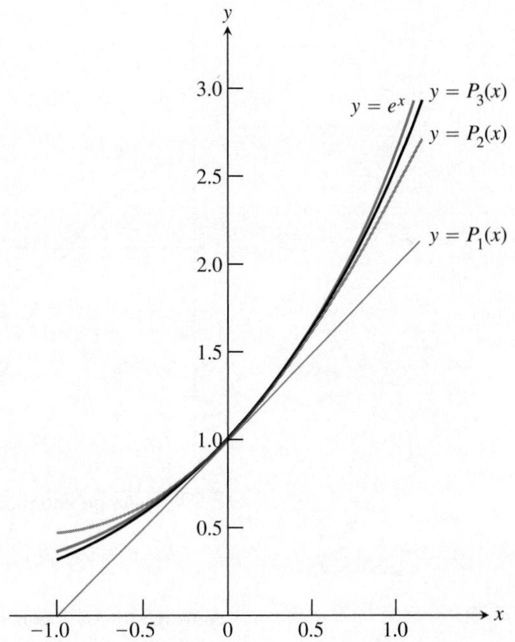

8.20 The graph of $f(x) = e^x$ and its Taylor polynomials $P_1(x) = 1 + x$, $P_2(x) = 1 + x + (x^2/2!)$, and $P_3(x) = 1 + x + (x^2/2!) + (x^3/3!)$. Notice the close agreement near the center $x = 0$.

Example 3 Find the Taylor polynomials generated by $f(x) = \cos x$ at $a = 0$.

Solution The cosine and its derivatives are

$$\begin{aligned} f(x) &= \cos x, & f'(x) &= -\sin x, \\ f''(x) &= -\cos x, & f^{(3)}(x) &= \sin x, \\ &\vdots & &\vdots \\ f^{(2n)}(x) &= (-1)^n \cos x, & f^{(2n+1)}(x) &= (-1)^{n+1}\sin x. \end{aligned}$$

When $x = 0$, the cosines are 1 and the sines are 0, so

$$f^{(2n)}(0) = (-1)^n, \qquad f^{(2n+1)}(0) = 0.$$

Notice that the Taylor polynomials of order $2n$ and of order $2n + 1$ are identical:

$$P_{2n}(x) = P_{2n+1}(x) = 1 - \frac{x^2}{2!} + \frac{x^4}{4!} - \cdots + (-1)^n \frac{x^{2n}}{(2n)!}.$$

Figure 8.21 shows how well these polynomials approximate $y = \cos x$ near $x = 0$. Only the right-hand portions of the graphs are shown because the graphs are symmetric about the y-axis.

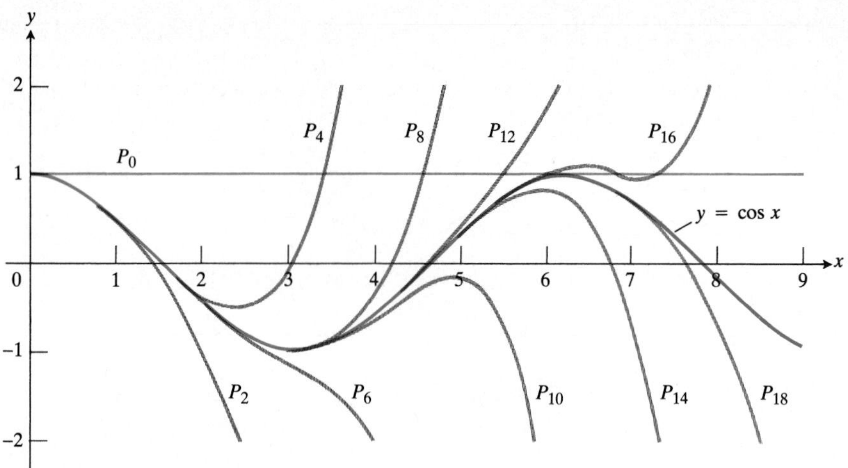

8.21 The polynomials

$$P_{2n}(x) = \sum_{k=0}^{n} [(-1)^k x^{2k}/(2k)!]$$

converge to $\cos x$ as $n \to \infty$. Notice how we can deduce the behavior of $\cos x$ arbitrarily far away solely from the values of the cosine and its derivatives at $a = 0$.

Taylor's Theorem with Remainder

The preceding examples involved Taylor polynomials of order n for $n = 0, 1, 2, \ldots$. As $n \to \infty$, the Taylor polynomials become the partial sums of the Taylor series generated by f at a. The next theorem helps us find out whether these sums converge to $f(x)$ in some interval $(a - h, a + h)$ or, perhaps, for all x.

THEOREM 15

Taylor's Theorem

If f and its first n derivatives $f', f'', \ldots, f^{(n)}$ are continuous on $[a, b]$ or on $[b, a]$, and $f^{(n)}$ is differentiable on (a, b) or on (b, a), then there exists a number c between a and b such that

$$f(b) = f(a) + f'(a)(b - a) + \frac{f''(a)}{2!}(b - a)^2 + \cdots$$

$$+ \frac{f^{(n)}(a)}{n!}(b - a)^n + \frac{f^{(n+1)}(c)}{(n + 1)!}(b - a)^{n+1}.$$

We prove the theorem assuming $a < b$. The proof for the case $a > b$ is nearly the same.

Proof (for $a < b$) The Taylor polynomial

$$P_n(x) = f(a) + f'(a)(x - a) + \frac{f''(a)}{2!}(x - a)^2 + \cdots + \frac{f^{(n)}(a)}{n!}(x - a)^n$$

and its first n derivatives match the function f and its first n derivatives at $x = a$. We do not disturb that matching if we add another term of the form $K(x - a)^{n+1}$, where K is any constant, because such a term and its first n derivatives are all equal to zero at $x = a$. The new function

$$\phi_n(x) = P_n(x) + K(x - a)^{n+1}$$

and its first n derivatives still agree with f and its first n derivatives at $x = a$.

We now choose the particular value of K that makes the curve $y = \phi_n(x)$ agree with the original curve $y = f(x)$ at $x = b$. In symbols,

$$f(b) = P_n(b) + K(b - a)^{n+1}, \qquad \text{or} \qquad K = \frac{f(b) - P_n(b)}{(b - a)^{n+1}}. \tag{5}$$

With K defined by Eq. (5), the function

$$F(x) = f(x) - \phi_n(x) \tag{6}$$

measures the difference between the original function f and the approximating function ϕ_n for each x in $[a, b]$.

We now use Rolle's theorem. First, because $F(a) = F(b) = 0$ and both F and F' are continuous on $[a, b]$, we know that

$$F'(c_1) = 0 \qquad \text{for some } c_1 \text{ in } (a, b).$$

Next, because $F'(a) = F'(c_1) = 0$ and both F' and F'' are continuous on $[a, c_1]$, we know that

$$F''(c_2) = 0 \qquad \text{for some } c_2 \text{ in } (a, c_1).$$

Rolle's theorem, applied successively to F'', F''', ..., $F^{(n-1)}$ implies the existence of

$$\begin{aligned} &c_3 \text{ in } (a, c_2) && \text{such that } F'''(c_3) = 0, \\ &c_4 \text{ in } (a, c_3) && \text{such that } F^{(4)}(c_4) = 0, \\ &\quad\vdots \\ &c_n \text{ in } (a, c_{n-1}) && \text{such that } F^{(n)}(c_n) = 0. \end{aligned}$$

Finally, because $F^{(n)}$ is continuous on $[a, c_n]$ and differentiable on (a, c_n), and $F^{(n)}(a) = F^{(n)}(c_n) = 0$, Rolle's theorem implies that there is a number c_{n+1} in (a, c_n) such that

$$F^{(n+1)}(c_{n+1}) = 0. \tag{7}$$

If we differentiate

$$F(x) = f(x) - P_n(x) - K(x - a)^{n+1}$$

$n + 1$ times, we get

$$F^{(n+1)}(x) = f^{(n+1)}(x) - 0 - (n + 1)!\,K. \tag{8}$$

Equations (7) and (8) together give

$$K = \frac{f^{(n+1)}(c)}{(n + 1)!} \qquad \text{for some number } c = c_{n+1} \text{ in } (a, b). \tag{9}$$

Equations (5) and (9) give

$$f(b) = P_n(b) + \frac{f^{(n+1)}(c)}{(n+1)!}(b-a)^{n+1}.$$

This concludes the proof. ∎

When we apply Taylor's theorem, we usually want to hold a fixed and treat b as an independent variable. Taylor's formula is easier to use in circumstances like these if we change b to x. Here is how the theorem reads with this change.

Taylor's Formula

If f has derivatives of all orders in an open interval I containing a, then for each positive integer n and for each x in I,

$$f(x) = f(a) + f'(a)(x-a) + \frac{f''(a)}{2!}(x-a)^2 + \cdots + \frac{f^{(n)}(a)}{n!}(x-a)^n + R_n(x), \tag{10}$$

where

$$R_n(x) = \frac{f^{(n+1)}(c)}{(n+1)!}(x-a)^{n+1} \quad \text{for some } c \text{ between } a \text{ and } x. \tag{11}$$

When we state Taylor's theorem this way, it says that for each x in I,

$$f(x) = P_n(x) + R_n(x). \tag{12}$$

Pause for a moment to think about how remarkable this equation is. For any value of n we want, the equation gives both a polynomial approximation of f of that order and a formula for the error involved in using that approximation over the interval I.

Equation (10) is called **Taylor's formula.** The function $R_n(x)$ is called the **remainder of order n** or the **error term** for the approximation of f by $P_n(x)$ over I. When $R_n(x) \to 0$ as $n \to \infty$ for all x in I, we say that the Taylor series generated by f at $x = a$ **converges** to f on I and we write

$$f(x) = \sum_{k=0}^{\infty} \frac{f^{(k)}(a)}{k!}(x-a)^k. \tag{13}$$

Example 4 *The Maclaurin Series for e^x.* Show that the Taylor series generated by $f(x) = e^x$ at $x = 0$ converges to $f(x)$ for every real value of x.

Solution The function has derivatives of all orders throughout the interval $-\infty < x < \infty$. Equations (10) and (11) with $f(x) = e^x$ and $a = 0$ give

$$e^x = 1 + x + \frac{x^2}{2!} + \cdots + \frac{x^n}{n!} + R_n(x), \qquad \text{(Polynomial from Example 2)}$$

and

$$R_n(x) = \frac{e^c}{(n+1)!}x^{n+1} \quad \text{for some } c \text{ between 0 and } x.$$

Since e^x is an increasing function of x, e^c lies between $e^0 = 1$ and e^x. When x is negative, so is c, and $e^c < 1$. When x is zero, $e^x = 1$ and $R_n(x) = 0$. When x is positive, so is c, and $e^c < e^x$. Thus,

$$|R_n(x)| \le \frac{|x|^{n+1}}{(n+1)!} \qquad \text{when } x \le 0,$$

and

$$|R_n(x)| < e^x \frac{x^{n+1}}{(n+1)!} \qquad \text{when } x > 0.$$

Finally, because

$$\lim_{n\to\infty} \frac{x^{n+1}}{(n+1)!} = 0 \qquad \text{for every } x, \qquad \text{(Section 8.1)}$$

$\lim_{n\to\infty} R_n(x) = 0$, and the series converges to e^x for every x.

$$e^x = \sum_{k=0}^{\infty} \frac{x^k}{k!} = 1 + x + \frac{x^2}{2!} + \cdots + \frac{x^k}{k!} + \cdots. \qquad (14)$$

Estimating the Remainder

It is often possible to estimate $R_n(x)$ as we did in Example 4. This method of estimation is so convenient that we state it as a theorem for future reference.

THEOREM 16

The Remainder Estimation Theorem

If there are positive constants M and r such that $|f^{(n+1)}(t)| \le Mr^{n+1}$ for all t between a and x, inclusive, then the remainder term $R_n(x)$ in Taylor's theorem satisfies the inequality

$$|R_n(x)| \le M \frac{r^{n+1}|x-a|^{n+1}}{(n+1)!}.$$

If these conditions hold for every n and all the other conditions of Taylor's theorem are satisfied by f, then the series converges to $f(x)$.

In the simplest examples, we can take $r = 1$ provided f and all its derivatives are bounded in magnitude by some constant M. But if $f(x) = 2\cos(3x)$, each time we differentiate we get a factor of 3 and r needs to be greater than 1. In this particular case, we can take $r = 3$ along with $M = 2$.

We are now ready to look at some examples of how the Remainder Estimation Theorem and Taylor's theorem can be used together to settle questions of convergence. As you will see, they can also be used to determine the accuracy with which a function is approximated by one of its Taylor polynomials.

Example 5 *The Maclaurin Series for* sin *x*. Show that the Maclaurin series for $\sin x$ converges to $\sin x$ for all x.

Solution The function and its derivatives are

$$\begin{aligned} f(x) &= \sin x, & f'(x) &= \cos x, \\ f''(x) &= -\sin x, & f'''(x) &= -\cos x, \\ &\vdots \\ f^{(2k)}(x) &= (-1)^k \sin x, & f^{(2k+1)}(x) &= (-1)^k \cos x, \end{aligned}$$

so

$$f^{(2k)}(0) = 0 \quad \text{and} \quad f^{(2k+1)}(0) = (-1)^k.$$

The series has only odd-power terms and, for $n = 2k + 1$, Taylor's theorem gives

$$\sin x = x - \frac{x^3}{3!} + \frac{x^5}{5!} - \cdots + \frac{(-1)^k x^{2k+1}}{(2k+1)!} + R_{2k+1}(x).$$

All the derivatives of $\sin x$ have absolute values less than or equal to 1, so we can apply the Remainder Estimation Theorem with $M = 1$ and $r = 1$ to obtain

$$|R_{2k+1}(x)| \le 1 \cdot \frac{|x|^{2k+2}}{(2k+2)!}.$$

Since $(|x|^{2k+2}/(2k+2)!) \to 0$ as $k \to \infty$, whatever the value of x, $R_{2k+1}(x) \to 0$, and the Maclaurin series for $\sin x$ converges to $\sin x$ for every x.

$$\sin x = \sum_{k=0}^{\infty} \frac{(-1)^k x^{2k+1}}{(2k+1)!} = x - \frac{x^3}{3!} + \frac{x^5}{5!} - \frac{x^7}{7!} + \cdots. \tag{15}$$

Example 6 *The Maclaurin Series for* cos *x*. Show that the Maclaurin series for $\cos x$ converges to $\cos x$ for every value of x.

Solution We add the remainder term to the Taylor polynomial for $\cos x$ in Example 3 to obtain Taylor's formula for $\cos x$ with $n = 2k$:

$$\cos x = 1 - \frac{x^2}{2!} + \frac{x^4}{4!} - \cdots + (-1)^k \frac{x^{2k}}{(2k)!} + R_{2k}(x).$$

Because the derivatives of the cosine have absolute value less than or equal to 1, the Remainder Estimation Theorem with $M = 1$ and $r = 1$ gives

$$|R_{2k}(x)| \le 1 \cdot \frac{|x|^{2k+1}}{(2k+1)!}.$$

For every value of x, $R_{2k} \to 0$ as $k \to \infty$. Therefore, the series converges to $\cos x$ for every value of x.

Who Invented Taylor Series?

Brook Taylor (1685–1731) did not invent Taylor series, and Maclaurin series were not developed by Colin Maclaurin (1698–1746). James Gregory was working with Taylor series when Taylor was only a few years old, and he published the Maclaurin series for tan x, sec x, $\tan^{-1}x$, and $\sec^{-1}x$ ten years before Maclaurin was born. At about the same time, Nicolaus Mercator discovered the Maclaurin series for $\ln(1 + x)$.

Taylor was not aware of Gregory's work when he published his book *Methodus incrementorum directa et inversa* in 1715, which contained what we now call Taylor series. Maclaurin quoted Taylor's work in a calculus book he wrote in 1742. The book popularized series representations of functions and although Maclaurin never claimed to have discovered them, Taylor series centered at $a = 0$ became known as Maclaurin series. History evened things up in the end. Maclaurin, a brilliant mathematician, was the original discoverer of the rule for solving systems of equations that we now call Cramer's rule.

$$\cos x = \sum_{k=0}^{\infty} \frac{(-1)^k x^{2k}}{(2k)!} = 1 - \frac{x^2}{2!} + \frac{x^4}{4!} - \frac{x^6}{6!} + \cdots. \tag{16}$$

Example 7 Find the Maclaurin series for cos $2x$ and show that it converges to cos $2x$ for every value of x.

Solution We replace the x in Eq. (16) by $2x$, to obtain

$$\cos 2x = \sum_{k=0}^{\infty} \frac{(-1)^k (2x)^{2k}}{(2k)!} = 1 - \frac{(2x)^2}{2!} + \frac{(2x)^4}{4!} - \frac{(2x)^6}{6!} + \cdots.$$

The Maclaurin series for cos x converges to cos x for every value of x and therefore converges for every value of $2x$.

Truncation Error

The Maclaurin series for e^x converges to e^x for all x. But we still need to decide how many terms to use to approximate e^x to a given degree of accuracy.

Here are some examples of how to use the Remainder Estimation Theorem to estimate truncation error.

Example 8 Calculate e with an error of less than 10^{-6}.

Solution We can use the result of Example 4 with $x = 1$ to write

$$e = 1 + 1 + \frac{1}{2!} + \cdots + \frac{1}{n!} + R_n(1),$$

with

$$R_n(1) = e^c \frac{1}{(n+1)!} \qquad \text{for some } c \text{ between 0 and 1.}$$

For the purposes of this example, we assume that we know that $e < 3$. Hence, we are certain that

$$\frac{1}{(n+1)!} < R_n(1) < \frac{3}{(n+1)!}$$

because $1 < e^c < 3$.

By experiment we find that $1/9! > 10^{-6}$, while $3/10! < 10^{-6}$. Thus we should take $(n + 1)$ to be at least 10, or n to be at least 9. With an error of less than 10^{-6},

$$e = 1 + 1 + \frac{1}{2} + \frac{1}{3!} + \cdots + \frac{1}{9!} = 2.7182\ 82. \qquad \text{(rounded)}$$

Example 9 For what values of x can we replace sin x by $x - (x^3/3!)$ with an error of magnitude no greater than 3×10^{-4}?

Solution Here we can take advantage of the fact that the Maclaurin series for $\sin x$ is an alternating series for every nonzero value of x. According to the Alternating Series Estimation Theorem (Section 8.5), the error in truncating

$$\sin x = x - \frac{x^3}{3!} \mathrel{\vdots} + \frac{x^5}{5!} - \cdots$$

after $(x^3/3!)$ is no greater than

$$\left|\frac{x^5}{5!}\right| = \frac{|x|^5}{120}.$$

Therefore the error will be less than or equal to 3×10^{-4} if

$$\frac{|x|^5}{120} < 3 \times 10^{-4} \quad \text{or} \quad |x| < \sqrt[5]{360 \times 10^{-4}} \approx 0.514. \quad \begin{pmatrix}\text{Rounded}\\ \text{down, to be}\\ \text{safe}\end{pmatrix}$$

The Alternating Series Estimation Theorem tells us something that the Remainder Estimation Theorem does not: namely, that the estimate $x - (x^3/3!)$ for $\sin x$ is an underestimate when x is positive because then $x^5/120$ is positive.

Figure 8.22 shows the graph of $\sin x$, along with the graphs of a number of its approximating Taylor polynomials. Notice that the graph of $P_3(x) = x - (x^3/3!)$ is almost indistinguishable from the sine curve when $-1 \leq x \leq 1$. However, it crosses the x-axis at $\pm\sqrt{6} \approx \pm 2.45$, whereas the sine curve crosses the axis at $\pm\pi \approx \pm 3.14$.

You might wonder how the estimate given by the Remainder Estimation Theorem compares with the one just obtained from the Alternating Series Estimation Theorem. If we write

$$\sin x = x - \frac{x^3}{3!} + R_3,$$

then the Remainder Estimation Theorem gives

$$|R_3| \leq 1 \cdot \frac{|x|^4}{4!} = \frac{|x|^4}{24},$$

which is not very good. But when we recognize that $x - (x^3/3!) = 0 + x + 0x^2 - (x^3/3!) + 0x^4$ is the Taylor polynomial of order 4 as well as of order

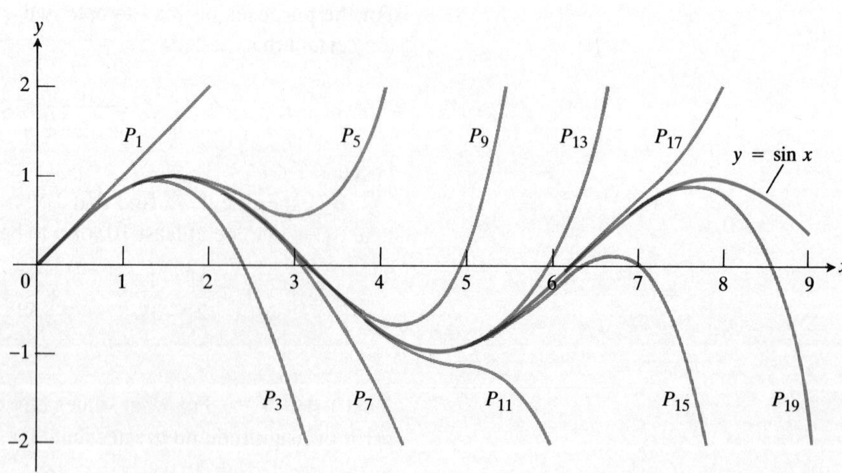

8.22 The polynomials

$$P_{2n+1}(x) = \sum_{k=0}^{n} [(-1)^k x^{2k+1}/(2k+1)!]$$

converge to $\sin x$ as $n \to \infty$.

3, then we have

$$\sin x = x - \frac{x^3}{3!} + 0 + R_4,$$

and the Remainder Estimation Theorem with $M = r = 1$ gives

$$|R_4| \leq 1 \cdot \frac{|x|^5}{5!} = \frac{|x|^5}{120}.$$

This is what we had from the Alternating Series Estimation Theorem.

Combining Taylor Series

On common intervals of convergence, Taylor series can be added, subtracted, and multiplied by constants just as other series can, and the results are once again Taylor series. The Taylor series for $f(x) + g(x)$ is the sum of the Taylor series for $f(x)$ and $g(x)$ because the nth derivative of $f + g$ is $f^{(n)} + g^{(n)}$, and so on. Thus we obtain the Maclaurin series for $(1 + \cos 2x)/2$ by adding 1 to the Maclaurin series for $\cos 2x$ and dividing the combined results by 2, and the Maclaurin series for $\sin x + \cos x$ is the term-by-term sum of the Maclaurin series for $\sin x$ and $\cos x$.

$e^{i\theta} = \cos\theta + i\sin\theta$

As you may recall, a complex number is a number of the form $a + bi$, where a and b are real numbers and $i = \sqrt{-1}$. If we substitute $x = i\theta$ (θ real) in the Maclaurin series for e^x and use the relations

$$i^2 = -1, \qquad i^3 = i^2 i = -i, \qquad i^4 = i^2 i^2 = 1, \qquad i^5 = i^4 i = i,$$

and so on, to simplify the result, we obtain

$$\begin{aligned} e^{i\theta} &= 1 + \frac{i\theta}{1!} + \frac{i^2\theta^2}{2!} + \frac{i^3\theta^3}{3!} + \frac{i^4\theta^4}{4!} + \frac{i^5\theta^5}{5!} + \frac{i^6\theta^6}{6!} + \cdots \\ &= \left(1 - \frac{\theta^2}{2!} + \frac{\theta^4}{4!} - \frac{\theta^6}{6!} + \cdots\right) + i\left(\theta - \frac{\theta^3}{3!} + \frac{\theta^5}{5!} - \cdots\right) = \cos\theta + i\sin\theta. \end{aligned} \tag{17}$$

This does not *prove* that $e^{i\theta} = \cos\theta + i\sin\theta$ because we have not yet defined what it means to raise e to an imaginary power. But it does say how we ought to define $e^{i\theta}$ to be consistent with other things we know.

DEFINITION

For any real number θ, $\quad e^{i\theta} = \cos\theta + i\sin\theta. \qquad (18)$

Equation (18), called **Euler's formula,** enables us to define e^{a+bi} to be $e^a \cdot e^{bi}$ for any complex number $a + bi$.

One of the amazing consequences of Euler's formula is the equation

$$e^{i\pi} = -1.$$

When written in the form $e^{i\pi} + 1 = 0$, this equation combines the five most important constants in mathematics.

EXERCISES 8.7

In Exercises 1–8, find the Taylor polynomials of orders 0, 1, 2, and 3 generated by f at a.

1. $f(x) = \ln x, \quad a = 1$
2. $f(x) = \ln(1 + x), \quad a = 0$
3. $f(x) = 1/x, \quad a = 2$
4. $f(x) = 1/(x + 2), \quad a = 0$
5. $f(x) = \sin x, \quad a = \pi/4$
6. $f(x) = \cos x, \quad a = \pi/4$
7. $f(x) = \sqrt{x}, \quad a = 4$
8. $f(x) = \sqrt{x + 4}, \quad a = 0$

Find the Maclaurin series for the functions in Exercises 9–18.

9. e^{-x}
10. $e^{x/2}$
11. $\sin 3x$
12. $5 \cos \pi x$
13. $\cos(-x)$
14. $x \sin x$
15. $\cosh x = (e^x + e^{-x})/2$
16. $\sinh x = (e^x - e^{-x})/2$
17. $(x^2/2) - 1 + \cos x$
18. $\cos^2 x$ (*Hint:* $\cos^2 x = (1 + \cos 2x)/2$)

Quadratic Approximations

Write out Taylor's formula (Eq. 10) with $n = 2$ and $a = 0$ for the functions in Exercises 19–24. This will give you the quadratic approximations of these functions at $x = 0$ and the associated error terms.

19. $e^{\tan x}$
20. $e^{\sin x}$
21. $\ln(\cos x)$
22. $\dfrac{1}{\sqrt{1 - x^2}}$
23. $\sinh x$
24. $\cosh x$
25. Use the Taylor series generated by e^x at $x = a$ to show that
$$e^x = e^a\left[1 + (x - a) + \frac{(x - a)^2}{2!} + \cdots\right].$$
26. Find the Taylor series generated by e^x at $a = 1$. Compare your answer with the formula in Exercise 25.
27. For approximately what values of x can you replace $\sin x$ by $x - (x^3/6)$ with an error of magnitude no greater than 5×10^{-4}?
28. If $\cos x$ is replaced by $1 - (x^2/2)$ and $|x| < 0.5$, what estimate can be made of the error? Does $1 - (x^2/2)$ tend to be too large, or too small?
29. How close is the approximation $\sin x = x$ when $|x| < 10^{-3}$? For which of these values of x is $x < \sin x$?
30. The estimate $\sqrt{1 + x} = 1 + (x/2)$ is used when x is small. Estimate the error when $|x| < 0.01$.
31. The approximation $e^x = 1 + x + (x^2/2)$ is used when x is small. Use the Remainder Estimation Theorem to estimate the error when $|x| < 0.1$.
32. When $x < 0$, the series for e^x is an alternating series. Use the Alternating Series Estimation Theorem to estimate the error that results from replacing e^x by $1 + x + (x^2/2)$ when $-0.1 < x < 0$. Compare with Exercise 31.
33. Estimate the error in the approximation $\sinh x = x + (x^3/3!)$ when $|x| < 0.5$. (*Hint:* Use R_4, not R_3.)
34. When $0 \le h \le 0.01$, show that e^h may be replaced by $1 + h$ with an error of magnitude no greater than 0.6% of h. Use $e^{0.01} = 1.01$.

Each of the series in Exercises 35 and 36 is the value of the Maclaurin series of a function $f(x)$ at some point. What function and what point? What is the sum of the series?

35. $(0.1) - \dfrac{(0.1)^3}{3!} + \dfrac{(0.1)^5}{5!} - \cdots + \dfrac{(-1)^k(0.1)^{2k+1}}{(2k + 1)!} + \cdots$
36. $1 - \dfrac{\pi^2}{4^2 \cdot 2!} + \dfrac{\pi^4}{4^4 \cdot 4!} - \cdots + \dfrac{(-1)^k(\pi)^{2k}}{4^{2k} \cdot (2k!)} - \cdots$
37. Differentiate the Maclaurin series for $\sin x$, $\cos x$, and e^x term by term and compare your results with the Maclaurin series for $\cos x$, $\sin x$, and e^x.
38. Integrate the Maclaurin series for $\sin x$, $\cos x$, and e^x term by term and compare your results with the Maclaurin series for $\cos x$, $\sin x$, and e^x.
39. Multiply the Maclaurin series for $2 \cos x$ and $\sin x$ together to find the first five nonzero terms of the product series. Confirm that this is the beginning of the Maclaurin series for $\sin 2x$.
40. Multiply the Maclaurin series for e^x and $\cos x$ together to find the first five nonzero terms of the Maclaurin series for $e^x \cos x$.
41. Use the Maclaurin series for $\cos x$ and the Alternating Series Estimation Theorem to show that
$$\frac{1}{2} - \frac{x^2}{24} < \frac{1 - \cos x}{x^2} < \frac{1}{2} \quad \text{for} \quad x \neq 0.$$
(This is the inequality in Section 1.9, Exercise 21.)

42. Use series to verify that $y = y_0 e^x$ is a solution of the equation $y' = y$ for any constant y_0.

43. Use the identity $\sin^2 x = (1 - \cos 2x)/2$ to obtain a series for $\sin^2 x$. Then differentiate this series to obtain a series for $2 \sin x \cos x$. Check that this is the series for $\sin 2x$.

44. *A convergent Taylor series that converges to its generating function only at its center.* It can be shown (though not simply) that the function f defined by the rule

$$f(x) = \begin{cases} 0 & \text{if } x = 0 \\ e^{-1/x^2} & \text{if } x \neq 0 \end{cases}$$

has derivatives of all orders at $x = 0$ (Fig. 8.23) and that $f^{(n)}(0) = 0$ for all n.

a) Find the Maclaurin series generated by f. At what values of x does the series converge? At what values of x does the series converge to $f(x)$?

b) Write out Taylor's formula (Eq. 10) for f, taking $a = 0$ and assuming $x \neq 0$. What does the formula tell you about the value of $R_n(x)$?

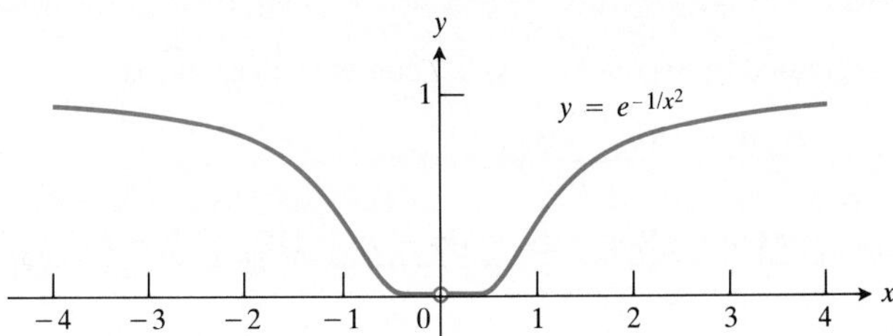

8.23 The graph of the continuous extension of $y = e^{-1/x^2}$ is so flat at the origin that all of its derivatives there are zero (Exercise 44).

45. Use Eq. (18) to write the following powers of e in the form $a + bi$.

a) $e^{-i\pi}$ b) $e^{i\pi/4}$ c) $e^{-i\pi/2}$

46. *Euler's identities.* Use Eq. (18) to show that

$$\cos \theta = \frac{e^{i\theta} + e^{-i\theta}}{2} \quad \text{and} \quad \sin \theta = \frac{e^{i\theta} - e^{-i\theta}}{2i}.$$

47. Establish the equations in Exercise 46 by combining the formal Maclaurin series for $e^{i\theta}$ and $e^{-i\theta}$.

48. Show that

a) $\cosh i\theta = \cos \theta$,

b) $\sinh i\theta = i \sin \theta$.

49. By multiplying the Maclaurin series for e^x and $\sin x$, find the terms through x^5 of the Maclaurin series for $e^x \sin x$. This series is the imaginary part of the series for

$$e^x \cdot e^{ix} = e^{(1+i)x}.$$

Use this fact to check your answer. For what values of x should the series for $e^x \sin x$ converge?

50. When a and b are real, we define $e^{(a+ib)x}$ to be $e^{ax}(\cos bx + i \sin bx)$. From this definition, show that

$$\frac{d}{dx} e^{(a+ib)x} = (a + ib)\, e^{(a+ib)x}.$$

51. Use the definition of $e^{i\theta}$ to show that for any real numbers θ, θ_1, and θ_2,

a) $e^{i\theta_1} e^{i\theta_2} = e^{i(\theta_1 + \theta_2)}$,

b) $e^{-i\theta} = 1/e^{i\theta}$.

52. Two complex numbers $a + ib$ and $c + id$ are equal if and only if $a = c$ and $b = d$. Use this fact to evaluate

$$\int e^{ax} \cos bx \, dx \quad \text{and} \quad \int e^{ax} \sin bx \, dx$$

from

$$\int e^{(a+ib)x} \, dx = \frac{a - ib}{a^2 + b^2} e^{(a+ib)x} + C,$$

where $C = C_1 + iC_2$ is a complex constant of integration.

53. *The Maclaurin series generated by $f(x) = \sum_{n=0}^{\infty} a_n x^n$ is $\sum_{n=0}^{\infty} a_n x^n$.* A function defined by a power series $\sum_{n=0}^{\infty} a_n x^n$ with a radius of convergence $c > 0$ has a Maclaurin series that converges to the function at every point of $(-c, c)$. Show this by showing that the Maclaurin series generated by $f(x) = \sum_{n=0}^{\infty} a_n x^n$ is the series $\sum_{n=0}^{\infty} a_n x^n$ itself.

An immediate consequence of this is that series like

$$x \sin x = x^2 - \frac{x^4}{3!} + \frac{x^6}{5!} - \frac{x^8}{7!} + \cdots$$

and

$$x^2 e^x = x^2 + x^3 + \frac{x^4}{2!} + \frac{x^5}{3!} + \cdots,$$

obtained by multiplying Maclaurin series by powers of x, as well as series obtained by integration and differentiation of convergent power series, are themselves the Maclaurin series generated by the functions they represent.

54. *Maclaurin series for even functions and odd functions (continuation of Section 8.6, Exercise 45).* Suppose that $f(x) = \sum_{n=0}^{\infty} a_n x^n$ converges for all x in an open interval $(-c, c)$. Show that

a) If f is even, then $a_1 = a_3 = a_5 = \cdots = 0$, i.e., the series for f contains only even powers of x.

b) If f is odd, then $a_0 = a_2 = a_4 = \cdots = 0$, i.e., the series for f contains only odd powers of x.

55. *Taylor polynomials of periodic functions*

a) Show that every continuous periodic function $f(x)$, $-\infty < x < \infty$, is bounded in magnitude by showing that there exists a positive constant M such that $|f(x)| \leq M$ for all x.

b) Show that the graph of every Taylor polynomial generated by $f(x) = \cos x$ must eventually move away from the graph of $\cos x$ as $|x|$ increases. You can see this happening in Fig. 8.21. The Taylor polynomials of sin x behave in a similar way (Fig. 8.22).

56. GRAPHER

a) Graph the curves $y = 1 - (x^2/6)$ and $y = (\sin x)/x$ together with the line $y = 1$.

b) Use the Maclaurin series for $\sin x$ and the Alternating Series Estimation Theorem to show that

$$1 - \frac{x^2}{6} < \frac{\sin x}{x} < 1 \quad \text{for} \quad x \neq 0.$$

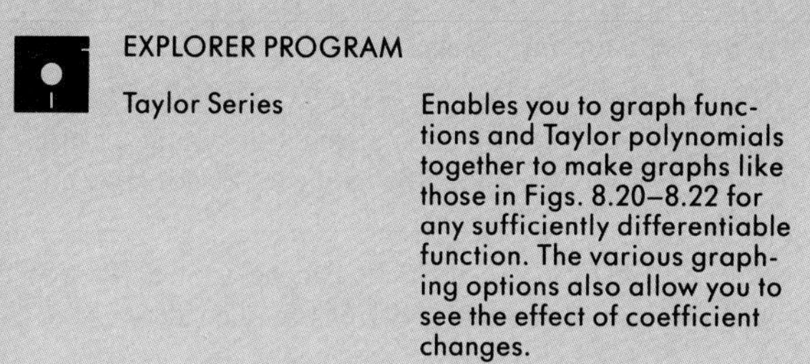

EXPLORER PROGRAM

Taylor Series — Enables you to graph functions and Taylor polynomials together to make graphs like those in Figs. 8.20–8.22 for any sufficiently differentiable function. The various graphing options also allow you to see the effect of coefficient changes.

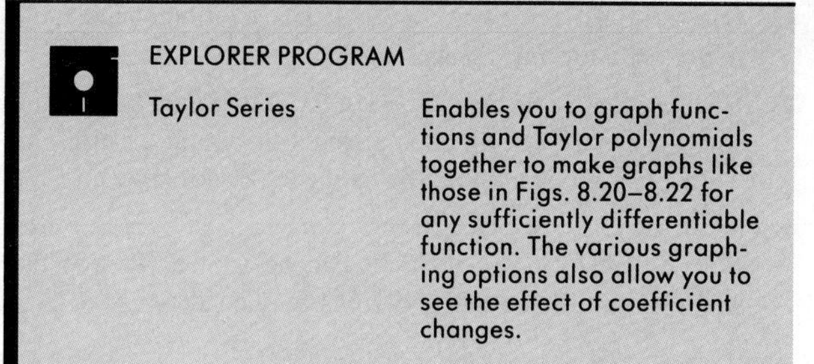

8.8 Calculations with Taylor Series

This section introduces the binomial series, discusses how to choose useful centers for Taylor series, and shows how series are sometimes used to evaluate nonelementary integrals. The section concludes with a table of frequently used Maclaurin series.

The Binomial Series

The Maclaurin series generated by $f(x) = (1 + x)^m$, when m is constant, is

$$1 + mx + \frac{m(m-1)}{2!}x^2 + \frac{m(m-1)(m-2)}{3!}x^3 + \cdots + \frac{m(m-1)(m-2)\cdots(m-k+1)}{k!}x^k + \cdots. \tag{1}$$

This series, called the **binomial series,** converges absolutely for $|x| < 1$. To derive the series, we first list the function and its derivatives:

$$\begin{aligned} f(x) &= (1+x)^m, \\ f'(x) &= m(1+x)^{m-1}, \\ f''(x) &= m(m-1)(1+x)^{m-2}, \\ f'''(x) &= m(m-1)(m-2)(1+x)^{m-3}, \\ &\vdots \\ f^{(k)}(x) &= m(m-1)(m-2)\cdots(m-k+1)(1+x)^{m-k}. \end{aligned}$$

We then evaluate these at $x = 0$ and substitute in the Maclaurin series formula to obtain

$$1 + mx + \frac{m(m-1)}{2!}x^2 + \cdots + \frac{m(m-1)(m-2)\cdots(m-k+1)}{k!}x^k + \cdots. \tag{2}$$

If m is an integer greater than or equal to zero, the series stops after $(m + 1)$ terms because the coefficients from $k = m + 1$ on are zero.

If m is not a positive integer or zero, the series is infinite and converges for $|x| < 1$. To see why, let u_k be the term involving x^k. Then apply the ratio test for absolute convergence to see that

$$\left|\frac{u_{k+1}}{u_k}\right| = \left|\frac{m-k}{k+1}x\right| \to |x| \qquad \text{as } k \to \infty.$$

Our derivation of the binomial series shows only that it is generated by $(1+x)^m$ and converges for $|x| < 1$. The derivation does not show that the series actually converges to $(1+x)^m$. The series does, but we shall assume that part without proof.

For $-1 < x < 1$,

$$(1+x)^m = 1 + \sum_{k=1}^{\infty} \binom{m}{k} x^k, \tag{3}$$

where

$$\binom{m}{1} = m, \qquad \binom{m}{2} = \frac{m(m-1)}{2!},$$

and

$$\binom{m}{k} = \frac{m(m-1)(m-2)\cdots(m-k+1)}{k!} \qquad \text{for } k \geq 3.$$

Example 1 Show that when $m = -1$, Eq. (3) gives the geometric series

$$\frac{1}{1+x} = 1 - x + x^2 - x^3 + \cdots + (-1)^k x^k + \cdots. \tag{4}$$

Solution When $m = -1$,

$$\binom{-1}{1} = -1, \qquad \binom{-1}{2} = \frac{-1(-2)}{2!} = 1,$$

and

$$\binom{-1}{k} = \frac{-1(-2)(-3)\cdots(-1-k+1)}{k!} = (-1)^k \left(\frac{k!}{k!}\right) = (-1)^k.$$

With these coefficient values, Eq. (3) becomes

$$(1+x)^{-1} = 1 + \sum_{k=1}^{\infty} (-1)^k x^k = 1 - x + x^2 - x^3 + \cdots + (-1)^k x^k + \cdots,$$

which is Eq. (4).

Choosing Centers for Taylor Series

Taylor's formula,

$$f(x) = f(a) + f'(a)(x-a) + \frac{f''(a)}{2!}(x-a)^2 + \cdots$$
$$+ \frac{f^{(n)}(a)}{n!}(x-a)^n + \frac{f^{(n+1)}(c)}{(n+1)!}(x-a)^{n+1}, \tag{5}$$

expresses the value of f at x in terms of f and its derivatives at a. In numerical computations, we therefore need a to be a point where we know the values of f and its derivatives. We also need a to be close enough to the values of x we are interested in to make $(x - a)^{n+1}$ so small we can neglect the remainder.

Example 2 What value of a might we choose in Taylor's formula (Eq. 5) to compute sin 35° efficiently?

Solution The radian measure for 35° is $35\pi/180$. We could choose $a = 0$ and use the series

$$\sin x = x - \frac{x^3}{3!} + \frac{x^5}{5!} - \cdots + (-1)^n \frac{x^{2n+1}}{(2n+1)!} + 0 \cdot x^{2n+2} + R_{2n+2}(x). \quad (6)$$

Alternatively, we could choose $a = \pi/6$ (which corresponds to 30°) and use the series

$$\sin x = \sin\frac{\pi}{6} + \cos\frac{\pi}{6}\left(x - \frac{\pi}{6}\right) - \sin\frac{\pi}{6}\frac{(x - \pi/6)^2}{2!} - \cos\frac{\pi}{6}\frac{(x - \pi/6)^3}{3!}$$

$$+ \cdots + \sin\left(\frac{\pi}{6} + n\frac{\pi}{2}\right)\frac{(x - \pi/6)^n}{n!} + R_n(x).$$

The remainder in Eq. (6) satisfies the inequality

$$|R_{2n+2}(x)| \le \frac{|x|^{2n+3}}{(2n+3)!},$$

which tends to zero as $n \to \infty$, no matter how large $|x|$ may be. We could therefore calculate sin 35° by placing

$$x = \frac{35\pi}{180} \approx 0.6108\ 652$$

in the approximation

$$\sin x \approx x - \frac{x^3}{6} + \frac{x^5}{120} - \frac{x^7}{5040}.$$

This gives a truncation error of magnitude no greater than 3.3×10^{-8}, since

$$\left|R_8\left(\frac{35\pi}{180}\right)\right| < \frac{(0.611)^9}{9!} < 3.3 \times 10^{-8}.$$

By using the series with $a = \pi/6$, we could obtain equal accuracy with a smaller exponent n, but at the expense of introducing $\cos \pi/6 = \sqrt{3}/2$ as one of the coefficients. In this series, with $a = \pi/6$, we would take $x = 35\pi/180$, and the quantity $(x - a)$ would be

$$x - \frac{\pi}{6} = \frac{35\pi}{180} - \frac{30\pi}{180}$$

$$= \frac{5\pi}{180} \approx 0.0872\ 665,$$

which decreases rapidly as it is raised to higher powers.

Evaluating Nonelementary Integrals

Maclaurin series are often used to express nonelementary integrals in terms of series.

Example 3 Express $\int \sin x^2\,dx$ as a power series.

Solution From the series for $\sin x$ we obtain

$$\sin x^2 = x^2 - \frac{x^6}{3!} + \frac{x^{10}}{5!} - \frac{x^{14}}{7!} + \frac{x^{18}}{9!} - \cdots.$$

Therefore,

$$\int \sin x^2\,dx = C + \frac{x^3}{3} - \frac{x^7}{7\cdot 3!} + \frac{x^{11}}{11\cdot 5!} - \frac{x^{15}}{15\cdot 7!} + \frac{x^{19}}{19\cdot 9!} - \cdots.$$

Example 4 Estimate $\int_0^1 \sin x^2\,dx$ with an error of less than 0.001.

Solution From the indefinite integral in Example 3,

$$\int_0^1 \sin x^2\,dx = \frac{1}{3} - \frac{1}{7\cdot 3!} + \frac{1}{11\cdot 5!} - \frac{1}{15\cdot 7!} + \frac{1}{19\cdot 9!} - \cdots.$$

The series alternates, and we find by experiment that

$$\frac{1}{11\cdot 5!} \approx 0.0007\ 6$$

is the first term to be numerically less than 0.001. The sum of the preceding two terms gives

$$\int_0^1 \sin x^2\,dx \approx \frac{1}{3} - \frac{1}{42} \approx 0.310.$$

With two more terms we could estimate

$$\int_0^1 \sin x^2\,dx \approx 0.3102\ 68$$

with an error of less than 10^{-6}, and with only one term beyond that we have

$$\int_0^1 \sin x^2\,dx \approx \frac{1}{3} - \frac{1}{42} + \frac{1}{1320} - \frac{1}{75600} + \frac{1}{6894720} \approx 0.3102\ 68303,$$

with an error of less than 10^{-9}. To guarantee this accuracy with the error formula for the trapezoidal rule would require using about 13,000 subintervals.

Arctangents

In Section 8.6, Example 4, we found a series for $\tan^{-1}x$ by differentiating to get

$$\frac{d}{dx}\tan^{-1}x = \frac{1}{1+x^2} = 1 - x^2 + x^4 - x^6 + \cdots$$

and integrating to get

$$\tan^{-1}x = x - \frac{x^3}{3} + \frac{x^5}{5} - \frac{x^7}{7} + \cdots.$$

However, we did not prove the term-by-term integration theorem on which this conclusion depended. We now derive the series again by integrating both sides of the finite formula

$$\frac{1}{1+t^2} = 1 - t^2 + t^4 - t^6 + \cdots + (-1)^n t^{2n} + \frac{(-1)^{n+1}t^{2n+2}}{1+t^2}, \tag{7}$$

in which the last term comes from adding the remaining terms as a geometric series with first term $a = t^{2n+2}$ and ratio $r = -t^2$. Integrating both sides of Eq. (7) from $t = 0$ to $t = x$ gives

$$\tan^{-1}x = x - \frac{x^3}{3} + \frac{x^5}{5} - \frac{x^7}{7} + \cdots + (-1)^n \frac{x^{2n+1}}{2n+1} + R(n, x),$$

where

$$R(n, x) = \int_0^x \frac{(-1)^{n+1}t^{2n+2}}{1+t^2}\,dt.$$

The denominator of the integrand is greater than or equal to 1; hence

$$|R(n, x)| \le \int_0^{|x|} t^{2n+2}\,dt = \frac{|x|^{2n+3}}{2n+3}.$$

If $|x| \le 1$, the right side of this inequality approaches zero as $n \to \infty$. Therefore $\lim_{n\to\infty} R(n, x) = 0$ if $|x| \le 1$ and

$$\tan^{-1}x = \sum_{n=0}^{\infty} \frac{(-1)^n x^{2n+1}}{2n+1}, \qquad |x| \le 1.$$

We take this route instead of finding the Maclaurin series directly because the formulas for the higher order derivatives of $\tan^{-1}x$ are unmanageable.

$$\tan^{-1}x = x - \frac{x^3}{3} + \frac{x^5}{5} - \frac{x^7}{7} + \cdots, \qquad |x| \le 1 \tag{8}$$

When we put $x = 1$ and $\tan^{-1}1 = \pi/4$ in Eq. (8), we get **Leibniz's formula:**

$$\frac{\pi}{4} = 1 - \frac{1}{3} + \frac{1}{5} - \frac{1}{7} + \frac{1}{9} - \cdots + \frac{(-1)^n}{2n+1} + \cdots.$$

This series converges too slowly to be a useful source of decimal approximations of π. It is better to use a formula like

$$\pi = 48 \tan^{-1}\frac{1}{18} + 32 \tan^{-1}\frac{1}{57} - 20 \tan^{-1}\frac{1}{239}, \tag{9}$$

which uses values of x closer to zero.

Frequently Used Maclaurin Series

$$\frac{1}{1-x} = 1 + x + x^2 + \cdots + x^n + \cdots = \sum_{n=0}^{\infty} x^n, \qquad |x| < 1$$

$$\frac{1}{1+x} = 1 - x + x^2 - \cdots + (-x)^n + \cdots = \sum_{n=0}^{\infty} (-1)^n x^n, \qquad |x| < 1$$

$$e^x = 1 + x + \frac{x^2}{2!} + \cdots + \frac{x^n}{n!} + \cdots = \sum_{n=0}^{\infty} \frac{x^n}{n!}, \qquad |x| < \infty$$

$$\sin x = x - \frac{x^3}{3!} + \frac{x^5}{5!} - \cdots + (-1)^n \frac{x^{2n+1}}{(2n+1)!} + \cdots = \sum_{n=0}^{\infty} \frac{(-1)^n x^{2n+1}}{(2n+1)!}, \qquad |x| < \infty$$

$$\cos x = 1 - \frac{x^2}{2!} + \frac{x^4}{4!} - \cdots + (-1)^n \frac{x^{2n}}{(2n)!} + \cdots = \sum_{n=0}^{\infty} \frac{(-1)^n x^{2n}}{(2n)!}, \qquad |x| < \infty$$

$$\ln(1+x) = x - \frac{x^2}{2} + \frac{x^3}{3} - \cdots + (-1)^{n-1} \frac{x^n}{n} + \cdots = \sum_{n=1}^{\infty} \frac{(-1)^{n-1} x^n}{n}, \qquad -1 < x \le 1$$

$$\ln \frac{1+x}{1-x} = 2 \tanh^{-1} x = 2\left(x + \frac{x^3}{3} + \frac{x^5}{5} + \cdots + \frac{x^{2n+1}}{2n+1} + \cdots\right) = 2\sum_{n=0}^{\infty} \frac{x^{2n+1}}{2n+1}, \qquad |x| < 1$$

$$\tan^{-1} x = x - \frac{x^3}{3} + \frac{x^5}{5} - \cdots + (-1)^n \frac{x^{2n+1}}{2n+1} + \cdots = \sum_{n=0}^{\infty} \frac{(-1)^n x^{2n+1}}{2n+1}, \qquad |x| \le 1$$

Binomial Series

$$(1+x)^m = 1 + mx + \frac{m(m-1)x^2}{2!} + \frac{m(m-1)(m-2)x^3}{3!} + \cdots + \frac{m(m-1)(m-2)\cdots(m-k+1)x^k}{k!} + \cdots$$

$$= 1 + \sum_{k=1}^{\infty} \binom{m}{k} x^k, \qquad |x| < 1,$$

where

$$\binom{m}{1} = m, \qquad \binom{m}{2} = \frac{m(m-1)}{2!}, \qquad \binom{m}{k} = \frac{m(m-1)\ldots(m-k+1)}{k!} \qquad \text{for } k \ge 3.$$

Note: It is customary to define $\binom{m}{0}$ to be 1 and to take $x^0 = 1$ (even in the usually excluded case where $x = 0$) to write the binomial series compactly as

$$(1+x)^m = \sum_{k=0}^{\infty} \binom{m}{k} x^k, \qquad |x| < 1.$$

If m is a *positive integer,* the series terminates at x^m, and the result converges for all x.

EXERCISES 8.8

What Taylor series would you use to represent the functions in Exercises 1–6 near the given values of x? (There may be more than one good answer.) Write out the first four nonzero terms of the series you choose.

1. $\cos x$ near $x = 1$

2. $\sin x$ near $x = 6.3$

3. e^x near $x = 0.4$

4. $\ln x$ near $x = 1.3$

5. $\cos x$ near $x = 69$

6. $\tan^{-1}x$ near $x = 2$

CALCULATOR In Exercises 7–14, use series and a calculator to estimate the value of each integral with an error of magnitude less than 10^{-3}.

7. $\displaystyle\int_0^{0.2} \sin x^2\, dx$

8. $\displaystyle\int_0^{0.1} \tan^{-1}x\, dx$

9. $\displaystyle\int_0^{0.1} x^2 e^{-x^2}\, dx$

10. $\displaystyle\int_0^{0.1} \frac{\tan^{-1}x}{x}\, dx$

11. $\displaystyle\int_0^{0.4} \frac{1-e^{-x}}{x}\, dx$

12. $\displaystyle\int_0^{0.1} \frac{\ln(1+x)}{x}\, dx$

13. $\displaystyle\int_0^{0.1} \frac{1}{\sqrt{1+x^4}}\, dx$

14. $\displaystyle\int_0^{0.25} \sqrt[3]{1+x^2}\, dx$

CALCULATOR Use series to evaluate the integrals in Exercises 15–18 as accurately as your calculator will allow.

15. $\displaystyle\int_0^{0.1} \frac{\sin x}{x}\, dx$

16. $\displaystyle\int_0^{0.1} e^{-x^2}\, dx$

17. $\displaystyle\int_0^{0.1} \sqrt{1+x^4}\, dx$

18. $\displaystyle\int_0^{1} \frac{1-\cos x}{x^2}\, dx$

19. Replace x by $-x$ in the Maclaurin series for $\ln(1+x)$ to obtain a series for $\ln(1-x)$. Then subtract this from the Maclaurin series for $\ln(1+x)$ to show that for $|x| < 1$,

$$\ln\frac{1+x}{1-x} = 2\left(x + \frac{x^3}{3} + \frac{x^5}{5} + \dots\right).$$

20. How many terms of the Maclaurin series for $\ln(1+x)$ should you add to be sure of calculating $\ln(1.1)$ with an error of magnitude less than 10^{-8}?

21. According to the Alternating Series Estimation Theorem, how many terms of the Maclaurin series for $\tan^{-1} 1$ would you have to add to be sure of finding $\pi/4$ with an error of magnitude less than 10^{-3}?

22. Show that the Maclaurin series for $\tan^{-1}x$ diverges for $|x| > 1$.

23. CALCULATOR About how many terms of the Maclaurin series for $\tan^{-1}x$ would you have to use to evaluate each term on the right-hand side of the equation

$$\pi = 48\tan^{-1}\frac{1}{18} + 32\tan^{-1}\frac{1}{57} - 20\tan^{-1}\frac{1}{239}$$

with an error of magnitude less than 10^{-6}? In contrast, the convergence of $\sum_{n=1}^{\infty} (1/n^2)$ to $\pi^2/6$ is so slow that even 50 terms will not yield two-place accuracy.

24. Integrate the binomial series for $(1-x^2)^{-1/2}$ to show that for $|x| < 1$

$$\sin^{-1}x = x + \sum_{n=1}^{\infty} \frac{1\cdot 3\cdot 5\cdot\dots\cdot(2n-1)}{2\cdot 4\cdot 6\cdot\dots\cdot(2n)}\frac{x^{2n+1}}{2n+1}.$$

25. a) Use the binomial series and the fact that

$$\frac{d}{dx}\sin^{-1}x = (1-x^2)^{-1/2}$$

to generate the first four nonzero terms of the Maclaurin series for $\sin^{-1}x$. What is the radius of convergence?

b) Use your result in (a) to find the first five nonzero terms of the Maclaurin series for $\cos^{-1}x$.

26. a) Find the first four nonzero terms of the Maclaurin series for

$$\sinh^{-1}x = \int_0^x \frac{dt}{\sqrt{1+t^2}}.$$

b) CALCULATOR Use the first three terms of the series in (a) to estimate $\sinh^{-1} 0.25$. Give an upper bound for the magnitude of the estimation error.

27. Obtain the Maclaurin series for $1/(1+x)^2$ from the series for $-1/(1+x)$.

28. Use the Maclaurin series for $1/(1-x^2)$ to obtain a series for $2x/(1-x^2)^2$.

29. Integrate the first three nonzero terms of the Maclaurin series for $\tan t$ from 0 to x to obtain the first three nonzero terms of the Maclaurin series for $\ln \sec x$.

30. *The series for $\tanh^{-1}(x) = (1/2)\ln((1+x)/(1-x))$.*

a) Show that

$$\int_0^x \frac{dt}{1-t^2} = \int_0^x \left(1 + t^2 + t^4 + \cdots + t^{2n} + \frac{t^{2n+2}}{1-t^2}\right) dt$$

or, put another way, that

$$\tanh^{-1}x = x + \frac{x^3}{3} + \frac{x^5}{5} + \cdots + \frac{x^{2n+1}}{2n+1} + R(n, x),$$

where

$$R(n, x) = \int_0^x \frac{t^{2n+2}}{1-t^2}\, dt.$$

b) Show that

$$|R(n, x)| \le \frac{1}{1-x^2}\frac{|x|^{2n+3}}{2n+3} \quad \text{if} \quad |x| < 1.$$

This shows that $\lim_{n\to\infty} R(n, x) = 0$ if $|x| < 1$ and hence that

$$\tanh^{-1}x = \sum_{n=0}^{\infty} \frac{x^{2n+1}}{2n+1}, \quad |x| < 1.$$

c) Show that the series for $\tanh^{-1}x$ diverges if $|x| \ge 1$.

✲ 8.9 A Computer Mystery

You may never have to actually use a series to compute the value of π, ln 2, e, sin 35°, or the like, because if you need any of these numbers you can either look up appropriate approximations in a table or (more likely) find them with a calculator. But it was an exciting new idea in mathematics (and still can be an exciting idea) to realize that we can get *all* the information about e^x, for example, from just knowing that $e^0 = 1$ and that the derivative of e^x is e^x. That is enough to generate the Maclaurin series for e^x, which converges to e^x for all x. By multiplying the series for e^x by the series for e^y, we can also show that $e^x \cdot e^y = e^{x+y}$; by differentiation, we can show that the derivative of the series for e^x is that same series, and so on.

Those of you who are interested in computer science may find the following "detective story" challenging. We wanted to compare, for purposes of illustration, three different ways of calculating e.

1. e is the root of the equation $\ln x = 1$. (Newton's method produced 2.7 1828 1828 on the third iteration, starting with $x_0 = 3$.)
2. e is the sum of the series

$$1 + \frac{1}{1} + \frac{1}{2!} + \frac{1}{3!} + \frac{1}{4!} + \frac{1}{5!} + \cdots.$$

(This produced the same degree of precision when we included 14 or more terms.)

3. $e = \lim_{n\to\infty} \left(1 + \frac{1}{n}\right)^n$

This is often taken as the definition of e. With a computer, one can just pick any large value of n and tell the computer to print the value of $a_n = [1 + (1/n)]^n$. Of course, it is also instructive to see what happens for various values of n along the way. When we did that, we got a surprise. Here is a portion of the table of values we got, starting at $n = 10^5$.

n	$\left(1 + \frac{1}{n}\right)^n$	
10^5	2.7 1826 8237	
2×10^5	2.7 1827 5305	
3×10^5	2.7 1827 6664	
4×10^5	2.7 1827 7479	
5×10^5	2.7 1827 7751	
6×10^5	2.7 1827 9110	
7×10^5	2.7 1827 9110	(What's this? A repetition?)
8×10^5	2.7 1828 1828	(Hooray!)
9×10^5	2.7 1827 9110	(The machine likes this number?)
10^6	2.7 1828 1828	(Well, that's better.)
10^{13}	1	(What's going on?)
⋮		
10^{19}	1	

We offer this as a puzzle. What is the machine doing? (Don't read on until you give it your best try!) Want a clue? Recall the definition of a^b.

A Computer Mystery Solved

We have just presented some data for $[1 + (1/n)]^n$ for selected values of n between 10^5 and 10^{19}. How can we account for the obvious errors in some of these answers?

Let's start with the definition of a^b as $\exp(b \ln a)$:

$$\left(1 + \frac{1}{n}\right)^n = \exp\left[n \ln\left(1 + \frac{1}{n}\right)\right]. \tag{1}$$

For large values of n, $1/n$ is small, and we look at the series for $\ln(1 + x)$ with $x = 1/n$. The result is

$$\ln\left(1 + \frac{1}{n}\right) = \frac{1}{n} - \frac{1}{2n^2} + \frac{1}{3n^3} - \cdots. \tag{2}$$

Clearly, this series approaches zero as $n \to \infty$. But Eq. (1) requires that we multiply this series by n. Doing so, we get

$$n \ln\left(1 + \frac{1}{n}\right) = 1 - \frac{1}{2n} + \frac{1}{3n^2} - \cdots. \tag{3}$$

For large values of n, the terms from $1/3n^2$ and beyond are very small compared to the first two terms. So, for a first approximation, we have

$$\left(1 + \frac{1}{n}\right)^n \approx \exp\left(1 - \frac{1}{2n}\right). \tag{4}$$

How about the irregular data that we got from the computer? We guessed that it might come from inaccurate values for $\ln[1 + (1/n)]^n$, so we programmed the computer to give those data. At the same time, we had the computer give the values of $1 - (1/2n)$ and $1 - (1/2n) + (1/3n^2)$. These latter values were the same for all values of n shown in the table. And they are certainly more accurate than the machine's values of $n \ln[1 + (1/n)]$.

n	$n \ln[1 + (1/n)]$	$1 - (1/2n) + (1/3n^2)$
10^5	0.99999 5	0.99999 5
2×10^5	0.99999 76	0.99999 75
3×10^5	0.99999 81	0.99999 8333
4×10^5	0.99999 84	0.99999 875
5×10^5	0.99999 85	0.99999 90
6×10^5	0.99999 90	0.99999 91667
7×10^5	0.99999 90	0.99999 92857
8×10^5	1	0.99999 94444
9×10^5	0.99999 9	0.99999 95
10^6	1	0.99999 95455
1.1×10^6	1.00000 01	0.99999 95833
1.2×10^6	0.99999 6	0.99999 96154
1.3×10^6	0.99999 0	0.99999 96429
1.4×10^6	1.00000 04	0.99999 96667
1.5×10^6	1.00000 05	0.99999 96875
2×10^6	1	0.99999 97500

Naturally, if the computer has wrong values for $n \ln[1 + (1/n)]$, it will give wrong answers for $\exp(n \ln[1 + (1/n)])$. So, we have tracked down the source of the trouble; the values of $\ln[1 + (1/n)]$ are not accurate enough to give right answers when multiplied by n, if $n \geq 3 \times 10^5$. For $n = 10^{13}$, the computer says $\ln(1.0000\ 00000\ 0001) = 0$. When it multiplies this by 10^{13}, it still says "zero" and $\exp(0) = 1$. We have gone beyond the machine's realm of reliability. It isn't the computer's fault—or ours. But we have gained some insight by solving the mystery.

One more remark: If we use the approximation (4) and the fact that

$$\exp\left(1 - \frac{1}{2n}\right) = \exp(1) \cdot \exp\left(-\frac{1}{2n}\right), \tag{5}$$

we can go a step further. Remember that we are talking about $n \geq 3 \times 10^5$, so $1/2n \leq (1/6) \times 10^{-5}$. If we put $h = -1/2n$ in the Maclaurin series for $\exp(h) = e^h$, we get

$$\exp\left(-\frac{1}{2n}\right) = 1 - \frac{1}{2n} \tag{6}$$

with an error less than $(1/2)(1/2n)^2$ by the Alternating Series Estimation Theorem. This error is less than 0.3×10^{-11} for $n \geq 3 \times 10^5$. Therefore, for large values of n, the combined results of Eqs. (4), (5), and (6) yield

$$\left(1 + \frac{1}{n}\right)^n \approx e \cdot \left(1 - \frac{1}{2n}\right) \approx e - \frac{e}{2n}.$$

For $n = 10^6$, the machine should *not* give 2.7 1828 1828 (which seems so very accurate), but 2.7 1828 0469 to nine decimals.

REVIEW QUESTIONS

1. Define infinite sequence (sequence), infinite series (series), and sequence of partial sums of a series.

2. Define convergence for (a) sequences, (b) series.

3. Describe Picard's method for finding roots. When can it be expected to work?

4. Which of the following statements are true, and which are false?

a) If a sequence does not converge, then it diverges.

b) If a sequence $\{a_n\}$ does not converge, then a_n tends to infinity as n does.

c) If a series does not converge, then its nth term does not approach zero as n tends to infinity.

d) If the nth term of a series does not approach zero as n tends to infinity, then the series diverges.

e) If a sequence $\{a_n\}$ converges, then there is a number L such that a_n lies within 1 unit of L (i) for all values of n, (ii) for all but a finite number of values of n.

f) If all partial sums of a series are less than some constant L, then the series converges.

g) If a series converges, then its partial sums s_n are bounded (that is, $m \leq s_n \leq M$ for some constants m and M).

5. What tests do you know for the convergence and divergence of infinite series?

6. Under what circumstances do you know for sure that a bounded sequence converges?

7. Define absolute convergence and conditional convergence for infinite series. Give examples of series that (a) converge absolutely, (b) converge conditionally, (c) diverge.

8. Under what circumstances can you guarantee that rearranging infinitely many terms of an infinite series will not alter the series' sum? When might it alter the series' sum? Give examples.

9. What are geometric series? Under what circumstances do they converge? diverge?

10. What test would you apply to decide whether an alternating series converges? Give examples of convergent and of divergent alternating series. How can you estimate the error involved in using a partial sum to estimate the sum of a convergent alternating series?

11. What is a power series? How do you test a power series for convergence? What kinds of results can you get? Give examples.

12. What are the basic facts about term-by-term differentiation and integration of power series? Give examples.
13. What is the Taylor series generated by a function $f(x)$ at a point $x = a$? What do you need to know about f to construct the series? Give an example.
14. What is the Maclaurin series generated by a function $f(x)$? Give an example.
15. What are Taylor polynomials and what good are they? Give examples.
16. What is Taylor's formula and what does Taylor's theorem say about it? When does the Taylor series generated by f at $x = a$ converge to f?
17. What does Taylor's formula tell us about the error in a linearization?
18. What are the Maclaurin series for e^x, $\sin x$, $\cos x$, $1/(1+x)$, $1/(1-x)$, $1/(1+x^2)$, $\tan^{-1}x$, and $\ln(1+x)$? How do you estimate the errors involved in replacing these series by their partial sums? Give examples.
19. What do you take into account when choosing a center for a Taylor series? Illustrate with examples.
20. How can you sometimes use series to evaluate nonelementary integrals? Give an example.

MISCELLANEOUS EXERCISES

Determine which of the sequences $\{a_n\}$ in Exercises 1–14 converge and which diverge. Find the limit of each convergent sequence.

1. $a_n = 1 + \dfrac{(-1)^n}{n}$
2. $a_n = \dfrac{1 - 2^n}{2^n}$
3. $a_n = \cos \dfrac{n\pi}{2}$
4. $a_n = \left(\dfrac{3}{n}\right)^{1/n}$
5. $a_n = \dfrac{n}{\ln(n^2)}$
6. $a_n = \left(\dfrac{n+5}{n}\right)^n$
7. $a_n = \sqrt[2n]{\dfrac{3^n}{n}}$
8. $a_n = \dfrac{1}{3^{2n-1}}$
9. $a_n = \dfrac{(-4)^n}{n!}$
10. $a_n = \dfrac{\ln(2n^3+1)}{n}$
11. $a_n = n(2^{1/n} - 1)$
12. $a_n = \sqrt[n]{2n+1}$
13. $a_n = \left(1 + \dfrac{1}{n}\right)^{-n}$
14. $a_n = \dfrac{\left(\dfrac{10}{11}\right)^n + \dfrac{1}{n}}{\left(\dfrac{9}{10}\right)^n + \left(\dfrac{11}{12}\right)^n}$

In Exercises 15–18, find the sums of the series that converge.

15. $\displaystyle\sum_{n=1}^{\infty} \ln\left(\frac{n}{n+1}\right)$
16. $\displaystyle\sum_{n=2}^{\infty} \frac{-2}{n(n+1)}$
17. $\displaystyle\sum_{n=0}^{\infty} e^{-n}$
18. $\displaystyle\sum_{n=1}^{\infty} (-1)^n \frac{3}{4^n}$

Which of the series in Exercises 19–40 converge absolutely, which converge conditionally, and which diverge? Give reasons for your answers.

19. $\displaystyle\sum_{n=1}^{\infty} \frac{1}{\sqrt{n}}$
20. $\displaystyle\sum_{n=1}^{\infty} \frac{-5}{n}$
21. $\displaystyle\sum_{n=1}^{\infty} \frac{(-1)^n}{\sqrt{n}}$
22. $\displaystyle\sum_{n=1}^{\infty} \frac{1}{2n^3}$
23. $\displaystyle\sum_{n=1}^{\infty} \frac{(-1)^n}{\ln(n+1)}$
24. $\displaystyle\sum_{n=2}^{\infty} \frac{1}{n(\ln n)^2}$
25. $\displaystyle\sum_{n=1}^{\infty} \frac{(-1)^n}{n\sqrt{n^2+1}}$
26. $\displaystyle\sum_{n=1}^{\infty} \frac{(-1)^n 3n^2}{n^3+1}$
27. $\displaystyle\sum_{n=1}^{\infty} \frac{n+1}{n!}$
28. $\displaystyle\sum_{n=1}^{\infty} \frac{(-1)^n (n^2+1)}{2n^2+n-1}$
29. $\displaystyle\sum_{n=1}^{\infty} \frac{(-3)^n}{n!}$
30. $\displaystyle\sum_{n=1}^{\infty} \frac{2^n 3^n}{n^n}$
31. $\displaystyle\sum_{n=1}^{\infty} \frac{1}{\sqrt{n(n+1)(n+2)}}$
32. $\displaystyle\sum_{n=2}^{\infty} \frac{1}{n\sqrt{n^2-1}}$
33. $\displaystyle\sum_{n=1}^{\infty} \frac{1}{(3n-2)^{n+(1/2)}}$
34. $\displaystyle\sum_{n=1}^{\infty} \frac{(\tan^{-1} n)^2}{n^2+1}$
35. $\displaystyle\sum_{n=1}^{\infty} (-1)^n \tanh n$
36. $\displaystyle\sum_{n=1}^{\infty} \tan^{-1}\left(\frac{\cos n\pi}{n}\right)$
37. $\displaystyle\sum_{n=1}^{\infty} (-1)^n \ln(1 + e^{-n})$
38. $\displaystyle\sum_{n=2}^{\infty} \frac{\log_n (n!)}{n^3}$ (*Hint:* First show that $\log_n (n!) < n$.)
39. $\displaystyle\sum_{n=1}^{\infty} \frac{\ln (n!)}{n^4}$ (*Hint:* First show that $n! \le n^n$.)
40. $\displaystyle\sum_{n=1}^{\infty} \frac{\Gamma(n)}{n^n}$ (See Chapter 7, Miscellaneous Exercise 161.)

Each of Exercises 41–50 gives a formula for the nth term of a series. (a) Find the series' radius and interval of convergence. For what values of x does the series converge (b) absolutely; (c) conditionally?

41. $\dfrac{(x+2)^n}{n3^n}$
42. $\dfrac{(x-1)^{2n-2}}{(2n-1)!}$
43. $\dfrac{x^n}{n^n}$
44. $\dfrac{n+1}{2n+1}\dfrac{(x-1)^n}{2^n}$
45. $\dfrac{(-1)^{n-1}(x-1)^n}{n^2}$
46. $\dfrac{x^n}{\sqrt{n}}$
47. $(\operatorname{csch} n)\, x^n$
48. $(\coth n)\, x^n$
49. $\dfrac{(n+1)\, x^{2n-1}}{3^n}$
50. $\dfrac{(\cos x)^n}{n}$

51. Find the radius of convergence of the series
$$\sum_{n=1}^{\infty} \frac{2 \cdot 5 \cdot 8 \cdot \cdots \cdot (3n-1)}{2 \cdot 4 \cdot 6 \cdot \cdots \cdot (2n)} x^n.$$

52. a) Show that the series
$$\sum_{n=1}^{\infty} \left(\sin \frac{1}{2n} - \sin \frac{1}{2n+1}\right)$$
converges.

b) CALCULATOR Estimate the error involved in using the sum of the first twenty terms of the series ($n = 20$) to estimate the sum of the series. Is the estimate too large, or too small?

Each of the series in Exercises 53–58 is the value of the Maclaurin series of a function $f(x)$ at a particular point. What function and what point? What is the sum of the series?

53. $1 - \frac{1}{4} + \frac{1}{16} - \cdots + (-1)^n \frac{1}{4^n} + \cdots$

54. $\frac{2}{3} - \frac{4}{18} + \frac{8}{81} - \cdots + (-1)^{n-1} \frac{2^n}{n3^n} + \cdots$

55. $\pi - \frac{\pi^3}{3!} + \frac{\pi^5}{5!} - \cdots + (-1)^n \frac{\pi^{2n+1}}{(2n+1)!} + \cdots$

56. $1 - \frac{\pi^2}{9 \cdot 2!} + \frac{\pi^4}{81 \cdot 4!} - \cdots + (-1)^n \frac{\pi^{2n}}{3^{2n}(2n)!} + \cdots$

57. $1 + \ln 2 + \frac{(\ln 2)^2}{2!} + \cdots + \frac{(\ln 2)^n}{n!} + \cdots$

58. $\frac{1}{\sqrt{3}} - \frac{1}{9\sqrt{3}} + \frac{1}{45\sqrt{3}} - \cdots + (-1)^{n-1} \frac{1}{(2n-1)(\sqrt{3})^{2n-1}} + \cdots$

In Exercises 59 and 60, find the first four nonzero terms of the Taylor series for the function $f(x)$ at the center $x = a$.

59. $f(x) = \sqrt{3 + x^2}$ at $x = -1$

60. $f(x) = 1/(1 - x)$ at $x = 2$

61. CALCULATOR Start with $x_0 = 1$ and press [sin] and [$\sqrt{x}$] alternately to start generating a sequence of terms with the recursion formula $x_{n+1} = \sqrt{\sin(x_n)}$. Does the sequence appear to be convergent? If so, what equation does the limit satisfy?

62. Figure 8.24 provides an informal proof that $\sum_{n=1}^{\infty} (1/n^2)$ is less than 2. Explain what is going on. (*Source:* "Convergence with Pictures" by P. J. Rippon, *American Mathematical Monthly,* Vol. 93, No. 6, 1986, pp. 476–78.)

63. Find a closed-form formula for the nth partial sum of the series $\sum_{n=2}^{\infty} \ln[1 - (1/n^2)]$ and use it to determine whether the series converges.

64. Evaluate $\sum_{k=2}^{\infty} 1/(k^2 - 1)$ by finding the limit as $n \to \infty$ of the nth partial sum.

65. Prove that the sequence $\{x_n\}$ and the series $\sum_{k=1}^{\infty} (x_{k+1} - x_k)$ both converge or both diverge.

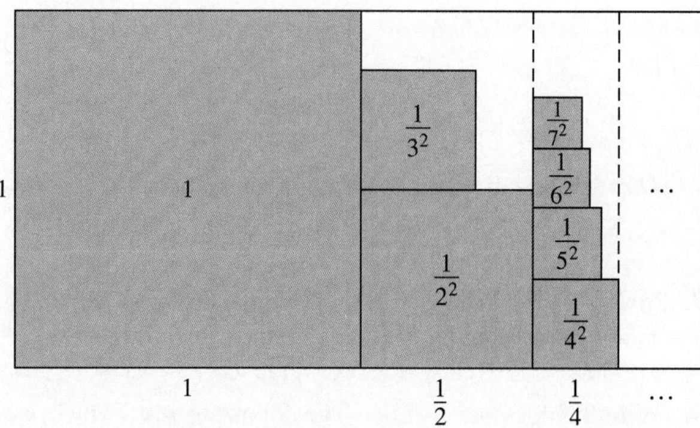

8.24 The figure for Exercise 62.

66. Assuming $|x| > 1$, show that
$$\frac{1}{1-x} = -\frac{1}{x} - \frac{1}{x^2} - \frac{1}{x^3} - \cdots.$$

67. *Generalizing Euler's constant.* Figure 8.25 shows the graph of a positive twice-differentiable decreasing function f whose second derivative is positive on $(0, \infty)$. For each n, the number A_n is the area of the lunar region between the curve and the line segment joining the points $(n, f(n))$ and $(n+1, f(n+1))$.

a) Use the figure to show that $\sum_{n=1}^{\infty} A_n < (1/2)(f(1) - f(2))$.

b) Then show the existence of
$$\lim_{n\to\infty} \left[\sum_{k=1}^{n} f(k) - \frac{1}{2}(f(1) + f(n)) - \int_1^n f(x)\,dx\right].$$

c) Then show the existence of
$$\lim_{n\to\infty} \left[\sum_{k=1}^{n} f(k) - \int_1^n f(x)\,dx\right].$$

If $f(x) = 1/x$, the limit in (c) is Euler's constant. (*Source:* "Convergence with Pictures" by P. J. Rippon, *American Mathematical Monthly,* Vol. 93, No. 6, 1986, pp. 476–78.)

8.25 The figure for Exercise 67.

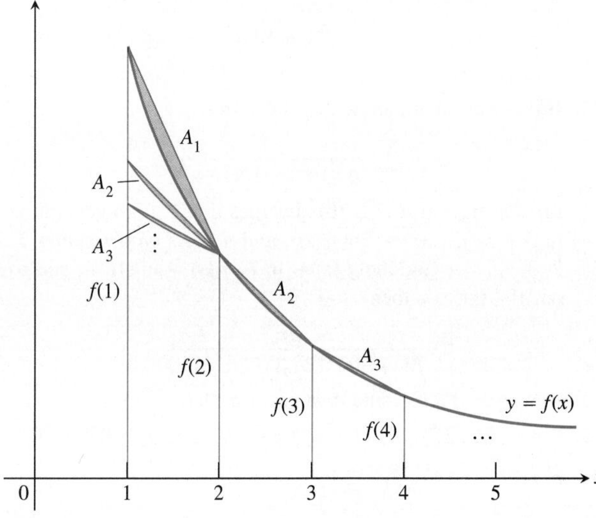

68. *Nicole Oresme's theorem.* Prove Nicole Oresme's theorem that

$$1+\frac{1}{2}\cdot 2+\frac{1}{4}\cdot 3+\cdots+\frac{n}{2^{n-1}}+\cdots=4.$$

$\left(\textit{Hint:}\text{ Differentiate both sides of the equation } 1/(1-x) = 1+\sum_{n=1}^{\infty} x^n.\right)$

69. Prove that the sequence of approximations $x_0, x_1, x_2, \ldots$ $(x_0 \neq 1)$ generated by Newton's method to find the zero of $f(x) = (x-1)^{40}$ really does converge to 1.

70. *Raabe's (or Gauss's) test.* The following test, which we state without proof, is an extension of the Ratio Test.

Raabe's test: If $\sum_{n=1}^{\infty} u_n$ is a series of positive constants and there exist constants C, K, and N such that

$$\frac{u_{n+1}}{u_n} = 1 - \frac{C}{n} + \frac{f(n)}{n^2}, \qquad (1)$$

where $|f(n)| < K$ for $n \geq N$, then $\sum_{n=1}^{\infty} u_n$ converges if $C > 1$ and diverges if $C \leq 1$.

Show that the results of Raabe's test agree with what you know about the series $\sum_{n=1}^{\infty} (1/n^2)$ and $\sum_{n=1}^{\infty} (1/n)$.

71. *(Continuation of Exercise 70.)* Suppose that the terms of $\sum_{n=1}^{\infty} u_n$ are defined recursively by the formulas

$$u_1 = 1, \quad u_{n+1} = \frac{(2n-1)^2}{(2n)(2n+1)} u_n.$$

Apply Raabe's test to determine whether the series converges.

72. a) Suppose $a_1, a_2, a_3, \ldots, a_n$ are positive numbers satisfying the following conditions:

i) $a_1 \geq a_2 \geq a_3 \geq \cdots$;

ii) the series $a_2 + a_4 + a_8 + a_{16} + \cdots$ diverges.

Show that the series

$$\frac{a_1}{1}+\frac{a_2}{2}+\frac{a_3}{3}+\cdots$$

diverges.

b) Use the result in (a) to show that

$$1+\sum_{n=2}^{\infty}\frac{1}{n \ln n}$$

diverges.

73. If p is a constant, show that the series

$$1+\sum_{n=3}^{\infty}\frac{1}{n\cdot \ln n \cdot [\ln(\ln n)]^p}$$

(a) converges if $p > 1$, (b) diverges if $p \leq 1$. In general, if $f_1(x) = x$, $f_{n+1}(x) = \ln(f_n(x))$, and n takes on the values 1, 2, 3, ..., we find that $f_2(x) = \ln x$, $f_3(x) = \ln(\ln x)$, and so on. If $f_n(a) > 1$, then

$$\int_a^{\infty} \frac{dx}{f_1(x) f_2(x) \ldots f_n(x)(f_{n+1}(x))^p}$$

converges if $p > 1$ and diverges if $p \leq 1$.

74. Prove that $\sum_{n=1}^{\infty} a_n/(1+a_n)$ converges if $a_n > 0$ for all n and $\sum_{n=1}^{\infty} a_n$ converges.

75. If $\sum_{n=1}^{\infty} a_n$ converges, and if $a_n \neq 1$ and $a_n > 0$ for all n,

a) Show that $\sum_{n=1}^{\infty} a_n^2$ converges.

b) Does $\sum_{n=1}^{\infty} a_n/(1-a_n)$ converge? Explain.

76. *(Continuation of Exercise 75.)* If $\sum_{n=1}^{\infty} a_n$ converges, and if $1 > a_n > 0$ for all n, show that $\sum_{n=1}^{\infty} \ln(1-a_n)$ converges. (*Hint:* First show that $|\ln(1-a_n)| \leq a_n/(1-a_n)$.)

77. a) Find the interval of convergence of the series

$$y = 1+\frac{1}{6}x^3+\frac{1}{180}x^6+\cdots+\frac{1\cdot 4\cdot 7\cdot \cdots \cdot (3n-2)}{(3n)!}x^{3n}+\cdots.$$

b) Show that the function defined by the series satisfies the differential equation of the form

$$\frac{d^2y}{dx^2} = x^a y + b$$

and find the values of the constants a and b.

78. a) Find the Maclaurin series for the function $x^2/(1+x)$.

b) Does the series converge at $x = 1$? Explain.

79. Find the first three nonzero terms of the Maclaurin series for $\ln(\cos x)$

a) by substituting the series for $y = 1 - \cos x$ into the series for $\ln(1-y)$.

b) by applying the definition to the function $f(x) = \ln(\cos x)$.

Which method is faster?

80. Find the first three nonzero terms of the expansion of $f(x) = \sqrt{1+x^2}$ in powers of $(x-1)$.

81. Expand $f(x) = 1/(1-x)$ in powers of $(x-2)$. Find the series' interval and radius of convergence.

82. Expand $f(x) = 1/(x+1)$ in powers of $(x-3)$. Find the series' interval and radius of convergence.

83. Expand $f(x) = \cos x$ in powers of $(x-\pi)$. Find the series' radius and interval of convergence.

84. Expand $f(x) = 1/x$ in powers of $(x-a)$. Find the interval of convergence if (a) $a > 0$, (b) $a < 0$.

85. *A fast estimate of $\pi/2$.* As you saw if you did Exercise 75 in Section 8.1, the sequence generated by starting with $x_0 = 1$ and applying the recursion formula $x_{n+1} = x_n + \cos x_n$ converges rapidly to $\pi/2$. To explain the speed of the convergence, let $\epsilon_n = (\pi/2) - x_n$ (Fig. 8.26). Then

$$\begin{aligned}\epsilon_{n+1} &= \frac{\pi}{2} - x_n - \cos x_n \\ &= \epsilon_n - \cos\left(\frac{\pi}{2} - \epsilon_n\right) \\ &= \epsilon_n - \sin \epsilon_n \\ &= \frac{1}{3!}(\epsilon_n)^3 - \frac{1}{5!}(\epsilon_n)^5 + \cdots.\end{aligned}$$

Use this equality to show that

$$0 < \epsilon_{n+1} < \frac{1}{6}(\epsilon_n)^3.$$

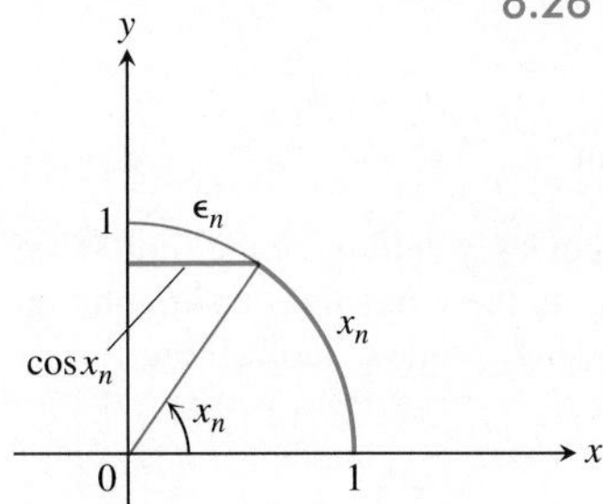

8.26 The figure for Exercise 85.

86. Suppose $\sum_{n=1}^{\infty} a_n$ is a convergent series of positive numbers. Does $\sum_{n=1}^{\infty} \ln(1 + a_n)$ converge? Explain.

CALCULATOR Estimate the values of the integrals in Exercises 87 and 88 with an error of magnitude less than 10^{-8}.

87. $\displaystyle\int_0^1 x \sin(x^3)\, dx$ **88.** $\displaystyle\int_0^{1/64} \frac{\tan^{-1} x}{\sqrt{x}}\, dx$

89. *Series for $\tan^{-1} x$ for $|x| > 1$.* Derive the series

$$\tan^{-1}x = \frac{\pi}{2} - \frac{1}{x} + \frac{1}{3x^3} - \frac{1}{5x^5} + \cdots, \quad x > 1,$$

$$\tan^{-1}x = -\frac{\pi}{2} - \frac{1}{x} + \frac{1}{3x^3} - \frac{1}{5x^5} + \cdots, \quad x < -1,$$

by integrating the series

$$\frac{1}{1+t^2} = \frac{1}{t^2} \cdot \frac{1}{1 + (1/t^2)} = \frac{1}{t^2} - \frac{1}{t^4} + \frac{1}{t^6} - \frac{1}{t^8} + \cdots$$

in the first case from x to ∞ and in the second case from $-\infty$ to x.

90. Use series multiplication to find the terms through x^5 of the Maclaurin series for (a) $(\tan^{-1}x)[\ln(1 + x)]$; (b) $(\tan^{-1}x)/(1 - x)$.

91. *An infinite product.* The infinite product

$$\prod_{n=1}^{\infty} (1 + a_n) = (1 + a_1)(1 + a_2)(1 + a_3)\ldots$$

is said to converge if the series

$$\sum_{n=1}^{\infty} \ln(1 + a_n),$$

obtained by taking the natural logarithm of the product, converges. Prove that the product converges if $a_n > -1$ for every n and if $\sum_{n=1}^{\infty} |a_n|$ converges. (*Hint:* Show that

$$|\ln(1 + a_n)| \le \frac{|a_n|}{1 - |a_n|} < 2|a_n|$$

when $|a_n| < 1/2$.)

92. *The value of $\sum_{n=1}^{\infty} \tan^{-1}(2/n^2)$.*

a) Use the formula for the tangent of the difference of two angles to show that

$$\tan(\tan^{-1}(n + 1) - \tan^{-1}(n - 1)) = \frac{2}{n^2}$$

and hence that

$$\tan^{-1}\frac{2}{n^2} = \tan^{-1}(n + 1) - \tan^{-1}(n - 1).$$

b) Show that

$$\sum_{n=1}^{N} \tan^{-1}\frac{2}{n^2} = \tan^{-1}(N + 1) + \tan^{-1}N - \frac{\pi}{4}.$$

c) Find the value of $\sum_{n=1}^{\infty} \tan^{-1}\dfrac{2}{n^2}$.

93. *Quality control*

a) Differentiate the series

$$\frac{1}{1-x} = 1 + x + x^2 + \cdots + x^n + \cdots$$

to obtain a series for $1/(1 - x)^2$.

b) In one throw of two dice, the probability of getting a roll of 7 is $p = 1/6$. If you throw the dice repeatedly, the probability that a 7 will appear for the first time at the nth throw is $q^{n-1}p$, where $q = 1 - p = 5/6$. The expected number of throws until a 7 first appears is $\sum_{n=1}^{\infty} n\, q^{n-1}p$. Find the sum of this series.

c) As an engineer applying statistical control to an industrial operation, you inspect items taken at random from the assembly line. You classify each sampled item as either "good" or "bad." If the probability of an item's being good is p and of an item being bad is $q = 1 - p$, the probability that the first bad item is found is the nth one inspected is $p^{n-1}q$. The average number inspected up to and including the first bad item found is $\sum_{n=1}^{\infty} n\, p^{n-1} q$. Evaluate this sum, assuming $0 < p < 1$.

94. *Expected value.* Suppose that a random variable X may assume the values 1, 2, 3, . . . , with probabilities p_1, p_2, $p_3, \ldots$, where p_k is the probability that X equals k ($k = 1, 2, 3, \ldots$). Suppose also that $p_k \ge 0$ and that $\sum_{k=1}^{\infty} p_k = 1$. The **expected value** of X, denoted by $E(X)$, is the number $\sum_{k=1}^{\infty} kp_k$, provided the series converges. In each of the following cases, show that $\sum_{k=1}^{\infty} p_k = 1$ and find $E(X)$ if it exists. (*Hint:* See Exercise 93.)

a) $p_k = 2^{-k}$ b) $p_k = \dfrac{5^{k-1}}{6^k}$

c) $p_k = \dfrac{1}{k(k+1)} = \dfrac{1}{k} - \dfrac{1}{k+1}$

95. a) Prove the following theorem: If $\{c_n\}$ is a sequence of numbers such that every sum $t_n = \sum_{k=1}^{n} c_k$ is bounded, then the series $\sum_{n=1}^{\infty} c_n/n$ converges and is equal to $\sum_{n=1}^{\infty} t_n/(n(n + 1))$.

Outline of proof: Replace c_1 by t_1 and c_n by $t_n - t_{n-1}$ for $n \ge 2$. If $s_{2n+1} = \sum_{k=1}^{2n+1} c_k/k$, show that

$$s_{2n+1} = t_1\left(1 - \frac{1}{2}\right) + t_2\left(\frac{1}{2} - \frac{1}{3}\right) + \cdots + t_{2n}\left(\frac{1}{2n} - \frac{1}{2n+1}\right) + \frac{t_{2n+1}}{2n+1}$$
$$= \sum_{k=1}^{2n} \frac{t_k}{k(k+1)} + \frac{t_{2n+1}}{2n+1}.$$

Because $|t_k| < M$ for some constant M, the series $\sum_{k=1}^{\infty} t_k/(k(k + 1))$ converges absolutely and s_{2n+1} has a limit as $n \to \infty$. Finally, if $s_{2n} = \sum_{k=1}^{2n} c_k/k$, then $s_{2n+1} - s_{2n} = c_{2n+1}/(2n + 1)$ approaches zero as

$n \to \infty$ because $|c_{2n+1}| = |t_{2n+1} - t_{2n}| < 2M$. Hence the sequence of partial sums of the series $\Sigma c_k/k$ converges and the limit is $\Sigma_{k=1}^{\infty} t_k/(k(k+1))$.

b) Show how the foregoing theorem applies to the alternating harmonic series

$$1 - \frac{1}{2} + \frac{1}{3} - \frac{1}{4} + \frac{1}{5} - \frac{1}{6} + \cdots.$$

c) Show that the series

$$1 - \frac{1}{2} - \frac{1}{3} + \frac{1}{4} + \frac{1}{5} - \frac{1}{6} - \frac{1}{7} + \cdots$$

converges. (After the first term, the signs are two negative, two positive, two negative, two positive, and so on in that pattern.)

96. *The convergence of* $\Sigma_{n=1}^{\infty}[(-1)^{n-1}x^n]/n$ *to* $\ln(1+x)$ *for* $-1 < x \le 1$.

a) Show by long division or otherwise that

$$\frac{1}{1+t} = 1 - t + t^2 - t^3 + \cdots + (-1)^n t^n + \frac{(-1)^{n+1}t^{n+1}}{1+t}.$$

b) By integrating the equation of part (a) with respect to t from 0 to x, show that

$$\ln(1+x) = x - \frac{x^2}{2} + \frac{x^3}{3} - \frac{x^4}{4} + \cdots + (-1)^n \frac{x^{n+1}}{n+1} + R_{n+1}$$

where

$$R_{n+1} = (-1)^{n+1}\int_0^x \frac{t^{n+1}}{1+t}\,dt.$$

c) If $x \ge 0$, show that

$$|R_{n+1}| \le \int_0^x t^{n+1}\,dt = \frac{x^{n+2}}{n+2}.$$

(*Hint:* As t varies from 0 to x,

$$1 + t \ge 1 \quad \text{and} \quad t^{n+1}/(1+t) \le t^{n+1},$$

and

$$\left|\int_0^x f(t)\,dt\right| \le \int_0^x |f(t)|\,dt.)$$

d) If $-1 < x < 0$, show that

$$|R_{n+1}| \le \left|\int_0^x \frac{t^{n+1}}{1-|x|}\,dt\right| = \frac{|x|^{n+2}}{(n+2)(1-|x|)}.$$

(*Hint:* If $x < t \le 0$, then $|1+t| \ge 1 - |x|$ and

$$\left|\frac{t^{n+1}}{1+t}\right| \le \frac{|t|^{n+1}}{1-|x|}.)$$

e) Use the foregoing results to prove that the series

$$x - \frac{x^2}{2} + \frac{x^3}{3} - \frac{x^4}{4} + \cdots + \frac{(-1)^n x^{n+1}}{n+1} + \cdots$$

converges to $\ln(1+x)$ for $-1 < x \le 1$.

Power Series and Indeterminate Forms

In considering

$$\lim_{x\to a}\frac{f(x)}{g(x)}$$

when $f(a) = g(a) = 0$, we can quickly determine the limit if we have Taylor series that converge to these functions on an interval centered at $x = a$. The following example shows how this is done.

EXAMPLE 1 Evaluate $\lim_{x\to 0}\frac{\sin x - \tan x}{x^3}$.

SOLUTION To terms in x^5, the Maclaurin series for $\sin x$ and $\tan x$ are

$$\sin x = x - \frac{x^3}{3!} + \frac{x^5}{5!} - \cdots, \qquad \tan x = x + \frac{x^3}{3} + \frac{2x^5}{15} + \cdots.$$

Hence,

$$\sin x - \tan x = -\frac{x^3}{2} - \frac{x^5}{8} - \cdots = x^3\left(-\frac{1}{2} - \frac{x^2}{8} - \cdots\right)$$

and

$$\lim_{x\to 0}\frac{\sin x - \tan x}{x^3} = \lim_{x\to 0}\left(-\frac{1}{2} - \frac{x^2}{8} - \cdots\right) = -\frac{1}{2}.$$

Use series to find the limits in Exercises 97–113.

97. $\lim_{x\to 0}\frac{e^x - (1+x)}{x^2}$

98. $\lim_{t\to 0}\frac{1 - \cos t - (t^2/2)}{t^4}$

99. $\lim_{\theta\to\infty}\theta\sin\frac{1}{\theta}$

100. $\lim_{x\to 0}\frac{1-\cos x}{\sin x}$

101. $\lim_{x\to 0}\frac{\sin x}{e^x - 1}$

102. $\lim_{y\to 0}\frac{\sin y - y + (y^3/6)}{y^5}$

103. $\lim_{u\to 0}\frac{e^u - e^{-u} - 2u}{u - \sin u}$

104. $\lim_{x\to 0}\frac{x - \tan^{-1}x}{x^3}$

105. $\lim_{\theta\to 0}\frac{\tan\theta - \sin\theta}{\theta^3\cos\theta}$

106. $\lim_{x\to\infty}x^2\left(e^{-1/x^2} - 1\right)$

107. $\lim_{h\to 0}\frac{\ln(1+h^2)}{1-\cos h}$

108. $\lim_{t\to 0}\frac{\tan 3t}{t}$

109. $\lim_{x\to 1}\frac{\ln x}{x-1}$

110. $\lim_{x\to 1}\frac{\ln x^2}{x-1}$

111. $\lim_{x\to 0}\left(\frac{1}{\sin x} - \frac{1}{x}\right)$

112. $\lim_{x\to 0}\left(\frac{1}{2 - 2\cos x} - \frac{1}{x^2}\right)$

113. $\lim_{x\to 0}\frac{\ln(1-x) - \sin x}{1 - \cos^2 x}$

114. Find values of r and s that make

$$\lim_{x\to 0}\left(\frac{\sin 3x}{x^3} + \frac{r}{x^2} + s\right) = 0.$$

CHAPTER

9

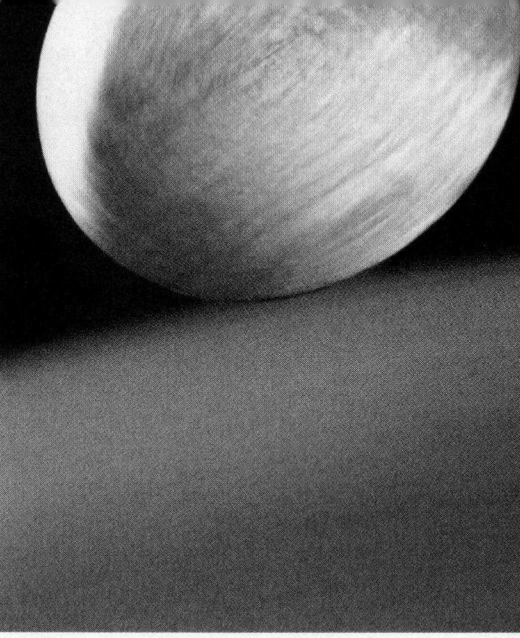

CONIC SECTIONS, PARAMETRIZED CURVES, AND POLAR COORDINATES

Overview The study of motion has been important since ancient times, and, as you know, calculus gives us the mathematics we need to describe motion. In this chapter, we extend our ability to analyze motion by showing how to keep track of the position of a moving body as a function of time. We do this with parametric equations. We study them in the coordinate plane here and then extend our work to three dimensions in Chapters 10 and 11. We begin our study by developing equations for conic sections, since these are the paths traveled by planets, satellites, and other bodies (even electrons) whose motions are driven by inverse square forces. Planetary motion is best described in polar coordinates (another of Newton's inventions, although James Bernoulli usually gets the credit because he published first), so we spend our remaining time finding out what curves, derivatives, and integrals look like in this new coordinate system.

9.1 Equations for Conic Sections

The conic sections that originated in Greek geometry are described today as the graphs of quadratic equations in the coordinate plane. The Greeks of Plato's time described these curves as the curves formed by cutting a double cone with a plane (Fig. 9.1); hence the name "conic section." We begin by reviewing briefly the equations for parabolas and circles and then continue on to ellipses and hyperbolas.

We use the mathematics of conic sections to describe the paths of planets, comets, moons, asteroids, and satellites that are moved through space by gravitational forces. Once we know that the path of a freely moving body is a conic section, we immediately have useful information about its velocity and the force that drives it, as we shall see in Chapter 11.

Equations from the Distance Formula: Circles and Parabolas

As you know, the distance between two points (x_1, y_1) and (x_2, y_2) in the coordinate plane is

$$d = \sqrt{(x_2 - x_1)^2 + (y_2 - y_1)^2}.$$

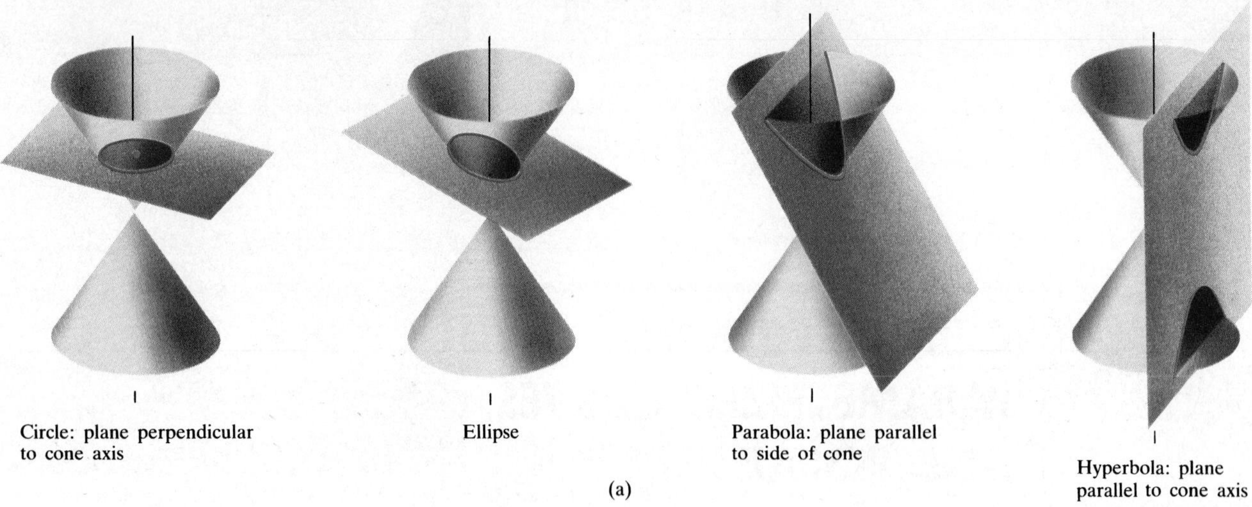

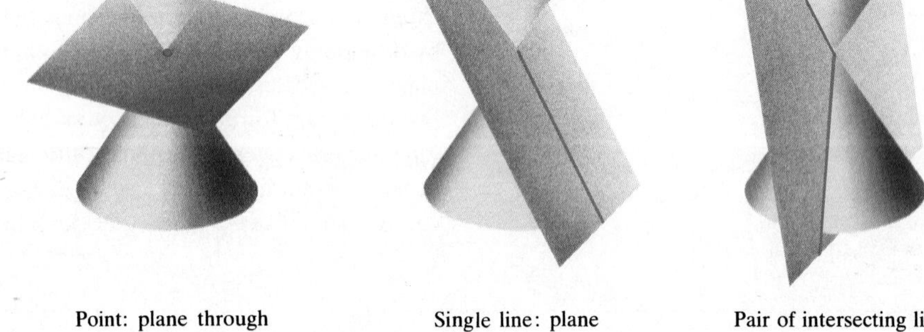

9.1 The standard conic sections (a) are the curves in which a plane cuts a double cone. Hyperbolas come in two parts, called *branches*. The point and lines obtained by passing the plane through the cone's vertex (b) are *degenerate* conic sections.

In Section 1.5, we used this formula to derive the standard equation for circles centered at the origin and the standard equations for parabolas with vertices at the origin (see Table 9.1). The horizontal and vertical shift formulas then gave us equations for circles and a variety of parabolas in other locations.

The Standard-Form Equation for a Circle of Radius a Centered at the Origin

$$x^2 + y^2 = a^2$$

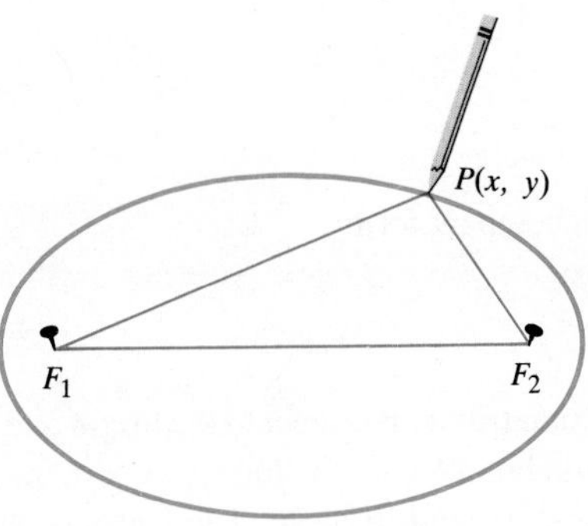

9.2 How to draw an ellipse.

TABLE 9.1
The standard-form equations for parabolas with vertices at the origin ($p > 0$)

Equation	Focus	Directrix	Axis	Direction
$x^2 = 4py$	$(0, p)$	$y = -p$	y-axis	Opens up
$x^2 = -4py$	$(0, -p)$	$y = p$	y-axis	Opens down
$y^2 = 4px$	$(p, 0)$	$x = -p$	x-axis	Opens to right
$y^2 = -4px$	$(-p, 0)$	$x = p$	x-axis	Opens to left

Ellipses

The equations we use for ellipses come from the distance formula, too.

DEFINITIONS

An **ellipse** is the set of points in a plane whose distances from two fixed points in the plane have a constant sum. The two fixed points are the **foci** of the ellipse.

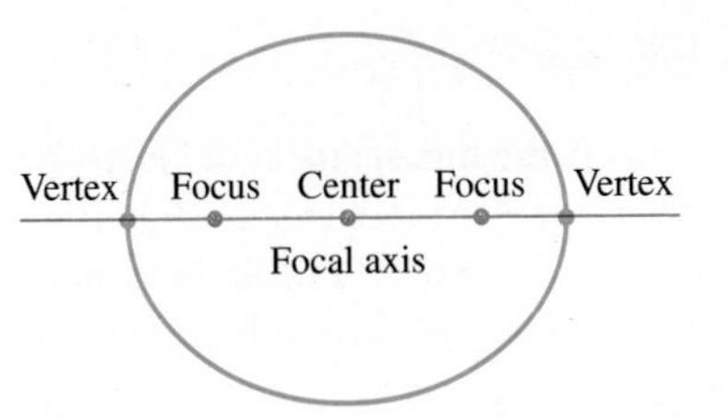

9.3 The center, vertices, and foci of an ellipse lie on the focal axis.

The quickest way to construct an ellipse uses the definition. Put a loop of string around two tacks F_1 and F_2, pull the string taut with a pencil point P, and move the pencil around to trace a closed curve (Fig. 9.2). The curve is an ellipse because the sum $PF_1 + PF_2$, being equal to the length of the loop minus the distance between the tacks, remains constant. The ellipse's foci lie at F_1 and F_2.

DEFINITIONS

The line through the foci of an ellipse is the ellipse's **focal axis.** The point on the axis halfway between the foci is the ellipse's **center.** The points where the focal axis crosses the ellipse are the ellipse's **vertices** (Fig. 9.3).

If the foci are $F_1(-c, 0)$ and $F_2(c, 0)$ (Fig. 9.4), and the sum of the distances $PF_1 + PF_2$ is denoted by $2a$, then the coordinates of a point P on the ellipse satisfy the equation

$$\sqrt{(x+c)^2 + y^2} + \sqrt{(x-c)^2 + y^2} = 2a. \tag{1}$$

To simplify this equation, we move the second radical to the right-hand side, square, isolate the remaining radical, and square again, obtaining

$$\frac{x^2}{a^2} + \frac{y^2}{a^2 - c^2} = 1. \tag{2}$$

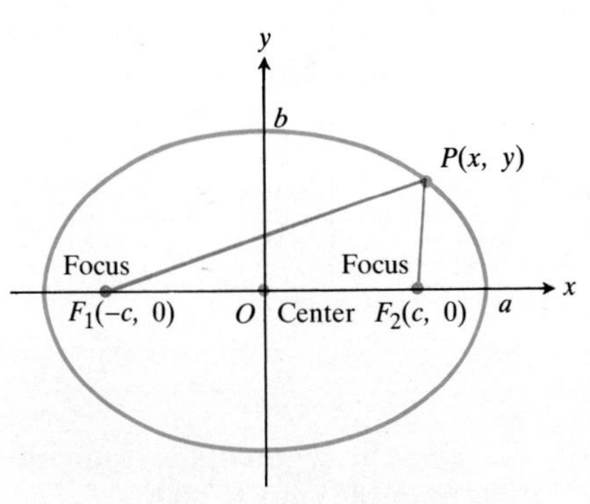

9.4 The ellipse defined by the equation $PF_1 + PF_2 = 2a$ is the graph of the equation $(x^2/a^2) + (y^2/b^2) = 1$.

Since the sum $PF_1 + PF_2$ is greater than the length F_1F_2 (triangle inequality for triangle PF_1F_2), the number $2a$ is greater than $2c$. Accordingly, a is greater than c and the number $a^2 - c^2$ in Eq. (2) is positive.

The algebraic steps taken to arrive at Eq. (2) can be reversed to show that every point P whose coordinates satisfy an equation of this form with $0 < c < a$ also satisfies the equation $PF_1 + PF_2 = 2a$. Thus, a point lies on the ellipse if and only if its coordinates satisfy Eq. (2).

If

$$b = \sqrt{a^2 - c^2}, \tag{3}$$

then $a^2 - c^2 = b^2$ and Eq. (2) takes the more compact form

$$\frac{x^2}{a^2} + \frac{y^2}{b^2} = 1. \tag{4}$$

Equation (4) reveals that this ellipse is symmetric with respect to the origin and both coordinate axes. It lies inside the rectangle bounded by the lines $x = \pm a$ and $y = \pm b$. It crosses the axes at the points $(\pm a, 0)$ and $(0, \pm b)$. The tangents at these points are perpendicular to the axes because the slope

$$\frac{dy}{dx} = -\frac{b^2 x}{a^2 y}$$

is zero when $x = 0$ and infinite when $y = 0$. These observations are the basis of the drawing lesson below.

The Major and Minor Axes of an Ellipse

The **major axis** of the ellipse described by Eq. (4) is the line segment of length $2a$ joining the points $(\pm a, 0)$. The **minor axis** is the line segment of length $2b$ joining the points $(0, \pm b)$. The number a itself is called the **semimajor axis** and the number b the **semiminor axis.** The number c, which can be found from Eq. (3) as

$$c = \sqrt{a^2 - b^2}, \tag{5}$$

is the **center-to-focus distance** of the ellipse.

DRAWING LESSON

How to Graph the Ellipse $\frac{x^2}{a^2} + \frac{y^2}{b^2} = 1$

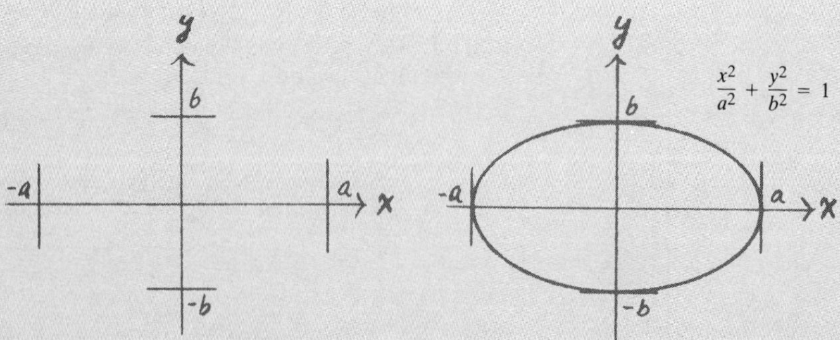

1. **Mark the points $(\pm a, 0)$ and $(0, \pm b)$ with line segments perpendicular to the coordinate axes.**

2. **Use the segments as tangent lines to guide your drawing.**

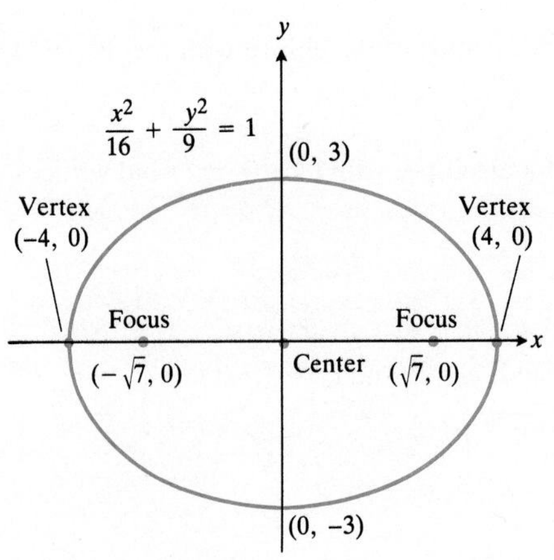

9.5 The major axis of $(x^2/16) + (y^2/9) = 1$ is horizontal (Example 1).

Example 1 *Major Axis Horizontal.* The ellipse

$$\frac{x^2}{16} + \frac{y^2}{9} = 1 \tag{6}$$

(Fig. 9.5) has

Semimajor axis: $a = \sqrt{16} = 4$, Semiminor axis: $b = \sqrt{9} = 3$,
Center-to-focus distance: $c = \sqrt{16 - 9} = \sqrt{7}$,
Foci: $(\pm c, 0) = (\pm\sqrt{7}, 0)$,
Vertices: $(\pm a, 0) = (\pm 4, 0)$,
Center: $(0, 0)$.

Example 2 *Major Axis Vertical.* The ellipse

$$\frac{x^2}{9} + \frac{y^2}{16} = 1, \tag{7}$$

obtained by interchanging x and y in Eq. (6), represents an ellipse with its major axis vertical instead of horizontal (Fig. 9.6). With a^2 still equal to 16 and b^2 equal to 9, we have

Semimajor axis: $a = \sqrt{16} = 4$, Semiminor axis: $b = \sqrt{9} = 3$,
Center-to-focus distance: $c = \sqrt{16 - 9} = \sqrt{7}$,
Foci: $(0, \pm c) = (0, \pm\sqrt{7})$,
Vertices: $(0, \pm a) = (0, \pm 4)$,
Center: $(0, 0)$.

There is never any cause for confusion in analyzing equations like (6) and (7). We simply find the intercepts on the coordinate axes; then we know which way the major axis runs because it is the longer of the two axes. The center always lies at the origin and the foci always lie on the major axis.

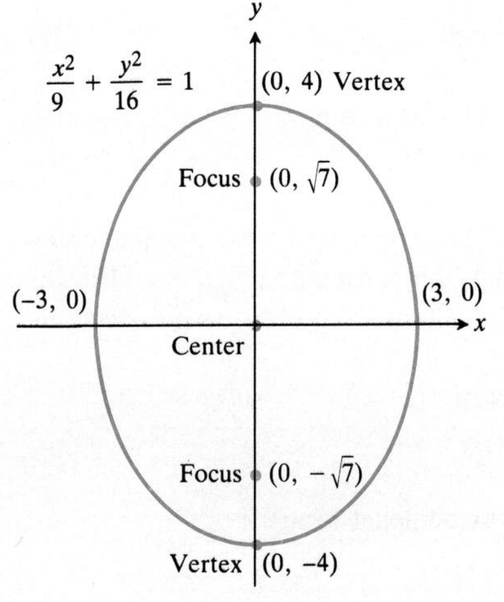

9.6 The major axis of $(x^2/9) + (y^2/16) = 1$ is vertical (Example 2).

Standard-Form Equations for Ellipses Centered at the Origin

Foci on the x-axis: $\frac{x^2}{a^2} + \frac{y^2}{b^2} = 1 \quad (a > b)$

Center-to-focus distance: $c = \sqrt{a^2 - b^2}$
Foci: $(\pm c, 0)$
Vertices: $(\pm a, 0)$

Foci on the y-axis: $\frac{x^2}{b^2} + \frac{y^2}{a^2} = 1 \quad (a > b)$

Center-to-focus distance: $c = \sqrt{a^2 - b^2}$
Foci: $(0, \pm c)$
Vertices: $(0, \pm a)$

In each case, a is the semimajor axis and b is the semiminor axis.

Example 3 Find the standard-form equation of the ellipse with foci $(0, \pm 3)$ and vertices $(0, \pm 4)$.

Solution The standard-form equation for an ellipse with foci $(0, \pm c)$ and vertices $(0, \pm a)$ is

$$\frac{x^2}{b^2} + \frac{y^2}{a^2} = 1,$$

where $c = \sqrt{a^2 - b^2}$. In the ellipse at hand, $c = 3$ and $a = 4$, so

$$3 = \sqrt{(4)^2 - b^2},$$

$$9 = 16 - b^2,$$

$$b^2 = 7.$$

The equation we seek is

$$\frac{x^2}{7} + \frac{y^2}{16} = 1.$$

Hyperbolas

DEFINITIONS

A **hyperbola** is the set of points in a plane whose distances from two fixed points in the plane have a constant difference. The two fixed points are the **foci** of the hyperbola.

If the foci are $F_1(-c, 0)$ and $F_2(c, 0)$ (Fig. 9.7) and the constant difference is $2a$, then a point (x, y) lies on the hyperbola if and only if

$$\sqrt{(x + c)^2 + y^2} - \sqrt{(x - c)^2 + y^2} = \pm 2a. \tag{8}$$

To simplify this equation, we move the second radical to the right-hand side, square, isolate the remaining radical, and square again, obtaining

$$\frac{x^2}{a^2} + \frac{y^2}{a^2 - c^2} = 1. \tag{9}$$

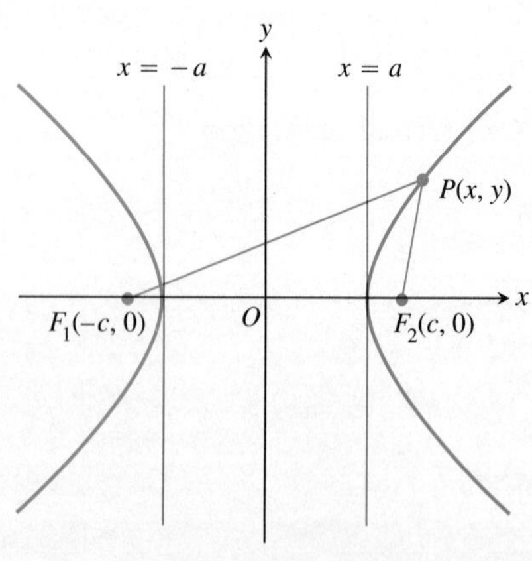

9.7 Hyperbolas have two branches. For points on the right-hand branch of the hyperbola shown here, $PF_1 - PF_2 = 2a$. For points on the left-hand branch, $PF_2 - PF_1 = 2a$.

So far, this looks just like the equation for an ellipse. But now $a^2 - c^2$ is negative because $2a$, being the difference of two sides of triangle PF_1F_2, is less than $2c$, the third side.

The algebraic steps taken to arrive at Eq. (9) can be reversed to show that every point P whose coordinates satisfy an equation of this form with $0 < a < c$ also satisfies Eq. (8). Thus, a point lies on the hyperbola if and only if its coordinates satisfy Eq. (9).

If we let b denote the positive square root of $c^2 - a^2$,

$$b = \sqrt{c^2 - a^2}, \tag{10}$$

then $a^2 - c^2 = -b^2$ and Eq. (9) takes the more compact form

$$\frac{x^2}{a^2} - \frac{y^2}{b^2} = 1. \tag{11}$$

The only difference between Eq. (11) and the equation for an ellipse (Eq. 4) is the minus sign in the equation and the new relation

$$c^2 = a^2 + b^2 \tag{12}$$

given by Eq. (10).

Like the ellipse, the hyperbola is symmetric with respect to the origin and both coordinate axes. It crosses the x-axis at the points $(\pm a, 0)$. The tangents at these points are vertical because the derivative

$$\frac{dy}{dx} = \frac{b^2x}{a^2y}$$

is infinite when $y = 0$. The hyperbola has no y-intercepts; in fact, no part of the curve lies between the lines $x = -a$ and $x = a$.

DEFINITIONS

> The line through the foci of a hyperbola is the hyperbola's **focal axis.** The point on the axis halfway between the foci is the hyperbola's **center.** The points where the focal axis crosses the hyperbola are the hyperbola's **vertices** (Fig. 9.8).

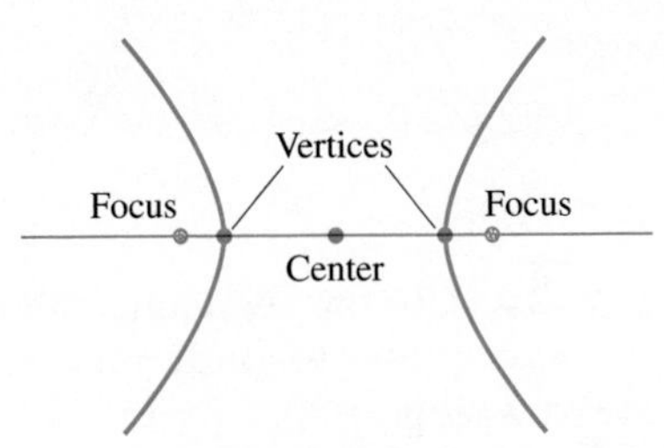

9.8 The center, vertices, and foci of a hyperbola lie on the focal axis.

Asymptotes of Hyperbolas—Graphing

As we saw in Section 3.4, the distance between a curve and some fixed line may approach zero as the curve moves farther and farther from the origin. If this happens, we call the line an **asymptote** of the curve. The hyperbola

$$\frac{x^2}{a^2} - \frac{y^2}{b^2} = 1 \tag{13}$$

has two asymptotes, the lines

$$y = \pm\frac{b}{a}x \tag{14}$$

(Exercise 67). The asymptotes give us the guidance we need to graph hyperbolas quickly. (See the drawing lesson on page 630.) The fastest way to find the equations of the asymptotes is to replace the 1 in Eq. (13) by 0 and solve the resulting equation for y:

$$\underbrace{\frac{x^2}{a^2} - \frac{y^2}{b^2} = 1}_{\text{hyperbola}} \quad\Rightarrow\quad \underbrace{\frac{x^2}{a^2} - \frac{y^2}{b^2} = 0}_{\text{0 for 1}} \quad\Rightarrow\quad \underbrace{y = \pm\frac{b}{a}x.}_{\text{asymptotes}} \tag{15}$$

Example 4 *Foci on the x-axis.* The equation

$$\frac{x^2}{4} - \frac{y^2}{5} = 1 \tag{16}$$

is Eq. (11) with $a^2 = 4$ and $b^2 = 5$ (Fig. 9.9a). We have

Center-to-focus distance: $c = \sqrt{a^2 + b^2} = \sqrt{4 + 5} = 3$,

Foci: $(\pm c, 0) = (\pm 3, 0)$, Vertices: $(\pm a, 0) = (\pm 2, 0)$,

Center: $(0, 0)$,

Asymptotes: $\dfrac{x^2}{4} - \dfrac{y^2}{5} = 0$ or $y = \pm\dfrac{\sqrt{5}}{2}x$.

DRAWING LESSON

How to Graph the Hyperbola $\frac{x^2}{a^2} - \frac{y^2}{b^2} = 1$

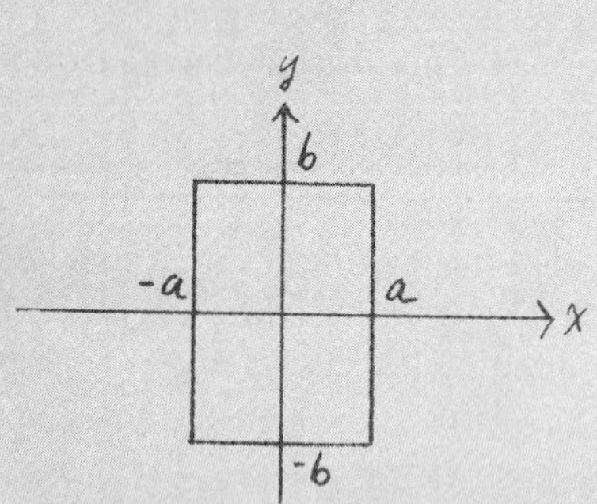

1. Mark the points $(\pm a, 0)$ and $(0, \pm b)$ with line segments and complete the rectangle they determine.

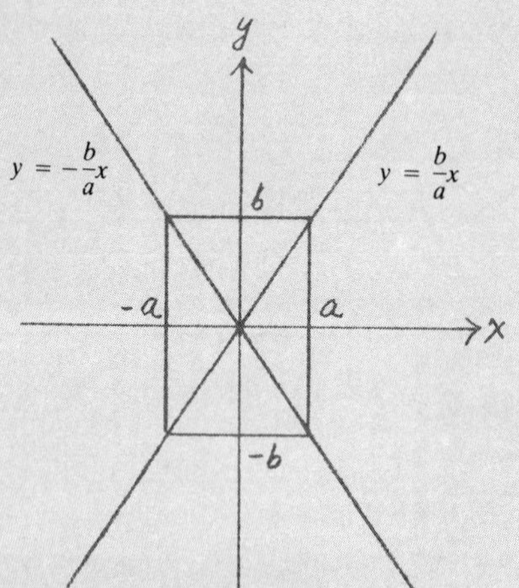

2. Sketch the asymptotes by extending the rectangle's diagonals.

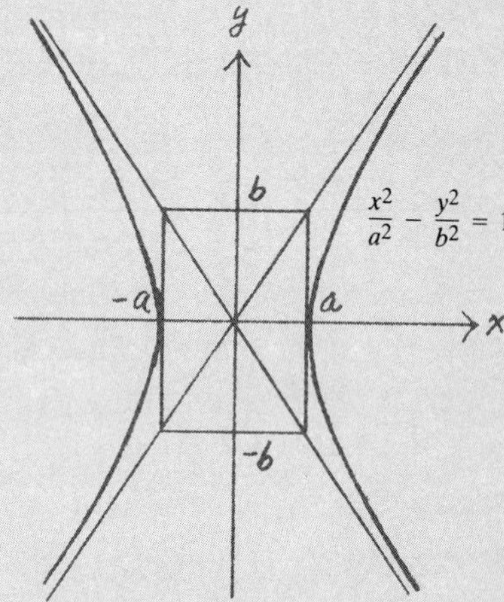

3. Use the rectangle and asymptotes to guide your drawing.

Standard-Form Equations for Hyperbolas Centered at the Origin

Foci on the x-axis: $\frac{x^2}{a^2} - \frac{y^2}{b^2} = 1$

Center-to-focus distance: $c = \sqrt{a^2 + b^2}$

Foci: $(\pm c, 0)$

Vertices: $(\pm a, 0)$

Asymptotes: $\frac{x^2}{a^2} - \frac{y^2}{b^2} = 0$ or $y = \pm\frac{b}{a}x$

Foci on the y-axis: $\frac{y^2}{a^2} - \frac{x^2}{b^2} = 1$

Center-to-focus distance: $c = \sqrt{a^2 + b^2}$

Foci: $(0, \pm c)$

Vertices: $(0, \pm a)$

Asymptotes: $\frac{y^2}{a^2} - \frac{x^2}{b^2} = 0$ or $y = \pm\frac{a}{b}x$

Notice the difference in the asymptote equations (b/a in the first, a/b in the second).

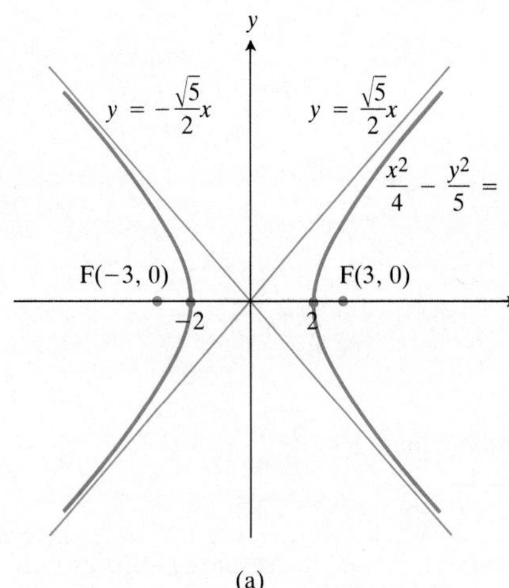

(a)

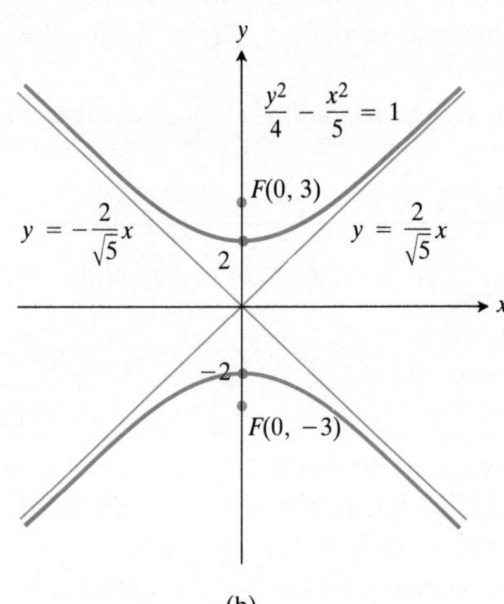

(b)

9.9 (a) The hyperbola in Example 4. (b) The hyperbola in Example 5.

Example 5 *Foci on the y-axis.* The hyperbola

$$\frac{y^2}{4} - \frac{x^2}{5} = 1, \tag{17}$$

obtained by interchanging x and y in Eq. (16), represents a hyperbola with vertices on the y-axis (Fig. 9.9b). With a^2 still equal to 4 and b^2 equal to 5, we have

Center-to-focus distance: $c = \sqrt{a^2 + b^2} = \sqrt{4 + 5} = 3$,

Foci: $(0, \pm c) = (0, \pm 3)$, Vertices: $(0, \pm a) = (0, \pm 2)$,

Center: $(0, 0)$,

Asymptotes: $\frac{y^2}{4} - \frac{x^2}{5} = 0$ or $y = \pm\frac{2}{\sqrt{5}}x$.

Example 6 Find an equation for the hyperbola with asymptotes $y = \pm(4/3)x$ and foci $(\pm 10, 0)$.

Solution The standard-form equation for a hyperbola with foci $(\pm c, 0)$ on the x-axis is

$$\frac{x^2}{a^2} - \frac{y^2}{b^2} = 1,$$

where $c = \sqrt{a^2 + b^2}$. From the asymptote equation $y = \pm(b/a)x$, we learn that

$$\frac{b}{a} = \frac{4}{3}, \quad \text{or} \quad b = \frac{4}{3}a.$$

Hence,

$$c^2 = a^2 + b^2 = a^2 + \frac{16}{9}a^2 = \frac{25}{9}a^2,$$

and we have

$$a^2 = \frac{9}{25}c^2 = \frac{9}{25}(10)^2 = 36,$$

$$b^2 = c^2 - a^2 = 100 - 36 = 64. \quad \left(\begin{array}{l}\text{The foci are } (\pm 10, 0),\\ \text{so } c = 10 \text{ and } c^2 = 100.\end{array}\right)$$

The equation we seek is

$$\frac{x^2}{36} - \frac{y^2}{64} = 1.$$

Classifying Conic Sections by Eccentricity: The Focus–Directrix Equation

Although the center-to-focus distance c does not appear in the standard equation for an ellipse,

$$\frac{x^2}{a^2} + \frac{y^2}{b^2} = 1 \quad (a > b),$$

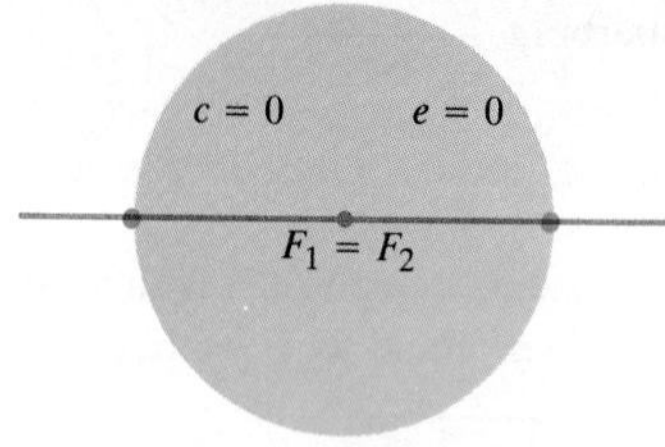

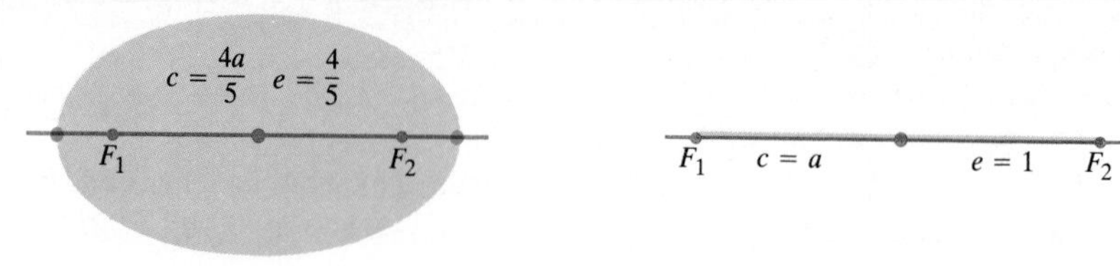

9.10 The ellipse changes from a circle to a line segment as c increases from 0 to a.

we may still determine the value of c from the equation

$$c = \sqrt{a^2 - b^2}.$$

If we keep a fixed and vary c over the interval $0 \le c \le a$, the resulting ellipses will vary in shape (Fig. 9.10). They are circles if $c = 0$ (so that $a = b$) and flatten as c increases. In the extreme case $c = a$, the foci and vertices overlap and the ellipse degenerates into a line segment.

We use the ratio of c to a to describe the various shapes the ellipse can take. We call this ratio the ellipse's eccentricity.

DEFINITION

The **eccentricity** of the ellipse $(x^2/a^2) + (y^2/b^2) = 1$ $(a > b)$ is the number

$$e = \frac{c}{a} = \frac{\sqrt{a^2 - b^2}}{a}. \tag{18}$$

TABLE 9.2
Eccentricities of planetary orbits

Mercury	0.21	Saturn	0.06
Venus	0.01	Uranus	0.05
Earth	0.02	Neptune	0.01
Mars	0.09	Pluto	0.25
Jupiter	0.05		

The planets in the solar system revolve around the sun in elliptical orbits with the sun at one focus. Most of the planets, including Earth, have orbits that are nearly circular, as can be seen from the eccentricities in Table 9.2. Pluto, however, has a fairly eccentric orbit, with $e = 0.25$, as does Mercury, with $e = 0.21$. Other members of the solar system have orbits that are even more eccentric. Icarus, an asteroid about 1 mile wide that revolves around the sun every 409 Earth days, has an orbital eccentricity of 0.83 (Fig. 9.11).

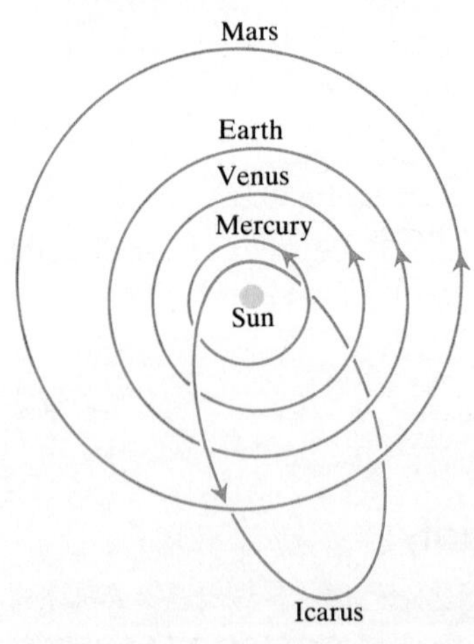

9.11 The orbit of the asteroid Icarus is highly eccentric. The earth's orbit is so nearly circular that both its foci lie inside the sun.

Example 7 The orbit of Halley's comet is an ellipse 36.18 astronomical units long by 9.12 astronomical units wide. (One *astronomical unit* [AU] is the semimajor axis of the earth's orbit, about 92,600,000 miles.) Its eccentricity is

$$e = \frac{\sqrt{a^2 - b^2}}{a} = \frac{\sqrt{(36.18/2)^2 - (9.12/2)^2}}{(36.18/2)} = \frac{\sqrt{(18.09)^2 - (4.56)^2}}{18.09}$$

$$= 0.97. \qquad \text{(Rounded, with a calculator)}$$

Whereas a parabola has one focus and one directrix, each ellipse has two foci and two directrices. These are the lines perpendicular to the major axis at distances $\pm a/e$ from the center. The parabola has the property that

$$PF = 1 \cdot PD \tag{19}$$

for any point P on it, where F is the focus and D is the point nearest P on the directrix. For an ellipse, it can be shown that the equations that replace (19) are

$$PF_1 = e \cdot PD_1, \qquad PF_2 = e \cdot PD_2. \tag{20}$$

Halley's Comet

Edmund Halley (1656–1742; pronounced *haw*-ley), British biologist, geologist, sea captain, pirate, spy, Antarctic voyager, astronomer, adviser on fortifications, company founder and director, and author of the first actuarial mortality tables, was also the mathematician who pushed and harried Newton into writing his *Principia*. Despite his accomplishments, Halley is known today chiefly as the man who calculated the orbit of the great comet of 1682: "wherefore if according to what we have already said [the comet] should return again about the year 1758, candid posterity will not refuse to acknowledge that this was first discovered by an Englishman." Indeed, candid posterity did not refuse—ever since the comet's return in 1758, it has been known as Halley's comet.

Last seen rounding the sun during the winter and spring of 1985–86, the comet is due to return again in the year 2062. A recent study indicates that the comet has made about 2000 cycles so far with about the same number to go before the sun erodes it away completely.

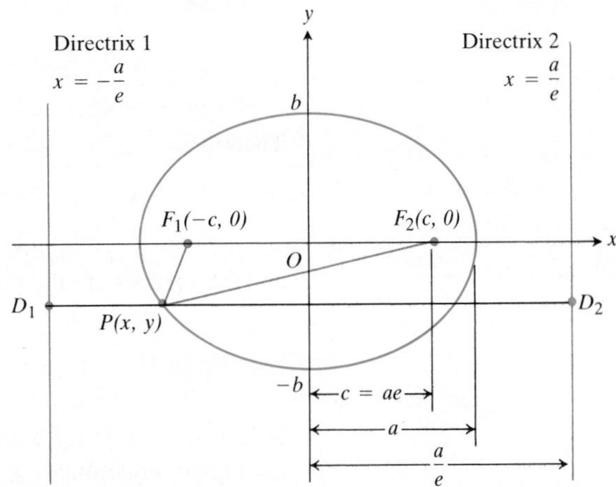

9.12 The foci and directrices of the ellipse $(x^2/a^2) + (y^2/b^2) = 1$. Directrix 1 corresponds to focus F_1, and directrix 2 to focus F_2.

Here, e is the eccentricity, P is any point on the ellipse, F_1 and F_2 are the foci, and D_1 and D_2 are the points on the directrices nearest P (Fig. 9.12).

In each equation in (20) the directrix and focus must correspond; that is, if we use the distance from P to F_1, we must also use the distance from P to the directrix at the same end of the ellipse. The directrix $x = -a/e$ corresponds to $F_1(-c, 0)$, and the directrix $x = a/e$ corresponds to $F_2(c, 0)$.

We define the eccentricity of a hyperbola with the same formula we use for the ellipse, $e = c/a$, only in this case c equals $\sqrt{a^2 + b^2}$ instead of $\sqrt{a^2 - b^2}$. In contrast to the eccentricity of an ellipse, the eccentricity of a hyperbola is always greater than 1.

DEFINITION

The **eccentricity** of the hyperbola $(x^2/a^2) - (y^2/b^2) = 1$ is the number

$$e = \frac{c}{a} = \frac{\sqrt{a^2 + b^2}}{a}. \qquad (21)$$

In both ellipse and hyperbola, the eccentricity is the ratio of the distance between the foci to the distance between the vertices (because $c/a = 2c/2a$).

$$\text{Eccentricity} = \frac{\text{distance between foci}}{\text{distance between vertices}} \qquad (22)$$

In an ellipse, the foci are closer together than the vertices and the ratio is less than 1. In a hyperbola, the foci are farther apart than the vertices and the ratio is greater than 1.

Example 8 Locate the vertices of an ellipse of eccentricity 0.8 whose foci lie at the points $(0, \pm 7)$.

Solution The vertices are the points $(0, \pm a)$, where

$$\frac{c}{a} = \frac{7}{a} = e = 0.8.$$

Therefore,

$$a = \frac{7}{0.8} = 8.75$$

and the vertices are the points $(0, \pm 8.75)$.

Example 9 Find the eccentricity of the hyperbola $9x^2 - 16y^2 = 144$.

Solution We divide both sides of the hyperbola's equation by 144 to put it in standard form, obtaining

$$\frac{9x^2}{144} - \frac{16y^2}{144} = 1 \quad \text{and thus} \quad \frac{x^2}{16} - \frac{y^2}{9} = 1.$$

With $a^2 = 16$ and $b^2 = 9$, we find that

$$c = \sqrt{a^2 + b^2} = \sqrt{16 + 9} = 5,$$

so

$$e = \frac{c}{a} = \frac{5}{4}.$$

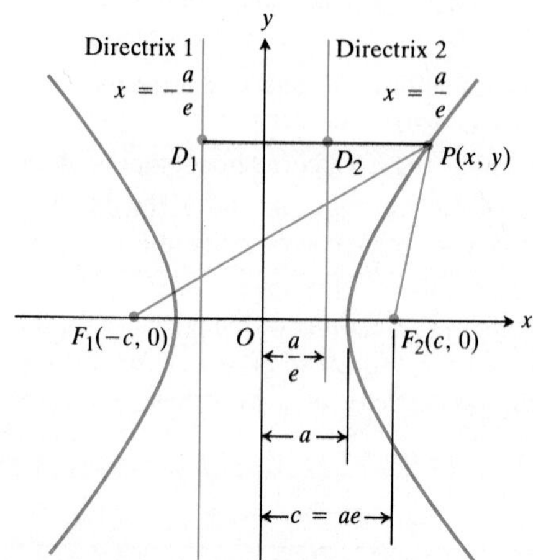

9.13 The foci and directrices of the hyperbola $(x^2/a^2) - (y^2/b^2) = 1$. No matter where P lies on the hyperbola, $PF_1 = e \cdot PD_1$ and $PF_2 = e \cdot PD_2$.

As with the ellipse, it can be shown that the lines $x = \pm a/e$ act as directrices for the hyperbola and that

$$PF_1 = e \cdot PD_1 \quad \text{and} \quad PF_2 = e \cdot PD_2. \tag{23}$$

Here P is any point on the hyperbola, F_1 and F_2 are the foci, and D_1 and D_2 are the points nearest P on the directrices (Fig. 9.13).

To complete the picture, we now define the eccentricity of a parabola to be $e = 1$. Equations (19), (20), and (23) then have the common form $PF = e \cdot PD$.

DEFINITION

The **eccentricity** of a parabola is $e = 1$.

The "focus–directrix" equation $PF = e \cdot PD$ unites the parabola, ellipse, and hyperbola in the following way. Suppose that a point P moves in a plane in such a way that its distance PF from a fixed point F in the plane (the focus) is a constant multiple of its distance PD from a fixed line in the plane (the directrix). That is, suppose that

$$PF = e \cdot PD, \tag{24}$$

where e is the constant of proportionality. Then the path traced by P is

a) a *parabola* if $e = 1$,

b) an *ellipse* of eccentricity e if $e < 1$, and

c) a *hyperbola* of eccentricity e if $e > 1$.

Equation (24) may not look like much to get excited about. There are no coordinates in it and when we try to translate it into coordinate form it translates in different ways, depending on the numerical size of e. At least, that is what happens in the Cartesian plane. However, in the polar coordinate plane, as we shall see in Section 9.7, the equation $PF = e \cdot PD$ translates into a single equation regardless of the size of e, an equation so simple that it has been the equation of choice for astronomers and space scientists for nearly three hundred years.

Given the focus and corresponding directrix of a hyperbola centered at the origin and with foci on the x-axis, we can use the dimensions shown in Fig. 9.13 to find e. Knowing e, we can derive a Cartesian equation for the hyperbola from the equation $PF = e \cdot PD$, as in the next example. We can find equations for ellipses centered at the origin and with foci on the x-axis in a similar way, using the dimensions shown in Fig. 9.12.

Example 10 Find the standard-form equation for the hyperbola centered at the origin that has a focus at (3, 0) and has the line $x = 1$ as the corresponding directrix.

Solution We first use the dimensions shown in Fig. 9.13 to find the hyperbola's eccentricity. The focus is

$$(c, 0) = (3, 0), \quad \text{so} \quad c = 3.$$

The directrix is the line

$$x = \frac{a}{e} = 1, \quad \text{so} \quad a = e.$$

When combined with the equation $e = c/a$ that defines eccentricity, these results give

$$e = \frac{c}{a} = \frac{3}{e}, \quad \text{so} \quad e^2 = 3 \quad \text{and} \quad e = \sqrt{3}.$$

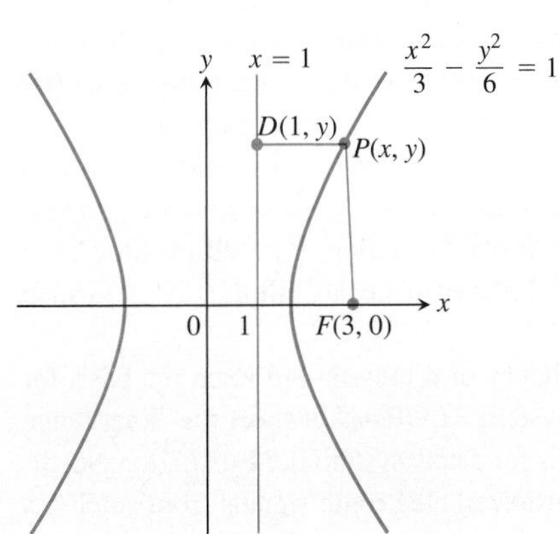

9.14 The hyperbola in Example 10.

Knowing e, we can now derive the equation we want from the equation $PF = e \cdot PD$. In the notation of Fig. 9.14, we have

$$\begin{aligned} PF &= e \cdot PD && \text{(Eq. (24))} \\ \sqrt{(x-3)^2 + (y-0)^2} &= \sqrt{3}\,|x-1| && (e = \sqrt{3}) \\ x^2 - 6x + 9 + y^2 &= 3(x^2 - 2x + 1) \\ 2x^2 - y^2 &= 6 \\ \frac{x^2}{3} - \frac{y^2}{6} &= 1. \end{aligned}$$

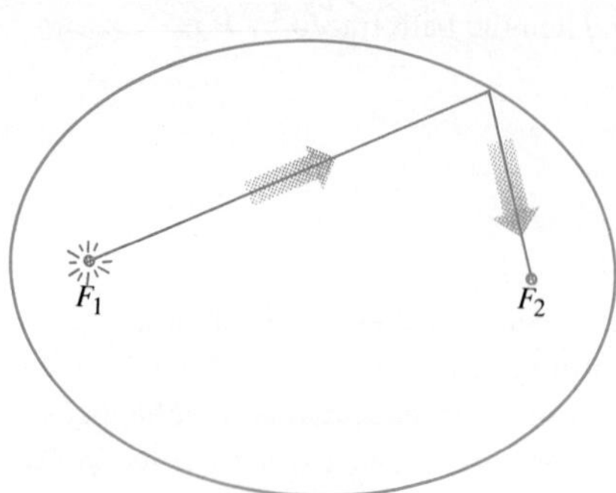

9.15 An elliptical mirror (shown here in profile) reflects light from one focus to the other.

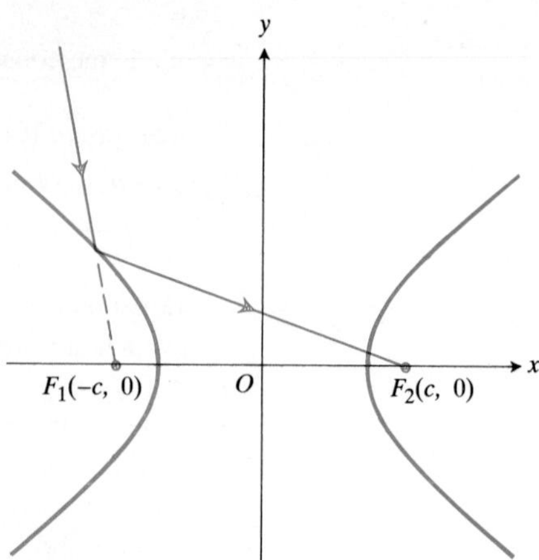

9.16 In this profile of a hyperbolic mirror, light coming toward focus F_1 is reflected toward focus F_2.

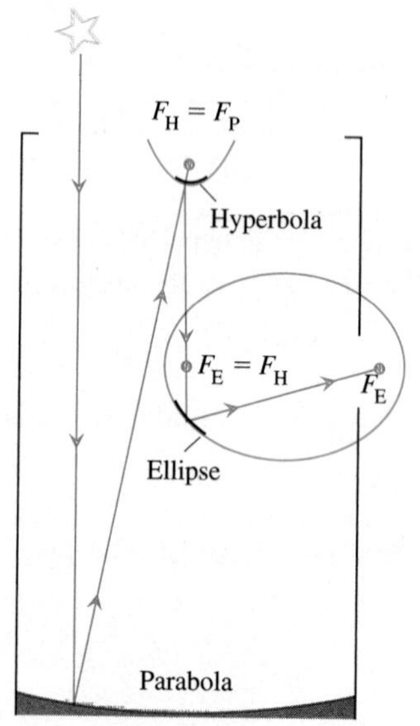

9.17 In this schematic drawing of a reflecting telescope, starlight reflects off a primary parabolic mirror toward the mirror's focus F_P. It is then reflected by a small hyperbolic mirror, whose focus is $F_H = F_P$, toward the second focus of the hyperbola, $F_E = F_H$. Since this focus is shared by an ellipse, the light is reflected by the elliptical mirror to the ellipse's second focus to be seen by an observer.

As recent experience with NASA's Hubble space telescope shows, the mirrors have to be nearly perfect to focus properly. The aberration causing the malfunction in Hubble's primary mirror amounts to about half a wavelength of visible light, no more than 1/50 the width of a human hair.

Reflective Properties

Like parabolas, ellipses and hyperbolas have reflective properties that are important in science and engineering. If an ellipse is revolved about its major axis to generate a surface (the surface is called an *ellipsoid*), and the interior is silvered to produce a mirror, light from one focus will be reflected to the other focus (Fig. 9.15). Ellipsoids reflect sound the same way, and this property is used to construct *whispering galleries,* rooms in which a person standing at one focus can hear a whisper from the other focus. Statuary Hall in the U.S. Capitol building is a whispering gallery. Ellipsoids also appear in instruments used to study aircraft noise in wind tunnels (sound at one focus can be received at the other focus with relatively little interference from other sources).

Light directed toward one focus of a hyperbolic mirror is reflected toward the other focus (Fig. 9.16). This property of hyperbolas is combined with the reflective properties of parabolas and ellipses in designing modern telescopes (Fig. 9.17).

Other Applications

Ellipses appear in airplane wings (British Spitfire) and sometimes in gears designed for racing bicycles. Stereo systems often have elliptical styli, and water pipes are sometimes designed with elliptical cross sections to allow for expansion when the water freezes. The triggering mechanisms in some lasers are elliptical, and stones on a beach become more and more elliptical as they are ground down by waves. There are also applications of ellipses to fossil formation. The ellipsolith, once thought to be a separate species, is now known to be an elliptically deformed nautilus.

Hyperbolic paths arise in Einstein's theory of relativity and form the basis for the (unrelated) LORAN radio navigation system. (LORAN is short for "long range navigation.") Hyperbolas also form the basis for a new system the Burlington Northern Railroad is developing for using synchronized electronic signals from satellites to track freight trains. A few years ago, computers aboard Burlington Northern locomotives in Minnesota were able to track trains to within one mile per hour of their speed and to within 150 feet of their actual location.

EXERCISES 9.1

Match the parabolas in Exercises 1–4 with the following equations:

$$x^2 = 2y, \quad x^2 = -6y, \quad y^2 = 8x, \quad y^2 = -4x.$$

Then find the parabola's focus and directrix.

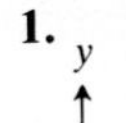

1.

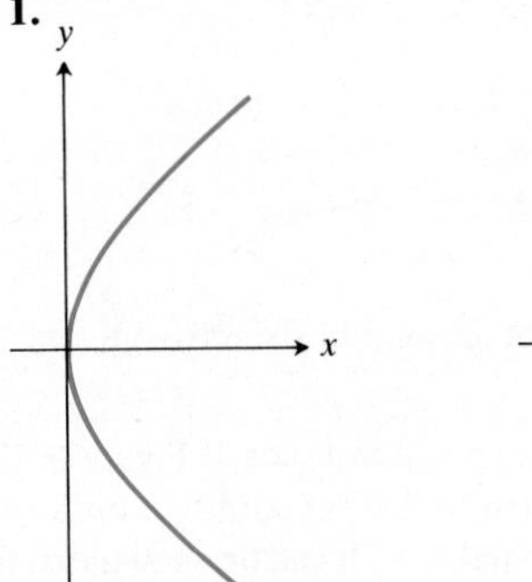

2.

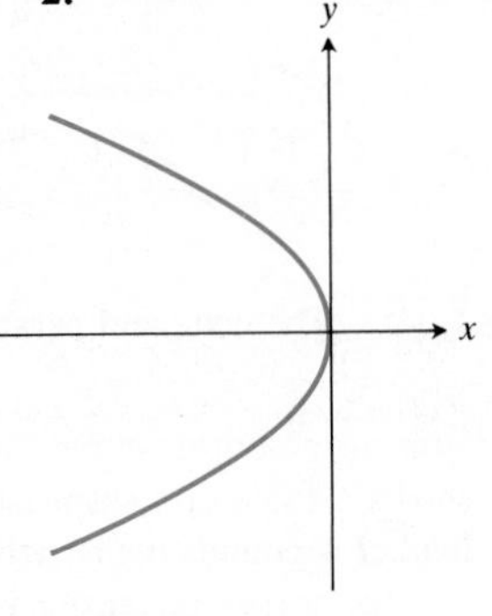

3.

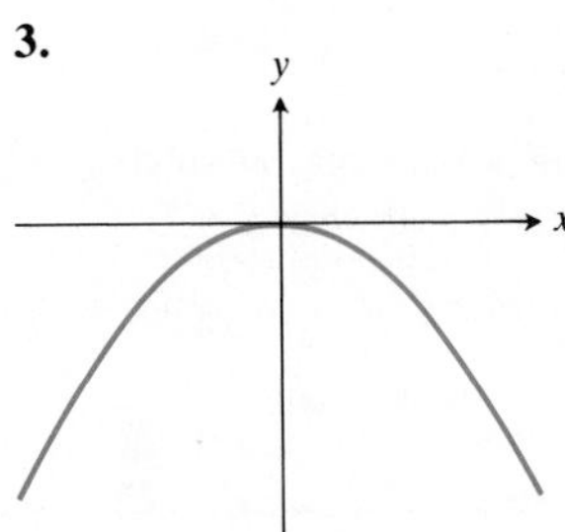

4.

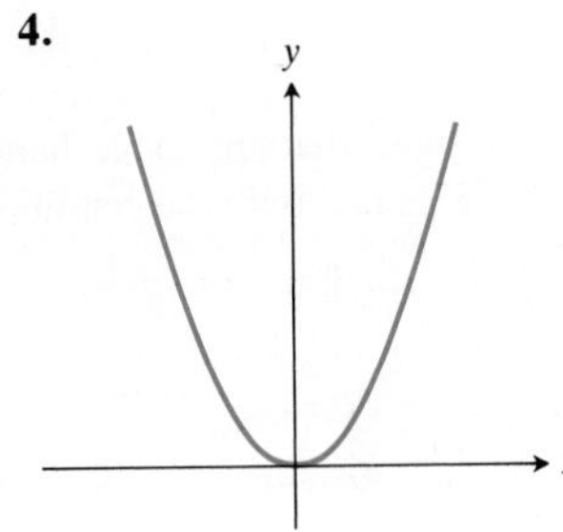

Match the conic sections in Exercises 5–8 with the following equations:

$$\frac{x^2}{4} + \frac{y^2}{9} = 1, \quad \frac{x^2}{2} + y^2 = 1, \quad \frac{y^2}{4} - x^2 = 1, \quad \frac{x^2}{4} - \frac{y^2}{9} = 1.$$

Then find the conic section's foci, eccentricity, and directrices. If the conic section is a hyperbola, find its asymptotes as well.

5.

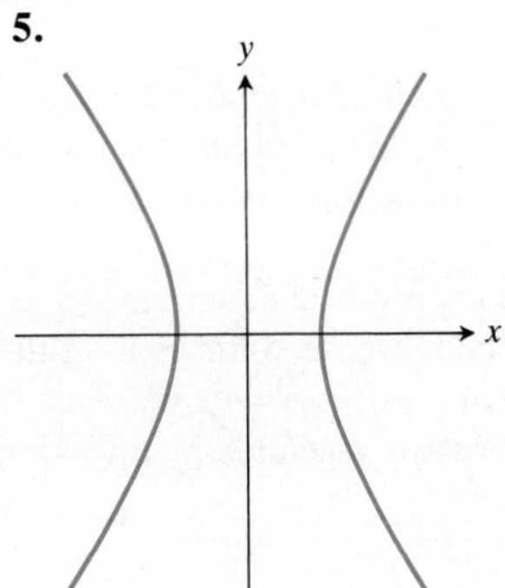

6.

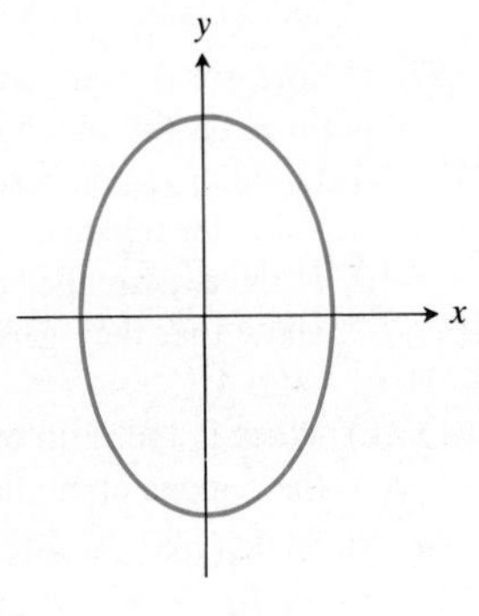

7.

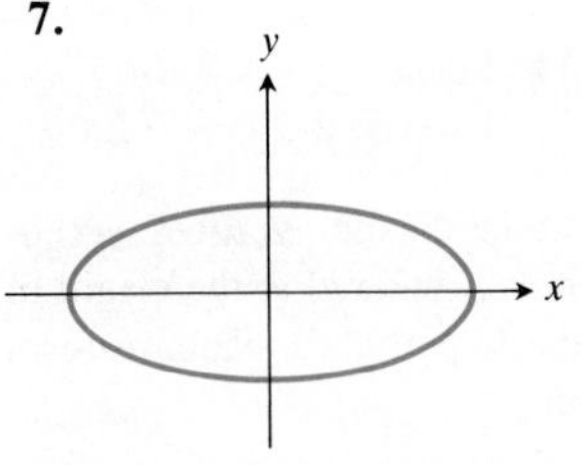

8.

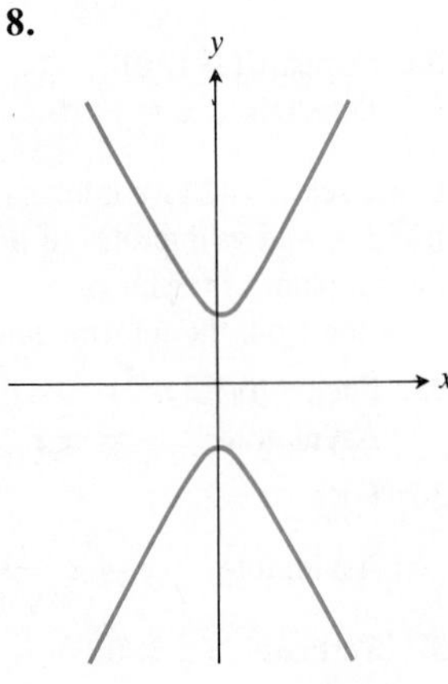

Exercises 9–16 give equations for ellipses. Put each equation in standard form and find the ellipse's eccentricity. Then sketch the ellipse. Include the foci in your sketch.

9. $16x^2 + 25y^2 = 400$

10. $7x^2 + 16y^2 = 112$

11. $2x^2 + y^2 = 2$

12. $2x^2 + y^2 = 4$

13. $3x^2 + 2y^2 = 6$

14. $9x^2 + 10y^2 = 90$

15. $6x^2 + 9y^2 = 54$

16. $169x^2 + 25y^2 = 4225$

Exercises 17–24 give equations for hyperbolas. Put each equation in standard form and find the hyperbola's eccentricity and asymptotes. Then sketch the hyperbola. Include the asymptotes and foci in your sketch.

17. $x^2 - y^2 = 1$

18. $9x^2 - 16y^2 = 144$

19. $y^2 - x^2 = 8$

20. $y^2 - x^2 = 4$

21. $8x^2 - 2y^2 = 16$

22. $y^2 - 3x^2 = 3$

23. $8y^2 - 2x^2 = 16$

24. $64x^2 - 36y^2 = 2304$

Exercises 25–30 give information about the foci, vertices, and eccentricity of ellipses centered at the origin of the xy-plane. In each case, find the ellipse's standard-form equation from the information given.

25. Foci: $(\pm\sqrt{2}, 0)$
Vertices: $(\pm 2, 0)$

26. Foci: $(0, \pm 4)$
Vertices: $(0, \pm 5)$

27. Foci: $(0, \pm 3)$
Eccentricity: 0.5

28. Foci: $(\pm 8, 0)$
Eccentricity: 0.2

29. Vertices: $(0, \pm 70)$
Eccentricity: 0.1

30. Vertices: $(\pm 10, 0)$
Eccentricity: 0.24

Exercises 31–34 give information about the foci and corresponding directrices of ellipses centered at the origin of the xy-plane. In each case, use the dimensions in Fig. 9.12 to find the eccentricity of the ellipse. Then use the equation $PF = e \cdot PD$ to find the ellipse's standard-form equation.

31. Focus: $(\sqrt{5}, 0)$
Directrix: $x = \dfrac{9}{\sqrt{5}}$

32. Focus: $(4, 0)$
Directrix: $x = \dfrac{16}{3}$

33. Focus: $(-4, 0)$
Directrix: $x = -16$

34. Focus: $(-\sqrt{2}, 0)$
Directrix: $x = -2\sqrt{2}$

Exercises 35–42 give information about the foci, vertices, eccentricities, and asymptotes of hyperbolas centered at the origin of the xy-plane. In each case, find the hyperbola's standard-form equation from the information given.

35. Foci: $(0, \pm\sqrt{2})$
Asymptotes: $y = \pm x$

36. Foci: $(\pm 2, 0)$
Asymptotes: $y = \pm\dfrac{1}{\sqrt{3}}x$

37. Vertices: $(\pm 3, 0)$
Asymptotes: $y = \pm\dfrac{4}{3}x$

38. Vertices: $(0, \pm 2)$
Asymptotes: $y = \pm\dfrac{1}{2}x$

39. Vertices: $(0, \pm 1)$
Eccentricity: 3

40. Vertices: $(\pm 2, 0)$
Eccentricity: 2

41. Foci: $(\pm 3, 0)$
Eccentricity: 3

42. Foci: $(0, \pm 5)$
Eccentricity: 1.25

Exercises 43–46 give information about the foci and corresponding directrices of hyperbolas centered at the origin of the xy-plane. In each case, use the dimensions in Fig. 9.13 to find the eccentricity of the hyperbola. Then use the equation $PF = e \cdot PD$ to find the hyperbola's standard-form equation.

43. Focus: $(4, 0)$
Directrix: $x = 2$

44. Focus: $(\sqrt{10}, 0)$
Directrix: $x = \sqrt{2}$

45. Focus: $(-2, 0)$
Directrix: $x = -\dfrac{1}{2}$

46. Focus: $(-6, 0)$
Directrix: $x = -2$

Sketch the regions whose points satisfy the inequalities or sets of inequalities in Exercises 47–52.

47. $9x^2 + 16y^2 \le 144$

48. $x^2 + y^2 \ge 1$ and $4x^2 + y^2 \le 4$

49. $x^2 + 4y^2 \ge 4$ and $4x^2 + 9y^2 \le 36$

50. $(x^2 + y^2 - 4)(x^2 + 9y^2 - 9) \le 0$

51. $4y^2 - x^2 \ge 4$

52. $|x^2 - y^2| \le 1$

53. *Archimedes' formula for the volume of a parabolic solid.* The region enclosed by the parabola $y = (4h/b^2)x^2$ and the line $y = h$ is revolved about the y-axis to generate a solid (Fig. 9.18). Show that the volume of the solid is 3/2 the volume of the corresponding cone.

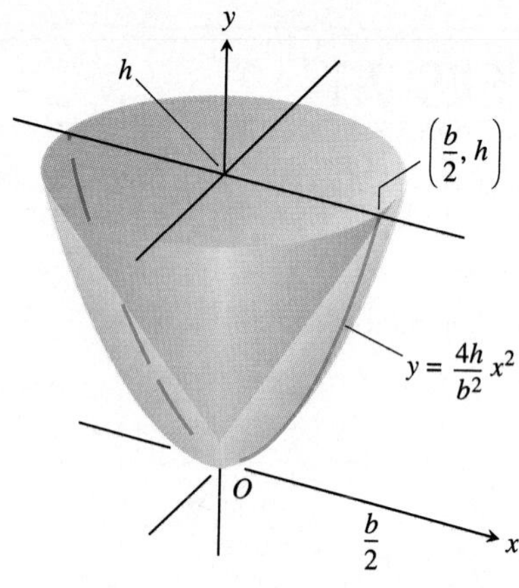

9.18 The cone and parabolic solid in Exercise 53.

54. *Suspension bridge cables hang in parabolas.* Figure 9.19 shows a cable of a suspension bridge supporting a uniform load of w pounds per horizontal foot. It can be shown that if H is the horizontal tension in the cable at the origin, then the curve of the cable satisfies the differential equation

$$\frac{dy}{dx} = \frac{w}{H}x.$$

Show that the cable hangs in a parabola by solving this equation with the condition that $y = 0$ when $x = 0$.

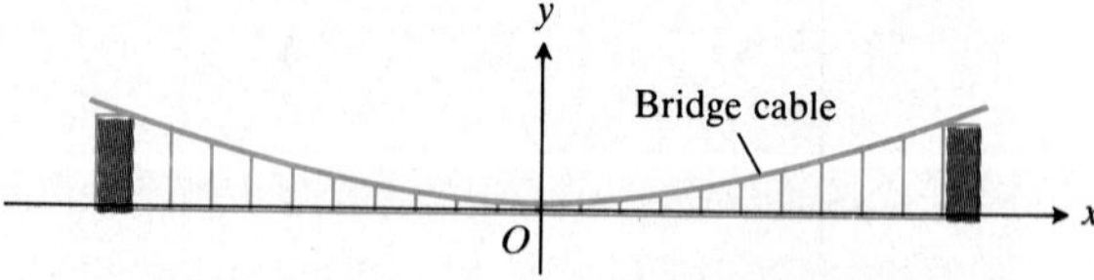

9.19 The suspension bridge cable in Exercise 54.

55. Find equations for the lines that are tangent to the circle $(x - 2)^2 + (y - 1)^2 = 5$ at the points where the circle crosses the coordinate axes. (*Hint:* Use implicit differentiation.)

56. Find an equation for the circle centered at $(-2, 1)$ that passes through the point $(1, 3)$. Is the point $(1.1, 2.8)$ inside, outside, or on the circle?

57. If lines are drawn parallel to the coordinate axes through a point P on the parabola $y^2 = kx$, the parabola divides the rectangular region bounded by these lines and the axes into two smaller regions.

a) If the two smaller regions are revolved about the y-axis, show that they generate solids whose volumes have the ratio 4:1.

b) What is the ratio of the volumes generated by revolving the regions about the x-axis?

58. Show that the tangents to the curve $y^2 = 4px$ from any point on the line $x = -p$ are perpendicular.

59. Find the dimensions of the rectangle of largest area that can be inscribed in the ellipse $x^2 + 4y^2 = 4$ with its sides parallel to the coordinate axes. What is the area of the rectangle?

60. Find the center of mass of a thin homogeneous plate that is bounded below by the x-axis and above by the ellipse $(x^2/9) + (y^2/16) = 1$.

61. Find the volume of the solid generated by revolving the region enclosed by the ellipse $9x^2 + 4y^2 = 36$ about the (a) x-axis, (b) y-axis.

62. The "triangular" region in the first quadrant bounded by the x-axis, the line $x = 4$, and the hyperbola $9x^2 - 4y^2 = 36$ is revolved about the x-axis to generate a solid. Find the volume of the solid.

63. The region bounded on the left by the y-axis, on the right by the hyperbola $x^2 - y^2 = 1$, and above and below by the lines $y = \pm 3$ is revolved about the y-axis to generate a solid. Find the volume of the solid.

64. The curve $y = \sqrt{x^2 + 1}$, $0 \le x \le \sqrt{2}$, which is part of the upper branch of the hyperbola $y^2 - x^2 = 1$, is revolved about the x-axis to generate a surface. Find the area of the surface.

65. The circular waves in Fig. 9.20 were made by touching the surface of the water in a ripple tank, first at A and then at B. As the circular waves expanded, their point of intersection seemed to trace a hyperbola. Did it really do that?

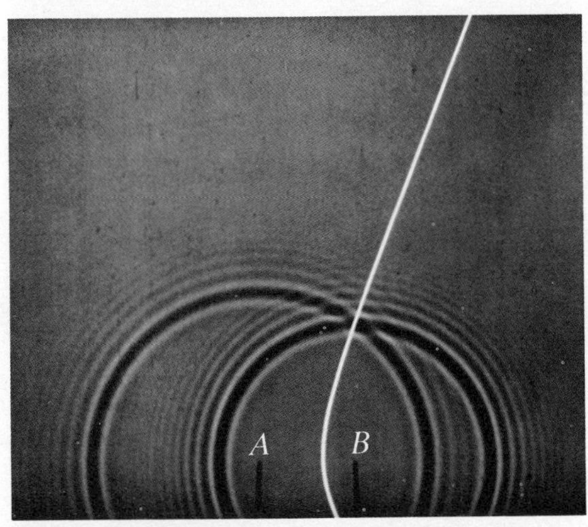

9.20 The expanding circles of ripples in Exercise 65. Photograph from *PSSC Physics,* Second Edition, 1965, D. C. Heath & Company with Education Development Center, Inc. *NCFMF Book of Film Notes,* 1974, The MIT Press with Education Development Center, Inc., Newton, Massachusetts.

To find out, we can model the waves with circles centered at A and B (Fig. 9.21). At time t, the point P is $r_A(t)$ units

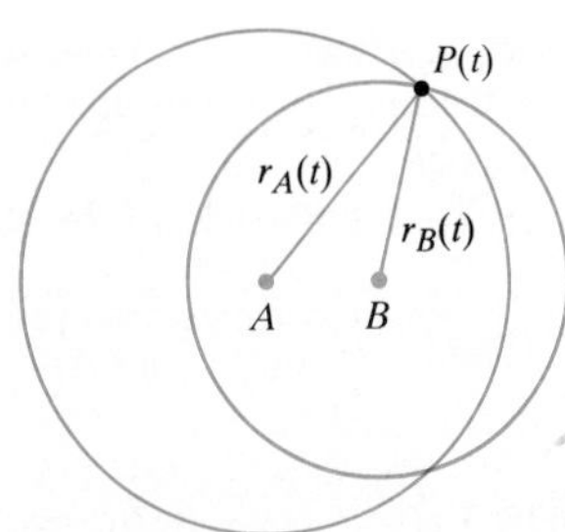

9.21 The model for the waves in Exercise 65.

from A and $r_B(t)$ units from B. Since the radii of the circles increase at a constant rate, the rate at which the waves are traveling is

$$\frac{dr_A}{dt} = \frac{dr_B}{dt}.$$

Conclude from this equation that $r_A - r_B$ has a constant value, so that P must lie on a hyperbola with foci at A and B.

66. *The reflective property of parabolas.* Figure 9.22 shows a typical point $P(x_0, y_0)$ on the parabola $y^2 = 4px$. The line L is tangent to the parabola at P. The parabola's focus lies at $F(p, 0)$. The ray L' extending from P to the right is parallel to the x-axis. We show that light from F to P will be reflected out along L' by showing that β equals α. Establish this equality by taking the following steps.

1. Show that $\tan \beta = 2p/y_0$.
2. Show that $\tan \phi = y_0/(x_0 - p)$.
3. Use the identity

$$\tan \alpha = \frac{\tan \phi - \tan \beta}{1 + \tan \phi \tan \beta}$$

to show that $\tan \alpha = 2p/y_0$.

Since the angles involved are both acute, $\tan \beta = \tan \alpha$ implies $\beta = \alpha$.

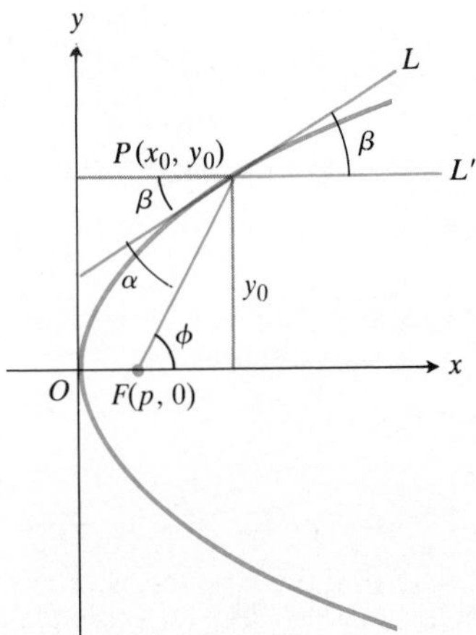

9.22 A parabolic reflector sends all light from the focus out parallel to the parabola's axis (Exercise 66).

67. *The asymptotes of $(x^2/a^2) - (y^2/b^2) = 1$.* Show that the vertical distance between the line $y = (b/a)x$ and the upper half of the right-hand branch $y = (b/a)\sqrt{x^2 - a^2}$ of the hyperbola $(x^2/a^2) - (y^2/b^2) = 1$ approaches 0 by showing that

$$\lim_{x\to\infty}\left(\frac{b}{a}x - \frac{b}{a}\sqrt{x^2 - a^2}\right) = \frac{b}{a}\lim_{x\to\infty}\left(x - \sqrt{x^2 - a^2}\right) = 0.$$

Similar results hold for the remaining portions of the hyperbola and the lines $y = \pm(b/a)x$.

9.2 The Graphs of Quadratic Equations in x and y; Rotations About the Origin

In this section, we establish one of the most amazing results in analytic geometry, which is that the Cartesian graph of any equation of the form

$$Ax^2 + Bxy + Cy^2 + Dx + Ey + F = 0, \tag{1}$$

in which A, B, and C are not all zero, is nearly always a conic section. The only exceptions are the case in which the graph consists of two parallel lines and the case in which there is no graph at all (Table 9.3). It is conventional to call all graphs of Eq. (1), curved or not, **quadratic curves.**

The Cross-Product Term

You may have noticed that the term Bxy did not appear in the equations for the conic sections in Section 9.1. This happened because the axes of the conic sections ran parallel to (in fact, coincided with) the coordinate axes.

To see what happens when the parallelism is absent, let us write an equation for a hyperbola with $a = 3$ and foci at $F_1(-3, -3)$ and $F_2(3, 3)$ (Fig. 9.23). The equa-

TABLE 9.3
Examples of quadratic curves

$Ax^2 + Bxy + Cy^2 + Dx + Ey + F = 0$								
	A	**B**	**C**	**D**	**E**	**F**	**Equation**	**Remarks**
Circle	1		1			−4	$x^2 + y^2 = 4$	$A = C$
Parabola			1	−9			$y^2 = 9x$	Quadratic in y, linear in x
Ellipse	4		9			−36	$4x^2 + 9y^2 = 36$	A, C have same sign, $A \neq C$
Hyperbola	1		−1			−1	$x^2 - y^2 = 1$	A, C have opposite signs
One line (still a conic section)	1						$x^2 = 0$	y-axis
Intersecting lines (still a conic section)		1		1	−1	−1	$xy + x - y - 1 = 0$	Factors to $(x-1)(y+1)=0$, so $x = 1$, $y = -1$
Parallel lines (not a conic section)	1			−3		2	$x^2 - 3x + 2 = 0$	Factors to $(x-1)(x-2) = 0$, so $x = 1$, $x = 2$
Point	1		1				$x^2 + y^2 = 0$	The origin
No graph	1					1	$x^2 = -1$	No graph

9.23 The focal axis of the hyperbola $2xy = 9$ makes an angle of $\pi/4$ radians with the positive x-axis.

tion $|PF_1 - PF_2| = 2a$ then becomes $|PF_1 - PF_2| = 2(3) = 6$ and

$$\sqrt{(x+3)^2 + (y+3)^2} - \sqrt{(x-3)^2 + (y-3)^2} = \pm 6.$$

When we transpose one radical, square, solve for the radical that still appears, and square again, this reduces to

$$2xy = 9, \tag{2}$$

which is a special case of Eq. (1) in which the cross-product term is present. The asymptotes of the hyperbola in Eq. (2) are the x- and y-axes, and the focal axis makes an angle of $\pi/4$ radians with the positive x-axis. As in this example, the cross-product term is present in Eq. (1) only when the axes of the conic are tilted.

Rotating the Coordinate Axes to Eliminate the Cross-Product Term

To eliminate the xy-term from the equation of a conic, we rotate the coordinate axes to eliminate the "tilt" in the axes of the conic. The equations for the rotations we use are derived in the following way. In the notation of Fig. 9.24, which shows a counterclockwise rotation about the origin through an angle α,

$$\begin{aligned} x &= OM = OP\cos(\theta+\alpha) = OP\cos\theta\cos\alpha - OP\sin\theta\sin\alpha, \\ y &= MP = OP\sin(\theta+\alpha) = OP\cos\theta\sin\alpha + OP\sin\theta\cos\alpha. \end{aligned} \tag{3}$$

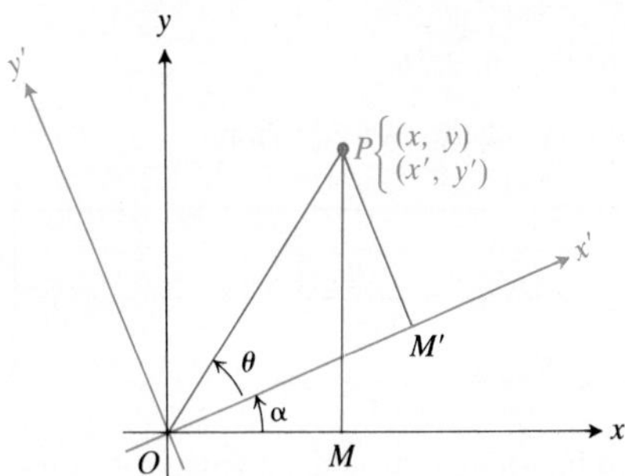

9.24 A counterclockwise rotation through angle α about the origin.

Since

$$OP\cos\theta = OM' = x' \quad \text{and} \quad OP\sin\theta = M'P = y', \tag{4}$$

Eqs. (3) reduce to the following.

Equations for Rotating the Coordinate Axes

$$\begin{aligned} x &= x'\cos\alpha - y'\sin\alpha, \\ y &= x'\sin\alpha + y'\cos\alpha \end{aligned} \tag{5}$$

Example 1 The x- and y-axes are rotated through an angle of $\pi/4$ radians about the origin. Find an equation for the hyperbola $2xy = 9$ in the new coordinates.

Solution Since $\cos\pi/4 = \sin\pi/4 = 1/\sqrt{2}$, we substitute

$$x = \frac{x' - y'}{\sqrt{2}}, \qquad y = \frac{x' + y'}{\sqrt{2}}$$

from Eqs. (5) into the equation $2xy = 9$ and obtain

$$2\left(\frac{x'-y'}{\sqrt{2}}\right)\left(\frac{x'+y'}{\sqrt{2}}\right) = 9$$

$$x'^2 - y'^2 = 9$$

$$\frac{x'^2}{9} - \frac{y'^2}{9} = 1.$$

See Fig. 9.25.

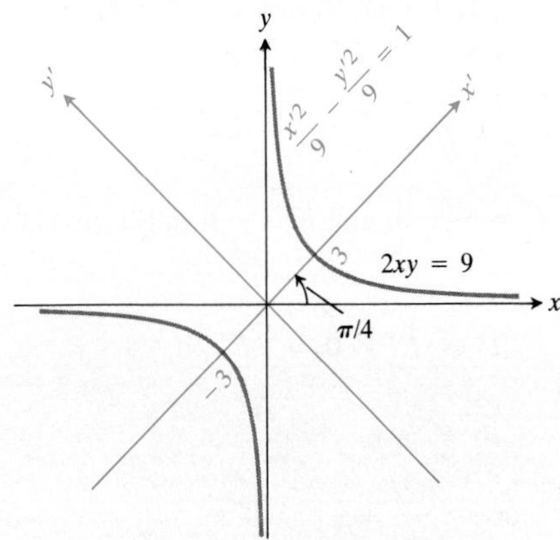

9.25 The hyperbola in Example 1.

If we apply the rotation equations in (5) to the general quadratic equation (1), we obtain a new quadratic equation

$$A'x'^2 + B'x'y' + C'y'^2 + D'x' + E'y' + F' = 0. \tag{6}$$

The new coefficients are related to the old ones by the equations

$$\begin{aligned} A' &= A\cos^2\alpha + B\cos\alpha\sin\alpha + C\sin^2\alpha, \\ B' &= B\cos 2\alpha + (C - A)\sin 2\alpha, \\ C' &= A\sin^2\alpha - B\sin\alpha\cos\alpha + C\cos^2\alpha, \\ D' &= D\cos\alpha + E\sin\alpha, \\ E' &= -D\sin\alpha + E\cos\alpha, \\ F' &= F. \end{aligned} \tag{7}$$

These equations show, among other things, that if we start with a quadratic equation for a curve in which the cross-product term is present ($B \neq 0$), we can find a rotation angle α that produces a quadratic equation in which no cross-product term appears ($B' = 0$). To find α, we put $B' = 0$ in the second equation in (7) and solve the resulting equation,

$$B\cos 2\alpha + (C - A)\sin 2\alpha = 0,$$

for α. In practice, this means finding α from one of the two equations

$$\cot 2\alpha = \frac{A - C}{B} \quad \text{or} \quad \tan 2\alpha = \frac{B}{A - C}. \tag{8}$$

Example 2 The coordinate axes are to be rotated through an angle α to produce an equation for the curve

$$x^2 + xy + y^2 - 6 = 0 \tag{9}$$

that has no cross-product term. Find a suitable value for α and the corresponding new equation.

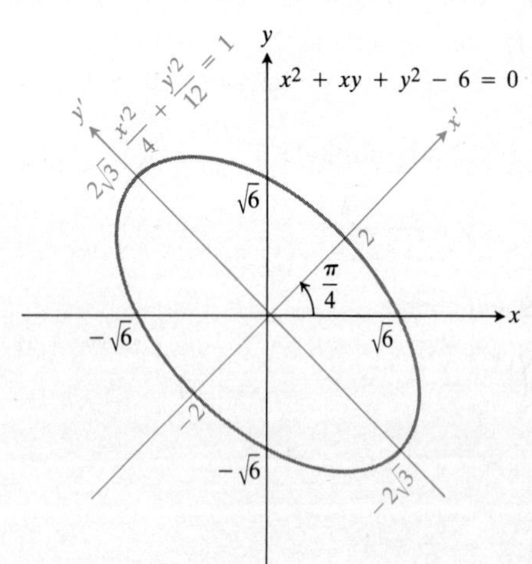

9.26 The ellipse in Example 2.

Solution 1 *Using Eqs. (8), (7), and (6).* Equation (9) has $A = B = C = 1$. We substitute these values into Eq. (8) to find α:

$$\cot 2\alpha = \frac{A - C}{B} = \frac{1 - 1}{1} = 0, \qquad 2\alpha = \frac{\pi}{2}, \qquad \alpha = \frac{\pi}{4}.$$

Substituting $\alpha = \pi/4$, $A = B = C = 1$, $D = E = 0$, and $F = -6$ into Eqs. (7) gives

$$A' = \frac{3}{2}, \qquad B' = 0, \qquad C' = \frac{1}{2}, \qquad D' = E' = 0, \qquad F' = -6.$$

Equation (6) then gives

$$\frac{3}{2}x'^2 + \frac{1}{2}y'^2 - 6 = 0, \qquad \text{or} \qquad \frac{x'^2}{4} + \frac{y'^2}{12} = 1.$$

This is the equation of an ellipse with foci on the new y'-axis (Fig. 9.26).

Solution 2 *Using the Rotation Equations Directly.* This method brings in fewer formulas but requires more arithmetic. We begin by finding α as in Solution 1 and substitute the value found, $\alpha = \pi/4$, into the rotation equations to get

$$x = x' \cos \alpha - y' \sin \alpha = \frac{\sqrt{2}}{2}x' - \frac{\sqrt{2}}{2}y',$$

$$y = x' \sin \alpha + y' \cos \alpha = \frac{\sqrt{2}}{2}x' + \frac{\sqrt{2}}{2}y'.$$

We then substitute the primed expressions for x and y in the original equation, $x^2 + xy + y^2 - 6 = 0$. This gives

$$\left(\frac{\sqrt{2}}{2}x' - \frac{\sqrt{2}}{2}y'\right)^2 + \left(\frac{\sqrt{2}}{2}x' - \frac{\sqrt{2}}{2}y'\right)\left(\frac{\sqrt{2}}{2}x' + \frac{\sqrt{2}}{2}y'\right) + \left(\frac{\sqrt{2}}{2}x' + \frac{\sqrt{2}}{2}y'\right)^2 - 6 = 0,$$

or, skipping over the arithmetic,

$$\frac{x'^2}{4} + \frac{y'^2}{12} = 1.$$

The Graphs of Quadratic Equations

We now return to our analysis of the graph of the general quadratic equation.

Since axes may always be rotated to eliminate the cross-product term, there is no loss of generality in assuming that this has been done, and our equation has the form

$$Ax^2 + Cy^2 + Dx + Ey + F = 0. \tag{10}$$

Equation (10) represents

a) a *circle* if $A = C \neq 0$ (special cases: the graph is a point or there is no graph at all);

b) a *parabola* if Eq. (10) is quadratic in one variable and linear in the other;

c) an *ellipse* if A and C are both positive or both negative (special cases: a single point or no graph at all);

d) a *hyperbola* if A and C have opposite signs (special case: a pair of intersecting lines);

e) a *straight line* if A and C are zero and at least one of D and E is different from zero;

f) *one or two straight lines* if the left-hand side of Eq. (10) can be factored into the product of two linear factors.

If $A = C \neq 0$, for example, we can complete the squares on the x-terms and y-terms to rewrite Eq. (10) in the form

$$A(x - h)^2 + A(y - k)^2 = F'' \tag{11}$$

and divide by A to obtain the equivalent equation

$$(x - h)^2 + (y - k)^2 = F''/A. \tag{12}$$

The form of Eq. (12) reveals the graph to be a circle of radius $\sqrt{F''/A}$ centered at (h, k). The exceptions (a point, no graph at all) arise when $F'' = 0$ or $F''/A < 0$.

Similarly, if A and C have opposite signs, completing the squares gives an equivalent equation of the form

$$\pm |A|(x - h)^2 \mp |C|(y - k)^2 = F''. \tag{13}$$

If $F'' \neq 0$, we may divide through by F'' and put the equation in the form

$$\pm \frac{(x-h)^2}{a^2} \mp \frac{(y-k)^2}{b^2} = 1. \tag{14}$$

The graph is a hyperbola centered at (h, k), i.e., a hyperbola that has been shifted to place its center at the point (h, k). The exception (two intersecting lines) occurs when $F'' = 0$, for then Eq. (13) reduces to

$$y = \pm \sqrt{\frac{|A|}{|C|}}(x - h) + k. \tag{15}$$

The Discriminant

There is a quick way to tell whether the graph of the equation

$$Ax^2 + Bxy + Cy^2 + Dx + Ey + F = 0 \tag{16}$$

is a parabola, an ellipse, or a hyperbola. The test does not require us to eliminate the xy-term first.

As we have seen, if B is not zero, then rotating the axes through the angle α determined by the equation

$$\cot 2\alpha = \frac{A - C}{B} \tag{17}$$

will change Eq. (16) into the equivalent form

$$A'x'^2 + C'y'^2 + D'x' + E'y' + F' = 0 \tag{18}$$

without a cross-product term.

Now, the graph of Eq. (18) is a (real or degenerate)

a) *parabola* if A' or $C' = 0$, that is, if $A'C' = 0$;

b) *ellipse* if A' and C' have the same sign, that is, if $A'C' > 0$;

c) *hyperbola* if A' and C' have opposite signs, that is, if $A'C' < 0$.

It can also be verified, by using Eqs. (7), that for any rotation of axes,

$$B^2 - 4AC = B'^2 - 4A'C'. \tag{19}$$

This means that the quantity $B^2 - 4AC$ is not changed by a rotation. But when we rotate through the angle α given by Eq. (17), B' becomes zero, so that

$$B^2 - 4AC = -4A'C'$$

(Exercise 46). Since the curve is a parabola if $A'C' = 0$, an ellipse if $A'C' > 0$, and a hyperbola if $A'C' < 0$, the curve must be

a) a *parabola* if $B^2 - 4AC = 0$,

b) an *ellipse* if $B^2 - 4AC < 0$, (20)

c) a *hyperbola* if $B^2 - 4AC > 0$.

The number $B^2 - 4AC$ is called the **discriminant** of Eq. (16). What we have just seen is that the graph of Eq. (16) is a parabola if the discriminant is zero, an ellipse if the discriminant is negative, and a hyperbola if the discriminant is positive (with the understanding that occasional degenerate cases may arise).

Example 3

a) $3x^2 - 6xy + 3y^2 + 2x - 7 = 0$ represents a parabola because

$$B^2 - 4AC = (-6)^2 - 4 \cdot 3 \cdot 3 = 36 - 36 = 0.$$

b) $x^2 + xy + y^2 - 1 = 0$ represents an ellipse because

$$B^2 - 4AC = (1)^2 - 4 \cdot 1 \cdot 1 = -3 < 0.$$

c) $xy - y^2 - 5y + 1 = 0$ represents a hyperbola because

$$B^2 - 4AC = (1)^2 - 4(0)(-1) = 1 > 0.$$

* How Calculators Use Rotations to Evaluate Sines and Cosines

Some calculators use rotations to calculate sines and cosines of arbitrary angles. The procedure goes something like this: The calculator has, stored,

1. ten angles or so, say

$$\alpha_1 = \sin^{-1}(10^{-1}), \quad \alpha_2 = \sin^{-1}(10^{-2}), \quad \ldots, \quad \alpha_{10} = \sin^{-1}(10^{-10}),$$

   and

2. twenty numbers, the sines and cosines of the angles $\alpha_1, \alpha_2, \ldots, \alpha_{10}$.

To calculate the sine and cosine of an arbitrary angle θ, we enter θ (in radians) into the calculator. The calculator subtracts or adds multiples of 2π to θ to replace θ by the angle between 0 and 2π that has the same sine and cosine as θ (we shall continue to call the angle θ). The calculator then "writes" θ as a sum of multiples of α_1 (as many as possible, without overshooting) plus multiples of α_2 (again, as many as possible), and so on, working its way to α_{10}. This gives

$$\theta \approx m_1\alpha_1 + m_2\alpha_2 + \cdots + m_{10}\alpha_{10}.$$

The calculator then rotates the point (1, 0) through m_1 copies of α_1 (through α_1, m_1 times in succession), plus m_2 copies of α_2, and so on, finishing off with m_{10} copies of α_{10} (Fig. 9.27). The coordinates of the final position of (1, 0) on the unit circle are the values the calculator gives for $(\cos\theta, \sin\theta)$.

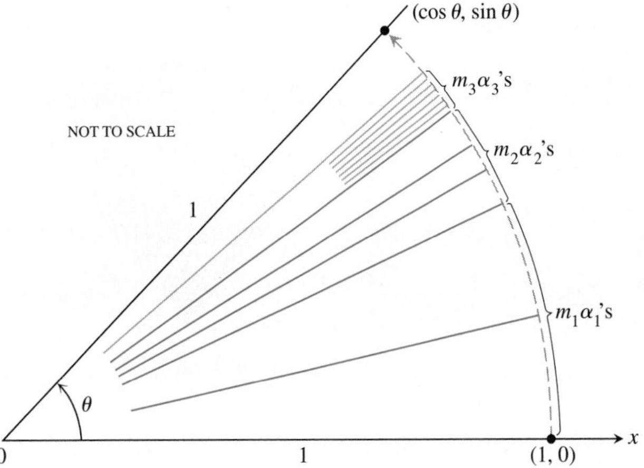

9.27 To calculate the sine and cosine of an angle θ between 0 and 2π, the calculator rotates the point (1, 0) to an appropriate location on the unit circle and displays the resulting coordinates.

* Sections of a Cone

We have defined ellipses, parabolas, and hyperbolas in the coordinate plane algebraically, with coordinate equations derived from the distance formula. We have also seen that plane curves whose points P satisfy equations of the form $PF = e \cdot PD$ for suitably chosen points F and D in the coordinate plane are "algebraic" parabolas, ellipses, and hyperbolas in this sense. In this section, we show that every nondegenerate curve obtained by slicing a double right circular cone with a plane that is not perpendicular to the axis of the cone satisfies an equation of the form $PF = e \cdot PD$ for suitably chosen points F and P in the cutting plane. Thus, every geometric ellipse, parabola, or hyperbola is also an algebraic ellipse, parabola, or hyperbola in its own plane.

As you know, circles do not satisfy a focus–directrix equation, but geometric circles are algebraic circles nonetheless. If we slice a right circular cone with a plane perpendicular to the cone's axis at a point O different from the cone's vertex, the points on the circle of intersection have a constant distance r from O. If we then coordinatize the plane with the origin at O, the circle is the graph of the equation $x^2 + y^2 = r^2$. Thus, every nondegenerate geometric conic section is an algebraic conic section in the coordinates of the plane that cuts the section from the cone.

The Basic Argument Suppose that the cutting plane makes an acute angle α with the axis of the cone and that the acute angle between the side and axis of the cone is β (Fig. 9.28). Then the section is

a) an ellipse if $\beta < \alpha < 90°$;

b) a parabola if $\alpha = \beta$;

c) a hyperbola if $0 \leq \alpha < \beta$.

9.28 The intersection of a plane and a right circular double cone in (a) an ellipse, (b) a parabola, (c) a hyperbola. The angle β is the angle between the side and axis of the cone. The angle α is the acute angle between the plane and the cone's axis.

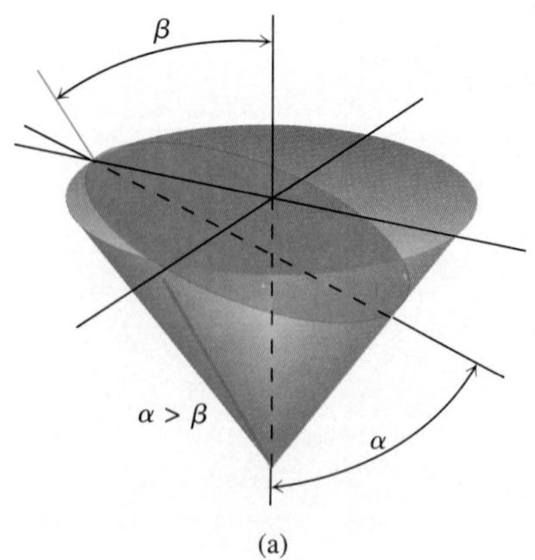

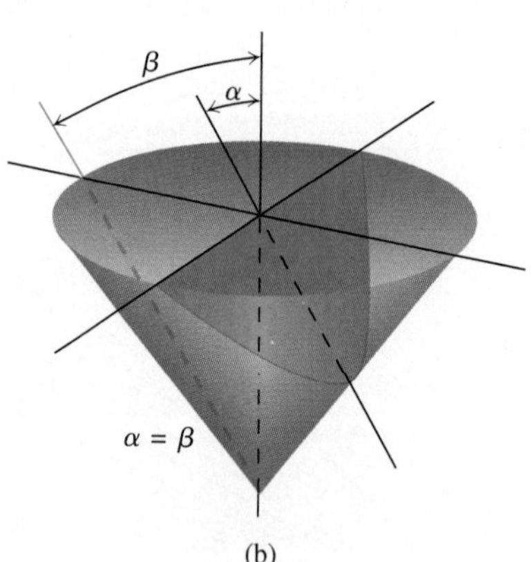

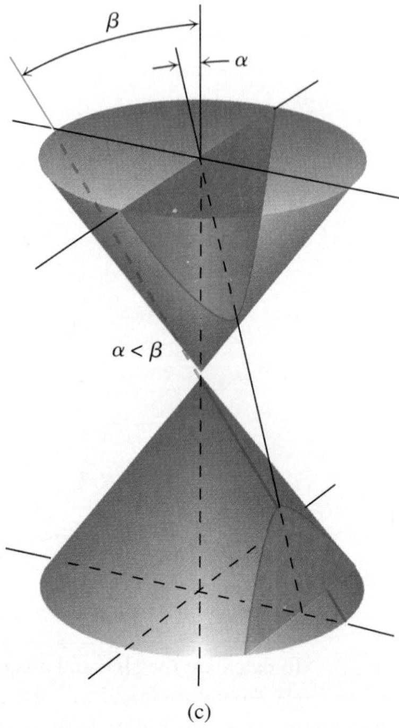

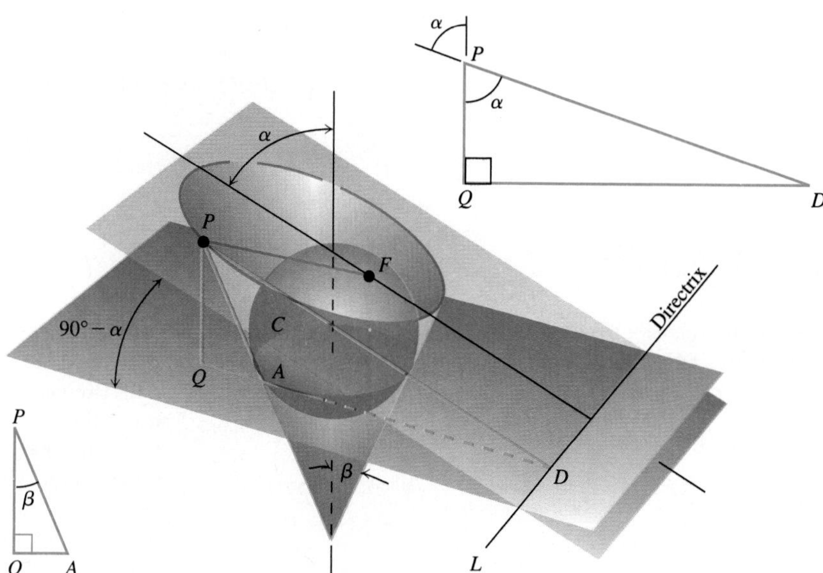

9.29 Every nondegenerate geometric ellipse satisfies a focus–directrix equation in the cutting plane. Line L is the directrix that corresponds to the focus F.

The construction we are about to make assumes we are working with an ellipse, but the argument works for the other two cases as well.

Inscribe a sphere that is tangent to the cone along a circle C and tangent to the cutting plane at a point F (Fig. 9.29). Let P be any point on the conic section. We shall see that F is a focus of the conic section and that the corresponding directrix is the line L in which the cutting plane intersects the plane of circle C.

To this end, let Q be the point where the line through P parallel to the axis of the cone intersects the plane of C. Let A be the point where the line joining P to the vertex of the cone intersects C. Let line PD be the perpendicular from P to L. Then $\overline{PA}$ and $\overline{PF}$, being tangents to the sphere from a common point P, have the same length:

$$PA = PF.$$

From right triangle PQA, we have

$$PQ = PA \cos \beta.$$

From right triangle PQD, we also have

$$PQ = PD \cos \alpha.$$

Equating the formulas for PQ gives, in sequence,

$$PA \cos \beta = PD \cos \alpha,$$

$$\frac{PA}{PD} = \frac{\cos \alpha}{\cos \beta},$$

$$\frac{PF}{PD} = \frac{\cos \alpha}{\cos \beta}, \qquad (PA = PF) \tag{21}$$

$$PF = \frac{\cos \alpha}{\cos \beta} \cdot PD,$$

$$PF = e \cdot PD. \qquad \left(e = \frac{\cos \alpha}{\cos \beta}\right)$$

The last equation in this sequence is the one we have been looking for. It characterizes P as lying on an algebraically defined ellipse, parabola, or hyperbola with focus F and directrix L, depending on whether $e < 1$, $e = 1$, or $e > 1$.

EXERCISES 9.2

Use the discriminant $B^2 - 4AC$ to decide whether the equations in Exercises 1–16 represent parabolas, ellipses, or hyperbolas.

1. $x^2 - 3xy + y^2 - x = 0$

2. $3x^2 - 18xy + 27y^2 - 5x + 7y = -4$

3. $3x^2 - 7xy + \sqrt{17}\, y^2 = 1$

4. $2x^2 - \sqrt{15}\, xy + 2y^2 + x + y = 0$

5. $x^2 + 2xy + y^2 + 2x - y + 2 = 0$

6. $2x^2 - y^2 + 4xy - 2x + 3y = 6$

7. $x^2 + 4xy + 4y^2 - 3x = 6$

8. $x^2 + y^2 + 3x - 2y = 10$

9. $xy + y^2 - 3x = 5$

10. $3x^2 + 6xy + 3y^2 - 4x + 5y = 12$

11. $3x^2 - 5xy + 2y^2 - 7x - 14y = -1$

12. $2x^2 - 4.9xy + 3y^2 - 4x = 7$

13. $x^2 - 3xy + 3y^2 + 6y = 7$

14. $25x^2 + 21xy + 4y^2 - 350x = 0$

15. $6x^2 + 3xy + 2y^2 + 17y + 2 = 0$

16. $3x^2 + 12xy + 12y^2 + 435x - 9y + 72 = 0$

In Exercises 17–26, rotate the coordinate axes to change the given equation into an equation that has no cross-product (xy) term. Then identify the graph of the equation. (The new equations will vary with the size and direction of the rotation you choose.)

17. $xy = 2$

18. $x^2 + xy + y^2 = 1$

19. $3x^2 + 2\sqrt{3}xy + y^2 - 8x + 8\sqrt{3}y = 0$

20. $x^2 - \sqrt{3}xy + 2y^2 = 1$

21. $x^2 - 2xy + y^2 = 2$

22. $x^2 - 3xy + y^2 = 5$

23. $\sqrt{2}x^2 + 2\sqrt{2}xy + \sqrt{2}y^2 - 8x + 8y = 0$

24. $xy - y - x + 1 = 0$

25. $3x^2 + 2xy + 3y^2 = 19$

26. $3x^2 + 4\sqrt{3}xy - y^2 = 7$

CALCULATOR The conic sections in Exercises 17–26 were chosen to have rotation angles that were "nice" in the sense that once we knew $\cot 2\alpha$ or $\tan 2\alpha$ we could identify 2α and find $\sin \alpha$ and $\cos \alpha$ from familiar triangles. The conic sections we encounter in practice may not have such nice rotation angles, and we may have to use a calculator to determine α from the value of $\cot 2\alpha$ or $\tan 2\alpha$.

In Exercises 27–32, use a calculator to find an angle α through which the coordinate axes can be rotated to change the given equation into a quadratic equation that has no cross-product term. Then find $\sin \alpha$ and $\cos \alpha$ to two decimal places and use Eqs. (7) to find the coefficients of the new equation to the nearest decimal place. In each case, say whether the conic section is an ellipse, a hyperbola, or a parabola.

27. $x^2 - xy + 3y^2 + x - y - 3 = 0$

28. $2x^2 + xy - 3y^2 + 3x - 7 = 0$

29. $x^2 - 4xy + 4y^2 - 5 = 0$

30. $2x^2 - 12xy + 18y^2 - 49 = 0$

31. $3x^2 + 5xy + 2y^2 - 8y - 1 = 0$

32. $2x^2 + 7xy + 9y^2 + 20x - 86 = 0$

33. What effect does a 90° rotation about the origin have on the equations of the following conic sections? Give the new equation in each case.
a) The ellipse $(x^2/a^2) + (y^2/b^2) = 1 \quad (a > b)$
b) The hyperbola $(x^2/a^2) - (y^2/b^2) = 1$
c) The circle $x^2 + y^2 = a^2$
d) The line $y = mx$
e) The line $y = mx + b$

34. What effect does a 180° rotation about the origin have on the equations of the following conic sections? Give the new equation in each case.
a) The ellipse $(x^2/a^2) + (y^2/b^2) = 1 \quad (a > b)$
b) The hyperbola $(x^2/a^2) - (y^2/b^2) = 1$
c) The circle $x^2 + y^2 = a^2$
d) The line $y = mx$
e) The line $y = mx + b$

35. *The hyperbola $xy = a$.* The hyperbola $xy = 1$ is one of many hyperbolas of the form $xy = a$ that appear in science and mathematics.
a) Rotate the coordinate axes through an angle of 45° to change the equation $xy = 1$ into an equation with no xy-term. What is the new equation?
b) Do the same for the equation $xy = a$.

36. Find the eccentricity of the hyperbola $xy = 2$.

37. Show that the equation $x^2 + y^2 = a^2$ becomes $x'^2 +$

$y'^2 = a^2$ for every choice of the angle α in the rotation equations.

38. Show that rotating the axes through an angle of $\pi/4$ radians will eliminate the xy-term from Eq. (1) whenever $A = C$.

39. a) What kind of conic section is the curve $xy + 2x - y = 0$?
b) Solve the equation $xy + 2x - y = 0$ for y and sketch the curve as the graph of a rational function of x.
c) Find equations for the lines parallel to the line $y = -2x$ that are normal to the curve. Add the lines to your sketch.

40. What values of a, b, and c make the ellipse

$$4x^2 + y^2 + ax + by + c = 0$$

tangent to the x-axis at the origin and also pass through the point $(-1, 2)$?

41. Starting with the equation

$$Ax^2 + Bxy + Cy^2 + Dx + Ey + F = 0,$$

find an equation for the conic section that has all of the following properties:

1. It is symmetric about the origin.
2. It passes through the point $(1, 0)$.
3. The line $y = 1$ is tangent to it at the point $(-2, 1)$.

What kind of conic section is it?

42. Find an equation for the curve $x^2 + 2xy + y^2 - 1 = 0$ after a rotation that eliminates the xy-term. Identify the graph.

43. Prove or find counterexamples to the following statements about the graph of the equation $Ax^2 + Bxy + Cy^2 + Dx + Ey + F = 0$.
a) If $AC > 0$, the graph is an ellipse.
b) If $AC > 0$, the graph is a hyperbola.
c) If $AC < 0$, the graph is a hyperbola.

44. *A nice area formula for ellipses.* When $B^2 - 4AC$ is negative, the equation

$$Ax^2 + Bxy + Cy^2 = 1$$

represents an ellipse. If the ellipse's semi-axes are a and b, its area is πab (a standard formula). Show that the area is also given by the formula $2\pi/\sqrt{4AC - B^2}$. (*Hint:* Rotate the coordinate axes to eliminate the xy-term and apply Eq. (19) to the new equation.)

45. *Other invariants.* We describe the fact that $B'^2 - 4A'C'$ equals $B^2 - 4AC$ after a rotation about the origin by saying that the discriminant of a quadratic equation is an **invariant** of the equation. Use Eqs. (7) to show that the numbers (a) $A + C$ and (b) $D^2 + E^2$ are also invariants, in the sense that

$$A' + C' = A + C \quad \text{and} \quad D'^2 + E'^2 = D^2 + E^2.$$

We can use these equalities to check against numerical errors when we rotate axes. They can also be helpful in shortening the work required to find values for the new coefficients.

46. *A proof that $B'^2 - 4A'C' = B^2 - 4AC$.* Use Eqs. (7) to show that $B'^2 - 4A'C' = B^2 - 4AC$ for any rotation of axes about the origin. The calculation works out nicely but requires patience.

***47.** Sketch a figure similar to Fig. 9.29 for the case in which the conic section is a parabola. Then derive Eq. (21) on the basis of your figure.

***48.** Sketch a figure similar to Fig. 9.29 for the case in which the conic section is a hyperbola. Then derive Eq. (21) on the basis of your figure.

***49.** Which parts of the geometric construction described in the last subsection ("Sections of a Cone") become impossible if the conic section is a circle?

EXPLORER PROGRAM

Conic Sections — Shows how the coefficients in the equation $Ax^2 + Bxy + Cy^2 + Dx + Ey + F = 0$ change as the conic section it describes is rotated in the xy-plane

9.3 Parametrizations of Curves

When the path of a particle moving in the plane looks like the curve in Fig. 9.30, we cannot hope to describe it with a Cartesian formula that expresses y directly in terms of x or x directly in terms of y. Instead, we express each of the particle's coordinates as a function of time t and describe the path with a pair of equations, $x = f(t)$ and $y = g(t)$. Indeed, for studying motion, equations like these are preferable to a Cartesian formula for the path because they immediately tell us the particle's position at any time t. They become equations for the motion as well as equations for the path along which the motion takes place. They also enable us to calculate the particle's velocity and acceleration at any time t, as we shall see in

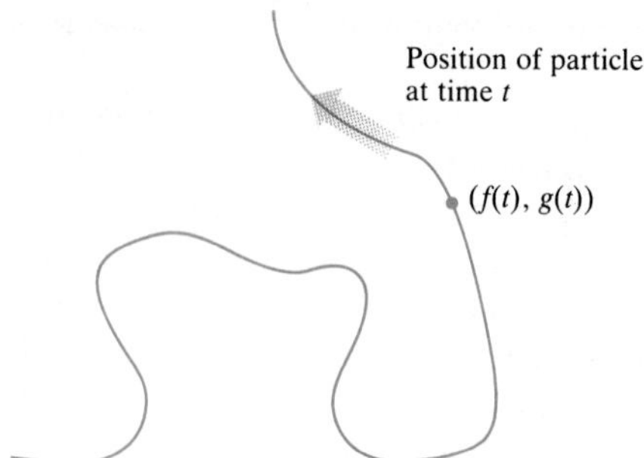

9.30 The path traced by a particle moving in the xy-plane is not always the graph of a function of x or a function of y.

Chapter 11. In the present section, we focus on the geometry of curves defined by parametric equations.

DEFINITIONS

If x and y are given as continuous functions

$$x = f(t), \qquad y = g(t) \tag{1}$$

over an interval of t-values, then the set of points $(x, y) = (f(t), g(t))$ defined by these equations is called a **curve** in the coordinate plane. The equations are **parametric equations** for the curve. The variable t is a **parameter** for the curve and its domain I is called the **parameter interval.** If I is a closed interval, $a \le t \le b$, the point $(f(a), g(a))$ is the **initial point** of the curve and $(f(b), g(b))$ is the **terminal point** of the curve. When we give parametric equations and a parameter interval for a curve in the plane, we say that we have **parametrized** the curve. The equations and interval constitute a **parametrization** of the curve.

In many applications, t denotes time, but it might instead denote an angle (as in some of the following examples) or the distance a particle has traveled along its path from its starting point (as it sometimes will when we later study motion).

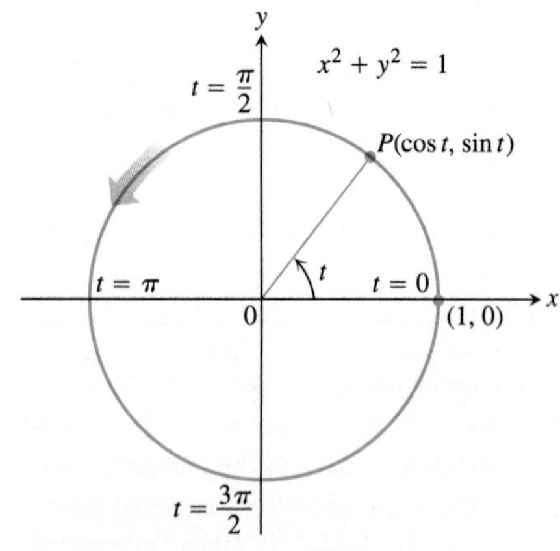

9.31 The equations $x = \cos t$ and $y = \sin t$ describe motion on the unit circle $x^2 + y^2 = 1$. The arrow shows the direction of increasing t (Example 1).

Example 1 *The Circle $x^2 + y^2 = 1$.* The equations and parameter interval

$$x = \cos t, \qquad y = \sin t, \qquad 0 \le t \le 2\pi,$$

describe the position $P(x, y)$ of a particle that moves counterclockwise around the circle $x^2 + y^2 = 1$ as t increases (Fig. 9.31).

We know that the point lies on this circle for every value of t because

$$x^2 + y^2 = \cos^2 t + \sin^2 t = 1.$$

But how much of the circle does the point $P(x, y)$ actually traverse? To find out, we track the motion as t runs from 0 to 2π. The parameter t is the radian measure of the angle that radius OP makes with the positive x-axis. The particle starts at $(1, 0)$, moves up and to the left as t approaches $\pi/2$, and continues around the circle to stop again at $(1, 0)$ when $t = 2\pi$. The particle traces the circle exactly once.

Example 2 *A Semicircle.* The equations and parameter interval

$$x = \cos t, \qquad y = -\sin t, \qquad 0 \le t \le \pi,$$

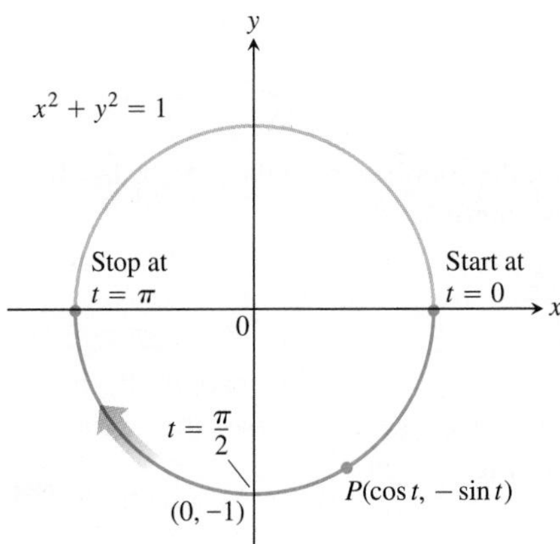

9.32 The point $P(\cos t, -\sin t)$ moves clockwise as t increases from 0 to π (Example 2).

describe the position $P(x, y)$ of a particle that moves clockwise around the circle $x^2 + y^2 = 1$ as t increases from 0 to π.

We know that the point P lies on this circle for all t because its coordinates satisfy the circle's equation. How much of the circle does the particle traverse? To find out, we track the motion as t runs from 0 to π. As in Example 1, the particle starts at (1, 0). But now as t increases, y becomes negative, decreasing to -1 when $t = \pi/2$ and then increasing back to 0 as t approaches π. The motion stops at $t = \pi$ with only the lower half of the circle covered (Fig. 9.32).

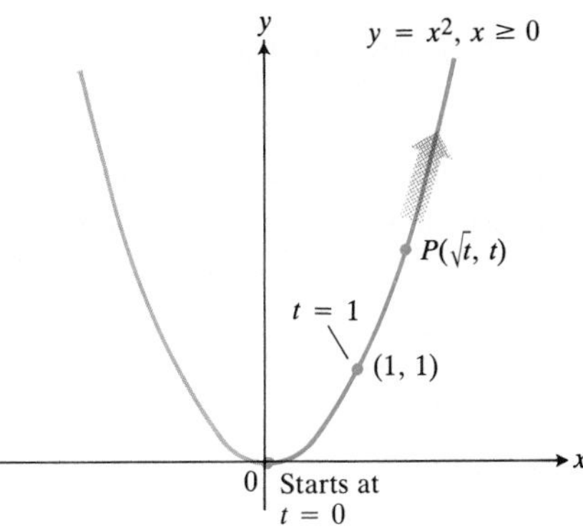

9.33 The equations $x = \sqrt{t}$, $y = t$ and interval $t \geq 0$ describe the motion of a particle that traces the right-hand half of the parabola $y = x^2$ (Example 3).

Example 3 *Half a Parabola.* The position $P(x, y)$ of a particle moving in the xy-plane is given by the equations and parameter interval

$$x = \sqrt{t}, \qquad y = t, \qquad t \geq 0.$$

Identify the path traced by the particle and describe the motion.

Solution We try to identify the path by eliminating t between the equations $x = \sqrt{t}$ and $y = t$. With any luck, this will produce a recognizable algebraic relation between x and y. We find that

$$y = t = (\sqrt{t})^2 = x^2.$$

This means the particle's position coordinates satisfy the equation $y = x^2$, so the particle moves along the parabola $y = x^2$.

It would be a mistake, however, to conclude that the particle's path is the entire parabola $y = x^2$—it is only half the parabola. The particle's x-coordinate is never negative. The particle starts at (0, 0) when $t = 0$ and rises into the first quadrant as t increases (Fig. 9.33).

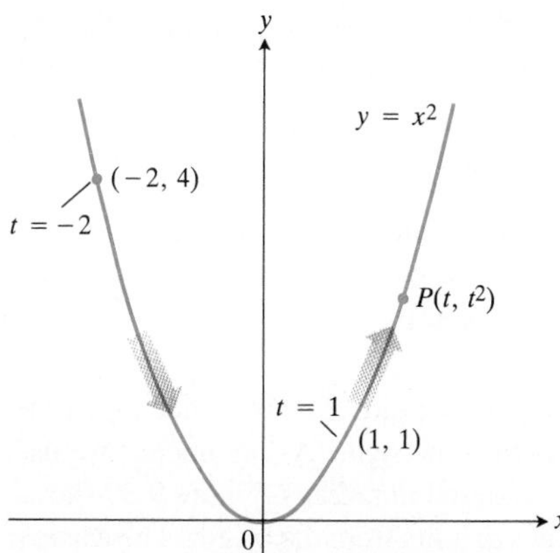

9.34 The path defined by the parametric equations $x = t$, $y = t^2$ and interval $-\infty < t < \infty$ (Example 4).

Example 4 *A Complete Parabola.* The position $P(x, y)$ of a particle moving in the xy-plane is given by the equations and parameter interval

$$x = t, \qquad y = t^2, \qquad -\infty < t < \infty.$$

Identify the particle's path and describe the motion.

Solution We identify the path by eliminating t between the equations $x = t$ and $y = t^2$, obtaining

$$y = (t)^2 = x^2.$$

The particle's position coordinates satisfy the equation $y = x^2$, so the particle moves along this curve.

In contrast to Example 3, the particle now traverses the entire parabola. As t increases from $-\infty$ to ∞, the particle comes down the left-hand side, passes through the origin, and moves up the right-hand side (Fig. 9.34).

As Example 4 illustrates, the graph of a function $y = f(x)$, $x \in I$, has the automatic parametrization $x = t$, $y = f(t)$, $t \in I$. This is so simple we usually do not use it, but the point of view is occasionally helpful.

Example 5 *A Parametrization of the Ellipse $x^2/a^2 + y^2/b^2 = 1$.* Describe the motion of a particle whose position $P(x, y)$ at time t is given by the

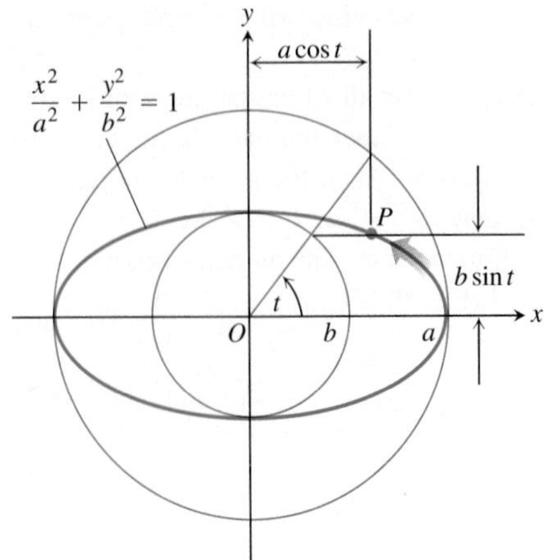

9.35 The coordinates of P are $x = a\cos t$, $y = b\sin t$ (Example 5).

equations and parameter interval

$$x = a\cos t, \qquad y = b\sin t, \qquad 0 \le t \le 2\pi.$$

Solution We find a Cartesian equation for the coordinates of the particle by eliminating t between the equations for x and y. Since

$$\frac{x^2}{a^2} + \frac{y^2}{b^2} = \frac{a^2\cos^2 t}{a^2} + \frac{b^2\sin^2 t}{b^2} = \cos^2 t + \sin^2 t = 1,$$

the motion takes place on the ellipse $x^2/a^2 + y^2/b^2 = 1$. The particle begins at $(a, 0)$ when t equals zero and moves counterclockwise around the ellipse, traversing it exactly once as t moves from 0 to 2π (Fig. 9.35).

Example 6 *A Parametrization of the Circle $x^2 + y^2 = a^2$.* The equations and parameter interval

$$x = a\cos t, \qquad y = a\sin t, \qquad 0 \le t \le 2\pi,$$

obtained by taking $b = a$ in Example 5, describe the circle $x^2 + y^2 = a^2$.

Example 7 *A Parametrization of the Right-hand Branch of the Hyperbola $x^2 - y^2 = 1$.* Describe the motion of the particle whose position $P(x, y)$ at time t is given by the equations and parameter interval

$$x = \sec t, \qquad y = \tan t, \qquad -\frac{\pi}{2} < t < \frac{\pi}{2}.$$

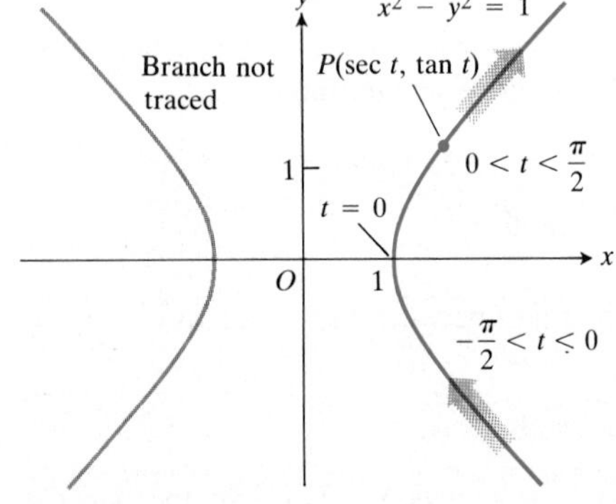

9.36 The equations $x = \sec t$, $y = \tan t$ and interval $-\pi/2 < t < \pi/2$ describe the right-hand branch of the hyperbola $x^2 - y^2 = 1$ (Example 7).

Solution We find a Cartesian equation for the coordinates of P by eliminating t between the equations for x and y. Since

$$x^2 - y^2 = \sec^2 t - \tan^2 t = 1,$$

we see that the motion takes place somewhere on the hyperbola $x^2 - y^2 = 1$. Since $x = \sec t$ is always positive for the parameter values $-\pi/2 < t < \pi/2$, the motion takes place on the hyperbola's right-hand branch. As t moves from $-\pi/2$ to $\pi/2$, the particle comes in along the lower half of the right-hand branch, reaching the vertex $(1, 0)$ at $t = 0$. It then moves into the first quadrant to complete the coverage of the right-hand branch as t approaches $\pi/2$ (Fig. 9.36).

Example 8 *Cycloids.* A wheel of radius a rolls along a horizontal straight line without slipping. Find parametric equations for the path traced by a point P on the wheel's circumference. The path is called a **cycloid.**

Solution We take the line to be the x-axis, mark a point P on the wheel, start the wheel with P at the origin, and roll the wheel to the right. As parameter, we use the angle t through which the wheel turns, measured in radians. Figure 9.37 shows the wheel a short while later, when its base lies at units from the origin. The wheel's center C lies at (at, a) and the coordinates of P are

$$x = at + a\cos\theta, \qquad y = a + a\sin\theta. \tag{2}$$

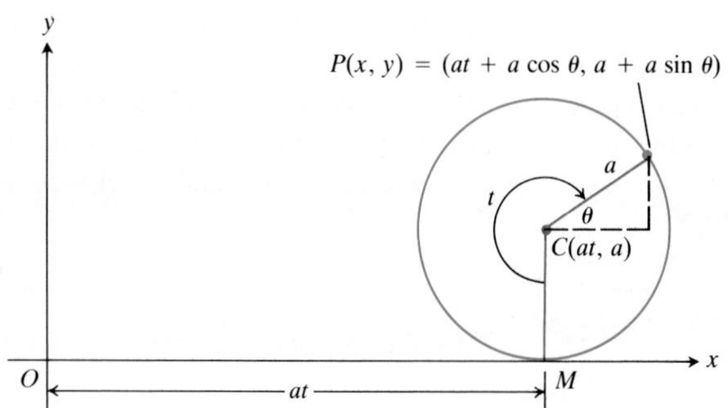

9.37 The position of $P(x, y)$ on the rolling wheel at time t (Example 8).

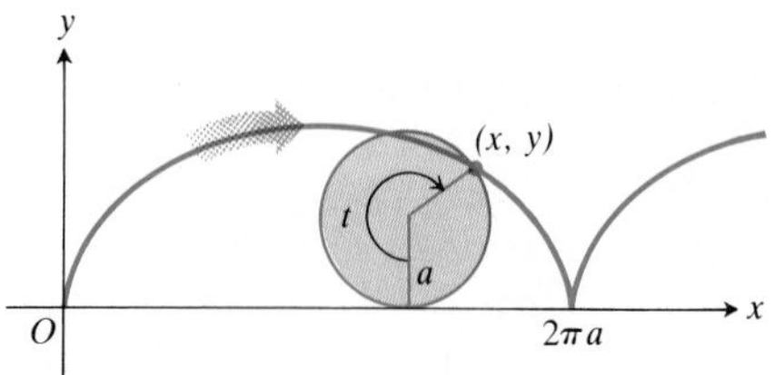

9.38 The cycloid $x = a(t - \sin t)$, $y = a(1 - \cos t)$, shown for $t \geq 0$.

To express θ in terms of t, we observe that $t + \theta = 3\pi/2$, so that

$$\theta = \frac{3\pi}{2} - t. \tag{3}$$

This makes

$$\cos\theta = \cos\left(\frac{3\pi}{2} - t\right) = -\sin t, \qquad \sin\theta = \sin\left(\frac{3\pi}{2} - t\right) = -\cos t. \tag{4}$$

The equations we seek are

$$x = at - a\sin t, \qquad y = a - a\cos t. \tag{5}$$

These are usually written with the a factored out:

$$x = a(t - \sin t), \qquad y = a(1 - \cos t). \tag{6}$$

Figure 9.38 shows the first arch of the cylcoid and part of the next.

✲ Brachistochrones and Tautochrones

If we turn Fig. 9.38 upside down, Eqs. (6) still apply and the resulting curve (Fig. 9.39) has two interesting physical properties. The first relates to the origin O and the point B at the bottom of the first arch. Among all smooth curves joining these points, the cycloid is the curve along which a frictionless bead, subject only to the force of gravity, will slide from O to B the fastest. This makes the cycloid a **brachistochrone** (brah-*kiss*-toe-krone), or shortest time curve for these points. The second property is that even if you start the bead partway down the curve toward B, it will still take the bead the same amount of time to reach B. This makes the cycloid a **tautochrone** (*taw*-toe-krone), or same-time curve for O and B.

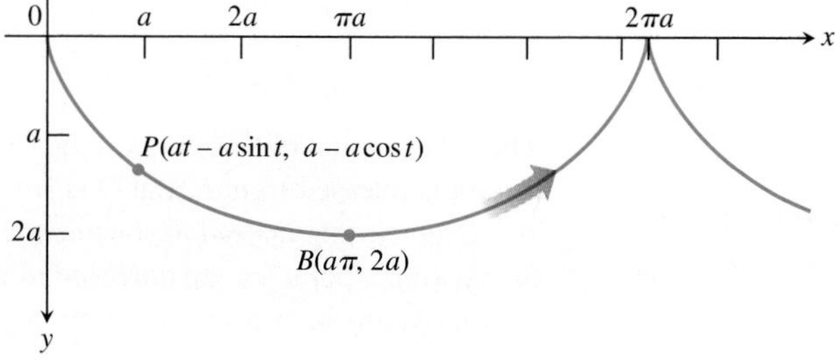

9.39 To study motion along an upside-down cycloid under the influence of gravity, we turn Fig. 9.38 upside down. This points the y-axis in the direction of the gravitational force and makes the downward y-coordinates positive. The equations and parameter interval for the cycloid are still

$$x = a(t - \sin t),$$
$$y = a(1 - \cos t), \quad t \geq 0.$$

The arrow shows the direction of increasing t.

Huygens's Clock

One problem with a pendulum clock whose bob swings in a circular arc is that the frequency of the swing depends on the amplitude of the swing. The wider the swing, the longer it takes the bob to return to center.

This does not happen if the bob can be made to swing in a cycloid, as you will see if you read about the cycloid's tautochrone property at the end of the section. In 1673, Christiaan Huygens (1629–1695), the Dutch mathematician, physicist, and astronomer who discovered the rings of Saturn, driven by a need to make accurate determinations of longitude at sea, designed a pendulum clock whose bob would swing in a cycloid. He hung the bob from a fine wire constrained by guards that caused it to draw up as it swung. How were the guards shaped? They were cycloids, too.

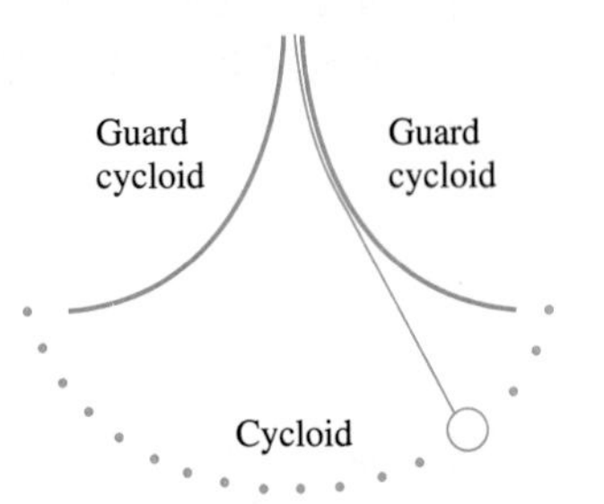

Are there any other brachistochrones joining O and B, or is the cycloid the only one? We can formulate this as a mathematical question in the following way. At the start, the kinetic energy of the bead is zero, since its velocity is zero. The work done by gravity in moving the bead from (0, 0) to any other point (x, y) in the plane is mgy, and this must equal the change in kinetic energy. That is,

$$mgy = \frac{1}{2}mv^2 - \frac{1}{2}m(0)^2. \tag{7}$$

Thus, the velocity of the bead when it reaches (x, y) has to be

$$v = \sqrt{2gy}. \tag{8}$$

That is,

$$\frac{ds}{dt} = \sqrt{2gy} \qquad (ds \text{ is the arc length differential along the bead's path.})$$

or

$$dt = \frac{ds}{\sqrt{2gy}} = \frac{\sqrt{1 + (dy/dx)^2}\,dx}{\sqrt{2gy}}. \tag{9}$$

The time T_f it takes the bead to slide along a particular path $y = f(x)$ from O to $B(a\pi, 2a)$ is

$$T_f = \int_{x=0}^{x=a\pi} \sqrt{\frac{1 + (dy/dx)^2}{2gy}}\,dx. \tag{10}$$

What curves $y = f(x)$, if any, minimize the value of this integral?

At first sight, we might guess that the straight line joining O and B would give the shortest time, but perhaps not. There might be some advantage in having the bead fall vertically at first to build up its velocity faster. With a higher velocity, the bead could travel a longer path and still reach B first. Indeed, this is the right idea. The solution, from a branch of mathematics known as the calculus of variations, is that the original cycloid from O to B is the one and only brachistochrone for O and B.

While the solution of the brachistochrone problem is beyond our present reach, we can still show why the cycloid is a tautochrone. For the cycloid, Eq. (10) takes the form

$$\begin{aligned} T_{\text{cycloid}} &= \int_{x=0}^{x=a\pi} \sqrt{\frac{dx^2 + dy^2}{2gy}} \\ &= \int_{t=0}^{t=\pi} \sqrt{\frac{a^2(2 - 2\cos t)}{2ga(1 - \cos t)}}\,dt \qquad \left(\begin{aligned}&\text{From Eqs. (6), } dx = a(1 - \cos t)\,dt, \\ &dy = a\sin t\,dt, \text{ and } y = a(1 - \cos t)\end{aligned}\right) \\ &= \int_0^{\pi} \sqrt{\frac{a}{g}}\,dt = \pi\sqrt{\frac{a}{g}}. \end{aligned} \tag{11}$$

Thus, the amount of time it takes the frictionless bead to slide down the cycloid to B after it is released from rest at O is $\pi\sqrt{a/g}$.

Suppose that instead of starting the bead at O we start it at some lower point on the cycloid, a point (x_0, y_0) corresponding to the parameter value $t_0 > 0$. The bead's velocity at any later point (x, y) on the cycloid is

$$v = \sqrt{2g(y - y_0)} = \sqrt{2ga(\cos t_0 - \cos t)}. \qquad (y = a(1 - \cos t))$$

The Witch of Agnesi

Although l'Hôpital wrote the first text on differential calculus, the first text to include differential and integral calculus along with analytic geometry, infinite series, and differential equations was written in the 1740s by the Italian mathematician Maria Gaetana Agnesi (1718–1799). Agnesi, a gifted scholar and linguist whose Latin essay defending higher education for women was published when she was only nine years old, was a well-published scientist by age 20 and an honorary faculty member of the University of Bologna by age 30.

Today, Agnesi is remembered chiefly for a bell-shaped curve called *the witch of Agnesi*. This name, found only in English texts, is the result of a mistranslation. Agnesi's own name for the curve was *versiera* or "turning curve." John Colson, a noted Cambridge mathematician who felt Agnesi's text so important that he learned Italian to translate it "for the benefit of British youth" (he particularly had in mind young women, for whom he hoped Agnesi would be a role model), probably confused versiera with *avversiera*, which means "wife of the devil" and translates into "witch." You can find out more about the witch by doing Exercise 28.

Accordingly, the time required for the bead to slide down to B is

$$T = \int_{t_0}^{\pi} \sqrt{\frac{a^2(2 - \cos t)}{2ag(\cos t_0 - \cos t)}}\, dt = \sqrt{\frac{a}{g}} \int_{t_0}^{\pi} \sqrt{\frac{1 - \cos t}{\cos t_0 - \cos t}}\, dt$$

$$= \sqrt{\frac{a}{g}} \int_{t_0}^{\pi} \sqrt{\frac{2\sin^2(t/2)}{[2\cos^2(t_0/2) - 1] - [2\cos^2(t/2) - 1]}}\, dt$$

$$= \sqrt{\frac{a}{g}} \int_{t_0}^{\pi} \frac{\sin(t/2)\, dt}{\sqrt{\cos^2(t_0/2) - \cos^2(t/2)}}$$

$$= \sqrt{\frac{a}{g}} \int_{t=t_0}^{t=\pi} \frac{-2\, du}{\sqrt{a^2 - u^2}} \qquad \begin{pmatrix} u = \cos(t/2) \\ -2du = \sin(t/2)\, dt \\ a = \cos(t_0/2) \end{pmatrix}$$

$$= 2\sqrt{\frac{a}{g}} \left[-\sin^{-1}\frac{u}{a} \right]_{t=t_0}^{t=\pi}$$

$$= 2\sqrt{\frac{a}{g}} \left[-\sin^{-1}\frac{\cos(t/2)}{\cos(t_0/2)} \right]_{t_0}^{\pi} = 2\sqrt{\frac{a}{g}}(-\sin^{-1}0 + \sin^{-1}1) = \pi\sqrt{\frac{a}{g}}.$$

This is precisely the time it takes the bead to slide to B from O. It takes the bead the same amount of time to reach B no matter where it starts. Beads starting simultaneously from O, A, and C in Fig. 9.40, for instance, will all reach B at the same time.

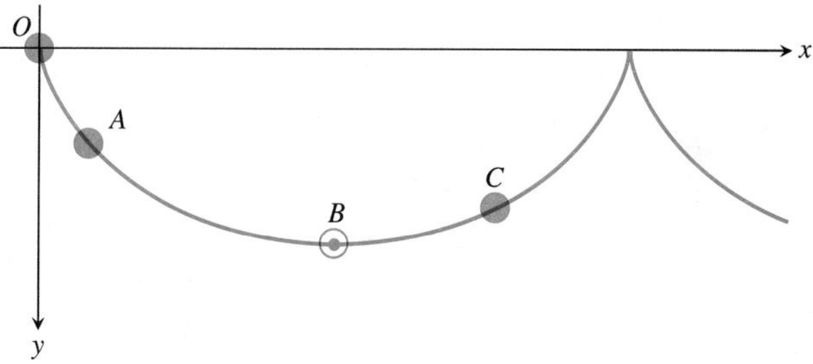

9.40 Beads released simultaneously on the cycloid at O, A, and C will reach B at the same time.

Standard Parametrizations

Circle $x^2 + y^2 = a^2$:

$x = a\cos t$

$y = a\sin t$

$0 \le t \le 2\pi$

Ellipse $\dfrac{x^2}{a^2} + \dfrac{y^2}{b^2} = 1$:

$x = a\cos t$

$y = b\sin t$

$0 \le t \le 2\pi$

Cycloid generated by a circle of radius a:

$$x = a(t - \sin t), \qquad y = a(1 - \cos t)$$

EXERCISES 9.3

Exercises 1–24 give parametric equations and parameter intervals for the motion of a particle in the xy-plane. Identify the particle's path by finding a Cartesian equation for it. Graph the Cartesian equation. Indicate the portion of the graph traced by the particle and the direction of motion.

1. $x = \cos t,\quad y = \sin t,\quad 0 \leq t \leq \pi$

2. $x = \cos 2t,\quad y = \sin 2t,\quad 0 \leq t \leq \pi$

3. $x = \sin 2\pi t,\quad y = \cos 2\pi t,\quad 0 \leq t \leq 1$

4. $x = \cos(\pi - t),\quad y = \sin(\pi - t),\quad 0 \leq t \leq \pi$

5. $x = 4 \cos t,\quad y = 2 \sin t,\quad 0 \leq t \leq 2\pi$

6. $x = 4 \sin t,\quad y = 2 \cos t,\quad 0 \leq t \leq \pi$

7. $x = 4 \cos t,\quad y = 5 \sin t,\quad 0 \leq t \leq \pi$

8. $x = 4 \sin t,\quad y = 5 \cos t,\quad 0 \leq t \leq 2\pi$

9. $x = 3t,\quad y = 9t^2,\quad -\infty < t < \infty$

10. $x = -\sqrt{t},\quad y = t,\quad t \geq 0$

11. $x = t,\quad y = \sqrt{t},\quad t \geq 0$

12. $x = \sec^2 t - 1,\quad y = \tan t,\quad -\pi/2 < t < \pi/2$

13. $x = -\sec t,\quad y = \tan t,\quad -\pi/2 < t < \pi/2$

14. $x = \csc t,\quad y = \cot t,\quad 0 < t < \pi$

15. $x = 2t - 5,\quad y = 4t - 7,\quad -\infty < t < \infty$

16. $x = 1 - t,\quad y = 1 + t,\quad -\infty < t < \infty$

17. $x = t,\quad y = 1 - t,\quad 0 \leq t \leq 1$

18. $x = 3t,\quad y = 2 - 2t,\quad 0 \leq t \leq 1$

19. $x = t,\quad y = \sqrt{1 - t^2},\quad -1 \leq t \leq 1$

20. $x = t,\quad y = \sqrt{4 - t^2},\quad 0 \leq t \leq 2$

21. $x = t^2,\quad y = \sqrt{t^4 + 1},\quad t \geq 0$

22. $x = \sqrt{t + 1},\quad y = \sqrt{t},\quad t \geq 0$

23. $x = \cosh t,\quad y = \sinh t,\quad -\infty < t < \infty$

24. $x = 2 \sinh t,\quad y = 2 \cosh t,\quad -\infty < t < \infty$

25. Find parametric equations and a parameter interval for the motion of a particle that starts at $(a, 0)$ and traces the circle $x^2 + y^2 = a^2$
a) once clockwise,
b) once counterclockwise,
c) twice clockwise,
d) twice counterclockwise.
(There are many correct ways to do these, so your answers may not be the same as the ones in the back of the book.)

26. Find parametric equations and a parameter interval for the motion of a particle that starts at $(a, 0)$ and traces the ellipse $(x^2/a^2) + (y^2/b^2) = 1$
a) once clockwise,
b) once counterclockwise,
c) twice clockwise,
d) twice counterclockwise.
(As in Exercise 25, there are many correct answers.)

27. *The involute of a circle.* If a string wound around a fixed circle is unwound while held taut in the plane of the circle, its end P traces an *involute* of the circle. In Fig. 9.41, the circle in question is the unit circle in the xy-plane and the initial position of the tracing point is the point $(1, 0)$ on the x-axis. The unwound portion of the string is tangent to the circle at Q, and t is the radian measure of the angle from the positive x-axis to segment OQ. Derive parametric equations for the involute by expressing the coordinates x and y of P in terms of t for $t \geq 0$.

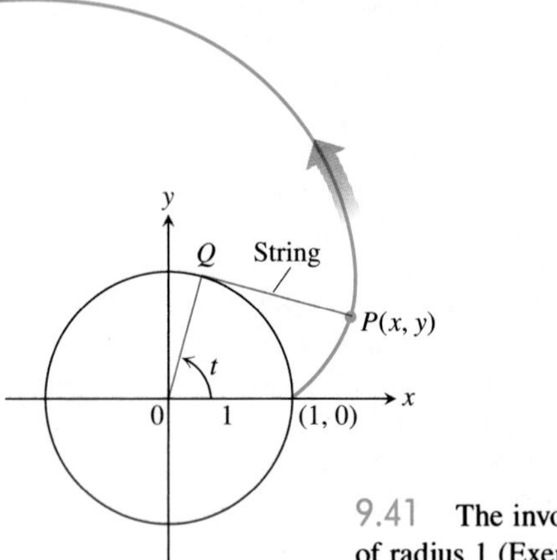

9.41 The involute of a circle of radius 1 (Exercise 27).

28. *The witch of Maria Agnesi.* The bell-shaped witch of Maria Agnesi can be constructed in the following way. Start with a circle of radius 1, centered at the point $(0, 1)$ on the y-axis (Fig. 9.42). Choose a point A on the line $y = 2$ and connect it to the origin with a line segment. Call the point where the segment crosses the circle B. Let P be the point where the vertical line through A crosses the horizontal line through B. The witch is the curve traced by P as A moves along the line $y = 2$. Find parametric equations and a parameter interval for the witch by expressing the coordi-

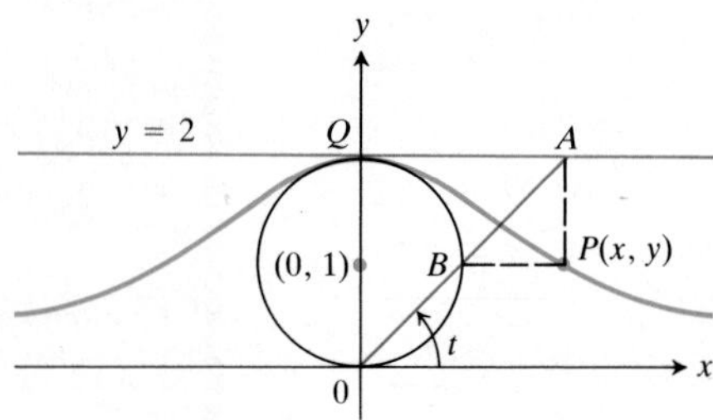

9.42 The witch of Maria Agnesi (Exercise 28).

nates of P in terms of t, the radian measure of the angle that segment OA makes with the positive x-axis. The following equalities (which you may assume) will help:

1. $x = AQ$
2. $y = 2 - AB \sin t$
3. $AB \cdot OA = (AQ)^2$

29. Find the point on the parabola $x = t$, $y = t^2$, $-\infty < t < \infty$, closest to the point (2, 1/2). (*Hint:* Minimize the square of the distance as a function of t.)

30. Find the point on the ellipse $x = 2 \cos t$, $y = \sin t$, $0 \leq t \leq 2\pi$ closest to the point (3/4, 0). (*Hint:* Minimize the square of the distance as a function of t.)

31. *Parametrizations of lines in the plane (Fig. 9.43)*

a) Show that the equations and parameter interval $x = x_0 + (x_1 - x_0)t$, $y = y_0 + (y_1 - y_0)t$, $-\infty < t < \infty$, describe the line through the points (x_0, y_0) and (x_1, y_1).

b) Using the same parameter interval, write parametric equations for the line through a point (x_1, y_1) and the origin.

c) Using the same parameter interval, write parametric equations for the line through $(-1, 0)$ and $(0, 1)$.

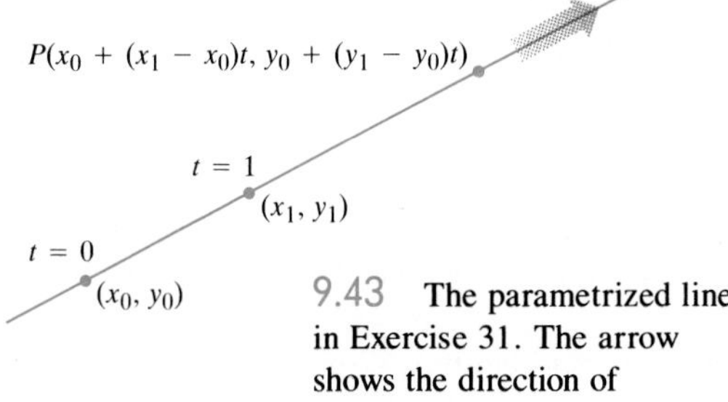

9.43 The parametrized line in Exercise 31. The arrow shows the direction of increasing t.

Computer Graphing Exercises

If you have access to a parametric equation grapher, graph the following equations over the given parameter intervals.

32. *Ellipse.* $x = 4 \cos t$, $y = 2 \sin t$, over

a) $0 \leq t \leq 2\pi$,

b) $0 \leq t \leq \pi$,

c) $-\pi/2 \leq t \leq \pi/2$.

33. *Hyperbola branch.* $x = \sec t$ (enter as 1/cos(t)), $y = \tan t$ (enter as (sin (t)/cos (t))), over

a) $-1.5 \leq t \leq 1.5$,

b) $-0.5 \leq t \leq 0.5$,

c) $-0.1 \leq t \leq 0.1$.

34. *Parabola.* $x = 2t + 3$, $y = t^2 - 1$, $-2 \leq t \leq 2$

35. *Cycloid.* $x = t - \sin t$, $y = 1 - \cos t$, over

a) $0 \leq t \leq 2\pi$,

b) $0 \leq t \leq 4\pi$,

c) $\pi \leq t \leq 3\pi$.

36. *Astroid.* $x = \cos^3 t$, $y = \sin^3 t$, over

a) $0 \leq t \leq 2\pi$,   b) $-\pi/2 \leq t \leq \pi/2$.

37. *A nice curve (a deltoid)*

$$x = 2 \cos t + \cos 2t,$$
$$y = 2 \sin t - \sin 2t,$$
$$0 \leq t \leq 2\pi$$

What happens if you replace 2 with -2 in the equations for x and y? Graph the new equations and find out.

38. *An even nicer curve*

$$x = 3 \cos t + \cos 3t,$$
$$y = 3 \sin t - \sin 3t,$$
$$0 \leq t \leq 2\pi$$

What happens if you replace 3 with -3 in the equations for x and y? Graph the new equations and find out.

39. *Projectile motion.* Graph

$$x = (64 \cos \alpha)t,$$
$$y = -16t^2 + (64 \sin \alpha)t,$$
$$0 \leq t \leq 4 \sin \alpha,$$

for the following firing angles:

a) $\alpha = \pi/4$,   b) $\alpha = \pi/6$,   c) $\alpha = \pi/3$,

d) $\alpha = \pi/2$ (watch out—here it comes!)

40. *Three beautiful curves*

a) *Epicycloid:*

$$x = 9 \cos t - \cos 9t$$
$$y = 9 \sin t - \sin 9t$$
$$0 \leq t \leq 2\pi$$

b) *Hypocycloid:*

$$x = 8 \cos t + 2 \cos 4t$$
$$y = 8 \sin t - 2 \sin 4t$$
$$0 \leq t \leq 2\pi$$

c) *Hypotrochoid:*

$$x = \cos t + 5 \cos 3t$$
$$y = 6 \cos t - 5 \sin 3t$$
$$0 \leq t \leq 2\pi$$

EXPLORER PROGRAM

PowerGrapher

Traces the curves for $x(t)$, $y(t)$, and $P(x, y)$ in side-by-side displays as t increases through the parameter interval. Also graphs $P(x, y)$ in a separate display.

Graphs different sets of parametric equations in a common display.

9.4 The Calculus of Parametrized Curves

This section shows how to find slopes, lengths, centroids, and surface areas associated with parametrized curves.

Slopes of Parametrized Curves

DEFINITIONS

A parametrized curve $x = f(t)$, $y = g(t)$ is said to be **differentiable at $t = t_0$** if f and g are differentiable at $t = t_0$. The curve is **differentiable** if it is differentiable at every parameter value. The curve is **smooth** if f and g are smooth, i.e., if f and g have continuous first derivatives.

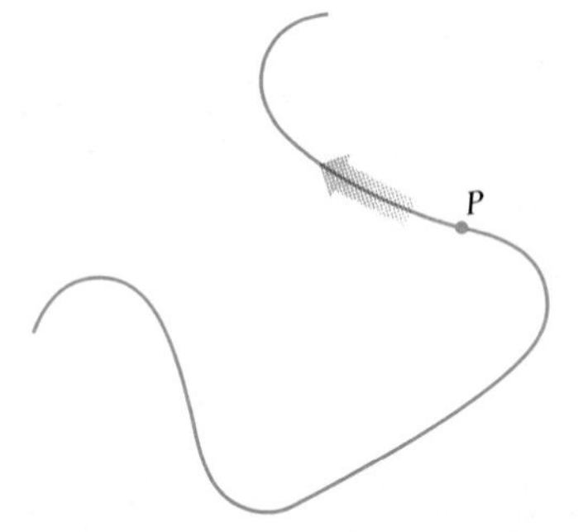

9.44 When the three first derivatives dx/dt, dy/dt, and dy/dx exist at a point P on a parametrized curve, they are related by the Chain Rule equation

$$\frac{dy}{dt} = \frac{dy}{dx}\frac{dx}{dt}.$$

At a point on a differentiable parametrized curve where y is also a differentiable function of x, the derivatives dx/dt, dy/dt, and dy/dx are related by the Chain Rule equation

$$\frac{dy}{dt} = \frac{dy}{dx}\frac{dx}{dt} \tag{1}$$

(Fig. 9.44). If $dx/dt \neq 0$, we may divide both sides of this equation by dx/dt to solve for dy/dx.

Formula for Finding dy/dx from dy/dt and dx/dt ($dx/dt \neq 0$)

$$\frac{dy}{dx} = \frac{dy/dt}{dx/dt} \tag{2}$$

Example 1 Find the tangent to the right-hand hyperbola branch

$$x = \sec t, \qquad y = \tan t, \qquad -\frac{\pi}{2} < t < \frac{\pi}{2},$$

at the point $(\sqrt{2}, 1)$, where $t = \pi/4$ (Fig. 9.45).

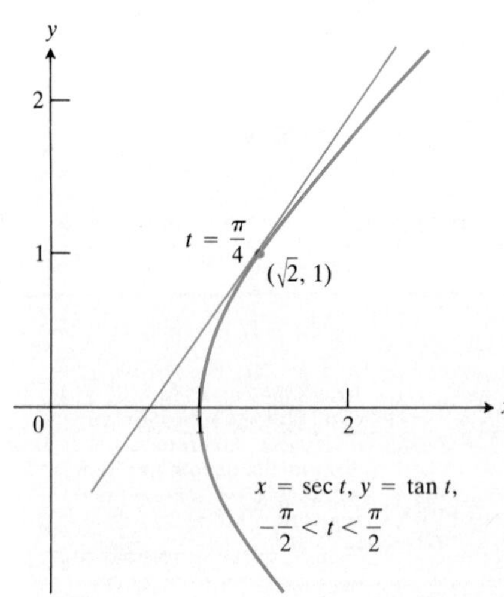

9.45 The hyperbola branch in Example 1.

Solution The slope of the curve at t is

$$\frac{dy}{dx} = \frac{dy/dt}{dx/dt} = \frac{\sec^2 t}{\sec t \tan t} = \frac{\sec t}{\tan t}. \qquad \text{(Eq.(2))}$$

Setting t equal to $\pi/4$ gives

$$\left.\frac{dy}{dx}\right|_{t=\pi/4} = \frac{\sec(\pi/4)}{\tan(\pi/4)} = \frac{\sqrt{2}}{1} = \sqrt{2}.$$

The point–slope equation of the tangent is

$$y - y_0 = m(x - x_0)$$
$$y - 1 = \sqrt{2}(x - \sqrt{2})$$
$$y = \sqrt{2}x - 2 + 1$$
$$y = \sqrt{2}x - 1.$$

The Parametric Formula for d^2y/dx^2

If the parametric equations for a curve define y as a twice-differentiable function of x, we may calculate d^2y/dx^2 as a function of t in the following way:

$$\frac{d^2y}{dx^2} = \frac{d}{dx}(y')$$

$$= \frac{dy'/dt}{dx/dt}. \qquad \text{(Eq. (2) with } y \text{ replaced by } y')$$

Formula for Finding d^2y/dx^2 from dx/dt and $y' = dy/dx$ $(dx/dt \neq 0)$

$$\frac{d^2y}{dx^2} = \frac{dy'/dt}{dx/dt} \tag{3}$$

Notice the lack of symmetry in Eq. (3). To find d^2y/dx^2, we divide the derivative of y' by the derivative of x, not by the derivative of x'.

Example 2 Find d^2y/dx^2 if $x = t - t^2$ and $y = t - t^3$.

To find d^2y/dx^2 in terms of t,
1. Express $y' = dy/dx$ in terms of t
2. Find dy'/dt
3. Divide dy'/dt by dx/dt.

Solution STEP 1: Express y' in terms of t:

$$y' = \frac{dy}{dx} = \frac{dy/dt}{dx/dt} = \frac{1 - 3t^2}{1 - 2t} \qquad \text{(Eq. (2) with } x = t - t^2,\ y = t - t^3)$$

STEP 2: Differentiate y' with respect to t:

$$\frac{dy'}{dt} = \frac{d}{dt}\left(\frac{1 - 3t^2}{1 - 2t}\right) = \frac{2 - 6t + 6t^2}{(1 - 2t)^2}$$

STEP 3: Divide dy'/dt by dx/dt:

$$\frac{dx}{dt} = \frac{d}{dt}(t - t^2) = 1 - 2t$$

$$\frac{dy'/dt}{dx/dt} = \frac{2 - 6t + 6t^2}{(1 - 2t)^2} \cdot \frac{1}{1 - 2t}$$

$$= \frac{2 - 6t + 6t^2}{(1 - 2t)^3}.$$

Lengths of Parametrized Curves; Centroids

We can find an integral for the length of a smooth curve

$$x = f(t), \qquad y = g(t), \qquad a \leq t \leq b,$$

by rewriting the integral $L = \int ds$ from Section 5.4 in the following way:

$$\text{Length} = \int_{t=a}^{t=b} ds = \int_a^b \sqrt{dx^2 + dy^2} \qquad (ds = \sqrt{dx^2 + dy^2})$$

$$= \int_a^b \sqrt{\left(\frac{dx^2}{dt^2} + \frac{dy^2}{dt^2}\right) dt^2} = \int_a^b \sqrt{\left(\frac{dx}{dt}\right)^2 + \left(\frac{dy}{dt}\right)^2}\, dt. \tag{4}$$

The only requirement besides the continuity of the integrand is that the point $P(x, y)$ not trace out any portion of the curve more than once as t moves from a to b.

> **Parametric Formula for Arc Length**
>
> If the functions $x = f(t)$ and $y = g(t)$ have continuous first derivatives with respect to t for $a \le t \le b$, and if the point $P(x, y)$ traces the curve defined by these equations exactly once as t moves from $t = a$ to $t = b$, then the length of the curve is given by the formula
>
> $$\text{Length} = \int_a^b \sqrt{\left(\frac{dx}{dt}\right)^2 + \left(\frac{dy}{dt}\right)^2}\, dt. \tag{5}$$

What if we have two different parametrizations for a curve whose length we want to find—does it matter which one we use? The answer, from advanced calculus, is no. As long as the parametrization we choose meets the conditions preceding Eq. (5), the formula gives the correct length.

Example 3 Find the length of the astroid (Fig. 9.46)

$$x = \cos^3 t, \qquad y = \sin^3 t, \qquad 0 \le t \le 2\pi.$$

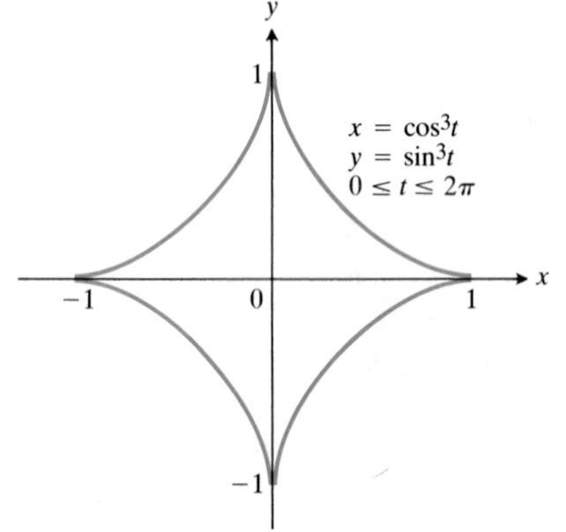

9.46 The astroid in Example 3.

Solution Because of the curve's symmetry with respect to the coordinate axes, its length is four times the length of the first-quadrant portion. We have

$$x = \cos^3 t, \qquad y = \sin^3 t,$$

$$\left(\frac{dx}{dt}\right)^2 = [3\cos^2 t(-\sin t)]^2 = 9\cos^4 t \sin^2 t,$$

$$\left(\frac{dy}{dt}\right)^2 = [3\sin^2 t(\cos t)]^2 = 9\sin^4 t \cos^2 t,$$

$$\sqrt{\left(\frac{dx}{dt}\right)^2 + \left(\frac{dy}{dt}\right)^2} = \sqrt{9\cos^2 t \sin^2 t\underbrace{(\cos^2 t + \sin^2 t)}_{1}},$$

$$= \sqrt{9\cos^2 t \sin^2 t}$$

$$= 3|\cos t \sin t|$$

$$= 3\cos t \sin t. \qquad (\cos t \sin t \ge 0 \text{ for } 0 \le t \le \pi/2)$$

Therefore,

$$\text{Length of first-quadrant portion} = \int_0^{\pi/2} 3\cos t \sin t\, dt$$

$$= \frac{3}{2}\int_0^{\pi/2} \sin 2t\, dt$$

$$= -\frac{3}{4}\cos 2t\Big]_0^{\pi/2} = \frac{3}{2}.$$

The length of the astroid is four times this: $4(3/2) = 6$.

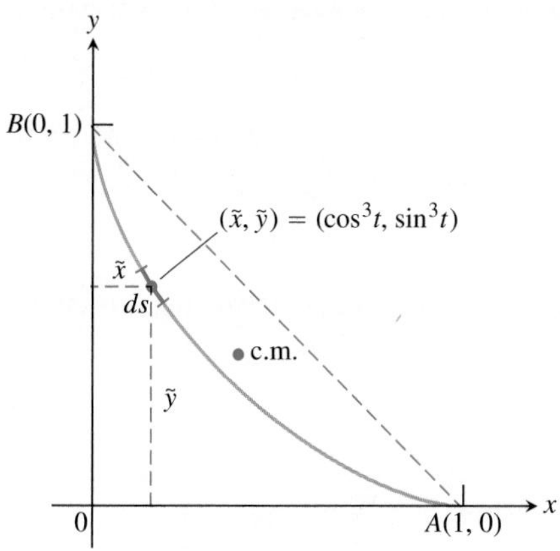

9.47 A typical segment of the arc in Example 4. The centroid (c.m.) of the curve lies about a third of the way toward chord AB.

Example 4 Find the centroid of the first-quadrant arc of the astroid in Example 3.

Solution We take the curve's density to be $\delta = 1$ and calculate the curve's mass and moments about the coordinate axes as we did at the end of Section 5.6.

The distribution of mass is symmetric about the line $y = x$, so $\bar{x} = \bar{y}$. A typical segment of the curve (Fig. 9.47) has mass

$$dm = 1 \cdot ds = \sqrt{\left(\frac{dx}{dt}\right)^2 + \left(\frac{dy}{dx}\right)^2}\, dt = 3 \cos t \sin t\, dt \qquad \text{(From Example 3)}$$

The curve's mass is

$$M = \int_0^{\pi/2} dm = \int_0^{\pi/2} 3 \cos t \sin t\, dt = \frac{3}{2}. \qquad \text{(Again from Example 3)}$$

The curve's moment about the x-axis is

$$M_x = \int \tilde{y}\, dm = \int_0^{\pi/2} \sin^3 t \cdot 3 \cos t \sin t\, dt$$

$$= 3 \int_0^{\pi/2} \sin^4 t \cos t\, dt = 3 \cdot \frac{\sin^5 t}{5}\bigg]_0^{\pi/2} = \frac{3}{5}.$$

Hence,

$$\bar{y} = \frac{M_x}{M} = \frac{3/5}{3/2} = \frac{2}{5}.$$

The centroid is the point (2/5, 2/5) (Fig. 9.47).

The Area of a Surface of Revolution

The formula $S = \int 2\pi \rho\, ds$ developed in Section 5.5 for the area of the surface swept out by revolving a smooth curve about an axis translates into $S = \int 2\pi\, y\, ds$ if the axis is the x-axis and into $S = \int 2\pi\, x\, ds$ if the axis is the y-axis. With $ds = \sqrt{(dx/dt)^2 + (dy/dt)^2}\, dt$, these lead to the following formulas.

Parametric Formulas for the Area of a Surface of Revolution

If the functions $x = f(t)$ and $y = g(t)$ are nonnegative and have continuous first derivatives with respect to t for $a \le t \le b$, and if the point $P(x, y)$ traces the curve defined by these equations exactly once as t moves from $t = a$ to $t = b$, then the areas of the surfaces generated by revolving the curve about the coordinate axes are as follows.

1. Revolution about the x-axis ($y \ge 0$)

$$\text{Surface area} = \int_a^b 2\pi\, y \sqrt{\left(\frac{dx}{dt}\right)^2 + \left(\frac{dy}{dt}\right)^2}\, dt \qquad (6)$$

2. Revolution about the y-axis ($x \ge 0$)

$$\text{Surface area} = \int_a^b 2\pi\, x \sqrt{\left(\frac{dx}{dt}\right)^2 + \left(\frac{dy}{dt}\right)^2}\, dt \qquad (7)$$

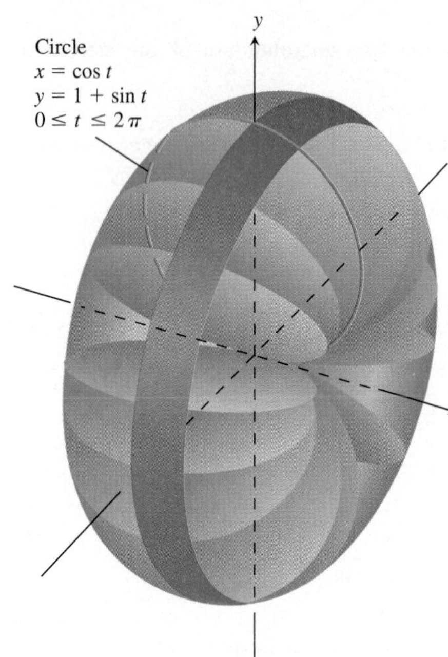

9.48 The surface in Example 5.

As with length, we can calculate surface area from any convenient parametrization that meets the criteria stated above.

Example 5 The standard parametrization of the circle of radius 1 centered at the point (0, 1) in the xy-plane is

$$x = \cos t, \quad y = 1 + \sin t, \quad 0 \le t \le 2\pi.$$

Use these equations to find the area of the surface swept out by revolving this circle about the x-axis (Fig. 9.48).

Solution We use Eq. (6) with

$$y = 1 + \sin t, \quad \frac{dx}{dt} = -\sin t, \quad \frac{dy}{dt} = \cos t$$

to obtain

$$\begin{aligned} \text{Area} &= \int_a^b 2\pi y \sqrt{\left(\frac{dx}{dt}\right)^2 + \left(\frac{dy}{dt}\right)^2}\, dt \qquad \text{(Eq.(6))} \\ &= \int_0^{2\pi} 2\pi(1 + \sin t)\underbrace{\sqrt{(-\sin t)^2 + (\cos t)^2}}_{1}\, dt \\ &= 2\pi \int_0^{2\pi} (1 + \sin t)\, dt = 2\pi\Big[t - \cos t\Big]_0^{2\pi} \\ &= 2\pi[(2\pi - 1) - (0 - 1)] = 4\pi^2. \end{aligned}$$

EXERCISES 9.4

In Exercises 1–12, find an equation for the line tangent to the curve at the point defined by the given value of t. Also, find the value of d^2y/dx^2 at this point.

1. $x = 2\cos t, \quad y = 2\sin t; \quad t = \pi/4$

2. $x = \sin 2\pi t, \quad y = \cos 2\pi t; \quad t = -1/6$

3. $x = 4\sin t, \quad y = 2\cos t; \quad t = \pi/4$

4. $x = \cos t, \quad y = \sqrt{3}\cos t; \quad t = 2\pi/3$

5. $x = t, \quad y = \sqrt{t}; \quad t = 1/4$

6. $x = \sec^2 t - 1, \quad y = \tan t; \quad t = -\pi/4$

7. $x = \sec t, \quad y = \tan t; \quad t = \pi/6$

8. $x = -\sqrt{t + 1}, \quad y = \sqrt{3t}; \quad t = 3$

9. $x = 2t^2 + 3, \quad y = t^4; \quad t = -1$

10. $x = 1/t, \quad y = -2 + \ln t; \quad t = 1$

11. $x = t - \sin t, \quad y = 1 - \cos t; \quad t = \pi/3$

12. $x = \cos t, \quad y = 1 + \sin t; \quad t = \pi/2$

Find the lengths of the curves in Exercises 13–18.

13. $x = \cos t, \quad y = t + \sin t, \quad 0 \le t \le \pi$

14. $x = t^3, \quad y = 3t^2/2, \quad 0 \le t \le \sqrt{3}$

15. $x = t^2/2, \quad y = (2t + 1)^{3/2}/3, \quad 0 \le t \le 4$

16. $x = (2t + 3)^{3/2}/3, \quad y = t + t^2/2, \quad 0 \le t \le 3$

17. $x = 8\cos t + 8t\sin t,$
$y = 8\sin t - 8t\cos t, \quad 0 \le t \le \pi/2$

18. $x = \ln(\sec t + \tan t) - \sin t,$
$y = \cos t, \quad 0 \le t \le \pi/3$ (See Fig. 9.49.)

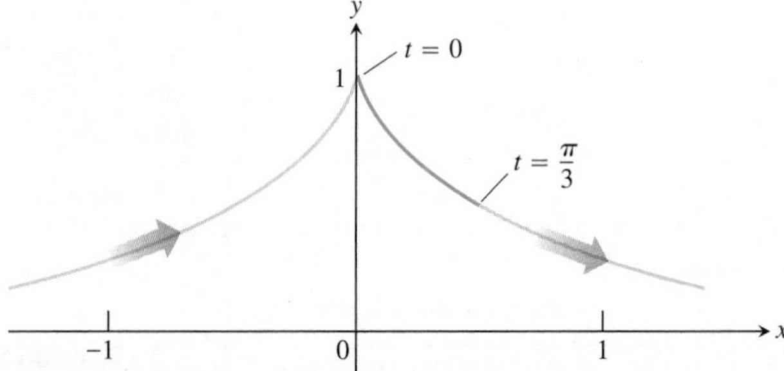

9.49 The curve in Exercise 18 is a portion of a curve that covers the entire x-axis as t runs between $-\pi/2$ and $\pi/2$. The arrows show the direction of increasing t.

Find the areas of the surfaces generated by revolving the curves in Exercises 19–22 about the indicated axes.

19. $x = \cos t,\quad y = 2 + \sin t,\quad 0 \le t \le 2\pi,$
about the x-axis

20. $x = (2/3)t^{3/2},\quad y = 2\sqrt{t},\quad 0 \le t \le \sqrt{3},$
about the y-axis

21. $x = t + \sqrt{2},\quad y = (t^2/2) + \sqrt{2}\,t,\quad -\sqrt{2} \le t \le \sqrt{2},$
about the y-axis

22. $x = \ln(\sec t + \tan t) - \sin t,$
$y = \cos t,\quad 0 \le t \le \pi/3,$ about the x-axis

23. *A cone frustum.* The line segment joining the points (0, 1) and (2, 2) is revolved about the x-axis to generate a frustum of a cone. Find the surface area of the frustum with the parametrization

$$x = 2t,\quad y = t + 1,\quad 0 \le t \le 1.$$

Check your result with the geometry formula

$$\text{Area} = \pi(r_1 + r_2)(\text{slant height}).$$

24. *A cone.* The line segment joining the origin to the point (h, r) is revolved about the x-axis to generate a cone of height h and base radius r. Find the cone's surface area with the parametrization

$$x = ht,\quad y = rt,\quad 0 \le t \le 1.$$

Check your result with the geometry formula

$$\text{Area} = \pi(r)(\text{slant height}).$$

25. a) Find the coordinates of the centroid of the curve

$$x = \cos t + t \sin t,\quad y = \sin t - t \cos t,\quad 0 \le t \le \pi/2.$$

b) CALCULATOR The curve is a portion of the involute in Fig. 9.41. Sketch the curve. Find the centroid's coordinates to the nearest tenth and add the centroid to your sketch.

26. a) Find the coordinates of the centroid of the curve

$$x = e^t \cos t,\quad y = e^t \sin t,\quad 0 \le t \le \pi.$$

b) CALCULATOR Sketch the curve. Find the centroid's coordinates to the nearest tenth and add the centroid to your sketch.

27. a) Find the coordinates of the centroid of the curve

$$x = \cos t,\quad y = t + \sin t,\quad 0 \le t \le \pi.$$

b) Sketch the curve and add the centroid to your sketch.

28. INTEGRAL EVALUATOR Most centroid calculations for curves are done with a calculator or computer that has an integral evaluation program. As a case in point, find, to the nearest hundredth, the coordinates of the centroid of the curve

$$x = t^3,\quad y = 3t^2/2,\quad 0 \le t \le \sqrt{3}.$$

29. *Length is independent of parametrization.* To illustrate the fact that the numbers we get for length do not depend on the way we parametrize our curves (except for the mild restrictions mentioned earlier), calculate the length of the semicircle $y = \sqrt{1 - x^2}$ with these two different parametrizations:

a) $x = \cos 2t,\quad y = \sin 2t,\quad 0 \le t \le \pi/2$
b) $x = \sin \pi t,\quad y = \cos \pi t,\quad -1/2 \le t \le 1/2$

30. *Elliptic integrals.* The length of the ellipse

$$x = a \cos t,\quad y = b \sin t,\quad 0 \le t \le 2\pi$$

turns out to be

$$\text{Length} = 4a \int_0^{\pi/2} \sqrt{1 - e^2 \cos^2 t}\, dt,$$

where e is the ellipse's eccentricity. The integral in this formula, called an *elliptic integral,* is nonelementary except when $e = 0$ or 1.

a) CALCULATOR Use the Trapezoidal Rule with $n = 10$ to estimate the length of the ellipse when $a = 1$ and $e = 1/2$.

b) Use the fact that the absolute value of the second derivative of $f(t) = \sqrt{1 - e^2\cos^2 t}$ is less than 1 to find an upper bound for the error in the estimate you obtained in (a).

The curves drawn by computer in Exercises 31 and 32 are called *Bowditch curves* or *Lissajous figures.* In each case, find the point in the first quadrant where the tangent to the curve is horizontal and find the equations of the two tangents at the origin.

31.

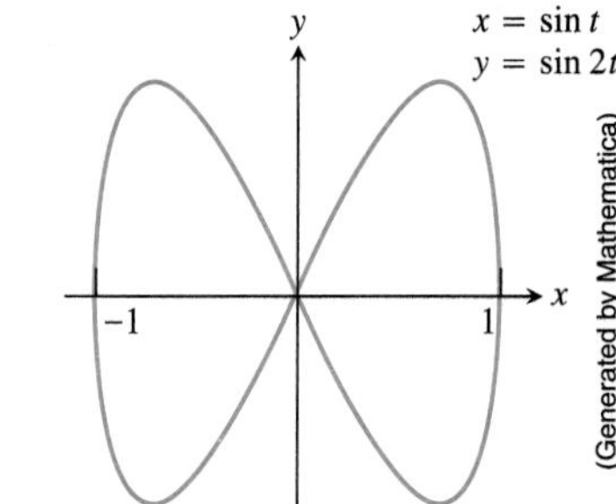

32.

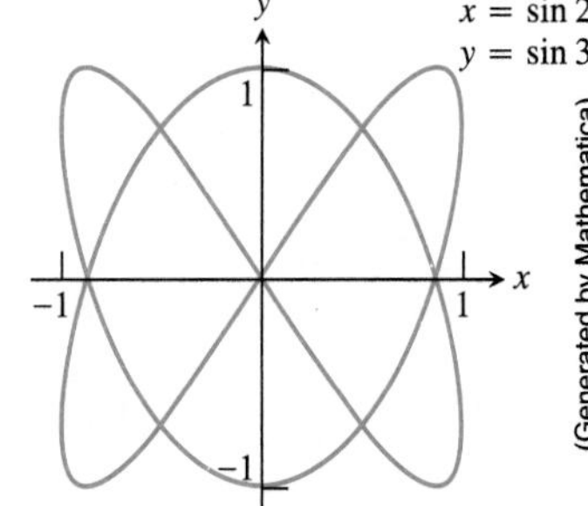

Computer Grapher

Graph the parametric curves in Exercises 33–39 over parameter intervals of your choice. The curves are all Bowditch curves (Lissajous figures), their general formula being

$$x = a \sin(mt + d),\qquad y = b \sin nt,$$

with m and n integers.

33. $x = \sin 2t, \quad y = \sin t$

34. $x = \sin 3t, \quad y = \sin 4t$

35. $x = \sin t, \quad y = \sin 4t$

36. $x = \sin t, \quad y = \sin 5t$

37. $x = \sin 3t, \quad y = \sin 5t$

38. $x = \sin(3t + \pi/2), \quad y = \sin 5t$

39. $x = \sin(3t + \pi/4), \quad y = \sin 5t$

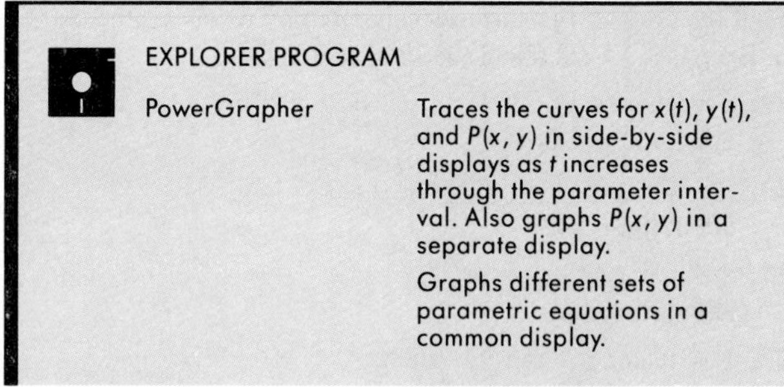

EXPLORER PROGRAM

PowerGrapher	Traces the curves for $x(t)$, $y(t)$, and $P(x, y)$ in side-by-side displays as t increases through the parameter interval. Also graphs $P(x, y)$ in a separate display.
	Graphs different sets of parametric equations in a common display.

9.5 Polar Coordinates

In this section, we define polar coordinates and study their relation to Cartesian coordinates. One of the distinctions between polar and Cartesian coordinates is that while a point in the plane has just one pair of Cartesian coordinates, it has infinitely many pairs of polar coordinates. This has interesting consequences for graphing, as we shall see in the next section. Polar coordinates enable us to describe all conic sections with a single equation, as we shall see in Section 9.7, and the calculus we have done in rectangular coordinates carries over to this new system as well.

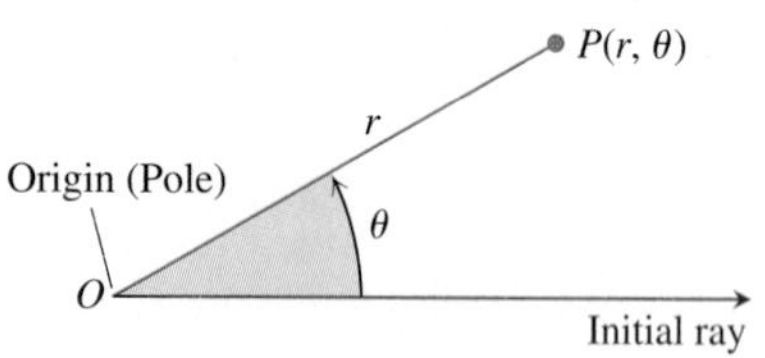

9.50 To define polar coordinates for the plane, we start with an origin called the pole, and an initial ray.

Definition of Polar Coordinates

To define polar coordinates, we first fix an **origin** O (called the **pole**) and an **initial ray** from O (Fig. 9.50). Then each point P can be located by assigning to it a **polar coordinate pair** (r, θ), in which the first number, r, gives the directed distance from O to P and the second number, θ, gives the directed angle from the initial ray to the segment OP:

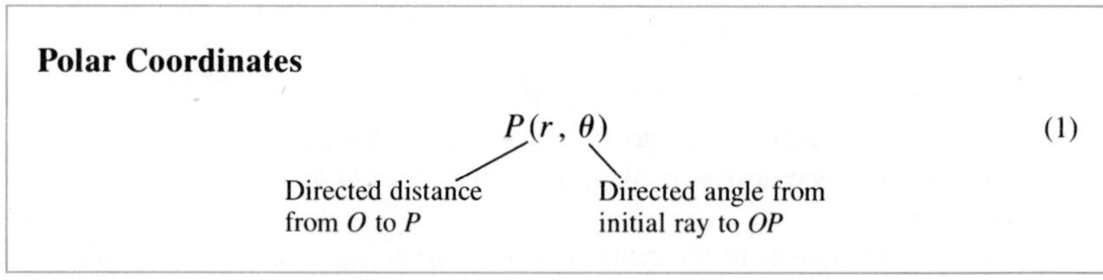

Polar Coordinates

$$P(r, \theta) \tag{1}$$

As in trigonometry, the angle θ is positive when measured counterclockwise and negative when measured clockwise. But the angle associated with a given point is not unique. For instance, the point 2 units from the origin along the ray $\theta = 30°$ has polar coordinates $r = 2$, $\theta = 30°$. It also has coordinates $r = 2$, $\theta = -330°$ (Fig. 9.51). As we shall see in Example 1, it has infinitely many coordinate pairs.

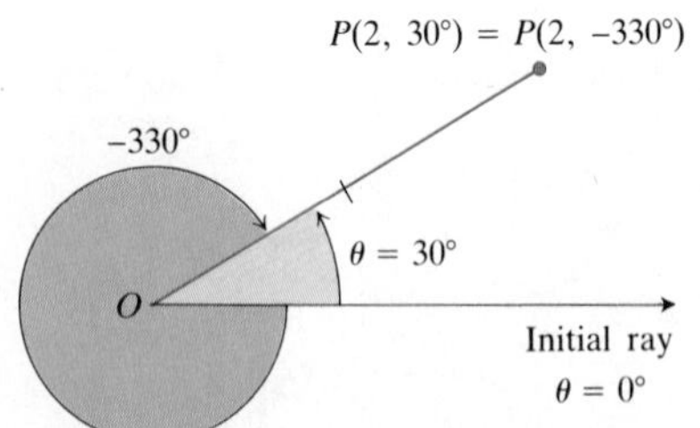

9.51 Polar coordinates are not unique. Since the ray $\theta = -330°$ is the same as the ray $\theta = 30°$, the point P defined by the coordinates $(2, 30°)$ also has the coordinates $(2, -330°)$.

Negative Values of r; Changing to Radian Measure

There are occasions when we wish to allow r to be negative. That is why we say "directed distance" in (1). The ray $\theta = 30°$ and the ray $\theta = 210°$ together make a complete line through O (Fig. 9.52). The point $P\ (2, 210°)$ 2 units from O on the ray

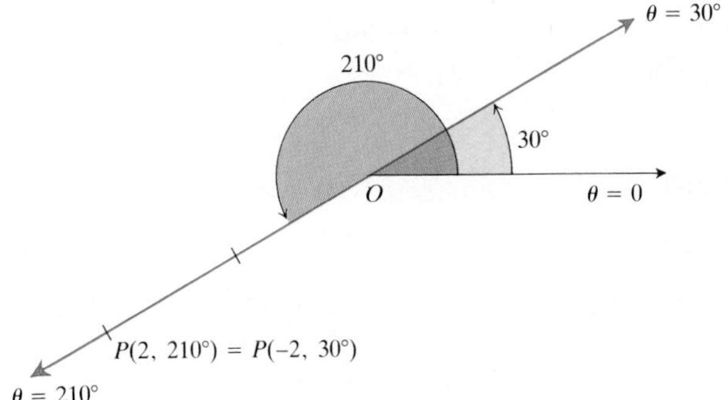

9.52 Points can have polar coordinates with negative r-values.

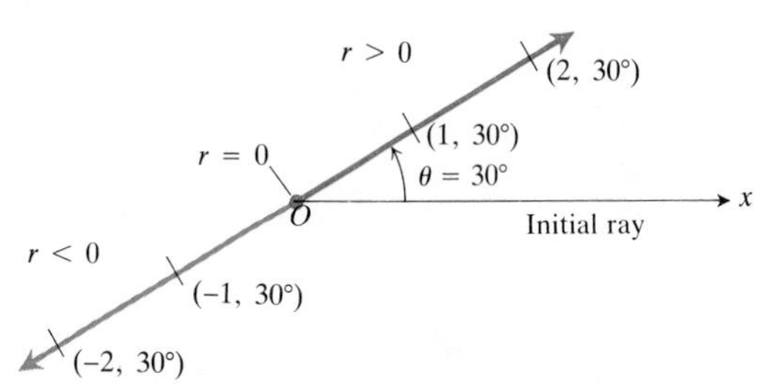

9.53 The ray $\theta = 30°$ and its opposite, the ray $\theta = 210°$, make a straight line.

$\theta = 210°$ has polar coordinates $r = 2$, $\theta = 210°$. It can be reached by turning 210° counterclockwise from the initial ray and going forward 2 units. It can also be reached by turning 30° counterclockwise from the initial ray and going *backward* two units. So we say that the point also has polar coordinates $r = -2$, $\theta = 30°$.

Whenever the angle between two rays is 180°, the rays make a straight line. We then say that each ray is the **opposite** of the other. Points on the ray $\theta = \alpha$ have polar coordinates (r, α) with $r \geq 0$. Points on the opposite ray, the ray $\theta = \alpha + 180°$, have coordinates (r, α) with $r \leq 0$ (Fig. 9.53).

Example 1 Find all the polar coordinates of the point (2, 30°). Express the angles in radians as well as in degrees.

Solution We sketch the initial ray of the coordinate system, draw the ray through the origin that makes a 30° angle with the initial ray, and mark the point (2, 30°) (Fig. 9.54). We then find formulas for the coordinate pairs in which $r = 2$ and $r = -2$ and convert the formulas to radian measure.

For $r = 2$: The angles

$$\begin{array}{ll} 30° + 1 \cdot 360° = 390° & 30° - 1 \cdot 360° = -330° \\ 30° + 2 \cdot 360° = 750° & 30° - 2 \cdot 360° = -690° \\ 30° + 3 \cdot 360° = 1110° & 30° - 3 \cdot 360° = -1050° \\ \vdots & \vdots \end{array} \tag{2}$$

all end in the same ray as the angle 30°. Thus, the polar coordinates

$$(2, 30° + n \cdot 360°), \qquad n = 0, \pm 1, \pm 2, \ldots \tag{3}$$

all identify the point (2, 30°).

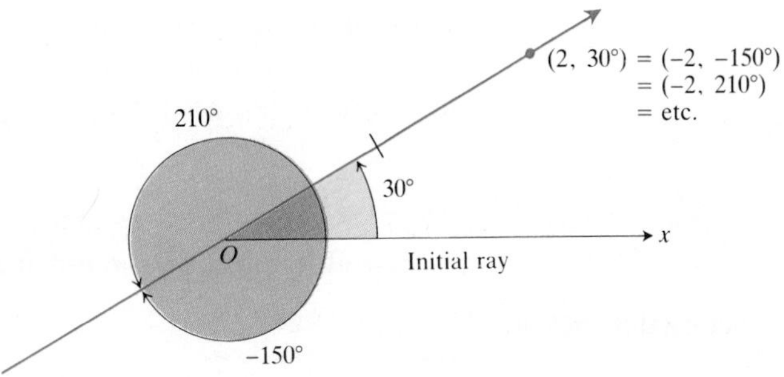

9.54 The point $P(2, 30°)$ has many different polar coordinates.

For $r = -2$: Numerous as they are, the coordinates in (3) are not the only polar coordinates of the point (2, 30°). The angles

$$\begin{array}{ll} -150^\circ & -150^\circ \\ -150^\circ + 360^\circ = 210^\circ & -150^\circ - 360^\circ = -510^\circ \\ -150^\circ + 720^\circ = 570^\circ & -150^\circ - 720^\circ = -870^\circ \\ \vdots & \vdots \end{array} \tag{4}$$

all define the ray opposite the ray $\theta = 30°$. Hence, the polar coordinates

$$(-2, -150° + n \cdot 360°), \qquad n = 0, \pm 1, \pm 2, \ldots \tag{5}$$

represent the point (2, 30°) as well.

Radian measure: The radian formulas that correspond to (3) and (5) are

$$\left(2, \frac{\pi}{6} + 2n\pi\right), \qquad n = 0, \pm 1, \pm 2, \ldots \tag{6}$$

and

$$\left(-2, -\frac{5\pi}{6} + 2n\pi\right), \qquad n = 0, \pm 1, \pm 2, \ldots. \tag{7}$$

When $n = 0$, these formulas give

$$(2, \pi/6) \quad \text{and} \quad (-2, -5\pi/6).$$

When $n = 1$, they give

$$(2, 13\pi/6) \quad \text{and} \quad (-2, 7\pi/6),$$

and so on.

The Use of Radian Measure

Although nothing in the definition of polar coordinates requires the use of radian measure, we shall need to have all angles in radians when we differentiate and integrate trigonometric functions of θ. We shall therefore use radian measure almost exclusively from now on.

Elementary Coordinate Equations and Inequalities

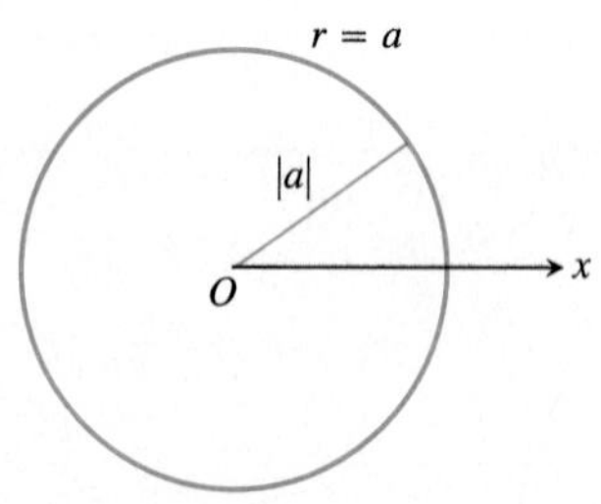

9.55 The polar equation for this circle is $r = a$.

If we hold r fixed at a constant nonzero value $r = a$, then the point $P(r, \theta)$ lies $|a|$ units from the origin. As θ varies over any interval of length 2π radians, P traces a circle of radius $|a|$ centered at the origin (Fig. 9.55).

Circle of Radius $|a|$ Centered at the Origin

$$r = a \tag{8}$$

(a)

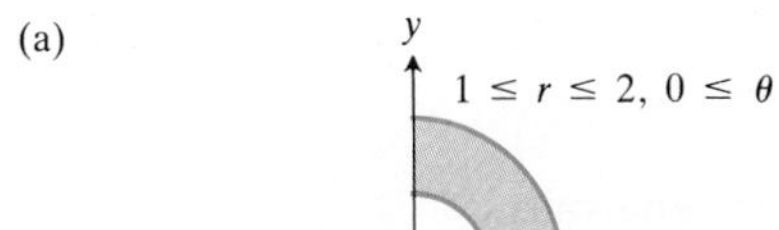

(b)

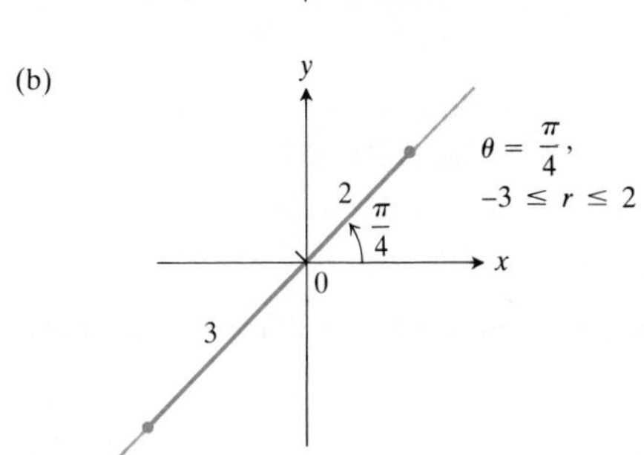

(c)

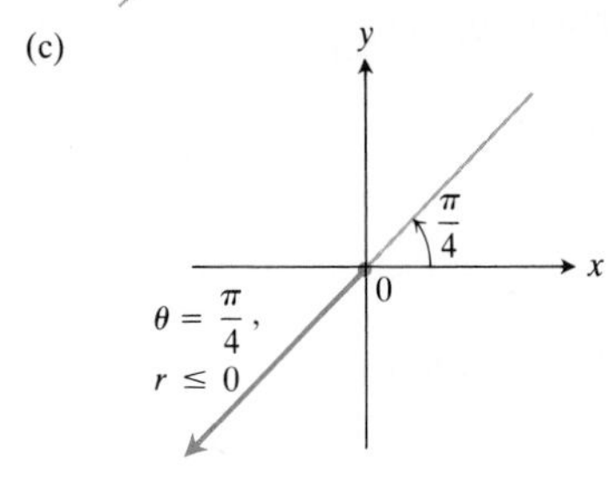

(d)

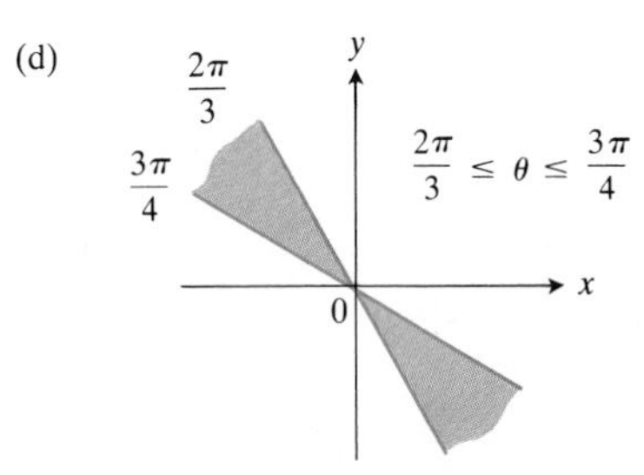

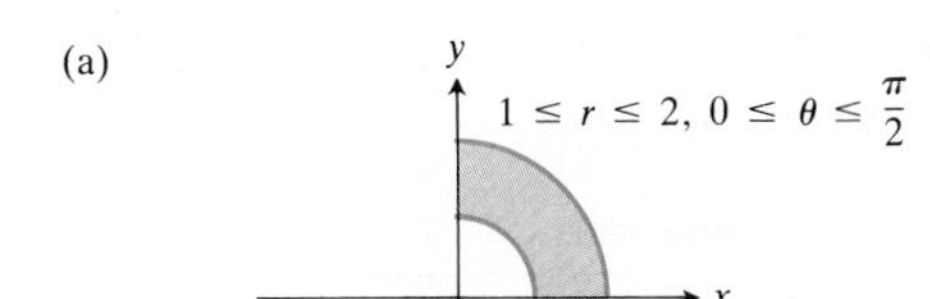

9.56 The graphs of typical inequalities in r and θ (Example 4).

Example 2 The equation $r = 1$ is an equation for the circle of radius 1 centered at the origin. So is the equation $r = -1$.

If we hold θ fixed at a constant value $\theta = \theta_0$ and let r run between $-\infty$ and ∞, the point $P(r, \theta)$ traces a line through the origin that makes an angle of measure θ_0 with the initial ray. The line therefore consists of all the points in the plane that have coordinates of the form (r, θ_0).

Equation for Lines Through the Origin

$$\theta = \theta_0 \tag{9}$$

Example 3 One equation for the line in Fig. 9.53 is $\theta = \pi/6$. The equations $\theta = 7\pi/6$ and $\theta = -5\pi/6$ are also equations for this line.

Equations of the form $r = a$ and $\theta = \theta_0$ can be combined to define regions, segments, and rays.

Example 4 Graph the sets of points whose polar coordinates satisfy the following conditions.

a) $1 \le r \le 2$ and $0 \le \theta \le \frac{\pi}{2}$

b) $-3 \le r \le 2$ and $\theta = \frac{\pi}{4}$

c) $r \le 0$ and $\theta = \frac{\pi}{4}$

d) $\frac{2\pi}{3} \le \theta \le \frac{3\pi}{4}$ (no restriction on r)

Solution The graphs are shown in Fig. 9.56.

Cartesian Versus Polar Coordinates

When we use both polar and Cartesian coordinates in a plane, we place the two origins together and take the initial polar ray to be the positive x-axis. The ray $\theta = \pi/2$, $r > 0$, becomes the positive y-axis (Fig. 9.57). The two sets of coordinates are then related by the following equations.

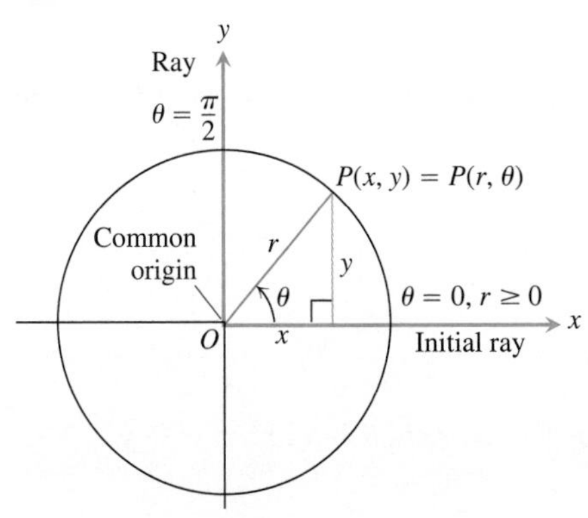

9.57 The usual way to relate polar and Cartesian coordinates.

Equations Relating Polar and Cartesian Coordinates

$$x = r\cos\theta, \quad y = r\sin\theta, \quad x^2 + y^2 = r^2, \quad \frac{y}{x} = \tan\theta \tag{10}$$

Equations (10) are the equations we use to rewrite polar equations in Cartesian form and vice versa.

Example 5

Polar equation	**Cartesian equivalent**
$r \cos \theta = 2$	$x = 2$
$r^2 \cos \theta \sin \theta = 4$	$xy = 4$
$r^2 \cos^2 \theta - r^2 \sin^2 \theta = 1$	$x^2 - y^2 = 1$
$r = 1 + 2r \cos \theta$	$y^2 - 3x^2 - 4x - 1 = 0$
$r = 1 - \cos \theta$	$x^4 + y^4 + 2x^2y^2 + 2x^3 + 2xy^2 - y^2 = 0$

With some curves, we are better off with polar coordinates; with others, we aren't.

Example 6 Find a Cartesian equation for the curve

$$r \cos\left(\theta - \frac{\pi}{3}\right) = 3.$$

Solution We use the identity

$$\cos(A - B) = \cos A \cos B + \sin A \sin B$$

with $A = \theta$ and $B = \pi/3$:

$$r \cos\left(\theta - \frac{\pi}{3}\right) = 3,$$

$$r\left(\cos \theta \cos \frac{\pi}{3} + \sin \theta \sin \frac{\pi}{3}\right) = 3,$$

$$r \cos \theta \cdot \frac{1}{2} + r \sin \theta \cdot \frac{\sqrt{3}}{2} = 3,$$

$$\frac{1}{2}x + \frac{\sqrt{3}}{2}y = 3,$$

$$x + \sqrt{3}y = 6.$$

Example 7 Replace the following polar equations by equivalent Cartesian equations, and identify their graphs.

a) $r \cos \theta = -4$

b) $r^2 = 4r \cos \theta$

c) $r = \dfrac{4}{2 \cos \theta - \sin \theta}$

Solution We use the substitutions $r \cos \theta = x$, $r \sin \theta = y$, $r^2 = x^2 + y^2$.

a) $r \cos \theta = -4$

The Cartesian equation: $r \cos \theta = -4$

$x = -4$

The graph: Vertical line through $x = -4$ on the x-axis

b) $r^2 = 4r\cos\theta$

The Cartesian equation:

$$r^2 = 4r\cos\theta$$
$$x^2 + y^2 = 4x$$
$$x^2 - 4x + y^2 = 0$$
$$x^2 - 4x + 4 + y^2 = 4 \qquad \text{(Completing the square)}$$
$$(x-2)^2 + y^2 = 4$$

The graph: Circle, radius 2, center $(h, k) = (2, 0)$

c) $r = \dfrac{4}{2\cos\theta - \sin\theta}$

The Cartesian equation:

$$r(2\cos\theta - \sin\theta) = 4$$
$$2r\cos\theta - r\sin\theta = 4$$
$$2x - y = 4$$
$$y = 2x - 4$$

The graph: Line, slope $m = 2$, y-intercept $b = -4$

EXERCISES 9.5

NOTE: *All angles are in radians.*

1. Pick out the polar coordinate pairs that label the same point.

a) $(3, 0)$ b) $(-3, 0)$
c) $(-3, \pi)$ d) $(-3, 2\pi)$
e) $(2, 2\pi/3)$ f) $(2, -2\pi/3)$
g) $(2, 7\pi/3)$ h) $(-2, \pi/3)$
i) $(2, -\pi/3)$ j) $(2, \pi/3)$
k) $(-2, -\pi/3)$ l) $(-2, 2\pi/3)$
m) (r, θ) n) $(r, \theta + \pi)$
o) $(-r, \theta + \pi)$ p) $(-r, \theta)$

2. Find the Cartesian coordinates of the points whose polar coordinates are given in parts (a)–(l) of Exercise 1.

3. Plot the following points (given in polar coordinates). Then find all the polar coordinates of each point.

a) $(2, \pi/2)$ b) $(2, 0)$
c) $(-2, \pi/2)$ d) $(-2, 0)$

4. Plot the following points (given in polar coordinates). Then find all the polar coordinates of each point.

a) $(3, \pi/4)$ b) $(-3, \pi/4)$
c) $(3, -\pi/4)$ d) $(-3, -\pi/4)$

5. Find the Cartesian coordinates of the following points (given in polar coordinates).

a) $(\sqrt{2}, \pi/4)$ b) $(1, 0)$
c) $(0, \pi/2)$ d) $(-\sqrt{2}, \pi/4)$
e) $(-3, 5\pi/6)$ f) $(5, \tan^{-1}(4/3))$
g) $(-1, 7\pi)$ h) $(2\sqrt{3}, 2\pi/3)$

6. Find all polar coordinates of the origin.

Graph the sets of points whose polar coordinates satisfy the equations and inequalities in Exercises 7–22.

7. $r = 2$
8. $0 \le r \le 2$
9. $r \ge 1$
10. $1 \le r \le 2$
11. $0 \le \theta \le \pi/6, \quad r \ge 0$
12. $\theta = 2\pi/3, \quad r \le -2$
13. $\theta = \pi/3, \quad -1 \le r \le 3$
14. $\theta = 11\pi/4, \quad r \ge -1$
15. $\theta = \pi/2, \quad r \ge 0$
16. $\theta = \pi/2, \quad r \le 0$
17. $0 \le \theta \le \pi, \quad r = 1$
18. $0 \le \theta \le \pi, \quad r = -1$
19. $\pi/4 \le \theta \le 3\pi/4, \quad 0 \le r \le 1$
20. $-\pi/4 \le \theta \le \pi/4, \quad -1 \le r \le 1$
21. $-\pi/2 \le \theta \le \pi/2, \quad 1 \le r \le 2$
22. $0 \le \theta \le \pi/2, \quad 1 \le |r| \le 2$

Replace the polar equations in Exercises 23–42 by equivalent Cartesian equations. Then identify the graph.

23. $r\cos\theta = 2$
24. $r\sin\theta = -1$
25. $r\sin\theta = 4$
26. $r\cos\theta = 0$
27. $r\sin\theta = 0$
28. $r\cos\theta = -3$
29. $r\cos\theta + r\sin\theta = 1$
30. $r\sin\theta = r\cos\theta$
31. $r^2 = 1$
32. $r^2 = 4r\sin\theta$

33. $r = \dfrac{5}{\sin\theta - 2\cos\theta}$

34. $r^2 \sin 2\theta = 2$

35. $r = \cot\theta \csc\theta$

36. $r = 4\tan\theta \sec\theta$

37. $r = \csc\theta\, e^{r\cos\theta}$

38. $r\sin\theta = \ln r + \ln\cos\theta$

39. $r^2 + 2r^2\cos\theta\sin\theta = 1$

40. $\cos^2\theta = \sin^2\theta$

41. $r = 2\cos\theta + 2\sin\theta$

42. $r = 2\cos\theta - \sin\theta$

Replace the Cartesian equations in Exercises 43–52 by equivalent polar equations.

43. $x = 7$

44. $y = 1$

45. $x = y$

46. $x - y = 3$

47. $x^2 + y^2 = 4$

48. $x^2 - y^2 = 1$

49. $\dfrac{x^2}{9} + \dfrac{y^2}{4} = 1$

50. $xy = 2$

51. $y^2 = 4x$

52. $x^2 - y^2 = 25\sqrt{x^2 + y^2}$

9.6 Graphing in Polar Coordinates

The graph of a polar coordinate equation $F(r, \theta) = 0$ consists of the points whose polar coordinates in some form satisfy the equation. We say "in some form" because some coordinate pairs of a point on the graph may not satisfy the equation even when others do.

To speed our work, we look for symmetries, for values of θ at which the curve passes through the origin, and for points at which r takes on extreme values. When the curve passes through the origin, we also try to calculate the curve's slope there. This section gives the details.

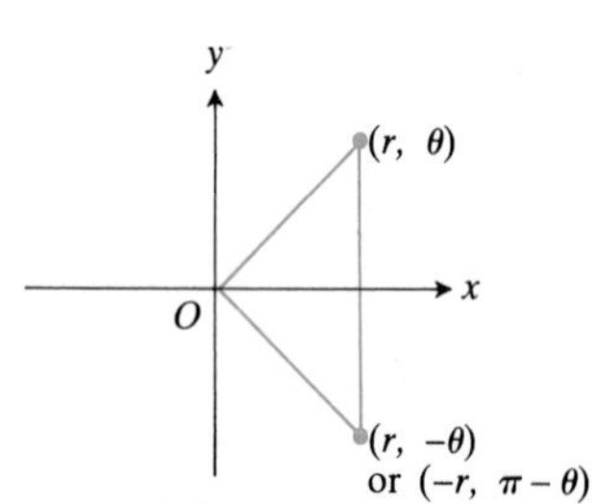

About the x-axis

(a)

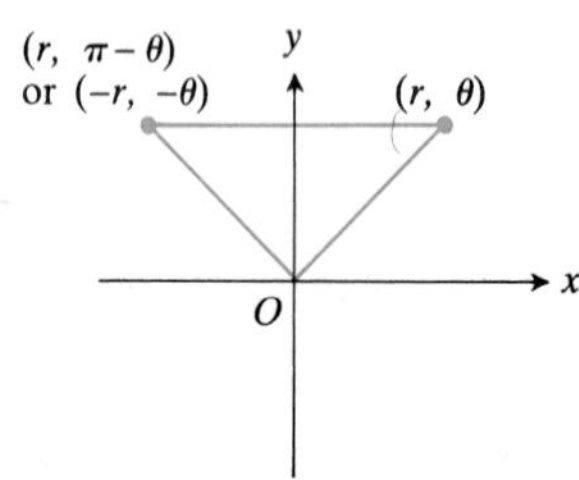

About the y-axis

(b)

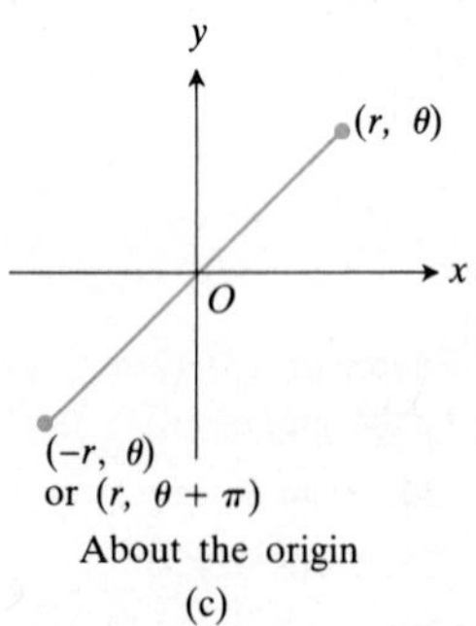

About the origin

(c)

9.58 Three tests for symmetry.

Symmetry

The three parts of Fig. 9.58 illustrate the standard polar coordinate tests for symmetry.

Symmetry Tests for Polar Graphs

1. *Symmetry about the x-axis:* If the point (r, θ) lies on the graph, the point $(r, -\theta)$ or $(-r, \pi - \theta)$ lies on the graph (Fig. 9.58a).
2. *Symmetry about the y-axis:* If the point (r, θ) lies on the graph, the point $(r, \pi - \theta)$ or $(-r, -\theta)$ lies on the graph (Fig. 9.58b).
3. *Symmetry about the origin:* If the point (r, θ) lies on the graph, the point $(-r, \theta)$ or $(r, \theta + \pi)$ lies on the graph (Fig. 9.58c).

If a curve has any two of the symmetries listed here, it also has the third (as you will be invited to show in Exercise 34). If two of the tests are positive there is no need to apply the third.

Slope

The slope of a polar curve $r = f(\theta)$ is not given by the derivative $r' = df/d\theta$, but by a different formula. To see why, and what the formula is, think of the graph of f as the graph of the parametric equations

$$x = r\cos\theta = f(\theta)\cos\theta, \qquad y = r\sin\theta = f(\theta)\sin\theta. \tag{1}$$

If f is a differentiable function of θ, then so are x and y and, when $dx/d\theta \neq 0$, we may calculate dy/dx from the parametric formula

$$\frac{dy}{dx} = \frac{dy/d\theta}{dx/d\theta} \qquad \text{(Section 9.4, Eq. (2) with } t = \theta\text{)}$$

$$= \frac{\frac{d}{d\theta}(f(\theta) \cdot \sin\theta)}{\frac{d}{d\theta}(f(\theta) \cdot \cos\theta)}$$

$$= \frac{\frac{df}{d\theta}\sin\theta + f(\theta)\cos\theta}{\frac{df}{d\theta}\cos\theta - f(\theta)\sin\theta} \qquad \text{(Product Rule for Derivatives)}$$

$$= \frac{r'\sin\theta + r\cos\theta}{r'\cos\theta - r\sin\theta}. \qquad (r = f,\ r' = df/d\theta)$$

Slope of a Polar Curve

If $r = f(\theta)$ is differentiable and $dx/d\theta \neq 0$, then the slope dy/dx at the point (r, θ) on the graph of f is given by the formula

$$\text{Slope at } (r, \theta) = \frac{r'\sin\theta + r\cos\theta}{r'\cos\theta - r\sin\theta}. \qquad (2)$$

If $r = 0$ when $\theta = \theta_0$, then Eq. (2) reduces to

$$\text{Slope at } (0, \theta_0) = \frac{r'\sin\theta_0}{r'\cos\theta_0} = \frac{\sin\theta_0}{\cos\theta_0} = \tan\theta_0. \qquad (3)$$

Slopes at the Origin

If the graph of $r = f(\theta)$ passes through the origin at the value $\theta = \theta_0$, the slope of the curve there is

$$\text{Slope at } (0, \theta_0) = \tan\theta_0. \qquad (4)$$

The reason we say "slope at $(0, \theta_0)$" and not just "slope at the origin" is that a polar curve may pass through the origin more than once, with different slopes at different θ-values. This is not the case in our first example, however.

Example 1 *A Cardioid.* Graph the curve

$$r = 1 - \cos\theta.$$

Solution The curve is symmetric about the x-axis because

$$\begin{aligned} (r, \theta) \text{ on the graph} \quad &\Rightarrow \quad r = 1 - \cos\theta \\ &\Rightarrow \quad r = 1 - \cos(-\theta) \qquad \begin{pmatrix}\cos\theta = \\ \cos(-\theta)\end{pmatrix} \\ &\Rightarrow \quad (r, -\theta) \text{ on the graph.} \end{aligned}$$

θ	$r = 1 - \cos\theta$
0	0
$\frac{\pi}{3}$	$\frac{1}{2}$
$\frac{\pi}{2}$	1
$\frac{2\pi}{3}$	$\frac{3}{2}$
π	2

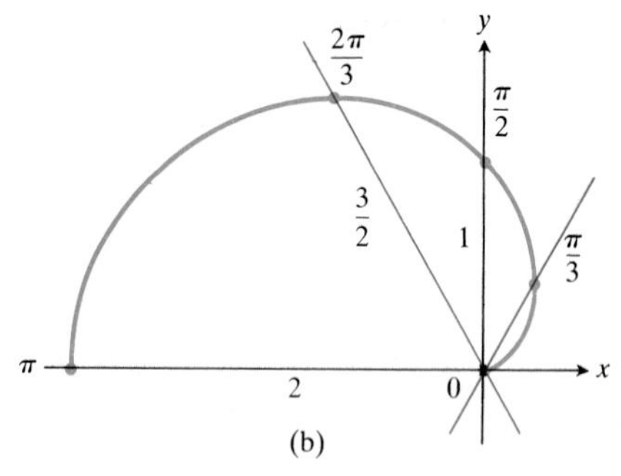

(a)

(b)

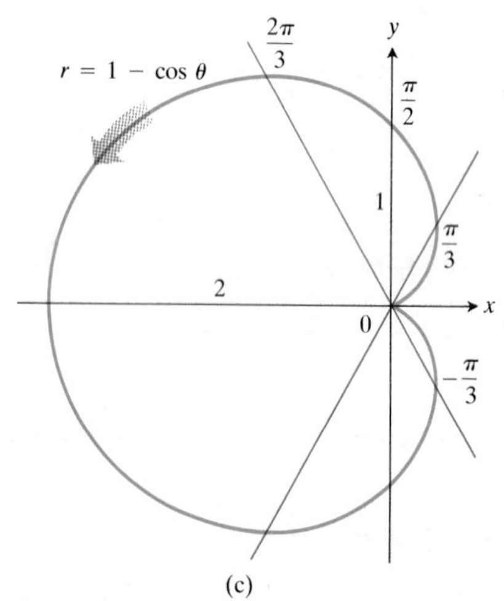

(c)

9.59 The steps in graphing the cardioid $r = 1 - \cos\theta$ (Example 1). The arrow shows the direction of increasing θ.

As θ increases from 0 to π, $\cos\theta$ decreases from 1 to -1, and $r = 1 - \cos\theta$ increases from a minimum value of 0 to a maximum value of 2. As θ continues on from π to 2π, $\cos\theta$ increases from -1 back to 1 and r decreases from 2 back to 0. The curve starts to repeat when $\theta = 2\pi$ because the cosine has period 2π.

The curve leaves the origin with slope $\tan(0) = 0$ and returns to the origin with slope $\tan(2\pi) = 0$.

We make a table of values from $\theta = 0$ to $\theta = \pi$, plot the points, draw a smooth curve through them with a horizontal tangent at the origin, and reflect the curve across the x-axis to complete the graph (Fig. 9.59). The curve is called a *cardioid* because of its heart shape.

Example 2

Graph the curve $r^2 = 4\cos\theta$.

Solution Although $\cos\theta$ has period 2π, the equation $r^2 = 4\cos\theta$ requires $\cos\theta \geq 0$, so we get the entire graph by running θ from $-\pi/2$ to $\pi/2$. The curve is symmetric about the x-axis because

$$
\begin{aligned}
(r, \theta) \text{ on the graph} \quad &\Rightarrow \quad r^2 = 4\cos\theta \\
&\Rightarrow \quad r^2 = 4\cos(-\theta) \qquad \begin{pmatrix}\cos\theta = \\ \cos(-\theta)\end{pmatrix} \\
&\Rightarrow \quad (r, -\theta) \text{ on the graph.}
\end{aligned}
$$

The curve is also symmetric about the origin because

$$
\begin{aligned}
(r, \theta) \text{ on the graph} \quad &\Rightarrow \quad r^2 = 4\cos\theta \\
&\Rightarrow \quad (-r)^2 = 4\cos\theta \\
&\Rightarrow \quad (-r, \theta) \text{ on the graph.}
\end{aligned}
$$

Together, these two symmetries imply symmetry about the y-axis.

The curve passes through the origin when $\theta = -\pi/2$ and $\theta = \pi/2$. It has a vertical tangent both times because $\tan\theta$ is infinite.

For each value of θ in the interval between $-\pi/2$ and $\pi/2$, the formula $r^2 = 4\cos\theta$ gives two values of r:

$$r = \pm 2\sqrt{\cos\theta}.$$

We make a short table of values, plot the corresponding points, and use information about symmetry and tangents to guide us in connecting the points with a smooth curve (Fig. 9.60).

θ	$\cos\theta$	$r = \pm 2\sqrt{\cos\theta}$
0	1	± 2
$\pm\frac{\pi}{6}$	$\frac{\sqrt{3}}{2}$	± 1.9
$\pm\frac{\pi}{4}$	$\frac{1}{\sqrt{2}}$	± 1.7
$\pm\frac{\pi}{3}$	$\frac{1}{2}$	± 1.4
$\pm\frac{\pi}{2}$	0	0

(a)

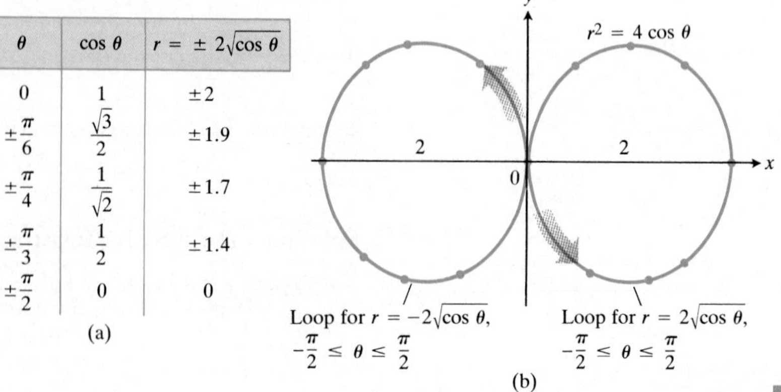

(b)

9.60 The graph of $r^2 = 4\cos\theta$ (Example 2). The arrows show the direction of increasing θ. The values of r in the table were found with a calculator and rounded.

Steps for Quick Graphing

STEP 1: First graph $r = f(\theta)$ in the *Cartesian* $r\theta$-plane (that is, plot the values of θ on a horizontal axis and the corresponding values of r along a vertical axis).

STEP 2: Then use the Cartesian graph as a "table" and guide to sketch the *polar* coordinate graph.

Faster Graphing

One way to graph a polar equation $r = f(\theta)$ is to make a table of (r, θ) values, plot the corresponding points, and connect them in order of increasing θ. This can work well if enough points have been plotted to reveal all the loops and dimples in the graph. In this section we describe another method of graphing that is usually quicker and more reliable. The steps in the new method are described in the sidelight.

This method is better than simple point plotting because the Cartesian graph, even when hastily drawn, shows at a glance where r is positive, negative, and nonexistent, as well as where r is increasing and decreasing. As examples, we graph $r = 1 + \cos(\theta/2)$ and $r^2 = \sin 2\theta$.

Example 3 Graph the curve

$$r = 1 + \cos\frac{\theta}{2}.$$

Solution We first graph r as a function of θ in the Cartesian $r\theta$-plane. Since the cosine has period 2π, we must let θ run from 0 to 4π to produce the entire graph (Fig. 9.61a). The arrows from the θ-axis to the curve give the radius vectors for graphing $r = 1 + \cos(\theta/2)$ in the polar plane (Fig. 9.61b).

Example 4 *A Lemniscate.* Graph the curve $r^2 = \sin 2\theta$.

Solution Here we begin by plotting r^2 (not r) as a function of θ in the Cartesian $r^2\theta$-plane, treating r^2 as a variable that may have negative as well as positive values (Fig. 9.62a). We pass from there to the graph of $r = \pm\sqrt{\sin 2\theta}$ in the $r\theta$-plane (Fig. 9.62b) and then draw the polar graph (Fig. 9.62c). The graph in Fig. 9.62(b) "covers" the final polar graph in Fig. 9.62(c) twice. We could have managed with either loop alone, with the two upper halves, or with the two lower halves. The double covering does no harm, however, and we actually learn a little more about the behavior of the function this way.

Finding the Points Where Curves Intersect

The fact that we can represent a point in different ways in polar coordinates makes extra care necessary in deciding when a point lies on the graph of a polar equation and in determining the points at which the graphs of polar equations intersect. The problem is that a point of intersection may satisfy the equation of one curve with polar coordinates that are different from the ones with which it satisfies the equation of another curve. Thus, solving the equations of two curves simultaneously may not identify all their points of intersection. The only sure way to identify all the points of intersection is to graph the equations.

Example 5 Show that the point $(2, \pi/2)$ lies on the curve $r = 2\cos 2\theta$.

Solution It may seem at first that the point $(2, \pi/2)$ does not lie on the curve because substituting the given coordinates into the equation gives

$$2 = 2\cos 2\left(\frac{\pi}{2}\right) = 2\cos\pi = -2,$$

How to Use Cartesian Graphs to Draw Polar Graphs

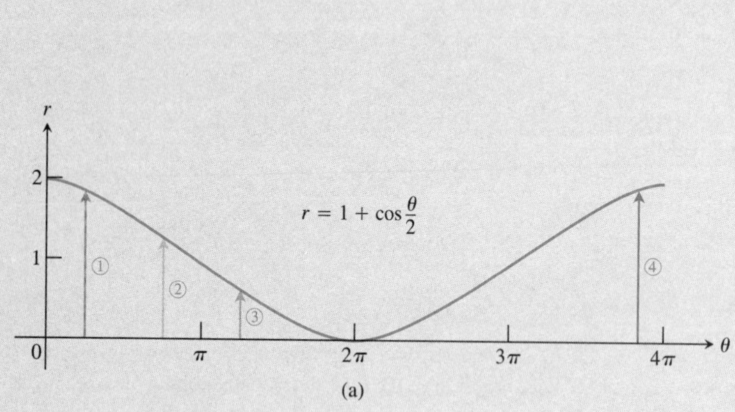

(a)

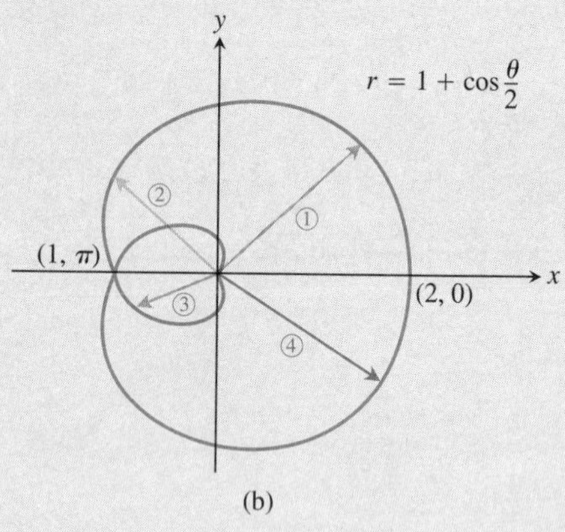

(b)

9.61 (a) The graph of $r = 1 + \cos(\theta/2)$ in the Cartesian $r\theta$-plane gives us radius vectors from the θ-axis to the curve. (b) The radius vectors help us draw the graph in the polar $r\theta$-plane.

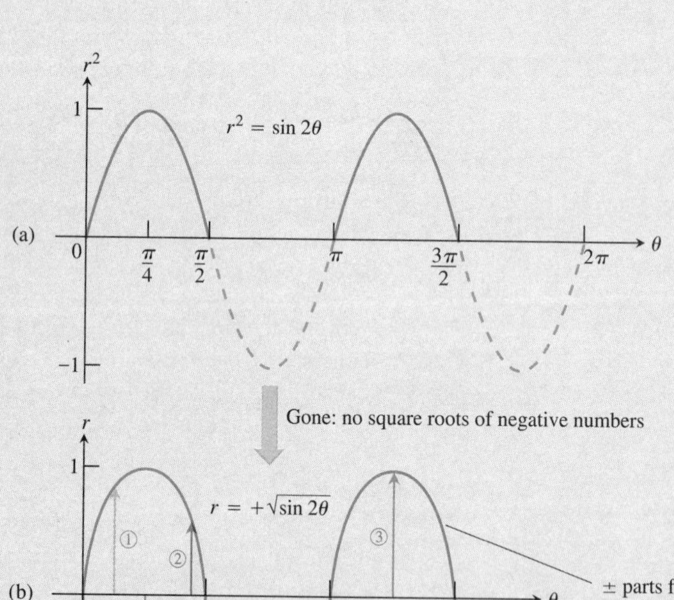

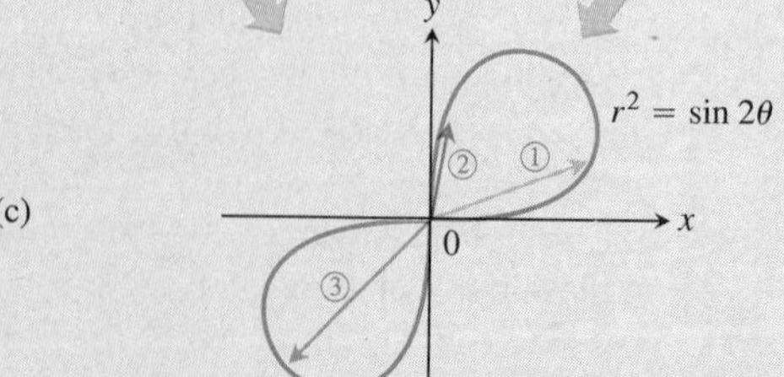

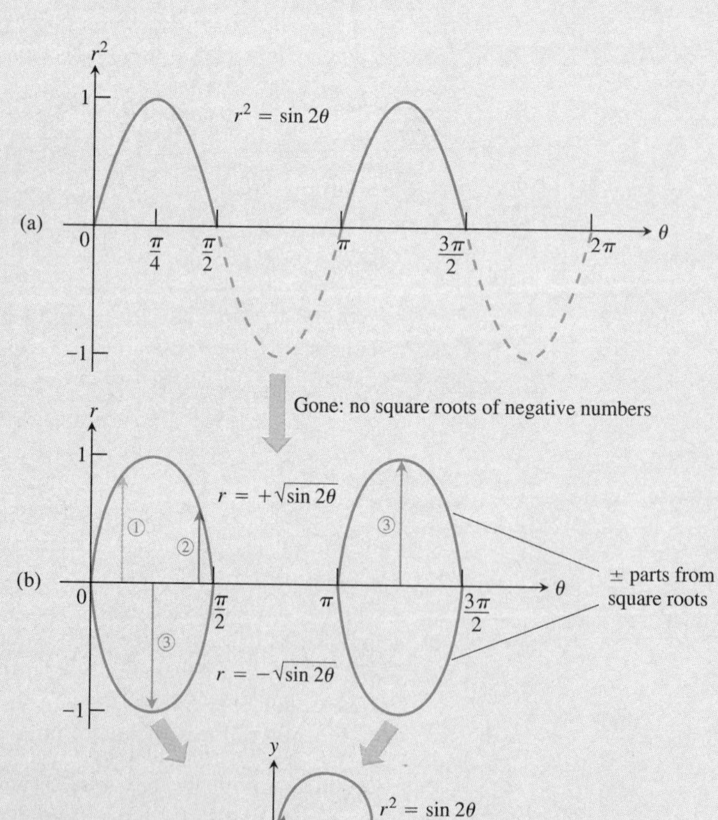

9.62 (a) The graph of $r^2 = \sin 2\theta$ in the Cartesian $r^2\theta$-plane includes negative values of the dependent variable r^2 as well as positive values. (b) When we graph r vs. θ in the Cartesian $r\theta$-plane, we ignore the points where r is imaginary but plot + and − parts from the points where r^2 is positive. (c) In the polar $r\theta$-plane, the radius vectors from the previous sketch cover the final graph twice.

which is not a true equality. The magnitude is right, but the sign is wrong. This suggests looking for a pair of coordinates for the given point in which r is negative, for example,

$$\left(-2, -\frac{\pi}{2}\right).$$

When we try these in the equation $r = 2 \cos 2\theta$, we find

$$-2 = 2 \cos 2\left(-\frac{\pi}{2}\right) = 2(-1) = -2,$$

and the equation is satisfied. The point $(2, \pi/2)$ does lie on the curve.

Example 6 Find the points of intersection of the curves

$$r^2 = 4 \cos \theta \qquad \text{and} \qquad r = 1 - \cos \theta.$$

Solution In Cartesian coordinates, we can always find the points where two curves cross by solving their equations simultaneously. In polar coordinates, the story is different. Simultaneous solution may reveal some intersection points without revealing others. In this example, simultaneous solution reveals only two of the four intersection points. The others may be found by graphing. (Also, see Exercise 33.)

If we substitute $\cos \theta = r^2/4$ in the equation $r = 1 - \cos \theta$, we get

$$r = 1 - \cos \theta = 1 - \frac{r^2}{4}$$

$$4r = 4 - r^2$$

$$r^2 + 4r - 4 = 0$$

$$r = -2 \pm 2\sqrt{2}. \qquad \text{(Quadratic Formula)}$$

The value $r = -2 - 2\sqrt{2}$ has too large an absolute value to belong to either curve. The values of θ corresponding to $r = -2 + 2\sqrt{2}$ are

$$\begin{aligned} \theta &= \cos^{-1}(1 - r) && \text{(From } r = 1 - \cos \theta) \\ &= \cos^{-1}(1 - (2\sqrt{2} - 2)) && \text{(Set } r = 2\sqrt{2} - 2) \\ &= \cos^{-1}(3 - 2\sqrt{2}) \\ &= \pm 80^\circ. && \text{(With a calculator, rounded)} \end{aligned}$$

We have thus identified two intersection points:

$$(r, \theta) = (2\sqrt{2} - 2, \pm 80^\circ).$$

If we graph the equations $r^2 = 4 \cos \theta$ and $r = 1 - \cos \theta$ together (Fig. 9.63), as we can now do by combining the graphs in Figs. 9.59 and 9.60, we see that the curves also intersect at the origin and at the point $(2, \pi)$. Why weren't the r-values of these points revealed by the simultaneous solution? The answer is that the points $(0, 0)$ and $(2, \pi)$ are not on the curves "simultaneously." They are not reached at the same value of θ. On the curve $r = 1 - \cos \theta$, the point $(2, \pi)$ is reached when $\theta = \pi$. On the curve $r^2 = 4 \cos \theta$, it is reached when $\theta = 0$, where it is identified not by the coordinates $(2, \pi)$, which do not satisfy the equation, but by the coordi-

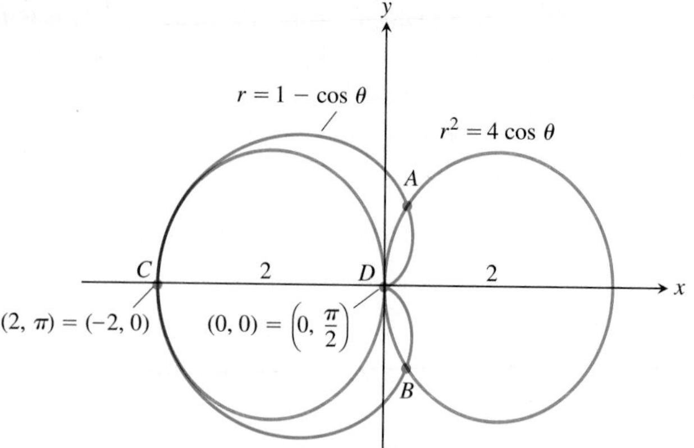

9.63 The four points of intersection of the curves $r = 1 - \cos\theta$ and $r^2 = 4\cos\theta$ (Example 6). Only A and B were found by simultaneous solution. The other two were disclosed by graphing.

nates $(-2, 0)$, which do. Similarly, the cardioid reaches the origin when $\theta = 0$, but the curve $r^2 = 4\cos\theta$ reaches the origin when $\theta = \pi/2$.

EXERCISES 9.6

Sketch the curves in Exercises 1–10.

1. $r = 1 + \cos\theta$

2. $r = 2 - 2\cos\theta$

3. $r = 1 - \sin\theta$

4. $r = 1 + \sin\theta$

5. $r = 2 + \sin\theta$

6. $r = 1 + 2\sin\theta$

7. $r^2 = 4\cos 2\theta$

8. $r^2 = 4\sin\theta$

9. $r = \theta$

10. $r = \sin(\theta/2)$

Use Eq. (2) to find the slopes of the curves in Exercises 11–14 at the given points. Sketch the curves along with their tangents at these points.

11. *Cardioid.* $r = -1 + \cos\theta$; $\theta = \pm\pi/2$

12. *Cardioid.* $r = -1 + \sin\theta$; $\theta = 0, \pi/2, \pi$

13. *Four-leaved rose.* $r = \sin 2\theta$; $\theta = \pm\pi/4, \pm 3\pi/4$, and the values of θ at which the curve passes through the origin

14. *Four-leaved rose.* $r = \cos 2\theta$; $\theta = 0, \pm\pi/2, \pi$, and the values of θ at which the curve passes through the origin

Graph the lemniscates in Exercises 15 and 16.

15. $r^2 = 4\cos 2\theta$

16. $r^2 = 4\sin 2\theta$

Graph the limaçons in Exercises 17–20. Limaçon ("*lee*-ma-sahn") is Old French for "snail." You will see why the name is appropriate when you graph the limaçons in Exercise 17. Equations for limaçons have the form $r = a \pm b\cos\theta$ or $r = a \pm b\sin\theta$. There are four basic shapes.

17. *Limaçons with an inner loop*

a) $r = \dfrac{1}{2} + \cos\theta$

b) $r = \dfrac{1}{2} + \sin\theta$

18. *Cardioids*

a) $r = 1 - \cos\theta$

b) $r = -1 + \sin\theta$

19. *Dimpled limaçons*

a) $r = \dfrac{3}{2} + \cos\theta$

b) $r = \dfrac{3}{2} - \sin\theta$

20. *Cardioid-like limaçons*

a) $r = 2 + \cos\theta$

b) $r = -2 + \sin\theta$

21. Sketch the region defined by the inequality $0 \le r \le 2 - 2\cos\theta$.

22. Sketch the region defined by the inequality $0 \le r^2 \le \cos\theta$.

23. Show that the point $(2, 3\pi/4)$ lies on the curve $r = 2\sin 2\theta$.

24. Show that the point $(1/2, 3\pi/2)$ lies on the curve $r = -\sin(\theta/3)$.

Find the points of intersection of the pairs of curves in Exercises 25–32.

25. $r = 1 + \cos\theta$, $r = 1 - \cos\theta$

26. $r = 1 + \sin\theta$, $r = 1 - \sin\theta$

27. $r = 1 - \sin\theta$, $r^2 = 4\sin\theta$

28. $r^2 = \sqrt{2}\sin\theta$, $r^2 = \sqrt{2}\cos\theta$

29. $r = 1$, $r = 2\sin 2\theta$

30. $r = 1$, $r^2 = 2\sin 2\theta$

31. $r = \sqrt{2}\cos 2\theta$, $r = \sqrt{2}\sin 2\theta$

32. $r^2 = \sqrt{2}\cos 2\theta$, $r^2 = \sqrt{2}\sin 2\theta$

33. *Continuation of Example 6.* The simultaneous solution of the equations

$$r^2 = 4\cos\theta, \quad (5)$$

$$r = 1 - \cos\theta, \quad (6)$$

in the text did not reveal the points $(0, 0)$ and $(2, \pi)$ in which their graphs intersected.

a) We could have found the point $(2, \pi)$, however, by replacing the (r, θ) in Eq. (5) by the equivalent $(-r, \theta + \pi)$ to obtain

$$\begin{aligned} r^2 &= 4 \cos \theta \\ (-r)^2 &= 4 \cos(\theta + \pi) \\ r^2 &= -4 \cos \theta . \end{aligned} \tag{7}$$

Solve Eqs. (6) and (7) simultaneously to show that $(2, \pi)$ is a common solution. (This will still not reveal that the graphs intersect at $(0, 0)$.)

b) The origin is still a special case. (It often is.) Here is one way to handle it: Set $r = 0$ in Eqs. (5) and (6) and solve each equation for a corresponding value of θ. Since $(0, \theta)$ is the origin for *any* θ, this will show that both curves pass through the origin even if they do so for different θ-values.

34. Show that a curve with any two of the symmetries listed at the beginning of the section automatically has the third.

Computer Grapher

Find the points of intersection of the pairs of curves in Exercises 35–38.

35. $r^2 = \sin 2\theta, \quad r^2 = \cos 2\theta$

36. $r = 1 + \cos \dfrac{\theta}{2}, \quad r = 1 - \sin \dfrac{\theta}{2}$

37. $r = 1, \quad r = 2 \sin 2\theta$

38. $r = 1, \quad r^2 = 2 \sin 2\theta$

39. *A rose within a rose.* Graph the equation $r = 1 - 2 \sin 3\theta$.

40. Graph the *nephroid of Freeth:*

$$r = 1 + 2 \sin \frac{\theta}{2}$$

41. Graph the *roses* $r = \cos m\, \theta$ for $m = 1/3, 2, 3,$ and 7.

42. *Spirals.* Polar coordinates are just the thing for defining spirals. Graph the following spirals:

a) *A logarithmic spiral:* $r = e^{\theta/10}$

b) *A hyperbolic spiral:* $r = 8/\theta$

c) *Equilateral hyperbola:* $r = \pm 10/\sqrt{\theta}$ (Try using different colors for the two branches.)

EXPLORER PROGRAM

PowerGrapher	Graphs equations of the form $r = f(\theta)$ singly or together in a common display

9.7 Polar Equations for Conic Sections

Polar coordinates are important in astronomy and astronautical engineering because the ellipses, parabolas, and hyperbolas along which satellites, moons, planets, and comets move can all be described with one general polar equation. In Cartesian coordinates, the equations of conics have different forms, but not so here. This section develops the general equation, along with special equations for lines and circles.

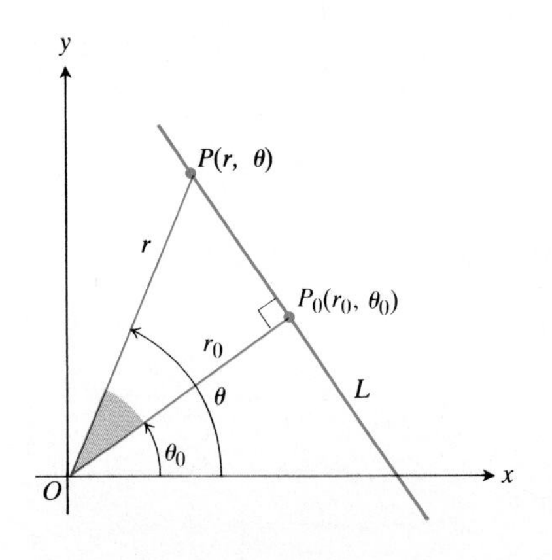

9.64 We can obtain a polar equation for line L by reading the relation $r_0/r = \cos(\theta - \theta_0)$ from triangle OP_0P.

Lines

Suppose the perpendicular from the origin to line L meets L at the point $P_0(r_0, \theta_0)$, with $r_0 \geq 0$ (Fig. 9.64). Then, if $P(r, \theta)$ is any other point on L, the points P, P_0, and O are the vertices of a right triangle, from which we can establish the trigonometric relation

$$\frac{r_0}{r} = \cos(\theta - \theta_0) \tag{1}$$

or

$$r \cos(\theta - \theta_0) = r_0. \tag{2}$$

The coordinates of P_0 satisfy this equation as well.

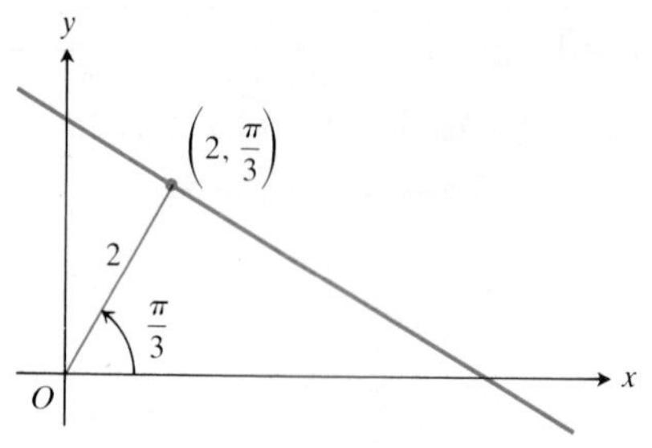

9.65 The standard polar equation for the line shown here is

$$r \cos\left(\theta - \frac{\pi}{3}\right) = 2$$

(Example 1).

The Standard Polar Equation for Lines

If the point $P_0(r_0, \theta_0)$ is the foot of the perpendicular from the origin to the line L, and $r_0 \geq 0$, then an equation for L is

$$r \cos(\theta - \theta_0) = r_0. \tag{3}$$

Example 1 Use the identity $\cos(A - B) = \cos A \cos B + \sin A \sin B$ to find a Cartesian equation for the line in Fig. 9.65.

Solution

$$r \cos\left(\theta - \frac{\pi}{3}\right) = 2$$

$$r\left(\cos\theta \cos\frac{\pi}{3} + \sin\theta \sin\frac{\pi}{3}\right) = 2$$

$$\frac{1}{2} r \cos\theta + \frac{\sqrt{3}}{2} r \sin\theta = 2$$

$$\frac{1}{2} x + \frac{\sqrt{3}}{2} y = 2$$

$$x + \sqrt{3} y = 4$$

Circles

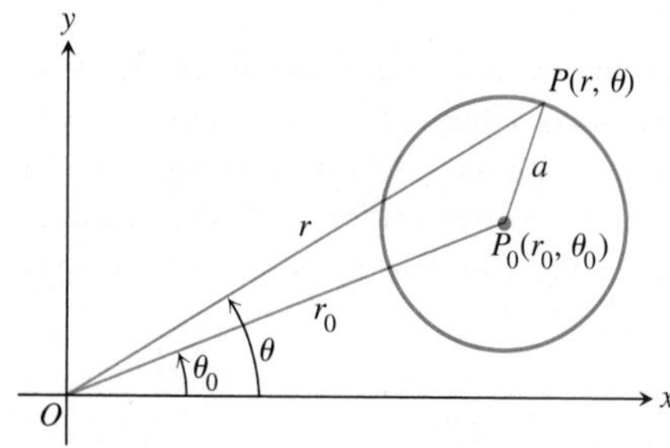

9.66 We can get a polar equation for this circle by applying the Law of Cosines to triangle OP_0P.

To find a polar equation for the circle of radius a centered at $P_0(r_0, \theta_0)$, we let $P(r, \theta)$ be a point on the circle and apply the Law of Cosines to triangle OP_0P (Fig. 9.66). This gives

$$a^2 = r_0^2 + r^2 - 2r_0 r \cos(\theta - \theta_0). \tag{4}$$

If the circle passes through the origin, then $r_0 = a$ and Eq. (4) simplifies somewhat:

$$a^2 = a^2 + r^2 - 2ar \cos(\theta - \theta_0) \qquad \text{(Eq.(4) with } r_0 = a\text{)}$$

$$r^2 = 2ar \cos(\theta - \theta_0)$$

$$r = 2a \cos(\theta - \theta_0). \tag{5}$$

If the circle's center lies on the positive x-axis, so that $\theta_0 = 0$, Eq. (5) becomes

$$r = 2a \cos\theta. \tag{6}$$

If, instead, the circle's center lies on the positive y-axis, $\theta = \pi/2$, $\cos(\theta - \pi/2) = \sin\theta$, and Eq. (5) becomes

$$r = 2a \sin\theta. \tag{7}$$

We obtain equations for circles through the origin centered on the negative x- and y-axes from Eqs. (6) and (7) by replacing r with $-r$.

Polar Equations for Circles Through the Origin Centered on the x- and y-Axes, Radius a

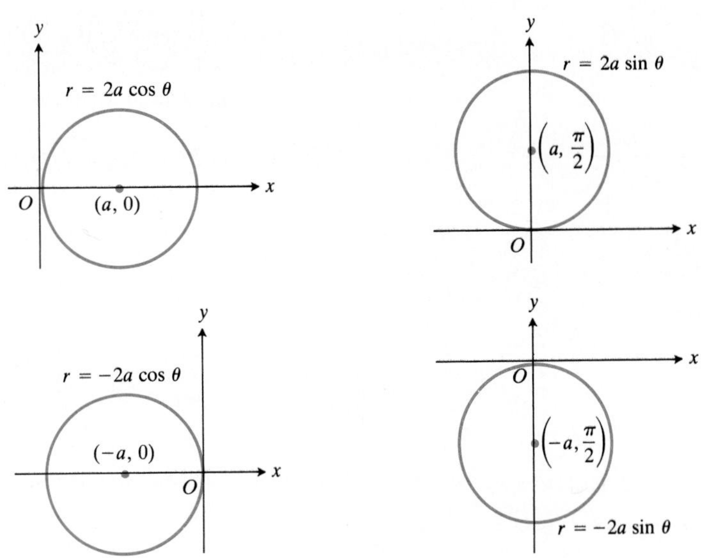

Example 2 *Circles Through the Origin*

Radius	Center (polar coordinates)	Equation
3	$(3, 0)$	$r = 6\cos\theta$
2	$(2, \pi/2)$	$r = 4\sin\theta$
1/2	$(-1/2, 0)$	$r = -\cos\theta$
1	$(-1, \pi/2)$	$r = -2\sin\theta$

9.67 If a conic section is put in this position, then $PF = r$ and $PD = k - r\cos\theta$.

Ellipses, Parabolas, and Hyperbolas

To find polar equations for ellipses, parabolas, and hyperbolas, we first assume that the conic has one focus at the origin (for the parabola, its only focus) and that the corresponding directrix is the vertical line $x = k$ lying to the right of the origin (Fig. 9.67). This makes

$$PF = r \tag{8}$$

and

$$PD = k - FB = k - r\cos\theta. \tag{9}$$

The conic's focus–directrix equation $PF = e \cdot PD$ then becomes

$$r = e(k - r\cos\theta), \tag{10}$$

TABLE 9.4
Equations for conic sections ($e > 0$)

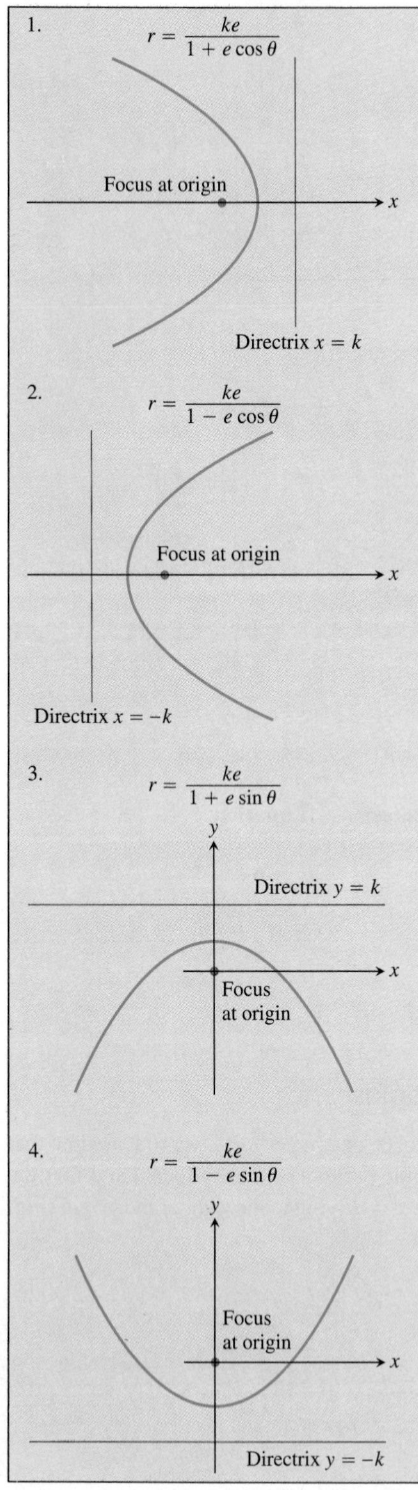

which can be solved for r to obtain

$$r = \frac{ke}{1 + e\cos\theta}. \tag{11}$$

This equation represents an ellipse if $0 < e < 1$, a parabola if $e = 1$, and a hyperbola if $e > 1$. And there we have it—ellipses, parabolas, and hyperbolas all with the same basic coordinate equation.

Example 3 *Typical Conics from Eq. (11)*

$e = \frac{1}{2}$: ellipse $r = \frac{k}{2 + \cos\theta}$

$e = 1$: parabola $r = \frac{k}{1 + \cos\theta}$

$e = 2$: hyperbola $r = \frac{2k}{1 + 2\cos\theta}$

From time to time, you may see variations of Eq. (11) that come from relocating the directrix. If the directrix is the line $x = -k$ to the left of the origin (the origin is still the focus), the equation you get in place of Eq. (11) is

$$r = \frac{ke}{1 - e\cos\theta}. \tag{12}$$

The denominator now has a $(-)$ instead of a $(+)$. If the directrix is either of the lines $y = k$ or $y = -k$, the equations have sines in them instead of cosines, as shown in Table 9.4.

Example 4 Find an equation for the hyperbola with eccentricity 3/2, directrix $x = 2$, and focus at the origin.

Solution We use Eq. (1) in Table 9.4 with $k = 2$ and $e = 3/2$ to get

$$r = \frac{2(3/2)}{1 + (3/2)\cos\theta}, \quad \text{or} \quad r = \frac{6}{2 + 3\cos\theta}.$$

Example 5 Find the directrix of the parabola

$$r = \frac{25}{10 + 10\cos\theta}.$$

Solution We divide the numerator and denominator by 10 to put the equation in standard form:

$$r = \frac{5/2}{1 + \cos\theta}.$$

This is the equation

$$r = \frac{ke}{1 + e\cos\theta}$$

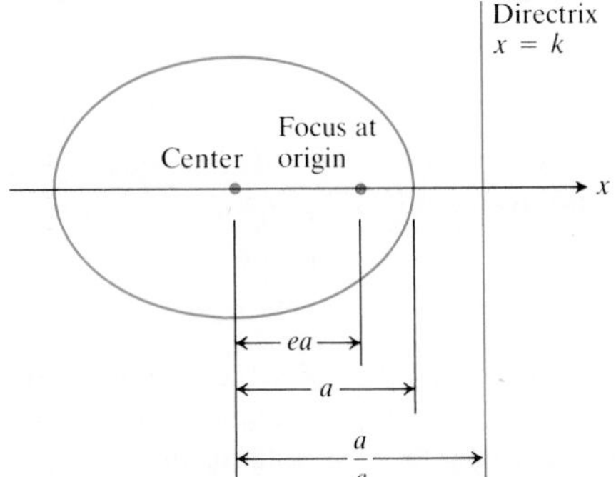

9.68 In an ellipse with semimajor axis a, the focus–directrix distance is $k = (a/e) - ea$, so $ke = a(1 - e^2)$.

with $k = 5/2$ and $e = 1$. The equation of the directrix is $x = 5/2$.

From the ellipse diagram in Fig. 9.68, we see that k is related to the eccentricity e and the semimajor axis a by the equation

$$k = \frac{a}{e} - ea. \tag{13}$$

From this, we find that $ke = a(1 - e^2)$. Replacing ke by $a(1 - e^2)$ in Eq. (11) gives the standard polar equation for an ellipse with eccentricity e and semimajor axis a.

Ellipse with Eccentricity *e* and Semimajor Axis *a*

$$r = \frac{a(1 - e^2)}{1 + e \cos \theta} \tag{14}$$

Notice that when $e = 0$, Eq. (14) becomes $r = a$, which represents a circle.

Equation (14) is the starting point for calculating planetary orbits in astronomy.

Example 6 Find a polar equation for an ellipse with semimajor axis 39.44 AU (astronomical units) and eccentricity 0.25. This is the approximate size of Pluto's orbit around the sun.

Solution We use Eq. (14) with $a = 39.44$ and $e = 0.25$ to find

$$r = \frac{39.44(1 - (0.25)^2)}{1 + 0.25 \cos \theta}$$

$$= \frac{147.9}{4 + \cos \theta}.$$

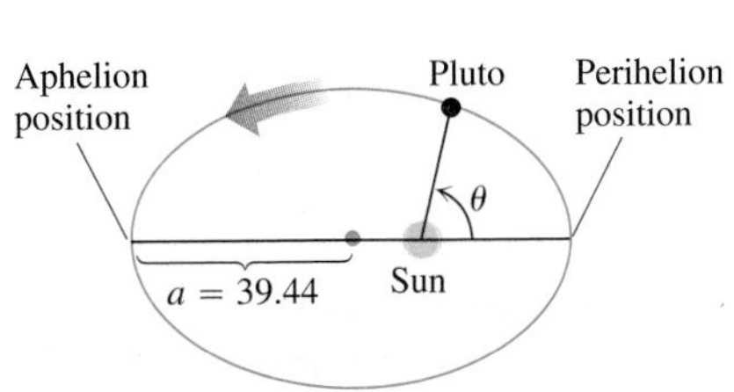

9.69 The orbit of Pluto (Example 6).

At its point of closest approach (perihelion), Pluto is

$$r_{\min} = \frac{147.9}{4 + 1} = 29.58 \text{ AU}$$

from the sun. At its most distant point (aphelion), Pluto is

$$r_{\max} = \frac{147.9}{4 - 1} = 49.3 \text{ AU}$$

from the sun (Fig. 9.69).

Example 7 Find the distance from one focus of the ellipse in Example 6 to the associated directrix.

Solution We use Eq. (13) with $a = 39.44$ and $e = 0.25$ to find

$$k = 39.44\left(\frac{1}{0.25} - 0.25\right) = 147.9 \text{ AU}.$$

EXERCISES 9.7

Find polar and Cartesian equations for the lines in Exercises 1–4.

1.

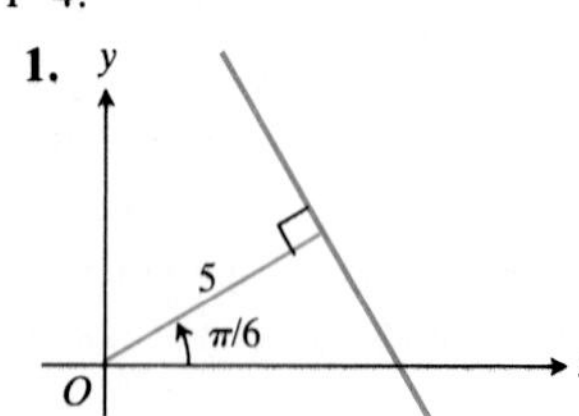

2.

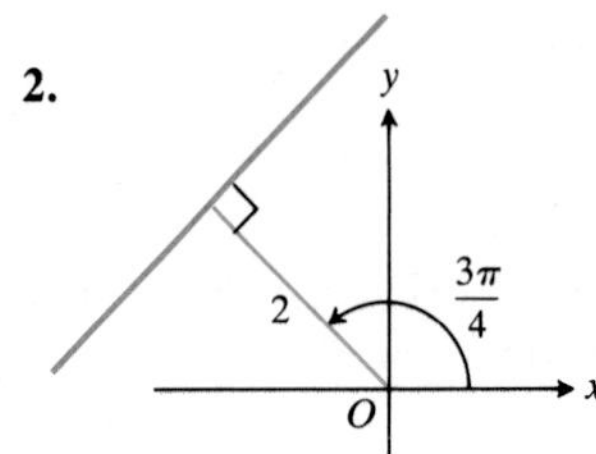

3. **4.**

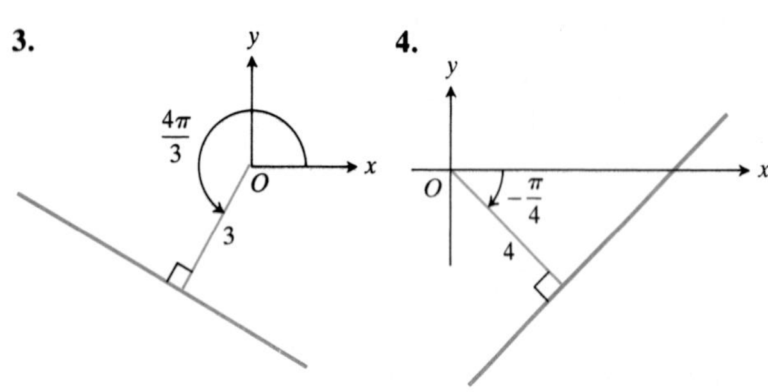

Sketch the lines in Exercises 5–8 and find Cartesian equations for them.

5. $r \cos\left(\theta - \frac{\pi}{4}\right) = \sqrt{2}$

6. $r \cos\left(\theta - \frac{2\pi}{3}\right) = 3$

7. $r \cos\left(\theta - \frac{3\pi}{2}\right) = 1$

8. $r \cos\left(\theta + \frac{\pi}{3}\right) = 2$

Find polar equations for the circles in Exercises 9–12.

9.

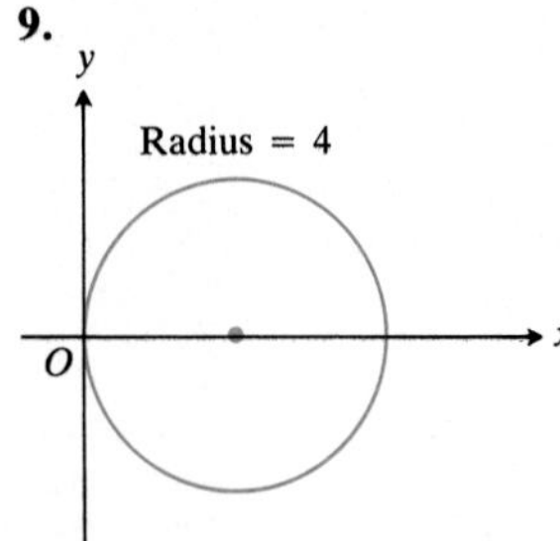

10.

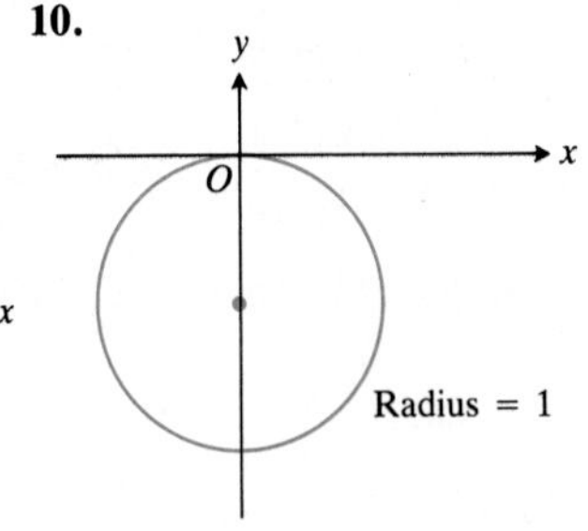

11.

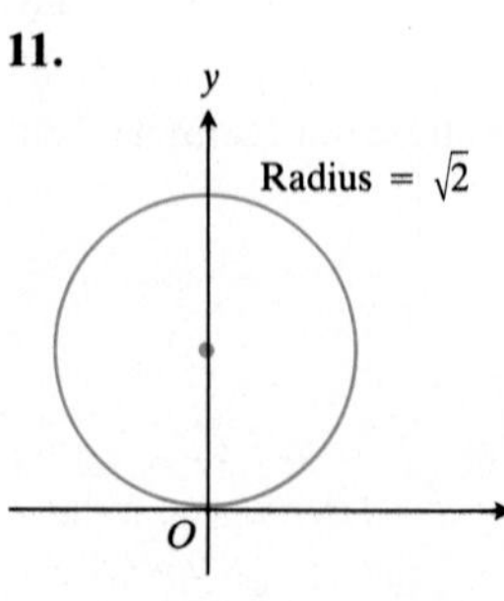

12.

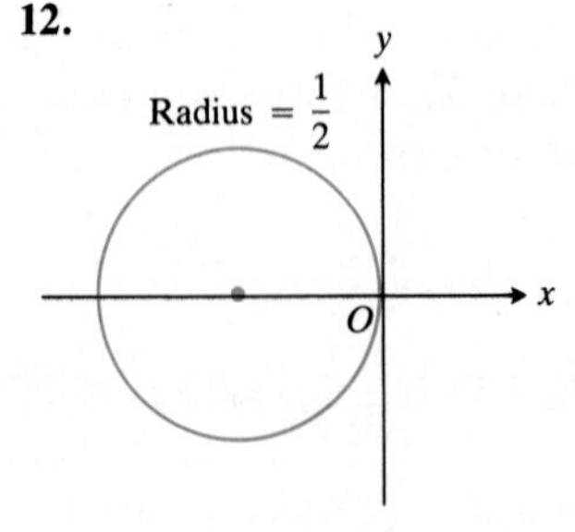

Sketch the circles in Exercises 13–16. Give polar coordinates for their centers and identify their radii.

13. $r = 4 \cos \theta$

14. $r = 6 \sin \theta$

15. $r = -2 \cos \theta$

16. $r = -8 \sin \theta$

Exercises 17–24 give the eccentricities of conic sections with one focus at the origin, along with the directrix corresponding to that focus. Find a polar equation for each conic section.

17. $e = 1, \quad x = 2$

18. $e = 1, \quad y = 2$

19. $e = 2, \quad x = 4$

20. $e = 5, \quad y = -6$

21. $e = 1/2, \quad x = 1$

22. $e = 1/4, \quad x = -2$

23. $e = 1/5, \quad y = -10$

24. $e = 1/3, \quad y = 6$

Sketch the parabolas and ellipses in Exercises 25–32. Include the directrix that corresponds to the focus at the origin. Label the vertices with appropriate polar coordinates. Label the centers of the ellipses as well.

25. $r = \dfrac{1}{1 + \cos \theta}$

26. $r = \dfrac{6}{2 + \cos \theta}$

27. $r = \dfrac{25}{10 - 5 \cos \theta}$

28. $r = \dfrac{4}{2 - 2 \cos \theta}$

29. $r = \dfrac{400}{16 + 8 \sin \theta}$

30. $r = \dfrac{12}{3 + 3 \sin \theta}$

31. $r = \dfrac{8}{2 - 2 \sin \theta}$

32. $r = \dfrac{4}{2 - \sin \theta}$

Sketch the regions defined by the inequalities in Exercises 33 and 34.

33. $0 \le r \le 2 \cos \theta$

34. $-3 \cos \theta \le r \le 0$

35. *Perihelion and aphelion.* Suppose that a planet travels about its sun in an ellipse whose semimajor axis has length a.

a) Show that $r = a(1 - e)$ when the planet is closest to the sun and that $r = a(1 + e)$ when the planet is farthest from the sun.

b) Use the data in Table 9.5 to find how close each planet in our solar system comes to the sun and how far away each planet gets from the sun.

36. *Planetary orbits.* In Example 6, we found a polar equation for the orbit of Pluto. Use the data in Table 9.5 to find polar equations for the orbits of the other planets.

37. a) Find Cartesian equations for the curves $r = 2 \sin \theta$ and $r = \csc \theta$.

b) Sketch the curves together and label their points of intersection in both Cartesian and polar coordinates.

38. Repeat Exercise 37 for $r = 2 \cos \theta$ and $r = \sec \theta$.

39. Find a polar equation for the parabola whose focus lies at the origin and whose directrix is the line $r \cos \theta = 4$.

TABLE 9.5
Semimajor axes and eccentricities of the planets in our solar system

Planet	Semimajor axis (astronomical units)	Eccentricity
Mercury	0.3871	0.2056
Venus	0.7233	0.0068
Earth	1.000	0.0167
Mars	1.524	0.0934
Jupiter	5.203	0.0484
Saturn	9.539	0.0543
Uranus	19.18	0.0460
Neptune	30.06	0.0082
Pluto	39.44	0.2481

40. Find a polar equation for the parabola whose focus lies at the origin and whose directrix is the line $r\cos(\theta - \pi/2) = 2$.

41. *The space engineer's formula for eccentricity.* The space engineer's formula for the eccentricity of an elliptical orbit is

$$e = \frac{r_{\max} - r_{\min}}{r_{\max} + r_{\min}},$$

where r is the distance from the space vehicle to the focus of the ellipse along which it travels. Why does the formula work? (*Hint:* You do not need calculus. Just think about the geometric description of eccentricity.)

42. *Halley's comet (Continuation of Example 7, Section 9.1)*

a) Write an equation for the orbit of Halley's comet in a coordinate system in which the sun lies at the origin and the other focus lies on the negative x-axis, scaled in astronomical units.

b) How close does the comet come to the sun in astronomical units? In miles?

c) What is the farthest the comet gets from the sun in astronomical units? In miles?

43. *Drawing ellipses with string (Continuation of Exercise 41).* You have a string with a knot in each end that can be pinned to a drawing board. The string is 10 in. long from the center of one knot to the center of the other. How far apart should the pins be to use the method illustrated in Fig. 9.2 (Section 9.1) to draw an ellipse of eccentricity 0.2? The resulting ellipse would resemble the orbit of Mercury.

44. a) Show that the equations $x = r\cos\theta$, $y = r\sin\theta$ transform the polar equation

$$r = \frac{k}{1 + e\cos\theta}$$

into the Cartesian equation

$$(1 - e^2)x^2 + y^2 + 2k\,ex - k^2 = 0.$$

b) Then apply the criteria of Section 9.2 to show that

$$\begin{aligned} e = 0 \quad &\Rightarrow \quad \text{circle} \\ 0 < e < 1 \quad &\Rightarrow \quad \text{ellipse} \\ e = 1 \quad &\Rightarrow \quad \text{parabola} \\ e > 1 \quad &\Rightarrow \quad \text{hyperbola.} \end{aligned}$$

Computer Grapher

Graph the lines and conic sections in Exercises 45–54.

45. $r = 3\sec(\theta - \pi/3)$
46. $r = 4\sec(\theta + \pi/6)$
47. $r = 4\sin\theta$
48. $r = -2\cos\theta$
49. $r = 8/(4 + \cos\theta)$
50. $r = 8/(4 + \sin\theta)$
51. $r = 1/(1 - \sin\theta)$
52. $r = 1/(1 + \cos\theta)$
53. $r = 1/(1 + 2\sin\theta)$
54. $r = 1/(1 + 2\cos\theta)$

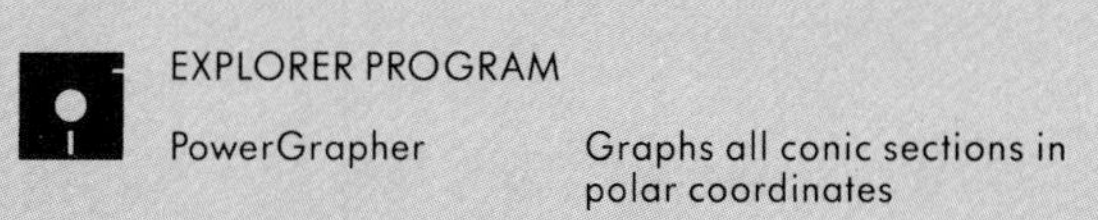

EXPLORER PROGRAM
PowerGrapher — Graphs all conic sections in polar coordinates

9.8 Integration in Polar Coordinates

This section shows how to calculate areas of plane regions, lengths of curves, and areas of surfaces of revolution in polar coordinates. The general methods for setting up the integrals are the same as for Cartesian coordinates but the resulting formulas are somewhat different. Formulas for finding the centroids of fan-shaped regions are derived at the end of the exercises.

Area in the Plane

The region TOS in Fig. 9.70 is bounded by the rays $\theta = \alpha$ and $\theta = \beta$ and the curve $r = f(\theta)$. We partition angle TOS into n subangles of measure $\Delta\theta$ and approximate a typical sector POQ by a circular sector of radius $r_k = f(\theta_k)$ and central angle $\Delta\theta$

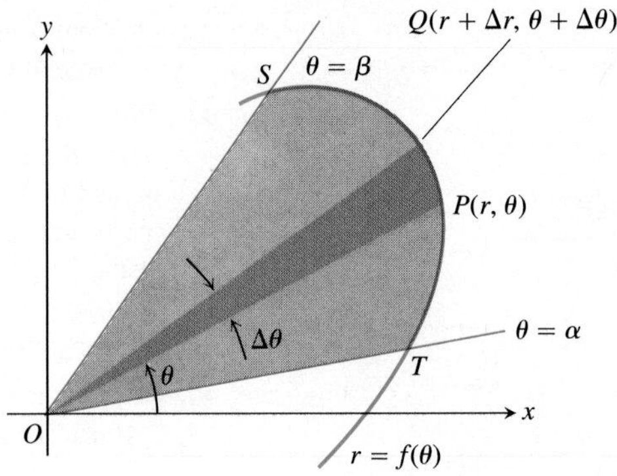

9.70 To derive a formula for the area swept out by the radius OP as P moves from T to S along the curve $r = f(\theta)$, we partition region TOS into sectors.

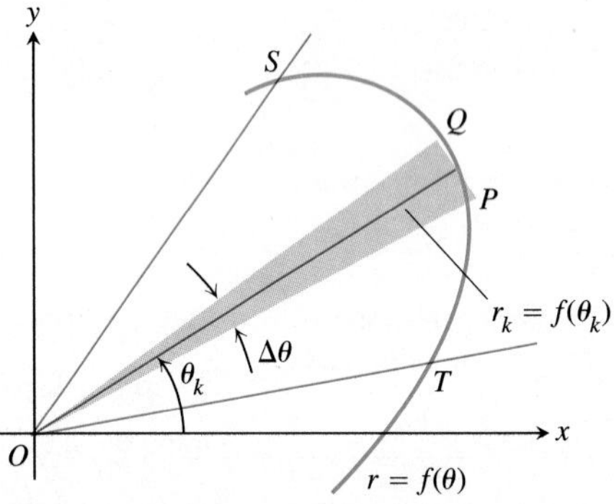

9.71 For some θ_k between θ and $\theta + \Delta\theta$, the area of the shaded circular sector just equals the area of the sector POQ bounded by the curve shown in Fig. 9.70.

(Fig. 9.71). If f is continuous for $\alpha \le \theta \le \beta$, we can choose θ_k to make the area $(1/2)r_k^2\,\Delta\theta$ of the circular sector equal the area of sector POQ. If we choose such a radius for each of the n sectors determined by the partition of angle TOS, the area of region TOS will equal the sum

$$A = \sum_{k=1}^{n} \frac{1}{2} r_k^2\,\Delta\theta = \sum_{k=1}^{n} \frac{1}{2}\left(f(\theta_k)\right)^2 \Delta\theta.$$

Letting $\Delta\theta \to 0$ then shows that

$$A = \lim_{\Delta\theta \to 0} \sum \frac{1}{2}(f(\theta_k))^2\,\Delta\theta = \int_\alpha^\beta \frac{1}{2} f^2(\theta)\,d\theta = \int_\alpha^\beta \frac{1}{2} r^2\,d\theta.$$

Area of the Fan-shaped Region Between the Origin and the Curve $r = f(\theta)$, $\alpha \le \theta \le \beta$

$$A = \int_\alpha^\beta \frac{1}{2} r^2\,d\theta. \qquad (1)$$

This is the integral of the **area differential** (Fig. 9.72)

$$dA = \frac{1}{2} r^2\,d\theta. \qquad (2)$$

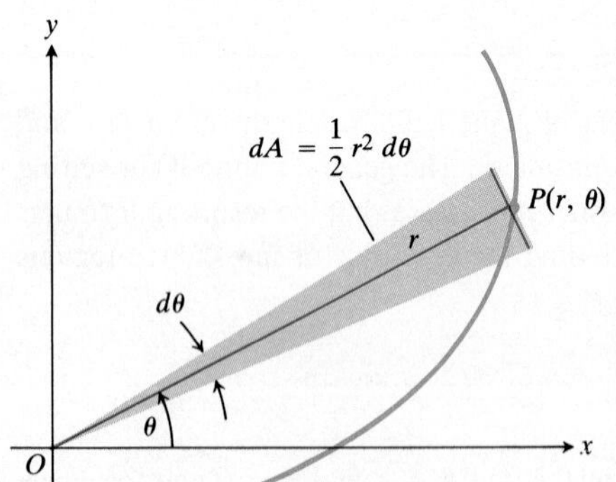

9.72 The area differential dA.

Example 1 Find the area of the region enclosed by the cardioid $r = 2(1 + \cos\theta)$.

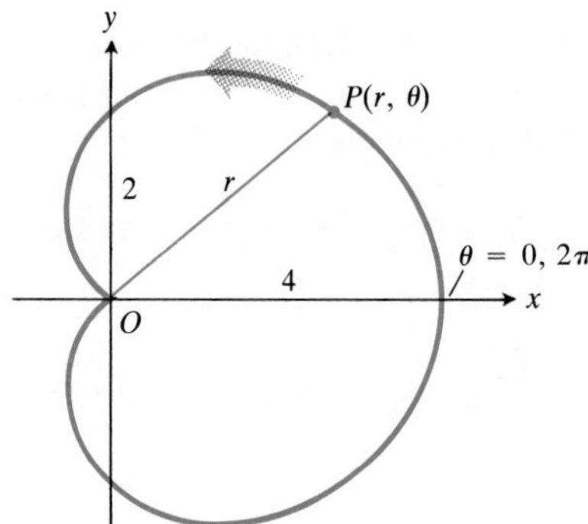

9.73 The cardioid $r = 2(1 + \cos\theta)$ (Example 1).

Solution We graph the cardioid (Fig. 9.73) and determine that the radius OP sweeps out the region exactly once as θ runs from 0 to 2π. The area is therefore

$$\begin{aligned}\int_{\theta=0}^{\theta=2\pi} \frac{1}{2}r^2\,d\theta &= \int_0^{2\pi} \frac{1}{2}\cdot 4(1+\cos\theta)^2\,d\theta\\ &= \int_0^{2\pi} 2(1 + 2\cos\theta + \cos^2\theta)\,d\theta\\ &= \int_0^{2\pi}\left(2 + 4\cos\theta + 2\frac{1+\cos 2\theta}{2}\right)d\theta\\ &= \int_0^{2\pi}(3 + 4\cos\theta + \cos 2\theta)\,d\theta\\ &= 3\theta + 4\sin\theta + \frac{\sin 2\theta}{2}\bigg]_0^{2\pi}\\ &= 6\pi - 0 = 6\pi.\end{aligned}$$

Example 2 Find the area inside the smaller loop of the limaçon

$$r = 2\cos\theta + 1.$$

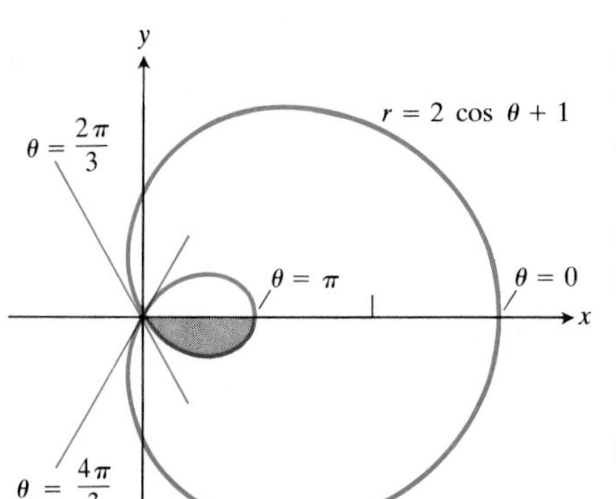

9.74 The limaçon in Example 2.

Solution After sketching the curve (Fig. 9.74), we see that the smaller loop is traced out by the point (r, θ) as θ increases from $\theta = 2\pi/3$ to $\theta = 4\pi/3$. Since the curve is symmetric about the x-axis (the equation is unaltered when we replace θ by $-\theta$), we may calculate the area of the shaded half of the inner loop by integrating from $\theta = 2\pi/3$ to $\theta = \pi$. The area A we seek will be twice the value of the resulting integral:

$$A = 2\int_{2\pi/3}^{\pi}\frac{1}{2}r^2\,d\theta = \int_{2\pi/3}^{\pi} r^2\,d\theta.$$

Since

$$\begin{aligned}r^2 &= (2\cos\theta + 1)^2\\ &= 4\cos^2\theta + 4\cos\theta + 1\\ &= 4\cdot\frac{1+\cos 2\theta}{2} + 4\cos\theta + 1\\ &= 2 + 2\cos 2\theta + 4\cos\theta + 1\\ &= 3 + 2\cos 2\theta + 4\cos\theta,\end{aligned}$$

we have

$$\begin{aligned}A &= \int_{2\pi/3}^{\pi}(3 + 2\cos 2\theta + 4\cos\theta)\,d\theta\\ &= \Big[3\theta + \sin 2\theta + 4\sin\theta\Big]_{2\pi/3}^{\pi}\\ &= (3\pi) - \left(2\pi - \frac{\sqrt{3}}{2} + 4\cdot\frac{\sqrt{3}}{2}\right)\\ &= \pi - \frac{3\sqrt{3}}{2}.\end{aligned}$$

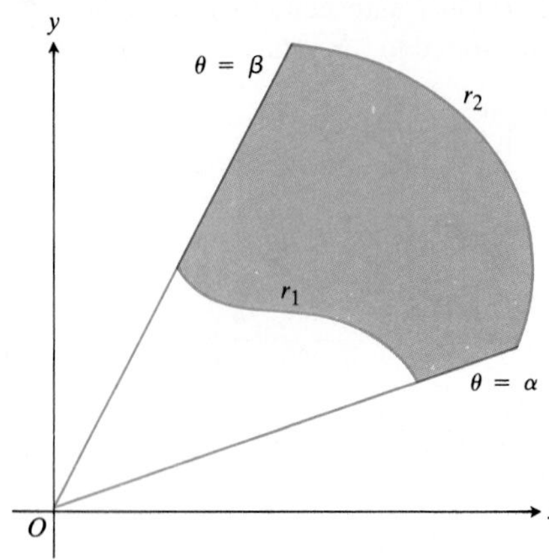

9.75 The area of the shaded region is calculated by subtracting the area of the region between r_1 and the origin from the area of the region between r_2 and the origin.

To find the area of a region like the one in Fig. 9.75, which lies between two polar curves from $\theta = \alpha$ to $\theta = \beta$, we subtract the integral of $(1/2)r_1^2\,d\theta$ from the integral of $(1/2)r_2^2\,d\theta$. This leads to the following formula.

Area of the Region $0 \le r_1(\theta) \le r \le r_2(\theta), \quad \alpha \le \theta \le \beta$

$$A = \int_\alpha^\beta \frac{1}{2} r_2^2\,d\theta - \int_\alpha^\beta \frac{1}{2} r_1^2\,d\theta = \int_\alpha^\beta \frac{1}{2}\left(r_2^2 - r_1^2\right) d\theta \tag{3}$$

Example 3 Find the area of the region that lies inside the circle $r = 1$ and outside the cardioid $r = 1 - \cos\theta$.

Solution We sketch the region to determine its boundaries and find the limits of integration (Fig. 9.76). The outer curve is $r_2 = 1$, the inner curve is $r_1 = 1 - \cos\theta$, and θ runs from $-\pi/2$ to $\pi/2$. The area, from Eq. (3), is

$$\begin{aligned} A &= \int_{-\pi/2}^{\pi/2} \frac{1}{2}\left(r_2^2 - r_1^2\right) d\theta \\ &= 2\int_0^{\pi/2} \frac{1}{2}\left(r_2^2 - r_1^2\right) d\theta \qquad \text{(Symmetry)} \\ &= \int_0^{\pi/2} \left(1 - (1 - 2\cos\theta + \cos^2\theta)\right) d\theta \\ &= \int_0^{\pi/2} \left(2\cos\theta - \cos^2\theta\right) d\theta = \int_0^{\pi/2} \left(2\cos\theta - \frac{\cos 2\theta + 1}{2}\right) d\theta \\ &= \left[2\sin\theta - \frac{\sin 2\theta}{4} - \frac{\theta}{2}\right]_0^{\pi/2} = 2 - \frac{\pi}{4}. \end{aligned}$$

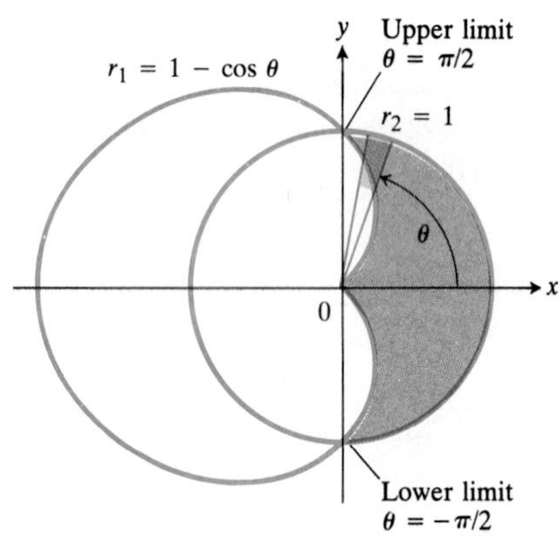

9.76 The region and limits of integration in Example 3.

The Length of a Curve

We calculate the length of a curve $r = f(\theta)$, $\alpha \le \theta \le \beta$, by expressing the differential $ds = \sqrt{dx^2 + dy^2}$ in terms of θ and integrating from α to β. To express ds in terms of θ, we first write dx and dy as

$$\begin{aligned} dx &= d(r\cos\theta) = -r\sin\theta\,d\theta + \cos\theta\,dr, \\ dy &= d(r\sin\theta) = r\cos\theta\,d\theta + \sin\theta\,dr. \end{aligned} \tag{4}$$

We then square and add (arithmetic omitted) to obtain

$$\begin{aligned} ds &= \sqrt{dx^2 + dy^2} \\ &= \sqrt{r^2\,d\theta^2 + dr^2}. \end{aligned} \tag{5}$$

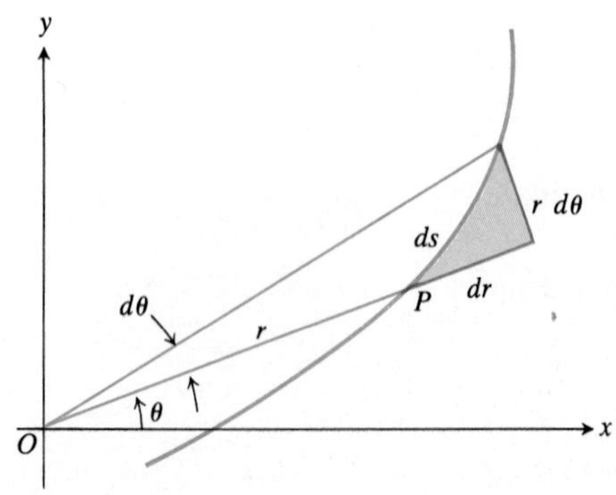

9.77 For arc length, $ds^2 = r^2\,d\theta^2 + dr^2$.

Think of ds as the hypotenuse of a right triangle whose sides are $r\,d\theta$ and dr (Fig. 9.77). For the purpose of evaluation, Eq. (5) is usually written with $d\theta$ factored out:

$$ds = \sqrt{r^2 + \left(\frac{dr}{d\theta}\right)^2}\,d\theta. \tag{6}$$

Length of a Curve

If $r = f(\theta)$ has a continuous first derivative for $\alpha \le \theta \le \beta$ and if the point $P(r, \theta)$ traces the curve $r = f(\theta)$ exactly once as θ runs from α to β, then the length of the curve is given by the formula

$$\text{Length} = \int_{\alpha}^{\beta} \sqrt{r^2 + \left(\frac{dr}{d\theta}\right)^2}\, d\theta. \tag{7}$$

Example 4 Find the length of the cardioid $r = 1 - \cos\theta$.

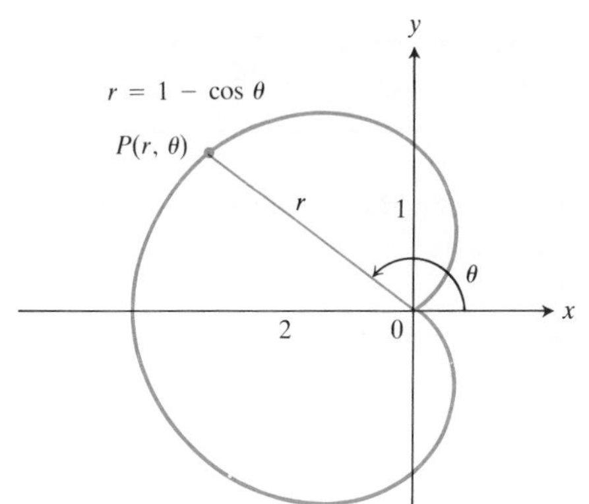

9.78 Example 4 calculates the length of this cardioid. The point P traces the curve once as θ runs from 0 to 2π.

Solution We sketch the cardioid to determine the limits of integration (Fig. 9.78). The point $P(r, \theta)$ starts at the origin and traces the curve once, counterclockwise as θ runs from 0 to 2π, so these are the values we take for α and β.

With

$$r = 1 - \cos\theta, \qquad \frac{dr}{d\theta} = \sin\theta,$$

we have

$$\begin{aligned} r^2 + \left(\frac{dr}{d\theta}\right)^2 &= (1 - \cos\theta)^2 + (\sin\theta)^2 \\ &= 1 - 2\cos\theta + \underbrace{\cos^2\theta + \sin^2\theta}_{1} \\ &= 2 - 2\cos\theta \end{aligned}$$

and

$$\begin{aligned} \text{Length} &= \int_{\alpha}^{\beta} \sqrt{r^2 + \left(\frac{dr}{d\theta}\right)^2}\, d\theta = \int_0^{2\pi} \sqrt{2 - 2\cos\theta}\, d\theta \\ &= \int_0^{2\pi} \sqrt{4\sin^2\frac{\theta}{2}}\, d\theta \qquad \left(1 - \cos\theta = 2\sin^2\frac{\theta}{2}\right) \\ &= \int_0^{2\pi} 2\left|\sin\frac{\theta}{2}\right| d\theta \\ &= \int_0^{2\pi} 2\sin\frac{\theta}{2}\, d\theta \qquad \left(\sin\frac{\theta}{2} \ge 0 \text{ for } 0 \le \theta \le 2\pi\right) \\ &= \left[-4\cos\frac{\theta}{2}\right]_0^{2\pi} = 4 + 4 = 8. \end{aligned}$$

The Area of a Surface of Revolution

The formula for the area of a surface of revolution is $S = \int 2\pi\rho\, ds$, just as in rectangular coordinates, but now we express the radius function ρ and the arc length differential ds in terms of r and θ.

Area of a Surface of Revolution

If $r = f(\theta)$ has a continuous first derivative for $\alpha \le \theta \le \beta$ and if the point $P(r, \theta)$ traces the curve $r = f(\theta)$ exactly once as θ runs from α to β, then the areas of the surfaces generated by revolving the curve about the x- and y-axes are given by the following formulas:

1. Revolution about the x-axis ($y \ge 0$)

$$\text{Surface area} = \int_{\alpha}^{\beta} 2\pi y \, ds = \int_{\alpha}^{\beta} 2\pi r \sin\theta \sqrt{r^2 + \left(\frac{dr}{d\theta}\right)^2}\, d\theta \tag{8}$$

2. Revolution about the y-axis ($x \ge 0$)

$$\text{Surface area} = \int_{\alpha}^{\beta} 2\pi x \, ds = \int_{\alpha}^{\beta} 2\pi r \cos\theta \sqrt{r^2 + \left(\frac{dr}{d\theta}\right)^2}\, d\theta \tag{9}$$

Example 5 Find the area of the surface generated by revolving the right-hand loop of the lemniscate $r^2 = \cos 2\theta$ about the y-axis.

Solution We sketch the loop to determine the limits of integration (Fig. 9.79). The point $P(r, \theta)$ traces the curve once, counterclockwise as θ runs from $-\pi/4$ to $\pi/4$, so these are the values we take for α and β.

We evaluate the surface area integrand

$$2\pi r \cos\theta \sqrt{r^2 + \left(\frac{dr}{d\theta}\right)^2} = 2\pi \cos\theta \sqrt{r^4 + \left(r\frac{dr}{d\theta}\right)^2} \tag{10}$$

in stages. First of all, $r^2 = \cos 2\theta$, so

$$2r\frac{dr}{d\theta} = -2\sin 2\theta$$

$$r\frac{dr}{d\theta} = -\sin 2\theta$$

$$\left(r\frac{dr}{d\theta}\right)^2 = \sin^2 2\theta.$$

9.79 The right-hand half of a lemniscate (a) is revolved about the y-axis to generate a surface (b), whose area is calculated in Example 5.

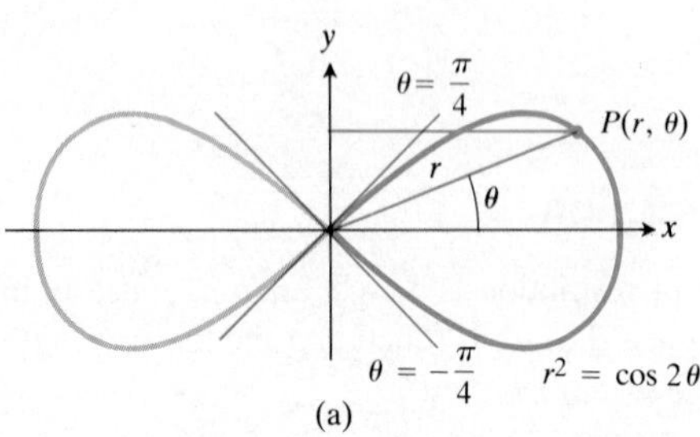

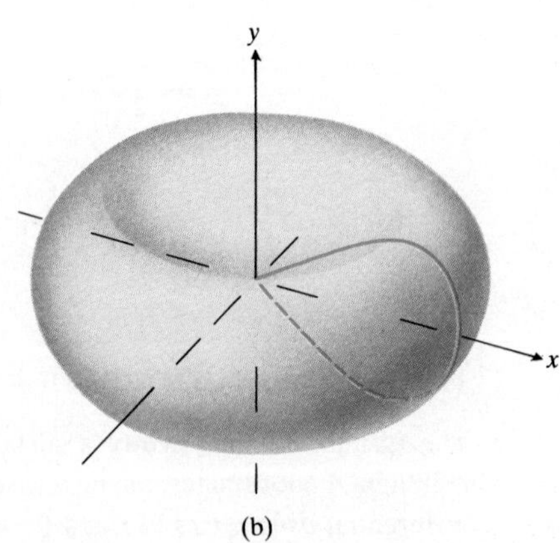

Since

$$r^4 = (r^2)^2 = \cos^2 2\theta,$$

the square root on the right-hand side of Eq. (10) simplifies to

$$\sqrt{r^4 + \left(r\frac{dr}{d\theta}\right)^2} = \sqrt{\cos^2 2\theta + \sin^2 2\theta} = 1.$$

Hence,

$$\begin{aligned}\text{Surface area} &= \int_\alpha^\beta 2\pi r \cos\theta \sqrt{r^2 + \left(\frac{dr}{d\theta}\right)^2}\, d\theta \\ &= \int_{-\pi/4}^{\pi/4} 2\pi \cos\theta \cdot (1)\, d\theta \\ &= 2\pi\Big[\sin\theta\Big]_{-\pi/4}^{\pi/4} = 2\pi\left[\frac{\sqrt{2}}{2} + \frac{\sqrt{2}}{2}\right] = 2\pi\sqrt{2}.\end{aligned}$$

EXERCISES 9.8

Find the areas of the regions in Exercises 1–16.

1. Inside the convex limaçon $r = 4 + 2\cos\theta$
2. Inside the cardioid $r = a(1 + \cos\theta)$, $a > 0$
3. Inside one leaf of the four-leaved rose $r = \cos 2\theta$
4. Inside the lemniscate $r^2 = 2a^2\cos 2\theta$, $a > 0$
5. Inside one loop of the lemniscate $r^2 = 4\sin 2\theta$
6. Inside the six-leaved rose $r^2 = 2\sin 3\theta$ (Fig. 9.80)

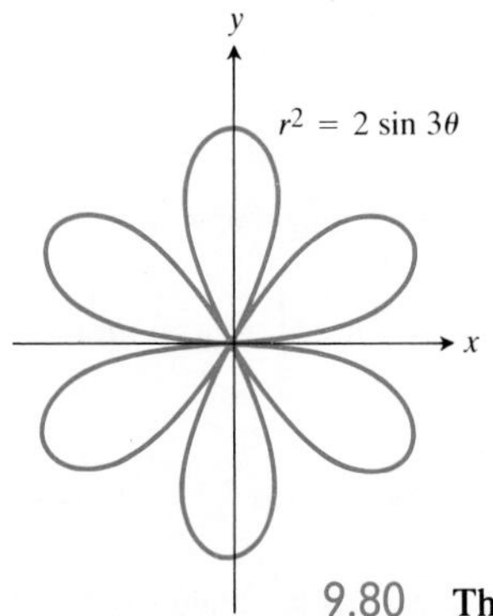

9.80 The rose in Exercise 6.

7. Shared by the circles $r = 2\cos\theta$ and $r = 2\sin\theta$
8. Shared by the circles $r = 1$ and $r = 2\sin\theta$
9. Shared by the circle $r = 2$ and the cardioid $r = 2(1 - \cos\theta)$
10. Shared by the cardioids $r = 2(1 + \cos\theta)$ and $r = 2(1 - \cos\theta)$
11. Inside the lemniscate $r^2 = 6\cos 2\theta$ and outside the circle $r = \sqrt{3}$
12. Inside the circle $r = 3a\cos\theta$ and outside the cardioid $r = a(1 + \cos\theta)$, $a > 0$
13. Inside the circle $r = -2\cos\theta$ and outside the circle $r = 1$
14. a) Inside the outer loop of the limaçon $r = 2\cos\theta + 1$ in Example 2
    b) Inside the outer loop and outside the inner loop
15. Inside the circle $r = 6$ above the line $r = 3\csc\theta$
16. Inside the lemniscate $r^2 = 6\cos 2\theta$ to the right of the line $r = (3/2)\sec\theta$.
17. a) Find the area of the shaded region in Fig. 9.81.
    b) It looks as if the graph of $r = \tan\theta$ could be asymptotic to the lines $x = 1$ and $x = -1$. Is it?

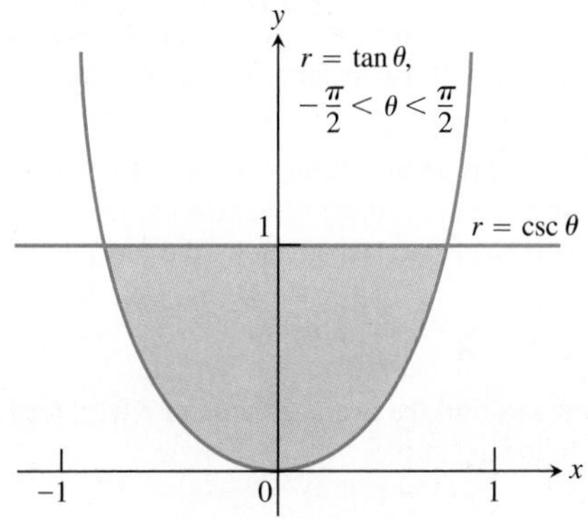

9.81 The region in Exercise 17.

18. The area of the region that lies inside the cardioid $r = \cos\theta + 1$ and outside the circle $r = \cos\theta$ is not

$$\frac{1}{2}\int_0^{2\pi} [(\cos\theta + 1)^2 - \cos^2\theta]\, d\theta = \pi.$$

Why not? What *is* the area?

Find the lengths of the curves in Exercises 19–25.

19. The spiral $r = \theta^2, \quad 0 \le \theta \le \sqrt{5}$

20. The spiral $r = e^{\theta}/\sqrt{2}, \quad 0 \le \theta \le \pi$

21. The cardioid $r = 1 + \cos\theta$

22. The curve $r = a\sin^2(\theta/2), \quad 0 \le \theta \le \pi, \quad a > 0$

23. The curve $r = \cos^3(\theta/3), \quad 0 \le \theta \le \pi/4$

24. The curve $r = \sqrt{1 + \sin 2\theta}, \quad 0 \le \theta \le \pi\sqrt{2}$

25. The curve $r = \sqrt{\cos 2\theta}, \quad 0 \le \theta \le \pi/6$

26. *Circumferences of circles.* As usual, when faced with a new formula, it is a good idea to try it out on familiar objects to be sure it gives results consistent with past experience. Use the length formula in Eq. (7) to calculate the circumferences of the following circles:
a) $r = a$ b) $r = a\cos\theta$ c) $r = a\sin\theta$

Find the areas of the surfaces generated by revolving the curves in Exercises 27–30 about the indicated axes.

27. $r = \sqrt{\cos 2\theta}, \quad 0 \le \theta \le \pi/4, \quad y$-axis

28. $r = \sqrt{2}e^{\theta/2}, \quad 0 \le \theta \le \pi/2, \quad x$-axis

29. $r^2 = \cos 2\theta, \quad x$-axis

30. $r = 2a\cos\theta, \quad y$-axis

31. $r = f(\theta)$ vs. $r = 2f(\theta)$. Suppose that the function $r = f(\theta)$ satisfies the criteria set forth for Eqs. (7) and (8) on the interval $\alpha \le \theta \le \beta$.

a) Show that the length of the curve $r = 2f(\theta)$, $\alpha \le \theta \le \beta$, is twice the length of the curve $r = f(\theta)$, $\alpha \le \theta \le \beta$.

b) Show that the area of the surface generated by revolving the curve $r = 2f(\theta)$, $\alpha \le \theta \le \beta$, about the x-axis is four times the area of the surface generated by revolving the curve $r = f(\theta)$, $\alpha \le \theta \le \beta$, about the x-axis.

32. *Average value.* If f is an integrable function of θ, the average value of the polar coordinate r over the curve $r = f(\theta)$, $\alpha \le \theta \le \beta$, with respect to θ is given by the formula

$$r_{\text{av}} = \frac{1}{\beta - \alpha}\int_{\alpha}^{\beta} f(\theta)\, d\theta.$$

Use this formula to find the average value of r with respect to θ over the following curves.

a) The cardioid $r = a(1 - \cos\theta)$

b) The circle $r = a$

c) The circle $r = a\cos\theta, \quad -\pi/2 \le \theta \le \pi/2$

Centroids of Fan-Shaped Regions

Since the centroid of a triangle is located on each median, two-thirds of the way from the vertex to the opposite base, the lever arm for the moment about the x-axis of the thin triangular region in Fig. 9.82 is about $(2/3)r\sin\theta$. Similarly, the lever arm for the moment of the triangular region about the y-axis is about $(2/3)r\cos\theta$. These approximations improve as $\Delta\theta \to 0$ and lead to the following formulas for the coordinates of the centroid of region AOB:

$$\bar{x} = \frac{\displaystyle\int \frac{2}{3}r\cos\theta \cdot \frac{1}{2}r^2\, d\theta}{\displaystyle\int \frac{1}{2}r^2\, d\theta} = \frac{\displaystyle\frac{2}{3}\int r^3\cos\theta\, d\theta}{\displaystyle\int r^2\, d\theta},$$

$$\bar{y} = \frac{\displaystyle\int \frac{2}{3}r\sin\theta \cdot \frac{1}{2}r^2\, d\theta}{\displaystyle\int \frac{1}{2}r^2\, d\theta} = \frac{\displaystyle\frac{2}{3}\int r^3\sin\theta\, d\theta}{\displaystyle\int r^2\, d\theta},$$

with limits $\theta = \alpha$ to $\theta = \beta$ on all integrals.

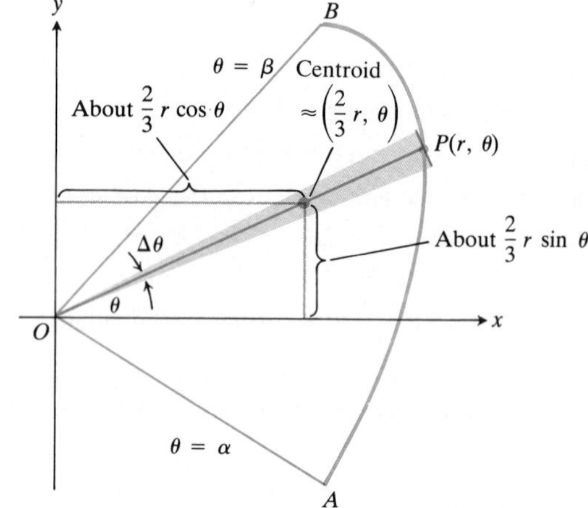

9.82 The moment of the thin triangular sector about the x-axis is approximately

$$\frac{2}{3}r\sin\theta\, dA = \frac{2}{3}r\sin\theta \cdot \frac{1}{2}r^2 d\theta = \frac{1}{3}r^3\sin\theta\, d\theta.$$

33. Find the centroid of the region enclosed by the cardioid $r = a(1 + \cos\theta)$.

34. Find the centroid of the semicircular region $0 \le r \le a$, $0 \le \theta \le \pi$.

EXPLORER PROGRAM

PowerGrapher — Graphs curves in polar coordinates in color

REVIEW QUESTIONS

1. What reflective properties do parabolas, ellipses, and hyperbolas have? What applications use these properties?

2. How are ellipses defined in terms of distance? Give typical equations for ellipses. Graph one of the equations and include the ellipse's vertices and foci. How is the eccentricity of an ellipse defined? Sketch an ellipse whose eccentricity is close to 1 and another whose eccentricity is close to 0.

3. How are hyperbolas defined in terms of distance? Give typical equations for hyperbolas. Graph one of the equations and include the hyperbola's vertices, foci, axes, and asymptotes. What values can the eccentricities of hyperbolas have?

4. Explain the equation $PF = e \cdot PD$.

5. What can be said about the graph of the equation
$$Ax^2 + Bxy + Cy^2 + Dx + Ey + F = 0$$
if A, B, and C are not all zero? What can you tell from the number $B^2 - 4AC$?

6. How do you find a coordinate system in which the new equation for a conic section has no xy-term? Give an example.

7. Give parametric equations for a circle, parabola, and ellipse. In each case, give a parameter domain that covers the conic exactly once.

8. Give parametric equations for one branch of a hyperbola. What is the appropriate domain for the parameter if the curve is to be traced out exactly once by your equations?

9. What is a cycloid? What are typical parametric equations for a cycloid? What physical problems have cycloids for solutions?

10. How do you find dy/dx and d^2y/dx^2 when a curve in the xy-plane is given by parametric equations? Give examples.

11. How do you find the length of a parametrized curve? Give an example.

12. How do you find the area of a surface of revolution generated by a parametrized curve? Give an example.

13. Make a diagram to show the standard relations between Cartesian coordinates (x, y) and polar coordinates (r, θ). Express each set of coordinates in terms of the other kind.

14. If a point has polar coordinates (r_0, θ_0), what other polar coordinates does the point have?

15. How do you test the graph of the equation $F(r, \theta) = 0$ for symmetry about the origin? About the x-axis? About the y-axis? Give examples.

16. Describe a technique for graphing the equation $r = f(\theta)$ that involves the Cartesian $r\theta$-plane. Give an example.

17. What are the standard equations for lines, circles, ellipses, parabolas, and hyperbolas in polar coordinates? Give examples.

18. How do you find the areas of plane regions in polar coordinates? Give examples.

19. How do you find the length of a curve in polar coordinates? Give an example.

20. How do you find the area of a surface of revolution in polar coordinates? Give an example.

MISCELLANEOUS EXERCISES

1. Find an equation for the parabola with focus (4, 0) and directrix $x = 3$. Sketch the parabola together with its vertex, focus, and directrix.

2. Find the vertex, focus, and directrix of the parabola
$$x^2 - 6x - 12y + 9 = 0.$$

3. Find an equation for the curve traced by the point $P(x, y)$ if the distance from P to the vertex of the parabola $x^2 = 4y$ is twice the distance from P to the focus. Identify the curve.

4. A line segment of length $a + b$ runs from the x-axis to the y-axis. The point P on the segment lies a units from one end and b units from the other end. Show that P traces an ellipse as the ends of the segment slide along the axes.

5. The vertices of an ellipse of eccentricity 0.5 lie at the points $(0, \pm 2)$. Where do the foci lie?

6. Find an equation for the ellipse of eccentricity 2/3 that has the line $x = 2$ as a directrix and the point (4, 0) as the corresponding focus.

7. One focus of a hyperbola lies at the point $(0, -7)$ and the corresponding directrix is the line $y = -1$. Find an equation for the hyperbola if its eccentricity is (a) 2; (b) 5.

8. Find an equation for the hyperbola with foci $(0, -2)$ and (0, 2) that passes through the point (12, 7).

Two curves are said to be *orthogonal* if their tangents cross at right angles at every point where the curves intersect. Exercises 9–12 are about orthogonal conic sections.

9. Sketch the curves $xy = 2$ and $x^2 - y^2 = 3$ together and show that they are orthogonal.

10. Sketch the curves $y^2 = 4x + 4$ and $y^2 = 64 - 16x$ together and show that they are orthogonal.

11. Show that the curves $2x^2 + 3y^2 = a^2$ and $ky^2 = x^3$ are orthogonal for all values of the constants a and $k (a \neq 0, k \neq 0)$. Sketch the four curves corresponding to $a = 2$, $a = 4$, $k = 1/2$, $k = -2$ in one diagram.

12. Show that the parabolas

$$y^2 = 4a(a - x)$$

and

$$y^2 = 4b(b + x),$$

$a > 0$ and $b > 0$, have a common focus, the same for any a and b. Show that the parabolas intersect at the points $(a - b, \pm 2\sqrt{ab})$ and that each a-parabola is orthogonal to every b-parabola. (By varying a and b, we obtain two families of confocal parabolas. Each family is said to be a set of *orthogonal trajectories* of the other family. See Fig. 9.83.)

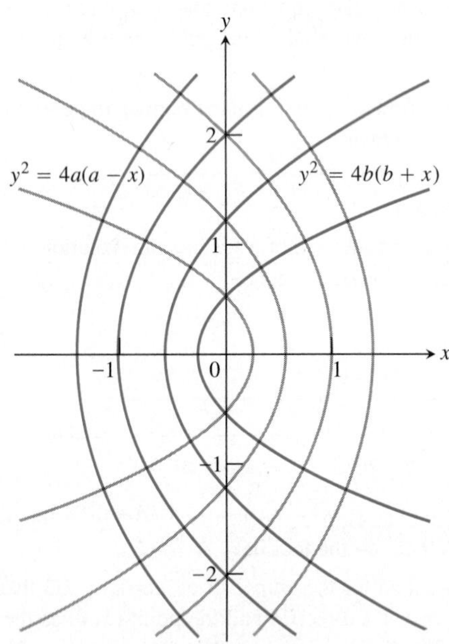

9.83 Some of the parabolas in Exercise 12.

13. A comet moves in a parabolic orbit with the sun at the focus. When the comet is 4×10^7 miles from the sun, the line from the comet to the sun makes a 60° angle with the orbit's axis (Fig. 9.84). How close will the comet come to the sun?

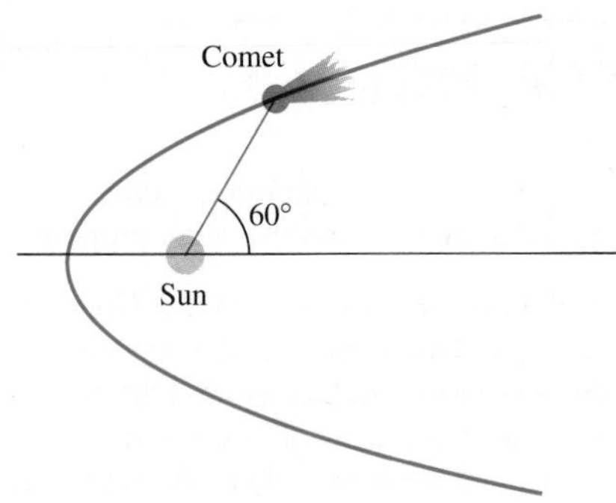

9.84 The comet in Exercise 13.

14. A ripple tank is made by bending a strip of tin around the perimeter of an ellipse for the wall of the tank and soldering a flat bottom onto this. An inch or two of water is put in the tank and the experimenter pokes a finger into it, right at one focus of the ellipse. Ripples radiate outward through the water, reflect from the strip around the edge of the tank, and in a short time a drop of water spurts up at the second focus. Why?

15. Set up the integrals that give (a) the area of a quadrant of the circle $x^2 + y^2 = a^2$, (b) the area of a quadrant of the ellipse $b^2x^2 + a^2y^2 = a^2b^2$. Show that the integral in (b) is b/a times the integral in (a), and deduce the area of the ellipse from the known area of the circle.

16. A rope with a ring in one end is looped over two pegs in a horizontal line. The free end, after being passed through the ring, has a weight suspended from it to make the rope hang taut. If the rope slips freely over the pegs and through the ring, the weight will descend as far as possible. Assume that the length of the rope is at least four times as great as the distance between the pegs and that the configuration of the rope is symmetric with respect to the line of the vertical part of the rope.

a) Find the angle formed at the bottom of the loop (Fig. 9.85).

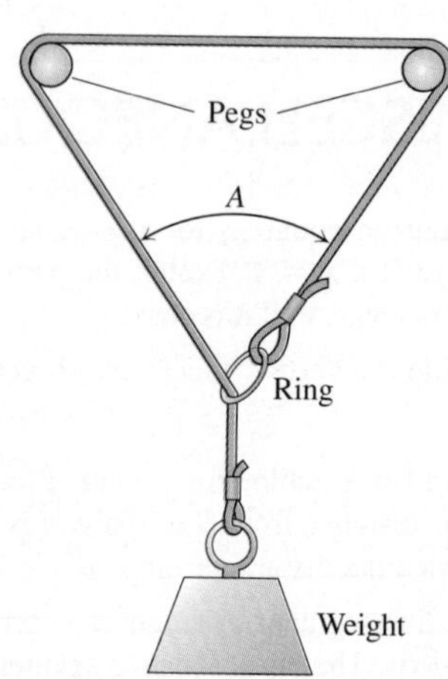

9.85 Exercise 16 asks how large the angle A will be when the frictionless rope shown here is pulled tight by the weight.

b) Show that for each fixed position of the ring on the rope, the possible locations of the ring in space lie on an ellipse with foci at the pegs.

c) Justify the original symmetry assumption by combining the result in (b) with the assumption that the rope and weight will take a rest position of minimal potential energy.

17. Two radar stations lie 20 km apart along an east–west line. A low-flying plane traveling from west to east is known to have a speed of v_0 km/sec. At $t = 0$ a signal is sent from the station at $(-10, 0)$, bounces off the plane, and is received at $(10, 0)$ $30/c$ seconds later (c is the velocity of the signal). When $t = 10/v_0$, another signal is sent out from the station at $(-10, 0)$, reflects off the plane, and is once again received $30/c$ seconds later by the other station. Find the position of the plane when it reflects the second signal under the assumption that v_0 is much less than c.

18. *LORAN.* A radio signal was sent simultaneously from towers A and B located several hundred miles apart on the California coast. A ship offshore received the signal from A 1400 microseconds before it received the signal from B.

a) Assume that the radio signals traveled at 980 ft per microsecond. What can be said about the approximate location of the ship relative to the two towers?

b) Find out what you can about LORAN and other hyperbolic radio navigation systems. (See, for example, *American Practical Navigator,* by Nathaniel Bowditch, Vol. I, U.S. Defense Mapping Agency Hydrographic Center, Publication No. 9, 1977, Chapter 43.)

19. On a level plane, at the same instant, you hear the sound of a rifle and that of the bullet hitting the target. What is your location in relation to the rifle and target?

20. Show that no tangent can be drawn from the origin to the hyperbola $x^2 - y^2 = 1$. (*Hint:* If the tangent to a curve at a point $P(x, y)$ on the curve passes through the origin, then the slope of the curve at P is y/x.)

21. *Constructing tangents to parabolas.* Show that the tangent to the parabola $y^2 = 4px$ at the point $P(x_1, y_1) \neq (0, 0)$ on the parabola meets the axis of symmetry x_1 units to the left of the vertex. This provides an accurate way to construct a tangent to the parabola at any point other than the origin (where we already have the y-axis): Mark the point $P(x_1, y_1,)$ in question, drop a perpendicular from P to the x-axis, measure $2x_1$ units to the left, mark that point, and draw a line from there through P.

22. *The reflective property of ellipses.* An ellipsoid is generated by rotating an ellipse about its major axis. The inside surface of the ellipsoid is silvered to produce a mirror. Show that a light ray emanating from one focus will be reflected to the other focus. (*Hint:* Put the ellipse in standard position in the plane, as in Fig. 9.86, and show that the lines from the foci to a point P on the ellipse make equal angles with the tangent at P.)

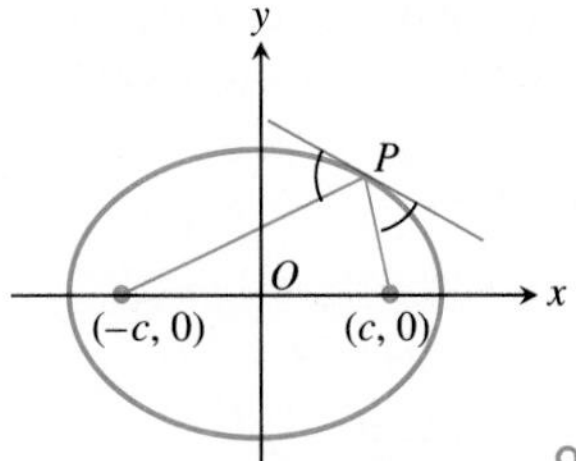

9.86 The ellipse in Exercise 22.

What points in the xy-plane satisfy the equations and inequalities in Exercises 23–30? Draw a figure for each exercise.

23. $(x^2 - y^2 - 1)(x^2 + y^2 - 25)(x^2 + 4y^2 - 4) = 0$

24. $(x + y)(x^2 + y^2 - 1) = 0$

25. $(x^2/9) + (y^2/16) \leq 1$

26. $(x^2/9) - (y^2/16) \leq 1$

27. $(9x^2 + 4y^2 - 36)(4x^2 + 9y^2 - 16) \leq 0$

28. $(9x^2 + 4y^2 - 36)(4x^2 + 9y^2 - 16) > 0$

29. $x^4 - (y^2 - 9)^2 = 0$

30. $x^2 + xy + y^2 < 3$

Use the discriminant to decide whether the equations in Exercises 31–34 represent parabolas, ellipses, or hyperbolas.

31. $x^2 + xy + y^2 + x + y + 1 = 0$

32. $x^2 + 3xy + 2y^2 + x + y + 1 = 0$

33. $x^2 + 4xy + 4y^2 + x + y + 1 = 0$

34. $x^2 + 2xy - 2y^2 + x + y + 1 = 0$

Identify the conic sections in Exercises 35 and 36. Then rotate the coordinate axes to find a new equation for the conic section that has no xy-term. (The new equations will vary with the rotations you choose.)

35. $2x^2 + xy + 2y^2 - 15 = 0$

36. $x^2 + 2\sqrt{3}xy - y^2 + 4 = 0$

37. Find the points on the parabola $x = 2t$, $y = t^2$, $-\infty < t < \infty$, closest to the point $(0, 3)$.

38. Is the curve $\sqrt{x} + \sqrt{y} = 1$ part of a conic section? If so, what kind of conic section? If not, why not?

39. CALCULATOR Find the eccentricity of the ellipse $x^2 + xy + y^2 = 1$ to the nearest hundredth.

40. Show that any tangent to the hyperbola $xy = a^2$ makes a triangle of area $2a^2$ with the hyperbola's asymptotes.

41. Find the eccentricity of the hyperbola $xy = 1$.

42. Show that the curve $2xy - \sqrt{2}y + 2 = 0$ is a hyperbola. Find the hyperbola's center, vertices, foci, axes, and asymptotes.

Exercises 43–48 give parametric equations and parameter intervals for the motion of a particle in the xy-plane. Identify the particle's path by finding a Cartesian equation for it. Graph the Cartesian equation and indicate the direction of motion and the portion traced by the particle.

43. $x = t/2, \quad y = t + 1; \quad -\infty < t < \infty$

44. $x = \sqrt{t}, \quad y = 1 - \sqrt{t}; \quad t \geq 0$

45. $x = (1/2)\tan t, \quad y = (1/2)\sec t; \quad -\pi/2 < t < \pi/2$

46. $x = -2\cos t, \quad y = 2\sin t; \quad 0 \leq t \leq \pi$

47. $x = -\cos t, \quad y = \cos^2 t; \quad 0 \leq t \leq \pi$

48. $x = 4\cos t, \quad y = 9\sin t; \quad 0 \leq t \leq 2\pi$

49. Find parametric equations and a parameter interval for the motion of a particle in the xy-plane that traces the ellipse $16x^2 + 9y^2 = 144$ once counterclockwise. (There are many ways to do this, so your answer may not be the same as the one in the back of the book.)

50. Find parametric equations and a parameter interval for the motion of a particle that starts at the point $(-2, 0)$ in the xy-plane and traces the circle $x^2 + y^2 = 4$ three times clockwise. (There are many ways to do this.)

51. *Epicycloids.* When a circle rolls externally along the circumference of a second, fixed circle, any point P on the circumference of the rolling circle describes an *epicycloid* (Fig. 9.87). Let the fixed circle have its center at the origin O and have radius a. Let the radius of the rolling circle be b and let the initial position of the tracing point P be $A(a, 0)$. Find parametric equations for the epicycloid, using as the parameter the angle θ from the positive x-axis to the line through the circles' centers.

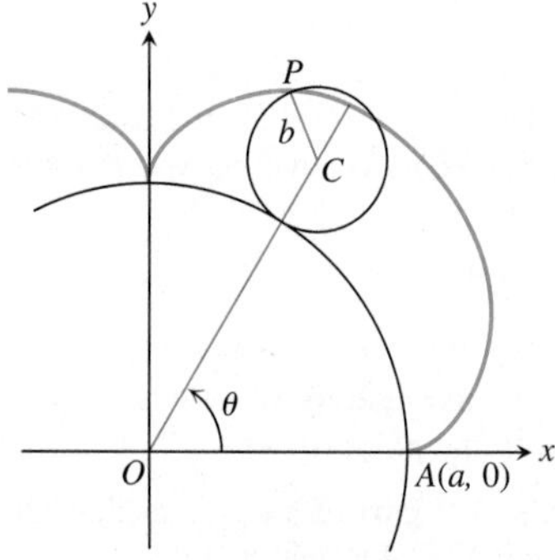

9.87 An epicycloid with $b = a/4$ (Exercise 51).

52. *Hypocycloids.* When a circle rolls on the inside of a fixed circle, any point P on the circumference of the rolling circle describes a *hypocycloid.* Let the fixed circle be $x^2 + y^2 = a^2$, let the radius of the rolling circle be b, and let the initial position of the tracing point P be $A(a, 0)$. Find parametric equations for the hypocycloid, using as the parameter the angle θ from the positive x-axis to the line joining the circles' centers. In particular, if $b = a/4$, as in Fig. 9.88, show that the hypocycloid is the astroid

$$x = a\cos^3\theta, \quad y = a\sin^3\theta.$$

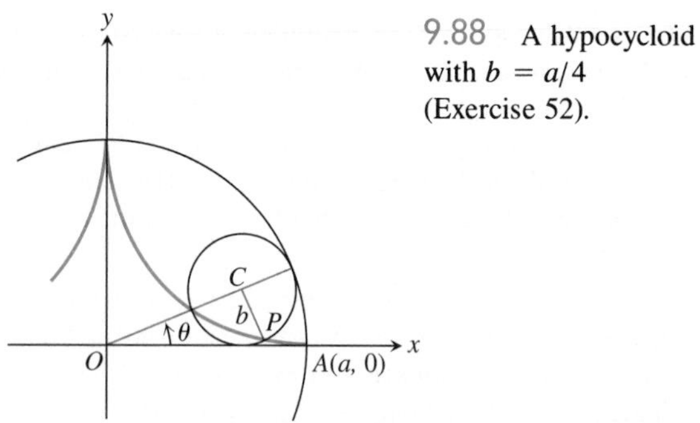

9.88 A hypocycloid with $b = a/4$ (Exercise 52).

53. *More about hypocycloids.* Figure 9.89 shows a circle of radius a tangent to the inside of a circle of radius $2a$. The point P, shown as the point of tangency in the figure, is attached to the smaller circle. What path does P trace as the smaller circle rolls around the inside of the larger circle?

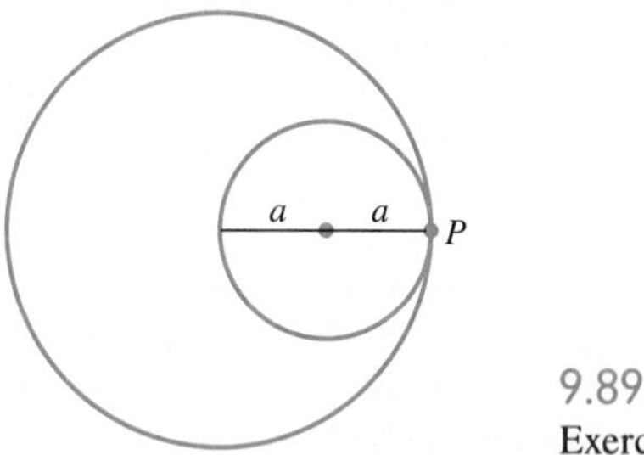

9.89 The circles in Exercise 53.

54. *Generating a cardioid with circles.* Cardioids are special epicycloids (Exercise 51). Show that if you roll a circle of radius a about another circle of radius a in the polar coordinate plane, as in Fig. 9.90, the original point of contact P will trace a cardioid. (*Hint:* Start by showing that angles OBC and PAD both have measure θ.)

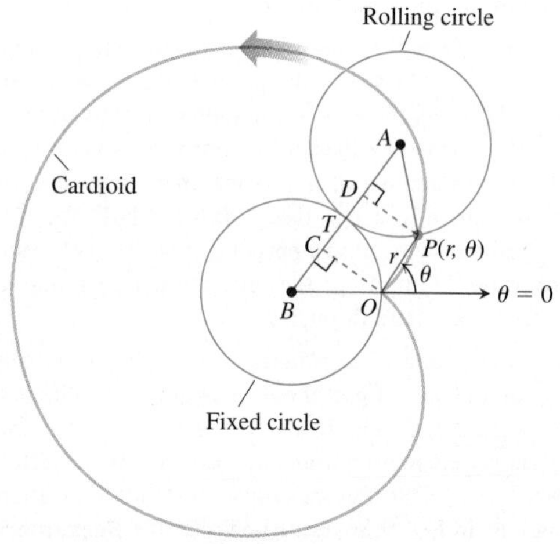

9.90 As the circle centered at A rolls around the circle centered at B, the point P traces a cardioid (Exercise 54).

In Exercises 55 and 56, find an equation for the line in the xy-plane that is tangent to the curve at the point corresponding to the given value of t. Also, find the value of d^2y/dx^2 at this point.

55. $x = (1/2)\tan t, \quad y = (1/2)\sec t; \quad t = \pi/3$

56. $x = 1 + 1/t^2, \quad y = 1 - 3/t; \quad t = 2$

Find the lengths of the curves in Exercises 57 and 58.

57. $x = e^{2t} - \dfrac{t}{8}, \quad y = e^t; \quad 0 \le t \le \ln 2$

58. The closed loop in Fig. 9.91.

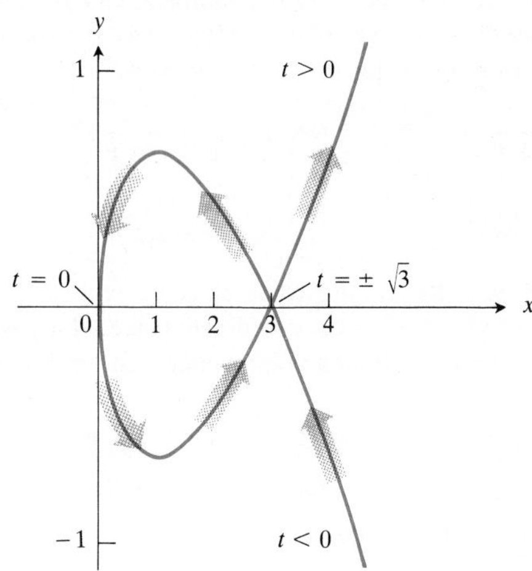

9.91 Exercise 58 refers to the curve $x = t^2$, $y = (t^3/3) - t$ shown here. The loop starts at $t = -\sqrt{3}$ and ends at $t = \sqrt{3}$.

Find the areas of the surfaces generated by revolving the curves in Exercises 59 and 60 about the axes indicated.

59. $x = t^2/2, \quad y = 2t, \quad 0 \le t \le \sqrt{5}; \quad x\text{-axis}$

60. $x = t^2 + 1/(2t), \quad y = 4\sqrt{t}, \quad 1/\sqrt{2} \le t \le 1; \quad y\text{-axis}$

61. Find the centroid of the region enclosed by the x-axis and the cycloid arch

$$x = a(t - \sin t), \quad y = a(1 - \cos t), \quad 0 \le t \le 2\pi.$$

(*Hint:* Express $\tilde{x}$, $\tilde{y}$, and dx in terms of t and dt.)

62. (Continuation of Exercise 61.) Use a theorem of Pappus to find the volume of the solid generated by revolving the region in Exercise 61 about the x-axis.

63. A point $P(x, y)$ moves in the plane in such a way that for $t \ge 0$,

$$\frac{dx}{dt} = \frac{1}{t+2} \quad \text{and} \quad \frac{dy}{dt} = 2t.$$

a) Express x and y as functions of t if $x = \ln 2$ and $y = 1$ when $t = 0$.

b) Express y in terms of x.

c) Express x in terms of y.

d) Find the average rate of change in y with respect to x as t varies from 0 to 2.

e) Find dy/dx when $t = 1$.

64. Find parametric equations and a Cartesian equation for the curve traced by the point $P(x, y)$ if its coordinates satisfy the differential equations

$$\frac{dx}{dt} = -2y, \quad \frac{dy}{dt} = \cos t,$$

subject to the conditions that $x = 3$ and $y = 0$ when $t = 0$. Identify the curve.

65. Find the first moments about the coordinate axes of the curve

$$x = (2/3)t^{3/2}, \quad y = 2\sqrt{t}, \quad 0 \le t \le \sqrt{3}.$$

66. *Pythagorean triples.* Suppose that the coordinates of a particle $P(x, y)$ moving in the plane are

$$x = \frac{1 - t^2}{1 + t^2} \quad \text{and} \quad y = \frac{2t}{1 + t^2}$$

for $-\infty < t < \infty$. Show that $x^2 + y^2 = 1$ and hence that the motion takes place on the unit circle. What one point of the circle is not covered by the motion? Sketch the circle and indicate the direction of motion for increasing t. For what values of t does $(x, y) = (0, -1)$? $(1, 0)$? $(0, 1)$?

From $x^2 + y^2 = 1$, we obtain

$$(t^2 - 1)^2 + (2t)^2 = (t^2 + 1)^2,$$

an equation of interest in number theory because it generates *Pythagorean triples* of integers. When t is an integer greater than 1, $a = t^2 - 1$, $b = 2t$, and $c = t^2 + 1$ are positive integers that satisfy the equation $a^2 + b^2 = c^2$.

Each of the graphs in Exercises 67–74 is the graph of one of the equations (a)–(l) listed below. Find the equation for each graph.

a) $r = \cos 2\theta$ b) $r \cos \theta = 1$

c) $r = \dfrac{6}{1 - 2\cos\theta}$ d) $r = \sin 2\theta$

e) $r = \theta$ f) $r^2 = \cos 2\theta$

g) $r = 1 + \cos\theta$ h) $r = 1 - \sin\theta$

i) $r = \dfrac{2}{1 - \cos\theta}$ j) $r^2 = \sin 2\theta$

k) $r = -\sin\theta$ l) $r = 2\cos\theta + 1$

67. Four-leaved rose

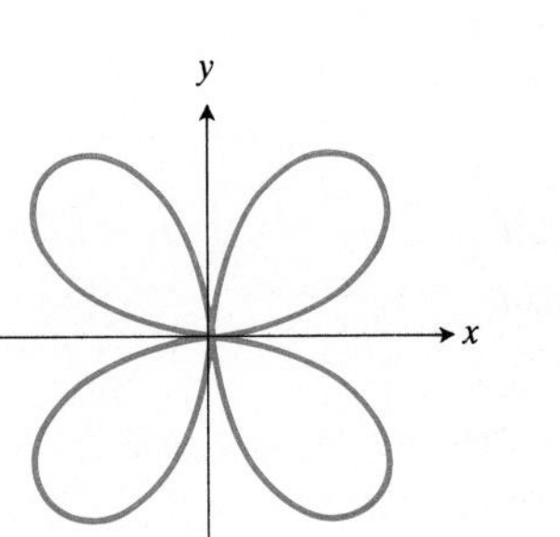

68. Spiral

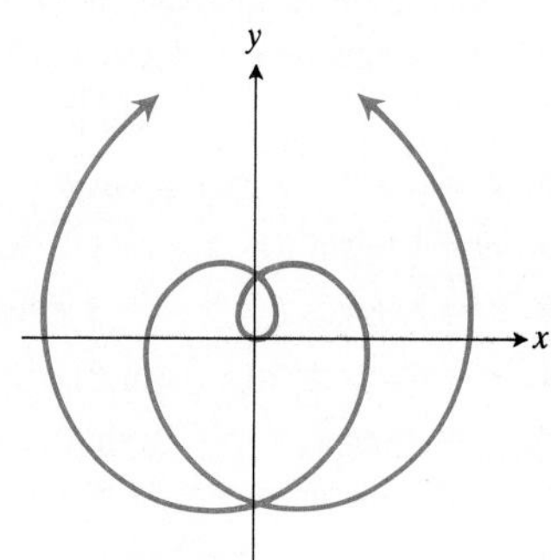

69. Limaçon

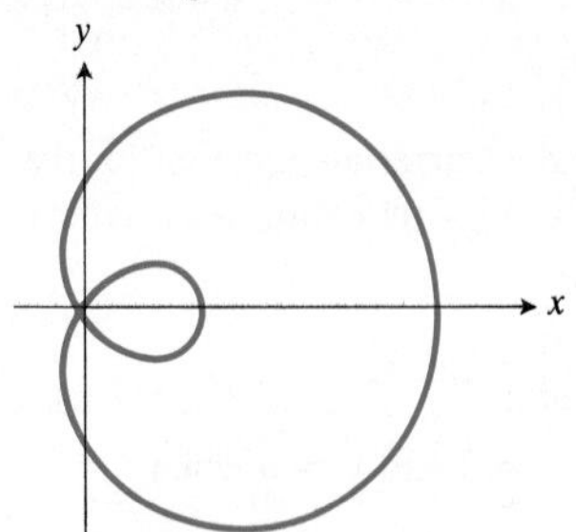

70. Lemniscate

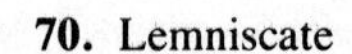

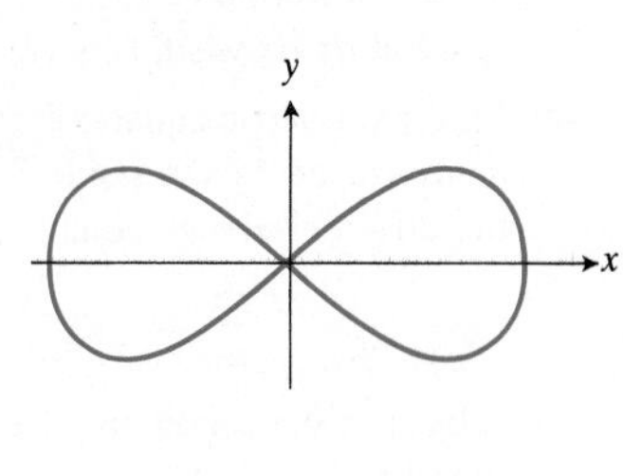

71. Circle

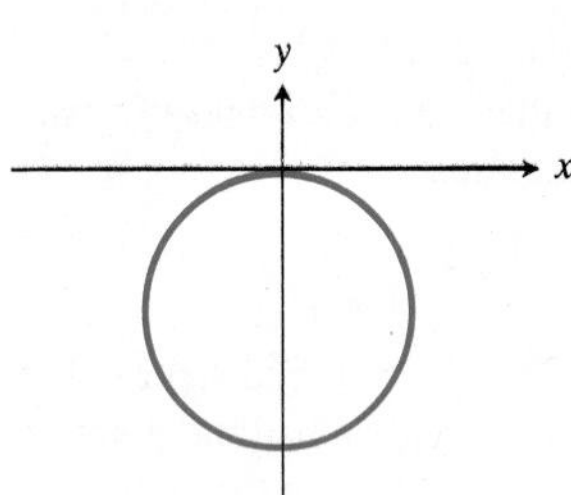

72. Cardioid

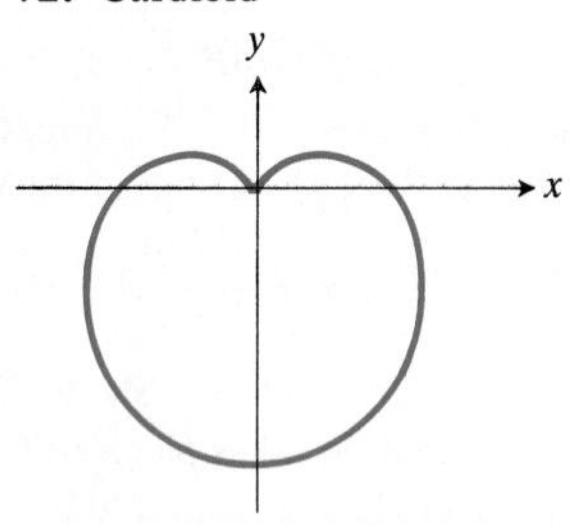

73. Parabola

74. Lemniscate

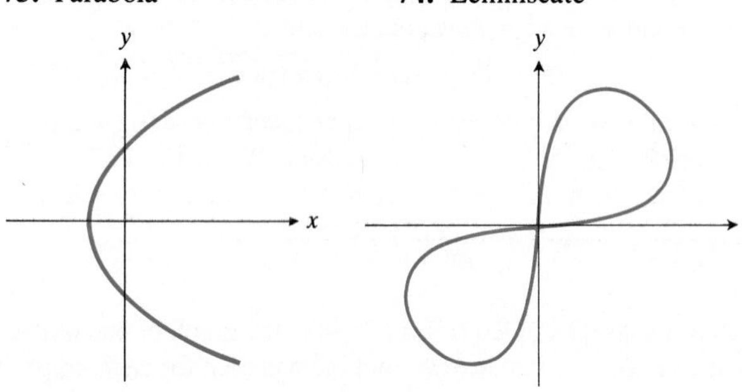

Sketch the regions defined by the polar coordinate inequalities in Exercises 75 and 76.

75. $0 \le r \le 6 \cos \theta$

76. $-4 \sin \theta \le r \le 0$

Find the points of intersection of the curves given by the polar coordinate equations in Exercises 77–84.

77. $r = \sin \theta, \quad r = 1 + \sin \theta$

78. $r = \cos \theta, \quad r = 1 - \cos \theta$

79. $r = 1 + \sin \theta, \quad r = -1 + \sin \theta$

80. $r = 1 + \cos \theta, \quad r = -1 - \cos \theta$

81. $r = a, \quad r = a(1 - \sin \theta), \quad a > 0$

82. $r = a \sec \theta, \quad r = 2a \sin \theta, \quad a > 0$

83. $r = a \cos \theta, \quad r = a(1 + \cos \theta), \quad a > 0$

84. $r = a(1 + \cos 2\theta), \quad r = a \cos 2\theta, \quad a > 0$

Find equations for the lines that are tangent to the polar coordinate curves in Exercises 85 and 86 at the origin.

85. The lemniscate $r^2 = \cos 2\theta$

86. The limaçon $r = 2 \cos \theta + 1$

87. Find polar coordinate equations for the lines that are tangent to the tips of the petals of the four-leaved rose $r = \sin 2\theta$.

88. Find polar coordinate equations for the lines that are tangent to the cardioid $r = 1 + \sin \theta$ at the points where it crosses the x-axis.

Sketch the conic sections whose polar coordinate equations are given in Exercises 89–92. Give polar coordinates for the vertices and, in the case of ellipses, for the centers as well.

89. $r = \dfrac{2}{1 + \cos \theta}$

90. $r = \dfrac{8}{2 + \cos \theta}$

91. $r = \dfrac{6}{1 - 2 \cos \theta}$

92. $r = \dfrac{12}{3 + \sin \theta}$

Exercises 93–96 give the eccentricities of conic sections with one focus at the origin of the polar coordinate plane, along with the directrix for that focus. Find a polar equation for each conic section.

93. $e = 2, \quad r \cos \theta = 2$

94. $e = 1, \quad r \cos \theta = -4$

95. $e = 1/2, \quad r \sin \theta = 2$

96. $e = 1/3, \quad r \sin \theta = -6$

97. Find a polar coordinate equation for

a) the parabola with focus at the origin and vertex at $(a, \pi/4)$;

b) the ellipse with foci at the origin and $(2, 0)$ and one vertex at $(4, 0)$;

c) the hyperbola with one focus at the origin, center at $(2, \pi/2)$, and a vertex at $(1, \pi/2)$.

98. *A satellite orbit.* A satellite is in an orbit that passes over the North and South Poles of the earth. When it is over the South Pole it is at the highest point of its orbit, 1000 miles above the earth's surface. Above the North Pole it is at the lowest point of its orbit, 300 miles above the earth's surface.

a) Assuming that the orbit (with reference to the earth) is an ellipse with one focus at the center of the earth, find its eccentricity. (Take the diameter of the earth to be 8000 miles.)

b) Using the north–south axis of the earth as the x-axis and the center of the earth as origin, find a polar coordinate equation for the orbit.

Find the areas of the regions in the polar coordinate plane described in Exercises 99–102.

99. Enclosed by the limaçon $r = 2 - \cos \theta$

100. Enclosed by one leaf of the three-leaved rose $r = \sin 3\theta$

101. Inside the two-leaved rose $r = 1 + \cos 2\theta$ and outside the circle $r = 1$

102. Inside the cardioid $r = 2(1 + \sin \theta)$ and outside the circle $r = 2 \sin \theta$

Find the areas of the surfaces generated by revolving the polar coordinate curves in Exercises 103 and 104 about the indicated axes.

103. $r = \sqrt{\cos 2\theta}$, $0 \le \theta \le \pi/4$, about the x-axis

104. $r^2 = \sin 2\theta$, about the y-axis

105. a) Find an equation in polar coordinates for the curve

$$x = e^{2t}\cos t,\quad y = e^{2t}\sin t,\quad -\infty < t < \infty.$$

b) Sketch the curve.

c) Find the length of the curve from $t = 0$ to $t = 2\pi$.

106. Find the length of the curve $r = 2\sin^3(\theta/3)$, $0 \le \theta \le 3\pi$, in the polar coordinate plane.

107. Find the area of the surface generated by revolving the first-quadrant portion of the cardioid $r = 1 + \cos\theta$ about the x-axis. (*Hint:* Use the identities $1 + \cos\theta = 2\cos^2(\theta/2)$ and $\sin\theta = 2\sin(\theta/2)\cos(\theta/2)$ to simplify the integral.)

108. Sketch the regions enclosed by the curves $r = 2a\cos^2(\theta/2)$ and $r = 2a\sin^2(\theta/2)$, $a > 0$, in the polar coordinate plane and find the area of the portion of the plane they have in common.

The Angle Between the Radius Vector and the Tangent Line to a Polar Coordinate Curve

In Cartesian coordinates, when we want to discuss the direction of a curve at a point, we use the angle ϕ measured counterclockwise from the positive x-axis to the tangent line. In polar coordinates, it is more convenient to calculate the angle ψ from the *radius vector* to the tangent line (Fig. 9.92). The angle ϕ can then be calculated from the relation

$$\phi = \theta + \psi, \tag{1}$$

which comes from applying the exterior angle theorem to the triangle in Fig. 9.92.

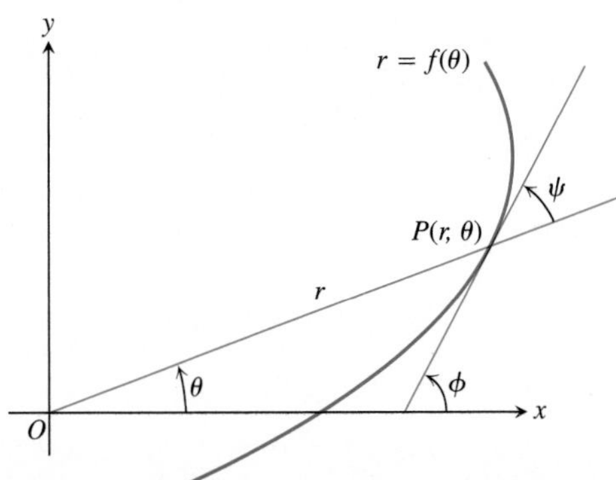

9.92 The angle ψ between the tangent line and the radius vector.

Suppose the equation of the curve is given in the form $r = f(\theta)$, where $f(\theta)$ is a differentiable function of θ. Then

$$x = r\cos\theta \quad\text{and}\quad y = r\sin\theta \tag{2}$$

are differentiable functions of θ with

$$\frac{dx}{d\theta} = -r\sin\theta + \cos\theta\,\frac{dr}{d\theta},\qquad \frac{dy}{d\theta} = r\cos\theta + \sin\theta\,\frac{dr}{d\theta}. \tag{3}$$

Since $\psi = \phi - \theta$ from (1),

$$\tan\psi = \tan(\phi - \theta) = \frac{\tan\phi - \tan\theta}{1 + \tan\phi\tan\theta}.$$

Furthermore,

$$\tan\phi = \frac{dy}{dx} = \frac{dy/d\theta}{dx/d\theta}$$

because $\tan\phi$ is the slope of the curve at P. Also,

$$\tan\theta = \frac{y}{x}.$$

Hence

$$\tan\psi = \frac{\dfrac{dy/d\theta}{dx/d\theta} - \dfrac{y}{x}}{1 + \dfrac{y}{x}\dfrac{dy/d\theta}{dx/d\theta}} = \frac{x\dfrac{dy}{d\theta} - y\dfrac{dx}{d\theta}}{x\dfrac{dx}{d\theta} + y\dfrac{dy}{d\theta}} \tag{4}$$

The numerator in the last expression in Eq. (4) is found from Eqs. (2) and (3) to be

$$x\frac{dy}{d\theta} - y\frac{dx}{d\theta} = r^2.$$

Similarly, the denominator is

$$x\frac{dx}{d\theta} + y\frac{dy}{d\theta} = r\frac{dr}{d\theta}.$$

When we substitute these into Eq. (4), we obtain

$$\tan\psi = \frac{r}{dr/d\theta}. \tag{5}$$

This is the equation we use for finding ψ.

109. Show, by reference to a figure, that the angle β between the tangents to two curves at a point of intersection may be found from the formula

$$\tan\beta = \frac{\tan\psi_2 - \tan\psi_1}{1 + \tan\psi_2\tan\psi_1}. \tag{6}$$

When will the two curves intersect at right angles?

110. Find the value of $\tan\psi$ for the curve $r = \sin^4(\theta/4)$.

111. Find the angle between the curve $r = 2a\sin 3\theta$ and its tangent when $\theta = \pi/3$.

112. For the hyperbolic spiral $r\theta = a$ show that $\psi = 3\pi/4$ when $\theta = 1$ radian, and that $\psi \to \pi/2$ as the spiral winds around the origin. Sketch the curve and indicate ψ for $\theta = 1$ radian.

113. The circles $r = \sqrt{3} \cos \theta$ and $r = \sin \theta$ intersect at the point $(\sqrt{3}/2, \pi/3)$. Show that their tangents are perpendicular there.

114. Sketch the cardioid $r = a(1 + \cos \theta)$ and circle $r = 3a \cos \theta$ in one diagram and find the angle between their tangents at the point of intersection that lies in the first quadrant.

115. Find the points of intersection of the parabolas

$$r = \frac{1}{1 - \cos \theta} \quad \text{and} \quad r = \frac{3}{1 + \cos \theta}$$

and the angles between their tangents at these points.

116. Find points on the cardioid $r = a(1 + \cos \theta)$ where the tangent line is (a) horizontal, (b) vertical.

117. Show that the parabolas $r = a/(1 + \cos \theta)$ and $r = b/(1 - \cos \theta)$ are orthogonal at each point of intersection $(ab \neq 0)$.

118. Find the angle at which the cardioid $r = a(1 - \cos \theta)$ crosses the ray $\theta = \pi/2$.

119. Find the angle between the line $r = 3 \sec \theta$ and the cardioid $r = 4(1 + \cos \theta)$ at one of their intersections.

120. Find the slope of the tangent line to the curve $r = a \tan(\theta/2)$ at $\theta = \pi/2$.

121. Find the angle at which the parabolas $r = 1/(1 - \cos \theta)$ and $r = 1/(1 - \sin \theta)$ intersect in the first quadrant.

122. The equation $r^2 = 2 \csc 2\theta$ represents a curve in polar coordinates.

a) Sketch the curve.

b) Find an equivalent Cartesian equation for the curve.

c) Find the angle at which the curve intersects the ray $\theta = \pi/4$.

123. Suppose that the angle ψ from the radius vector to the tangent line of the curve $r = f(\theta)$ has the constant value α.

a) Show that the area bounded by the curve and two rays $\theta = \theta_1$, $\theta = \theta_2$, is proportional to $r_2^2 - r_1^2$, where (r_1, θ_1) and (r_2, θ_2) are polar coordinates of the ends of the arc of the curve between these rays. Find the factor of proportionality.

b) Show that the length of the arc of the curve in part (a) is proportional to $r_2 - r_1$ and find the proportionality constant.

124. Let P be a point on the hyperbola $r^2 \sin 2\theta = 2a^2$. Show that the triangle formed by OP, the tangent at P, and the initial line is isosceles.

CHAPTER

10

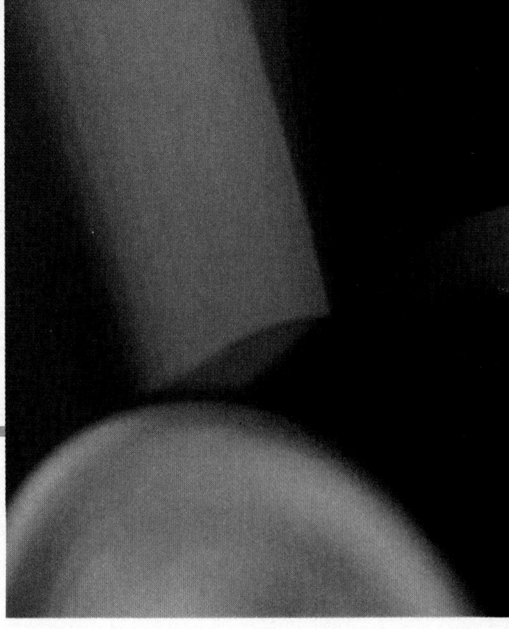

VECTORS AND ANALYTIC GEOMETRY IN SPACE

Overview This chapter introduces vectors and three-dimensional coordinate systems. Just as the coordinate plane is the natural place to study functions of a single variable, coordinate space is the place to study functions of two variables (or more). We establish coordinates in space by adding a third axis that measures distance above and below the xy-plane. This builds on what we already know without forcing us to start over again.

Equations in three variables define surfaces in space the way equations in two variables define curves in the plane. We use these surfaces to graph functions of two variables (not in this chapter, but later), define regions, bound solids, describe walls of containers, and so on. In short, we use them to do all the things we do in the plane, but stepped up one dimension.

Once in space, we can model motion in three dimensions and track the positions of moving bodies with vectors. We can also calculate the directions and magnitudes of their velocities and accelerations and predict the effects of the forces that are driving them. As we shall see in Chapter 11, coordinates and vectors make a powerful combination. Coordinates tell us where moving bodies are, and vectors tell us what is happening to them as they go along.

10.1 Vectors in the Plane

Some of the things we measure are completely determined by their magnitudes. To record mass, length, or time, for example, we need only write down a number and name an appropriate unit of measure. But we need more than that to describe a force, displacement, or velocity, for these quantities have direction as well as magnitude. To describe a force, we need to record the direction in which it acts as well as how large it is. To describe a body's displacement, we have to say in what direction it moves as well as how far. To describe a body's velocity at any given time, we have to know where the body is headed as well as how fast it is going.

Quantities that have direction as well as magnitude are usually represented by arrows that point in the direction of the action and whose lengths give the magnitude of the action in terms of a suitably chosen unit. Arrows with the same length and direction are regarded as equivalent. We can draw the arrow that represents a force or velocity wherever we want as long as it has the right length and direction.

When we work with arrows in mathematics, we think of them as directed line segments and we call the sets of equivalent segments *vectors.*

DEFINITIONS

Directed line segments (arrows) in the plane are **equivalent** if they have the same length and direction. Each directed line segment is also equivalent to itself. The collection of all directed line segments equivalent to a given directed line segment is the **vector** determined by the segment. Each directed line segment in the vector **represents** the vector. When we draw a directed line segment as an arrow, we say that we **draw** the vector it represents. The **length (magnitude)** of a vector is the common length of the directed line segments that represent it. The **direction** of a vector is the common direction of the directed line segments that represent it.

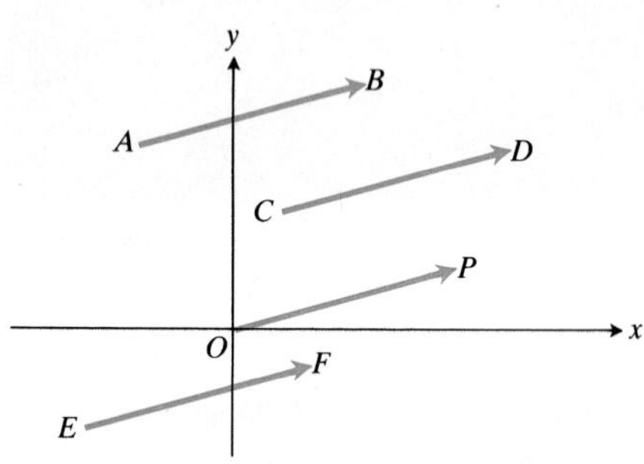

10.1 Arrows with the same length and direction represent the same vector (Example 1).

Thus, the arrows we use when we draw vectors are understood to represent the same vector if they have the same length, are parallel, and point the same way.

In print, vectors are usually described with single boldface roman letters, as in **v** ("vector vee"). The vector defined by the directed line segment from point A to point B is written as $\overrightarrow{AB}$ ("vector *ab*"). In handwritten work it is customary to draw small arrows above letters that represent vectors. Thus the equation appearing as $\mathbf{v} = \overrightarrow{AB}$ in print would appear as $\vec{v} = \overrightarrow{AB}$ when written by hand.

Example 1 The four arrows in Fig. 10.1 have the same length and direction. They therefore represent the same vector, and we write

$$\overrightarrow{AB} = \overrightarrow{CD} = \overrightarrow{OP} = \overrightarrow{EF}.$$

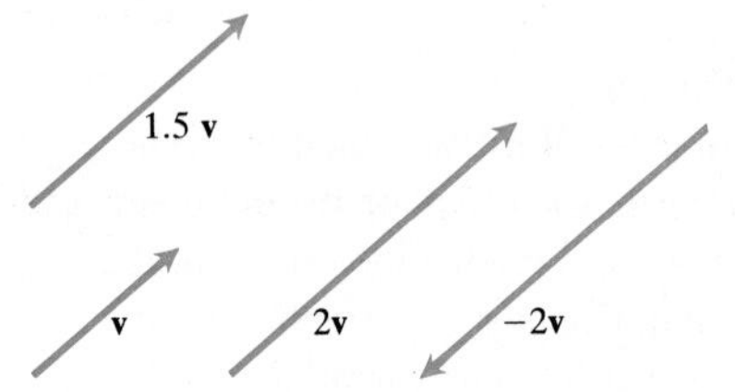

10.2 Scalar multiples of **v**.

Scalars and Scalar Multiples

We multiply a vector by a positive real number by multiplying its length by the number (Fig. 10.2). To multiply a vector by 2, we double its length. To multiply a vector by 1.5, we increase its length by 50%, and so on. We multiply a vector by a negative number by reversing the vector's direction and multiplying the length by the number's absolute value. To multiply a vector by -2, we reverse the vector's direction and double its length.

If c is a nonzero real number and **v** a vector, the direction of $c\mathbf{v}$ agrees with that of **v** if c is positive and is opposite to that of **v** if c is negative. Since real numbers work like scaling factors in this context, we call them **scalars** and call multiples like $c\mathbf{v}$ scalar multiples of **v**.

To include multiplication by zero, we adopt the convention that multiplying a vector by zero produces the **zero vector 0**, consisting of points that are degenerate line segments of zero length. Unlike other vectors, the vector **0** has no direction.

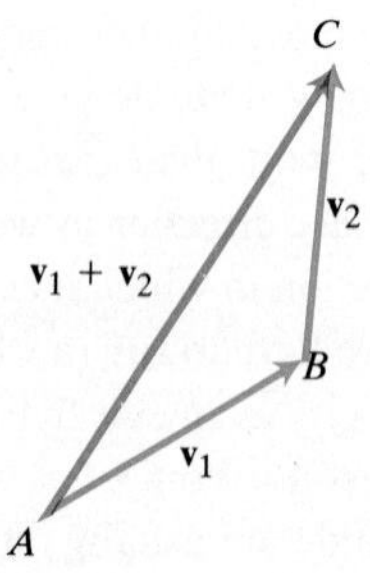

10.3 The sum of $\mathbf{v}_1$ and $\mathbf{v}_2$.

Geometric Addition: The Parallelogram Law

Two vectors $\mathbf{v}_1$ and $\mathbf{v}_2$ may be added geometrically by drawing a representative of $\mathbf{v}_1$, say from A to B as in Fig. 10.3, and then a representative of $\mathbf{v}_2$ starting from the

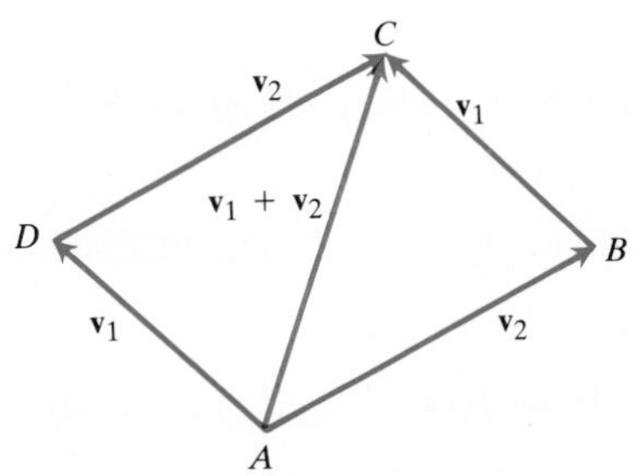

10.4 The Parallelogram Law of Addition. Quadrilateral $ABCD$ is a parallelogram because both pairs of opposite sides have equal lengths. The law was used by Aristotle to describe the combined action of two forces.

terminal point B of $\mathbf{v}_1$. In Fig. 10.3, $\mathbf{v}_2 = \overrightarrow{BC}$. The sum $\mathbf{v}_1 + \mathbf{v}_2$ is then the vector represented by the arrow from the initial point A of $\mathbf{v}_1$ to the terminal point C of $\mathbf{v}_2$. That is, if

$$\mathbf{v}_1 = \overrightarrow{AB} \qquad \text{and} \qquad \mathbf{v}_2 = \overrightarrow{BC},$$

then

$$\begin{aligned} \mathbf{v}_1 + \mathbf{v}_2 &= \overrightarrow{AB} + \overrightarrow{BC} \\ &= \overrightarrow{AC}. \end{aligned}$$

This description of addition is sometimes called the **Parallelogram Law** of addition because $\mathbf{v}_1 + \mathbf{v}_2$ is given by the diagonal of the parallelogram determined by $\mathbf{v}_1$ and $\mathbf{v}_2$ (Fig. 10.4).

Components

Whenever a vector $\mathbf{v}$ can be written as a sum

$$\mathbf{v} = \mathbf{v}_1 + \mathbf{v}_2,$$

the vectors $\mathbf{v}_1$ and $\mathbf{v}_2$ are said to be **components** of $\mathbf{v}$. We also say that we have **represented** or **resolved** $\mathbf{v}$ in terms of $\mathbf{v}_1$ and $\mathbf{v}_2$.

The most common algebra of vectors is based on representing each vector in terms of components parallel to the Cartesian coordinate axes and writing each component as an appropriate multiple of a **basic** vector of unit length. The basic vector in the positive x-direction is the vector $\mathbf{i}$ determined by the directed line segment that runs from (0, 0) to (1, 0). The basic vector in the positive y-direction is the vector $\mathbf{j}$ determined by the directed line segment from (0, 0) to (0, 1). Then $a\mathbf{i}$, a being a scalar, represents a vector of length $|a|$ parallel to the x-axis, pointing to the right if a is positive and to the left if a is negative. Similarly, $b\mathbf{j}$ is a vector of length $|b|$ parallel to the y-axis, pointing up if b is positive and down if b is negative. Figure 10.5 shows a vector $\mathbf{v} = \overrightarrow{AC}$ resolved into its $\mathbf{i}$- and $\mathbf{j}$-components as the sum

$$\mathbf{v} = a\mathbf{i} + b\mathbf{j}.$$

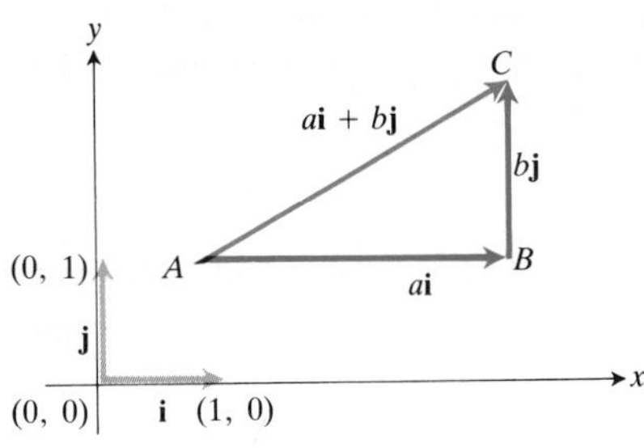

10.5 The basic vectors $\mathbf{i}$ and $\mathbf{j}$. Any vector $\overrightarrow{AC}$ in the plane can be expressed as a scalar multiple of $\mathbf{i}$ plus a scalar multiple of $\mathbf{j}$.

DEFINITION

If $\mathbf{v} = a\mathbf{i} + b\mathbf{j}$, the vectors $a\mathbf{i}$ and $b\mathbf{j}$ are the **vector components of v in the directions of i and j.** The numbers a and b are the **scalar components of v in the directions of i and j.**

Components give us a way to define the equality of vectors algebraically.

DEFINITION

Equality of Vectors (Algebraic Definition)

$$a\mathbf{i} + b\mathbf{j} = a'\mathbf{i} + b'\mathbf{j} \quad \Leftrightarrow \quad a = a' \text{ and } b = b' \tag{1}$$

That is, two vectors are equal if and only if their scalar components in the directions of $\mathbf{i}$ and $\mathbf{j}$ are identical.

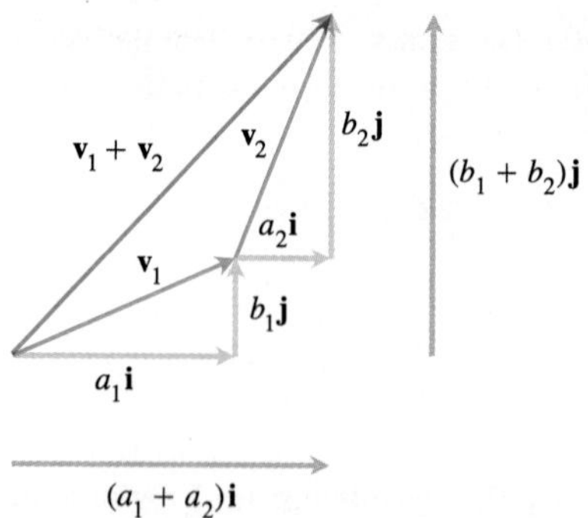

10.6 If $\mathbf{v}_1 = a_1\,\mathbf{i} + b_1\,\mathbf{j}$ and $\mathbf{v}_2 = a_2\,\mathbf{i} + b_2\,\mathbf{j}$, then $\mathbf{v}_1 + \mathbf{v}_2 = (a_1 + a_2)\,\mathbf{i} + (b_1 + b_2)\,\mathbf{j}$.

Algebraic Addition

Two vectors may be added algebraically by adding their corresponding scalar components, as shown in Fig. 10.6.

> If $\mathbf{v}_1 = a_1\,\mathbf{i} + b_1\,\mathbf{j}$, and $\mathbf{v}_2 = a_2\,\mathbf{i} + b_2\,\mathbf{j}$, then
>
> $$\mathbf{v}_1 + \mathbf{v}_2 = (a_1 + a_2)\,\mathbf{i} + (b_1 + b_2)\,\mathbf{j}. \tag{2}$$

Example 2

$$(2\,\mathbf{i} - 4\,\mathbf{j}) + (5\,\mathbf{i} + 3\,\mathbf{j}) = (2 + 5)\,\mathbf{i} + (-4 + 3)\,\mathbf{j} = 7\,\mathbf{i} - \mathbf{j}.$$

Subtraction

The negative of a vector $\mathbf{v}$ is the vector $-\mathbf{v}$ that has the same length as $\mathbf{v}$ but points in the opposite direction. To subtract a vector $\mathbf{v}_2$ from a vector $\mathbf{v}_1$, we add $-\mathbf{v}_2$ to $\mathbf{v}_1$. This may be done geometrically by drawing $-\mathbf{v}_2$ from the tip of $\mathbf{v}_1$ and then drawing the vector from the initial point of $\mathbf{v}_1$ to the tip of $-\mathbf{v}_2$, as shown in Fig. 10.7(a), where

$$\overrightarrow{AD} = \overrightarrow{AB} + \overrightarrow{BD} = \mathbf{v}_1 + (-\mathbf{v}_2) = \mathbf{v}_1 - \mathbf{v}_2.$$

Another way to draw $\mathbf{v}_1 - \mathbf{v}_2$ is to draw $\mathbf{v}_1$ and $\mathbf{v}_2$ with a common initial point and then draw $\mathbf{v}_1 - \mathbf{v}_2$ as the vector from the tip of $\mathbf{v}_2$ to the tip of $\mathbf{v}_1$. This is illustrated in Fig. 10.7(b), where

$$\overrightarrow{CB} = \overrightarrow{CA} + \overrightarrow{AB} = -\mathbf{v}_2 + \mathbf{v}_1 = \mathbf{v}_1 - \mathbf{v}_2.$$

Thus, $\overrightarrow{CB}$ is the vector that when added to $\mathbf{v}_2$ gives $\mathbf{v}_1$:

$$\overrightarrow{CB} + \mathbf{v}_2 = (\mathbf{v}_1 - \mathbf{v}_2) + \mathbf{v}_2 = \mathbf{v}_1.$$

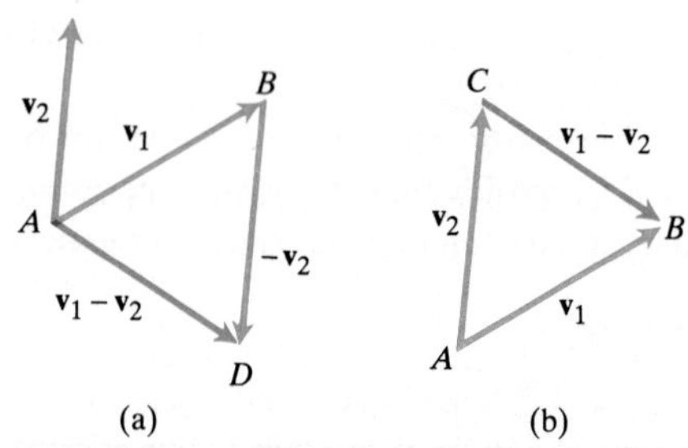

10.7 Two ways to draw $\mathbf{v}_1 - \mathbf{v}_2$: (a) as $\mathbf{v}_1 + (-\mathbf{v}_2)$ and (b) as the vector from the tip of $\mathbf{v}_2$ to the tip of $\mathbf{v}_1$.

In terms of components, vector subtraction follows the algebraic law

> $$\mathbf{v}_1 - \mathbf{v}_2 = (a_1 - a_2)\,\mathbf{i} + (b_1 - b_2)\,\mathbf{j}, \tag{3}$$

which says that corresponding scalar components are subtracted.

Example 3

$$(6\,\mathbf{i} + 2\,\mathbf{j}) - (3\,\mathbf{i} - 5\,\mathbf{j}) = (6 - 3)\,\mathbf{i} + (2 - (-5))\,\mathbf{j} = 3\,\mathbf{i} + 7\,\mathbf{j}.$$

We find the components of the vector from a point $P_1(x_1, y_1)$ to a point $P_2(x_2, y_2)$ by subtracting the components of $\overrightarrow{OP_1} = x_1\,\mathbf{i} + y_1\,\mathbf{j}$ from the components of $\overrightarrow{OP_2} = x_2\,\mathbf{i} + y_2\,\mathbf{j}$.

> The vector from $P_1(x_1, y_1)$ to $P_2\,(x_2, y_2)$ is
>
> $$\overrightarrow{P_1P_2} = (x_2 - x_1)\,\mathbf{i} + (y_2 - y_1)\,\mathbf{j}. \tag{4}$$

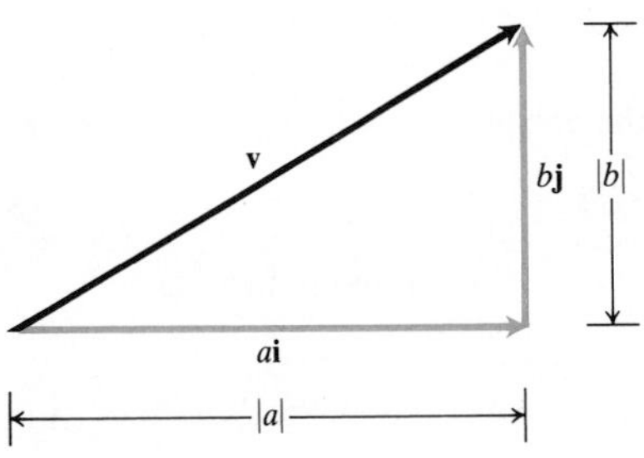

10.8 The length of $\mathbf{v}$ is $\sqrt{|a|^2 + |b|^2} = \sqrt{a^2 + b^2}$.

Example 4 The vector from P_1 (3, 4) to P_2 (5, 1) is

$$\overrightarrow{P_1P_2} = (5 - 3)\,\mathbf{i} + (1 - 4)\,\mathbf{j} = 2\,\mathbf{i} - 3\,\mathbf{j}.$$

Length

We calculate the length of $\mathbf{v} = a\mathbf{i} + b\mathbf{j}$ by representing $\mathbf{v}$ as the hypotenuse of a right triangle with sides $|a|$ and $|b|$ (Fig. 10.8) and applying the Pythagorean theorem to get

$$\text{Length of } \mathbf{v} = \sqrt{|a|^2 + |b|^2} = \sqrt{a^2 + b^2}.$$

The usual symbol for the length is $|\mathbf{v}|$, which is read "the length of $\mathbf{v}$" or "the magnitude of $\mathbf{v}$," the latter being more common in applied fields. The bars are the same as the ones we use for absolute values.

The length or magnitude of $\mathbf{v} = a\mathbf{i} + b\mathbf{j}$ is

$$|\mathbf{v}| = \sqrt{a^2 + b^2}. \qquad (5)$$

Scalar Multiplication

Scalar multiplication can be accomplished component by component.

If c is a scalar and $\mathbf{v} = a\mathbf{i} + b\mathbf{j}$ is a vector, then

$$c\mathbf{v} = c(a\mathbf{i} + b\mathbf{j}) = (ca)\,\mathbf{i} + (cb)\,\mathbf{j}. \qquad (6)$$

To check that the length of $c\mathbf{v}$ is still $|c|$ times the length of $\mathbf{v}$ when we do scalar multiplication this way, we can calculate the length with Eq. (5):

$$\begin{aligned}
|c\mathbf{v}| &= |(ca)\,\mathbf{i} + (cb)\,\mathbf{j}| && \text{(Eq. 6)} \\
&= \sqrt{(ca)^2 + (cb)^2} && \text{(Eq. (5) with } ca \text{ and } cb \text{ in place of } a \text{ and } b) \\
&= \sqrt{c^2(a^2 + b^2)} \\
&= \sqrt{c^2}\,\sqrt{a^2 + b^2} \\
&= |c|\,|\mathbf{v}|.
\end{aligned}$$

If c is a scalar and $\mathbf{v}$ is a vector, then

$$|c\mathbf{v}| = |c|\,|\mathbf{v}|. \qquad (7)$$

Example 5 If $c = -2$ and $\mathbf{v} = -3\,\mathbf{i} + 4\,\mathbf{j}$, then

$$|\mathbf{v}| = |-3\,\mathbf{i} + 4\,\mathbf{j}| = \sqrt{(-3)^2 + (4)^2} = \sqrt{9 + 16} = \sqrt{25} = 5$$

$$|-2\mathbf{v}| = |(-2)(-3\,\mathbf{i} + 4\,\mathbf{j})| = |6\,\mathbf{i} - 8\,\mathbf{j}| = \sqrt{(6)^2 + (-8)^2} = \sqrt{36 + 64}$$

$$= \sqrt{100} = 10 = |-2|5 = |c|\,|\mathbf{v}|.$$

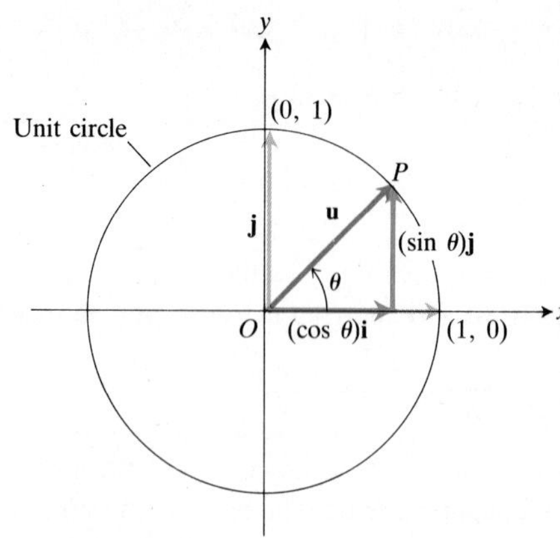

10.9 The unit vector that makes an angle of measure θ with the positive x-axis. Every unit vector has the form

$$\mathbf{u} = (\cos\theta)\,\mathbf{i} + (\sin\theta)\,\mathbf{j}$$

for some θ.

The Zero Vector

In terms of components, the zero vector is the vector

$$\mathbf{0} = 0\,\mathbf{i} + 0\,\mathbf{j}. \tag{8}$$

It is the only vector whose length is zero, as we can see from the fact that

$$|a\mathbf{i} + b\mathbf{j}| = \sqrt{a^2 + b^2} = 0 \quad \Leftrightarrow \quad a = b = 0.$$

Unit Vectors

Any vector $\mathbf{u}$ whose length is equal to the unit of length used along the coordinate axes is called a **unit vector.** The vectors $\mathbf{i}$ and $\mathbf{j}$ are unit vectors:

$$|\,\mathbf{i}\,| = |1\,\mathbf{i} + 0\,\mathbf{j}| = \sqrt{1^2 + 0^2} = 1, \qquad |\,\mathbf{j}\,| = |0\,\mathbf{i} + 1\,\mathbf{j}| = \sqrt{0^2 + 1^2} = 1.$$

If $\mathbf{u}$ is the unit vector obtained by rotating $\mathbf{i}$ through an angle θ in the positive direction, then $\mathbf{u}$ has a horizontal component $\cos\theta$ and a vertical component $\sin\theta$ (Fig. 10.9), so that

$$\mathbf{u} = (\cos\theta)\,\mathbf{i} + (\sin\theta)\,\mathbf{j}. \tag{9}$$

If we allow the angle θ in Eq. (9) to vary from 0 to 2π, the point P in Fig. 10.9 traces the unit circle $x^2 + y^2 = 1$ once counterclockwise. Since this includes all possible directions, every unit vector in the plane is given by Eq. (9) for some value of θ.

In handwritten work it is common to denote unit vectors with small "hats," as in $\hat{\mathrm{u}}$ (pronounced "u hat"). In hat notation, $\mathbf{i}$ and $\mathbf{j}$ become $\hat{\imath}$ and $\hat{\jmath}$.

Direction as a Vector

It is common in subjects like classical electricity and magnetism, which use vectors a great deal, to define the direction of a nonzero vector $\mathbf{v}$ to be the unit vector obtained by dividing $\mathbf{v}$ by its own length.

DEFINITION

If $\mathbf{v} \neq \mathbf{0}$, the **direction** of $\mathbf{v}$ is the vector $\dfrac{\mathbf{v}}{|\mathbf{v}|}$. (10)

Notice that instead of just saying that $\mathbf{v}/|\mathbf{v}|$ *represents* the direction of $\mathbf{v}$, we say that it *is* the direction of $\mathbf{v}$.

To see that $\mathbf{v}/|\mathbf{v}|$ really is a unit vector, we can calculate its length directly:

$$\begin{aligned}\text{Length of } \frac{\mathbf{v}}{|\mathbf{v}|} &= \left|\frac{\mathbf{v}}{|\mathbf{v}|}\right| \\ &= \frac{1}{|\mathbf{v}|}\,|\mathbf{v}| \qquad \left(\text{Eq. (7) with } c = \frac{1}{|\mathbf{v}|}\right) \\ &= 1.\end{aligned}$$

Any nonzero vector can be expressed in terms of its length and direction by using the equation

$$\mathbf{v} = |\mathbf{v}| \cdot \frac{\mathbf{v}}{|\mathbf{v}|} = (\text{length of } \mathbf{v}) \cdot (\text{direction of } \mathbf{v}). \tag{11}$$

Example 6 Express $\mathbf{v} = 3\,\mathbf{i} - 4\,\mathbf{j}$ in terms of its length and direction.

Solution Length of $\mathbf{v}$: $|\mathbf{v}| = \sqrt{(3)^2 + (-4)^2} = \sqrt{9 + 16} = 5$

Direction of $\mathbf{v}$: $\dfrac{\mathbf{v}}{|\mathbf{v}|} = \dfrac{3\,\mathbf{i} - 4\,\mathbf{j}}{5} = \dfrac{3}{5}\mathbf{i} - \dfrac{4}{5}\mathbf{j}$

$$\mathbf{v} = 3\,\mathbf{i} - 4\,\mathbf{j} = \underbrace{5}_{\text{length of } \mathbf{v}}\underbrace{\left(\frac{3}{5}\mathbf{i} - \frac{4}{5}\mathbf{j}\right)}_{\text{direction of } \mathbf{v}}$$

It follows from the definition of direction as a vector that vectors **A** and **B** have the same direction if

$$\frac{\mathbf{A}}{|\mathbf{A}|} = \frac{\mathbf{B}}{|\mathbf{B}|} \quad \text{or} \quad \mathbf{A} = \frac{|\mathbf{A}|}{|\mathbf{B}|}\mathbf{B}. \tag{12}$$

Thus, if **A** and **B** have the same direction, **A** is a positive scalar multiple of **B**. Conversely, if $\mathbf{A} = k\mathbf{B}$, $k > 0$, then

$$\frac{\mathbf{A}}{|\mathbf{A}|} = \frac{k\mathbf{B}}{|k\mathbf{B}|} = \frac{k}{|k|}\frac{\mathbf{B}}{|\mathbf{B}|} = \frac{k}{k}\frac{\mathbf{B}}{|\mathbf{B}|} = \frac{\mathbf{B}}{|\mathbf{B}|}. \tag{13}$$

Therefore, two nonzero vectors **A** and **B** have the same direction if and only if **A** is a positive scalar multiple of **B**.

We say that two nonzero vectors **A** and **B** have *opposite* directions if their directions are opposite in sign:

$$\frac{\mathbf{A}}{|\mathbf{A}|} = -\frac{\mathbf{B}}{|\mathbf{B}|}. \tag{14}$$

From this it follows that **A** and **B** have opposite directions if and only if **A** is a negative scalar multiple of **B**.

Example 7

a) Same direction: $\mathbf{A} = 3\,\mathbf{i} - 4\,\mathbf{j}$ and $\mathbf{B} = \frac{3}{2}\mathbf{i} - 2\,\mathbf{j} = \frac{1}{2}\mathbf{A}$

(**B** is a positive scalar multiple of **A**.)

b) Opposite directions: $\mathbf{A} = 3\,\mathbf{i} - 4\,\mathbf{j}$ and $\mathbf{B} = -9\,\mathbf{i} + 12\,\mathbf{j} = -3\mathbf{A}$

(**B** is a negative scalar multiple of **A**.)

Slopes, Tangents, and Normals

Two vectors are said to be **parallel** if they are either positive or negative scalar multiples of one another or, equivalently, if the line segments representing them are parallel. Similarly, a vector is parallel to a line if the segments that represent the vector are parallel to the line. The **slope** of a vector that is not parallel to the y-axis is the slope shared by the lines parallel to the vector. Thus, when $a \neq 0$, the vector

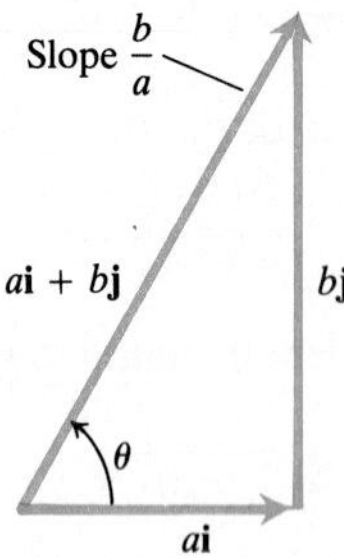

10.10 If $a \neq 0$, the vector $a\mathbf{i} + b\mathbf{j}$ has slope $b/a = \tan\theta$.

$\mathbf{v} = a\mathbf{i} + b\mathbf{j}$ has a well-defined slope, which can be calculated from the components of $\mathbf{v}$ as the number b/a (Fig. 10.10).

When we say that a vector is **tangent** or **normal** to a curve at a point, we mean that the vector is parallel or normal to the line that is tangent to the curve at the point. The next example shows how to find such vectors.

Example 8 Find unit vectors tangent and normal to the curve

$$y = \frac{x^3}{2} + \frac{1}{2}$$

at the point (1, 1).

Solution We find the unit vectors that are parallel and normal to the curve's tangent line at the point (1, 1), shown in Fig. 10.11.

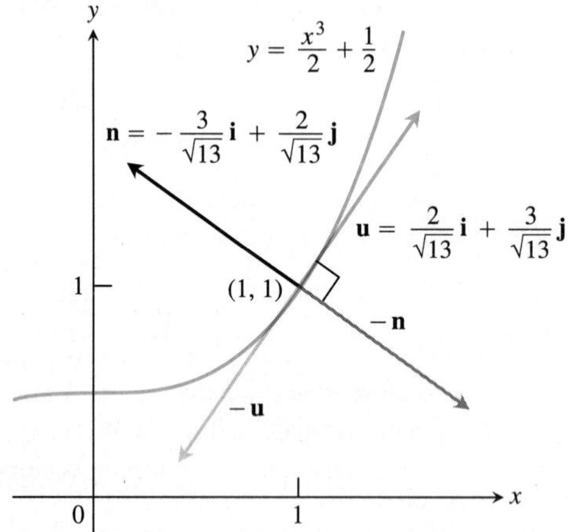

10.11 The unit tangent and normal vectors at the point (1, 1) on the curve $y = (x^3/2) + 1/2$.

The slope of the line tangent to the curve at (1, 1) is

$$y' = \left.\frac{3x^2}{2}\right|_{x=1} = \frac{3}{2}.$$

We find a unit vector with this slope. The vector $\mathbf{v} = 2\,\mathbf{i} + 3\,\mathbf{j}$ has slope 3/2, as does every nonzero multiple of $\mathbf{v}$. To find a multiple of $\mathbf{v}$ that is a unit vector, we divide $\mathbf{v}$ by its length,

$$|\mathbf{v}| = \sqrt{2^2 + 3^2} = \sqrt{13}.$$

This produces the unit vector

$$\mathbf{u} = \frac{\mathbf{v}}{|\mathbf{v}|} = \frac{2}{\sqrt{13}}\mathbf{i} + \frac{3}{\sqrt{13}}\mathbf{j}.$$

The vector $\mathbf{u}$ is tangent to the curve at (1, 1) because it has the same direction as $\mathbf{v}$. Of course, the vector

$$-\mathbf{u} = -\frac{2}{\sqrt{13}}\mathbf{i} - \frac{3}{\sqrt{13}}\mathbf{j},$$

which points in the opposite direction, is also tangent to the curve at (1, 1). Without some additional requirement, there is no reason to prefer one of these vectors to the other.

If $\mathbf{v} = a\mathbf{i} + b\mathbf{j}$, then $\mathbf{p} = -b\mathbf{i} + a\mathbf{j}$ and $\mathbf{q} = b\mathbf{i} - a\mathbf{j}$ are perpendicular to $\mathbf{v}$ because their slopes are both $-a/b$, the negative reciprocal of $\mathbf{v}$'s slope.

To find unit vectors normal to the curve at (1, 1), we look for unit vectors whose slopes are the negative reciprocal of the slope of $\mathbf{u}$. This is quickly done by interchanging the scalar components of $\mathbf{u}$ and changing the sign of one of them. We obtain

$$\mathbf{n} = -\frac{3}{\sqrt{13}}\mathbf{i} + \frac{2}{\sqrt{13}}\mathbf{j} \quad \text{and} \quad -\mathbf{n} = \frac{3}{\sqrt{13}}\mathbf{i} - \frac{2}{\sqrt{13}}\mathbf{j}.$$

Again, either one will do. The vectors have opposite directions but both are normal to the curve at the point (1, 1).

EXERCISES 10.1

1. The vectors **A**, **B**, and **C** in Fig. 10.12 lie in a plane. Copy them on a sheet of paper. Then, by arranging the vectors head to tail, as in Figs. 10.3 and 10.6, sketch

a) **A** + **B** b) **A** + **B** + **C**
c) **A** − **B** d) **A** − **C**

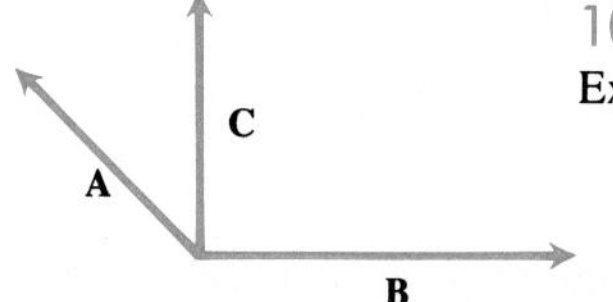

10.12 The vectors for Exercise 1.

2. The vectors **A**, **B**, and **C** in Fig. 10.13 lie in a plane. Copy them on a sheet of paper. Then, by arranging the vectors head to tail, as in Figs. 10.3 and 10.6, sketch

a) **A** − **B** b) **A** − **B** + **C**
c) 2**A** − **B** d) **A** + **B** + **C**

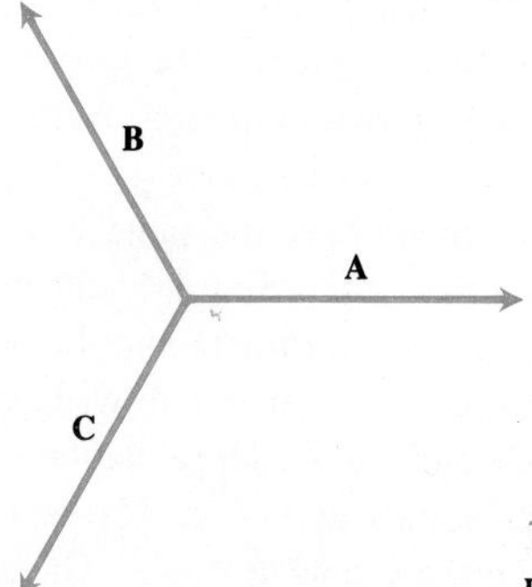

10.13 The vectors for Exercise 2.

Write the sums and differences in Exercises 3–8 in the form $a\,\mathbf{i} + b\,\mathbf{j}$.

3. $(2\,\mathbf{i} - 7\,\mathbf{j}) + (\mathbf{i} + 6\,\mathbf{j})$

4. $(\sqrt{3}\,\mathbf{i} - 3\,\mathbf{j}) + 6\,\mathbf{i}$

5. $(-2\,\mathbf{i} + 6\,\mathbf{j}) - 2(\mathbf{i} + \mathbf{j}) + 3\,\mathbf{i} - 4\,\mathbf{j}$

6. $3\left(\frac{\mathbf{i}}{2} - \frac{\mathbf{j}}{3}\right) - \frac{1}{2}(5\,\mathbf{i} - 2\,\mathbf{j})$

7. $2((\ln 2)\,\mathbf{i} + \mathbf{j}) - ((\ln 8)\,\mathbf{i} + \pi\mathbf{j})$

8. $(\mathbf{i} + \sqrt{2}\,\mathbf{j}) - 7(\mathbf{i} - \mathbf{j}) - \sqrt{2}\,(2\,\mathbf{i} + \mathbf{j})$

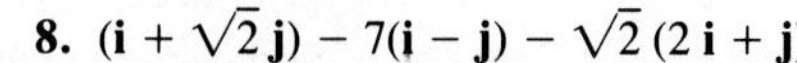

Express the vectors in Exercises 9–16 in the form $a\mathbf{i} + b\mathbf{j}$ and sketch them as arrows in the coordinate plane.

9. $\overrightarrow{P_1P_2}$ if P_1 is the point (1, 3) and P_2 is the point (2, −1)

10. $\overrightarrow{OP_3}$ if O is the origin and P_3 is the midpoint of the vector $\overrightarrow{P_1P_2}$ joining $P_1(2, -1)$ and $P_2(-4, 3)$

11. The vector from the point $A(2, 3)$ to the origin

12. The sum of the vectors $\overrightarrow{AB}$ and $\overrightarrow{CD}$, given the four points $A(1, -1)$, $B(2, 0)$, $C(-1, 3)$, and $D(-2, 2)$

13. The unit vectors $\mathbf{u} = (\cos\theta)\,\mathbf{i} + (\sin\theta)\,\mathbf{j}$ for $\theta = \pi/6$ and $\theta = 2\pi/3$. Include the circle $x^2 + y^2 = 1$ in your sketch.

14. The unit vectors $\mathbf{u} = (\cos\theta)\,\mathbf{i} + (\sin\theta)\,\mathbf{j}$ for $\theta = -\pi/4$ and $\theta = -3\pi/4$. Include the circle $x^2 + y^2 = 1$ in your sketch.

15. The unit vector obtained by rotating **j** 120° clockwise about the origin

16. The unit vector obtained by rotating **i** 135° counterclockwise about the origin

In Exercises 17–20, find the unit vectors that are tangent and normal to the curve at the given point (four vectors in all). Then sketch the vectors and curve together.

17. $y = x^2$, (2, 4)

18. $x^2 + 2y^2 = 6$, (2, 1)

19. $y = \tan^{-1}x$, $(1, \pi/4)$

20. $y = \sum_{n=0}^{\infty} \frac{x^n}{n!}$, (0, 1)

In Exercises 21–24, find the unit vectors that are tangent and normal to the curve at the given point (four vectors in all).

21. $3x^2 + 8xy + 2y^2 - 3 = 0$, (1, 0)

22. $x^2 - 6xy + 8y^2 - 2x - 1 = 0$, (1, 1)

23. $y = \int_0^x \sqrt{3 + t^4}\,dt$, (0, 0)

24. $y = \int_e^x \ln(\ln t)\,dt$, $(e, 0)$

Use Eq. (11) to express the vectors in Exercises 25–30 as products of their lengths and directions.

25. $\mathbf{i} + \mathbf{j}$

26. $2\,\mathbf{i} - 3\,\mathbf{j}$

27. $\sqrt{3}\,\mathbf{i} + \mathbf{j}$

28. $-2\,\mathbf{i} + 3\,\mathbf{j}$

29. $5\,\mathbf{i} + 12\,\mathbf{j}$

30. $-5\,\mathbf{i} - 12\,\mathbf{j}$

31. Show that $\mathbf{A} = 3\,\mathbf{i} + 6\,\mathbf{j}$ and $\mathbf{B} = -\mathbf{i} - 2\,\mathbf{j}$ have opposite directions. Sketch **A** and **B** together.

32. Show that $\mathbf{A} = 3\,\mathbf{i} + 6\,\mathbf{j}$ and $\mathbf{B} = (1/2)\,\mathbf{i} + \mathbf{j}$ have the same direction.

33. Find a vector 2 units long in the direction of $\mathbf{A} = -\mathbf{i} - \mathbf{j}$. How many such vectors are there?

34. Find a vector 5 units long in the direction opposite to the direction of $\mathbf{A} = (3/5)\,\mathbf{i} + (4/5)\,\mathbf{j}$. How many such vectors are there?

35. Let $\mathbf{v}$ be a vector in the plane not parallel to the y-axis. Is the slope of $-\mathbf{v}$ the same as the slope of $\mathbf{v}$, or is it the negative of the slope of $\mathbf{v}$? Explain.

10.2 Cartesian (Rectangular) Coordinates and Vectors in Space

Our goal now is to describe the three-dimensional Cartesian coordinate system and learn our way around in space. This means defining distance, practicing with the arithmetic of vectors (the rules are the same as in the plane but with an extra term), and making connections between sets of points in space and equations and inequalities. Everyone we know finds it harder to draw in three dimensions than in two, so we have included some drawing tips as well (pages 714–715). The Cartesian coordinates for space are often called *rectangular coordinates* because the axes that define them meet at right angles.

Cartesian Coordinates

To locate points in space, we use three mutually perpendicular coordinate axes, arranged as in Fig. 10.14. The axes Ox, Oy, and Oz shown there make what is known as a *right-handed* coordinate frame. When you hold your right hand so that the fingers curl from the positive x-axis toward the positive y-axis, your thumb points along the positive z-axis.

The Cartesian coordinates (x, y, z) of a point P in space are the numbers at which the planes through P perpendicular to the axes cut the axes.

Points that lie on the x-axis have y- and z-coordinates equal to zero. That is, they have coordinates of the form $(x, 0, 0)$. Similarly, points on the y-axis have coordinates of the form $(0, y, 0)$. Points on the z-axis have coordinates of the form $(0, 0, z)$.

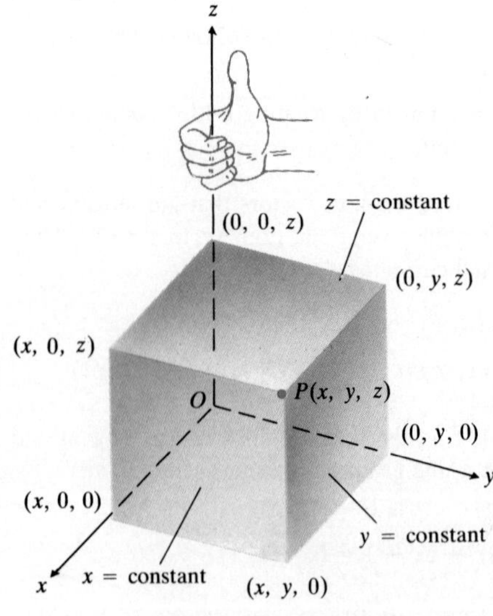

10.14 The Cartesian coordinate system is right-handed.

The points in a plane perpendicular to the x-axis all have the same x-coordinate, this being the number at which that plane cuts the x-axis. Similarly, the points in a plane perpendicular to the y-axis have a common y-coordinate and the points in a plane perpendicular to the z-axis have a common z-coordinate. To write equations for these planes, we name the common coordinate's value. The equation $x = 2$ is an equation for the plane perpendicular to the x-axis at $x = 2$. The equation $y = 3$ is an equation for the plane perpendicular to the y-axis at $y = 3$. The equation $z = 5$ is an equation for the plane perpendicular to the z-axis at $z = 5$. Figure 10.15 shows the planes $x = 2$, $y = 3$, and $z = 5$, together with their intersection point $(2, 3, 5)$.

The planes $x = 2$ and $y = 3$ in Fig. 10.15 intersect in a line that runs parallel to the z-axis. This line is described by the *pair* of equations $x = 2$, $y = 3$. A point (x, y, z) lies on the line if and only if $x = 2$ and $y = 3$. Similarly, the line of intersection of the planes $y = 3$ and $z = 5$ is described by the equation pair $y = 3$, $z = 5$. This line runs parallel to the x-axis. The line of intersection of the

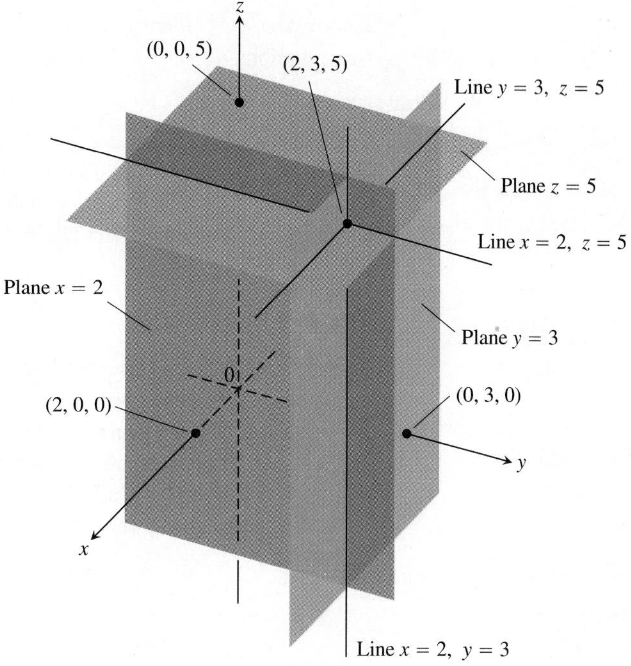

10.15 The planes $x = 2$, $y = 3$, and $z = 5$ determine three lines through the point (2, 3, 5).

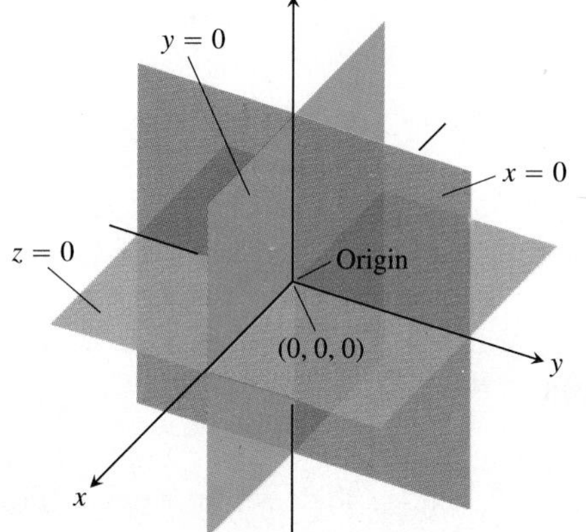

10.16 The planes $x = 0$, $y = 0$, and $z = 0$ are the planes determined by the coordinate axes. They divide space into eight cells called octants.

planes $x = 2$ and $z = 5$, parallel to the y-axis, is described by the equation pair $x = 2$, $z = 5$.

The planes determined by the three coordinate axes are the ***xy*-plane,** whose standard equation is $z = 0$; the ***yz*-plane,** whose standard equation is $x = 0$; and the ***xz*-plane,** whose standard equation is $y = 0$. They meet in the point (0, 0, 0), which is called the **origin** of the coordinate system (Fig. 10.16).

The three **coordinate planes** $x = 0$, $y = 0$, and $z = 0$ divide space into eight cells called **octants.** The octant in which the point coordinates are all positive is called the **first octant,** but there is no conventional numbering for the remaining seven octants.

In the following examples, we match coordinate equations and inequalities with the sets of points they define in space.

Example 1

Defining equations and inequalities	Verbal description
$z \geq 0$	The half-space consisting of the points on and above the xy-plane.
$x = -3$	The plane perpendicular to the x-axis at $x = -3$. This plane lies parallel to the yz-plane and 3 units behind it.
$z = 0,\ x \leq 0,\ y \geq 0$	The second quadrant of the xy-plane.
$x \geq 0,\ y \geq 0,\ z \geq 0$	The first octant.
$-1 \leq y \leq 1$	The slab between the planes $y = -1$ and $y = 1$ (planes included).
$y = -2,\ z = 2$	The line in which the planes $y = -2$ and $z = 2$ intersect. Alternatively, the line through the point $(0, -2, 2)$ parallel to the x-axis.

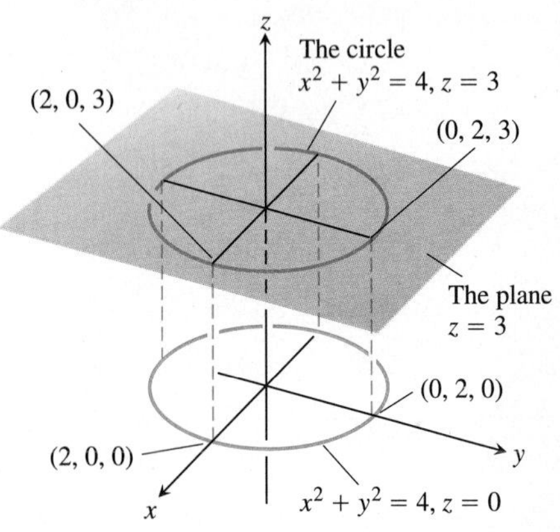

10.17 The circle $x^2 + y^2 = 4$, $z = 3$.

Example 2 Identify the set of points $P(x, y, z)$ whose coordinates satisfy the two equations

$$x^2 + y^2 = 4 \quad \text{and} \quad z = 3.$$

Solution The points lie in the horizontal plane $z = 3$ and, in this plane, make up the circle $x^2 + y^2 = 4$. We call this set of points "the circle $x^2 + y^2 = 4$ in the plane $z = 3$" or, more simply, "the circle $x^2 + y^2 = 4$, $z = 3$" (Fig. 10.17).

Vectors in Space

The sets of equivalent directed line segments we use to represent forces, displacements, and velocities in space are called vectors, just as they are in the plane. The same rules of addition, subtraction, and scalar multiplication apply.

The vectors represented by the directed line segments from the origin to the points (1, 0, 0), (0, 1, 0), and (0, 0, 1) are the **basic vectors.** We denote them by **i**, **j**, and **k**. The **position vector r** from the origin O to the typical point $P(x, y, z)$ is

$$\mathbf{r} = \overrightarrow{OP} = x\mathbf{i} + y\mathbf{j} + z\mathbf{k}. \tag{1}$$

Addition, Subtraction, and Scalar Multiplication for Vectors in Space

For any vectors $\mathbf{A} = a_1\,\mathbf{i} + a_2\,\mathbf{j} + a_3\,\mathbf{k}$ and $\mathbf{B} = b_1\,\mathbf{i} + b_2\,\mathbf{j} + b_3\,\mathbf{k}$, and for any scalar c,

$$\mathbf{A} + \mathbf{B} = (a_1 + b_1)\,\mathbf{i} + (a_2 + b_2)\,\mathbf{j} + (a_3 + b_3)\,\mathbf{k}$$

$$\mathbf{A} - \mathbf{B} = (a_1 - b_1)\,\mathbf{i} + (a_2 - b_2)\,\mathbf{j} + (a_3 - b_3)\,\mathbf{k}$$

$$c\mathbf{A} = (ca_1)\,\mathbf{i} + (ca_2)\,\mathbf{j} + (ca_3)\,\mathbf{k}.$$

The Vector Between Two Points

Often we want to express the vector $\overrightarrow{P_1P_2}$ from the point $P_1(x_1, y_1, z_1)$ to the point $P_2(x_2, y_2, z_2)$ in terms of the coordinates of P_1 and P_2. To do so, we observe (Fig. 10.18) that

$$\begin{aligned}\overrightarrow{P_1P_2} &= \overrightarrow{OP_2} - \overrightarrow{OP_1}\\ &= (x_2\,\mathbf{i} + y_2\,\mathbf{j} + z_2\,\mathbf{k}) - (x_1\,\mathbf{i} + y_1\,\mathbf{j} + z_1\,\mathbf{k})\\ &= (x_2 - x_1)\,\mathbf{i} + (y_2 - y_1)\,\mathbf{j} + (z_2 - z_1)\,\mathbf{k}.\end{aligned} \tag{2}$$

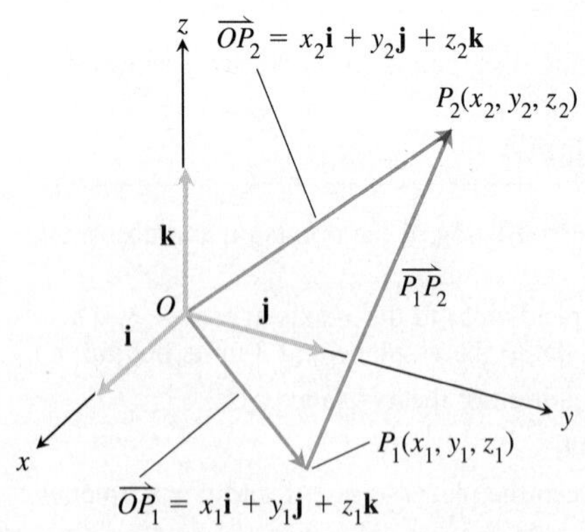

10.18 The vector from P_1 to P_2 is $\overrightarrow{P_1P_2} = (x_2 - x_1)\,\mathbf{i} + (y_2 - y_1)\,\mathbf{j} + (z_2 - z_1)\,\mathbf{k}$.

The vector from $P_1(x_1, y_1, z_1)$ to $P_2(x_2, y_2, z_2)$ is

$$\overrightarrow{P_1P_2} = (x_2 - x_1)\,\mathbf{i} + (y_2 - y_1)\,\mathbf{j} + (z_2 - z_1)\,\mathbf{k}. \tag{3}$$

Length and Direction

As always, the important features of a vector are its length and direction. The length of a vector $a\mathbf{i} + b\mathbf{j} + c\mathbf{k}$ is calculated by applying the Pythagorean theorem twice.

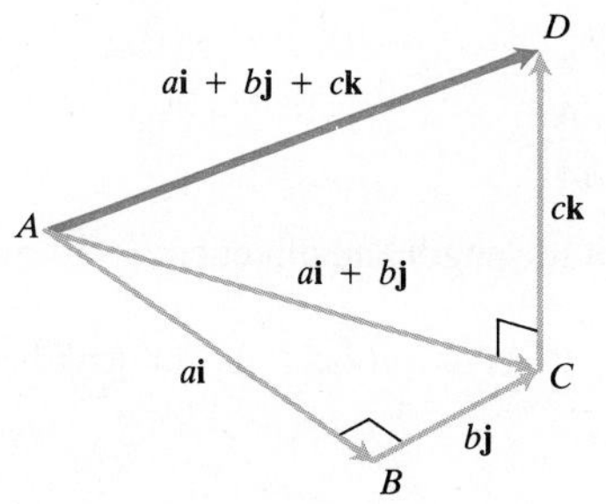

10.19 The length of the vector $\overrightarrow{AD}$ can be determined by applying the Pythagorean theorem to the right triangles ABC and ACD.

In the notation of Fig. 10.19,

$$|\overrightarrow{AC}| = |a\mathbf{i} + b\mathbf{j}| = \sqrt{a^2 + b^2},$$

from triangle ABC, and

$$|a\mathbf{i} + b\mathbf{j} + c\mathbf{k}| = |\overrightarrow{AD}| = \sqrt{|\overrightarrow{AC}|^2 + |\overrightarrow{CD}|^2}$$
$$= \sqrt{a^2 + b^2 + c^2},$$

from triangle ACD.

The **length** or **magnitude** of $\mathbf{A} = a\mathbf{i} + b\mathbf{j} + c\mathbf{k}$ is

$$|\mathbf{A}| = |a\mathbf{i} + b\mathbf{j} + c\mathbf{k}| = \sqrt{a^2 + b^2 + c^2}. \tag{4}$$

Example 3 The length of $\mathbf{A} = \mathbf{i} - 2\,\mathbf{j} + 3\,\mathbf{k}$ is

$$|\mathbf{A}| = \sqrt{(1)^2 + (-2)^2 + (3)^2} = \sqrt{1 + 4 + 9} = \sqrt{14}.$$

If we multiply a vector $\mathbf{A} = a_1\,\mathbf{i} + a_2\,\mathbf{j} + a_3\,\mathbf{k}$ by a scalar c, the length of $c\mathbf{A}$ is $|c|$ times the length of $\mathbf{A}$, just as in the plane. The reason is the same, as well:

$$c\mathbf{A} = ca_1\,\mathbf{i} + ca_2\,\mathbf{j} + ca_3\,\mathbf{k},$$
$$|c\mathbf{A}| = \sqrt{(ca_1)^2 + (ca_2)^2 + (ca_3)^2} = \sqrt{c^2a_1^2 + c^2a_2^2 + c^2a_3^2} \tag{5}$$
$$= |c|\sqrt{a_1^2 + a_2^2 + a_3^2} = |c||\mathbf{A}|.$$

Example 4 If $\mathbf{A}$ is the vector of Example 3, then the length of

$$2\mathbf{A} = 2(\,\mathbf{i} - 2\,\mathbf{j} + 3\,\mathbf{k}) = 2\,\mathbf{i} - 4\,\mathbf{j} + 6\,\mathbf{k}$$

is

$$\sqrt{(2)^2 + (-4)^2 + (6)^2} = \sqrt{4 + 16 + 36} = \sqrt{56}$$
$$= \sqrt{4 \cdot 14} = 2\sqrt{14} = 2|\mathbf{A}|.$$

As with vectors in the plane, vectors of unit length are called **unit vectors.** The vectors **i**, **j**, and **k** are unit vectors because

$$|\mathbf{i}| = |1\,\mathbf{i} + 0\,\mathbf{j} + 0\,\mathbf{k}| = \sqrt{1^2 + 0^2 + 0^2} = 1,$$
$$|\mathbf{j}| = |0\,\mathbf{i} + 1\,\mathbf{j} + 0\,\mathbf{k}| = \sqrt{0^2 + 1^2 + 0^2} = 1,$$
$$|\mathbf{k}| = |0\,\mathbf{i} + 0\,\mathbf{j} + 1\,\mathbf{k}| = \sqrt{0^2 + 0^2 + 1^2} = 1.$$

The direction of a nonzero vector $\mathbf{A}$ is the unit vector obtained by dividing $\mathbf{A}$ by its length $|\mathbf{A}|$.

The **direction** of $\mathbf{A}$ is $\dfrac{\mathbf{A}}{|\mathbf{A}|}$. (6)

As in the plane, we can use the equation

$$\mathbf{A} = |\mathbf{A}| \cdot \frac{\mathbf{A}}{|\mathbf{A}|} \tag{7}$$

to express any nonzero vector as a product of its length and direction.

Example 5 Express $\mathbf{A} = \mathbf{i} - 2\,\mathbf{j} + 3\,\mathbf{k}$ as a product of its length and direction.

Solution

$$\begin{aligned} \mathbf{A} &= |\mathbf{A}| \cdot \frac{\mathbf{A}}{|\mathbf{A}|} && \text{(Eq. 7)} \\ &= \sqrt{14} \cdot \frac{\mathbf{i} - 2\,\mathbf{j} + 3\,\mathbf{k}}{\sqrt{14}} && \text{(Values from Example 3)} \\ &= \sqrt{14}\left(\frac{1}{\sqrt{14}}\,\mathbf{i} - \frac{2}{\sqrt{14}}\,\mathbf{j} + \frac{3}{\sqrt{14}}\,\mathbf{k}\right) = (\text{length of } \mathbf{A}) \cdot (\text{direction of } \mathbf{A}) \end{aligned}$$

Example 6 Find a unit vector $\mathbf{u}$ in the direction of the vector from $P_1(1, 0, 1)$ to $P_2(3, 2, 0)$.

Solution The vector we want is the direction of $\overrightarrow{P_1P_2}$. To find it, we divide $\overrightarrow{P_1P_2}$ by its own length:

$$\overrightarrow{P_1P_2} = (3 - 1)\,\mathbf{i} + (2 - 0)\,\mathbf{j} + (0 - 1)\,\mathbf{k} = 2\,\mathbf{i} + 2\,\mathbf{j} - \mathbf{k},$$

$$|\overrightarrow{P_1P_2}| = \sqrt{(2)^2 + (2)^2 + (-1)^2} = \sqrt{4 + 4 + 1} = \sqrt{9} = 3,$$

$$\mathbf{u} = \frac{\overrightarrow{P_1P_2}}{|\overrightarrow{P_1P_2}|} = \frac{2\,\mathbf{i} + 2\,\mathbf{j} - \mathbf{k}}{3} = \frac{2}{3}\,\mathbf{i} + \frac{2}{3}\,\mathbf{j} - \frac{1}{3}\,\mathbf{k}.$$

Example 7 Find a vector 6 units long in the direction of $\mathbf{A} = 2\,\mathbf{i} + 2\,\mathbf{j} - \mathbf{k}$.

Solution The vector we want is

$$6\frac{\mathbf{A}}{|\mathbf{A}|} = 6\,\frac{2\,\mathbf{i} + 2\,\mathbf{j} - \mathbf{k}}{\sqrt{2^2 + 2^2 + (-1)^2}} = 6\,\frac{2\,\mathbf{i} + 2\,\mathbf{j} - \mathbf{k}}{3} = 4\,\mathbf{i} + 4\,\mathbf{j} - 2\,\mathbf{k}.$$

Distance in Space

To find the distance between two points P_1 and P_2 in space, we find the length of $\overrightarrow{P_1P_2}$. Equation (8) gives the resulting formula.

The Distance Between $P_1(x_1, y_1, z_1)$ and $P_2(x_2, y_2, z_2)$

$$|\overrightarrow{P_1P_2}| = \sqrt{(x_2 - x_1)^2 + (y_2 - y_1)^2 + (z_2 - z_1)^2} \tag{8}$$

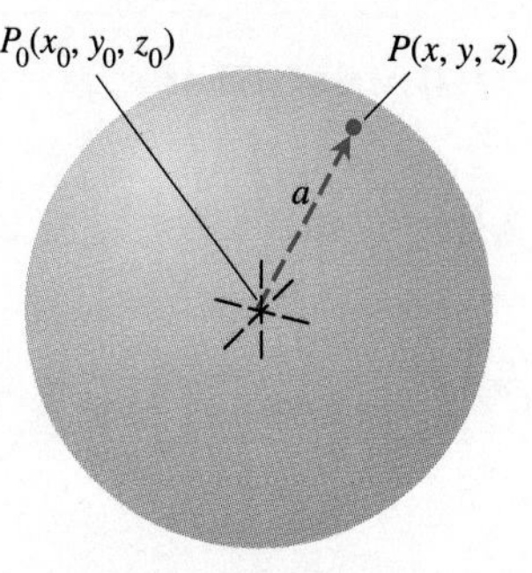

10.20 The standard equation of the sphere of radius a centered at the point (x_0, y_0, z_0) is

$$(x - x_0)^2 + (y - y_0)^2 + (z - z_0)^2 = a^2.$$

Example 8 The distance between $P_1(2, 1, 5)$ and $P_2(-2, 3, 0)$ is

$$\begin{aligned} |\overrightarrow{P_1P_2}| &= \sqrt{(-2-2)^2 + (3-1)^2 + (0-5)^2} \\ &= \sqrt{16 + 4 + 25} \\ &= \sqrt{45} = 3\sqrt{5}. \end{aligned}$$

Spheres

We use Eq. (8) to write equations for spheres (Fig. 10.20). Since a point $P(x, y, z)$ lies on the sphere of radius a centered at $P_0(x_0, y_0, z_0)$ if and only if it lies a units from P_0, it lies on the sphere if and only if

$$|\overrightarrow{P_0P}| = a$$

or

$$(x - x_0)^2 + (y - y_0)^2 + (z - z_0)^2 = a^2.$$

> **The Standard Equation for the Sphere of Radius a and Center (x_0, y_0, z_0)**
>
> $$(x - x_0)^2 + (y - y_0)^2 + (z - z_0)^2 = a^2 \tag{9}$$

Example 9 Find the center and radius of the sphere

$$x^2 + y^2 + z^2 + 2x - 4y = 0.$$

Solution Complete the squares on the x-terms and y-terms to obtain

$$x^2 + 2x + 1 + y^2 - 4y + 4 + z^2 = 0 + 1 + 4$$

$$(x + 1)^2 + (y - 2)^2 + z^2 = 5.$$

This is Eq. (9) with $x_0 = -1$, $y_0 = 2$, $z_0 = 0$, and $a = \sqrt{5}$. The center is $(-1, 2, 0)$ and the radius is $\sqrt{5}$.

Example 10 *Sets Bounded by Spheres or Portions of Spheres*

Defining equations and inequalities	Description
a) $x^2 + y^2 + z^2 < 4$	The interior of the sphere $x^2 + y^2 + z^2 = 4$.
b) $x^2 + y^2 + z^2 \leq 4$	The solid ball bounded by the sphere $x^2 + y^2 + z^2 = 4$. Alternatively, the sphere $x^2 + y^2 + z^2 = 4$ together with its interior.
c) $x^2 + y^2 + z^2 > 4$	The exterior of the sphere $x^2 + y^2 + z^2 = 4$.
d) $x^2 + y^2 + z^2 = 4, \quad z \leq 0$	The lower hemisphere cut from the sphere $x^2 + y^2 + z^2 = 4$ by the xy-plane (the plane $z = 0$).

How to Draw Three-dimensional Objects to Look Three-dimensional

1. Break lines. When one line passes behind another, break it to show that it doesn't touch and that part of it is hidden.

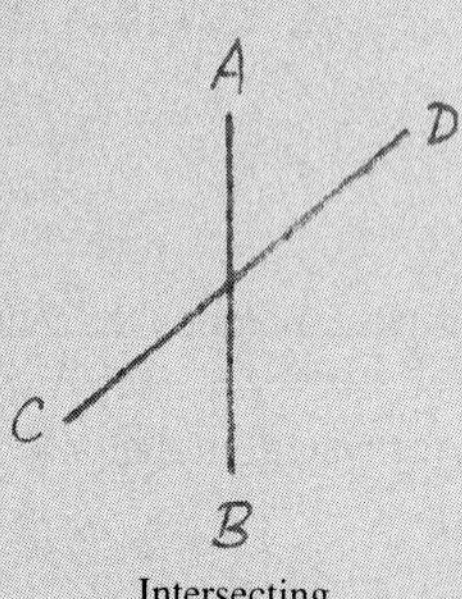

Intersecting

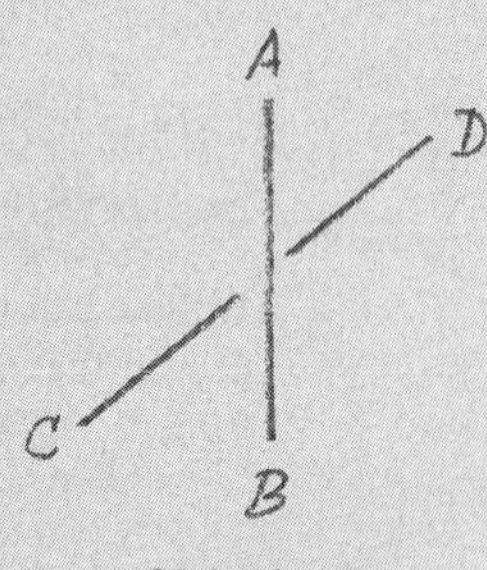

CD behind *AB*

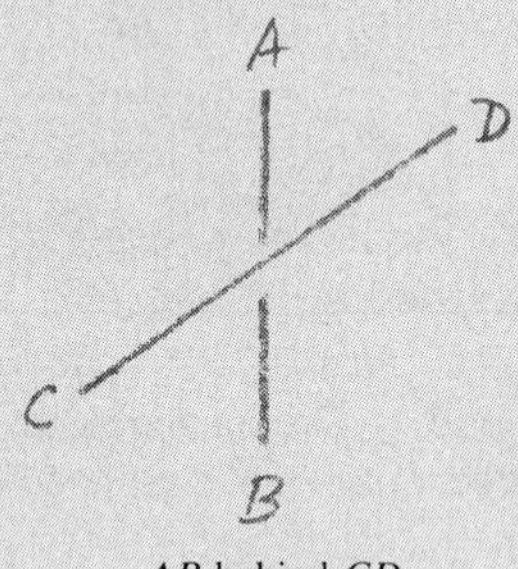

AB behind *CD*

2. Make the angle between the positive x-axis and the positive y-axis large enough.

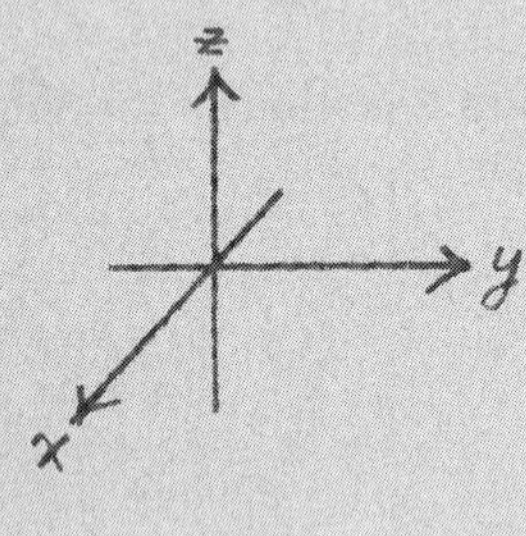

This

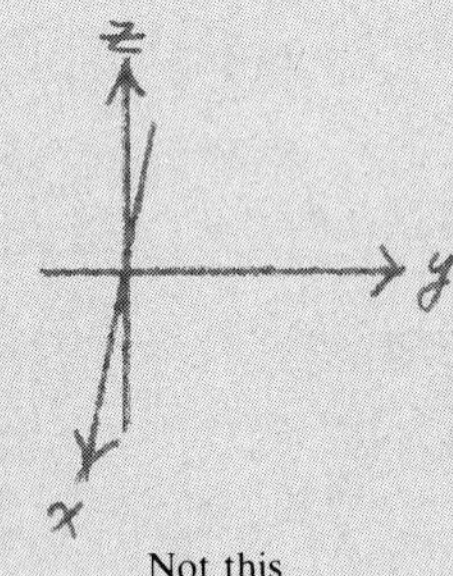

Not this

3. Draw planes parallel to the coordinate planes as if they were rectangles with sides parallel to the coordinate axes.

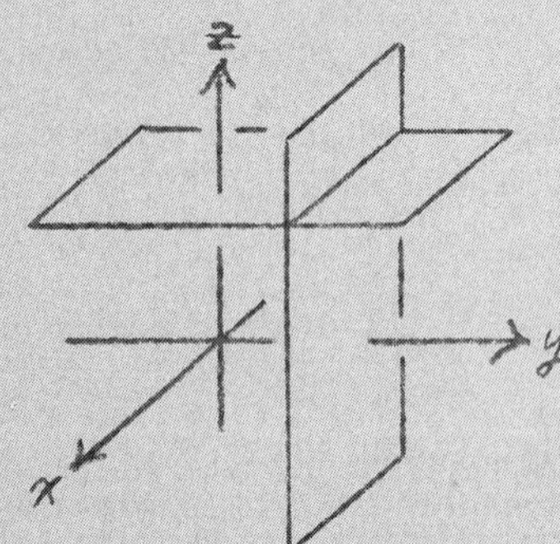

4. Dash or omit hidden portions of lines. Don't let the line touch the boundary of the parallelogram that represents the plane, unless the line lies in the plane.

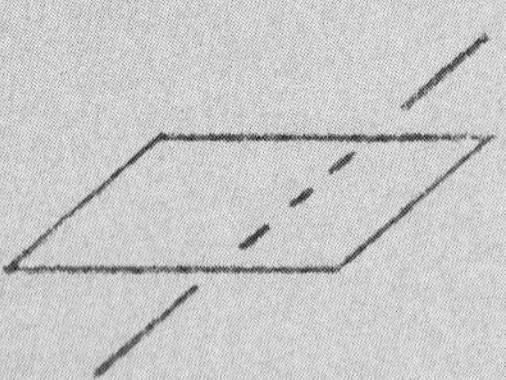
Line below plane

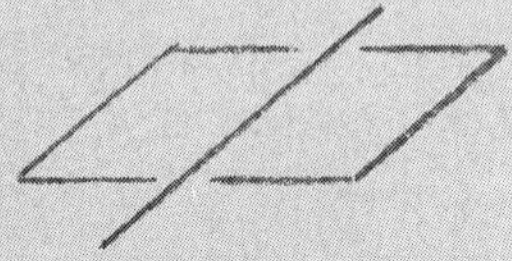
Line above plane

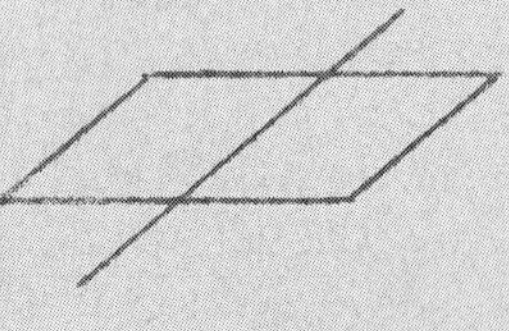
Line *in* plane

5. Spheres: Draw the sphere first (outline and equator); draw axes, if any, later. Use line breaks and dashed lines.

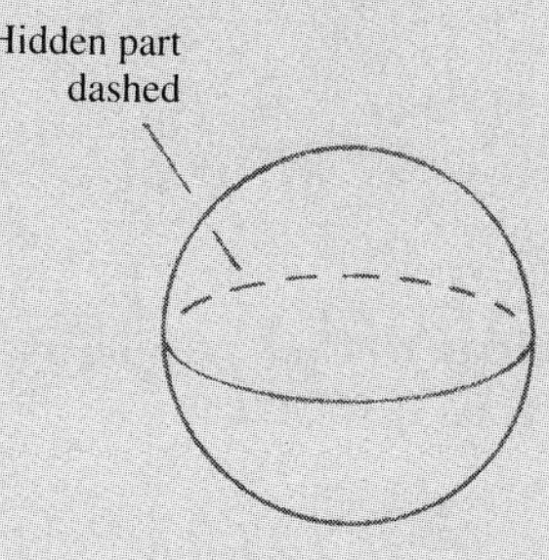

Sphere first

Axes later

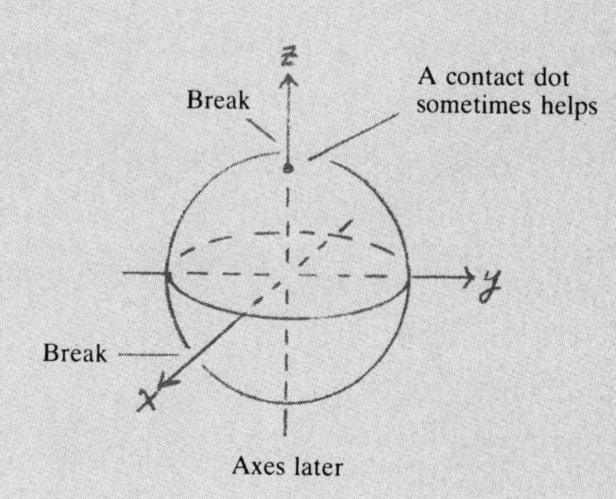

6. A general rule for perspective: Draw the object as if it lies some distance away, below, and to the left.

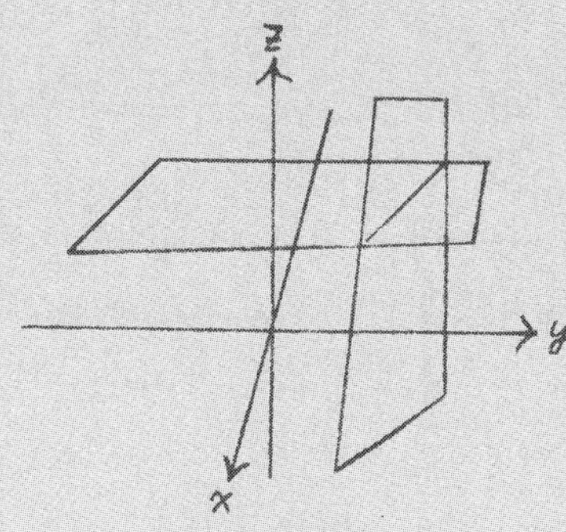

Advice ignored

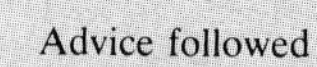

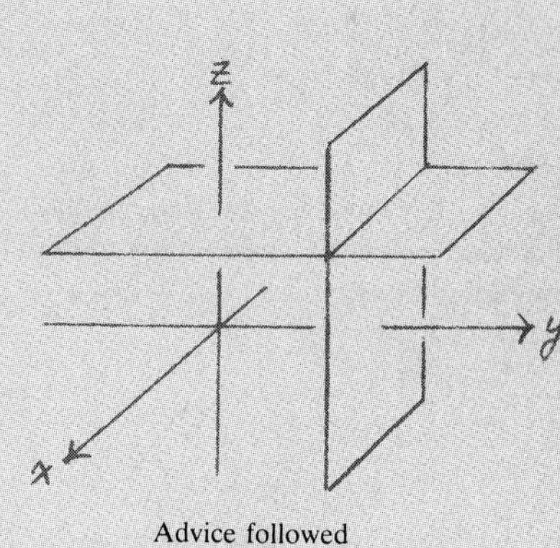

Advice followed

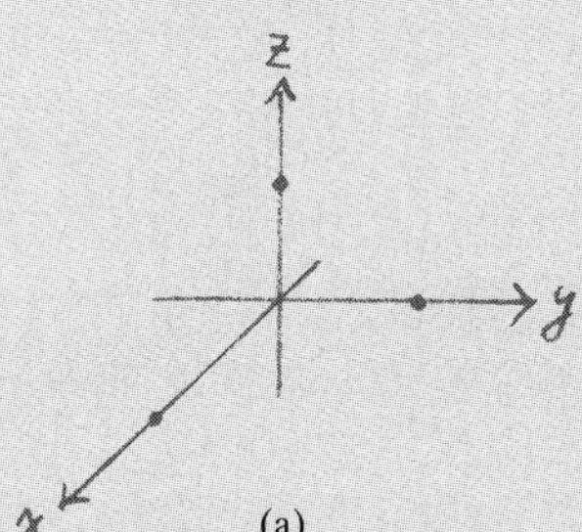

(a)

(b)

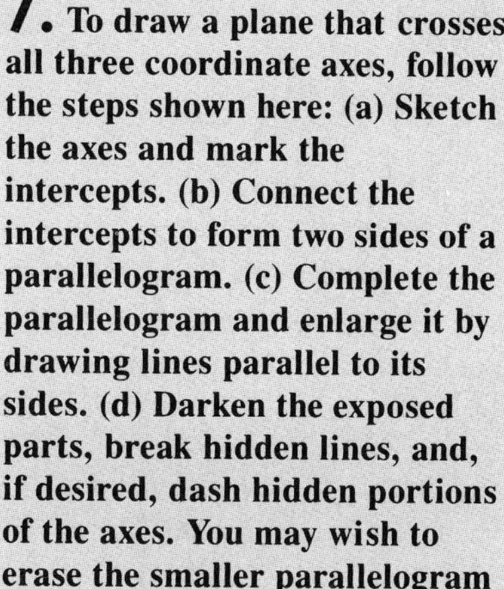

7. To draw a plane that crosses all three coordinate axes, follow the steps shown here: (a) Sketch the axes and mark the intercepts. (b) Connect the intercepts to form two sides of a parallelogram. (c) Complete the parallelogram and enlarge it by drawing lines parallel to its sides. (d) Darken the exposed parts, break hidden lines, and, if desired, dash hidden portions of the axes. You may wish to erase the smaller parallelogram at this point.

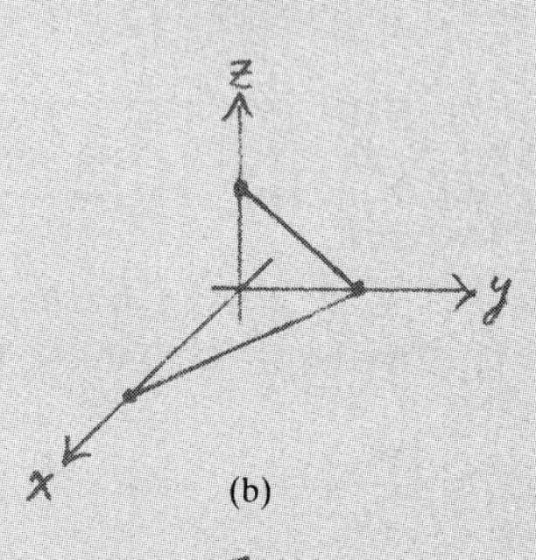

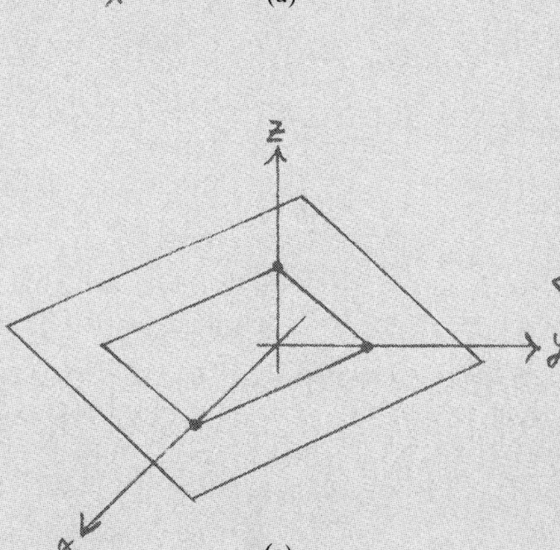

(c)

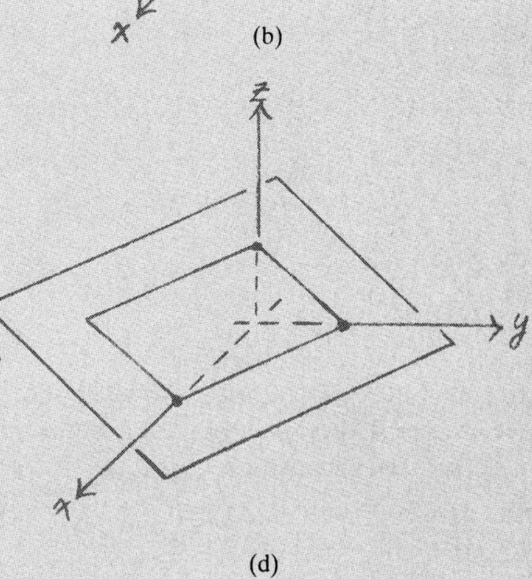

(d)

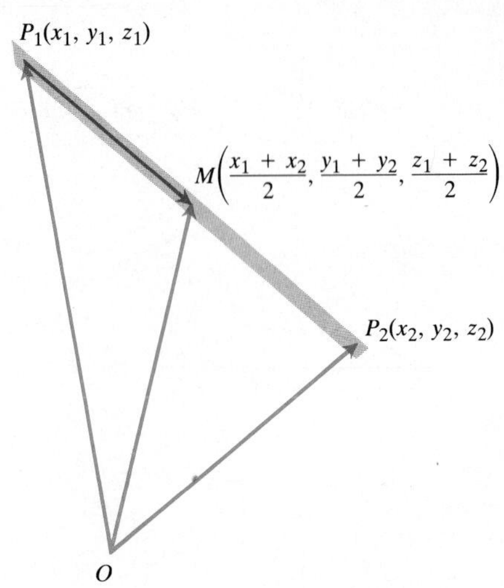

10.21 The coordinates of the point halfway between P_1 and P_2 are found by averaging the coordinates of P_1 and P_2.

Midpoints of Line Segments

The coordinates of the midpoint M of the line segment joining two points $P_1(x_1, y_1, z_1)$ and $P_2(x_2, y_2, z_2)$ are found by averaging the coordinates of P_1 and P_2:

$$M = \left(\frac{x_1 + x_2}{2}, \frac{y_1 + y_2}{2}, \frac{z_1 + z_2}{2}\right). \tag{10}$$

To see why, observe that these coordinates are the scalar components of the position vector $\overrightarrow{OM}$ (Fig. 10.21) and that

$$\begin{aligned}\overrightarrow{OM} &= \overrightarrow{OP_1} + \frac{1}{2}(\overrightarrow{P_1P_2}) = \overrightarrow{OP_1} + \frac{1}{2}(\overrightarrow{OP_2} - \overrightarrow{OP_1}) \\ &= \frac{1}{2}(\overrightarrow{OP_1} + \overrightarrow{OP_2}) \\ &= \frac{x_1 + x_2}{2}\mathbf{i} + \frac{y_1 + y_2}{2}\mathbf{j} + \frac{z_1 + z_2}{2}\mathbf{k}.\end{aligned}$$

Example 11 The midpoint of the segment joining $P_1(3, -2, 0)$ and $P_2(7, 4, 4)$ is

$$\left(\frac{3+7}{2}, \frac{-2+4}{2}, \frac{0+4}{2}\right) = (5, 1, 2).$$

EXERCISES 10.2

In Exercises 1–12, give a geometric description of the set of points in space whose coordinates satisfy the given pairs of equations.

1. $x = 2, \quad y = 3$

2. $x = -1, \quad z = 0$

3. $y = 0, \quad z = 0$

4. $x = 1, \quad y = 0$

5. $x^2 + y^2 = 4, \quad z = 0$

6. $x^2 + y^2 = 4, \quad z = -2$

7. $x^2 + z^2 = 4, \quad y = 0$

8. $y^2 + z^2 = 1, \quad x = 0$

9. $x^2 + y^2 + z^2 = 1, \quad x = 0$

10. $x^2 + y^2 + z^2 = 25, \quad y = -4$

11. $x^2 + y^2 + (z + 3)^2 = 25, \quad z = 0$

12. $x^2 + (y - 1)^2 + z^2 = 4, \quad y = 0$

In Exercises 13–18, describe the sets of points in space whose coordinates satisfy the given inequalities or combinations of equations and inequalities.

13. a) $x \geq 0, \quad y \geq 0, \quad z = 0$
b) $x \geq 0, \quad y \leq 0, \quad z = 0$

14. a) $0 \leq x \leq 1$
b) $0 \leq x \leq 1, \quad 0 \leq y \leq 1$
c) $0 \leq x \leq 1, \quad 0 \leq y \leq 1, \quad 0 \leq z \leq 1$

15. a) $x^2 + y^2 + z^2 \leq 1$
b) $x^2 + y^2 + z^2 > 1$

16. a) $x^2 + y^2 \leq 1, \quad z = 0$
b) $x^2 + y^2 \leq 1, \quad z = 3$
c) $x^2 + y^2 \leq 1$, no restriction on z

17. a) $x^2 + y^2 + z^2 = 1, \quad z \geq 0$
b) $x^2 + y^2 + z^2 \leq 1, \quad z \geq 0$

18. a) $x = y, \quad z = 0$
b) $x = y$, no restriction on z

In Exercises 19–28, describe the given set with a single equation or with a pair of equations.

19. The plane perpendicular to the
a) x-axis at $(3, 0, 0)$
b) y-axis at $(0, -1, 0)$
c) z-axis at $(0, 0, -2)$

20. The plane through the point $(3, -1, 2)$ perpendicular to the
a) x-axis b) y-axis c) z-axis

21. The plane through the point $(3, -1, 1)$ parallel to the
a) xy-plane b) yz-plane c) xz-plane

22. The circle of radius 2 centered at (0, 0, 0) and lying in the
a) xy-plane b) yz-plane c) xz-plane

23. The circle of radius 2 centered at (0, 2, 0) and lying in the
a) xy-plane b) yz-plane c) plane $y = 2$

24. The circle of radius 1 centered at $(-3, 4, 1)$ and lying in a plane parallel to the
a) xy-plane b) yz-plane c) xz-plane

25. The line through the point $(1, 3, -1)$ parallel to the
a) x-axis b) y-axis c) z-axis

26. The set of points in space equidistant from the origin and the point (0, 2, 0)

27. The circle in which the plane through the point (1, 1, 3) perpendicular to the z-axis meets the sphere of radius 5 centered at the origin

28. The set of points in space that lie 2 units from the point (0, 0, 1) and, at the same time, 2 units from the point $(0, 0, -1)$

Write inequalities to describe the sets in Exercises 29–34.

29. The slab bounded by the planes $z = 0$ and $z = 1$ (planes included)

30. The solid cube in the first octant bounded by the planes $x = 2$, $y = 2$, and $z = 2$

31. The half-space consisting of the points on and below the xy-plane

32. The upper hemisphere of the sphere of radius 1 centered at the origin

33. The (a) interior and (b) exterior of the sphere of radius 1 centered at the point (1, 1, 1)

34. The closed region bounded by the spheres of radius 1 and radius 2 centered at the origin. (*Closed* means the spheres are to be included. Had we wanted the spheres left out, we would have asked for the *open* region bounded by the spheres. This is analogous to the way we use "closed" and "open" to describe intervals: "closed" means endpoints included, "open" means endpoints left out. Closed sets include boundaries; open sets leave them out.)

In Exercises 35–46, express each vector as a product of its length and direction.

35. $2\mathbf{i} + \mathbf{j} - 2\mathbf{k}$

36. $3\mathbf{i} - 6\mathbf{j} + 2\mathbf{k}$

37. $\mathbf{i} + 4\mathbf{j} - 8\mathbf{k}$

38. $9\mathbf{i} - 2\mathbf{j} + 6\mathbf{k}$

39. $5\mathbf{k}$

40. $6\mathbf{i}$

41. $-4\mathbf{j}$

42. $\frac{3}{5}\mathbf{i} + \frac{4}{5}\mathbf{k}$

43. $-\frac{1}{3}\mathbf{j} + \frac{1}{4}\mathbf{k}$

44. $\frac{1}{\sqrt{2}}\mathbf{i} - \frac{1}{\sqrt{2}}\mathbf{k}$

45. $\frac{1}{\sqrt{6}}\mathbf{i} - \frac{1}{\sqrt{6}}\mathbf{j} - \frac{1}{\sqrt{6}}\mathbf{k}$

46. $\frac{\mathbf{i}}{\sqrt{3}} + \frac{\mathbf{j}}{\sqrt{3}} + \frac{\mathbf{k}}{\sqrt{3}}$

In Exercises 47–52, find the length and direction of the vector from point P_1 to point P_2. Also find the midpoint of the line segment joining P_1 and P_2.

47. $P_1(1, 1, 1)$, $P_2(3, 3, 0)$

48. $P_1(-1, 1, 5)$, $P_2(2, 5, 0)$

49. $P_1(1, 4, 5)$, $P_2(4, -2, 7)$

50. $P_1(3, 4, 5)$, $P_2(2, 3, 4)$

51. $P_1(0, 0, 0)$, $P_2(2, -2, -2)$

52. $P_1(5, 3, -2)$, $P_2(0, 0, 0)$

53. Find the vectors whose lengths and directions are given. Try to answer without writing anything down.

Length	Direction
a) 2	$\mathbf{i}$
b) 4	$\mathbf{j}$
c) $\sqrt{3}$	$\mathbf{k}$
d) $\frac{1}{2}$	$\frac{3}{5}\mathbf{j} + \frac{4}{5}\mathbf{k}$
e) 7	$\frac{6}{7}\mathbf{i} - \frac{2}{7}\mathbf{j} + \frac{3}{7}\mathbf{k}$
f) a	$u_1\mathbf{i} + u_2\mathbf{j} + u_3\mathbf{k}$

54. Find a vector of magnitude 7 in the direction of $\mathbf{A} = 12\mathbf{i} - 5\mathbf{k}$.

55. Find a vector $\sqrt{5}$ units long in the direction of $\mathbf{A} = \mathbf{i} + \mathbf{j} + \mathbf{k}$.

56. Find a vector 1/2 unit long in the direction of $\mathbf{A} = \mathbf{i} - \mathbf{j}$.

Spheres and Distance

57. Find the centers and radii of the following spheres.
a) $(x + 2)^2 + y^2 + (z - 2)^2 = 8$
b) $\left(x + \frac{1}{2}\right)^2 + \left(y + \frac{1}{2}\right)^2 + \left(z + \frac{1}{2}\right)^2 = \frac{21}{4}$
c) $(x - \sqrt{2})^2 + (y - \sqrt{2})^2 + (z + \sqrt{2})^2 = 2$
d) $x^2 + \left(y + \frac{1}{3}\right)^2 + \left(z - \frac{1}{3}\right)^2 = \frac{29}{9}$

58. Find equations for the spheres whose centers and radii are given here.

Center	Radius
a) (1, 2, 3)	$\sqrt{14}$
b) $(0, -1, 5)$	2
c) $(-2, 0, 0)$	$\sqrt{3}$
d) $(0, -7, 0)$	7

Find the centers and radii of the spheres in Exercises 59–62.

59. $x^2 + y^2 + z^2 + 4x - 4z = 0$

60. $2x^2 + 2y^2 + 2z^2 + x + y + z = 9$

61. $x^2 + y^2 + z^2 - 2az = 0$

62. $3x^2 + 3y^2 + 3z^2 + 2y - 2z = 9$

63. Find a formula for the distance from the point $P(x, y, z)$ to the
a) x-axis, b) y-axis, c) z-axis.

64. Find a formula for the distance from the point $P(x, y, z)$ to the
a) xy-plane, b) yz-plane, c) xz-plane.

Geometry with Vectors

65. Suppose that A, B, and C are the corner points of the thin triangular plate of constant density shown in Fig. 10.22.
a) Find the vector from C to the midpoint M of side AB.
b) Find the vector from C to the point that lies two-thirds of the way from C to M on the median CM.
c) Find the coordinates of the point in which the medians of ΔABC intersect. According to Exercise 22, Section 5.6, this point is the plate's center of mass.

66. Find the vector from the origin to the point of intersection of the medians of the triangle whose vertices are

$$A(1, -1, 2),\quad B(2, 1, 3),\quad \text{and}\quad C(-1, 2, -1).$$

67. Let $ABCD$ be a general, not necessarily planar, quadrilateral in space. Show that the two segments joining the midpoints of opposite sides of $ABCD$ bisect each other. (*Hint:* Show that the segments have the same midpoint.)

68. Vectors are drawn from the center of a regular n-sided polygon in the plane to the vertices of the polygon. Show that the sum of the vectors is zero. (*Hint:* What happens to the sum if you rotate the polygon about its center?)

69. Suppose that A, B, and C are vertices of a triangle and that a, b, and c are, respectively, the midpoints of the opposite sides. Show that $\overrightarrow{Aa} + \overrightarrow{Bb} + \overrightarrow{Cc} = 0$.

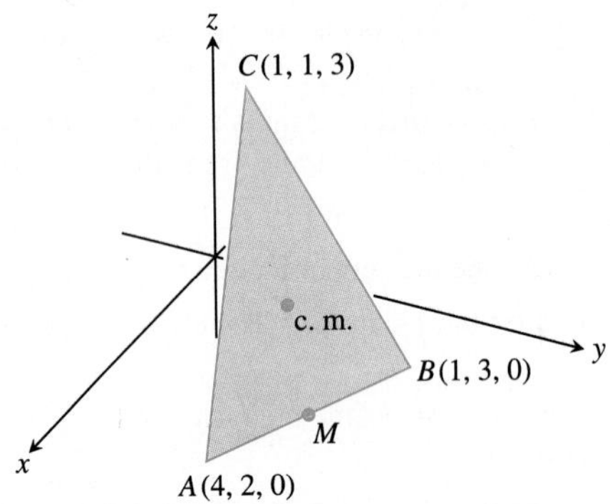

10.22 The triangular plate in Exercise 65.

10.3 Dot Products

We now introduce the dot product of two vectors, the first of two methods we shall learn for multiplying vectors together. Our motivation is the need to calculate the work done by a constant force in displacing a mass. If we can represent the force and displacement as vectors, we can calculate the work by finding their dot product. Dot products are also called *scalar products* because the result of the multiplication is a scalar and not a vector. Cross products, which we shall study in the next section, are always vectors.

Scalar products have applications in mathematics as well as in engineering and physics. This section describes the geometric and algebraic properties on which many of these applications depend.

DEFINITION

The **scalar product** $\mathbf{A} \cdot \mathbf{B}$ ("**A** dot **B**") or **dot product** of two vectors **A** and **B** is the number

$$\mathbf{A} \cdot \mathbf{B} = |\mathbf{A}||\mathbf{B}| \cos \theta, \qquad (1)$$

where θ measures the smaller angle made by **A** and **B** when their initial points coincide (as in Fig. 10.23).

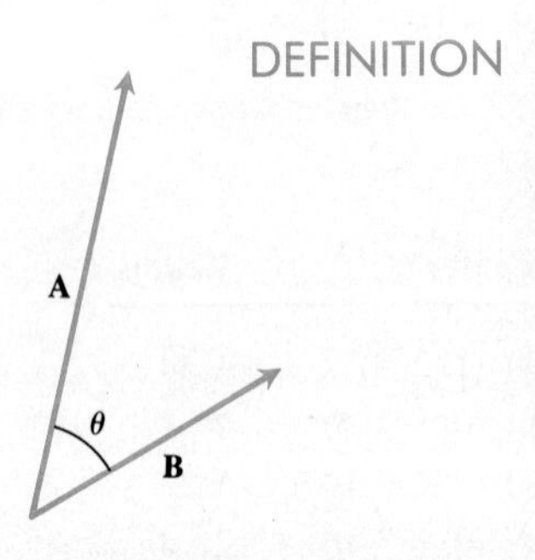

10.23 $\mathbf{A} \cdot \mathbf{B}$ is the number $|\mathbf{A}||\mathbf{B}| \cos \theta$.

In words, the scalar product of **A** and **B** is the length of **A** times the length of **B** times the cosine of the angle between **A** and **B**. The product is a scalar, not a vector. It is called the dot product because of the dot in the notation $\mathbf{A} \cdot \mathbf{B}$.

From Eq. (1) we see that the scalar product of two vectors is positive when the angle between them is acute, negative when the angle is obtuse.

Since the angle a vector **A** makes with itself is zero, and the cosine of zero is 1,

$$\mathbf{A}\cdot\mathbf{A} = |\mathbf{A}||\mathbf{A}|\cos 0 = |\mathbf{A}||\mathbf{A}|(1) = |\mathbf{A}|^2, \qquad \text{or} \qquad |\mathbf{A}| = \sqrt{\mathbf{A}\cdot\mathbf{A}}. \tag{2}$$

As we shall see, this provides a handy way to calculate a vector's length.

Calculation

To calculate $\mathbf{A}\cdot\mathbf{B}$ from the components of **A** and **B**, we let

$$\mathbf{A} = a_1\,\mathbf{i} + a_2\,\mathbf{j} + a_3\,\mathbf{k},$$

$$\mathbf{B} = b_1\,\mathbf{i} + b_2\,\mathbf{j} + b_3\,\mathbf{k},$$

and

$$\mathbf{C} = \mathbf{B} - \mathbf{A} = (b_1 - a_1)\,\mathbf{i} + (b_2 - a_2)\,\mathbf{j} + (b_3 - a_3)\,\mathbf{k}.$$

Then we apply the law of cosines to a triangle whose sides represent the vectors **A**, **B**, and **C** (Fig. 10.24) and obtain

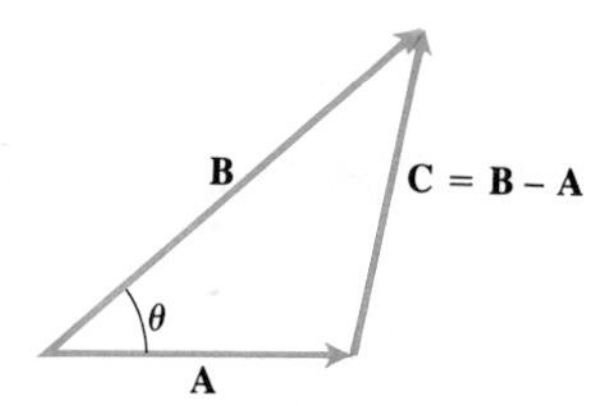

10.24 Equation (3) is obtained by applying the law of cosines to a triangle whose sides represent **A**, **B**, and **C** = **B** − **A**.

$$|\mathbf{C}|^2 = |\mathbf{A}|^2 + |\mathbf{B}|^2 - 2|\mathbf{A}||\mathbf{B}|\cos\theta,$$

$$|\mathbf{A}||\mathbf{B}|\cos\theta = \frac{|\mathbf{A}|^2 + |\mathbf{B}|^2 - |\mathbf{C}|^2}{2}.$$

The left side of this equation is $\mathbf{A}\cdot\mathbf{B}$, and we may evaluate the right side by applying Eq. (4) of Section 10.2 to find the lengths of **A**, **B**, and **C**. The result of this algebra is the formula

$$\mathbf{A}\cdot\mathbf{B} = a_1b_1 + a_2\,b_2 + a_3\,b_3. \tag{3}$$

Thus, to find the scalar product of two given vectors we multiply their corresponding **i**, **j**, and **k** components and add the results. In particular, from Eq. (2) we have

$$|\mathbf{A}| = \sqrt{\mathbf{A}\cdot\mathbf{A}} = \sqrt{a_1^2 + a_2^2 + a_3^2}. \tag{4}$$

When we solve Eq. (1) for θ, we get a formula for finding angles between nonzero vectors.

The Angle Between Two Nonzero Vectors

The angle between two nonzero vectors **A** and **B** is

$$\theta = \cos^{-1}\left(\frac{\mathbf{A}\cdot\mathbf{B}}{|\mathbf{A}||\mathbf{B}|}\right). \tag{5}$$

Since the values of the arc cosine lie in $[0, \pi]$, Eq. (5) automatically gives the smaller of the two angles made by **A** and **B**.

Example 1 Find the angle between $\mathbf{A} = \mathbf{i} - 2\,\mathbf{j} - 2\,\mathbf{k}$ and $\mathbf{B} = 6\,\mathbf{i} + 3\,\mathbf{j} + 2\,\mathbf{k}$.

Solution We use Eq. (5):

$$\mathbf{A} \cdot \mathbf{B} = (1)(6) + (-2)(3) + (-2)(2) = 6 - 6 - 4 = -4,$$

$$|\mathbf{A}| = \sqrt{\mathbf{A} \cdot \mathbf{A}} = \sqrt{(1)^2 + (-2)^2 + (-2)^2} = \sqrt{9} = 3,$$

$$|\mathbf{B}| = \sqrt{\mathbf{B} \cdot \mathbf{B}} = \sqrt{(6)^2 + (3)^2 + (2)^2} = \sqrt{49} = 7,$$

$$\theta = \cos^{-1}\left(\frac{\mathbf{A} \cdot \mathbf{B}}{|\mathbf{A}||\mathbf{B}|}\right) \qquad \text{(Eq. (5))}$$

$$= \cos^{-1}\left(\frac{-4}{(3)(7)}\right) = \cos^{-1}\left(-\frac{4}{21}\right)$$

$$= 101°. \qquad \text{(Calculator, rounded)}$$

Laws of Multiplication

From the equation $\mathbf{A} \cdot \mathbf{B} = a_1b_1 + a_2b_2 + a_3b_3$, we can see right away that

$$\mathbf{A} \cdot \mathbf{B} = \mathbf{B} \cdot \mathbf{A}. \tag{6}$$

In other words, scalar multiplication is commutative. We can also see from Eq. (3) that if c is any number, then

$$(c\mathbf{A}) \cdot \mathbf{B} = \mathbf{A} \cdot (c\mathbf{B}) = c(\mathbf{A} \cdot \mathbf{B}). \tag{7}$$

If $\mathbf{C} = c_1\,\mathbf{i} + c_2\,\mathbf{j} + c_3\,\mathbf{k}$ is any third vector, then

$$\mathbf{A} \cdot (\mathbf{B} + \mathbf{C}) = a_1(b_1 + c_1) + a_2(b_2 + c_2) + a_3(b_3 + c_3)$$

$$= (a_1b_1 + a_2b_2 + a_3b_3) + (a_1c_1 + a_2c_2 + a_3c_3)$$

$$= \mathbf{A} \cdot \mathbf{B} + \mathbf{A} \cdot \mathbf{C}.$$

Hence scalar products obey the distributive law:

$$\mathbf{A} \cdot (\mathbf{B} + \mathbf{C}) = \mathbf{A} \cdot \mathbf{B} + \mathbf{A} \cdot \mathbf{C}. \tag{8}$$

If we combine this with the commutative law, Eq. (6), it is also evident that

$$(\mathbf{A} + \mathbf{B}) \cdot \mathbf{C} = \mathbf{A} \cdot \mathbf{C} + \mathbf{B} \cdot \mathbf{C}. \tag{9}$$

Equations (8) and (9) together permit us to multiply sums of vectors by the familiar laws of algebra. For example,

$$(\mathbf{A} + \mathbf{B}) \cdot (\mathbf{C} + \mathbf{D}) = \mathbf{A} \cdot \mathbf{C} + \mathbf{A} \cdot \mathbf{D} + \mathbf{B} \cdot \mathbf{C} + \mathbf{B} \cdot \mathbf{D}. \tag{10}$$

Orthogonal Vectors

Two vectors whose scalar product is zero are said to be *orthogonal*. From the equation $\mathbf{A} \cdot \mathbf{B} = |\mathbf{A}||\mathbf{B}| \cos \theta$, we can see that when neither $|\mathbf{A}|$ nor $|\mathbf{B}|$ is zero, $\mathbf{A} \cdot \mathbf{B}$ is zero if and only if $\cos\theta$ is zero, or θ equals 90°. Hence, for the vectors we are dealing with, "orthogonal" means the same as "perpendicular." In other scientific contexts in which the word is used, there may be no such geometric interpretation.

The zero vector $\mathbf{0} = 0\,\mathbf{i} + 0\,\mathbf{j} + 0\,\mathbf{k}$ is orthogonal to every vector because its dot product with every vector is zero.

DEFINITION

Vectors $\mathbf{A}$ and $\mathbf{B}$ are **orthogonal** if $\mathbf{A}\cdot\mathbf{B} = 0$. (11)

Example 2 The vectors $\mathbf{A} = 3\,\mathbf{i} - 2\,\mathbf{j} + \mathbf{k}$ and $\mathbf{B} = 2\,\mathbf{j} + 4\,\mathbf{k}$ are orthogonal because

$$\mathbf{A}\cdot\mathbf{B} = (3)(0) + (-2)(2) + (1)(4) = 0.$$

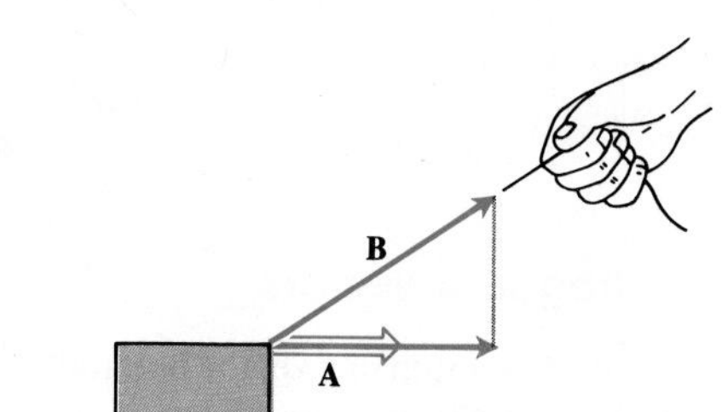

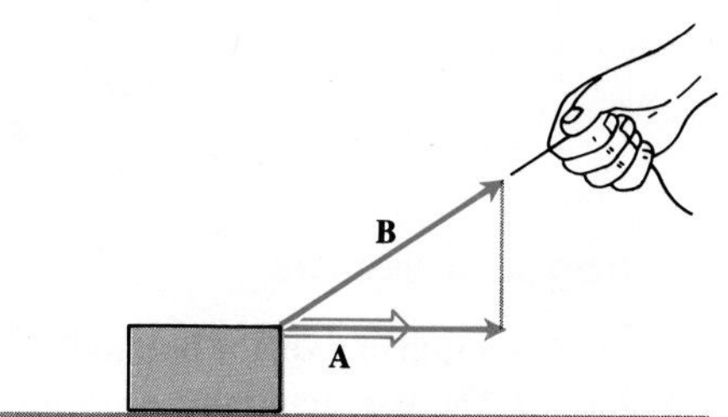

10.25 We may pull on the box with force $\mathbf{B}$ to move the box in the direction of $\mathbf{A}$. The effective force in this direction is represented by the vector projection of $\mathbf{B}$ onto $\mathbf{A}$.

Vector Projections and Scalar Components

The vector we get by projecting a vector $\mathbf{B}$ onto the line through a vector $\mathbf{A}$ is called the **vector projection of B onto A,** sometimes denoted

$$\text{proj}_{\mathbf{A}}\mathbf{B} \qquad (\text{"the vector projection of } \mathbf{B} \text{ onto } \mathbf{A}\text{"}).$$

If $\mathbf{B}$ represents a force, then the vector projection of $\mathbf{B}$ onto $\mathbf{A}$ represents the effective force in the direction of $\mathbf{A}$ (Fig. 10.25).

If the angle between $\mathbf{B}$ and $\mathbf{A}$ is acute, the length of the vector projection of $\mathbf{B}$ onto $\mathbf{A}$ is $|\mathbf{B}|\cos\theta$. If the angle is obtuse, its cosine is negative and the length of the vector projection of $\mathbf{B}$ onto $\mathbf{A}$ is $-|\mathbf{B}|\cos\theta$. In either case, the number $|\mathbf{B}|\cos\theta$ itself is called the **scalar component of B in the direction of A** (Fig. 10.26).

The scalar component of $\mathbf{B}$ in the direction of $\mathbf{A}$ can be found by dividing both sides of the equation $\mathbf{A}\cdot\mathbf{B} = |\mathbf{A}||\mathbf{B}|\cos\theta$ by $|\mathbf{A}|$. This gives

$$|\mathbf{B}|\cos\theta = \frac{\mathbf{A}\cdot\mathbf{B}}{|\mathbf{A}|} = \mathbf{B}\cdot\frac{\mathbf{A}}{|\mathbf{A}|}. \qquad (12)$$

Equation (12) says that the scalar component of $\mathbf{B}$ in the direction of $\mathbf{A}$ can be obtained by "dotting" $\mathbf{B}$ with the direction of $\mathbf{A}$.

The vector projection of $\mathbf{B}$ onto $\mathbf{A}$ is the scalar component of $\mathbf{B}$ in the direction of $\mathbf{A}$ times the direction of $\mathbf{A}$. If the angle between $\mathbf{A}$ and $\mathbf{B}$ is acute, the vector projection has length $|\mathbf{B}|\cos\theta$ and direction $\mathbf{A}/|\mathbf{A}|$. If the angle is obtuse, the vector projection has length $-|\mathbf{B}|\cos\theta$ and direction $-\mathbf{A}/|\mathbf{A}|$. In either case,

$$\text{proj}_{\mathbf{A}}\mathbf{B} = (|\mathbf{B}|\cos\theta)\frac{\mathbf{A}}{|\mathbf{A}|}. \qquad (13)$$

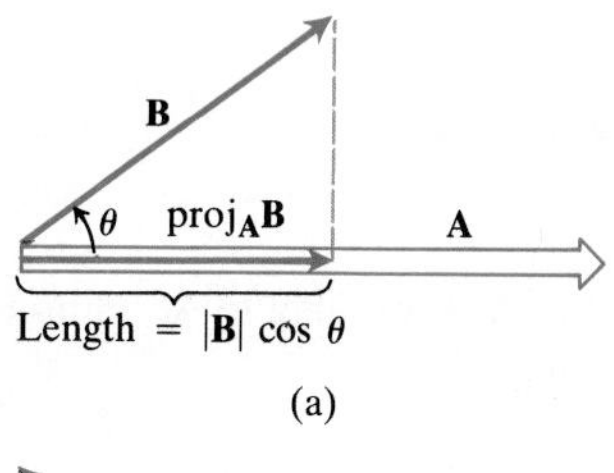

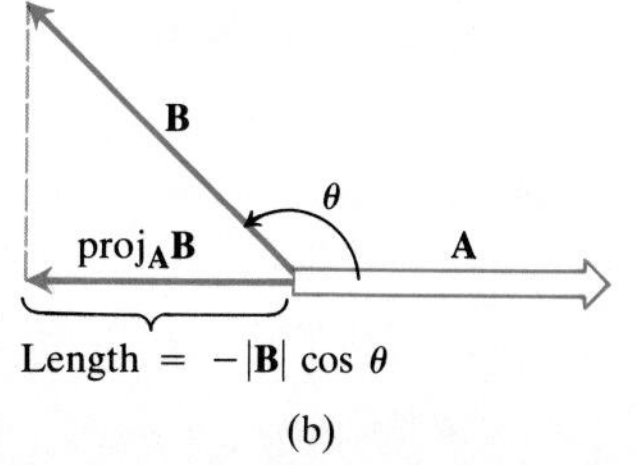

10.26 (a) When θ is acute, the length of the vector projection of $\mathbf{B}$ onto $\mathbf{A}$ is $|\mathbf{B}|\cos\theta$, the scalar component of $\mathbf{B}$ in the direction of $\mathbf{A}$. (b) When θ is obtuse, the scalar component of $\mathbf{B}$ in the direction of $\mathbf{A}$ is negative and the length of the vector projection of $\mathbf{B}$ onto $\mathbf{A}$ is $-|\mathbf{B}|\cos\theta$.

Equations (12) and (13) together give a useful way to calculate $\text{proj}_{\mathbf{A}}\mathbf{B}$:

$$\begin{aligned} \text{proj}_{\mathbf{A}}\mathbf{B} &= (|\mathbf{B}|\cos\theta)\frac{\mathbf{A}}{|\mathbf{A}|} && \text{(Eq. (13))} \\ &= \left(\frac{\mathbf{A}\cdot\mathbf{B}}{|\mathbf{A}|}\right)\frac{\mathbf{A}}{|\mathbf{A}|} && \text{(Eq. (12))} \\ &= \frac{\mathbf{A}\cdot\mathbf{B}}{\mathbf{A}\cdot\mathbf{A}}\mathbf{A}. && (|\mathbf{A}|^2 = \mathbf{A}\cdot\mathbf{A}) \end{aligned} \qquad (14)$$

Example 3 Find the vector projection of $\mathbf{B} = 6\mathbf{i} + 3\mathbf{j} + 2\mathbf{k}$ onto $\mathbf{A} = \mathbf{i} - 2\mathbf{j} - 2\mathbf{k}$ and the scalar component of $\mathbf{B}$ in the direction of $\mathbf{A}$.

Solution We find $\text{proj}_{\mathbf{A}}\mathbf{B}$ from Eq. (14):

$$\begin{aligned}\text{proj}_{\mathbf{A}}\mathbf{B} &= \frac{\mathbf{A}\cdot\mathbf{B}}{\mathbf{A}\cdot\mathbf{A}}\mathbf{A} = \frac{6-6-4}{1+4+4}(\mathbf{i} - 2\mathbf{j} - 2\mathbf{k}) \\ &= -\frac{4}{9}(\mathbf{i} - 2\mathbf{j} - 2\mathbf{k}) = -\frac{4}{9}\mathbf{i} + \frac{8}{9}\mathbf{j} + \frac{8}{9}\mathbf{k}.\end{aligned}$$

We find the scalar component of $\mathbf{B}$ in the direction of $\mathbf{A}$ from Eq. (12):

$$\begin{aligned}|\mathbf{B}|\cos\theta &= \mathbf{B}\cdot\frac{\mathbf{A}}{|\mathbf{A}|} = (6\mathbf{i} + 3\mathbf{j} + 2\mathbf{k})\cdot\left(\frac{1}{3}\mathbf{i} - \frac{2}{3}\mathbf{j} - \frac{2}{3}\mathbf{k}\right) \\ &= 2 - 2 - \frac{4}{3} = -\frac{4}{3}.\end{aligned}$$

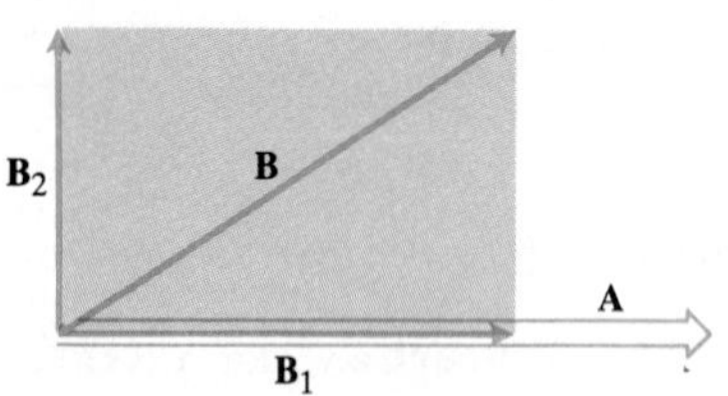

10.27 The vector $\mathbf{B}$ as the sum of vectors parallel and orthogonal to $\mathbf{A}$.

Writing a Vector as a Sum of Orthogonal Vectors

In mechanics, we often want to express a vector $\mathbf{B}$ as a sum of a vector $\mathbf{B}_1$ parallel to a vector $\mathbf{A}$ and a vector $\mathbf{B}_2$ orthogonal to $\mathbf{A}$ (Fig. 10.27). We can do this by writing $\mathbf{B}$ as a sum of its vector projection on $\mathbf{A}$ plus whatever is left over, because the left-over part will automatically be orthogonal to $\mathbf{A}$.

Formula for Writing B as a Vector Parallel to A Plus a Vector Orthogonal to A

$$\mathbf{B} = \underbrace{\frac{\mathbf{A}\cdot\mathbf{B}}{\mathbf{A}\cdot\mathbf{A}}\mathbf{A}}_{\substack{\mathbf{B}_1 \\ \text{parallel to } \mathbf{A}}} + \underbrace{\left(\mathbf{B} - \frac{\mathbf{A}\cdot\mathbf{B}}{\mathbf{A}\cdot\mathbf{A}}\mathbf{A}\right)}_{\substack{\mathbf{B}_2 \\ \text{orthogonal to } \mathbf{A}}}. \tag{15}$$

The vector $\mathbf{B}_1$, being the vector projection of $\mathbf{B}$ onto $\mathbf{A}$, is parallel to $\mathbf{A}$, while $\mathbf{B}_2$ can be seen to be orthogonal to $\mathbf{A}$ because $\mathbf{A}\cdot\mathbf{B}_2$ is zero:

$$\mathbf{A}\cdot\mathbf{B}_2 = \mathbf{A}\cdot\left(\mathbf{B} - \frac{\mathbf{A}\cdot\mathbf{B}}{\mathbf{A}\cdot\mathbf{A}}\mathbf{A}\right) = \mathbf{A}\cdot\mathbf{B} - \frac{\mathbf{A}\cdot\mathbf{B}}{\mathbf{A}\cdot\mathbf{A}}\mathbf{A}\cdot\mathbf{A} = \mathbf{A}\cdot\mathbf{B} - \mathbf{A}\cdot\mathbf{B} = 0. \tag{16}$$

Example 4 Express $\mathbf{B} = 2\mathbf{i} + \mathbf{j} - 3\mathbf{k}$ as the sum of a vector parallel to $\mathbf{A} = 3\mathbf{i} - \mathbf{j}$ and a vector orthogonal to $\mathbf{A}$.

Solution We use Eq. (15). With

$$\mathbf{A}\cdot\mathbf{B} = 6 - 1 = 5 \quad \text{and} \quad \mathbf{A}\cdot\mathbf{A} = 9 + 1 = 10,$$

Eq. (15) gives

$$\begin{aligned}\mathbf{B} &= \frac{\mathbf{A}\cdot\mathbf{B}}{\mathbf{A}\cdot\mathbf{A}}\mathbf{A} + \left(\mathbf{B} - \frac{\mathbf{A}\cdot\mathbf{B}}{\mathbf{A}\cdot\mathbf{A}}\mathbf{A}\right) = \frac{5}{10}(3\mathbf{i} - \mathbf{j}) + \left(2\mathbf{i} + \mathbf{j} - 3\mathbf{k} - \frac{5}{10}(3\mathbf{i} - \mathbf{j})\right) \\ &= \left(\frac{3}{2}\mathbf{i} - \frac{1}{2}\mathbf{j}\right) + \left(\frac{1}{2}\mathbf{i} + \frac{3}{2}\mathbf{j} - 3\mathbf{k}\right).\end{aligned}$$

Where Vectors Came From

Although Aristotle used vectors to describe the effects of forces, the idea of resolving vectors into geometric components parallel to the coordinate axes came from Descartes. The algebra of vectors we use today was developed simultaneously and independently in the 1870s by Josiah Willard Gibbs (1839–1903), a mathematical physicist at Yale University, and by the English mathematical physicist Oliver Heaviside (1850–1925), the Heaviside of Heaviside layer fame. The works of Gibbs and Heaviside grew out of more complicated mathematical theories developed some years earlier by the Irish mathematician William Hamilton (1805–1865) and the German linguist, physicist, and geometer Hermann Grassman (1809–1877). Hamilton's quaternions and Grassman's algebraic forms are still in use but tend to appear in more theoretical work.

Check: The first vector in the sum is parallel to **A** because it is $(1/2)\mathbf{A}$. The second vector in the sum is orthogonal to **A** because

$$\left(\frac{1}{2}\mathbf{i} + \frac{3}{2}\mathbf{j} - 3\mathbf{k}\right) \cdot (3\mathbf{i} - \mathbf{j}) = \frac{3}{2} - \frac{3}{2} = 0.$$

Lines in the Plane and Distances from Points to Lines

Dot products give a new understanding of the equations we write for lines in the plane and a quick way to calculate distances from points to lines.

Suppose L is a line through the point $P_0(x_0, y_0)$ perpendicular to a vector $\mathbf{N} = A\mathbf{i} + B\mathbf{j}$ (Fig. 10.28). Then a point $P(x, y)$ lies on L if and only if

$$\mathbf{N} \cdot \overrightarrow{P_0P} = 0 \quad \text{or} \quad A(x - x_0) + B(y - y_0) = 0. \tag{17}$$

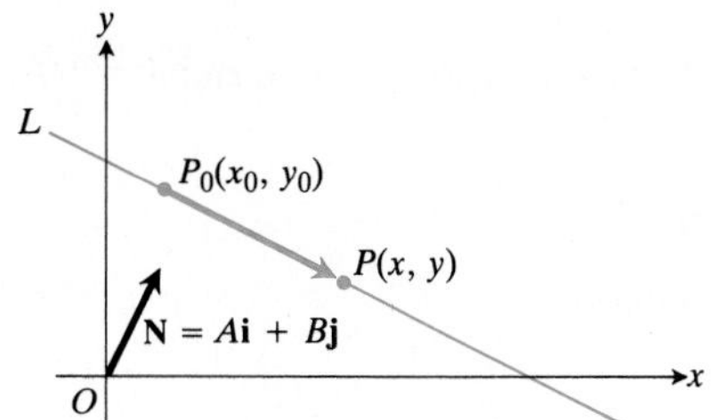

10.28 A point $P(x, y)$ lies on the line through P_0 perpendicular to **N** if and only if $\mathbf{N} \cdot \overrightarrow{P_0P} = 0$.

When rearranged, the second equation becomes

$$Ax + By = Ax_0 + By_0. \tag{18}$$

> **Line Through $P(x_0, y_0)$ Perpendicular to $\mathbf{N} = A\mathbf{i} + B\mathbf{j}$**
>
> $$Ax + By = C, \qquad C = Ax_0 + By_0 \tag{19}$$

Notice how the components of **N** become the coefficients in the equation $Ax + By = C$.

Example 5 Find an equation for the line through $P_0(3, 5)$ perpendicular to $\mathbf{N} = \mathbf{i} + 2\mathbf{j}$.

Solution We use Eqs. (19) with $A = 1$, $B = 2$, and $C = (1)(3) + (2)(5) = 13$ to get

$$x + 2y = 13.$$

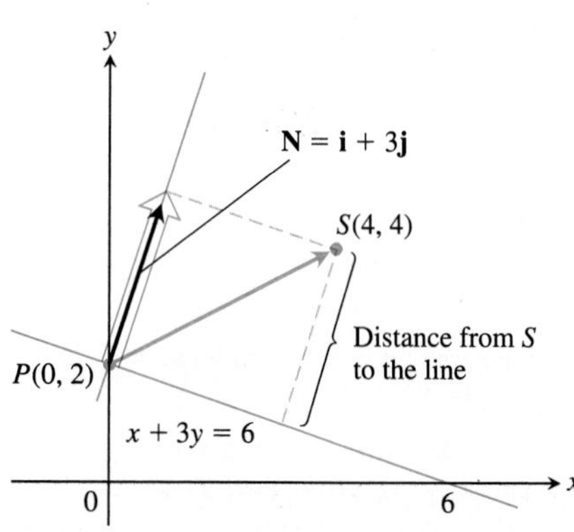

10.29 The distance from point S to the line is the length of the vector projection of $\overrightarrow{PS}$ onto **N** (Example 6).

Example 6 Find the distance from the point $S(4, 4)$ to the line L: $x + 3y = 6$.

Solution We find a point P on the line and calculate the distance as the length of the vector projection of $\overrightarrow{PS}$ onto a vector **N** perpendicular to the line (Fig. 10.29). Any point on the line will do for P and we can find **N** from the coefficients of $x + 3y = 6$ as $\mathbf{N} = \mathbf{i} + 3\mathbf{j}$. With $P = (0, 2)$, say, we then have

$$\overrightarrow{PS} = (4 - 0)\mathbf{i} + (4 - 2)\mathbf{j} = 4\mathbf{i} + 2\mathbf{j},$$

$$\text{distance from } S \text{ to } L = |\text{proj}_{\mathbf{N}}\, \overrightarrow{PS}| = \left|\overrightarrow{PS} \cdot \frac{\mathbf{N}}{|\mathbf{N}|}\right|$$

$$= \left|(4\mathbf{i} + 2\mathbf{j}) \cdot \frac{\mathbf{i} + 3\mathbf{j}}{\sqrt{(1)^2 + (3)^2}}\right| = \frac{4 + 6}{\sqrt{10}} = \sqrt{10}.$$

To find the distance from a point S to the line L: $Ax + By = C$, find

1. a point P on L,
2. $\overrightarrow{PS}$,
3. the direction of $\mathbf{N} = A\mathbf{i} + B\mathbf{j}$.

Then, calculate the distance as

$$\left|\overrightarrow{PS} \cdot \frac{\mathbf{N}}{|\mathbf{N}|}\right|.$$

Work

As we mentioned at the beginning of the section, the work done by a constant force **F** when the point of application undergoes a displacement $\overrightarrow{PQ}$ (Fig. 10.30) is defined to be the dot product of **F** with $\overrightarrow{PQ}$.

DEFINITION

The **work** done by a constant force **F** acting through a displacement $\overrightarrow{PQ}$ is

$$\text{Work} = \mathbf{F} \cdot \overrightarrow{PQ} = |\mathbf{F}||\overrightarrow{PQ}| \cos\theta. \tag{20}$$

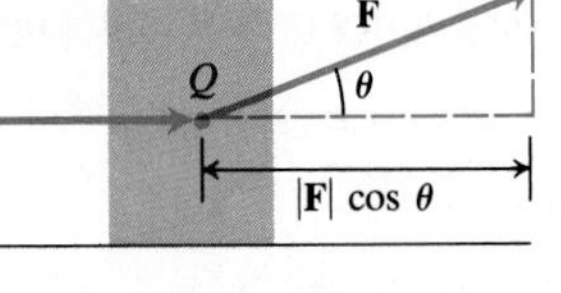

10.30 The work done by a constant force **F** during a displacement $\overrightarrow{PQ}$ is $(|\mathbf{F}| \cos\theta)|\overrightarrow{PQ}|$.

Example 7 If $|\mathbf{F}| = 40$ newtons (about 9 pounds), $|\overrightarrow{PQ}| = 3$ m, and $\theta = 60°$, the work done by **F** in acting from P to Q is

$$\begin{aligned} \text{Work} &= |\mathbf{F}||\overrightarrow{PQ}| \cos\theta && \text{(Eq. 20)} \\ &= (40)(3) \cos 60° && \text{(Given values)} \\ &= (120)(1/2) \\ &= 60 \text{ newton-meters.} \end{aligned}$$

We shall encounter work problems of greater interest in Chapter 14, where we show how to use integration to calculate the work done in moving a mass or a charged particle along a curve in space against a variable force.

EXERCISES 10.3

In Exercises 1–12, find $\mathbf{A} \cdot \mathbf{B}$, $|\mathbf{A}|$, $|\mathbf{B}|$, the cosine of the angle between **A** and **B**, the scalar component of **B** in the direction of **A**, and the vector projection of **B** onto **A**.

1. $\mathbf{A} = 3\mathbf{i} + 2\mathbf{j}$, $\mathbf{B} = 5\mathbf{j} + \mathbf{k}$
2. $\mathbf{A} = \mathbf{i}$, $\mathbf{B} = 5\mathbf{j} - 3\mathbf{k}$
3. $\mathbf{A} = 3\mathbf{i} - 2\mathbf{j} - \mathbf{k}$, $\mathbf{B} = -2\mathbf{j}$
4. $\mathbf{A} = -2\mathbf{i} + 7\mathbf{j}$, $\mathbf{B} = \mathbf{k}$
5. $\mathbf{A} = 5\mathbf{j} - 3\mathbf{k}$, $\mathbf{B} = \mathbf{i} + \mathbf{j} + \mathbf{k}$
6. $\mathbf{A} = \frac{1}{\sqrt{2}}\mathbf{i} + \frac{1}{\sqrt{3}}\mathbf{j} + \frac{1}{\sqrt{6}}\mathbf{k}$, $\mathbf{B} = \frac{1}{\sqrt{2}}\mathbf{j} - \mathbf{k}$
7. $\mathbf{A} = -\mathbf{i} + \mathbf{j}$, $\mathbf{B} = \sqrt{2}\,\mathbf{i} + \sqrt{3}\,\mathbf{j} + 2\mathbf{k}$
8. $\mathbf{A} = \mathbf{i} + \mathbf{k}$, $\mathbf{B} = \mathbf{i} + \mathbf{j} + \mathbf{k}$
9. $\mathbf{A} = 2\mathbf{i} - 4\mathbf{j} + \sqrt{5}\,\mathbf{k}$, $\mathbf{B} = -2\mathbf{i} + 4\mathbf{j} - \sqrt{5}\,\mathbf{k}$
10. $\mathbf{A} = -5\mathbf{i} + \mathbf{j}$, $\mathbf{B} = 2\mathbf{i} + \sqrt{17}\,\mathbf{j} + 10\mathbf{k}$
11. $\mathbf{A} = 10\mathbf{i} + 11\mathbf{j} - 2\mathbf{k}$, $\mathbf{B} = 3\mathbf{j} + 4\mathbf{k}$
12. $\mathbf{A} = 2\mathbf{i} + 10\mathbf{j} - 11\mathbf{k}$, $\mathbf{B} = 2\mathbf{i} + 2\mathbf{j} + \mathbf{k}$
13. Write $\mathbf{B} = 3\mathbf{j} + 4\mathbf{k}$ as the sum of a vector parallel to $\mathbf{A} = \mathbf{i} + \mathbf{j}$ and a vector orthogonal to **A**.
14. Write $\mathbf{B} = \mathbf{j} + \mathbf{k}$ as the sum of a vector parallel to $\mathbf{A} = \mathbf{i} + \mathbf{j}$ and a vector orthogonal to **A**.
15. Write $\mathbf{B} = 8\mathbf{i} + 4\mathbf{j} - 12\mathbf{k}$ as the sum of a vector parallel to $\mathbf{A} = \mathbf{i} + 2\mathbf{j} - \mathbf{k}$ and a vector orthogonal to **A**.
16. $\mathbf{B} = \mathbf{i} + (\mathbf{j} + \mathbf{k})$ is already the sum of a vector parallel to **i** and a vector orthogonal to **i**. If you use Eq. (15) with $\mathbf{A} = \mathbf{i}$, do you get $\mathbf{B}_1 = \mathbf{i}$ and $\mathbf{B}_2 = \mathbf{j} + \mathbf{k}$? Try it and find out.

In Exercises 17–20, find an equation for the line in the xy-plane that passes through the given point perpendicular to the given vector. Then sketch the line. Include the vector in your sketch as a vector starting at the origin (as in Fig. 10.28).

17. $P(2, 1)$, $\mathbf{i} + 2\mathbf{j}$
18. $P(-2, 1)$, $\mathbf{i} - \mathbf{j}$
19. $P(-1, 2)$, $-2\mathbf{i} - \mathbf{j}$
20. $P(-1, 2)$, $2\mathbf{i} - 3\mathbf{j}$

In Exercises 21–24, find the distance in the xy-plane from the point to the line.

21. $S(2, 8)$, $x + 3y = 6$
22. $S(0, 0)$, $x + 3y = 6$
23. $S(2, 1)$, $x + y = 1$
24. $S(1, 3)$, $y = -2x$

25. Show that the vectors

$$\mathbf{A} = \frac{1}{\sqrt{3}}(\mathbf{i} - \mathbf{j} + \mathbf{k}), \quad \mathbf{B} = \frac{1}{\sqrt{2}}(\mathbf{j} + \mathbf{k}),$$

$$\mathbf{C} = \frac{1}{\sqrt{6}}(-2\,\mathbf{i} - \mathbf{j} + \mathbf{k})$$

are orthogonal to one another.

26. Find the vector projections of $\mathbf{D} = \mathbf{i} + \mathbf{j} + \mathbf{k}$ on the vectors $\mathbf{A}$, $\mathbf{B}$, and $\mathbf{C}$ of Exercise 25. Then show that $\mathbf{D}$ is the sum of these vector projections.

27. *Cancellation in dot products is not valid.* In real-number multiplication, if $ab_1 = ab_2$ and a is not zero, we can safely cancel the a and conclude that $b_1 = b_2$. Not so for vector multiplication: If $\mathbf{A} \cdot \mathbf{B}_1 = \mathbf{A} \cdot \mathbf{B}_2$ and $\mathbf{A} \neq \mathbf{0}$, it is not safe to conclude that $\mathbf{B}_1 = \mathbf{B}_2$. See if you can come up with an example. Keep it simple: Experiment with $\mathbf{i}$, $\mathbf{j}$, and $\mathbf{k}$.

28. *Sums and differences.* In Fig. 10.31, it looks as if $\mathbf{v}_1 + \mathbf{v}_2$ and $\mathbf{v}_1 - \mathbf{v}_2$ are orthogonal. Is this mere coincidence, or are there circumstances under which we may expect the sum of two vectors to be orthogonal to their difference? Find out by expanding the left-hand side of the equation

$$(\mathbf{v}_1 + \mathbf{v}_2) \cdot (\mathbf{v}_1 - \mathbf{v}_2) = 0.$$

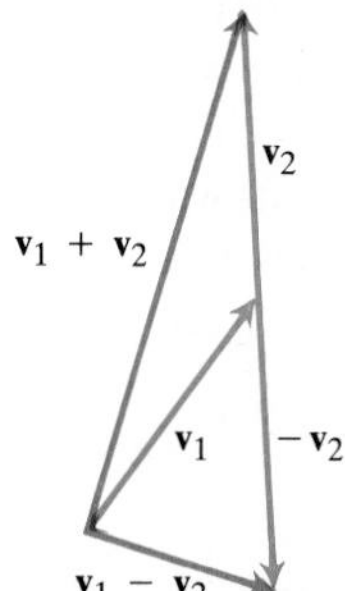

10.31 The vectors in Exercise 28.

29. Show that the diagonals of a rhombus are perpendicular.

30. Show that squares are the only rectangles with perpendicular diagonals.

CALCULATOR Use a calculator to find the angles in Exercises 31–34 to the nearest tenth of a degree.

31. Find the interior angles of the triangle ABC whose vertices are $A(-1, 0, 2)$, $B(2, 1, -1)$, and $C(1, -2, 2)$.

32. Find the angle between $\mathbf{A} = 2\,\mathbf{i} + 2\,\mathbf{j} + \mathbf{k}$ and $\mathbf{B} = 2\,\mathbf{i} + 10\,\mathbf{j} - 11\,\mathbf{k}$.

33. Find the angle between the diagonal of a cube and the diagonal of one of its faces. (*Hint:* Use a cube whose edges represent $\mathbf{i}$, $\mathbf{j}$, and $\mathbf{k}$.)

34. Find the angle between the diagonal of a cube and one of the edges it meets at a vertex.

Work

35. Find the work done by a force $\mathbf{F} = -5\,\mathbf{k}$ (magnitude 5 newtons) in moving an object along the line from the origin to the point $(1, 1, 1)$ (distance in meters).

36. A locomotive exerted a constant force of 60,000 newtons on a freight train while pulling it 1 km along a straight track. How much work did the locomotive do?

37. How much work does it take to slide a crate 20 m along a loading dock by pulling on it with a 200-newton force at an angle of 30° from the horizontal?

38. The wind passing over a boat's sail exerted a 1000-lb-magnitude force $\mathbf{F}$ as shown in Fig. 10.32. If the force vector made a 60° angle with the line of the boat's forward motion, how much work did the wind perform in moving the boat forward 1 mi? Answer in foot-pounds.

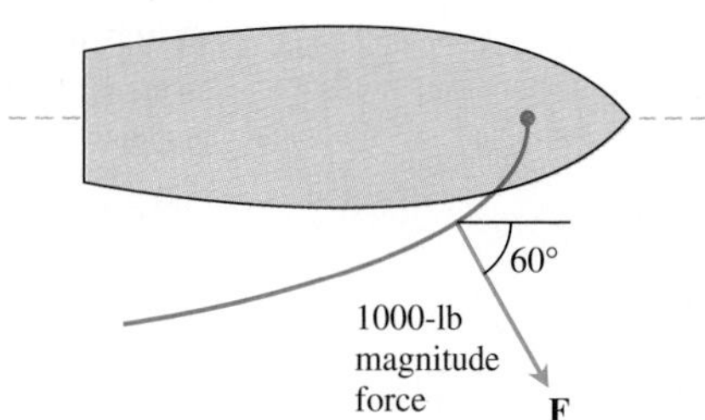

10.32 The boat in Exercise 38.

Angles Between Curves

The angles between two differentiable curves at a point of intersection are the angles between the curves' tangents at these points. Find the angles between the curves in Exercises 39–43 at their points of intersection. (You will not need a calculator.)

39. $3x + y = 5, \quad 2x - y = 4$

40. $y = \sqrt{3}x - 1, \quad y = -\sqrt{3}x + 2$

41. $y = \sqrt{(3/4) + x}, \quad y = \sqrt{(3/4) - x}$

42. $y = x^3, \quad y = \sqrt{x}$ (two points of intersection)

43. $y = x^2, \quad y = \sqrt[3]{x}$ (two points of intersection)

10.4 Cross Products

When we turn a bolt by applying a force to a wrench (Fig. 10.33), the torque we produce acts along the axis of the bolt to drive the bolt forward. The magnitude of the torque depends on how far out on the wrench the force is applied and on how much of the force is actually perpendicular to the wrench at that point. The number we use to measure the torque's magnitude is a product made up of the length of the lever arm **A** and the scalar component of **B** perpendicular to **A**. In the notation of Fig. 10.33,

$$\text{magnitude of the torque vector} = |\mathbf{A}||\mathbf{B}|\sin\theta.$$

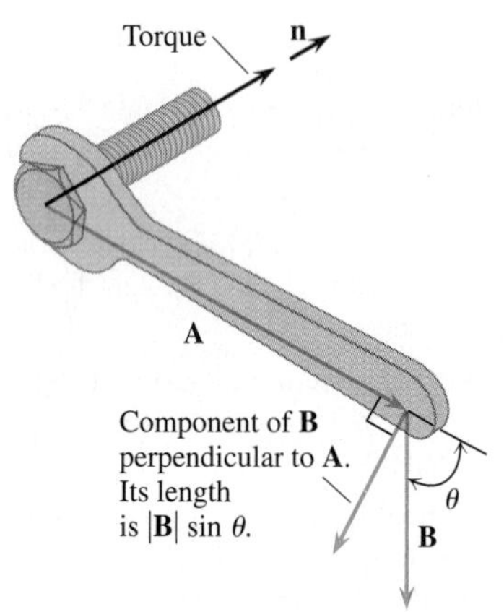

10.33 The torque vector describes the tendency of the force **B** to drive the bolt forward.

If we let **n** be a unit vector along the axis of the bolt in the direction of the torque, then the complete description of the torque vector is

$$\text{Torque vector} = \mathbf{n}|\mathbf{A}||\mathbf{B}|\sin\theta.$$

In mathematics, we call the vector $\mathbf{n}|\mathbf{A}||\mathbf{B}|\sin\theta$ the vector product of **A** and **B**.

Vector products are widely used to describe the effects of forces in studies of electricity, magnetism, fluid mechanics, and planetary motion. The goal of this section is to present the mathematical properties of vector products that account for their use in these fields. We shall use these properties ourselves in Chapter 11, where we study motion in space, and in Chapter 14, where we study the vector integrals of fluid flow. In the next section we shall also see how to combine vector products with scalar products to produce equations for lines and planes in space.

The Vector Product of Two Vectors in Space

When we define vector products in mathematics, we start with two nonzero vectors **A** and **B** in space without requiring them to have any particular physical interpretation. If **A** and **B** are not parallel, they determine a plane. We select a unit vector **n** perpendicular to the plane by the **right-hand rule.** This means we choose **n** to be the unit (normal) vector that points the way your right thumb points when your fingers curl through the angle θ from **A** to **B** (Fig. 10.34). We then define the **vector product A × B** ("**A** cross **B**") of **A** and **B** to be the vector

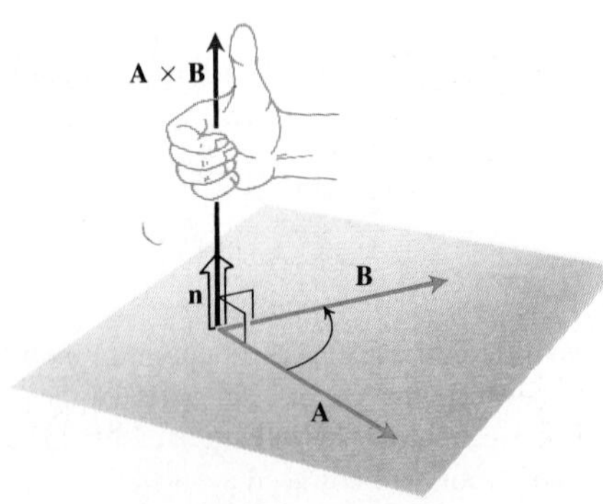

10.34 The construction of **A × B**.

$$\mathbf{A} \times \mathbf{B} = \mathbf{n}|\mathbf{A}||\mathbf{B}|\sin\theta. \tag{1}$$

Since $\mathbf{A} \times \mathbf{B}$ is a scalar multiple of **n**, it is perpendicular to both **A** and **B**.

If θ approaches 0° or 180° in Eq. (1), the length of $\mathbf{A} \times \mathbf{B}$ approaches zero. We therefore define $\mathbf{A} \times \mathbf{B}$ to be **0** if **A** and **B** are parallel (and fail to determine a plane). This is consistent with our torque interpretation as well. If the force **B** in Fig. 10.33 is parallel to the wrench, meaning that we are trying to turn the bolt by pushing or pulling straight along the handle of the wrench, the torque produced is **0**.

If one or both of **A** and **B** is zero, we define $\mathbf{A} \times \mathbf{B}$ to be zero. This way, the cross product of two vectors **A** and **B** is zero if and only if **A** and **B** are parallel or one or both of them is the zero vector.

The vector product of **A** and **B** is often called the **cross product** of **A** and **B** because of the cross in the notation $\mathbf{A} \times \mathbf{B}$. In contrast to the dot product $\mathbf{A} \cdot \mathbf{B}$, which is a scalar, the cross product $\mathbf{A} \times \mathbf{B}$ is a vector.

A × B Versus B × A

Reversing the order of the factors in a nonzero cross product reverses the direction of the resulting vector. When the fingers of our right hand curl through the angle θ from **B** to **A**, our thumb points the opposite way and the unit vector we choose in

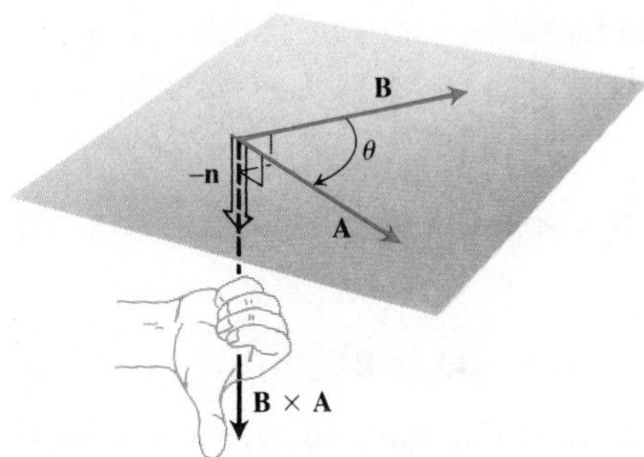

10.35 The construction of $\mathbf{B} \times \mathbf{A}$.

forming $\mathbf{B} \times \mathbf{A}$ is the negative of the one we choose in forming $\mathbf{A} \times \mathbf{B}$ (Fig. 10.35). Thus,

$$\mathbf{B} \times \mathbf{A} = -(\mathbf{A} \times \mathbf{B}) \tag{2}$$

for all vectors $\mathbf{A}$ and $\mathbf{B}$. Unlike the dot product, the cross product is not commutative.

When we apply the definition to calculate the pairwise cross products of the unit vectors $\mathbf{i}$, $\mathbf{j}$, and $\mathbf{k}$, we find (Fig. 10.36)

$$\begin{aligned} \mathbf{i} \times \mathbf{j} &= -(\mathbf{j} \times \mathbf{i}) = \mathbf{k}, \\ \mathbf{j} \times \mathbf{k} &= -(\mathbf{k} \times \mathbf{j}) = \mathbf{i}, \\ \mathbf{k} \times \mathbf{i} &= -(\mathbf{i} \times \mathbf{k}) = \mathbf{j}, \end{aligned} \tag{3}$$

and

$$\mathbf{i} \times \mathbf{i} = \mathbf{j} \times \mathbf{j} = \mathbf{k} \times \mathbf{k} = \mathbf{0}.$$

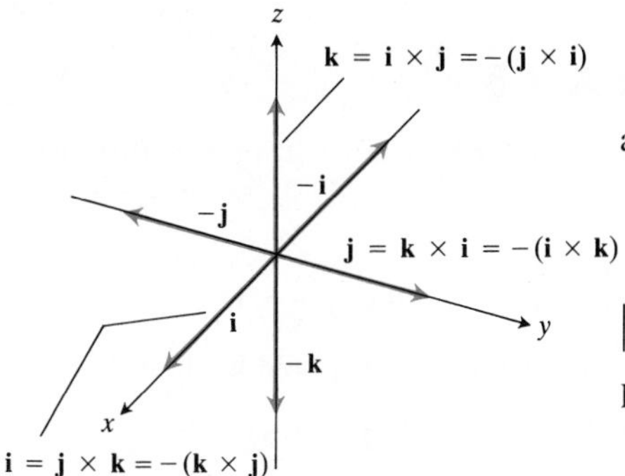

10.36 The pairwise cross products of $\mathbf{i}$, $\mathbf{j}$, and $\mathbf{k}$ can be read from this diagram.

$|\mathbf{A} \times \mathbf{B}|$ Is the Area of a Parallelogram

Because $\mathbf{n}$ is a unit vector, the magnitude of $\mathbf{A} \times \mathbf{B}$ is

$$|\mathbf{A} \times \mathbf{B}| = |\mathbf{n}||\mathbf{A}||\mathbf{B}||\sin\theta| = |\mathbf{A}||\mathbf{B}||\sin\theta|. \tag{4}$$

This is the area of the parallelogram determined by $\mathbf{A}$ and $\mathbf{B}$ (Fig. 10.37), $|\mathbf{A}|$ being the base of the parallelogram and $|\mathbf{B}||\sin\theta|$ the height.

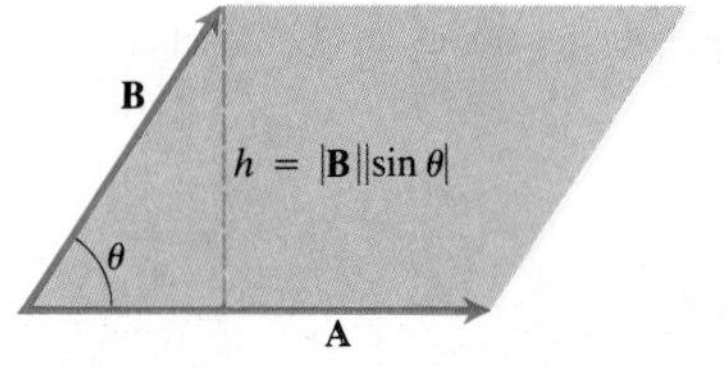

10.37 The parallelogram determined by $\mathbf{A}$ and $\mathbf{B}$. Its area is $|\mathbf{A} \times \mathbf{B}|$.

The Magnitude of a Torque

Equation (4) is the equation we use to calculate magnitudes of torques.

Example 1 The magnitude of the torque exerted by force $\mathbf{F}$ about the pivot point P in Fig. 10.38 is

$$\begin{aligned} |\overrightarrow{PQ} \times \mathbf{F}| &= |\overrightarrow{PQ}||\mathbf{F}| \sin 70° && \text{(Eq. (4))} \\ &= (3)(20)(0.94) && \text{(Calculator, rounded)} \\ &= 56.4 \text{ ft-lb.} \end{aligned}$$

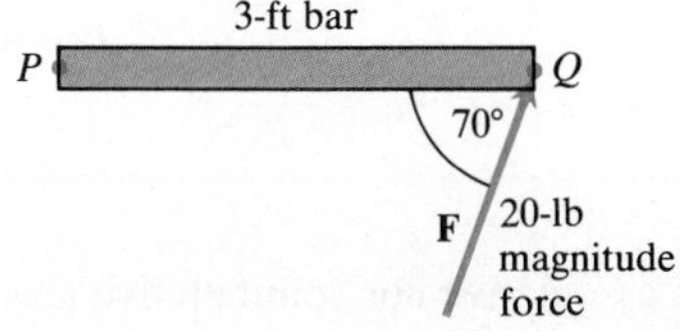

10.38 The magnitude of the torque exerted by $\mathbf{F}$ about P is about 56.4 ft · lb (Example 1).

The Associative and Distributive Laws

As a rule, cross-product multiplication is not associative because $(\mathbf{A} \times \mathbf{B}) \times \mathbf{C}$ lies in the plane of $\mathbf{A}$ and $\mathbf{B}$ whereas $\mathbf{A} \times (\mathbf{B} \times \mathbf{C})$ lies in the plane of $\mathbf{B}$ and $\mathbf{C}$. However, the **Scalar Distributive Law**

$$(r\mathbf{A}) \times (s\mathbf{B}) = (rs)\mathbf{A} \times \mathbf{B} \tag{5}$$

does hold, as do the **Vector Distributive Laws**

$$\mathbf{A} \times (\mathbf{B} + \mathbf{C}) = \mathbf{A} \times \mathbf{B} + \mathbf{A} \times \mathbf{C} \tag{6}$$

and

$$(\mathbf{B} + \mathbf{C}) \times \mathbf{A} = \mathbf{B} \times \mathbf{A} + \mathbf{C} \times \mathbf{A}. \tag{7}$$

As a special case of (5) we have

$$(-\mathbf{A}) \times \mathbf{B} = \mathbf{A} \times (-\mathbf{B}) = -(\mathbf{A} \times \mathbf{B}).$$

The Scalar Distributive Law can be verified by applying Eq. (1) to the products on both sides of Eq. (5) and comparing the results. The Vector Distributive Law in Eq. (6) is not easy to prove, however. We shall assume it here and leave the proof to Appendix 9. Equation (7) follows from Eq. (6): multiply both sides of Eq. (6) by -1 and reverse the orders of the products.

Determinant Formulas

(See Appendix 10.)

$$\begin{vmatrix} a & b \\ c & d \end{vmatrix} = ad - bc$$

Example

$$\begin{vmatrix} 2 & 1 \\ -4 & 3 \end{vmatrix} = (2)(3) - (-4)(1) = 6 + 4 = 10$$

$$\begin{vmatrix} a_{11} & a_{12} & a_{13} \\ a_{21} & a_{22} & a_{23} \\ a_{31} & a_{32} & a_{33} \end{vmatrix} = a_{11}\begin{vmatrix} a_{22} & a_{23} \\ a_{32} & a_{33} \end{vmatrix} - a_{12}\begin{vmatrix} a_{21} & a_{23} \\ a_{31} & a_{33} \end{vmatrix} + a_{13}\begin{vmatrix} a_{21} & a_{22} \\ a_{31} & a_{32} \end{vmatrix}$$

Example

$$\begin{vmatrix} -5 & 3 & 1 \\ 2 & 1 & 1 \\ -4 & 3 & 1 \end{vmatrix} = (-5)\begin{vmatrix} 1 & 1 \\ 3 & 1 \end{vmatrix} - (3)\begin{vmatrix} 2 & 1 \\ -4 & 1 \end{vmatrix} + (1)\begin{vmatrix} 2 & 1 \\ -4 & 3 \end{vmatrix}$$

$$= -5(1 - 3) - 3(2 + 4) + 1(6 + 4)$$

$$= 10 - 18 + 10 = 2$$

The Determinant Formula for A × B

Our next objective is to show how to calculate the components of $\mathbf{A} \times \mathbf{B}$ from the components of $\mathbf{A}$ and $\mathbf{B}$.

Suppose

$$\mathbf{A} = a_1\mathbf{i} + a_2\mathbf{j} + a_3\mathbf{k}, \qquad \mathbf{B} = b_1\mathbf{i} + b_2\mathbf{j} + b_3\mathbf{k}.$$

Then the distributive laws and the rules for multiplying $\mathbf{i}$, $\mathbf{j}$, and $\mathbf{k}$ tell us that

$$\begin{aligned} \mathbf{A} \times \mathbf{B} &= (a_1\mathbf{i} + a_2\mathbf{j} + a_3\mathbf{k}) \times (b_1\mathbf{i} + b_2\mathbf{j} + b_3\mathbf{k}) \\ &= a_1b_1\,\mathbf{i} \times \mathbf{i} + a_1b_2\,\mathbf{i} \times \mathbf{j} + a_1b_3\,\mathbf{i} \times \mathbf{k} \\ &\quad + a_2b_1\,\mathbf{j} \times \mathbf{i} + a_2b_2\,\mathbf{j} \times \mathbf{j} + a_2b_3\,\mathbf{j} \times \mathbf{k} \\ &\quad + a_3b_1\,\mathbf{k} \times \mathbf{i} + a_3b_2\,\mathbf{k} \times \mathbf{j} + a_3b_3\,\mathbf{k} \times \mathbf{k} \\ &= (a_2b_3 - a_3b_2)\,\mathbf{i} + (a_3b_1 - a_1b_3)\,\mathbf{j} + (a_1b_2 - a_2b_1)\,\mathbf{k}. \end{aligned} \tag{8}$$

The terms at the end of Eq. (8) are the same as the terms in the expansion of the symbolic determinant

$$\begin{vmatrix} \mathbf{i} & \mathbf{j} & \mathbf{k} \\ a_1 & a_2 & a_3 \\ b_1 & b_2 & b_3 \end{vmatrix}.$$

We therefore have the following rule.

> If $\mathbf{A} = a_1\mathbf{i} + a_2\mathbf{j} + a_3\mathbf{k}$ and $\mathbf{B} = b_1\mathbf{i} + b_2\mathbf{j} + b_3\mathbf{k}$, then
>
> $$\mathbf{A} \times \mathbf{B} = \begin{vmatrix} \mathbf{i} & \mathbf{j} & \mathbf{k} \\ a_1 & a_2 & a_3 \\ b_1 & b_2 & b_3 \end{vmatrix}. \tag{9}$$

Equation (9) is remarkable, given that neither the associative nor commutative law holds for cross-product multiplication.

Example 2 Find $\mathbf{A} \times \mathbf{B}$ and $\mathbf{B} \times \mathbf{A}$ if

$$\mathbf{A} = 2\mathbf{i} + \mathbf{j} + \mathbf{k}, \qquad \mathbf{B} = -4\mathbf{i} + 3\mathbf{j} + \mathbf{k}.$$

Solution We use Eq. (9) to find $\mathbf{A} \times \mathbf{B}$:

$$\mathbf{A} \times \mathbf{B} = \begin{vmatrix} \mathbf{i} & \mathbf{j} & \mathbf{k} \\ 2 & 1 & 1 \\ -4 & 3 & 1 \end{vmatrix} = \begin{vmatrix} 1 & 1 \\ 3 & 1 \end{vmatrix} \mathbf{i} - \begin{vmatrix} 2 & 1 \\ -4 & 1 \end{vmatrix} \mathbf{j} + \begin{vmatrix} 2 & 1 \\ -4 & 3 \end{vmatrix} \mathbf{k}$$

$$= -2\,\mathbf{i} - 6\,\mathbf{j} + 10\,\mathbf{k}.$$

Equation (2) then gives

$$\mathbf{B} \times \mathbf{A} = -\mathbf{A} \times \mathbf{B} = 2\,\mathbf{i} + 6\,\mathbf{j} - 10\,\mathbf{k}.$$

Example 3 Find a vector perpendicular to the plane of $P(1, -1, 0)$, $Q(2, 1, -1)$, and $R(-1, 1, 2)$.

Solution The vector $\overrightarrow{PQ} \times \overrightarrow{PR}$ is perpendicular to the plane because it is perpendicular to both vectors. In terms of components,

$$\overrightarrow{PQ} = (2-1)\,\mathbf{i} + (1+1)\,\mathbf{j} + (-1-0)\,\mathbf{k} = \mathbf{i} + 2\,\mathbf{j} - \mathbf{k},$$

$$\overrightarrow{PR} = (-1-1)\,\mathbf{i} + (1+1)\,\mathbf{j} + (2-0)\,\mathbf{k} = -2\,\mathbf{i} + 2\,\mathbf{j} + 2\,\mathbf{k},$$

$$\overrightarrow{PQ} \times \overrightarrow{PR} = \begin{vmatrix} \mathbf{i} & \mathbf{j} & \mathbf{k} \\ 1 & 2 & -1 \\ -2 & 2 & 2 \end{vmatrix}$$

$$= \begin{vmatrix} 2 & -1 \\ 2 & 2 \end{vmatrix} \mathbf{i} - \begin{vmatrix} 1 & -1 \\ -2 & 2 \end{vmatrix} \mathbf{j} + \begin{vmatrix} 1 & 2 \\ -2 & 2 \end{vmatrix} \mathbf{k}$$

$$= 6\,\mathbf{i} + 6\,\mathbf{k}.$$

Example 4 Find the area of the triangle with vertices $P(1, -1, 0)$, $Q(2, 1, -1)$, and $R(-1, 1, 2)$ (Fig. 10.39).

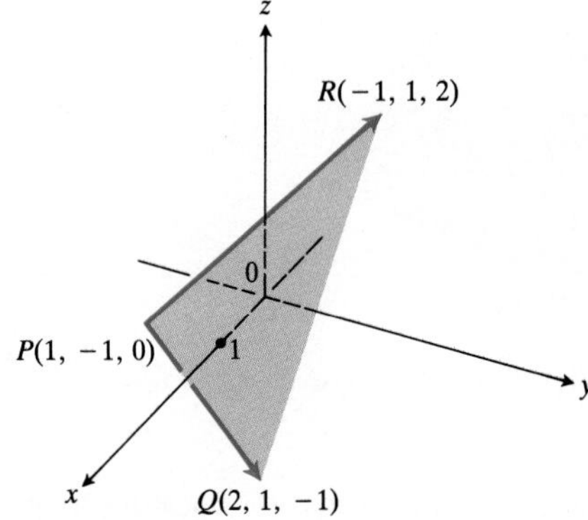

10.39 The area of triangle PQR is half of $|\overrightarrow{PQ} \times \overrightarrow{PR}|$ (Example 4).

Solution The area of the parallelogram determined by P, Q, and R is

$$|\overrightarrow{PQ} \times \overrightarrow{PR}| = |6\,\mathbf{i} + 6\,\mathbf{k}| \qquad \text{(Values from Example 3)}$$

$$= \sqrt{(6)^2 + (6)^2} = \sqrt{2 \cdot 36} = 6\sqrt{2}.$$

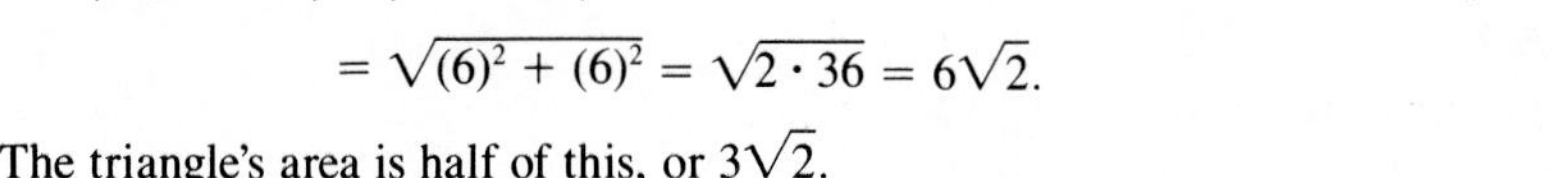

The triangle's area is half of this, or $3\sqrt{2}$.

Example 5 Find a unit vector perpendicular to the plane of $P(1, -1, 0)$, $Q(2, 1, -1)$, and $R(-1, 1, 2)$.

Solution Since $\overrightarrow{PQ} \times \overrightarrow{PR}$ is perpendicular to the plane, its direction $\mathbf{n}$ is a unit vector perpendicular to the plane. In component form,

$$\mathbf{n} = \frac{\overrightarrow{PQ} \times \overrightarrow{PR}}{|\overrightarrow{PQ} \times \overrightarrow{PR}|}$$

$$= \frac{6\,\mathbf{i} + 6\,\mathbf{k}}{6\sqrt{2}} \qquad \text{(Values from Examples 3 and 4)}$$

$$= \frac{1}{\sqrt{2}}\,\mathbf{i} + \frac{1}{\sqrt{2}}\,\mathbf{k}.$$

EXERCISES 10.4

In Exercises 1–8, find the length and direction (when defined) of $\mathbf{A} \times \mathbf{B}$ and $\mathbf{B} \times \mathbf{A}$.

1. $\mathbf{A} = 2\mathbf{i} - 2\mathbf{j} - \mathbf{k}, \quad \mathbf{B} = \mathbf{i} - \mathbf{k}$

2. $\mathbf{A} = 2\mathbf{i} + 3\mathbf{j}, \quad \mathbf{B} = -\mathbf{i} + \mathbf{j}$

3. $\mathbf{A} = 2\mathbf{i} - 2\mathbf{j} + 4\mathbf{k}, \quad \mathbf{B} = -\mathbf{i} + \mathbf{j} - 2\mathbf{k}$

4. $\mathbf{A} = \mathbf{i} + \mathbf{j} - \mathbf{k}, \quad \mathbf{B} = \mathbf{0}$

5. $\mathbf{A} = 2\mathbf{i}, \quad \mathbf{B} = -3\mathbf{j}$

6. $\mathbf{A} = \mathbf{i} \times \mathbf{j}, \quad \mathbf{B} = \mathbf{j} \times \mathbf{k}$

7. $\mathbf{A} = -8\mathbf{i} - 2\mathbf{j} - 4\mathbf{k}, \quad \mathbf{B} = 2\mathbf{i} + 2\mathbf{j} + \mathbf{k}$

8. $\mathbf{A} = \frac{3}{2}\mathbf{i} - \frac{1}{2}\mathbf{j} + \mathbf{k}, \quad \mathbf{B} = \mathbf{i} + \mathbf{j} + 2\mathbf{k}$

In Exercises 9–14, sketch the coordinate axes and then include the vectors $\mathbf{A}$, $\mathbf{B}$, and $\mathbf{A} \times \mathbf{B}$ as vectors coming out from the origin.

9. $\mathbf{A} = \mathbf{i}, \quad \mathbf{B} = \mathbf{j}$

10. $\mathbf{A} = \mathbf{i} + \mathbf{k}, \quad \mathbf{B} = \mathbf{j}$

11. $\mathbf{A} = \mathbf{i} - \mathbf{k}, \quad \mathbf{B} = \mathbf{j} + \mathbf{k}$

12. $\mathbf{A} = 2\mathbf{i} - \mathbf{j}, \quad \mathbf{B} = \mathbf{i} + 2\mathbf{j}$

13. $\mathbf{A} = \mathbf{i} + 3\mathbf{j} + 2\mathbf{k}, \quad \mathbf{B} = \mathbf{k}$

14. $\mathbf{A} = \mathbf{i} + 2\mathbf{j}, \quad \mathbf{B} = 2\mathbf{j} + \mathbf{k}$

In Exercises 15–18:

a) Find a vector $\mathbf{N}$ perpendicular to the plane of the points P, Q, and R.

b) Find the area of triangle PQR.

c) Find a unit vector perpendicular to plane PQR.

15. $P(1, -1, 2), \quad Q(2, 0, -1), \quad R(0, 2, 1)$

16. $P(1, 1, 1), \quad Q(2, 1, 3), \quad R(3, -1, 1)$

17. $P(2, -2, 1), \quad Q(3, -1, 2), \quad R(3, -1, 1)$

18. $P(-2, 2, 0), \quad Q(0, 1, -1), \quad R(-1, 2, -2)$

19. Let $\mathbf{A} = 5\mathbf{i} - \mathbf{j} + \mathbf{k}$, $\mathbf{B} = \mathbf{j} - 5\mathbf{k}$, $\mathbf{C} = -15\mathbf{i} + 3\mathbf{j} - 3\mathbf{k}$. Which vectors, if any, are (a) perpendicular, (b) parallel?

20. Let $\mathbf{A} = \mathbf{i} + 2\mathbf{j} - \mathbf{k}$, $\mathbf{B} = -\mathbf{i} + \mathbf{j} + \mathbf{k}$, $\mathbf{C} = \mathbf{i} + \mathbf{k}$, $\mathbf{D} = -(\pi/2)\mathbf{i} - \pi\mathbf{j} + (\pi/2)\mathbf{k}$. Which vectors, if any, are (a) perpendicular, (b) parallel?

In Exercises 21 and 22, find the magnitude of the torque exerted by $\mathbf{F}$ on the bolt at P if $|\overrightarrow{PQ}| = 8$ in. and $|\mathbf{F}| = 30$ lb. Answer in foot-pounds.

21. **22.**

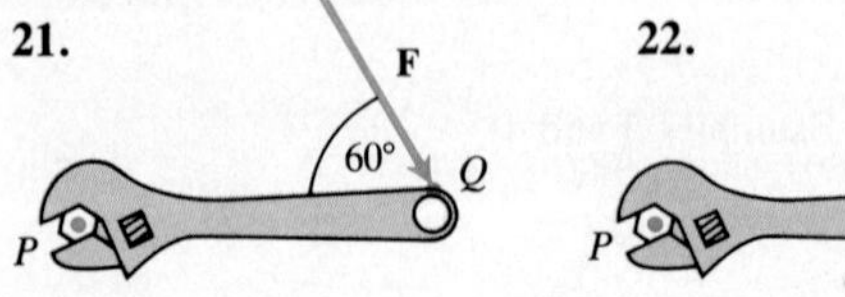

23. If $\mathbf{A} = 2\mathbf{i} - \mathbf{j}$ and $\mathbf{B} = \mathbf{i} + 3\mathbf{j} - 2\mathbf{k}$, find $\mathbf{A} \times \mathbf{B}$. Then calculate $(\mathbf{A} \times \mathbf{B}) \cdot \mathbf{A}$ and $(\mathbf{A} \times \mathbf{B}) \cdot \mathbf{B}$.

24. Is $(\mathbf{A} \times \mathbf{B}) \cdot \mathbf{A}$ always zero? Explain. What about $(\mathbf{A} \times \mathbf{B}) \cdot \mathbf{B}$?

25. Given vectors $\mathbf{A}$, $\mathbf{B}$, and $\mathbf{C}$, use dot-product and cross-product notation, as appropriate, to describe the following:

a) The vector projection of $\mathbf{A}$ onto $\mathbf{B}$

b) A vector orthogonal to $\mathbf{A}$ and $\mathbf{B}$

c) A vector with the length of $\mathbf{A}$ and the direction of $\mathbf{B}$

d) A vector orthogonal to $\mathbf{A} \times \mathbf{B}$ and $\mathbf{C}$

e) A vector in the plane of $\mathbf{B}$ and $\mathbf{C}$ perpendicular to $\mathbf{A}$

26. *Cancellation is not valid in cross products either.* Find an example to show that $\mathbf{A} \times \mathbf{B} = \mathbf{A} \times \mathbf{C}$ need not imply that $\mathbf{B}$ equals $\mathbf{C}$ even if $\mathbf{A} \neq \mathbf{0}$.

27. If $\overrightarrow{AB} \times \overrightarrow{AC} = 2\mathbf{i} - 4\mathbf{j} + 4\mathbf{k}$, find the area of triangle ABC.

28. a) Suppose that on each triangular face of a tetrahedron $ABCD$ (Fig. 10.40) we construct an outward normal vector whose magnitude equals the area of the face. Show that the sum of these four vectors is $\mathbf{0}$.

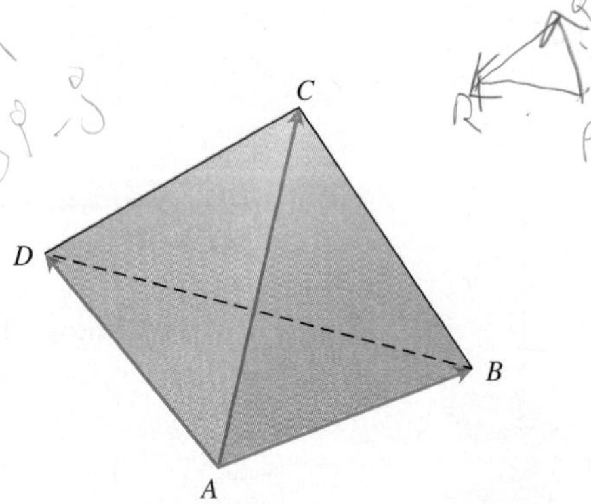

10.40 The tetrahedron in Exercise 28.

b) Formulate and prove an analogous statement for triangles in the coordinate plane.

c) The solid in (a) need not be a tetrahedron. Show that the conclusion holds for any starlike polyhedron with triangular faces. (A polyhedron is *starlike* if it has an interior point from which it is possible to view all the faces simultaneously without other faces getting in the way.) Do not give a technical argument. Just describe the basic idea behind the proof.

The statement in (a) is actually true for any polyhedron in space. It need not be starlike and its faces need not be triangular.

10.5 Lines and Planes in Space

This section shows how to use scalar and vector products to write equations for lines, line segments, and planes in space.

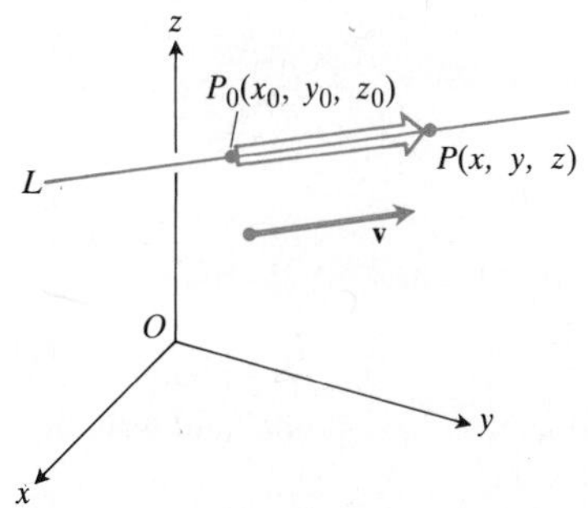

10.41 A point P lies on the line through P_0 parallel to $\mathbf{v}$ if and only if $\overrightarrow{P_0P}$ is a scalar multiple of $\mathbf{v}$.

Equations for Lines and Line Segments

Suppose L is a line in space that passes through a point $P_0(x_0, y_0, z_0)$ and lies parallel to a vector $\mathbf{v} = A\mathbf{i} + B\mathbf{j} + C\mathbf{k}$. Then L is the set of all points $P(x, y, z)$ for which the vector P_0P is parallel to $\mathbf{v}$ (Fig. 10.41). That is, P lies on L if and only if

$$\overrightarrow{P_0P} = t\mathbf{v} \tag{1}$$

for some number t, $-\infty < t < \infty$.

When we write Eq. (1) in terms of **i**-, **j**-, and **k**-components and equate the corresponding components of the two sides, we get three equations involving the parameter t:

$$(x - x_0)\,\mathbf{i} + (y - y_0)\,\mathbf{j} + (z - z_0)\,\mathbf{k} = t(A\mathbf{i} + B\mathbf{j} + C\mathbf{k}), \quad \begin{pmatrix}\text{Eq. (1) spelled}\\ \text{out}\end{pmatrix}$$

$$x - x_0 = tA, \qquad y - y_0 = tB, \qquad z - z_0 = tC. \quad \text{(Components equated)}$$

When rearranged, these last three equations give us the standard parametrization of the line for the parameter interval $-\infty < t < \infty$.

Standard Parametrization of the Line Through $P_0(x_0, y_0, z_0)$ Parallel to $\mathbf{v} = A\mathbf{i} + B\mathbf{j} + C\mathbf{k}$

$$x = x_0 + tA, \qquad y = y_0 + tB, \qquad z = z_0 + tC, \qquad -\infty < t < \infty \tag{2}$$

We rarely use any t-interval other than $(-\infty, \infty)$ when we parametrize a line, so we call the equations in (2) the **standard parametric equations** for the line through P_0 parallel to $\mathbf{v}$. When we want parametric equations for this line, these are the ones we produce, the interval $(-\infty, \infty)$ being understood.

Example 1 Find parametric equations for the line through $(-2, 0, 4)$ parallel to $\mathbf{v} = 2\mathbf{i} + 4\mathbf{j} - 2\mathbf{k}$ (Fig. 10.42).

10.42 Selected points and parameter values on the line $x = -2 + 2t$, $y = 4t$, $z = 4 - 2t$. The arrows show the direction of increasing t (Example 1).

Solution With $P_0(x_0, y_0, z_0)$ equal to $(-2, 0, 4)$ and $A\mathbf{i} + B\mathbf{j} + C\mathbf{k}$ equal to $2\mathbf{i} + 4\mathbf{j} - 2\mathbf{k}$, Eqs. (2) become

$$x = -2 + 2t, \qquad y = 4t, \qquad z = 4 - 2t.$$

Example 2 Find parametric equations for the line through the points $P(-3, 2, -3)$ and $Q(1, -1, 4)$.

Solution The vector

$$\overrightarrow{PQ} = 4\mathbf{i} - 3\mathbf{j} + 7\mathbf{k}$$

is parallel to the line, and Eqs. (2) with $(x_0, y_0, z_0) = (-3, 2, -3)$ give

$$x = -3 + 4t, \qquad y = 2 - 3t, \qquad z = -3 + 7t. \tag{3}$$

We could equally well have chosen $Q(1, -1, 4)$ as the "base point" and written

$$x = 1 + 4t, \qquad y = -1 - 3t, \qquad z = 4 + 7t \tag{4}$$

as equations for the line. The parametrizations in (3) and (4) place you at different points for a given value of t, but each set of equations covers the line completely as t runs from $-\infty$ to ∞.

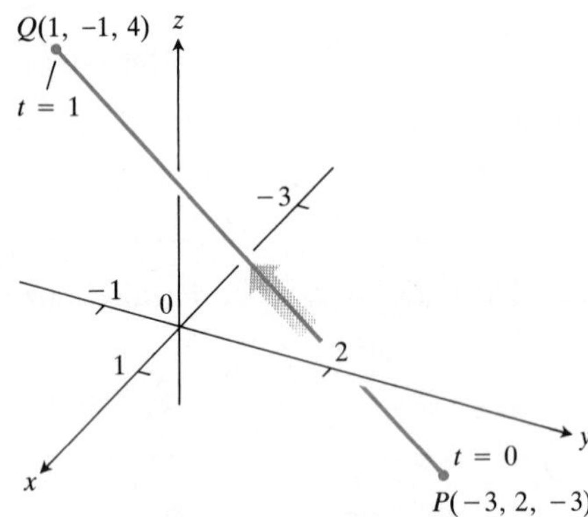

10.43 Example 3 derives a parametrization of the line segment joining P and Q. The arrow shows the direction of increasing t for this parametrization.

To parametrize a line segment joining two points, we first parametrize the line through the points. We then find the t-values for the points and restrict t to lie in the closed interval bounded by these values. The line equations together with this added restriction give a parametrization for the segment.

Example 3 Parametrize the line segment joining the points $P(-3, 2, -3)$ and $Q(1, -1, 4)$ (Fig. 10.43).

Solution We begin with equations for the line through P and Q, taking them, in this case, from Example 2:

$$x = -3 + 4t, \qquad y = 2 - 3t, \qquad z = -3 + 7t. \tag{5}$$

We observe that the point

$$(x, y, z) = (-3 + 4t, 2 - 3t, -3 + 7t)$$

passes through $P(-3, 2, -3)$ at $t = 0$ and $Q(1, -1, 4)$ at $t = 1$. We add the restriction $0 \le t \le 1$ to Eqs. (5) to parametrize the segment:

$$x = -3 + 4t, \qquad y = 2 - 3t, \qquad z = -3 + 7t, \qquad 0 \le t \le 1.$$

The Distance from a Point to a Line in Space

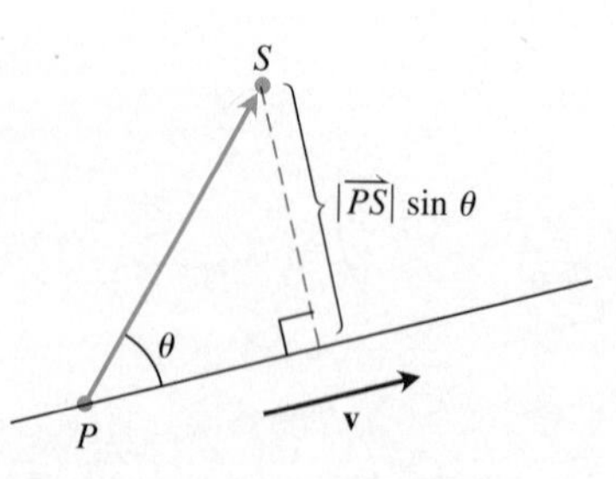

10.44 The distance from S to the line through P parallel to $\mathbf{v}$ is $|\overrightarrow{PS}| \sin\theta$, where θ is the angle between $\overrightarrow{PS}$ and $\mathbf{v}$.

To find the distance from a point S to a line L that passes through a point P parallel to a vector $\mathbf{v}$, we find the length of the component of $\overrightarrow{PS}$ normal to the line (Fig. 10.44). In the notation of Fig. 10.44, the length is

$$\begin{aligned} |\overrightarrow{PS}| \sin\theta &= |\overrightarrow{PS}| \frac{|\overrightarrow{PS} \times \mathbf{v}|}{|\overrightarrow{PS}|\,|\mathbf{v}|} \qquad \left(|\overrightarrow{PS} \times \mathbf{v}| = |\overrightarrow{PS}|\,|\mathbf{v}| \sin\theta\right) \\ &= \frac{|\overrightarrow{PS} \times \mathbf{v}|}{|\mathbf{v}|}. \end{aligned} \tag{6}$$

Distance from a Point S to a Line Through P Parallel to v

$$\text{Distance} = \frac{|\overrightarrow{PS} \times \mathbf{v}|}{|\mathbf{v}|} \tag{7}$$

Example 4 Find the distance from the point $S(1, 1, 5)$ to the line

$$L: \quad x = 1 + t, \quad y = 3 - t, \quad z = 2t.$$

Solution We see from the equations for L that L passes through $P(1, 3, 0)$ parallel to $\mathbf{v} = \mathbf{i} - \mathbf{j} + 2\,\mathbf{k}$. With

$$\overrightarrow{PS} = (1 - 1)\mathbf{i} + (1 - 3)\mathbf{j} + (5 - 0)\mathbf{k} = -2\mathbf{j} + 5\mathbf{k}$$

and

$$\overrightarrow{PS} \times \mathbf{v} = \begin{vmatrix} \mathbf{i} & \mathbf{j} & \mathbf{k} \\ 0 & -2 & 5 \\ 1 & -1 & 2 \end{vmatrix} = \mathbf{i} + 5\,\mathbf{j} + 2\,\mathbf{k},$$

Eq. (7) gives

$$\text{Distance from } S \text{ to } L = \frac{|\overrightarrow{PS} \times \mathbf{v}|}{|\mathbf{v}|} = \frac{\sqrt{1 + 25 + 4}}{\sqrt{1 + 1 + 4}} = \frac{\sqrt{30}}{\sqrt{6}} = \sqrt{5}.$$

Equations for Planes

Suppose M is a plane in space that passes through a point $P_0(x_0, y_0, z_0)$ and is normal (perpendicular) to the nonzero vector $\mathbf{N} = A\mathbf{i} + B\mathbf{j} + C\mathbf{k}$. Then M consists of all points $P(x, y, z)$ for which the vector P_0P is orthogonal to $\mathbf{N}$ (Fig. 10.45). That is, P lies on M if and only if

$$\mathbf{N} \cdot \overrightarrow{P_0P} = 0 \tag{8}$$

or

$$A(x - x_0) + B(y - y_0) + C(z - z_0) = 0.$$

When rearranged, this becomes

$$Ax + By + Cz = Ax_0 + By_0 + Cz_0.$$

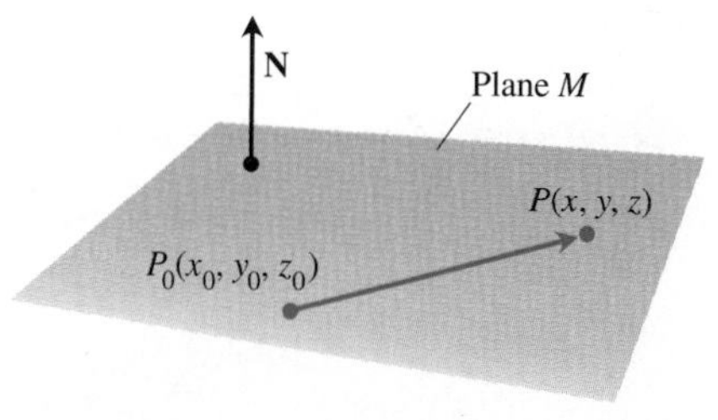

10.45 The standard equation for a plane in space is defined in terms of a vector normal to the plane: A point P lies in the plane through P_0 normal to $\mathbf{N}$ if and only if $\mathbf{N} \cdot \overrightarrow{P_0P} = 0$.

Standard Equation for the Plane Through $P_0(x_0, y_0, z_0)$ Normal to $\mathbf{N} = A\mathbf{i} + B\mathbf{j} + C\mathbf{k}$.

$$Ax + By + Cz = D, \qquad D = Ax_0 + By_0 + Cz_0 \tag{9}$$

Notice how the components of $\mathbf{N}$ become coefficients in the equation

$$Ax + By + Cz = D,$$

just the way they did for lines in two dimensions.

Example 5 Find an equation for the plane through $P_0(-3, 0, 7)$ perpendicular to $\mathbf{N} = 5\,\mathbf{i} + 2\,\mathbf{j} - \mathbf{k}$.

Solution We use Eqs. (9) to get

$$D = 5(-3) + 2(0) - 1(7) = -15 - 7 = -22$$

and

$$5x + 2y - z = -22.$$

Example 6 Find an equation for the plane through the points $A(0, 0, 1)$, $B(2, 0, 0)$, and $C(0, 3, 0)$.

Solution We find a vector normal to the plane and use it with one of the points (it does not matter which one) to write an equation for the plane.

The cross product

$$\overrightarrow{AB} \times \overrightarrow{AC} = \begin{vmatrix} \mathbf{i} & \mathbf{j} & \mathbf{k} \\ 2 & 0 & -1 \\ 0 & 3 & -1 \end{vmatrix} = 3\,\mathbf{i} + 2\,\mathbf{j} + 6\,\mathbf{k}$$

is normal to the plane. We substitute the components of this vector and the coordinates of the point $(0, 0, 1)$ into Eqs. (9) to get $D = 3(0) + 2(0) + 6(1) = 6$ and

$$3x + 2y + 6z = 6$$

as an equation for the plane.

Example 7 Find the point in which the line

$$x = \frac{8}{3} + 2t, \qquad y = -2t, \qquad z = 1 + t$$

meets the plane $3x + 2y + 6z = 6$.

Solution The point

$$\left(\frac{8}{3} + 2t,\ -2t,\ 1 + t\right)$$

lies in the plane if its coordinates satisfy the equation of the plane; that is, if

$$3\left(\frac{8}{3} + 2t\right) + 2(-2t) + 6(1 + t) = 6$$

$$8 + 6t - 4t + 6 + 6t = 6$$

$$8t = -8$$

$$t = -1.$$

The point of intersection is

$$(x, y, z)\Big|_{t=-1} = \left(\frac{8}{3} - 2,\ 2,\ 1 - 1\right) = \left(\frac{2}{3},\ 2,\ 0\right).$$

To find the distance from a point S to a plane $Ax + By + Cz = D$, find

1. a point P on the plane,
2. $\overrightarrow{PS}$,
3. the direction of $\mathbf{N} = A\mathbf{i} + B\mathbf{j} + C\mathbf{k}$.

Then calculate the distance as

$$\left| \overrightarrow{PS} \cdot \frac{\mathbf{N}}{|\mathbf{N}|} \right|.$$

Example 8 Find the distance from the point $S(1, 1, 3)$ to the plane $3x + 2y + 6z = 6$.

Solution We use the same approach we used in Section 10.3 to find the distance from a point to a line: We find a point P in the plane and calculate the length of the vector projection of $\overrightarrow{PS}$ onto a vector $\mathbf{N}$ normal to the plane (Fig. 10.46).

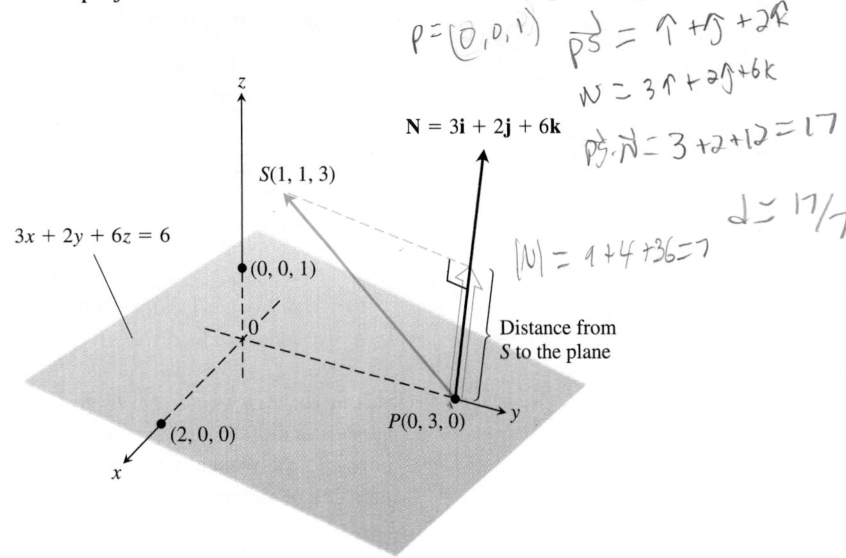

10.46 The distance from S to the plane is the length of the vector projection of $\overrightarrow{PS}$ onto $\mathbf{N}$ (Example 8).

The coefficients in the equation $3x + 2y + 6z = 6$ give

$$\mathbf{N} = 3\,\mathbf{i} + 2\,\mathbf{j} + 6\,\mathbf{k}.$$

The points on the plane easiest to find from the plane's equation are the intercepts. If we take P to be the y-intercept $(0, 3, 0)$, then

$$\begin{aligned}
\overrightarrow{PS} &= (1 - 0)\,\mathbf{i} + (1 - 3)\,\mathbf{j} + (3 - 0)\,\mathbf{k} \\
&= \mathbf{i} - 2\,\mathbf{j} + 3\,\mathbf{k}, \\
|\mathbf{N}| &= \sqrt{(3)^2 + (2)^2 + (6)^2} \\
&= \sqrt{49} = 7,
\end{aligned}$$

$$\begin{aligned}
\text{Distance from } S \text{ to the plane} &= \left| \overrightarrow{PS} \cdot \frac{\mathbf{N}}{|\mathbf{N}|} \right| \\
&= \left| (\mathbf{i} - 2\,\mathbf{j} + 3\,\mathbf{k}) \cdot \left(\frac{3}{7}\,\mathbf{i} + \frac{2}{7}\,\mathbf{j} + \frac{6}{7}\,\mathbf{k} \right) \right| \\
&= \left| \frac{3}{7} - \frac{4}{7} + \frac{18}{7} \right| = \frac{17}{7}.
\end{aligned}$$

Angles Between Planes; Lines of Intersection

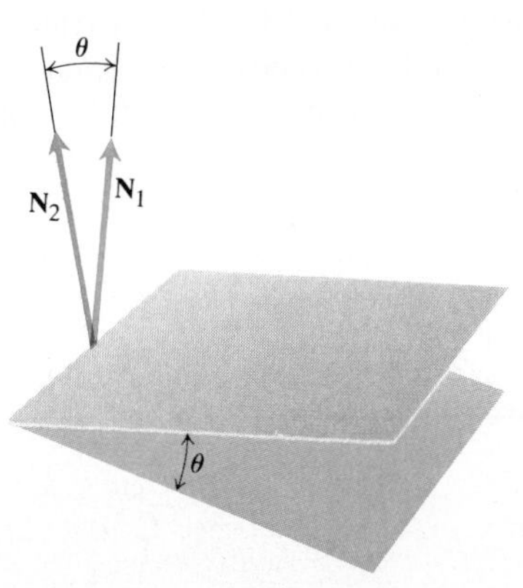

10.47 The angle between two planes is obtained from the angle between their normals.

The angle between two intersecting planes is defined to be the (acute) angle made by their normal vectors (Fig. 10.47).

Example 9 Find the angle between the planes $3x - 6y - 2z = 15$ and $2x + y - 2z = 5$.

Solution The vectors

$$\mathbf{N}_1 = 3\,\mathbf{i} - 6\,\mathbf{j} - 2\,\mathbf{k}, \qquad \mathbf{N}_2 = 2\,\mathbf{i} + \mathbf{j} - 2\,\mathbf{k}$$

are normals to the planes. The angle between them is

$$\begin{aligned} \theta &= \cos^{-1}\left(\frac{\mathbf{N}_1 \cdot \mathbf{N}_2}{|\mathbf{N}_1||\mathbf{N}_2|}\right) && \text{(Eq. (5), Section 10.3)} \\ &= \cos^{-1}\frac{4}{21} \\ &= 79°. && \text{(Calculator, rounded)} \end{aligned}$$

Example 10 Find a vector parallel to the line of intersection of the planes $3x - 6y - 2z = 15$ and $2x + y - 2z = 5$.

Solution Any vector parallel to the line of intersection will be parallel to both planes and therefore perpendicular to their normals. Conversely, any vector perpendicular to the planes' normals will be parallel to both planes and hence parallel to their line of intersection. These requirements are met by the vector

$$\mathbf{v} = \mathbf{N}_1 \times \mathbf{N}_2 = \begin{vmatrix} \mathbf{i} & \mathbf{j} & \mathbf{k} \\ 3 & -6 & -2 \\ 2 & 1 & -2 \end{vmatrix} = 14\,\mathbf{i} + 2\,\mathbf{j} + 15\,\mathbf{k}.$$

Any nonzero scalar multiple of $\mathbf{v}$ will do as well.

Example 11 Find parametric equations for the line in which the planes $3x - 6y - 2z = 15$ and $2x + y - 2z = 5$ intersect.

Solution We find a vector parallel to the line and a point on the line and use Eqs. (2).

Example 10 gives a vector parallel to the line, $\mathbf{v} = 14\,\mathbf{i} + 2\,\mathbf{j} + 15\,\mathbf{k}$. To find a point on the line, we can take any point common to the two planes. Substituting $z = 0$ in the plane equations and solving for x and y simultaneously gives the point $(3, -1, 0)$. The line is

$$x = 3 + 14t, \qquad y = -1 + 2t, \qquad z = 15t.$$

EXERCISES 10.5

Find parametric equations for the lines in Exercises 1–12.

1. The line through $P(3, -4, -1)$ parallel to the vector $\mathbf{i} + \mathbf{j} + \mathbf{k}$
2. The line through $P(1, 2, -1)$ and $Q(-1, 0, 1)$
3. The line through $P(-2, 0, 3)$ and $Q(3, 5, -2)$
4. The line through $P(1, 2, 0)$ and $Q(1, 1, -1)$
5. The line through the origin parallel to the vector $2\,\mathbf{j} + \mathbf{k}$

6. The line through the point $(3, -2, 1)$ parallel to the line $x = 1 + 2t$, $y = 2 - t$, $z = 3t$
7. The line through $(1, 1, 1)$ parallel to the z-axis
8. The line through $(2, 4, 5)$ perpendicular to the plane $3x + 7y - 5z = 21$
9. The line through $(0, -7, 0)$ perpendicular to the plane $x + 2y + 2z = 13$
10. The line through $(2, 3, 0)$ perpendicular to the vectors $\mathbf{A} = \mathbf{i} + 2\mathbf{j} + 3\mathbf{k}$ and $\mathbf{B} = 3\mathbf{i} + 4\mathbf{j} + 5\mathbf{k}$
11. The x-axis
12. The z-axis

Find parametrizations for the line segments joining the points in Exercises 13–20. Draw coordinate axes and sketch each segment, indicating the direction of increasing t for your parametrization.

13. $(0, 0, 0)$, $(1, 1, 1)$
14. $(0, 0, 0)$, $(1, 0, 0)$
15. $(1, 0, 0)$, $(1, 1, 0)$
16. $(1, 1, 0)$, $(1, 1, 1)$
17. $(0, -1, 1)$, $(0, 1, 1)$
18. $(3, 0, 0)$, $(0, 2, 0)$
19. $(2, 2, 0)$, $(1, 2, -2)$
20. $(1, -1, -2)$, $(0, 2, 1)$

Find equations for the planes in Exercises 21–26.

21. The plane through $P_0(0, 2, -1)$ normal to $\mathbf{N} = 3\mathbf{i} - 2\mathbf{j} - \mathbf{k}$
22. The plane through $(1, -1, 3)$ parallel to the plane $3x + y + z = 7$.
23. The plane through $(1, 1, -1)$, $(2, 0, 2)$, and $(0, -2, 1)$
24. The plane through $(2, 4, 5)$, $(1, 5, 7)$, and $(-1, 6, 8)$
25. The plane through $P_0(2, 4, 5)$ perpendicular to the line
$$x = 5 + t, \quad y = 1 + 3t, \quad z = 4t$$
26. The plane through $A(1, -2, 1)$ perpendicular to the vector from the origin to A
27. Use Eqs. (2) to generate a parametrization of the line through $P(2, -4, 7)$ parallel to $\mathbf{v}_1 = 2\mathbf{i} - \mathbf{j} + 3\mathbf{k}$. Then generate another parametrization of the line using the point $P_2(-2, -2, 1)$ and the vector $\mathbf{v}_2 = -\mathbf{i} + (1/2)\mathbf{j} - (3/2)\mathbf{k}$.
28. Use Eqs. (9) to generate an equation for the plane through $P_1(4, 1, 5)$ normal to $\mathbf{N}_1 = \mathbf{i} - 2\mathbf{j} + \mathbf{k}$. Then generate another equation for the plane using the point $P_2(3, -2, 0)$ and the normal vector $\mathbf{N} = -\sqrt{2}\,\mathbf{i} + 2\sqrt{2}\,\mathbf{j} - \sqrt{2}\,\mathbf{k}$.

In Exercises 29–34, find the distance from the point to the line.

29. $x = 4t$, $y = -2t$, $z = 2t$; $S(0, 0, 12)$
30. $x = 5 + 3t$, $y = 5 + 4t$, $z = -3 - 5t$; $S(0, 0, 0)$
31. $x = 2 + 2t$, $y = 1 + 6t$, $z = 3$; $S(2, 1, 3)$
32. $x = 2t$, $y = 1 + 2t$, $z = 2t$; $S(2, 1, -1)$
33. $x = 4 - t$, $y = 3 + 2t$, $z = -5 + 3t$; $S(3, -1, 4)$
34. $x = 10 + 4t$, $y = -3$, $z = 4t$; $S(-1, 4, 3)$

In Exercises 35–40, find the distance from the point to the plane.

35. $(2, -3, 4)$, $x + 2y + 2z = 13$
36. $(0, 0, 0)$, $3x + 2y + 6z = 6$
37. $(0, 1, 1)$, $4y + 3z = -12$
38. $(2, 2, 3)$, $2x + y + 2z = 4$
39. $(0, -1, 0)$, $2x + y + 2z = 4$
40. $(1, 0, -1)$, $-4x + y + z = 4$

In Exercises 41–44, find the point in which the line meets the plane.

41. $x = 1 - t$, $y = 3t$, $z = 1 + t$; $2x - y + 3z = 6$
42. $x = 2$, $y = 3 + 2t$, $z = -2 - 2t$; $6x + 3y - 4z = -12$
43. $x = 1 + 2t$, $y = 1 + 5t$, $z = 3t$; $x + y + z = 2$
44. $x = -1 + 3t$, $y = -2$, $z = 5t$; $2x - 3z = 7$

Find the angles between the planes in Exercises 45 and 46 (calculator not needed).

45. $x + y = 1$, $2x + y - 2z = 2$
46. $5x + y - z = 10$, $x - 2y + 3z = -1$

CALCULATOR Use a calculator to find the angles between the planes in Exercises 47–50 to the nearest tenth of a degree.

47. $2x + 2y + 2z = 3$, $2x - 2y - z = 5$
48. $x + y + z = 1$, $z = 0$ (the xy-plane)
49. $2x + 2y - z = 3$, $x + 2y + z = 2$
50. $4y + 3z = -12$, $3x + 2y + 6z = 6$

Find parametrizations for the lines in which the planes in Exercises 51–54 intersect.

51. $x + y + z = 1$, $x + y = 2$
52. $3x - 6y - 2z = 3$, $2x + y - 2z = 2$
53. $x - 2y + 4z = 2$, $x + y - 2z = 5$
54. $5x - 2y = 11$, $4y - 5z = -17$
55. Find the points in which the line $x = 1 + 2t$, $y = -1 - t$, $z = 3t$ meets the coordinate planes.
56. Find equations for the line in the plane $z = 3$ that makes a 30°-angle with $\mathbf{i}$ and a 60°-angle with $\mathbf{j}$.
57. *Perspective in computer graphics.* In computer graphics and perspective drawing we need to represent objects seen by the eye in space as images on a two-dimensional plane. Suppose the eye is at $E(x_0, 0, 0)$ (Fig. 10.48) and that we want to represent a point $P_1(x_1, y_1, z_1)$ as a point on the yz-plane. We do this by projecting P_1 onto the plane with a ray from E. The point P_1 will be portrayed as the point $P(0, y, z)$. The problem for us as graphics designers is to find y and z given E and P_1.

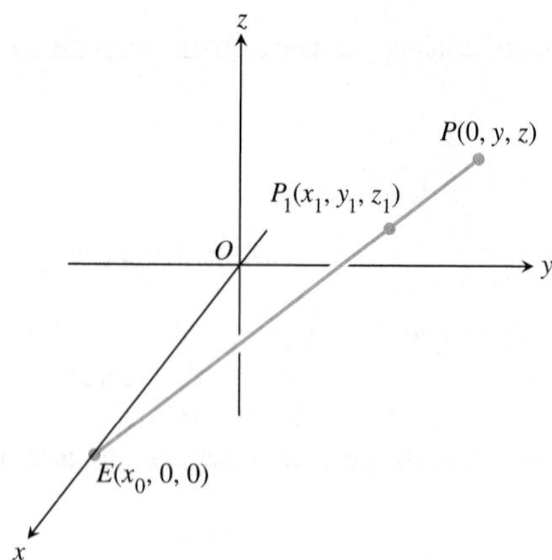

10.48 The figure for Exercise 57.

a) Write a vector equation that holds between $\overrightarrow{EP}$ and $\overrightarrow{EP_1}$. Use the equation to express y and z in terms of x_0, x_1, y_1, and z_1.

b) Test the formulas obtained for y and z in part (a) by investigating their behavior at $x_1 = 0$ and $x_1 = x_0$ and by seeing what happens as $x_0 \to \infty$.

58. *Hidden lines.* Here is another typical problem in computer graphics. Your eye is at (4, 0, 0). You are looking at a triangular plate whose vertices are at (1, 0, 1), (1, 1, 0), and (−2, 2, 2). The line segment from (1, 0, 0) to (0, 2, 2) passes through the plate. What portion of the line segment is hidden from your view by the plate? (This is an exercise in finding intersections of lines and planes.)

COMPUTER Exercises 59–62 give the coordinates of points A, B, C, and S. What outputs do you get when you use them as inputs for the vector utility program listed in this section?

	A	B	C	S
59.	(1, 0, −1)	(0, 2, 0)	(−1, 0, 1)	(0, 0, 3)
60.	(1, 0, 1)	(0, 1, 1)	(1, 1, 0)	(2, 2, 2)
61.	(−2, 0, 4)	(2, 8, 0)	(2, 4, −2)	(0, 4, 2)
62.	(1, 3, 2)	(2, −1, 4)	(3, 0, −1)	(0, 0, 0)

COMPUTER PROGRAM: VECTOR OPERATIONS

The following BASIC program enables you to calculate the results of most of the vector operations you have seen so far. If you enter the coordinates of points A, B, and C, and later, the coordinates of a point S, the program will display the following information on the screen.

- The lengths of $\overrightarrow{OA}$, $\overrightarrow{OB}$, and $\overrightarrow{AB}$
- The coordinates of the midpoint of segment AB
- $\overrightarrow{OA} \cdot \overrightarrow{OB}$
- The angle between $\overrightarrow{OA}$ and $\overrightarrow{OB}$ in both radians and degrees (when the vectors are not parallel)
- The components of $\overrightarrow{OA} \times \overrightarrow{OB}$
- Parametric equations for line AB
- Parametric equations for the line through C parallel to $\overrightarrow{AB}$
- The distance from C to line AB
- The components of a vector **N** normal to plane ABC
- An equation for plane ABC
- The distance from S to plane ABC

Most of the computations follow the formulas in the text. The program uses subscripted variables called "arrays" to hold the components of vectors. In this way, the variable $A(I)$ can refer to any of the components of vector $\overrightarrow{OA}$ by setting $I = 1$, 2, or 3. This allows FOR loops to compute component values without using several statements.

Another use of FOR loops with subscripted variables is to get a sum, as in getting the dot product. The SUM must be set to zero before the loop begins; the successive values are added into SUM within the loop.

BASIC does not have an arc cosine function. Its only inverse trigonometric function is the arc tangent, ATN(X). To get the arc cosine, we used the identity

$$\cos^{-1}(x) = \frac{\pi}{2} - \tan^{-1}\left(\frac{x}{\sqrt{1-x^2}}\right).$$

This relation holds as long as x is different from 1, which it will be if vectors $\overrightarrow{OA}$ and $\overrightarrow{OB}$ are not parallel.

To find the distance from a point S to a plane, the program finds a line through S normal to the plane, calculates the parameter value of the point of intersection, finds the coordinates of the point, and calculates its distance from S. If $N_1 x + N_2 y + N_3 z = D$ is the plane and $\mathbf{N}$ is the normal vector, the normal line through (S_1, S_2, S_3) is

$$x = S_1 + N_1 t, \qquad y = S_2 + N_2 t, \qquad z = S_3 + N_3 t.$$

At the intersection of this line with the plane, the values of x, y, and z satisfy the plane's equation, so

$$N_1(S_1 + N_1 t) + N_2(S_2 + N_2 t) + N_3(S_3 + N_3 t) = D.$$

Solving for t (the program calls the solution T) gives

$$T = \frac{D - (N_1S_1 + N_2S_2 + N_3S_3)}{N_1^2 + N_2^2 + N_3^2} = \frac{D - \mathbf{N}\cdot\overrightarrow{OS}}{|\mathbf{N}|}.$$

The distance from the intersection to S is

$$\begin{aligned}\text{Distance} &= \sqrt{(S_1 - (S_1 + N_1T))^2 + (S_2 - (S_2 + N_2T))^2 + (S_3 - (S_3 + N_3T))^2} \\ &= \sqrt{N_1^2T^2 + N_2^2T^2 + N_3^2T^2} = |T||\mathbf{N}|.\end{aligned}$$

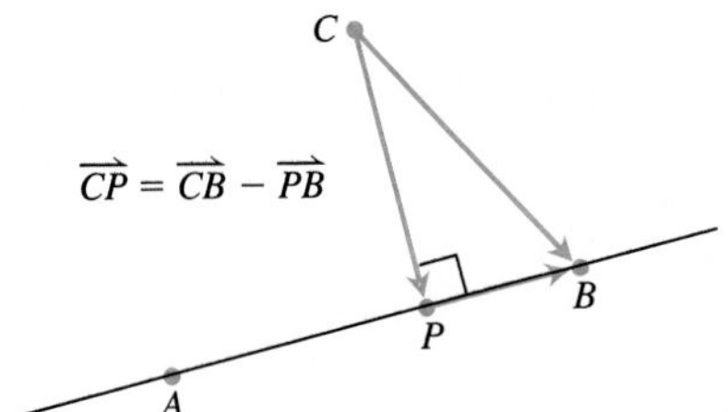

10.49 Another way to find the distance from a point C to a line AB is to calculate it as $|\overrightarrow{CB} - \overrightarrow{PB}|$.

To find the distance from C to line AB, the program finds the vector projection $\overrightarrow{PB}$ of $\overrightarrow{CB}$ onto $\overrightarrow{AB}$ and calculates $|\overrightarrow{CB} - \overrightarrow{PB}|$ (Fig. 10.49).

PROGRAM	COMMENTS
10 DIM A(3), B(3), C(3), AB(3), AC(3), XPAB(3), Q(3), N(3)	Define arrays (subscripted variables) for vectors.
20 PI = 3.141593	Define π.
30 PRINT "ENTER THE COORDINATES OF THREE POINTS"	Give instructions to the user.
40 PRINT "RESPOND TO EACH PROMPT WITH 3 NUMBERS,"	
50 PRINT "USE A COMMA BETWEEN THE VALUES"	
60 PRINT	
70 INPUT "ENTER COMPONENTS OF POINT A "; A(1), A(2), A(3)	Get components of $\overrightarrow{OA}$,
80 INPUT "ENTER COMPONENTS OF POINT B "; B(1), B(2), B(3)	and of $\overrightarrow{OB}$.
90 INPUT "ENTER COORDINATES OF POINT C "; C(1), C(2), C(3)	Get coordinates of point C.
100 PRINT	
110 SUM = 0: FOR I = 1 TO 3: SUM = SUM + A(I) ^ 2: NEXT I	Get sum for length of $\overrightarrow{OA}$.
120 LA = SQR(SUM)	Get square root.
130 PRINT "LENGTH OF VECTOR OA IS "; LA	Print length of $\overrightarrow{OA}$.
140 SUM = 0: FOR I = 1 TO 3: SUM = SUM + B(I) ^ 2: NEXT I	Get sum for length of $\overrightarrow{OB}$.
150 LB = SQR(SUM)	Get square root.
160 PRINT "LENGTH OF VECTOR OB IS "; LB	Print length of $\overrightarrow{OB}$.
170 FOR I = 1 TO 3: AB(I) = B(I) − A(I): NEXT I	Get components of $\overrightarrow{AB}$.
180 SUM = 0: FOR I = 1 TO 3: SUM = SUM + AB(I) ^ 2: NEXT I	Get sum for length of $\overrightarrow{AB}$.

	PROGRAM	COMMENTS
190	LAB = SQR(SUM)	Get square root.
200	PRINT "LENGTH OF VECTOR AB IS "; LAB	Print length of $\overrightarrow{AB}$.
210	PRINT "MIDPOINT OF AB IS "; (A(1) + B(1))/2, (A(2) + B(2))/2, (A(3) + B(3))/2	Compute and print coordinates of midpoint of segment AB.
220	DPAB = 0: FOR I = 1 TO 3: DPAB = DPAB + A(I) * B(I): NEXT I	Get sum for dot product of $\overrightarrow{OA}$ and $\overrightarrow{OB}$.
230	PRINT "THE DOT PRODUCT OF OA AND OB IS "; DPAB	Print dot product.
240	X = DPAB/ABS(LA * LB)	Put a value into temporary variable x.
250	ANGAB = 0: IF ABS(X) < 1 THEN ANGAB = PI/2 − ATN(X/SQR(1 − X ^ 2))	Get angle between $\overrightarrow{OA}$ and $\overrightarrow{OB}$—use trigonometric identity.
260	PRINT "ANGLE BETWEEN OA AND OB IS "; ANGAB; " RADIANS"	Print angle in radian measure.
270	PRINT " THIS IS EQUAL TO "; ANGAB/PI * 180; " DEGREES"	Print angle in degrees.
280	XPAB(1) = A(2) * B(3) − A(3) * B(2)	Get components of $\overrightarrow{OA} \times \overrightarrow{OB}$ using the determinant rule.
290	XPAB(2) = A(3) * B(1) − A(1) * B(3)	
300	XPAB(3) = A(1) * B(2) − A(2) * B(1)	
310	PRINT "COMPONENTS OF CROSS PRODUCT, OA X OB: "; XPAB(1); XPAB(2); XPAB(3)	Print components of $\overrightarrow{OA} \times \overrightarrow{OB}$.
320	PRINT "LINE THROUGH A AND B (PARAMETRIC FORM):"	Print first part of output for line through A and B,
330	PRINT " X = "; A(1);" + ";AB(1);"(t)"	then the equation for x,
340	PRINT " Y = "; A(2); " + ";AB(2);"(t)"	next the equation for y,
350	PRINT " Z = "; A(3); " + ";AB(3);"(t)"	and lastly the equation for z.
360	PRINT "LINE THROUGH C PARALLEL TO AB:"	Print the first part of output for parallel line through C,
370	PRINT " X = "; C(1); " + ";AB(1);"(t)"	then the equation for x,
380	PRINT " Y = "; C(2); " + ";AB(2);"(t)"	next the equation for y,
390	PRINT " Z = "; C(3); " + ";AB(3);"(t)"	and lastly the equation for z.
400	FOR I = 1 TO 3: CB(I) = B(I) − C(I): NEXT I	Compute components of $\overrightarrow{CB}$
410	DPABCB = 0: FOR I = 1 TO 3: DPABCB = DPABCB + AB(I) * CB(I): NEXT I	$\overrightarrow{AB} \cdot \overrightarrow{CB}$.
420	FOR I = 1 TO 3: P(I) = DPABCB/LAB ^ 2* AB(I): NEXT I	Components of $\overrightarrow{PB} = \text{proj}_{\overrightarrow{AB}} \overrightarrow{CB}$
430	SUM = 0: FOR I = 1 TO 3: SUM = SUM + (CB(I) − P(I)) ^ 2: NEXT I	$\overrightarrow{CP} \cdot \overrightarrow{CP}$
440	PRINT "DISTANCE FROM C TO LINE AB IS ": SQR(SUM)	The distance from C to line AB is $\sqrt{\overrightarrow{CP} \cdot \overrightarrow{CP}}$.
450	FOR I = 1 TO 3: AC(I) = A(I) − C(I): NEXT I	The components of $\overrightarrow{AC}$
460	N(1) = AB(2) * AC(3) − AB(3) * AC(2)	Get components of a vector **N** normal to plane through A, B, and C using the determinant rule.
470	N(2) = AB(3) * AC(1) − AB(1) * AC(3)	
480	N(3) = AB(1) * AC(2) − AB(2) * AC(1)	
490	PRINT "COMPONENTS OF A NORMAL TO PLANE ABC ARE";	Print first part of output for normal vector,
500	PRINT " "; N(1); N(2); N(3)	and then the components.
510	SUM = 0: FOR I = 1 TO 3: SUM = SUM + N(I) * C(I): NEXT I	Get product of components of **N** and coordinates of C.
520	PRINT "EQUATION OF PLANE ABC IS "	Print first part of output for equation of plane,
530	PRINT " "; N(1);"X + ";N(2);"Y + ";N(3);"Z = ";SUM	and then the equation.
540	PRINT "TO FIND THE DISTANCE FROM POINT S TO THE PLANE"	Prompt user and get coordinates of
550	INPUT " ENTER COORDINATES OF S: ", S(1),S(2),S(3)	point S.
560	S2 = 0: FOR I = 1 TO 3: S2 = S2 + (S(I) − C(I)) * N(I): NEXT I	Find $\mathbf{N} \cdot \overrightarrow{OS}$.
570	S3 = 0: FOR I = 1 TO 3: S3 = S3 + N(I) ^ 2: NEXT I	Add squares of components of **N**.
580	DIST = S2/SQR(S3)	Compute distance to the plane.
590	PRINT "DISTANCE IS ";ABS(DIST)	Print the distance.
600	END	

Here is the output for

$$A = (2, 0, 0), \quad B = (0, 3, 0), \quad C = (0, 0, 1), \quad S = (1, 1, 3).$$

```
LENGTH OF VECTOR OA IS  2
LENGTH OF VECTOR OB IS  3
LENGTH OF VECTOR AB IS  3.605551
MIDPOINT OF AB IS  1    1.5    0
THE DOT PRODUCT OF OA AND OB IS 0
ANGLE BETWEEN OA AND OB IS 1.570797 RADIANS
  THIS IS EQUAL TO 90 DEGREES
COMPONENTS OF CROSS PRODUCT, OA X OB: 0 0 6
LINE THROUGH A AND B (PARAMETRIC FORM):
  X = 2 - 2(t)
  Y = 0 + 3(t)
  Z = 0 + 0(t)
LINE THROUGH C PARALLEL TO AB:
  X = 0 +  - 2(t)
  Y = 0 + 3(t)
  Z = 1 + 0(t)
DISTANCE FROM C TO LINE AB IS 1.941451
COMPONENTS OF A NORMAL TO PLANE ABC ARE  - 3  - 2  - 6
EQUATION OF PLANE ABC IS  - 3X +  - 2Y +  - 6Z = -6
TO FIND THE DISTANCE FROM POINT S TO THE PLANE
  ENTER COORDINATES OF S:  1, 1, 3
DISTANCE IS  2.428572
```

10.6 Products of Three Vectors or More

The triple products $(\mathbf{A} \times \mathbf{B}) \cdot \mathbf{C}$ and $(\mathbf{A} \times \mathbf{B}) \times \mathbf{C}$ appear frequently in engineering and physics in the calculations associated with electricity, magnetism, and fluid flow. In this section, we study the three-dimensional geometry of the triple scalar product $(\mathbf{A} \times \mathbf{B}) \cdot \mathbf{C}$ (we shall use it to derive the surface area formulas in Section 14.4) and derive the identities that are used to evaluate triple products from the components of the vectors being multiplied.

The Triple Scalar or Box Product

The product $(\mathbf{A} \times \mathbf{B}) \cdot \mathbf{C}$ is called the **triple scalar product** of **A**, **B**, and **C** (in that order). As you can see from the formula

$$|(\mathbf{A} \times \mathbf{B}) \cdot \mathbf{C}| = |\mathbf{A} \times \mathbf{B}||\mathbf{C}||\cos\theta|, \tag{1}$$

the absolute value of the product is the volume of the parallelepiped (parallelogram-sided box) determined by **A**, **B**, and **C** (Fig. 10.50). The number $|\mathbf{A} \times \mathbf{B}|$ is the area of the base parallelogram. The number $|\mathbf{C}||\cos\theta|$ is the parallelepiped's height. Because of the geometry here, $(\mathbf{A} \times \mathbf{B}) \cdot \mathbf{C}$ is often called the **box product** of **A**, **B**, and **C**.

By treating the planes of **B** and **C** and of **C** and **A** as the base planes of the parallelepiped determined by **A**, **B**, and **C**, we see that

$$(\mathbf{A} \times \mathbf{B}) \cdot \mathbf{C} = (\mathbf{B} \times \mathbf{C}) \cdot \mathbf{A} = (\mathbf{C} \times \mathbf{A}) \cdot \mathbf{B}. \tag{2}$$

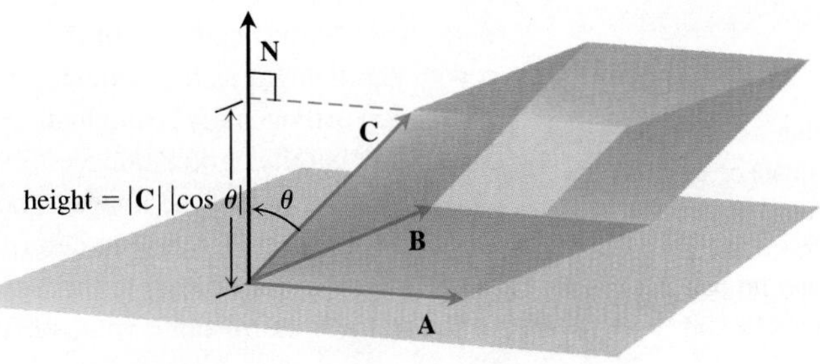

10.50 The volume of the parallelepiped shown here is $|(\mathbf{A} \times \mathbf{B}) \cdot \mathbf{C}|$. The absolute value bars correct the sign when θ is greater than 90°.

Since the dot product is commutative, Eq. (2) also gives

$$(\mathbf{A} \times \mathbf{B}) \cdot \mathbf{C} = \mathbf{A} \cdot (\mathbf{B} \times \mathbf{C}). \tag{3}$$

The dot and cross may be interchanged in a triple scalar product without altering its value.

The triple scalar product can be evaluated as a determinant in the following way:

$$\begin{aligned}\mathbf{A} \cdot (\mathbf{B} \times \mathbf{C}) &= \mathbf{A} \cdot \left[\begin{vmatrix} b_2 & b_3 \\ c_2 & c_3 \end{vmatrix} \mathbf{i} - \begin{vmatrix} b_1 & b_3 \\ c_1 & c_3 \end{vmatrix} \mathbf{j} + \begin{vmatrix} b_1 & b_2 \\ c_1 & c_2 \end{vmatrix} \mathbf{k} \right] \\ &= a_1 \begin{vmatrix} b_2 & b_3 \\ c_2 & c_3 \end{vmatrix} - a_2 \begin{vmatrix} b_1 & b_3 \\ c_1 & c_3 \end{vmatrix} + a_3 \begin{vmatrix} b_1 & b_2 \\ c_1 & c_2 \end{vmatrix} \\ &= \begin{vmatrix} a_1 & a_2 & a_3 \\ b_1 & b_2 & b_3 \\ c_1 & c_2 & c_3 \end{vmatrix}.\end{aligned}$$

The number $\mathbf{A} \cdot (\mathbf{B} \times \mathbf{C})$ can be calculated with the formula

$$\mathbf{A} \cdot (\mathbf{B} \times \mathbf{C}) = (\mathbf{A} \times \mathbf{B}) \cdot \mathbf{C} = \begin{vmatrix} a_1 & a_2 & a_3 \\ b_1 & b_2 & b_3 \\ c_1 & c_2 & c_3 \end{vmatrix}. \tag{4}$$

Example 1 Find the volume of the box (parallelepiped) determined by $\mathbf{A} = \mathbf{i} + 2\mathbf{j} - \mathbf{k}$, $\mathbf{B} = -2\mathbf{i} + 3\mathbf{k}$, and $\mathbf{C} = 7\mathbf{j} - 4\mathbf{k}$.

Solution The volume is the absolute value of

$$\begin{aligned}\mathbf{A} \cdot (\mathbf{B} \times \mathbf{C}) &= \begin{vmatrix} 1 & 2 & -1 \\ -2 & 0 & 3 \\ 0 & 7 & -4 \end{vmatrix} \\ &= \begin{vmatrix} 0 & 3 \\ 7 & -4 \end{vmatrix} - 2\begin{vmatrix} -2 & 3 \\ 0 & -4 \end{vmatrix} - \begin{vmatrix} -2 & 0 \\ 0 & 7 \end{vmatrix} \\ &= -21 - 16 + 14 = -23,\end{aligned}$$

or 23.

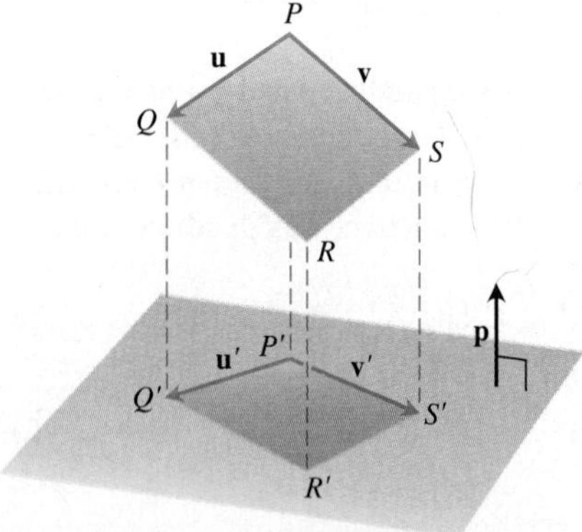

10.51 The parallelogram determined by two vectors **u** and **v** in space and the orthogonal projection of the parallelogram onto a plane. The projection lines, orthogonal to the plane, lie parallel to the unit normal vector **p**.

The Area of a Parallelogram's Projection on a Plane

Our next goal is to establish a result we shall need in Section 14.4 when we derive the standard integral formula for the area of a curved surface lying above a coordinate plane. At the beginning of the derivation, we approximate the surface by tiling it with parallelograms (in context, a natural thing to do). We then project each parallelogram orthogonally (straight down) onto the plane and work backward from the area of the parallelogram's image to find the area of the parallelogram itself. How do we find the area of the image? The answer, as we shall see in a moment, is that if the parallelogram's sides are determined by the vectors **u** and **v** (Fig. 10.51) and if **p** is a unit vector normal to the plane, then the area of the parallelogram's image in the plane is the absolute value of $(\mathbf{u} \times \mathbf{v}) \cdot \mathbf{p}$.

THEOREM 1

> The area of the orthogonal projection of the parallelogram determined by two vectors **u** and **v** in space onto a plane with unit normal vector **p** is given by the formula
>
> $$\text{Area} = |(\mathbf{u} \times \mathbf{v}) \cdot \mathbf{p}|. \tag{5}$$

Proof In the notation of Fig. 10.51, which shows a typical parallelogram determined by vectors **u** and **v** and its orthogonal projection onto a plane with unit normal vector **p**,

$$\begin{aligned} \mathbf{u} &= \overrightarrow{PP'} + \mathbf{u}' + \overrightarrow{Q'Q} \\ &= \mathbf{u}' + \overrightarrow{PP'} - \overrightarrow{QQ'} \qquad \left(\overrightarrow{Q'Q} = -\overrightarrow{QQ'}\right) \\ &= \mathbf{u}' + s\mathbf{p}. \qquad \text{(For some scalar } s \text{ because } \overrightarrow{PP'} - \overrightarrow{QQ'} \text{ is parallel to } \mathbf{p}) \end{aligned} \tag{6}$$

Similarly,

$$\mathbf{v} = \mathbf{v}' + t\mathbf{p} \tag{7}$$

for some scalar t. Hence,

$$\begin{aligned} \mathbf{u} \times \mathbf{v} &= (\mathbf{u}' + s\mathbf{p}) \times (\mathbf{v}' + t\mathbf{p}) \\ &= (\mathbf{u}' \times \mathbf{v}') + s(\mathbf{p} \times \mathbf{v}') + t(\mathbf{u}' \times \mathbf{p}) + st\underbrace{(\mathbf{p} \times \mathbf{p})}_{\mathbf{0}}. \end{aligned} \tag{8}$$

The vectors $\mathbf{p} \times \mathbf{v}'$ and $\mathbf{u}' \times \mathbf{p}$ are both orthogonal to **p**. Hence, when we dot both sides of Eq. (8) with **p**, the only surviving term on the right is $(\mathbf{u}' \times \mathbf{v}') \cdot \mathbf{p}$. We are left with

$$(\mathbf{u} \times \mathbf{v}) \cdot \mathbf{p} = (\mathbf{u}' \times \mathbf{v}') \cdot \mathbf{p}. \tag{9}$$

In particular,

$$|(\mathbf{u} \times \mathbf{v}) \cdot \mathbf{p}| = |(\mathbf{u}' \times \mathbf{v}') \cdot \mathbf{p}|. \tag{10}$$

The absolute value on the right is the volume of the box determined by $\mathbf{u}'$, $\mathbf{v}'$, and **p**. The height of this particular box is $|\mathbf{p}| = 1$, so the box's volume is the same as its base area, the area of parallelogram $P'Q'R'S'$. Combining this observation with Eq. (10) gives

$$\text{Area of } P'Q'R'S' = |(\mathbf{u}' \times \mathbf{v}') \cdot \mathbf{p}| = |(\mathbf{u} \times \mathbf{v}) \cdot \mathbf{p}|. \tag{11}$$

This says that the area of the orthogonal projection of the parallelogram determined by **u** and **v** onto a plane with unit normal vector **p** is $|(\mathbf{u} \times \mathbf{v}) \cdot \mathbf{p}|$, which is what we set out to prove. ∎

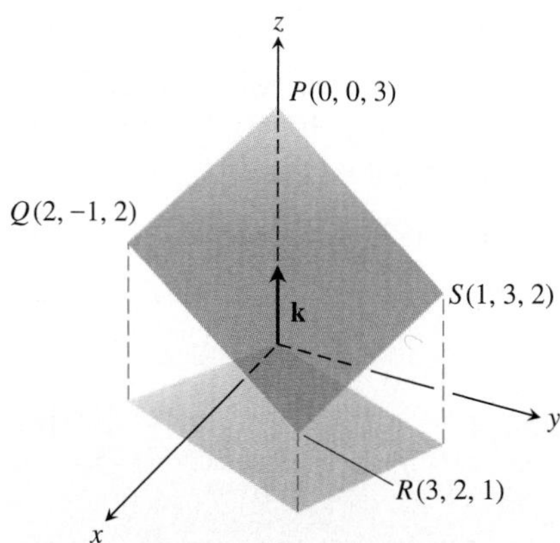

10.52 Example 2 calculates the area of the orthogonal projection of parallelogram *PQRS* on the *xy*-plane.

Example 2 Find the area of the orthogonal projection onto the xy-plane of the parallelogram determined by the points $P(0, 0, 3)$, $Q(2, -1, 2)$, $R(3, 2, 1)$, and $S(1, 3, 2)$ (Fig. 10.52).

Solution With

$$\mathbf{u} = \overrightarrow{PQ} = 2\mathbf{i} - \mathbf{j} - \mathbf{k}, \qquad \mathbf{v} = \overrightarrow{PS} = \mathbf{i} + 3\mathbf{j} - \mathbf{k}, \qquad \text{and} \qquad \mathbf{p} = \mathbf{k},$$

Eq. (5) gives

$$\text{Area} = (\mathbf{u} \times \mathbf{v}) \cdot \mathbf{p} = \begin{vmatrix} 2 & -1 & -1 \\ 1 & 3 & -1 \\ 0 & 0 & 1 \end{vmatrix} = \begin{vmatrix} 2 & -1 \\ 1 & 3 \end{vmatrix} = 7.$$

Triple Vector Products

The **triple vector products** $(\mathbf{A} \times \mathbf{B}) \times \mathbf{C}$ and $\mathbf{A} \times (\mathbf{B} \times \mathbf{C})$ are usually not equal, although the formulas for evaluating them from components are similar.

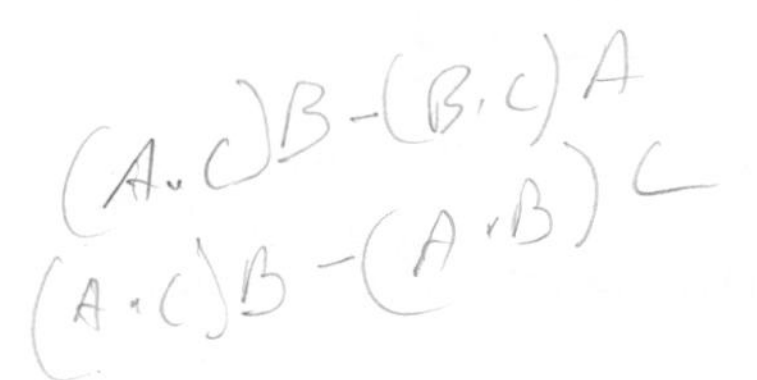

$$(\mathbf{A} \times \mathbf{B}) \times \mathbf{C} = (\mathbf{A} \cdot \mathbf{C})\mathbf{B} - (\mathbf{B} \cdot \mathbf{C})\mathbf{A} \tag{12}$$

$$\mathbf{A} \times (\mathbf{B} \times \mathbf{C}) = (\mathbf{A} \cdot \mathbf{C})\mathbf{B} - (\mathbf{A} \cdot \mathbf{B})\mathbf{C} \tag{13}$$

The first product is a multiple of **B** minus a multiple of **A**. The second is the same multiple of **B** minus a multiple of **C**. Take a moment to compare the two formulas.

Equation (13) follows from (12) by permuting the letters **A**, **B**, and **C**:

$$(\mathbf{A} \times \mathbf{B}) \times \mathbf{C} = (\mathbf{A} \cdot \mathbf{C})\mathbf{B} - (\mathbf{B} \cdot \mathbf{C})\mathbf{A} \qquad \text{(Eq. (12))}$$

$$(\mathbf{B} \times \mathbf{C}) \times \mathbf{A} = (\mathbf{B} \cdot \mathbf{A})\mathbf{C} - (\mathbf{C} \cdot \mathbf{A})\mathbf{B} \qquad (\mathbf{A} \to \mathbf{B} \to \mathbf{C} \to \mathbf{A})$$

$$= (\mathbf{A} \cdot \mathbf{B})\mathbf{C} - (\mathbf{A} \cdot \mathbf{C})\mathbf{B} \qquad \left(\text{Dot multiplication is commutative.}\right)$$

$$\mathbf{A} \times (\mathbf{B} \times \mathbf{C}) = (\mathbf{A} \cdot \mathbf{C})\mathbf{B} - (\mathbf{A} \cdot \mathbf{B})\mathbf{C} \qquad \left(\begin{array}{l}\text{Reverse the order of the} \\ \text{product on the left; change} \\ \text{the signs on the right.}\end{array}\right)$$

To derive Eq. (12), we first eliminate two degenerate cases. If any of the vectors is zero, Eq. (12) holds because both sides are zero. If **B** is a scalar multiple of **A**, both sides of Eq. (12) are zero again (see Exercise 12).

Suppose now that none of the vectors involved is zero and that **A** and **B** are not parallel. The vector $(\mathbf{A} \times \mathbf{B}) \times \mathbf{C}$ is orthogonal to $\mathbf{A} \times \mathbf{B}$ and consequently parallel to the plane of **A** and **B**. This means that

$$(\mathbf{A} \times \mathbf{B}) \times \mathbf{C} = m\mathbf{A} + n\mathbf{B} \tag{14}$$

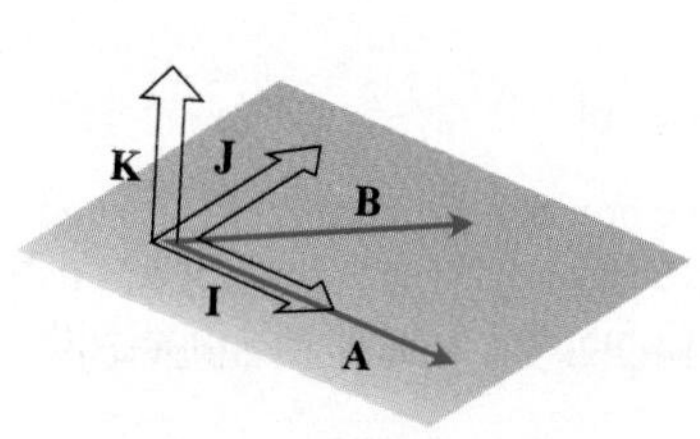

10.53 The orthogonal vectors **I** and **J** in the plane of **A** and **B**, together with their cross product **K**.

for some scalars m and n. To evaluate m and n, we let **I** equal $\mathbf{A}/|\mathbf{A}|$ and let **J** be a unit vector orthogonal to **I** in the plane of **A** and **B** (Fig. 10.53). We then let **K** denote $\mathbf{I} \times \mathbf{J}$ and write **A**, **B**, and **C** in terms of the unit vectors **I**, **J**, and **K**:

$$\mathbf{A} = a_1\,\mathbf{I}, \qquad \mathbf{B} = b_1\,\mathbf{I} + b_2\,\mathbf{J}, \qquad \mathbf{C} = c_1\,\mathbf{I} + c_2\,\mathbf{J} + c_3\,\mathbf{K}. \tag{15}$$

Then

$$\mathbf{A} \times \mathbf{B} = a_1 b_2\,\mathbf{K},$$

$$(\mathbf{A} \times \mathbf{B}) \times \mathbf{C} = a_1 b_2 c_1\,\mathbf{J} - a_1 b_2 c_2\,\mathbf{I}, \tag{16}$$

$$m(a_1\,\mathbf{I}) + n(b_1\,\mathbf{I} + b_2\,\mathbf{J}) = a_1 b_2 c_1\,\mathbf{J} - a_1 b_2 c_2\,\mathbf{I}. \qquad \text{(Eq. (14))}$$

This final equation is equivalent to the scalar equations

$$ma_1 + nb_1 = -a_1 b_2 c_2,$$

$$nb_2 = a_1 b_2 c_1.$$

If b_2 were zero, **A** and **B** would be parallel, contrary to our assumption. Hence b_2 is not zero and we may divide both sides of the equation $nb_2 = a_1b_2c_1$ by b_2 to obtain

$$n = a_1c_1 = \mathbf{A} \cdot \mathbf{C}.$$

Then, by substitution,

$$\begin{aligned} ma_1 &= -nb_1 - a_1b_2c_2 \\ &= -a_1c_1b_1 - a_1b_2c_2. \end{aligned}$$

Since $|\mathbf{A}| = a_1 \neq 0$, we may divide by a_1 to obtain

$$m = -(b_1c_1 + b_2c_2) = -(\mathbf{B} \cdot \mathbf{C}).$$

Substituting these values for m and n in Eq. (14) now gives Eq. (12). This concludes the derivation.

Example 3 Verify Eq. (12) for

$$\mathbf{A} = \mathbf{i} - \mathbf{j} + 2\,\mathbf{k}, \qquad \mathbf{B} = 2\,\mathbf{i} + \mathbf{j} + \mathbf{k}, \qquad \mathbf{C} = \mathbf{i} + 2\,\mathbf{j} - \mathbf{k}.$$

Solution Since

$$\mathbf{A} \cdot \mathbf{C} = -3 \qquad \text{and} \qquad \mathbf{B} \cdot \mathbf{C} = 3,$$

the right-hand side of Eq. (12) is

$$(\mathbf{A} \cdot \mathbf{C})\mathbf{B} - (\mathbf{B} \cdot \mathbf{C})\mathbf{A} = -3\mathbf{B} - 3\mathbf{A} = -3(3\,\mathbf{i} + 3\,\mathbf{k}) = -9\,\mathbf{i} - 9\,\mathbf{k}.$$

As for the left-hand side,

$$\mathbf{A} \times \mathbf{B} = \begin{vmatrix} \mathbf{i} & \mathbf{j} & \mathbf{k} \\ 1 & -1 & 2 \\ 2 & 1 & 1 \end{vmatrix} = -3\,\mathbf{i} + 3\,\mathbf{j} + 3\,\mathbf{k},$$

so

$$(\mathbf{A} \times \mathbf{B}) \times \mathbf{C} = \begin{vmatrix} \mathbf{i} & \mathbf{j} & \mathbf{k} \\ -3 & 3 & 3 \\ 1 & 2 & -1 \end{vmatrix} = -9\,\mathbf{i} - 9\,\mathbf{k}.$$

We can sometimes use Eq. (12) to replace a string of cross multiplications with an expression that is less cumbersome to evaluate.

Example 4 Express $(\mathbf{A} \times \mathbf{B}) \times (\mathbf{C} \times \mathbf{D})$ in terms of cross products with fewer factors.

Solution For convenience, set $\mathbf{C} \times \mathbf{D}$ equal to **V**. Then,

$$(\mathbf{A} \times \mathbf{B}) \times \mathbf{V} = (\mathbf{A} \cdot \mathbf{V})\mathbf{B} - (\mathbf{B} \cdot \mathbf{V})\mathbf{A} \qquad \text{(Eq. 12)}$$

and

$$(\mathbf{A} \times \mathbf{B}) \times (\mathbf{C} \times \mathbf{D}) = (\mathbf{A} \cdot \mathbf{C} \times \mathbf{D})\mathbf{B} - (\mathbf{B} \cdot \mathbf{C} \times \mathbf{D})\mathbf{A}. \qquad (\mathbf{V} = \mathbf{C} \times \mathbf{D})$$

This expresses the original product as a scalar multiple of **B** minus a scalar multiple of **A**. Had we wanted to do so, we could have used Eq. (13) instead to express it as a scalar multiple of **C** minus a scalar multiple of **D** (Exercise 10).

Notice that we tried to keep the answer as uncluttered as possible by omitting the usual parentheses from the expressions $\mathbf{A} \cdot \mathbf{C} \times \mathbf{D}$ and $\mathbf{B} \cdot \mathbf{C} \times \mathbf{D}$. There is only one way these products make sense, so we do not need parentheses to tell us what to do.

EXERCISES 10.6

In Exercises 1–4, verify that $(\mathbf{A} \times \mathbf{B}) \cdot \mathbf{C} = (\mathbf{B} \times \mathbf{C}) \cdot \mathbf{A} = (\mathbf{C} \times \mathbf{A}) \cdot \mathbf{B}$. Find the volume of the parallelepiped determined by **A**, **B**, and **C**. Also, find $(\mathbf{A} \times \mathbf{B}) \times \mathbf{C}$ and $\mathbf{A} \times (\mathbf{B} \times \mathbf{C})$.

	A	B	C
1.	$2\mathbf{i}$	$2\mathbf{j}$	$2\mathbf{k}$
2.	$\mathbf{i} - \mathbf{j} + \mathbf{k}$	$2\mathbf{i} + \mathbf{j} - 2\mathbf{k}$	$-\mathbf{i} + 2\mathbf{j} - \mathbf{k}$
3.	$2\mathbf{i} + \mathbf{j}$	$2\mathbf{i} - \mathbf{j} + \mathbf{k}$	$\mathbf{i} + 2\mathbf{k}$
4.	$\mathbf{i} + \mathbf{j} - 2\mathbf{k}$	$-\mathbf{i} - \mathbf{k}$	$2\mathbf{i} + 4\mathbf{j} - 2\mathbf{k}$

5. Which of the following are always true and which are not always true?

a) $|\mathbf{A}| = \sqrt{\mathbf{A} \cdot \mathbf{A}}$ b) $\mathbf{A} \cdot \mathbf{A} = |\mathbf{A}|$
c) $\mathbf{A} \times \mathbf{0} = \mathbf{0} \times \mathbf{A} = \mathbf{0}$ d) $\mathbf{A} \times (-\mathbf{A}) = \mathbf{0}$
e) $\mathbf{A} \times \mathbf{B} = \mathbf{B} \times \mathbf{A}$
f) $\mathbf{A} \times (\mathbf{B} + \mathbf{C}) = \mathbf{A} \times \mathbf{B} + \mathbf{A} \times \mathbf{C}$
g) $(\mathbf{A} \times \mathbf{B}) \cdot \mathbf{B} = 0$ h) $(\mathbf{A} \times \mathbf{B}) \cdot \mathbf{C} = \mathbf{A} \cdot (\mathbf{B} \times \mathbf{C})$
i) $(\mathbf{A} \times \mathbf{B}) \times \mathbf{C} = \mathbf{A} \times (\mathbf{B} \times \mathbf{C})$

6. Which of the following are always true and which are not always true?

a) $\mathbf{A} \cdot \mathbf{B} = \mathbf{B} \cdot \mathbf{A}$ b) $(\mathbf{A} \times \mathbf{B}) = -(\mathbf{B} \times \mathbf{A})$
c) $(-\mathbf{A}) \times \mathbf{B} = -(\mathbf{A} \times \mathbf{B})$
d) $(c\mathbf{A}) \cdot \mathbf{B} = \mathbf{A} \cdot (c\mathbf{B}) = c(\mathbf{A} \cdot \mathbf{B})$ (Any number c)
e) $(c\mathbf{A}) \times \mathbf{B} = \mathbf{A} \times (c\mathbf{B}) = c(\mathbf{A} \times \mathbf{B})$ (Any number c)
f) $\mathbf{A} \cdot \mathbf{A} = |\mathbf{A}|^2$ g) $(\mathbf{A} \times \mathbf{A}) \cdot \mathbf{A} = 0$
h) $(\mathbf{A} \times \mathbf{B}) \cdot \mathbf{A} = \mathbf{B} \cdot (\mathbf{A} \times \mathbf{B})$
i) $(\mathbf{A} \times \mathbf{B}) \times (\mathbf{C} \times \mathbf{A}) = k\mathbf{A}$ for some scalar k

7. Suppose

$$\mathbf{A} \cdot \mathbf{A} = 4, \quad \mathbf{B} \cdot \mathbf{B} = 4, \quad \mathbf{A} \cdot \mathbf{B} = 0,$$
$$(\mathbf{A} \times \mathbf{B}) \times \mathbf{C} = \mathbf{0}, \quad (\mathbf{A} \times \mathbf{B}) \cdot \mathbf{C} = 8.$$

Find

a) $\mathbf{A} \cdot \mathbf{C}$ b) $|\mathbf{C}|$ c) $|\mathbf{B} \times \mathbf{C}|$

(*Hint:* Picture the vectors and think geometrically. Use basic, coordinate-free definitions. Avoid long calculations.)

8. Show that any vector **A** satisfies the identity

$$\mathbf{i} \times (\mathbf{A} \times \mathbf{i}) + \mathbf{j} \times (\mathbf{A} \times \mathbf{j}) + \mathbf{k} \times (\mathbf{A} \times \mathbf{k}) = 2\mathbf{A}.$$

9. Use Eq. (4) and facts about determinants (Appendix 10) to show that

a) $\mathbf{A} \cdot (\mathbf{C} \times \mathbf{B}) = -\mathbf{A} \cdot (\mathbf{B} \times \mathbf{C})$
b) $\mathbf{A} \cdot (\mathbf{A} \times \mathbf{B}) = 0$
c) $(\mathbf{A} + \mathbf{D}) \cdot (\mathbf{B} \times \mathbf{C}) = \mathbf{A} \cdot (\mathbf{B} \times \mathbf{C}) + \mathbf{D} \cdot (\mathbf{B} \times \mathbf{C})$

10. Show that

$(\mathbf{A} \times \mathbf{B}) \times (\mathbf{C} \times \mathbf{D}) = (\mathbf{A} \times \mathbf{B} \cdot \mathbf{D})\mathbf{C} - (\mathbf{A} \times \mathbf{B} \cdot \mathbf{C})\mathbf{D}$.

11. Let $P(1, 2, -1)$, $Q(3, -1, 4)$, and $R(2, 6, 2)$ be three vertices of a parallelogram $PQRS$.

a) Find the coordinates of S.
b) Find the area of $PQRS$.
c) Find the area of the orthogonal projection of $PQRS$ on each coordinate plane.

12. Show that if **B** is a scalar multiple of **A**, then both sides of the equation

$$(\mathbf{A} \times \mathbf{B}) \times \mathbf{C} = (\mathbf{A} \cdot \mathbf{C})\mathbf{B} - (\mathbf{B} \cdot \mathbf{C})\mathbf{A}$$

are zero.

13. Show that the area of a parallelogram in space equals the square root of the sum of the squares of the areas of the parallelogram's orthogonal projections on the coordinate planes.

14. Show that $(\mathbf{A} \times \mathbf{B}) \times (\mathbf{C} \times \mathbf{D}) = \mathbf{0}$ if **A**, **B**, **C**, and **D** are coplanar.

10.7 Surfaces in Space

Just as we call the graph of an equation $F(x, y) = 0$ in the plane a curve, we call the graph of an equation $F(x, y, z) = 0$ in space a **surface.** We use surfaces to describe boundaries of solids, to model membranes across which fluids flow, to describe plates over which electrical charges are distributed, and to define the walls of containers that are subjected to pressures of various kinds. We shall see all this and more in the chapters to come.

Our goal in the present section is to become acquainted with the surfaces most commonly used in the theory and application of the calculus of functions of more than one variable. This means finding out what the surfaces look like, what their equations are, and how to draw them.

Cylinders

The simplest surfaces to draw and write equations for, besides planes, are cylinders.

DEFINITIONS

A **cylinder** is a surface composed of all the lines parallel to a given line that pass through a given plane curve. The curve is a **generating curve** for the cylinder.

In solid geometry, *cylinder* usually means *circular cylinder,* but that is not the case here. We allow our cylinders to have cross sections of any kind. The cross sections of the cylinder in our first example are parabolas, not circles. Cylinders, as we now define them, need not even be closed.

Example 1 *The Parabolic Cylinder $y = x^2$.* Find an equation for the cylinder made by the lines parallel to the z-axis that pass through the parabola $y = x^2$, $z = 0$ (Fig. 10.54).

Solution Suppose that the point $P_0(x_0, x_0^2, 0)$ lies on the parabola $y = x^2$ in the xy-plane. Then, for any value of z, the point $Q(x_0, x_0^2, z)$ will lie on the cylinder because it lies on the line $x = x_0$, $y = x_0^2$ through P_0 parallel to the z-axis. Conversely, any point $Q(x_0, x_0^2, z)$ whose y-coordinate is the square of its x-coordinate lies on the cylinder because it lies on the line $x = x_0$, $y = x_0^2$ through P_0 parallel to the z-axis (Fig. 10.55).

Regardless of the value of z, therefore, the points on the surface are the points whose coordinates satisfy the equation $y = x^2$. This makes $y = x^2$ an equation for the cylinder. Because of this, we call the cylinder "the cylinder $y = x^2$."

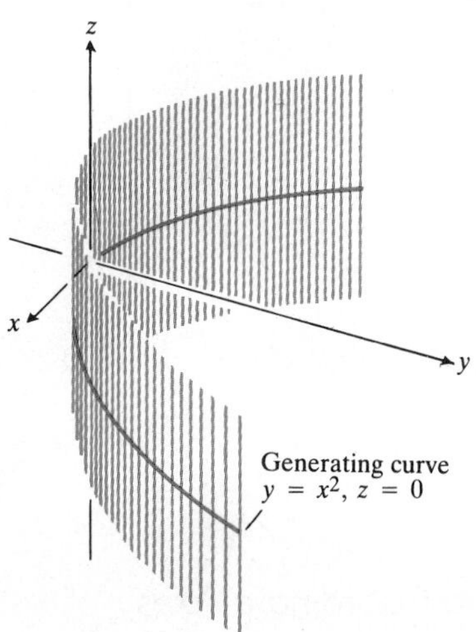

10.54 The cylinder generated by lines parallel to the z-axis and passing through the parabola $y = x^2$ in the xy-plane (Example 1).

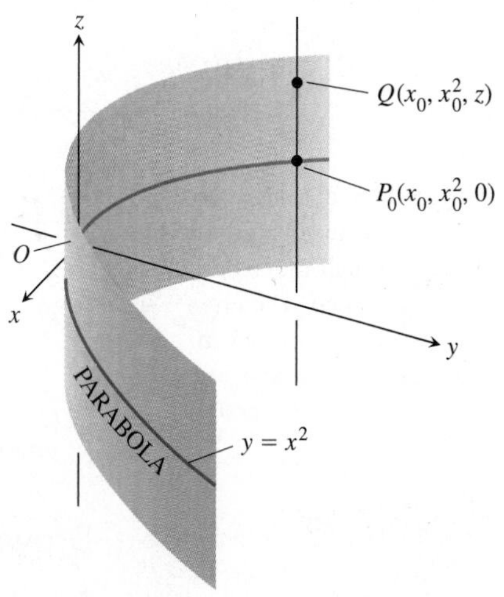

10.55 Every point of the cylinder in Fig. 10.54 has coordinates of the form (x_0, x_0^2, z). We call the cylinder "the cylinder $y = x^2$."

As Example 1 suggests, any curve $f(x, y) = c$ in the xy-plane defines a cylinder parallel to the z-axis whose equation is also $f(x, y) = c$. The equation $x^2 + y^2 = 1$ defines the circular cylinder made by the lines parallel to the z-axis that pass through the circle $x^2 + y^2 = 1$ in the xy-plane. The equation $x^2 + 4y^2 = 9$ defines the elliptical cylinder made by the lines parallel to the z-axis that pass through the ellipse $x^2 + 4y^2 = 9$ in the xy-plane.

In a similar way, any curve $g(x, z) = c$ in the xz-plane defines a cylinder parallel to the y-axis whose space equation is also $g(x, z) = c$. Any curve $h(y, z) = c$ defines a cylinder parallel to the x-axis whose space equation is also $h(y, z) = c$. An equation in any two of the three Cartesian coordinates defines a cylinder parallel to the axis of the third coordinate. See Figs. 10.56 and 10.57.

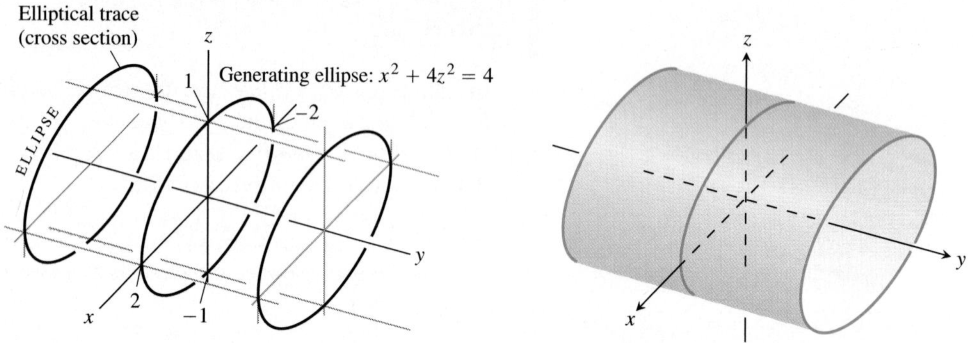

10.56 The elliptic cylinder $x^2 + 4z^2 = 4$ is made of lines parallel to the y-axis and passing through the ellipse $x^2 + 4z^2 = 4$ in the xz-plane. The cross sections or "traces" of the cylinder in planes perpendicular to the y-axis are ellipses congruent to the generating ellipse. The cylinder extends along the entire y-axis but we can draw only a finite portion of it.

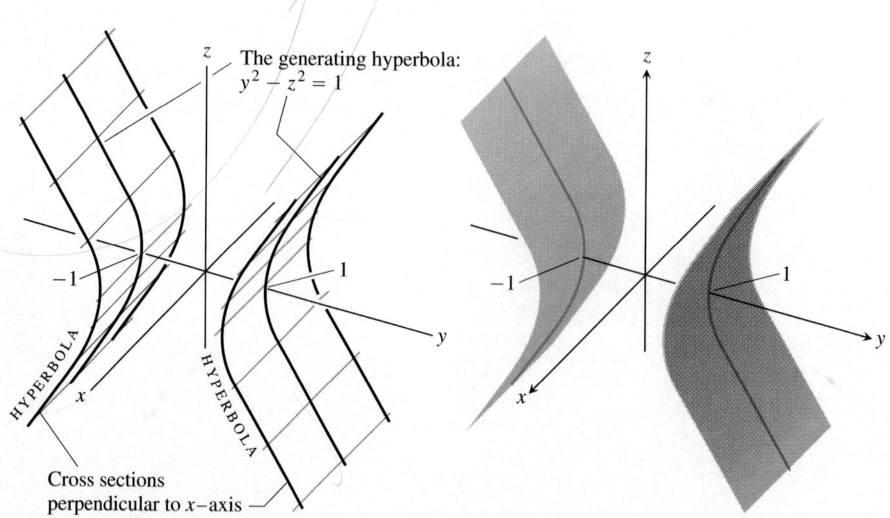

10.57 The hyperbolic cylinder $y^2 - z^2 = 1$ is made of lines parallel to the x-axis and passing through the hyperbola $y^2 - z^2 = 1$ in the yz-plane. The cross sections or "traces" of the cylinder in planes perpendicular to the x-axis are hyperbolas congruent to the generating hyperbola.

Drawing Cylinders Parallel to Coordinate Axes

See page 749 for some advice about drawing cylinders. As always, pencil is safer than pen because you can erase, but the advice applies no matter what medium you choose. Just determine which axes the cylinder is parallel to and carry out the steps shown.

How to Draw Cylinders Parallel to the Coordinate Axes

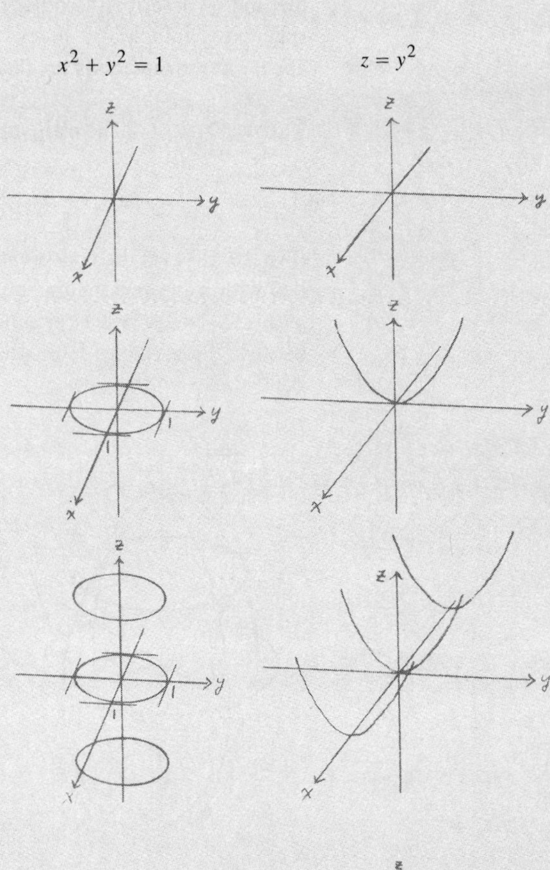

1. **Sketch all three coordinate axes** ***very lightly.***

2. **Sketch the trace of the cylinder in the coordinate plane of the two variables that appear in the cylinder's equation. Sketch** ***very lightly.***

3. **Sketch traces in parallel planes on either side (again, lightly).**

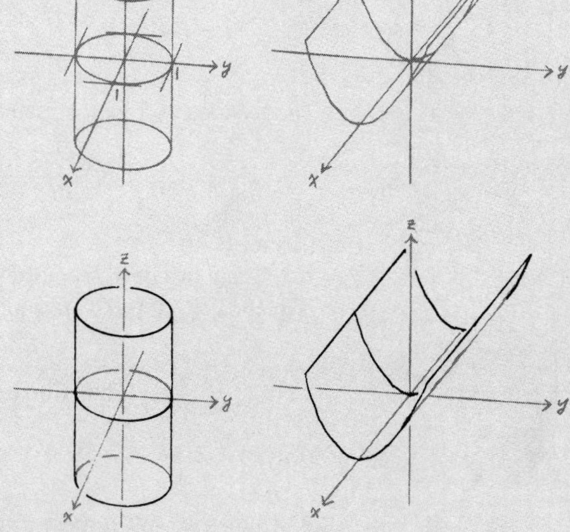

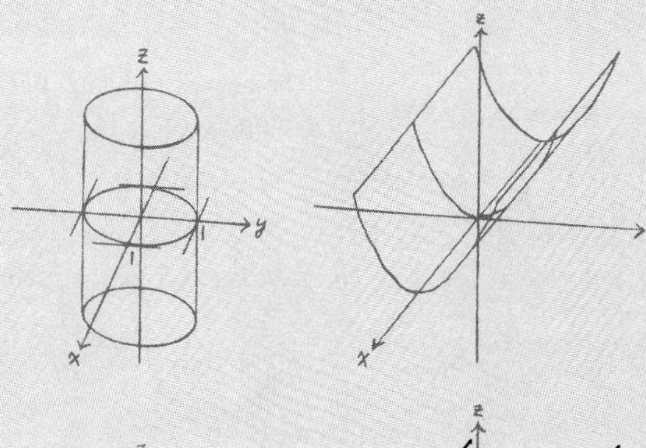

4. **Add parallel outer edges to give the shape definition.**

5. **If more definition is required, darken the parts of the lines that are exposed to view. Leave the hidden parts light. Use line breaks when you can.**

Quadric Surfaces

The cylinders in Figs. 10.55–10.57 are all **quadric surfaces,** surfaces whose equations combine quadratic terms with linear terms and constants. The examples that follow describe the other important quadric surfaces: ellipsoids (these include spheres), paraboloids, cones, and hyperboloids. The numbers a, b, and c that appear in the equations for these surfaces are assumed to be positive.

Example 2 The **ellipsoid**

$$\frac{x^2}{a^2} + \frac{y^2}{b^2} + \frac{z^2}{c^2} = 1 \tag{1}$$

(Fig. 10.58) cuts the coordinate axes at $(\pm a, 0, 0)$, $(0, \pm b, 0)$, and $(0, 0, \pm c)$. It lies within the rectangular box defined by the inequalities $|x| \leq a$, $|y| \leq b$, and $|z| \leq c$. The surface is symmetric with respect to each of the coordinate planes because the variables in the defining equation are squared.

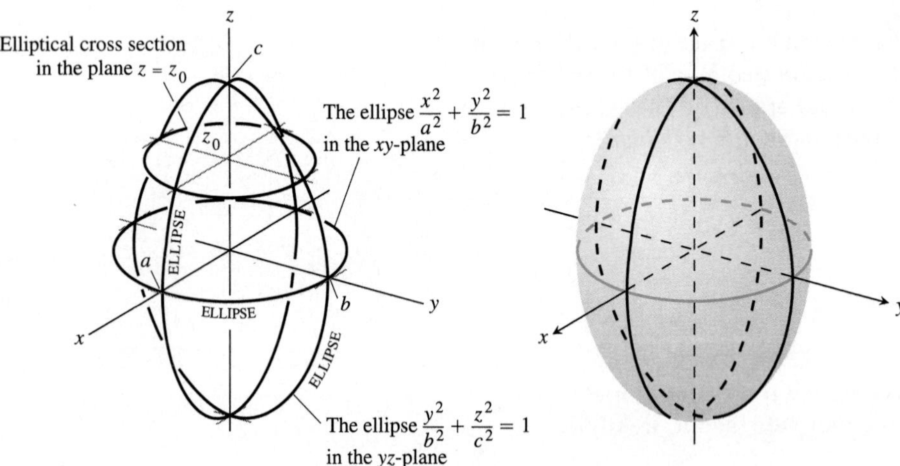

10.58 The ellipsoid

$$\frac{x^2}{a^2} + \frac{y^2}{b^2} + \frac{z^2}{c^2} = 1$$

in Example 2.

The curves in which the three coordinate planes cut the surface are ellipses. For example,

$$\frac{x^2}{a^2} + \frac{y^2}{b^2} = 1 \qquad \text{when} \qquad z = 0.$$

The section cut from the surface by the plane $z = z_0$, $|z_0| < c$, is the ellipse

$$\frac{x^2}{a^2(1 - (z_0/c)^2)} + \frac{y^2}{b^2(1 - (z_0/c)^2)} = 1 \tag{2}$$

(Exercise 70).

If any two of the semiaxes a, b, and c are equal, the surface is an ellipsoid of revolution. If all three are equal, the surface is a sphere.

Example 3 The **elliptic paraboloid**

$$\frac{x^2}{a^2} + \frac{y^2}{b^2} = \frac{z}{c} \tag{3}$$

is symmetric with respect to the planes $x = 0$ and $y = 0$ (Fig. 10.59). The only

The parabola $z = \frac{c}{a^2}x^2$ in the xz-plane

The ellipse $\frac{x^2}{a^2} + \frac{y^2}{b^2} = 1$ in the plane $z = c$

$z = c$

a

b

PARABOLA

ELLIPSE

The parabola $z = \frac{c}{b^2}y^2$ in the yz-plane

z x y

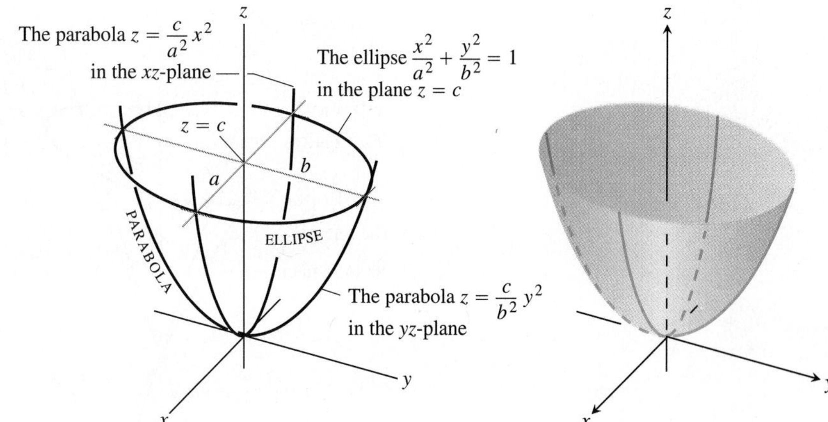

10.59 The elliptic paraboloid $(x^2/a^2) + (y^2/b^2) = z/c$ in Example 3. The cross sections perpendicular to the z-axis above the xy-plane are ellipses. The cross sections in the planes that contain the z-axis are parabolas.

intercept on the axes is the origin. Except for this point, the surface lies above the xy-plane because z is positive whenever either x or y is different from zero. The sections cut by the coordinate planes are

$$
\begin{aligned}
x = 0: &\quad \text{the parabola } z = \frac{c}{b^2}y^2, \\
y = 0: &\quad \text{the parabola } z = \frac{c}{a^2}x^2, \qquad (4) \\
z = 0: &\quad \text{the point } (0, 0, 0).
\end{aligned}
$$

Each plane $z = z_0$ above the xy-plane cuts the surface in the ellipse

$$\frac{x^2}{a^2} + \frac{y^2}{b^2} = \frac{z_0}{c}.$$

Example 4 The **circular paraboloid** or **paraboloid of revolution**

$$\frac{x^2}{a^2} + \frac{y^2}{a^2} = \frac{z}{c} \qquad (5)$$

is obtained by taking $b = a$ in Eq. (3) for the elliptic paraboloid. The cross sections of the surface by planes perpendicular to the z-axis are circles centered on the z-axis. The cross sections by planes containing the z-axis are congruent parabolas with a common focus at the point $(0, 0, a^2/4c)$.

Shapes cut from circular paraboloids are used for antennas in radio telescopes, satellite trackers, and microwave radio links (Fig. 10.60).

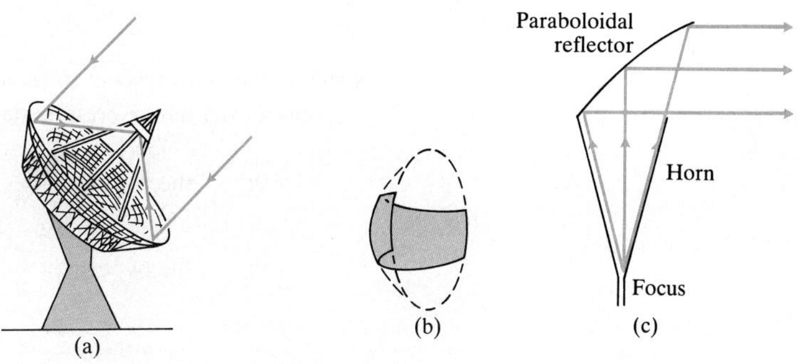

10.60 Many antennas are shaped like pieces of paraboloids of revolution. (a) Radio telescopes use the same principles as optical telescopes. (b) A "rectangular-cut" radar reflector. (c) The profile of a horn antenna in a microwave radio link.

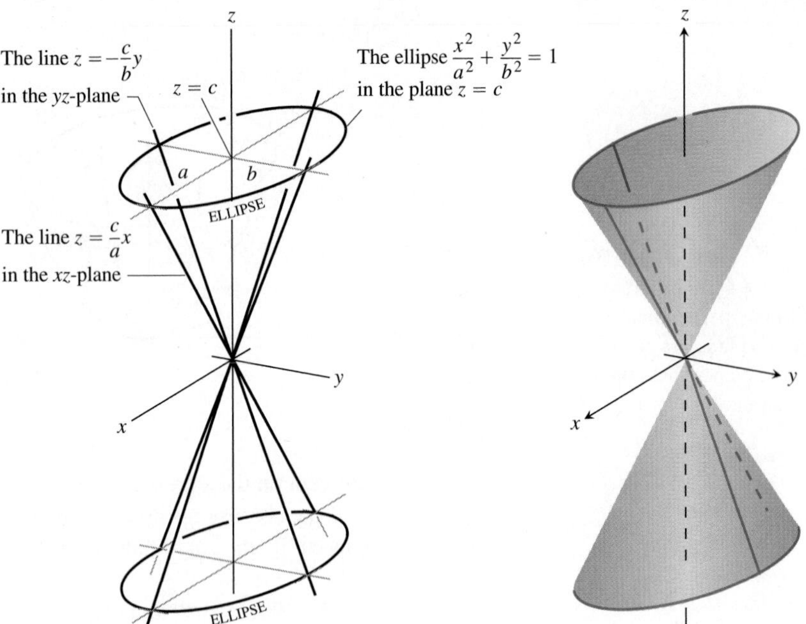

10.61 The elliptic cone $(x^2/a^2) + (y^2/b^2) = z^2/c^2$ in Example 5. Planes perpendicular to the z-axis cut the cone in ellipses above and below the xy-plane. Vertical planes that contain the z-axis cut it in pairs of intersecting lines.

Example 5 The **elliptic cone**

$$\frac{x^2}{a^2} + \frac{y^2}{b^2} = \frac{z^2}{c^2} \tag{6}$$

is symmetric with respect to the three coordinate planes (Fig. 10.61). The sections cut by the coordinate planes are

$x = 0$: the lines $z = \pm\frac{c}{b}y$, (7)

$y = 0$: the lines $z = \pm\frac{c}{a}x$, (8)

$z = 0$: the point (0, 0, 0).

The sections cut by planes $z = z_0$ above and below the xy-plane are ellipses whose centers lie on the z-axis and whose vertices lie on the lines in Eqs. (7) and (8).

If $a = b$, the cone is a right circular cone.

Example 6 The **hyperboloid of one sheet**

$$\frac{x^2}{a^2} + \frac{y^2}{b^2} - \frac{z^2}{c^2} = 1 \tag{9}$$

is symmetric with respect to each of the three coördinate planes (Fig. 10.62). The sections cut out by the coordinate planes are

$x = 0$: the hyperbola $\frac{y^2}{b^2} - \frac{z^2}{c^2} = 1$,

$y = 0$: the hyperbola $\frac{x^2}{a^2} - \frac{z^2}{c^2} = 1$, (10)

$z = 0$: the ellipse $\frac{x^2}{a^2} + \frac{y^2}{b^2} = 1$.

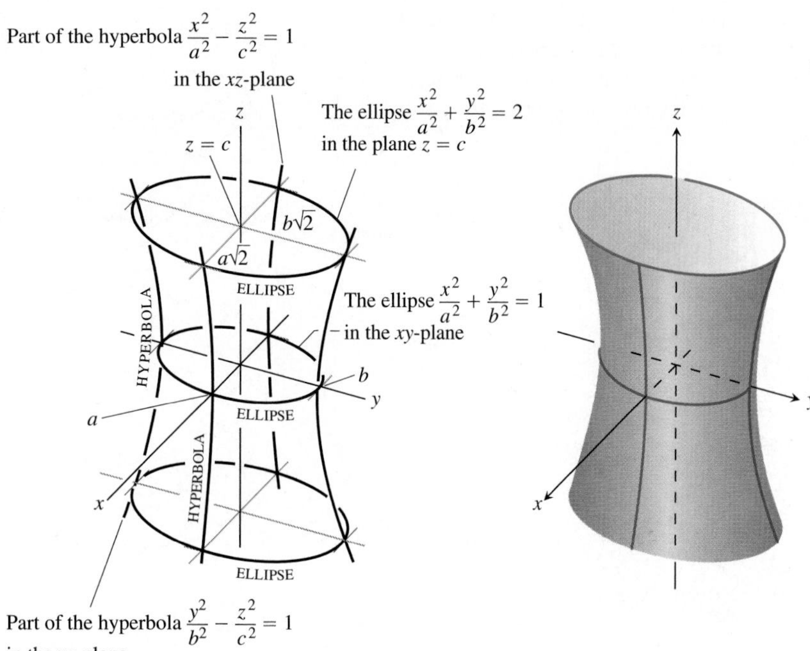

10.62 The hyperboloid $(x^2/a^2) + (y^2/b^2) - (z^2/c^2) = 1$ in Example 6. Planes perpendicular to the z-axis cut it in ellipses. Vertical planes containing the z-axis cut it in hyperbolas.

The plane $z = z_0$ cuts the surface in an ellipse with center on the z-axis and vertices on one of the hyperbolas in (10).

The surface is connected, meaning that it is possible to travel from one point on it to any other without leaving the surface. For this reason, it is said to have *one* sheet, in contrast to the hyperboloid in the next example, which has two sheets.

If $a = b$, the hyperboloid is a surface of revolution.

Example 7 The **hyperboloid of two sheets**

$$\frac{z^2}{c^2} - \frac{x^2}{a^2} - \frac{y^2}{b^2} = 1 \tag{11}$$

is symmetric with respect to the three coordinate planes (Fig. 10.63). The plane $z = 0$ does not intersect the surface; in fact, for a horizontal plane to intersect the surface, we must have $|z| \geq c$. The hyperbolic sections

$$x = 0: \quad \frac{z^2}{c^2} - \frac{y^2}{b^2} = 1, \qquad y = 0: \quad \frac{z^2}{c^2} - \frac{x^2}{a^2} = 1,$$

have their vertices and foci on the z-axis. The surface is separated into two portions, one above the plane $z = c$ and the other below the plane $z = -c$. This accounts for its name.

Equations (9) and (11) have different numbers of negative terms. The number in each case is the same as the number of sheets of the hyperboloid. If we compare with Eq. (6), we see that replacing the 1 on the right side of either Eq. (9) or (11) by zero gives the equation of a cone. If we replace the 1 on the right side of either Eq. (9) or Eq. (11) by 0, we obtain the equation

$$\frac{x^2}{a^2} + \frac{y^2}{b^2} - \frac{z^2}{c^2} = 0$$

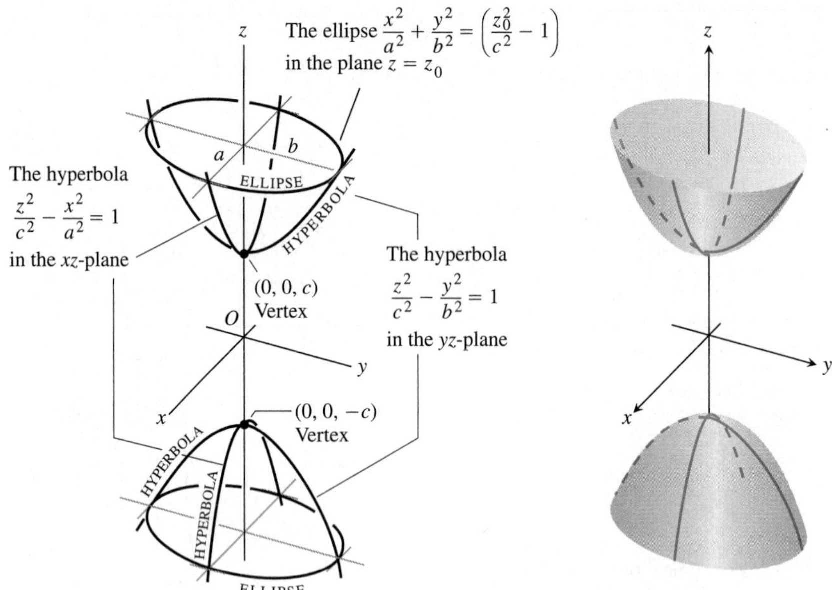

10.63 The hyperboloid $(z^2/c^2) - (x^2/a^2) - (y^2/b^2) = 1$ in Example 7. Planes perpendicular to the z-axis above and below the vertices cut it in ellipses. Vertical planes containing the z-axis cut it in hyperbolas.

for an elliptic cone (Eq. 6). The hyperboloids are asymptotic to this cone (Fig. 10.64) in the same way that the hyperbolas

$$\frac{x^2}{a^2} - \frac{y^2}{b^2} = \pm 1$$

are asymptotic to the lines

$$\frac{x^2}{a^2} - \frac{y^2}{b^2} = 0$$

in the xy-plane.

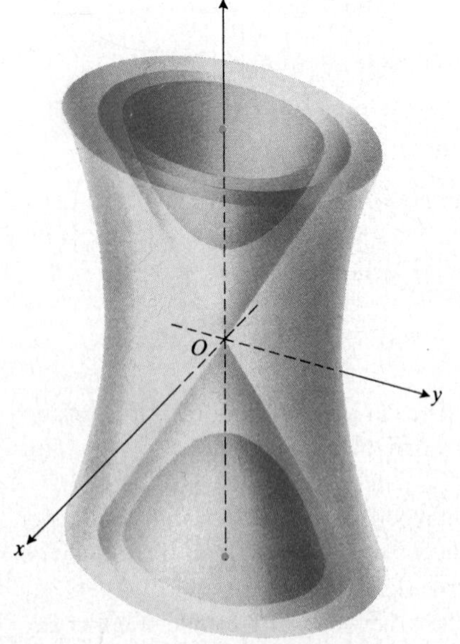

10.64 Both hyperboloids are asymptotic to the cone (Example 7).

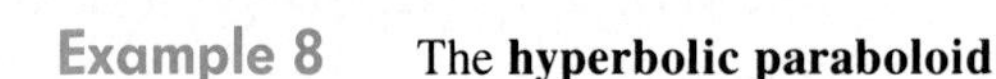

Example 8 The **hyperbolic paraboloid**

$$\frac{y^2}{b^2} - \frac{x^2}{a^2} = \frac{z}{c} \tag{12}$$

has symmetry with respect to the planes $x = 0$ and $y = 0$ (Fig. 10.65). The sections in these planes are

$$x = 0: \quad \text{the parabola } z = \frac{c}{b^2} y^2, \tag{13}$$

$$y = 0: \quad \text{the parabola } z = -\frac{c}{a^2} x^2. \tag{14}$$

In the plane $x = 0$, the parabola opens upward from the origin. The parabola in the plane $y = 0$ opens downward.

If we cut the surface by a plane $z = z_0 > 0$, the section is a hyperbola,

$$\frac{y^2}{b^2} - \frac{x^2}{a^2} = \frac{z_0}{c}, \tag{15}$$

with its focal axis parallel to the y-axis and its vertices on the parabola in (13). If z_0

DRAWING LESSON

How to Draw Quadric Surfaces

$x^2 + \frac{y^2}{4} + z^2 = 1$ $\qquad$ $z = 4 - x^2 - y^2$

1. Lightly sketch the three coordinate axes.

2. Decide on a scale and mark the intercepts on the axes.

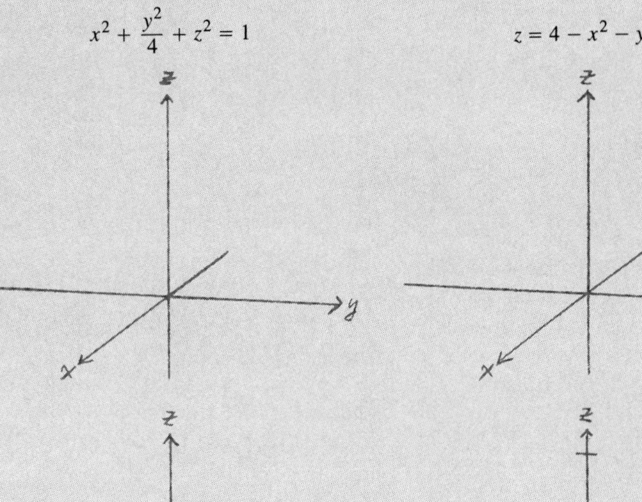

3. Sketch cross sections in the coordinate planes and in a few parallel planes, but don't clutter the picture. Use tangent lines as guides.

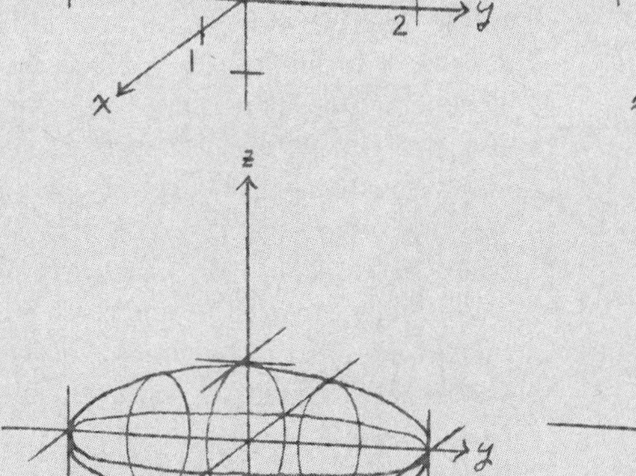

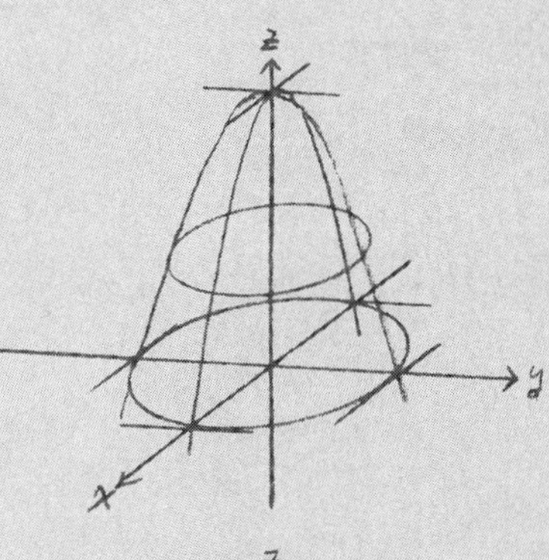

4. If more is required, darken the parts exposed to view. Leave the rest light. Use line breaks when you can.

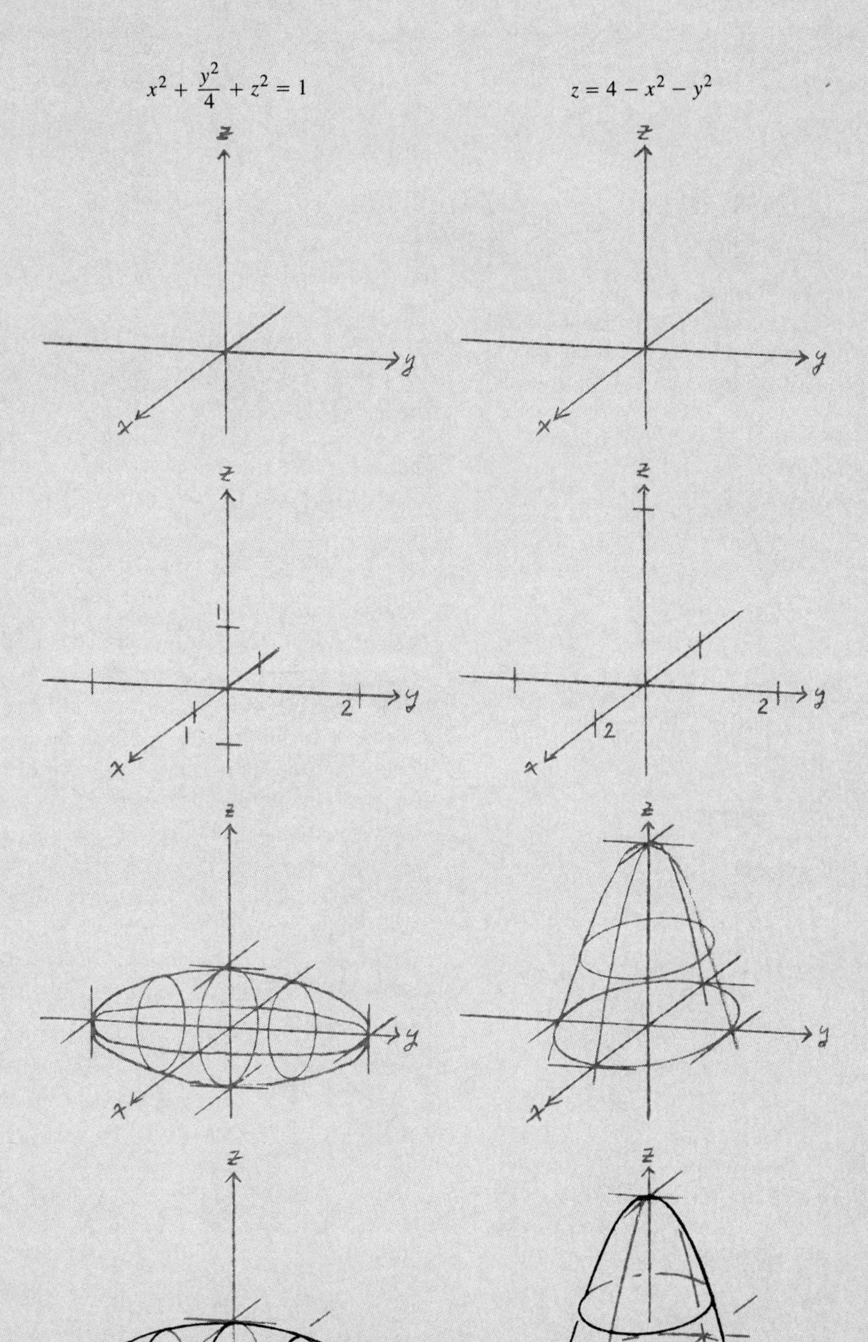

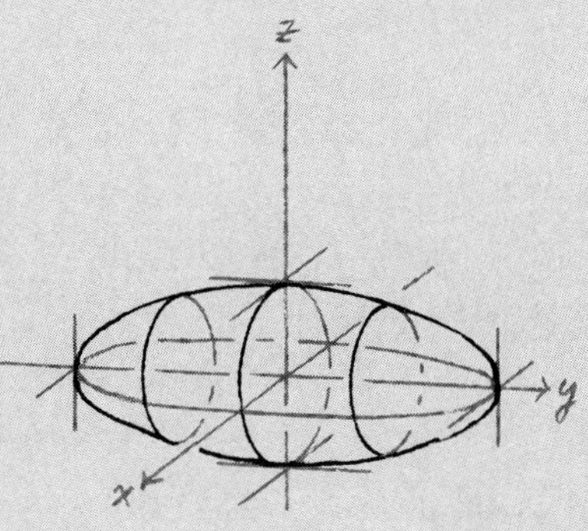

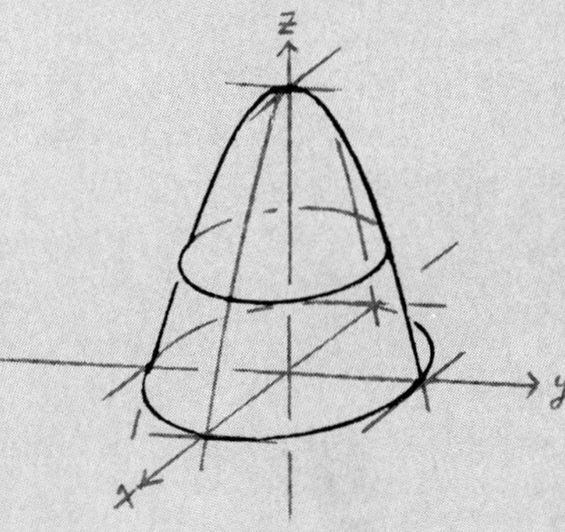

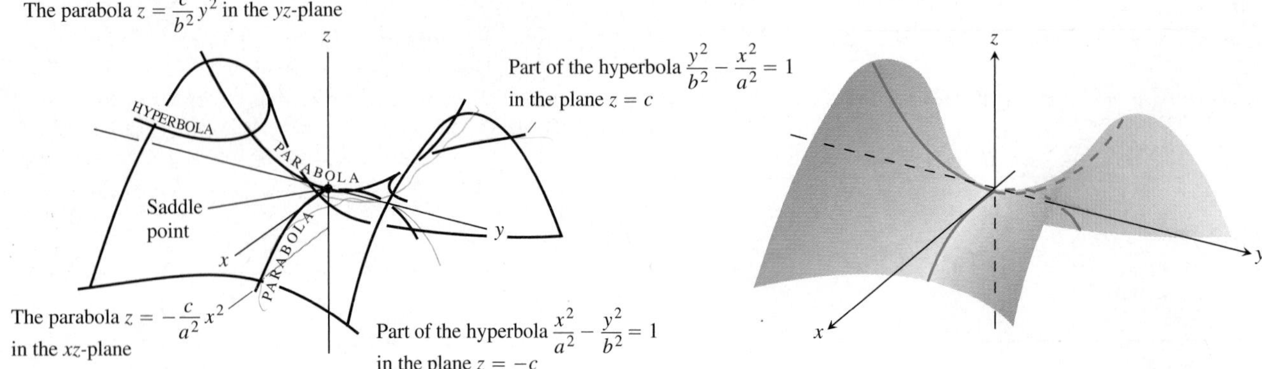

10.65 The hyperbolic paraboloid $(y^2/b^2) - (x^2/a^2) = z/c$ in Example 8. The cross sections in planes perpendicular to the z-axis above and below the xy-plane are hyperbolas. The cross sections in planes perpendicular to the other axes are parabolas.

is negative, the focal axis is parallel to the x-axis and the vertices lie on the parabola in (14).

Near the origin, the surface is shaped like a saddle. To a person traveling along the surface in the yz-plane, the origin looks like a minimum. To a person traveling in the xz-plane, the origin looks like a maximum. Such a point is called a **minimax** or **saddle point** of a surface. We shall discuss maximum and minimum points on surfaces in Chapter 12.

Liquid Mirror Telescopes

When a circular pan of liquid is rotated about its vertical axis, the surface of the liquid does not stay flat. Instead, it assumes the shape of a paraboloid of revolution, exactly what is needed for the primary mirror of a reflecting telescope. At the turn of the century, attempts to make reliable mirrors with revolving mercury failed because of surface ripples and focus losses caused by variations in the speed of rotation. Today, these difficulties can be overcome with synchronous motors driven by oscillator-stabilized power supplies and checked against constant clocks.

Using the same idea, astronomers at the Steward Observatory's Mirror Laboratory in Tucson, Arizona, have used a large heated spinning turntable to cast borosilicate glass blanks for lightweight mirrors. Spincast mirrors cost less than traditionally cast mirrors and can be made larger. Their shorter focal lengths also allow them to be installed in compact telescope frames that are less expensive to house and less likely to flex in strong winds.

EXERCISES 10.7

Sketch the surfaces in Exercises 1–64.

Cylinders

1. $x^2 + y^2 = 4$

2. $x^2 + z^2 = 4$

3. $z = y^2 - 1$

4. $x = y^2$

5. $x^2 + 4z^2 = 16$

6. $4x^2 + y^2 = 36$

7. $z^2 - y^2 = 1$

8. $yz = 1$

Ellipsoids

9. $9x^2 + y^2 + z^2 = 9$

10. $4x^2 + 4y^2 + z^2 = 16$

11. $4x^2 + 9y^2 + 4z^2 = 36$

12. $9x^2 + 4y^2 + 36z^2 = 36$

Paraboloids

13. $z = x^2 + 4y^2$

14. $z = x^2 + 9y^2$

15. $z = 8 - x^2 - y^2$

16. $z = 18 - x^2 - 9y^2$

17. $x = 4 - 4y^2 - z^2$

18. $y = 1 - x^2 - z^2$

Cones

19. $x^2 + y^2 = z^2$

20. $y^2 + z^2 = x^2$

21. $4x^2 + 9z^2 = 9y^2$

22. $9x^2 + 4y^2 = 36z^2$

Hyperboloids

23. $x^2 + y^2 - z^2 = 1$

24. $y^2 + z^2 - x^2 = 1$

25. $(y^2/4) + (z^2/9) - (x^2/4) = 1$

26. $(x^2/4) + (y^2/4) - (z^2/9) = 1$

27. $z^2 - x^2 - y^2 = 1$

28. $(y^2/4) - (x^2/4) - z^2 = 1$

29. $x^2 - y^2 - (z^2/4) = 1$

30. $(x^2/4) - y^2 - (z^2/4) = 1$

Hyperbolic Paraboloids

31. $y^2 - x^2 = z$

32. $x^2 - y^2 = z$

Assorted

33. $x^2 + y^2 + z^2 = 4$

34. $4x^2 + 4y^2 = z^2$

35. $z = 1 + y^2 - x^2$

36. $y^2 - z^2 = 4$

37. $y = -(x^2 + z^2)$

38. $z^2 - 4x^2 - 4y^2 = 4$

39. $16x^2 + 4y^2 = 1$

40. $z = x^2 + y^2 + 1$

41. $x^2 + y^2 - z^2 = 4$

42. $x = 4 - y^2$

43. $x^2 + z^2 = y$

44. $z^2 - (x^2/4) - y^2 = 1$

45. $x^2 + z^2 = 1$

46. $4x^2 + 4y^2 + z^2 = 4$

47. $16y^2 + 9z^2 = 4x^2$

48. $z = x^2 - y^2 - 1$

49. $9x^2 + 4y^2 + z^2 = 36$

50. $4x^2 + 9z^2 = y^2$

51. $x^2 + y^2 - 16z^2 = 16$

52. $z^2 + 4y^2 = 9$

53. $z = -(x^2 + y^2)$

54. $y^2 - x^2 - z^2 = 1$

55. $x^2 - 4y^2 = 1$

56. $z = 4x^2 + y^2 - 4$

57. $4y^2 + z^2 - 4x^2 = 4$

58. $z = 1 - x^2$

59. $x^2 + y^2 = z$

60. $(x^2/4) + y^2 - z^2 = 1$

61. $yz = -1$

62. $36x^2 + 9y^2 + 4z^2 = 36$

63. $9x^2 + 16y^2 = 4z^2$

64. $4z^2 - x^2 - y^2 = 4$

65. a) Express the area A of the cross section cut from the ellipsoid

$$x^2 + \frac{y^2}{4} + \frac{z^2}{9} = 1$$

by the plane $z = c$ as a function of c. (The area of an ellipse with semiaxes a and b is πab.)

b) Use slices perpendicular to the z-axis to find the volume of the ellipsoid in (a).

c) Now find the volume of the ellipsoid

$$\frac{x^2}{a^2} + \frac{y^2}{b^2} + \frac{z^2}{c^2} = 1.$$

Does your formula give the volume of a sphere of radius a if $a = b = c$?

66. A barrel has the shape of an ellipsoid with equal pieces cut from the ends by planes perpendicular to the z-axis (Fig. 10.66). The barrel is $2h$ units high, its midsection radius is R, and its end radii are both r. Find a formula for the barrel's volume. Then check two things. First, suppose the sides of the barrel are straightened to turn the barrel into a cylinder of radius R and height $2h$. Does your formula give the cylinder's volume? Second, suppose $r = 0$ and $h = R$ so the barrel is a sphere. Does your formula give the sphere's volume?

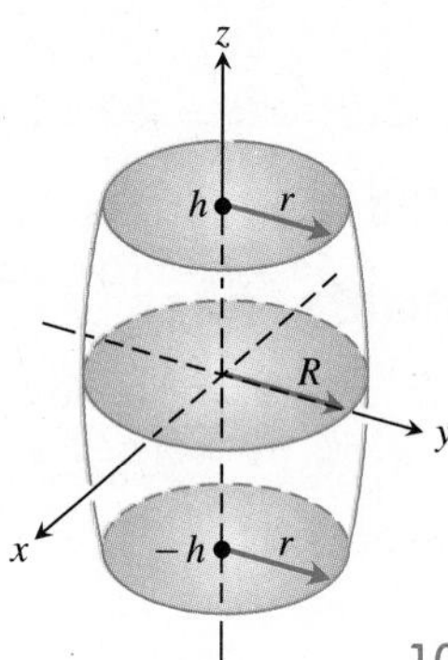

10.66 The barrel in Exercise 66.

67. Show that the volume of the segment cut from the paraboloid

$$\frac{x^2}{a^2} + \frac{y^2}{b^2} = \frac{z}{c}$$

by the plane $z = h$ equals half the segment's base times its altitude. (Figure 10.59 shows the segment for the special case $h = c$.)

68. a) Find the volume of the solid bounded by the hyperboloid

$$\frac{x^2}{a^2} + \frac{y^2}{b^2} - \frac{z^2}{c^2} = 1$$

and the planes $z = 0$ and $z = h$, $h > 0$.

b) Express your answer in (a) in terms of h and the areas A_0 and A_h of the regions cut by the hyperboloid from the planes $z = 0$ and $z = h$.

c) Show that the volume in (a) is also given by the formula

$$V = \frac{h}{6}\left(A_0 + 4A_m + A_h\right),$$

where A_m is the area of the region cut by the hyperboloid from the plane $z = h/2$.

69. *Cross sections of quadric surfaces cut by planes perpendicular to the coordinate axes are always conic sections.* The general equation for a quadric surface is

$$Gx^2 + Hy^2 + Iz^2 + Jxy + Kyz + Lxz + Mx + Ny + Pz + Q = 0. \quad (16)$$

Use this equation together with information from Section 9.2 to show that the cross section of any quadric surface by a plane perpendicular to one of the coordinate axes is a conic section.

70. Verify Eq. (2).

Computer Grapher

If you have access to a 3-D computer grapher, try graphing the surfaces in Exercises 71–77.

71. $z = y^2$ **72.** $z = 1 - y^2$

73. $z = x^2 + y^2$ **74.** $z = x^2 + 2y^2$

75. $z = \sqrt{1 - x^2}$ (upper half of a circular cylinder)

76. $z = \sqrt{1 - (y^2/4)}$ (upper half of an elliptical cylinder)

77. $z = \sqrt{x^2 + 2y^2 + 4}$ (one sheet of an elliptic hyperboloid)

EXPLORER PROGRAM

3D Grapher — Graphs equations of the form $z = f(x, y)$

10.8 Cylindrical and Spherical Coordinates

In this section we introduce two new systems of coordinates for space: the cylindrical coordinate system and the spherical coordinate system. In the cylindrical coordinate system, cylinders whose axes lie along the z-axis and planes that contain the z-axis have especially simple equations. In the spherical coordinate system, spheres centered at the origin and single cones at the origin whose axes lie along the z-axis have especially simple equations. When your work involves these shapes, these may be the best coordinate systems to use, as we shall see in later chapters.

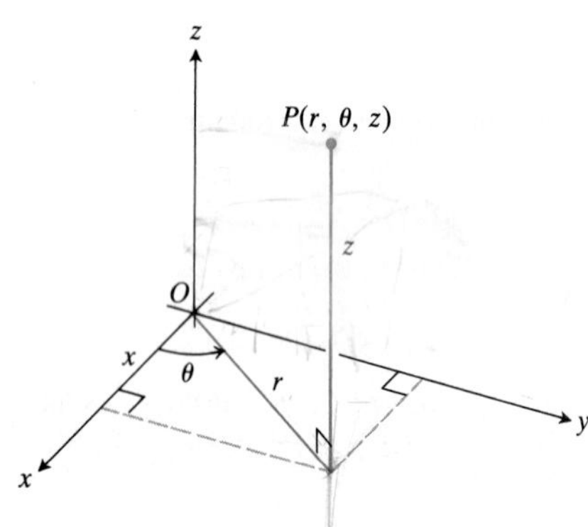

10.67 The cylindrical coordinates of a point in space are r, θ, and z.

Cylindrical Coordinates

We obtain cylindrical coordinates for space by combining polar coordinates in the xy-plane with the usual z-axis. This assigns to every point in space one or more coordinate triples of the form (r, θ, z), as shown in Fig. 10.67.

The values of x, y, r, and θ in cylindrical coordinates are related by the usual equations:

$$x = r\cos\theta, \qquad r^2 = x^2 + y^2, \qquad y = r\sin\theta, \qquad \tan\theta = y/x. \tag{1}$$

We shall use cylindrical coordinates to study planetary motion in Section 11.5.

In cylindrical coordinates, the equation $r = a$ describes not just a circle in the xy-plane but an entire cylinder about the z-axis (Fig. 10.68). The z-axis itself is given by the equation $r = 0$. The equation $\theta = \theta_0$ describes the plane that contains the z-axis and makes an angle of θ_0 radians with the positive x-axis.

Example 1 Describe the points in space whose cylindrical coordinates satisfy the equations

$$r = 2, \qquad \theta = \frac{\pi}{4}.$$

Solution These points make up the line in which the cylinder $r = 2$ cuts the portion of the plane $\theta = \pi/4$ in which r is positive (Fig. 10.69). This is the line through the point $(2, \pi/4, 0)$ parallel to the z-axis.

Example 2 Sketch the surface $r = 1 + \cos\theta$.

Solution The equation involves only r and θ; the coordinate variable z is missing. Therefore, the surface is a cylinder of lines that pass through the cardioid $r = 1 + \cos\theta$ in the $r\theta$-plane and lie parallel to the z-axis. The rules for sketching

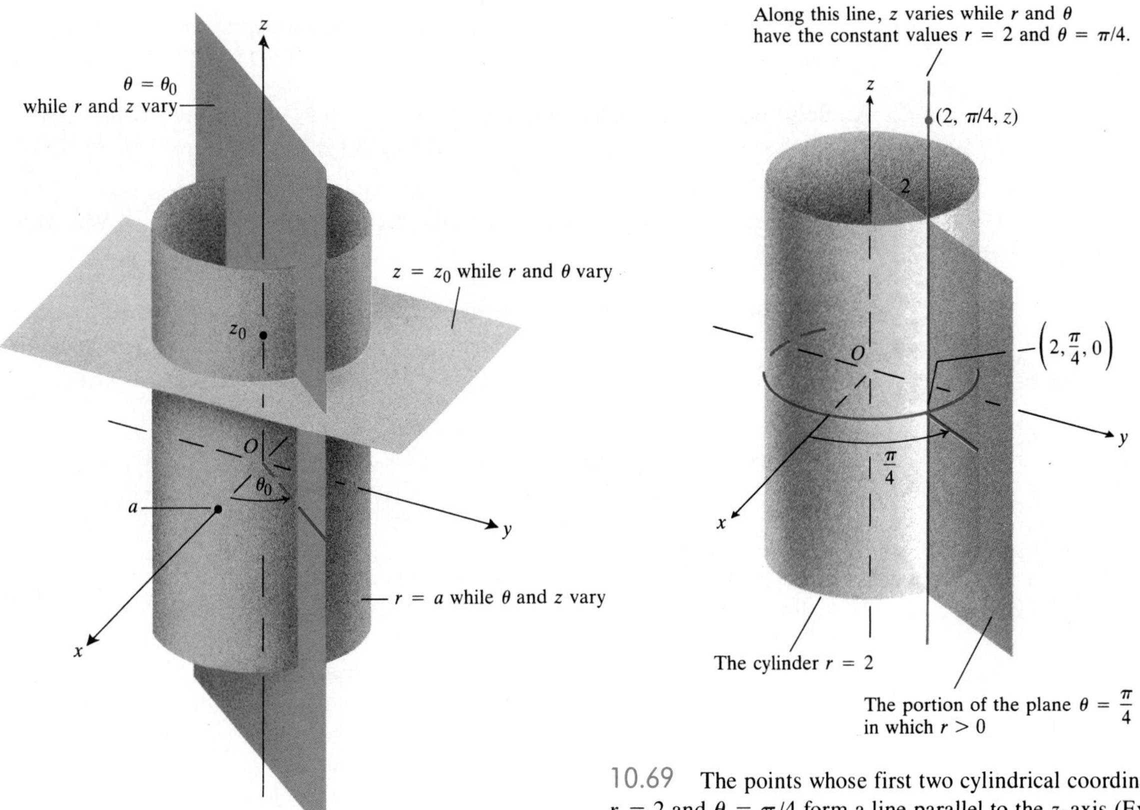

10.69 The points whose first two cylindrical coordinates are $r = 2$ and $\theta = \pi/4$ form a line parallel to the z-axis (Example 1).

10.68 Planes and cylinders that have constant-coordinate equations in cylindrical coordinates.

the cylinder are the same as always: sketch the x-, y-, and z-axes, draw a few perpendicular cross sections, connect the cross sections with parallel lines, and darken the exposed parts (Fig. 10.70).

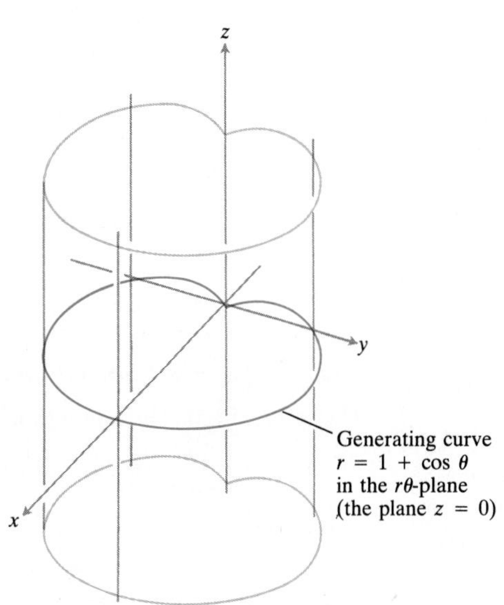

10.70 The cylindrical-coordinate equation $r = 1 + \cos\theta$ defines a cylinder in space whose cross sections perpendicular to the z-axis are cardioids (Example 2).

Example 3 Find a Cartesian equation for the surface $z = r^2$ and identify the surface.

Solution From Eqs. (1) we have $z = r^2 = x^2 + y^2$. The surface is the circular paraboloid $x^2 + y^2 = z$.

Example 4 Find an equation for the circular cylinder $4x^2 + 4y^2 = 9$ in cylindrical coordinates.

Solution The cylinder consists of the points whose distance from the z-axis is $\sqrt{x^2 + y^2} = 3/2$. The corresponding equation in cylindrical coordinates is $r = 3/2$.

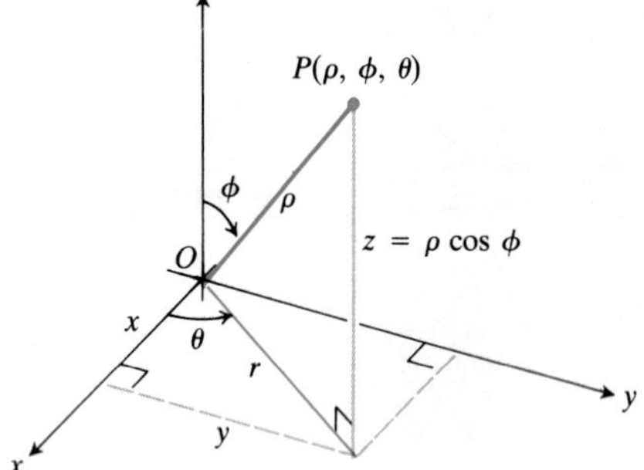

10.71 The spherical coordinates ρ, ϕ, and θ and their relation to x, y, z, and r.

Spherical Coordinates

Spherical coordinates locate points in space with two angles and a distance, as shown in Fig. 10.71.

The first coordinate, $\rho = |\overrightarrow{OP}|$, is the point's distance from the origin. Unlike r, the variable ρ is never negative. The second coordinate, ϕ, is the angle the vector $\overrightarrow{OP}$ makes with the positive z-axis. It is required to lie in the interval from 0 to π. The third coordinate is the angle θ from cylindrical coordinates.

The equation $\rho = a$ describes the sphere of radius a centered at the origin (Fig. 10.72). The equation $\phi = \phi_0$ describes a single cone whose vertex lies at the origin and whose axis lies along the z-axis. (We have to broaden our interpretation here to include the xy-plane as the cone $\phi = \pi/2$.) If ϕ_0 is greater than $\pi/2$, the cone $\phi = \phi_0$ opens downward.

A few books give spherical coordinates in the order (ρ, θ, ϕ), with the θ and ϕ reversed. Watch out for this when you read elsewhere.

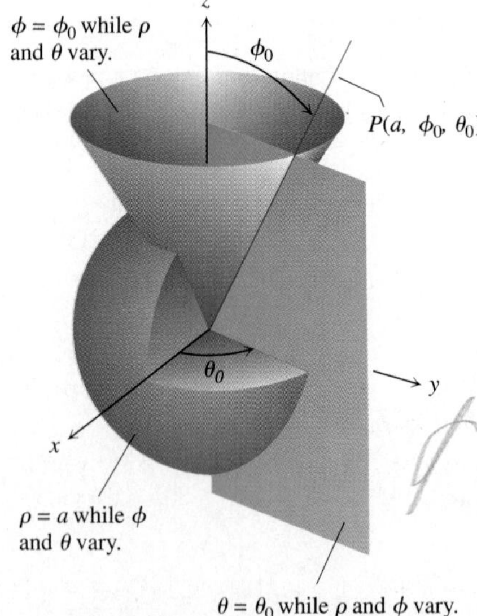

10.72 Spheres whose centers are at the origin, single cones at the origin whose axes lie along the z-axis, and half-planes "hinged" along the z-axis have constant-coordinate equations in spherical coordinates.

Selected Equations Relating Cartesian (Rectangular), Cylindrical, and Spherical Coordinates

$$r = \rho \sin \phi, \qquad x = r \cos \theta = \rho \sin \phi \cos \theta,$$
$$z = \rho \cos \phi, \qquad y = r \sin \theta = \rho \sin \phi \sin \theta, \tag{2}$$
$$\rho = \sqrt{x^2 + y^2 + z^2} = \sqrt{r^2 + z^2}$$

Example 5 Find a spherical coordinate equation for the sphere

$$x^2 + y^2 + (z - 1)^2 = 1.$$

Solution From Eqs. (2) we find that the left side of the equation is

$$\rho^2 \sin^2\phi(\cos^2\theta + \sin^2\theta) + \rho^2\cos^2\phi - 2\rho \cos \phi + 1 = \rho^2 - 2\rho \cos \phi + 1.$$

Hence the original equation transforms into

$$\rho^2 - 2\rho \cos \phi + 1 = 1,$$
$$\rho^2 = 2\rho \cos \phi,$$
$$\rho = 2 \cos \phi.$$

See Fig. 10.73.

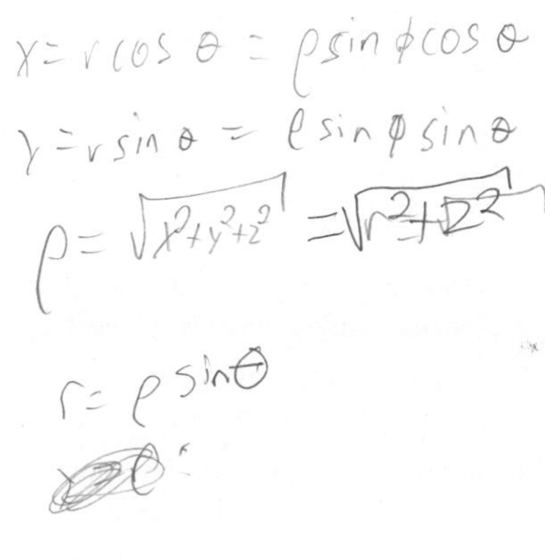

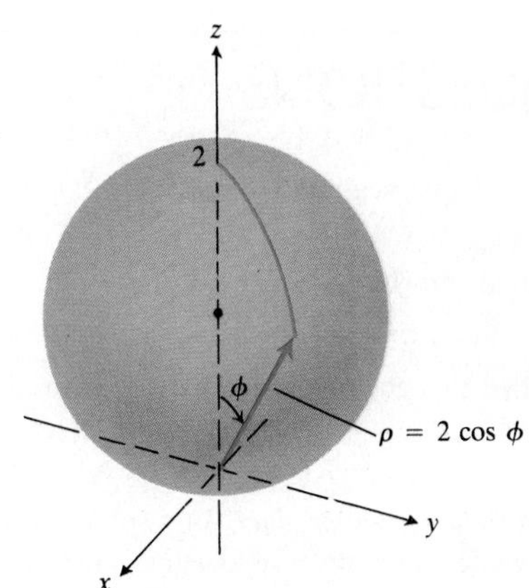

10.73 The sphere $\rho = 2\cos\phi$ in Example 5. Notice that the equation restricts ϕ to lie in the interval $0 \le \phi \le \pi/2$ because the spherical coordinate ρ is not allowed to be negative.

EXERCISES 10.8

The following table gives the coordinates of specific points in space in one of three coordinate systems. In Exercises 1–10, find coordinates for each point in the other two systems. There may be more than one right answer because points in cylindrical and spherical coordinates may have more than one coordinate triple.

	Rectangular (x, y, z)	Cylindrical (r, θ, z)	Spherical (ρ, ϕ, θ)
1.	(0, 0, 0)		
2.	(1, 0, 0)		
3.	(0, 1, 0)		
4.	(0, 0, 1)		
5.		(1, 0, 0)	
6.		$(\sqrt{2}, 0, 1)$	
7.		$(1, \pi/2, 1)$	
8.			$(\sqrt{3}, \pi/3, -\pi/2)$
9.			$(2\sqrt{2}, \pi/2, 3\pi/2)$
10.			$(\sqrt{2}, \pi, 3\pi/2)$

In Exercises 11–26, translate the equations from the given coordinate system (rectangular, cylindrical, spherical) into equations in the other two systems. Also, identify the set of points defined by the equation.

11. $r = 0$

12. $x^2 + y^2 = 5$

13. $z = 0$

14. $z = -2$

15. $\rho\cos\phi = 3$

16. $\sqrt{x^2 + y^2} = z$

17. $\rho\sin\phi\cos\theta = 0$

18. $\tan^2\phi = 1$

19. $x^2 + y^2 + z^2 = 4$

20. $\rho = 6\cos\phi$

21. $z = r^2\cos 2\theta$

22. $z^2 - r^2 = 1$

23. $r = \csc\theta$

24. $z = x^2 + y^2$

25. $3\tan^2\phi = 1$

26. $\rho^2\cos 2\phi = -1$

In Exercises 27–34, describe the sets of points in space whose cylindrical coordinates satisfy the given equations or pairs of equations. Sketch.

27. $r = 4$

28. $r^2 + z^2 = 1$

29. $r = 1 - \cos\theta$

30. $r = 2\cos\theta$

31. $r = 2, \quad z = 3$

32. $\theta = \pi/6, \quad z = r$

33. $r = 3, \quad z = \theta/2$

34. $r^2 = \cos 2\theta$

35. Find the rectangular coordinates of the center of the sphere
$$r^2 + z^2 = 4r\cos\theta + 6r\sin\theta + 2z.$$

36. What symmetry will you find in a surface whose spherical coordinate equation has the form $\rho = f(\phi)$ (independent of θ)?

In Exercises 37–44, describe the sets of points in space whose spherical coordinates satisfy the given equations or pairs of equations. Sketch.

37. $\phi = \pi/6$

38. $\rho = 6, \quad \phi = \pi/6$

39. $\rho = 5, \quad \theta = \pi/4$

40. $\theta = \pi/4, \quad \phi = \pi/4$

41. $\rho = \cos\phi$

42. $\rho = 1 - \cos\phi$

43. $\rho = \sin\phi$

44. $\theta = \pi/2, \quad \rho = 4\sin\phi$

REVIEW QUESTIONS

1. When do two directed line segments represent the same vector?
2. How are vectors added and subtracted?
3. How are the length and direction of a vector calculated?
4. If a vector is multiplied by a positive scalar, how is the result related to the original vector? What if the scalar is zero? Negative?
5. Define the *scalar product (dot product)* of two vectors. Which algebraic laws (commutative, associative, distributive) are satisfied by dot products and which, if any, are not? Give examples. When is the scalar product of two vectors equal to zero?
6. What is the vector projection of a vector **B** onto a vector **A**? How do you write **B** as the sum of a vector parallel to **A** and a vector orthogonal to **A**?
7. Define the *vector product (cross product)* of two vectors. Which algebraic laws (commutative, associative, distributive) are satisfied by cross products and which are not? Give examples. When is the vector product of two vectors equal to zero?
8. What is the determinant formula for evaluating the cross product of two vectors? Use it in an example.
9. How are vector and scalar products used to find equations for lines, line segments, planes? Give examples.
10. How can vectors be used to calculate the distance between a point and a line? A point and a plane? Give examples.
11. What is the geometric interpretation of $|(\mathbf{A} \times \mathbf{B}) \cdot \mathbf{C}|$ as a volume? How may the product be calculated from the components of **A**, **B**, and **C**? Give an example.
12. What theorem do you know about the areas of orthogonal projections of parallelograms on planes? Give an example.
13. What formulas are available for expressing $(\mathbf{A} \times \mathbf{B}) \times \mathbf{C}$ and $\mathbf{A} \times (\mathbf{B} \times \mathbf{C})$ as scalar multiples of **A**, **B**, and **C**? Give examples.
14. What is a cylinder? Give examples of equations that define cylinders in Cartesian coordinates; in cylindrical coordinates. What advice can you give about drawing cylinders?
15. Give examples of ellipsoids, paraboloids, cones, and hyperboloids (equations and sketches). What advice can you give about sketching these surfaces?
16. How are cylindrical and spherical coordinates defined? Draw diagrams that show how cylindrical and spherical coordinates are related to rectangular coordinates. What sets have constant-coordinate equations (like $x = 1$, $r = 1$, or $\phi = \pi/3$) in the three coordinate systems?

MISCELLANEOUS EXERCISES

1. Draw the unit vectors $\mathbf{u} = (\cos\theta)\,\mathbf{i} + (\sin\theta)\,\mathbf{j}$ for $\theta = 0$, $\pi/2$, $2\pi/3$, $5\pi/4$, and $5\pi/3$, together with the coordinate axes and unit circle.
2. Find the unit vector obtained by rotating
   a) **i** clockwise 45°
   b) **j** counterclockwise 120°

In Exercises 3 and 4, find the unit vectors that are tangent and normal to the curve at point P.

3. $y = \tan x$, $P(\pi/4, 1)$
4. $x^2 + y^2 = 25$, $P(3, 4)$

Express the vectors in Exercises 5–8 in terms of their lengths and directions.

5. $\sqrt{2}\,\mathbf{i} + \sqrt{2}\,\mathbf{j}$
6. $-\mathbf{i} - \mathbf{j}$
7. $2\,\mathbf{i} - 3\,\mathbf{j} + 6\,\mathbf{k}$
8. $\mathbf{i} + 2\,\mathbf{j} - \mathbf{k}$
9. Use vectors to prove that
$$(a^2 + b^2)(c^2 + d^2) \geq (ac + bd)^2$$
for any four numbers a, b, c, and d. (*Hint:* Let $\mathbf{A} = a\,\mathbf{i} + b\,\mathbf{j}$ and $\mathbf{B} = c\,\mathbf{i} + d\,\mathbf{j}$.)
10. In Fig. 10.74, D is the midpoint of side AB of triangle ABC, and E is one-third of the way between C and B. Use vectors to prove that F is the midpoint of line segment CD.

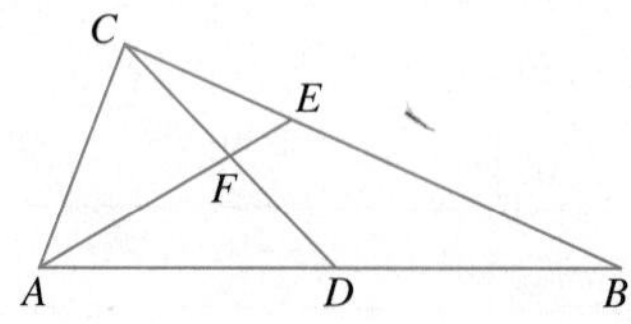

10.74 The triangle in Exercise 10.

11. The initial points of the vectors $2\,\mathbf{i} + 3\,\mathbf{j}$, $4\,\mathbf{i} + \mathbf{j}$, and $5\,\mathbf{i} + y\,\mathbf{j}$ lie at the origin. Find the value of y that makes them terminate on a common straight line.
12. Suppose that x and y are positive numbers whose sum is 1 and that $\mathbf{A} = \overrightarrow{OA}$ and $\mathbf{B} = \overrightarrow{OB}$ are vectors from the origin to

points A and B. Show that the point P determined by the vector $\overrightarrow{OP} = x\mathbf{A} + y\mathbf{B}$ lies on the line segment AB.

In Exercises 13 and 14, find $|\mathbf{A}|$, $|\mathbf{B}|$, $\mathbf{A} \cdot \mathbf{B}$, $\mathbf{B} \cdot \mathbf{A}$, $\mathbf{A} \times \mathbf{B}$, $\mathbf{B} \times \mathbf{A}$, $|\mathbf{A} \times \mathbf{B}|$, the acute angle between $\mathbf{A}$ and $\mathbf{B}$, the scalar component of $\mathbf{B}$ in the direction of $\mathbf{A}$, and the vector projection of $\mathbf{B}$ onto $\mathbf{A}$.

13. $\mathbf{A} = \mathbf{i} + \mathbf{j}$,
$\mathbf{B} = 2\mathbf{i} + \mathbf{j} - 2\mathbf{k}$

14. $\mathbf{A} = 5\mathbf{i} + \mathbf{j} + \mathbf{k}$,
$\mathbf{B} = \mathbf{i} - 2\mathbf{j} + 3\mathbf{k}$

In Exercises 15 and 16, write $\mathbf{B}$ as the sum of a vector parallel to $\mathbf{A}$ and a vector orthogonal to $\mathbf{A}$.

15. $\mathbf{A} = 2\mathbf{i} + \mathbf{j} - \mathbf{k}$,
$\mathbf{B} = \mathbf{i} + \mathbf{j} - 5\mathbf{k}$

16. $\mathbf{A} = \mathbf{i} - 2\mathbf{j}$,
$\mathbf{B} = \mathbf{i} + \mathbf{j} + \mathbf{k}$

In Exercises 17 and 18, draw coordinate axes and then sketch $\mathbf{A}$, $\mathbf{B}$, and $\mathbf{A} \times \mathbf{B}$ as vectors at the origin.

17. $\mathbf{A} = \mathbf{i}$, $\quad \mathbf{B} = \mathbf{i} + \mathbf{j}$

18. $\mathbf{A} = \mathbf{i} - \mathbf{j}$, $\quad \mathbf{B} = \mathbf{i} + \mathbf{j}$

19. *Work.* Find the work done in pushing a car 800 ft with a force of magnitude 70 lb directed 30° downward from the horizontal against the back of the car.

20. *Torque.* The operator's manual for the Toro 21-in. lawn mower says "tighten the spark plug to 15 ft · lb (20.4 N · m)." If you are installing the plug with an 11-in. socket wrench that places the center of your hand 9 in. from the axis of the spark plug, about how hard should you pull? Answer in pounds.

21. Show that $|\mathbf{A} + \mathbf{B}| \leq |\mathbf{A}| + |\mathbf{B}|$ for any vectors $\mathbf{A}$ and $\mathbf{B}$.

22. Suppose that vectors $\mathbf{A}$ and $\mathbf{B}$ are not parallel and that $\mathbf{A} = \mathbf{C} + \mathbf{D}$, where $\mathbf{C}$ is parallel to $\mathbf{B}$ and $\mathbf{D}$ is orthogonal to $\mathbf{B}$. Express $\mathbf{C}$ and $\mathbf{D}$ in terms of $\mathbf{A}$ and $\mathbf{B}$.

23. Show that $\mathbf{C} = |\mathbf{B}|\mathbf{A} + |\mathbf{A}|\mathbf{B}$ bisects the angle between $\mathbf{A}$ and $\mathbf{B}$.

24. Show that $|\mathbf{B}|\mathbf{A} + |\mathbf{A}|\mathbf{B}$ and $|\mathbf{B}|\mathbf{A} - |\mathbf{A}|\mathbf{B}$ are orthogonal.

25. *Dot multiplication is positive definite.* Show that dot multiplication of vectors is *positive definite;* that is, show that $\mathbf{A} \cdot \mathbf{A} \geq 0$ for every vector $\mathbf{A}$ and that $\mathbf{A} \cdot \mathbf{A} = 0$ if and only if $\mathbf{A} = \mathbf{0}$.

26. By forming the cross product of two appropriate vectors, derive the trigonometric identity

$$\sin(A - B) = \sin A \cos B - \cos A \sin B.$$

In Exercises 27 and 28, find the distance from the point to the line in the xy-plane.

27. $(3, 2)$, $\quad 3x + 4y = 2$

28. $(-1, 1)$, $\quad 5x - 12y = 9$

In Exercises 29 and 30, find the distance from the point to the line.

29. $(2, 2, 0)$; $\quad x = -t, \quad y = t, \quad z = -1 + t$

30. $(0, 4, 1)$; $\quad x = 2 + t, \quad y = 2 + t, \quad z = t$

31. Find parametric equations for the line that passes through the point $(1, 2, 3)$ parallel to the vector $\mathbf{v} = -3\mathbf{i} + 7\mathbf{k}$.

32. Find parametric equations for the line segment joining the points $P(1, 2, 0)$ and $Q(1, 3, -1)$.

In Exercises 33 and 34, find the distance from the point to the plane.

33. $(6, 0, -6)$, $\quad x - y = 4$

34. $(3, 0, 10)$, $\quad 2x + 3y + z = 2$

35. Find an equation for the plane that passes through the point $(3, -2, 1)$ normal to $\mathbf{N} = 2\mathbf{i} + \mathbf{j} - \mathbf{k}$.

36. Find an equation for the plane that passes through the point $(-1, 6, 0)$ perpendicular to the line $x = -1 + t$, $y = 6 - 2t$, $z = 3t$.

37. The equation $\mathbf{N} \cdot \overrightarrow{P_0P} = 0$ represents the plane through P_0 normal to $\mathbf{N}$. What set does the inequality $\mathbf{N} \cdot \overrightarrow{P_0P} > 0$ represent?

38. Find the angle between the planes $x = 7$ and $x + y + \sqrt{2}z = -3$.

39. Find the angle between the planes $x + y = 1$ and $y + z = 1$.

40. Find the distance from the point $P(1, 4, 0)$ to the plane through the points $A(0, 0, 0)$, $B(2, 0, -1)$, and $C(2, -1, 0)$.

41. Find the distance from the point $(2, 2, 3)$ to the plane $2x + 3y + 5z = 0$.

42. Find a vector parallel to the plane $2x - y - z = 4$ and orthogonal to the vector $\mathbf{i} + \mathbf{j} + \mathbf{k}$.

43. Find a unit vector orthogonal to $\mathbf{A}$ in the plane of $\mathbf{B}$ and $\mathbf{C}$ if $\mathbf{A} = 2\mathbf{i} - \mathbf{j} + \mathbf{k}$, $\mathbf{B} = \mathbf{i} + 2\mathbf{j} + \mathbf{k}$, and $\mathbf{C} = \mathbf{i} + \mathbf{j} - 2\mathbf{k}$.

44. Find a vector of magnitude 2 parallel to the line of intersection of the planes $x + 2y + z - 1 = 0$ and $x - y + 2z + 7 = 0$.

45. Find the point in which the line through the origin perpendicular to the plane $2x - y - z = 4$ meets the plane $3x + 5y + 2z = 6$.

46. Find parametric equations for the line in which the planes $x + 2y + z = 1$ and $x - y + 2z = -8$ intersect.

47. Show that the line in which the planes

$$x + 2y - 2z = 5 \quad \text{and} \quad 5x - 2y - z = 0$$

intersect is parallel to the line

$$x = -3 + 2t, \quad y = 3t, \quad z = 1 + 4t.$$

48. Find the point in which the line through $P(3, 2, 1)$ normal to the plane $2x - y + 2z = -2$ meets the plane.

49. What angle does the line of intersection of the planes $2x + y - z = 0$ and $x + y + 2z = 0$ make with the positive x-axis?

50. The line

$$L: \quad x = 3 + 2t, \quad y = 2t, \quad z = t$$

intersects the plane $x + 3y - z = -4$ in a point P. Find the coordinates of P and find equations for the line through P perpendicular to L.

51. Show that for every real number k the plane

$$x - 2y + z + 3 + k(2x - y - z + 1) = 0$$

contains the line of intersection of the planes

$$x - 2y + z + 3 = 0 \quad \text{and} \quad 2x - y - z + 1 = 0.$$

52. Find an equation for the plane through $A(-2, 0, -3)$ and $B(1, -2, 1)$ that lies parallel to the line through $C(-2, -13/5, 26/5)$ and $D(16/5, -13/5, 0)$.

53. *Submarine hunting.* Two surface ships on maneuvers are trying to determine a submarine's course and speed to prepare for an aircraft intercept. As shown in Fig. 10.75, ship A is located at (4, 0, 0) while ship B is located at (0, 5, 0). All coordinates are given in thousands of feet. Ship A locates the submarine in the direction of the vector $2\mathbf{i} + 3\mathbf{j} - (1/3)\mathbf{k}$, and ship B locates it in the direction of the vector $18\mathbf{i} - 6\mathbf{j} - \mathbf{k}$. Four minutes ago, the submarine was located at $(2, -1, -1/3)$. The aircraft is due in 20 min. Assuming the submarine moves in a straight line at a constant speed, to what position should the surface ships direct the aircraft?

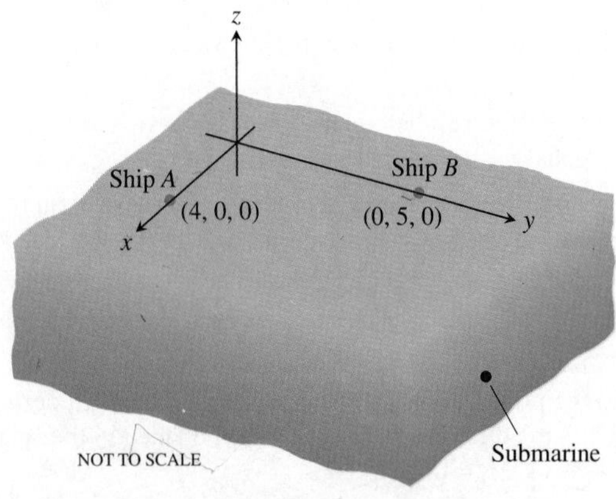

10.75 The ships and submarine in Exercise 53.

54. Two helicopters, H_1 and H_2, are traveling together. At time $t = 0$ hours, they separate and follow different straight-line paths given by

$$H_1: \quad x = 6 + 40t, \quad y = -3 + 10t, \quad z = -3 + 2t$$

$$H_2: \quad x = 6 + 110t, \quad y = -3 + 4t, \quad z = -3 + t,$$

all coordinates measured in miles. Due to system malfunctions, H_2 stops its flight at (446, 13, 1) and, in a negligible amount of time, lands at (446, 13, 0). Two hours later, H_1 is advised of this fact and heads toward H_2 at 150 mph. How long will it take H_1 to reach H_2?

55. *The distance between two lines.* Find the distance between the line L_1 through the points $A(1, 0, -1)$ and $B(-1, 1, 0)$ and the line L_2 through the points $C(3, 1, -1)$ and $D(4, 5, -2)$. The distance is to be measured along the line perpendicular to the two lines. First find a vector $\mathbf{N}$ perpendicular to both lines. Then project $\overrightarrow{AC}$ onto $\mathbf{N}$.

56. *(Continuation of Exercise 55.)* Find the distance between the line through $A(4, 0, 2)$ and $B(2, 4, 1)$ and the line through $C(1, 3, 2)$ and $D(2, 2, 4)$.

57. The parallelogram in Fig. 10.76 has vertices at $A(2, -1, 4)$, $B(1, 0, -1)$, $C(1, 2, 3)$, and D. Find
a) the coordinates of D,
b) the cosine of the interior angle at B,
c) the vector projection of $\overrightarrow{BA}$ onto $\overrightarrow{BC}$,
d) the area of the parallelogram,
e) an equation for the plane of the parallelogram,
f) the areas of the orthogonal projections of the parallelogram on the three coordinate planes.

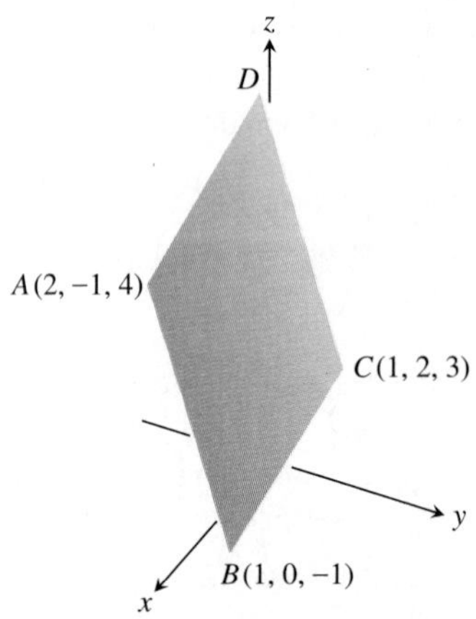

10.76 The parallelogram in Exercise 57.

58. Which of the following are equations for the plane through the points $P(1, 1, -1)$, $Q(3, 0, 2)$, and $R(-2, 1, 0)$?
a) $(2\mathbf{i} - 3\mathbf{j} + 3\mathbf{k}) \cdot ((x + 2)\mathbf{i} + (y - 1)\mathbf{j} + z\mathbf{k}) = 0$
b) $x = 3 - t, \quad y = -11t, \quad z = 2 - 3t$
c) $(x + 2) + 11(y - 1) = 3z$
d) $(2\mathbf{i} - 3\mathbf{j} + 3\mathbf{k}) \times ((x + 2)\mathbf{i} + (y - 1)\mathbf{j} + z\mathbf{k}) = \mathbf{0}$
e) $(2\mathbf{i} - \mathbf{j} + 3\mathbf{k}) \times (-3\mathbf{i} + \mathbf{k}) \cdot ((x + 2)\mathbf{i} + (y - 1)\mathbf{j} + z\mathbf{k}) = 0$

59. a) Show that

$$\begin{vmatrix} x_1 - x & y_1 - y & z_1 - z \\ x_2 - x & y_2 - y & z_2 - z \\ x_3 - x & y_3 - y & z_3 - z \end{vmatrix} = 0$$

is an equation for the plane through the three noncollinear points $P_1(x_1, y_1, z_1)$, $P_2(x_2, y_2, z_2)$, and $P_3(x_3, y_3, z_3)$.

b) What set of points in space is described by the equation

$$\begin{vmatrix} x & y & z & 1 \\ x_1 & y_1 & z_1 & 1 \\ x_2 & y_2 & z_2 & 1 \\ x_3 & y_3 & z_3 & 1 \end{vmatrix} = 0?$$

60. Show that the lines

$$x = a_1 s + b_1, \quad y = a_2 s + b_2, \quad z = a_3 s + b_3, \quad -\infty < s < \infty,$$

and

$$x = c_1 t + d_1, \quad y = c_2 t + d_2, \quad z = c_3 t + d_3, \quad -\infty < t < \infty,$$

intersect or are parallel if and only if

$$\begin{vmatrix} a_1 & c_1 & b_1 - d_1 \\ a_2 & c_2 & b_2 - d_2 \\ a_3 & c_3 & b_3 - d_3 \end{vmatrix} = 0.$$

61. Use vectors to show that the distance from $P_1(x_1, y_1)$ to the line $ax + by = c$ is

$$d = \frac{|ax_1 + by_1 - c|}{\sqrt{a^2 + b^2}}.$$

62. a) Use vectors to show that the distance from $P_1(x_1, y_1, z_1)$ to the plane $Ax + By + Cz = D$ is

$$d = \frac{|Ax_1 + By_1 + Cz_1 - D|}{\sqrt{A^2 + B^2 + C^2}}.$$

b) Find an equation for the sphere that is tangent to the planes $x + y + z = 3$ and $x + y + z = 9$ if the planes $2x - y = 0$ and $3x - z = 0$ pass through the center of the sphere.

63. a) Show that the distance between the parallel planes $Ax + By + Cz = D_1$ and $Ax + By + Cz = D_2$ is

$$d = \frac{|D_1 - D_2|}{|A\mathbf{i} + B\mathbf{j} + C\mathbf{k}|}. \qquad (1)$$

b) Use Eq. (1) to find the distance between the planes $2x + 3y - z = 6$ and $2x + 3y - z = 12$.

c) Find an equation for the plane parallel to the plane $2x - y + 2z = -4$ if the point $(3, 2, -1)$ is equidistant from the two planes.

d) Write equations for the planes that lie parallel to and 5 units away from the plane $x - 2y + z = 3$.

64. Show that if **A**, **B**, **C**, and **D** are any vectors, then

a) $\mathbf{A} \times (\mathbf{B} \times \mathbf{C}) + \mathbf{B} \times (\mathbf{C} \times \mathbf{A}) + \mathbf{C} \times (\mathbf{A} \times \mathbf{B}) = \mathbf{0}$,

b) $\mathbf{A} \times \mathbf{B} = (\mathbf{A} \cdot \mathbf{B} \times \mathbf{i})\mathbf{i} + (\mathbf{A} \cdot \mathbf{B} \times \mathbf{j})\mathbf{j} + (\mathbf{A} \cdot \mathbf{B} \times \mathbf{k})\mathbf{k}$,

c) $(\mathbf{A} \times \mathbf{B}) \cdot (\mathbf{C} \times \mathbf{D}) = \begin{vmatrix} \mathbf{A} \cdot \mathbf{C} & \mathbf{B} \cdot \mathbf{C} \\ \mathbf{A} \cdot \mathbf{D} & \mathbf{B} \cdot \mathbf{D} \end{vmatrix}$.

65. Prove that four points A, B, C, and D are coplanar (lie in a common plane) if and only if $\overrightarrow{AD} \cdot (\overrightarrow{AB} \times \overrightarrow{BC}) = 0$.

66. Prove or disprove the formula

$$\mathbf{A} \times [\mathbf{A} \times (\mathbf{A} \times \mathbf{B})] \cdot \mathbf{C} = -|\mathbf{A}|^2 \mathbf{A} \cdot \mathbf{B} \times \mathbf{C}.$$

The equations in Exercises 67–76 define sets both in the plane and in three-dimensional space. Identify the sets.

Rectangular Coordinates

67. $x = 0$

68. $x + y = 1$

69. $x^2 + y^2 = 4$

70. $x^2 + 4y^2 = 16$

71. $x = y^2$

72. $y^2 - x^2 = 1$

Cylindrical Coordinates

73. $r = 1 - \cos\theta$

74. $r = \sin\theta$

75. $r^2 = 2\cos 2\theta$

76. $r = \cos 2\theta$

Spherical Coordinates

Describe the sets defined by the spherical-coordinate equations and inequalities in Exercises 77–82.

77. $\rho = 2$

78. $\theta = \pi/4$

79. $\phi = \pi/6$

80. $\rho = 1, \quad \phi = \pi/2$

81. $\rho = 1, \quad 0 \le \phi \le \pi/2$

82. $1 \le \rho \le 2$

In Exercises 83–88, translate the equations from the given coordinate system (rectangular, cylindrical, spherical) into the other two systems. Identify the set of points defined by the equation.

Rectangular

83. $z = 2$

84. $y^2 = x^2 + z^2$

Cylindrical

85. $z = r^2\cos^2 2\theta$

86. $r = \cos\theta$

Spherical

87. $\rho = 4\sec\phi$

88. $\rho\tan\phi\sin\phi = -1$

Areas in the Plane

To find the area of a parallelogram in a plane, picture the plane as the xy-plane in space. Then write vectors for a pair of adjacent sides (their **k**-components will be zero) and calculate the magnitude of their cross product.

Find the areas of the parallelograms whose vertices are given in Exercises 89–92.

89. $A(1, 0), \quad B(0, 1), \quad C(-1, 0), \quad D(0, -1)$

90. $A(0, 0), \quad B(7, 3), \quad C(9, 8), \quad D(2, 5)$

91. $A(-1, 2), \quad B(2, 0), \quad C(7, 1), \quad D(4, 3)$

92. $A(-6, 0), \quad B(1, -4), \quad C(3, 1), \quad D(-4, 5)$

Find the areas of the triangles whose vertices are given in Exercises 93–96.

93. $A(0, 0),\quad B(-2, 3),\quad C(3, 1)$

94. $A(-1, -1),\quad B(3, 3),\quad C(2, 1)$

95. $A(-5, 3),\quad B(1, -2),\quad C(6, -2)$

96. $A(-6, 0),\quad B(10, -5),\quad C(-2, 4)$

97. Show that the area of a triangle in the xy-plane with one vertex at the origin and the other two vertices at (a_1, a_2) and (b_1, b_2) is

$$\pm\frac{1}{2}\begin{vmatrix} a_1 & a_2 \\ b_1 & b_2 \end{vmatrix}.$$

What controls the sign?

CHAPTER

11

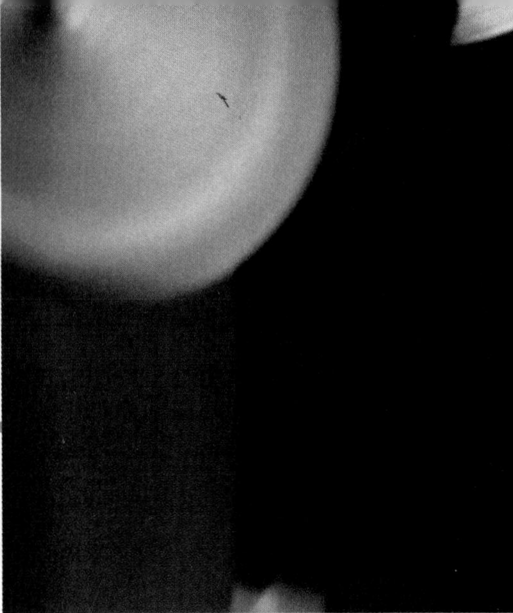

VECTOR-VALUED FUNCTIONS AND MOTION IN SPACE

Overview When a body travels through space, the three equations $x = f(t)$, $y = g(t)$, and $z = h(t)$ that give the body's coordinates as functions of time serve as parametric equations for the body's motion and path. With vector notation, we can condense these three equations into a single equation $\mathbf{r} = f(t)\,\mathbf{i} + g(t)\,\mathbf{j} + h(t)\,\mathbf{k}$ that gives the body's position as a vector function of time.

In this chapter, we show how to use calculus to differentiate and integrate vector functions to study the paths, velocities, and accelerations of moving bodies. As we go along, we shall see how our work answers the standard questions about the paths and motions of projectiles, planets, and satellites. In the final section, we combine our new vector calculus with what we know about vector algebra and geometry, derivatives, the solutions of differential equations and initial value problems, and the equations of conic sections in polar coordinates to derive Kepler's laws of planetary motion from Newton's laws of motion and gravitation.

11.1 Vector-Valued Functions and Space Curves

To track a particle moving in space, we run a vector $\mathbf{r}$ from the origin to the particle (Fig. 11.1) and study the changes in $\mathbf{r}$. If the particle's position coordinates are twice-differentiable functions of time, then so is $\mathbf{r}$ and we can find the particle's velocity and acceleration vectors at any time by differentiating $\mathbf{r}$. Conversely, if we know either the particle's velocity vector or acceleration vector as a continuous function of time, and if we have enough information about the particle's initial velocity and initial position, we can find $\mathbf{r}$ as a function of time by integration. This section shows how the calculations go.

Definitions

When a particle moves through space during a time interval I, we think of the particle's coordinates as functions defined on I:

$$x = f(t), \qquad y = g(t), \qquad z = h(t), \qquad t \in I. \tag{1}$$

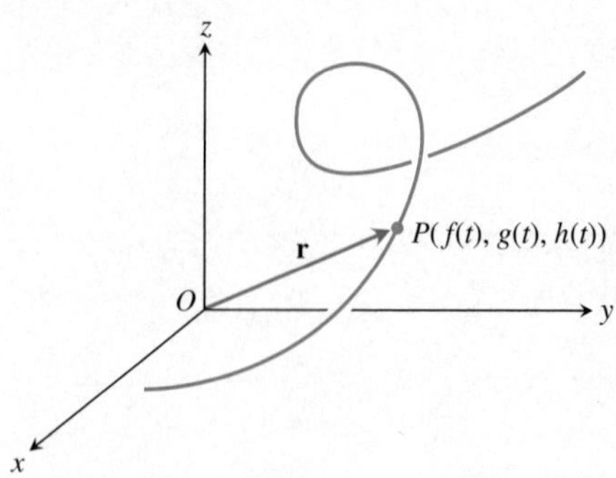

11.1 The position vector $\mathbf{r} = \overrightarrow{OP}$ of a particle moving through space is a function of time.

The points $(x, y, z) = (f(t), g(t), h(t))$, $t \in I$, make up the **curve** in space that we call the particle's path. The equations and interval in (1) **parametrize** the curve. The vector

$$\mathbf{r} = \overrightarrow{OP} = f(t)\,\mathbf{i} + g(t)\,\mathbf{j} + h(t)\,\mathbf{k} \tag{2}$$

from the origin to the particle's **position** $P(f(t), g(t), h(t))$ at time t is the particle's **position vector.** The functions f, g, and h are the **component functions (components)** of the position vector. We think of the particle's path as the **curve traced by r** during the time interval I.

Equation (2) defines $\mathbf{r}$ as a vector function of the real variable t on the interval I. More generally, a **vector function** or **vector-valued function** on a domain set D is a rule that assigns a vector in space to each element in D. For now, the domains will be intervals of real numbers. Later, in Chapter 14, the domains will be regions in the plane or in space. Vector functions will then be called "vector fields."

When we need to distinguish them from vector functions, we refer to real-valued functions as **scalar functions.** The components of $\mathbf{r}$ are scalar functions of t. When we define a vector-valued function by giving its component functions, we assume the vector function's domain to be the common domain of its component functions.

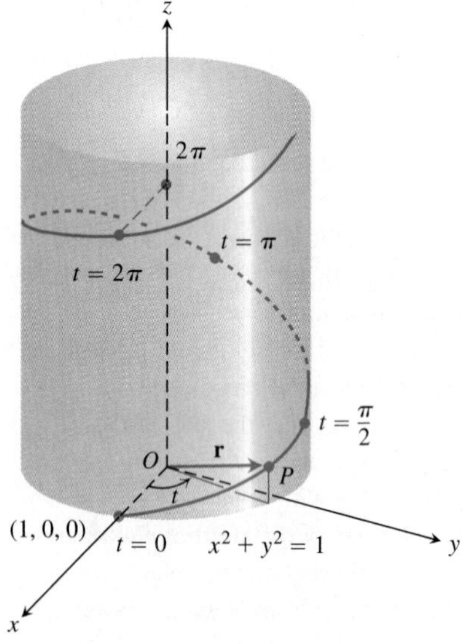

11.2 The upper half of the helix $\mathbf{r} = (\cos t)\,\mathbf{i} + (\sin t)\,\mathbf{j} + t\,\mathbf{k}$.

Example 1 *A Helix.* The vector function

$$\mathbf{r} = (\cos t)\,\mathbf{i} + (\sin t)\,\mathbf{j} + t\mathbf{k}$$

is defined for all real values of t. The curve traced by $\mathbf{r}$ is a helix (from an old Greek word for spiral) that winds around the circular cylinder $x^2 + y^2 = 1$ (Fig. 11.2). The curve lies on the cylinder because the **i**- and **j**-components of $\mathbf{r}$, being the x- and y-coordinates of the tip of $\mathbf{r}$, satisfy the cylinder's equation:

$$x^2 + y^2 = (\cos t)^2 + (\sin t)^2 = 1.$$

The curve rises as the **k**-component $z = t$ increases. Each time t increases by 2π, the curve completes one turn around the cylinder. The equations

$$x = \cos t, \qquad y = \sin t, \qquad z = t$$

parametrize the helix, the interval $-\infty < t < \infty$ being understood. You will find more helices in Fig. 11.3.

11.3 Helices drawn by computer.

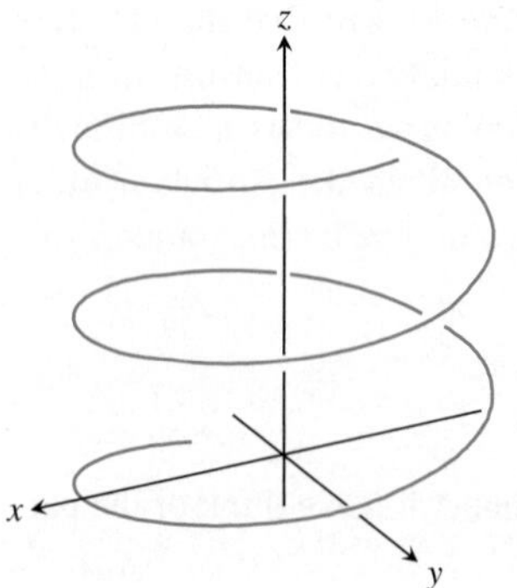

$\mathbf{r} = (\cos t)\mathbf{i} + (\sin t)\mathbf{j} + t\mathbf{k}$

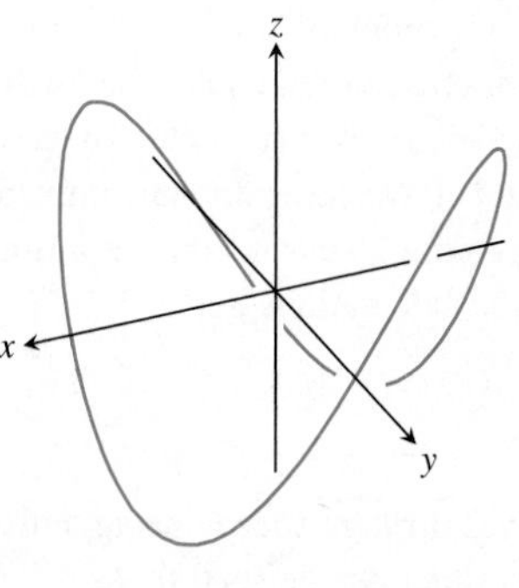

$\mathbf{r} = (\cos t)\mathbf{i} + (\sin t)\mathbf{j} + (\sin 2t)\mathbf{k}$

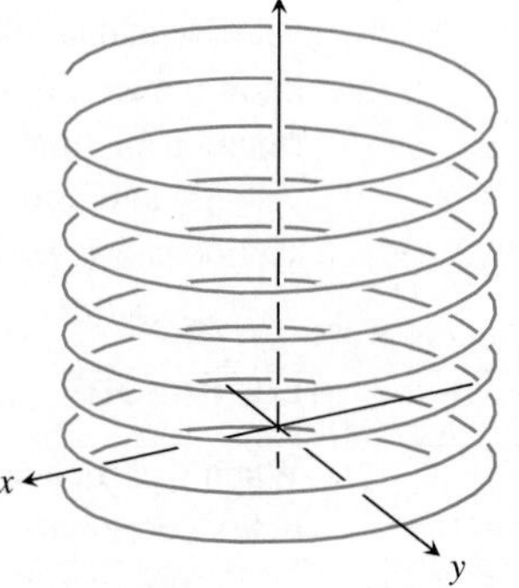

$\mathbf{r} = (\cos t)\mathbf{i} + (\sin t)\mathbf{j} + 0.3t\mathbf{k}$

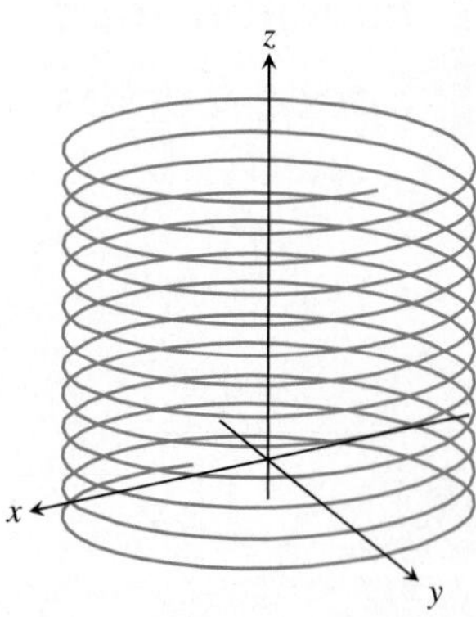

$\mathbf{r} = (\cos 5t)\mathbf{i} + (\sin 5t)\mathbf{j} + t\mathbf{k}$

(Generated by Mathematica)

Limits and Continuity

We define limits of vector functions in terms of components.

DEFINITION

A vector function **f** defined by the rule $\mathbf{f}(t) = f(t)\,\mathbf{i} + g(t)\,\mathbf{j} + h(t)\,\mathbf{k}$ has **limit** $\mathbf{L} = L_1\,\mathbf{i} + L_2\,\mathbf{j} + L_3\,\mathbf{k}$ **as** t **approaches** t_0 if

$$\lim_{t\to t_0} f(t) = L_1, \qquad \lim_{t\to t_0} g(t) = L_2, \qquad \text{and} \qquad \lim_{t\to t_0} h(t) = L_3. \tag{3}$$

We denote the existence of the limit in the usual way by writing

$$\lim_{t\to t_0} \mathbf{f}(t) = \mathbf{L}.$$

Example 2 If $\mathbf{f}(t) = (\cos t)\,\mathbf{i} + (\sin t)\,\mathbf{j} + t\mathbf{k}$, then

$$\begin{aligned}\lim_{t\to \pi/4} \mathbf{f}(t) &= \left(\lim_{t\to \pi/4} \cos t\right)\mathbf{i} + \left(\lim_{t\to \pi/4} \sin t\right)\mathbf{j} + \left(\lim_{t\to \pi/4} t\right)\mathbf{k} \\ &= \frac{\sqrt{2}}{2}\,\mathbf{i} + \frac{\sqrt{2}}{2}\,\mathbf{j} + \frac{\pi}{4}\,\mathbf{k}.\end{aligned}$$

We define continuity for vector functions the same way we define continuity for scalar functions.

DEFINITION

A vector function **f** is **continuous at a point** $t = t_0$ in its domain if $\lim_{t\to t_0}\mathbf{f}(t) = \mathbf{f}(t_0)$. The function is **continuous** if it is continuous at every point in its domain.

Since limits are defined in terms of components, we can test vector functions for continuity by examining their components (Exercise 36).

Component Test for Continuity at a Point

A vector function **f** defined by the rule $\mathbf{f}(t) = f(t)\,\mathbf{i} + g(t)\,\mathbf{j} + h(t)\,\mathbf{k}$ is continuous at $t = t_0$ if and only if f, g, and h are continuous at t_0.

Example 3 a) The function

$$\mathbf{r} = (\cos t)\,\mathbf{i} + (\sin t)\,\mathbf{j} + t\mathbf{k}$$

is continuous because $\cos t$, $\sin t$, and t are continuous.

b) The function

$$\mathbf{r} = (\cos t)\,\mathbf{i} + (\sin t)\,\mathbf{j} + \left\lfloor \frac{t}{2\pi} \right\rfloor \mathbf{k}$$

is discontinuous at every integer multiple of 2π.

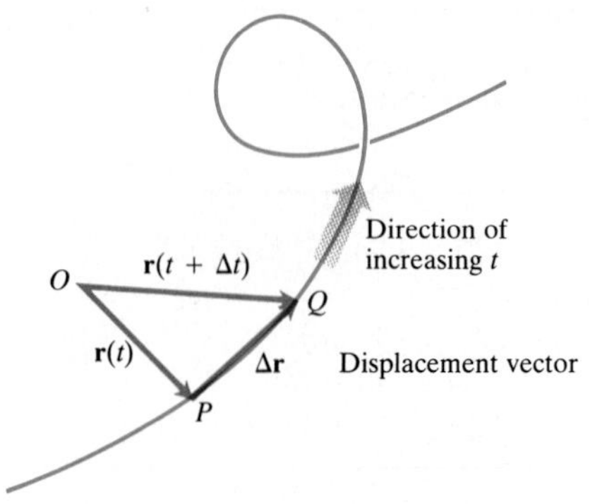

11.4 Between time t and time $t + \Delta t$, the particle moving along the path shown here undergoes the displacement $\overrightarrow{PQ} = \Delta\mathbf{r}$. The vector sum $\mathbf{r}(t) + \Delta\mathbf{r}$ gives the new position, $\mathbf{r}(t + \Delta t)$.

Derivatives and Motion

Suppose that $\mathbf{r} = f(t)\,\mathbf{i} + g(t)\,\mathbf{j} + h(t)\,\mathbf{k}$ is the position vector of a particle moving along a curve in space and that f, g, and h are differentiable functions of t. Then the difference between the particle's positions at time t and a nearby time $t + \Delta t$ can be expressed as the vector difference

$$\Delta\mathbf{r} = \mathbf{r}(t + \Delta t) - \mathbf{r}(t) \tag{4}$$

(Fig. 11.4). In terms of components,

$$\begin{aligned}\Delta\mathbf{r} &= \mathbf{r}(t + \Delta t) - \mathbf{r}(t)\\ &= [f(t + \Delta t)\,\mathbf{i} + g(t + \Delta t)\,\mathbf{j} + h(t + \Delta t)\,\mathbf{k}]\\ &\quad - [f(t)\,\mathbf{i} + g(t)\,\mathbf{j} + h(t)\,\mathbf{k}]\\ &= [f(t + \Delta t) - f(t)]\,\mathbf{i} + [g(t + \Delta t) - g(t)]\,\mathbf{j} + [h(t + \Delta t) - h(t)]\,\mathbf{k}.\end{aligned} \tag{5}$$

As Δt approaches zero, three things seem to happen simultaneously. First, Q approaches P along the curve. Second, the secant line PQ seems to approach a limiting position tangent to the curve at P. Third, the quotient $\Delta\mathbf{r}/\Delta t$ approaches the limit

$$\begin{aligned}\lim_{\Delta t\to 0}\frac{\Delta\mathbf{r}}{\Delta t} &= \left[\lim_{\Delta t\to 0}\frac{f(t + \Delta t) - f(t)}{\Delta t}\right]\mathbf{i} + \left[\lim_{\Delta t\to 0}\frac{g(t + \Delta t) - g(t)}{\Delta t}\right]\mathbf{j}\\ &\quad + \left[\lim_{\Delta t\to 0}\frac{h(t + \Delta t) - h(t)}{\Delta t}\right]\mathbf{k}\\ &= \left[\frac{df}{dt}\right]\mathbf{i} + \left[\frac{dg}{dt}\right]\mathbf{j} + \left[\frac{dh}{dt}\right]\mathbf{k}.\end{aligned} \tag{6}$$

We are therefore led by past experience to the following definition.

DEFINITION

The vector function $\mathbf{r} = f(t)\,\mathbf{i} + g(t)\,\mathbf{j} + h(t)\,\mathbf{k}$ is **differentiable at $t = t_0$** if f, g, and h are differentiable at t_0. Also, $\mathbf{r}$ is said to be **differentiable** if it is differentiable at every point of its domain. At any point t at which $\mathbf{r}$ is differentiable, its **derivative** is the vector

$$\frac{d\mathbf{r}}{dt} = \lim_{\Delta t\to 0}\frac{\mathbf{r}(t + \Delta t) - \mathbf{r}(t)}{\Delta t} = \frac{df}{dt}\mathbf{i} + \frac{dg}{dt}\mathbf{j} + \frac{dh}{dt}\mathbf{k}.$$

The curve traced by $\mathbf{r}$ is **smooth** if $d\mathbf{r}/dt$ is continuous, i.e., if f, g, and h have continuous first derivatives.

Look once again at Fig. 11.4. We drew the figure for Δt positive, so $\Delta\mathbf{r}$ points forward, in the direction of the motion. The vector $\Delta\mathbf{r}/\Delta t$, having the same direction as $\Delta\mathbf{r}$, points forward too. Had Δt been negative, $\Delta\mathbf{r}$ would have pointed backward, against the direction of motion. The quotient $\Delta\mathbf{r}/\Delta t$, however, being a negative scalar multiple of $\Delta\mathbf{r}$, would once again have pointed forward. No matter how $\Delta\mathbf{r}$ points, $\Delta\mathbf{r}/\Delta t$ points forward and we expect the vector $d\mathbf{r}/dt = \lim_{\Delta t\to 0}\Delta\mathbf{r}/\Delta t$, when different from $\mathbf{0}$, to do the same. This means that the derivative $d\mathbf{r}/dt$ is just what we want for modeling a particle's velocity. It points in the direction of motion and gives the rate of change of position with respect to time.

DEFINITIONS

If the position vector $\mathbf{r}$ of a particle moving in space is a differentiable function of time t, then the vector

$$\mathbf{v} = \frac{d\mathbf{r}}{dt}$$

is the particle's **velocity vector.** At any time t, the direction of $\mathbf{v}$ is the **direction of motion,** the magnitude of $\mathbf{v}$ is the particle's **speed,** and the derivative $\mathbf{a} = d\mathbf{v}/dt$, when it exists, is the particle's **acceleration vector.** In short,

1. Velocity is the derivative of position: $\mathbf{v} = \dfrac{d\mathbf{r}}{dt}$
2. Speed is the magnitude of velocity: Speed $= |\mathbf{v}|$
3. Acceleration is the derivative of velocity: $\mathbf{a} = \dfrac{d\mathbf{v}}{dt} = \dfrac{d^2\mathbf{r}}{dt^2}$
4. The vector $\mathbf{v}/|\mathbf{v}|$ is the direction of motion at time t.

We use the formula $\mathbf{A} = |\mathbf{A}| \cdot (\mathbf{A}/|\mathbf{A}|)$ to express the velocity of a moving particle as the product of its speed and direction.

$$\text{Velocity} = |\mathbf{v}| \cdot \frac{\mathbf{v}}{|\mathbf{v}|} = (\text{speed}) \cdot (\text{direction}) \tag{7}$$

Example 4 The vector

$$\mathbf{r} = (3\cos t)\,\mathbf{i} + (3\sin t)\,\mathbf{j} + t^2\,\mathbf{k}$$

gives the position of a moving body at time t. Find the body's speed and direction when $t = 2$. At what times, if any, are the body's velocity and acceleration orthogonal?

Solution

$$\mathbf{r} = (3\cos t)\,\mathbf{i} + (3\sin t)\,\mathbf{j} + t^2\,\mathbf{k}$$

$$\mathbf{v} = \frac{d\mathbf{r}}{dt} = -(3\sin t)\,\mathbf{i} + (3\cos t)\,\mathbf{j} + 2t\mathbf{k}$$

$$\mathbf{a} = \frac{d^2\mathbf{r}}{dt^2} = -(3\cos t)\,\mathbf{i} - (3\sin t)\,\mathbf{j} + 2\,\mathbf{k}$$

At $t = 2$, the body's speed and direction are

Speed: $|\mathbf{v}(2)| = \sqrt{(-3\sin 2)^2 + (3\cos 2)^2 + (2\cdot 2)^2} = 5$

Direction: $\dfrac{\mathbf{v}(2)}{|\mathbf{v}(2)|} = -\left(\dfrac{3}{5}\sin 2\right)\mathbf{i} + \left(\dfrac{3}{5}\cos 2\right)\mathbf{j} + \dfrac{4}{5}\mathbf{k}.$

To find when $\mathbf{v}$ and $\mathbf{a}$ are orthogonal, we look for values of t for which

$$\mathbf{v}\cdot\mathbf{a} = 9\sin t\cos t - 9\cos t\sin t + 4t = 4t = 0.$$

The only value is $t = 0$.

Differentiation Rules

Because the derivatives of vector functions are defined component by component, the rules for differentiating vector functions have the same form as the rules for differentiating scalar functions.

Differentiation Rules for Vector Functions

Constant Function Rule: $\frac{d}{dt}\mathbf{C} = \mathbf{0}$ (any constant vector $\mathbf{C}$)

If $\mathbf{u}$ and $\mathbf{v}$ are differentiable functions of t, then

Scalar Multiple Rule: $\frac{d}{dt}(c\mathbf{u}) = c\frac{d\mathbf{u}}{dt}$ (any number c)

Sum Rule: $\frac{d}{dt}(\mathbf{u} + \mathbf{v}) = \frac{d\mathbf{u}}{dt} + \frac{d\mathbf{v}}{dt}$

Difference Rule: $\frac{d}{dt}(\mathbf{u} - \mathbf{v}) = \frac{d\mathbf{u}}{dt} - \frac{d\mathbf{v}}{dt}$

Dot-Product Rule: $\frac{d}{dt}(\mathbf{u} \cdot \mathbf{v}) = \frac{d\mathbf{u}}{dt} \cdot \mathbf{v} + \mathbf{u} \cdot \frac{d\mathbf{v}}{dt}$

Cross-Product Rule: $\frac{d}{dt}(\mathbf{u} \times \mathbf{v}) = \frac{d\mathbf{u}}{dt} \times \mathbf{v} + \mathbf{u} \times \frac{d\mathbf{v}}{dt}$

Chain Rule (Short Form): If $\mathbf{r}$ is a differentiable function of t and t is a differentiable function of s, then

$$\frac{d\mathbf{r}}{ds} = \frac{d\mathbf{r}}{dt}\frac{dt}{ds}.$$

When using the Cross-Product Rule, remember to preserve the order of the factors. If $\mathbf{u}$ comes first on the left, it must come first on the right or the signs will be wrong.

We shall prove the product rules and the Chain Rule but leave the rules for constants, scalar multiples, sums, and differences as exercises.

Proof of the Dot-Product Rule

Suppose that $\quad \mathbf{u} = u_1(t)\,\mathbf{i} + u_2(t)\,\mathbf{j} + u_3(t)\,\mathbf{k}$

and $\quad \mathbf{v} = v_1(t)\,\mathbf{i} + v_2(t)\,\mathbf{j} + v_3(t)\,\mathbf{k}.$

Then

$$\begin{aligned}\frac{d}{dt}(\mathbf{u} \cdot \mathbf{v}) &= \frac{d}{dt}(u_1v_1 + u_2v_2 + u_3v_3) \\ &= \underbrace{u_1'v_1 + u_2'v_2 + u_3'v_3}_{\mathbf{u}' \cdot \mathbf{v}} + \underbrace{u_1v_1' + u_2v_2' + u_3v_3'}_{\mathbf{u} \cdot \mathbf{v}'}.\end{aligned}$$

■

Proof of the Cross-Product Rule We model the proof after the proof of the product rule for scalar functions. According to the definition of derivative,

$$\frac{d}{dt}(\mathbf{u} \times \mathbf{v}) = \lim_{h \to 0} \frac{\mathbf{u}(t+h) \times \mathbf{v}(t+h) - \mathbf{u}(t) \times \mathbf{v}(t)}{h}.$$

To change this fraction into an equivalent one that contains the difference quotients for the derivatives of $\mathbf{u}$ and $\mathbf{v}$, we subtract and add $\mathbf{u}(t) \times \mathbf{v}(t+h)$ in the numerator. Then

$$\begin{aligned}
&\frac{d}{dt}(\mathbf{u} \times \mathbf{v}) \\
&\quad = \lim_{h \to 0} \frac{\mathbf{u}(t+h) \times \mathbf{v}(t+h) - \mathbf{u}(t) \times \mathbf{v}(t+h) + \mathbf{u}(t) \times \mathbf{v}(t+h) - \mathbf{u}(t) \times \mathbf{v}(t)}{h} \\
&\quad = \lim_{h \to 0} \left[\frac{\mathbf{u}(t+h) - \mathbf{u}(t)}{h} \times \mathbf{v}(t+h) + \mathbf{u}(t) \times \frac{\mathbf{v}(t+h) - \mathbf{v}(t)}{h} \right] \qquad (8) \\
&\quad = \lim_{h \to 0} \frac{\mathbf{u}(t+h) - \mathbf{u}(t)}{h} \times \lim_{h \to 0} \mathbf{v}(t+h) + \lim_{h \to 0} \mathbf{u}(t) \times \lim_{h \to 0} \frac{\mathbf{v}(t+h) - \mathbf{v}(t)}{h}.
\end{aligned}$$

The last of these equalities holds because the limit of the cross product of two vector functions is the cross product of their limits if the latter exist (Exercise 37). As h approaches zero, $\mathbf{v}(t+h)$ approaches $\mathbf{v}(t)$ because $\mathbf{v}$, being differentiable at t, is continuous at t (Exercise 38). The two fractions approach the values of $d\mathbf{u}/dt$ and $d\mathbf{v}/dt$ at t. In short,

$$\frac{d}{dt}(\mathbf{u} \times \mathbf{v}) = \frac{d\mathbf{u}}{dt} \times \mathbf{v} + \mathbf{u} \times \frac{d\mathbf{v}}{dt}.$$

Proof of the Chain Rule Suppose that $\mathbf{r} = f(t)\,\mathbf{i} + g(t)\,\mathbf{j} + h(t)\,\mathbf{k}$ is a differentiable vector function of t and that t is a differentiable scalar function of some other variable s. Then f, g, and h are differentiable functions of s, and the Chain Rule for differentiable real-valued functions gives

$$\begin{aligned}
\frac{d\mathbf{r}}{ds} &= \frac{df}{ds}\mathbf{i} + \frac{dg}{ds}\mathbf{j} + \frac{dh}{ds}\mathbf{k} \\
&= \frac{df}{dt}\frac{dt}{ds}\mathbf{i} + \frac{dg}{dt}\frac{dt}{ds}\mathbf{j} + \frac{dh}{dt}\frac{dt}{ds}\mathbf{k} \\
&= \left(\frac{df}{dt}\mathbf{i} + \frac{dg}{dt}\mathbf{j} + \frac{dh}{dt}\mathbf{k}\right)\frac{dt}{ds} \\
&= \frac{d\mathbf{r}}{dt}\frac{dt}{ds}.
\end{aligned}$$

Derivatives of Vectors of Constant Length

We might at first think that a vector whose length remains constant as time passes has to have a zero derivative, but this need not be the case. Think of a moving clock hand. Its length remains constant as time passes, but its direction still changes. What we *can* say about the derivative of a vector of constant length is that it is always orthogonal to the vector. Direction changes take place at right angles.

If $\mathbf{u}$ is a differentiable vector function of t of constant length, then

$$\mathbf{u} \cdot \frac{d\mathbf{u}}{dt} = 0. \tag{9}$$

To see why Eq. (9) holds, suppose $\mathbf{u}$ is a differentiable function of t and that $|\mathbf{u}|$ is constant. Then $\mathbf{u} \cdot \mathbf{u} = |\mathbf{u}|^2$ is constant and we may differentiate both sides of this equation to get

$$\frac{d}{dt}(\mathbf{u} \cdot \mathbf{u}) = 0$$

$$\frac{d\mathbf{u}}{dt} \cdot \mathbf{u} + \mathbf{u} \cdot \frac{d\mathbf{u}}{dt} = 0 \qquad \text{(Dot-Product Rule with } \mathbf{v} = \mathbf{u}\text{)}$$

$$2\mathbf{u} \cdot \frac{d\mathbf{u}}{dt} = 0 \qquad \text{(Dot multiplication is commutative.)}$$

$$\mathbf{u} \cdot \frac{d\mathbf{u}}{dt} = 0.$$

Example 5 Show that

$$\mathbf{u} = (\sin t)\,\mathbf{i} + (\cos t)\,\mathbf{j} + \sqrt{3}\,\mathbf{k}$$

has constant length and is orthogonal to its derivative.

Solution

$$\mathbf{u} = (\sin t)\,\mathbf{i} + (\cos t)\,\mathbf{j} + \sqrt{3}\,\mathbf{k}$$

$$|\mathbf{u}| = \sqrt{(\sin t)^2 + (\cos t)^2 + (\sqrt{3})^2} = \sqrt{1 + 3} = 2$$

$$\frac{d\mathbf{u}}{dt} = (\cos t)\,\mathbf{i} - (\sin t)\,\mathbf{j}$$

$$\mathbf{u} \cdot \frac{d\mathbf{u}}{dt} = \sin t \cos t - \cos t \sin t = 0$$

Integrals of Vector Functions

A differentiable vector function $\mathbf{F}(t)$ is an **antiderivative** of a vector function $\mathbf{f}(t)$ on an interval I if $d\mathbf{F}/dt = \mathbf{f}$ at each point of I. If $\mathbf{F}$ is an antiderivative of $\mathbf{f}$ on I, it can be shown, working one component at a time, that every antiderivative of $\mathbf{f}$ on I has the form $\mathbf{F} + \mathbf{C}$ for some constant vector $\mathbf{C}$ (Exercise 41). The set of all antiderivatives of $\mathbf{f}$ on I is the **indefinite integral** of $\mathbf{f}$ on I; we denote it in the usual way and write

$$\int \mathbf{f}(t)\,dt = \mathbf{F}(t) + \mathbf{C}. \tag{10}$$

The usual arithmetic rules for indefinite integrals apply.

Example 6

$$\int((\cos t)\,\mathbf{i}+\mathbf{j}-2t\mathbf{k})\,dt$$

$$=\left(\int \cos t\,dt\right)\mathbf{i}+\left(\int dt\right)\mathbf{j}-\left(\int 2t\,dt\right)\mathbf{k} \qquad (11)$$

$$=(\sin t+C_1)\,\mathbf{i}+(t+C_2)\,\mathbf{j}-(t^2+C_3)\,\mathbf{k} \qquad (12)$$

$$=(\sin t)\,\mathbf{i}+t\mathbf{j}-t^2\mathbf{k}+\mathbf{C} \qquad (\mathbf{C}=C_1\mathbf{i}+C_2\mathbf{j}+C_3\mathbf{k})$$

As in the integration of scalar functions, we recommend that you skip the steps in (11) and (12) and go directly to the final form. Find an antiderivative for each component and add a constant vector at the end.

Definite integrals of vector functions are defined in terms of components.

DEFINITION

If the components of $\mathbf{f}(t)=f(t)\,\mathbf{i}+g(t)\,\mathbf{j}+h(t)\,\mathbf{k}$ are integrable over the interval $a\le t\le b$, then $\mathbf{f}$ is **integrable** over $[a, b]$ and the **definite integral** of $\mathbf{f}$ from a to b is

$$\int_a^b \mathbf{f}(t)\,dt=\left(\int_a^b f(t)\,dt\right)\mathbf{i}+\left(\int_a^b g(t)\,dt\right)\mathbf{j}+\left(\int_a^b h(t)\,dt\right)\mathbf{k}. \qquad (13)$$

All of the usual arithmetic rules for definite integrals apply (Exercise 39).

Example 7

$$\int_0^\pi((\cos t)\,\mathbf{i}+\mathbf{j}-2t\mathbf{k})\,dt=\left(\int_0^\pi \cos t\,dt\right)\mathbf{i}+\left(\int_0^\pi dt\right)\mathbf{j}-\left(\int_0^\pi 2t\,dt\right)\mathbf{k}$$

$$=\Big[\sin t\Big]_0^\pi\mathbf{i}+\Big[t\Big]_0^\pi\mathbf{j}-\Big[t^2\Big]_0^\pi\mathbf{k}$$

$$=[0-0]\,\mathbf{i}+[\pi-0]\,\mathbf{j}-[\pi^2-0^2]\,\mathbf{k}$$

$$=\pi\,\mathbf{j}-\pi^2\,\mathbf{k}$$

Example 8 *Finding a Particle's Position Function from its Velocity Function and Initial Position.* The velocity of a particle moving in space is

$$\frac{d\mathbf{r}}{dt}=(\cos t)\,\mathbf{i}-(\sin t)\,\mathbf{j}+\mathbf{k}.$$

Find the particle's position as a function of t if $\mathbf{r}=2\,\mathbf{i}+\mathbf{k}$ when $t=0$.

Solution Our goal is to solve the initial value problem that consists of

The differential equation: $\dfrac{d\mathbf{r}}{dt}=(\cos t)\,\mathbf{i}-(\sin t)\,\mathbf{j}+\mathbf{k}$,

The initial condition: $\mathbf{r}=2\,\mathbf{i}+\mathbf{k}$ when $t=0$.

To solve it, we first use what we know about antiderivatives to find the general solution of the differential equation. Integrating both sides with respect to t gives

$$\mathbf{r} = (\sin t)\,\mathbf{i} + (\cos t)\,\mathbf{j} + t\mathbf{k} + \mathbf{C}.$$

We then use the initial condition to find the right value for $\mathbf{C}$:

$$(\sin 0)\,\mathbf{i} + (\cos 0)\,\mathbf{j} + (0)\,\mathbf{k} + \mathbf{C} = 2\,\mathbf{i} + \mathbf{k} \qquad \begin{pmatrix}\mathbf{r} = 2\,\mathbf{i} + \mathbf{k} \\ \text{when } t = 0\end{pmatrix}$$

$$\mathbf{j} + \mathbf{C} = 2\,\mathbf{i} + \mathbf{k}$$

$$\mathbf{C} = 2\,\mathbf{i} - \mathbf{j} + \mathbf{k}$$

The particle's position as a function of t is

$$\mathbf{r} = (\sin t + 2)\,\mathbf{i} + (\cos t - 1)\,\mathbf{j} + (t + 1)\,\mathbf{k}.$$

To check (always a good idea), we can see from this formula that

$$\frac{d\mathbf{r}}{dt} = (\cos t + 0)\,\mathbf{i} + (-\sin t - 0)\,\mathbf{j} + (1 + 0)\,\mathbf{k} = (\cos t)\,\mathbf{i} - (\sin t)\,\mathbf{j} + \mathbf{k}$$

and that when $t = 0$,

$$\mathbf{r} = (\sin 0 + 2)\,\mathbf{i} + (\cos 0 - 1)\,\mathbf{j} + (0 + 1)\,\mathbf{k} = 2\,\mathbf{i} + \mathbf{k}.$$

Motion in the Plane

When a particle moves in the xy-plane, its z-coordinate is zero and its position vector reduces to

$$\mathbf{r} = f(t)\,\mathbf{i} + g(t)\,\mathbf{j}. \tag{14}$$

Example 9 The position vector of a particle moving along the parabola $y = x^2$ in the xy-plane is

$$\mathbf{r} = t\mathbf{i} + t^2\,\mathbf{j}.$$

Find the particle's velocity and acceleration vectors. Sketch them on the parabola together with the position vector at time $t = 1$.

Solution

$$\mathbf{r} = t\mathbf{i} + t^2\,\mathbf{j}$$

$$\mathbf{v} = \frac{d\mathbf{r}}{dt} = \mathbf{i} + 2t\mathbf{j}$$

$$\mathbf{a} = \frac{d\mathbf{v}}{dt} = 0\,\mathbf{i} + 2\,\mathbf{j} = 2\,\mathbf{j}.$$

At time $t = 1$,

$$\mathbf{r} = \mathbf{i} + \mathbf{j}, \qquad \mathbf{v} = \mathbf{i} + 2\,\mathbf{j}, \qquad \mathbf{a} = 2\,\mathbf{j}.$$

We sketch the parabola in the xy-plane and draw in the vectors (Fig. 11.5).

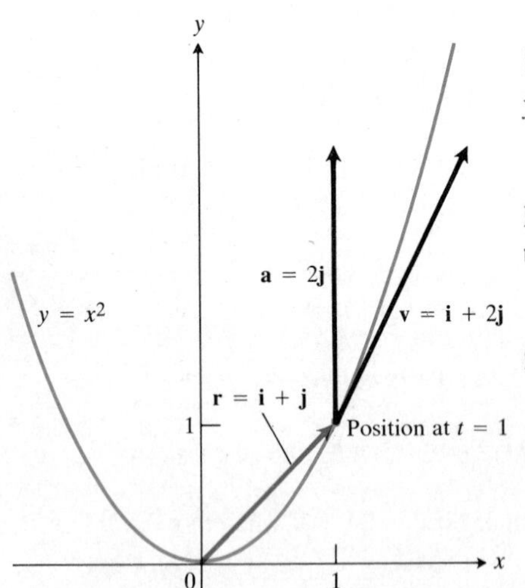

11.5 The position, velocity, and acceleration of the particle in Example 9 at $t = 1$. Notice the convention of drawing the velocity and acceleration as vectors at the point on the curve instead of at the origin.

EXERCISES 11.1

In Exercises 1–6, $\mathbf{r}$ is the position of a particle in space at time t. Find the particle's velocity and acceleration vectors. Then find the particle's speed and direction of motion at the given value of t. Write the velocity at that time as the product of its speed and direction.

1. $\mathbf{r} = (2\cos t)\,\mathbf{i} + (3\sin t)\,\mathbf{j} + 4t\mathbf{k}, \quad t = \pi/2$
2. $\mathbf{r} = (t+1)\,\mathbf{i} + (t^2-1)\,\mathbf{j} + 2t\mathbf{k}, \quad t = 1$
3. $\mathbf{r} = (\cos 2t)\,\mathbf{j} + (2\sin t)\,\mathbf{k}, \quad t = 0$
4. $\mathbf{r} = e^t\mathbf{i} + \frac{2}{9}e^{2t}\mathbf{j}, \quad t = \ln 3$
5. $\mathbf{r} = (\sec t)\,\mathbf{i} + (\tan t)\,\mathbf{j} + \frac{4}{3}t\mathbf{k}, \quad t = \pi/6$
6. $\mathbf{r} = (2\ln(t+1))\,\mathbf{i} + t^2\mathbf{j} + \frac{t^2}{2}\mathbf{k}, \quad t = 1$

In Exercises 7–10, $\mathbf{r}$ is the position of a particle in space at time t. Find the angle between the velocity and acceleration vectors at time $t = 0$.

7. $\mathbf{r} = (3t+1)\,\mathbf{i} + \sqrt{3}t\mathbf{j} + t^2\mathbf{k}$
8. $\mathbf{r} = \left(\frac{\sqrt{2}}{2}t\right)\mathbf{i} + \left(\frac{\sqrt{2}}{2}t - 16t^2\right)\mathbf{j}$
9. $\mathbf{r} = (\ln(t^2+1))\,\mathbf{i} + (\tan^{-1}t)\,\mathbf{j} + \sqrt{t^2+1}\,\mathbf{k}$
10. $\mathbf{r} = \frac{4}{9}(1+t)^{3/2}\,\mathbf{i} + \frac{4}{9}(1-t)^{3/2}\,\mathbf{j} + \frac{1}{3}t\mathbf{k}$

In Exercises 11 and 12, $\mathbf{r}$ is the position vector of a particle moving in space. Find the time or times in the given time interval when the velocity and acceleration vectors are orthogonal.

11. $\mathbf{r} = (t - \sin t)\,\mathbf{i} + (1 - \cos t)\,\mathbf{j}, \quad 0 \le t \le 2\pi$
12. $\mathbf{r} = (\sin t)\,\mathbf{i} + t\mathbf{j} + (\cos t)\,\mathbf{k}, \quad t \ge 0$

Evaluate the integrals in Exercises 13–16.

13. $\displaystyle\int_0^1 (t^3\mathbf{i} + 7\mathbf{j} + (t+1)\,\mathbf{k})\,dt$
14. $\displaystyle\int_1^4 \left(\frac{1}{t}\mathbf{i} + \frac{1}{5-t}\mathbf{j} + \frac{1}{2t}\mathbf{k}\right)dt$
15. $\displaystyle\int_{-\pi/4}^{\pi/4} ((\sin t)\,\mathbf{i} + (1+\cos t)\,\mathbf{j} + (\sec^2 t)\,\mathbf{k})\,dt$
16. $\displaystyle\int_0^{\pi/3} ((\sec t\tan t)\,\mathbf{i} + (\tan t)\,\mathbf{j} + (2\sin t\cos t)\,\mathbf{k})\,dt$

Exercises 17–20 give the position vectors of particles moving in the xy-plane. In each case, find the particle's velocity and acceleration vectors at the given times. Sketch them as vectors on the curve, together with the position vector.

17. *Motion on the circle* $x^2 + y^2 = 1$

$$\mathbf{r} = (\sin t)\,\mathbf{i} + (\cos t)\,\mathbf{j}; \quad t = \pi/4 \text{ and } \pi/2$$

18. *Motion on the circle* $x^2 + y^2 = 16$

$$\mathbf{r} = \left(4\cos\frac{t}{2}\right)\mathbf{i} + \left(4\sin\frac{t}{2}\right)\mathbf{j}; \quad t = \pi \text{ and } 3\pi/2$$

19. *Motion on the cycloid* $x = t - \sin t$, $y = 1 - \cos t$

$$\mathbf{r} = (t - \sin t)\,\mathbf{i} + (1 - \cos t)\,\mathbf{j}; \quad t = \pi \text{ and } 3\pi/2$$

20. *Motion on the parabola* $y = x^2 + 1$

$$\mathbf{r} = t\mathbf{i} + (t^2+1)\,\mathbf{j}; \quad t = -1, 0, \text{ and } 1$$

Exercises 21–24 give the velocity function and initial position of a particle moving in space. Find the particle's position function.

21. Velocity function: $\dfrac{d\mathbf{r}}{dt} = -t\mathbf{i} - t\mathbf{j} - t\mathbf{k}$

    Initial position: $\mathbf{r} = \mathbf{i} + 2\mathbf{j} + 3\mathbf{k}$ when $t = 0$

22. Velocity function: $\dfrac{d\mathbf{r}}{dt} = (180t)\,\mathbf{i} + (180t - 16t^2)\,\mathbf{j}$

    Initial position: $\mathbf{r} = 100\,\mathbf{j}$ when $t = 0$

23. Velocity function: $\dfrac{d\mathbf{r}}{dt} = \dfrac{3}{2}(t+1)^{1/2}\,\mathbf{i} + e^{-t}\mathbf{j} + \dfrac{1}{t+1}\mathbf{k}$

    Initial position: $\mathbf{r} = \mathbf{k}$ when $t = 0$

24. Velocity function: $\dfrac{d\mathbf{r}}{dt} = (t^3 + 4t)\,\mathbf{i} + t\mathbf{j} + 2t^2\,\mathbf{k}$

    Initial position: $\mathbf{r} = \mathbf{i} + \mathbf{j}$ when $t = 0$

Exercises 25 and 26 give the acceleration function and the initial position and velocity of a particle moving in space. Find the particle's position function.

25. Acceleration function: $\dfrac{d^2\mathbf{r}}{dt^2} = -32\,\mathbf{k}$

    Initial position and velocity: $\mathbf{r} = 100\,\mathbf{k}$ and $\dfrac{d\mathbf{r}}{dt} = 8\,\mathbf{i} + 8\,\mathbf{j}$ when $t = 0$

26. Acceleration function: $\dfrac{d^2\mathbf{r}}{dt^2} = -(\mathbf{i} + \mathbf{j} + \mathbf{k})$

    Initial position and velocity: $\mathbf{r} = 10\,\mathbf{i} + 10\,\mathbf{j} + 10\,\mathbf{k}$ and $\dfrac{d\mathbf{r}}{dt} = 0$ when $t = 0$

27. A particle moves on a cycloid in the xy-plane in such a way that its position at time t is

$$\mathbf{r} = (t - \sin t)\,\mathbf{i} + (1 - \cos t)\,\mathbf{j}.$$

Find the maximum and minimum values of $|\mathbf{v}|$ and $|\mathbf{a}|$. (*Hint:* Find the extreme values of $|\mathbf{v}|^2$ and $|\mathbf{a}|^2$ first and take square roots later.)

28. A particle moves around the ellipse $(y/3)^2 + (z/2)^2 = 1$ in the yz-plane in such a way that its position at time t is

$$\mathbf{r} = (3\cos t)\,\mathbf{j} + (2\sin t)\,\mathbf{k}.$$

Find the maximum and minimum values of $|\mathbf{v}|$ and $|\mathbf{a}|$. (See the hint in Exercise 27.)

29. Suppose a particle moves on a sphere centered at the origin in such a way that its position vector $\mathbf{r}$ is a differentiable function of time (Fig. 11.6). Show that the particle's velocity is always orthogonal to $\mathbf{r}$.

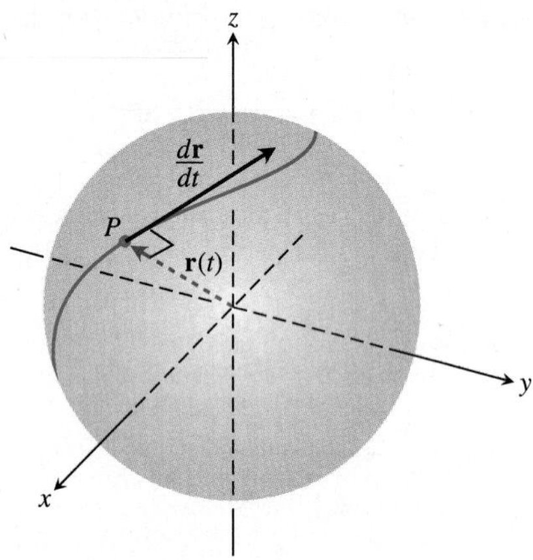

11.6 The velocity vector of a particle P that moves on the surface of a sphere is tangent to the sphere (Exercise 29).

30. Let $\mathbf{v}$ be a differentiable vector function of t. Show that if $\mathbf{v}\cdot(d\mathbf{v}/dt) = 0$ for all t, then $|\mathbf{v}|$ is constant.

31. *Derivatives of triple scalar products*

a) Show that if $\mathbf{u}$, $\mathbf{v}$, and $\mathbf{w}$ are differentiable vector functions of t, then

$$\frac{d}{dt}(\mathbf{u}\cdot\mathbf{v}\times\mathbf{w}) \qquad (15)$$

$$= \frac{d\mathbf{u}}{dt}\cdot\mathbf{v}\times\mathbf{w} + \mathbf{u}\cdot\frac{d\mathbf{v}}{dt}\times\mathbf{w} + \mathbf{u}\cdot\mathbf{v}\times\frac{d\mathbf{w}}{dt}.$$

b) Show that Eq. (15) is equivalent to

$$\frac{d}{dt}\begin{vmatrix} u_1 & u_2 & u_3 \\ v_1 & v_2 & v_3 \\ w_1 & w_2 & w_3 \end{vmatrix} = \begin{vmatrix} \frac{du_1}{dt} & \frac{du_2}{dt} & \frac{du_3}{dt} \\ v_1 & v_2 & v_3 \\ w_1 & w_2 & w_3 \end{vmatrix} + \begin{vmatrix} u_1 & u_2 & u_3 \\ \frac{dv_1}{dt} & \frac{dv_2}{dt} & \frac{dv_3}{dt} \\ w_1 & w_2 & w_3 \end{vmatrix}$$

$$+ \begin{vmatrix} u_1 & u_2 & u_3 \\ v_1 & v_2 & v_3 \\ \frac{dw_1}{dt} & \frac{dw_2}{dt} & \frac{dw_3}{dt} \end{vmatrix} \qquad (16)$$

Equation (16) says that the derivative of a 3 by 3 determinant of differentiable functions is the sum of the three determinants obtained from the original by differentiating one row at a time. The result extends to determinants of any order.

32. *(Continuation of Exercise 31)* Suppose that $\mathbf{r} = f(t)\,\mathbf{i} + g(t)\,\mathbf{j} + h(t)\,\mathbf{k}$ and that f, g, and h have derivatives through order three. Use Eq. (15) or (16) to show that

$$\frac{d}{dt}\left(\mathbf{r}\cdot\frac{d\mathbf{r}}{dt}\times\frac{d^2\mathbf{r}}{dt^2}\right) = \mathbf{r}\cdot\left(\frac{d\mathbf{r}}{dt}\times\frac{d^3\mathbf{r}}{dt^3}\right). \qquad (17)$$

(*Hint:* Differentiate on the left and look for vectors whose products are zero.)

33. *The Constant Function Rule.* Prove that if $\mathbf{f}$ is the vector function with the constant value $\mathbf{C}$, then $d\mathbf{f}/dt = \mathbf{0}$.

34. *The Scalar Multiple Rule.* Prove that if $\mathbf{f}$ is a differentiable function of t and c is any real number, then

$$\frac{d(c\mathbf{f})}{dt} = c\frac{d\mathbf{f}}{dt}.$$

35. *The Sum and Difference Rules.* Prove that if $\mathbf{u}$ and $\mathbf{v}$ are differentiable functions of t, then

$$\frac{d}{dt}(\mathbf{u}+\mathbf{v}) = \frac{d\mathbf{u}}{dt} + \frac{d\mathbf{v}}{dt}$$

and

$$\frac{d}{dt}(\mathbf{u}-\mathbf{v}) = \frac{d\mathbf{u}}{dt} - \frac{d\mathbf{v}}{dt}.$$

36. *The component test for continuity at a point.* Show that the vector function $\mathbf{f}$ defined by the rule $\mathbf{f}(t) = f(t)\mathbf{i} + g(t)\,\mathbf{j} + h(t)\,\mathbf{k}$ is continuous at $t = t_0$ if and only if f, g, and h are continuous at t_0.

37. *Limits of cross products of vector functions.* Suppose that $\mathbf{f}(t) = f_1(t)\,\mathbf{i} + f_2(t)\,\mathbf{j} + f_3(t)\,\mathbf{k}$, $\mathbf{g}(t) = g_1(t)\,\mathbf{i} + g_2(t)\,\mathbf{j} + g_3(t)\,\mathbf{k}$, $\lim_{t\to t_0}\mathbf{f}(t) = \mathbf{A}$, and $\lim_{t\to t_0}\mathbf{g}(t) = \mathbf{B}$. Use the determinant formula for cross products and the Limit Product Rule for scalar functions to show that

$$\lim_{t\to t_0}(\mathbf{f}(t)\times\mathbf{g}(t)) = \left(\lim_{t\to t_0}\mathbf{f}(t)\right)\times\left(\lim_{t\to t_0}\mathbf{g}(t)\right).$$

38. *Differentiable vector functions are continuous.* Show that if $\mathbf{f}(t) = f(t)\,\mathbf{i} + g(t)\,\mathbf{j} + h(t)\,\mathbf{k}$ is differentiable at $t = t_0$, then it is continuous at t_0 as well.

39. Establish the following properties of integrable vector functions.

a) The *Constant Scalar Multiple Rule:*

$$\int_a^b k\mathbf{f}(t)\,dt = k\int_a^b \mathbf{f}(t)\,dt \quad \text{(any scalar } k\text{)}$$

The *Rule for Negatives,*

$$\int_a^b (-\mathbf{f}(t))\,dt = -\int_a^b \mathbf{f}(t)\,dt,$$

is obtained by taking $k = -1$.

b) The *Sum and Difference Rules:*

$$\int_a^b (\mathbf{f}(t)\pm\mathbf{g}(t))\,dt = \int_a^b \mathbf{f}(t)\,dt \pm \int_a^b \mathbf{g}(t)\,dt$$

c) The *Constant Vector Multiple Rules:*

$$\int_a^b \mathbf{C} \cdot \mathbf{f}(t)\, dt = \mathbf{C} \cdot \int_a^b \mathbf{f}(t)\, dt \qquad \text{(any constant vector } \mathbf{C}\text{)}$$

and

$$\int_a^b \mathbf{C} \times \mathbf{f}(t)\, dt = \mathbf{C} \times \int_a^b \mathbf{f}(t)\, dt \qquad \text{(any constant vector } \mathbf{C}\text{)}$$

40. *Products of scalar and vector functions.* Suppose that the scalar function $u(t)$ and the vector function $\mathbf{f}(t)$ are both defined for $a \le t \le b$.

a) Show that $u\mathbf{f}$ is continuous on $[a, b]$ if u and $\mathbf{f}$ are continuous on $[a, b]$.

b) If u and $\mathbf{f}$ are both differentiable on $[a, b]$, show that $u\,\mathbf{f}$ is differentiable on $[a, b]$ and that

$$\frac{d}{dt}(u\mathbf{f}) = u\frac{d\mathbf{f}}{dt} + \mathbf{f}\frac{du}{dt}.$$

41. *Antiderivatives of vector functions*

a) Use Corollary 3 of the Mean Value Theorem for scalar functions to show that if two vector functions $\mathbf{F}_1(t)$ and $\mathbf{F}_2(t)$ have identical derivatives on an interval I, then the functions differ by a constant vector value throughout I.

b) Use the result in (a) to show that if $\mathbf{F}(t)$ is any antiderivative of $\mathbf{f}(t)$ on I, then every other antiderivative of $\mathbf{f}$ on I equals $\mathbf{F}(t) + \mathbf{C}$ for some constant vector $\mathbf{C}$.

42. *The Fundamental Theorem of Calculus.* The Fundamental Theorem of Calculus for scalar functions of a real variable holds for vector functions of a real variable as well. Prove this by using the theorem for scalar functions to show first that if a vector function $\mathbf{f}(t)$ is continuous for $a \le t \le b$, then

$$\frac{d}{dt}\int_a^t \mathbf{f}(\tau)\, d\tau = \mathbf{f}(t)$$

at every point t of $[a, b]$. Then use the conclusion in part (b) of Exercise 41 to show that if $\mathbf{F}$ is any antiderivative of $\mathbf{f}$ on $[a, b]$ then

$$\int_a^b \mathbf{f}(t)\, dt = \mathbf{F}(b) - \mathbf{F}(a).$$

11.2 Modeling Projectile Motion

When we shoot a projectile into the air we usually want to know beforehand how far it will go (will it reach the target?), how high it will rise (will it clear the hill?), and when it will land (when do we get results?). We get this information from equations that calculate the answers we want from the direction and magnitude of the projectile's initial velocity vector, described in terms of the angle and speed at which the projectile is fired. The equations come from combining calculus and Newton's second law of motion in vector form. In this section, we derive these equations and show how to use them to get the information we want about projectile motion.

The Vector and Parametric Equations for Ideal Projectile Motion

To derive equations for projectile motion, we assume that the projectile behaves like a particle moving in a vertical coordinate plane and that the only force acting on the projectile during its flight is the constant force of gravity, which always points straight down. In practice, none of these assumptions really holds. The ground moves beneath the projectile as the earth turns, the air creates a frictional force that varies with the projectile's speed and altitude, and the force of gravity changes as the projectile moves along. All this must be taken into account by applying corrections to the predictions of the ideal equations we are about to derive. The corrections, however, are not the subject of this section.

We assume that our projectile is launched from the origin at time $t = 0$ into the first quadrant with an initial velocity $\mathbf{v}_0$ (Fig. 11.7). If $\mathbf{v}_0$ makes an angle α with the horizontal, then

$$\mathbf{v}_0 = (|\mathbf{v}_0| \cos\alpha)\,\mathbf{i} + (|\mathbf{v}_0| \sin\alpha)\,\mathbf{j}. \qquad (1)$$

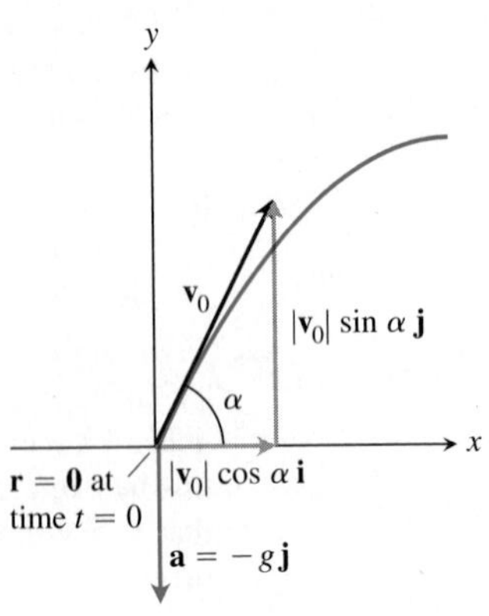

(a) Position, velocity, acceleration, and launch angle at $t = 0$.

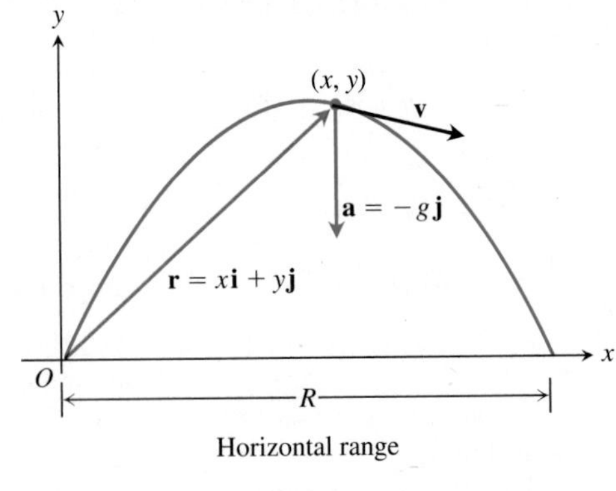

(b) Position, velocity, and acceleration at a later time t.

11.7 The flight of an ideal projectile.

If we use the simpler notation v_0 for the initial speed $|\mathbf{v}_0|$, then

$$\mathbf{v}_0 = (v_0 \cos \alpha)\,\mathbf{i} + (v_0 \sin \alpha)\,\mathbf{j}. \tag{2}$$

The projectile's initial position is

$$\mathbf{r}_0 = 0\,\mathbf{i} + 0\,\mathbf{j} = \mathbf{0}. \tag{3}$$

If the only force acting on the projectile during its flight is the constant downward acceleration of gravity of magnitude g, then

$$\frac{d^2\mathbf{r}}{dt^2} = -g\mathbf{j}. \tag{4}$$

We find the projectile's position as a function of time t by solving the following initial value problem:

Differential equation: $\dfrac{d^2\mathbf{r}}{dt^2} = -g\mathbf{j}$

Initial conditions: $\mathbf{r} = \mathbf{0}$ and $\dfrac{d\mathbf{r}}{dt} = \mathbf{v}_0$ when $t = 0$

The first integration gives

$$\frac{d\mathbf{r}}{dt} = -(gt)\,\mathbf{j} + \mathbf{v}_0. \tag{5}$$

A second integration gives

$$\mathbf{r} = -\frac{1}{2}gt^2\,\mathbf{j} + \mathbf{v}_0 t + \mathbf{r}_0. \tag{6}$$

Substituting the values of $\mathbf{v}_0$ and $\mathbf{r}_0$ from Eqs. (2) and (3) gives

$$\begin{aligned}\mathbf{r} &= -\frac{1}{2}gt^2\,\mathbf{j} + \underbrace{(v_0 \cos \alpha)t\mathbf{i} + (v_0 \sin \alpha)t\mathbf{j}}_{\mathbf{v}_0 t} + \mathbf{0} \\ &= (v_0 \cos \alpha)t\;\mathbf{i} + \left((v_0 \sin \alpha)t - \frac{1}{2}gt^2\right)\mathbf{j}.\end{aligned} \tag{7}$$

Equations of Motion for an Ideal Projectile Launched from the Origin at $t = 0$

Vector Form: $$\mathbf{r} = (v_0 \cos \alpha)t\ \mathbf{i} + \left((v_0 \sin \alpha)t - \frac{1}{2}gt^2\right)\mathbf{j} \tag{8}$$

Parametric Form: $$x = (v_0 \cos \alpha)t, \quad y = (v_0 \sin \alpha)t - \frac{1}{2}gt^2 \tag{9}$$

where α is the **launch angle (firing angle, angle of elevation)** and v_0 is the **initial speed.**

If we measure time in seconds and distance in meters, g is 9.8 m/sec^2 and the equations in (9) give x and y in meters. If we measure time in seconds and distance in feet, then g is 32 ft/sec^2 and the equations in (9) give x and y in feet.

Example 1 A projectile is fired from the origin over horizontal ground at an initial speed of 500 m/sec at a launch angle of 60°. Where will the projectile be 10 sec later?

Solution We use Eqs. (9) with $v_0 = 500$, $\alpha = 60°$, $g = 9.8$, and $t = 10$ to find the projectile's coordinates to the nearest meter 10 sec after firing:

$$x = (v_0 \cos \alpha)t = 500 \cdot \frac{1}{2} \cdot 10 = 2500 \text{ m}$$

$$\begin{aligned} y &= (v_0 \sin \alpha)t - \frac{1}{2}gt^2 \\ &= 500 \cdot \frac{\sqrt{3}}{2} \cdot 10 - \frac{1}{2} \cdot 9.8 \cdot (10)^2 \\ &= 2500\sqrt{3} - 490 \\ &= 3840 \text{ m.} \qquad \text{(Calculator, rounded)} \end{aligned}$$

Ten seconds after firing, the projectile is 3840 m in the air and 2500 m downrange.

Height, Flight Time, and Range

Equations (9) enable us to answer most questions about an ideal projectile fired from the origin.

The projectile reaches its highest point when its vertical velocity component is zero, that is, when

$$\frac{dy}{dt} = v_0 \sin \alpha - gt = 0, \qquad \text{or} \qquad t = \frac{v_0 \sin \alpha}{g}.$$

For this value of t, the value of y is

$$y_{\max} = (v_0 \sin \alpha)\left(\frac{v_0 \sin \alpha}{g}\right) - \frac{1}{2}g\left(\frac{v_0 \sin \alpha}{g}\right)^2 = \frac{(v_0 \sin \alpha)^2}{2g}. \tag{10}$$

To find when the projectile lands, when fired over horizontal ground, we set y equal to zero in Eqs. (9) and solve for t:

$$(v_0 \sin \alpha)t - \frac{1}{2}gt^2 = 0$$

$$t\left(v_0 \sin \alpha - \frac{1}{2}gt\right) = 0 \tag{11}$$

$$t = 0, \quad t = \frac{2v_0 \sin \alpha}{g}.$$

Since 0 is the time the projectile is fired, $(2v_0 \sin \alpha)/g$ must be the time when the projectile strikes the ground.

To find the projectile's range R, the distance from the origin to the point of impact on horizontal ground, we find the value of x when $t = (2v_0 \sin \alpha)/g$:

$$x = (v_0 \cos \alpha)t$$

$$R = (v_0 \cos \alpha)\left(\frac{2v_0 \sin \alpha}{g}\right)$$

$$= \frac{v_0^2}{g}(2 \sin \alpha \cos \alpha) = \frac{v_0^2}{g}\sin 2\alpha. \tag{12}$$

The range is largest when $\sin 2\alpha = 1$ or $\alpha = 45°$.

Height, Flight Time, and Range

For an ideal projectile fired from the origin over horizontal ground:

Maximum height: $$y_{max} = \frac{(v_0 \sin \alpha)^2}{2g} \tag{13}$$

Flight time (time to impact): $$t = \frac{2v_0 \sin \alpha}{g} \tag{14}$$

Range (distance to point of impact): $$R = \frac{v_0^2}{g}\sin 2\alpha \tag{15}$$

Example 2 Find the maximum height, flight time, and range of a projectile fired from the origin over horizontal ground at an initial speed of 500 m/sec and a launch angle of 60° (same projectile as in Example 1).

Solution (With a calculator, to the nearest meter and second)

Maximum height (Eq. 13): $$y_{max} = \frac{(v_0 \sin \alpha)^2}{2g} = \frac{(500 \sin 60°)^2}{2(9.8)} = 9566 \text{ m}$$

Flight time (Eq. 14): $$t = \frac{2v_0 \sin \alpha}{g} = \frac{2(500) \sin 60°}{9.8} = 88 \text{ sec}$$

Range (Eq. 15): $$R = \frac{v_0^2}{g}\sin 2\alpha = \frac{(500)^2 \sin 120°}{9.8} = 22{,}092 \text{ m}$$

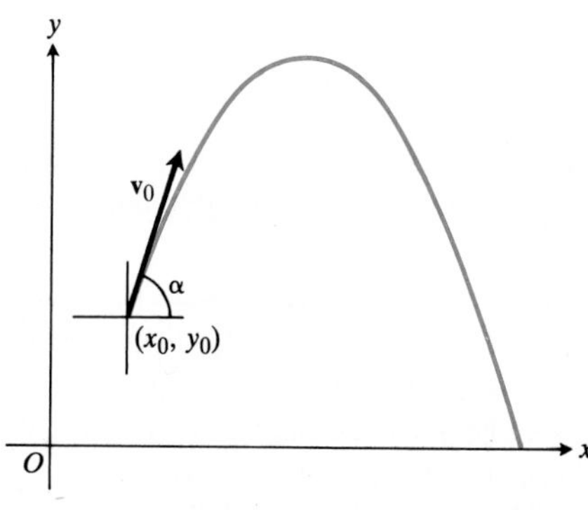

11.8 The path of a projectile fired from (x_0, y_0) with an initial velocity $\mathbf{v}_0$ at an angle of α degrees with the horizontal.

Firing from (x_0, y_0)

If we fire our ideal projectile from the point (x_0, y_0) instead of the origin (Fig. 11.8), the equations that replace Eqs. (9) are

$$x = x_0 + (v_0 \cos \alpha)t, \qquad y = y_0 + (v_0 \sin \alpha)t - \frac{1}{2} gt^2, \tag{16}$$

as you will be invited to show in Exercise 20.

Example 3 An athlete throws a 16-lb shot at an angle of 45° to the horizontal from 8 ft above the ground at an initial speed of 44 ft/sec (Fig. 11.9). How far from the inner edge of the stopboard does the shot land?

Solution We use a coordinate system in which the x-axis lies along the ground and the shot's coordinates at the time of launch ($t = 0$) are

$$x_0 = 0, \qquad y_0 = 8.$$

With these values for x_0 and y_0, and with $\alpha = 45°$ and $g = 32$ ft/sec², Eqs. (16) tell us that

$$x = 0 + (44 \cos 45°)t = 22\sqrt{2}t,$$

$$y = 8 + (44 \sin 45°)t - 16t^2 = 8 + 22\sqrt{2}t - 16t^2.$$

We find *when* the shot lands by setting y equal to zero and using the quadratic formula to solve for t, obtaining the positive solution

$$t = \frac{22\sqrt{2} + \sqrt{968 + 512}}{32} = 2.174 \text{ sec.} \qquad \text{(Calculator, rounded)}$$

We then find the value of x at this time from the equation $x = 22\sqrt{2}t$:

$$x = 22\sqrt{2}(2.174) = 67.64 \text{ ft.} \qquad \text{(Calculator, rounded)}$$

The shot lands about 67 ft 8 in. from the stopboard. Air resistance, variations in gravity, and the rotation of the earth have little effect on a dense object, like a shot, that moves relatively slowly and for only a few seconds. In such cases, we expect good results from the ideal model.

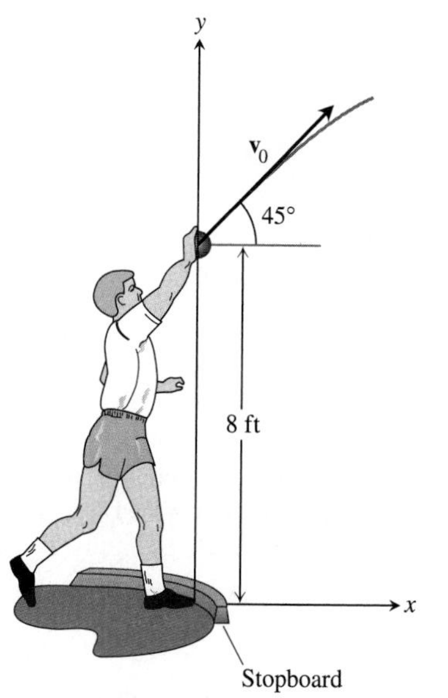

11.9 The shot put in Example 3. How far will the shot go?

Ideal Trajectories Are Parabolic

It is often claimed that water from a hose traces a parabola in the air, but anyone who looks closely enough will see this is not so. The air slows the water down and its forward progress is too slow at the end to match the rate at which it falls.

What is really being claimed is that ideal projectiles move along parabolas, and this we can see from Eqs. (9). If we substitute $t = x/(v_0 \cos \alpha)$ from the first equation into the second, we obtain the Cartesian coordinate equation

$$y = -\left(\frac{g}{2v_0^2 \cos^2\alpha}\right) x^2 + (\tan \alpha)\, x. \tag{17}$$

This equation is quadratic in x and linear in y, so its graph is a parabola.

EXERCISES 11.2

CALCULATOR The projectiles in the following exercises are to be treated as ideal projectiles whose behavior is faithfully portrayed by the equations derived in the text. Most of the arithmetic, however, is realistic and is best done with a calculator. All launch angles are assumed to be measured from the horizontal. All projectiles are assumed to be fired from the origin over horizontal ground, unless stated otherwise.

1. A projectile is fired at a speed of 840 m/sec at an angle of 60°. How long will it take to get 21 km downrange?

2. Find the muzzle speed of a gun whose maximum range is 24.5 km.

3. A projectile is fired over level ground with an initial speed of 500 m/sec at an angle of elevation of 45°.
   a) When and how far away will the projectile strike?
   b) How high overhead will the projectile be when it is 5 km downrange?
   c) What is the highest the projectile will go?

4. A baseball is thrown from the stands 32 ft above the field at an angle of 30°. When and how far away will the ball strike the ground if its initial speed is 32 ft/sec?

5. Show that a projectile fired at an angle of α degrees, $0<\alpha<90$, has the same range as a projectile fired at the same speed at an angle of $(90-\alpha)$ degrees. (In models that take air resistance into account, this symmetry is lost.)

6. What two angles of elevation will enable a projectile to reach a target 16 km downrange on the same level as the gun if the projectile's initial speed is 400 m/sec?

7. A spring gun at ground level fires a golf ball at an angle of 45°. The ball lands 10 m away. What was the ball's initial speed? For the same initial speed, find the two firing angles that make the range 6 m.

8. An electron in a TV tube is beamed horizontally at a speed of 5×10^6 m/sec toward the face of the tube 40 cm away. About how far will the electron drop before it hits?

9. Laboratory tests designed to find how far golf balls of different hardness go when hit with a driver showed that a 100-compression, Surlyn-covered, two-piece ball hit with a club-head speed of 100 mph at a launch angle of 9° carried 248.8 yd.
   a) What was the launch speed of the ball? (It was more than 100 mph. At the same time the club head was moving forward, the compressed ball was kicking away from the club face, adding to the ball's forward speed. The data are from "Does Compression Really Matter?" by Lew Fishman, *Golf Digest*, August 1986, pp. 35–37.)
   b) How much work was done on the ball getting it into the air? (Golf balls weigh 1.6 oz.)

10. Show that doubling a projectile's initial speed at a given launch angle multiplies its range by 4. By about what percentage should you increase the initial speed to double the height and range?

11. Show that a projectile attains three-quarters of its maximum height in half the time it takes to reach the maximum height.

12. A human cannonball is to be fired with an initial speed of $v_0 = 80\sqrt{10/3}$ ft/sec. The circus performer (of the right caliber, naturally) hopes to land on a special cushion located 200 ft downrange. The circus is being held in a large room with a flat ceiling 75 ft high. Can the performer be fired to the cushion without striking the ceiling? If so, what should the cannon's angle of elevation be?

13. A golf ball leaves the ground at a 30° angle at a speed of 90 ft/sec. Will it clear the top of a 30-ft tree 135 ft away?

14. A golf ball is hit from the tee to an elevated green with an initial speed of 116 ft/sec at an angle of elevation of 45°, as shown in Fig. 11.10. Will the ball reach the pin?

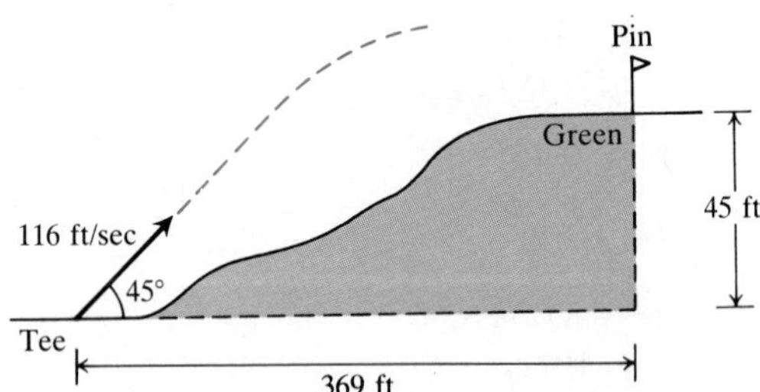

11.10 The tee and green in Exercise 14.

15. All other things being equal, the shot in Example 3 would have gone slightly farther if it had been launched at a 40° angle. How much farther? Answer in inches.

16. In Moscow, in 1987, Natalya Lisouskaya of the USSR set a women's world record by putting an 8-lb 13-oz shot 73 ft 10 in.
   a) Assuming that she launched the shot at a 40° angle to the horizontal 7 ft above the ground, what was the shot's initial speed?
   b) How much work was done on the shot to launch it at that speed? (Do not include the work it took to lift the shot 7 ft off the ground.)

17. A baseball hit by a Boston Red Sox player at a 20° angle from 3 ft above the ground just cleared the left end of the "Green Monster," the left-field wall in Fenway Park (Fig. 11.11). The wall there is 37 ft high and 315 ft from home plate. About how fast was the ball going? How long did it take the ball to reach the wall?

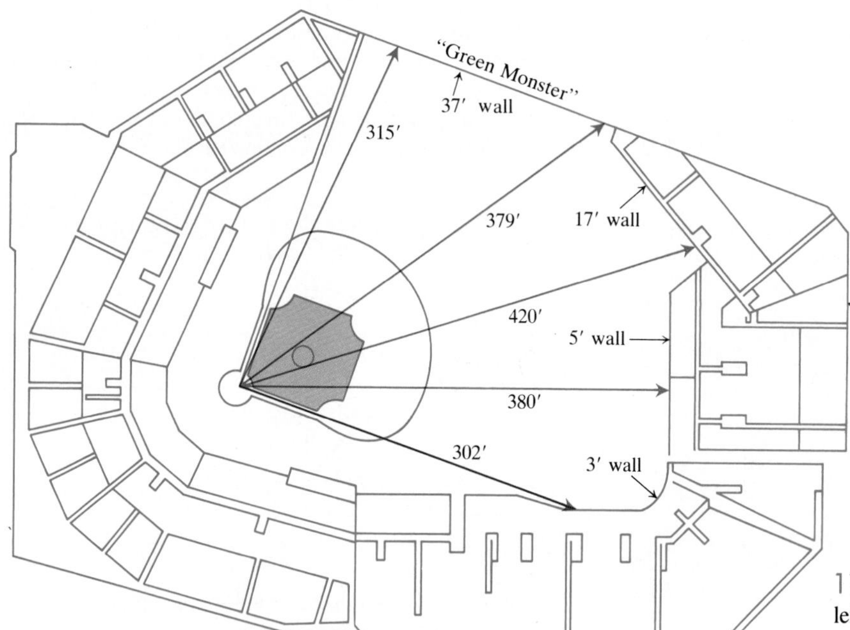

11.11 The "Green Monster," the left-field wall at Fenway Park in Boston (Exercise 17).

18. The multiflash photograph in Fig. 11.12 shows a model train engine moving at a constant speed on a straight track. As the engine moved along, a marble was fired into the air by a spring in the engine's smokestack. The marble, which continued to move with the same forward speed as the engine, rejoined the engine 1 sec after it was fired. Measure the angle the marble's path made with the horizontal and use the information to find how high the marble went and how fast the engine was moving.

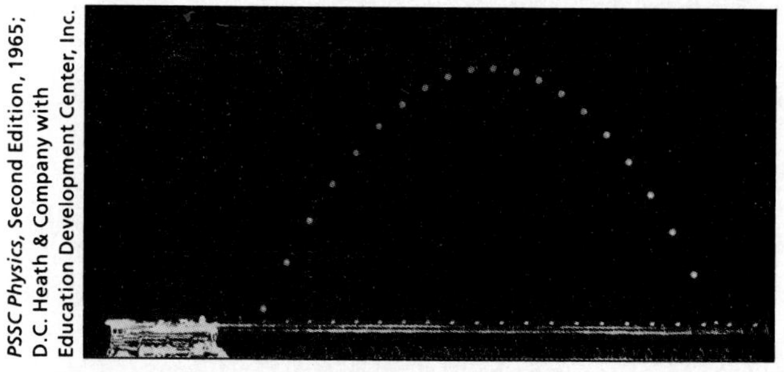

11.12 The train in Exercise 18.

19. Figure 11.13 shows an experiment with two marbles. Marble A was launched toward marble B with launch angle α and initial speed v_0. At the same instant, marble B was released to fall from rest $R \tan \alpha$ units directly above a spot R units downrange from A. The marbles were found to collide regardless of the value of v_0. Was this mere coincidence, or must this happen? How do you know?

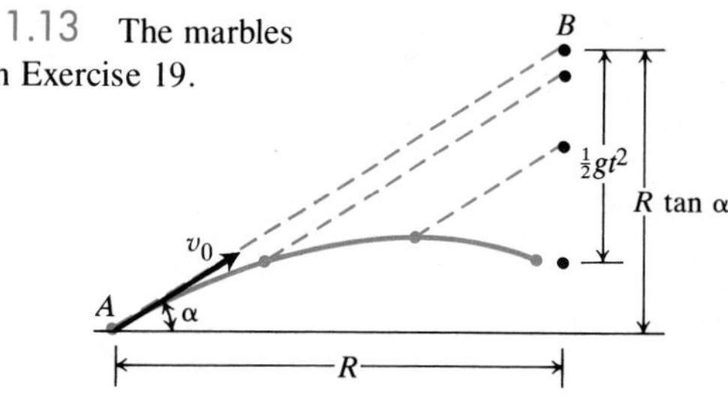

11.13 The marbles in Exercise 19.

20. Derive the equations

$$x = x_0 + (v_0 \cos \alpha)t,$$

$$y = y_0 + (v_0 \sin \alpha)t - \frac{1}{2} gt^2$$

(Eqs. 16 in the text) by solving the following initial value problem for a vector $\mathbf{r}$ in the plane.

Differential equation: $\dfrac{d^2\mathbf{r}}{dt^2} = -g\mathbf{j}$

Initial conditions: $\mathbf{r} = x_0\,\mathbf{i} + y_0\,\mathbf{j}$ and

$$\frac{d\mathbf{r}}{dt} = (v_0 \cos \alpha)\,\mathbf{i} + (v_0 \sin \alpha)\,\mathbf{j}$$

when $t = 0$

21. An ideal projectile, launched from the origin into the first octant at time $t = 0$ with initial velocity $\mathbf{v}_0$ (Fig. 11.14), experiences a constant downward acceleration $\mathbf{a} = -g\mathbf{k}$ from gravity. Find the projectile's velocity and position as functions of t.

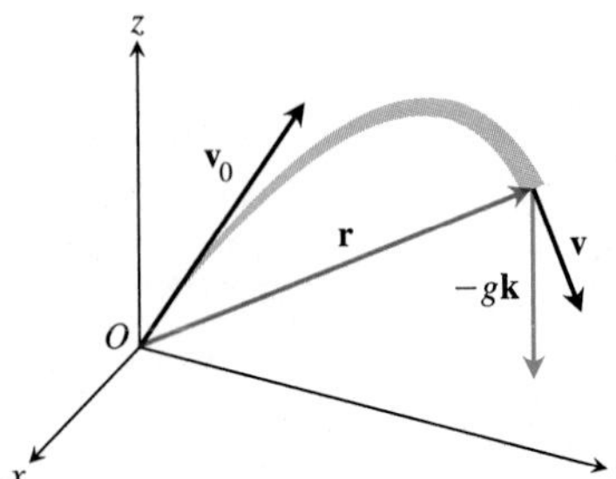

11.14 The projectile in Exercise 21.

22. *Air resistance proportional to velocity.* If a projectile of mass m launched with initial velocity $\mathbf{v}_0$ encounters an air resistance proportional to its velocity, the total force $m(d^2\mathbf{r}/dt^2)$ on the projectile satisfies the equation

$$m\frac{d^2\mathbf{r}}{dt^2} = -mg\,\mathbf{j} - k\frac{d\mathbf{r}}{dt},$$

where k is the proportionality constant. Show that one integration of this equation gives

$$\frac{d\mathbf{r}}{dt} + \frac{k}{m}\mathbf{r} = \mathbf{v}_0 - gt\,\mathbf{j}.$$

Solve this equation. To do so, multiply both sides of the equation by $e^{(k/m)t}$. The left-hand side will then be the derivative of a product, with the result that both sides of the equation can now be integrated.

(The function $e^{(k/m)t}$ is called an *integrating factor* for the differential equation because multiplying the equation by it makes the equation integrable. We shall say more about integrating factors in Chapter 15.)

23. An ideal projectile is launched straight down an inclined plane, as shown in profile in Fig. 11.15.

a) Show that the greatest downhill range is achieved when the initial velocity vector bisects angle AOR.

b) If the projectile were fired uphill instead of down, what launch angle would maximize its range?

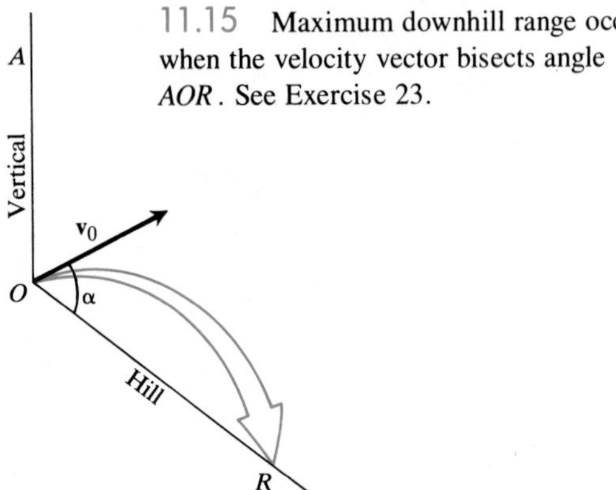

11.15 Maximum downhill range occurs when the velocity vector bisects angle AOR. See Exercise 23.

24. GRAPHER If you have access to a parametric equation grapher and have not yet done Exercise 39 in Section 9.3, do it now. It is about ideal projectile motion.

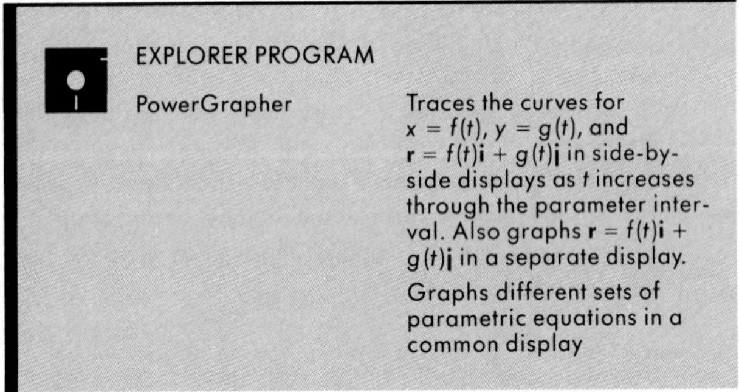

EXPLORER PROGRAM

PowerGrapher	Traces the curves for $x = f(t)$, $y = g(t)$, and $\mathbf{r} = f(t)\mathbf{i} + g(t)\mathbf{j}$ in side-by-side displays as t increases through the parameter interval. Also graphs $\mathbf{r} = f(t)\mathbf{i} + g(t)\mathbf{j}$ in a separate display. Graphs different sets of parametric equations in a common display

11.3 Directed Distance and the Unit Tangent Vector T

As you can imagine, differentiable curves, especially those with continuous first and second derivatives, have been subjects of intense study, for their mathematical interest as well as for their applications to motion in space. In this section and the next, we describe some of the features that account for the importance of these curves.

Distance Along a Curve

One of the special features of space curves whose coordinate functions have continuous first derivatives is that, like plane curves, they are smooth enough to have a measurable length. This enables us to locate points along these curves by giving their directed distance s along the curve from some **base point,** the way we locate points on coordinate axes by giving their directed distance from the origin

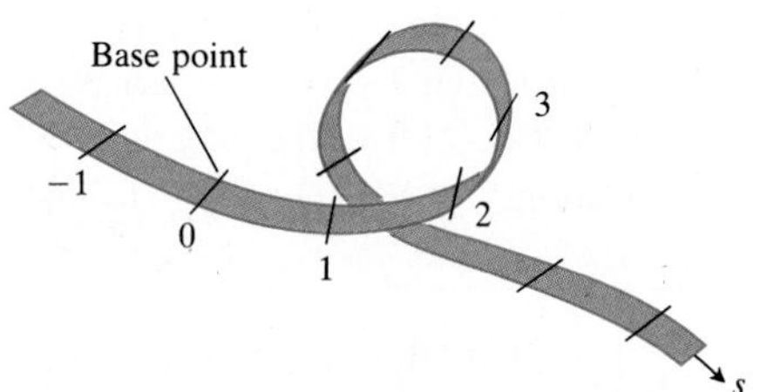

11.16 Smooth curves can be scaled like coordinate axes or like tape measures that include negative numbers as well as positive numbers. The coordinate given to each point is its directed "highway" distance from the base point.

(Fig. 11.16). Although time is the natural parameter for describing a moving body's velocity and acceleration, the **directed distance coordinate** s is the natural parameter for studying a curve's geometry. The relationships between these parameters play an important role in calculations of space flight.

To calculate the distances along parametrized curves in space, we add a z-term to the length formula we use for curves in the plane.

DEFINITION

If $x = f(t)$, $y = g(t)$, and $z = h(t)$ have continuous first derivatives with respect to t for $a \le t \le b$, and the position vector $\mathbf{r} = f(t)\,\mathbf{i} + g(t)\,\mathbf{j} + h(t)\,\mathbf{k}$ defined by these equations traces its curve exactly once as t moves from $t = a$ to $t = b$, then the **length** of the curve is given by the formula

$$\text{Length} = \int_a^b \sqrt{\left(\frac{dx}{dt}\right)^2 + \left(\frac{dy}{dt}\right)^2 + \left(\frac{dz}{dt}\right)^2}\, dt. \tag{1}$$

Just as for plane curves, we can calculate the length of a curve in space from any convenient parametrization that meets the stated conditions. Again, we shall omit the proof.

Notice that the square root in Eq. (1) is $|\mathbf{v}|$, the length of the velocity vector $d\mathbf{r}/dt$. This lets us write the formula for length a shorter way.

Length Formula (Short Form)

$$\text{Length} = \int_a^b |\mathbf{v}|\, dt \tag{2}$$

As in linear motion, distance traveled is the integral of speed.

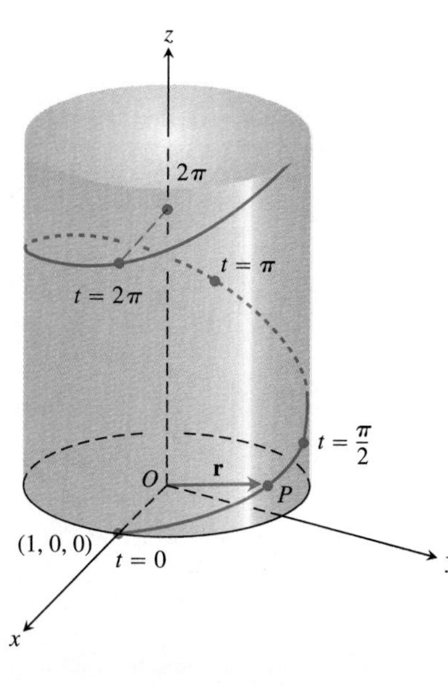

11.17 The helix $\mathbf{r} = (\cos t)\,\mathbf{i} + (\sin t)\,\mathbf{j} + t\,\mathbf{k}$ makes a complete turn during the time interval from $t = 0$ to $t = 2\pi$. We calculate the distance traveled by P by integrating $|\mathbf{v}|$ from $t = 0$ to $t = 2\pi$ (Example 1).

Example 1 Find the length of one turn of the helix

$$\mathbf{r} = (\cos t)\,\mathbf{i} + (\sin t)\,\mathbf{j} + t\mathbf{k}.$$

Solution The helix makes one full turn as t runs from 0 to 2π (Fig. 11.17). We find the length of this portion of the curve from Eq. (2):

$$\text{Length} = \int_a^b |\mathbf{v}|\, dt = \int_0^{2\pi} \sqrt{(-\sin t)^2 + (\cos t)^2 + (1)^2}\, dt$$

$$= \int_0^{2\pi} \sqrt{2}\, dt = 2\pi\sqrt{2}.$$

This is $\sqrt{2}$ times the length of the circle in the xy-plane over which the helix stands.

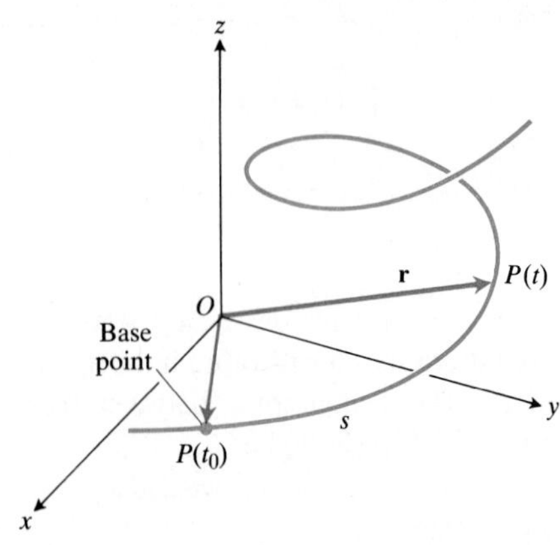

11.18 The directed distance along the curve from $P(t_0)$ to any point $P(t)$ is
$$s = \int_{t_0}^{t} |\mathbf{v}(\tau)|\, d\tau.$$

If $x = f(t)$, $y = g(t)$, and $z = h(t)$ have continuous first derivatives with respect to t and we choose a base point $P(t_0)$ on the curve $\mathbf{r} = f(t)\,\mathbf{i} + g(t)\,\mathbf{j} + h(t)\,\mathbf{k}$ (Fig. 11.18), the integral of $|\mathbf{v}|$ from t_0 to t gives the directed distance along the curve from $P(t_0)$ to $P(t)$. The distance is a function of t, and we denote it by s.

Directed Distance Along a Curve from t_0 to t

$$s = \int_{t_0}^{t} \sqrt{x'(\tau)^2 + y'(\tau)^2 + z'(\tau)^2}\, d\tau = \int_{t_0}^{t} |\mathbf{v}(\tau)|\, d\tau \qquad (3)$$

The value of s is positive if t is greater than t_0 and negative if t is less than t_0.

Example 2 If $t_0 = 0$, the directed distance along the helix
$$\mathbf{r} = (\cos t)\,\mathbf{i} + (\sin t)\,\mathbf{j} + t\mathbf{k}$$
from t_0 to t is
$$\begin{aligned} s &= \int_{t_0}^{t} |\mathbf{v}(\tau)|\, d\tau && \text{(Eq. (3))} \\ &= \int_{0}^{t} \sqrt{2}\, d\tau && \text{(Value from Example 1)} \\ &= \sqrt{2}t. \end{aligned}$$

Thus, $s = 2\pi\sqrt{2}$ when $t = 2\pi$, $s = -2\pi\sqrt{2}$ when $t = -2\pi$, and so on (Fig. 11.19).

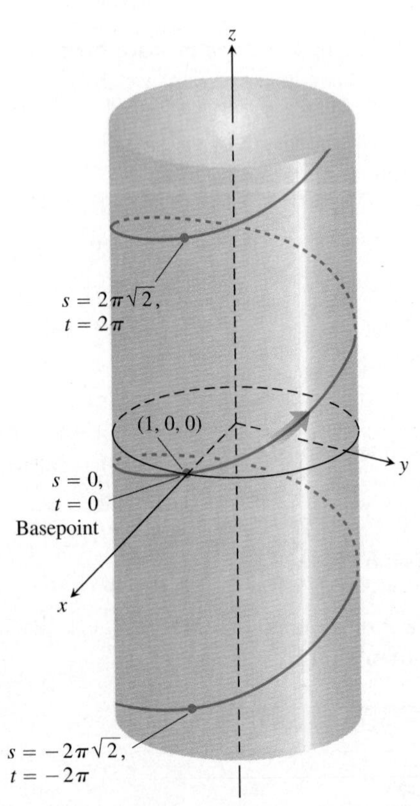

11.19 Directed distances along the helix $\mathbf{r} = (\cos t)\,\mathbf{i} + (\sin t)\,\mathbf{j} + t\,\mathbf{k}$ from the base point (1, 0, 0) (Example 2).

Example 3 *Distance Along a Line.* Show that if $\mathbf{u} = u_1\,\mathbf{i} + u_2\,\mathbf{j} + u_3\,\mathbf{k}$ is a unit vector, then the directed distance along the line
$$\mathbf{r} = (x_0 + tu_1)\,\mathbf{i} + (y_0 + tu_2)\,\mathbf{j} + (z_0 + tu_3)\,\mathbf{k}$$
from the point $P_0(x_0, y_0, z_0)$ where $t = 0$ is t itself.

Solution

$$\mathbf{v} = \frac{d}{dt}(x_0 + tu_1)\,\mathbf{i} + \frac{d}{dt}(y_0 + tu_2)\,\mathbf{j} + \frac{d}{dt}(z_0 + tu_3)\,\mathbf{k} = u_1\,\mathbf{i} + u_2\,\mathbf{j} + u_3\,\mathbf{k} = \mathbf{u},$$

so

$$s(t) = \int_0^t |\mathbf{v}|\, d\tau = \int_0^t |\mathbf{u}|\, d\tau = \int_0^t 1\, d\tau = t.$$

If the derivatives beneath the radical in Eq. (3) are continuous, the Fundamental Theorem of Calculus tells us that s is a differentiable function of t whose derivative is

$$\frac{ds}{dt} = |\mathbf{v}(t)|. \qquad (4)$$

As we expect, the speed with which a particle moves along a smooth path is the magnitude of $\mathbf{v}$.

Notice that while the base point $P(t_0)$ plays a role in defining s in Eq. (3) it plays no role in Eq. (4). The rate at which a moving particle covers the distance along its path has nothing to do with how far away the base point is.

Notice also that as long as $|\mathbf{v}|$ is different from zero, *as we shall assume it to be in all examples from now on, ds/dt* is positive and s is an increasing function of t.

The Unit Tangent Vector T

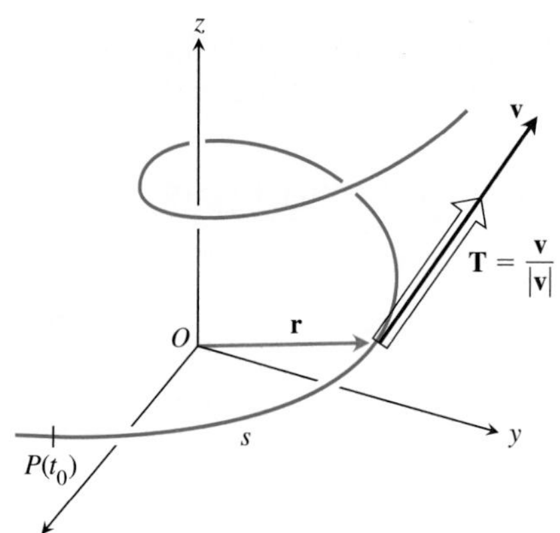

11.20 We find the unit tangent vector **T** by dividing **v** by $|\mathbf{v}|$.

Since ds/dt is positive for the curves we are considering from now on, s is one-to-one and has an inverse that gives t as a differentiable function of s (Section 6.1). The derivative of the inverse is

$$\frac{dt}{ds} = \frac{1}{ds/dt} = \frac{1}{|\mathbf{v}|}. \tag{5}$$

This makes **r** a differentiable function of s whose derivative can be calculated with the Chain Rule for vector-valued functions to be

$$\frac{d\mathbf{r}}{ds} = \frac{d\mathbf{r}}{dt}\frac{dt}{ds} = \mathbf{v}\cdot\frac{1}{|\mathbf{v}|} = \frac{\mathbf{v}}{|\mathbf{v}|}. \tag{6}$$

This tells us that $d\mathbf{r}/ds$ has the constant length

$$\left|\frac{d\mathbf{r}}{ds}\right| = \frac{1}{|\mathbf{v}|}|\mathbf{v}| = 1. \tag{7}$$

Together, Eqs. (6) and (7) say that $d\mathbf{r}/ds$ is a unit vector that points in the direction of **v**. We call $d\mathbf{r}/ds$ the unit tangent vector of the curve traced by **r** and denote it by **T** (Fig. 11.20).

DEFINITION

The **unit tangent vector** of a differentiable curve $\mathbf{r} = \mathbf{f}(t)$ is

$$\mathbf{T} = \frac{d\mathbf{r}}{ds} = \frac{d\mathbf{r}/dt}{ds/dt} = \frac{\mathbf{v}}{|\mathbf{v}|}. \tag{8}$$

The unit tangent vector **T** is a differentiable function of t whenever **v** is a differentiable function of t. As we shall see in the next section, **T** is one of three unit vectors in a traveling reference frame that is used to describe the motion of space vehicles and other bodies moving in three dimensions.

Example 4 Find the unit tangent vector of the helix

$$\mathbf{r} = (\cos t)\,\mathbf{i} + (\sin t)\,\mathbf{j} + t\mathbf{k}.$$

Solution

$$\mathbf{v} = (-\sin t)\,\mathbf{i} + (\cos t)\,\mathbf{j} + \mathbf{k}$$

$$|\mathbf{v}| = \sqrt{(-\sin t)^2 + (\cos t)^2 + (1)^2} = \sqrt{2}$$

$$\mathbf{T} = \frac{\mathbf{v}}{|\mathbf{v}|} = -\frac{\sin t}{\sqrt{2}}\mathbf{i} + \frac{\cos t}{\sqrt{2}}\mathbf{j} + \frac{1}{\sqrt{2}}\mathbf{k}$$

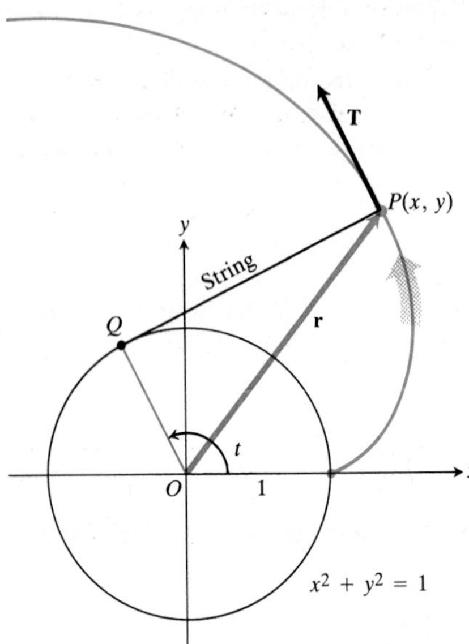

11.21 The *involute* of a circle is the path traced by the endpoint P of a string unwinding from a circle, here the unit circle in the xy-plane. The position vector of P can be shown to be

$\mathbf{r} = (\cos t + t\sin t)\,\mathbf{i} + (\sin t - t\cos t)\,\mathbf{j}$,

where t is the angle from the positive x-axis to Q. Example 5 derives a formula for the curve's unit tangent vector $\mathbf{T}$.

Example 5 *The Involute of a Circle (Fig. 11.21).* Find the unit tangent vector of the curve

$$\mathbf{r} = (\cos t + t\sin t)\,\mathbf{i} + (\sin t - t\cos t)\,\mathbf{j}, \qquad t > 0.$$

Solution

$$\mathbf{v} = \frac{d\mathbf{r}}{dt} = (-\sin t + \sin t + t\cos t)\,\mathbf{i} + (\cos t - \cos t + t\sin t)\,\mathbf{j}$$

$$= (t\cos t)\,\mathbf{i} + (t\sin t)\,\mathbf{j}$$

$$|\mathbf{v}| = \sqrt{t^2\cos^2 t + t^2\sin^2 t} = \sqrt{t^2} = |t| = t \qquad (|t| = t \text{ because } t > 0)$$

$$\mathbf{T} = \frac{\mathbf{v}}{|\mathbf{v}|} = \frac{\mathbf{v}}{t} = (\cos t)\,\mathbf{i} + (\sin t)\,\mathbf{j}$$

Example 6 For the counterclockwise motion

$$\mathbf{r} = (\cos t)\,\mathbf{i} + (\sin t)\,\mathbf{j}$$

around the unit circle,

$$\mathbf{v} = (-\sin t)\,\mathbf{i} + (\cos t)\,\mathbf{j}$$

is already a unit vector, so $\mathbf{T} = \mathbf{v}$ (Fig. 11.22).

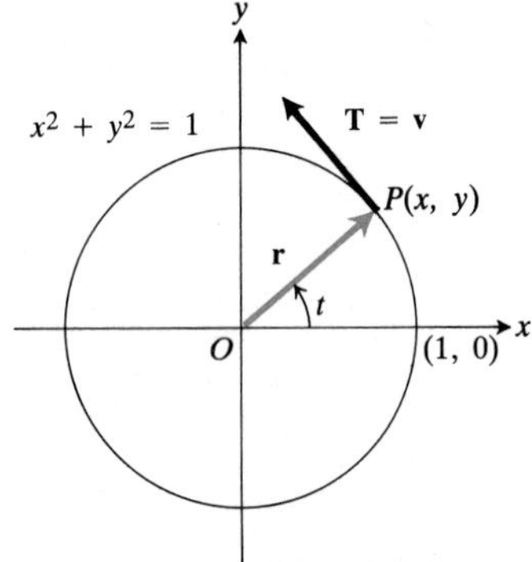

11.22 For the counterclockwise motion $\mathbf{r} = (\cos t)\,\mathbf{i} + (\sin t)\,\mathbf{j}$ about the unit circle, the unit tangent vector is $\mathbf{v}$ itself (Example 6).

EXERCISES 11.3

In Exercises 1–8, find the curve's unit tangent vector. Also, find the length of the indicated portion of the curve.

1. $\mathbf{r} = (2\cos t)\,\mathbf{i} + (2\sin t)\,\mathbf{j} + \sqrt{5}t\,\mathbf{k}$, from $t = 0$ to $t = \pi$
2. $\mathbf{r} = (6\sin 2t)\,\mathbf{i} + (6\cos 2t)\,\mathbf{j} + 5t\,\mathbf{k}$, from $t = 0$ to $t = \pi$
3. $\mathbf{r} = t\mathbf{i} + (2/3)t^{3/2}\,\mathbf{k}$, from $t = 0$ to $t = 8$
4. $\mathbf{r} = (\cos^3 t)\,\mathbf{j} + (\sin^3 t)\,\mathbf{k}$, from $t = 0$ to $t = \pi/2$
5. $\mathbf{r} = (2 + t)\,\mathbf{i} - (t + 1)\,\mathbf{j} + t\mathbf{k}$, from $t = 0$ to $t = 3$
6. $\mathbf{r} = 6t^3\,\mathbf{i} - 2t^3\,\mathbf{j} - 3t^3\,\mathbf{k}$, from $t = -1$ to $t = 1$
7. $\mathbf{r} = (t\cos t)\,\mathbf{i} + (t\sin t)\,\mathbf{j} + (2\sqrt{2}/3)t^{3/2}\,\mathbf{k}$, from $t = 0$ to $t = \pi$
8. $\mathbf{r} = (t\sin t + \cos t)\,\mathbf{i} + (t\cos t - \sin t)\,\mathbf{j}$, from $t = 0$ to $t = \sqrt{2}$

In Exercises 9–12, find the directed distance along the curve from the point where $t = 0$ by evaluating the integral

$$s = \int_0^t |\mathbf{v}(\tau)|\,d\tau$$

from Eq. (3). Then find the length of the indicated portion of the curve.

9. $\mathbf{r} = (4\cos t)\,\mathbf{i} + (4\sin t)\,\mathbf{j} + 3t\mathbf{k}$, from $t = 0$ to $t = \pi/2$
10. $\mathbf{r} = (\cos t + t\sin t)\,\mathbf{i} + (\sin t - t\cos t)\,\mathbf{j}$, from $t = \pi/2$ to $t = \pi$
11. $\mathbf{r} = (e^t\cos t)\,\mathbf{i} + (e^t\sin t)\,\mathbf{j} + e^t\,\mathbf{k}$, from $t = 0$ to $t = -\ln 4$
12. $\mathbf{r} = (1 + 2t)\,\mathbf{i} + (1 + 3t)\,\mathbf{j} + (6 - 6t)\,\mathbf{k}$, from $t = 0$ to $t = -1$
13. Find the length of the curve
$$\mathbf{r} = (\sqrt{2}t)\,\mathbf{i} + (\sqrt{2}t)\,\mathbf{j} + (1 - t^2)\,\mathbf{k}$$
from (0, 0, 1) to $(\sqrt{2}, \sqrt{2}, 0)$.
14. The length $2\pi\sqrt{2}$ of the turn of the helix in Example 1 is

also the length of the diagonal of a square 2π units on a side. Show how to obtain this square by cutting away and flattening a portion of the cylinder around which the helix winds.

15. a) Show that the curve

$$\mathbf{r} = (\cos t)\,\mathbf{i} + (\sin t)\,\mathbf{j} + (1 - \cos t)\,\mathbf{k},\ 0 \le t \le 2\pi$$

is an ellipse by showing that it is the curve of intersection of a cylinder and a plane. Find equations for the cylinder and plane.

b) Sketch the ellipse on the cylinder. Add to your sketch the unit tangent vectors at $t = 0,\ \pi/2,\ \pi$, and $3\pi/2$.

c) Show that the acceleration vector always lies parallel to the plane (orthogonal to a vector normal to the plane). This means that if you draw the acceleration as a vector attached to the ellipse, it will lie in the plane of the ellipse. Add the acceleration vectors for $t = 0,\ \pi/2,\ \pi$, and $3\pi/2$ to your sketch.

d) Write an integral for the length of the ellipse. Do not try to evaluate the integral—it is nonelementary.

e) NUMERICAL INTEGRATOR Estimate the length of the ellipse to two decimal places.

16. *Length is independent of parametrization.* To illustrate the fact that the length of a smooth space curve does not depend on the parametrization we use to compute it as long as the parametrization meets the conditions given with Eq. (1), calculate the length of one turn of the helix in Example 1 with the following parametrizations.

a) $\mathbf{r} = (\cos 4t)\,\mathbf{i} + (\sin 4t)\,\mathbf{j} + 4t\mathbf{k},\quad 0 \le t \le \pi/2$

b) $\mathbf{r} = (\cos(t/2))\,\mathbf{i} + (\sin(t/2))\,\mathbf{j} + (t/2)\,\mathbf{k},\quad 0 \le t \le 4\pi$

c) $\mathbf{r} = (\cos t)\,\mathbf{i} - (\sin t)\,\mathbf{j} - t\mathbf{k},\quad -2\pi \le t \le 0$

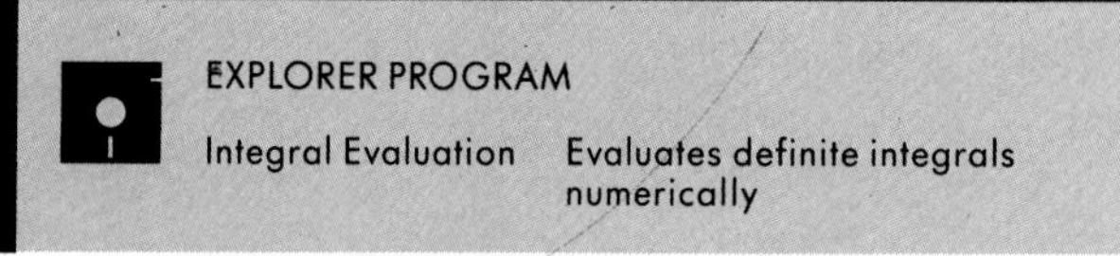

11.4 Curvature, Torsion, and the Frenet Frame

SKIP

In this section we define a frame of mutually orthogonal unit vectors that always travels with a body moving along a curve in space (Fig. 11.23). The frame has three vectors. The first is $\mathbf{T}$, the unit tangent vector. The second is $\mathbf{N}$, the unit vector that gives the direction of $d\mathbf{T}/ds$. The third is $\mathbf{B} = \mathbf{T} \times \mathbf{N}$. These vectors and their derivatives, when available, give useful information about a vehicle's orientation in space and about how the vehicle's path turns and twists as the vehicle moves along.

For example, the magnitude of the derivative $d\mathbf{T}/ds$ tells how much a vehicle's path turns to the left or right as it moves along; it is called the *curvature* of the vehicle's path. The number $-(d\mathbf{B}/ds) \cdot \mathbf{N}$ tells how much a vehicle's path rotates or twists as the vehicle moves along; it is called the *torsion* of the vehicle's path. Look at Fig. 11.23 again. If P were a train climbing a curved banked track, the rate at which the headlight turned from side to side per unit distance would be the curvature of the track. The rate at which the engine rotated about its longitudinal axis (the line of $\mathbf{T}$) would be the torsion.

We begin with curves in the plane and then move into space.

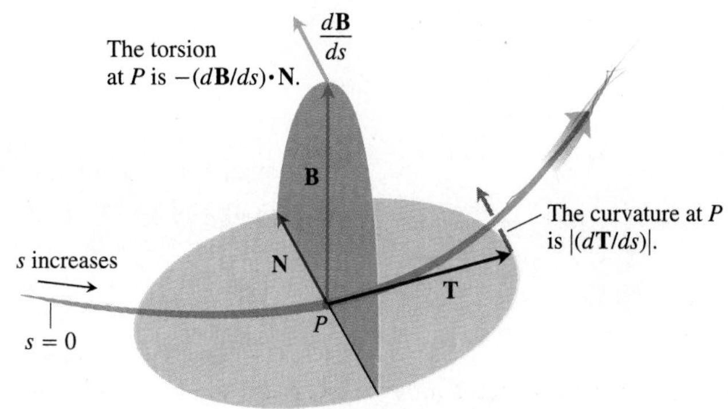

11.23 Every moving body travels with a frame of mutually orthogonal unit vectors (**T**, **N**, and **B**) that describes how the body moves.

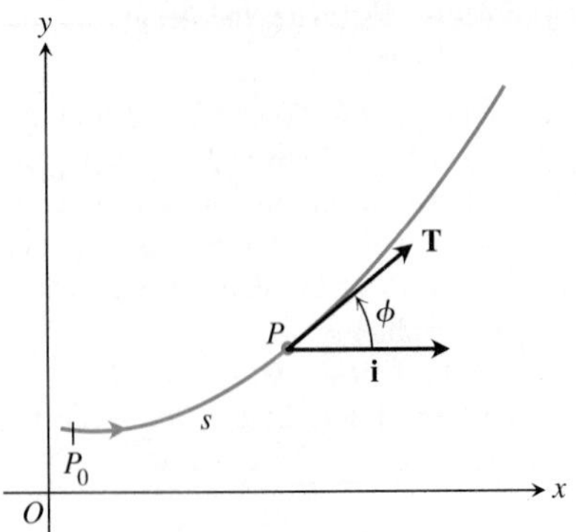

11.24 The value of $|d\phi/ds|$ at P is the curvature of the curve at P.

The Curvature of a Plane Curve

As we move along a differentiable curve in the plane, the unit tangent vector $\mathbf{T}$ turns as the curve bends. We measure the rate at which $\mathbf{T}$ turns by measuring the change in the angle ϕ that $\mathbf{T}$ makes with $\mathbf{i}$ (Fig. 11.24). At each point P, the absolute value of $d\phi/ds$, stated in radians per unit of length along the curve, is called the **curvature** at P. If $|d\phi/ds|$ is large, $\mathbf{T}$ turns sharply as we pass through P and the curvature at P is large. If $|d\phi/ds|$ is close to zero, $\mathbf{T}$ turns more slowly and the curvature at P is small. On circles and lines, the curvature is constant, as we shall see in Examples 1 and 2. On other curves, the curvature can vary from place to place. The traditional symbol for the curvature function is the Greek letter κ (kappa).

If $x = f(t)$ and $y = g(t)$ are twice-differentiable functions of t, we can derive a formula for the curvature of the curve $\mathbf{r} = f(t)\mathbf{i} + g(t)\mathbf{j}$ in the following way. In Newton's dot notation, in which $\dot{y}$ ("y dot") means dy/dt, $\ddot{y}$ ("y double dot") means d^2y/dt^2, and so on, we have

$$\tan\phi = \frac{dy}{dx} = \frac{dy/dt}{dx/dt} = \frac{\dot{y}}{\dot{x}} \quad \text{and} \quad \phi = \tan^{-1}\left(\frac{\dot{y}}{\dot{x}}\right). \tag{1}$$

Hence,

$$\begin{aligned}\frac{d\phi}{ds} = \frac{d\phi}{dt}\cdot\frac{dt}{ds} &= \frac{1}{1+(\dot{y}/\dot{x})^2}\,\frac{d}{dt}\left(\frac{\dot{y}}{\dot{x}}\right)\cdot\frac{1}{(\dot{x}^2+\dot{y}^2)^{1/2}} \qquad \left(\frac{dt}{ds} = \frac{1}{|\mathbf{v}|}\right)\\ &= \frac{\dot{x}^2}{(\dot{x}^2+\dot{y}^2)^{3/2}}\,\frac{\dot{x}\ddot{y}-\dot{y}\ddot{x}}{\dot{x}^2} \qquad (2)\\ &= \frac{\dot{x}\ddot{y}-\dot{y}\ddot{x}}{|\mathbf{v}|^3}.\end{aligned}$$

The curvature, therefore, is

$$\left|\frac{d\phi}{ds}\right| = \frac{|\dot{x}\ddot{y}-\dot{y}\ddot{x}|}{|\mathbf{v}|^3}. \tag{3}$$

The observation that $|\dot{x}\ddot{y}-\dot{y}\ddot{x}|$ is the magnitude of

$$\mathbf{v}\times\mathbf{a} = \begin{vmatrix}\mathbf{i} & \mathbf{j} & \mathbf{k}\\ \dot{x} & \dot{y} & 0\\ \ddot{x} & \ddot{y} & 0\end{vmatrix} \tag{4}$$

enables us to write Eq. (3) in a compact vector form.

Curvature

$$\kappa = \frac{|\mathbf{v}\times\mathbf{a}|}{|\mathbf{v}|^3} \tag{5}$$

Equation (5) calculates the curvature, a geometric property of the curve, from the velocity and acceleration of any vector representation of the curve in which $|\mathbf{v}|$ is different from zero. Take a moment to think about how remarkable this really is: From any formula for motion along a curve, no matter how variable the motion may be (as long as $\mathbf{v}$ is never zero), we can calculate a physical property of the curve that seems to have nothing to do with the way the curve is traversed.

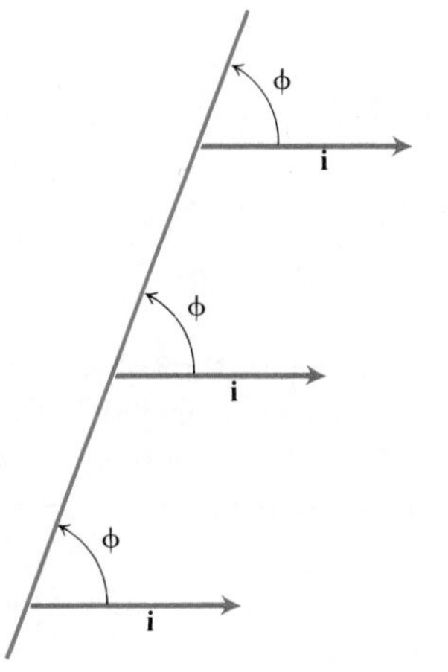

11.25 Along a line, the angle ϕ stays the same from point to point and the curvature, $d\phi/ds$, is zero (Example 1).

Example 1 *The Curvature of a Straight Line Is Zero.* On a straight line, ϕ is constant (Fig. 11.25), so $d\phi/ds = 0$.

Example 2 *The Curvature of a Circle of Radius a Is 1/a.* To see why, parametrize the circle with the equation

$$\mathbf{r} = (a \cos t)\,\mathbf{i} + (a \sin t)\,\mathbf{j}.$$

Then

$$\mathbf{v} = -(a \sin t)\,\mathbf{i} + (a \cos t)\,\mathbf{j},$$

$$\mathbf{a} = -(a \cos t)\,\mathbf{i} - (a \sin t)\,\mathbf{j}.$$

Hence

$$\mathbf{v} \times \mathbf{a} = \begin{vmatrix} \mathbf{i} & \mathbf{j} & \mathbf{k} \\ -a \sin t & a \cos t & 0 \\ -a \cos t & -a \sin t & 0 \end{vmatrix} = (a^2\sin^2 t + a^2\cos^2 t)\,\mathbf{k} = a^2\,\mathbf{k},$$

$$|\mathbf{v}|^3 = \left[\sqrt{(-a \sin t)^2 + (a \cos t)^2}\right]^3 = a^3,$$

and

$$\kappa = \frac{|\mathbf{v} \times \mathbf{a}|}{|\mathbf{v}|^3} = \frac{|a^2\,\mathbf{k}|}{a^3} = \frac{a^2|\mathbf{k}|}{a^3} = \frac{1}{a}.$$

The larger the circle, the more gradually it curves.

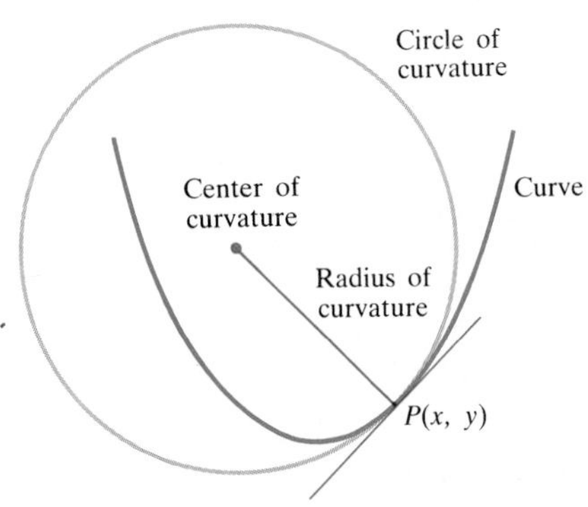

11.26 The osculating circle or circle of curvature at $P(x, y)$ lies toward the inner side of the curve.

Circle of Curvature and Radius of Curvature

The **circle of curvature** or **osculating circle** at a point P on a plane curve where $\kappa \neq 0$ is the circle in the plane of the curve that

1. is tangent to the curve at P (has the same tangent the curve has);
2. has the same curvature the curve has at P; and
3. lies toward the concave or inner side of the curve (as in Fig. 11.26).

The **radius of curvature** of the curve at P is the radius of the circle of curvature, which, according to Example 2, is

$$\text{Radius of curvature} = \rho = \frac{1}{\kappa}. \tag{6}$$

To calculate ρ, we calculate κ and take its reciprocal.

The **center of curvature** of the curve at P is the center of the circle of curvature.

The Principal Unit Normal Vector for Curves in the Plane

The vectors $d\mathbf{T}/ds$ and $d\mathbf{T}/d\phi$ are related by the Chain Rule equation

$$\frac{d\mathbf{T}}{ds} = \frac{d\mathbf{T}}{d\phi}\frac{d\phi}{ds}. \tag{7}$$

Furthermore, since $\mathbf{T}$ is a vector of constant length, $d\mathbf{T}/ds$ and $d\mathbf{T}/d\phi$ are both orthogonal to $\mathbf{T}$ (Section 11.1). They have the same direction if $d\phi/ds$ is positive

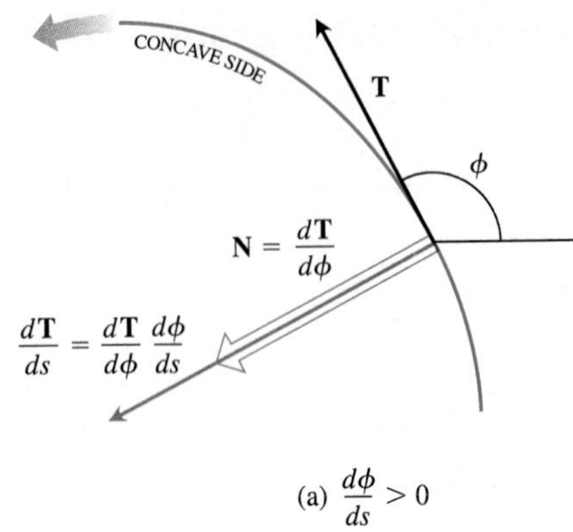

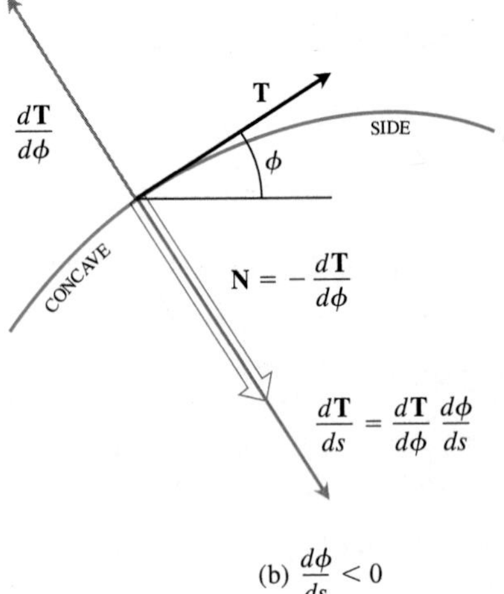

11.27 The vector $d\mathbf{T}/ds$, normal to the curve, always points inward. The principal unit normal vector $\mathbf{N}$ is the direction of $d\mathbf{T}/ds$.

and opposite directions if $d\phi/ds$ is negative. Now

$$\mathbf{T} = (\cos\phi)\,\mathbf{i} + (\sin\phi)\,\mathbf{j}, \tag{8}$$

and

$$\frac{d\mathbf{T}}{d\phi} = -(\sin\phi)\,\mathbf{i} + (\cos\phi)\,\mathbf{j} = \cos\left(\phi + \frac{\pi}{2}\right)\mathbf{i} + \sin\left(\phi + \frac{\pi}{2}\right)\mathbf{j} \tag{9}$$

is the unit vector obtained by rotating $\mathbf{T}$ counterclockwise through $\pi/2$ radians. Therefore, if we stand at a point on the curve facing in the direction of $\mathbf{T}$, the vector $d\mathbf{T}/ds = (d\mathbf{T}/d\phi)(d\phi/ds)$ will point toward the left if $d\phi/ds$ is positive and toward the right if $d\phi/ds$ is negative. In other words, $d\mathbf{T}/ds$ will point toward the concave side of the curve (Fig. 11.27).

Since $d\mathbf{T}/d\phi$ is a unit vector, the magnitude of $d\mathbf{T}/ds$ at any point on the curve is the curvature at that point, as we can see from the equation

$$\left|\frac{d\mathbf{T}}{ds}\right| = \left|\frac{d\mathbf{T}}{d\phi}\right|\left|\frac{d\phi}{ds}\right| = (1)(\kappa) = \kappa. \tag{10}$$

When $d\mathbf{T}/ds \neq \mathbf{0}$, its direction is given by the unit vector

$$\mathbf{N} = \frac{d\mathbf{T}/ds}{|d\mathbf{T}/ds|} = \frac{1}{\kappa}\frac{d\mathbf{T}}{ds}. \tag{11}$$

Since $\mathbf{N}$ points the same way $d\mathbf{T}/ds$ does, $\mathbf{N}$ is always orthogonal to $\mathbf{T}$ and directed toward the concave side of the curve. The vector $\mathbf{N}$ is the **principal unit normal vector** of the curve.

Because the directed distance is defined with ds/dt positive, dt/ds is positive and the Chain Rule gives

$$\begin{aligned}\mathbf{N} &= \frac{d\mathbf{T}/ds}{|d\mathbf{T}/ds|} = \frac{(d\mathbf{T}/dt)(dt/ds)}{|d\mathbf{T}/dt|\,|dt/ds|} \\ &= \frac{d\mathbf{T}/dt}{|d\mathbf{T}/dt|}.\end{aligned} \tag{12}$$

This formula enables us to find $\mathbf{N}$ without having to find ϕ, κ, or s first.

Example 3 Find $\mathbf{T}$ and $\mathbf{N}$ for the circular motion

$$\mathbf{r} = (\cos 2t)\,\mathbf{i} + (\sin 2t)\,\mathbf{j}.$$

Solution We first find $\mathbf{T}$:

$$\mathbf{v} = -(2\sin 2t)\,\mathbf{i} + (2\cos 2t)\,\mathbf{j},$$

$$|\mathbf{v}| = \sqrt{4\sin^2 2t + 4\cos^2 2t} = 2,$$

$$\mathbf{T} = \frac{\mathbf{v}}{|\mathbf{v}|} = -(\sin 2t)\,\mathbf{i} + (\cos 2t)\,\mathbf{j}.$$

From this we find

$$\frac{d\mathbf{T}}{dt} = -(2\cos 2t)\,\mathbf{i} - (2\sin 2t)\,\mathbf{j},$$

$$\left|\frac{d\mathbf{T}}{dt}\right| = \sqrt{4\cos^2 2t + 4\sin^2 2t} = 2,$$

and

$$\mathbf{N} = \frac{d\mathbf{T}/dt}{|d\mathbf{T}/dt|} = -(\cos 2t)\,\mathbf{i} - (\sin 2t)\,\mathbf{j}. \qquad \text{(Eq. (12))}$$

Curvature for Curves in Space

In space there is no natural way to define an angle like ϕ with which to measure the change in $\mathbf{T}$ along a differentiable curve. But we still have s, the directed distance along the curve, and can define the curvature to be

$$\kappa = \left|\frac{d\mathbf{T}}{ds}\right|,$$

as it worked out to be for curves in the plane. The formula

$$\kappa = \frac{|\mathbf{v} \times \mathbf{a}|}{|\mathbf{v}|^3} \tag{13}$$

still holds, as you will see if you do the calculation in Exercise 26.

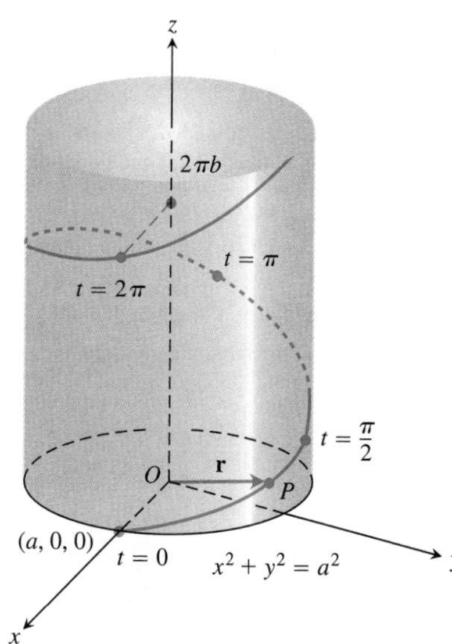

11.28 The helix $\mathbf{r} = (a \cos t)\,\mathbf{i} + (a \sin t)\,\mathbf{j} + bt\,\mathbf{k}$, drawn with a and b positive (Example 4).

Example 4 How do the values of a and b control the curvature of the helix (Fig. 11.28)

$$\mathbf{r} = (a \cos t)\,\mathbf{i} + (a \sin t)\,\mathbf{j} + bt\mathbf{k}? \qquad (a, b \geq 0)$$

Solution We calculate the curvature with Eq. (13),

$$\mathbf{v} = -(a \sin t)\,\mathbf{i} + (a \cos t)\,\mathbf{j} + b\mathbf{k},$$

$$\mathbf{a} = -(a \cos t)\,\mathbf{i} - (a \sin t)\,\mathbf{j},$$

$$\mathbf{v} \times \mathbf{a} = \begin{vmatrix} \mathbf{i} & \mathbf{j} & \mathbf{k} \\ -a \sin t & a \cos t & b \\ -a \cos t & -a \sin t & 0 \end{vmatrix} = (ab \sin t)\,\mathbf{i} - (ab \cos t)\,\mathbf{j} + a^2\,\mathbf{k},$$

$$\kappa = \frac{|\mathbf{v} \times \mathbf{a}|}{|\mathbf{v}|^3} = \frac{\sqrt{a^2b^2 + a^4}}{(a^2 + b^2)^{3/2}} = \frac{a\sqrt{a^2 + b^2}}{(a^2 + b^2)^{3/2}} = \frac{a}{a^2 + b^2}. \tag{14}$$

From Eq. (14) we see that increasing b for a fixed a decreases the curvature. Decreasing a for a fixed b eventually decreases the curvature as well. In other words, stretching a spring tends to straighten it.

If $b = 0$, the helix reduces to a circle of radius a and its curvature reduces to $1/a$, as it should. If $a = 0$, the helix becomes the z-axis, and its curvature reduces to 0, again as it should.

N for Curves in Space

To define the principal unit normal vector of a curve in space, we use the same definition we use for a curve in the plane:

$$\mathbf{N} = \frac{d\mathbf{T}/ds}{|d\mathbf{T}/ds|} = \frac{1}{\kappa}\frac{d\mathbf{T}}{ds} = \frac{d\mathbf{T}/dt}{|d\mathbf{T}/dt|}. \tag{15}$$

As before, $d\mathbf{T}/ds$ and its unit vector direction $\mathbf{N}$ are orthogonal to $\mathbf{T}$ because $\mathbf{T}$ has constant length, in this case 1.

Example 5 Find $\mathbf{N}$ for the helix in Example 4.

Solution Using values from Example 4, we have

$$\mathbf{T} = \frac{\mathbf{v}}{|\mathbf{v}|} = \frac{-(a \sin t)\,\mathbf{i} + (a \cos t)\,\mathbf{j} + b\mathbf{k}}{\sqrt{a^2 + b^2}}$$

$$\frac{d\mathbf{T}}{dt} = -\frac{a}{\sqrt{a^2 + b^2}}\left((\cos t)\,\mathbf{i} + (\sin t)\,\mathbf{j}\right)$$

$$\left|\frac{d\mathbf{T}}{dt}\right| = \frac{a}{\sqrt{a^2 + b^2}}\sqrt{\cos^2 t + \sin^2 t} = \frac{a}{\sqrt{a^2 + b^2}}$$

$$\mathbf{N} = \frac{d\mathbf{T}/dt}{|d\mathbf{T}/dt|} = -(\cos t)\,\mathbf{i} - (\sin t)\,\mathbf{j}.$$

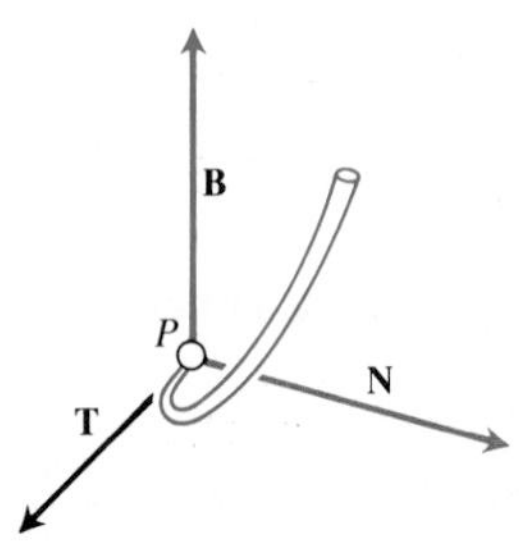

11.29 The vectors **T**, **N**, and **B** (in that order) make a right-handed frame of mutually orthogonal unit vectors in space. We call it the **Frenet** (fre-*nay*) **frame** (after Jean-Frédéric Frenet, 1816–1900).

Torsion and the Binormal Vector

The **binormal vector** of a curve in space is the vector $\mathbf{B} = \mathbf{T} \times \mathbf{N}$, a unit vector orthogonal to both $\mathbf{T}$ and $\mathbf{N}$ (Fig. 11.29). Together, $\mathbf{T}$, $\mathbf{N}$, and $\mathbf{B}$ define a moving right-handed vector frame that plays a significant role in calculating the flight paths of space vehicles.

The three planes determined by $\mathbf{T}$, $\mathbf{N}$, and $\mathbf{B}$ are shown in Fig. 11.30. The curvature $\kappa = |d\mathbf{T}/ds|$ can be thought of as the rate at which the normal plane turns as the point P moves along the curve. Similarly, the **torsion** $\tau = -(d\mathbf{B}/ds)\cdot\mathbf{N}$ is the rate at which the osculating plane twists about $\mathbf{T}$ as P moves along the curve. It gives a measure of how much the curve twists to the left or right.

The most widely used formula for torsion, derived in more advanced texts, is

$$\tau = \frac{\begin{vmatrix} \dot{x} & \dot{y} & \dot{z} \\ \ddot{x} & \ddot{y} & \ddot{z} \\ \dddot{x} & \dddot{y} & \dddot{z} \end{vmatrix}}{|\mathbf{v} \times \mathbf{a}|^2} \qquad (\mathbf{v} \times \mathbf{a} \neq \mathbf{0}). \tag{16}$$

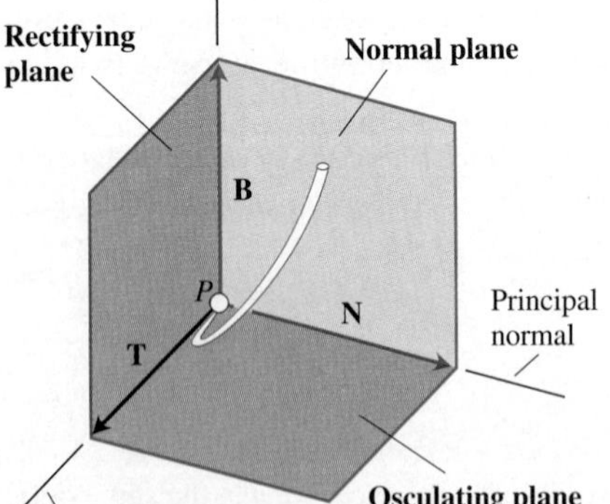

11.30 The three planes determined by **T**, **N**, and **B** have special names.

This formula calculates the torsion directly from the derivatives of the component functions $x = f(t)$, $y = g(t)$, $z = h(t)$, that make up $\mathbf{r}$. The determinant's first row comes from $\mathbf{v}$, the second row comes from $\mathbf{a}$, and the third row comes from $\dot{\mathbf{a}}$.

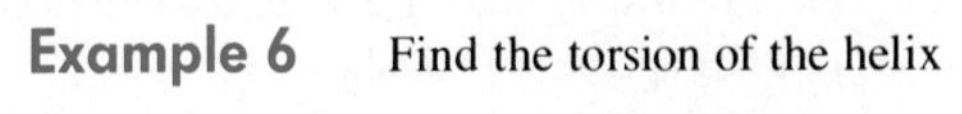

Example 6 Find the torsion of the helix

$$\mathbf{r} = (\cos t)\,\mathbf{i} + (\sin t)\,\mathbf{j} + t\mathbf{k}.$$

Solution We evaluate Eq. (16). We find the entries in the determinant by differentiating $\mathbf{r}$:

$$\mathbf{v} = -(\sin t)\,\mathbf{i} + (\cos t)\,\mathbf{j} + \mathbf{k}$$

$$\mathbf{a} = -(\cos t)\,\mathbf{i} - (\sin t)\,\mathbf{j}$$

$$\dot{\mathbf{a}} = (\sin t)\,\mathbf{i} - (\cos t)\,\mathbf{j}$$

Then,

$$\tau = \frac{\begin{vmatrix} \dot{x} & \dot{y} & \dot{z} \\ \ddot{x} & \ddot{y} & \ddot{z} \\ \dddot{x} & \dddot{y} & \dddot{z} \end{vmatrix}}{|\mathbf{v} \times \mathbf{a}|^2} = \frac{\begin{vmatrix} -\sin t & \cos t & 1 \\ -\cos t & -\sin t & 0 \\ \sin t & -\cos t & 0 \end{vmatrix}}{\left\| \begin{matrix} \mathbf{i} & \mathbf{j} & \mathbf{k} \\ -\sin t & \cos t & 1 \\ -\cos t & -\sin t & 0 \end{matrix} \right\|^2}$$

$$= \frac{\cos^2 t + \sin^2 t}{|(\sin t)\,\mathbf{i} - (\cos t)\,\mathbf{j} + \mathbf{k}|^2} = \frac{1}{2}.$$

The Tangential and Normal Components of Acceleration

When a moving body is accelerated by gravity, brakes, a combination of rocket motors, or whatever, we usually want to know how much of the acceleration acts to move the body straight ahead in the direction of motion, that is, in the tangential direction **T**. We can find out if we use the Chain Rule to rewrite **v** as

$$\mathbf{v} = \frac{d\mathbf{r}}{dt} = \frac{d\mathbf{r}}{ds}\frac{ds}{dt} = \mathbf{T}\frac{ds}{dt} \tag{17}$$

and differentiate both ends of this string of equalities to get

$$\begin{aligned} \mathbf{a} &= \frac{d\mathbf{v}}{dt} = \frac{d}{dt}\left(\mathbf{T}\frac{ds}{dt}\right) = \frac{d^2s}{dt^2}\mathbf{T} + \frac{ds}{dt}\frac{d\mathbf{T}}{dt} \\ &= \frac{d^2s}{dt^2}\mathbf{T} + \frac{ds}{dt}\left(\frac{d\mathbf{T}}{ds}\frac{ds}{dt}\right) = \frac{d^2s}{dt^2}\mathbf{T} + \frac{ds}{dt}\left(\kappa\mathbf{N}\frac{ds}{dt}\right) \\ &= \frac{d^2s}{dt^2}\mathbf{T} + \kappa\left(\frac{ds}{dt}\right)^2\mathbf{N}. \end{aligned} \tag{18}$$

This is a remarkable equation. There is no **B** in it. No matter how the path of the moving body we are watching may appear to twist and turn, the acceleration **a** always lies in the plane of **T** and **N** orthogonal to **B**. The equation also tells us exactly how much of the acceleration takes place tangent to the motion (d^2s/dt^2) and how much takes place normal to the motion ($\kappa(ds/dt)^2$).

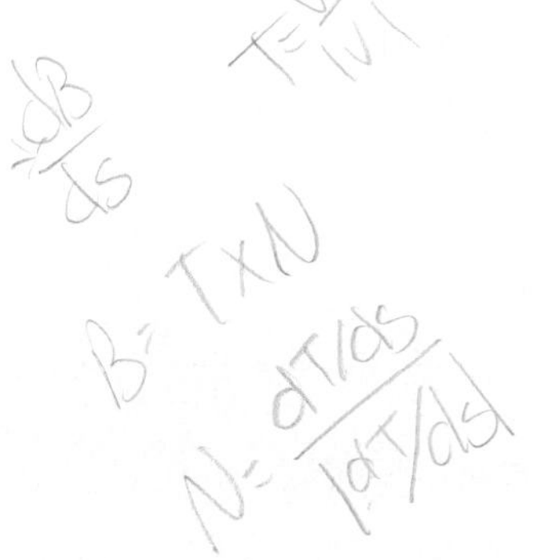

The **tangential** and **normal** scalar components of acceleration are

$$a_{\mathrm{T}} = \frac{d^2s}{dt^2} = \frac{d}{dt}|\mathbf{v}| \quad \text{and} \quad a_{\mathrm{N}} = \kappa\left(\frac{ds}{dt}\right)^2 = \kappa|\mathbf{v}|^2. \tag{19}$$

That is,

$$\mathbf{a} = \frac{d^2s}{dt^2}\mathbf{T} + \kappa\left(\frac{ds}{dt}\right)^2\mathbf{N}. \tag{20}$$

Notice that the normal scalar component of the acceleration is the curvature times the square of the speed. This explains why you have to hold on when your car makes a sharp (large κ), high-speed (large $|\mathbf{v}|$) turn.

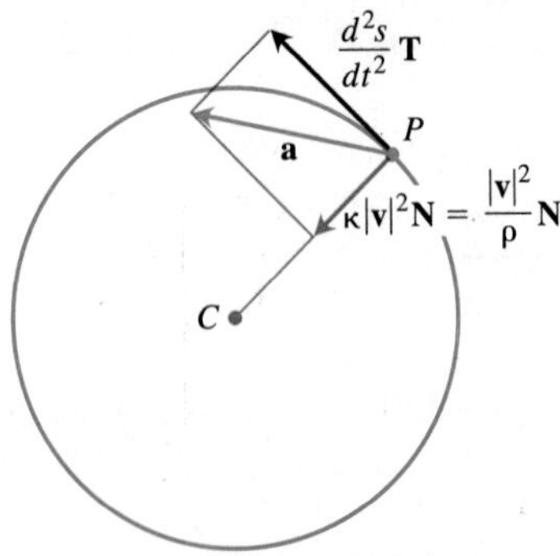

11.31 The tangential and normal components of the acceleration of a body speeding up as it moves counterclockwise around a circle of radius ρ.

If a body moves in a circle at a constant speed, d^2s/dt^2 is zero and all the acceleration points along **N** toward the circle's center. If the body is speeding up or slowing down, **a** has a nonzero tangential component (Fig. 11.31).

To calculate a_N we usually use the formula

$$a_N = \sqrt{|\mathbf{a}|^2 - a_T^2}, \tag{21}$$

which comes from solving the equation $|\mathbf{a}|^2 = \mathbf{a} \cdot \mathbf{a} = a_T^2 + a_N^2$ for a_N. With this formula we can find a_N without having to calculate κ first.

Example 7 Without finding **T** and **N**, write the acceleration of the motion

$$\mathbf{r} = (\cos t + t \sin t)\,\mathbf{i} + (\sin t - t \cos t)\,\mathbf{j}, \qquad t > 0$$

in the form $\mathbf{a} = a_T\mathbf{T} + a_N\mathbf{N}$. (The path of the motion is the involute of the circle in Fig. 11.32.)

Solution We use the first of Eqs. (19) to find a_T:

$$\mathbf{v} = (t \cos t)\,\mathbf{i} + (t \sin t)\,\mathbf{j} \qquad \begin{pmatrix}\text{Value from Section 11.3,}\\ \text{Example 5}\end{pmatrix}$$

$$|\mathbf{v}| = \sqrt{t^2\cos^2 t + t^2\sin^2 t} = \sqrt{t^2} = |t| = t \qquad (t > 0)$$

$$a_T = \frac{d}{dt}|\mathbf{v}| = \frac{d}{dt}(t) = 1 \qquad \text{(Eq. 19)}$$

Knowing a_T, we use Eq. (21) to find a_N:

$$\mathbf{a} = (\cos t - t \sin t)\,\mathbf{i} + (\sin t + t \cos t)\,\mathbf{j}$$

$$|\mathbf{a}|^2 = t^2 + 1 \qquad \begin{pmatrix}\text{after some}\\ \text{algebra}\end{pmatrix}$$

$$a_N = \sqrt{|\mathbf{a}|^2 - a_T^2} = \sqrt{(t^2 + 1) - (1)} = \sqrt{t^2} = t.$$

We then use Eq. (20) to find **a**:

$$\mathbf{a} = a_T\mathbf{T} + a_N\mathbf{N} = (1)\mathbf{T} + (t)\mathbf{N} = \mathbf{T} + t\mathbf{N}.$$

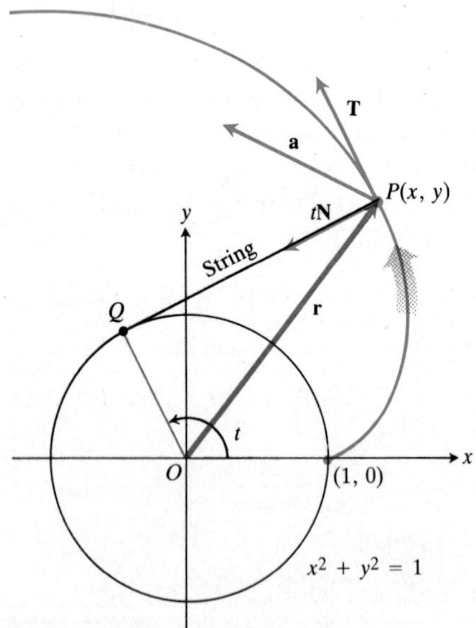

11.32 The tangential and normal acceleration components of the motion $\mathbf{r} = (\cos t + t \sin t)\,\mathbf{i} + (\sin t - t \cos t)\,\mathbf{j}$ shown here are **T** and t**N** (Example 7).

Motion Formulas ($|\mathbf{v}| \neq 0$)

Unit tangent vector: $\mathbf{T} = \dfrac{\mathbf{v}}{|\mathbf{v}|}$

Principal unit normal vector: $\mathbf{N} = \dfrac{d\mathbf{T}/dt}{|d\mathbf{T}/dt|}$

Binormal vector: $\mathbf{B} = \mathbf{T} \times \mathbf{N}$

Curvature: $\kappa = \left|\dfrac{d\mathbf{T}}{ds}\right| = \dfrac{|\mathbf{v} \times \mathbf{a}|}{|\mathbf{v}|^3}$

Torsion: $\tau = -\dfrac{d\mathbf{B}}{ds} \cdot \mathbf{N} = \dfrac{\begin{vmatrix} \dot{x} & \dot{y} & \dot{z} \\ \ddot{x} & \ddot{y} & \ddot{z} \\ \dddot{x} & \dddot{y} & \dddot{z} \end{vmatrix}}{|\mathbf{v} \times \mathbf{a}|^2}$

Tangential and normal scalar components of acceleration:

$$\mathbf{a} = a_T\mathbf{T} + a_N\mathbf{N}$$

$$a_T = \frac{d}{dt}|\mathbf{v}|$$

$$a_N = \sqrt{|\mathbf{a}|^2 - a_T^2}$$

EXERCISES 11.4

Find **T**, **N**, and κ for the plane curves in Exercises 1–4.

1. $\mathbf{r} = t\mathbf{i} + (\ln \cos t)\mathbf{j}, \quad -\pi/2 < t < \pi/2$

2. $\mathbf{r} = (\ln \sec t)\mathbf{i} + t\mathbf{j}, \quad -\pi/2 < t < \pi/2$

3. $\mathbf{r} = (2t + 3)\mathbf{i} + (5 - t^2)\mathbf{j}$

4. $\mathbf{r} = (\cos t + t \sin t)\mathbf{i} + (\sin t - t \cos t)\mathbf{j}, \quad t > 0$

Find **T**, **N**, **B**, κ, and τ for the space curves in Exercises 5–12.

5. $\mathbf{r} = (3 \sin t)\mathbf{i} + (3 \cos t)\mathbf{j} + 4t\mathbf{k}$

6. $\mathbf{r} = (\cos t + t \sin t)\mathbf{i} + (\sin t - t \cos t)\mathbf{j} + 3\mathbf{k}$

7. $\mathbf{r} = (e^t \cos t)\mathbf{i} + (e^t \sin t)\mathbf{j} + 2\mathbf{k}$

8. $\mathbf{r} = (6 \sin 2t)\mathbf{i} + (6 \cos 2t)\mathbf{j} + 5t\mathbf{k}$

9. $\mathbf{r} = (t^3/3)\mathbf{i} + (t^2/2)\mathbf{j}, \quad t > 0$

10. $\mathbf{r} = (\cos^3 t)\mathbf{i} + (\sin^3 t)\mathbf{j}, \quad 0 < t < \pi/2$

11. $\mathbf{r} = t\mathbf{i} + a(\cosh(t/a))\mathbf{j}, \quad a > 0$

12. $\mathbf{r} = (\cosh t)\mathbf{i} - (\sinh t)\mathbf{j} + t\mathbf{k}$

In Exercises 13–16, write **a** in the form $\mathbf{a} = a_\mathrm{T}\mathbf{T} + a_\mathrm{N}\mathbf{N}$ without finding **T** and **N**.

13. $\mathbf{r} = (2t + 3)\mathbf{i} + (t^2 - 1)\mathbf{j}$

14. $\mathbf{r} = \ln(t^2 + 1)\mathbf{i} + (t - 2 \tan^{-1} t)\mathbf{j}$

15. $\mathbf{r} = (a \cos t)\mathbf{i} + (a \sin t)\mathbf{j} + bt\mathbf{k}$

16. $\mathbf{r} = (1 + 3t)\mathbf{i} + (t - 2)\mathbf{j} - 3t\mathbf{k}$

In Exercises 17–20, write **a** in the form $\mathbf{a} = a_\mathrm{T}\mathbf{T} + a_\mathrm{N}\mathbf{N}$ at the given value of t without finding **T** and **N**. You can save yourself some work by evaluating the vectors **v** and **a** at the given value of t *before* finding their lengths.

17. $\mathbf{r} = (t + 1)\mathbf{i} + 2t\mathbf{j} + t^2\mathbf{k}, \quad t = 1$

18. $\mathbf{r} = (t \cos t)\mathbf{i} + (t \sin t)\mathbf{j} + t^2\mathbf{k}, \quad t = 0$

19. $\mathbf{r} = t^2\mathbf{i} + (t + (1/3)t^3)\mathbf{j} + (t - (1/3)t^3)\mathbf{k}, \quad t = 0$

20. $\mathbf{r} = (e^t \cos t)\mathbf{i} + (e^t \sin t)\mathbf{j} + \sqrt{2}e^t\mathbf{k}, \quad t = 0$

In Exercises 21 and 22, find **r**, **T**, **N**, and **B** at the given value of t. Then find equations for the osculating, normal, and rectifying planes at that value of t.

21. $\mathbf{r} = (\cos t)\mathbf{i} + (\sin t)\mathbf{j} - \mathbf{k}, \quad t = \pi/4$

22. $\mathbf{r} = (\cos t)\mathbf{i} + (\sin t)\mathbf{j} + t\mathbf{k}, \quad t = 0$

23. The speedometer on your car reads a steady 35 mph. Could you be accelerating? Explain.

24. Show that if a particle's speed is constant its acceleration is either zero or normal to its path.

25. Show that a particle's speed will be constant if the acceleration is always perpendicular to the velocity.

26. *How to derive the formula* $\kappa = |\mathbf{v} \times \mathbf{a}|/|\mathbf{v}|^3$ *for space curves.* To derive the formula, carry out these steps.

STEP 1: Use equations $\mathbf{v} = \mathbf{T}(ds/dt)$ and $\mathbf{a} = (d^2s/dt^2)\mathbf{T} + \kappa(ds/dt)^2\,\mathbf{N}$ to find a formula for $|\mathbf{v} \times \mathbf{a}|$.

STEP 2: Solve the resulting equation for κ, assuming $\mathbf{v} \times \mathbf{a} \neq 0$.

27. If a_N and $|\mathbf{v}|$ are known, the equation $a_\mathrm{N} = \kappa|\mathbf{v}|^2$ gives a convenient way to find curvature. Use it to find the curvature and radius of curvature of the curve

$$\mathbf{r} = (\cos t + t \sin t)\mathbf{i} + (\sin t - t \cos t)\mathbf{j}, \quad t > 0$$

in Example 7.

28. Show that the torsion of any sufficiently differentiable plane curve is zero when defined.

29. Show that a moving particle will move in a straight line if the normal component of its acceleration is zero.

30. Show that κ and τ are both zero for the line

$$\mathbf{r} = (x_0 + At)\mathbf{i} + (y_0 + Bt)\mathbf{j} + (z_0 + Ct)\mathbf{k}.$$

31. *Maximizing the curvature of a helix.* In Example 4, we found the curvature of the helix $\mathbf{r} = (a \cos t)\mathbf{i} + (a \sin t)\mathbf{j} + bt\mathbf{k}$ $(a,\ b \geq 0)$ to be $\kappa = a/(a^2 + b^2)$. What is the largest value κ can have for a given value of b?

32. *The torsion of a helix.* Find the torsion of the helix

$$\mathbf{r} = (a \cos t)\mathbf{i} + (a \sin t)\mathbf{j} + bt\mathbf{k} \quad (a, b > 0).$$

What is the largest value τ can have for a given value of a?

33. *Differentiable curves with zero torsion lie in planes.* That a sufficiently differentiable curve with zero torsion lies in a plane is a special case of the fact that a particle whose velocity remains normal to a fixed vector **C** moves in a plane perpendicular to **C**. This, in turn, can be viewed as the solution of the following problem in calculus.

Suppose $\mathbf{r} = f(t)\mathbf{i} + g(t)\mathbf{j} + h(t)\mathbf{k}$ is twice differentiable for all t in an interval $[a, b]$, that $\mathbf{r} = 0$ when $t = a$, and that $\mathbf{v} \cdot \mathbf{k} = 0$ for all t in $[a, b]$. Then $h(t) = 0$ for all t in $[a, b]$.

Solve this problem. (*Hint:* Start with $\mathbf{a} = d^2\mathbf{r}/dt^2$ and apply the initial conditions in reverse order.)

34. *A formula that calculates torsion from* **B** *and* **v**. If we start with the definition $\tau = -(d\mathbf{B}/ds) \cdot \mathbf{N}$ and apply the Chain Rule to rewrite $d\mathbf{B}/ds$ as

$$\frac{d\mathbf{B}}{ds} = \frac{d\mathbf{B}}{dt}\frac{dt}{ds} = \frac{d\mathbf{B}}{dt}\frac{1}{|\mathbf{v}|},$$

we arrive at the formula

$$\tau = -\frac{1}{|\mathbf{v}|}\left(\frac{d\mathbf{B}}{dt} \cdot \mathbf{N}\right).$$

The advantage of this formula over the one in Eq. (16) is that it is easier to derive and state. The disadvantage is that it can take a lot of work to evaluate without a computer. Use the new formula to find the torsion of the helix in Example 6.

35. Find an equation for the circle of curvature of the curve $\mathbf{r} = t\mathbf{i} + (\sin t)\mathbf{j}$ at the point $(\pi/2, 1)$. (The curve parametrizes the graph of $y = \sin x$ in the xy-plane.)

36. a) GRAPHER Graph the curve $\mathbf{r} = (2 \ln t)\mathbf{i} - (t + (1/t))\mathbf{j}$ for $0 < t < 10$.
b) Find an equation for the circle of curvature at the point $(0, -2)$.

37. Equation (2) says that the curvature of a plane curve $\mathbf{r} = f(t)\mathbf{i} + g(t)\mathbf{j}$ defined by twice differentiable functions $x = f(t)$ and $y = g(t)$ is given by the formula

$$\kappa = \frac{|\dot{x}\ddot{y} - \dot{y}\ddot{x}|}{(\dot{x}^2 + \dot{y}^2)^{3/2}}.$$

Apply the formula to find the curvatures of the following curves.
a) $\mathbf{r} = t\mathbf{i} + (\ln \sin t)\mathbf{j}, \quad 0 < t < \pi$
b) $\mathbf{r} = (\tan^{-1}(\sinh t))\mathbf{i} + (\ln \cosh t)\mathbf{j}$.

38. *A formula for the curvature of the graph of a function in the xy-plane.*
a) The graph $y = f(x)$ in the xy-plane automatically has the parametrization $x = x$, $y = f(x)$, and the vector formula $\mathbf{r} = x\mathbf{i} + f(x)\mathbf{j}$. Use this formula to show that if f is a twice-differentiable function of x, then

$$\kappa = \frac{|f''(x)|}{(1 + (f'(x))^2)^{3/2}}.$$

b) Use the formula in (a) to find the curvature of $y = \ln(\cos x)$, $-\pi/2 < x < \pi/2$. Compare your answer with the answer in Exercise 1.

39. a) Use the formula in Exercise 38 to find the curvature of the curve $y = e^x$ at the point $(0, 1)$. Then find an equation for the osculating circle at this point. Sketch the curve and circle together.
b) Use the circle's equation to find the values of dy/dx and d^2y/dx^2 for the circle at the point $(1, 0)$. Show that these derivatives have the same values as the corresponding derivatives of the function $y = e^x$ at this point.

40. *Smooth curves in the plane with zero curvature are linear.* Suppose that $x = f(t)$ and $y = g(t)$ have continuous first derivatives with respect to t and that

$$\frac{d\phi}{ds} = \frac{d}{ds}\tan^{-1}\left(\frac{\dot{y}}{\dot{x}}\right) = 0$$

for all t in some interval I. Show that $y = C_1x + C_2$ for some constants C_1 and C_2. Thus, the curve $\mathbf{r} = f(t)\mathbf{i} + g(t)\mathbf{j}$, $t \in I$, lies along a straight line if its curvature is zero.

✲ COMPUTER PROGRAM: v, a, |v|, T, N, B, AND κ

The following BASIC program computes the velocity, acceleration, speed, the unit tangent, normal, and binormal vectors, and the curvature of the curve

$$\mathbf{r} = f(t)\mathbf{i} + g(t)\mathbf{j} + h(t)\mathbf{k} \text{ at any value of } t \text{ you select.}$$

There is a major difference between a numerical method like the one implemented here and the analytic technique by which we find the functions that define the quantities evaluated by the method. When we solve a problem analytically, we find a function that we can evaluate for any value of the parameter. The numerical method implemented by the program listed below must be run again with each new parameter value. The analytical result has another advantage in that we can often recognize critical values of the parameter where important changes will occur.

Another important difference between the two approaches is that a numerical solution almost always involves approximations. For example, a derivative is found from the ratio of two finite differences. It is not found as the limiting value of this ratio that is the definition of the derivative. The accuracy of such derivative approximations depends on the proper choice of the size of Δt. You may want to vary the value for DT in line 60 to investigate this point.

	PROGRAM	COMMENT
10	DIM F(3), G(3), H(3), DR(3), DDR(3)	Define arrays (subscripted variables)
20	DIM TV(3), NV(3), VXA(3), BV(3)	for vectors
30	DEF FNF(T) = 3*COS(T)	Define parameter function for x
40	DEF FNG(T) = 3*SIN(T)	and for y
50	DEF FNH(T) = T^2	and for z

	PROGRAM	COMMENT
60	DT = .01	Δt for derivative computations
70	INPUT "ENTER A VALUE FOR T ", TT	Get a t-value from the user in variable TT
80	T = TT	Save this in variable T
90	FOR I = −1 TO 1: F(I + 2) = FNF(T + I*DT): NEXT I	Compute three values for $x = f(t)$
100	FOR I = −1 TO 1: G(I + 2) = FNG(T + I*DT): NEXT I	and three for $y = g(t)$
110	FOR I = −1 TO 1: H(I + 2) = FNH(T + I*DT): NEXT I	and three for $z = h(t)$
120	DR(1) = (F(3) − F(1))/2/DT	Compute first component of velocity
130	DR(2) = (G(3) − G(1))/2/DT	and the second
140	DR(3) = (H(3) − H(1))/2/DT	and the third
150	DDR(1) = (F(3) − 2*F(2) + F(1))/DT^2	Do the same for
160	DDR(2) = (G(3) − 2*G(2) + G(1))/DT^2	the second
170	DDR(3) = (H(3) − 2*H(2) + H(1))/DT^2	derivative
180	PRINT "AT T = ";T;" THE COMPONENTS OF V AND A ARE"	Begin the output line for **V** and **A**,
190	PRINT DR(1); DR(2); DR(3)	now the velocity
200	PRINT DDR(1); DDR(2); DDR(3)	and the acceleration
210	SUM = 0: FOR I = 1 TO 3: SUM = SUM + DR(I)^2: NEXT I	Get sum of squares components of **V**
220	SPD = SQR(SUM)	and compute the speed
230	PRINT "SPEED = "; SPD	Print the speed
240	FOR I = 1 TO 3: TV(I) = DR(I)/SPD: NEXT I	Compute components of **T**
250	PRINT "COMPONENTS OF T VECTOR: "; TV(1); TV(2); TV(3)	and print them
260	SUM = 0: FOR I = 1 TO 3: SUM = SUM + DDR(I)^2: NEXT I	Sum squares of components of **A**
270	MAGNA = SQR(SUM)	and find $\lvert\mathbf{A}\rvert$
280	FOR I = 1 TO 3: NV(I) = DDR(I)/MAGNA: NEXT I	Get components of **N**
290	PRINT "COMPONENTS OF N VECTOR: "; NV(1); NV(2); NV(3)	and print them
300	VXA(1) = DR(2)*DDR(3) − DR(3)*DDR(2)	Get first component of $\mathbf{V} \times \mathbf{A}$
310	VXA(2) = DR(3)*DDR(1) − DR(1)*DDR(3)	and then the second
320	VXA(3) = DR(1)*DDR(2) − DR(2)*DDR(1)	and the third
330	FOR I = 1 TO 3: BV (1) = VXA(I)/SPD/MAGNA: NEXT I	Get components of **B**
340	PRINT "COMPONENTS OF B VECTOR: "; BV(1); BV(2); BV(3)	and print them
350	SUM = 0: FOR I = 1 TO 3: SUM = SUM + VXA(I)^2: NEXT I	Sum squares of components of $\mathbf{V} \times \mathbf{A}$
360	PRINT "CURVATURE = "; SQR(SUM)/SPD^3	and compute and print the curvature
370	END	

Here is the output for $t = 2$.

```
AT T = 2 COMPONENTS OF V AND A ARE:
 - 2.727842   -1.248431   4.000008
   1.249313   -2.737045   2.000332
SPEED = 4.999977
COMPONENTS OF T VECTOR:  - 0.545571    - 0.2496874   0.8000053
COMPONENTS OF N VECTOR:    0.345786    - 0.7575618   0.5536536
COMPONENTS OF B VECTOR:    0.4678131     0.578688    0.4996421
CURVATURE = 0.1295347
```

Computer

Use the computer program just listed (or any similar program) to find **v**, **a**, |**v**|, **T**, **N**, **B**, and κ for the curves in Exercises 41–44 at the given values of t. (Change lines 30, 40, and 50 for each exercise.)

41. $\mathbf{r} = (t \cos t)\,\mathbf{i} + (t \sin t)\,\mathbf{j} + t\mathbf{k}, \quad t = \sqrt{3}$

42. $\mathbf{r} = (e^t \cos t)\,\mathbf{i} + (e^t \sin t)\,\mathbf{j} + e^t\,\mathbf{k}, \quad t = \ln 2$

43. $\mathbf{r} = (t - \sin t)\,\mathbf{i} + (1 - \cos t)\,\mathbf{j} + \sqrt{-t}\,\mathbf{k}, \quad t = -3\pi$

44. $\mathbf{r} = (3t - t^2)\,\mathbf{i} + (3t^2)\,\mathbf{j} + (3t + t^3)\,\mathbf{k}, \quad t = 1$

11.5 Planetary Motion and Satellites

SKIP

In this section, we derive Kepler's laws of planetary motion from Newton's laws of motion and gravitation and discuss the orbits of Earth satellites. The derivation of Kepler's laws from Newton's is one of the triumphs of mathematical modeling with calculus. It draws on almost everything we have studied so far, including the algebra and geometry of vectors in space, the calculus of vector functions, the solutions of differential equations and initial value problems, and the polar coordinate description of conic sections.

Vector Equations for Motion in Polar and Cylindrical Coordinates

When a particle moves along a curve in the polar coordinate plane, we express its position, velocity, and acceleration in terms of the moving unit vectors

$$\mathbf{u}_r = (\cos\theta)\,\mathbf{i} + (\sin\theta)\,\mathbf{j}, \qquad \mathbf{u}_\theta = -(\sin\theta)\,\mathbf{i} + (\cos\theta)\,\mathbf{j} \tag{1}$$

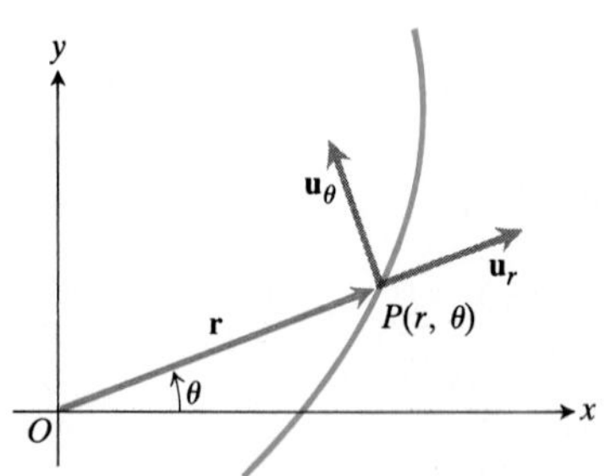

11.33 The length of the position vector $\mathbf{r}$ is the positive polar coordinate r of the point P. Thus, $\mathbf{u}_r$, which is $\mathbf{r}/|\mathbf{r}|$, is also $\mathbf{r}/r$. Equations (1) express $\mathbf{u}_r$ and $\mathbf{u}_\theta$ in terms of $\mathbf{i}$ and $\mathbf{j}$.

(Fig. 11.33). The vector $\mathbf{u}_r$ is the direction of the particle's position vector $\mathbf{r} = \overrightarrow{OP}$ and is related to $\mathbf{r}$ by the equations

$$\mathbf{u}_r = \frac{\mathbf{r}}{|\mathbf{r}|} = \frac{\mathbf{r}}{r} \qquad \text{and} \qquad \mathbf{r} = r\mathbf{u}_r, \tag{2}$$

where r, the length of the position vector, is the positive polar distance coordinate of P. The vector $\mathbf{u}_r$ points in the direction of increasing r. The vector $\mathbf{u}_\theta$, orthogonal to $\mathbf{u}_r$, points in the direction of increasing θ. We find from (1) that

$$\begin{aligned} \frac{d\mathbf{u}_r}{d\theta} &= -(\sin\theta)\,\mathbf{i} + (\cos\theta)\,\mathbf{j} = \mathbf{u}_\theta, \\ \frac{d\mathbf{u}_\theta}{d\theta} &= -(\cos\theta)\,\mathbf{i} - (\sin\theta)\,\mathbf{j} = -\mathbf{u}_r. \end{aligned} \tag{3}$$

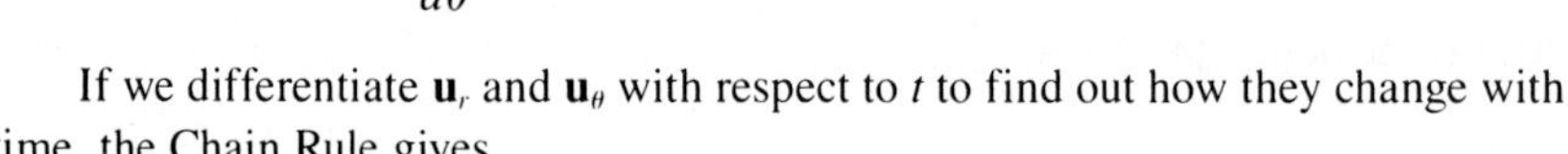

If we differentiate $\mathbf{u}_r$ and $\mathbf{u}_\theta$ with respect to t to find out how they change with time, the Chain Rule gives

$$\dot{\mathbf{u}}_r = \frac{d\mathbf{u}_r}{d\theta}\,\dot\theta = \dot\theta\,\mathbf{u}_\theta, \qquad \dot{\mathbf{u}}_\theta = \frac{d\mathbf{u}_\theta}{d\theta}\,\dot\theta = -\dot\theta\,\mathbf{u}_r. \tag{4}$$

As in the previous section, we use Newton's dot notation for time derivatives to keep the formulas as simple as we can: $\dot{\mathbf{u}}_r$ means $d\mathbf{u}_r/dt$, $\dot\theta$ means $d\theta/dt$, and so on.

When we express the particle's velocity vector in terms of coordinate unit vectors (Fig. 11.34), we find that

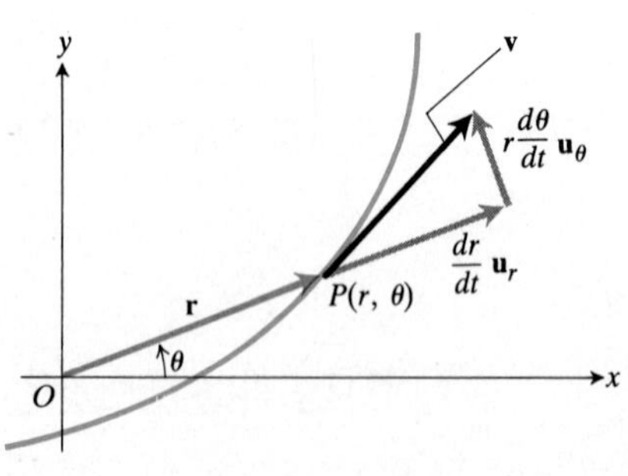

11.34 In polar coordinates, the velocity vector is
$\mathbf{v} = (dr/dt)\mathbf{u}_r + r(d\theta/dt)\mathbf{u}_\theta$.

$$\mathbf{v} = \frac{d\mathbf{r}}{dt} = \frac{d}{dt}\left(r\mathbf{u}_r\right) = \dot r\,\mathbf{u}_r + r\,\dot{\mathbf{u}}_r = \dot r\,\mathbf{u}_r + r\dot\theta\,\mathbf{u}_\theta. \tag{5}$$

The acceleration is

$$\mathbf{a} = \dot{\mathbf{v}} = (\ddot r\,\mathbf{u}_r + \dot r\,\dot{\mathbf{u}}_r) + (r\ddot\theta\,\mathbf{u}_\theta + \dot r\dot\theta\,\mathbf{u}_\theta + r\dot\theta\,\dot{\mathbf{u}}_\theta). \tag{6}$$

When we use Eqs. (4) to evaluate $\dot{\mathbf{u}}_r$ and $\dot{\mathbf{u}}_\theta$ and group the components, Eq. (6) becomes

$$\mathbf{a} = (\ddot r - r\dot\theta^2)\mathbf{u}_r + (r\ddot\theta + 2\dot r\dot\theta)\mathbf{u}_\theta. \tag{7}$$

Planets Move in Planes

Let us set our new coordinate system aside for a moment and turn our attention to the physics of a planet moving about a sun.

Newton's law of gravitation says that if $\mathbf{r}$ is the radius vector from the center of a sun of mass M to the center of a planet of mass m, then the force $\mathbf{F}$ of the gravitational attraction between the planet and sun is given by the equation

$$\mathbf{F} = -\frac{GmM}{|\mathbf{r}|^2}\frac{\mathbf{r}}{|\mathbf{r}|} \tag{8}$$

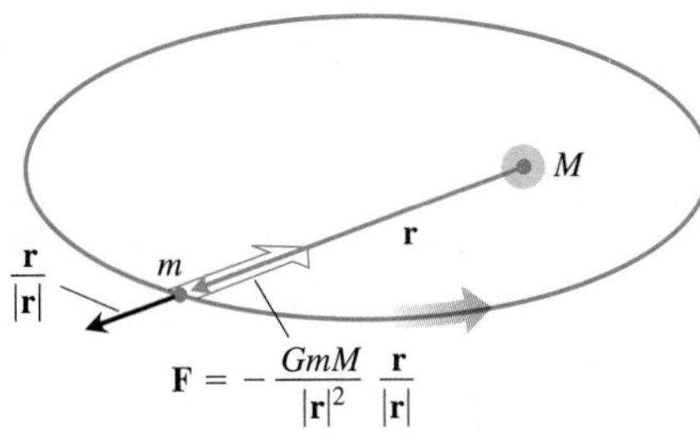

11.35 The force of gravity is directed along the line joining the centers of mass.

(Fig. 11.35). The number G is the (universal) **gravitational constant.** If we measure mass in kilograms, force in newtons, and distance in meters, G is about 6.6720×10^{-11} Nm^{2}kg^{-2}.

Combining Eq. (8) with the equation $\mathbf{F} = m\ddot{\mathbf{r}}$, Newton's second law of motion, gives

$$m\ddot{\mathbf{r}} = -\frac{GmM}{|\mathbf{r}|^2}\frac{\mathbf{r}}{|\mathbf{r}|},$$

$$\ddot{\mathbf{r}} = -\frac{GM}{|\mathbf{r}|^2}\frac{\mathbf{r}}{|\mathbf{r}|}. \tag{9}$$

The planet is accelerated toward the sun's center at all times.

Equation (9) says that $\ddot{\mathbf{r}}$ is a scalar multiple of $\mathbf{r}$ and hence that

$$\mathbf{r} \times \ddot{\mathbf{r}} = \mathbf{0}. \tag{10}$$

A routine vector calculation shows $\mathbf{r} \times \ddot{\mathbf{r}}$ to be the derivative of $\mathbf{r} \times \dot{\mathbf{r}}$:

$$\frac{d}{dt}(\mathbf{r} \times \dot{\mathbf{r}}) = \underbrace{\dot{\mathbf{r}} \times \dot{\mathbf{r}}}_{\mathbf{0}} + \mathbf{r} \times \ddot{\mathbf{r}} = \mathbf{r} \times \ddot{\mathbf{r}}. \tag{11}$$

Hence Eq. (10) is equivalent to

$$\frac{d}{dt}(\mathbf{r} \times \dot{\mathbf{r}}) = \mathbf{0}, \tag{12}$$

which integrates to

$$\mathbf{r} \times \dot{\mathbf{r}} = \mathbf{C} \tag{13}$$

for some constant vector $\mathbf{C}$.

The mathematical significance of Eq. (13) is that it tells us that $\mathbf{r}$ and $\dot{\mathbf{r}}$ always lie in a plane normal to a fixed vector $\mathbf{C}$. The real-world interpretation is that the center of mass of the planet tracked by $\mathbf{r}$ moves in a plane through the center of the sun (Fig. 11.36).

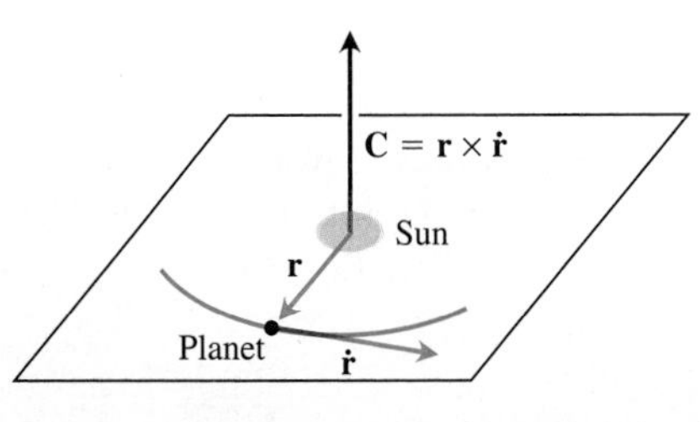

11.36 A planet that obeys Newton's laws of gravitation and motion travels in the plane through the sun's center of mass normal to $\mathbf{C} = \mathbf{r} \times \dot{\mathbf{r}}$.

Coordinates and Initial Conditions

We now introduce cylindrical coordinates in a way that places the origin at the sun's center of mass and makes the plane of the planet's motion the polar coordinate plane. This makes $\mathbf{r}$ the planet's polar coordinate position vector and makes $|\mathbf{r}|$ equal to r and $\mathbf{r}/|\mathbf{r}|$ equal to $\mathbf{u}_r$. We also position the z-axis in a way that makes $\mathbf{k}$ the direction of $\mathbf{C}$. Thus, $\mathbf{k}$ has the same right-hand relation to $\mathbf{r} \times \dot{\mathbf{r}}$ that $\mathbf{C}$ does, and the planet's motion is counterclockwise when viewed from the positive z-axis. This

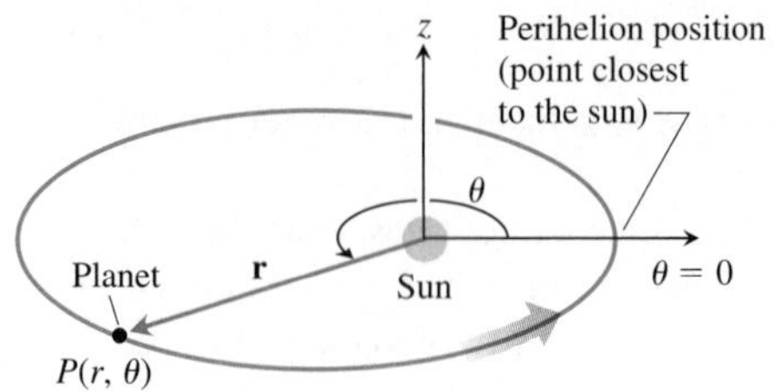

11.37 The coordinate system for planetary motion. The motion is counterclockwise when viewed from above, as it is here, and $d\theta/dt$ is positive.

makes θ increase with t, so that $d\theta/dt$ is positive for all t. Finally, we rotate the polar coordinate plane about the z-axis, if necessary, to make the initial ray coincide with the direction $\mathbf{r}$ has when the planet is closest to the sun. This runs the ray through the planet's **perihelion** position (Fig. 11.37).

If we now measure time so that $t = 0$ at perihelion, we have the following initial conditions for the planet's motion.

1. $r = r_0$, the minimum radius, when $t = 0$
2. $\dot{r} = 0$ when $t = 0$ (because r has a minimum value then)
3. $\theta = 0$ when $t = 0$
4. $|\mathbf{v}| = v_0$ when $t = 0$

Since

$$\begin{aligned} v_0 &= |\mathbf{v}|_{t=0} && \text{(Standard notation)} \\ &= \left|\dot{r}\mathbf{u}_r + r\dot{\theta}\mathbf{u}_\theta\right|_{t=0} && \text{(Eq. (5))} \\ &= \left|r\dot{\theta}\mathbf{u}_\theta\right|_{t=0} && (\dot{r} = 0 \text{ when } t = 0) \\ &= (|r\dot{\theta}||\mathbf{u}_\theta|)_{t=0} \\ &= |r\dot{\theta}|_{t=0} && (|\mathbf{u}_\theta| = 1) \\ &= (r\dot{\theta})_{t=0}, && (r \text{ and } \dot{\theta} \text{ are both positive}) \end{aligned}$$

we also know that

5. $r\dot{\theta} = v_0$ when $t = 0$.

Statement of Kepler's First Law (The Conic Section Law)

Kepler's first law says that a planet's path is a conic section with the sun at one focus. The eccentricity of the conic is

$$e = \frac{r_0 v_0^2}{GM} - 1 \tag{14}$$

and the polar equation is

$$r = \frac{(1 + e)r_0}{1 + e\cos\theta}. \tag{15}$$

The derivation uses Kepler's second law, so we shall state and prove the second law before proving the first law.

Kepler's Second Law (The Equal Area Law)

Kepler's second law says that the radius vector from the sun to a planet (the vector $\mathbf{r}$ in our model) sweeps out equal areas in equal times (Fig. 11.38). To derive the law, we use Eq. (5) to evaluate the cross product $\mathbf{C} = \mathbf{r} \times \dot{\mathbf{r}}$ from Eq. (13):

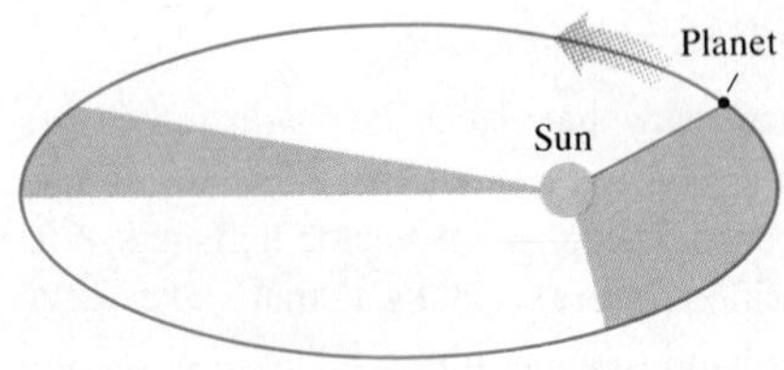

11.38 The line joining a planet to its sun sweeps over equal areas in equal times.

$$\begin{aligned} \mathbf{C} &= \mathbf{r} \times \dot{\mathbf{r}} = \mathbf{r} \times \mathbf{v} \\ &= r\mathbf{u}_r \times (\dot{r}\mathbf{u}_r + r\dot{\theta}\mathbf{u}_\theta) && \text{(Eq. (5))} \\ &= r\dot{r}\underbrace{(\mathbf{u}_r \times \mathbf{u}_r)}_{\mathbf{0}} + r(r\dot{\theta})\underbrace{(\mathbf{u}_r \times \mathbf{u}_\theta)}_{\mathbf{k}} && (16) \\ &= r(r\dot{\theta})\,\mathbf{k}. \end{aligned}$$

The German astronomer, mathematician, and physicist Johannes Kepler (1571–1630) was the first, and until Descartes the only, scientist to demand physical (as opposed to theological) explanations of celestial phenomena. His three laws of motion, the results of a lifetime of work, changed the course of astronomy forever and played a crucial role in the development of Newton's physics.

Setting t equal to zero shows that

$$\mathbf{C} = [r(r\dot{\theta})]_{t=0}\,\mathbf{k} = r_0 v_0\,\mathbf{k}. \tag{17}$$

Substituting this value for $\mathbf{C}$ in Eq. (16) gives

$$r_0 v_0\,\mathbf{k} = r^2\dot{\theta}\,\mathbf{k}, \qquad \text{or} \qquad r^2\dot{\theta} = r_0 v_0. \tag{18}$$

This is where the area comes in. The area differential in polar coordinates is

$$dA = \frac{1}{2} r^2\, d\theta$$

(Section 9.8). Accordingly, dA/dt has the constant value

$$\frac{dA}{dt} = \frac{1}{2} r^2 \dot{\theta} = \frac{1}{2} r_0 v_0, \tag{19}$$

which is Kepler's second law.

For Earth, r_0 is about 150,000,000 km, v_0 is about 30 km/sec, and dA/dt is about 2,250,000,000 km^2/sec. Every time your heart beats, Earth advances 30 km along its orbit and the radius joining Earth to the sun sweeps out 2,250,000,000 km^2 of area.

Proof of Kepler's First Law

To prove that a planet moves along a conic section with one focus at its sun, we need to express the planet's radius r as a function of θ. This requires a long sequence of calculations and some substitutions that are not altogether obvious.

We begin with the equation that comes from equating the coefficients of $\mathbf{u}_r = \mathbf{r}/|\mathbf{r}|$ in Eqs. (7) and (9):

$$\ddot{r} - r\dot{\theta}^2 = -\frac{GM}{r^2}. \tag{20}$$

We eliminate $\dot{\theta}$ temporarily by replacing it with $r_0 v_0 / r^2$ from Eq. (18) and rearrange the resulting equation to get

$$\ddot{r} = \frac{r_0^2 v_0^2}{r^3} - \frac{GM}{r^2}. \tag{21}$$

We change this into a first-order equation by a change of variable. With

$$p = \frac{dr}{dt}, \qquad \frac{d^2r}{dt^2} = \frac{dp}{dt} = \frac{dp}{dr}\frac{dr}{dt} = p\frac{dp}{dr}, \qquad \text{(Chain Rule)}$$

Eq. (21) becomes

$$p\frac{dp}{dr} = \frac{r_0^2 v_0^2}{r^3} - \frac{GM}{r^2}. \tag{22}$$

Multiplying through by 2 and integrating with respect to r gives

$$p^2 = (\dot{r})^2 = -\frac{r_0^2 v_0^2}{r^2} + \frac{2GM}{r} + C_1. \tag{23}$$

The initial conditions that $r = r_0$ and $\dot{r} = 0$ when $t = 0$ determine the value of C_1 to be

$$C_1 = v_0^2 - \frac{2GM}{r_0}.$$

Accordingly, Eq. (23), after a suitable rearrangement, becomes

$$\dot{r}^2 = v_0^2\left(1 - \frac{r_0^2}{r^2}\right) + 2GM\left(\frac{1}{r} - \frac{1}{r_0}\right). \tag{24}$$

The effect of going from Eq. (20) to Eq. (24) has been to replace a second order differential equation in r by a first order differential equation in r. Our goal is still to express r in terms of θ, so we now bring θ back into the picture. To accomplish this, we divide both sides of Eq. (24) by the squares of the corresponding sides of the equation $r^2\dot{\theta} = r_0 v_0$ (Eq. 18) and use the fact that $\dot{r}/\dot{\theta} = (dr/dt)/(d\theta/dt) = dr/d\theta$ to get

$$\begin{aligned}\frac{1}{r^4}\left(\frac{dr}{d\theta}\right)^2 &= \frac{1}{r_0^2} - \frac{1}{r^2} + \frac{2GM}{r_0^2 v_0^2}\left(\frac{1}{r} - \frac{1}{r_0}\right) \\ &= \frac{1}{r_0^2} - \frac{1}{r^2} + 2h\left(\frac{1}{r} - \frac{1}{r_0}\right). \qquad \left(h = \frac{GM}{r_0^2 v_0^2}\right)\end{aligned} \tag{25}$$

To simplify further, we substitute

$$u = \frac{1}{r}, \qquad u_0 = \frac{1}{r_0}, \qquad \frac{du}{d\theta} = -\frac{1}{r^2}\frac{dr}{d\theta}, \qquad \left(\frac{du}{d\theta}\right)^2 = \frac{1}{r^4}\left(\frac{dr}{d\theta}\right)^2,$$

obtaining

$$\left(\frac{du}{d\theta}\right)^2 = u_0^2 - u^2 + 2\,hu - 2\,hu_0 = (u_0 - h)^2 - (u - h)^2, \tag{26}$$

$$\frac{du}{d\theta} = \pm\sqrt{(u_0 - h)^2 - (u - h)^2}. \tag{27}$$

Which sign do we take? We know that $\dot{\theta} = r_0 v_0/r^2$ is positive. Also, r starts from a minimum value at $t = 0$, so it cannot immediately decrease, and $\dot{r} \geq 0$, at least for early positive values of t. Therefore,

$$\frac{dr}{d\theta} = \frac{\dot{r}}{\dot{\theta}} \geq 0 \qquad \text{and} \qquad \frac{du}{d\theta} = -\frac{1}{r^2}\frac{dr}{d\theta} \leq 0.$$

The correct sign for Eq. (27) is the negative sign. With this determined, we rearrange Eq. (27) and integrate both sides with respect to θ:

$$\begin{aligned}\frac{-1}{\sqrt{(u_0 - h)^2 - (u - h)^2}}\frac{du}{d\theta} &= 1 \\ \cos^{-1}\left(\frac{u - h}{u_0 - h}\right) &= \theta + C_2.\end{aligned} \tag{28}$$

The constant of integration is zero because $u = u_0$ when $\theta = 0$ and because $\cos^{-1}(1) = 0$. Therefore,

$$\frac{u - h}{u_0 - h} = \cos\theta$$

and

$$\frac{1}{r} = u = h + (u_0 - h)\cos\theta. \tag{29}$$

A few more algebraic maneuvers produce the final equation

$$r = \frac{(1 + e)r_0}{1 + e \cos \theta}, \tag{30}$$

where

$$e = \frac{1}{r_0 h} - 1 = \frac{r_0 v_0^2}{GM} - 1. \tag{31}$$

Together, Eqs. (30) and (31) say that the path of the planet is a conic section with one focus at the sun and with eccentricity $(r_0 v_0^2/GM) - 1$. This is the modern formulation of Kepler's first law.

Statement of Kepler's Third Law (The Time–Distance Law)

The time T it takes a planet to go around its sun once is the planet's **orbital period.** The semimajor axis a of the planet's orbit is the planet's **mean distance** from the sun. *Kepler's third law* says that T and a are related by the equation

$$\frac{T^2}{a^3} = \frac{4\pi^2}{GM}. \tag{32}$$

Since the right-hand side of this equation is constant within a given solar system, the ratio of T^2 to a^3 is the same for every planet in the system.

Kepler's third law is the starting point for working out the size of our solar system. It allows the semimajor axis of each planetary orbit to be expressed in astronomical units, Earth's semimajor axis being one unit. The distance between any two planets at any time can then be predicted in astronomical units and all that remains is to find one of these distances in kilometers. This can be done by bouncing radar waves off Venus, for example. The astronomical unit is now known, after a series of such measurements, to be 149,597,870 km.

We derive Kepler's third law by combining two different formulas for the area enclosed by the planet's elliptical orbit:

Formula 1: $\quad \text{Area} = \pi ab \quad$ (The geometry formula in which a is the semimajor axis and b is the semiminor axis)

Formula 2:
$$\begin{aligned} \text{Area} &= \int_0^T dA \\ &= \int_0^T \frac{1}{2} r_0 v_0 \, dt \qquad \text{(Eq. (19))} \\ &= \frac{1}{2} T r_0 v_0. \end{aligned}$$

Equating these gives

$$T = \frac{2\pi ab}{r_0 v_0} = \frac{2\pi a^2}{r_0 v_0}\sqrt{1 - e^2}. \qquad \text{(For any ellipse, } b = a\sqrt{1 - e^2}.) \tag{33}$$

It remains only to express a and e in terms of r_0, v_0, G, and M. Equation (31) does this for e. For a, we observe that setting θ equal to π in Eq. (30) gives

$$r_{\max} = r_0 \frac{1 + e}{1 - e}.$$

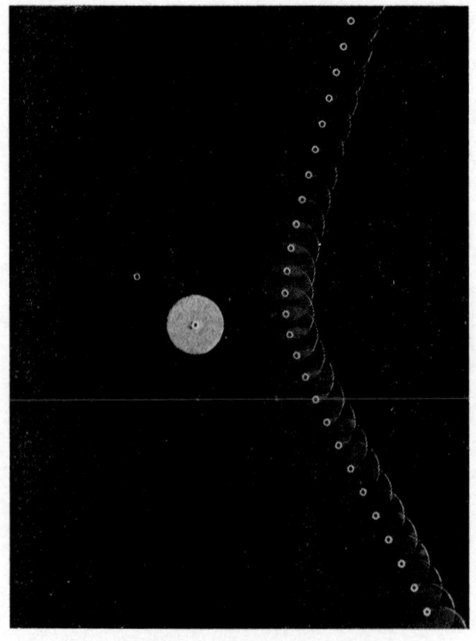

PSSC Physics, Second Edition, 1965; D.C. Heath & Company with Education Development Center, Inc.

11.39 This multiflash photograph shows a body being deflected by an inverse square law force. It moves along a hyperbola.

Hence,

$$2a = r_0 + r_{max} = \frac{2r_0}{1-e} = \frac{2r_0 GM}{2GM - r_0 v_0^2}. \tag{34}$$

Squaring both sides of Eq. (33) and substituting the results of Eqs. (31) and (34) now produces Kepler's third law, as you will see if you do Exercise 14.

Orbit Data

Although Kepler discovered his laws empirically and stated them only for the six planets known at the time, the modern derivations of Kepler's laws show that they apply to any body driven by a force that obeys an inverse square law. They apply to Halley's comet and the asteroid Icarus. They apply to the moon's orbit about Earth, and they applied to the orbit of the spacecraft Apollo 8 about the moon. They also applied to the air puck shown in Fig. 11.39 being deflected by an inverse square law force—its path is a hyperbola. Charged particles fired at the nuclei of atoms scatter along hyperbolic paths.

Tables 11.1–11.3 give additional data for planetary orbits and for the orbits of seven of Earth's artificial satellites (Fig. 11.40). Vanguard 1 sent back data that revealed differences between the levels of Earth's oceans and provided the first determination of the precise locations of some of the more isolated Pacific islands. The data also verified that the gravitation of the sun and moon would affect the orbits of Earth's satellites and that solar light could exert enough pressure to deform an orbit.

Syncom 3 is one of a series of U.S. Department of Defense telecommunications satellites. Tiros 11 (for "television infrared observation satellite") is one of a series of weather satellites. GOES 4 (for "geostationary operational environmental satellite") is one of a series of satellites designed to gather information about Earth's atmosphere. Its orbital period, 1436.2 minutes, is nearly the same as Earth's rotational period of 1436.1 minutes, and its orbit is nearly circular ($e = 0.0003$). Intelsat 5 is a heavy-capacity commercial telecommunications satellite.

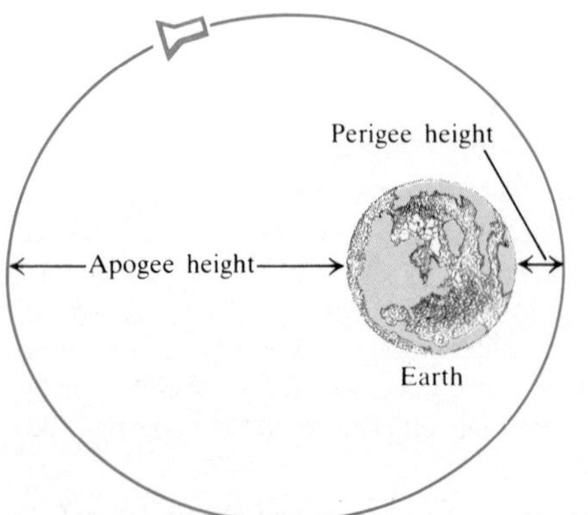

11.40 The orbit of an Earth satellite: $2a$ = diameter of Earth + perigee height + apogee height.

Circular Orbits

For circular orbits, e is zero, $r = r_0$ is a constant, and Eq. (14) gives

$$r = r_0 = \frac{GM}{v_0^2}, \tag{35}$$

which reduces to

$$r = \frac{GM}{v^2} \tag{36}$$

because v is constant as well (Exercise 12). Kepler's third law becomes

$$\frac{T^2}{r^3} = \frac{4\pi^2}{GM} \tag{37}$$

because $a = r$.

TABLE 11.1
Values of a, e, and T for the major planets

Planet	Semimajor axis a†	Eccentricity e	Period T
Mercury	57.95	0.2056	87.967 days
Venus	108.11	0.0068	224.701 days
Earth	149.57	0.0167	365.256 days
Mars	227.84	0.0934	1.8808 years
Jupiter	778.14	0.0484	11.8613 years
Saturn	1427.0	0.0543	29.4568 years
Uranus	2870.3	0.0460	84.0081 years
Neptune	4499.9	0.0082	164.784 years
Pluto	5909	0.2481	248.35 years

†Millions of kilometers

TABLE 11.3
Numerical data

Gravitational constant: $G = 6.6720 \times 10^{-11}\ \text{Nm}^2\text{kg}^{-2}$
(When you use this value of G in a calculation, remember to express force in newtons, distance in meters, mass in kilograms, and time in seconds.)
Sun's mass: 1.99×10^{30} kg
Earth's mass: 5.975×10^{24} kg
Equatorial radius of Earth: 6378.533 km
Polar radius of Earth: 6356.912 km
Earth's rotational period: 1436.1 min
Earth's orbital period: 1 year = 365.256 days

TABLE 11.2
Data on Earth's satellites

Name	Launch date	Time or expected time aloft	Weight at launch (kg)	Period (min)	Perigee height (km)	Apogee height (km)	Semimajor axis a (km)	Eccentricity
Sputnik 1	Oct. 1957	57.6 days	83.6	96.2	215	939	6,955	0.052
Vanguard 1	March 1958	300 years	1.47	138.5	649	4,340	8,872	0.208
Syncom 3	Aug. 1964	$> 10^6$ years	39	1436.2	35,718	35,903	42,189	0.002
Skylab 4	Nov. 1973	84.06 days	13,980	93.11	422	437	6,808	0.001
Tiros 11	Oct. 1978	500 years	734	102.12	850	866	7,236	0.001
GOES 4	Sept. 1980	$> 10^6$ years	627	1436.2	35,776	35,800	42,166	0.0003
Intelsat 5	Dec. 1980	$> 10^6$ years	1,928	1417.67	35,143	35,707	41,803	0.007

EXERCISES 11.5

Reminder: When a calculation involves the gravitational constant G, express distance in meters, mass in kilograms, and time in seconds.

Calculator Exercises

1. Since the orbit of Skylab 4 had a semimajor axis of $a = 6808$ km, Kepler's third law with M equal to the earth's mass should give the period. Calculate it. Compare your result with the value in Table 11.2.

2. Earth's distance from the sun at perihelion is approximately 149,577,000 km, and the eccentricity of the earth's orbit about the sun is 0.0167. Compute the velocity v_0 of the earth in its orbit at perihelion. (Use Eq. 14.)

3. In July 1965, the USSR launched Proton 1, weighing 12,200 kg (at launch), with a perigee height of 183 km, an apogee height of 589 km, and a period of 92.25 minutes. Using the relevant data for the mass of the earth and the gravitational constant G, compute the semimajor axis a of the orbit from Eq. (32). Compare your answer with the number you get by adding the perigee and apogee heights to the diameter of the earth.

4. a) The Viking 1 orbiter, which surveyed Mars from August 1975 to June 1976, had a period of 1639 min. Use this and the fact that the mass of Mars is 6.418×10^{23} kg to find the semimajor axis of the Viking 1 orbit.

b) The Viking 1 orbiter was 1499 km from the surface of

Mars at its closest point and 35,800 km from the surface at its farthest point. Use this information together with the value you obtained in part (a) to estimate the average diameter of Mars.

5. The Viking 2 orbiter, which surveyed Mars from September 1975 to August 1976, moved in an ellipse whose semimajor axis was 22,030 km. What was the orbital period? (Express your answer in minutes.)

6. If a satellite is to hold a geostationary orbit, what must the semimajor axis of its orbit be? Compare your result with the semimajor axes of the satellites in Table 11.2.

7. The mass of Mars is 6.418×10^{23} kg. If a satellite revolving about Mars is to hold a stationary orbit (have the same period as the period of Mars's rotation, which is 1477.4 min), what must the semimajor axis of its orbit be?

8. The period of the moon's rotation about the earth is 2.36055×10^6 sec. About how far away is the moon?

9. A satellite moves around the earth in a circular orbit. Express the satellite's speed as a function of the orbit's radius.

10. If T is measured in seconds and a in meters, what is the value of T^2/a^3 for planets in our solar system? For satellites orbiting the earth? For satellites orbiting the moon? (The moon's mass is 7.354×10^{22} kg.)

Noncalculator Exercises

11. For what values of v_0 in Eq. (14) is the orbit in Eq. (15) a circle? An ellipse? A hyperbola?

12. Show that a planet in a circular orbit moves with a constant speed. (*Hint:* This is a consequence of one of Kepler's laws.)

13. Suppose that $\mathbf{r}$ is the position vector of a particle moving along a plane curve and dA/dt is the rate at which the vector sweeps out area. Without introducing coordinates, and assuming the necessary derivatives exist, give a geometric argument based on increments and limits for the validity of the equation

$$\frac{dA}{dt} = \frac{1}{2}|\mathbf{r} \times \dot{\mathbf{r}}|.$$

14. Complete the derivation of Kepler's third law (the part following Eq. 33).

Motion in the Polar Coordinate Plane

15. A particle moves in the plane in such a way that its polar coordinates at time t are $r = t$, $\theta = t$. Find $\mathbf{v}$, $\mathbf{a}$, and the curvature of the particle's path as functions of t.

16. At time t, the polar coordinates of a particle moving in the plane are

$$r = e^{\omega t} + e^{-\omega t}, \qquad \theta = t,$$

where ω ("omega") is a constant. Find the acceleration vector at time $t = 0$.

17. A ball is placed in a long frictionless tube that is pivoted at one end and rotates with constant angular velocity $d\theta/dt = 2$. The position of the ball at time t in polar coordinates is $r = \cosh 2t$, $\theta = 2t$. Show that the $\mathbf{u}_r$-component of the acceleration is always zero.

REVIEW QUESTIONS

1. State the rules for differentiating and integrating vector functions. Give examples.

2. How do you define and calculate the velocity, speed, direction of motion, and acceleration of a body moving along a sufficiently differentiable space curve? Give an example.

3. What is special about the derivatives of vectors of constant length? Give an example.

4. What are the vector and parametric equations for ideal projectile motion? How do you find a projectile's maximum height, flight time, and range? Give examples.

5. How do you define and calculate the length of a segment of a space curve? Give an example. What mathematical assumptions are involved in the definition?

6. How do you measure distance along a smooth curve in space from a preselected base point? Give an example of a directed-distance function.

7. What is a differentiable curve's unit tangent vector? Give an example.

8. Define curvature, circle of curvature (osculating circle), center of curvature, and radius of curvature for twice-differentiable curves in the plane. Give examples. What curves have zero curvature? constant curvature?

9. What is a plane curve's principal normal vector? Which way does it point? Give an example.

10. How do you define $\mathbf{N}$ and κ for curves in space? How are these quantities related? Give examples.

11. What is a curve's binormal vector? Give an example. What is the relation of this vector to the curve's torsion? Give an example.

12. What formulas are available for writing a moving body's acceleration as a sum of its tangential and normal components? Give an example. Why might one want to write the acceleration this way? What if the body moves at a constant speed? At a constant speed around a circle?

13. State Kepler's laws. To what do they apply?

MISCELLANEOUS EXERCISES

Graph the curves in Exercises 1 and 2, and sketch their velocity and acceleration vectors at the given values of t.

1. $\mathbf{r} = (4 \cos t)\,\mathbf{i} + (\sqrt{2} \sin t)\,\mathbf{j}, \quad t = 0 \text{ and } \pi/4$

2. $\mathbf{r} = (\sqrt{3} \sec t)\,\mathbf{i} + (\sqrt{3} \tan t)\,\mathbf{j}, \quad t = 0 \text{ and } \pi/6$

Evaluate the integrals in Exercises 3 and 4.

3. $\displaystyle\int_0^1 [(3 + 6t)\,\mathbf{i} + (4 + 8t)\,\mathbf{j} + (6\pi \cos \pi t)\,\mathbf{k}]\,dt$

4. $\displaystyle\int_e^{e^2} \left[\frac{2 \ln t}{t}\,\mathbf{i} + \frac{1}{t \ln t}\,\mathbf{j} + \frac{1}{t}\,\mathbf{k}\right] dt$

Exercises 5 and 6 give the velocity and initial position of a particle moving in space. Find the particle's position function.

5. $\dfrac{d\mathbf{r}}{dt} = -(\sin t)\,\mathbf{i} + (\cos t)\,\mathbf{j} + \mathbf{k}, \quad \mathbf{r} = \mathbf{j} \text{ when } t = 0$

6. $\dfrac{d\mathbf{r}}{dt} = \dfrac{1}{t^2 + 1}\,\mathbf{i} - \dfrac{1}{\sqrt{1 - t^2}}\,\mathbf{j} + \dfrac{1}{\sqrt{t^2 + 1}}\,\mathbf{k}, \quad \mathbf{r} = \mathbf{j} + \mathbf{k}$ when $t = 0$.

Exercises 7 and 8 give the acceleration and the initial position and velocity of a particle moving in space. Find the particle's position function.

7. $\dfrac{d^2\mathbf{r}}{dt^2} = 2\,\mathbf{j}, \quad \mathbf{r} = \mathbf{i} \text{ and } \dfrac{d\mathbf{r}}{dt} = \mathbf{k} \text{ when } t = 0$

8. $\dfrac{d^2\mathbf{r}}{dt^2} = -2\,\mathbf{i} - 4\,\mathbf{j}, \quad \mathbf{r} = 3\,\mathbf{i} + 3\,\mathbf{j} \text{ and } \dfrac{d\mathbf{r}}{dt} = 4\,\mathbf{i} \text{ when } t = 1$

Find the lengths of the curves in Exercises 9 and 10.

9. $\mathbf{r} = (2 \cos t)\,\mathbf{i} + (2 \sin t)\,\mathbf{j} + t^2\,\mathbf{k}, \quad 0 \le t \le \pi/4$

10. $\mathbf{r} = (3 \cos t)\,\mathbf{i} + (3 \sin t)\,\mathbf{j} + 2t^{3/2}\,\mathbf{k}, \quad 0 \le t \le 3$

In Exercises 11–14, find $\mathbf{T}$, $\mathbf{N}$, $\mathbf{B}$, κ, and τ at the given value of t. Then find equations for the osculating, normal, and rectifying planes at that value of t.

11. $\mathbf{r} = \dfrac{4}{9}(1 + t)^{3/2}\,\mathbf{i} + \dfrac{4}{9}(1 - t)^{3/2}\,\mathbf{j} + \dfrac{1}{3}t\,\mathbf{k}, \quad t = 0$

12. $\mathbf{r} = (e^t \sin 2t)\,\mathbf{i} + (e^t \cos 2t)\,\mathbf{j} + 2e^t\,\mathbf{k}, \quad t = 0$

13. $\mathbf{r} = t\,\mathbf{i} + \dfrac{1}{2}e^{2t}\,\mathbf{j}, \quad t = \ln 2$

14. $\mathbf{r} = (3 \cosh 2t)\,\mathbf{i} + (3 \sinh 2t)\,\mathbf{j} + 6t\,\mathbf{k}, \quad t = \ln 2$

In Exercises 15 and 16, write $\mathbf{a}$ in the form $\mathbf{a} = a_{\mathrm{T}}\mathbf{T} + a_{\mathrm{N}}\mathbf{N}$ at $t = 0$ without finding $\mathbf{T}$ and $\mathbf{N}$.

15. $\mathbf{r} = (2 + 3t + 3t^2)\,\mathbf{i} + (4t + 4t^2)\,\mathbf{j} - (6 \cos t)\,\mathbf{k}$

16. $\mathbf{r} = (2 + t)\,\mathbf{i} + (t + 2t^2)\,\mathbf{j} + (1 + t^2)\,\mathbf{k}$

17. Find $\mathbf{T}$, $\mathbf{N}$, $\mathbf{B}$, κ, and τ as functions of t if $\mathbf{r} = (\sin t)\,\mathbf{i} + (\sqrt{2} \cos t)\,\mathbf{j} + (\sin t)\,\mathbf{k}$.

18. The position of a particle in the plane at time t is

$$\mathbf{r} = \frac{1}{\sqrt{1 + t^2}}\,\mathbf{i} + \frac{t}{\sqrt{1 + t^2}}\,\mathbf{j}.$$

Find the particle's highest speed.

19. Suppose $\mathbf{r} = (e^t \cos t)\,\mathbf{i} + (e^t \sin t)\,\mathbf{j}$. Show that the angle between $\mathbf{r}$ and $\mathbf{a}$ never changes. What *is* the angle?

20. At what times in the interval $0 \le t \le \pi$ are the velocity and acceleration vectors of the motion $\mathbf{r} = \mathbf{i} + (5 \cos t)\,\mathbf{j} + (3 \sin t)\,\mathbf{k}$ orthogonal?

21. The position of a particle moving in space at time $t \ge 0$ is

$$\mathbf{r} = 2\,\mathbf{i} + \left(4 \sin \frac{t}{2}\right)\mathbf{j} + \left(3 - \frac{t}{\pi}\right)\mathbf{k}.$$

Find the first time $\mathbf{r}$ is perpendicular to the vector $\mathbf{i} - \mathbf{j}$.

22. A particle moves around the unit circle in the xy-plane. Its position at time t is $\mathbf{r} = x\mathbf{i} + y\,\mathbf{j}$, where x and y are differentiable functions of t. Find dy/dt if $\mathbf{v} \cdot \mathbf{i} = y$. Is the motion clockwise, or counterclockwise?

23. You send a message through a pneumatic tube that follows the curve $9y = x^3$ (distance in meters). At the point (3, 3), $\mathbf{v} \cdot \mathbf{i} = 4$ and $\mathbf{a} \cdot \mathbf{i} = -2$. Find the values of $\mathbf{v} \cdot \mathbf{j}$ and $\mathbf{a} \cdot \mathbf{j}$ at (3, 3).

24. A particle moves in the plane so that its velocity and position vectors are always orthogonal. Show that the particle moves in a circle centered at the origin.

25. A circular wheel with radius 1 ft and center C rolls to the right along the x-axis at a half turn per second (Fig. 11.41). At time t seconds, the position vector of the point P on the wheel's circumference is

$$\mathbf{r} = (\pi t - \sin \pi t)\,\mathbf{i} + (1 - \cos \pi t)\,\mathbf{j}.$$

a) Sketch the curve traced by P during the interval $0 \le t \le 3$.

b) Find $\mathbf{v}$ and $\mathbf{a}$ at $t = 0$, 1, 2, and 3 and add these vectors to your sketch.

c) At any given time, what is the forward speed of the topmost point of the wheel? of C?

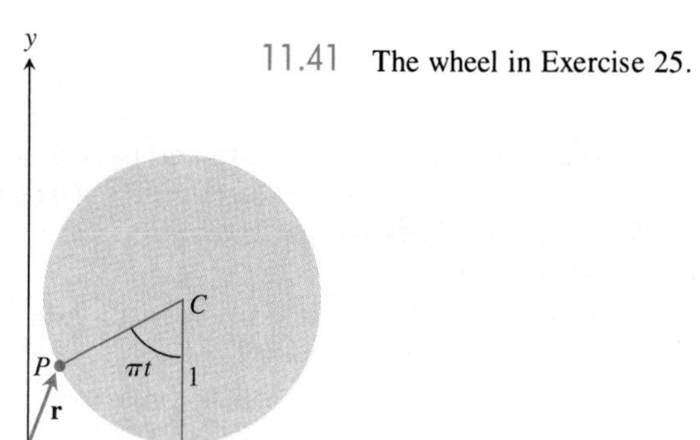

11.41 The wheel in Exercise 25.

26. The curve

$$\mathbf{r} = (2\sqrt{t}\cos t)\,\mathbf{i} + (3\sqrt{t}\sin t)\,\mathbf{j} + \sqrt{1-t}\,\mathbf{k}, \quad 0 \le t \le 1,$$

lies on a quadric surface. Describe the surface and find an equation for it.

27. Using the spherical coordinate θ as parameter, find a formula $\mathbf{r} = f(\theta)\,\mathbf{i} + g(\theta)\,\mathbf{j} + h(\theta)\,\mathbf{k}$ for the curve in which the plane $y + z = 0$ cuts the sphere $\rho = a$. Then find the length of the curve.

28. The line through the origin and the point $A(1, 1, 1)$ is the axis of rotation of a rigid body rotating with a constant angular speed of 6 rad/sec. The rotation appears to be clockwise when we look toward the origin from A. Find the velocity of the point of the body that is at the position $(1, 3, 2)$.

29. CALCULATOR *Javelin.* In Potsdam in 1988, Petra Felke of (then) East Germany set a women's world record by throwing a javelin 262 ft 5 in.

a) Assuming that Felke launched the javelin at a 40° angle to the horizontal 7 ft above the ground, what was the javelin's initial speed?

b) How high did the javelin go?

c) A women's regulation javelin is 7 ft 2 1/2 in. (220 cm) long and weighs 1.32 lb (600 gm). How much work did Felke do on the javelin to launch it at that speed? (Do not include the work it took to lift the javelin the 7 ft.)

30. A golf ball is hit with an initial speed v_0 at an angle α to the horizontal from a point that lies at the foot of a straight-sided hill that is inclined at an angle ϕ to the horizontal, where

$$0 < \phi < \alpha < \frac{\pi}{2}.$$

Show that the ball lands at a distance

$$\frac{2v_0^2 \cos\alpha}{g\cos^2\phi}\sin(\alpha - \phi),$$

measured up the face of the hill. Hence, show that the greatest range that can be achieved for a given v_0 occurs when $\alpha = (\phi/2) + (\pi/4)$, i.e., when the initial velocity vector bisects the angle between the vertical and the hill.

31. *Synchronous curves.* By eliminating α from the ideal projectile equations

$$x = v_0(\cos\alpha)t, \quad y = (v_0\sin\alpha)t - \frac{1}{2}gt^2,$$

show that $x^2 + (y + gt^2/2)^2 = v_0^2t^2$. This shows that projectiles launched simultaneously from the origin at the same initial speed will, at any given instant, all lie on the circle of radius v_0t centered at $(0, -gt^2/2)$, regardless of their launch angle. These circles are the *synchronous curves* of the launching.

32. At point P, the velocity and acceleration of a particle moving in the plane are $\mathbf{v} = 3\,\mathbf{i} + 4\,\mathbf{j}$ and $\mathbf{a} = 5\,\mathbf{i} + 15\,\mathbf{j}$. Find the curvature of the particle's path at P.

33. Show that the radius of curvature of a twice-differentiable plane curve $\mathbf{r} = f(t)\,\mathbf{i} + g(t)\,\mathbf{j}$ is given by the formula

$$\rho = \frac{\dot{x}^2 + \dot{y}^2}{\sqrt{\ddot{x}^2 + \ddot{y}^2 - \ddot{s}^2}}, \quad \text{where } \ddot{s} = \frac{d}{dt}\sqrt{\dot{x}^2 + \dot{y}^2}.$$

34. Find the point on the curve $y = e^x$ where the curvature is greatest.

35. Express the curvature of the curve

$$\mathbf{r} = \left(\int_0^t \cos\left(\frac{1}{2}\pi\theta^2\right)d\theta\right)\mathbf{i} + \left(\int_0^t \sin\left(\frac{1}{2}\pi\theta^2\right)d\theta\right)\mathbf{j}$$

as a function of the directed distance s measured along the curve from the origin (Fig. 11.42).

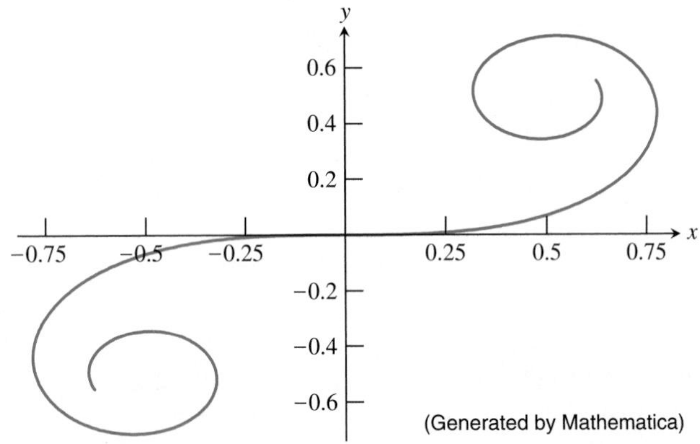

11.42 The curve in Exercise 35.

36. Find equations for the osculating, normal, and rectifying planes of the curve $\mathbf{r} = t\,\mathbf{i} + t^2\,\mathbf{j} + t^3\,\mathbf{k}$ at the point $(1, 1, 1)$.

37. Figure 11.43 shows the distance s measured counterclockwise around the circle $x^2 + y^2 = a^2$ from the point $(a, 0)$ to a point P. It also shows the angle ϕ that the tangent at P makes with the x-axis. Use the equations $s = a\theta$ and $\phi = \theta + \pi/2$ to calculate the circle's curvature directly from the definition $\kappa = |d\phi/ds|$.

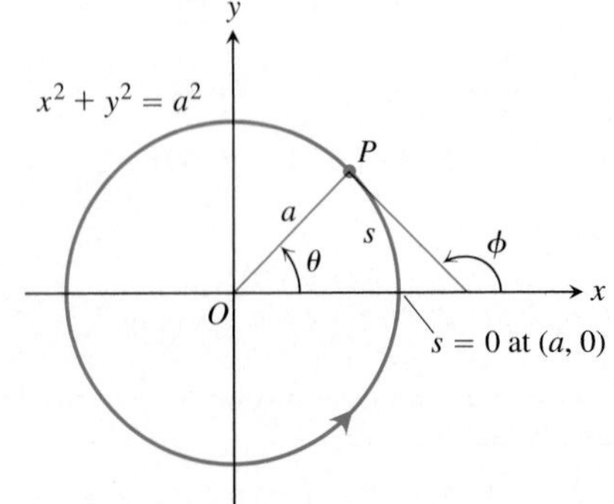

11.43 The circle in Exercise 37.

38. A frictionless particle P, starting from rest at time $t = 0$ at the point $(a, 0, 0)$, slides down the helix

$$\mathbf{r} = (a \cos \theta)\,\mathbf{i} + (a \sin \theta)\,\mathbf{j} + b\theta\,\mathbf{k} \qquad (a, b > 0)$$

under the influence of gravity, as in Fig. 11.44. The θ in this equation is the cylindrical coordinate θ and the helix is the curve $r = a$, $z = b\theta$, $\theta \geq 0$, in cylindrical coordinates. We assume θ to be a differentiable function of t for the motion. The law of conservation of energy tells us that the particle's speed after it has fallen a distance z is $\sqrt{2gz}$, where g is the constant acceleration of gravity.

a) Find the angular velocity $d\theta/dt$ when $\theta = 2\pi$.
b) Express the particle's θ- and z-coordinates as functions of t.
c) Express the tangential and normal components of the velocity $d\mathbf{r}/dt$ and acceleration $d^2\mathbf{r}/dt^2$ as functions of t. Does the acceleration have any nonzero component in the direction of the binormal vector $\mathbf{B}$?

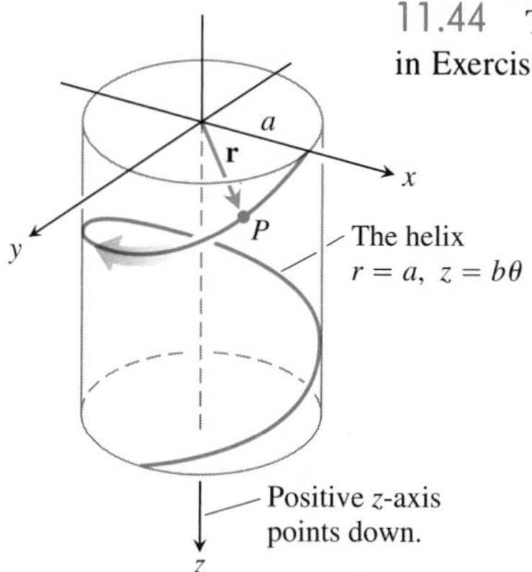

11.44 The circular helix in Exercise 38.

39. Suppose the curve in Exercise 38 is replaced by the conical helix $r = a\theta$, $z = b\theta$ shown in Fig. 11.45.

a) Express the angular velocity $d\theta/dt$ as a function of θ.
b) Express the distance the particle travels along the helix as a function of θ.

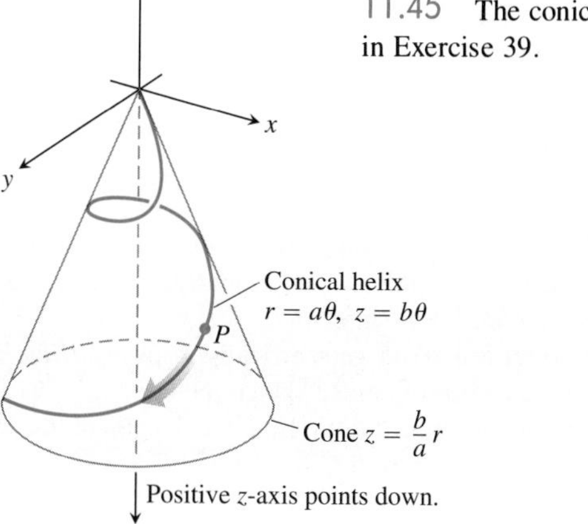

11.45 The conical helix in Exercise 39.

40. *The view from Skylab 4.* What percentage of Earth's surface area could the astronauts see when Skylab 4 was at its apogee height, 437 km above the surface? To find out, model the visible surface as the surface generated by revolving the circular arc GT in Fig. 11.46 about the y-axis. Then carry out these steps:

1. Use similar triangles to show that $y_0/6380 = 6380/(6380 + 437)$. Solve for y_0.
2. Calculate the visible area as
$$VA = \int_{y_0}^{6380} 2\pi x \, ds\,.$$
3. Express the result as a percentage of Earth's surface area.

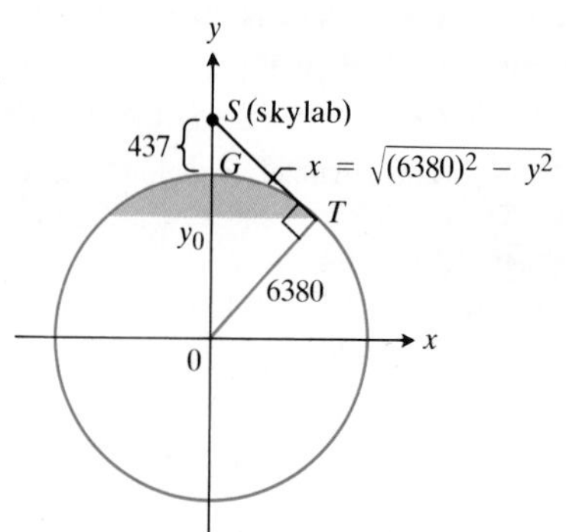

11.46 How much of the earth can you see from space at any one time? Exercise 40 shows how to find out.

41. Deduce from the orbit equation

$$r = \frac{(1 + e)r_0}{1 + e \cos \theta}$$

that a planet is closest to its sun when $\theta = 0$ and show that $r = r_0$ at that time.

42. GRAPHER or ROOT FINDER *A Kepler equation.* The problem of locating a planet in its orbit at a given time and date eventually leads to solving "Kepler" equations of the form

$$f(x) = x - 1 - \frac{1}{2} \sin x = 0.$$

a) Show that this particular equation has a solution between $x = 0$ and $x = 2$.
b) Find the solution to as many places as you can.

43. Express the curvature of a twice-differentiable curve $r = f(\theta)$ in the polar coordinate plane in terms of f and its derivatives.

44. A slender rod through the origin of the polar coordinate plane rotates (in the plane) about the origin at the rate of 3 rad/min. A beetle starting from the point (2, 0) crawls along the rod toward the origin at the rate of 1 in./min.

a) Find the beetle's acceleration and velocity in polar form when it is halfway to (1 in. from) the origin.

b) CALCULATOR To the nearest tenth of an inch, what will be the length of the path the beetle has traveled by the time it reaches the origin?

45. In Section 11.5, we found the velocity of a particle moving in the plane to be

$$\mathbf{v} = \dot{x}\,\mathbf{i} + \dot{y}\,\mathbf{j} = \dot{r}\mathbf{u}_r + r\dot{\theta}\mathbf{u}_\theta.$$

a) Express $\dot{x}$ and $\dot{y}$ in terms of $\dot{r}$ and $r\dot{\theta}$ by evaluating the dot products $\mathbf{v}\cdot\mathbf{i}$ and $\mathbf{v}\cdot\mathbf{j}$.

b) Express $\dot{r}$ and $r\dot{\theta}$ in terms of $\dot{x}$ and $\dot{y}$ by evaluating the dot products $\mathbf{v}\cdot\mathbf{u}_r$ and $\mathbf{v}\cdot\mathbf{u}_\theta$.

46. *Unit vectors for position and motion in cylindrical coordinates.* When the position of a particle moving in space is given in cylindrical coordinates, the unit vectors we use to describe its position and motion are

$$\mathbf{u}_r = (\cos\theta)\,\mathbf{i} + (\sin\theta)\,\mathbf{j}, \qquad \mathbf{u}_\theta = -(\sin\theta)\,\mathbf{i} + (\cos\theta)\,\mathbf{j},$$

and $\mathbf{k}$ (Fig. 11.47). The particle's position vector is then $\mathbf{r} = r\mathbf{u}_r + z\,\mathbf{k}$, where r is the positive polar distance coordinate of the particle's position.

a) Show that $\mathbf{u}_r$, $\mathbf{u}_\theta$, and $\mathbf{k}$, in this order, form a right-handed frame of unit vectors.

b) Show that

$$\frac{d\mathbf{u}_r}{d\theta} = \mathbf{u}_\theta \quad\text{and}\quad \frac{d\mathbf{u}_\theta}{d\theta} = -\mathbf{u}_r.$$

c) Assuming that the necessary derivatives with respect to t exist, express $\mathbf{v} = \dot{\mathbf{r}}$ and $\mathbf{a} = \ddot{\mathbf{r}}$ in terms of $\mathbf{u}_r$, $\mathbf{u}_\theta$, $\mathbf{k}$, $\dot{r}$, and $\dot{\theta}$. (The dots indicate derivatives with respect to t: $\dot{\mathbf{r}}$ means $d\mathbf{r}/dt$, $\ddot{\mathbf{r}}$ means $d^2\mathbf{r}/dt^2$, and so on.) Section 11.5 shows how the vectors mentioned here are used in describing planetary motion.

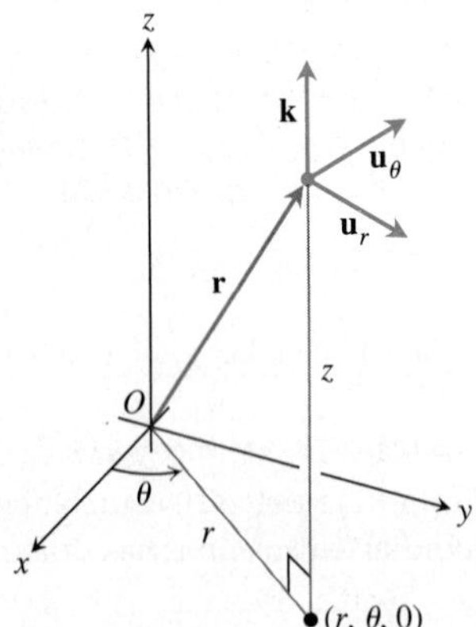

11.47 The unit vectors for describing motion in cylindrical coordinates (Exercise 46).

47. *Unit vectors for position and motion in spherical coordinates.* Hold two of the three spherical coordinates ρ, ϕ, θ of a point P in space constant while letting the remaining coordinate increase. Let $\mathbf{u}$, with a subscript corresponding to the increasing coordinate, be the unit vector that points in the direction in which P then starts to move. The three resulting unit vectors, $\mathbf{u}_\rho$, $\mathbf{u}_\phi$, and $\mathbf{u}_\theta$ at P are shown in Fig. 11.48.

a) Express $\mathbf{u}_\rho$, $\mathbf{u}_\phi$, and $\mathbf{u}_\theta$ in terms of $\mathbf{i}$, $\mathbf{j}$, and $\mathbf{k}$.

b) Show that $\mathbf{u}_\rho \cdot \mathbf{u}_\phi = 0$.

c) Show that $\mathbf{u}_\theta = \mathbf{u}_\rho \times \mathbf{u}_\phi$.

d) Show that $\mathbf{u}_\rho$, $\mathbf{u}_\phi$, and $\mathbf{u}_\theta$, in that order, make a right-handed frame of mutually orthogonal vectors.

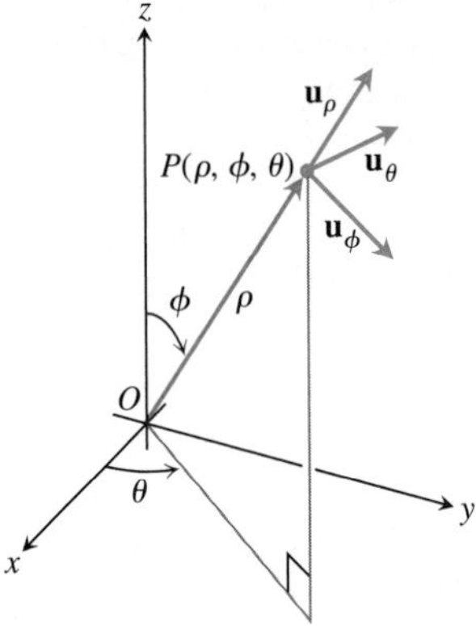

11.48 The unit vectors for describing motion in spherical coordinates (Exercise 47).

48. *(Continuation of Exercise 47)* A particle P moves in space in such a way that its spherical coordinates are differentiable functions of time t. Express the particle's position and velocity vectors in terms of ρ, ϕ, and θ, the derivatives of ρ, ϕ, and θ, and the vectors $\mathbf{u}_\rho$, $\mathbf{u}_\phi$, and $\mathbf{u}_\theta$.

49. *Arc length in cylindrical coordinates*

a) Show that when you express $ds^2 = dx^2 + dy^2 + dz^2$ in terms of cylindrical coordinates, you get $ds^2 = dr^2 + r^2d\theta^2 + dz^2$.

b) Interpret this result geometrically in terms of the edges and a diagonal of a box. Sketch the box.

c) Use the result in (a) to find the length of the curve $r = e^\theta$, $z = e^\theta$, $0 \le \theta \le \ln 8$.

50. *Arc length in spherical coordinates*

a) Show that when you express $ds^2 = dx^2 + dy^2 + dz^2$ in terms of spherical coordinates, you get $ds^2 = d\rho^2 + \rho^2 d\phi^2 + \rho^2\sin^2\phi\, d\theta^2$.

b) Interpret this result geometrically in terms of the edges and a diagonal of a box. Sketch the box.

c) Use the result in (a) to find the length of the curve $\rho = 2e^\theta$, $\phi = \pi/6$, $0 \le \theta \le \ln 8$.

CHAPTER

12

FUNCTIONS OF TWO OR MORE VARIABLES AND THEIR DERIVATIVES

Overview We now begin our study of functions that have more than one independent variable. Functions of two or more variables appear more often in science than functions of a single variable, and their calculus is richer. Their derivatives are more varied because of the different ways in which the variables can interact. Their integrals lead to a greater variety of applications. The studies of probability, statistics, fluid dynamics, and electricity, to mention only a few, all lead in natural ways to functions of more than one variable.

In the present chapter, we introduce real-valued functions of two or more variables, define and calculate their derivatives, learn how to write the Chain Rule, and find out how to calculate the rates at which functions change as we move in different directions through their domains. We also learn how to estimate values of functions of two or more variables and how to calculate their extreme values in a given region.

The remaining chapters of the book build on what we do here. Chapter 13 deals with integrals of functions of two and three variables. These are the integrals with which we calculate the volumes and centers of mass of three-dimensional solids. In Chapter 14, we once more combine calculus with vectors, this time to derive the integral theorems that provide the mathematical foundation for the studies of fluid flow and electricity in physics and engineering. Finally, in Chapter 15, we examine various ways to solve equations of the form $dy/dx = f(x, y)$, equations in which the derivative of an unknown function y of x is given as a function of the two variables x and y.

12.1 Functions of Two or More Independent Variables

The values of many functions are determined by more than one independent variable. For example, the function $V = \pi r^2 h$ calculates the volume of a circular cylinder from its radius and height. The function $f(x, y) = x^2 + y^2$ calculates the height of the paraboloid $z = x^2 + y^2$ above the point (x, y) in the xy-plane. The function

$$w = \cos(1.7 \times 10^{-2}t - 0.2x)\, e^{-0.2x},$$

in Example 9, calculates the temperature x feet belowground on the tth day of the year as a fraction of the surface temperature on that day.

In this section, we define functions of more than one independent variable and discuss two ways to display their values graphically.

Functions and Variables

Real-valued functions of two or more real variables are defined much the way you would imagine from the single-variable case. The domains are sets of ordered pairs (triples, quadruples, whatever) of real numbers, and the ranges are sets of real numbers.

DEFINITIONS

Suppose D is a set of n-tuples of real numbers $(x_1, x_2, \ldots, x_n)$. A **real-valued function** f on D is a rule that assigns a real number

$$w = f(x_1, x_2, \ldots, x_n)$$

to each element in D. The set D is the function's **domain.** The set of w-values taken on by f is the function's **range.** The symbol w is the **dependent variable** of f and f is said to be a function of the n **independent variables** x_1 to x_n. We call the x's the function's **input variables** and call w the function's **output variable.**

If f is a function of two independent variables, we usually call the variables x and y and picture the domain of f as a region in the xy-plane. If f is a function of three independent variables, we call the variables x, y, and z and picture the domain of f as a region in space.

When functions of more than one variable arise in applications, we tend to use letters that remind us of what the variables stand for. To say that the volume of a right circular cylinder is a function of its radius and height, we might write $V = f(r, h)$. To be even more specific, we might replace the notation $f(r, h)$ by the formula that calculates the value of V from the values of r and h and write $V = \pi r^2 h$. In either case, r and h would be the independent variables and V the dependent variable of the function.

As usual, we evaluate functions defined by formulas by substituting the values of the independent variables in the formula and calculating the corresponding value of the dependent variable.

Example 1 The value of the function $f(x, y, z) = \sqrt{x^2 + y^2 + z^2}$ at the point $(3, 0, 4)$ is

$$f(3, 0, 4) = \sqrt{(3)^2 + (0)^2 + (4)^2} = \sqrt{25} = 5.$$

Domains

In defining functions of more than one variable, we continue the practice of excluding inputs that lead to complex numbers or division by zero. If $f(x, y) = \sqrt{y - x^2}$, we do not allow y to be less than x^2. If $f(x, y) = 1/xy$, we do not allow the product xy to be zero.

These are the only restrictions, however, and the domains of functions are otherwise assumed to be the largest sets for which the defining rules generate real numbers.

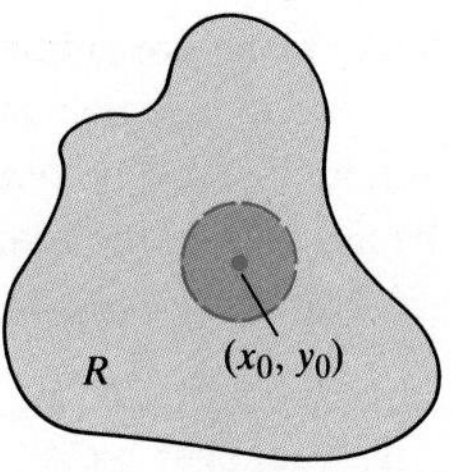

(a) Interior point

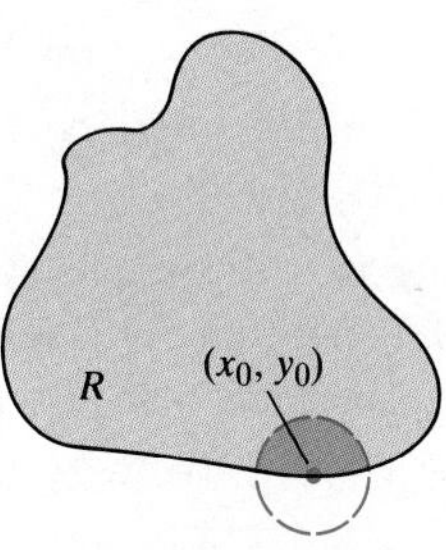

(b) Boundary point

12.1 Interior points and boundary points of a plane region R. An interior point is necessarily a point of R. A boundary point of R need not belong to R.

Example 2 *Functions of Two Variables*

Function	Domain	Range
$w = \sqrt{y - x^2}$	$y \geq x^2$	$w \geq 0$
$w = \dfrac{1}{xy}$	$xy \neq 0$	$w \neq 0$
$w = \sin xy$	Entire plane	$-1 \leq w \leq 1$
$w = -\dfrac{1}{x^2 + y^2}$	$(x, y) \neq (0, 0)$	$-\infty < w < 0$

Example 3 *Functions of Three Variables*

Function	Domain	Range
$w = \sqrt{x^2 + y^2 + z^2}$	Entire space	$w \geq 0$
$w = \dfrac{1}{x^2 + y^2 + z^2}$	$(x, y, z) \neq (0, 0, 0)$	$0 < w < \infty$
$w = xy \ln z$	Half-space $z > 0$	$-\infty < w < \infty$

The domains of functions defined on portions of the plane can have interior points and boundary points just as the domains of functions defined on intervals of the real line can.

DEFINITIONS

A point (x_0, y_0) in a region (set) R in the xy-plane is an **interior point** of R if it is the center of a disk that lies entirely in R (Fig. 12.1). A point (x_0, y_0) is a **boundary point** of R if every disk centered at (x_0, y_0) contains points that lie outside of R as well as points that lie in R. (The boundary point itself need not belong to R.)

The interior points of a region, as a set, make up the **interior** of the region. The region's boundary points make up its **boundary.** A region is **open** if it consists entirely of interior points. A region is **closed** if it contains all of its boundary points (Fig. 12.2).

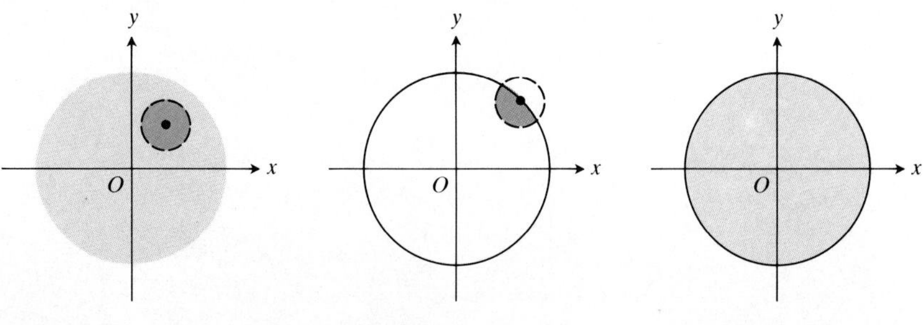

$\{(x, y) \mid x^2 + y^2 < 1\}$ Open unit disk. Every point an interior point.

$\{(x, y) \mid x^2 + y^2 = 1\}$ Boundary of unit disk. (The unit circle.)

$\{(x, y) \mid x^2 + y^2 \leq 1\}$ Closed unit disk. Contains all boundary points.

12.2 Interior points and boundary points of the unit disk in the plane.

As with intervals of real numbers, some regions in the plane are neither open nor closed. If you start with the open disk in Fig. 12.2 and add to it some but not all of its boundary points, the resulting set is neither open nor closed. The boundary points that *are* there keep the set from being open. The absence of the remaining boundary points keeps the set from being closed.

DEFINITIONS

A region in the plane is **bounded** if it lies inside a disk of fixed radius. A region is **unbounded** if it is not bounded.

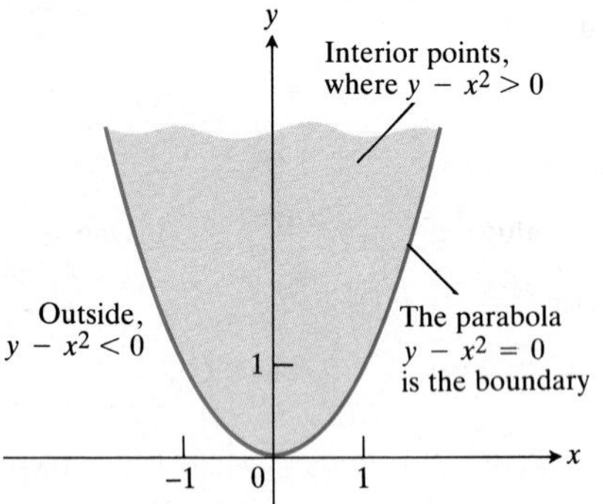

12.3 The domain of $f(x, y) = \sqrt{y - x^2}$ consists of the shaded region and its bounding parabola $y = x^2$.

Example 4

Bounded sets in the plane:	Line segments, triangles, interiors of triangles, rectangles, disks
Unbounded sets in the plane:	Lines, coordinate axes, the graphs of functions defined on infinite intervals, quadrants, half-planes, the plane itself

Example 5 The domain of the function $f(x, y) = \sqrt{y - x^2}$ is closed and unbounded (Fig. 12.3). The parabola $y = x^2$ is the boundary of the domain. The points inside the parabola make up the domain's interior.

The definitions of interior, boundary, open, closed, bounded, and unbounded for regions in space are similar to those for regions in the plane. To accommodate the extra dimension, we use solid spheres instead of disks.

DEFINITIONS

A point (x_0, y_0, z_0) in a region D in space is an **interior point** of D if it is the center of a solid sphere that lies entirely in D (Fig. 12.4). A point (x_0, y_0, z_0) is a **boundary point** of D if every sphere centered at (x_0, y_0, z_0) encloses points that lie outside of D as well as points that lie inside D. The **interior** of D is the set of interior points of D. The **boundary** of D is the set of boundary points of D.

A region D is **open** if it consists entirely of interior points. A region is **closed** if it contains its boundary.

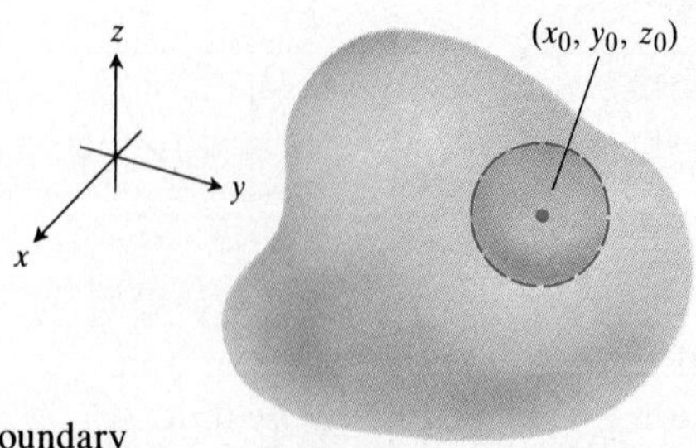

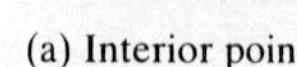

(a) Interior point

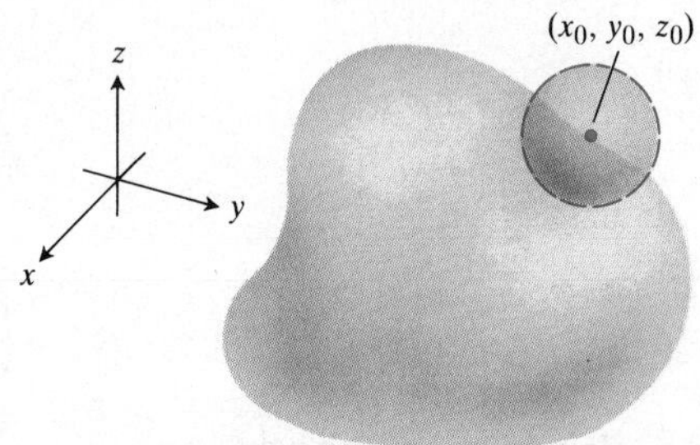

(b) Boundary point

12.4 Interior points and boundary points of a region in space.

Example 6

Open sets in space:	Interior of a solid sphere; the open half-space $z > 0$; the first octant (bounding planes absent); space itself
Closed sets in space:	Lines; planes; the closed half-space $z \geq 0$; the first octant together with its bounding planes; space itself
Neither open nor closed:	A solid sphere with part of its boundary removed; a solid cube with a missing face, edge, or corner point

Graphs and Level Curves of Functions of Two Variables

There are two standard ways to picture the values of a function $f(x, y)$. One is to draw and label curves in the domain on which f has a constant value. The other is to sketch the surface $z = f(x, y)$ in space.

DEFINITIONS

The set of points in the plane where a function $f(x, y)$ has a constant value $f(x, y) = c$ is called a **level curve** of f. The set of all points $(x, y, f(x, y))$ in space, for (x, y) in the domain of f, is called the **graph** of f. The graph of f is also called the **surface** $z = f(x, y)$.

Example 7

Graph the function

$$f(x, y) = 100 - x^2 - y^2$$

in space and plot the level curves $f(x, y) = 0$, $f(x, y) = 51$, and $f(x, y) = 75$, in the domain of f in the plane.

Solution The domain of f is the entire xy-plane and the range of f is the set of real numbers less than or equal to 100. The graph is the paraboloid $z = 100 - x^2 - y^2$, a portion of which is shown in Fig. 12.5.

The level curve $f(x, y) = 0$ is the set of points in the xy-plane at which

$$f(x, y) = 100 - x^2 - y^2 = 0, \quad \text{or} \quad x^2 + y^2 = 100,$$

which is the circle of radius 10 centered at the origin. Similarly, the level curves $f(x, y) = 51$ and $f(x, y) = 75$ are the circles

$$f(x, y) = 100 - x^2 - y^2 = 75, \quad \text{or} \quad x^2 + y^2 = 25,$$

$$f(x, y) = 100 - x^2 - y^2 = 51, \quad \text{or} \quad x^2 + y^2 = 49.$$

The level curve $f(x, y) = 100$ consists of the origin alone. (It is still called a level curve.)

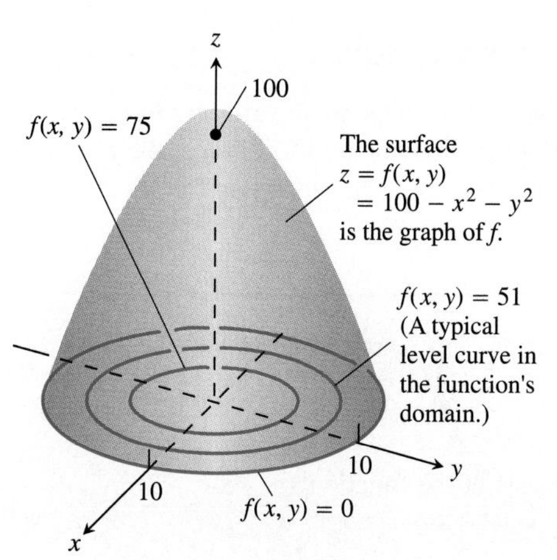

12.5 The graph and selected level curves of the function $f(x, y) = 100 - x^2 - y^2$.

Contour Lines

The curve in space in which the plane $z = c$ cuts a surface $z = f(x, y)$ consists of the points that represent the function value $f(x, y) = c$. It is called the **contour**

The contour line $f(x, y) = 100 - x^2 - y^2 = 75$ is the circle $x^2 + y^2 = 25$ in the plane $z = 75$.

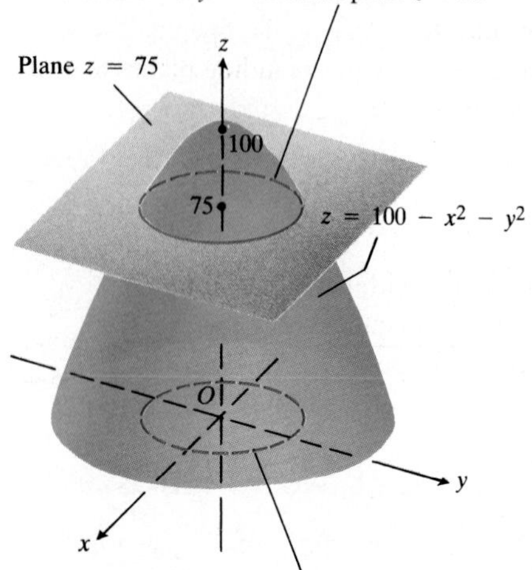

The level curve $f(x, y) = 100 - x^2 - y^2 = 75$ is the circle $x^2 + y^2 = 25$ in the xy-plane.

12.6 The graph of $f(x, y) = 100 - x^2 - y^2$ and its intersection with the plane $z = 75$.

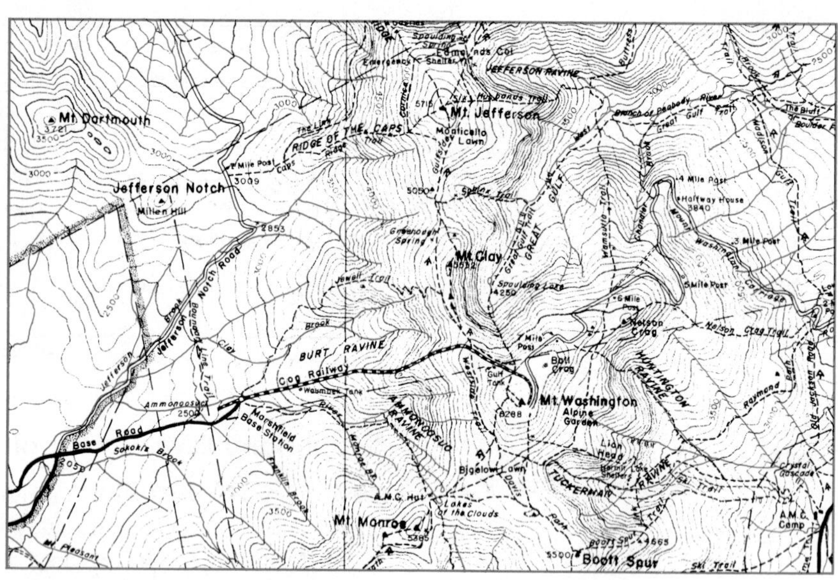

12.7 Contours on Mt. Washington in north-central New Hampshire. The streams, which follow paths of steepest descent, run perpendicular to the contours. So does the Cog Railway. (Based on a map by Louis F. Cutter, © 1987 Appalachian Mountain Club, Boston, Mass., by permission.)

line $f(x, y) = c$ to distinguish it from the level curve $f(x, y) = c$ in the domain of f. Figure 12.6 shows the contour line $f(x, y) = 75$ on the surface $z = 100 - x^2 - y^2$ defined by the function $f(x, y) = 100 - x^2 - y^2$. The contour line lies directly above the circle $x^2 + y^2 = 25$, which is the level curve $f(x, y) = 75$ in the function's domain.

Not everyone makes this distinction, however, and you may wish to call both kinds of curves by a single name and rely on context to convey which one you have in mind. On most maps, for example, the curves that represent constant elevation (height above sea level) are called contours, not level curves (Fig. 12.7).

Level Surfaces of Functions of Three Variables

In the plane, the points at which a function of two independent variables has a constant value $f(x, y) = c$ make a curve in the function's domain. In space, the points where a function of three independent variables has a constant value $f(x, y, z) = c$ make a surface in the function's domain.

DEFINITION

> The set of points in space where a function of three independent variables has a constant value $f(x, y, z) = c$ is called a **level surface** of f.

Example 8 Describe the level surfaces of the function

$$f(x, y, z) = \sqrt{x^2 + y^2 + z^2}.$$

Solution The value of f is the distance from the origin to the point (x, y, z). Each level surface

$$\sqrt{x^2 + y^2 + z^2} = c, \qquad c > 0,$$

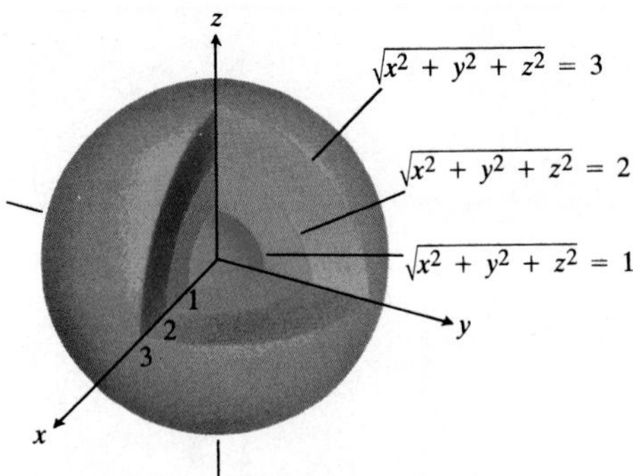

12.8 The level surfaces of the function $f(x, y, z) = \sqrt{x^2 + y^2 + z^2}$ are concentric spheres.

is a sphere of radius c centered at the origin (Fig. 12.8). The level surface $\sqrt{x^2 + y^2 + z^2} = 0$ consists of the origin alone.

We are not graphing the function here. The graph of the function, made up of the points $\left(x, y, z, \sqrt{x^2 + y^2 + z^2}\right)$, lies in a four-variable space. Instead, we are looking at level surfaces in the function's domain.

The function's level surfaces show how the function's values change as we move around in its domain. If we remain on a sphere of radius c centered at the origin, the function maintains a constant value, namely c. If we move from one sphere to another, the function's value changes. It increases if we move away from the origin and decreases if we move toward the origin. The way the function's values change depends on the direction we take. The dependence of change on direction is important. We shall return to it in Section 12.7.

Computer Graphing

The three-dimensional graphing programs for computers make it possible to graph functions of two variables with only a few keystrokes. We can often get information more quickly from a graph than we can from a formula.

Example 9 Figure 12.9 shows a computer-generated graph of the function

$$w = \cos(1.7 \times 10^{-2}t - 0.2x)\, e^{-0.2x},$$

where t is in days and x is in feet. The graph shows how the temperature beneath the earth's surface varies with time. The variation is given as a fraction of the variation at the surface. At a depth of 15 ft, the variation (change in vertical amplitude in the figure) is about 5 percent of the surface variation. At 30 ft, there is almost no variation in temperature during the year.

Another thing the graph shows that is not immediately apparent from the equation for f is that the temperature 15 ft below the surface is about half a year out of phase with the surface temperature. When the temperature is lowest on the surface (late January, say) it is at its highest 15 ft below. Fifteen feet below the ground, the seasons are reversed.

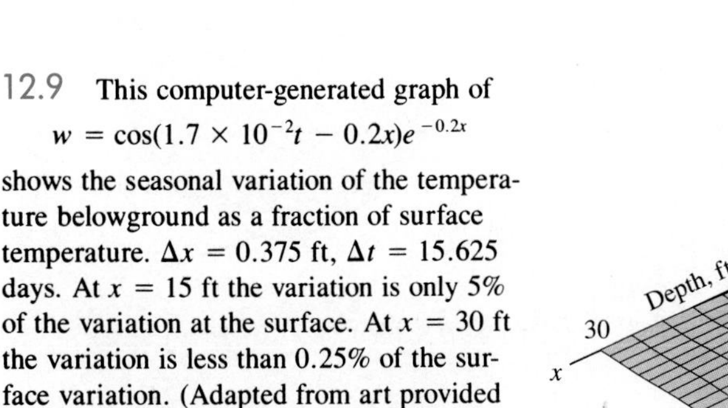

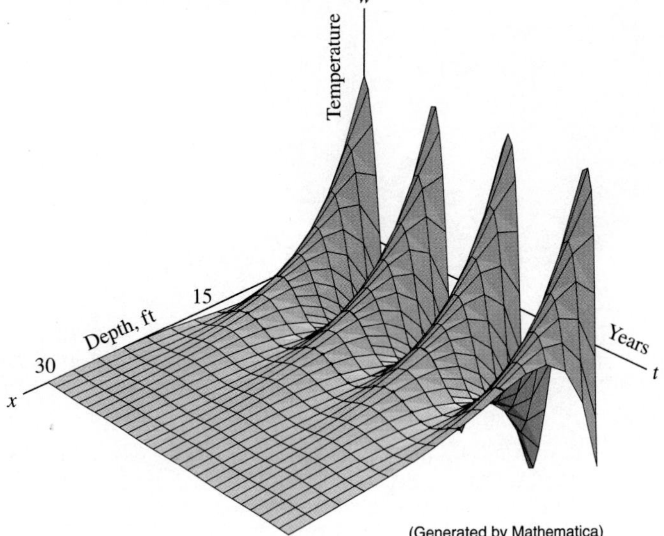

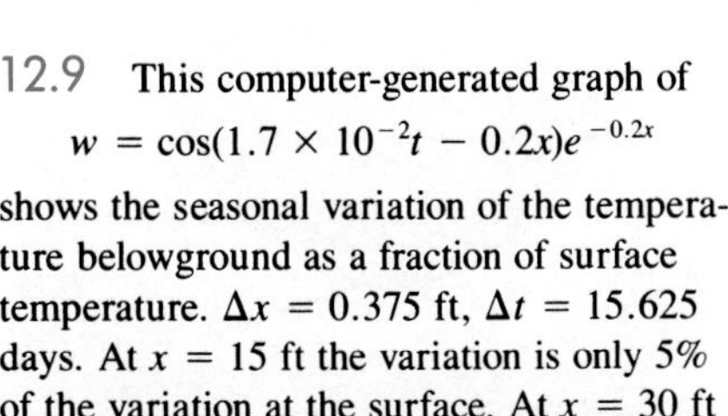

12.9 This computer-generated graph of

$$w = \cos(1.7 \times 10^{-2}t - 0.2x)e^{-0.2x}$$

shows the seasonal variation of the temperature belowground as a fraction of surface temperature. $\Delta x = 0.375$ ft, $\Delta t = 15.625$ days. At $x = 15$ ft the variation is only 5% of the variation at the surface. At $x = 30$ ft the variation is less than 0.25% of the surface variation. (Adapted from art provided by Norton Starr for G. C. Berresford's "Differential Equations and Root Cellars," *The UMAP Journal,* Vol. 2, No. 3, (1981), pp. 53–75.)

Surfaces Defined by Functions of Two Variables

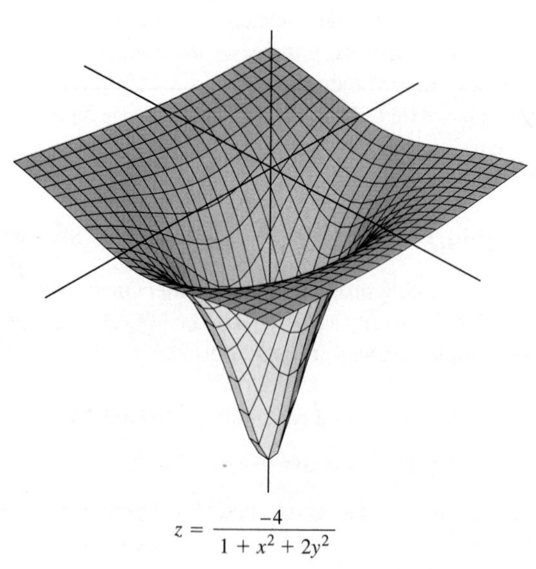

$z = \dfrac{-4}{1 + x^2 + 2y^2}$

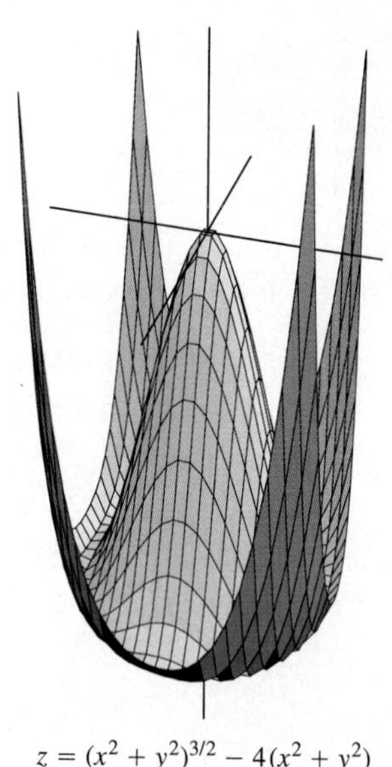

$z = (x^2 + y^2)^{3/2} - 4(x^2 + y^2)$

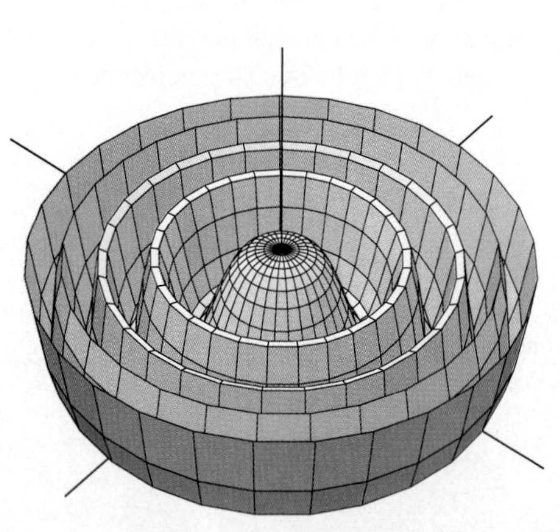

$z = 1 + \cos(x^2 + y^2)$

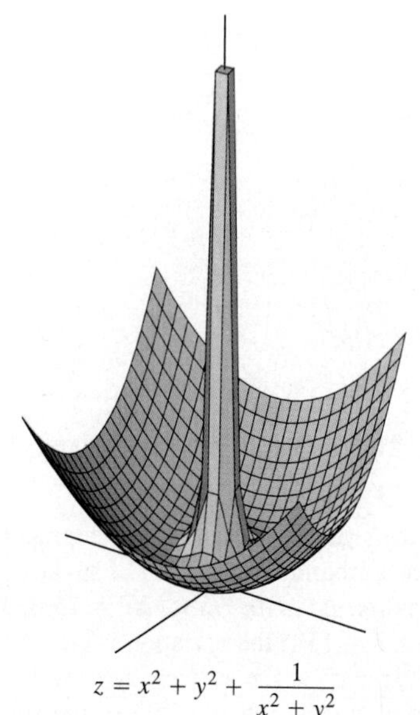

$z = x^2 + y^2 + \dfrac{1}{x^2 + y^2}$

(Generated by Mathematica)

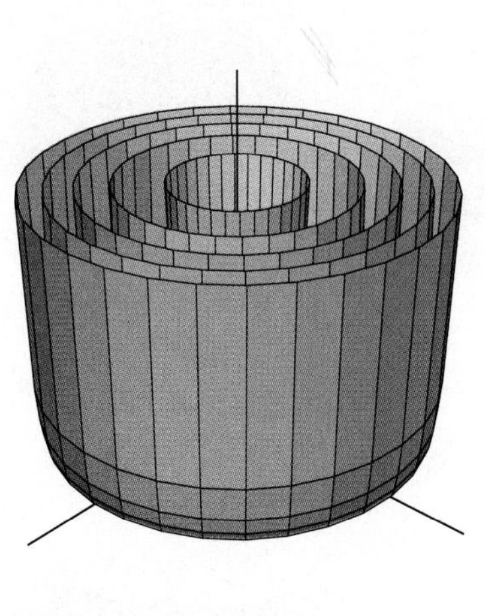

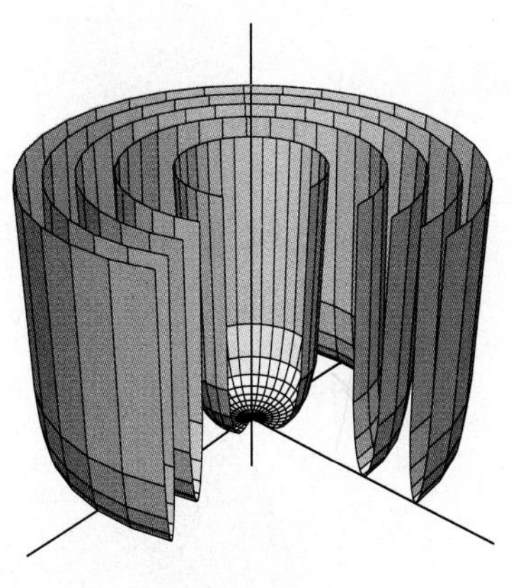

$z = \ln(\sec(x^2 + y^2))$

$z = e^{-y} \cos x$

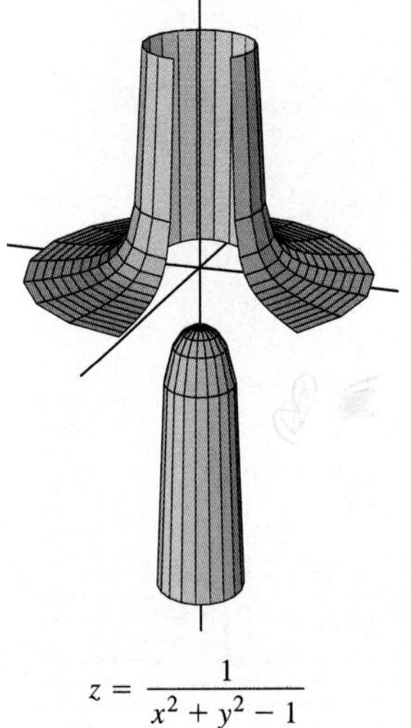

$$z = \frac{1}{x^2 + y^2 - 1}$$

(Generated by Mathematica)

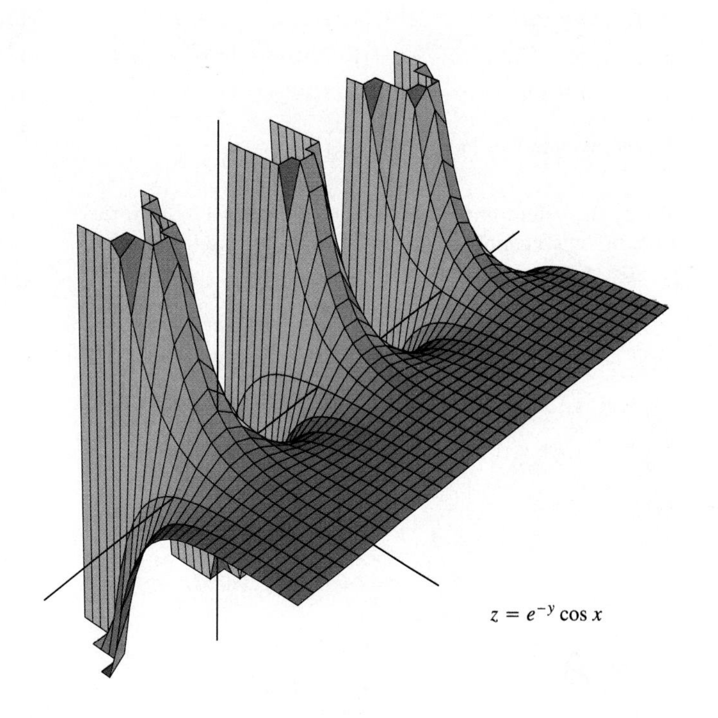

EXERCISES 12.1

In Exercises 1–10, find the function's domain and range and describe the function's level curves.

1. $f(x, y) = \sqrt{y - x}$

2. $f(x, y) = 4x^2 + 9y^2$

3. $f(x, y) = y/x^2$

4. $f(x, y) = x^2 - y^2$

5. $f(x, y) = e^{\cos xy}$

6. $f(x, y) = \ln(x^2 + y^2)$

7. $f(x, y) = \sin^{-1}(y - x)$

8. $f(x, y) = \tan^{-1}(y/x)$

9. $f(x, y) = \ln x - \ln y$

10. $f(x, y) = \sqrt{\dfrac{x - y}{x + y}}$

Display the values of the functions in Exercises 11–20 in two ways: (a) by sketching the surface $z = f(x, y)$, and (b) by drawing an assortment of level curves in the function's domain. Label each level curve with the appropriate function value.

11. $f(x, y) = y^2$

12. $f(x, y) = 4 - y^2$

13. $f(x, y) = 1 + x^2 + y^2$

14. $f(x, y) = 4 - x^2 - y^2$

15. $f(x, y) = \sqrt{x^2 + y^2}$

16. $f(x, y) = 4x^2 + y^2$

17. $f(x, y) = 1 - |y|$

18. $f(x, y) = 1 - |x| - |y|$

19. $f(x, y) = e^{-(x^2 + y^2)}$

20. $f(x, y) = 1/(x^2 + y^2)$

Exercises 21–26 show level curves for the functions graphed in (a)–(f). Match each set of level curves with the appropriate function.

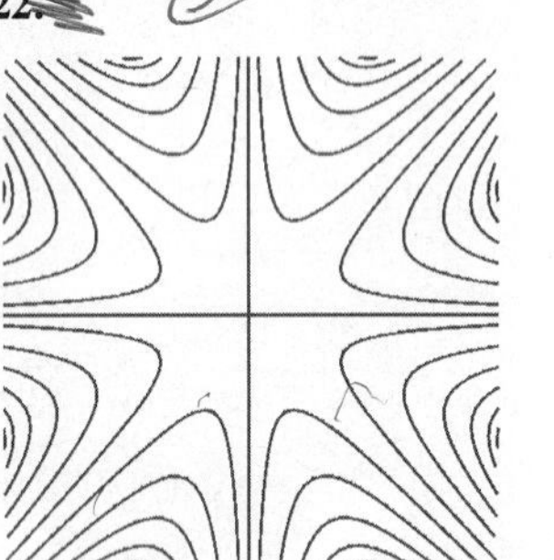

21. **22.**

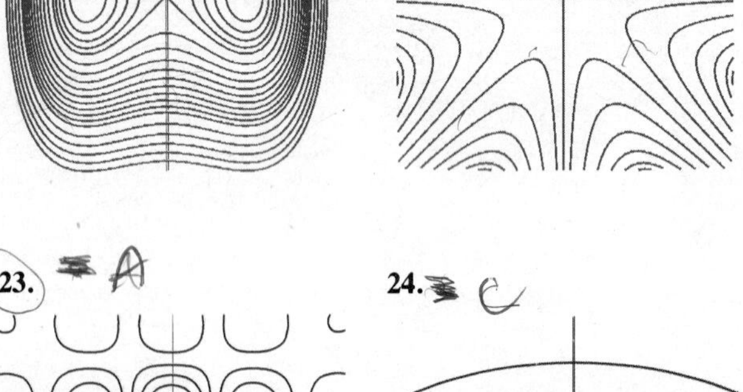

23. **24.**

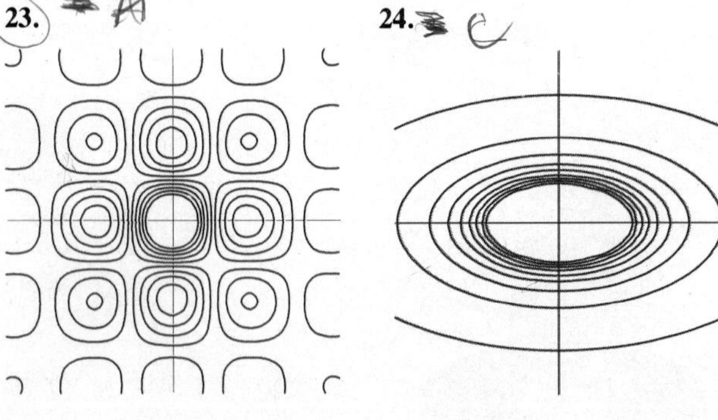

25.

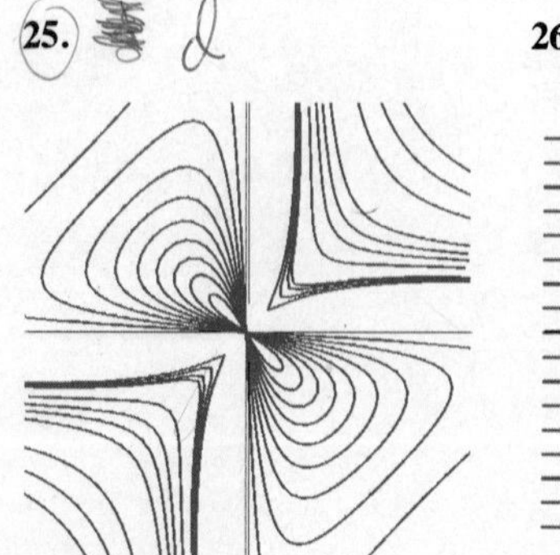

26.

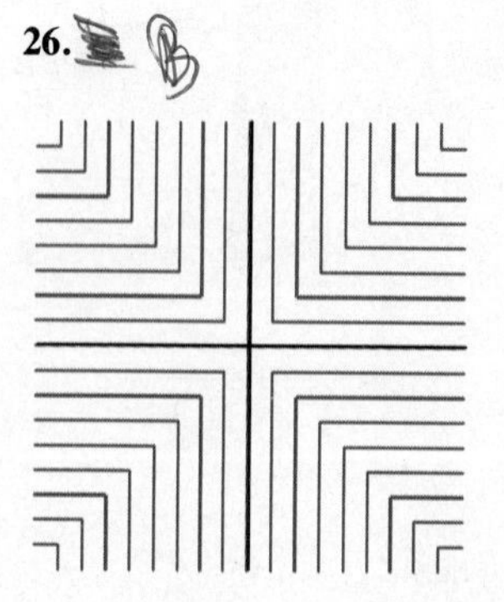

(a) (b)

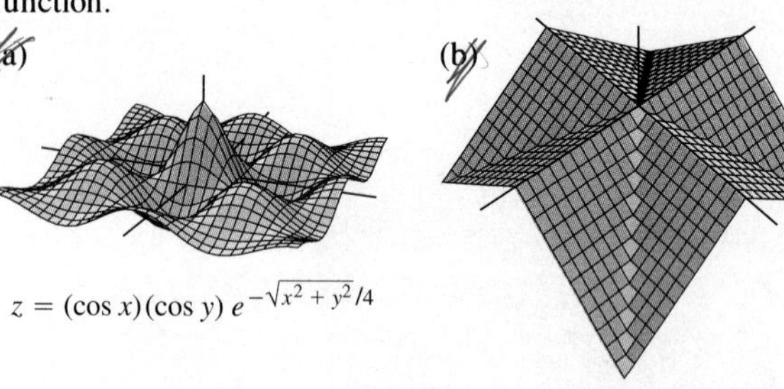

$z = (\cos x)(\cos y)\, e^{-\sqrt{x^2 + y^2}/4}$

$z = \dfrac{||x| - |y||}{2} - \dfrac{(|x| + |y|)}{2}$

(c)

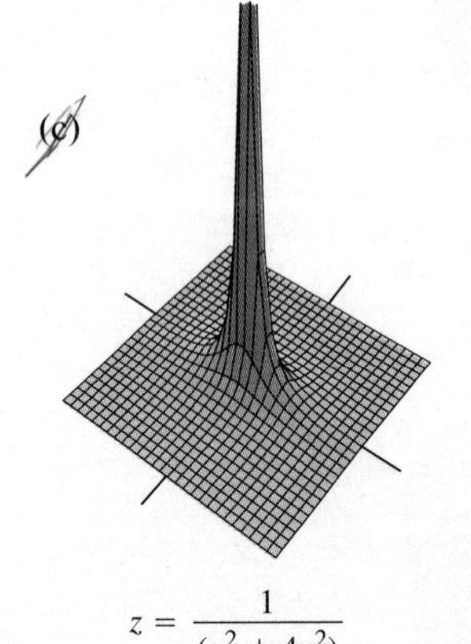

$z = \dfrac{1}{(x^2 + 4y^2)}$

(d)

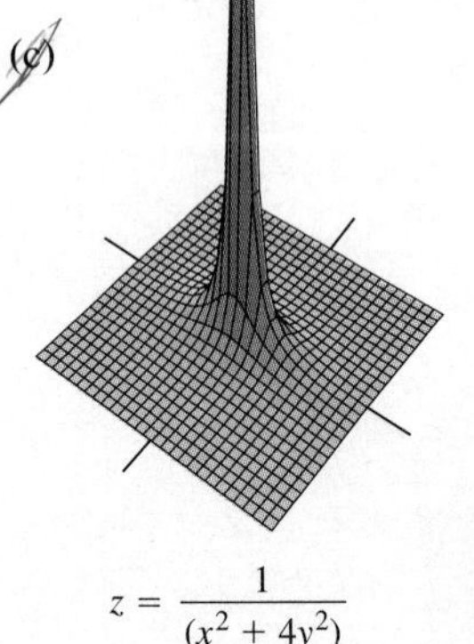

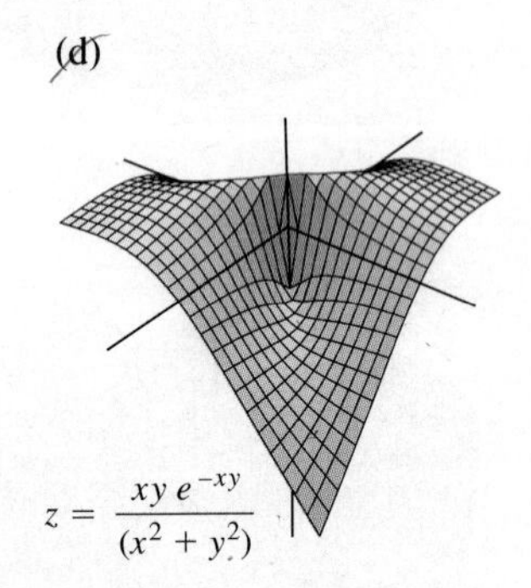

$z = \dfrac{xy\, e^{-xy}}{(x^2 + y^2)}$

(e)

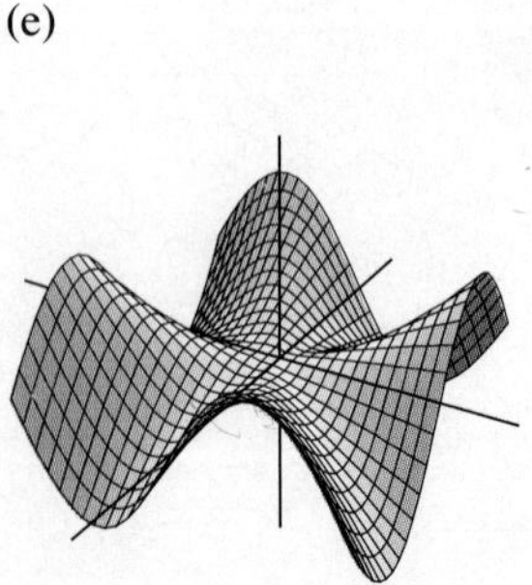

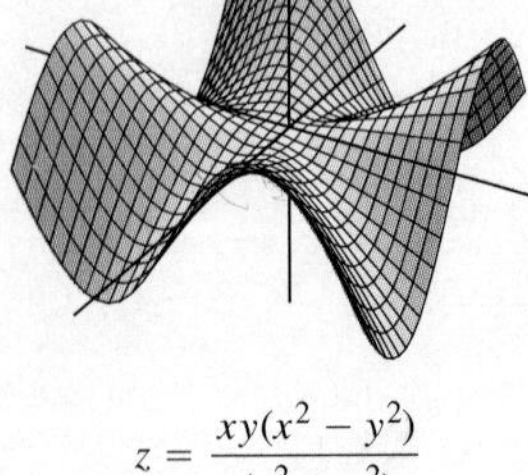

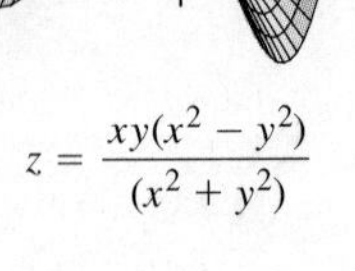

$z = \dfrac{xy(x^2 - y^2)}{(x^2 + y^2)}$

(f)

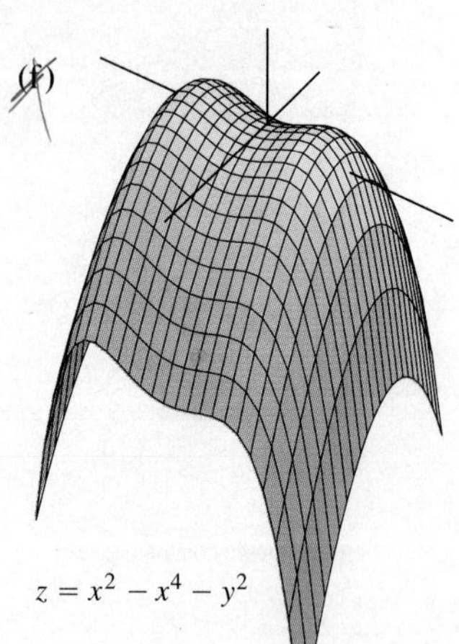

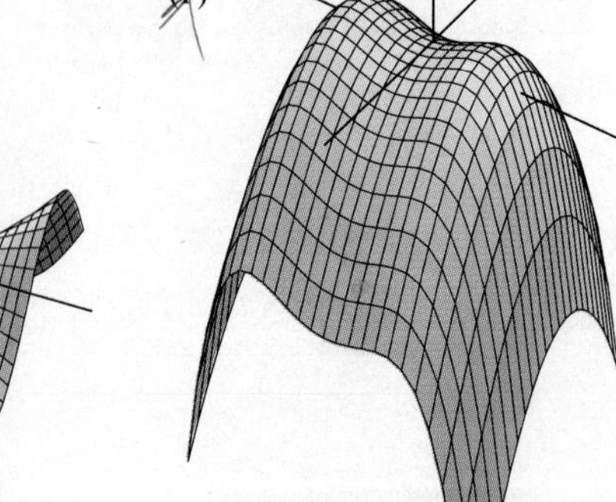

$z = x^2 - x^4 - y^2$

Describe the domains, ranges, and level surfaces of the functions in Exercises 27–34. Sketch a typical level surface for each function and label the surface with its function value.

27. $f(x, y, z) = \ln(x^2 + y^2 + z^2)$

28. $f(x, y, z) = \cosh(x^2 + z^2)$

29. $f(x, y, z) = \tan^{-1}(z - x^2 - y^2)$

30. $f(x, y, z) = \sin^{-1} z$

31. $f(x, y, z) = e^{-|z|}$

32. $f(x, y, z) = (x^2/25) + (y^2/16) + (z^2/9)$

33. $f(x, y, z) = \log_2\left(z - \sqrt{x^2 + y^2}\right)$

34. $f(x, y, z) = |x| + |y| + |z|$

In Exercises 35–38, find an equation for the level curve of the function $f(x, y)$ that passes through the given point.

35. $f(x, y) = 16 - x^2 - y^2, \quad \left(2\sqrt{2}, \sqrt{2}\right)$

36. $f(x, y) = \sqrt{x^2 - 1}, \quad (1, 0)$

37. $f(x, y) = \displaystyle\int_x^y \frac{dt}{1 + t^2}, \quad \left(-\sqrt{2}, \sqrt{2}\right)$

38. $f(x, y) = \displaystyle\sum_{n=0}^{\infty} \left(\frac{x}{y}\right)^n, \quad (1, 2)$

In Exercises 39–42, find an equation for the level surface of the function $f(x, y, z)$ that passes through the given point.

39. $f(x, y, z) = \sqrt{x - y} - \ln z, \quad (3, -1, 1)$

40. $f(x, y, z) = \cos(x^2 - y + z^2), \quad (-1, 2, 1)$

41. $f(x, y, z) = \displaystyle\sum_{n=0}^{\infty} \frac{(x + y)^n}{n!\, z^n}, \quad (\ln 2, \ln 4, 3)$

42. $f(x, y, z) = \displaystyle\int_x^y \frac{d\theta}{\sqrt{1 - \theta^2}} + \int_{\sqrt{2}}^z \frac{dt}{t\sqrt{t^2 - 1}}, \quad (0, 1/2, 2)$

43. *The Concorde's sonic booms.* The width w of the region in which people on the ground hear the Concorde's sonic boom directly, not reflected from a layer in the atmosphere, is a function of

T = air temperature at ground level (in degrees Kelvin),

h = the Concorde's altitude (in km),

d = the vertical temperature gradient (temperature drop in degrees Kelvin per km).

The formula for w is

$$w = 4(Th/d)^{1/2}.$$

See Fig. 12.10.

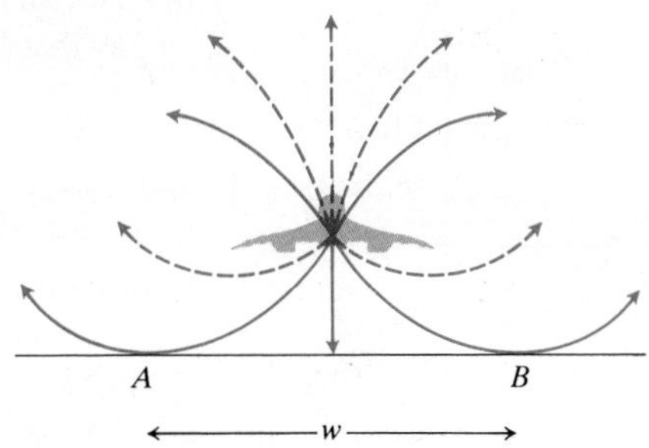

Sonic boom carpet

12.10 Sound waves from the Concorde bend as the temperature changes above and below the altitude at which the plane flies. The sonic boom carpet is the region on the ground that receives shock waves directly from the plane, not reflected from the atmosphere or diffracted along the ground. The carpet is determined by the grazing rays striking the ground from the point directly under the plane (Exercise 43).

The Washington-bound Concorde approaches the United States from Europe on a course that takes it south of Nantucket Island at an altitude of 16.8 km. If the surface temperature is 290 K and the vertical temperature gradient is 5 K/km, how far south of Nantucket must the plane be flown to keep its sonic boom carpet away from the island? (From "Concorde Sonic Booms as an Atmospheric Probe" by N. K. Balachandra, W. L. Donn, and D. H. Rind, *Science*, July 1, 1977, Vol. 197, p. 47.)

44. *The maximum value of a function on a line in space.* Does the function $f(x, y, z) = xyz$ have a maximum value on the line $x = 20 - t$, $y = t$, $z = 20$? If so, what is it? (*Hint:* Along the line, $w = f(x, y, z)$ is a differentiable function of t.)

EXPLORER PROGRAM

3D Grapher — Graphs surfaces defined by equations of the form $z = f(x, y)$ over rectangular regions in the xy-plane

12.2 Limits and Continuity

This section deals with limits and continuity of functions of more than one variable. The definitions are analogous to those for functions of a single variable, except that there are now more variables to watch.

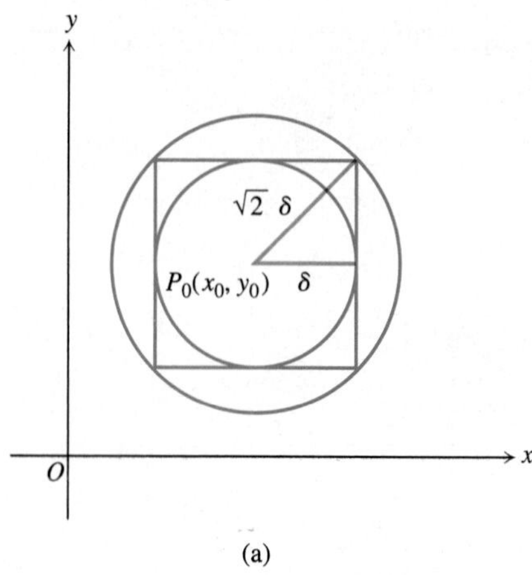

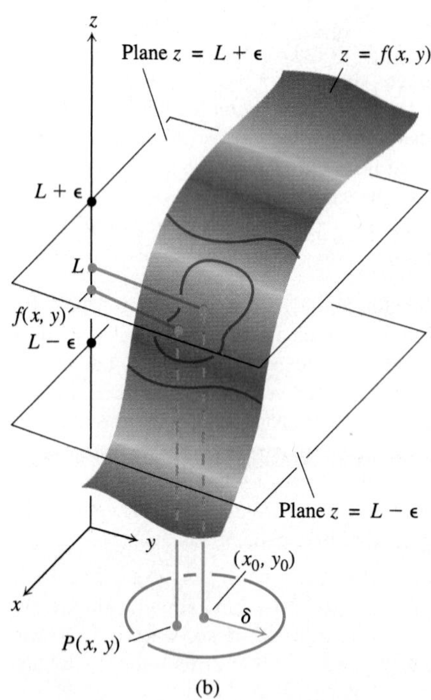

12.11 (a) The open square $|x - x_0| < \delta$, $|y - y_0| < \delta$ lies inside the open disk $\sqrt{(x - x_0)^2 + (y - y_0)^2} < \sqrt{2}\delta$ and contains the open disk $\sqrt{(x - x_0)^2 + (y - y_0)^2} < \delta$. Thus, we can make both $|x - x_0|$ and $|y - y_0|$ small by making $\sqrt{(x - x_0)^2 + (y - y_0)^2}$ small, and conversely. (b) For every point (x, y) within δ of (x_0, y_0), the value of $f(x, y)$ lies within ϵ of L.

Limits

If the values of a real-valued function $f(x, y)$ lie arbitrarily close to a fixed real number L for all points (x, y) sufficiently close to a point (x_0, y_0) but not equal to (x_0, y_0), we say that L is the limit of f as (x, y) approaches (x_0, y_0). In symbols, we write

$$\lim_{(x, y)\to(x_0, y_0)} f(x, y) = L, \tag{1}$$

and we say "the limit of f as (x, y) approaches (x_0, y_0) equals L." This is like the limit of a function of one variable, except that two independent variables are involved instead of one. In addition, if (x_0, y_0) is an interior point of f's domain, (x, y) can approach (x_0, y_0) from any direction. The direction of approach may turn out to be an issue, as some of the following examples show.

For (x, y) to be "close" to (x_0, y_0) means that the Cartesian distance $\sqrt{(x - x_0)^2 + (y - y_0)^2}$ is small in some sense. Since

$$|x - x_0| = \sqrt{(x - x_0)^2} \le \sqrt{(x - x_0)^2 + (y - y_0)^2} \tag{2}$$

and

$$|y - y_0| = \sqrt{(y - y_0)^2} \le \sqrt{(x - x_0)^2 + (y - y_0)^2}, \tag{3}$$

the inequality

$$\sqrt{(x - x_0)^2 + (y - y_0)^2} < \delta$$

for any value of δ implies that

$$|x - x_0| < \delta \quad \text{and} \quad |y - y_0| < \delta.$$

Conversely, if for some $\delta > 0$ both $|x - x_0| < \delta$ and $|y - y_0| < \delta$, then

$$\sqrt{(x - x_0)^2 + (y - y_0)^2} < \sqrt{\delta^2 + \delta^2} = \sqrt{2}\delta,$$

which is small if δ is small (Fig. 12.11). Thus, in calculating limits we may think either in terms of the distance in the plane or in terms of differences in individual coordinates.

> The **limit** of $f(x, y)$ as $(x, y)\to(x_0, y_0)$ is the number L if for any $\epsilon > 0$ there exists a $\delta > 0$ such that for all points $(x, y) \neq (x_0, y_0)$ in the domain of f, either
>
> 1. $\sqrt{(x - x_0)^2 + (y - y_0)^2} < \delta \quad \Rightarrow \quad |f(x, y) - L| < \epsilon$ or
> 2. $|x - x_0| < \delta$ and $|y - y_0| < \delta \quad \Rightarrow \quad |f(x, y) - L| < \epsilon$.

These equivalent definitions apply to boundary points as well as interior points of the domain of f. The only requirement is that the point (x, y) remain in the domain at all times.

It can be shown, as for functions of a single variable, that

$$\lim_{(x, y)\to(x_0, y_0)} x = x_0,$$
$$\lim_{(x, y)\to(x_0, y_0)} y = y_0, \tag{4}$$
$$\lim_{(x, y)\to(x_0, y_0)} k = k. \qquad \text{(Any number } k\text{)}$$

It can also be shown that the limit of the sum of two functions is the sum of their limits (when the latter exist), with similar results for the limits of the differences, products, constant multiples, and quotients of functions.

THEOREM 1

Properties of Limits of Functions of Two Variables

The following rules hold if $\lim_{(x, y)\to(x_0, y_0)} f(x, y) = L_1$ and $\lim_{(x, y)\to(x_0, y_0)} g(x, y) = L_2$.

1. *Sum Rule:* $\lim [f(x, y) + g(x, y)] = L_1 + L_2$
2. *Difference Rule:* $\lim [f(x, y) - g(x, y)] = L_1 - L_2$
3. *Product Rule:* $\lim f(x, y) \cdot g(x, y) = L_1 \cdot L_2$
4. *Constant Multiple Rule:* $\lim kg(x, y) = kL_2$ (Any number k)
5. *Quotient Rule:* $\lim \dfrac{f(x, y)}{g(x, y)} = \dfrac{L_1}{L_2}$ if $L_2 \neq 0$.

All limits are to be taken as $(x, y) \to (x_0, y_0)$, and L_1 and L_2 are to be real numbers.

When we combine Theorem 1 with the functions and limits in (4) we obtain the useful result that the limits of polynomials and rational functions as $(x, y) \to (x_0, y_0)$ may be calculated by evaluating the functions at the point (x_0, y_0). The only requirement is that the functions be defined at (x_0, y_0).

Example 1

a) $\lim_{(x, y)\to(3, -4)} (x^2 + y^2) = (3)^2 + (-4)^2 = 9 + 16 = 25$

b) $\lim_{(x, y)\to(0, 1)} \dfrac{x - xy + 3}{x^2y + 5xy - y^3} = \dfrac{0 - 0(1) + 3}{(0)^2(1) + 5(0)(1) - (1)^3} = -3$

Continuity

The definition of continuity for functions of two variables is analogous to the definition for functions of a single variable.

DEFINITIONS

A function $f(x, y)$ is said to be **continuous at the point (x_0, y_0)** if

1. f is defined at (x_0, y_0),
2. $\lim_{(x, y)\to(x_0, y_0)} f(x, y)$ exists,
3. $\lim_{(x, y)\to(x_0, y_0)} f(x, y) = f(x_0, y_0)$.

A function is said to be **continuous** if it is continuous at every point of its domain.

As with the definition of limit, the definition of continuity applies at boundary points as well as interior points of the domain of f. The only requirement is that the point (x, y) remain in the domain at all times.

As you may have guessed, one of the consequences of Theorem 1 is that algebraic combinations of continuous functions are continuous at every point at which all the functions involved are defined. This means that sums, differences, products, and constant multiples of continuous functions are continuous at any shared domain point. Also, the quotient of two continuous functions is continuous at any point at which the quotient is defined. In particular, polynomials and rational functions of two variables are continuous at every point at which they are defined.

If $z = f(x, y)$ is a continuous function of x and y, and $w = g(z)$ is a continuous function of z, then the composite $w = g(f(x, y))$ is continuous. Thus, composites like

$$e^{x-y}, \qquad \cos\frac{xy}{x^2+1}, \qquad \ln(1 + x^2y^2)$$

are continuous at every point (x, y).

As with functions of a single variable, the general rule is that composites of continuous functions are continuous. The only requirement is that each function be continuous where it is applied.

Example 2 Show that the function

$$f(x, y) = \begin{cases} \dfrac{2xy}{x^2+y^2}, & (x, y) \neq (0, 0), \\ 0, & (x, y) = (0, 0), \end{cases}$$

is continuous at every point except the origin (Fig. 12.12).

Solution The function f is continuous at any point $(x, y) \neq (0, 0)$ because its values are then given by a rational function of x and y.

At $(0, 0)$ the value of f is defined, but f, we claim, has no limit as $(x, y) \to (0, 0)$. The reason is that different paths of approach to the origin can lead to different results, as we shall now see.

For every value of m, the line $y = mx$ is a level curve of f because the value

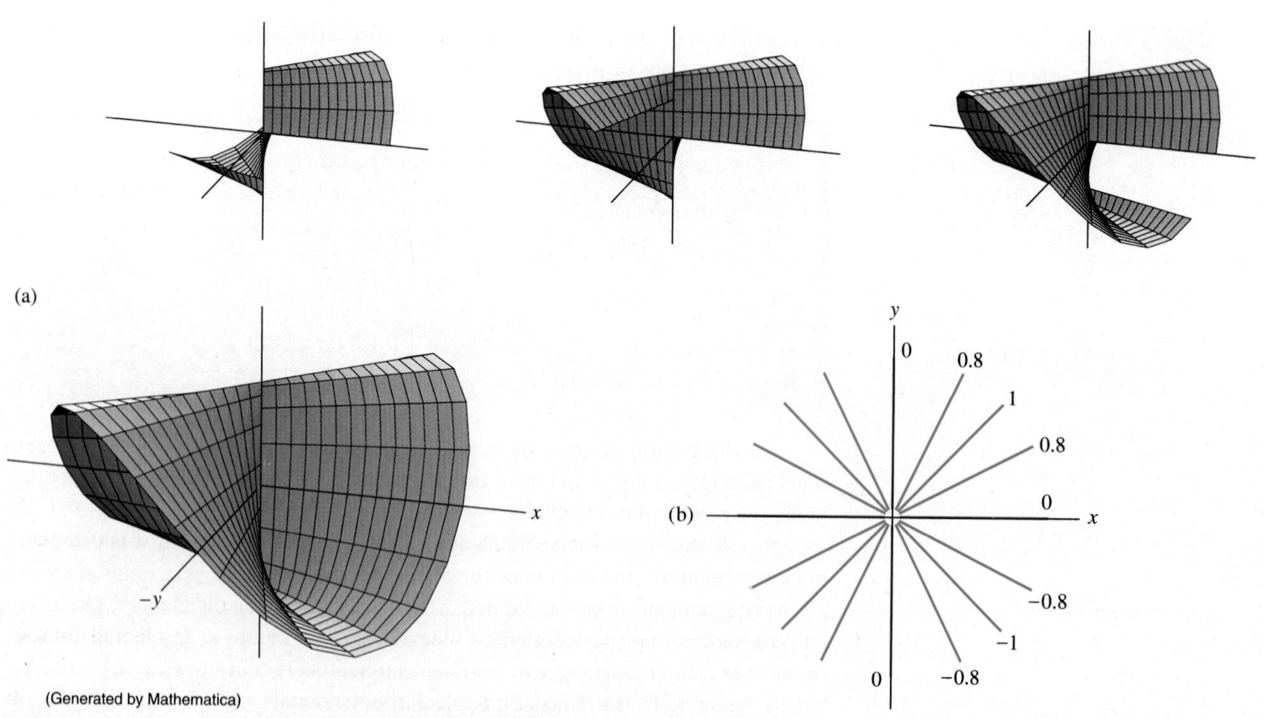

12.12 a) Stages in a computer's construction of the graph of

$$f(x, y) = \begin{cases} \dfrac{2xy}{x^2+y^2}, & (x, y) \neq (0, 0), \\ 0, & (x, y) = (0, 0). \end{cases}$$

The function is continuous at every point except the origin. b) The level curves of f are the open rays from the origin together with the origin itself (as a degenerate curve). The rays in quadrants I and IV have been labeled with function values.

of f on the line has the constant value

$$f(x, y)\Big|_{y=mx} = \frac{2xy}{x^2+y^2}\Big|_{y=mx} = \frac{2x(mx)}{x^2+(mx)^2} = \frac{2mx^2}{x^2+m^2x^2} = \frac{2m}{1+m^2}.$$

Therefore, f has this number as its limit as (x, y) approaches the origin along the line:

$$\lim_{\substack{(x, y)\to(0, 0)\\ \text{along } y=mx}} f(x, y) = \lim_{(x, y)\to(0, 0)} \left[f(x, y)\Big|_{y=mx} \right] = \frac{2m}{1+m^2}.$$

This limit changes with m. There is therefore no single number we may call the limit of f as (x, y) approaches the origin. The limit fails to exist, and the function is not continuous.

Example 2 illustrates an important point about limits of functions of two variables (or even more variables, for that matter). For a limit to exist at a point, the limit must be the same along every approach path. Therefore, if we ever find paths with different limits, we know the function has no limit at the point they approach.

The Two-Path Test for Discontinuity

If a function $f(x, y)$ has different limits along two different paths as (x, y) approaches (x_0, y_0), then $\lim_{(x, y)\to(x_0, y_0)} f(x, y)$ does not exist.

Example 3 Show that the function

$$f(x, y) = \frac{2x^2y}{x^4+y^2}$$

(Fig. 12.13) has no limit as (x, y) approaches $(0, 0)$.

Solution Along the curve $y = kx^2$, $x \neq 0$, the function has the constant value

$$f(x, y)\Big|_{y=kx^2} = \frac{2x^2y}{x^4+y^2}\Big|_{y=kx^2} = \frac{2x^2(kx^2)}{x^4+(kx^2)^2} = \frac{2kx^4}{x^4+k^2x^4} = \frac{2k}{1+k^2}.$$

Therefore,

$$\lim_{\substack{(x, y)\to(0, 0)\\ \text{along } y=kx^2}} f(x, y) = \lim_{(x, y)\to(0, 0)} \left[f(x, y)\Big|_{y=kx^2} \right] = \frac{2k}{1+k^2}.$$

This limit varies with the path of approach. If (x, y) approaches $(0, 0)$ along the parabola $y = x^2$, for instance, $k = 1$ and the limit is 1. If (x, y) approaches $(0, 0)$ along the x-axis, $k = 0$ and the limit is 0. By the two-path test, f has no limit as (x, y) approaches $(0, 0)$.

The language we use here may seem contradictory. You might well ask, "What do you mean f has no limit as (x, y) approaches the origin—it has lots of limits." But that is the point. There is no path-independent limit, and therefore, by our definition, $\lim_{(x, y)\to(0, 0)} f(x, y)$ does not exist. It is our translating this formal statement

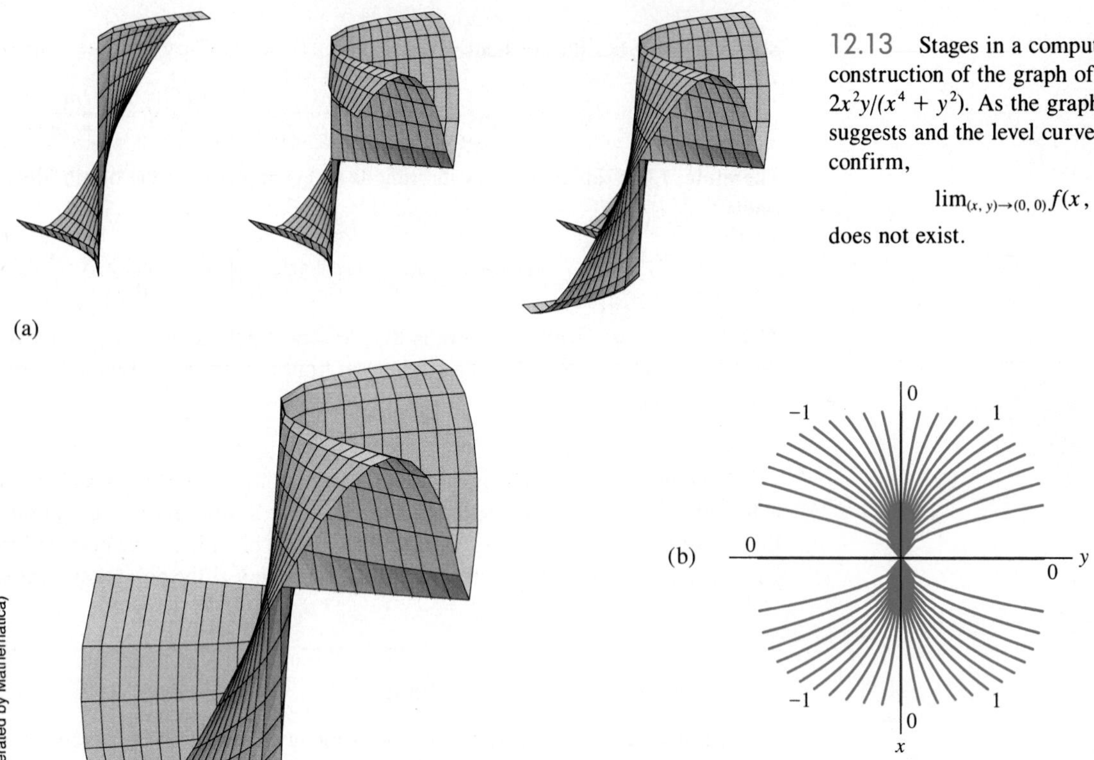

12.13 Stages in a computer's construction of the graph of $f(x, y) = 2x^2y/(x^4 + y^2)$. As the graph in (a) suggests and the level curve values in (b) confirm,

$$\lim_{(x,\, y)\to(0,\, 0)} f(x, y)$$

does not exist.

into the more colloquial "has no limit" that creates the apparent contradiction. The mathematics is fine. The problem arises in how we tend to talk about it. In the crunch, we need formality to keep things straight.

Functions of More Than Two Variables

The definitions of limit and continuity for functions of two variables and the conclusions about limits and continuity for sums, products, quotients, and composites of functions of two variables all extend to functions of three or more variables. Thus, functions like

$$\ln(x + y + z) \qquad \text{and} \qquad \frac{y \sin z}{x - 1}$$

are continuous throughout their domains and limits like

$$\lim_{P\to(1,\,0,\,-1)} \frac{e^{x+z}}{z^2 + \cos\sqrt{xy}} = \frac{e^{1-1}}{(-1)^2 + \cos 0} = \frac{1}{2},$$

where P denotes the point (x, y, z), may be found by direct substitution.

Continuous Functions Defined on Closed, Bounded Regions

As we know, a function of a single variable that is continuous throughout a closed, bounded interval $[a, b]$ takes on an absolute maximum value and an absolute minimum value at least once in $[a, b]$. The same is true of a function $z = f(x, y)$ that is continuous on a closed, bounded set R in the plane (like a line segment, disk, or filled-in triangle). The function takes on an absolute maximum value at some point in R and an absolute minimum value at some point in R.

Similar theorems hold for functions of three or more variables. A continuous function $w = f(x, y, z)$, for example, must take on absolute maximum and minimum values on any closed, bounded set (solid ball or cube, spherical shell, rectangular plate) on which it is defined.

It is important to know how to find these extreme values and we shall show how to do so in Section 12.8. But first, we need to know about derivatives, and that will be the topic of the next section.

EXERCISES 12.2

Find the limits in Exercises 1–12.

1. $\lim_{(x,y)\to(0,0)} \dfrac{3x^2 - y^2 + 5}{x^2 + y^2 + 2}$

2. $\lim_{(x,y)\to(1,1)} \ln|1 + x^2y^2|$

3. $\lim_{(x,y)\to(0,\ln 2)} e^{x-y}$

4. $\lim_{(x,y)\to(0,4)} \dfrac{x}{\sqrt{y}}$

5. $\lim_{P\to(1,3,4)} \sqrt{x^2 + y^2 + z^2 - 1}$

6. $\lim_{P\to(1,2,6)} \left(\dfrac{1}{x} + \dfrac{1}{y} + \dfrac{1}{z}\right)$

7. $\lim_{(x,y)\to(0,\pi/4)} \sec x \tan y$

8. $\lim_{(x,y)\to(0,0)} \cos \dfrac{x^2 + y^2}{x + y + 1}$

9. $\lim_{(x,y)\to(1,1)} \cos \sqrt[3]{|xy| - 1}$

10. $\lim_{(x,y)\to(1,0)} \dfrac{x \sin y}{x^2 + 1}$

11. $\lim_{(x,y)\to(0,0)} \dfrac{e^y \sin x}{x}$

12. $\lim_{(x,y)\to(0,0)} \tan^{-1}\left(1/\sqrt{x^2 + y^2}\right)$

Find the limits in Exercises 13–16 by rewriting the fractions first.

13. $\lim_{\substack{(x,y)\to(1,1)\\ x\neq y}} \dfrac{x^2 - 2xy + y^2}{x - y}$

14. $\lim_{\substack{(x,y)\to(1,1)\\ x\neq y}} \dfrac{x^2 - y^2}{x - y}$

15. $\lim_{\substack{(x,y)\to(1,1)\\ x\neq 1}} \dfrac{xy - y - 2x + 2}{x - 1}$

16. $\lim_{\substack{(x,y)\to(2,-4)\\ y\neq -4,\ x\neq x^2}} \dfrac{y + 4}{x^2y - xy + 4x^2 - 4x}$

Find the limits in Exercises 17–22. In each exercise, P stands for the point (x, y, z).

17. $\lim_{P\to(2,3,-6)} \sqrt{x^2 + y^2 + z^2}$

18. $\lim_{P\to(0,-2,0)} \ln\sqrt{x^2 + y^2 + z^2}$

19. $\lim_{P\to(3,3,0)} (\sin^2 x + \cos^2 y + \sec^2 z)$

20. $\lim_{P\to(\pi,0,3)} ze^{-2y} \cos 2x$

21. $\lim_{P\to(-1/4,\pi/2,2)} \tan^{-1} xyz$

22. $\lim_{P\to(1,-1,-1)} \dfrac{2xy + yz}{x^2 + z^2}$

At what points (x, y) in the plane are the functions given by the formulas in Exercises 23–26 continuous?

23. a) $\sin(x + y)$ b) $\ln(x^2 + y^2)$

24. a) $\dfrac{x + y}{x - y}$ b) $\dfrac{y}{x^2 + 1}$

25. a) $\sin \dfrac{1}{xy}$ b) $\dfrac{x + y}{2 + \cos x}$

26. a) $\dfrac{x^2 + y^2}{x^2 - 3x + 2}$ b) $\dfrac{1}{x^2 - y}$

At what points (x, y, z) in space are the functions given by the formulas in Exercises 27–30 continuous?

27. a) $x^2 + y^2 - 2z^2$ b) $\sqrt{x^2 + y^2 - 1}$

28. a) $\ln xyz$ b) $e^{x+y} \cos z$

29. a) $xy \sin \dfrac{1}{z}$ b) $\dfrac{1}{x^2 + y^2 + z^2 - 1}$

30. a) $\dfrac{1}{|x| + |y| + |z|}$ b) $\dfrac{1}{|xy| + |z|}$

By considering different paths of approach, show that the functions in Exercises 31–38 have no limit as $(x, y)\to(0, 0)$.

31. $f(x, y) = \dfrac{x}{\sqrt{x^2 + y^2}}$

32. $f(x, y) = \dfrac{x^4}{x^4 + y^2}$

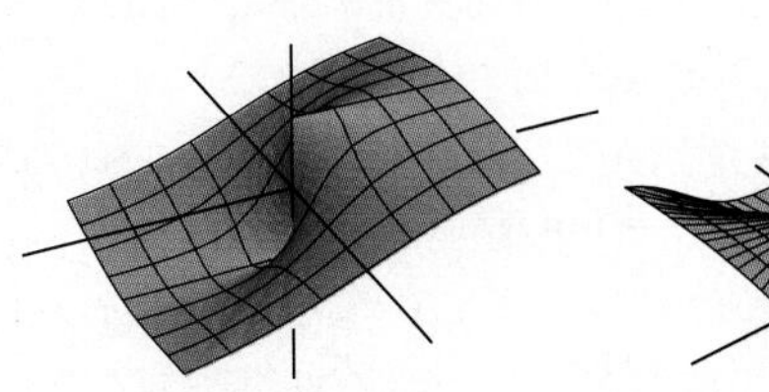

(Generated by Mathematica) (Generated by Mathematica)

33. $f(x, y) = \dfrac{x^4 - y^2}{x^4 + y^2}$

34. $f(x, y) = \dfrac{xy}{|xy|}$

35. $f(x, y) = \dfrac{x - y}{x + y}$

36. $f(x, y) = \dfrac{x + y}{x - y}$

37. $f(x, y) = \dfrac{x^2 + y}{y}$

38. $f(x, y) = \dfrac{x^2}{x^2 - y}$

39. Continuation of Example 2

a) Reread Example 2. Then substitute $m = \tan\theta$ into the formula

$$f(x, y)\Big|_{y = mx} = \frac{2m}{1 + m^2}$$

and simplify the result to show how the value of f varies with the line's angle of inclination.

b) Use the formula you obtained in (a) to show that the limit of f as $(x, y) \to (0, 0)$ along the line $y = mx$ varies from -1 to 1 depending on the angle of approach.

40. Define $f(0, 0)$ in a way that extends

$$f(x, y) = xy\frac{x^2 - y^2}{x^2 + y^2}$$

to be continuous at the origin.

The Sandwich Theorem for functions of two variables states that if $g(x, y) \le f(x, y) \le h(x, y)$ for all $(x, y) \ne (x_0, y_0)$ in a disk centered at (x_0, y_0) and if g and h have the same finite limit L as $(x, y) \to (x_0, y_0)$, then $\lim_{(x, y)\to(x_0, y_0)} f(x, y)$ exists and equals L as well. Use this result to find the limits in Exercises 41 and 42.

41. Find $\displaystyle\lim_{(x, y)\to(0, 0)} \frac{\tan^{-1} xy}{xy}$ if

$$1 - \frac{x^2y^2}{3} < \frac{\tan^{-1} xy}{xy} < 1.$$

42. Find $\displaystyle\lim_{(x, y)\to(0, 0)} \frac{4 - 4\cos\sqrt{|xy|}}{|xy|}$ if

$$2|xy| - \frac{x^2y^2}{6} < 4 - 4\cos\sqrt{|xy|} < 2|xy|.$$

Changing to Polar Coordinates

If you cannot make any headway with $\lim_{(x, y)\to(0, 0)} f(x, y)$ in rectangular coordinates, try changing to polar coordinates. Substitute $x = r\cos\theta$, $y = r\sin\theta$, and investigate the limit of the resulting expression as $r \to 0$. In other words, try to decide whether there exists a number L satisfying the following criterion:

Given $\epsilon > 0$, there exists a $\delta > 0$ such that for all r and θ,

$$|r| < \delta \;\Rightarrow\; |f(r, \theta) - L| < \epsilon. \tag{5}$$

If such an L exists, then

$$\lim_{(x, y)\to(0, 0)} f(x, y) = \lim_{r\to 0} f(r, \theta) = L.$$

For instance,

$$\lim_{(x, y)\to(0, 0)} \frac{x^3}{x^2 + y^2} = \lim_{r\to 0}\frac{r^3\cos^3\theta}{r^2} = \lim_{r\to 0} r\cos^3\theta = 0.$$

To verify the last of these equalities, we need to show that implication (5) in the criterion for limit is satisfied with $f(r, \theta) = r\cos^3\theta$ and $L = 0$. That is, we need to show that given any $\epsilon > 0$ there exists a $\delta > 0$ such that for all r and θ,

$$|r| < \delta \;\Rightarrow\; |r\cos^3\theta - 0| < \epsilon.$$

Since

$$|r\cos^3\theta| = |r||\cos^3\theta| \le |r| \cdot 1 = |r|,$$

the implication holds for all r and θ if we take $\delta = \epsilon$.

In contrast,

$$\frac{x^2}{x^2 + y^2} = \frac{r^2\cos^2\theta}{r^2} = \cos^2\theta$$

takes on all values from 0 to 1 regardless of how small $|r|$ is, so that $\lim_{(x, y)\to(0, 0)} x^2/(x^2 + y^2)$ does not exist.

In each of these instances, the existence or nonexistence of the limit as $r \to 0$ is fairly clear. Shifting to polar coordinates does not always help, however, and may even tempt us to false conclusions. For example, the limit may exist along every straight line (or ray) $\theta =$ constant and yet fail to exist in the broader sense. Example 3 illustrates this point. In polar coordinates, $f(x, y) = (2x^2y)/(x^4 + y^2)$ becomes

$$f(r\cos\theta, r\sin\theta) = \frac{r\cos\theta\sin 2\theta}{r^2\cos^4\theta + \sin^2\theta}$$

for $r \ne 0$. If we hold θ constant and let $r \to 0$, the limit is 0. On the path $y = x^2$, however, we have $r\sin\theta = r^2\cos^2\theta$ and

$$f(r\cos\theta, r\sin\theta) = \frac{r\cos\theta\sin 2\theta}{r^2\cos^4\theta + (r\cos^2\theta)^2}$$

$$= \frac{2r\cos^2\theta\sin\theta}{2r^2\cos^4\theta} = \frac{r\sin\theta}{r^2\cos^2\theta} = 1.$$

In Exercises 43–48, find the limit of f as $(x, y) \to (0, 0)$ or show that the limit does not exist.

43. $f(x, y) = \dfrac{x^3 - xy^2}{x^2 + y^2}$

44. $f(x, y) = \cos\left(\dfrac{x^3 - y^3}{x^2 + y^2}\right)$

45. $f(x, y) = \dfrac{y^2}{x^2 + y^2}$

46. $f(x, y) = \dfrac{2x}{x^2 + x + y^2}$

47. $f(x, y) = \tan^{-1}\left(\dfrac{|x| + |y|}{x^2 + y^2}\right)$

48. $f(x, y) = \dfrac{x^2 - y^2}{x^2 + y^2}$

In Exercises 49 and 50, define $f(0, 0)$ in a way that extends f to be continuous at the origin.

49. $f(x, y) = \ln\left(\dfrac{y^2 + 3x}{3x^2 + x}\right)$

50. $f(x, y) = \dfrac{xy - x - y}{x + y}$

The Definitions of Limit and Continuity

Each of Exercises 51–54 gives a function $f(x, y)$ and a positive number ϵ. In each exercise, either show that there exists a $\delta > 0$ such that for all (x, y),

$$\sqrt{x^2 + y^2} < \delta \;\Rightarrow\; |f(x, y) - f(0, 0)| < \epsilon$$

or show that there exists a $\delta > 0$ such that for all (x, y),

$$|x| < \delta \quad \text{and} \quad |y| < \delta \;\Rightarrow\; |f(x, y) - f(0, 0)| < \epsilon.$$

Do either one or the other, whichever seems more convenient. There is no need to do both.

51. $f(x, y) = x^2 + y^2, \quad \epsilon = 0.01$

52. $f(x, y) = y/(x^2 + 1), \quad \epsilon = 0.05$

53. $f(x, y) = (x + y)/(x^2 + 1), \quad \epsilon = 0.01$

54. $f(x, y) = (x + y)/(2 + \cos x), \quad \epsilon = 0.02$

Each of Exercises 55–58 gives a function $f(x, y, z)$ and a positive number ϵ. In each exercise, either show that there exists a $\delta > 0$ such that for all (x, y, z),

$$\sqrt{x^2 + y^2 + z^2} < \delta \;\Rightarrow\; |f(x, y, z) - f(0, 0, 0)| < \epsilon$$

or show that there exists a $\delta > 0$ such that for all (x, y, z),

$$|x| < \delta, \quad |y| < \delta, \quad \text{and}$$
$$|z| < \delta \;\Rightarrow\; |f(x, y, z) - f(0, 0, 0)| < \epsilon.$$

Do either one or the other, whichever seems more convenient. There is no need to do both.

55. $f(x, y, z) = x^2 + y^2 + z^2, \quad \epsilon = 0.015$

56. $f(x, y, z) = xyz, \quad \epsilon = 0.008$

57. $f(x, y, z) = \dfrac{x + y + z}{x^2 + y^2 + z^2 + 1}, \quad \epsilon = 0.015$

58. $f(x, y, z) = \tan^2 x + \tan^2 y + \tan^2 z, \quad \epsilon = 0.03$

59. Show that $f(x, y, z) = x + y - z$ is continuous at every point (x_0, y_0, z_0).

60. Show that $f(x, y, z) = x^2 + y^2 + z^2$ is continuous at the origin.

12.3 Partial Derivatives

Partial derivatives are what we get when we hold all but one of the independent variables of a function constant and differentiate with respect to that one variable. In this section, we show how partial derivatives arise geometrically and how we can calculate them by applying the rules for differentiating functions of a single variable. We begin with functions of two independent variables.

Definitions and Notation

If (x_0, y_0) is a point in the domain of a function $f(x, y)$, the vertical plane $y = y_0$ will cut the surface $z = f(x, y)$ in the curve $z = f(x, y_0)$ (Fig. 12.14). This curve is the graph of the function $z = f(x, y_0)$ in the plane $y = y_0$. The horizontal coordinate in this plane is x; the vertical coordinate is z.

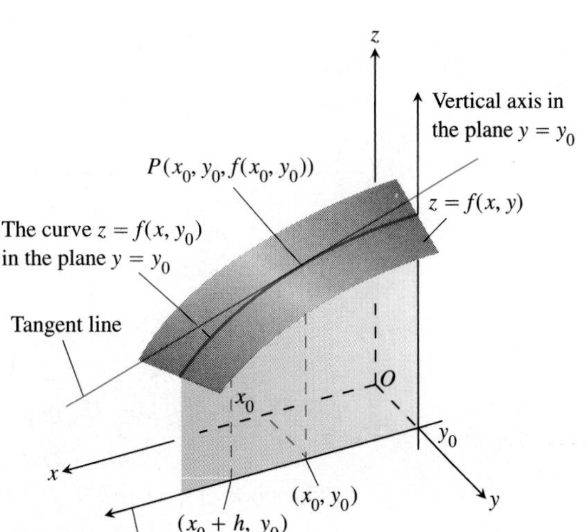

12.14 The intersection of the plane $y = y_0$ with the surface $z = f(x, y)$, viewed from a point above the first quadrant of the xy-plane.

We define the partial derivative of f with respect to x at the point (x_0, y_0) as the ordinary derivative of $f(x, y_0)$ with respect to x at the point $x = x_0$.

DEFINITION

Partial Derivative with Respect to x

The partial derivative of $f(x, y)$ with respect to x at the point (x_0, y_0) is

$$\left.\frac{\partial f}{\partial x}\right|_{(x_0, y_0)} = \left.\frac{d}{dx} f(x, y_0)\right|_{x=x_0} = \lim_{h \to 0} \frac{f(x_0 + h, y_0) - f(x_0, y_0)}{h}, \tag{1}$$

provided the limit exists. (The letter ∂ is a special round d.)

The slope of the curve $z = f(x, y_0)$ at the point $P(x_0, y_0, f(x_0, y_0))$ in the plane $y = y_0$ is the value of the partial derivative of f with respect to x at (x_0, y_0). The tangent to the curve at P is the line in the plane $y = y_0$ that passes through P with this slope. The partial derivative $\partial f/\partial x$ at (x_0, y_0) gives the rate of change of f with respect to x when y is held fixed at the value y_0.

The notation we use for the partial derivative of f with respect to x depends on what aspect of the derivative we want to emphasize. The usual list looks like this:

$\frac{\partial f}{\partial x}(x_0, y_0)$ or $f_x(x_0, y_0)$	"Partial derivative of f with respect to x at (x_0, y_0)" or "f sub x at (x_0, y_0)." Convenient for stressing the point (x_0, y_0).
$\left.\frac{\partial z}{\partial x}\right\|_{(x_0, y_0)}$ or $\left(\frac{\partial z}{\partial x}\right)_{(x_0, y_0)}$	"Partial derivative of z with respect to x at (x_0, y_0)." Common in science and engineering when you are dealing with variables and do not mention the function explicitly.
f_x, $\frac{\partial f}{\partial x}$, z_x, or $\frac{\partial z}{\partial x}$	"Partial derivative of f (or z) with respect to x." Convenient when you regard the partial derivative as a function in its own right.

The definition of the partial derivative of $f(x, y)$ with respect to y at a point (x_0, y_0) is similar to the definition of the partial derivative of f with respect to x. We hold x fixed at the value x_0 and take the ordinary derivative of $f(x_0, y)$ with respect to y at y_0.

DEFINITION

Partial Derivative with Respect to y

The partial derivative of $f(x, y)$ with respect to y at the point (x_0, y_0) is

$$\left.\frac{\partial f}{\partial y}\right|_{(x_0, y_0)} = \left.\frac{d}{dy} f(x_0, y)\right|_{y=y_0} = \lim_{k \to 0} \frac{f(x_0, y_0 + k) - f(x_0, y_0)}{k}, \tag{2}$$

provided the limit exists.

The slope of the curve $z = f(x_0, y)$ at the point $P(x_0, y_0, f(x_0, y_0))$ in the vertical plane $x = x_0$ (Fig. 12.15) is the partial derivative of f with respect to y at (x_0, y_0). The tangent to the curve at P is the line in the plane $x = x_0$ that passes through P

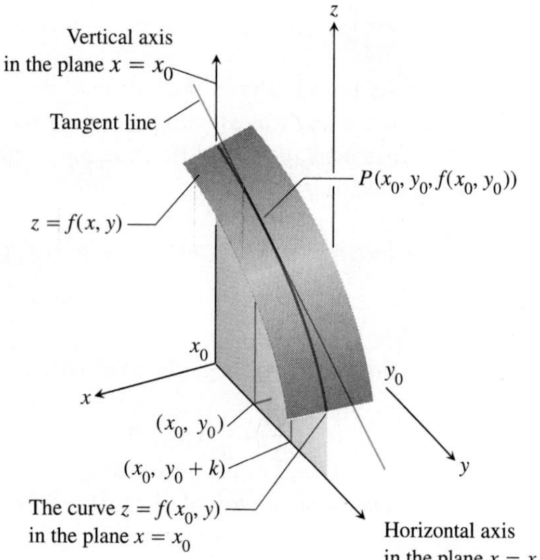

12.15 The intersection of the plane $x = x_0$ with the surface $z = f(x, y)$, viewed from above the first quadrant of the xy-plane.

with this slope. The partial derivative gives the rate of change of f with respect to y at (x_0, y_0) when x is held fixed at the value x_0.

The partial derivative with respect to y is denoted the same way as the partial derivative with respect to x. We have

$$\frac{\partial f}{\partial y}(x_0, y_0), \qquad f_y(x_0, y_0), \qquad \frac{\partial f}{\partial y}, \qquad f_y, \tag{3}$$

and so on.

Notice that we now have two tangent lines associated with the surface $z = f(x, y)$ at the point $P(x_0, y_0, f(x_0, y_0))$ (Fig. 12.16). Is the plane they determine tangent to the surface at P? It would be nice if it were, but we shall have to learn more about partial derivatives to find out.

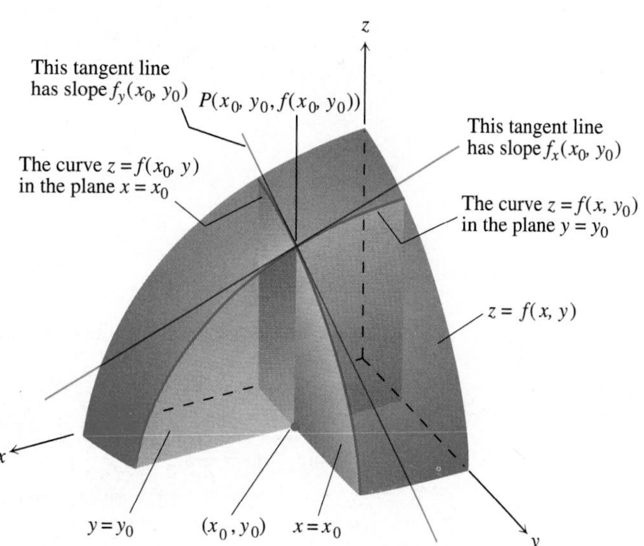

12.16 Figures 12.14 and 12.15 combined. The tangent lines at $(x_0, y_0, f(x_0, y_0))$ determine a plane that, in this picture at least, appears to be tangent to the surface.

Calculations

As Eq. (1) shows, we calculate $\partial f/\partial x$ by differentiating f with respect to x in the usual way while treating y as a constant. Similarly, from Eq. (2) we see that we can calculate $\partial f/\partial y$ by differentiating f with respect to y in the usual way while holding x constant.

Example 1 Find the value of $\partial f/\partial x$ at the point $(4, -5)$ if

$$f(x, y) = x^2 + 3xy + y - 1.$$

Solution We regard y as a constant and differentiate with respect to x:

$$\frac{\partial f}{\partial x} = \frac{\partial}{\partial x}(x^2 + 3xy + y - 1) = 2x + 3y + 0 - 0 = 2x + 3y.$$

The value of $\partial f/\partial x$ at $(4, -5)$ is $2(4) + 3(-5) = -7$.

Example 2 Find the value of $\partial f/\partial y$ at the point $(4, -5)$ if

$$f(x, y) = x^2 + 3xy + y - 1.$$

Solution We regard x as a constant and differentiate with respect to y:

$$\frac{\partial f}{\partial y} = \frac{\partial}{\partial y}(x^2 + 3xy + y - 1) = 0 + 3x + 1 - 0 = 3x + 1.$$

The value of $\partial f/\partial y$ at $(4, -5)$ is $3(4) + 1 = 13$.

Example 3 Find $\partial f/\partial y$ if

$$f(x, y) = y \sin xy.$$

Solution We treat x as a constant and f as a product of y and $\sin xy$:

$$\frac{\partial f}{\partial y} = \frac{\partial}{\partial y}(y \sin xy) = y \frac{\partial}{\partial y} \sin xy + \sin xy \frac{\partial}{\partial y}(y)$$

$$= y \cos xy \frac{\partial}{\partial y}(xy) + \sin xy = xy \cos xy + \sin xy.$$

Example 4 Find f_x if

$$f(x, y) = \frac{2y}{y + \cos x}.$$

Solution We treat f as the quotient of $2y$ divided by $(y + \cos x)$. With y held constant, we get

$$f_x = \frac{\partial}{\partial x}\left(\frac{2y}{y + \cos x}\right) = \frac{(y + \cos x)\frac{\partial}{\partial x}(2y) - 2y\frac{\partial}{\partial x}(y + \cos x)}{(y + \cos x)^2}$$

$$= \frac{(y + \cos x)(0) - 2y(-\sin x)}{(y + \cos x)^2} = \frac{2y \sin x}{(y + \cos x)^2}.$$

Implicit differentiation works for partial derivatives the way it works for ordinary derivatives.

Example 5 Find $\partial z/\partial x$ if the equation

$$yz - \ln z = x + y$$

defines z as a function of the two independent variables x and y and the partial derivative exists.

Solution We differentiate both sides of the equation with respect to x, holding y fixed and treating z as a differentiable function of x:

$$\frac{\partial}{\partial x}(yz) - \frac{\partial}{\partial x}\ln z = \frac{\partial x}{\partial x} + \frac{\partial y}{\partial x}$$

$$y\frac{\partial z}{\partial x} - \frac{1}{z}\frac{\partial z}{\partial x} = 1 + 0$$

$$\left(y - \frac{1}{z}\right)\frac{\partial z}{\partial x} = 1$$

$$\frac{\partial z}{\partial x} = \frac{z}{yz - 1}.$$

Functions of More Than Two Variables

The definitions of the partial derivatives of functions of more than two independent variables are like the definitions for functions of two variables. They are ordinary derivatives with respect to one variable, taken while the other independent variables are held constant.

Example 6 If x, y, and z are independent variables and

$$f(x, y, z) = x\sin(y + 3z),$$

then

$$\frac{\partial f}{\partial z} = \frac{\partial}{\partial z}(x\sin(y + 3z)) = x\frac{\partial}{\partial z}\sin(y + 3z)$$

$$= x\cos(y + 3z)\frac{\partial}{\partial z}(y + 3z) = 3x\cos(y + 3z).$$

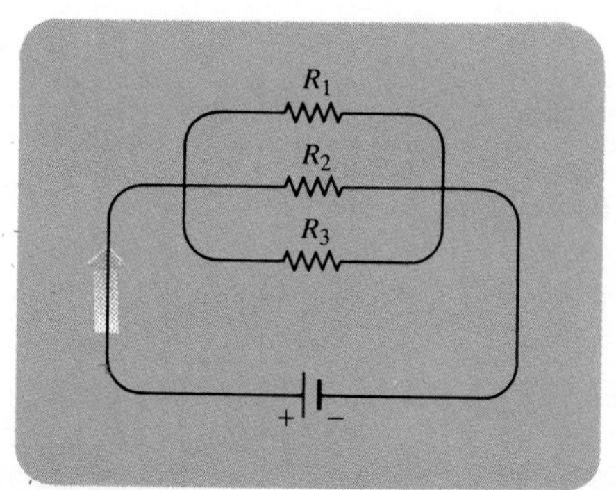

12.17 Resistors arranged this way are said to be connected in parallel (Example 7). Each resistor lets a portion of the current through. Their combined resistance R is calculated with the formula

$$\frac{1}{R} = \frac{1}{R_1} + \frac{1}{R_2} + \frac{1}{R_3}.$$

Example 7 *Electrical Resistors in Parallel.* If resistors of R_1, R_2, and R_3 ohms are connected in parallel to make an R-ohm resistor, the value of R can be found from the equation

$$\frac{1}{R} = \frac{1}{R_1} + \frac{1}{R_2} + \frac{1}{R_3} \qquad (4)$$

(Fig. 12.17). Find the value of $\partial R/\partial R_2$ when $R_1 = 30$, $R_2 = 45$, and $R_3 = 90$ ohms.

Solution To find $\partial R/\partial R_2$, we treat R_1 and R_3 as constants and differentiate both sides of Eq. (4) with respect to R_2:

$$\frac{\partial}{\partial R_2}\left(\frac{1}{R}\right) = \frac{\partial}{\partial R_2}\left(\frac{1}{R_1} + \frac{1}{R_2} + \frac{1}{R_3}\right)$$

$$-\frac{1}{R^2}\frac{\partial R}{\partial R_2} = 0 - \frac{1}{R_2^2} + 0$$

$$\frac{\partial R}{\partial R_2} = \frac{R^2}{R_2^2} = \left(\frac{R}{R_2}\right)^2.$$

When $R_1 = 30$, $R_2 = 45$, and $R_3 = 90$,

$$\frac{1}{R} = \frac{1}{30} + \frac{1}{45} + \frac{1}{90} = \frac{3 + 2 + 1}{90} = \frac{6}{90} = \frac{1}{15},$$

so $R = 15$ and

$$\frac{\partial R}{\partial R_2} = \left(\frac{15}{45}\right)^2 = \left(\frac{1}{3}\right)^2 = \frac{1}{9}.$$

Second Order Partial Derivatives

When we differentiate a function $f(x, y)$ twice, we produce its second order derivatives. These derivatives are usually denoted by

$\dfrac{\partial^2 f}{\partial x^2}$	"d squared f d x squared"	or	f_{xx}	"f sub x x"
$\dfrac{\partial^2 f}{\partial y^2}$	"d squared f d y squared"		f_{yy}	"f sub y y"
$\dfrac{\partial^2 f}{\partial x\,\partial y}$	"d squared f d x d y"		f_{yx}	"f sub y x"
$\dfrac{\partial^2 f}{\partial y\,\partial x}$	"d squared f d y d x"		f_{xy}	"f sub x y"

The defining equations are

$$\frac{\partial^2 f}{\partial x^2} = \frac{\partial}{\partial x}\left(\frac{\partial f}{\partial x}\right), \qquad \frac{\partial^2 f}{\partial x\,\partial y} = \frac{\partial}{\partial x}\left(\frac{\partial f}{\partial y}\right),$$

and so on. Notice the order in which the derivatives are taken:

$\dfrac{\partial^2 f}{\partial x\,\partial y}$ (Differentiate first with respect to y, then with respect to x.)

$f_{yx} = (f_y)_x$ (Means the same thing.)

Example 8 If $f(x, y) = x \cos y + ye^x$, then

$$\frac{\partial f}{\partial x} = \cos y + ye^x,$$

$$\frac{\partial^2 f}{\partial y\,\partial x} = \frac{\partial}{\partial y}\left(\frac{\partial f}{\partial x}\right) = -\sin y + e^x,$$

$$\frac{\partial^2 f}{\partial x^2} = \frac{\partial}{\partial x}\left(\frac{\partial f}{\partial x}\right) = ye^x.$$

Also,

$$\frac{\partial f}{\partial y} = -x \sin y + e^x$$

$$\frac{\partial^2 f}{\partial x\, \partial y} = \frac{\partial}{\partial x}\left(\frac{\partial f}{\partial y}\right) = -\sin y + e^x$$

$$\frac{\partial^2 f}{\partial y^2} = \frac{\partial}{\partial y}\left(\frac{\partial f}{\partial y}\right) = -x \cos y.$$

The Mixed Derivative Theorem

You may have noticed that the "mixed" second order derivatives

$$\frac{\partial^2 f}{\partial y\, \partial x} \quad \text{and} \quad \frac{\partial^2 f}{\partial x\, \partial y}$$

in Example 8 were equal. This was no mere coincidence. They have to be equal whenever f, f_x, f_y, f_{xy}, and f_{yx} are continuous.

THEOREM 2

The Mixed Derivative Theorem

If $f(x, y)$ and its partial derivatives f_x, f_y, f_{xy}, and f_{yx} are defined throughout an open region containing a point (a, b) and are all continuous at (a, b), then

$$f_{xy}(a, b) = f_{yx}(a, b). \tag{5}$$

You can find a proof of Theorem 2 in Appendix 11.

Theorem 2 says that when we want to calculate a mixed second order derivative we may differentiate in either order. This can work to our advantage, as the next example shows.

Example 9 Find $\partial^2 w/\partial x \partial y$ if

$$w = xy + \frac{e^y}{y^2 + 1}.$$

Solution The symbol $\partial^2 w/\partial x \partial y$ tells us to differentiate first with respect to y and then with respect to x. However, if we postpone the differentiation with respect to y and differentiate first with respect to x, we can get the answer more quickly. In two steps,

$$\frac{\partial w}{\partial x} = y \quad \text{and} \quad \frac{\partial^2 w}{\partial y\, \partial x} = 1.$$

We are in for more work if we differentiate first with respect to y. (Just try it.)

Partial Derivatives of Still Higher Order

Although we shall deal mostly with first and second order partial derivatives because these appear most frequently in applications, there is no theoretical limit to how many times we can differentiate a function as long as the derivatives involved exist. Thus we get third and fourth order derivatives denoted by symbols like

$$\frac{\partial^3 f}{\partial x\, \partial y^2} = f_{yyx},$$

$$\frac{\partial^4 f}{\partial x^2\, \partial y^2} = f_{yyxx},$$

and so on. As with second order derivatives, the order of differentiation is immaterial as long as the function and its derivatives through the order in question are all defined throughout an open region containing the point at which the derivatives are taken and are continuous at that point.

EXERCISES 12.3

In Exercises 1–22, find $\partial f/\partial x$ and $\partial f/\partial y$.

1. $f(x, y) = 2x^2 - 3y - 4$
2. $f(x, y) = x^2 - xy + y^2$
3. $f(x, y) = (x^2 - 1)(y + 2)$
4. $f(x, y) = 5xy - 7x^2 - y^2 + 3x - 6y + 2$
5. $f(x, y) = (xy - 1)^2$
6. $f(x, y) = (2x - 3y)^3$
7. $f(x, y) = \sqrt{x^2 + y^2}$
8. $f(x, y) = (x^3 + (y/2))^{2/3}$
9. $f(x, y) = 1/(x + y)$
10. $f(x, y) = x/(x^2 + y^2)$
11. $f(x, y) = (x + y)/(xy - 1)$
12. $f(x, y) = \tan^{-1}(y/x)$
13. $f(x, y) = \ln (x + y)$
14. $f(x, y) = e^{-x} \sin(x + y)$
15. $f(x, y) = e^{xy} \ln y$
16. $f(x, y) = e^{(x + y + 1)}$
17. $f(x, y) = \cos^2(3x - y^2)$
18. $f(x, y) = \sin^{-1} \sqrt{xy}$
19. $f(x, y) = x^y$
20. $f(x, y) = \log_y x$
21. $f(x, y) = \int_x^y f(t)\, dt$ (f continuous for all t)
22. $f(x, y) = \sum_{n=0}^{\infty} (xy)^n \quad (|xy| < 1)$

In Exercises 23–34, find f_x, f_y, and f_z.

23. $f(x, y, z) = 1 + xy^2 - 2z^2$
24. $f(x, y, z) = xy + yz + xz$
25. $f(x, y, z) = x - \sqrt{y^2 + z^2}$
26. $f(x, y, z) = (x^2 + y^2 + z^2)^{-1/2}$
27. $f(x, y, z) = \sin^{-1} (xyz)$
28. $f(x, y, z) = \sec^{-1} (x + yz)$
29. $f(x, y, z) = \ln(x + 2y + 3z)$
30. $f(x, y, z) = yz \ln(xy)$
31. $f(x, y, z) = e^{-(x^2 + y^2 + z^2)}$
32. $f(x, y, z) = e^{-xyz}$
33. $f(x, y, z) = \tanh(x + 2y + 3z)$
34. $f(x, y, z) = \sinh(xy - z^2)$

In Exercises 35–40, find the partial derivative of the function with respect to each variable.

35. $f(t, \alpha) = \cos(2\pi t - \alpha)$
36. $g(u, v) = v^2 e^{(2u/v)}$
37. $h(\rho, \phi, \theta) = \rho \sin \phi \cos \theta$
38. $g(r, \theta, z) = r(1 - \cos \theta) - z$
39. *Work done by the heart* (Section 2.6, Exercise 48)

$$W(P, V, p, v, g) = PV + \frac{Vpv^2}{2g}$$

40. *Wilson lot size formula* (Section 3.5, Exercise 49)

$$A(c, h, k, m, q) = \frac{km}{q} + cm + \frac{hq}{2}$$

Find all the second order partial derivatives of the functions in Exercises 41–46.

41. $f(x, y) = x + y + xy$
42. $f(x, y) = \sin xy$
43. $g(x, y) = x^2 y + \cos y + y \sin x$
44. $h(x, y) = xe^y + y + 1$
45. $r(x, y) = \ln(x + y)$
46. $s(x, y) = \tan^{-1}(y/x)$

In Exercises 47–50, verify that $w_{xy} = w_{yx}$.

47. $w = \ln(2x + 3y)$
48. $w = e^x + x \ln y + y \ln x$
49. $w = xy^2 + x^2y^3 + x^3y^4$
50. $w = x \sin y + y \sin x + xy$

51. Which order of differentiation will calculate f_{xy} faster: x first, or y first? Try to answer without writing anything down.

a) $f(x, y) = x \sin y + e^y$

b) $f(x, y) = 1/x$
c) $f(x, y) = y + (x/y)$
d) $f(x, y) = y + x^2y + 4y^3 - \ln(y^2 + 1)$
e) $f(x, y) = x^2 + 5xy + \sin x + 7e^x$
f) $f(x, y) = x \ln xy$

52. The fifth order partial derivative $\partial^5 f/\partial x^2 \partial y^3$ is zero for each of the following functions. To show this as quickly as possible, which variable would you differentiate with respect to first: x, or y? Try to answer without writing anything down.
a) $f(x, y) = y^2x^4e^x + 2$
b) $f(x, y) = y^2 + y(\sin x - x^4)$
c) $f(x, y) = x^2 + 5xy + \sin x + 7e^x$
d) $f(x, y) = xe^{y^2/2}$

53. Find the value of $\partial z/\partial x$ at the point (1, 1, 1) if the equation
$$xy + z^3x - 2yz = 0$$
defines z as a function of the two independent variables x and y and the partial derivative exists.

54. Find the value of $\partial x/\partial z$ at the point $(1, -1, -3)$ if the equation
$$xz + y \ln x - x^2 + 4 = 0$$
defines x as a function of the two independent variables y and z and the partial derivative exists.

Exercises 55 and 56 are about the angles and sides of the triangle in Fig. 12.18.

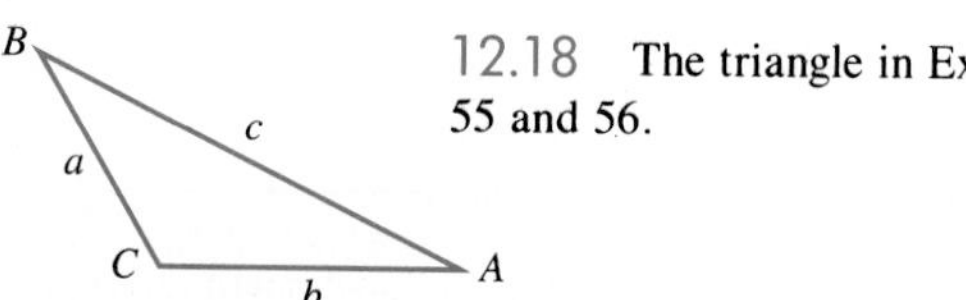

12.18 The triangle in Exercises 55 and 56.

55. Express A implicitly as a function of a, b, and c and calculate $\partial A/\partial a$ and $\partial A/\partial b$.

56. Express a implicitly as a function of A, b, and B and calculate $\partial a/\partial A$ and $\partial a/\partial B$.

57. Express v_x in terms of u and v if the equations $x = v \ln u$ and $y = u \ln v$ define u and v as functions of the independent variables x and y, and if v_x exists. (*Hint:* Differentiate both equations with respect to x and solve for v_x with Cramer's rule.)

58. Find $\partial x/\partial u$ and $\partial y/\partial u$ if the equations $u = x^2 - y^2$ and $v = x^2 - y$ define x and y as functions of the independent variables u and v, and the partial derivatives exist. (See the hint in Exercise 57.) Then let $s = x^2 + y^2$ and find $\partial s/\partial u$.

Laplace Equations

The *three-dimensional Laplace equation*
$$\frac{\partial^2 f}{\partial x^2} + \frac{\partial^2 f}{\partial y^2} + \frac{\partial^2 f}{\partial z^2} = 0 \tag{6}$$
is satisfied by steady-state temperature distributions $T = f(x, y, z)$ in space, by gravitational potentials, and by electrostatic potentials. The *two-dimensional Laplace equation*
$$\frac{\partial^2 f}{\partial x^2} + \frac{\partial^2 f}{\partial y^2} = 0, \tag{7}$$
obtained by dropping the $\partial^2 f/\partial z^2$ term from (6), describes potentials and steady-state temperature distributions in a plane (Fig. 12.19).

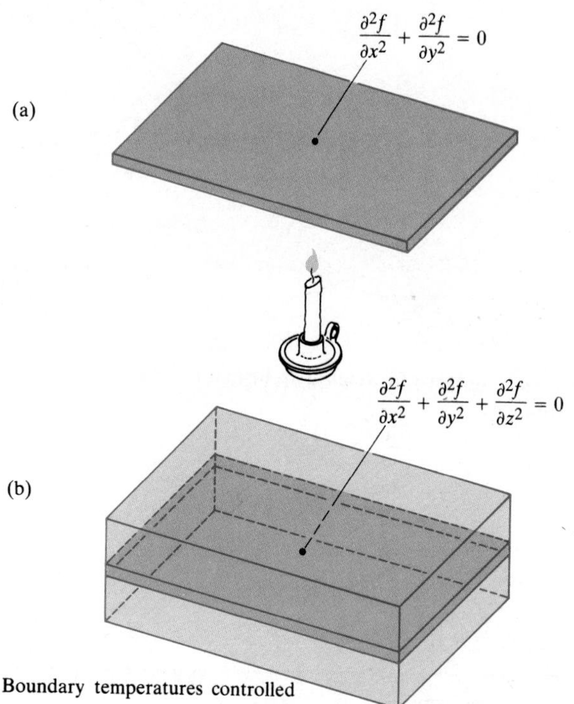

12.19 Steady-state temperature distributions in planes and solids satisfy Laplace equations. The plane (a) may be treated as a thin slice of the solid (b) perpendicular to the z-axis.

Show that each function in Exercises 59–64 satisfies a Laplace equation.

59. $f(x, y, z) = x^2 + y^2 - 2z^2$

60. $f(x, y, z) = e^{-2y} \cos 2x$

61. $f(x, y, z) = \ln\sqrt{x^2 + y^2}$

62. $f(x, y, z) = (x^2 + y^2 + z^2)^{-1/2}$

63. $f(x, y, z) = e^{3x + 4y} \cos 5z$

64. $f(x, y, z) = \cos 3x \cos 4y \sinh 5z$

65. For what values of n does
$$f(x, y, z) = (x^2 + y^2 + z^2)^n$$
satisfy the three-dimensional Laplace equation?

66. Find all solutions of the two-dimensional Laplace equation of the form
a) $f(x, y) = ax^2 + bxy + cy^2$
b) $f(x, y) = ax^3 + bx^2y + cxy^2 + dy^3$

The Wave Equation

If we stand on an ocean shore and take a snapshot of the waves, the picture shows a regular pattern of peaks and valleys in an instant of time (Fig. 12.20). We see periodic vertical motion in space, with respect to distance. If we stand in the water, we can feel the rise and fall of the water as the waves go by. We see periodic vertical motion in time. In physics, this beautiful symmetry is expressed by the *one-dimensional wave equation*

$$\frac{\partial^2 w}{\partial t^2} = c^2 \frac{\partial^2 w}{\partial x^2}, \tag{8}$$

where w is the wave height, x is the distance variable, t is the time variable, and c is the velocity with which the waves are propagated.

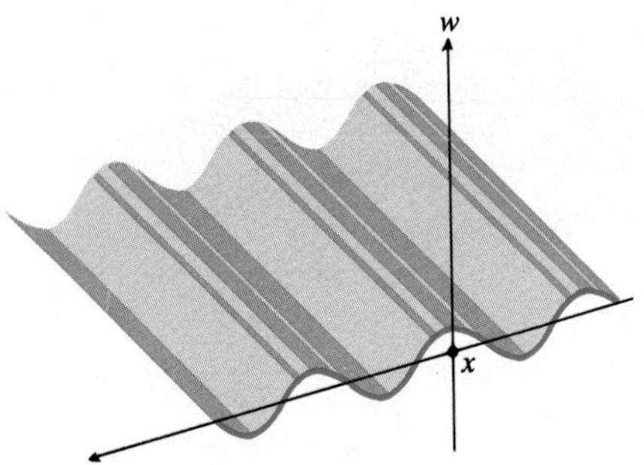

12.20 Waves in water at an instant in time. As time passes,

$$\frac{\partial^2 w}{\partial t^2} = c^2 \frac{\partial^2 w}{\partial x^2}.$$

In our example, x is the distance across the ocean's surface, but in other applications x might be the distance along a vibrating string, distance through air (sound waves), or distance through space (light waves). The number c varies with the medium and type of wave.

Equations with partial derivatives are normally harder to solve than equations with ordinary derivatives because the solutions of a partial differential equation may have so many different forms. As a case in point, show that the functions in Exercises 67–73 are all solutions of the wave equation.

67. $w = \sin(x + ct)$

68. $w = \cos(2x + 2ct)$

69. $w = \sin(x + ct) + \cos(2x + 2ct)$

70. $w = \ln(2x + 2ct)$

71. $w = \tan(2x - 2ct)$

72. $w = 5\cos(3x + 3ct) + e^{x+ct}$

73. $w = f(u)$, where f is a differentiable function of u, $u = a(x + ct)$, and a is a constant.

12.4 Differentiability, Linearization, and Differentials

We know what it means for a function of two variables to have partial derivatives at a point, but we have yet to say what it means for such a function to be differentiable at a point. In this section, we define differentiability and proceed from there to linearizations and differentials. The mathematical results of the section stem from the Increment Theorem. As we shall see in the next section, this theorem also underlies the Chain Rule for functions of two variables.

Differentiability

Surprising as it may seem, the starting point for differentiability is not Fermat's difference quotient but rather the idea of increment. You may recall from our work with functions of a single variable that if $y = f(x)$ is differentiable at $x = x_0$, then the change in the value of f that results from changing x from x_0 to $x_0 + \Delta x$ is given by an equation of the form

$$\Delta y = f'(x_0)\Delta x + \epsilon\,\Delta x \tag{1}$$

in which $\epsilon \to 0$ as $\Delta x \to 0$. We used this property in Section 2.6 to derive the Chain Rule for functions of a single variable. For functions of two variables, the analogous property becomes the definition of differentiability. The Increment Theorem tells us when we may expect the property to hold.

THEOREM 3

The Increment Theorem for Functions of Two Variables

Suppose that the first partial derivatives of $f(x, y)$ are defined throughout an open region R containing the point (x_0, y_0) and that f_x and f_y are continuous at (x_0, y_0). Then the change $\Delta z = f(x_0 + \Delta x, y_0 + \Delta y) - f(x_0, y_0)$ in the value of f that results from moving from (x_0, y_0) to another point $(x_0 + \Delta x, y_0 + \Delta y)$ in R satisfies an equation of the form

$$\Delta z = f_x(x_0, y_0)\Delta x + f_y(x_0, y_0)\Delta y + \epsilon_1 \Delta x + \epsilon_2 \Delta y, \tag{2}$$

in which $\epsilon_1, \epsilon_2 \to 0$ as $\Delta x, \Delta y \to 0$.

You will see where the epsilons come from if you read the proof in Appendix 11. You will also see that analogous theorems hold for functions of more than two variables.

The definition of differentiability for functions of two variables is based on Eq. (2).

DEFINITION

A function $f(x, y)$ is **differentiable at (x_0, y_0)** if $f_x(x_0, y_0)$ and $f_y(x_0, y_0)$ exist and Eq. (2) holds for f at (x_0, y_0). We call f **differentiable** if it is differentiable at every point in its domain.

In light of this definition, we have the immediate corollary of Theorem 3 that a function is differentiable if its first partial derivatives are continuous.

COROLLARY OF THEOREM 3

If the partial derivatives f_x and f_y of a function $f(x, y)$ are continuous throughout an open region R, then f is differentiable at every point of R.

If we replace the Δz in Eq. (2) by the expression $f(x, y) - f(x_0, y_0)$ and rewrite the equation as

$$f(x, y) = f(x_0, y_0) + f_x(x_0, y_0)\Delta x + f_y(x_0, y_0)\Delta y + \epsilon_1 \Delta x + \epsilon_2 \Delta y, \tag{3}$$

we see that the right-hand side of the new equation approaches $f(x_0, y_0)$ as Δx and Δy approach 0. This tells us that a function $f(x, y)$ is continuous at every point where it is differentiable.

THEOREM 4

If a function $f(x, y)$ is differentiable at (x_0, y_0), then f is continuous at (x_0, y_0).

We can see from Theorems 3 and 4 that a function $f(x, y)$ is continuous at a point (x_0, y_0) if its first partial derivatives are continuous throughout some open region R containing the point. Our experience with functions of a single variable might tempt us to think that the mere existence of f_x and f_y at (x_0, y_0) would be enough to ensure the continuity of f at (x_0, y_0). After all, the existence of the derivative of a single-variable function at a point does ensure continuity. But here we have a departure from the single-variable theory. A function of two variables can be

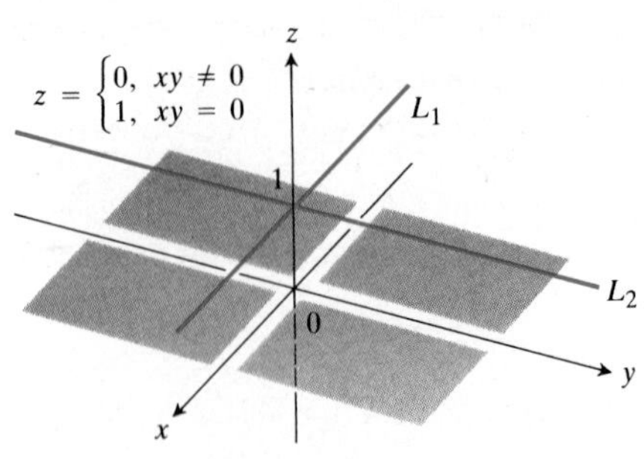

12.21 The function

$$f(x, y) = \begin{cases} 0, & xy \neq 0, \\ 1, & xy = 0, \end{cases}$$

is not continuous at (0, 0). The limit of f as (x, y) approaches (0, 0) along the line $y = x$ is 0, but $f(0, 0)$ itself is 1. The first partial derivatives of f at (0, 0) both exist, $f_x(0, 0) = 0$ being the slope of line L_1 and $f_y(0, 0) = 0$ being the slope of line L_2.

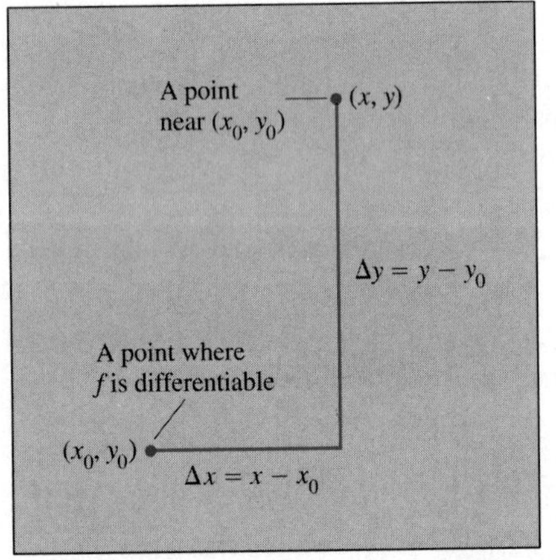

12.22 If f is differentiable at (x_0, y_0), then the value of f at any point (x, y) nearby is given by Eq. (5).

discontinuous at a point where its first partial derivatives exist (Fig. 12.21). Existence alone is not enough.

The Differentiability of Functions of More Than Two Variables

A function $w = f(x, y, z)$ of three independent variables is differentiable at a point $P_0(x_0, y_0, z_0)$ if f_x, f_y, and f_z exist there and f satisfies an equation of the form

$$\Delta w = f_x\Delta x + f_y\Delta y + f_z\Delta z + \epsilon_1\Delta x + \epsilon_2\Delta y + \epsilon_3\Delta z \tag{4}$$

in which f_x, f_y, and f_z are evaluated at P_0 and $\epsilon_1, \epsilon_2, \epsilon_3 \to 0$ as $\Delta x, \Delta y, \Delta z \to 0$. A generalization of the Increment Theorem tells us that f is differentiable at a point P_0 if its first partial derivatives are continuous throughout an open region containing P_0. Similar results hold for functions of any finite number of independent variables.

Linearization

The functions of two variables that arise in science and mathematics can be complicated and we sometimes need to replace them with simpler functions that give the accuracy required for specific applications without being so hard to work with. We do this in a way that is similar to the way we find linear replacements for functions of a single variable (Section 2.6). The source of the linear replacements, or linearizations as we again call them, is Eq. (3).

Suppose the function we wish to replace is $z = f(x, y)$ and that we want the replacement to be effective near a point (x_0, y_0) at which we know the values of f, f_x, and f_y and at which f is differentiable. Since f is differentiable, Eq. (3) holds for f at (x_0, y_0). Therefore, if we move from (x_0, y_0) to any point (x, y) by increments $\Delta x = x - x_0$ and $\Delta y = y - y_0$ (Fig. 12.22), the new value of f will be

$$\begin{aligned} f(x, y) = f(x_0, y_0) &+ f_x(x_0, y_0)(x - x_0) \\ &+ f_y(x_0, y_0)(y - y_0) + \epsilon_1\Delta x + \epsilon_2\Delta y, \end{aligned} \qquad \left(\begin{array}{l}\text{Eq. (3), with} \\ \Delta x = x - x_0 \\ \text{and } \Delta y = y - y_0\end{array}\right) \tag{5}$$

where $\epsilon_1, \epsilon_2 \to 0$ as $\Delta x, \Delta y \to 0$. If the increments Δx and Δy are small, the products $\epsilon_1\Delta x$ and $\epsilon_2\Delta y$ will be smaller still and we will have

$$f(x, y) \approx \underbrace{f(x_0, y_0) + f_x(x_0, y_0)(x - x_0) + f_y(x_0, y_0)(y - y_0)}_{L(x, y)}. \tag{6}$$

In other words, as long as Δx and Δy are small, f will have approximately the same value as the linear function L. When f is too hard to use, and our work can tolerate the error involved, we can safely replace f by L.

DEFINITIONS

The **linearization** of a function $f(x, y)$ at a point (x_0, y_0) where f is differentiable is the function

$$L(x, y) = f(x_0, y_0) + f_x(x_0, y_0)(x - x_0) + f_y(x_0, y_0)(y - y_0). \tag{7}$$

The approximation

$$L(x, y) \approx f(x, y) \tag{8}$$

is the **standard linear approximation** of f at (x_0, y_0).

In Section 12.7 we shall see that the plane $z = L(x, y)$ is tangent to the surface $z = f(x, y)$ at the point (x_0, y_0). Thus, the linearization of a function of two variables is a tangent-plane approximation to the function in the same way that the linearization of a function of a single variable is a tangent-line approximation. In the meantime, here is an example of a linearization.

Example 1 Find the linearization of

$$f(x, y) = x^2 - xy + \frac{1}{2}y^2 + 3$$

at the point (3, 2).

Solution We evaluate Eq. (7) with

$$f(x_0, y_0) = \left(x^2 - xy + \frac{1}{2}y^2 + 3\right)_{(3,\,2)} = 8,$$

$$f_x(x_0, y_0) = \frac{\partial}{\partial x}\left(x^2 - xy + \frac{1}{2}y^2 + 3\right)_{(3,\,2)} = (2x - y)_{(3,\,2)} = 4,$$

$$f_y(x_0, y_0) = \frac{\partial}{\partial y}\left(x^2 - xy + \frac{1}{2}y^2 + 3\right)_{(3,\,2)} = (-x + y)_{(3,\,2)} = -1,$$

getting

$$\begin{aligned} L(x, y) &= f(x_0, y_0) + f_x(x_0, y_0)(x - x_0) + f_y(x_0, y_0)(y - y_0) \qquad \text{(Eq. (7))} \\ &= 8 + (4)(x - 3) + (-1)(y - 2) = 4x - y - 2. \end{aligned}$$

The linearization of f at (3, 2) is $L(x, y) = 4x - y - 2$.

How Accurate Is the Standard Linear Approximation?

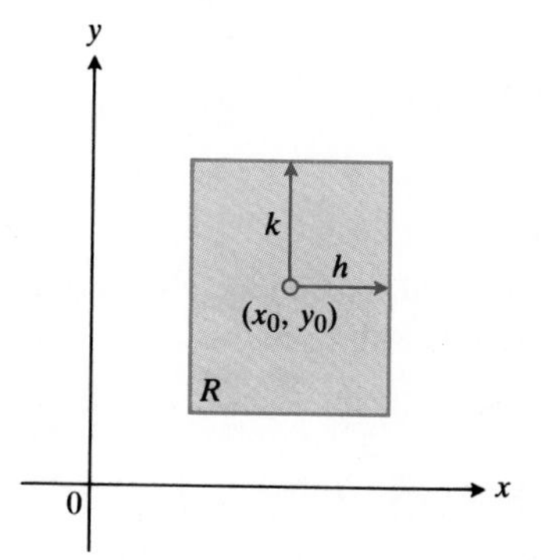

12.23 The rectangular region R: $|x - x_0| \leq h$, $|y - y_0| \leq k$ in the xy-plane. On this kind of region, we can find error bounds for our approximations.

To find the error in the approximation $f(x, y) \approx L(x, y)$, we need the second order partial derivatives of f. Suppose that the first and second order partial derivatives of f are continuous throughout an open set containing a closed rectangular region R centered at (x_0, y_0) and given by the inequalities

$$|x - x_0| \leq h, \qquad |y - y_0| \leq k \tag{9}$$

(Fig. 12.23). Since R is closed and bounded, the second partial derivatives all have maximum values on R. If B is the largest of these values, then, as explained in Section 12.10, the error $E(x, y) = f(x, y) - L(x, y)$ in the standard linear approximation satisfies the inequality

$$|E(x, y)| \leq \frac{1}{2}B\left(|x - x_0| + |y - y_0|\right)^2 \tag{10}$$

throughout R.

When we use this inequality to estimate E, we usually cannot find the values of f_{xx}, f_{yy}, and f_{xy} that determine B and we have to replace B itself with an upper-bound or "worst-case" value instead. If M is any common upper bound for $|f_{xx}|$, $|f_{yy}|$, and $|f_{xy}|$ on R, then B will be less than or equal to M and we will know that

$$|E(x, y)| \leq \frac{1}{2}M\left(|x - x_0| + |y - y_0|\right)^2. \tag{11}$$

This is the inequality we normally use in estimating E. When we need to make $|E(x, y)|$ small for a given M, we just make $|x - x_0|$ and $|y - y_0|$ small.

The Error in the Standard Linear Approximation of $f(x, y)$ near (x_0, y_0)

If f has continuous first and second partial derivatives throughout an open set containing a rectangle R centered at (x_0, y_0) and if M is any upper bound for the values of $|f_{xx}|$, $|f_{yy}|$, and $|f_{xy}|$ on R, then the error $E(x, y)$ incurred in replacing $f(x, y)$ on R by its linearization

$$L(x, y) = f(x_0, y_0) + f_x(x_0, y_0)(x - x_0) + f_y(x_0, y_0)(y - y_0) \tag{12}$$

satisfies the inequality

$$|E(x, y)| \le \frac{1}{2} M\left(|x - x_0| + |y - y_0|\right)^2. \tag{13}$$

Example 2 In Example 1, we found the linearization of

$$f(x, y) = x^2 - xy + \frac{1}{2}y^2 + 3$$

at (3, 2) to be

$$L(x, y) = 4x - y - 2.$$

Find an upper bound for the error in the approximation $f(x, y) \approx L(x, y)$ over the rectangle

$$R: \quad |x - 3| \le 0.1, \quad |y - 2| \le 0.1.$$

Express the upper bound as a percentage of $f(3, 2)$, the value of f at the center of the rectangle.

Solution We use the inequality

$$|E(x, y)| \le \frac{1}{2} M\left(|x - x_0| + |y - y_0|\right)^2. \qquad \text{(Eq. (13))}$$

To find a suitable value for M, we calculate f_{xx}, f_{xy}, and f_{yy}, finding, after a routine differentiation, that all three derivatives are constant, with values

$$|f_{xx}| = |2| = 2, \qquad |f_{xy}| = |-1| = 1, \qquad |f_{yy}| = |1| = 1.$$

The largest of these is 2, so we may safely take M to be 2. With $(x_0, y_0) = (3, 2)$, we then know that, throughout R,

$$|E(x, y)| \le \frac{1}{2}(2)\left(|x - 3| + |y - 2|\right)^2 = \left(|x - 3| + |y - 2|\right)^2.$$

Finally, since $|x - 3| \le 0.1$ and $|y - 2| \le 0.1$ on R, we have

$$|E(x, y)| \le (0.1 + 0.1)^2 = 0.04.$$

As a percentage of $f(3, 2) = 8$, the error is no greater than

$$\frac{0.04}{8} \times 100 = 0.5\%.$$

As long as (x, y) stays in R, the approximation $f(x, y) \approx L(x, y)$ will be in error by no more than 0.04, which is 1/2% of the value of f at the center of R.

Predicting Change with Differentials

Suppose we know the values of a differentiable function $f(x, y)$ and its first partial derivatives at a point (x_0, y_0) and we want to predict how much this value will change if we move to a point $(x_0 + \Delta x, y_0 + \Delta y)$ nearby. If Δx and Δy are small, f and its linearization will change by nearly the same amount, so calculating the change in L gives a practical way to estimate the change in f.

The change in f is

$$\Delta f = f(x_0 + \Delta x, y_0 + \Delta y) - f(x_0, y_0).$$

A straightforward calculation with Eq. (7), using the notation $x - x_0 = \Delta x$ and $y - y_0 = \Delta y$, shows that the corresponding change in L is

$$\begin{aligned} \Delta L &= L(x_0 + \Delta x, y_0 + \Delta y) - L(x_0, y_0) \\ &= f_x(x_0, y_0)\Delta x + f_y(x_0, y_0)\Delta y. \end{aligned} \tag{14}$$

The formula for Δf is usually as hard to work with as the formula for f. The change in L, however, is just a known constant times Δx plus a known constant times Δy.

The change ΔL is usually described in the more suggestive notation

$$df = f_x(x_0, y_0)\, dx + f_y(x_0, y_0)\, dy, \tag{15}$$

in which df denotes the change in the linearization that results from the changes dx and dy in x and y. As usual, we call dx and dy differentials of x and y, and we call df the corresponding differential of f.

DEFINITION

If we move (x_0, y_0) to a nearby point $(x_0 + dx, y_0 + dy)$, the resulting differential in f is

$$df = f_x(x_0, y_0)\, dx + f_y(x_0, y_0)\, dy. \tag{16}$$

This change in the linearization of f is called the **total differential of f.**

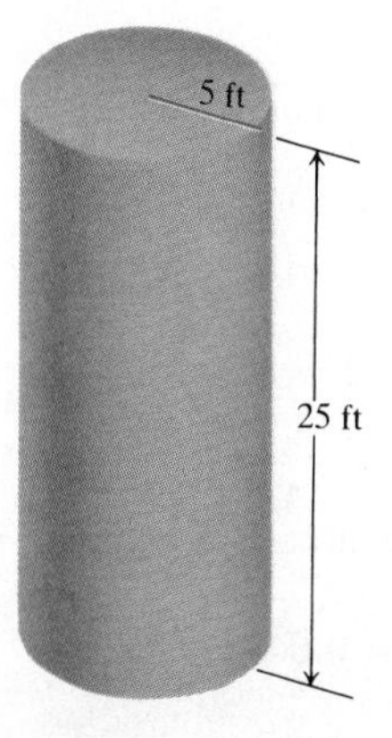

12.24 A small change in this tank's radius will change the volume 10 times as much as a change of the same size in the tank's height (Example 3).

Example 3 *Sensitivity to Change.* Your company manufactures right circular cylindrical molasses storage tanks that are 25 ft high with a radius of 5 ft (Fig. 12.24). How sensitive are the tanks' volumes to small variations in height and radius?

Solution As a function of radius r and height h, the typical tank's volume is

$$V = \pi r^2 h.$$

The change in volume caused by small changes dr and dh in radius and height is approximately

$$\begin{aligned} dV &= V_r(5, 25)\, dr + V_h(5, 25)\, dh \qquad \begin{pmatrix}\text{Eq. (16) with } f = V \text{ and} \\ (x_0, y_0) = (5, 25)\end{pmatrix} \\ &= (2\pi rh)_{(5, 25)}\, dr + (\pi r^2)_{(5, 25)}\, dh \\ &= 250\pi\, dr + 25\pi\, dh. \end{aligned}$$

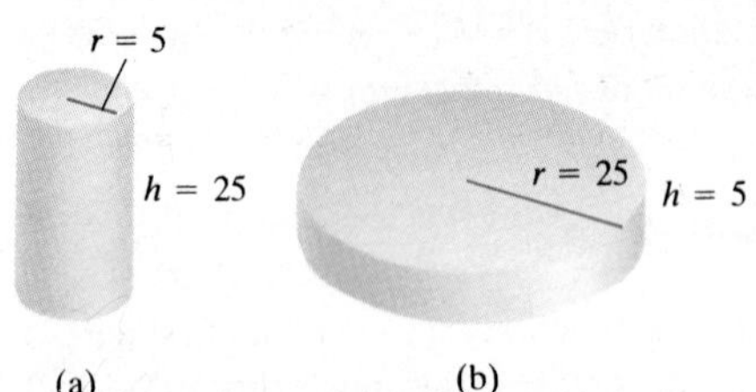

12.25 The volume of cylinder (a) is more sensitive to a small change in r than it is to an equally small change in h. The volume of cylinder (b) is more sensitive to small changes in h than it is to small changes in r.

Thus, a 1-unit change in r will change V by about 250π units. A 1-unit change in h will change V by about 25π units. The tank's volume is 10 times more sensitive to a small change in r than it is to a small change of equal size in h. As a quality control engineer concerned with being sure the tanks have the correct volume, you would want to pay special attention to their radii.

In contrast, if the values of r and h are reversed to make $r = 25$ and $h = 5$, then the total differential in V becomes

$$dV = (2\pi rh)_{(25,\,5)}\,dr + (\pi r^2)_{(25,\,5)}\,dh = 250\pi\,dr + 625\pi\,dh.$$

Now the volume is more sensitive to changes in h than to changes in r (Fig. 12.25).

The general rule to be learned from this example is that functions are most sensitive to small changes in the variables that generate the largest partial derivatives.

Absolute, Relative, and Percentage Change

When we move from (x_0, y_0) to a nearby point, we can describe the corresponding change in the value of a function $f(x, y)$ in three different ways.

	True	**Estimate**	
Absolute change:	Δf	df	
Relative change:	$\dfrac{\Delta f}{f(x_0, y_0)}$	$\dfrac{df}{f(x_0, y_0)}$	(17)
Percentage change:	$\dfrac{\Delta f}{f(x_0, y_0)} \times 100$	$\dfrac{df}{f(x_0, y_0)} \times 100$	

Example 4 Suppose that the variables r and h change from the initial values of $(r_0, h_0) = (1, 5)$ by the amounts $dr = 0.03$ and $dh = -0.1$. Estimate the resulting absolute, relative, and percentage changes in the values of the function $V = \pi r^2 h$.

Solution To estimate the absolute change in V, we evaluate

$$dV = V_r(r_0, h_0)\,dr + V_h(r_0, h_0)\,dh \tag{18}$$

to get

$$dV = 2\pi r_0 h_0\,dr + \pi r_0^2\,dh = 2\pi(1)(5)(0.03) + \pi(1)^2(-0.1)$$

$$= 0.3\pi - 0.1\pi = 0.2\pi.$$

We divide this by $V(r_0, h_0)$ to estimate the relative change:

$$\frac{dV}{V(r_0, h_0)} = \frac{0.2\pi}{\pi r_0^2 h_0} = \frac{0.2\pi}{\pi(1)^2(5)} = 0.04.$$

We multiply this by 100 to estimate the percentage change:

$$\frac{dV}{V(r_0, h_0)} \times 100 = 0.04 \times 100 = 4\%.$$

Example 5 The volume $V = \pi r^2 h$ of a right circular cylinder is to be calculated from measured values of r and h. Suppose that r is measured with an error of no more than 2% and h with an error of no more than 0.5%. Estimate the resulting possible percentage error in the calculation of V.

Solution We are told that

$$\left|\frac{dr}{r} \times 100\right| \leq 2 \quad \text{and} \quad \left|\frac{dh}{h} \times 100\right| \leq 0.5.$$

Since

$$\frac{dV}{V} = \frac{2\pi rh\, dr + \pi r^2\, dh}{\pi r^2 h} = \frac{2\, dr}{r} + \frac{dh}{h},$$

we have

$$\begin{aligned}\left|\frac{dV}{V} \times 100\right| &= \left|2\frac{dr}{r} \times 100 + \frac{dh}{h} \times 100\right| \\ &\leq 2\left|\frac{dr}{r} \times 100\right| + \left|\frac{dh}{h} \times 100\right| \\ &\leq 2(2) + 0.5 = 4.5.\end{aligned}$$

We estimate the error in the volume calculation to be at most 4.5%.

How accurately do we have to measure r and h to have a reasonable chance of calculating $V = \pi r^2 h$ with an error, say, of less than 2%? Questions like this are hard to answer for functions of more than one variable because there is usually no single right answer. Since

$$\frac{dV}{V} = 2\frac{dr}{r} + \frac{dh}{h},$$

we see that dV/V is controlled by a combination of dr/r and dh/h. If we can measure h with great accuracy, we might come out all right even if we are sloppy about measuring r. On the other hand, our measurement of h might have so large a dh that the resulting dV/V would be too crude an estimate of $\Delta V/V$ to be useful even if dr were zero.

What we do in such cases is look for a reasonable square about the measured values (r_0, h_0) in which V will not vary by more than the allowed amount from $V_0 = \pi r_0^2 h_0$. The next example shows how this is done.

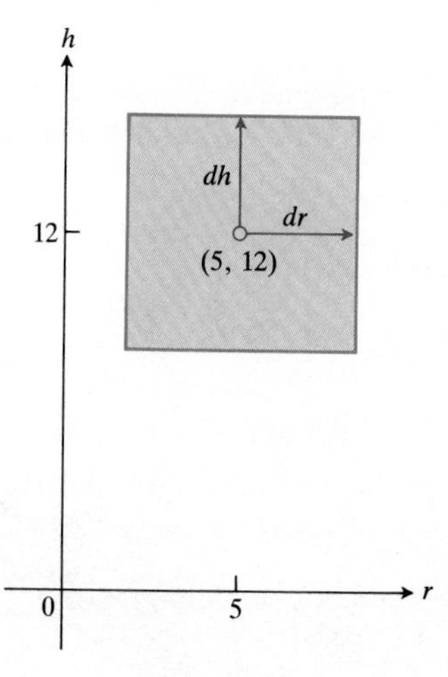

12.26 A small square about the point (5, 12) in the rh-plane (Example 6).

Example 6 Find a reasonable square about the point $(r_0, h_0) = (5, 12)$ in which the value of $V = \pi r^2 h$ will not vary by more than ± 0.1.

Solution We approximate the variation ΔV by the differential

$$dV = 2\pi r_0 h_0\, dr + \pi r_0^2\, dh = 2\pi(5)(12)\, dr + \pi(5)^2\, dh = 120\pi\, dr + 25\pi\, dh.$$

Since the region to which we are restricting our attention is a square (Fig. 12.26), we may set $dh = dr$ to get

$$dV = 120\pi\, dr + 25\pi\, dr = 145\pi\, dr.$$

We then ask, How small must we take dr to be sure that $|dV|$ is no larger than 0.1? To answer, we start with the inequality

$$|dV| \leq 0.1,$$

express dV in terms of dr,

$$|145\pi\, dr| \leq 0.1,$$

and find a corresponding upper bound for dr:

$$|dr| \leq \frac{0.1}{145\pi} \approx 2.1 \times 10^{-4}. \qquad \left(\begin{array}{l}\text{Rounding down to make sure} \\ dr \text{ won't accidentally be too big}\end{array}\right)$$

With $dh = dr$, then, the square we want is described by the inequalities

$$|r - 5| \leq 2.1 \times 10^{-4}, \qquad |h - 12| \leq 2.1 \times 10^{-4}.$$

As long as (r, h) stays in this square, we may expect $|dV|$ to be less than or equal to 0.1 and we may expect $|\Delta V|$ to be approximately the same size.

Functions of More Than Two Variables

Analogous results hold for differentiable functions of more than two variables.

1. The **linearization** of a differentiable function $f(x, y, z)$ at a point $P_0(x_0, y_0, z_0)$ is

$$L(x, y, z) = f(P_0) + f_x(P_0)(x - x_0) + f_y(P_0)(y - y_0) + f_z(P_0)(z - z_0). \tag{19}$$

2. Suppose that R is a closed rectangular solid centered at P_0 and lying in an open region on which the second partial derivatives of f are continuous. Suppose also that $|f_{xx}|$, $|f_{yy}|$, $|f_{zz}|$, $|f_{xy}|$, $|f_{xz}|$, and $|f_{yz}|$ are all less than or equal to M throughout R. Then the **error** $E(x, y, z) = f(x, y, z) - L(x, y, z)$ in the approximation of f by L is bounded throughout R by the inequality

$$|E| \leq \frac{1}{2} M\left(|x - x_0| + |y - y_0| + |z - z_0|\right)^2. \tag{20}$$

3. If the second partial derivatives of f are continuous and if x, y, and z change from x_0, y_0, and z_0 by small amounts dx, dy, and dz, the **total differential**

$$df = f_x(P_0)\, dx + f_y(P_0)\, dy + f_z(P_0)\, dz \tag{21}$$

gives a good approximation of the resulting change in f.

Example 7 Find the linearization $L(x, y, z)$ of

$$f(x, y, z) = x^2 - xy + 3 \sin z$$

at the point $(x_0, y_0, z_0) = (2, 1, 0)$. Find an upper bound for the error incurred in replacing f by L on the rectangle

$$R: \quad |x - 2| \leq 0.01, \quad |y - 1| \leq 0.02, \quad |z| \leq 0.01.$$

Solution A routine evaluation gives

$$f(2, 1, 0) = 2, \qquad f_x(2, 1, 0) = 3, \qquad f_y(2, 1, 0) = -2, \qquad f_z(2, 1, 0) = 3.$$

With these values, Eq. (19) becomes

$$L(x, y, z) = 2 + 3(x - 2) + (-2)(y - 1) + 3(z - 0) = 3x - 2y + 3z - 2.$$

Equation (20) gives an upper bound for the error incurred by replacing f by L on R. Since

$$f_{xx} = 2, \qquad f_{yy} = 0, \qquad f_{zz} = -3 \sin z,$$
$$f_{xy} = -1, \qquad f_{xz} = 0, \qquad f_{yz} = 0,$$

we may safely take M to be $\max|-3 \sin z| = 3$. Hence

$$|E| \le \frac{1}{2}(3)(0.01 + 0.02 + 0.01)^2 = 0.0024.$$

The error will be no greater than 0.0024.

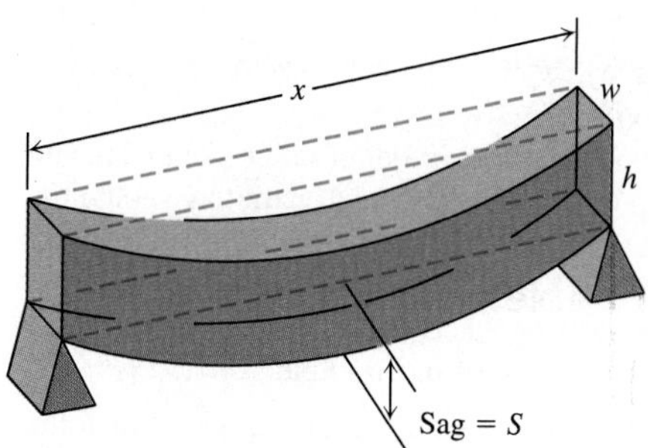

12.27 A beam supported at its two ends before and after loading. Example 8 shows how the sag S is related to the weight of the load and the dimensions of the beam.

Example 8 *Controlling Sag in Uniformly Loaded Beams.* A rectangular beam, supported at both ends, will sag when subjected to a uniform load (constant weight per linear foot). The amount S of sag (Fig. 12.27) is calculated with the formula

$$S = C\frac{px^4}{wh^3}.$$

In this equation,

p = the load (newtons per meter of beam length),

x = the length between supports (m),

w = the width of the beam (m),

h = the height of the beam (m),

C = a constant that depends on the units of measurement and on the material from which the beam is made.

Find dS for a beam 4 m long, 10 cm wide, and 20 cm high that is subjected to a load of 100 N/m (Fig. 12.28). What conclusions can be drawn about the beam from the expression for dS?

Solution Since S is a function of the four independent variables p, x, w, and h, its differential dS is given by the equation

$$dS = S_p\, dp + S_x\, dx + S_w\, dw + S_h\, dh.$$

When we write this out for a particular set of values p_0, x_0, w_0, and h_0 and simplify the result, we find that

$$dS = S_0\left(\frac{dp}{p_0} + \frac{4dx}{x_0} - \frac{dw}{w_0} - \frac{3dh}{h_0}\right), \tag{22}$$

where $S_0 = S(p_0, x_0, w_0, h_0) = Cp_0x_0^4/w_0h_0^3$.

If $p_0 = 100$ N/m, $x_0 = 4$ m, $w_0 = 0.1$ m, and $h_0 = 0.2$ m, then

$$dS = S_0\left(\frac{dp}{100} + dx - 10dw - 15dh\right). \tag{23}$$

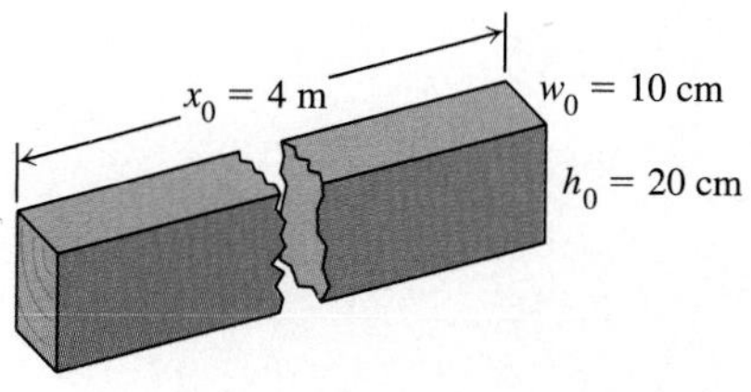

12.28 The dimensions of the beam in Example 8.

Here is what we can learn from Eq. (23). Since dp and dx appear with positive coefficients, increases in p and x will increase the sag. But dw and dh appear with negative coefficients, so increases in w and h will decrease the sag (make the beam stiffer). The sag is not very sensitive to changes in load because the coefficient of dp is $1/100$. The magnitude of the coefficient of dh is greater than the magnitude of the coefficient of dw. Making the beam 1 cm higher will therefore decrease the sag more than making the beam 1 cm wider.

EXERCISES 12.4

In Exercises 1–6, find the linearization $L(x, y)$ of the given function at each point.

1. $f(x, y) = x^2 + y^2 + 1$ at (a) $(0, 0)$, (b) $(1, 1)$
2. $f(x, y) = x^3y^4$ at (a) $(1, 1)$, (b) $(0, 0)$
3. $f(x, y) = e^x \cos y$ at (a) $(0, 0)$, (b) $(0, \pi/2)$
4. $f(x, y) = (x + y + 2)^2$ at (a) $(0, 0)$, (b) $(1, 2)$
5. $f(x, y) = 3x - 4y + 5$ at (a) $(0, 0)$, (b) $(1, 1)$
6. $f(x, y) = e^{2y-x}$ at (a) $(0, 0)$, (b) $(1, 2)$

In Exercises 7–12, find the linearization $L(x, y)$ of the function $f(x, y)$ at P_0. Then use Eq. (13) to find an upper bound for the magnitude $|E|$ of the error in the approximation $f(x, y) \approx L(x, y)$ over the rectangle R.

7. $f(x, y) = x^2 - 3xy + 5$ at $P_0(2, 1)$,
   R: $|x - 2| \le 0.1$, $|y - 1| \le 0.1$
8. $f(x, y) = (1/2)x^2 + xy + (1/4)y^2 + 3x - 3y + 4$ at $P_0(2, 2)$,
   R: $|x - 2| \le 0.1$, $|y - 2| \le 0.1$
9. $f(x, y) = 1 + y + x \cos y$ at $P_0(0, 0)$,
   R: $|x| \le 0.2$, $|y| \le 0.2$
   (Use $|\cos y| \le 1$ and $|\sin y| \le 1$ in estimating E.)
10. $f(x, y) = \ln x + \ln y$ at $P_0(1, 1)$,
    R: $|x - 1| \le 0.2$, $|y - 1| \le 0.2$
11. $f(x, y) = e^x \cos y$ at $P_0(0, 0)$,
    R: $|x| \le 0.1$, $|y| \le 0.1$
    (Use $e^x \le 1.11$ and $|\cos y| \le 1$ in estimating E.)
12. $f(x, y) = xy^2 + y \cos(x - 1)$ at $P_0(1, 2)$,
    R: $|x - 1| \le 0.1$, $|y - 2| \le 0.1$
13. You plan to calculate the area of a long, thin rectangle from measurements of its length and width. Which measurement should you be more careful with? Why?
14. a) Around the point $(1, 0)$, is $f(x, y) = x^2(y + 1)$ more sensitive to changes in x, or to changes in y?
    b) What ratio of dx to dy will make df equal zero at $(1, 0)$?
15. Suppose T is to be calculated from the formula $T = x(e^y + e^{-y})$ and that x and y are known to be 2 and $\ln 2$ with maximum possible errors of $|dx| = 0.1$ and $|dy| = 0.02$. Estimate the maximum possible error in the computed value of T.
16. About how accurately may $V = \pi r^2 h$ be calculated from measurements of r and h that are in error by 1%?
17. If $r = 5.0$ cm and $h = 12.0$ cm to the nearest millimeter, what should we expect the maximum percentage error in calculating $V = \pi r^2 h$ to be?
18. To estimate the volume of a cylinder of radius about 2 m and height about 3 m, approximately how accurately should the radius and height be measured so that the error in the volume estimate will not exceed 0.1 m^3? Assume that the possible error dr in measuring r is equal to the possible error dh in measuring h.
19. Give a reasonable square centered at $(1, 1)$ over which the value of $f(x, y) = x^3y^4$ will not vary by more than ± 0.1.
20. *Variation in electrical resistance.* The resistance R produced by wiring resistors of R_1 and R_2 ohms in parallel (Fig. 12.29) can be calculated from the formula

$$\frac{1}{R} = \frac{1}{R_1} + \frac{1}{R_2}.$$

a) Show that

$$dR = \left(\frac{R}{R_1}\right)^2 dR_1 + \left(\frac{R}{R_2}\right)^2 dR_2.$$

b) You have designed a two-resistor circuit like the one in Fig. 12.29 to have resistances of $R_1 = 100$ ohms and $R_2 = 400$ ohms, but there is always some variation in manufacturing and the resistors received by your firm will probably not have these exact values. Will the value of R be more sensitive to variation in R_1, or to variation in R_2?

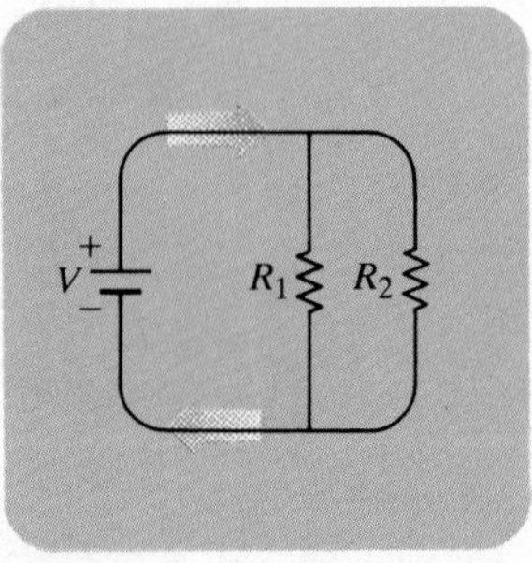

12.29 The circuit in Exercises 20 and 21.

21. *Continuation of Exercise 20.* In another circuit like the one in Fig. 12.29, you plan to change R_1 from 20 to 20.1 ohms and R_2 from 25 to 24.9 ohms. By about what percentage will this change R?

22. *Error carry-over in coordinate changes*

a) If $x = 3 \pm 0.01$ and $y = 4 \pm 0.01$ (Fig. 12.30), with approximately what accuracy can you calculate the polar coordinates r and θ of the point $P(x, y)$ from the formulas $r^2 = x^2 + y^2$ and $\theta = \tan^{-1}(y/x)$? Express your estimates as percentage changes of the values that r and θ have at the point $(x_0, y_0) = (3, 4)$.

b) At the point $(x_0, y_0) = (3, 4)$, are the values of r and θ more sensitive to changes in x, or to changes in y?

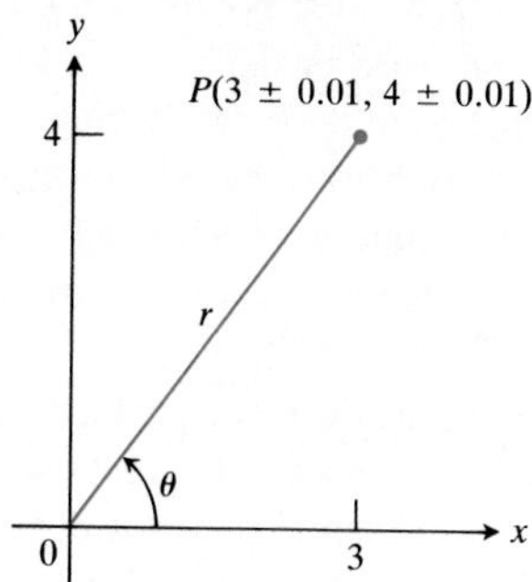

12.30 How much do the errors in the Cartesian coordinates of P affect the polar coordinates of P? See Exercise 22.

23. Does a function $f(x, y)$ with continuous first partial derivatives throughout an open region R have to be continuous on R? Explain.

24. If a function $f(x, y)$ has continuous second partial derivatives throughout an open region R, must the first order partial derivatives of f be continuous on R? Explain.

Functions of More Than Two Variables

Find the linearizations $L(x, y, z)$ of the functions in Exercises 25–30 at the given points.

25. $f(x, y, z) = xy + yz + xz$ at
a) $(1, 1, 1)$ b) $(1, 0, 0)$ c) $(0, 0, 0)$

26. $f(x, y, z) = x^2 + y^2 + z^2$ at
a) $(1, 1, 1)$ b) $(0, 1, 0)$ c) $(1, 0, 0)$

27. $f(x, y, z) = \sqrt{x^2 + y^2 + z^2}$ at
a) $(1, 0, 0)$ b) $(1, 1, 0)$ c) $(1, 2, 2)$

28. $f(x, y, z) = (\sin xy)/z$ at
a) $(\pi/2, 1, 1)$ b) $(2, 0, 1)$

29. $f(x, y, z) = e^x + \cos(y + z)$ at
a) $(0, 0, 0)$ b) $\left(0, \frac{\pi}{2}, 0\right)$ c) $\left(0, \frac{\pi}{4}, \frac{\pi}{4}\right)$

30. $f(x, y, z) = \tan^{-1}(xyz)$ at
a) $(1, 0, 0)$ b) $(1, 1, 0)$ c) $(1, 1, 1)$

In Exercises 31–34, find the linearization $L(x, y, z)$ of the function $f(x, y, z)$ at P_0. Then use Eq. (20) to find an upper bound for the magnitude of the error E in the approximation $f(x, y, z) \approx L(x, y, z)$ over the region R.

31. $f(x, y, z) = xz - 3yz + 2$ at $P_0(1, 1, 2)$
R: $|x - 1| \le 0.01$, $|y - 1| \le 0.01$, $|z - 2| \le 0.02$

32. $f(x, y, z) = x^2 + xy + yz + (1/4)z^2$ at $P_0(1, 1, 2)$
R: $|x - 1| \le 0.01$, $|y - 1| \le 0.01$, $|z - 2| \le 0.08$

33. $f(x, y, z) = xy + 2yz - 3xz$ at $P_0(1, 1, 0)$
R: $|x - 1| \le 0.01$, $|y - 1| \le 0.01$, $|z| \le 0.01$

34. $f(x, y, z) = \sqrt{2} \cos x \sin(y + z)$ at $P_0(0, 0, \pi/4)$
R: $|x| \le 0.01$, $|y| \le 0.01$, $|z - \pi/4| \le 0.01$

35. The beam of Example 8 is tipped on its side so that $h = 0.1$ m and $w = 0.2$ m.
a) What is the value of dS now?
b) Compare the sensitivity of the newly positioned beam to a small change in height with its sensitivity to an equally small change in width.

36. If $|a|$ is much greater than $|b|$, $|c|$, and $|d|$, to which of a, b, c, and d is the value of the determinant

$$f(a, b, c, d) = \begin{vmatrix} a & b \\ c & d \end{vmatrix}$$

most sensitive?

37. Estimate how strongly simultaneous errors of 2% in a, b, and c might affect the calculation of the product

$$p(a, b, c) = abc.$$

38. Estimate how much wood it takes to make a hollow rectangular box whose inside measurements are 5 ft long by 3 ft wide by 2 ft deep if the box is made of lumber 1/2-in. thick and the box has no top.

39. The area of a triangle is $(1/2)ab \sin C$, where a and b are the lengths of two sides of the triangle and C is the measure of the included angle. In surveying a triangular plot, you have measured a, b, and C to be 150 ft, 200 ft, and 60°, respectively. By about how much might your area calculation be in error if your values of a and b are off by half a foot each and your measurement of C is off by 2°? See Fig. 12.31. Remember to use radians.

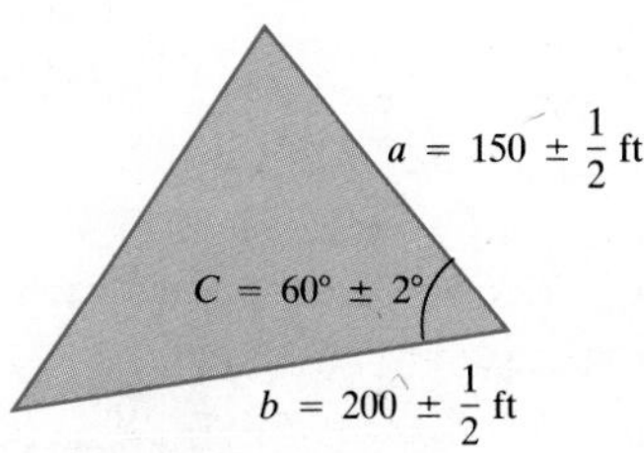

12.31 How accurately can you calculate the triangle's area from the measurement shown? See Exercise 39.

40. Suppose that $u = xe^y + y \sin z$ and that x, y, and z can be measured with maximum possible errors of ± 0.2, ± 0.6, and $\pm \pi/180$, respectively. Estimate the maximum possible error in calculating u from the measured values $x = 2$, $y = \ln 3$, $z = \pi/2$.

41. *The Wilson lot size formula.* The Wilson lot size formula in economics says that the most economical quantity Q of goods (radios, shoes, brooms, whatever) for a store to order is given by the formula $Q = \sqrt{2KM/h}$, where K is the cost of placing the order, M is the number of items sold per week, and h is the weekly holding cost for each item (cost of space, utilities, security, and so on). To which of the variables K, M, and h is Q most sensitive near the point $(K_0, M_0, h_0) = (2, 20, 0.05)$?

12.5 The Chain Rule

When we are interested in the temperature $w = f(x, y, z)$ at points along a curve $x = g(t)$, $y = h(t)$, $z = k(t)$ in space, or in the pressure or density along a path through a gas or fluid, we may think of f as a function of the single variable t. For each value of t, the temperature at the point $(g(t), h(t), k(t))$ is the value of the composite $f(g(t), h(t), k(t))$. If we then wish to know the rate at which f changes with respect to t along the path, we have only to differentiate this composite with respect to t, provided, of course, the derivative exists.

Sometimes we can find the derivative by substituting the formulas for g, h, and k into the formula for f and differentiating directly with respect to t. But we often have to work with functions whose formulas are too complicated for convenient substitution or for which formulas are not readily available. To find a function's derivatives under circumstances like these, we use the Chain Rule. The form the Chain Rule takes depends on how many variables are involved but, except for the presence of additional variables, it works just like the Chain Rule from Chapter 2. The basic idea remains the same: Composites of differentiable functions are differentiable.

The Chain Rule for Functions of Two Variables

In Chapter 2, we used the Chain Rule when $w = f(x)$ was a differentiable function of x and $x = g(t)$ was a differentiable function of t. This made w a differentiable function of t and the Chain Rule said that dw/dt could be calculated with the formula

$$\frac{dw}{dt} = \frac{dw}{dx}\frac{dx}{dt}.$$

The analogous formula for a function $w = f(x, y)$ of two independent variables is given in Theorem 5.

THEOREM 5

Chain Rule for Functions of Two Independent Variables

If $w = f(x, y)$ is differentiable and x and y are differentiable functions of t, then w is a differentiable function of t and

$$\frac{dw}{dt} = \frac{\partial f}{\partial x}\frac{dx}{dt} + \frac{\partial f}{\partial y}\frac{dy}{dt}. \qquad (1)$$

The way to remember the Chain Rule is to picture the diagram below. To find dw/dt, start at w and read down each route to t, multiplying derivatives along the way. Then add the products.

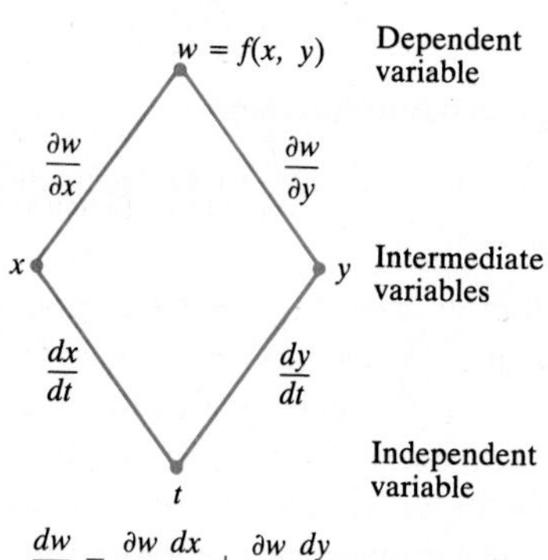

$$\frac{dw}{dt} = \frac{\partial w}{\partial x}\frac{dx}{dt} + \frac{\partial w}{\partial y}\frac{dy}{dt}$$

Proof The proof consists of showing that if x and y are differentiable at $t = t_0$, then w is differentiable at t_0 and

$$\left(\frac{dw}{dt}\right)_{t_0} = \left(\frac{\partial w}{\partial x}\right)_{t_0}\left(\frac{dx}{dt}\right)_{t_0} + \left(\frac{\partial w}{\partial y}\right)_{t_0}\left(\frac{dy}{dt}\right)_{t_0}. \tag{2}$$

Let Δx, Δy, and Δw be the increments that result from changing t from t_0 to $t_0 + \Delta t$. Since f is differentiable,

$$\Delta w = \left(\frac{\partial w}{\partial x}\right)_{t_0}\Delta x + \left(\frac{\partial w}{\partial y}\right)_{t_0}\Delta y + \epsilon_1\Delta x + \epsilon_2\Delta y, \tag{3}$$

where $\epsilon_1, \epsilon_2 \to 0$ as $\Delta x, \Delta y \to 0$. We divide Eq. (3) through by Δt and let Δt approach zero. The division gives

$$\frac{\Delta w}{\Delta t} = \left(\frac{\partial w}{\partial x}\right)_{t_0}\frac{\Delta x}{\Delta t} + \left(\frac{\partial w}{\partial y}\right)_{t_0}\frac{\Delta y}{\Delta t} + \epsilon_1\frac{\Delta x}{\Delta t} + \epsilon_2\frac{\Delta y}{\Delta t}, \tag{4}$$

and letting Δt approach zero gives

$$\begin{aligned}\left(\frac{dw}{dt}\right)_{t_0} &= \lim_{\Delta t\to 0}\frac{\Delta w}{\Delta t}\\ &= \left(\frac{\partial w}{\partial x}\right)_{t_0}\left(\frac{dx}{dt}\right)_{t_0} + \left(\frac{\partial w}{\partial y}\right)_{t_0}\left(\frac{dy}{dt}\right)_{t_0} + 0\cdot\left(\frac{dx}{dt}\right)_{t_0} + 0\cdot\left(\frac{dy}{dt}\right)_{t_0}.\end{aligned} \tag{5}$$

This completes the proof. ∎

Example 1 Use the Chain Rule to find the derivative of

$$w = xy$$

with respect to t along the path $x = \cos t$, $y = \sin t$. What is the derivative's value at $t = \pi/2$?

Solution We evaluate the right-hand side of Eq. (1) with $w = xy$, $x = \cos t$, and $y = \sin t$:

$$\frac{\partial w}{\partial x} = y = \sin t, \quad \frac{\partial w}{\partial y} = x = \cos t, \quad \frac{dx}{dt} = -\sin t, \quad \frac{dy}{dt} = \cos t,$$

$$\begin{aligned}\frac{dw}{dt} &= \frac{\partial w}{\partial x}\frac{dx}{dt} + \frac{\partial w}{\partial y}\frac{dy}{dt} = (\sin t)(-\sin t) + (\cos t)(\cos t) \qquad \left(\text{Eq. (1) with values from above}\right)\\ &= -\sin^2 t + \cos^2 t = \cos 2t.\end{aligned}$$

In this case we can check the result with a more direct calculation. As a function of t,

$$w = xy = \cos t \sin t = \frac{1}{2}\sin 2t,$$

so

$$\frac{dw}{dt} = \frac{d}{dt}\left(\frac{1}{2}\sin 2t\right) = \frac{1}{2}\cdot 2\cos 2t = \cos 2t.$$

Here we have three routes from w to t instead of two. But the rule for evaluating dw/dt is still the same. Read down each route, multiplying derivatives along the way; then add.

Chain Rule

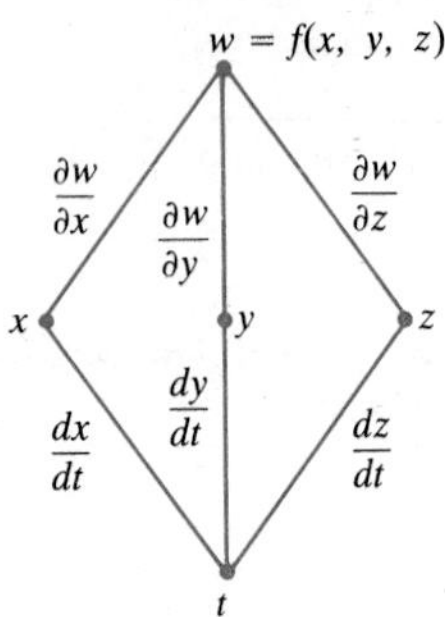

$$\frac{dw}{dt} = \frac{\partial w}{\partial x}\frac{dx}{dt} + \frac{\partial w}{\partial y}\frac{dy}{dt} + \frac{\partial w}{\partial z}\frac{dz}{dt}$$

In either case,

$$\left(\frac{dw}{dt}\right)_{t=\pi/2} = \cos\left(2 \cdot \frac{\pi}{2}\right) = \cos \pi = -1.$$

The Chain Rule for Functions of Three Variables

To get the Chain Rule for functions of three variables, we add a term to Eq. (1).

Chain Rule for Functions of Three Independent Variables

If $w = f(x, y, z)$ is differentiable and x, y, and z are differentiable functions of t, then w is a differentiable function of t and

$$\frac{dw}{dt} = \frac{\partial f}{\partial x}\frac{dx}{dt} + \frac{\partial f}{\partial y}\frac{dy}{dt} + \frac{\partial f}{\partial z}\frac{dz}{dt}. \qquad (6)$$

The derivation is identical with the derivation of Eq. (1) except that there are now three intermediate variables instead of two. The diagram we use for remembering the new equation is similar as well.

Example 2 *Changes in a Function's Values Along a Helix.* Find dw/dt if

$$w = xy + z, \qquad x = \cos t, \qquad y = \sin t, \qquad z = t$$

(Fig. 12.32). What is the derivative's value at $t = 0$?

Solution

$$\begin{aligned}\frac{dw}{dt} &= \frac{\partial w}{\partial x}\frac{dx}{dt} + \frac{\partial w}{\partial y}\frac{dy}{dt} + \frac{\partial w}{\partial z}\frac{dz}{dt} \qquad \text{(Eq. (6))}\\ &= (y)(-\sin t) + (x)(\cos t) + (1)(1)\\ &= (\sin t)(-\sin t) + (\cos t)(\cos t) + 1\\ &= -\sin^2 t + \cos^2 t + 1 = 1 + \cos 2t,\end{aligned}$$

$$\left(\frac{dw}{dt}\right)_{t=0} = 1 + \cos(0) = 2.$$

The Chain Rule for Functions Defined on Surfaces

If we are interested in the temperature $w = f(x, y, z)$ at points (x, y, z) on a globe in space, we might prefer to think of x, y, and z as functions of the variables r and s that give the points' longitudes and latitudes. If $x = g(r, s)$, $y = h(r, s)$, and $z = k(r, s)$, we could then express the temperature as a function of r and s with the composite function

$$w = f(g(r, s), h(r, s), k(r, s)).$$

Under the right conditions, w would have partial derivatives with respect to both r and s that could be calculated in the following way.

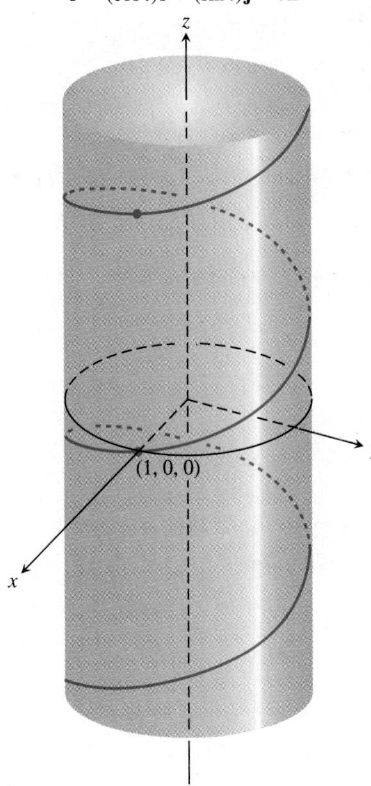

12.32 Example 2 shows how the values of the function $w = xy + z$ vary with t along this helix.

Chain Rule for Two Independent Variables and Three Intermediate Variables

If $w = f(x, y, z)$, $x = g(r, s)$, $y = h(r, s)$, and $z = k(r, s)$ are differentiable, then w is a differentiable function of r and s and its partial derivatives are given by the equations

$$\frac{\partial w}{\partial r} = \frac{\partial w}{\partial x}\frac{\partial x}{\partial r} + \frac{\partial w}{\partial y}\frac{\partial y}{\partial r} + \frac{\partial w}{\partial z}\frac{\partial z}{\partial r}, \tag{7}$$

$$\frac{\partial w}{\partial s} = \frac{\partial w}{\partial x}\frac{\partial x}{\partial s} + \frac{\partial w}{\partial y}\frac{\partial y}{\partial s} + \frac{\partial w}{\partial z}\frac{\partial z}{\partial s}. \tag{8}$$

Equation (7) can be derived from Eq. (6) by holding s fixed and setting r equal to t. Similarly, Eq. (8) can be derived from Eq. (6) by holding r fixed and setting s equal to t. The tree diagrams are shown in Fig. 12.33.

Example 3 Express $\partial w/\partial r$ and $\partial w/\partial s$ in terms of r and s if

$$w = x + 2y + z^2, \qquad x = \frac{r}{s}, \qquad y = r^2 + \ln s, \qquad z = 2r.$$

Solution

$$\frac{\partial w}{\partial r} = \frac{\partial w}{\partial x}\frac{\partial x}{\partial r} + \frac{\partial w}{\partial y}\frac{\partial y}{\partial r} + \frac{\partial w}{\partial z}\frac{\partial z}{\partial r} \qquad \text{(Eq. (7))}$$

$$= (1)\left(\frac{1}{s}\right) + (2)(2r) + (2z)(2) = \frac{1}{s} + 4r + (4r)(2) = \frac{1}{s} + 12r,$$

$$\frac{\partial w}{\partial s} = \frac{\partial w}{\partial x}\frac{\partial x}{\partial s} + \frac{\partial w}{\partial y}\frac{\partial y}{\partial s} + \frac{\partial w}{\partial z}\frac{\partial z}{\partial s} \qquad \text{(Eq. (8))}$$

$$= (1)\left(-\frac{r}{s^2}\right) + 2\left(\frac{1}{s}\right) + (2z)(0) = \frac{2}{s} - \frac{r}{s^2}.$$

If f is a function of two variables instead of three, Eqs. (7) and (8) become one term shorter.

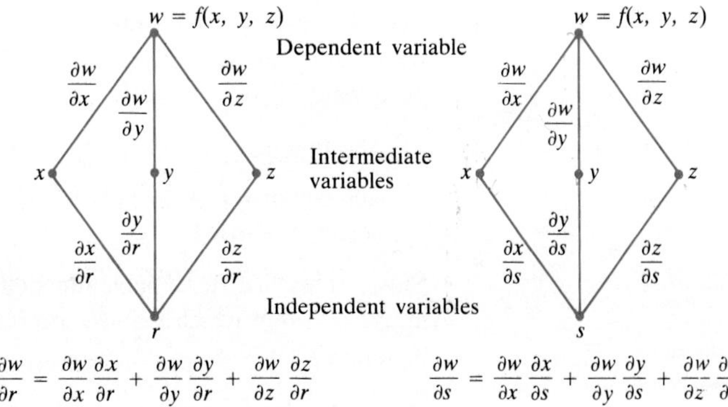

12.33 Tree diagrams for Eqs. (7) and (8).

If $w = f(x, y)$, $x = g(r, s)$, and $y = h(r, s)$, then

$$\frac{\partial w}{\partial r} = \frac{\partial w}{\partial x}\frac{\partial x}{\partial r} + \frac{\partial w}{\partial y}\frac{\partial y}{\partial r} \quad \text{and} \quad \frac{\partial w}{\partial s} = \frac{\partial w}{\partial x}\frac{\partial x}{\partial s} + \frac{\partial w}{\partial y}\frac{\partial y}{\partial s}. \tag{9}$$

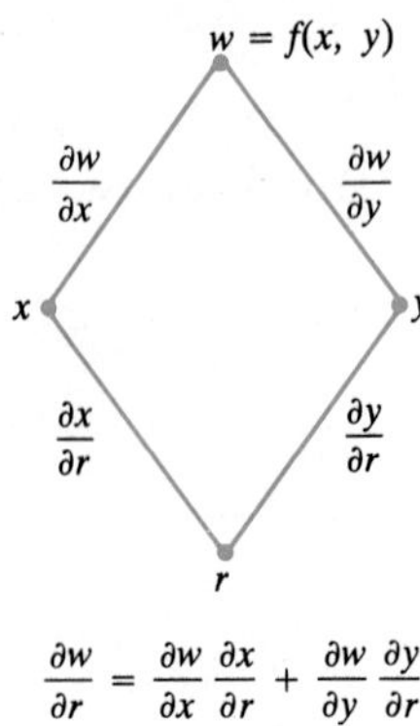

12.34 Tree diagram for Eqs. (9).

Figure 12.34 shows a tree diagram for the first equation in (9). The diagram for the second equation in (9) is similar—just replace r with s.

Example 4 Express $\partial w/\partial r$ and $\partial w/\partial s$ in terms of r and s if

$$w = x^2 + y^2, \qquad x = r - s, \qquad y = r + s.$$

Solution We use Eqs. (9):

$$\begin{aligned}\frac{\partial w}{\partial r} &= \frac{\partial w}{\partial x}\frac{\partial x}{\partial r} + \frac{\partial w}{\partial y}\frac{\partial y}{\partial r} \\ &= (2x)(1) + (2y)(1) \\ &= 2(r - s) + 2(r + s) \\ &= 4r,\end{aligned} \qquad \begin{aligned}\frac{\partial w}{\partial s} &= \frac{\partial w}{\partial x}\frac{\partial x}{\partial s} + \frac{\partial w}{\partial y}\frac{\partial y}{\partial s} \\ &= (2x)(-1) + (2y)(1) \\ &= -2(r - s) + 2(r + s) \\ &= 4s.\end{aligned}$$

If f is a function of x alone, Eqs. (7) and (8) simplify still further.

If $w = f(x)$ and $x = g(r, s)$, then

$$\frac{\partial w}{\partial r} = \frac{dw}{dx}\frac{\partial x}{\partial r} \quad \text{and} \quad \frac{\partial w}{\partial s} = \frac{dw}{dx}\frac{\partial x}{\partial s}. \tag{10}$$

Here dw/dx is the ordinary (single-variable) derivative (Fig. 12.35).

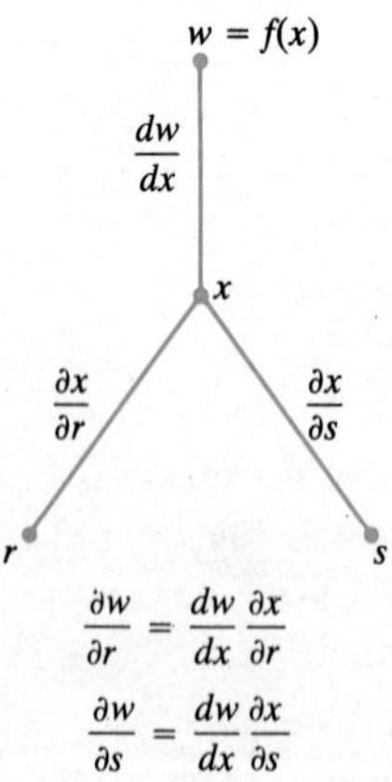

12.35 Tree diagram for Eqs. (10).

Implicit Differentiation (Continued from Chapter 2)

The two-variable Chain Rule in Eq. (1) leads to a formula that takes most of the work out of implicit differentiation. The formula arises in the following way. Suppose that

1. the function $F(x, y)$ is differentiable and
2. the equation $F(x, y) = 0$ defines y implicitly as a differentiable function of x, say $y = h(x)$.

Now, x is also a differentiable function of x, so what we have, in effect, is a function $w = F(u, v)$ in which u and v are both differentiable functions of x. By Eq. (1), therefore,

$$\frac{dw}{dx} = \frac{\partial w}{\partial u}\frac{du}{dx} + \frac{\partial w}{\partial v}\frac{dv}{dx}. \qquad \text{(Eq. (1) with } x = u,\ y = v,\ t = x\text{)}$$

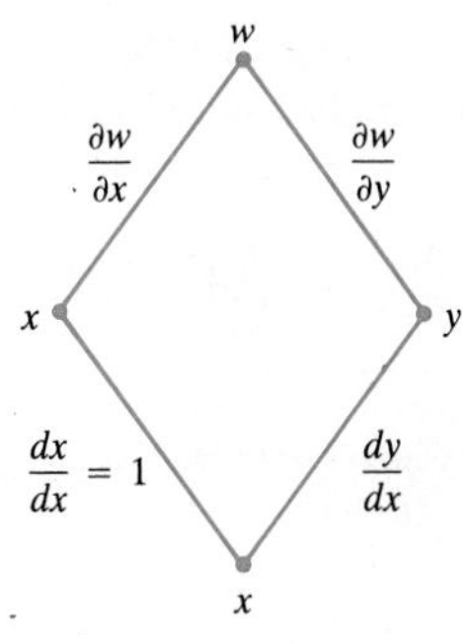

12.36 Tree diagram for Eq. (11).

With $u = x$, and $v = y = h(x)$,

$$\frac{dw}{dx} = \frac{\partial w}{\partial x}\frac{dx}{dx} + \frac{\partial w}{\partial y}\frac{dy}{dx} = \frac{\partial w}{\partial x} \cdot 1 + \frac{\partial w}{\partial y}\frac{dy}{dx} \tag{11}$$

(Fig. 12.36). Since $w = F(x, h(x)) = 0$ for all values of x (because h is defined by setting $F(x, y) = 0$), $dw/dx = 0$ and Eq. (11) reduces still further to

$$0 = \frac{\partial w}{\partial x} + \frac{\partial w}{\partial y}\frac{dy}{dx}. \tag{12}$$

If $\partial w/\partial y \neq 0$, we can solve Eq. (12) for dy/dx to get

$$\frac{dy}{dx} = -\frac{\partial w/\partial x}{\partial w/\partial y} = -\frac{F_x}{F_y}. \tag{13}$$

Suppose that $F(x, y)$ is differentiable and that the equation $F(x, y) = 0$ defines y as a differentiable function of x. Then, at any point where $F_y \neq 0$,

$$\frac{dy}{dx} = -\frac{F_x}{F_y}. \tag{14}$$

Example 5 Find dy/dx if $x^2 + \sin y - 2y = 0$.

Solution Take $F(x, y) = x^2 + \sin y - 2y$. Then

$$\frac{dy}{dx} = -\frac{F_x}{F_y} = -\frac{2x}{\cos y - 2}. \qquad \text{(Eq. (14))}$$

This calculation is significantly shorter than the single-variable calculation with which we found dy/dx in Section 2.5, Example 3.

Remembering the Different Forms of the Chain Rule

How are we to remember all the different forms of the Chain Rule? The answer is that there is no need to remember them all. The best thing to do is to remember a few key equations, say (1), (6), and (7). Construct the others, when you need them, from tree diagrams. You can always draw the appropriate diagram by placing the dependent variable on top, the intermediate variables in the middle, and the selected independent variable at the bottom.

The Chain Rule for Functions of Many Variables

Suppose $w = f(x, y, \ldots, v)$ is a differentiable function of the variables $x, y, \ldots, v$ (a finite set) and $x, y, \ldots, v$ are themselves differentiable functions of the variables $p, q, \ldots, t$ (another finite set). Then w is a differentiable function of the variables p through t and the partial derivatives of w with respect to these variables are given by equations of the form

$$\frac{\partial w}{\partial p} = \frac{\partial w}{\partial x}\frac{\partial x}{\partial p} + \frac{\partial w}{\partial y}\frac{\partial y}{\partial p} + \cdots + \frac{\partial w}{\partial v}\frac{\partial v}{\partial p}. \tag{15}$$

The other equations are obtained by replacing p by $q, \ldots, t$, one at a time.

One way to remember Eq. (15) is to think of the right-hand side as the dot product of two vectors with components

$$\underbrace{\left(\frac{\partial w}{\partial x}, \frac{\partial w}{\partial y}, \ldots, \frac{\partial w}{\partial v}\right)}_{\substack{\text{derivatives of } w \text{ with} \\ \text{respect to the} \\ \text{intermediate variables}}} \quad \text{and} \quad \underbrace{\left(\frac{\partial x}{\partial p}, \frac{\partial y}{\partial p}, \ldots, \frac{\partial v}{\partial p}\right)}_{\substack{\text{derivatives of the intermediate} \\ \text{variables with respect to the} \\ \text{selected independent variable}}}. \tag{16}$$

EXERCISES 12.5

In Exercises 1–6, (a) express dw/dt as a function of t, both by using the Chain Rule and by expressing w in terms of t and differentiating directly with respect to t. Then (b) evaluate dw/dt at the given value of t.

1. $w = x^2 + y^2, \quad x = \cos t, \quad y = \sin t; \quad t = \pi$
2. $w = x^2 + y^2, \quad x = \cos t + \sin t, \quad y = \cos t - \sin t;$ $t = 0$
3. $w = \dfrac{x}{z} + \dfrac{y}{z}, \quad x = \cos^2 t, \quad y = \sin^2 t, \quad z = 1/t; \quad t = 3$
4. $w = \ln(x^2 + y^2 + z^2), \quad x = \cos t, \quad y = \sin t,$ $z = 4\sqrt{t}; \quad t = 3$
5. $w = 2ye^x - \ln z, \quad x = \ln(t^2 + 1), \quad y = \tan^{-1} t,$ $z = e^t; \quad t = 1$
6. $w = z - \sin xy, \quad x = t, \quad y = \ln t, \quad z = e^{t-1}; \quad t = 1$

In Exercises 7 and 8, (a) express $\partial z/\partial r$ and $\partial z/\partial \theta$ as functions of r and θ both by using the Chain Rule and by expressing z directly in terms of r and θ before differentiating. Then (b) evaluate $\partial z/\partial r$ and $\partial z/\partial \theta$ at the given point (r, θ).

7. $z = 4\,e^x \ln y, \quad x = \ln(r \cos \theta), \quad y = r \sin \theta;$ $(r, \theta) = (2, \pi/4)$
8. $z = \tan^{-1}(x/y), \quad x = r \cos \theta, \quad y = r \sin \theta;$ $(r, \theta) = (1.3, \pi/6)$

In Exercises 9 and 10, (a) express $\partial w/\partial u$ and $\partial w/\partial v$ as functions of u and v both by using the Chain Rule and by expressing w directly in terms of u and v before differentiating. Then (b) evaluate $\partial w/\partial u$ and $\partial w/\partial v$ at the given point (u, v).

9. $w = xy + yz + xz, \quad x = u + v, \quad y = u - v, \quad z = uv;$ $(u, v) = (1/2, 1)$
10. $w = \ln(x^2 + y^2 + z^2), \quad x = ue^v \sin u, \quad y = ue^v \cos u,$ $z = ue^v; \quad (u, v) = (-2, 0)$

In Exercises 11 and 12, (a) express $\partial u/\partial x$, $\partial u/\partial y$, and $\partial u/\partial z$ as functions of x, y, and z both by using the Chain Rule and by expressing u directly in terms of p, q, and r before differentiating. Then (b) evaluate $\partial u/\partial x$, $\partial u/\partial y$, and $\partial u/\partial z$ at the given point (x, y, z).

11. $u = \dfrac{p - q}{q - r}, \quad p = x + y + z, \quad q = x - y + z,$ $r = x + y - z; \quad (x, y, z) = (\sqrt{3}, 2, 1)$
12. $u = e^{qr} \sin^{-1} p, \quad p = \sin x, \quad q = z^2 \ln y, \quad r = 1/z;$ $(x, y, z) = (\pi/4, 1/2, -1/2)$

In Exercises 13–24, draw a tree diagram and write a Chain Rule formula for each derivative.

13. $\dfrac{dz}{dt}$ for $z = f(x, y), \quad x = g(t), \quad y = h(t)$
14. $\dfrac{dz}{dt}$ for $z = f(u, v, w), \quad u = g(t), \quad v = h(t), \quad w = k(t)$
15. $\dfrac{\partial w}{\partial u}$ and $\dfrac{\partial w}{\partial v}$ for $w = h(x, y, z), \quad x = f(u, v), \quad y = g(u, v),$ $z = k(u, v)$
16. $\dfrac{\partial w}{\partial x}$ and $\dfrac{\partial w}{\partial y}$ for $w = f(r, s, t), \quad r = g(x, y), \quad s = h(x, y),$ $t = k(x, y)$
17. $\dfrac{\partial w}{\partial u}$ and $\dfrac{\partial w}{\partial v}$ for $w = g(x, y), \quad x = h(u, v), \quad y = k(u, v)$
18. $\dfrac{\partial w}{\partial x}$ and $\dfrac{\partial w}{\partial y}$ for $w = g(u, v), \quad u = h(x, y), \quad v = k(x, y)$
19. $\dfrac{\partial z}{\partial t}$ and $\dfrac{\partial z}{\partial s}$ for $z = f(x, y), \quad x = g(t, s), \quad y = h(t, s)$
20. $\dfrac{\partial y}{\partial r}$ for $y = f(u), \quad u = g(r, s)$
21. $\dfrac{\partial w}{\partial s}$ and $\dfrac{\partial w}{\partial t}$ for $w = g(u), \quad u = h(s, t)$
22. $\dfrac{\partial w}{\partial p}$ for $w = f(x, y, z, v), \quad x = g(p, q), \quad y = h(p, q),$ $z = j(p, q), \quad v = k(p, q)$
23. $\dfrac{\partial w}{\partial r}$ and $\dfrac{\partial w}{\partial s}$ for $w = f(x, y), \quad x = g(r), \quad y = h(s)$
24. $\dfrac{\partial w}{\partial s}$ for $w = g(x, y), \quad x = h(r, s, t), \quad y = k(r, s, t)$

Assuming that the equations in Exercises 25–28 define y as a differentiable function of x, use Eq. (14) to find the value of dy/dx at the given point.

25. $x^3 - 2y^2 + xy = 0, \quad (1, 1)$

26. $xy + y^2 - 3x - 3 = 0, \quad (-1, 1)$

27. $x^2 + xy + y^2 - 7 = 0, \quad (1, 2)$

28. $xe^y + \sin xy + y - \ln 2 = 0, \quad (0, \ln 2)$

Equation (14) can be generalized to functions of three variables and even more. The three-variable version goes like this:

If the equation $F(x, y, z) = 0$ determines z as a differentiable function of x and y, then, at points where $F_z \neq 0$,

$$\frac{\partial z}{\partial x} = -\frac{F_x}{F_z} \quad \text{and} \quad \frac{\partial z}{\partial y} = -\frac{F_y}{F_z}. \tag{17}$$

Use these equations to find the values of $\partial z/\partial x$ and $\partial z/\partial y$ at the points in Exercises 29–32.

29. $z^3 - xy + yz + y^3 - 2 = 0, \quad (1, 1, 1)$

30. $\frac{1}{x} + \frac{1}{y} + \frac{1}{z} - 1 = 0, \quad (2, 3, 6)$

31. $\sin(x + y) + \sin(y + z) + \sin(x + z) = 0, \quad (\pi, \pi, \pi)$

32. $xe^y + ye^z + 2\ln x - 2 - 3\ln 2 = 0, \quad (1, \ln 2, \ln 3)$

33. Find $\partial w/\partial r$ when $r = 1$, $s = -1$ if $w = (x + y + z)^2$, $x = r - s$, $y = \cos(r + s)$, $z = \sin(r + s)$.

34. Find $\partial w/\partial v$ when $u = -1$, $v = 2$ if $w = xy + \ln z$, $x = v^2/u$, $y = u + v$, $z = \sin u$.

35. Find $\partial w/\partial v$ when $u = 0$, $v = 0$ if $w = x^2 + (y/x)$, $x = u - 2v + 1$, $y = 2u + v - 2$.

36. Find $\partial z/\partial u$ when $u = 0$, $v = 1$ if $z = \sin xy + x \sin y$, $x = u^2 + v^2$, $y = uv$.

37. Find $\partial z/\partial u$ and $\partial z/\partial v$ when $u = \ln 2$, $v = 1$ if $z = 5\tan^{-1}x$ and $x = e^u + \ln v$.

38. Find $\partial w/\partial q$ when $p = 1$, $q = 4$ if $w = xy + yz + zu + uv$, $x = pq$, $y = q\tan^{-1}p$, $z = \ln(q - 3p)$, $u = \sqrt{q + 5p}$, $v = p + q + 1$.

39. If $f(u, v, w)$ is differentiable and $u = x - y$, $v = y - z$, and $w = z - x$, show that

$$\frac{\partial f}{\partial x} + \frac{\partial f}{\partial y} + \frac{\partial f}{\partial z} = 0.$$

40. a) Show that if we substitute polar coordinates $x = r\cos\theta$ and $y = r\sin\theta$ in a differentiable function $w = f(x, y)$, then

$$\frac{\partial w}{\partial r} = f_x \cos\theta + f_y \sin\theta$$

and

$$\frac{1}{r}\frac{\partial w}{\partial \theta} = -f_x \sin\theta + f_y \cos\theta.$$

b) Solve the equations in (a) to express f_x and f_y in terms of $\partial w/\partial r$ and $\partial w/\partial \theta$.

c) Show that

$$(f_x)^2 + (f_y)^2 = \left(\frac{\partial w}{\partial r}\right)^2 + \frac{1}{r^2}\left(\frac{\partial w}{\partial \theta}\right)^2.$$

41. *Changing voltage in a circuit.* The voltage V in a circuit that satisfies the law $V = IR$ (Fig. 12.37) is slowly dropping as the battery wears out. At the same time, the resistance R is increasing as the resistor heats up. Use the equation

$$\frac{dV}{dt} = \frac{\partial V}{\partial I}\frac{dI}{dt} + \frac{\partial V}{\partial R}\frac{dR}{dt}$$

to find how the current is changing at the instant when $R = 600$ ohms, $I = 0.04$ amp, $dR/dt = 0.5$ ohm/sec, and $dV/dt = -0.01$ volt/sec.

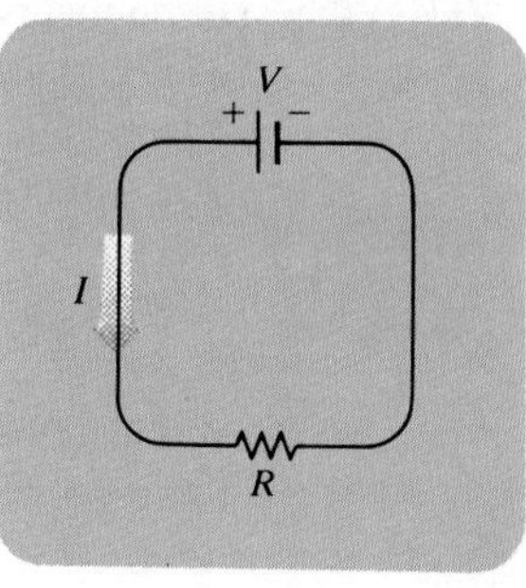

12.37 The battery-driven circuit in Exercise 41.

42. *Changing dimensions in a box.* The dimensions a, b, and c of the box in Fig. 12.38 are changing with time. At the instant in question, $a = 13$ cm, $b = 9$ cm, $c = 5$ cm, $da/dt = dc/dt = 2$ cm/sec, and $db/dt = -5$ cm/sec.

a) How fast are the volume $V = abc$ and the surface area $S = 2ab + 2ac + 2bc$ changing at that instant?

b) Is the box's diagonal increasing, or decreasing?

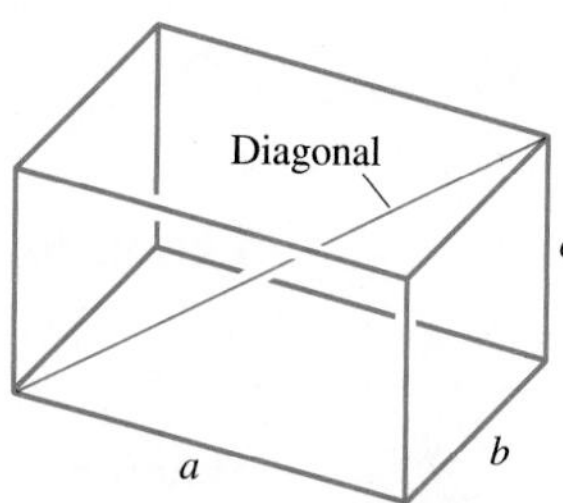

12.38 The box in Exercise 42.

43. Show that if $w = f(u, v)$ satisfies the Laplace equation $f_{uu} + f_{vv} = 0$, and if $u = (x^2 - y^2)/2$ and $v = xy$, then w satisfies the Laplace equation $w_{xx} + w_{yy} = 0$.

44. Let $w = f(u) + g(v)$, where $u = x + iy$ and $v = x - iy$ and $i = \sqrt{-1}$. Show that w satisfies the Laplace equation $w_{xx} + w_{yy} = 0$ if all the necessary functions are differentiable.

Changes in Functions Along Curves

45. Suppose that the partial derivatives of a function $f(x, y, z)$ at points on the helix $x = \cos t$, $y = \sin t$, $z = t$ are

$$f_x = \cos t, \quad f_y = \sin t, \quad f_z = t^2 + t - 2.$$

At what points on the curve, if any, can f take on extreme values?

46. Let $w = x^2 e^{2y} \cos 3z$. Find the value of dw/dt at the point $(1, \ln 2, 0)$ on the curve $x = \cos t$, $y = \ln(t + 2)$, $z = t$.

47. Let $T = f(x, y)$ be the temperature at the point (x, y) on the circle $x = \cos t$, $y = \sin t$ and suppose that

$$\frac{aT}{ax} = 8x - 4y, \quad \frac{dT}{ay} = 8y - 4x.$$

a) Find where the maximum and minimum temperatures on the circle occur by examining the derivatives dT/dt and d^2T/dt^2.

b) Suppose $T = 4x^2 - 4xy + 4y^2$. Find the maximum and minimum values of T on the circle.

12.6 Partial Derivatives with Constrained Variables*

In finding partial derivatives of functions like $w = f(x, y)$, we have assumed x and y to be independent. But in many applications this is not the case. For example, the internal energy U of a gas may be expressed as a function $U = f(p, v, T)$ of pressure, volume, and temperature. If the individual molecules of the gas do not interact, however, p, v, and T obey the ideal gas law

$$pv = nRT \qquad (n \text{ and } R \text{ constant})$$

and therefore fail to be independent. Finding partial derivatives in situations like these can be complicated. But it is better to face the complication now than to meet it for the first time while you are also trying to learn economics, engineering, or physics.

Decide Which Variables Are Dependent and Which Are Independent

If the variables in a function $w = f(x, y, z)$ are constrained by a relation like the one imposed on x, y, and z by the equation $z = x^2 + y^2$, the geometric meanings and the numerical values of the partial derivatives of f will depend on which variables are chosen to be dependent and which are chosen to be independent. To see how this choice can affect the outcome, we consider the calculation of $\partial w/\partial x$ when $w = x^2 + y^2 + z^2$ and $z = x^2 + y^2$.

Example 1 Find $\partial w/\partial x$ if

$$w = x^2 + y^2 + z^2 \qquad \text{and} \qquad z = x^2 + y^2.$$

Solution We are given two equations in the four unknowns x, y, z, and w. Like many such systems, this one can be solved for two of the unknowns (the dependent variables) in terms of the others (the independent variables). In being asked for $\partial w/\partial x$, we are told that w is to be a dependent variable and x an independent variable. The possible choices for the other variables come down to

Dependent	*Independent*
w, z	x, y
w, y	x, z

*This section is based on notes written for MIT by Arthur P. Mattuck.

In either case, we can express w explicitly in terms of the selected independent variables. We do this by using the second equation to eliminate the remaining dependent variable in the first equation.

In the first case, the remaining dependent variable is z. We eliminate it from the first equation by replacing it by $x^2 + y^2$. The resulting expression for w is

$$\begin{aligned} w &= x^2 + y^2 + z^2 = x^2 + y^2 + (x^2 + y^2)^2 \\ &= x^2 + y^2 + x^4 + 2x^2y^2 + y^4 \end{aligned}$$

and

$$\frac{\partial w}{\partial x} = 2x + 4x^3 + 4xy^2. \tag{1}$$

This is the formula for $\partial w/\partial x$ when x and y are the independent variables.

In the second case, where the independent variables are x and z and the remaining dependent variable is y, we eliminate the dependent variable y in the expression for w by replacing y^2 by $z - x^2$. This gives

$$w = x^2 + y^2 + z^2 = x^2 + (z - x^2) + z^2 = z + z^2$$

and

$$\frac{\partial w}{\partial x} = 0. \tag{2}$$

This is the formula for $\partial w/\partial x$ when x and z are the independent variables.

The formulas for $\partial w/\partial x$ in Eqs. (1) and (2) are genuinely different. We cannot change either formula into the other by using the relation $z = x^2 + y^2$. There is not just one $\partial w/\partial x$, there are two, and we see that the original instruction to find $\partial w/\partial x$ was incomplete. *Which* $\partial w/\partial x$? we ask.

The geometric interpretations of Eqs. (1) and (2) help to explain why the equations differ. The function $w = x^2 + y^2 + z^2$ measures the square of the distance from the point (x, y, z) to the origin. The condition $z = x^2 + y^2$ says that the point (x, y, z) lies on the surface of the paraboloid of revolution shown in Fig. 12.39. What does it mean to calculate $\partial w/\partial x$ at a point $P(x, y, z)$ that can move only on this surface? What is the value of $\partial w/\partial x$ when the coordinates of P are, say, $(1, 0, 1)$?

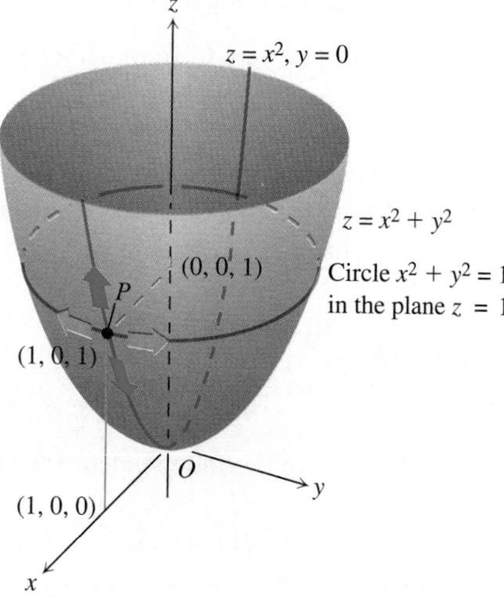

12.39 If P is constrained to lie on the paraboloid $z = x^2 + y^2$, the value of the partial derivative of $w = x^2 + y^2 + z^2$ with respect to x at P depends on the direction of motion (Example 1). (a) As x changes, with $y = 0$, P moves up or down the surface on the parabola $z = x^2$ in the xz-plane and $\partial w/\partial x = 2x + 4x^3$. (b) As x changes, with $z = 1$, P moves on the circle $x^2 + y^2 = 1$, $z = 1$, and $\partial w/\partial x = 0$.

If we take x and y to be independent, then we find $\partial w/\partial x$ by holding y fixed (at $y = 0$ in this case) and letting x vary. This means that P moves along the parabola $z = x^2$ in the xz-plane. As P moves on this parabola, w, which is the square of the distance from P to the origin, changes. We calculate $\partial w/\partial x$ in this case (our first solution above) to be

$$\frac{\partial w}{\partial x} = 2x + 4x^3 + 4xy^2.$$

At the point $P(1, 0, 1)$, the value of this derivative is

$$\frac{\partial w}{\partial x} = 2 + 4 + 0 = 6.$$

If we take x and z to be independent, then we find $\partial w/\partial x$ by holding z fixed while x varies. Since the z-coordinate of P is 1, varying x moves P along a circle in the plane $z = 1$. As P moves along this circle, its distance from the origin remains constant, and w, being the square of this distance, does not change. That is,

$$\frac{\partial w}{\partial x} = 0,$$

as we found in our second solution.

How to Find $\partial w/\partial x$ When the Variables in $w = f(x, y, z)$ Are Constrained by Another Equation

As we saw in Example 1, a typical routine for finding $\partial w/\partial x$ when the variables in the function $w = f(x, y, z)$ are related by another equation has three steps.

STEP 1: Decide which variables are to be dependent and which are to be independent. (In practice, the decision is based on the physical or theoretical context of our work. In the exercises at the end of this section, we say which variables are which.)

STEP 2: Eliminate the other dependent variable(s) in the expression for w.

STEP 3: Differentiate as usual.

These steps apply to finding $\partial w/\partial y$ and $\partial w/\partial z$ as well.

If we cannot carry out step 2 after deciding which variables are dependent, we differentiate the equations as they are and try to solve for $\partial w/\partial x$ afterward. The next example shows how this is done.

Example 2 Find $\partial w/\partial x$ at the point $(x, y, z) = (2, -1, 1)$ if

$$w = x^2 + y^2 + z^2, \qquad z^3 - xy + yz + y^3 = 0,$$

and x and y are the independent variables.

Solution It is not convenient to eliminate z in the expression for w. We therefore differentiate both equations implicitly with respect to x, treating x and y as independent variables and w and z as dependent variables. This gives

$$\frac{\partial w}{\partial x} = 2x + 2z\frac{\partial z}{\partial x} \tag{3}$$

and

$$3z^2\frac{\partial z}{\partial x} - y + y\frac{\partial z}{\partial x} + 0 = 0. \tag{4}$$

These equations may now be combined to express $\partial w/\partial x$ in terms of x, y, and z. We solve Eq. (4) for $\partial z/\partial x$ to get

$$\frac{\partial z}{\partial x} = \frac{y}{y + 3z^2}$$

and substitute into Eq. (3) to get

$$\frac{\partial w}{\partial x} = 2x + \frac{2yz}{y + 3z^2}.$$

The value of this derivative at $(x, y, z) = (2, -1, 1)$ is

$$\left(\frac{\partial w}{\partial x}\right)_{(2,\,-1,\,1)} = 2(2) + \frac{2(-1)(1)}{-1 + 3(1)^2}$$

$$= 4 + \frac{-2}{2} = 3.$$

To show what variables are assumed to be independent in calculating a derivative, we can use the following notation:

$$\left(\frac{\partial w}{\partial x}\right)_y \qquad \partial w/\partial x \text{ with } x \text{ and } y \text{ independent}, \tag{5}$$

$$\left(\frac{\partial w}{\partial x}\right)_z \qquad \partial w/\partial x \text{ with } x \text{ and } z \text{ independent}, \tag{6}$$

$$\left(\frac{\partial f}{\partial y}\right)_{x,\,t} \qquad \partial f/\partial y \text{ with } y,\ x, \text{ and } t \text{ independent}. \tag{7}$$

Example 3 Find

a) $\left(\frac{\partial w}{\partial x}\right)_{y,\,z}$ and b) $\left(\frac{\partial w}{\partial x}\right)_{t,\,z}$

if $w = x^2 + y - z + \sin t$ and $x + y = t$.

Solution

a) With x, y, z independent, we have

$$t = x + y, \qquad w = x^2 + y - z + \sin(x + y),$$

$$\left(\frac{\partial w}{\partial x}\right)_{y,\,z} = 2x + 0 - 0 + \cos(x + y)\frac{\partial}{\partial x}(x + y)$$

$$= 2x + \cos(x + y).$$

b) With x, t, z independent, we have

$$y = t - x, \qquad w = x^2 + (t - x) - z + \sin t,$$

$$\left(\frac{\partial w}{\partial x}\right)_{t,\,z} = 2x - 1 - 0 + 0 = 2x - 1.$$

Arrow Diagrams

In solving problems like the one in Example 3, it often helps to start with an arrow diagram that shows how the variables and functions are related. If

$$w = x^2 + y - z + \sin t \qquad \text{and} \qquad x + y = t$$

and we are asked to find $\partial w/\partial x$ when x, y, and z are independent, the appropriate diagram is one like this:

$$\begin{pmatrix} x \\ y \\ z \end{pmatrix} \quad \rightarrow \quad \begin{pmatrix} x \\ y \\ z \\ t \end{pmatrix} \quad \rightarrow \quad w \tag{8}$$

independent variables — intermediate variables and relations — dependent variable

$x = x$
$y = y$
$z = z$
$t = x + y$

The diagram shows the independent variables on the left, the intermediate variables and their relation to the independent variables in the middle, and the dependent variable on the right.

To find $\partial w/\partial x$, we first apply the four-variable form of the Chain Rule to w, getting

$$\frac{\partial w}{\partial x} = \frac{\partial w}{\partial x}\frac{\partial x}{\partial x} + \frac{\partial w}{\partial y}\frac{\partial y}{\partial x} + \frac{\partial w}{\partial z}\frac{\partial z}{\partial x} + \frac{\partial w}{\partial t}\frac{\partial t}{\partial x}. \tag{9}$$

We then use the formula for w,

$$w = x^2 + y - z + \sin t,$$

to evaluate the partial derivatives of w that appear on the right-hand side of Eq. (9). This gives

$$\begin{aligned} \frac{\partial w}{\partial x} &= 2x\frac{\partial x}{\partial x} + (1)\frac{\partial y}{\partial x} + (-1)\frac{\partial z}{\partial x} + \cos t\frac{\partial t}{\partial x} \\ &= 2x\frac{\partial x}{\partial x} + \frac{\partial y}{\partial x} - \frac{\partial z}{\partial x} + \cos t\frac{\partial t}{\partial x}. \end{aligned} \tag{10}$$

To calculate the remaining partial derivatives, we apply what we know about the dependence and independence of the variables involved. As shown in the diagram (8), the variables x, y, and z are independent and $t = x + y$. Hence,

$$\frac{\partial x}{\partial x} = 1, \quad \frac{\partial y}{\partial x} = 0, \quad \frac{\partial z}{\partial x} = 0, \quad \frac{\partial t}{\partial x} = \frac{\partial}{\partial x}(x + y) = (1 + 0) = 1. \tag{11}$$

We substitute these values into Eq. (10) to find $\partial w/\partial x$:

$$\left(\frac{\partial w}{\partial x}\right)_{y,z} = 2x(1) + 0 - 0 + (\cos t)(1)$$

$$= 2x + \cos t$$

$$= 2x + \cos(x + y). \qquad \left(\text{In terms of the independent variables}\right)$$

Working Without Specific Formulas

In applications, we often have to find an expression for the derivative of a function when neither the function nor the relation among its variables is given explicitly. Two observations can help us.

First, if a differentiable function $F(r, s)$ has a constant value throughout some region R of the rs-plane, and r and s are independent variables, then

$$\frac{\partial F}{\partial r} = \frac{\partial F}{\partial s} = 0 \tag{12}$$

throughout R. In particular, Eq. (12) holds if $F(r, s) = 0$ throughout R.

Second, suppose that

$$f(x, y, z) = f(x(r, s), y(r, s), z(r, s)) = F(r, s) \tag{13}$$

throughout some region R of the rs-plane, where all the functions are differentiable and r and s are independent variables. To display this information, we draw an arrow diagram:

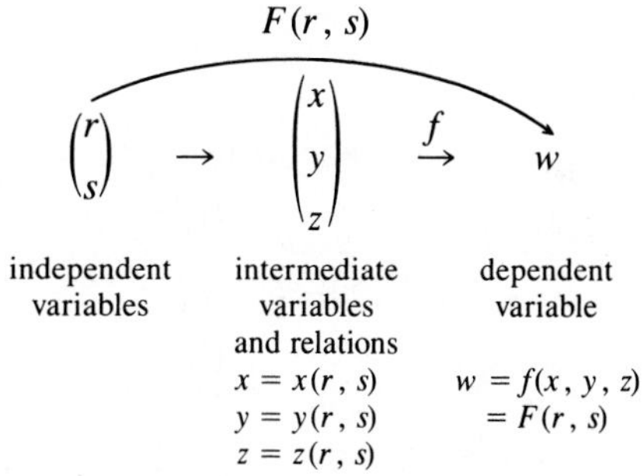

$$\begin{pmatrix} r \\ s \end{pmatrix} \rightarrow \begin{pmatrix} x \\ y \\ z \end{pmatrix} \xrightarrow{f} w \quad \text{(curved arrow labelled } F(r, s) \text{ from } (r, s) \text{ to } w) \tag{14}$$

independent variables	intermediate variables and relations	dependent variable
	$x = x(r, s)$ $y = y(r, s)$ $z = z(r, s)$	$w = f(x, y, z)$ $= F(r, s)$

Suppose further that $F(r, s) = 0$ throughout R. Then

$$\frac{\partial f}{\partial x}\frac{\partial x}{\partial r} + \frac{\partial f}{\partial y}\frac{\partial y}{\partial r} + \frac{\partial f}{\partial z}\frac{\partial z}{\partial r} = 0 \tag{15}$$

and

$$\frac{\partial f}{\partial x}\frac{\partial x}{\partial s} + \frac{\partial f}{\partial y}\frac{\partial y}{\partial s} + \frac{\partial f}{\partial z}\frac{\partial z}{\partial s} = 0. \tag{16}$$

The left-hand sides of these equations are $\partial F/\partial r$ and $\partial F/\partial s$, and these derivatives are zero because $F(r, s)$ has a constant value throughout R.

Example 4 Suppose that the equation $f(x, y, z) = 0$ determines z as a differentiable function of the independent variables x and y and that $\partial f/\partial z \neq 0$. Show that

$$\left(\frac{\partial z}{\partial x}\right)_y = -\frac{\partial f/\partial x}{\partial f/\partial z}.$$

Solution We regard all three of x, y, and z to be functions of the two independent variables x and y as in the diagram

$$\begin{pmatrix} x \\ y \end{pmatrix} \to \begin{pmatrix} x \\ y \\ z \end{pmatrix} \xrightarrow{f} w \qquad (17)$$

$$x = x$$
$$y = y$$
$$z = z(x, y)$$

This is the diagram in (14) with $r = x$, $s = y$, and a few details left out. From Eq. (15) with $r = x$, we have

$$\frac{\partial f}{\partial x}\frac{\partial x}{\partial x} + \frac{\partial f}{\partial y}\frac{\partial y}{\partial x} + \frac{\partial f}{\partial z}\frac{\partial z}{\partial x} = 0$$

$$\frac{\partial f}{\partial x} \cdot 1 + \frac{\partial f}{\partial y} \cdot 0 + \frac{\partial f}{\partial z}\frac{\partial z}{\partial x} = 0$$

$$\frac{\partial z}{\partial x} = -\frac{\partial f/\partial x}{\partial f/\partial z}.$$

EXERCISES 12.6

In Exercises 1–3, begin by drawing a diagram that shows the relations among the variables.

1. If $w = x^2 + y^2 + z^2$ and $z = x^2 + y^2$, find
   a) $\left(\frac{\partial w}{\partial y}\right)_z$ b) $\left(\frac{\partial w}{\partial z}\right)_x$ c) $\left(\frac{\partial w}{\partial z}\right)_y$

2. If $w = x^2 + y - z + \sin t$ and $x + y = t$, find
   a) $\left(\frac{\partial w}{\partial y}\right)_{x,z}$ b) $\left(\frac{\partial w}{\partial y}\right)_{z,t}$ c) $\left(\frac{\partial w}{\partial z}\right)_{x,y}$
   d) $\left(\frac{\partial w}{\partial z}\right)_{y,t}$ e) $\left(\frac{\partial w}{\partial t}\right)_{x,z}$ f) $\left(\frac{\partial w}{\partial t}\right)_{y,z}$

3. Let $U = f(p, v, T)$ be the internal energy of a gas that obeys the ideal gas law $pv = nRT$ (n and R constant). Find
   a) $\left(\frac{\partial U}{\partial p}\right)_v$ b) $\left(\frac{\partial U}{\partial T}\right)_v$

4. Find
   a) $\left(\frac{\partial w}{\partial x}\right)_y$ b) $\left(\frac{\partial w}{\partial z}\right)_y$
   at the point $(x, y, z) = (0, 1, \pi)$ if
   $$w = x^2 + y^2 + z^2 \quad \text{and} \quad y \sin z + z \sin x = 1.$$

5. Find
   a) $\left(\frac{\partial w}{\partial y}\right)_x$ b) $\left(\frac{\partial w}{\partial y}\right)_z$
   at the point $(w, x, y, z) = (4, 2, 1, -1)$ if
   $$w = x^2y^2 + yz - z^3 \quad \text{and} \quad x^2 + y^2 + z^2 = 6.$$

6. Find $\left(\frac{\partial u}{\partial y}\right)_x$ at the point $(u, v) = (\sqrt{2}, 1)$ if $x = u^2 + v^2$ and $y = uv$.

7. Suppose that $x^2 + y^2 = r^2$ and $x = r \cos \theta$, as in polar coordinates. Find
   $$\left(\frac{\partial x}{\partial r}\right)_\theta \quad \text{and} \quad \left(\frac{\partial r}{\partial x}\right)_y.$$

8. Suppose that the equation $g(x, y, z) = 0$ determines z as a differentiable function of the independent variables x and y and that $g_z \neq 0$. Show that
   $$\left(\frac{\partial z}{\partial y}\right)_x = -\frac{\partial g/\partial y}{\partial g/\partial z}.$$

9. Establish the fact, widely used in hydrodynamics, that if $f(x, y, z) = 0$, then
   $$\left(\frac{\partial x}{\partial y}\right)_z \left(\frac{\partial y}{\partial z}\right)_x \left(\frac{\partial z}{\partial x}\right)_y = -1.$$
   (*Hint:* Express all the derivatives in terms of the formal partial derivatives $\partial f/\partial x$, $\partial f/\partial y$, and $\partial f/\partial z$.)

10. If $z = x + f(u)$, where $u = xy$, show that
   $$x\frac{\partial z}{\partial x} - y\frac{\partial z}{\partial y} = x.$$

11. Suppose
   $$w = x^2 - y^2 + 4z + t \quad \text{and} \quad x + 2z + t = 25.$$

Show that the equations

$$\frac{\partial w}{\partial x} = 2x - 1 \quad \text{and} \quad \frac{\partial w}{\partial x} = 2x - 2$$

each give $\partial w/\partial x$, depending on which variables are chosen to be dependent and which are chosen to be independent. Identify the independent variables in each case.

12.7 Directional Derivatives, Gradient Vectors, and Tangent Planes

We saw in Section 12.4 that if a function $f(x, y, z)$ is differentiable, the rate at which the values of f change with respect to t along a differentiable curve $x = g(t)$, $y = h(t)$, $z = k(t)$ can be found from the formula

$$\frac{df}{dt} = \frac{\partial f}{\partial x}\frac{dx}{dt} + \frac{\partial f}{\partial y}\frac{dy}{dt} + \frac{\partial f}{\partial z}\frac{dz}{dt}. \tag{1}$$

At any particular point $P_0(g(t_0), h(t_0), k(t_0))$, this formula gives the rate of change of f with respect to increasing t and therefore depends, among other things, on the direction of motion along the curve. This observation is particularly important when the curve is a straight line and the parameter t is the distance along the line measured from P_0 in the direction of a given unit vector $\mathbf{u}$. For then df/dt is the rate of change of f with respect to distance in its domain in the direction of $\mathbf{u}$. By varying $\mathbf{u}$, we can find the rates at which f changes with respect to distance as we move through P_0 in different directions. These "directional derivatives" are extremely useful in science and engineering as well as in mathematics. The first goal of this section is to develop a formula for calculating them. The second goal is to show how to use vectors called gradients to define tangent planes and normal lines at points on level surfaces of functions of three variables.

Directional Derivatives of Functions of Three Variables

Suppose that the function $f(x, y, z)$ is differentiable throughout some region D in space, that $P_0(x_0, y_0, z_0)$ is a point in D, and that $\mathbf{u} = u_1\mathbf{i} + u_2\mathbf{j} + u_3\mathbf{k}$ is a unit vector (Fig. 12.40). Then

$$x = x_0 + su_1, \qquad y = y_0 + su_2, \qquad z = z_0 + su_3 \tag{2}$$

are parametric equations for the line through P_0 parallel to $\mathbf{u}$, and the parameter s measures the directed distance along this line from P_0 to the point $P_s(x_0 + su_1, y_0 + su_2, z_0 + su_3)$, as we saw in Example 3, Section 11.3. To calculate df/ds, the rate at which f changes with respect to distance in the direction of $\mathbf{u}$ at P_0, we substitute the derivatives

$$\frac{dx}{ds} = u_1, \qquad \frac{dy}{ds} = u_2, \qquad \frac{dz}{ds} = u_3$$

into the equation

$$\frac{df}{ds} = \frac{\partial f}{\partial x}\frac{dx}{ds} + \frac{\partial f}{\partial y}\frac{dy}{ds} + \frac{\partial f}{\partial z}\frac{dz}{ds}$$

and evaluate the partial derivatives of f at P_0. This gives

$$\frac{df}{ds} = \left(\frac{\partial f}{\partial x}\right)_{P_0} u_1 + \left(\frac{\partial f}{\partial y}\right)_{P_0} u_2 + \left(\frac{\partial f}{\partial z}\right)_{P_0} u_3. \tag{3}$$

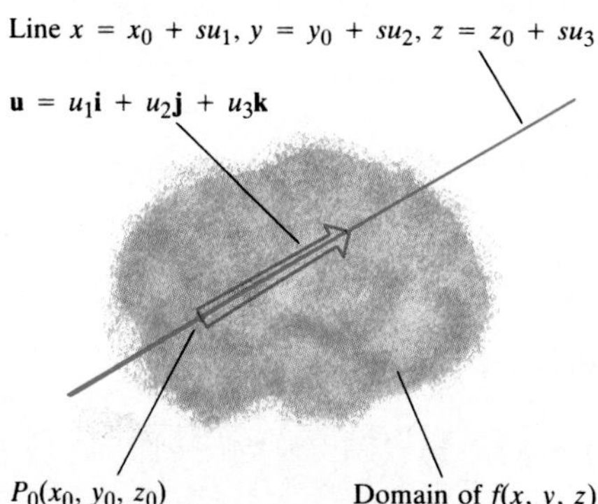

12.40 The derivative of f in the direction of the unit vector $\mathbf{u}$ at a point P_0 in its domain is the value of df/ds at P_0.

The expression on the right-hand side of Eq. (3) is the dot product of $\mathbf{u}$ and the

The notation ∇f is read "grad f" as well as "gradient of f" and "del f." The symbol ∇ by itself is read "del." Another notation for the gradient is grad f, read the way it is written.

vector

$$\nabla f = \left(\frac{\partial f}{\partial x}\right)_{P_0} \mathbf{i} + \left(\frac{\partial f}{\partial y}\right)_{P_0} \mathbf{j} + \left(\frac{\partial f}{\partial z}\right)_{P_0} \mathbf{k}.$$

This vector is called the *gradient* of f at the point P_0.

It is customary to picture the gradient as a vector in the domain of f. Its components are calculated by evaluating the three partial derivatives of f at (x_0, y_0, z_0). The derivative on the left-hand side of Eq. (3) is called the *derivative of f at the point P_0 in the direction of* $\mathbf{u}$. It is often denoted by

$$(D_{\mathbf{u}} f)_{P_0} \qquad \text{or} \qquad \left(\frac{df}{ds}\right)_{\mathbf{u}, P_0}$$

DEFINITIONS

If the partial derivatives of $f(x, y, z)$ are defined at $P_0(x_0, y_0, z_0)$, then the **gradient** of f at P_0 is the vector

$$\nabla f = \frac{\partial f}{\partial x} \mathbf{i} + \frac{\partial f}{\partial y} \mathbf{j} + \frac{\partial f}{\partial z} \mathbf{k} \tag{4}$$

obtained by evaluating the partial derivatives of f at P_0.

If $\mathbf{u}$ is a unit vector, then the **derivative of f at P_0 in the direction of u** is the number

$$(D_{\mathbf{u}} f)_{P_0} = (\nabla f)_{P_0} \cdot \mathbf{u}, \tag{5}$$

which is the scalar product of the gradient f at P_0 and $\mathbf{u}$.

Example 1 Find the derivative of

$$f(x, y, z) = x^3 - xy^2 - z$$

at $P_0(1, 1, 0)$ in the direction of the vector $\mathbf{A} = 2\,\mathbf{i} - 3\,\mathbf{j} + 6\,\mathbf{k}$.

Solution The direction of $\mathbf{A}$ is obtained by dividing $\mathbf{A}$ by its length:

$$|\mathbf{A}| = \sqrt{(2)^2 + (-3)^2 + (6)^2} = \sqrt{49} = 7,$$

$$\mathbf{u} = \frac{\mathbf{A}}{|\mathbf{A}|} = \frac{2}{7}\mathbf{i} - \frac{3}{7}\mathbf{j} + \frac{6}{7}\mathbf{k}.$$

The partial derivatives of f at P_0 are

$$f_x = 3x^2 - y^2|_{(1,1,0)} = 2, \qquad f_y = -2xy|_{(1,1,0)} = -2, \qquad f_z = -1|_{(1,1,0)} = -1.$$

The gradient of f at P_0 is

$$\nabla f|_{(1,1,0)} = 2\,\mathbf{i} - 2\,\mathbf{j} - \mathbf{k}$$

(Fig. 12.41). The derivative of f at P_0 in the direction of $\mathbf{A}$ is therefore

$$(D_{\mathbf{u}} f)|_{(1,1,0)} = \nabla f|_{(1,1,0)} \cdot \mathbf{u} = (2\,\mathbf{i} - 2\,\mathbf{j} - \mathbf{k}) \cdot \left(\frac{2}{7}\mathbf{i} - \frac{3}{7}\mathbf{j} + \frac{6}{7}\mathbf{k}\right)$$

$$= \frac{4}{7} + \frac{6}{7} - \frac{6}{7} = \frac{4}{7}.$$

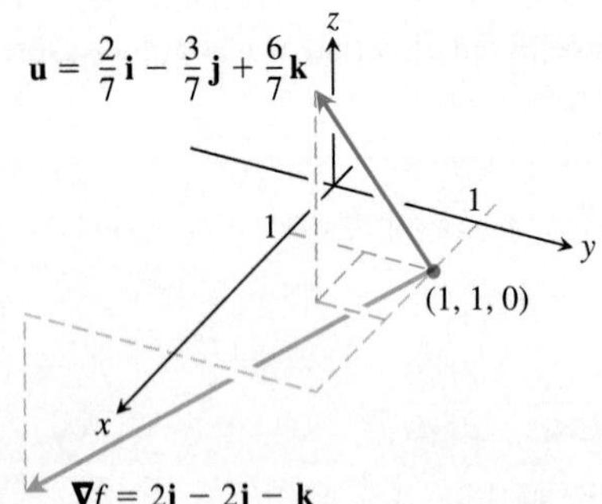

12.41 The change in f in the direction of $\mathbf{u}$ is $\nabla f \cdot \mathbf{u} = 4/7$ (Example 1).

Increments and Distance

The directional derivative plays the role of an ordinary derivative when we want to estimate how much a function f changes if we move a small distance ds from a point P_0 to a nearby point. If f were a function of a single variable, we would have

$$df = f'(P_0)\,ds. \qquad \text{(Ordinary derivative} \times \text{increment)}$$

For a function of two or more variables, we use the formula

$$df = (\nabla f|_{P_0} \cdot \mathbf{u})\,ds, \qquad \text{(Directional derivative} \times \text{increment)}$$

where $\mathbf{u}$ is the direction of motion away from P_0. The product on the right-hand side of this formula is the total differential of f in disguise:

$$\begin{aligned} df &= f_x\,dx + f_y\,dy + f_z\,dz & \text{(Total differential of } f\text{)} \\ &= \nabla f \cdot (dx\mathbf{i} + dy\mathbf{j} + dz\mathbf{k}) \\ &= \nabla f \cdot \underbrace{\frac{dx\mathbf{i} + dy\mathbf{j} + dz\mathbf{k}}{\sqrt{dx^2 + dy^2 + dz^2}}}_{\substack{\text{unit vector } \mathbf{u} \text{ in the} \\ \text{direction of change}}} \underbrace{\sqrt{dx^2 + dy^2 + dz^2}}_{ds} & (6) \\ &= (\nabla f \cdot \mathbf{u})\,ds. \end{aligned}$$

Estimating the Change in f in a Direction u

To estimate the change in the value of a function f when we move a small distance ds from a point P_0 in a particular direction $\mathbf{u}$, use the formula

$$df = \underbrace{(\nabla f|_{P_0} \cdot \mathbf{u})}_{\substack{\text{directional} \\ \text{derivative}}} \cdot \underbrace{ds}_{\substack{\text{distance} \\ \text{increment}}}. \qquad (7)$$

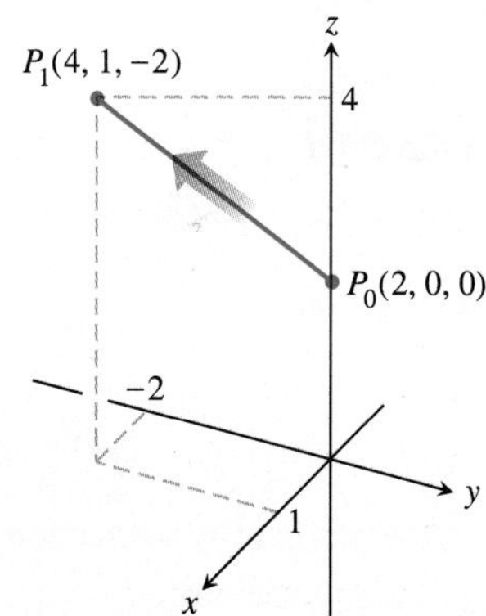

12.42 Example 2 shows how to estimate the change in the value of $f(x, y, z) = xe^y + yz$ as (x, y, z) moves a short distance $ds = 0.1$ units from P_0 along the line toward P_1.

Example 2 Estimate how much the value of

$$f(x, y, z) = xe^y + yz$$

will change if the point $P(x, y, z)$ is moved from $P_0(2, 0, 0)$ straight toward $P_1(4, 1, -2)$ a distance of $ds = 0.1$ units (Fig. 12.42).

Solution We first find the derivative of f at P_0 in the direction of the vector

$$\overrightarrow{P_0P_1} = 2\,\mathbf{i} + \mathbf{j} - 2\,\mathbf{k}.$$

The direction of this vector is

$$\mathbf{u} = \frac{\overrightarrow{P_0P_1}}{|\overrightarrow{P_0P_1}|} = \frac{\overrightarrow{P_0P_1}}{3} = \frac{2}{3}\mathbf{i} + \frac{1}{3}\mathbf{j} - \frac{2}{3}\mathbf{k}.$$

The gradient of f at P_0 is

$$\nabla f|_{(2,0,0)} = (e^y\,\mathbf{i} + (xe^y + z)\,\mathbf{j} + y\mathbf{k})|_{(2,0,0)} = \mathbf{i} + 2\,\mathbf{j}.$$

Therefore,

$$\nabla f|_{P_0} \cdot \mathbf{u} = (\mathbf{i} + 2\,\mathbf{j}) \cdot \left(\frac{2}{3}\mathbf{i} + \frac{1}{3}\mathbf{j} - \frac{2}{3}\mathbf{k}\right) = \frac{2}{3} + \frac{2}{3} = \frac{4}{3}.$$

The change df in f that results from moving $ds = 0.1$ units away from P_0 in the direction of $\mathbf{u}$ is approximately

$$df = \nabla f|_{P_0} \cdot \mathbf{u}\, ds = \left(\frac{4}{3}\right)(0.1) \approx 0.13.$$

If we evaluate the dot product in the formula for the directional derivative, to obtain

$$D_{\mathbf{u}} f = \nabla f \cdot \mathbf{u} = |\nabla f|\, |\mathbf{u}| \cos\theta = |\nabla f| \cos\theta, \tag{8}$$

the following facts come to light.

Properties of the Directional Derivative $D_{\mathbf{u}} f = \nabla f \cdot \mathbf{u} = |\nabla f| \cos\theta$

1. The directional derivative has its largest positive value when $\cos\theta = 1$, or when $\mathbf{u}$ is the direction of the gradient. That is, f increases most rapidly at any point in its domain in the direction of ∇f. The derivative in this direction is $D_{\mathbf{u}} f = |\nabla f| \cos(0) = |\nabla f|$.
2. Similarly, f decreases most rapidly in the direction of $-\nabla f$. The derivative in this direction is $D_{\mathbf{u}} f = |\nabla f| \cos(\pi) = -|\nabla f|$.
3. $D_{-\mathbf{u}} f = \nabla f \cdot (-\mathbf{u}) = -\nabla f \cdot \mathbf{u} = -D_{\mathbf{u}} f$. That is, the derivative of f in the direction of $-\mathbf{u}$ is the negative of the derivative of f in the direction of $\mathbf{u}$.
4. The relationships of the partial derivatives of f to the directional derivative are

   $$D_{\mathbf{i}} f = \nabla f \cdot \mathbf{i} = f_x, \qquad D_{\mathbf{j}} f = \nabla f \cdot \mathbf{j} = f_y, \qquad D_{\mathbf{k}} f = \nabla f \cdot \mathbf{k} = f_z.$$

   Thus,

   $$f_x = \text{derivative of } f \text{ in the } \mathbf{i}\text{-direction},$$
   $$f_y = \text{derivative of } f \text{ in the } \mathbf{j}\text{-direction},$$
   $$f_z = \text{derivative of } f \text{ in the } \mathbf{k}\text{-direction}.$$

   Combining these results with property 3 gives

   $$D_{-\mathbf{i}} f = -f_x, \qquad D_{-\mathbf{j}} f = -f_y, \qquad D_{-\mathbf{k}} f = -f_z.$$

5. Any direction $\mathbf{u}$ normal (perpendicular) to the gradient is a direction of zero change in f because

   $$D_{\mathbf{u}} f = |\nabla f| \cos(\pi/2) = |\nabla f| \cdot 0 = 0.$$

Example 3 Find the directions in which $f(x, y, z) = x^2 + y^2 + z^2$ increases and decreases most rapidly at the point $(1, 1, 1)$ in its domain. At what rates does f change in these directions?

Solution The most rapid increase at (x, y, z) is in the direction of

$$\nabla f = 2x\mathbf{i} + 2y\mathbf{j} + 2z\mathbf{k}.$$

At (1, 1, 1) we have

$$\nabla f = 2\,\mathbf{i} + 2\,\mathbf{j} + 2\,\mathbf{k}, \qquad |\nabla f| = \sqrt{4+4+4} = 2\sqrt{3}.$$

The direction of ∇f is

$$\mathbf{u} = \frac{\nabla f}{|\nabla f|} = \frac{1}{\sqrt{3}}(\mathbf{i} + \mathbf{j} + \mathbf{k}).$$

The derivative of f in this direction is $|\nabla f| = 2\sqrt{3}$.

The direction of most rapid decrease at (1, 1, 1) is

$$-\mathbf{u} = -\frac{1}{\sqrt{3}}(\mathbf{i} + \mathbf{j} + \mathbf{k})$$

and the derivative of f in this direction is $-\,|\nabla f| = -2\sqrt{3}$.

Results for Functions of Two Variables

For functions of two variables, we get results much like the ones for functions of three variables. We obtain the two-variable formulas by dropping the z-terms from the three-variable formulas. Thus, for a function $f(x, y)$ and a unit vector $\mathbf{u} = u_1\,\mathbf{i} + u_2\,\mathbf{j}$,

$$\nabla f = \frac{\partial f}{\partial x}\,\mathbf{i} + \frac{\partial f}{\partial y}\,\mathbf{j}, \tag{9}$$

and

$$D_{\mathbf{u}} f = \nabla f \cdot \mathbf{u} = \frac{\partial f}{\partial x} u_1 + \frac{\partial f}{\partial y} u_2. \tag{10}$$

The five properties listed earlier for directional derivatives of functions of three variables hold equally well in the two-variable case. In addition, property 5, which says that any direction normal to the gradient is a direction of zero change for the function, can now be turned around to say that the gradient of a function $f(x, y)$ is always normal to the function's level curves (Fig. 12.43). (See Exercise 51 for additional details.)

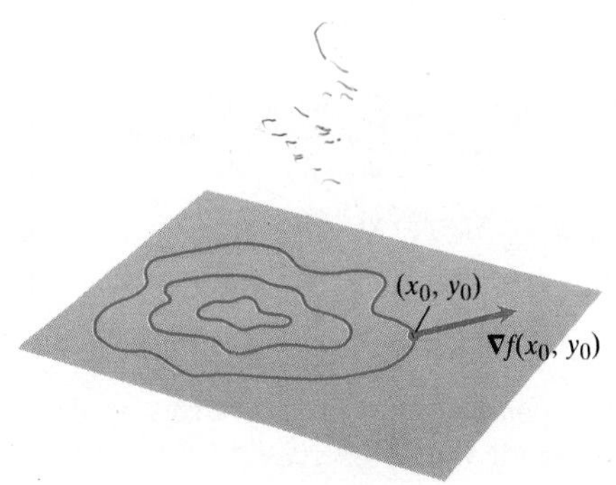

12.43 The gradient of a differentiable function of two variables is always normal to the function's level curves.

> At every point (x_0, y_0) in the domain of $f(x, y)$, the gradient of f is normal to the level curve through (x_0, y_0).

Example 4

a) Find the derivative of

$$f(x, y) = x^2 + y^2$$

at the point $P_0(1, 1)$ in the direction of the unit vector $\mathbf{u} = u_1\,\mathbf{i} + u_2\,\mathbf{j}$.

b) In what direction in its domain (the xy-plane) does f increase most rapidly at (1, 1)? What is the derivative of f in this direction?

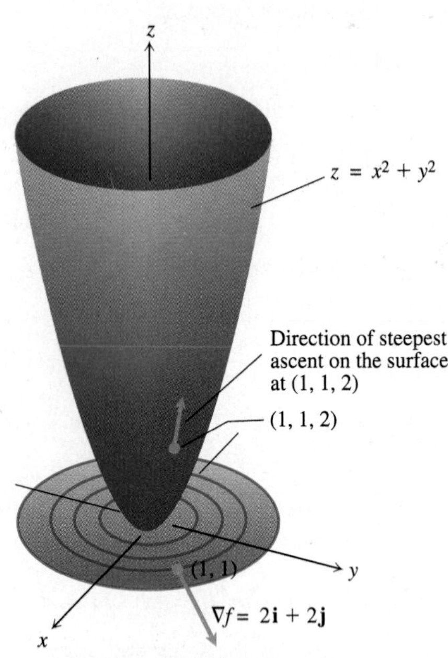

12.44 The gradient of $f(x, y) = x^2 + y^2$ at the point $(1, 1)$ is the vector $\nabla f = 2\mathbf{i} + 2\mathbf{j}$ in the xy-plane. This vector points in the direction in which f increases most rapidly at the point $(1, 1)$ and is normal to the level curve there. The corresponding direction on the surface $z = f(x, y)$ is the direction of steepest ascent at the point $(1, 1, 2)$ (Example 4).

Solution

a) We have

$$f(x, y) = x^2 + y^2, \qquad f_x(1, 1) = 2x\big|_{(1,1)} = 2, \qquad f_y(1, 1) = 2y\big|_{(1,1)} = 2,$$

$$(\nabla f)_{(1,1)} = f_x(1, 1)\,\mathbf{i} + f_y(1, 1)\,\mathbf{j} = 2\,\mathbf{i} + 2\,\mathbf{j},$$

and

$$(D_{\mathbf{u}} f)_{(1,1)} = (\nabla f)_{(1,1)} \cdot \mathbf{u} = (2\,\mathbf{i} + 2\,\mathbf{j}) \cdot \mathbf{u} = 2u_1 + 2u_2.$$

The derivative of f at $(1, 1)$ in the direction of $\mathbf{u}$ is $2u_1 + 2u_2$.

b) The function increases most rapidly at $(1, 1)$ in the direction of ∇f at $(1, 1)$, which is

$$\left(\frac{\nabla f}{|\nabla f|}\right)_{(1,1)} = \frac{2\,\mathbf{i} + 2\,\mathbf{j}}{\sqrt{2^2 + 2^2}} = \frac{2}{\sqrt{8}}\mathbf{i} + \frac{2}{\sqrt{8}}\mathbf{j} = \frac{1}{\sqrt{2}}(\mathbf{i} + \mathbf{j})$$

(Fig. 12.44). The derivative of f in this direction at $(1, 1)$ is

$$|\nabla f|_{(1,1)} = 2\sqrt{2}.$$

Algebra Rules for Gradients

If we know the gradients of two functions f and g, we automatically know the gradients of their constant multiples, sum, difference, and product.

Algebra Rules for Gradients

1. *Constant Multiple Rule:* $\nabla(kf) = k\nabla f$ (Any number k)
2. *Sum Rule:* $\nabla(f + g) = \nabla f + \nabla g$
3. *Difference Rule:* $\nabla(f - g) = \nabla f - \nabla g$
4. *Product Rule:* $\nabla(fg) = f\nabla g + g\nabla f$

These rules have the same form as the corresponding rules for derivatives, as they should. You will see where the rules come from if you do Exercise 52.

Example 5 With

$$f(x, y, z) = e^x, \qquad g(x, y, z) = y - z,$$
$$\nabla f = e^x\,\mathbf{i}, \qquad \nabla g = \mathbf{j} - \mathbf{k},$$

we find

1. $\nabla(2f) = \nabla(2e^x) = 2e^x\,\mathbf{i} = 2\nabla f,$
2. $\nabla(f + g) = \nabla(e^x + y - z) = e^x\,\mathbf{i} + \mathbf{j} - \mathbf{k} = (e^x\,\mathbf{i}) + (\mathbf{j} - \mathbf{k}) = \nabla f + \nabla g,$
3. $\nabla(f - g) = \nabla(e^x - y + z) = e^x\,\mathbf{i} - \mathbf{j} + \mathbf{k} = (e^x\,\mathbf{i}) - (\mathbf{j} - \mathbf{k}) = \nabla f - \nabla g,$
4. $\nabla(fg) = \nabla(ye^x - ze^x) = (ye^x - ze^x)\,\mathbf{i} + e^x\,\mathbf{j} - e^x\,\mathbf{k}$
$\quad = (y - z)(e^x\,\mathbf{i}) + e^x(\mathbf{j} - \mathbf{k}) = g\nabla f + f\nabla g.$

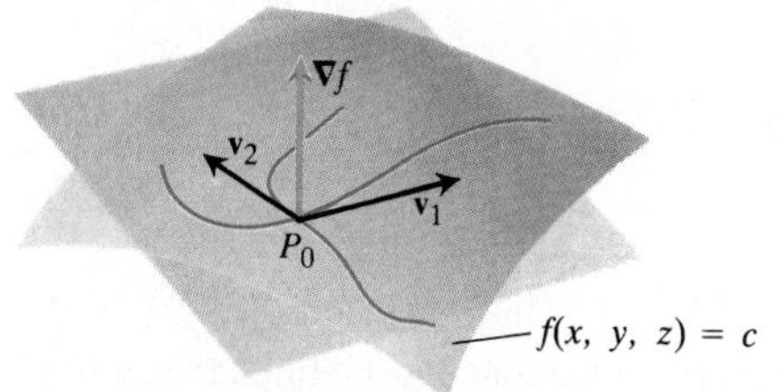

12.45 ∇f is orthogonal to the velocity vector of every differentiable curve in the surface through P_0. The velocity vectors at P_0 therefore lie in a common plane, which we call the tangent plane at P_0.

Equations for Tangent Planes and Normal Lines to Level Surfaces

To find equations for tangent planes and normal lines, we suppose that $P_0(x_0, y_0, z_0)$ is a point on the level surface $f(x, y, z) = c$ of a differentiable function f. If $x = g(t)$, $y = h(t)$, $z = k(t)$ is any differentiable curve on the surface, then

$$f(g(t), h(t), k(t)) = c \tag{11}$$

for every value of t. Differentiating both sides of this equation with respect to t gives

$$\frac{\partial f}{\partial x}\frac{dx}{dt} + \frac{\partial f}{\partial y}\frac{dy}{dt} + \frac{\partial f}{\partial z}\frac{dz}{dt} = 0. \tag{12}$$

The left-hand side of this equation is the dot product of ∇f with the curve's velocity vector $\mathbf{v}$, so the equation can be written in the form

$$\nabla f \cdot \mathbf{v} = 0. \tag{13}$$

What Eq. (13) shows more clearly than Eq. (12) is that the gradient of f is orthogonal to the velocity vector at every point of the curve.

Now let us restrict our attention to the differentiable curves that lie on the surface $f(x, y, z) = c$ and pass through P_0. All the velocity vectors at P_0 are orthogonal to ∇f at P_0, and hence all the tangent lines to these curves lie in the plane through P_0 normal to ∇f. We call this plane the tangent plane of the surface at P_0 (Fig. 12.45). We call the line through P_0 perpendicular to the plane the surface's normal line at P_0.

DEFINITIONS

The **tangent plane** at the point $P_0(x_0, y_0, z_0)$ on the level surface $f(x, y, z) = c$ is the plane

$$f_x(P_0)(x - x_0) + f_y(P_0)(y - y_0) + f_z(P_0)(z - z_0) = 0. \tag{14}$$

The **normal line** of the surface at P_0 is the line

$$x = x_0 + f_x(P_0)t, \qquad y = y_0 + f_y(P_0)t, \qquad z = z_0 + f_z(P_0)t. \tag{15}$$

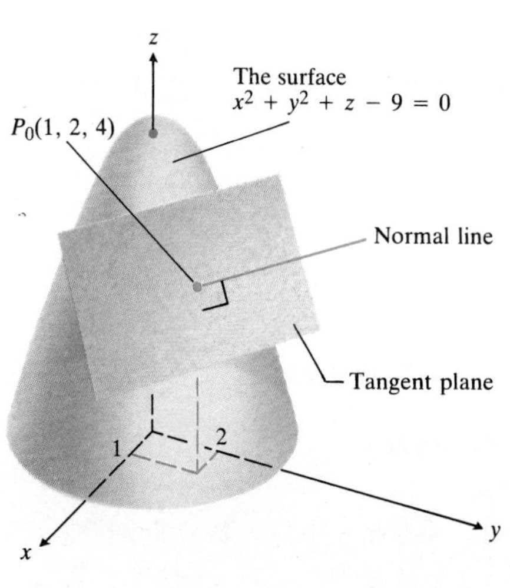

12.46 The tangent plane and normal line to the surface $x^2 + y^2 + z - 9 = 0$ at $P_0(1, 2, 4)$.

Example 6 Find equations for the tangent plane and normal line of the surface

$$f(x, y, z) = x^2 + y^2 + z - 9 = 0 \qquad \text{(A circular paraboloid)}$$

at the point $P_0(1, 2, 4)$.

Solution The surface is shown in Fig. 12.46.

The tangent plane is the plane through P_0 perpendicular to the gradient of f at P_0. The gradient is

$$\nabla f|_{P_0} = (2x\mathbf{i} + 2y\mathbf{j} + \mathbf{k})_{(1,2,4)} = 2\,\mathbf{i} + 4\,\mathbf{j} + \mathbf{k}.$$

The plane is therefore the plane

$$2(x - 1) + 4(y - 2) + (z - 4) = 0 \qquad \text{or} \qquad 2x + 4y + z = 14.$$

The line normal to the surface at P_0 is

$$x = 1 + 2t, \qquad y = 2 + 4t, \qquad z = 4 + t.$$

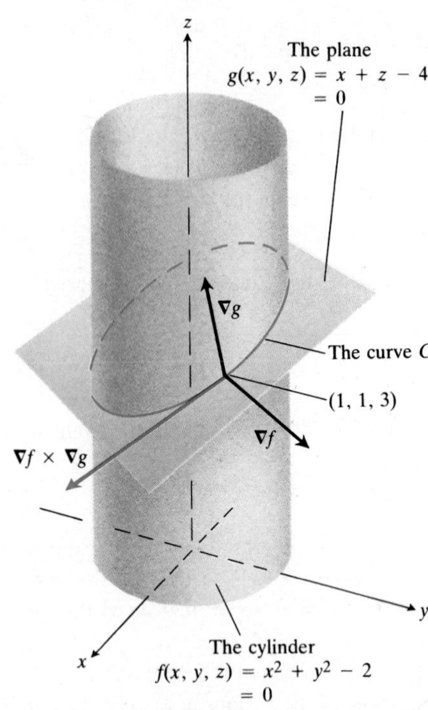

12.47 The cylinder $f(x, y, z) = x^2 + y^2 - 2 = 0$ and the plane $g(x, y, z) = x + z - 4 = 0$ intersect in a curve C. Any vector tangent to C will automatically be orthogonal to ∇f and ∇g and hence parallel to $\nabla f \times \nabla g$. This information enables us to write an equation for the tangent line in Example 7.

Example 7 The surfaces

$$f(x, y, z) = x^2 + y^2 - 2 = 0 \qquad \text{(A cylinder)}$$

and

$$g(x, y, z) = x + z - 4 = 0 \qquad \text{(A plane)}$$

meet in a curve C (Fig. 12.47). Find parametric equations for the line tangent to C at the point $P_0(1, 1, 3)$.

Solution The tangent line is perpendicular to both ∇f and ∇g at P_0, so $\mathbf{v} = \nabla f \times \nabla g$ is parallel to the line. The components of $\mathbf{v}$ and the coordinates of P_0 give us equations for the line. We have

$$\nabla f_{(1,1,3)} = (2x\mathbf{i} + 2y\mathbf{j})_{(1,1,3)} = 2\,\mathbf{i} + 2\,\mathbf{j}$$

$$\nabla g_{(1,1,3)} = (\mathbf{i} + \mathbf{k})_{(1,1,3)} = \mathbf{i} + \mathbf{k}$$

$$\mathbf{v} = (2\,\mathbf{i} + 2\,\mathbf{j}) \times (\mathbf{i} + \mathbf{k}) = \begin{vmatrix} \mathbf{i} & \mathbf{j} & \mathbf{k} \\ 2 & 2 & 0 \\ 1 & 0 & 1 \end{vmatrix} = 2\,\mathbf{i} - 2\,\mathbf{j} - 2\,\mathbf{k}.$$

The line is

$$x = 1 + 2t, \qquad y = 1 - 2t, \qquad z = 3 - 2t.$$

Equations for Lines Tangent to Level Curves

The fact that the gradient of a function $f(x, y)$ is normal to the function's level curves in the plane enables us to use ∇f to write equations for tangent lines to level curves.

> The tangent to the level curve $f(x, y) = c$ at the point (x_0, y_0) is the line
>
> $$f_x(x_0, y_0)(x - x_0) + f_y(x_0, y_0)(y - y_0) = 0. \qquad (16)$$

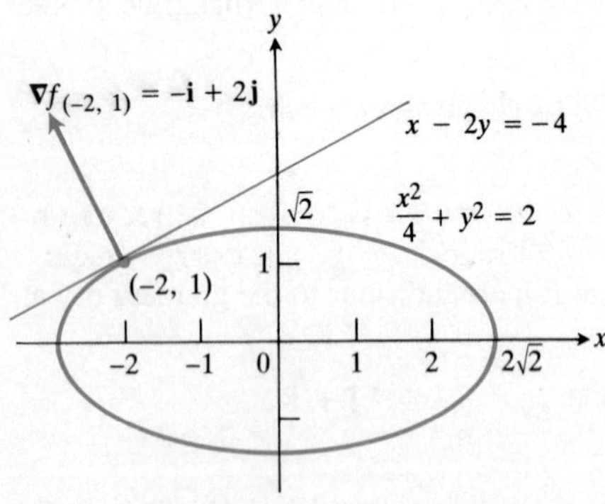

12.48 We can find the tangent to the ellipse $(x^2/4) + y^2 = 2$ by treating the ellipse as a level curve of the function $f(x, y) = (x^2/4) + y^2$ (Example 8).

Example 8 Find an equation for the tangent to the ellipse

$$\frac{x^2}{4} + y^2 = 2$$

(Fig. 12.48) at the point $(-2, 1)$.

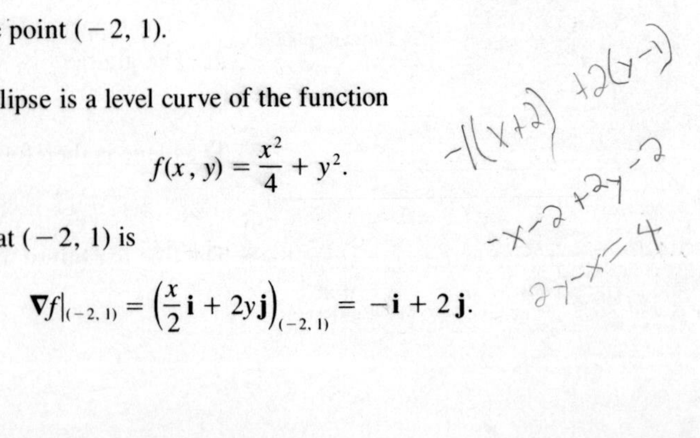

Solution The ellipse is a level curve of the function

$$f(x, y) = \frac{x^2}{4} + y^2.$$

The gradient of f at $(-2, 1)$ is

$$\nabla f|_{(-2,1)} = \left(\frac{x}{2}\mathbf{i} + 2y\mathbf{j}\right)_{(-2,1)} = -\mathbf{i} + 2\,\mathbf{j}.$$

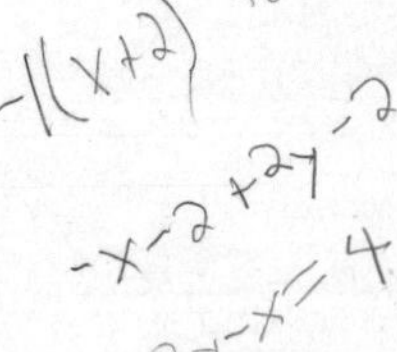

The tangent is the line

$$(-1)(x+2)+(2)(y-1)=0 \qquad \text{(Eq. (16))}$$

$$x-2y=-4.$$

EXERCISES 12.7

In Exercises 1–4, find ∇f at the given point.

1. $f(x, y, z) = x^2 + y^2 - 2z^2 + z \ln x, \quad (1, 1, 1)$
2. $f(x, y, z) = 2z^3 - 3(x^2 + y^2)z + \tan^{-1} xz, \quad (1, 1, 1)$
3. $f(x, y, z) = (x^2 + y^2 + z^2)^{-1/2} + \ln(xyz), \quad (-1, 2, -2)$
4. $f(x, y, z) = e^{x+y} \cos z + (y + 1)\sin^{-1} x, \quad (0, 0, \pi/6)$

In Exercises 5–8, sketch the surface $f(x, y, z) = c$ together with ∇f at the given point.

5. $x^2 + y^2 - z = 0, \quad (1, 1, 2)$
6. $(y^2/2) + z^2 = 2, \quad (0, \sqrt{2}, 1)$
7. $(x^2/2) + (y^2/2) - (z^2/2) = 0, \quad (1, 1, -\sqrt{2})$
8. $z^2 - y^2 = 3, \quad (0, 1, 2)$

In Exercises 9–12, find ∇f at the given point. Then sketch ∇f together with the level curve through the point.

9. $f(x, y) = y - x, \quad (2, 1)$
10. $f(x, y) = \ln(x^2 + y^2), \quad (1, 1)$
11. $f(x, y) = y - x^2, \quad (-1, 0)$
12. $f(x, y) = x^2 - y^2, \quad (2, \sqrt{3})$

In Exercises 13–18, find the derivative of f at P_0 in the direction of **A**.

13. $f(x, y, z) = xy + yz + zx, \quad P_0(1, -1, 2)$
    $\mathbf{A} = 3\mathbf{i} + 6\mathbf{j} - 2\mathbf{k}$
14. $f(x, y, z) = x^2 + 2y^2 - 3z^2, \quad P_0(1, 1, 1)$
    $\mathbf{A} = \mathbf{i} + \mathbf{j} + \mathbf{k}$
15. $f(x, y, z) = 3e^x \cos yz, \quad P_0(0, 0, 0)$
    $\mathbf{A} = -2\mathbf{i} + \mathbf{j} + 2\mathbf{k}$
16. $f(x, y, z) = \cos xy + e^{yz} + \ln zx, \quad P_0(1, 0, 1/2)$
    $\mathbf{A} = \mathbf{i} + \mathbf{j} - \sqrt{2}\mathbf{k}$
17. $f(x, y) = x - (y^2/x) + \sqrt{3} \sec^{-1}(2xy), \quad P_0(1, 1),$
    $\mathbf{A} = 12\mathbf{i} + 5\mathbf{j}$
18. $f(x, y) = \tan^{-1}(y/x) + \sqrt{3} \sin^{-1}(xy/2), \quad P_0(1, 1),$
    $\mathbf{A} = 3\mathbf{i} - 2\mathbf{j}$

In Exercises 19–24, find the directions in which f increases and decreases most rapidly at P_0. Then find the derivatives of f in these directions.

19. $f(x, y) = x^2 y + e^{xy} \sin y, \quad P_0(1, 0)$
20. $f(x, y) = x^{2/3} + xy + y^{2/3}, \quad P_0(-1, 1)$
21. $f(x, y, z) = xe^y + 4z^2 \cos^{-1}(\ln x), \quad P_0(1, \ln 2, 1/2)$
22. $f(x, y, z) = (x/y) - yz - 4\sqrt{xyz}, \quad P_0(4, 1, 1)$
23. $f(x, y, z) = \ln xy + \ln yz + \ln xz, \quad P_0(1, 1, 1)$
24. $f(x, y, z) = \ln(x^2 + y^2 - 1) + \sinh(xyz), \quad P_0(1, 1, 0)$

In Exercises 25–30, find equations for (a) the tangent plane and (b) the normal line at the point P_0 on the surface.

25. $2z - x^2 = 0, \quad P_0(2, 0, 2)$
26. $z - \ln(x^2 + y^2) = 0, \quad P_0(1, 0, 0)$
27. $x^2 + 2xy - y^2 + z^2 = 7, \quad P_0(1, -1, 3)$
28. $x - y^3 + zy - 2z^2 = 0, \quad P_0(-4, -2, 1)$
29. $\cos \pi x - x^2 y + e^{xz} + yz = 4, \quad P_0(0, 1, 2)$
30. $x^2 + y^2 - 2xy - x + 3y - z = -4, \quad P_0(2, -3, 18)$

In Exercises 31–34, find parametric equations for the line that is tangent to the curve of intersection of the surfaces at the given point.

31. Surfaces: $x + y^2 + 2z = 4, \quad x = 1$
    Point: $(1, 1, 1)$
32. Surfaces: $xyz = 1, \quad x^2 + 2y^2 + 3z^2 = 6$
    Point: $(1, 1, 1)$
33. Surfaces: $x^3 + 3x^2y^2 + y^3 + 4xy - z^2 = 0,$
    $x^2 + y^2 + z^2 = 11$
    Point: $(1, 1, 3)$
34. Surfaces: $x^2 + y^2 = 4, \quad x^2 + y^2 - z = 0$
    Point: $(\sqrt{2}, \sqrt{2}, 4)$

In Exercises 35–38, sketch the curve $f(x, y) = c$ together with ∇f and the tangent line at the given point. Then write an equation for the tangent line.

35. $x^2 + y^2 = 4, \quad (\sqrt{2}, \sqrt{2})$
36. $x^2 - y = 1, \quad (\sqrt{2}, 1)$
37. $xy = -4, \quad (2, -2)$
38. $x^2 - xy + y^2 = 7, \quad (-1, 2)$ (This is the curve in Section 2.5, Example 4.)
39. By about how much will the value of $\cos \pi xy + xy^2$ change if (x, y) is moved a distance of $ds = 0.1$ unit from $(-1, -1)$ in the direction of $\mathbf{i} + \mathbf{j}$?

40. By about how much will the value of $\ln\sqrt{x^2+y^2+z^2}$ change if the point (x, y, z) is moved a distance of $ds = 0.1$ unit from $(3, 4, 12)$ in the direction of $3\mathbf{i} + 6\mathbf{j} - 2\mathbf{k}$?

41. By about how much will

$$f(x, y, z) = e^x \cos yz$$

change if the point $P(x, y, z)$ moves a distance of $ds = 0.1$ unit from the origin toward the point $(2, 1, -2)$?

42. By about how much will

$$f(x, y, z) = x + x\cos z - y\sin z + y$$

change if the point $P(x, y, z)$ moves from $P_0(2, -1, 0)$ a distance of $ds = 0.2$ unit toward the point $P_1(0, 1, 2)$?

43. In what two directions is the derivative of $f(x, y) = xy + y^2$ equal to zero at the point $(3, 2)$? (Fig. 12.49.)

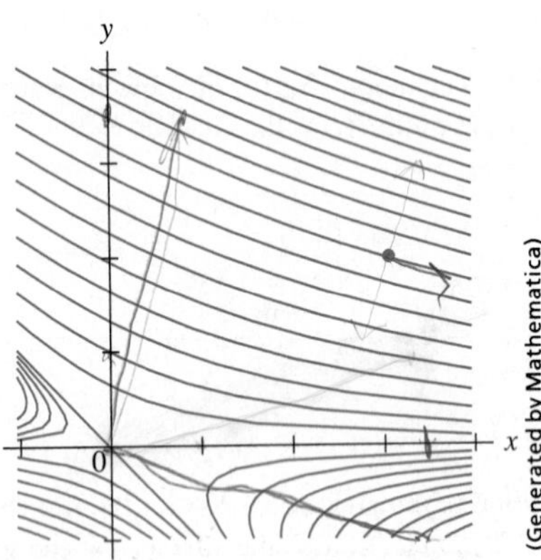

12.49 Level curves of $f(x, y) = xy + y^2$ (Exercise 43).

44. In what two directions is the derivative of $f(x, y) = (x^2 - y^2)/(x^2 + y^2)$ equal to zero at the point $(1, 1)$? (Fig. 12.50.)

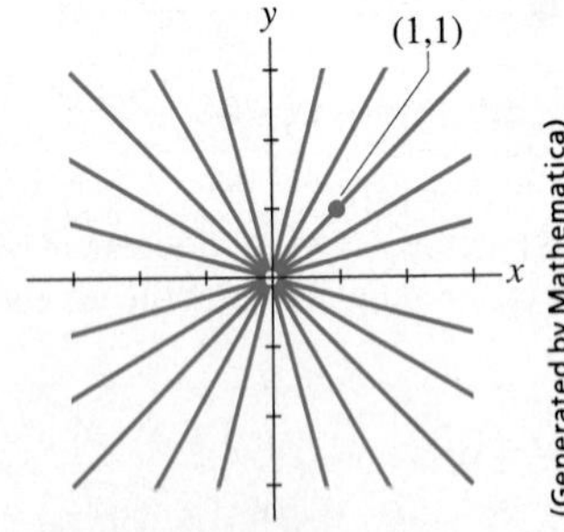

12.50 Level curves of $f(x, y) = (x^2 - y^2)/(x^2 + y^2)$ (Exercise 44).

45. *Change along a helix.* Find the derivative of $f(x, y, z) = x^2 + y^2 + z^2$ in the direction of the unit tangent vector of the helix $\mathbf{r} = (\cos t)\mathbf{i} + (\sin t)\mathbf{j} + t\mathbf{k}$ at the points where $t = -\pi/4$, 0, and $\pi/4$. The function f gives the square of the distance from a point $P(x, y, z)$ on the helix to the origin. The derivatives calculated here give the rates at which the square of the distance is changing with respect to t as P moves through the points where $t = -\pi/4$, 0, and $\pi/4$.

46. *Change along the involute of a circle.* Find the derivative of $f(x, y) = x^2 + y^2$ in the direction of the unit tangent vector of the curve

$$\mathbf{r} = (\cos t + t\sin t)\mathbf{i} + (\sin t - t\cos t)\mathbf{j}, \quad t > 0$$

(Fig. 12.51).

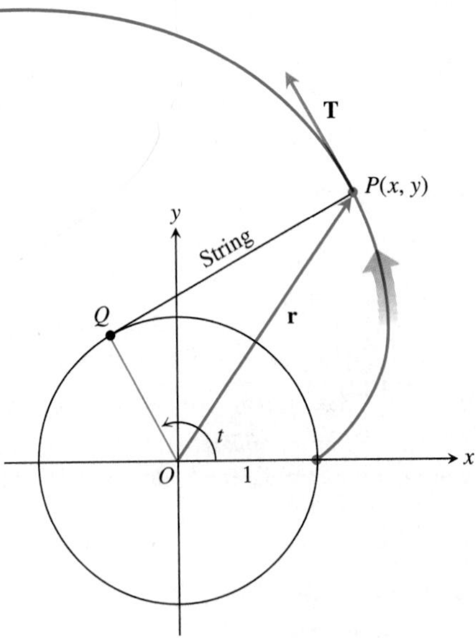

12.51 The involute of the unit circle from Section 11.3, Example 5. If you move out along the involute, covering distance along the curve at a steady rate, your distance from the origin will increase at a constant rate as well. (This is how to interpret the result of your calculation in Exercise 46.)

47. The derivative of $f(x, y)$ at $P_0(1, 2)$ in the direction of $\mathbf{i} + \mathbf{j}$ is $2\sqrt{2}$ and in the direction of $-2\mathbf{j}$ is -3. What is the derivative of f in the direction of $-\mathbf{i} - 2\mathbf{j}$?

48. The derivative of $f(x, y, z)$ at a point P is greatest in the direction of $\mathbf{A} = \mathbf{i} + \mathbf{j} - \mathbf{k}$. In this direction the value of the derivative is $2\sqrt{3}$.

a) Find ∇f at P.
b) Find the derivative of f at P in the direction of $\mathbf{i} + \mathbf{j}$.

49. *Normal curves and tangent curves.* A curve is **normal** to a surface $f(x, y, z) = c$ at a point of intersection if the curve's velocity vector is a scalar multiple of ∇f at the point. The curve is **tangent** to the surface at a point of intersection if its velocity vector is orthogonal to ∇f there.

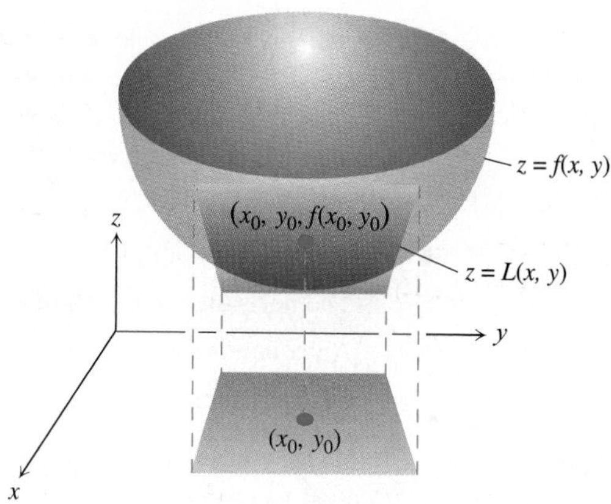

12.52 The graph of a function $z = f(x, y)$ and its linearization at a point (x_0, y_0). The plane defined by L is tangent to the surface at the point above the point (x_0, y_0). This furnishes a geometric explanation of why the values of L lie close to those of f in the immediate neighborhood of (x_0, y_0) (Exercise 50).

a) Show that the curve

$$\mathbf{r} = \sqrt{t}\,\mathbf{i} + \sqrt{t}\,\mathbf{j} - \frac{1}{4}(t+3)\,\mathbf{k}$$

is normal to the surface $x^2 + y^2 - z = 3$ when $t = 1$.

b) Show that the curve

$$\mathbf{r} = \sqrt{t}\,\mathbf{i} + \sqrt{t}\,\mathbf{j} + (2t-1)\,\mathbf{k}$$

is tangent to the surface $x^2 + y^2 - z = 1$ when $t = 1$.

50. *The linearization of f(x, y) is a tangent–plane approximation.* Show that the tangent plane at the point $P_0(x_0, y_0, f(x_0, y_0))$ on the surface $z = f(x, y)$ defined by a differentiable function f is the plane

$$f_x(x_0, y_0)(x - x_0) + f_y(x_0, y_0)(y - y_0) - (z - f(x_0, y_0)) = 0$$

or

$$z = f(x_0, y_0) + f_x(x_0, y_0)(x - x_0) + f_y(x_0, y_0)(y - y_0).$$

Thus the tangent plane at P_0 is the graph of the linearization of f at P_0 (Fig. 12.52).

51. *Another way to see why gradients are normal to level curves.* Suppose that a differentiable function $f(x, y)$ has a constant value c along the differentiable curve $x = g(t)$, $y = h(t)$ for all values of t. Differentiate both sides of the equation $f(g(t), h(t)) = c$ with respect to t to show that ∇f is normal to the curve's tangent vector at every point.

52. *The algebra rules for gradients.* Given a constant k and the gradients

$$\nabla f = \frac{\partial f}{\partial x}\mathbf{i} + \frac{\partial f}{\partial y}\mathbf{j} + \frac{\partial f}{\partial z}\mathbf{k} \quad \text{and} \quad \nabla g = \frac{\partial g}{\partial x}\mathbf{i} + \frac{\partial g}{\partial y}\mathbf{j} + \frac{\partial g}{\partial z}\mathbf{k},$$

use the facts that

$$\frac{\partial}{\partial x}(kf) = k\frac{\partial f}{\partial x}, \qquad \frac{\partial}{\partial x}(f \pm g) = \frac{\partial f}{\partial x} \pm \frac{\partial g}{\partial x},$$

and so on, to establish the following rules:

a) $\nabla(kf) = k\nabla f$

b) $\nabla(f + g) = \nabla f + \nabla g$

c) $\nabla(f - g) = \nabla f - \nabla g$

d) $\nabla(fg) = f\nabla g + g\nabla f$

COMPUTER Exercises 53–56 each give a function $f(x, y, z)$, a point P_0, and a vector $\mathbf{v}$. Use the gradient utility program at the end of the section to find ∇f at P_0, the derivative of f in the direction of $\mathbf{v}$, and equations for the tangent plane and normal line at P_0 on the level surface of f through P_0.

	$f(x, y, z)$	P_0	$\mathbf{v}$
53.	$4x^2y + y^2z$	$(1, 0, 1)$	$\mathbf{i} + \mathbf{j} + \mathbf{k}$
54.	$xyz + y^3 - z$	$(1, 2, -1)$	$2\mathbf{i} - \mathbf{j} + 2\mathbf{k}$
55.	$z\cos(x^2 + y^2)$	$(1, 1, 1)$	$2\mathbf{i} + 2\mathbf{j} - 2\mathbf{k}$
56.	$x^3z + x^2y^2 + \sin yz$	$(1, 0, -1)$	$3\mathbf{i} - \mathbf{j} + 5\mathbf{k}$

✲ COMPUTER PROGRAM

A Gradient Utility Program

For versions of BASIC that accept multivariable functions, the following BASIC program carries out the standard calculations associated with gradients. Key in the program, including a formula for a differentiable function $f(x, y, z)$ in line 20. (To apply the program to a new function, you have only to change this one line.) Then run the program and enter, as prompted, the coordinates of a point P_0 and the components of a vector $\mathbf{v}$. In response to your entries, the program will display the following:

The gradient of f at P_0

The derivative of f at P_0 in the direction of $\mathbf{v}$

An equation for the plane tangent to the level surface of f through P_0

Equations for the line normal to the surface at P_0

The vector you enter for the direction need not be a unit vector. Any vector with the direction you want will do.

Here is a listing of the program with $f(x, y, z) = x^2 + y^2 + z$ in line 20.

	PROGRAM	COMMENTS
10	`DIM P(3), V(3), DF(3)`	Define arrays to hold components of vectors
20	`DEF FNF(X,Y, Z) = X^2 + Y^2 + Z`	Define $f(x, y, z)$
30	`PRINT "ENTER X-, Y-, Z-COORDINATES OF THE POINT P0."`	Prompt the user for coordinates of a point
40	`INPUT " USE COMMAS BETWEEN ", P(1),P(2),P(3)`	and accept the input
50	`PRINT "ENTER COMPONENTS OF A VECTOR THAT"`	Prompt the user for components of a vector
60	`INPUT" DEFINES THE DIRECTION ", V(1),V(2),V(3)`	and accept the input
70	`DEL = .01`	Define delta value for computing derivatives
80	`DF(1) = (FNF(P(1) + DEL,P(2),P(3) ) - FNF(P(1) -DEL,P(2),P(3) ) )/DEL/2`	Compute $\partial f/\partial x$
90	`DF(2) = (FNF(P(1),P(2)+DEL,P(3) )-FNF(P(1),P(2) -DEL,P(3) ) )/DEL/2`	then $\partial y/\partial x$
100	`DF(3) = (FNF(P(1),P(2),P(3)+DEL)-FNF(P(1),P(2),P(3) -DEL) )/DEL/2`	and $\partial z/\partial x$
110	`PRINT "COMPONENTS OF GRADIENT AT THE POINT"`	Output line for components of gradient vector
120	`PRINT " ARE: ";DF(1);DF(2);DF(3)`	
130	`PRINT`	Print a blank line, cosmetic only
140	`SUM = 0: FOR I = 1 TO 3: SUM = SUM + V(I)^2: NEXT I`	Sum squares for
150	`LENV = SQR(SUM)`	the length of the vector
160	`SUM = 0: FOR I = 1 TO 3: SUM = SUM + DF(I)*V(I)/LENV: NEXT I`	Dot gradient with unit vector
170	`PRINT "DERIVATIVE OF F AT P0 IN"`	Output line for
180	`PRINT " THE DIRECTION OF YOUR VECTOR IS ";SUM`	directional derivative
190	`PRINT`	Print a blank line
200	`SUM = 0: FOR I = 1 TO 3: SUM = SUM + P(I)*DF(I): NEXT I`	Sum products of gradient components and coordinates
210	`PRINT "THE PLANE TANGENT TO THE LEVEL"`	
220	`PRINT "SURFACE THROUGH P0 AT POINT PO IS"`	
230	`PRINT " ";DF(1);"X + ";DF(2);"Y + ";DF(3);"Z = ";SUM`	
240	`PRINT "THE LINE NORMAL TO THE LEVEL"`	
250	`PRINT "SURFACE THROUGH P0 AT POINT PO IS"`	
260	`PRINT " X = ";P(1);" + ";DF(1);"T"`	
270	`PRINT " Y = ";P(2);" + ";DF(2);"T"`	
280	`PRINT " Z = ";P(3);" + ";DF(3);"T"`	
290	`END`	

With $f(x, y, z) = x^2 + y^2 + z$ in line 20, here is the screen output for $P_0 = (1, 2, 4)$, $\mathbf{v} = 2\mathbf{i} - 3\mathbf{j} + 5\mathbf{k}$ and $c = f(1, 2, 4) = (1)^2 + (2)^2 + 4 = 9$. The inputs have been highlighted.

```
ENTER X-, Y-, Z-COORDINATES OF THE POINT, P0
   USE COMMAS BETWEEN        1, 2, 4
ENTER COMPONENTS OF A VECTOR THAT
   DEFINES THE DIRECTION      2, -3, 5
COMPONENTS OF GRADIENT AT THE POINT
   ARE:        1.99995   3.999996   1.000023
DERIVATIVE OF F AT P0 IN
   THE DIRECTION OF YOUR VECTOR IS  -.4866601
TO FIND THE NORMAL LINE AND TANGENT PLANE
   AT P0 ON THE LEVEL SURFACE THROUGH P0,
   ENTER THE VALUE C OF F AT P0      9
THE EQUATION FOR THE PLANE IS
   1.99995 X + 3.999996 Y + 1.000023 Z = 14.00003
THE NORMAL LINE IS
   X = 1 + 1.99995 T
   Y = 2 + 3.999996 T
   Z = 4 + 1.000023 T
```

12.8 Maxima, Minima, and Saddle Points

As we mentioned at the end of Section 12.2, continuous functions defined on closed bounded regions in the xy-plane take on absolute maximum and minimum values on their domains (Figs. 12.53 and 12.54). It is important to be able to find these values and to know where they occur. For example, what is the highest temperature on a heated metal plate and where is it taken on? Where does a given surface attain its highest point above a given patch of the xy-plane? As we shall see in a moment, we can often answer questions like these by examining the partial derivatives of some function.

(Generated by Mathematica)

12.53 The function

$$z = (\cos x)(\cos y)e^{-\sqrt{x^2 + y^2}}$$

has a maximum value of 1 and a minimum value of about -0.067 on the square region $|x| \le 3\pi/2$, $|y| \le 3\pi/2$.

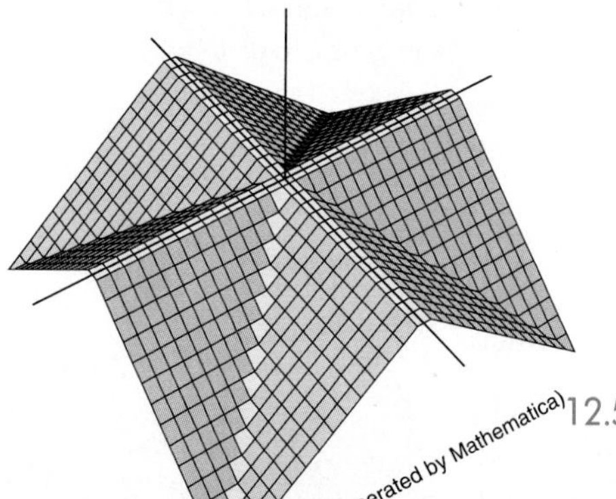

12.54 The "ridge" surface

$$z = \frac{1}{2}\left(\big||x| - |y|\big| - |x| - |y|\right)$$

viewed from the point (10, 15, 20). The defining function has a maximum value of 0 and a minimum value of $-a$ on the square region $|x| \le a$, $|y| \le a$.

Our goal in this section is to show how to use partial derivatives to find the local maximum and minimum values of a continuous function $f(x, y)$ on a region R in the xy-plane. The first step is to use the function's first partial derivatives to make a (usually) short and comprehensive list of points where f can assume its local extreme values. What we do then depends on whether R is closed and bounded. If it is, we appeal to a theorem from advanced calculus that a continuous function on a closed, bounded region R assumes an absolute maximum value and an absolute minimum value on R, and we look through the list to find what these values are. If R is not closed and bounded, the function may not have absolute maximum and minimum values on R. However, we can still try to use the function's second partial derivatives to tell which points on the list, if any, give local maximum and minimum values. The routine is much the same as the routine for functions of a single variable. The main difference is that the first and second derivative tests now involve more derivatives.

The Derivative Tests

To find the extreme values of a continuous function of a single variable, we first look for points where the graph has a horizontal tangent line. At such a point we then look for a local maximum, a local minimum, or a point of inflection. We also examine the values of the function at the boundary points of its domain and at any points where the first derivative does not exist.

For a continuous function $f(x, y)$ of two independent variables, we look for points where the surface $z = f(x, y)$ has a horizontal tangent plane. At such points we then look for a local maximum, a local minimum, or a **saddle point** (Fig. 12.55). We also examine the values of the function at the boundary points in its domain and at any points where either f_x or f_y fails to exist.

We organize the search into three steps:

STEP 1: *Make a list that includes the points where f has its local maxima and minima and evaluate f at all the points on the list.*

As we shall see later, the local maxima and minima of f can occur only at

i) boundary points of R,

ii) interior points of R where $f_x = f_y = 0$ (**the first derivative test**) and points where f_x or f_y fails to exist. (We call these the **critical points** of f.)

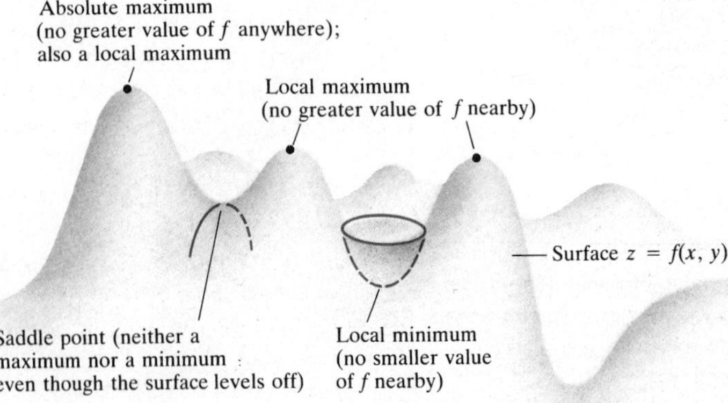

12.55 How values of $f(x, y)$ are classified. Saddle points are the three-dimensional analogues of points of inflection. Local values are often called *relative* values.

Therefore, to make the list, we find the maximum and minimum values of f on the boundary of R and find the values of f at the critical points.

In theory, such a list could be long and might even include infinitely many points, but that is usually not the case in practice and definitely not the case in this book. The lists in the examples and exercises will be short.

STEP 2: *If R is closed and bounded, look through the list for the maximum and minimum values of f. These will be the absolute maximum and minimum values of f on R.*

As we mentioned in the introduction, a function that is continuous on a closed bounded region of the plane has an absolute maximum value on the region and an absolute minimum value on the region. Since absolute maxima and minima are also local maxima and minima, the absolute maximum and minimum values of f already appear somewhere in the list we made in step 1. We have only to glance at the list to see what they are. If we then wish to learn which of the remaining values, if any, are local maxima and minima, we can go on to step 3.

STEP 3: *If R is not closed and bounded, try the following second derivative test.*

The fact that $f_x = f_y = 0$ at an interior point (a, b) of R does not in itself guarantee that f will have an extreme value there. However, if f and its first and second partial derivatives are continuous on R, there is a second derivative test that may identify the behavior of f at (a, b). The **second derivative test** goes like this:

If $f_x(a, b) = f_y(a, b) = 0$, then

i) f has a **local maximum** at (a, b) if $f_{xx} < 0$ and $f_{xx}f_{yy} - f_{xy}^2 > 0$ at (a, b);

ii) f has a **local minimum** at (a, b) if $f_{xx} > 0$ and $f_{xx}f_{yy} - f_{xy}^2 > 0$ at (a, b);

iii) f has a **saddle point** at (a, b) if $f_{xx}f_{yy} - f_{xy}^2 < 0$ at (a, b).

iv) The test is *inconclusive* at (a, b) if $f_{xx}f_{yy} - f_{xy}^2 = 0$ at (a, b). We must find some other way to determine the behavior of f at (a, b).

The expression $f_{xx}f_{yy} - f_{xy}^2$ is called the **discriminant** of f. It is sometimes easier to remember in the determinate form

$$f_{xx}f_{yy} - f_{xy}^2 = \begin{vmatrix} f_{xx} & f_{xy} \\ f_{xy} & f_{yy} \end{vmatrix}.$$

The second derivative test is derived in Section 12.10.

We now look at examples that show these tests at work. After that, we show why the condition $f_x = f_y = 0$ is a necessary condition for having an extreme value at an interior point of the domain of a differentiable function.

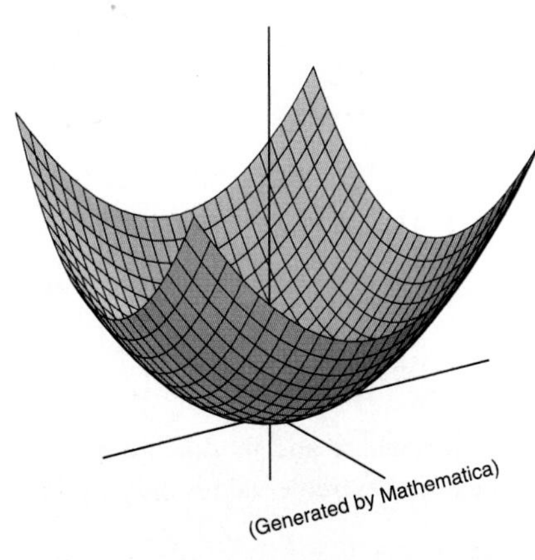

12.56 The graph of the function $f(x, y) = x^2 + y^2$ is the paraboloid $z = x^2 + y^2$. The function has only one extreme value, an absolute minimum value of 0 at the origin (Example 1).

The Tests at Work

In the first example we look at the function $f(x, y) = x^2 + y^2$, whose behavior we already know from looking at the formula: Its value is zero at the origin and increases steadily as (x, y) moves away from the origin. The point of Example 1 is to show how the derivative tests reveal this behavior.

Example 1 Find the extreme values of $f(x, y) = x^2 + y^2$.

Solution The domain of f has no boundary points, for it is the entire plane (Fig. 12.56). The derivatives $f_x = 2x$ and $f_y = 2y$ exist everywhere. Therefore, local maxima and minima can occur only where

$$f_x = 2x = 0 \quad \text{and} \quad f_y = 2y = 0.$$

The only possibility is the origin, where the value of f is zero. Since f is never negative, we see that zero is an absolute minimum.

We have not needed the second derivative test at all. Had we used it, we would have found

$$f_{xx} = 2, \qquad f_{yy} = 2, \qquad f_{xy} = 0$$

and

$$f_{xx} f_{yy} - f_{xy}^2 = (2)(2) - (0)^2 = 4 > 0,$$

identifying (0, 0) as a local minimum. This in itself does not identify (0, 0) as an absolute minimum. It takes more information to do that.

One virtue of the procedure we have described for finding extreme values is that it applies even to functions whose graphs are too complicated to draw. The next example illustrates this point.

Example 2 Find the extreme values of the function

$$f(x, y) = xy - x^2 - y^2 - 2x - 2y + 4.$$

Solution The function is defined and differentiable for all x and y and its domain has no boundary points. The function therefore has extreme values only at the points where f_x and f_y are simultaneously zero. This leads to

$$f_x = y - 2x - 2 = 0, \qquad f_y = x - 2y - 2 = 0,$$

or

$$x = y = -2.$$

Therefore, the point $(-2, -2)$ is the only point where f may take on an extreme value. To see if it does so, we calculate

$$f_{xx} = -2, \qquad f_{yy} = -2, \qquad f_{xy} = 1.$$

The discriminant of f at $(a, b) = (-2, -2)$ is

$$f_{xx} f_{yy} - f_{xy}^2 = (-2)(-2) - (1)^2 = 4 - 1 = 3.$$

The combination

$$f_{xx} < 0 \quad \text{and} \quad f_{xx} f_{yy} - f_{xy}^2 > 0$$

tells us that f has a local maximum at $(-2, -2)$. The value of f at this point is $f(-2, -2) = 8$.

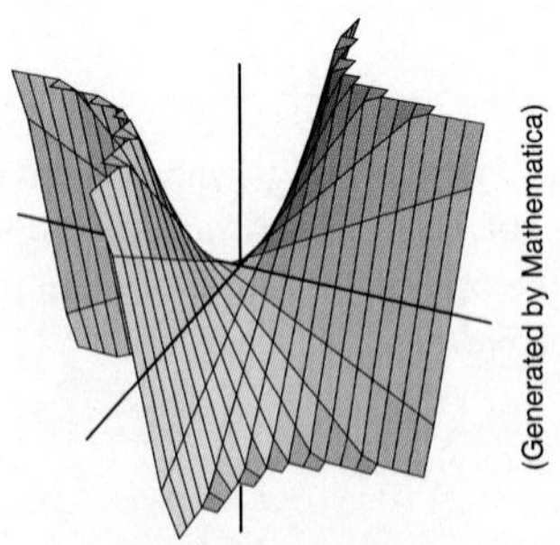

12.57 The surface $z = xy$ has a saddle point at the origin (Example 3).

Example 3 Find the extreme values of $f(x, y) = xy$.

Solution Since the function is differentiable everywhere and its domain has no boundary points (Fig. 12.57), the function can assume extreme values only where

$$f_x = y = 0 \quad \text{and} \quad f_y = x = 0.$$

Thus, the origin is the only point where f might have an extreme value. To see what happens there, we calculate

$$f_{xx} = 0, \qquad f_{yy} = 0, \qquad f_{xy} = 1.$$

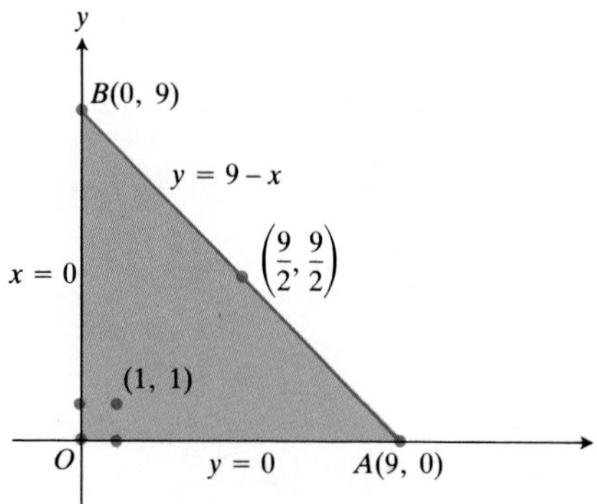

12.58 This triangular plate is the domain of the function in Example 4.

The discriminant,

$$f_{xx}f_{yy} - f_{xy}^2 = -1,$$

is negative. Therefore the function has a saddle point at (0, 0). We conclude that $f(x, y) = xy$ assumes no extreme values at all.

Example 4 Find the absolute maximum and minimum values of

$$f(x, y) = 2 + 2x + 2y - x^2 - y^2$$

on the triangular plate in the first quadrant bounded by the lines $x = 0$, $y = 0$, $y = 9 - x$.

Solution Since f is differentiable, the only places where f can assume extreme values are points on the boundary of the triangle (Fig. 12.58) and points inside the triangle where $f_x = f_y = 0$.

Boundary points. We take the triangle one side at a time:

1. On the segment OA, $y = 0$. The function

   $$f(x, y) = f(x, 0) = 2 + 2x - x^2$$

   may now be regarded as a function of x defined on the closed interval $0 \leq x \leq 9$. Its extreme values (we know from Chapter 3) may occur at the endpoints.

   $$x = 0 \qquad \text{where } f(0, 0) = 2,$$

   $$x = 9 \qquad \text{where } f(9, 0) = 2 + 18 - 81 = -61,$$

   and at the interior points where $f'(x, 0) = 2 - 2x = 0$. The only interior point where $f'(x, 0) = 0$ is $x = 1$, where

   $$f(x, 0) = f(1, 0) = 3.$$

2. On the segment OB, $x = 0$ and

   $$f(x, y) = f(0, y) = 2 + 2y - y^2.$$

   We know from the symmetry of f in x and y and from the analysis we just carried out that the candidates on this segment are

   $$f(0, 0) = 2, \qquad f(0, 9) = -61, \qquad f(0, 1) = 3.$$

3. We have already accounted for the values of f at the endpoints of AB, so we have only to look at the interior points of AB. With

   $$y = 9 - x,$$

   we have

   $$f(x, y) = 2 + 2x + 2(9 - x) - x^2 - (9 - x)^2 = -61 + 18x - 2x^2.$$

   Setting $f'(x, 9 - x) = 18 - 4x = 0$ gives

   $$x = \frac{18}{4} = \frac{9}{2}.$$

At this value of x,

$$y = 9 - \frac{9}{2} = \frac{9}{2}, \qquad \text{and} \qquad f(x, y) = f\left(\frac{9}{2}, \frac{9}{2}\right) = -\frac{41}{2}.$$

Interior points. For these we have

$$f_x = 2 - 2x = 0, \qquad f_y = 2 - 2y = 0,$$

or

$$(x, y) = (1, 1),$$

where

$$f(1, 1) = 4.$$

Summary. We list all the candidates:

$$4, \quad 2, \quad -61, \quad 3, \quad -\frac{41}{2}.$$

The maximum is 4, which f assumes at $(1, 1)$. The minimum is -61, which f assumes at $(0, 9)$ and $(9, 0)$.

The Condition $f_x(a, b) = f_y(a, b) = 0$

The assertion that a function $f(x, y)$ with defined first partial derivatives can have an extreme value at an interior point of its domain only if f_x and f_y are both zero at that point is called the first derivative test for local extreme values.

THEOREM 6

The First Derivative Test for Local Extreme Values

If $f(x, y)$ has a local maximum or local minimum value at an interior point (a, b) of its domain where f_x and f_y are both defined, then f_x and f_y are both zero at (a, b).

Proof Suppose the value of f at (a, b) is a local maximum. Then

1. $x = a$ is an interior point of the domain of the curve $z = f(x, b)$ in which the plane $y = b$ cuts the surface $z = f(x, y)$ (Fig. 12.59);
2. the function $z = f(x, b)$ has a local maximum value at $x = a$;
3. the value of the derivative of $z = f(x, b)$ at $x = a$ is therefore zero (Theorem 1, Section 3.2).

Since this derivative is precisely $f_x(a, b)$, we conclude that $f_x(a, b) = 0$.

A similar argument with the function $z = f(a, y)$ shows that $f_y(a, b) = 0$.

This proves the theorem for local maximum values. To prove it for local minimum values, replace f by $-f$ and run through the argument again.

Despite the power of Theorem 6, we urge you to remember its limitations. It does not apply to points where either f_x or f_y fails to exist. It does not apply to boundary points of a function's domain. At a boundary point, a function can have an extreme value even if one or both of f_x and f_y are different from zero.

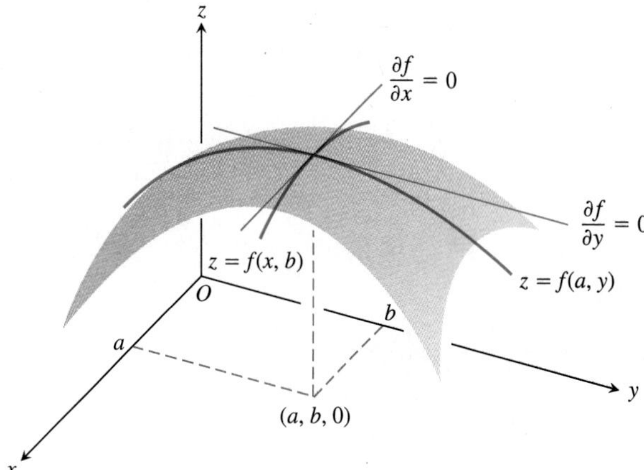

12.59 The maximum of f occurs at $x = a$, $y = b$.

Summary of Max–Min Tests

The extreme values of $f(x, y)$ can occur only at

i) **boundary points** of the domain of f,

ii) **critical points** (interior points where $f_x = f_y = 0$ or points where f_x or f_y fails to exist).

If the first and second order partial derivatives of f are continuous throughout an open region containing a point (a, b) and $f_x(a, b) = f_y(a, b) = 0$, you may be able to classify (a, b) with the **second derivative test:**

i) $f_{xx} < 0$ and $f_{xx}f_{yy} - f_{xy}^2 > 0$ at (a, b) $\Rightarrow$ *local maximum,*

ii) $f_{xx} > 0$ and $f_{xx}f_{yy} - f_{xy}^2 > 0$ at (a, b) $\Rightarrow$ *local minimum,*

iii) $f_{xx}f_{yy} - f_{xy}^2 < 0$ at (a, b) $\Rightarrow$ *saddle point,*

iv) $f_{xx}f_{yy} - f_{xy}^2 = 0$ at (a, b) $\Rightarrow$ test is *inconclusive.*

EXERCISES 12.8

Find all maxima, minima, and saddle points of the functions in Exercises 1–30. Which, if any, of the maxima and minima are absolute?

1. $f(x, y) = x^2 + xy + y^2 + 3x - 3y + 4$

2. $f(x, y) = x^2 + 3xy + 3y^2 - 6x + 3y - 6$

3. $f(x, y) = 5xy - 7x^2 + 3x - 6y + 2$

4. $f(x, y) = 2xy - 5x^2 - 2y^2 + 4x + 4y - 4$

5. $f(x, y) = x^2 + xy + 3x + 2y + 5$

6. $f(x, y) = y^2 + xy - 2x - 2y + 2$

7. $f(x, y) = 2xy - 5x^2 - 2y^2 + 4x - 4$

8. $f(x, y) = 2xy - x^2 - 2y^2 + 3x + 4$

9. $f(x, y) = x^2 - 4xy + y^2 + 6y + 2$

10. $f(x, y) = 3x^2 + 6xy + 7y^2 - 2x + 4y$

11. $f(x, y) = 2x^2 + 3xy + 4y^2 - 5x + 2y$

12. $f(x, y) = 4x^2 - 6xy + 5y^2 - 20x + 26y$

13. $f(x, y) = x^2 - 4xy + y^2 + 5x - 2y$

14. $f(x, y) = x^2 - y^2 - 2x + 4y + 6$

15. $f(x, y) = x^2 - 2xy + 2y^2 - 2x + 2y + 1$

16. $f(x, y) = x^2 + 2xy$

17. $f(x, y) = 3 + 2x + 2y - 2x^2 - 2xy - y^2$

18. $f(x, y) = x^2 + xy + y^2 + x - 4y + 5$

19. $f(x, y) = x^3 - y^3 - 2xy + 6$

20. $f(x, y) = x^3 + y^3 + 3x^2 - 3y^2 - 8$

21. $f(x, y) = 6x^2 - 2x^3 + 3y^2 + 6xy$

22. $f(x, y) = 9x^3 + y^3/3 - 4xy$

23. $f(x, y) = x^3 + 3xy + y^3$

24. $f(x, y) = 4xy - x^4 - y^4$

25. $f(x, y) = \dfrac{1}{x^2 + y^2 - 1}$ **26.** $f(x, y) = \dfrac{1}{x} + xy + \dfrac{1}{y}$

27. $f(x, y) = \dfrac{-4}{1 + x^2 + y^2}$ **28.** $f(x, y) = \displaystyle\int_{x^2+2}^{y^2+5} \sqrt{t - 1}\, dt$

29. $f(x, y) = y \sin x$ **30.** $f(x, y) = e^{2x} \cos y$

In Exercises 31–38, find the absolute maxima and minima of the functions on the given domains.

31. $f(x, y) = 2x^2 - 4x + y^2 - 4y + 1$ on the closed triangular plate bounded by the lines $x = 0$, $y = 2$, $y = 2x$ in the first quadrant

32. $D(x, y) = x^2 - xy + y^2 + 1$ on the closed triangular plate in the first quadrant bounded by the lines $x = 0$, $y = 4$, $y = x$

33. $f(x, y) = x^2 + y^2$ on the closed triangular plate bounded by the lines $x = 0$, $y = 0$, $y + 2x = 2$ in the first quadrant

34. $T(x, y) = x^2 + xy + y^2 - 6x$ on the rectangular plate $0 \leq x \leq 5$, $-3 \leq y \leq 3$

35. $T(x, y) = x^2 + xy + y^2 - 6x + 2$ on the rectangular plate $0 \leq x \leq 5$, $-3 \leq y \leq 0$

36. $f(x, y) = 48xy - 32x^3 - 24y^2$ on the rectangular plate $0 \leq x \leq 1$, $0 \leq y \leq 1$

37. $f(x, y) = (4x - x^2)\cos y$ on the rectangular plate $1 \leq x \leq 3$, $-\pi/4 \leq y \leq \pi/4$ (Fig. 12.60)

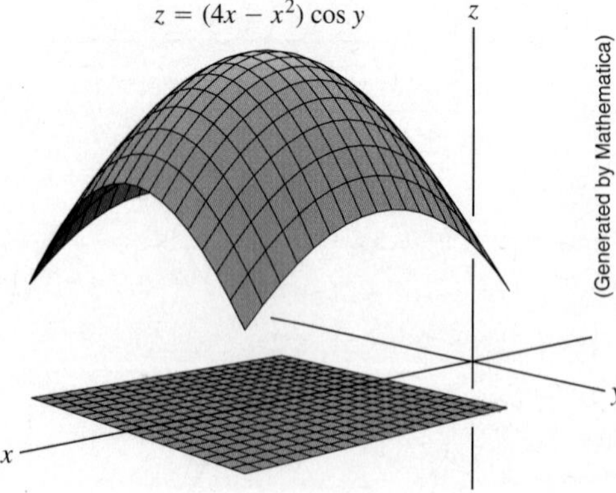

12.60 The function and domain in Exercise 37.

38. $f(x, y) = 4x - 8xy + 2y + 1$ on the triangular plate bounded by the lines $x = 0$, $y = 0$, $x + y = 1$ in the first quadrant

39. *Temperatures.* A flat circular plate has the shape of the region $x^2 + y^2 \leq 1$. The plate, including the boundary where $x^2 + y^2 = 1$, is heated so that the temperature at the point (x, y) is

$$T(x, y) = x^2 + 2y^2 - x.$$

Find the temperatures at the hottest and coldest points on the plate (Fig. 12.61).

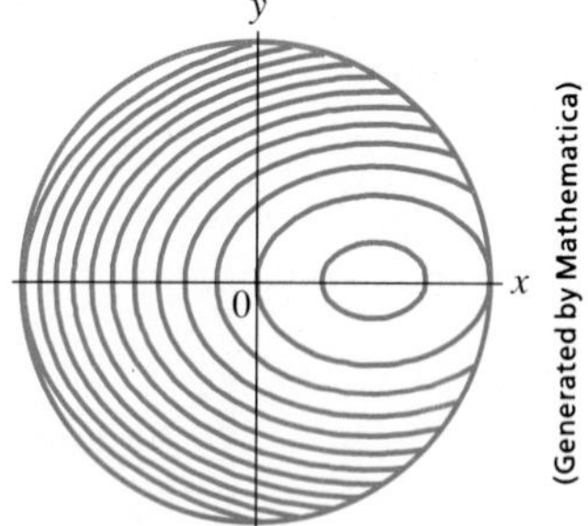

12.61 Curves of constant temperature are called isotherms. The figure shows isotherms of the temperature function $T(x, y) = x^2 + 2y^2 - x$ on the disk $x^2 + y^2 \leq 1$ in the xy-plane. Exercise 39 asks you to locate the extreme temperatures.

40. Find the critical point of

$$f(x, y) = xy + 2x - \ln x^2 y$$

in the open first quadrant ($x > 0$, $y > 0$) and show that f takes on a minimum there (Fig. 12.62).

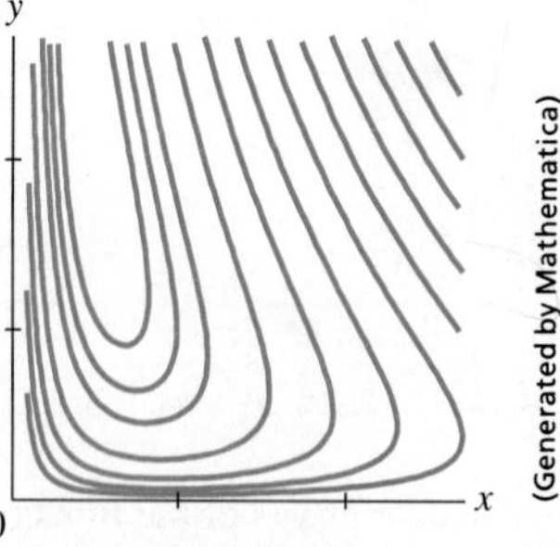

12.62 The function $f(x, y) = xy + 2x - \ln x^2 y$ takes on a minimum value somewhere in the open first quadrant $x > 0$, $y > 0$. Exercise 40 asks you to find where.

41. Find the maxima, minima, and saddle points of $f(x, y)$, if any, given that

a) $f_x = 2x - 4y$ and $f_y = 2y - 4x$

b) $f_x = 2x - 2$ and $f_y = 2y - 4$

c) $f_x = 9x^2 - 9$ and $f_y = 2y + 4$

42. The discriminant $f_{xx}f_{yy} - f_{xy}^2$ is zero at the origin for each of the following functions, so the second derivative test fails there. Determine whether the function has a maximum, a minimum, or neither at the origin by imagining what the surface $z = f(x, y)$ looks like.

a) $f(x, y) = x^2y^2$ b) $f(x, y) = 1 - x^2y^2$
c) $f(x, y) = xy^2$ d) $f(x, y) = x^3y^2$
e) $f(x, y) = x^3y^3$ f) $f(x, y) = x^4y^4$

Extreme Values on Parametrized Curves

To find the extreme values of a function $f(x, y)$ on a curve $x = g(t)$, $y = h(t)$, we treat f as a function of the single variable t and use the Chain Rule to find where df/dt is zero. As in any other single-variable case, the extreme values of f are then found among the values at the

a) critical points (points where df/dt is zero or fails to exist), and
b) endpoints of the parameter domain.

43. Find the absolute maximum and minimum values of the following functions on

i) the quarter-circle $x^2 + y^2 = 4$ in the first quadrant,
ii) the half-circle $x^2 + y^2 = 4$, $y \geq 0$, and
iii) the full circle $x^2 + y^2 = 4$.

a) $f(x, y) = xy$ b) $f(x, y) = x + y$
c) $f(x, y) = 2x^2 + y^2$

Use the parametrization $x = 2\cos t$, $y = 2\sin t$.

44. Find the absolute maximum and minimum values, if any, of the function $f(x, y) = xy$ on the following.

a) The line $x = 2t$, $y = t + 1$
b) The line segment $x = 2t$, $y = t + 1$, $-1 \leq t \leq 0$
c) The line segment $x = 2t$, $y = t + 1$, $0 \leq t \leq 1$

45. Find the absolute maximum and minimum values of the following functions on

i) the quarter-ellipse $(x^2/9) + (y^2/4) = 1$ in the first quadrant,
ii) the half-ellipse $(x^2/9) + (y^2/4) = 1$, $y \geq 0$, and
iii) the full ellipse $(x^2/9) + (y^2/4) = 1$.

a) $f(x, y) = x^2 + 3y^2$ b) $f(x, y) = 2x + 3y$

46. Find the absolute maximum and minimum values of $f(x, y) = xy$ on the ellipse $x^2 + 4y^2 = 8$.

Least Squares and Regression Lines

When we try to fit a line $y = mx + b$ to a set of numerical data points $(x_1, y_1), (x_2, y_2), \ldots, (x_n, y_n)$ (Fig. 12.63), we usually choose the line that minimizes the sum of the squares of the vertical distances from the points to the line. In theory, this means finding the values of m and b that minimize the value of the function

$$w = (mx_1 + b - y_1)^2 + \cdots + (mx_n + b - y_n)^2. \qquad (1)$$

The values of m and b that do this are found with the first and second derivative tests to be

$$m = \frac{\left(\sum x_k\right)\left(\sum y_k\right) - n\sum x_k y_k}{\left(\sum x_k\right)^2 - n\sum x_k^2}, \qquad (2)$$

$$b = \frac{1}{n}\left(\sum y_k - m\sum x_k\right), \qquad (3)$$

with all sums running from $k = 1$ to $k = n$. Many scientific calculators have these formulas built in, enabling you to find m and b with only a few key presses after you have entered the data.

The line $y = mx + b$ determined by these values of m and b is called the **least squares line** or **regression line** for the data under study. Finding a least squares line lets you

1. summarize data with a simple expression,
2. predict values of y for other, experimentally untried values of x,
3. handle data analytically.

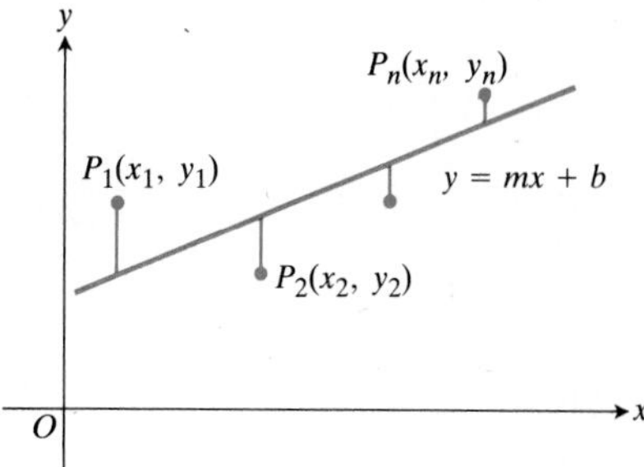

12.63 To fit a line to noncollinear points, we choose the line that minimizes the sum of the squares of the deviations.

EXAMPLE Find the least squares line for the points (0, 1), (1, 3), (2, 2), (3, 4), (4, 5).

Solution We organize the calculations in a table:

k	x_k	y_k	x_k^2	$x_k y_k$
1	0	1	0	0
2	1	3	1	3
3	2	2	4	4
4	3	4	9	12
5	4	5	16	20
Σ	10	15	30	39

Then we find

$$m = \frac{(10)(15) - 5(39)}{(10)^2 - 5(30)} = 0.9 \qquad \left(\text{Eq.(2) with } n = 5 \text{ and data from the table}\right)$$

and use the value of m to find

$$b = \frac{1}{5}(15 - (0.9)(10)) = 1.2. \qquad \left(\text{Eq. (3) with } n = 5, m = 0.9\right)$$

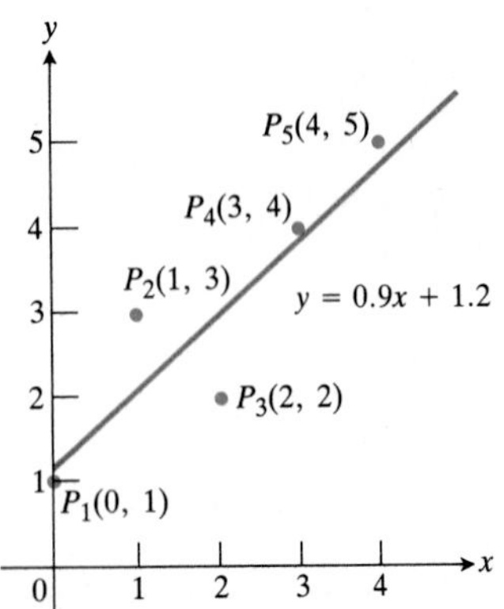

12.64 The least squares line for the data in the example.

The least squares line is $y = 0.9x + 1.2$ (Fig. 12.64).

In Exercises 47–50, use Eqs. (2) and (3) to find the least squares line for each set of data points. Then use the linear equation you obtain to predict the value of y that would correspond to $x = 4$.

47. $(-1, 2)$, $(0, 1)$, $(3, -4)$

48. $(-2, 0)$, $(0, 2)$, $(2, 3)$

49. $(0, 0)$, $(1, 2)$, $(2, 3)$

50. $(0, 1)$, $(2, 2)$, $(3, 2)$

51. CALCULATOR Write a linear equation for the effect of irrigation on the yield of alfalfa by fitting a least squares line to the data in Table 12.1 (from the University of California Experimental Station, *Bulletin* No. 450, p. 8). Plot the data and draw the line.

52. CALCULATOR *Craters on Mars.* One theory of crater formation suggests that the frequency of large craters should fall off as the square of the diameter (Marcus, *Science,* June 21, 1968, p. 1334). Pictures from Mariner IV show the frequencies listed in Table 12.2. Fit a line of the form $F = m(1/D^2) + b$ to the data. Plot the data and draw the line.

TABLE 12.1
Growth of alfalfa

x (total seasonal depth of water applied, in.)	y (average alfalfa yield, tons/acre)
12	5.27
18	5.68
24	6.25
30	7.21
36	8.20
42	8.71

TABLE 12.2
Crater sizes on Mars

Diameter in km, D	$1/D^2$ (for left value of class interval)	Frequency, F
32–45	0.001	53
45–64	0.0005	22
64–90	0.00025	14
90–128	0.000125	3

53. CALCULATOR *Köchel numbers.* In 1862, the German musicologist Ludwig von Köchel made a chronological list of the musical works of Wolfgang Amadeus Mozart. This list is the source of the Köchel numbers, or "K numbers," that now accompany the titles of Mozart's pieces (Sinfonia Concertante in E-flat major, K.364, for example). Table 12.3 gives the Köchel numbers and composition dates (y) of ten of Mozart's works.

a) Plot y vs. K to show that y is close to being a linear function of K.

b) Find a least squares line $y = m\text{K} + b$ for the data and add the line to your plot in (a).

c) Use the least squares line to estimate the year in which the Sinfonia Concertante was composed.

TABLE 12.3
Compositions by Mozart

Köchel number K	Year Composed y
1	1761
75	1771
155	1772
219	1775
271	1777
351	1780
425	1783
503	1786
575	1789
626	1791

54. CALCULATOR *Submarine sinkings.* The data in Table 12.4 show the results of an historical study of German submarines sunk by the U.S. Navy during 16 consecutive months of World War II. The data given for each month are the number of reported sinkings and the number of actual sinkings. The number of submarines sunk was slightly greater than the Navy's reports implied. Find a least squares line for estimating the number of actual sinkings from the number of reported sinkings.

TABLE 12.4
Sinkings of German submarines by U.S. during 16 consecutive months of WWII

Month	**Guesses by U.S. (reported sinkings)** x	**Actual number** y
1	3	3
2	2	2
3	4	6
4	2	3
5	5	4
6	5	3
7	9	11
8	12	9
9	8	10
10	13	16
11	14	13
12	3	5
13	4	6
14	13	19
15	10	15
16	16	15
	123	140

12.9 Lagrange Multipliers

Skip

As we saw in Section 12.8, we sometimes need to find the maximum and minimum values of functions whose domains are constrained to lie within some particular subset of the plane—a disk, for example, or a closed triangular region. But, as Fig. 12.65 suggests, functions may be subject to other kinds of constraints as well.

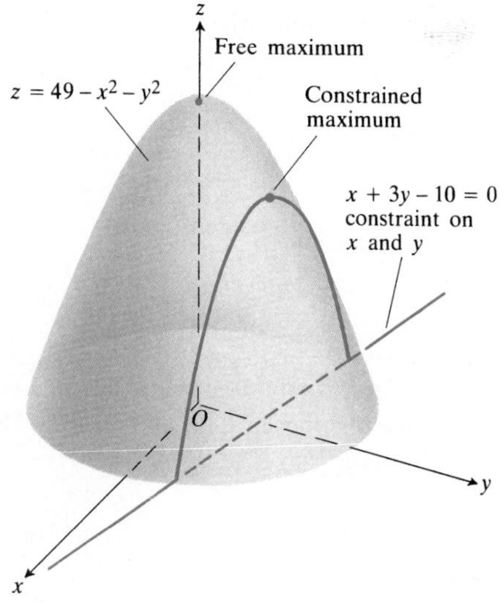

12.65 The function $f(x, y) = 49 - x^2 - y^2$, subject to the constraint $g(x, y) = x + 3y - 10 = 0$.

In this section, we explore a powerful method for finding the maxima and minima of constrained functions: the method of **Lagrange multipliers.** Lagrange developed the method in 1755 to solve sophisticated max–min problems in geometry. Today the method is important in economics, in engineering (where it is used in designing multistage rockets, for example), and in mathematics itself.

We begin with two examples and then describe the method in general terms and look at more examples.

Constrained Maxima and Minima

Example 1 Find the point $P(x, y, z)$ on the plane

$$2x + y - z - 5 = 0$$

that lies closest to the origin.

Solution The problem asks us to find the minimum value of the function

$$|\overrightarrow{OP}| = \sqrt{(x-0)^2 + (y-0)^2 + (z-0)^2} = \sqrt{x^2 + y^2 + z^2}$$

subject to the constraint that

$$2x + y - z - 5 = 0.$$

Since $|\overrightarrow{OP}|$ has a minimum value wherever the function

$$f(x, y, z) = x^2 + y^2 + z^2$$

has a minimum value, we may solve the problem by finding the minimum value of $f(x, y, z)$ subject to the constraint $2x + y - z - 5 = 0$. If we regard x and y as the independent variables in this equation and write z as

$$z = 2x + y - 5,$$

our problem reduces to one of finding the points (x, y) at which the function

$$h(x, y) = f(x, y, 2x + y - 5) = x^2 + y^2 + (2x + y - 5)^2$$

has its minimum value or values. Since the domain of h is the entire xy-plane, the first derivative test of Section 12.8 tells us that any minima that h might have must occur at points where

$$h_x = 2x + 2(2x + y - 5)(2) = 0, \qquad h_y = 2y + 2(2x + y - 5) = 0.$$

This leads to

$$10x + 4y = 20, \qquad 4x + 4y = 10,$$

and the solution

$$x = \frac{5}{3}, \qquad y = \frac{5}{6}.$$

We may apply a geometric argument together with the second derivative test to show that these values minimize h. The z-coordinate of the corresponding point on the plane $z = 2x + y - 5$ is

$$z = 2\left(\frac{5}{3}\right) + \frac{5}{6} - 5 = -\frac{5}{6}.$$

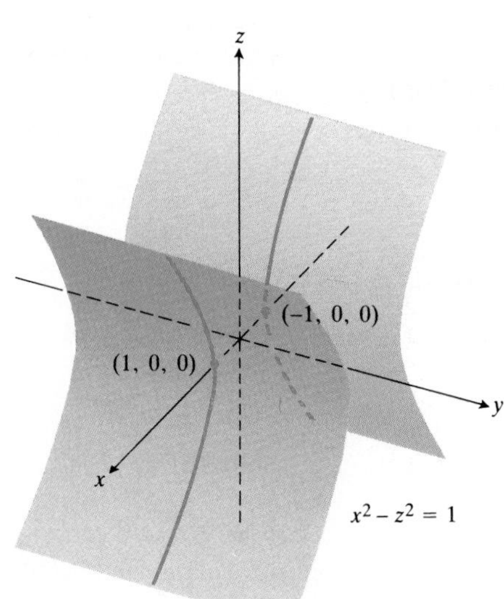

12.66 The hyperbolic cylinder $x^2 - z^2 - 1 = 0$ in Example 2.

Therefore, the point we seek is

$$\text{Closest point:} \qquad P\left(\frac{5}{3}, \frac{5}{6}, -\frac{5}{6}\right).$$

Attempts to solve a constrained maximum or minimum problem by substitution, as we might call the method of Example 1, do not always go smoothly, as the next example shows. This is one of the reasons for learning the new method of this section, which does not require us to decide in advance which of the constrained variables to regard as independent.

Example 2 Find the points on the hyperbolic cylinder

$$x^2 - z^2 - 1 = 0$$

closest to the origin.

Solution 1 The cylinder is shown in Fig. 12.66. We seek the points on the cylinder closest to the origin. These are the points whose coordinates minimize the value of the function

$$f(x, y, z) = x^2 + y^2 + z^2 \qquad \text{(Square of the distance)}$$

subject to the constraint that $x^2 - z^2 - 1 = 0$. If we regard x and y as independent variables in the constraint equation, then

$$z^2 = x^2 - 1$$

and the values of $f(x, y, z) = x^2 + y^2 + z^2$ on the cylinder are given by the function

$$h(x, y) = x^2 + y^2 + (x^2 - 1) = 2x^2 + y^2 - 1.$$

To find the points on the cylinder whose coordinates minimize f, we look for the points in the xy-plane whose coordinates minimize h. The only extreme value of h occurs where

$$h_x = 4x = 0 \qquad \text{and} \qquad h_y = 2y = 0,$$

that is, at the point (0, 0). But now we're in trouble—there are no points on the cylinder where both x and y are zero. What went wrong?

What happened was that the first derivative test found (as it should have) the point *in the domain of h* where h has a minimum value. We, on the other hand, want the points *on the cylinder* where h has a minimum value. While the domain of h is the entire xy-plane, the domain from which we can select the first two coordinates of the points (x, y, z) on the cylinder is restricted to the "shadow" of the cylinder on the xy-plane; it does not include the band between the lines $x = -1$ and $x = 1$ (Fig. 12.67).

We can avoid this problem if we treat y and z as independent variables (instead of x and y) and express x in terms of y and z as

$$x^2 = z^2 + 1.$$

With this substitution, $f(x, y, z) = x^2 + y^2 + z^2$ becomes

$$k(y, z) = (z^2 + 1) + y^2 + z^2 = 1 + y^2 + 2z^2$$

and we look for the points where k takes on its smallest value. The domain of k in

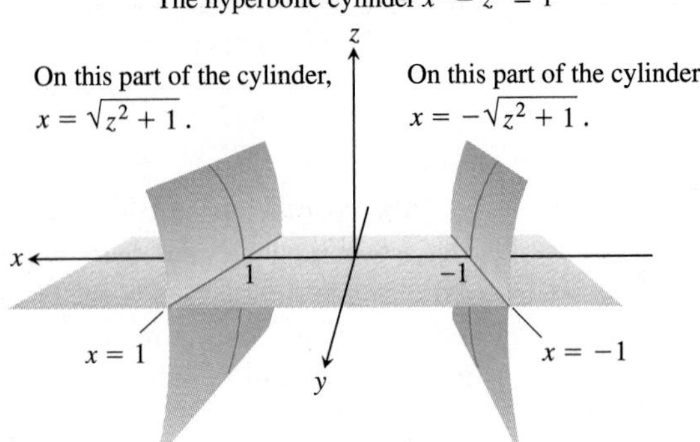

12.67 The region in the xy-plane from which the first two coordinates of the points (x, y, z) on the hyperbolic cylinder $x^2 - z^2 = 1$ are selected excludes the band $-1 < x < 1$ in the xy-plane.

the yz-plane now matches the domain from which we select the y- and z-coordinates of the points (x, y, z) on the cylinder. Hence, the points that minimize k in the plane will have corresponding points on the cylinder. The smallest values of k occur where

$$k_y = 2y = 0 \qquad \text{and} \qquad k_z = 4z = 0,$$

or where $y = z = 0$. This leads to

$$x^2 = z^2 + 1 = 1, \qquad x = \pm 1.$$

The corresponding points on the cylinder are $(\pm 1, 0, 0)$. We can see from the inequality

$$k(y, z) = 1 + y^2 + 2z^2 \geq 1$$

that the points $(\pm 1, 0, 0)$ give a minimum value for k. We can also see that the minimum distance from the origin to a point on the cylinder is 1 unit.

Solution 2 Another way to find the points on the cylinder closest to the origin is to imagine a small sphere centered at the origin expanding like a soap bubble until it just touches the cylinder (Fig. 12.68). At each point of contact, the cylinder and sphere have the same tangent plane and normal line. Therefore, if the sphere and cylinder are represented as the level surfaces obtained by setting

$$f(x, y, z) = x^2 + y^2 + z^2 - a^2 \qquad \text{and} \qquad g(x, y, z) = x^2 - z^2 - 1$$

equal to 0, then the gradients ∇f and ∇g will be parallel where the surfaces touch. At any point of contact we will therefore be able to find a scalar λ ("lambda") such that

$$\nabla f = \lambda \nabla g,$$

or

$$2x\mathbf{i} + 2y\mathbf{j} + 2z\mathbf{k} = \lambda(2x\mathbf{i} - 2z\mathbf{k}).$$

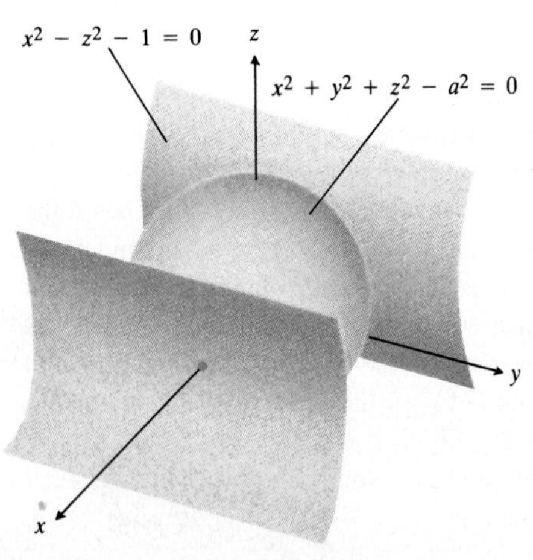

12.68 The sphere obtained by expanding a soap bubble centered at the origin until it just touches the hyperbolic cylinder $x^2 - z^2 - 1 = 0$. See Solution 2 of Example 2.

Thus, the coordinates x, y, and z of any point of tangency will have to satisfy the three scalar equations

$$2x = 2\lambda x, \qquad 2y = 0, \qquad 2z = -2\lambda z. \tag{1}$$

For what values of λ will a point (x, y, z) whose coordinates satisfy the equations in (1) also lie on the surface $x^2 - z^2 - 1 = 0$? To answer this question, we use the fact that no point on the surface has a zero x-coordinate to conclude that $x \neq 0$ in the first equation in (1). This means that $2x = 2\lambda x$ only if

$$2 = 2\lambda \qquad \text{or} \qquad \lambda = 1.$$

For $\lambda = 1$, the equation $2z = -2\lambda z$ becomes $2z = -2z$. If this equation is to be satisfied as well, z must be zero. Since $y = 0$ also (from the equation $2y = 0$), we conclude that the points we seek all have coordinates of the form

$$(x, 0, 0).$$

What points on the surface $x^2 - z^2 = 1$ have coordinates of this form? The points $(x, 0, 0)$ for which

$$x^2 - (0)^2 = 1, \qquad x^2 = 1, \qquad \text{or} \qquad x = \pm 1.$$

The points on the cylinder closest to the origin are the points $(\pm 1, 0, 0)$.

The Method of Lagrange Multipliers

In Solution 2 of Example 2, we solved the problem by the **method of Lagrange multipliers.** In general terms, the method says that the extreme values of a function $f(x, y, z)$ whose variables are subject to a constraint $g(x, y, z) = 0$ are to be found on the surface $g = 0$ at the points where

$$\nabla f = \lambda \nabla g$$

for some scalar λ (called a **Lagrange multiplier**).

To explore the method further and see why it works, we first make the following observation, which we state as a theorem.

THEOREM 7

The Orthogonal Gradient Theorem

Suppose that $f(x, y, z)$ is differentiable in a region whose interior contains a differentiable curve

$$C: \quad \mathbf{r} = g(t)\,\mathbf{i} + h(t)\,\mathbf{j} + k(t)\,\mathbf{k}.$$

If P_0 is a point on C where f has a local maximum or minimum relative to its values on C, then ∇f is orthogonal to C at P_0.

Proof We show that ∇f is orthogonal to the curve's velocity vector at P_0. The values of f on C are given by the composite $f(g(t), h(t), k(t))$, whose derivative with respect to t is

$$\frac{df}{dt} = \frac{\partial f}{\partial x}\frac{dg}{dt} + \frac{\partial f}{\partial y}\frac{dh}{dt} + \frac{\partial f}{\partial z}\frac{dk}{dt} = \nabla f \cdot \mathbf{v}.$$

At any point P_0 where f has a local maximum or minimum relative to its values on the curve, $df/dt = 0$, so

$$\nabla f \cdot \mathbf{v} = 0.$$

By dropping the z-terms in Theorem 7, we obtain a similar result for functions of two variables.

COROLLARY OF THEOREM 7

At the points on a differentiable curve $\mathbf{r} = g(t)\,\mathbf{i} + h(t)\,\mathbf{k}$ where a differentiable function $f(x, y)$ takes on its local maxima and minima relative to its values on the curve, $\nabla f \cdot \mathbf{v} = 0$.

Theorem 7 is the key to why the method of Lagrange multipliers works, as we shall now see. Suppose that $f(x, y, z)$ and $g(x, y, z)$ are differentiable and that P_0 is a point on the surface $g(x, y, z) = 0$ where f has a local maximum or minimum value relative to its other values on the surface. Then f takes on a local maximum or minimum at P_0 relative to its values on every differentiable curve through P_0 on the surface $g(x, y, z) = 0$. Therefore, ∇f is orthogonal to the velocity vector of every such differentiable curve through P_0. But so is ∇g (because ∇g is orthogonal to the level surface $g = 0$, as we saw in Section 12.7). Therefore, at P_0, ∇f is some scalar multiple λ of ∇g.

The Method of Lagrange Multipliers

Suppose that $f(x, y, z)$ and $g(x, y, z)$ are differentiable. To find the local maximum and minimum values of f subject to the constraint $g(x, y, z) = 0$, find the values of x, y, z, and λ that simultaneously satisfy the equations

$$\nabla f = \lambda \nabla g \quad \text{and} \quad g(x, y, z) = 0. \tag{2}$$

For functions of two independent variables, the appropriate equations are

$$\nabla f = \lambda \nabla g \quad \text{and} \quad g(x, y) = 0. \tag{3}$$

Example 3 Find the greatest and smallest values that the function

$$f(x, y) = xy$$

takes on the ellipse (Fig. 12.69)

$$\frac{x^2}{8} + \frac{y^2}{2} = 1.$$

Solution We are asked to find the extreme values of $f(x, y) = xy$ subject to the constraint

$$g(x, y) = \frac{x^2}{8} + \frac{y^2}{2} - 1 = 0.$$

To do so, we first find the values of x, y, and λ for which

$$\nabla f = \lambda \nabla g \quad \text{and} \quad g(x, y) = 0.$$

The gradient equation gives

$$y\mathbf{i} + x\mathbf{j} = \frac{\lambda}{4}x\mathbf{i} + \lambda y\mathbf{j},$$

from which we find

$$y = \frac{\lambda}{4}x, \qquad x = \lambda y, \qquad \text{and} \qquad y = \frac{\lambda}{4}(\lambda y) = \frac{\lambda^2}{4}y,$$

so that $y = 0$ or $\lambda = \pm 2$. We now consider these two cases.

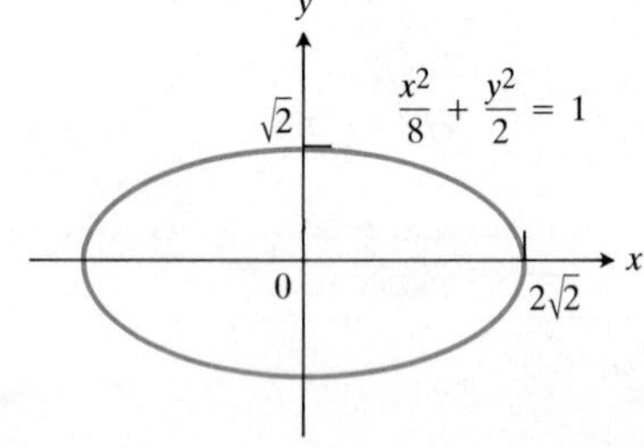

12.69 Example 3 shows how to find the largest and smallest values of the product xy on this ellipse.

CASE 1: If $y = 0$, then $x = y = 0$, but the point $(0, 0)$ is not on the ellipse. Hence, $y \neq 0$.

CASE 2: If $y \neq 0$, then $\lambda = \pm 2$ and $x = \pm 2y$. Substituting this in the equation $g(x, y) = 0$ gives

$$\frac{(\pm 2y)^2}{8} + \frac{y^2}{2} = 1, \qquad 4y^2 + 4y^2 = 8, \qquad \text{and} \qquad y = \pm 1.$$

The function $f(x, y) = xy$ therefore takes on its extreme values on the ellipse at the four points $(\pm 2, 1)$, $(\pm 2, -1)$. The extreme values are $xy = 2$ and $xy = -2$.

The Geometry of the Solution The level curves of the function $f(x, y) = xy$ are the hyperbolas $xy = c$ (Fig. 12.70). The farther the hyperbolas lie from the origin, the larger the absolute value of f. We want to find the extreme values of $f(x, y)$, given that the point (x, y) also lies on the ellipse $x^2 + 4y^2 = 8$. Which hyperbolas intersecting the ellipse lie farthest from the origin? The hyperbolas that just graze the ellipse, the ones that are tangent to it. At these points, any vector normal to the hyperbola is normal to the ellipse, so the gradient $\nabla f = y\mathbf{i} + x\mathbf{j}$ is a multiple $(\lambda = \pm 2)$ of the gradient $\nabla g = (x/4)\,\mathbf{i} + y\mathbf{j}$. At the point $(2, 1)$, for example,

$$\nabla f = \mathbf{i} + 2\,\mathbf{j}, \qquad \nabla g = \frac{1}{2}\mathbf{i} + \mathbf{j}, \qquad \text{and} \qquad \nabla f = 2\nabla g.$$

At the point $(-2, 1)$,

$$\nabla f = \mathbf{i} - 2\,\mathbf{j}, \qquad \nabla g = -\frac{1}{2}\mathbf{i} + \mathbf{j}, \qquad \text{and} \qquad \nabla f = -2\nabla g.$$

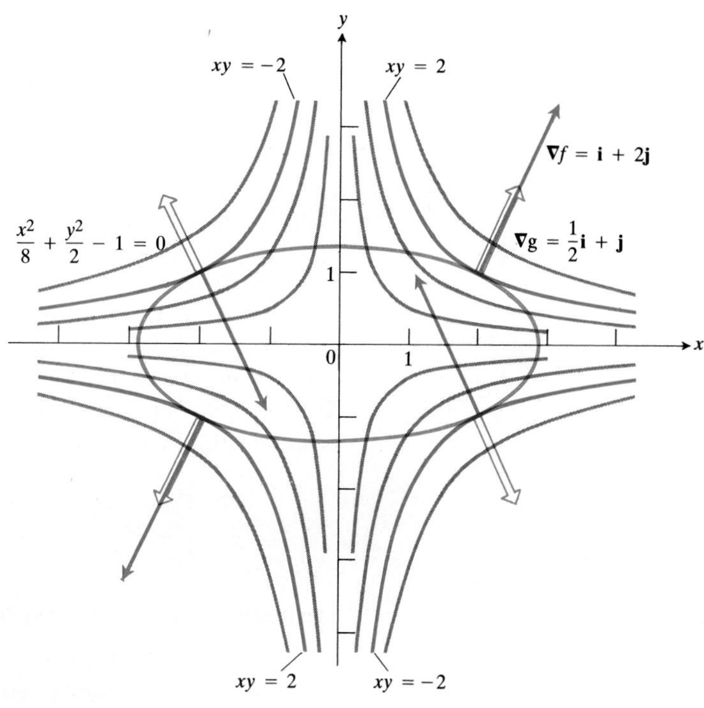

12.70 When subjected to the constraint $g(x, y) = x^2/8 + y^2/2 - 1 = 0$, the function $f(x, y) = xy$ takes on extreme values at the four points $(\pm 2, \pm 1)$. These are the points on the ellipse where ∇f (red) is a scalar multiple of ∇g (blue).

Example 4 Find the maximum and minimum values of the function $f(x, y) = 3x + 4y$ on the circle $x^2 + y^2 = 1$.

Solution We model this as a Lagrange multiplier problem with

$$f(x, y) = 3x + 4y, \qquad g(x, y) = x^2 + y^2 - 1$$

and look for the values of x, y, and λ that satisfy the equations

$$\begin{aligned} \nabla f = \lambda \nabla g: &\quad 3\mathbf{i} + 4\mathbf{j} = 2x\lambda\mathbf{i} + 2y\lambda\mathbf{j}, \\ g(x, y) = 0: &\quad x^2 + y^2 - 1 = 0. \end{aligned} \tag{4}$$

The gradient equation implies that $\lambda \neq 0$ and gives

$$x = \frac{3}{2\lambda}, \qquad y = \frac{2}{\lambda}.$$

These equations tell us, among other things, that x and y have the same sign. With these values for x and y, the equation $g(x, y) = 0$ gives

$$\left(\frac{3}{2\lambda}\right)^2 + \left(\frac{2}{\lambda}\right)^2 - 1 = 0,$$

so

$$\frac{9}{4\lambda^2} + \frac{4}{\lambda^2} = 1, \qquad 9 + 16 = 4\lambda^2, \qquad 4\lambda^2 = 25, \qquad \text{and} \qquad \lambda = \pm\frac{5}{2}.$$

Thus,

$$x = \frac{3}{2\lambda} = \pm\frac{3}{5}, \qquad y = \frac{2}{\lambda} = \pm\frac{4}{5},$$

and $f(x, y) = 3x + 4y$ has extreme values at $(x, y) = \pm(3/5, 4/5)$. (There are two points instead of four because x and y have the same sign.)

By calculating the value of $3x + 4y$ at the points $\pm(3/5, 4/5)$, we see that its maximum and minimum values on the circle $x^2 + y^2 = 1$ are

$$3\left(\frac{3}{5}\right) + 4\left(\frac{4}{5}\right) = \frac{25}{5} = 5 \qquad \text{and} \qquad 3\left(-\frac{3}{5}\right) + 4\left(-\frac{4}{5}\right) = -\frac{25}{5} = -5.$$

The Geometry of the Solution (Fig. 12.71) The level curves of $f(x, y) = 3x + 4y$ are the lines $3x + 4y = c$. The farther the lines lie from the origin, the larger the absolute value of f. We want to find the extreme values of $f(x, y)$ given that the point (x, y) also lies on the circle $x^2 + y^2 = 1$. Which lines intersecting the circle lie farthest from the origin? The lines tangent to the circle. At the points of tangency, any vector normal to the line is normal to the circle, so the gradient $\nabla f = 3\mathbf{i} + 4\mathbf{j}$ is a multiple ($\lambda = \pm 5/2$) of the gradient $\nabla g = 2x\mathbf{i} + 2y\mathbf{j}$. At the point $(3/5, 4/5)$, for example,

$$\nabla f = 3\mathbf{i} + 4\mathbf{j}, \qquad \nabla g = \frac{6}{5}\mathbf{i} + \frac{8}{5}\mathbf{j}, \qquad \text{and} \qquad \nabla f = \frac{5}{2}\nabla g.$$

Lagrange Multipliers with Two Constraints

In many applied problems we need to find the extreme values of a differentiable function $f(x, y, z)$ whose variables are subject to two constraints.

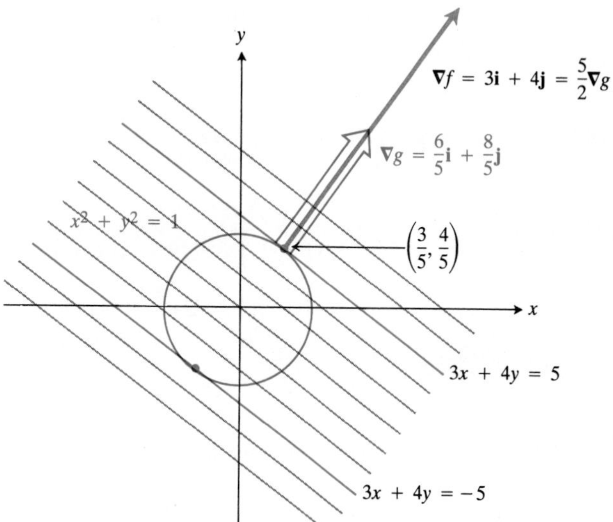

12.71 The function $f(x, y) = 3x + 4y$ takes on its largest value on the unit circle $g(x, y) = x^2 + y^2 - 1 = 0$ at the point $(3/5, 4/5)$ and its smallest value at the point $(-3/5, -4/5)$. At each of these points, ∇f is a scalar multiple of ∇g. The figure shows the gradients at the first point but not the second.

If the constraints on x, y, and z are

$$g_1(x, y, z) = 0 \qquad \text{and} \qquad g_2(x, y, z) = 0$$

and g_1 and g_2 are differentiable with ∇g_1 not parallel to ∇g_2, we find the constrained local maxima and minima of f by introducing two Lagrange multipliers λ and μ (mu, pronounced "mew"). That is, we locate the points $P(x, y, z)$ where f takes on its constrained extreme values by finding the values of x, y, z, λ, and μ that simultaneously satisfy the equations

$$\nabla f = \lambda \nabla g_1 + \mu \nabla g_2, \qquad g_1(x, y, z) = 0, \qquad g_2(x, y, z) = 0. \tag{5}$$

The equations in (5) have a nice geometric interpretation. The surfaces $g_1 = 0$ and $g_2 = 0$ (usually) intersect in a differentiable curve, say C (Fig. 12.72), and along this curve we seek the points where f has local maximum and minimum values relative to its other values on the curve. These are the points where ∇f is normal to C, as we saw in Theorem 7. But ∇g_1 and ∇g_2 are also normal to C at these points because C lies in the surfaces $g_1 = 0$ and $g_2 = 0$. Therefore ∇f lies in the plane determined by ∇g_1 and ∇g_2, which means that

$$\nabla f = \lambda \nabla g_1 + \mu \nabla g_2$$

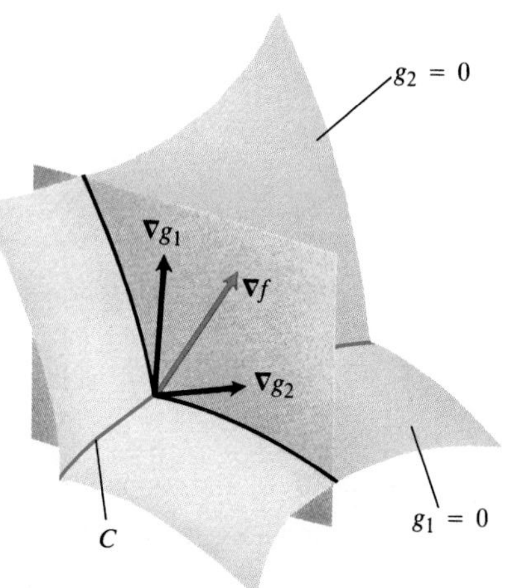

12.72 The vectors ∇g_1 and ∇g_2 lie in a plane perpendicular to the curve C because ∇g_1 is normal to the surface $g_1 = 0$ and ∇g_2 is normal to the surface $g_2 = 0$.

for some λ and μ. Since the points we seek also lie in both surfaces, their coordinates must satisfy the equations $g_1(x, y, z) = 0$ and $g_2(x, y, z) = 0$, which are the remaining requirements in Eq. (5).

Example 5 The plane $x + y + z = 1$ cuts the cylinder $x^2 + y^2 = 1$ in an ellipse (Fig. 12.73). Find the points on the ellipse that lie closest to and farthest from the origin.

Solution We model this as a Lagrange multiplier problem in which we find the extreme values of

$$f(x, y, z) = x^2 + y^2 + z^2$$

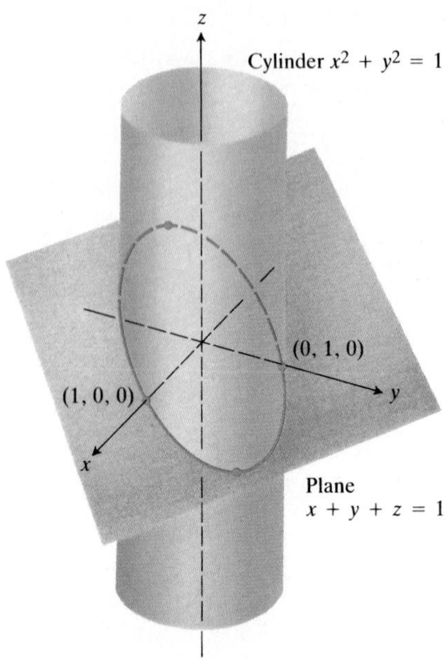

12.73 On the ellipse where the plane and cylinder meet, what are the points closest to and farthest from the origin (Example 5)?

(the square of the distance from (x, y, z) to the origin) subject to the two constraints

$$g_1(x, y, z) = x^2 + y^2 - 1 = 0, \tag{6}$$

$$g_2(x, y, z) = x + y + z - 1 = 0. \tag{7}$$

The gradient equation in (5) then gives

$$\nabla f = \lambda \nabla g_1 + \mu \nabla g_2 \qquad \text{(Eq. (5))}$$

$$2x\mathbf{i} + 2y\mathbf{j} + 2z\mathbf{k} = \lambda(2x\mathbf{i} + 2y\mathbf{j}) + \mu(\mathbf{i} + \mathbf{j} + \mathbf{k}) \tag{8}$$

$$2x\mathbf{i} + 2y\mathbf{j} + 2z\mathbf{k} = (2\lambda x + \mu)\mathbf{i} + (2\lambda y + \mu)\mathbf{j} + \mu\mathbf{k} \tag{9}$$

or

$$2x = 2\lambda x + \mu, \qquad 2y = 2\lambda y + \mu, \qquad 2z = \mu. \tag{10}$$

The scalar equations in (10) yield

$$\begin{aligned} 2x &= 2\lambda x + 2z \quad \Rightarrow \quad (1 - \lambda)x = z, \\ 2y &= 2\lambda y + 2z \quad \Rightarrow \quad (1 - \lambda)y = z. \end{aligned} \tag{11}$$

Equations (11) are satisfied simultaneously if either $\lambda = 1$ and $z = 0$ or if $\lambda \neq 1$ and $x = y = z/(1 - \lambda)$.

If $z = 0$, then solving Eqs. (6) and (7) simultaneously to find the corresponding points on the ellipse gives the two points (1, 0, 0) and (0, 1, 0). This makes sense when you look at Fig. 12.73.

If $x = y$, then Eqs. (6) and (7) give

$$\begin{aligned} x^2 + x^2 - 1 &= 0 & x + x + z - 1 &= 0 \\ 2x^2 &= 1 & z &= 1 - 2x \\ x &= \pm\frac{\sqrt{2}}{2} & z &= 1 \mp \sqrt{2}. \end{aligned} \tag{12}$$

The corresponding points on the ellipse are

$$\left(\frac{\sqrt{2}}{2}, \frac{\sqrt{2}}{2}, 1 - \sqrt{2}\right) \quad \text{and} \quad \left(-\frac{\sqrt{2}}{2}, -\frac{\sqrt{2}}{2}, 1 + \sqrt{2}\right). \tag{13}$$

Again this makes sense when you look at Fig. 12.73.

The points on the ellipse closest to the origin are (1, 0, 0) and (0, 1, 0). The points on the ellipse farthest from the origin are the two points displayed in (13).

EXERCISES 12.9

1. Find the points on the ellipse $x^2 + 2y^2 = 1$ where $f(x, y) = xy$ has its extreme values.
2. Find the extreme values of $f(x, y) = xy$ subject to the constraint $g(x, y) = x^2 + y^2 - 10 = 0$.
3. Find the maximum value of $f(x, y) = 49 - x^2 - y^2$ on the line $x + 3y = 10$ (Fig. 12.65).
4. How close does the line $y = x + 1$ come to the parabola $y^2 = x$?
5. Find the extreme values of $f(x, y) = x^2y$ on the line $x + y = 3$.
6. Find the points on the curve $x^2y = 2$ nearest the origin.
7. Use the method of Lagrange multipliers to find

a) the minimum value of $x + y$, subject to the constraints $xy = 16$, $x > 0$, $y > 0$;

b) the maximum value of xy, subject to the constraint $x + y = 16$.

Comment on the geometry of each solution.

8. Find the points on the curve $x^2 + xy + y^2 = 1$ in the xy-plane that are nearest to and farthest from the origin.

9. Find the dimensions of the closed right circular cylindrical can of smallest surface area whose volume is 16π cm^3.

10. Use the method of Lagrange multipliers to find the dimensions of the rectangle of greatest area that can be inscribed in the ellipse $x^2/16 + y^2/9 = 1$ with sides parallel to the coordinate axes.

11. The temperature at a point (x, y) on a metal plate is $T(x, y) = 4x^2 - 4xy + y^2$. An ant on the plate walks around the circle of radius 5 centered at the origin. What are the highest and lowest temperatures encountered by the ant?

12. Your firm has been asked to design a storage tank for liquid petroleum gas. The customer's specifications call for a cylindrical tank with hemispherical ends, and the tank is to hold 8000 m^3 of gas. The customer also wants to use the smallest amount of material possible in building the tank. What radius and height do you recommend for the cylindrical portion of the tank?

13. Find the maximum and minimum values of $x^2 + y^2$ subject to the constraint $x^2 - 2x + y^2 - 4y = 0$.

14. Find the point on the plane $x + 2y + 3z = 13$ closest to the point $(1, 1, 1)$.

15. Find the maximum and minimum values of

$$f(x, y, z) = x - 2y + 5z$$

on the sphere

$$x^2 + y^2 + z^2 = 30.$$

16. Find the minimum distance from the surface $x^2 + y^2 - z^2 = 1$ to the origin.

17. Find the point on the surface $z = xy + 1$ nearest the origin.

18. Find the points on the surface $z^2 = xy + 4$ closest to the origin.

19. Find the points on the sphere $x^2 + y^2 + z^2 = 25$ where $f(x, y, z) = x + 2y + 3z$ has its maximum and minimum values.

20. Find three real numbers whose sum is 9 and the sum of whose squares is as small as possible.

21. Find the largest product the positive numbers x, y, and z can have if $x + y + z^2 = 16$.

22. A space probe in the shape of the ellipsoid

$$4x^2 + y^2 + 4z^2 = 16$$

enters the earth's atmosphere and its surface begins to heat. After one hour, the temperature at the point (x, y, z) on the probe's surface is

$$T(x, y, z) = 8x^2 + 4yz - 16z + 600.$$

Find the hottest point on the probe's surface.

23. *An example from economics.* In economics, the usefulness or *utility* of amounts x and y of two capital goods G_1 and G_2 is sometimes measured by a function $U(x, y)$. For example, G_1 and G_2 might be two chemicals a pharmaceutical company needs to have on hand and $U(x, y)$ the gain from manufacturing a product whose synthesis requires different amounts of the chemicals depending on the process used. If G_1 costs a dollars per kilogram, G_2 costs b dollars per kilogram, and the total amount allocated for the purchase of G_1 and G_2 together is c dollars, then the company's managers want to maximize $U(x, y)$ given that $ax + by = c$. Thus, they need to solve a typical Lagrange multiplier problem.

Suppose that

$$U(x, y) = xy + 2x$$

and that the equation $ax + by = c$ becomes

$$2x + y = 30$$

when reduced to lowest terms. Find the maximum value of U and the corresponding values of x and y subject to this latter constraint.

24. You are in charge of erecting a radio telescope on a newly discovered planet. To minimize interference, you want to place it where the magnetic field of the planet is weakest. The planet is spherical, with a radius of 6 units. Based on a coordinate system whose origin is at the center of the planet, the strength of the magnetic field is given by $M(x, y, z) = 6x - y^2 + xz + 60$. Where should you locate the radio telescope?

25. *The condition $\nabla f = \lambda \nabla g$ is not sufficient.* While $\nabla f = \lambda \nabla g$ is a necessary condition for the occurrence of an extreme value of $f(x, y)$ subject to the condition $g(x, y) = 0$, it does not in itself guarantee that one exists. As a case in point, try using the method of Lagrange multipliers to find a maximum value of $f(x, y) = x + y$ subject to the constraint that $xy = 16$. The method will identify the two points $(4, 4)$ and $(-4, -4)$ as candidates for the location of extreme values. Yet the sum $(x + y)$ has no maximum value on the hyperbola $xy = 16$. The farther you go from the origin on this hyperbola in the first quadrant, the larger the sum $f(x, y) = x + y$ becomes.

26. *A least squares plane.* The plane

$$z = Ax + By + C$$

is to be "fitted" to the following points (x_k, y_k, z_k):

$$(0, 0, 0), \quad (0, 1, 1), \quad (1, 1, 1), \quad (1, 0, -1).$$

Find the values of A, B, and C that minimize the sum of squares of the deviations

$$\sum_{k=1}^{4} (Ax_k + By_k + C - z_k)^2.$$

Lagrange Multipliers with Two Constraints

27. Find the point closest to the origin on the line of intersection of the planes $y + 2z = 12$ and $x + y = 6$.

28. Find the maximum value that $f(x, y, z) = x^2 + 2y - z^2$ can have on the line of intersection of the planes $2x - y = 0$ and $y + z = 0$.

29. Find the extreme values of $f(x, y, z) = x^2yz + 1$ on the intersection of the plane $z = 1$ with the sphere $x^2 + y^2 + z^2 = 10$.

30. a) Find the maximum value of $w = xyz$ on the line of intersection of the two planes $x + y + z = 40$ and $x + y - z = 0$.
b) Give a geometric argument to support your claim that you have found a maximum, and not a minimum, value of w.

31. Find the extreme values of the function $f(x, y, z) = xy + z^2$ on the circle in which the plane $y - x = 0$ intersects the sphere $x^2 + y^2 + z^2 = 4$.

12.10 Taylor's Formula, Second Derivatives, and Error Estimates

Skip

In this section, we derive the second derivative test for extreme values and the formula for estimating errors in linearizations of differentiable functions of two variables. The derivations are based on Taylor's formula from Section 8.7. The use of Taylor's formula in this context leads to a remarkable extension of the formula to functions of two variables. The extended formula provides polynomial approximations of all orders for functions of two variables, along with explicit formulas for the approximation errors.

The Derivation of the Second Derivative Test

Our first application of Taylor's formula is the derivation of the second derivative test for extreme values of a twice-differentiable function $f(x, y)$ at a point $P(a, b)$ where $f_x = f_y = 0$. We assume that the first and second order derivatives of f are continuous in an open region R about P (Fig. 12.74). We let h and k be increments small enough to put the point $S(a + h, b + k)$ and the line segment joining it to P inside R. We parametrize the segment PS with the equations and parameter interval

$$x = a + th, \qquad y = b + tk, \qquad 0 \le t \le 1.$$

This makes the value of f along the segment a function of t. We denote the function by F. Its defining formula is

$$F(t) = f(a + th, b + tk).$$

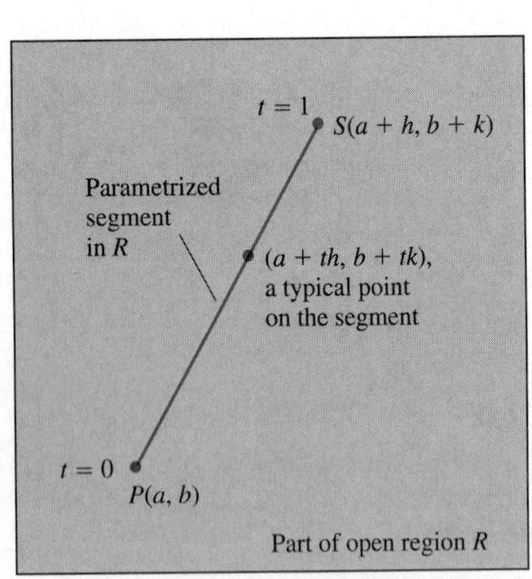

12.74 We begin the derivation of the second derivative test at $P(a, b)$ by parametrizing a typical line segment from P to a point S nearby.

The function F is a differentiable function of t because f is differentiable and x and y are differentiable functions of t. The Chain Rule gives the derivative of F with respect to t as

$$F'(t) = f_x \frac{dx}{dt} + f_y \frac{dy}{dt} = hf_x + kf_y. \tag{1}$$

Since f_x and f_y are continuous throughout R, the function F' is continuous on the closed interval $0 \le t \le 1$.

Since f_x and f_y are differentiable (they have continuous partial derivatives), F' is a differentiable function of t. The Chain Rule gives the derivative of F' with respect to t as

$$\begin{aligned} F'' &= \frac{\partial F'}{\partial x}\frac{dx}{dt} + \frac{\partial F'}{\partial y}\frac{dy}{dt} \\ &= \frac{\partial}{\partial x}\left(hf_x + kf_y\right) \cdot h + \frac{\partial}{\partial y}\left(hf_x + kf_y\right) \cdot k && \text{(From Eq. (1))} \qquad (2) \\ &= h^2 f_{xx} + 2hkf_{xy} + k^2 f_{yy}. && (f_{xy} = f_{yx}) \end{aligned}$$

Since F and F' are continuous on $[0, 1]$ and F' is differentiable on $(0, 1)$, we can apply Taylor's formula with $n = 2$ and $a = 0$ to obtain

$$F(1) = F(0) + F'(0)(1 - 0) + F''(c)\frac{(1-0)^2}{2}$$

$$F(1) = F(0) + F'(0) + \frac{1}{2}F''(c) \tag{3}$$

for some parameter value c between 0 and 1.

In terms of f,

$$\begin{aligned} F(1) &= f(a+h, b+k), \\ F(0) &= f(a, b), \\ F'(0) &= hf_x(a, b) + kf_y(a, b), \\ F''(c) &= (h^2 f_{xx} + 2hkf_{xy} + k^2 f_{yy})\big|_{(a+ch,\, b+ck)}. \end{aligned} \tag{4}$$

With these values, Eq. (3) becomes

$$\begin{aligned} f(a+h, b+k) = f(a, b) &+ hf_x(a, b) + kf_y(a, b) \\ &+ \frac{1}{2}\left(h^2 f_{xx} + 2hkf_{xy} + k^2 f_{yy}\right)\Big|_{(a+ch,\, b+ck)}, \end{aligned} \tag{5}$$

which reduces to

$$f(a+h, b+k) - f(a, b) = \frac{1}{2}\left(h^2 f_{xx} + 2hkf_{xy} + k^2 f_{yy}\right)\Big|_{(a+ch,\, b+ck)} \tag{6}$$

since $f_x(a, b) = f_y(a, b) = 0$. We derive the second derivative test from this equation.

The presence of a maximum or minimum value of f at (a, b) is determined by the sign of $f(a+h, b+k) - f(a, b)$. By Eq. (6), this is the same as the sign of the quadratic form

$$Q(c) = (h^2 f_{xx} + 2hkf_{xy} + k^2 f_{yy})\big|_{(a+ch,\, b+ck)}. \tag{7}$$

The derivatives in Eq. (7) are evaluated at a point on segment PQ. Since these derivatives are continuous throughout R, their values at $(a+ch, b+ck)$ will be nearly the same as their values at (a, b) when h and k are small. In particular, if $Q(0) \neq 0$, the sign of $Q(c)$ will be the same as the sign of $Q(0)$ for sufficiently small values of h and k.

We can predict the sign of

$$Q(0) = h^2 f_{xx}(a, b) + 2hkf_{xy}(a, b) + k^2 f_{yy}(a, b) \tag{8}$$

and hence the sign of $Q(c)$ from the signs of f_{xx} and $f_{xx}f_{yy} - f_{xy}^2$ at (a, b). To do so, we multiply both sides of Eq. (8) by f_{xx} and rearrange the right-hand side to get

$$f_{xx}Q(0) = (hf_{xx} + kf_{xy})^2 + (f_{xx}f_{yy} - f_{xy}^2)k^2. \tag{9}$$

If $f_{xx} \neq 0$, the sign of $Q(0)$ can be determined from this equation, and we are led to the following criteria for the behavior of $f(x, y)$ at (a, b).

1. If $f_{xx} < 0$ and $f_{xx}f_{yy} - f_{xy}^2 > 0$ at (a, b), then $Q(0) < 0$ for all sufficiently small nonzero values of h and k, and f has a *local maximum* value at (a, b).

2. If $f_{xx} > 0$ and $f_{xx}f_{yy} - f_{xy}^2 > 0$ at (a, b), then $Q(0) > 0$ for all sufficiently small nonzero values of h and k, and f has a *local minimum* value at (a, b).
3. If $f_{xx}f_{yy} - f_{xy}^2 < 0$ at (a, b), then it can be shown that there are combinations of arbitrarily small nonzero values of h and k for which $Q(0) > 0$, and other values for which $Q(0) < 0$. Thus, arbitrarily close to the point $P_0(a, b, f(a, b))$ on the surface $z = f(x, y)$ there are points above P_0 and points below P_0. The function f has a *saddle point* at (a, b).
4. Finally, if $f_{xx}f_{yy} - f_{xy}^2 = 0$, we can draw *no conclusion* about the sign of $Q(c)$. Another test is needed. It does follow from Eq. (9) that $f_{xx}Q(0) \geq 0$ for all choices of h and k but the possibility that $Q(0)$ equals zero prevents us from drawing any reliable conclusion about the sign of $Q(c)$.

This completes the derivation of the second derivative test.

The Error Formula for Linear Approximations

Our second application of Taylor's formula is the verification that the difference $E(x, y)$ between the values of a function $f(x, y)$ and those of its linearization $L(x, y)$ at (x_0, y_0) satisfies the inequality

$$|E(x, y)| \leq \frac{1}{2} B\left(|x - x_0| + |y - y_0|\right)^2. \tag{10}$$

The function f is assumed to have continuous first and second partial derivatives throughout an open set containing a closed rectangular region R centered at (x_0, y_0). The number B is the largest value that any of $|f_{xx}|$, $|f_{yy}|$, and $|f_{xy}|$ take on R.

The inequality we want comes from Eq. (5). We substitute x_0 and y_0 for a and b, and $x - x_0$ and $y - y_0$ for h and k, respectively, and rearrange the equation in the form

$$f(x, y) = \underbrace{f(x_0, y_0) + f_x(x_0, y_0)(x - x_0) + f_y(x_0, y_0)(y - y_0)}_{\text{linearization } L(x, y)} \tag{11}$$

$$\underbrace{+ \frac{1}{2}\left((x - x_0)^2 f_{xx} + 2(x - x_0)(y - y_0)f_{xy} + (y - y_0)^2 f_{yy}\right)\Big|_{(x_0 + c(x - x_0),\, y_0 + c(y - y_0))}}_{\text{error } E(x, y)}.$$

This remarkable equation shows how to calculate E directly from the derivatives of f. In particular, it tells us that

$$|E| \leq \frac{1}{2}\left((x - x_0)^2|f_{xx}| + 2|x - x_0|\,|y - y_0|\,|f_{xy}| + (y - y_0)^2|f_{yy}|\right). \tag{12}$$

Hence, if B is an upper bound for the values of $|f_{xx}|$, $|f_{xy}|$, and $|f_{yy}|$ on R,

$$\begin{aligned} |E| &\leq \frac{1}{2}\left((x - x_0)^2 B + 2|x - x_0|\,|y - y_0| B + (y - y_0)^2 B\right) \\ &\leq \frac{1}{2} B\left((x - x_0)^2 + 2|x - x_0|\,|y - y_0| + (y - y_0)^2\right) \qquad (13) \\ &\leq \frac{1}{2} B\left(|x - x_0| + |y - y_0|\right)^2. \end{aligned}$$

Combining these last inequalities gives the inequality

$$|E| \leq \frac{1}{2} B\left(|x - x_0| + |y - y_0|\right)^2 \tag{14}$$

that we set out to derive.

Taylor's Formula for Functions of Two Variables

The formula

$$F'(t) = \left(h \frac{\partial f}{\partial x} + k \frac{\partial f}{\partial y}\right)$$

from Eq. (1) says that applying the operator d/dt to $F(t) = f(a + ht, b + kt)$ gives the same result as applying the operator

$$\left(h \frac{\partial}{\partial x} + k \frac{\partial}{\partial y}\right)$$

to $f(x, y)$. Similarly, the formula

$$F''(t) = h^2 \frac{\partial^2 f}{\partial x^2} + 2hk \frac{\partial^2 f}{\partial x\, \partial y} + k^2 \frac{\partial^2 f}{\partial y^2}$$

from Eq. (2) says that applying d^2/dt^2 to $F(t)$ gives the same result as applying

$$\left(h \frac{\partial}{\partial x} + k \frac{\partial}{\partial y}\right)^2 = h^2 \frac{\partial^2}{\partial x^2} + 2hk \frac{\partial^2}{\partial x\, \partial y} + k^2 \frac{\partial^2}{\partial y^2}$$

to $f(x, y)$. These are the first two instances of a more general formula,

$$F^{(n)}(t) = \frac{d^n}{dt^n} F(t) = \left(h \frac{\partial}{\partial x} + k \frac{\partial}{\partial y}\right)^n f(x, y), \tag{15}$$

which says that applying d^n/dt^n to $F(t)$ gives the same result as applying the operator

$$\left(h \frac{\partial}{\partial x} + k \frac{\partial}{\partial y}\right)^n,$$

after expanding it by the binomial theorem, to $f(x, y)$.

Suppose now that $f(x, y)$ and its partial derivatives through order $n + 1$ are continuous throughout a rectangular region centered at the point (a, b). Then we may extend the Taylor formula for $F(t)$ to more terms,

$$F(t) = F(0) + F'(0) \cdot t + \frac{F''(0)}{2!} t^2 + \cdots + \frac{F^{(n)}(0)}{n!} t^n + \text{remainder},$$

and take $t = 1$ to obtain

$$F(1) = F(0) + F'(0) + \frac{F''(0)}{2!} + \cdots + \frac{F^{(n)}(0)}{n!} + \text{remainder}.$$

When we replace the first n derivatives on the right of this last series by their equivalent expressions from Eq. (15) evaluated at $t = 0$ and add the appropriate remainder term, we arrive at the following formula.

Taylor's Formula for $f(x, y)$ at the Point (a, b)

Suppose $f(x, y)$ and its partial derivatives through order $n + 1$ are continuous throughout an open rectangular region R centered at a point (a, b). Then, throughout R,

$$\begin{aligned} f(a+h, b+k) = {} & f(a, b) + (h f_x + k f_y)\big|_{(a, b)} \\ & + \frac{1}{2!}(h^2 f_{xx} + 2hk f_{xy} + k^2 f_{yy})\big|_{(a, b)} \\ & + \frac{1}{3!}(h^3 f_{xxx} + 3h^2 k f_{xxy} + 3hk^2 f_{xyy} + k^3 f_{yyy})\big|_{(a, b)} \\ & + \cdots + \frac{1}{n!}\left(h\frac{\partial}{\partial x} + k\frac{\partial}{\partial y}\right)^n f\bigg|_{(a, b)} \\ & + \frac{1}{(n+1)!}\left(h\frac{\partial}{\partial x} + k\frac{\partial}{\partial y}\right)^{n+1} f\bigg|_{(a+ch, b+ck)} \end{aligned} \qquad (16)$$

The derivatives in the first n derivative terms are evaluated at (a, b). The derivatives in the last term are evaluated at some point $(a + ch, b + ck)$ on the line segment joining (a, b) and $(a + h, b + k)$.

If $(a, b) = (0, 0)$ and we treat h and k as independent variables (denoting them now by x and y), then Eq. (16) assumes the following simpler form.

Taylor's Formula for $f(x, y)$ at the Origin

$$\begin{aligned} f(x, y) = {} & f(0, 0) + x f_x + y f_y \\ & + \frac{1}{2!}(x^2 f_{xx} + 2xy f_{xy} + y^2 f_{yy}) \\ & + \frac{1}{3!}(x^3 f_{xxx} + 3x^2 y f_{xxy} + 3xy^2 f_{xyy} + y^3 f_{yyy}) \\ & + \cdots + \frac{1}{n!}\left(x\frac{\partial}{\partial x} + y\frac{\partial}{\partial y}\right)^n f \\ & + \frac{1}{(n+1)!}\left(x\frac{\partial}{\partial x} + y\frac{\partial}{\partial y}\right)^{n+1} f\bigg|_{(cx, cy)} \end{aligned} \qquad (17)$$

The partial derivatives in the first n derivative terms are evaluated at $(0, 0)$. The partial derivatives in the final term are evaluated at (cy, cx), a point on the line segment joining the origin and (x, y).

Taylor's formula is the standard source of approximations of two-variable functions by polynomials in x and y. The first n derivative terms give the approximating polynomial; the last term gives the approximation error. The first three terms of Taylor's formula give the function's linearization. To improve on the linearization, we add higher power terms, the way we do for functions of a single variable.

Example 1 Use Taylor's formula to find a quadratic polynomial that approximates

$$f(x, y) = \sin x \sin y$$

near the origin. How accurate is the approximation if $|x| \leq 0.1$ and $|y| \leq 0.1$?

Solution We take $n = 2$ in Eq. (17):

$$\begin{aligned} f(x, y) = f(0, 0) &+ (x f_x + y f_y) \\ &+ \frac{1}{2}(x^2 f_{xx} + 2xy f_{xy} + y^2 f_{yy}) \\ &+ \frac{1}{6}(x^3 f_{xxx} + 3x^2 y f_{xxy} + 3xy^2 f_{xyy} + y^3 f_{yyy})_{(cx,\, cy)} \end{aligned}$$

The first and second partial derivatives are evaluated at the origin and the third partial derivatives are evaluated at (cx, cy). With

$$\begin{aligned} f(x, y) &= \sin x \sin y, \\ f(0, 0) &= \sin x \sin y|_{(0,0)} = 0, \\ f_x(0, 0) &= \cos x \sin y|_{(0,0)} = 0, \\ f_y(0, 0) &= \sin x \cos y|_{(0,0)} = 0, \\ f_{xx}(0, 0) &= -\sin x \sin y|_{(0,0)} = 0, \\ f_{xy}(0, 0) &= \cos x \cos y|_{(0,0)} = 1, \\ f_{yy}(0, 0) &= -\sin x \sin y|_{(0,0)} = 0, \end{aligned}$$

we have

$$\sin x \sin y \approx 0 + 0 + 0 + \frac{1}{2}(x^2(0) + 2xy(1) + y^2(0))$$

$$\sin x \sin y \approx xy.$$

This is the standard quadratic approximation of $\sin x \sin y$ at the origin.

The error in the approximation is

$$E(x, y) = \frac{1}{6}(x^3 f_{xxx} + 3x^2 y f_{xxy} + 3xy^2 f_{xyy} + y^3 f_{yyy})|_{(cx,\, cy)}.$$

The third derivatives never exceed 1 in absolute value because they are products of sines and cosines. Also, $|x| \leq 0.1$ and $|y| \leq 0.1$. Hence

$$\begin{aligned} |E(x, y)| &\leq \frac{1}{6}((0.1)^3 + 3(0.1)^3 + 3(0.1)^3 + (0.1)^3) \\ &\leq \frac{8}{6}(0.1)^3 \\ &\leq 0.00134. \qquad \text{(Rounded up, for safety)} \end{aligned}$$

The error in the approximation will not exceed 0.00134 as long as $|x| \leq 0.1$ and $|y| \leq 0.1$.

EXERCISES 12.10

In Exercises 1–6, use Taylor's formula for $f(x, y)$ at the origin to find quadratic and cubic polynomial approximations of f near the origin.

1. $f(x, y) = e^x \cos y$
2. $f(x, y) = e^x \ln(1 + y)$
3. $f(x, y) = \sin(x^2 + y^2)$
4. $f(x, y) = \cos(x^2 + y^2)$
5. $f(x, y) = \dfrac{1}{1 - x - y}$
6. $f(x, y) = \dfrac{1}{1 - x - y + xy}$
7. Use Taylor's formula to find a quadratic approximation of $f(x, y) = \cos x \cos y$ at the origin. Estimate the error in the approximation if $|x| \leq 0.1$ and $|y| \leq 0.1$.
8. Use Taylor's formula to find a quadratic approximation of $e^x \sin y$ at the origin. Estimate the error in the approximation if $|x| \leq 0.1$ and $|y| \leq 0.1$.
9. *Error bounds for linear approximations of functions of three variables*

   a) Suppose that R is a rectangular solid centered at $P(x_0, y_0, z_0)$ and lying in an open region on which $f(x, y, z)$ and its first and second order derivatives are continuous. Suppose also that h, k, and m are increments small enough for the line segment joining P to the point $S(x_0 + h, y_0 + k, z_0 + m)$ to lie in R. Parametrize PS with the equations and interval

   $$x = x_0 + th, \quad y = y_0 + tk, \quad z = z_0 + tm, \quad 0 \leq t \leq 1.$$

   Apply Taylor's formula for functions of a single variable to the function $F(t)$ whose defining formula is

   $$F(t) = f(x_0 + th, y_0 + tk, z_0 + tm)$$

   to show that

   $$\begin{aligned} f(x_0 + h, y_0 + k, z_0 + m) &= f(x_0, y_0, z_0) + hf_x(x_0, y_0, z_0) + kf_y(x_0, y_0, z_0) \\ &\quad + mf_z(x_0, y_0, z_0) \\ &\quad + \frac{1}{2}\left(h^2 f_{xx} + k^2 f_{yy} + m^2 f_{zz} + 2hk f_{xy} \right. \\ &\quad \left. + 2hm f_{xz} + 2km f_{yz}\right)\Big|_{(x_0 + ch,\, y_0 + ck,\, z_0 + cm)} \end{aligned} \qquad (5')$$

   for some parameter value c between 0 and 1. This is the three-variable version of Eq. (5) in the text.

   b) Use Eq. (5′) to derive the inequality

   $$|E| \leq \frac{1}{2} M\left(|x - x_0| + |y - y_0| + |z - z_0|\right)^2$$

   in Section 12.4, Eq. (20).

REVIEW QUESTIONS

1. Give examples of functions of two and three variables, and describe their domains and ranges.
2. Describe two ways to display the values of a function $f(x, y)$ graphically. Give examples.
3. What is a level surface of a function $f(x, y, z)$ of three independent variables? Give an example.
4. Give two equivalent definitions of

   $$\lim_{(x, y) \to (x_0, y_0)} f(x, y) = L.$$

   What is the basic theorem for calculating limits of sums, differences, products, constant multiples, and quotients of functions of two or more variables?
5. When is a function of two (three) variables continuous at a point in its domain? Give examples of functions that are continuous at some points but not continuous at others.
6. How are the partial derivatives $\partial f/\partial x$ and $\partial f/\partial y$ of a function $f(x, y)$ defined? What is their geometric meaning? What rates do they describe? How are they calculated? Give examples.
7. Give examples of second order partial derivatives of functions of two variables. What is the basic theorem about mixed second order derivatives? What about mixed partial derivatives of higher order? Give examples.
8. What does it mean for a function $f(x, y)$ to be differentiable? The Increment Theorem states conditions under which f will be differentiable. What are they?
9. Why are differentiable functions of two independent variables continuous?
10. What is the linearization $L(x, y)$ of a function $f(x, y)$ at a point (x_0, y_0)? How are linearizations used? Give an upper bound for the error incurred by approximating f by L in the neighborhood of (x_0, y_0). Give examples.
11. If (x, y) moves from (x_0, y_0) to a point $(x_0 + dx, y_0 + dy)$ nearby, how do we estimate the corresponding change in a differentiable function $f(x, y)$? How do we estimate the relative change? The percentage change? Give an example.
12. Define the gradient of a function $f(x, y, z)$. Describe the role the gradient plays in defining directional derivatives, tangent planes, and normal lines. What is the relation between ∇f and the directions in which f changes most

rapidly? least rapidly? What are the analogous results for functions of two variables? Give examples.

13. Describe how to find the extreme values (if any) of a function $f(x, y)$ that has continuous first and second order partial derivatives. Give examples.

14. How do you find the extreme values of a differentiable function along a differentiable curve in its domain? Give an example.

15. Describe the method of Lagrange multipliers and its geometric interpretation as it applies to a problem of maximizing or minimizing (a) a differentiable function $f(x, y)$ subject to a differentiable constraint $g(x, y) = 0$, (b) a differentiable function $f(x, y, z)$ subject to simultaneous differentiable constraints $g_1(x, y, z) = 0$ and $g_2(x, y, z) = 0$ when ∇g_1 is not parallel to ∇g_2. Give examples.

16. State Taylor's formula for a (sufficiently) differentiable function $f(x, y)$. Show how to use it to generate a quadratic polynomial approximation together with an appropriate error estimate.

MISCELLANEOUS EXERCISES

In Exercises 1–4, find the domain and range of f and identify the level curves. Sketch a typical level curve.

1. $f(x, y) = 9x^2 + y^2$
2. $f(x, y) = \sin(y - x)$
3. $f(x, y) = 1/xy$
4. $f(x, y) = \sqrt{x^2 - y}$

In Exercises 5–8, find the domain and range of f and identify the level surfaces. Sketch a typical level surface.

5. $f(x, y, z) = x^2 + y^2 - z$
6. $f(x, y, z) = x^2 + 4y^2 + 9z^2$
7. $f(x, y, z) = \dfrac{1}{x^2 + y^2 + z^2}$
8. $f(x, y, z) = \dfrac{1}{x^2 + y^2 + z^2 + 1}$

Find the limits in Exercises 9–14.

9. $\lim_{(x, y)\to(\pi, \ln 2)} e^y \cos x$
10. $\lim_{(x, y)\to(0, 0)} \dfrac{2 + y}{x + \cos y}$
11. $\lim_{\substack{(x, y)\to(1, 1) \\ x \neq y}} \dfrac{x^2 - y^2}{x - y}$
12. $\lim_{P\to(1, -1, e)} \ln |x + y + z|$
13. $\lim_{(x, y)\to(0, 0)} y \sin \dfrac{1}{x}$
14. $\lim_{(x, y)\to(1, 1)} \dfrac{x^3y^3 - 1}{xy - 1}$

Show that the limits in Exercises 15 and 16 do not exist.

15. $\lim_{(x, y)\to(0, 0)} \dfrac{y}{x^2 - y}$
16. $\lim_{(x, y)\to(0, 0)} \dfrac{x^2 + y^2}{xy}$

17. a) Let $f(x, y) = (x^2 - y^2)/(x^2 + y^2)$ for $(x, y) \neq (0, 0)$. Is it possible to define $f(0, 0)$ in a way that makes f continuous at the origin? Why?
b) Let

$$f(x, y) = \begin{cases} \dfrac{\sin(x - y)}{|x| + |y|}, & |x| + |y| \neq 0, \\ 0, & (x, y) = (0, 0). \end{cases}$$

Is f continuous at the origin? Why?

18. Let

$$f(r, \theta) = \begin{cases} \dfrac{\sin 6r}{6r}, & r \neq 0, \\ 1, & r = 0 \end{cases}$$

(Fig. 12.75). Find
a) $\lim_{r\to 0} f(r, \theta)$ b) $f_r(0, 0)$ c) $f_\theta(r, \theta)$, $r \neq 0$

12.75 The graph of the function in Exercise 18.

In Exercises 19–24, find the partial derivative of the function with respect to each variable.

19. $g(r, \theta) = r \cos \theta + r \sin \theta$
20. $f(x, y) = \dfrac{1}{2} \ln(x^2 + y^2) + \tan^{-1} \dfrac{y}{x}$
21. $f(R_1, R_2, R_3) = \dfrac{1}{R_1} + \dfrac{1}{R_2} + \dfrac{1}{R_3}$
22. $h(x, y, z) = \sin(2\pi x + y - 3z)$
23. $P(n, R, T, V) = \dfrac{nRT}{V}$ (the Ideal Gas Law)
24. $f(r, l, T, d) = \dfrac{1}{2rl}\sqrt{\dfrac{T}{\pi d}}$

(the frequency of a struck piano string, Chapter 2, Miscellaneous Exercise 50)

Find the second order partial derivatives of the functions in Exercises 25–28.

25. $f(x, y) = y + \dfrac{x}{y}$
26. $f(x, y) = e^x + y \sin x$
27. $f(x, y) = x + xy - 5x^3 + \ln(x^2 + 1)$

28. $f(x, y) = y^2 - 3xy + \cos y + 7e^y$

29. If you did Exercise 40 in Section 12.2, you know that the function

$$f(x, y) = \begin{cases} xy\dfrac{x^2 - y^2}{x^2 + y^2}, & (x, y) \neq (0, 0), \\ 0, & (x, y) = (0, 0) \end{cases}$$

is continuous at the origin (Fig. 12.76). Find $f_{xy}(0, 0)$ and $f_{yx}(0, 0)$.

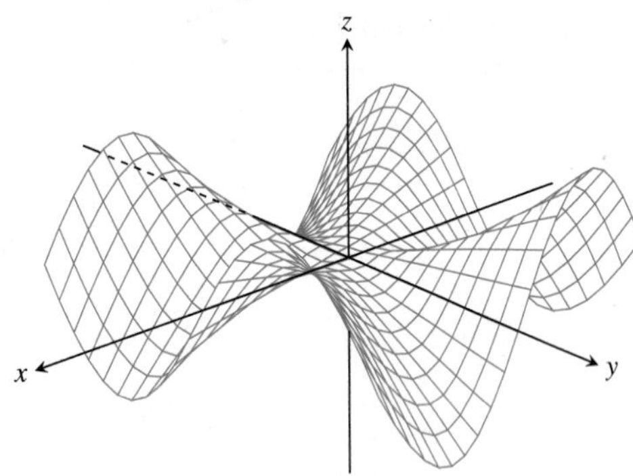

12.76 The graph of the function in Exercise 29.

30. Find a function $w = f(x, y)$ whose first partial derivatives are $\partial w/\partial x = 1 + e^x\cos y$ and $\partial w/\partial y = 2y - e^x\sin y$ and whose value at $(\ln 2, 0)$ is $\ln 2$.

In Exercises 31 and 32, find the linearization $L(x, y)$ of the function $f(x, y)$ at the point P_0. Then find an upper bound for the magnitude of the error E in the approximation $f(x, y) \approx L(x, y)$ over the rectangle R.

31. $f(x, y) = \sin x \cos y, \quad P_0(\pi/4, \pi/4)$

$R: \quad \left|x - \dfrac{\pi}{4}\right| \leq 0.1, \quad \left|y - \dfrac{\pi}{4}\right| \leq 0.1$

32. $f(x, y) = xy - 3y^2 + 2, \quad P_0(1, 1)$

$R: \quad |x - 1| \leq 0.1, \quad |y - 1| \leq 0.2$

Find the linearizations of the functions in Exercises 33 and 34 at the given points.

33. $f(x, y, z) = xy + 2yz - 3xz$ at $(1, 0, 0)$ and $(1, 1, 0)$

34. $f(x, y, z) = \sqrt{2} \cos x \sin(y + z)$ at $(0, 0, \pi/4)$ and $(\pi/4, \pi/4, 0)$.

35. You plan to calculate the volume inside a stretch of pipeline that is about 36 in. in diameter and 1 mi long. With which measurement should you be more careful—the length, or the diameter? Why?

36. Near the point $(1, 2)$, is $f(x, y) = x^2 - xy + y^2 - 3$ more sensitive to changes in x, or to changes in y? How do you know?

37. Suppose that the current I (amperes) in an electrical circuit is related to the voltage V (volts) and the resistance R (ohms) by the equation $I = V/R$. If the voltage drops from 24 to 23 volts and the resistance drops from 100 to 80 ohms, will I increase, or decrease? By about how much? Express the changes in V and R and the estimated change in I as percentages of their original values.

38. If $a = 10$ cm and $b = 16$ cm to the nearest millimeter, what should you expect the maximum percentage error to be in the calculated area $A = \pi ab$ of the ellipse $x^2/a^2 + y^2/b^2 = 1$?

39. Let $y = uv$ and $z = u + v$, where u and v are positive independent variables.

a) If u is measured with an error of 2% and v with an error of 3%, about what is the percentage error in the calculated value of y?

b) Show that the percentage error in the calculated value of z is less than the percentage error in the value of y.

40. CALCULATOR *Cardiac index.* To make different people comparable in studies of cardiac output (Section 3.1, Exercise 19), researchers divide the measured cardiac output by the body surface area to find the *cardiac index C*:

$$C = \frac{\text{cardiac output}}{\text{body surface area}}.$$

The body surface area B is calculated with the formula

$$B = 71.84w^{0.425}h^{0.725},$$

which gives B in square centimeters when w is measured in kilograms and h in centimeters. You are about to calculate the cardiac index of a person with the following measurements:

Cardiac output:	7 L/min
Weight:	70 kg
Height:	180 cm

Which will have a greater effect on the calculation, a 1-kg error in measuring the weight, or a 1-cm error in measuring the height?

41. Find dw/dt at $t = 0$ if $w = \sin(xy + \pi)$, $x = e^t$, $y = \ln(t + 1)$.

42. Find dw/dt at $t = 1$ if $w = xe^y + y \sin z - \cos z$, $x = 2\sqrt{t}$, $y = t - 1 + \ln t$, $z = \pi t$.

43. Find $\partial w/\partial r$ and $\partial w/\partial s$ when $r = \pi$ and $s = 0$ if $w = \sin(2x - y)$, $x = r + \sin s$, $y = rs$.

44. Find $\partial w/\partial u$ and $\partial w/\partial v$ when $u = v = 0$ if $w = \ln\sqrt{1 + x^2} - \tan^{-1}x$ and $x = 2e^u\cos v$.

Assuming that the equations in Exercises 45 and 46 define y as a differentiable function of x, find the value of dy/dx at the given point.

45. $1 - x - y^2 - \sin xy = 0, \quad (0, 1)$

46. $2xy + e^{x+y} - 2 = 0, \quad (0, \ln 2)$

47. Find the value of the derivative of $f(x, y, z) = xy + yz + xz$ with respect to t on the curve $x = \cos t$, $y = \sin t$, $z = \cos 2t$ at the point where $t = 1$ (Fig. 12.77).

(Generated by Mathematica)

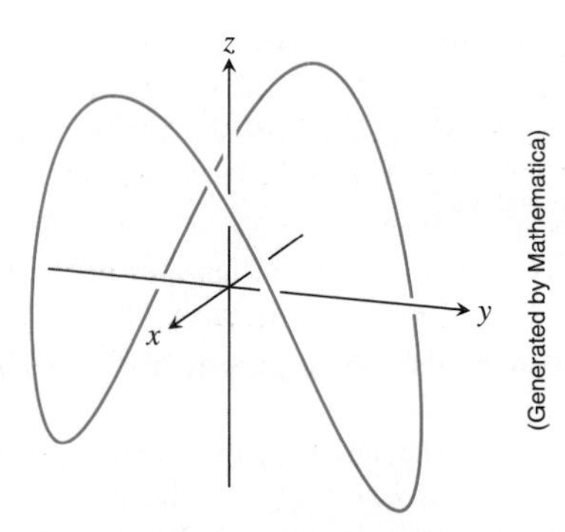

12.77 The curve in Exercise 47.

48. Show that if $w = f(s)$ is any differentiable function of s whatever, and $s = y + 5x$, then

$$\frac{\partial w}{\partial x} - 5\frac{\partial w}{\partial y} = 0.$$

49. Let $w = f(r, \theta)$, $r = \sqrt{x^2 + y^2}$, and $\theta = \tan^{-1}(y/x)$. Find $\partial w/\partial x$ and $\partial w/\partial y$ and express your answers in terms of r and θ.

50. Let $z = f(u, v)$, $u = ax + by$, and $v = ax - by$. Express z_x and z_y in terms of f_u, f_v, and the constants a and b.

51. If a and b are constants, $w = u^3 + \tanh u + \cos u$, and $u = ax + by$, show that

$$a\frac{\partial w}{\partial y} = b\frac{\partial w}{\partial x}.$$

52. If $w = \ln(x^2 + y^2 + 2z)$, $x = r + s$, $y = r - s$, and $z = 2rs$, find w_r and w_s by the Chain Rule. Then check your answer another way.

53. The equations $e^u \cos v - x = 0$ and $e^u \sin v - y = 0$ define u and v as differentiable functions of x and y. Show that the angle between the vectors

$$\frac{\partial u}{\partial x}\mathbf{i} + \frac{\partial u}{\partial y}\mathbf{j} \quad \text{and} \quad \frac{\partial v}{\partial x}\mathbf{i} + \frac{\partial v}{\partial y}\mathbf{j}$$

is constant.

54. Introducing polar coordinates $x = r\cos\theta$ and $y = r\sin\theta$ changes $f(x, y)$ to $g(r, \theta)$. Find the value of $\partial^2 g/\partial\theta^2$ at the point $(r, \theta) = (2, \pi/2)$, given that

$$\frac{\partial f}{\partial x} = \frac{\partial f}{\partial y} = \frac{\partial^2 f}{\partial x^2} = \frac{\partial^2 f}{\partial y^2} = 1$$

at that point.

55. *A proof of Leibniz's rule.* Leibniz's rule says that if f is continuous on $[a, b]$ and if $u(x)$ and $v(x)$ are differentiable functions of x whose values lie in $[a, b]$, then

$$\frac{d}{dx}\int_{u(x)}^{v(x)} f(t)\,dt = f(v(x))\frac{dv}{dx} - f(u(x))\frac{du}{dx}.$$

Prove the rule by setting

$$g(u, v) = \int_u^v f(t)\,dt, \qquad u = u(x), \qquad v = v(x)$$

and calculating dg/dx with the Chain Rule.

56. Suppose that f is a twice-differentiable function of r, that $r = \sqrt{x^2 + y^2 + z^2}$, and that

$$f_{xx} + f_{yy} + f_{zz} = 0.$$

Show that for some constants a and b,

$$f(r) = \frac{a}{r} + b.$$

57. A function $f(x, y)$ is *homogeneous of degree n* (n a nonnegative integer) if $f(tx, ty) = t^n f(x, y)$ for all t, x, and y. For such a function (sufficiently differentiable), prove that

a) $x\dfrac{\partial f}{\partial x} + y\dfrac{\partial f}{\partial y} = nf(x, y)$,

b) $x^2\left(\dfrac{\partial^2 f}{\partial x^2}\right) + 2xy\left(\dfrac{\partial^2 f}{\partial x\,\partial y}\right) + y^2\left(\dfrac{\partial^2 f}{\partial y^2}\right) = n(n-1)f.$

58. Let $\mathbf{r} = x\mathbf{i} + y\mathbf{j} + z\mathbf{k}$. Express x, y, and z as functions of the spherical coordinates ρ, ϕ, and θ and calculate $\partial\mathbf{r}/\partial\rho$, $\partial\mathbf{r}/\partial\phi$, and $\partial\mathbf{r}/\partial\theta$. Then express these derivatives in terms of the unit vectors

$$\mathbf{u}_\rho = (\sin\phi\cos\theta)\,\mathbf{i} + (\sin\phi\sin\theta)\,\mathbf{j} + (\cos\phi)\,\mathbf{k}$$
$$\mathbf{u}_\phi = (\cos\phi\cos\theta)\,\mathbf{i} + (\cos\phi\sin\theta)\,\mathbf{j} - (\sin\phi)\,\mathbf{k}$$
$$\mathbf{u}_\theta = -(\sin\theta)\,\mathbf{i} + (\cos\theta)\,\mathbf{j}.$$

In Exercises 59–62, find the directions in which f increases and decreases most rapidly at P_0 and find the derivative of f in each direction. Also, find the derivative of f at P_0 in the direction of the vector $\mathbf{A}$.

59. $f(x, y) = \cos x \cos y$, $P_0(\pi/4, \pi/4)$, $\mathbf{A} = 3\mathbf{i} + 4\mathbf{j}$

60. $f(x, y) = x^2 e^{-2y}$, $P_0(1, 0)$, $\mathbf{A} = \mathbf{i} + \mathbf{j}$

61. $f(x, y, z) = \ln(2x + 3y + 6z)$, $P_0(-1, -1, 1)$, $\mathbf{A} = 2\mathbf{i} + 3\mathbf{j} + 6\mathbf{k}$

62. $f(x, y, z) = x^2 + 3xy - z^2 + 2y + z + 4$, $P_0(0, 0, 0)$, $\mathbf{A} = \mathbf{i} + \mathbf{j} + \mathbf{k}$

In Exercises 63 and 64, sketch the surface $f(x, y, z) = c$ together with ∇f at the given points.

63. $x^2 + y + z^2 = 0$; $(0, -1, \pm 1)$, $(0, 0, 0)$

64. $y^2 + z^2 = 4$; $(2, \pm 2, 0)$, $(2, 0, \pm 2)$

In Exercises 65 and 66, find an equation for the plane tangent to the level surface $f(x, y, z) = c$ at the point P_0. Also find parametric equations for the line that is normal to the surface at P_0.

65. $x^2 - y - 5z = 0$, $P_0(2, -1, 1)$

66. $x^2 + y^2 + z = 4$, $P_0(1, 1, 2)$

In Exercises 67 and 68, find equations for the lines that are tangent and normal to the level curve $f(x, y) = c$ at the point P_0. Then sketch the lines and level curve together with ∇f at P_0.

67. $y - \sin x = 1$, $P_0(\pi, 1)$

68. $\dfrac{y^2}{2} - \dfrac{x^2}{2} = \dfrac{3}{2}$, $P_0(1, 2)$

In Exercises 69 and 70, find parametric equations for the line that is tangent to the curve of intersection of the surfaces at the given point.

69. Surfaces: $x^2 + 2y + 2z = 4, \quad y = 1$
Point: $(1, 1, 1/2)$

70. Surfaces: $x + y^2 + z = 2, \quad y = 1$
Point: $(1/2, 1, 1/2)$

71. Find the points on the surface
$$(y + z)^2 + (z - x)^2 = 16$$
where the normal line is parallel to the yz-plane.

72. Find the points on the surface
$$xy + yz + zx - x - z^2 = 0$$
where the tangent plane is parallel to the xy-plane.

73. Find the derivative of $f(x, y, z) = xyz$ in the direction of the velocity vector of the helix
$$\mathbf{r} = (\cos 3t)\,\mathbf{i} + (\sin 3t)\,\mathbf{j} + 3t\mathbf{k}$$
at time $t = \pi/3$.

74. What is the largest value that the directional derivative of $f(x, y, z) = xyz$ can have at the point $(1, 1, 1)$?

75. At the point $(1, 2)$ the function $f(x, y)$ has a derivative of 2 in the direction toward $(2, 2)$ and a derivative of -2 in the direction toward $(1, 1)$. Find the derivative in the direction toward $(4, 6)$.

76. Suppose that $\nabla f(x, y, z)$ is always parallel to the position vector $x\mathbf{i} + y\mathbf{j} + z\mathbf{k}$. Show that $f(0, 0, a) = f(0, 0, -a)$ for any a.

77. Show that the directional derivative of $f(x, y, z) = \sqrt{x^2 + y^2 + z^2}$ at the origin equals 1 in any direction but that f has no gradient vector at the origin.

78. Let $\mathbf{r} = x\mathbf{i} + y\mathbf{j} + z\mathbf{k}$ and let $r = |\mathbf{r}|$.
a) Show that $\nabla r = \mathbf{r}/r$.
b) Show that $\nabla(r^n) = nr^{n-2}\mathbf{r}$.
c) Find a function whose gradient equals $\mathbf{r}$.
d) Show that $\mathbf{r} \cdot d\mathbf{r} = r\,dr$.
e) Show that $\nabla(\mathbf{A} \cdot \mathbf{r}) = \mathbf{A}$ for any constant vector $\mathbf{A}$.

79. Show that the line normal to the surface $xy + z = 2$ at the point $(1, 1, 1)$ passes through the origin.

80. a) Sketch the surface $x^2 - y^2 + z^2 = 4$.
b) Find a vector normal to the surface at $(2, -3, 3)$. Add the vector to your sketch.
c) Find equations for the tangent plane and normal line at $(2, -3, 3)$.

81. Show that the curve
$$\mathbf{r} = (\ln t)\,\mathbf{i} + (t \ln t)\,\mathbf{j} + t\mathbf{k}$$
is tangent to the surface
$$xz^2 - yz + \cos xy = 1$$
at $(0, 0, 1)$.

82. Show that the curve
$$\mathbf{r} = \left(\frac{t^3}{4} - 2\right)\mathbf{i} + \left(\frac{4}{t} - 3\right)\mathbf{j} + \cos(t - 2)\,\mathbf{k}$$
is tangent to the surface
$$x^3 + y^3 + z^3 - xyz = 0$$
at $(0, -1, 1)$.

83. Which of the following statements are true if $f(x, y)$ is differentiable at (x_0, y_0)?
a) If $\mathbf{u}$ is a unit vector, the derivative of f at (x_0, y_0) in the direction of $\mathbf{u}$ is $(f_x(x_0, y_0)\,\mathbf{i} + f_y(x_0, y_0)\,\mathbf{j}) \cdot \mathbf{u}$.
b) The derivative of f at (x_0, y_0) in the direction of $\mathbf{u}$ is a vector.
c) The directional derivative of f at (x_0, y_0) has its greatest value in the direction of ∇f.
d) At (x_0, y_0), the gradient of f is normal to the curve $f(x, y) = f(x_0, y_0)$.

84. Suppose that a differentiable function $f(x, y)$ has the constant value c along the differentiable curve $x = g(t)$, $y = h(t)$; that is,
$$f(g(t), h(t)) = c$$
for all values of t. Differentiate both sides of this equation with respect to t to show that ∇f is orthogonal to the curve's tangent vector at every point on the curve.

85. Suppose cylindrical coordinates r, θ, z are introduced into a function $w = f(x, y, z)$ to yield $w = F(r, \theta, z)$. Show that
$$\nabla w = \frac{\partial w}{\partial r}\mathbf{u}_r + \frac{1}{r}\frac{\partial w}{\partial \theta}\mathbf{u}_\theta + \frac{\partial w}{\partial z}\mathbf{k}, \tag{1}$$
where
$$\mathbf{u}_r = (\cos\theta)\,\mathbf{i} + (\sin\theta)\,\mathbf{j}$$
$$\mathbf{u}_\theta = -(\sin\theta)\,\mathbf{i} + (\cos\theta)\,\mathbf{j}.$$
(*Hint:* Express the right-hand side of Eq. (1) in terms of **i**, **j**, and **k** and use the Chain Rule to express the components of **i**, **j**, and **k** in rectangular coordinates.)

86. Suppose spherical coordinates ρ, ϕ, θ are introduced into a function $w = f(x, y, z)$ to yield a function $w = F(\rho, \phi, \theta)$. Show that
$$\nabla w = \frac{\partial w}{\partial \rho}\mathbf{u}_\rho + \frac{1}{\rho}\frac{\partial w}{\partial \phi}\mathbf{u}_\phi + \frac{1}{\rho \sin\phi}\frac{\partial w}{\partial \theta}\mathbf{u}_\theta, \tag{2}$$
where
$$\mathbf{u}_\rho = (\sin\phi\cos\theta)\,\mathbf{i} + (\sin\phi\sin\theta)\,\mathbf{j} + (\cos\phi)\,\mathbf{k}$$
$$\mathbf{u}_\phi = (\cos\phi\cos\theta)\,\mathbf{i} + (\cos\phi\sin\theta)\,\mathbf{j} - (\sin\phi)\,\mathbf{k}$$
$$\mathbf{u}_\theta = -(\sin\theta)\,\mathbf{i} + (\cos\theta)\,\mathbf{j}.$$
(*Hint:* Express the right-hand side of Eq. (2) in terms of **i**, **j**, and **k** and use the Chain Rule to express the components of **i**, **j**, and **k** in rectangular coordinates.)

Test the functions in Exercises 87–92 for maxima, minima, and saddle points. Find the functions' values at these points.

87. $f(x, y) = x^2 - xy + y^2 + 2x + 2y - 4$

88. $f(x, y) = 5x^2 + 4xy - 2y^2 + 4x - 4y$

89. $f(x, y) = 6xy - x^3 - y^2$

90. $f(x, y) = 2x^3 + 3xy + y^3$

91. $f(x, y) = x^3 + y^3 - 3xy + 15$

92. $f(x, y) = x^3 + y^3 + 3x^2 - 3y^2$

In Exercises 93–100, find the absolute maximum and minimum values of f on the region R.

93. $f(x, y) = x^2 + xy + y^2 - 3x + 3y$
R: The triangular region cut from the first quadrant by the line $x + y = 4$

94. $f(x, y) = x^2 - y^2 - 2x + 4y + 1$
R: The rectangular region in the first quadrant bounded by the coordinate axes and the lines $x = 4$ and $y = 2$

95. $f(x, y) = y^2 - xy - 3y + 2x$
R: The square region enclosed by the lines $x = \pm 2$ and $y = \pm 2$

96. $f(x, y) = 2x + 2y - x^2 - y^2$
R: The square bounded by the coordinate axes and the lines $x = 2$, $y = 2$ in the first quadrant

97. $f(x, y) = x^2 - y^2 - 2x + 4y$
R: The triangular region bounded below by the x-axis, above by the line $y = x + 2$, and on the right by the line $x = 2$

98. $f(x, y) = 4xy - x^4 - y^4 + 16$
R: The triangular region bounded below by the line $y = -2$, above by the line $y = x$, and on the right by the line $x = 2$

99. $f(x, y) = x^3 + y^3 + 3x^2 - 3y^2$
R: The square region enclosed by the lines $x = \pm 1$ and $y = \pm 1$

100. $f(x, y) = x^3 + 3xy + y^3 + 1$
R: The square region enclosed by the lines $x = \pm 1$ and $y = \pm 1$

101. Show that the only possible maxima and minima of z on the surface $z = x^3 + y^3 - 9xy + 27$ occur at $(0, 0)$ and $(3, 3)$. Show that neither a maximum nor a minimum occurs at $(0, 0)$. Determine whether z has a maximum or a minimum at $(3, 3)$.

102. Find the maximum value of $f(x, y) = 6xye^{-(2x + 3y)}$ in the closed first quadrant (includes the nonnegative axes).

103. Find the points nearest the origin on the curve $xy^2 = 54$.

104. Find the extreme values of $f(x, y) = x^3 + y^2$ on the cylinder $x^2 + y^2 = 1$.

105. Suppose that the temperature T (degrees) at the point (x, y, z) on the sphere $x^2 + y^2 + z^2 = 1$ is $T = 400\,xyz^2$. Locate the highest and lowest temperatures on the sphere.

106. Find the extreme values of $f(x, y) = x^2 + 3y^2 + 2y$ on the unit disk $x^2 + y^2 \leq 1$.

107. Find the point(s) on the surface $xyz = 1$ closest to the origin.

108. A closed rectangular box is to have a volume of V in^3. The cost of the material used in the box is a cents/in^2 for top and bottom, b cents/in^2 for front and back, and c cents/in^2 for the remaining two sides. What dimensions minimize the total cost of materials?

109. Find the minimum volume for a region bounded by the planes $x = 0$, $y = 0$, $z = 0$ and a plane tangent to the ellipsoid

$$\frac{x^2}{a^2} + \frac{y^2}{b^2} + \frac{z^2}{c^2} = 1$$

at a point in the first octant.

110. Prove the following theorem: If $f(x, y)$ is defined in a region R, and f_x and f_y exist and are bounded in R, then $f(x, y)$ is continuous in R. (The assumption of boundedness is essential.)

The One-Dimensional Heat Equation

If $w(x, t)$ represents the temperature at position x at time t in a uniform conducting rod with perfectly insulated sides (Fig. 12.78), then the partial derivatives w_{xx} and w_t satisfy a differential equation of the form

$$w_{xx} = \frac{1}{c^2} w_t. \tag{3}$$

This equation is called the **one-dimensional heat equation.** The value of the positive constant c^2 is determined by the material from which the rod is made. It has been determined experimentally for a broad range of materials. For a given application one finds the appropriate value in a table. For dry soil, for example, $c^2 = 0.19$ ft^2/day.

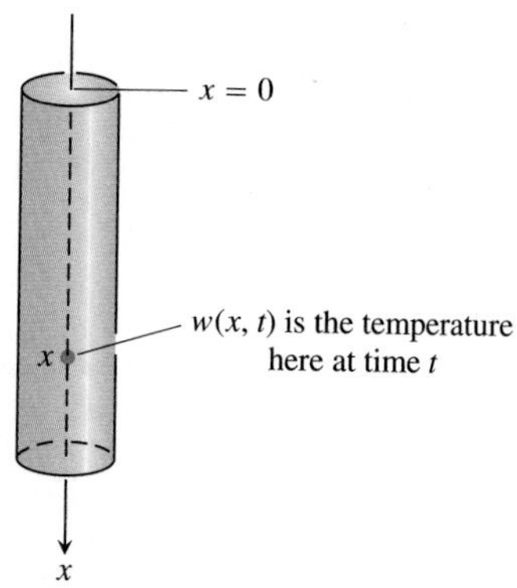

12.78 The temperature distribution in an insulated heat-conducting rod satisfies the equation

$$w_{xx} = \frac{1}{c^2} w_t.$$

In chemistry and biochemistry, the heat equation is known as the **diffusion equation.** In this context, $w(x, t)$ represents the concentration of a dissolved substance, a salt for instance, diffusing along a tube filled with liquid. The value of $w(x, t)$ is the concentration at point x at time t. In other applications, $w(x, t)$ represents the diffusion of a gas down a long, thin pipe.

In electrical engineering, the heat equation appears in the forms

$$v_{xx} = RCv_t \tag{4}$$

and

$$i_{xx} = RCi_t, \tag{5}$$

which are known as the **telegraph equations.** These equations describe the voltage v and the flow of current i in a coaxial cable or in any other cable in which leakage and inductance are negligible. The functions and constants in these equations are

$v(x, t)$ = voltage at point x at time t,

R = resistance per unit length,

C = capacitance to ground per unit of cable length,

$i(x, t)$ = current at point x at time t.

111. Find all solutions of the one-dimensional heat equation of the form $w = e^{rt} \sin \pi x$, r a constant.

112. Find all solutions of the one-dimensional heat equation that have the form $w = e^{rt} \sin kx$ and satisfy the conditions that $w(0, t) = 0$ and $w(L, t) = 0$. What happens to these solutions as $t \to \infty$?

The Cover Photograph

113. You plan to shape a test piece from a solid cylindrical blank of high-speed M-2 steel. The blank is 3.25 in. long, with a radius of 0.375 in. You will cut the piece by machining a disk from one end of the blank, shaping the other end into a right circular cone, and removing two cylindrical bands (of different depths) from the sides. Figure 12.79 shows the dimensions involved. You can control the radius and the horizontal length of your cuts with an accuracy of 1/2000 in. About how much error should you expect there to be in the mass of the finished piece? Express your answer as a percentage of the piece's ideal mass.

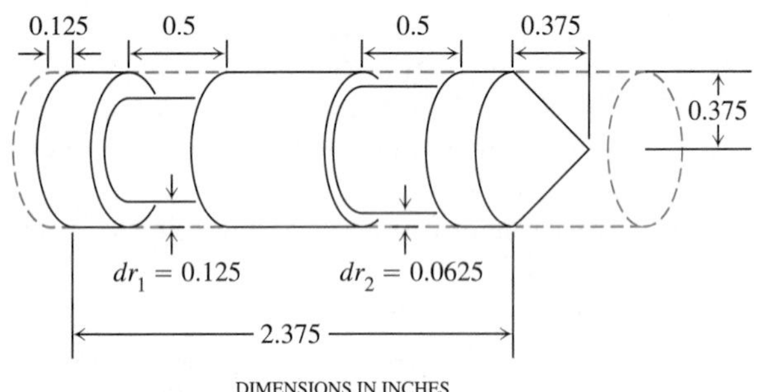

12.79 The cylindrical blank (outline) and machined piece in Exercise 113.

CHAPTER 13

MULTIPLE INTEGRALS

Overview This chapter introduces multiple integration, the complement of partial differentiation. The problems we can solve by integrating functions of two and three variables are similar to, but more general than, the problems we solve by integrating functions of a single variable. In Chapter 5, for example, we calculated the volumes of solids of revolution. Now we shall be able to calculate volumes of asymmetric solids in a variety of coordinate systems. Earlier, we calculated moments and centers of mass of rods and thin plates. Now we shall be able to handle plates more easily, work with more general density functions, and treat solids as well. As in the previous chapter, we perform the necessary calculations by drawing on our experience with functions of a single variable.

In the next chapter, we shall combine multiple integrals with vector functions to derive the magnificent vector integral theorems with which we calculate surface area and fluid flow. Multiple integrals also play an important part in statistics, electrical engineering, and the theory of electromagnetism.

13.1 Double Integrals

We now show how to integrate a continuous function $f(x, y)$ over a bounded region in the xy-plane. We begin with rectangular regions and then proceed to bounded regions of a more general nature. There are many similarities between the "double" integrals we define here and the "single" integrals we defined in Chapter 4 for functions of a single variable. Indeed, the connection is very strong. The basic theorem for evaluating double integrals says that every double integral can be evaluated in stages, using the single-integration methods already at our command.

Double Integrals Over Rectangles

Suppose that $f(x, y)$ is defined on a rectangular region R given by

$$R: \quad a \leq x \leq b, \quad c \leq y \leq d.$$

We imagine R to be covered by a network of lines parallel to the x- and y-axes

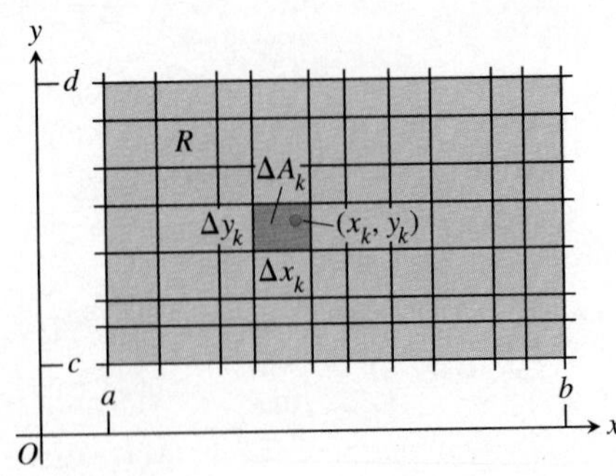

13.1 Rectangular grid partitioning the region R into small rectangles of area $\Delta A_k = \Delta x_k \Delta y_k$.

(Fig. 13.1). These lines partition R into small pieces of area

$$\Delta A = \Delta x\, \Delta y.$$

We number these in some order $\Delta A_1, \Delta A_2, \ldots, \Delta A_n$, choose a point (x_k, y_k) in each piece ΔA_k, and form the sum

$$S_n = \sum_{k=1}^{n} f(x_k, y_k)\, \Delta A_k. \tag{1}$$

If f is continuous throughout R, then, as we refine the mesh width to make both Δx and Δy go to zero, the sums in (1) approach a limit called the **double integral** of f over R. The notation for it is

$$\iint_R f(x, y)\, dA.$$

Thus,

$$\iint_R f(x, y)\, dA = \lim_{\Delta A \to 0} \sum_{k=1}^{n} f(x_k, y_k)\, \Delta A_k. \tag{2}$$

As with continuous functions of a single variable, the sums approach this limit no matter how the intervals $[a, b]$ and $[c, d]$ that determine R are partitioned, as long as the norms of the partitions both go to zero. The limit in (2) is also independent of the order in which the areas ΔA_k are numbered and independent of the choice of the point (x_k, y_k) within each ΔA_k. The values of the individual approximating sums S_n depend on these choices, but the sums approach the same limit in the end. The proof of the existence and uniqueness of this limit for a continuous function f is given in more advanced texts. The continuity of f is a sufficient, but not a necessary, condition for the existence of the double integral, and the limit in question exists for many discontinuous functions as well.

Properties of Double Integrals

Like single integrals, double integrals of continuous functions have algebraic properties that are useful in computations and applications. Among these properties are the following.

1. $\displaystyle \iint_R k f(x, y)\, dA = k \iint_R f(x, y)\, dA \qquad \text{(any number } k\text{)}$

2. $\displaystyle \iint_R (f(x, y) + g(x, y))\, dA = \iint_R f(x, y)\, dA + \iint_R g(x, y)\, dA$

3. $\displaystyle \iint_R (f(x, y) - g(x, y))\, dA = \iint_R f(x, y)\, dA - \iint_R g(x, y)\, dA$

4. $\displaystyle \iint_R f(x, y)\, dA \ge 0 \quad \text{if} \quad f(x, y) \ge 0 \text{ on } R$

5. $\displaystyle \iint_R f(x, y)\, dA \ge \iint_R g(x, y)\, dA \quad \text{if} \quad f(x, y) \ge g(x, y) \text{ on } R$

These are like the single-integral properties in Section 4.4. In addition, double inte-

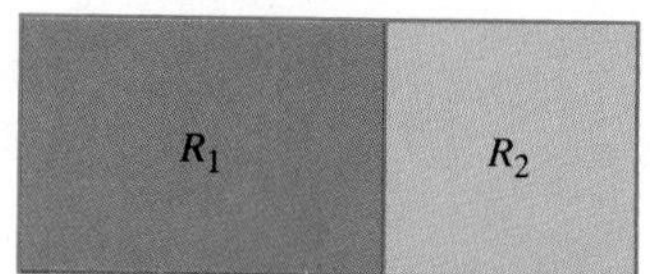

$$\iint\limits_{R_1 \cup R_2} f(x, y)\, dA = \iint\limits_{R_1} f(x, y)\, dA + \iint\limits_{R_2} f(x, y)\, dA$$

13.2 Double integrals have the same kind of domain additivity property that single integrals have.

grals have a "domain additivity" property:

6. $$\iint\limits_{R} f(x, y)\, dA = \iint\limits_{R_1} f(x, y)\, dA + \iint\limits_{R_2} f(x, y)\, dA.$$

The property holds when R is the union of two nonoverlapping rectangles R_1 and R_2 (Fig. 13.2). We shall omit the proofs.

Interpreting Double Integrals as Volumes

When $f(x, y)$ is positive, we may interpret the double integral of f over a rectangular region R as the volume of the solid prism bounded below by R and above by the surface $z = f(x, y)$ (Fig. 13.3). Each term $f(x_k, y_k)\,\Delta A_k$ in the sum $S_n = \Sigma f(x_k, y_k)\,\Delta A_k$ is the volume of a vertical rectangular prism that approximates the volume of the portion of the solid that stands directly above the base ΔA_k. The sum S_n thus approximates what we want to call the volume of the solid. We *define* this volume to be

$$\text{Volume} = \lim S_n = \iint\limits_{R} f(x, y)\, dA. \tag{3}$$

As you might expect, this more general method of calculating volume agrees with the methods in Chapter 5 but we shall not prove this here.

Fubini's Theorem for Evaluating Double Integrals

We are now ready to evaluate our first double integral.

Suppose we wish to calculate the volume under the plane $z = 4 - x - y$ over the rectangular region R: $0 \leq x \leq 2$, $0 \leq y \leq 1$ in the xy-plane. If we apply the method of slicing from Section 5.2, with slices perpendicular to the x-axis

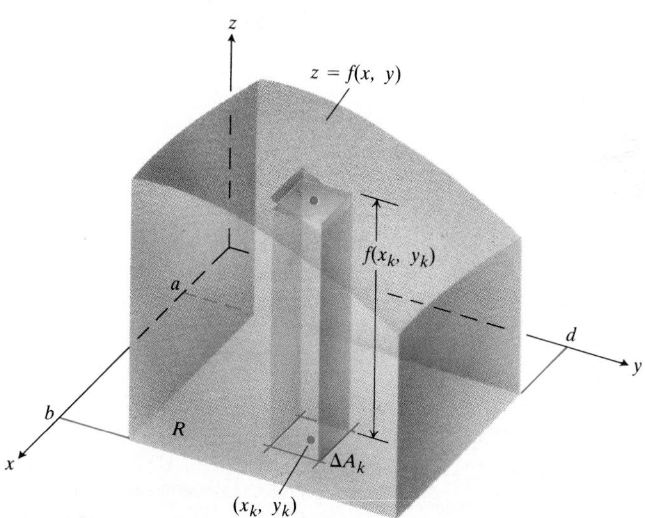

13.3 Approximating solids with rectangular prisms leads us to define the volumes of more general prisms as double integrals. The volume of the prism shown here is the double integral of $f(x, y)$ over the base region R.

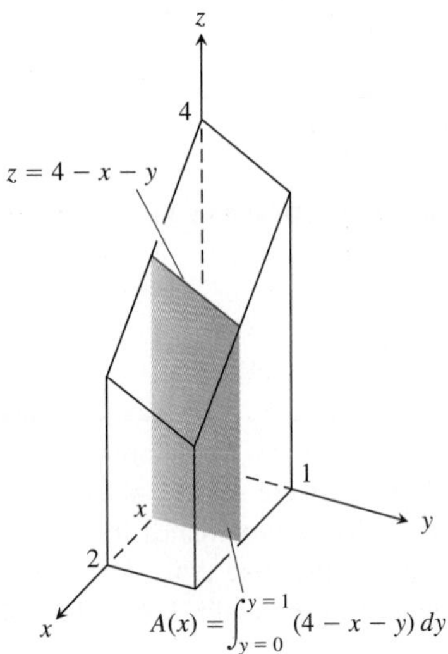

13.4 To obtain the cross-section area $A(x)$, we hold x fixed and integrate with respect to y.

(Fig. 13.4), then the volume is

$$\int_{x=0}^{x=2} A(x)\,dx, \tag{4}$$

where $A(x)$ is the cross-section area at x. For each value of x we may calculate $A(x)$ as the integral

$$A(x) = \int_{y=0}^{y=1} (4 - x - y)\,dy, \tag{5}$$

which is the area under the curve $z = 4 - x - y$ in the plane of the cross section at x. In calculating $A(x)$, x is held fixed and the integration takes place with respect to y. Combining (4) and (5), we see that the volume of the entire solid is

$$\begin{aligned} \text{Volume} &= \int_{x=0}^{x=2} A(x)\,dx = \int_{x=0}^{x=2} \left(\int_{y=0}^{y=1} (4 - x - y) dy \right) dx \\ &= \int_{x=0}^{x=2} \left[4y - xy - \frac{y^2}{2} \right]_{y=0}^{y=1} dx = \int_{x=0}^{x=2} \left(\frac{7}{2} - x \right) dx = \left[\frac{7}{2}x - \frac{x^2}{2} \right]_0^2 = 5. \end{aligned} \tag{6}$$

If we had just wanted to write instructions for calculating the volume, without carrying out any of the integrations, we could have written

$$\text{Volume} = \int_0^2 \int_0^1 (4 - x - y)\,dy\,dx.$$

The expression on the right, called an **iterated** or **repeated integral,** says that the volume is obtained by integrating $4 - x - y$ with respect to y from $y = 0$ to $y = 1$, holding x fixed, and then by integrating the resulting expression in x with respect to x from $x = 0$ to $x = 2$.

What would have happened if we had calculated the volume by slicing with planes perpendicular to the y-axis (Fig. 13.5)? As a function of y, the typical cross-section area is

$$A(y) = \int_{x=0}^{x=2} (4 - x - y)\,dx = \left[4x - \frac{x^2}{2} - xy \right]_{x=0}^{x=2} = 6 - 2y. \tag{7}$$

The volume of the entire solid is therefore

$$\text{Volume} = \int_{y=0}^{y=1} A(y)\,dy = \int_{y=0}^{y=1} (6 - 2y)\,dy = \left[6y - y^2 \right]_0^1 = 5,$$

in agreement with our earlier calculation.

Again, we may give instructions for calculating the volume as an iterated integral by writing

$$\text{Volume} = \int_0^1 \int_0^2 (4 - x - y)\,dx\,dy.$$

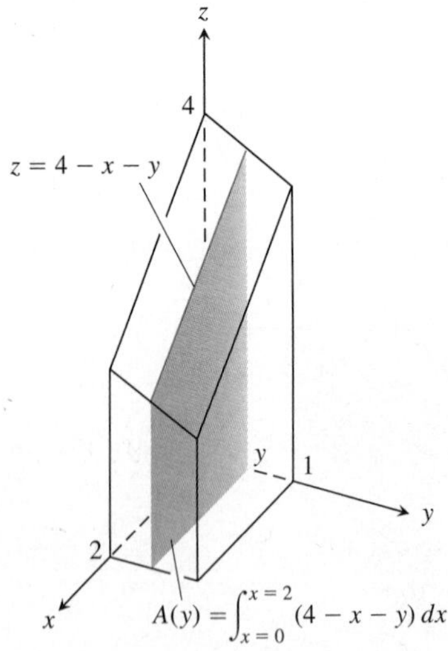

13.5 To obtain the cross-section area $A(y)$, we hold y fixed and integrate with respect to x.

The expression on the right says that the volume may be obtained by integrating $4 - x - y$ with respect to x from $x = 0$ to $x = 2$ (as we did in Eq. 7) and by integrating the result with respect to y from $y = 0$ to $y = 1$. In this iterated integral the order of integration is first x and then y, the reverse of the order we used in Eq. (6).

What do these two volume calculations with iterated integrals have to do with the double integral

$$\iint_R (4 - x - y)\, dA$$

over the rectangle $R: 0 \le x \le 2,\ 0 \le y \le 1$? The answer is that they both give the value of the double integral. A theorem proved by Guido Fubini (1879–1943) and published in 1907 says that the double integral of any continuous function over a rectangle can always be calculated as an iterated integral in either order of integration. (Fubini proved his theorem in greater generality, but this is how it translates into what we are doing now.)

THEOREM 1

Fubini's Theorem (First Form)

If $f(x, y)$ is continuous on the rectangular region $R: a \le x \le b,\ c \le y \le d$, then

$$\iint_R f(x, y)\, dA = \int_c^d \int_a^b f(x, y)\, dx\, dy = \int_a^b \int_c^d f(x, y)\, dy\, dx.$$

Fubini's theorem says that double integrals over rectangles can always be calculated as iterated integrals. This means that we can evaluate a double integral by integrating one variable at a time, using the integration techniques we already know for functions of a single variable.

Fubini's theorem also says that we may calculate the double integral by integrating in *either* order, a genuine convenience, as we shall see in Example 3. In particular, when we calculate a volume by slicing, we may use either planes perpendicular to the x-axis or planes perpendicular to the y-axis. We get the same answer both ways.

Even more important is the fact that Fubini's theorem holds for *any* continuous function $f(x, y)$. In particular, f may have negative values as well as positive values on R, and, as we shall see later, the integrals we evaluate with Fubini's theorem may represent other things besides volumes.

Example 1 Calculate $\iint_R f(x, y)\, dA$ for

$$f(x, y) = 1 - 6x^2y \quad \text{and} \quad R:\ 0 \le x \le 2, \quad -1 \le y \le 1.$$

Solution By Fubini's theorem,

$$\iint_R f(x, y)\, dA = \int_{-1}^{1} \int_0^2 (1 - 6x^2y)\, dx\, dy = \int_{-1}^{1} \Big[x - 2x^3y\Big]_{x=0}^{x=2} dy$$

$$= \int_{-1}^{1} (2 - 16y)\, dy = \Big[2y - 8y^2\Big]_{-1}^{1} = 4.$$

Reversing the order of integration gives the same answer:

$$\int_0^2 \int_{-1}^{1} (1 - 6x^2y)\, dy\, dx = \int_0^2 \Big[y - 3x^2y^2\Big]_{y=-1}^{y=1} dx$$

$$= \int_0^2 \Big[(1 - 3x^2) - (-1 - 3x^2)\Big]\, dx = \int_0^2 2\, dx = 4.$$

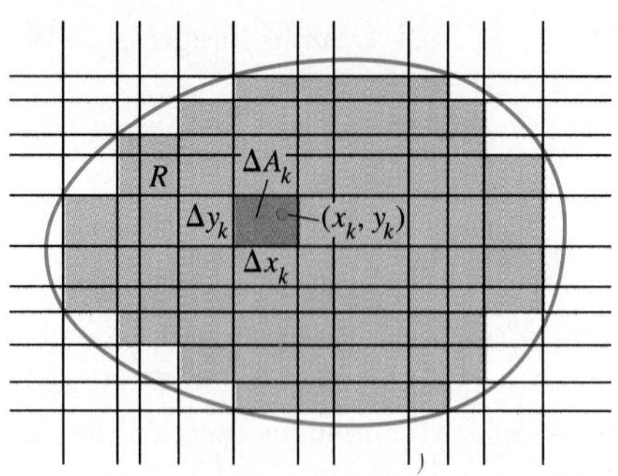

13.6 A rectangular grid partitioning a bounded nonrectangular region into cells.

Double Integrals over Bounded Nonrectangular Regions

To define the double integral of a function $f(x, y)$ over a bounded nonrectangular region, like the one shown in Fig. 13.6, we again imagine R to be covered by a rectangular grid, but we include in the partial sum only the small pieces of area $\Delta A = \Delta x \Delta y$ that lie entirely within the region (shaded in the figure). We number the pieces in some order, choose an arbitrary point (x_k, y_k) in each ΔA_k, and form the sum

$$S_n = \sum_{k=1}^{n} f(x_k, y_k)\, \Delta A_k .$$

The only difference between this sum and the one in Eq. (1) for rectangular regions is that now the areas ΔA_k may not cover all of R. But as the mesh becomes increasingly fine and the number of terms in S_n increases, more and more of R is included. If f is continuous and the boundary of R is made from the graphs of a finite number of continuous functions of x and/or continuous functions of y joined end to end, then the sums S_n will have a limit as the norms of the partitions that define the rectangular grid independently approach zero. We call the limit the **double integral** of f over R:

$$\iint_R f(x, y)\, dA = \lim_{\Delta A \to 0} \sum f(x_k, y_k)\, \Delta A_k .$$

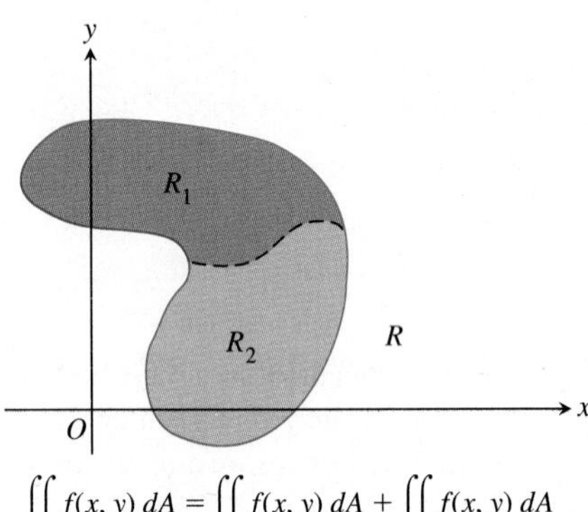

$$\iint_R f(x, y)\, dA = \iint_{R_1} f(x, y)\, dA + \iint_{R_2} f(x, y)\, dA$$

13.7 The domain additivity property stated earlier for rectangular regions also holds for regions bounded by continuous curves.

This limit may also exist under less restrictive circumstances, but we shall not pursue the point here.

Double integrals of continuous functions over nonrectangular regions have all the algebraic properties listed earlier for integrals over rectangular regions. The domain additivity property corresponding to property 6 says that if R is decomposed into nonoverlapping regions R_1 and R_2 with boundaries that are again made of a finite number of line segments or smooth curves (see Fig. 13.7), then

6′. $$\iint_R f(x, y)\, dA = \iint_{R_1} f(x, y)\, dA + \iint_{R_2} f(x, y)\, dA .$$

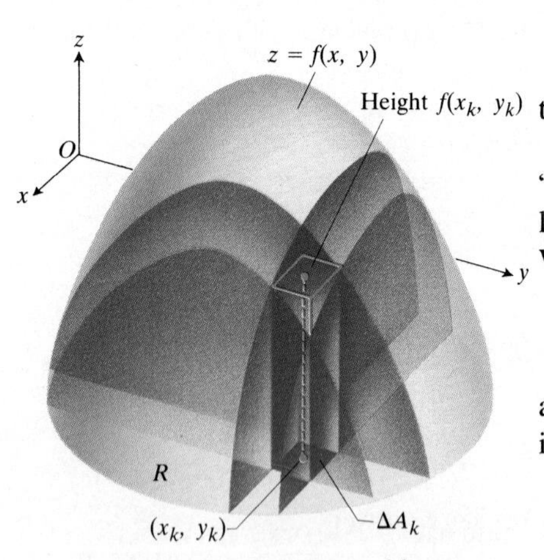

Volume $= \lim \Sigma f(x_k, y_k)\Delta A_k = \iint_R f(x, y)\, dA$

13.8 We define the volumes of solids with curved bases the same way we define the volumes of solids with rectangular bases.

If $f(x, y)$ is positive and continuous over R (Fig. 13.8), we define the volume of the solid region between R and the surface $z = f(x, y)$ to be $\iint_R f(x, y)\, dA$, as before.

If R is a region like the one shown in the xy-plane in Fig. 13.9, bounded "above" and "below" by the curves $y = g_2(x)$ and $y = g_1(x)$ and on the sides by the lines $x = a$, $x = b$, we may again calculate the volume by the method of slicing. We first calculate the cross-section area

$$A(x) = \int_{y=g_1(x)}^{y=g_2(x)} f(x, y)\, dy$$

and then integrate $A(x)$ from $x = a$ to $x = b$ to get the volume as an iterated integral:

$$V = \int_a^b A(x)\, dx = \int_a^b \int_{g_1(x)}^{g_2(x)} f(x, y)\, dy\, dx . \tag{8}$$

Similarly, if R is a region like the one shown in Fig. 13.10, bounded by the curves $x = h_2(y)$ and $x = h_1(y)$ and the lines $y = c$ and $y = d$, then the volume calculated by slicing is given by the iterated integral

$$\text{Volume} = \int_c^d \int_{h_1(y)}^{h_2(y)} f(x, y)\, dx\, dy . \tag{9}$$

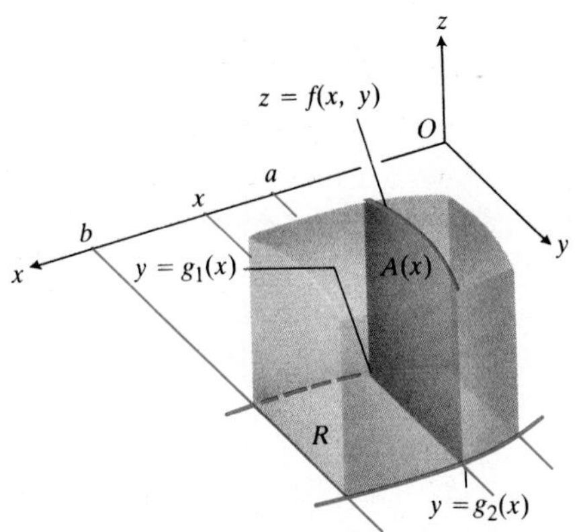

13.9 The area of the vertical slice shown here is

$$A(x) = \int_{g_1(x)}^{g_2(x)} f(x, y)\, dy.$$

To calculate the volume of the solid we integrate this area from $x = a$ to $x = b$.

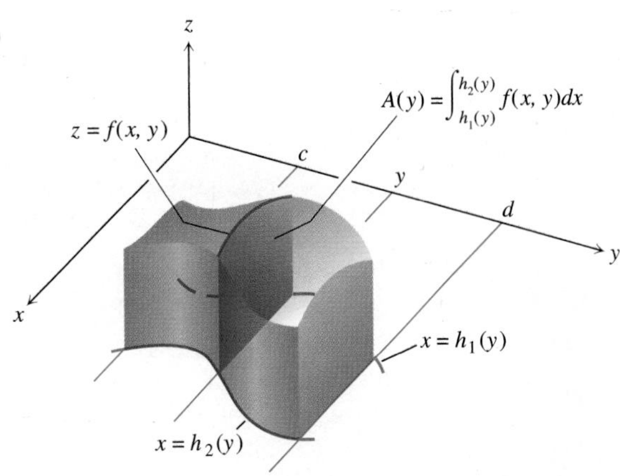

13.10 The volume of the solid shown here is

$$\int_c^d A(y)\, dy = \int_c^d \int_{h_1(y)}^{h_2(y)} f(x, y)\, dx\, dy.$$

The fact that the iterated integrals in Eqs. (8) and (9) both give the volume that we defined to be the double integral of f over R is a consequence of the following stronger form of Fubini's theorem.

THEOREM 2

Fubini's Theorem (Stronger Form)

Let $f(x, y)$ be continuous on a region R.

1. If R is defined by $a \le x \le b$, $g_1(x) \le y \le g_2(x)$, with g_1 and g_2 continuous on $[a, b]$, then
$$\iint_R f(x, y)\, dA = \int_a^b \int_{g_1(x)}^{g_2(x)} f(x, y)\, dy\, dx.$$
2. If R is defined by $c \le y \le d$, $h_1(y) \le x \le h_2(y)$, with h_1 and h_2 continuous on $[c, d]$, then
$$\iint_R f(x, y)\, dA = \int_c^d \int_{h_1(y)}^{h_2(y)} f(x, y)\, dx\, dy.$$

Example 2 Find the volume of the prism whose base is the triangle in the xy-plane bounded by the x-axis and the lines $y = x$ and $x = 1$ and whose top lies in the plane

$$z = f(x, y) = 3 - x - y.$$

Solution See Fig. 13.11(a). For any x between 0 and 1, y may vary from $y = 0$ to $y = x$ (Fig. 13.11b). Hence,

$$V = \int_0^1 \int_0^x (3 - x - y)\, dy\, dx = \int_0^1 \left[3y - xy - \frac{y^2}{2}\right]_{y=0}^{y=x} dx$$

$$= \int_0^1 \left(3x - \frac{3x^2}{2}\right) dx = \left[\frac{3x^2}{2} - \frac{x^3}{2}\right]_{x=0}^{x=1} = 1.$$

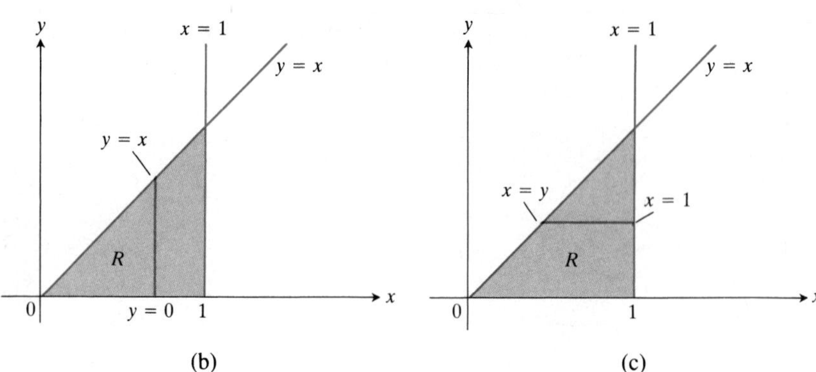

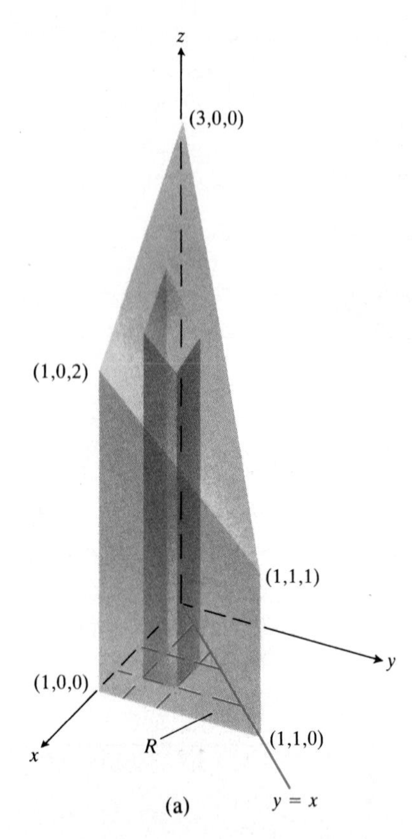

13.11 (a) Prism with a triangular base in the xy-plane. The volume of this prism is defined as a double integral over R. To evaluate it as an iterated integral, we may integrate first with respect to y and then with respect to x, or the other way around (Example 2).
(b) Integration limits of

$$\int_{x=0}^{x=1}\int_{y=0}^{y=x} f(x, y)\, dy\, dx.$$

If we integrate first with respect to y, we integrate along a vertical line through R and then integrate from left to right to include all the vertical lines in R.
(c) Integration limits of

$$\int_{y=0}^{y=1}\int_{x=y}^{x=1} f(x, y)\, dx\, dy.$$

If we integrate first with respect to x, we integrate along a horizontal line through R and then integrate from bottom to top to include all the horizontal lines in R.

When the order of integration is reversed (Fig. 13.11c), the integral for the volume is

$$\begin{aligned}
V &= \int_0^1\int_y^1 (3 - x - y)\, dx\, dy = \int_0^1 \left[3x - \frac{x^2}{2} - xy\right]_{x=y}^{x=1} dy \\
&= \int_0^1 \left(3 - \frac{1}{2} - y - 3y + \frac{y^2}{2} + y^2\right) dy \\
&= \int_0^1 \left(\frac{5}{2} - 4y + \frac{3}{2}y^2\right) dy \\
&= \left[\frac{5}{2}y - 2y^2 + \frac{y^3}{2}\right]_{y=0}^{y=1} = 1.
\end{aligned}$$

The two integrals are equal, as they should be.

While Fubini's theorem assures us that a double integral may be calculated as an iterated integral in either order of integration, the value of one integral may be easier to find than the value of the other. The next example shows how this can happen.

Example 3 Calculate

$$\iint_R \frac{\sin x}{x}\, dA,$$

where R is the triangle in the xy-plane bounded by the x-axis, the line $y = x$, and the line $x = 1$.

Solution The region of integration is shown in Fig. 13.12. If we integrate first with respect to y and then with respect to x, we find

$$\begin{aligned}
\int_0^1 \left(\int_0^x \frac{\sin x}{x}\, dy\right) dx &= \int_0^1 \left(y \frac{\sin x}{x}\right]_{y=0}^{y=x}\right) dx \\
&= \int_0^1 \sin x\, dx = -\cos(1) + 1 \approx 0.46.
\end{aligned}$$

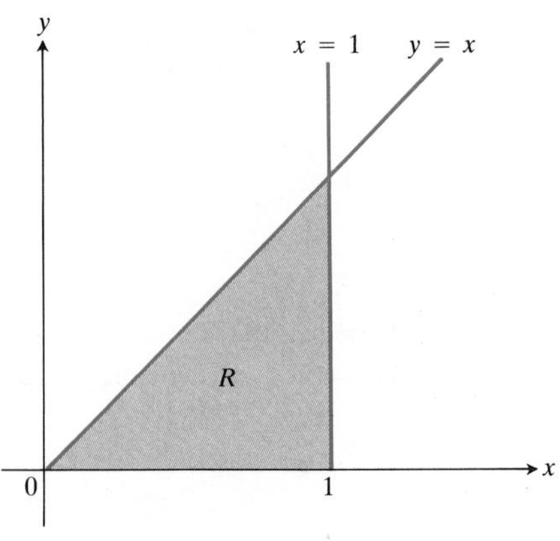

13.12 The region of integration in Example 3.

If we reverse the order of integration and attempt to calculate

$$\int_0^1 \int_y^1 \frac{\sin x}{x}\, dx\, dy,$$

we are stopped by the fact that $\int((\sin x)/x)\, dx$ cannot be expressed in terms of elementary functions.

There is no general rule for predicting which order of integration (if either) will be a good one in circumstances like these, so do not worry about how to start your integrations. Just forge ahead and if the order you choose first does not work, try the other.

Finding the Limits of Integration

The hardest part of evaluating a double integral can be finding the limits of integration. Fortunately, there is a good procedure to follow.

Procedure for Finding Limits of Integration

A. To evaluate $\iint_R f(x, y)\, dA$ over a region R, integrating first with respect to y and then with respect to x, take the following steps:

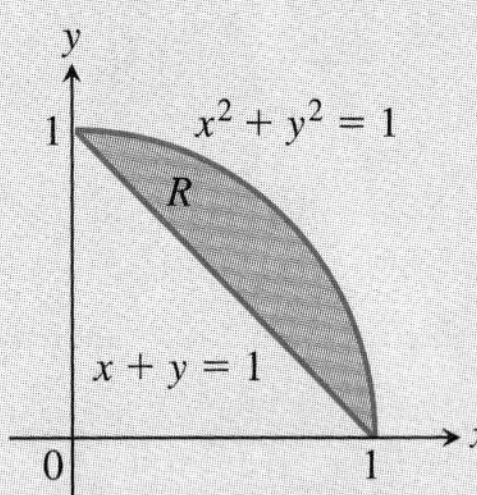

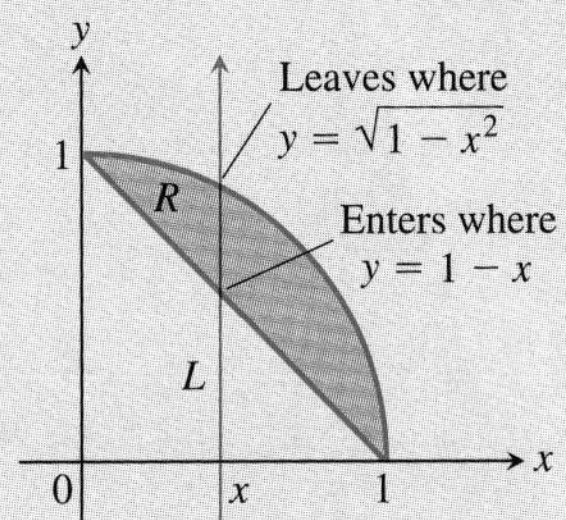

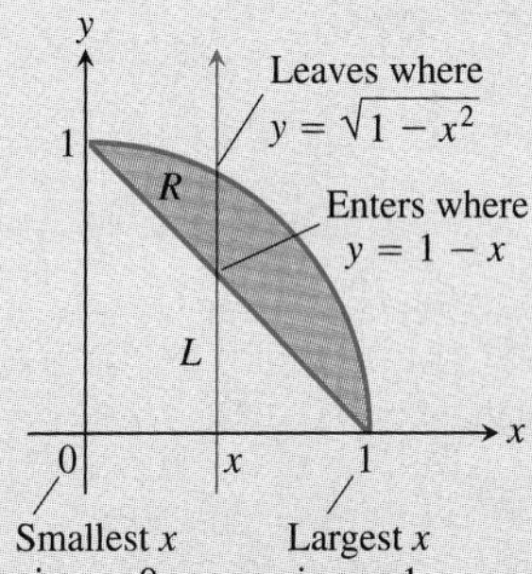

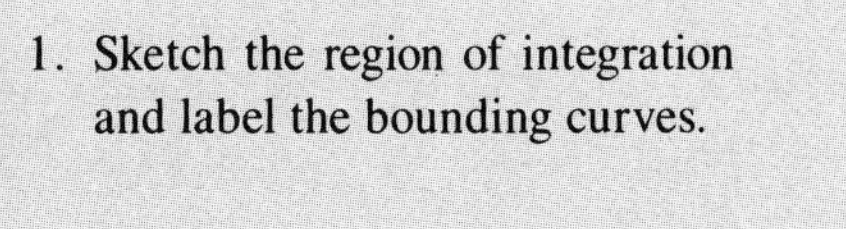

1. Sketch the region of integration and label the bounding curves.

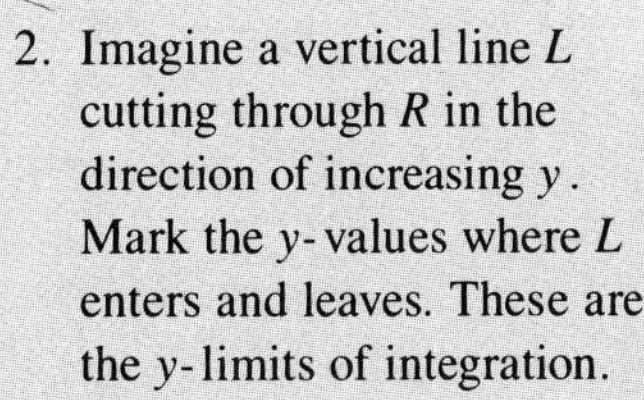

2. Imagine a vertical line L cutting through R in the direction of increasing y. Mark the y-values where L enters and leaves. These are the y-limits of integration.

3. Choose x-limits that include all the vertical lines through R. The integral is

$$\iint_R f(x, y)\, dA = \int_{x=0}^{x=1} \int_{y=1-x}^{y=\sqrt{1-x^2}} f(x, y)\, dy\, dx.$$

B. To evaluate the same double integral as an iterated integral with the order of integration reversed, use horizontal lines instead of vertical lines. The integral is

$$\iint_R f(x, y)\, dA = \int_0^1 \int_{1-y}^{\sqrt{1-y^2}} f(x, y)\, dx\, dy.$$

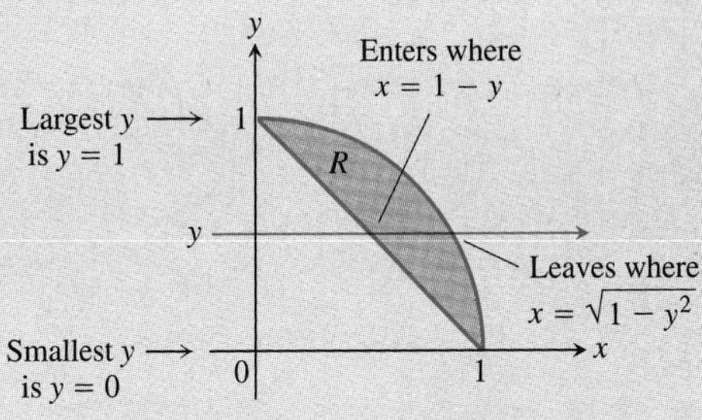

Example 4 Sketch the region over which the integration

$$\int_0^2 \int_{x^2}^{2x} (4x + 2)\, dy\, dx$$

takes place and write an equivalent integral with the order of integration reversed.

Solution The region of integration is given by the inequalities $x^2 \leq y \leq 2x$ and $0 \leq x \leq 2$. It is therefore the region bounded by the curves $y = x^2$ and $y = 2x$ between $x = 0$ and $x = 2$ (Fig. 13.13a).

To find the limits for integrating in the reverse order, we imagine a horizontal line passing from left to right through the region. It enters at $x = y/2$ and leaves at $x = \sqrt{y}$. To include all such lines, we let y run from $y = 0$ to $y = 4$ (Fig. 13.13b). The integral is

$$\int_0^4 \int_{y/2}^{\sqrt{y}} (4x + 2)\, dx\, dy.$$

The common value of these integrals is 8.

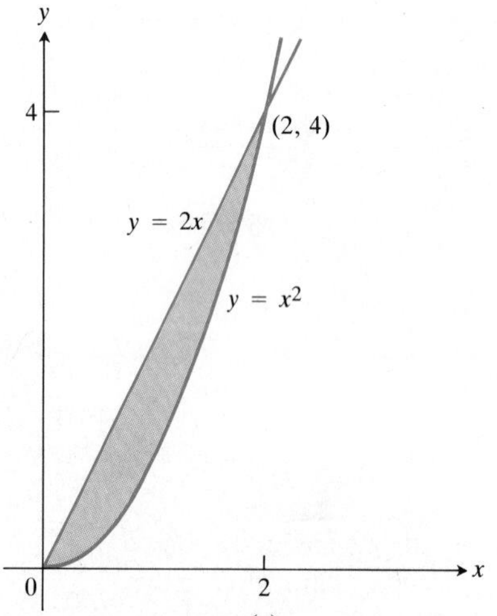

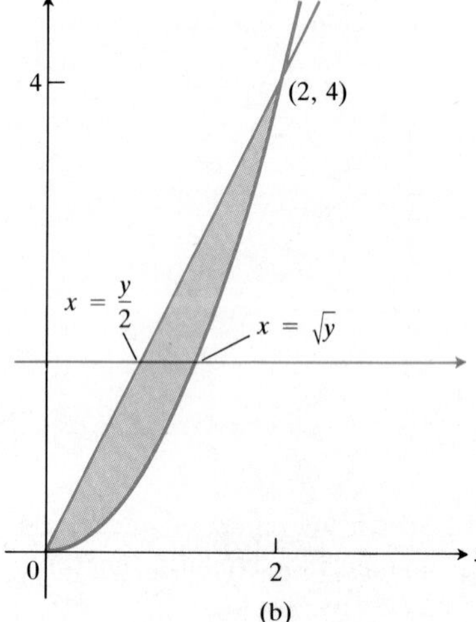

13.13 To write

$$\int_0^2 \int_{x^2}^{2x} (4x + 2)\, dy\, dx$$

as a double integral with the order of integration reversed, we (a) sketch the region of integration and (b) work out what the new limits of integration have to be. Example 4 gives the details.

EXERCISES 13.1

In Exercises 1–10 sketch the region of integration and then evaluate the integral.

1. $\int_0^3 \int_0^2 (4 - y^2)\, dy\, dx$

2. $\int_0^3 \int_{-2}^0 (x^2y - 2xy)\, dy\, dx$

3. $\int_{-1}^0 \int_{-1}^1 (x + y + 1)\, dx\, dy$

4. $\int_{\pi}^{2\pi} \int_0^{\pi} (\sin x + \cos y)\, dx\, dy$

5. $\int_0^{\pi} \int_0^x x \sin y\, dy\, dx$

6. $\int_0^{\pi} \int_0^{\sin x} y\, dy\, dx$

7. $\int_1^{\ln 8} \int_0^{\ln y} e^{x+y}\, dx\, dy$

8. $\int_1^2 \int_y^{y^2} dx\, dy$

9. $\int_{10}^{1} \int_{0}^{1/y} ye^{xy}\, dx\, dy$

10. $\int_{0}^{1} \int_{0}^{x^3} e^{y/x}\, dy\, dx$

In Exercises 11–16, integrate f over the given region.

11. $f(x, y) = x/y$ over the region in the first quadrant bounded by the lines $y = x$, $y = 2x$, $x = 1$, $x = 2$

12. $f(x, y) = x^2 + y^2$ over the triangular region with vertices $(0, 0)$, $(1, 0)$, and $(0, 1)$

13. $f(x, y) = 1/xy$ over the square $1 \le x \le 2$, $1 \le y \le 2$

14. $f(x, y) = y \cos xy$ over the rectangle $0 \le x \le \pi$, $0 \le y \le 1$

15. $f(u, v) = v - \sqrt{u}$ over the triangular region cut from the first quadrant of the uv-plane by the line $u + v = 1$

16. $f(s, t) = e^s \ln t$ over the region in the first quadrant of the st-plane that lies under the curve $s = \ln t$ from $t = 1$ to $t = 2$

Each of Exercises 17–20 gives an integral over a region in a Cartesian coordinate plane. Sketch the region and evaluate the integral.

17. $\int_{-2}^{0} \int_{v}^{-v} 2\, dp\, dv$ (the pv-plane)

18. $\int_{0}^{1} \int_{0}^{\sqrt{1-s^2}} 8t\, dt\, ds$ (the st-plane)

19. $\int_{-\pi/3}^{\pi/3} \int_{0}^{\sec t} 3 \cos t\, du\, dt$ (the tu-plane)

20. $\int_{0}^{3} \int_{-2}^{4-2u} \frac{4-2u}{v^2}\, dv\, du$ (the uv-plane)

In Exercises 21–26, sketch the region of integration and write an equivalent integral with the order of integration reversed. Then evaluate both integrals to confirm their equality.

21. $\int_{0}^{1} \int_{2}^{4-2x} dy\, dx$

22. $\int_{0}^{1} \int_{y}^{\sqrt{y}} dx\, dy$

23. $\int_{0}^{1} \int_{\sqrt{y}}^{1} dx\, dy$

24. $\int_{0}^{1} \int_{1}^{e^x} dy\, dx$

25. $\int_{0}^{3/2} \int_{0}^{9-4x^2} 16x\, dy\, dx$

26. $\int_{0}^{1} \int_{-\sqrt{1-y^2}}^{\sqrt{1-y^2}} 3y\, dx\, dy$

Evaluate the improper integrals in Exercises 27–30 as iterated integrals.

27. $\int_{1}^{\infty} \int_{e^{-x}}^{1} \frac{1}{x^3 y}\, dy\, dx$

28. $\int_{-1}^{1} \int_{-1/\sqrt{1-x^2}}^{1/\sqrt{1-x^2}} (2y + 1)\, dy\, dx$

29. $\int_{-\infty}^{\infty} \int_{-\infty}^{\infty} \frac{1}{(x^2+1)(y^2+1)}\, dx\, dy$

30. $\int_{0}^{\infty} \int_{0}^{\infty} xe^{-(x+2y)}\, dx\, dy$

In Exercises 31–38, sketch the region of integration and evaluate the integral.

31. $\int_{0}^{\pi} \int_{x}^{\pi} \frac{\sin y}{y}\, dy\, dx$

32. $\int_{0}^{1} \int_{y}^{1} x^2 e^{xy}\, dx\, dy$

33. $\int_{0}^{2} \int_{x}^{2} 2y^2 \sin xy\, dy\, dx$

34. $\int_{0}^{2} \int_{0}^{4-x^2} \frac{xe^{2y}}{4-y}\, dy\, dx$

35. $\int_{0}^{2\sqrt{\ln 3}} \int_{y/2}^{\sqrt{\ln 3}} e^{x^2}\, dx\, dy$

36. $\int_{0}^{1/16} \int_{y^{1/4}}^{1/2} \cos(16\pi x^5)\, dx\, dy$

37. $\int_{0}^{3} \int_{\sqrt{x/3}}^{1} e^{y^3}\, dy\, dx$

38. $\int_{0}^{8} \int_{\sqrt[3]{x}}^{2} \frac{dy\, dx}{y^4 + 1}$

39. Find the volume of the region that lies under the paraboloid $z = x^2 + y^2$ and above the triangle enclosed by the lines $y = x$, $x = 0$, and $x + y = 2$ in the xy-plane.

40. Find the volume of the solid that is bounded above by the cylinder $z = x^2$ and below by the region enclosed by the parabola $y = 4 - x^2$ and the line $y = x$ in the xy-plane.

41. Find the volume of the solid whose base is the region in the xy-plane that is bounded by the parabola $y = 4 - x^2$ and the line $y = 3x$, while the top of the solid is bounded by the plane $z = x + 4$.

42. Find the volume of the solid in the first octant bounded by the coordinate planes, the cylinder $x^2 + y^2 = 4$, and the plane $z + y = 3$.

43. Find the volume of the solid in the first octant bounded by the coordinate planes, the plane $x = 3$, and the parabolic cylinder $z = 4 - y^2$.

44. Find the volume of the solid cut from the first octant by the surface $z = 4 - x^2 - y$.

45. Find the volume of the wedge cut from the first octant by the cylinder $z = 12 - 3y^2$ and the plane $x + y = 2$.

46. Find the volume of the solid cut from the square column $|x| + |y| \le 1$ by the planes $z = 0$ and $3x + z = 3$.

47. Find the volume of the solid that is bounded on the front and back by the planes $x = 2$ and $x = 1$, on the sides by the cylinders $y = \pm 1/x$, and above and below by the planes $z = x + 1$ and $z = 0$.

48. Find the volume of the solid that is bounded on the front and back by the planes $x = \pm\pi/3$, on the sides by the cylinders $y = \pm\sec x$, above by the cylinder $z = 1 + y^2$, and below by the xy-plane.

49. Integrate $f(x, y) = \sqrt{4 - x^2}$ over the smaller sector cut from the disk $x^2 + y^2 \leq 4$ by the rays $\theta = \pi/6$ and $\theta = \pi/2$.

50. Integrate $f(x, y) = 1/(x^2 - x)(y - 1)^{2/3}$ over the infinite rectangle $2 \leq x < \infty$, $0 \leq y \leq 2$.

51. A solid right (noncircular) cylinder has its base R in the xy-plane and is bounded from above by the paraboloid $z = x^2 + y^2$. The cylinder's volume is

$$V = \int_0^1 \int_0^y (x^2 + y^2)\, dx\, dy + \int_1^2 \int_0^{2-y} (x^2 + y^2)\, dx\, dy.$$

Sketch the base region R and express the cylinder's volume as a single iterated integral with the order of integration reversed. Then evaluate the integral to find the volume.

52. Evaluate the integral

$$\int_0^2 (\tan^{-1}\pi x - \tan^{-1}x)\, dx.$$

(*Hint:* Write the integrand as an integral.)

53. Over what region R in the xy-plane does

$$\iint_R (4 - x^2 - 2y^2)\, dA$$

have its maximum value?

54. Over what region R in the xy-plane does

$$\iint_R \left[(x^2 + y^2)^2 - 5(x^2 + y^2) + 4\right] dA$$

have its minimum value?

Numerical Evaluation

Just as there are methods for evaluating single integrals numerically, there are corresponding methods for the numerical evaluation of double integrals. You will find a computer program for the two-variable trapezoidal rule immediately following this exercise set. If you can implement this program or if you have access to a double-integral evaluator of some other kind, estimate the values of the integrals in Exercises 55–58.

55. $\displaystyle\int_1^3 \int_1^x \frac{1}{xy}\, dy\, dx$

56. $\displaystyle\int_0^1 \int_0^1 e^{-(x^2 + y^2)}\, dy\, dx$

57. $\displaystyle\int_0^1 \int_0^1 \tan^{-1}xy\, dy\, dx$

58. $\displaystyle\int_{-1}^1 \int_0^{\sqrt{1-x^2}} 3\sqrt{1 - x^2 - y^2}\, dy\, dx$

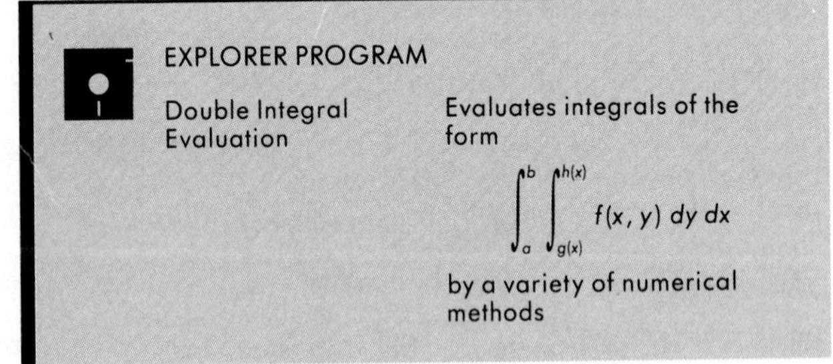

✲ A COMPUTER PROGRAM FOR EVALUATING DOUBLE INTEGRALS NUMERICALLY

The BASIC program listed here uses the trapezoidal rule to generate numerical approximations of integrals of the form

$$\int_a^b \int_{g_1(x)}^{g_2(x)} f(x, y)\, dy\, dx.$$

In the notation of the program, a is X0, b is XF (X final), and the interval $[a, b]$ is partitioned into NX subintervals of equal length. The vertical line segments that run from the curve $y = g_1(x)$ to the curve $y = g_2(x)$ above the partition points are each partitioned into NY subintervals of equal length (Fig. 13.14). These latter lengths vary with the length of the segment. The program calculates a trapezoidal sum of f-values along each vertical segment (as indicated in red in the figure) and then calculates a trapezoidal sum of these sums (indicated in black) to give the approximation for the region.

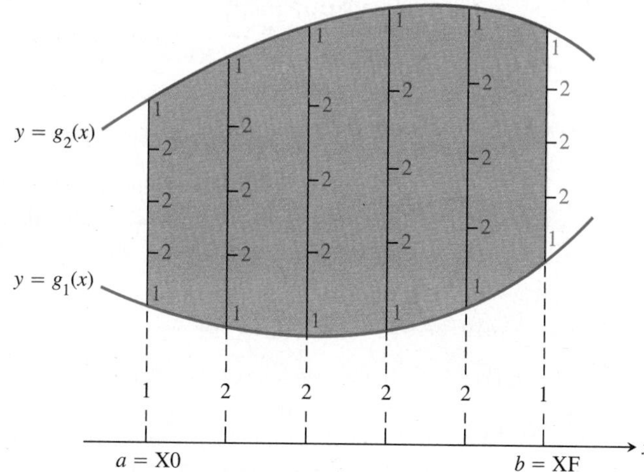

13.14 The partitions used in a typical trapezoidal rule estimate of the value of

$$\int_a^b \int_{y=g_1(x)}^{y=g_2(x)} f(x, y)\, dy\, dx.$$

For more accuracy, you can use Simpson's rule in place of the trapezoidal rule. You can find detailed descriptions of these and other methods in any standard text on numerical methods.

Program Listing

Here is a listing of the program with $f(x, y) = 4x + 2$, $g_1(x) = x^2$, and $g_2(x) = 2x$.

PROGRAM	COMMENT
10 NX = 20	NX is the number of x-intervals
20 NY = 20	NY is the number of y-intervals
30 DIM WX(NX + 1), WY(NY + 1)	Dimension arrays to hold weighting factors
40 DEF FN F(X, Y) = 4*X + 2	Define the function to be integrated
50 DEF FN G1(X) = X^2	Function G1(x) is the lower boundary
60 DEF FN G2(X) = 2*X	and G2(x) is the upper boundary
70 INPUT "ENTER X0, XF ", X0, XF	The left and right limits for x
80 DX = (XF − X0)/NX	Compute the value for delta x
90 FOR I = 2 TO NX: WX(I) = 2: NEXT I	Put the weights for x into
100 WX(1) = 1: WX(NX + 1) = 1	the WX array
110 FOR I = 2 TO NY: WY(I) = 2: NEXT I	Do the same for
120 WY(1) = 1: WY(NY + 1) = 1	the WY array
130 SUM = 0	Initialize the sum at zero
140 FOR I = 1 TO NX + 1	Begin a loop through all points in the region
150 X = X0 + (I − 1)*DX	Compute the x-value
160 Y0 = FN G1(X)	Compute the lower y-value
170 DY = (FN G2(X) − FN G1(X))/NY	and the delta-y at this x-value
180 FOR J = 1 TO NY + 1	Begin an inner loop at a constant value for x
190 Y = Y0 + (J − 1)*DY	moving up for the y's and
200 F = FN F(X, Y)	compute a function value at each point
210 SUM = SUM + WX(I)*WY(J)*F*DX/2*DY/2	and add to the sum after weighting it
220 NEXT J	End the inner loop
230 NEXT I	End the outer loop
240 PRINT "THE VALUE OF THE INTEGRAL IS ABOUT ";SUM	Print the answer
250 END	

Computation

With X0 = 0 and XF = 2, the program returns the value 7.980003 for

$$\int_0^2 \int_{x^2}^{2x} (4x + 2)\, dy\, dx.$$

The exact value of the integral is 8. With NX and NY equal to 50 instead of 20, the program gives the more accurate answer of 7.996799. With NX and NY equal to 100, the program gives 7.999367, but the calculation takes considerably longer.

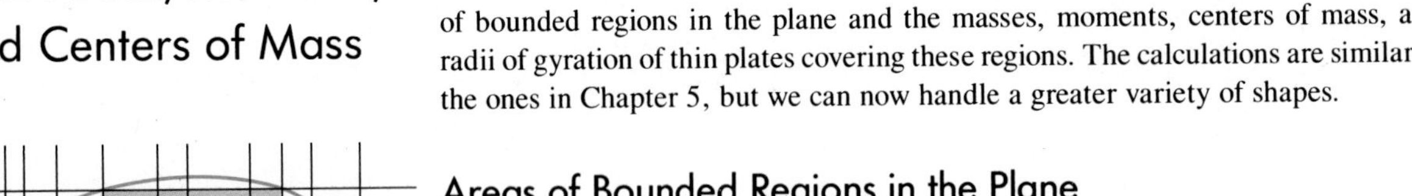

13.2 Areas, Moments, and Centers of Mass

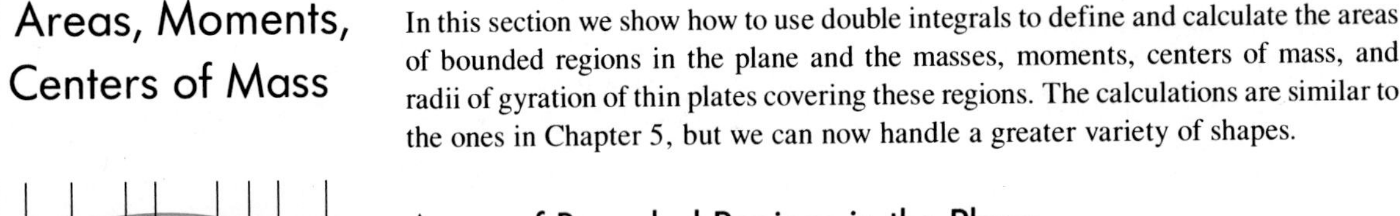

In this section we show how to use double integrals to define and calculate the areas of bounded regions in the plane and the masses, moments, centers of mass, and radii of gyration of thin plates covering these regions. The calculations are similar to the ones in Chapter 5, but we can now handle a greater variety of shapes.

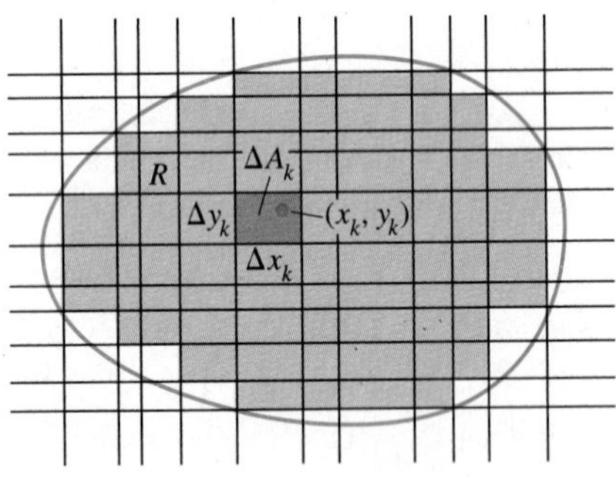

13.15 The first step in defining the area of a region is to partition the interior of the region into cells.

Areas of Bounded Regions in the Plane

If we take $f(x, y) = 1$ in the definition of the double integral over a region R in the preceding section, the partial sums reduce to

$$S_n = \sum_{k=1}^{n} f(x_k, y_k)\Delta A_k = \sum_{k=1}^{n} \Delta A_k. \tag{1}$$

This approximates what we would like to call the area of R. As the increments in x and y independently approach zero, the coverage of R by the ΔA_k's (Fig. 13.15) becomes more nearly complete, and we define the area of R to be the limit

$$\text{Area} = \lim \sum \Delta A_k = \iint_R dA. \tag{2}$$

DEFINITION

The **area** of a closed bounded plane region R is the value of the integral

$$\text{Area} = \iint_R dA. \tag{3}$$

As with the other definitions in this chapter, the definition of area applies to a greater variety of regions than does the earlier single-variable definition of area, but it agrees with the earlier definition on regions to which they both apply.

To evaluate the area integral in (3), we integrate the constant function $f(x, y) = 1$ over R.

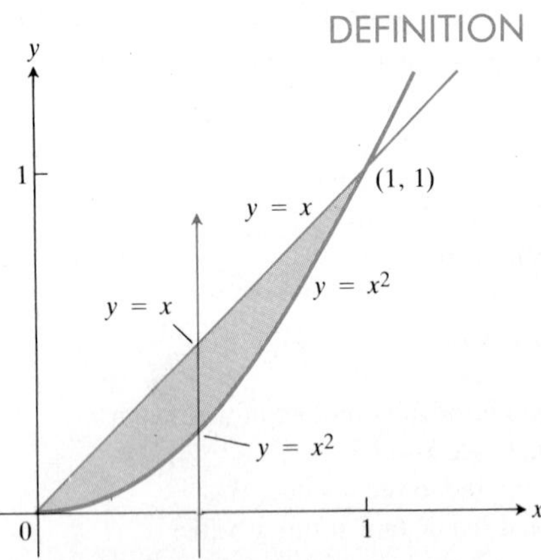

13.16 The area of the region between the parabola and the line in Example 1 is given by the double integral

$$\int_0^1 \int_{x^2}^{x} dy\, dx.$$

Example 1 Find the area of the region R bounded by $y = x$ and $y = x^2$ in the first quadrant.

Solution We sketch the region (Fig. 13.16) and calculate the area as

$$A = \int_0^1 \int_{x^2}^{x} dy\, dx = \int_0^1 (x - x^2)\, dx = \left[\frac{x^2}{2} - \frac{x^3}{3}\right]_0^1 = \frac{1}{6}.$$

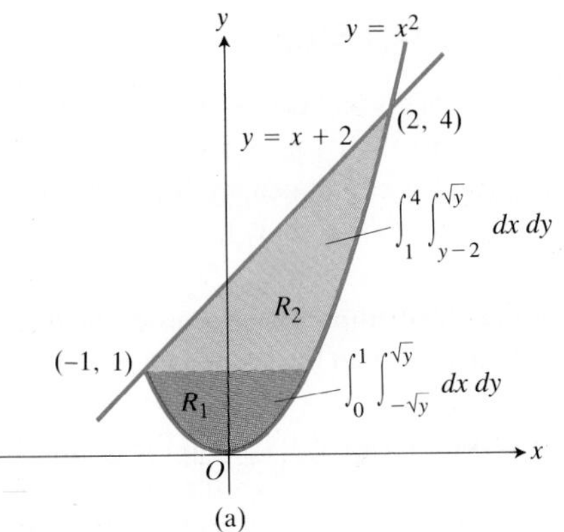

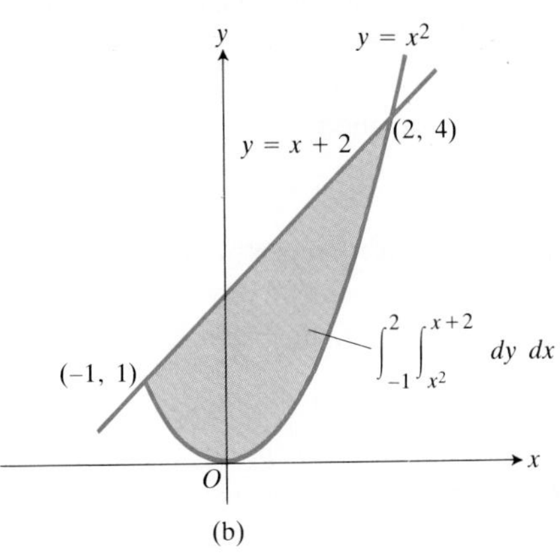

13.17 Calculating the area shown here takes (a) two integrals if the first integration is with respect to x, but (b) only one if the first integration is with respect to y (Example 2).

Example 2 Find the area of the region R enclosed by the parabola $y = x^2$ and the line $y = x + 2$.

Solution If we divide R into the regions R_1 and R_2 shown in Fig. 13.17(a), we may calculate the area as

$$A = \iint_{R_1} dA + \iint_{R_2} dA = \int_0^1 \int_{-\sqrt{y}}^{\sqrt{y}} dx\, dy + \int_1^4 \int_{y-2}^{\sqrt{y}} dx\, dy.$$

On the other hand, reversing the order of integration (Fig. 13.17b) gives

$$A = \int_{-1}^{2} \int_{x^2}^{x+2} dy\, dx.$$

Clearly, this result is simpler and is the only one we would bother to write down in practice. Evaluation of this integral leads to the result

$$A = \int_{-1}^{2} \Big[y \Big]_{x^2}^{x+2} dx = \int_{-1}^{2} (x + 2 - x^2)\, dx = \left[\frac{x^2}{2} + 2x - \frac{x^3}{3}\right]_{-1}^{2} = \frac{9}{2}.$$

Average Value

The average value of an integrable function of a single variable on a closed interval is the integral of the function over the interval divided by the length of the interval. For an integrable function of two variables defined on a closed and bounded region that has a measurable area, the average value is the integral over the region divided by the area of the region. If f is the function and R the region, then

$$\text{Average value of } f \text{ over } R = \frac{1}{\text{area of } R} \iint_R f\, dA. \tag{4}$$

If f is the density of a thin plate covering R, then the double integral of f over R divided by the area of R is the plate's average density in units of mass per unit area. If $f(x, y)$ is the distance from the point (x, y) to a fixed point P, then the average value of f over R is the average distance of points in R from P.

Example 3 Find the average value of

$$f(x, y) = x \cos xy$$

over the rectangle $R: 0 \le x \le \pi, 0 \le y \le 1$.

Solution The value of the integral of f over R is

$$\begin{aligned}\int_0^{\pi} \int_0^1 x \cos xy\, dy\, dx &= \int_0^{\pi} \Big[\sin xy\Big]_{y=0}^{y=1} dx \\ &= \int_0^{\pi} (\sin x - 0)\, dx \\ &= -\cos x \Big]_0^{\pi} \\ &= 1 + 1 = 2.\end{aligned}$$

The area of R is π. The average value of f over R is $2/\pi$.

First and Second Moments and Centers of Mass

To find the moments and centers of mass of thin sheets and plates, we use formulas similar to those in Chapter 5. The main difference is that now, with double integrals, we can accommodate a greater variety of shapes and density functions. The formulas are given in Table 13.1 The examples that follow show how the formulas are used.

The mathematical difference between the **first moments** M_x and M_y and the **moments of inertia,** or **second moments,** I_x and I_y is that the second moments use the *squares* of the "lever-arm" distances x and y.

The moment I_0 is also called the **polar moment** of inertia about the origin. It is calculated by integrating the density $\delta(x, y)$ times $r^2 = x^2 + y^2$, the square of the distance from a representative point (x, y) to the origin. Notice that $I_0 = I_x + I_y$; once we find two, we get the third automatically. (The moment I_0 is sometimes called I_z, for moment of inertia about the z-axis. The identity $I_z = I_x + I_y$ is then called the **Perpendicular Axis Theorem.**)

The **radius of gyration** R_x is defined by the equation

$$I_x = MR_x^2.$$

It tells how far from the x-axis the entire mass of the plate might be concentrated to give the same I_x. The radius of gyration gives a convenient way to express the moment of inertia in terms of a mass and a length. The radii R_y and R_0 are defined

TABLE 13.1
Mass and moment formulas for thin plates covering regions in the xy-plane

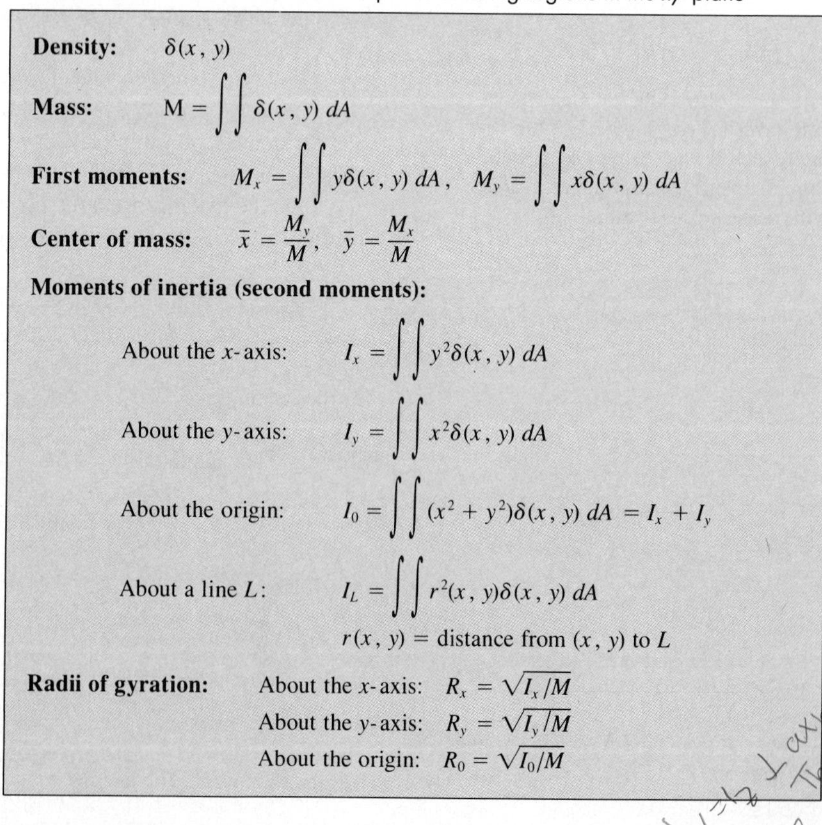

Density: $\delta(x, y)$

Mass: $M = \iint \delta(x, y)\, dA$

First moments: $M_x = \iint y\delta(x, y)\, dA, \quad M_y = \iint x\delta(x, y)\, dA$

Center of mass: $\bar{x} = \dfrac{M_y}{M}, \quad \bar{y} = \dfrac{M_x}{M}$

Moments of inertia (second moments):

About the x-axis: $I_x = \iint y^2\delta(x, y)\, dA$

About the y-axis: $I_y = \iint x^2\delta(x, y)\, dA$

About the origin: $I_0 = \iint (x^2 + y^2)\delta(x, y)\, dA = I_x + I_y$

About a line L: $I_L = \iint r^2(x, y)\delta(x, y)\, dA$

$r(x, y) =$ distance from (x, y) to L

Radii of gyration:
About the x-axis: $R_x = \sqrt{I_x/M}$
About the y-axis: $R_y = \sqrt{I_y/M}$
About the origin: $R_0 = \sqrt{I_0/M}$

in a similar way, with

$$I_y = MR_y^2 \qquad \text{and} \qquad I_0 = MR_0^2.$$

We take square roots to get the formulas in Table 13.1.

Why the interest in moments of inertia? A body's first moments tell us about balance and about the torque the body exerts about different axes in a gravitational field. But if the body is a rotating shaft, we are more likely to be interested in how much energy is stored in the shaft or about how much energy it will take to accelerate the shaft to a particular angular velocity. This is where the second moment or moment of inertia comes in.

Think of partitioning the shaft into small blocks of mass Δm_k and let r_k denote the distance from the kth block's center of mass to the axis of rotation (Fig. 13.18). If the shaft rotates at an angular velocity of $\omega = d\theta/dt$ radians per second, the block's center of mass will trace its orbit at a linear speed of

$$v_k = \frac{d}{dt}\left(r_k\theta\right) = r_k\frac{d\theta}{dt} = r_k\,\omega. \tag{5}$$

The block's kinetic energy will be approximately

$$\frac{1}{2}\Delta m_k\, v_k^2 = \frac{1}{2}\Delta m_k\left(r_k\,\omega\right)^2 = \frac{1}{2}\omega^2 r_k^2\,\Delta m_k. \tag{6}$$

The kinetic energy of the shaft will be approximately

$$\sum \frac{1}{2}\omega^2 r_k^2 \Delta m_k. \tag{7}$$

The integral approached by these sums as the shaft is partitioned into smaller and smaller blocks gives the shaft's kinetic energy:

$$\text{KE}_{\text{shaft}} = \int \frac{1}{2}\omega^2 r^2\, dm = \frac{1}{2}\omega^2 \int r^2\, dm. \tag{8}$$

The factor

$$I = \int r^2\, dm \tag{9}$$

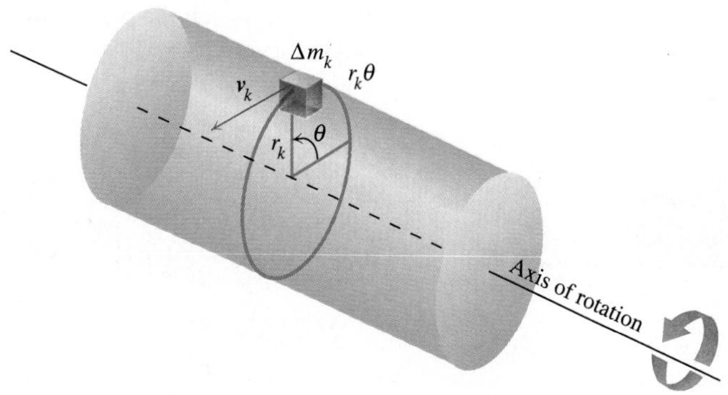

13.18 To find an integral for the amount of energy stored in a rotating shaft, we first imagine the shaft to be partitioned into small blocks. Each block has its own kinetic energy. We add the contributions of the individual blocks to find the kinetic energy of the shaft.

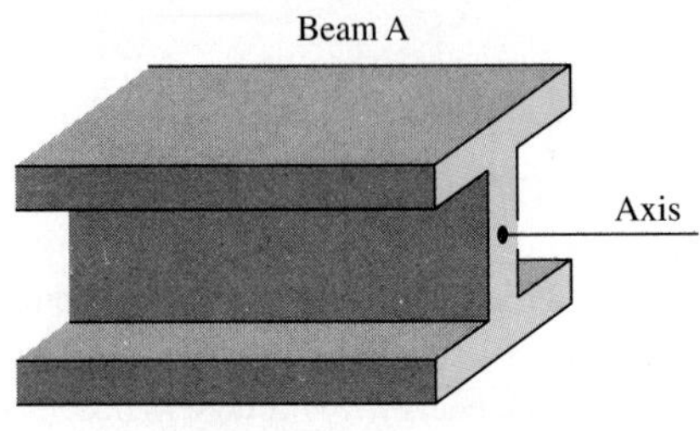

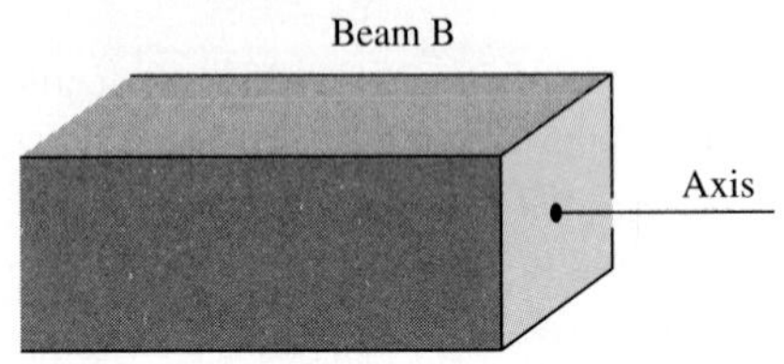

13.19 The greater the polar moment of inertia of the cross section of a beam about the beam's longitudinal axis, the stiffer the beam. Beams A and B have the same cross-section area, but beam A is stiffer.

is the moment of inertia of the shaft about its axis of rotation, and we see from Eq. (8) that the shaft's kinetic energy is

$$\mathrm{KE}_{\text{shaft}} = \frac{1}{2} I\omega^2. \tag{10}$$

To start a shaft of inertial moment I rotating at an angular velocity ω, we need to provide a kinetic energy of $\mathrm{KE} = (1/2)I\omega^2$. To stop the shaft, we have to take this amount of energy back out. To start a locomotive with mass m moving at a linear velocity v, we need to provide a kinetic energy of $\mathrm{KE} = (1/2)mv^2$. To stop the locomotive, we have to remove this amount of energy. The shaft's moment of inertia is analogous to the locomotive's mass. What makes the locomotive hard to start or stop is its mass. What makes the shaft hard to start or stop is its moment of inertia. The moment of inertia takes into account not only the mass but also its distribution.

The moment of inertia also plays a role in determining how much a horizontal metal beam will bend under a load. The stiffness of the beam is a constant times I, the polar moment of inertia of a typical cross section of the beam perpendicular to the beam's longitudinal axis. The greater the value of I, the stiffer the beam and the less it will bend under a given load. That is why we use I-beams instead of beams whose cross sections are square. The flanges at the top and bottom of the beam hold most of the beam's mass away from the longitudinal axis to maximize the value of I (Fig. 13.19).

If you want to see the moment of inertia at work, try the following experiment. Tape two coins to the ends of a pencil and twiddle the pencil about the center of mass. The moment of inertia accounts for the resistance you feel each time you change the direction of motion. Now move the coins an equal distance toward the center of mass and twiddle the pencil again. The system has the same mass and the same center of mass but now offers less resistance to the changes in motion. The moment of inertia has been reduced. The moment of inertia is what gives a baseball bat, golf club, or tennis racket its "feel." Tennis rackets that weigh the same, look the same, and have identical centers of mass will feel different and behave differently if their masses are not distributed the same way.

First moments are "balancing" moments. Second moments are "turning" moments.

Example 4 A thin plate covers the triangular region bounded by the x-axis and the lines $x = 1$ and $y = 2x$ in the first quadrant. The plate's density at the point (x, y) is $\delta(x, y) = 6x + 6y + 6$. Find the plate's mass, first moments, center of mass, moments of inertia, and radii of gyration about the coordinate axes.

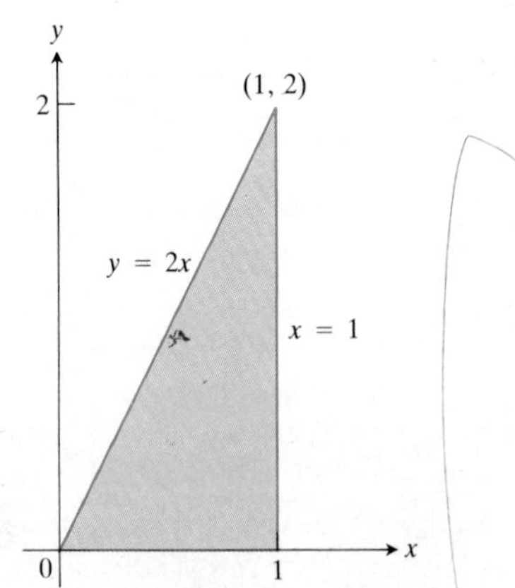

13.20 The triangular region covered by the plate in Example 4.

Solution We sketch the plate and put in enough detail to determine the limits of integration for the integrals we have to evaluate (Fig. 13.20).

The plate's mass is

$$\begin{aligned} M &= \int_0^1 \int_0^{2x} \delta(x, y)\, dy\, dx = \int_0^1 \int_0^{2x} (6x + 6y + 6)\, dy\, dx \\ &= \int_0^1 \left[6xy + 3y^2 + 6y \right]_{y=0}^{y=2x} dx \\ &= \int_0^1 (24x^2 + 12x)\, dx = \left[8x^3 + 6x^2 \right]_0^1 = 14. \end{aligned}$$

The first moment about the x-axis is

$$M_x = \int_0^1 \int_0^{2x} y\delta(x, y)\, dy\, dx = \int_0^1 \int_0^{2x} (6xy + 6y^2 + 6y)\, dy\, dx$$

$$= \int_0^1 \left[3xy^2 + 2y^3 + 3y^2\right]_{y=0}^{y=2x} dx = \int_0^1 (28x^3 + 12x^2)\, dx$$

$$= \left[7x^4 + 4x^3\right]_0^1 = 11.$$

A similar calculation gives

$$M_y = \int_0^1 \int_0^{2x} x\delta(x, y)\, dy\, dx = 10.$$

The coordinates of the center of mass are therefore

$$\bar{x} = \frac{M_y}{M} = \frac{10}{14} = \frac{5}{7}, \qquad \bar{y} = \frac{M_x}{M} = \frac{11}{14}.$$

The moment of inertia about the x-axis is

$$I_x = \int_0^1 \int_0^{2x} y^2\delta(x, y)\, dy\, dx = \int_0^1 \int_0^{2x} (6xy^2 + 6y^3 + 6y^2)\, dy\, dx$$

$$= \int_0^1 \left[2xy^3 + \frac{3}{2}y^4 + 2y^3\right]_{y=0}^{y=2x} dx = \int_0^1 (40x^4 + 16x^3)\, dx = \left[8x^5 + 4x^4\right]_0^1 = 12.$$

Similarly, the moment of inertia about the y-axis is

$$I_y = \int_0^1 \int_0^{2x} x^2\delta(x, y)\, dy\, dx = \frac{39}{5}.$$

Since we know I_x and I_y, we do not need to evaluate an integral to find I_0; we can use the equation $I_0 = I_x + I_y$ instead:

$$I_0 = 12 + \frac{39}{5} = \frac{60 + 39}{5} = \frac{99}{5}.$$

The three radii of gyration are

$$R_x = \sqrt{I_x/M} = \sqrt{12/14} = \sqrt{6/7},$$

$$R_y = \sqrt{I_y/M} = \sqrt{\left(\frac{39}{5}\right)/14} = \sqrt{39/70},$$

$$R_0 = \sqrt{I_0/M} = \sqrt{\left(\frac{99}{5}\right)/14} = \sqrt{99/70}.$$

Centroids of Geometric Figures

When the density of an object is constant, it cancels out of the numerator and denominator of the formulas for $\bar{x}$ and $\bar{y}$. As far as $\bar{x}$ and $\bar{y}$ are concerned, δ might as well be 1. Thus, when δ is constant, the location of the center of mass becomes a feature of the object's shape and not of the material of which it is made. In such cases, engineers may call the center of mass the **centroid** of the shape. To find a centroid, we set δ equal to 1 and proceed to find $\bar{x}$ and $\bar{y}$ as before, by dividing first moments by masses.

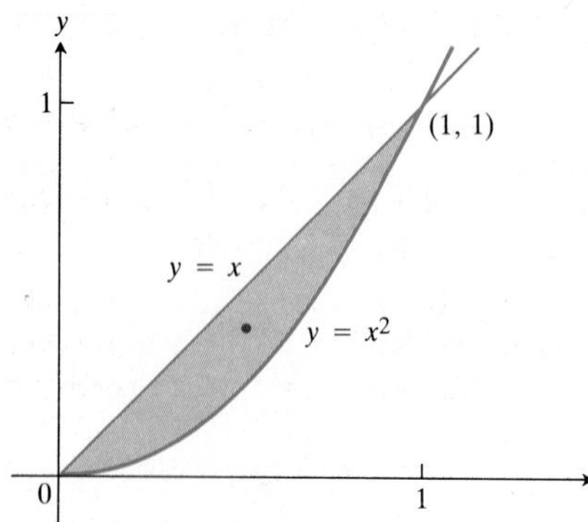

13.21 Example 5 calculates the coordinates of the centroid of the region shown here.

Example 5 Find the centroid of the region in the first quadrant that is bounded above by the line $y = x$ and below by the parabola $y = x^2$.

Solution We sketch the region and include enough detail to determine the limits of integration (Fig. 13.21). We then set δ equal to 1 and evaluate the appropriate formulas from Table 13.1:

$$M = \int_0^1 \int_{x^2}^{x} 1\, dy\, dx = \int_0^1 \left[y\right]_{y=x^2}^{y=x} dx = \int_0^1 (x - x^2)\, dx = \left[\frac{x^2}{2} - \frac{x^3}{3}\right]_0^1 = \frac{1}{6},$$

$$M_x = \int_0^1 \int_{x^2}^{x} y\, dy\, dx = \int_0^1 \left[\frac{y^2}{2}\right]_{y=x^2}^{y=x} dx = \int_0^1 \left(\frac{x^2}{2} - \frac{x^4}{2}\right) dx = \left[\frac{x^3}{6} - \frac{x^5}{10}\right]_0^1 = \frac{1}{15},$$

$$M_y = \int_0^1 \int_{x^2}^{x} x\, dy\, dx = \int_0^1 \left[xy\right]_{y=x^2}^{y=x} dx = \int_0^1 (x^2 - x^3)\, dx = \left[\frac{x^3}{3} - \frac{x^4}{4}\right]_0^1 = \frac{1}{12}.$$

From these values of M, M_x, and M_y, we find

$$\bar{x} = \frac{M_y}{M} = \frac{1/12}{1/6} = \frac{1}{2} \quad \text{and} \quad \bar{y} = \frac{M_x}{M} = \frac{1/15}{1/6} = \frac{2}{5}.$$

The centroid is the point $\left(\frac{1}{2}, \frac{2}{5}\right)$.

EXERCISES 13.2

In Exercises 1–8, sketch the region bounded by the given lines and curves. Then find the region's area by double integration.

1. The coordinate axes and the line $x + y = 2$
2. The x-axis, the curve $y = e^x$, and the lines $x = 0$, $x = \ln 2$
3. The y-axis and the lines $y = 2x$, $y = 4$
4. The parabola $x = -y^2$ and the line $y = x + 2$
5. The parabolas $x = y^2$ and $x = 2y - y^2$
6. The parabola $x = y - y^2$ and the line $x + y = 0$
7. The semiellipse $y = 2\sqrt{1 - x^2}$ and the lines $x = \pm 1$, $y = -1$
8. Above by $y = x^2$, below by $y = -1$, on the left by $x = -2$, and on the right by $y = 2x - 1$

The integrals and sums of integrals in Exercises 9–14 give the areas of regions in the xy-plane. Sketch each region, label each bounding curve with its equation, and give the coordinates of the points where the curves intersect. Then find the area of the region.

9. $\displaystyle\int_0^6 \int_{y^2/3}^{2y} dx\, dy$

10. $\displaystyle\int_0^3 \int_{-x}^{x(2-x)} dy\, dx$

11. $\displaystyle\int_0^{\pi/4} \int_{\sin x}^{\cos x} dy\, dx$

12. $\displaystyle\int_{-1}^2 \int_{y^2}^{y+2} dx\, dy$

13. $\displaystyle\int_{-1}^0 \int_{-2x}^{1-x} dy\, dx + \int_0^2 \int_{-x/2}^{1-x} dy\, dx$

14. $\displaystyle\int_0^2 \int_{x^2-4}^{0} dy\, dx + \int_0^4 \int_0^{\sqrt{x}} dy\, dx$

Constant Density

15. Find the center of mass of a thin plate of density $\delta = 3$ bounded by the lines $x = 0$, $y = x$, and the parabola $y = 2 - x^2$ in the first quadrant.
16. Find the moments of inertia and radii of gyration about the coordinate axes of a thin rectangular plate of constant density δ bounded by the coordinate axes and the lines $x = 3$ and $y = 3$ in the first quadrant.

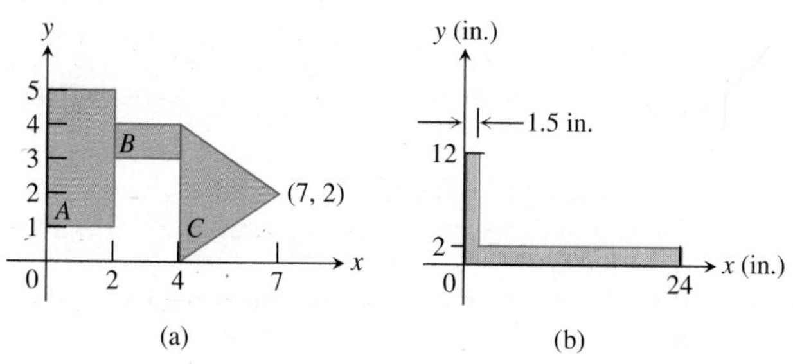

13.23 The figures for Exercises 51 and 52.

53. An isosceles triangle T has base $2a$ and altitude h. The base lies along the diameter of a semicircular disk D of radius a so that the two together make a shape resembling an ice cream cone. What relation must hold between a and h to place the centroid of $T \cup D$ on the common boundary of T and D? inside T?

54. An isosceles triangle T of altitude h has as its base one side of a square Q whose edges have length s. (The square and triangle do not overlap.) What relation must hold between h and s to place the centroid of $T \cup Q$ on the base of the triangle? Compare your answer with the answer to Exercise 53.

13.3 Double Integrals in Polar Form

Integrals are sometimes easier to evaluate if we change to polar coordinates. This section shows how to accomplish the change and how to evaluate integrals over regions whose boundaries are given by polar equations.

Integrals in Polar Coordinates

When we defined the double integral of a function over a region R in the xy-plane, we began by cutting R into rectangles whose sides were parallel to the coordinate axes. These were the natural shapes to use because their sides have either constant x-values or constant y-values. In polar coordinates, the natural shape is a "polar rectangle" whose sides have constant r- and θ values.

Suppose that a function $f(r, \theta)$ is defined over a region R that is bounded by the rays $\theta = \alpha$ and $\theta = \beta$ and by the continuous curves $r = g_1(\theta)$ and $r = g_2(\theta)$. Suppose also that $0 \le g_1(\theta) \le g_2(\theta) \le a$ for every value of θ between α and β. Then R lies in a fan-shaped region Q defined by the inequalities $0 \le r \le a$ and $\alpha \le \theta \le \beta$. See Fig. 13.24.

We cover Q by a grid of circular arcs and rays. The arcs are cut from circles centered at the origin, with radii $\Delta r, 2\Delta r, \ldots, m\Delta r$, where $\Delta r = a/m$. The rays

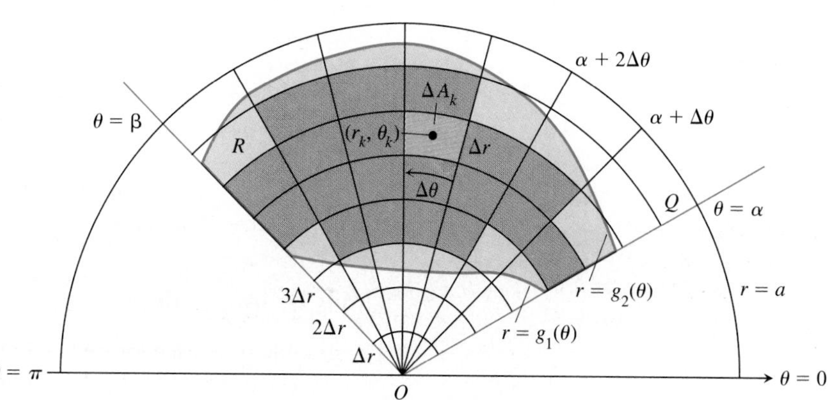

13.24 The region R: $g_1(\theta) \le r \le g_2(\theta)$, $\alpha \le \theta \le \beta$ is contained in the fan-shaped region Q: $0 \le r \le a$, $\alpha \le \theta \le \beta$. The partition of Q by circular arcs and rays induces a partition of R.

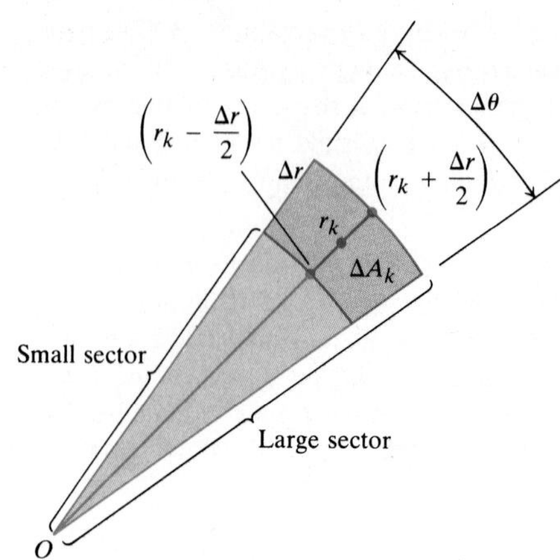

13.25 The observation that

$$\Delta A_k = \begin{pmatrix}\text{area of} \\ \text{large sector}\end{pmatrix} - \begin{pmatrix}\text{area of} \\ \text{small sector}\end{pmatrix}$$

leads to the formula $\Delta A_k = r_k \Delta r \Delta\theta$. The text explains why.

are given by

$$\theta = \alpha, \quad \theta = \alpha + \Delta\theta, \quad \theta = \alpha + 2\Delta\theta, \quad \ldots, \quad \theta = \alpha + m'\Delta\theta = \beta.$$

The arcs and rays divide Q into small patches called "polar rectangles."

We number the polar rectangles that lie inside R (the order does not matter), calling their areas

$$\Delta A_1, \quad \Delta A_2, \quad \ldots, \quad \Delta A_n.$$

We let (r_k, θ_k) be the center of the polar rectangle whose area is ΔA_k. By "center" we mean the point that lies halfway between the circular arcs on the ray that bisects the arcs. We then form the sum

$$S_n = \sum_{k=1}^{n} f(r_k, \theta_k)\Delta A_k. \tag{1}$$

If f is continuous throughout R, this sum will approach a limit as we refine the grid to make Δr and $\Delta\theta$ go to zero. The limit is called the double integral of f over R. In symbols,

$$\lim S_n = \iint_R f(r, \theta)\, dA.$$

To evaluate this limit, we first have to write the sum S_n in a way that expresses ΔA_k in terms of Δr and $\Delta\theta$. The radius of the inner arc bounding ΔA_k is $r_k - (\Delta r/2)$ (Fig. 13.25). The area of the circular sector subtended by this arc at the origin is therefore

$$\frac{1}{2}\left(r_k - \frac{\Delta r}{2}\right)^2 \Delta\theta. \tag{2}$$

Similarly, the radius of the outer boundary of ΔA_k is $r_k + (\Delta r/2)$. The area of the sector it subtends is

$$\frac{1}{2}\left(r_k + \frac{\Delta r}{2}\right)^2 \Delta\theta. \tag{3}$$

Therefore,

$$\begin{aligned}\Delta A_k &= \text{area of large sector} - \text{area of small sector} \\ &= \frac{\Delta\theta}{2}\left[\left(r_k + \frac{\Delta r}{2}\right)^2 - \left(r_k - \frac{\Delta r}{2}\right)^2\right] = \frac{\Delta\theta}{2}\,[2r_k\,\Delta r] = r_k \Delta r\, \Delta\theta.\end{aligned}$$

Combining this result with Eq. (1) gives

$$S_n = \sum_{k=1}^{n} f(r_k, \theta_k) r_k \Delta r \Delta\theta. \tag{4}$$

A version of Fubini's theorem now says that the limit approached by these sums can be evaluated by repeated single integrations with respect to r and θ as

$$\iint_R f(r, \theta)\, dA = \int_{\theta=\alpha}^{\theta=\beta} \int_{r=g_1(\theta)}^{r=g_2(\theta)} f(r, \theta)\, r\, dr\, d\theta. \tag{5}$$

Limits of Integration

The procedure we used for finding limits of integration for integrals in rectangular coordinates also works for polar coordinates.

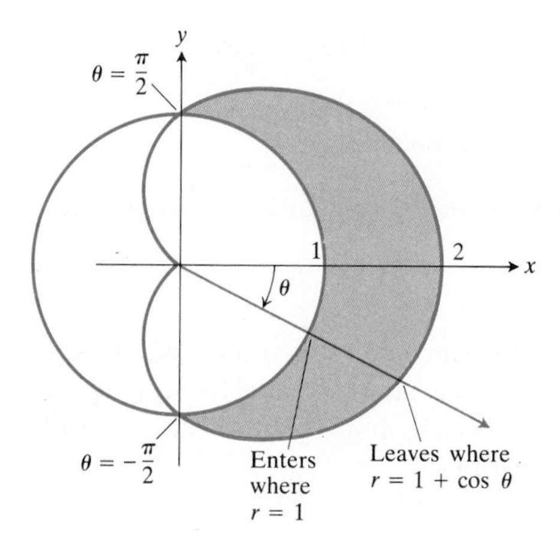

13.26 Example 1 describes the process by which we find limits of integration for calculating the area of the shaded region between the circle $r = 1$ and the cardioid $r = 1 + \cos\theta$.

Example 1 *How to Find Limits of Integration.* Find the limits of integration for integrating a function $f(r, \theta)$ over the region R that lies inside the cardioid $r = 1 + \cos\theta$ and outside the circle $r = 1$.

Solution We graph the cardioid and circle (Fig. 13.26) and carry out the following steps:

STEP 1: Holding θ fixed, let r increase to trace a ray out from the origin.

STEP 2: Integrate from the r-value where the ray enters R to the r-value where the ray leaves R.

STEP 3: Choose θ-limits to include all the rays from the origin that intersect R.

The result is the integral

$$\int_{-\pi/2}^{\pi/2} \int_{r=1}^{r=1+\cos\theta} f(r, \theta)\, r\, dr\, d\theta.$$

If $f(r, \theta)$ is the constant function whose value is 1, then the value of the integral of f over a region R is the area of R:

Area in Polar Coordinates

The area of a closed and bounded region R in the polar coordinate plane is given by the formula

$$\text{Area of } R = \iint_R r\, dr\, d\theta. \tag{6}$$

As you might expect, this formula for area is consistent with all earlier formulas, although we shall not prove the fact.

Example 2 Find the area enclosed by the lemniscate $r^2 = 4\cos 2\theta$.

Solution We graph the lemniscate to determine the limits of integration (Fig. 13.27) and see that the total area is four times the first-quadrant portion.

$$\text{Area} = 4\int_0^{\pi/4} \int_0^{\sqrt{4\cos 2\theta}} r\, dr\, d\theta = 4\int_0^{\pi/4} \left[\frac{r^2}{2}\right]_{r=0}^{r=\sqrt{4\cos 2\theta}} d\theta$$

$$= 4\int_0^{\pi/4} 2\cos 2\theta\, d\theta = 4\sin 2\theta\Big]_0^{\pi/4} = 4.$$

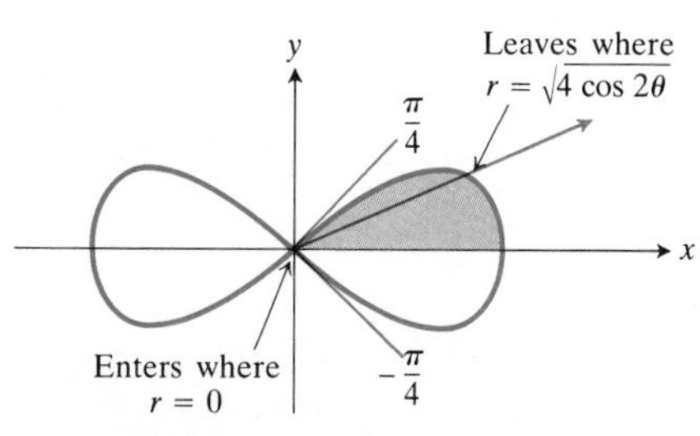

13.27 To integrate over the shaded region bounded by the lemniscate $r^2 = 4\cos 2\theta$, we run r from 0 to $\sqrt{4\cos 2\theta}$ and θ from 0 to $\pi/4$ (Example 2).

Changing Cartesian Integrals into Polar Integrals

The procedure for changing a Cartesian integral

$$\iint_R f(x, y)\, dx\, dy$$

into a polar integral has two steps:

STEP 1: Substitute $x = r\cos\theta$ and $y = r\sin\theta$, and replace $dx\,dy$ by $r\,dr\,d\theta$ in the Cartesian integral.

STEP 2: Supply polar limits of integration for the boundary of R.

The Cartesian integral then becomes

$$\iint_R f(x, y)\,dx\,dy = \iint_G f(r\cos\theta, r\sin\theta)\,r\,dr\,d\theta, \tag{7}$$

where G denotes the region of integration in polar coordinates. This is like the substitution method in Chapter 4 except that there are now two variables to substitute for instead of one. Notice that $dx\,dy$ is not replaced by $dr\,d\theta$ but by $r\,dr\,d\theta$. We shall go into the reasons for that briefly in Section 13.7.

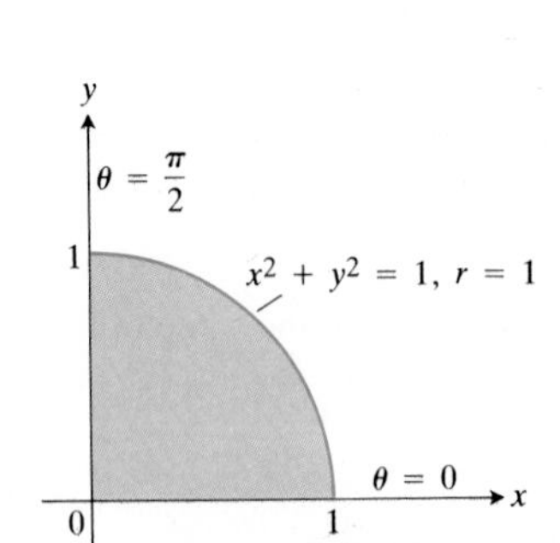

13.28 Example 3 evaluates the integral of a function over the region enclosed by the quarter circle $x^2 + y^2 = 1$ in the first quadrant. In polar coordinates, this region is described by simple inequalities:

$0 \le r \le 1$ and $0 \le \theta \le \pi/2$.

Example 3 Find the polar moment of inertia about the origin of a thin plate of density $\delta(x, y) = 1$ bounded by the quarter circle $x^2 + y^2 = 1$ in the first quadrant.

Solution We sketch the region of integration to determine the limits of integration (Fig. 13.28).

In Cartesian coordinates, the polar moment is the value of the integral

$$\int_0^1 \int_0^{\sqrt{1-x^2}} (x^2 + y^2)\,dy\,dx.$$

Integration with respect to y gives

$$\int_0^1 \left(x^2\sqrt{1-x^2} + \frac{(1-x^2)^{3/2}}{3}\right) dx,$$

an integral difficult to evaluate without tables.

Things go better if we change the original integral to polar coordinates. Substituting $x = r\cos\theta$, $y = r\sin\theta$, and replacing $dx\,dy$ by $r\,dr\,d\theta$, we get

$$\begin{aligned}\int_0^1 \int_0^{\sqrt{1-x^2}} (x^2 + y^2)\,dy\,dx &= \int_0^{\pi/2} \int_0^1 (r^2)\,r\,dr\,d\theta \\ &= \int_0^{\pi/2} \left[\frac{r^4}{4}\right]_{r=0}^{r=1} d\theta \\ &= \int_0^{\pi/2} \frac{1}{4}\,d\theta = \frac{\pi}{8}.\end{aligned}$$

Why was the polar coordinate transformation so effective? One reason is that $x^2 + y^2$ was simplified to r^2. Another is that the limits of integration became constants.

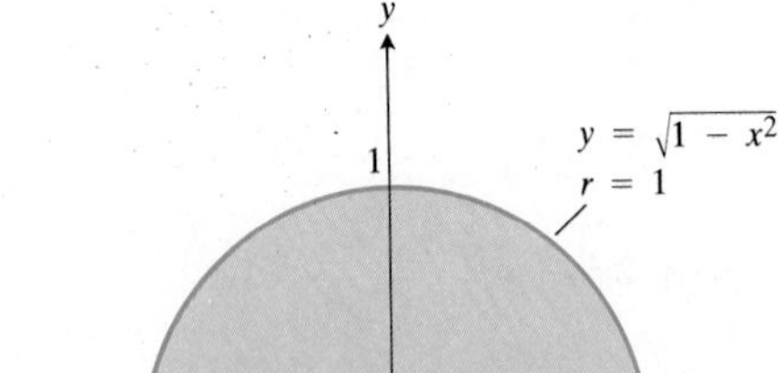

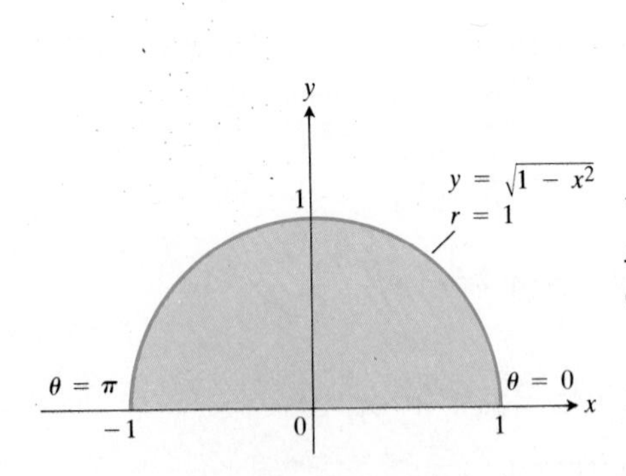

13.29 The semicircular region in Example 4 is described by the polar coordinate inequalities

$0 \le r \le 1$ and $0 \le \theta \le \pi$.

Example 4 Evaluate

$$\iint_R e^{x^2+y^2}\,dy\,dx,$$

where R is the semicircular region bounded by the x-axis and the curve $y = \sqrt{1-x^2}$ (Fig. 13.29).

Solution In Cartesian coordinates, the integral in question is a nonelementary integral and there is no direct way to integrate $e^{x^2+y^2}$ with respect to either x or y. Yet this integral and others like it are important in mathematics—in statistics, for example—and we must find a way to evaluate it. Polar coordinates save the day. Substituting $x = r\cos\theta$, $y = r\sin\theta$, and replacing $dx\,dy$ by $r\,dr\,d\theta$ enables us to evaluate the integral as

$$\iint_R e^{x^2+y^2}\,dy\,dx = \int_0^{\pi}\int_0^1 e^{r^2}\,r\,dr\,d\theta$$

$$= \int_0^{\pi}\left[\frac{1}{2}e^{r^2}\right]_0^1 d\theta$$

$$= \int_0^{\pi}\frac{1}{2}(e-1)\,d\theta = \frac{\pi}{2}(e-1).$$

The r in the $r\,dr\,d\theta$ was just what we needed to integrate e^{r^2}. Without it we would have been stuck, as we were at the beginning.

When we work in polar coordinates, we often have to integrate powers of sines and cosines. We accomplish this with reduction formulas, as in the next example. Reduction formulas were discussed in Section 7.6. You will find a number of useful reduction formulas in the integral table at the end of the book.

Example 5 Find the moment of inertia about the y-axis of a thin plate of constant density $\delta(x, y) = 1$ covering the region R enclosed by the cardioid $r = 1 - \cos\theta$.

Solution The required moment is

$$I_y = \iint_R x^2\,\delta(x, y)\,dA = \iint_R x^2\,dA.$$

We substitute $x = r\cos\theta$, replace dA by $r\,dr\,d\theta$, and sketch the region (Fig. 13.30) to determine the limits of integration:

$$I_y = \int_0^{2\pi}\int_0^{1-\cos\theta} r^3\cos^2\theta\,dr\,d\theta$$

$$= \int_0^{2\pi}\frac{1}{4}(1-\cos\theta)^4\cos^2\theta\,d\theta.$$

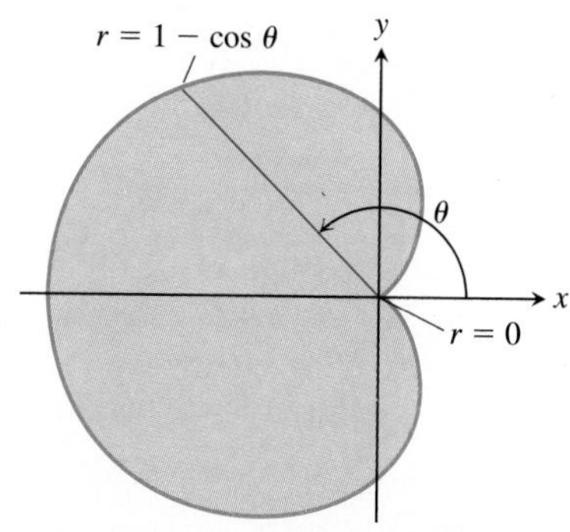

13.30 Example 5 shows how to find the moment of inertia about the y-axis of a thin plate covering the shaded region. To integrate over the region, we first run r from 0 to $1 - \cos\theta$ and then run θ from 0 to 2π.

We use the following reduction formula (Formula 61 in the integral table at the end of the book)

$$\int_0^{2\pi}\cos^n\theta\,d\theta = \frac{\cos^{n-1}\theta\sin\theta}{n}\Bigg]_0^{2\pi} + \frac{n-1}{n}\int_0^{2\pi}\cos^{n-2}\theta\,d\theta$$

$$= \frac{n-1}{n}\int_0^{2\pi}\cos^{n-2}\theta\,d\theta$$

to evaluate the powers of cos θ that arise when we expand the integrand:

$$\int_0^{2\pi} \cos^2\theta\, d\theta = \frac{1}{2}\int_0^{2\pi} d\theta = \pi,$$

$$\int_0^{2\pi} \cos^3\theta\, d\theta = \frac{2}{3}\int_0^{2\pi} \cos\theta\, d\theta = \frac{2}{3}\sin\theta\Big]_0^{2\pi} = 0,$$

$$\int_0^{2\pi} \cos^4\theta\, d\theta = \frac{3}{4}\int_0^{2\pi} \cos^2\theta\, d\theta = \frac{3\pi}{4},$$

$$\int_0^{2\pi} \cos^5\theta\, d\theta = \frac{4}{5}\int_0^{2\pi} \cos^3\theta\, d\theta = 0,$$

$$\int_0^{2\pi} \cos^6\theta\, d\theta = \frac{5}{6}\int_0^{2\pi} \cos^4\theta\, d\theta = \frac{5}{6}\frac{3\pi}{4} = \frac{5\pi}{8}.$$

Therefore

$$I_y = \frac{1}{4}\int_0^{2\pi} (\cos^2\theta - 4\cos^3\theta + 6\cos^4\theta - 4\cos^5\theta + \cos^6\theta)\, d\theta$$

$$= \frac{1}{4}\left[1 + \frac{18}{4} + \frac{5}{8}\right]\pi = \frac{49\pi}{32}.$$

EXERCISES 13.3

In Exercises 1–15, change the Cartesian integral into an equivalent polar integral and evaluate the polar integral.

1. $\int_{-1}^{1}\int_{0}^{\sqrt{1-x^2}} dy\, dx$

2. $\int_{-1}^{1}\int_{-\sqrt{1-x^2}}^{\sqrt{1-x^2}} dy\, dx$

3. $\int_{0}^{1}\int_{0}^{\sqrt{1-y^2}} (x^2+y^2)\, dx\, dy$

4. $\int_{-1}^{1}\int_{-\sqrt{1-y^2}}^{\sqrt{1-y^2}} (x^2+y^2)\, dx\, dy$

5. $\int_{-a}^{a}\int_{-\sqrt{a^2-x^2}}^{\sqrt{a^2-x^2}} dy\, dx$

6. $\int_{0}^{2}\int_{0}^{\sqrt{4-y^2}} (x^2+y^2)\, dx\, dy$

7. $\int_{0}^{1}\int_{y}^{\sqrt{2-y^2}} x\, dx\, dy$

8. $\int_{0}^{2}\int_{0}^{x} y\, dy\, dx$

9. $\int_{0}^{3}\int_{0}^{\sqrt{3}x} \frac{dy\, dx}{\sqrt{x^2+y^2}}$

10. $\int_{0}^{2}\int_{0}^{\sqrt{4-x^2}} \frac{xy}{\sqrt{x^2+y^2}}\, dy\, dx$

11. $\int_{0}^{1}\int_{0}^{\sqrt{1-x^2}} 5\sqrt{x^2+y^2}\, dy\, dx$

12. $\int_{0}^{1}\int_{0}^{\sqrt{1-x^2}} e^{-(x^2+y^2)}\, dy\, dx$

13. $\int_{0}^{2}\int_{0}^{\sqrt{1-(x-1)^2}} \frac{x+y}{x^2+y^2}\, dy\, dx$

14. $\int_{-1}^{1}\int_{-\sqrt{1-y^2}}^{\sqrt{1-y^2}} \ln(x^2+y^2+1)\, dx\, dy$

15. $\int_{-1}^{1}\int_{-\sqrt{1-x^2}}^{\sqrt{1-x^2}} \frac{2\, dy\, dx}{(1+x^2+y^2)^2}$

16. a) The usual way to evaluate the improper integral $I = \int_0^\infty e^{-x^2}\, dx$ is first to calculate its square:

$$I^2 = \left(\int_0^\infty e^{-x^2}\, dx\right)\left(\int_0^\infty e^{-y^2}\, dy\right) = \int_0^\infty\int_0^\infty e^{-(x^2+y^2)}\, dx\, dy.$$

Evaluate the last integral using polar coordinates and solve the resulting equation to show that $I = \sqrt{\pi}/2$.

b) Evaluate the integral

$$\int_0^\infty \frac{e^{-x}}{\sqrt{x}}\,dx.$$

17. Find the area of the region cut from the first quadrant by the curve $r = 2(2 - \sin 2\theta)^{1/2}$.

18. Find the area of the region that lies inside the cardioid $r = 1 + \cos\theta$ and outside the circle $r = 1$.

19. Find the area enclosed by one leaf of the rose $r = 12\cos 3\theta$.

20. Find the area of the region enclosed by the positive x-axis and spiral $r = 4\theta/3,\ 0 \le \theta \le 2\pi$. The region looks like a snail shell.

21. Find the area of the region cut from the first quadrant by the cardioid $r = 1 + \sin\theta$.

22. Find the area of the region common to the interiors of the cardioids $r = 1 + \cos\theta$ and $r = 1 - \cos\theta$.

23. Find the area of the region that is bounded on the left by the y-axis and on the right by the parabola $r = \sec^2(\theta/2)$.

24. Find the area of the region inside the cardioid $r = 2 + 2\cos\theta$ that lies to the right of the parabola $r = \sec^2(\theta/2)$.

25. Integrate the function $f(x, y) = 1/(1 - x^2 - y^2)$ over the disk $x^2 + y^2 \le 3/4$.

26. Integrate the function $f(x, y) = (\ln(x^2 + y^2))/(x^2 + y^2)$ over the region between the circles $x^2 + y^2 = 1$ and $x^2 + y^2 = e^2$.

27. Find the first moment about the x-axis of a thin plate of constant density $\delta(x, y) = 3$, bounded below by the x-axis and above by the cardioid $r = 1 - \cos\theta$.

28. Find the moment of inertia about the x-axis and the polar moment of inertia about the origin of a thin disk bounded by the circle $x^2 + y^2 = a^2$ if the density is $\delta(x, y) = k(x^2 + y^2)$, k a constant.

29. Find the centroid of the region enclosed by the cardioid $r = 1 + \cos\theta$.

30. Find the polar moment of inertia about the origin of a thin plate bounded by the cardioid $r = 1 + \cos\theta$ if the density is $\delta(x, y) = 1$.

31. Find the mass of a thin plate that covers that region that lies outside the curve $r = a$ and inside the curve $r = 2a\sin\theta$ if the density is inversely proportional to the distance from the origin.

32. Find the polar moment of inertia about the origin of a thin plate of density $\delta(x, y) = 1$ covering the region that lies inside the cardioid $r = 1 - \cos\theta$ and outside the circle $r = 1$.

33. The region that lies inside the cardioid $r = 1 + \cos\theta$ and outside the circle $r = 1$ is the base of a solid right cylinder. The top of the cylinder lies in the plane $z = x$. Find the cylinder's volume.

34. The region enclosed by the lemniscate $r^2 = 2\cos 2\theta$ is the base of a solid right cylinder whose top is bounded by the sphere $z = \sqrt{2 - r^2}$. Find the cylinder's volume.

35. Find the average height of the hemisphere $z = \sqrt{1 - x^2 - y^2}$ above the disk $x^2 + y^2 \le 1$.

36. Find the average value of the square of the distance from the points inside a unit disk to a point on the boundary.

37. Let P_0 be a point inside a circle of radius a and let h denote the distance from P_0 to the center of the circle. Let d denote the distance from an arbitrary point P to P_0. Find the average value of d^2 over the region enclosed by the circle. (*Hint:* Simplify your work by placing the center of the circle at the origin and P_0 on the x-axis.)

38. Suppose that the area of a region in the polar coordinate plane is

$$A = \int_{\pi/4}^{3\pi/4} \int_{\csc\theta}^{2\sin\theta} r\,dr\,d\theta.$$

a) Sketch the region and find its area.

b) Use one of Pappus's theorems together with the centroid information in Exercise 22 of Section 5.9 to find the volume of the solid generated by revolving the region about the x-axis.

13.4 Triple Integrals in Rectangular Coordinates: Volumes and Average Values

Triple integrals are used to calculate the volumes of irregular three-dimensional shapes and the average values of functions over three-dimensional regions. They are also the integrals we use to calculate masses and moments of three-dimensional solids. When combined with vectors, as they will be in Chapter 14, they are the integrals we use to describe many of the phenomena associated with fluid flow and electromagnetism.

In the present section, we define triple integrals and use them to calculate volumes and average values.

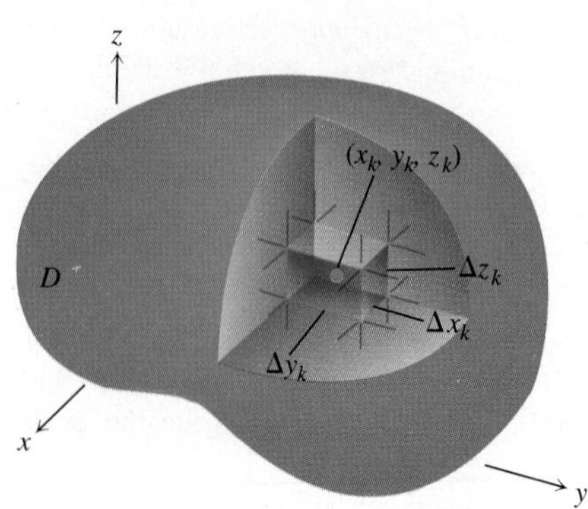

13.31 Partitioning a solid with rectangular cells of volume ΔV_k.

Triple Integrals

If $F(x, y, z)$ is a function defined on a closed bounded region D in space—the region occupied by a solid ball, for example, or a lump of clay—then the integral of F over D may be defined in the following way. We partition a rectangular region about D into rectangular cells by planes parallel to the coordinate planes (Fig. 13.31). We number the cells that lie inside D from 1 to n in some order, a typical cell having dimensions Δx_k by Δy_k by Δz_k and volume ΔV_k. We choose a point (x_k, y_k, z_k) in each cell and form the sum

$$S_n = \sum_{k=1}^{n} F(x_k, y_k, z_k)\,\Delta V_k. \tag{1}$$

If F is continuous and the bounding surface of D is made of smooth surfaces joined along continuous curves, then as Δx_k, Δy_k, and Δz_k approach zero independently the sums S_n approach a limit

$$\lim_{n\to\infty} S_n = \iiint_D F(x, y, z)\,dV. \tag{2}$$

We call this limit the **triple integral of F over D.** The limit also exists for some discontinuous functions.

Properties of Triple Integrals

Triple integrals have the same algebraic properties as double integrals and single integrals. If $F = F(x, y, z)$ and $G = G(x, y, z)$ are both integrable, then

1. $\displaystyle\iiint_D kF\,dV = k\iiint_D F\,dV$ (any number k),

2. $\displaystyle\iiint_D (F + G)\,dV = \iiint_D F\,dV + \iiint_D G\,dV,$

3. $\displaystyle\iiint_D (F - G)\,dV = \iiint_D F\,dV - \iiint_D G\,dV,$

4. $\displaystyle\iiint_D F\,dV \ge 0$ if $F \ge 0$ on D,

5. $\displaystyle\iiint_D F\,dV \ge \iiint_D G\,dV$ if $F \ge G$ on D.

Triple integrals also have a domain additivity property that proves useful in physics and engineering as well as in mathematics. If the domain D of a continuous function F is partitioned by smooth surfaces into a finite number of nonoverlapping cells $D_1, D_2, \ldots, D_n$, then

6. $\displaystyle\iiint_D F\,dV = \iiint_{D_1} F\,dV + \iiint_{D_2} F\,dV + \cdots + \iiint_{D_n} F\,dV.$

Volume of a Region in Space

If F is the constant function whose value is 1, then the sums in Eq. (1) reduce to

$$S_n = \sum F(x_k, y_k, z_k)\Delta V_k = \sum 1 \cdot \Delta V_k = \sum \Delta V_k. \tag{3}$$

As Δx, Δy, and Δz all approach zero, the cells ΔV_k become smaller and more numerous and fill up more and more of D. We therefore define the volume of D to be the triple integral

$$\lim \sum_{k=1}^{n} \Delta V_k = \iiint_D dV.$$

DEFINITION

The **volume** of a closed bounded region D in space is the value of the integral

$$\text{Volume of } D = \iiint_D dV. \tag{4}$$

As we shall see in a moment, this integral enables us to calculate the volumes of solids enclosed by curved surfaces.

Evaluation of Triple Integrals

The triple integral is seldom evaluated directly from its definition as a limit. Instead, we apply a three-dimensional version of Fubini's theorem to evaluate the integral by repeated single integrations.

For example, suppose we want to integrate a continuous function $F(x, y, z)$ over a region D that is bounded below by a surface $z = f_1(x, y)$, above by the surface $z = f_2(x, y)$, and on the side by a cylinder C parallel to the z-axis (Fig. 13.32). Let R denote the vertical projection of D onto the xy-plane, which is the region in the xy-plane enclosed by C. The integral of F over D is then evaluated as

$$\iiint_D F(x, y, z)\, dV = \iint_R \left(\int_{z=f_1(x,y)}^{z=f_2(x,y)} F(x, y, z)\, dz \right) dy\, dx,$$

or

$$\iiint_D F(x, y, z)\, dV = \iint_R \int_{z=f_1(x,y)}^{z=f_2(x,y)} F(x, y, z)\, dz\, dy\, dx. \tag{5}$$

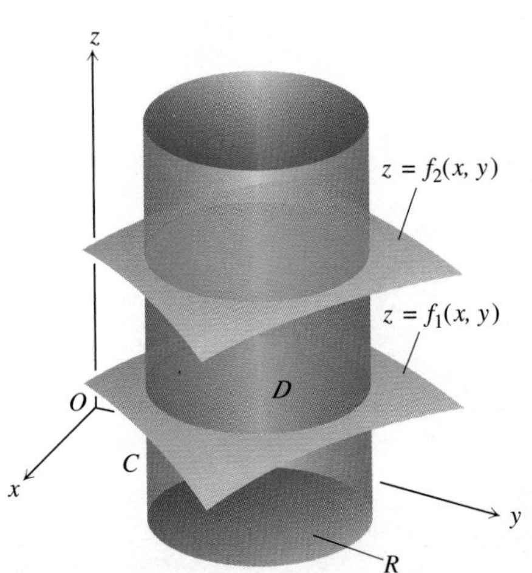

13.32 The enclosed volume can be found by evaluating

$$V = \iint_R \int_{z=f_1(x,y)}^{z=f_2(x,y)} dz\, dy\, dx.$$

The z-limits of integration indicate that for every (x, y) in the region R, z may extend from the lower surface $z = f_1(x, y)$ to the upper surface $z = f_2(x, y)$. The y- and x-limits of integration have not been given explicitly in Eq. (5) but are to be determined in the usual way from the boundaries of R.

In case the lateral surface of the cylinder reduces to zero, as in Fig. 13.33 and Example 1, we may find the equation of the boundary of R by eliminating z between the two equations $z = f_1(x, y)$ and $z = f_2(x, y)$. This gives

$$f_1(x, y) = f_2(x, y),$$

an equation that contains no z and that defines the boundary of R in the xy-plane.

To supply the z-limits of integration in any particular instance we may use a procedure like the one for double integrals. We imagine a line L through a point

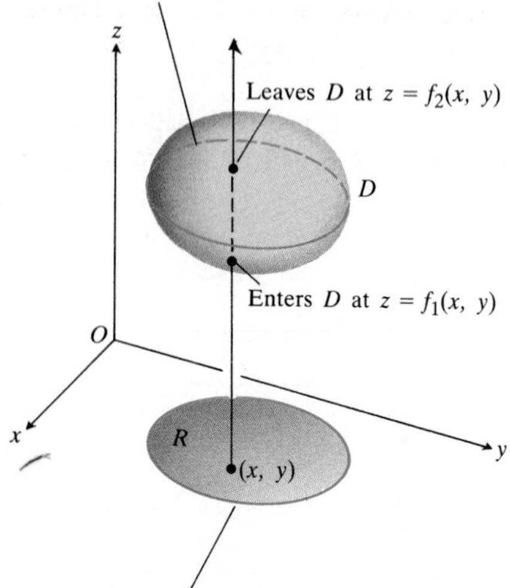

13.33 A schematic diagram for finding the limits of integration for a triple integral of a function F over a three-dimensional region D enclosed by two surfaces. For the region here,

$$\iiint_D F\,dV = \iint_R \int_{z=f_1(x,y)}^{z=f_2(x,y)} F\,dz\,dy\,dx.$$

(x, y) in R and parallel to the z-axis. As z increases, the line enters D at $z = f_1(x, y)$ and leaves D at $z = f_2(x, y)$. These give the lower and upper limits of the integration with respect to z. The result of this integration is now a function of x and y alone, which we integrate over R, supplying limits in the usual way.

Example 1 Find the volume of the three-dimensional region enclosed by the surfaces $z = x^2 + 3y^2$ and $z = 8 - x^2 - y^2$.

Solution The two surfaces (Fig. 13.34) intersect on the elliptical cylinder

$$x^2 + 3y^2 = 8 - x^2 - y^2, \qquad \text{or} \qquad x^2 + 2y^2 = 4.$$

The three-dimensional region projects into the two-dimensional region R in the xy-plane that is enclosed by the ellipse having this same equation. In the double integral with respect to y and x over R, if we integrate first with respect to y, holding x fixed, y varies from $-\sqrt{(4 - x^2)/2}$ to $+\sqrt{(4 - x^2)/2}$. Then x varies from -2 to $+2$. Thus we have

$$V = \int_{-2}^{2} \int_{-\sqrt{(4-x^2)/2}}^{\sqrt{(4-x^2)/2}} \int_{x^2+3y^2}^{8-x^2-y^2} dz\,dy\,dx = \int_{-2}^{2} \int_{-\sqrt{(4-x^2)/2}}^{\sqrt{(4-x^2)/2}} (8 - 2x^2 - 4y^2)\,dy\,dx$$

$$= \int_{-2}^{2} \left[(8 - 2x^2)y - \frac{4}{3}y^3\right]_{y=-\sqrt{(4-x^2)/2}}^{y=\sqrt{(4-x^2)/2}} dx$$

$$= \int_{-2}^{2} \left(2(8 - 2x^2)\sqrt{\frac{4 - x^2}{2}} - \frac{8}{3}\left(\frac{4 - x^2}{2}\right)^{3/2}\right) dx$$

$$= \int_{-2}^{2} \left(8\left(\frac{4 - x^2}{2}\right)^{3/2} - \frac{8}{3}\left(\frac{4 - x^2}{2}\right)^{3/2}\right) dx = \frac{4\sqrt{2}}{3}\int_{-2}^{2} (4 - x^2)^{3/2}\,dx$$

$$= 8\pi\sqrt{2}. \qquad \left(\begin{array}{l}\text{After integration with the} \\ \text{substitution } x = 2 \sin u\end{array}\right)$$

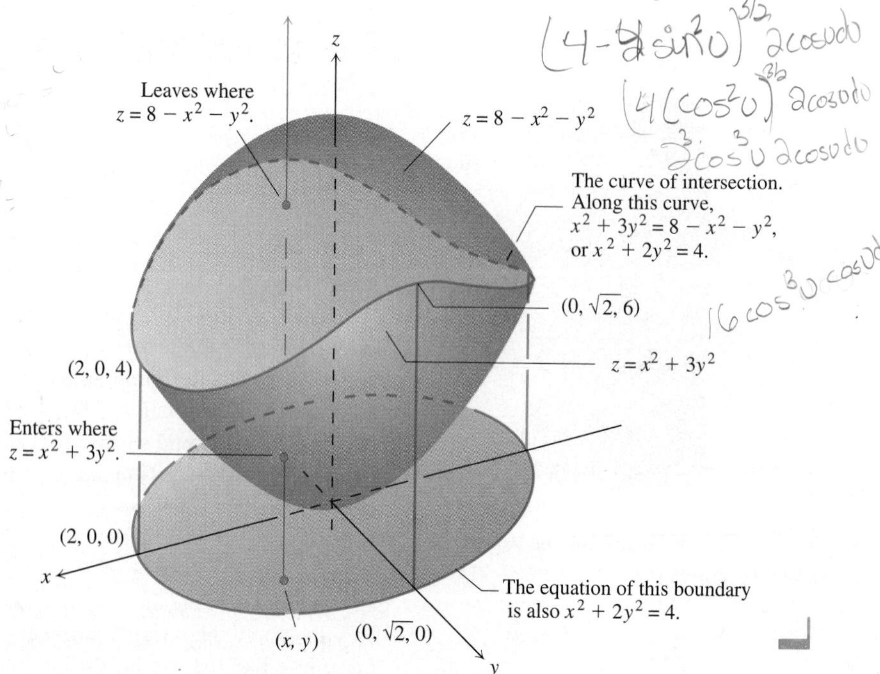

13.34 The volume of the region enclosed by these two paraboloids is calculated in Example 1.

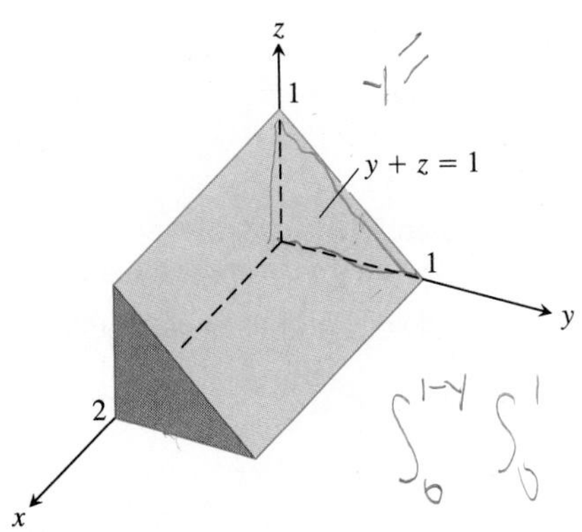

13.35 Example 2 shows how to calculate the volume of this prism with six different iterated triple integrals.

As we know, there are sometimes (but not always) two different orders in which the single integrations that evaluate a double integral may be worked. For triple integrals, there are sometimes (but not always) as many as *six* workable orders of integration. In the next example, all six are workable.

Example 2 Each of the following integrals gives the volume of the solid shown in Fig. 13.35.

a) $\displaystyle\int_0^1\int_0^{1-z}\int_0^2 dx\,dy\,dz$ b) $\displaystyle\int_0^1\int_0^{1-y}\int_0^2 dx\,dz\,dy$

c) $\displaystyle\int_0^1\int_0^2\int_0^{1-z} dy\,dx\,dz$ d) $\displaystyle\int_0^2\int_0^1\int_0^{1-z} dy\,dz\,dx$

e) $\displaystyle\int_0^1\int_0^2\int_0^{1-y} dz\,dx\,dy$ f) $\displaystyle\int_0^2\int_0^1\int_0^{1-y} dz\,dy\,dx$

Average Value of a Function in Space

The average value of a function F over a region D in space is defined by the formula

$$\text{Average value of } F \text{ over } D = \frac{1}{\text{volume of } D}\iiint_D F\,dV. \tag{6}$$

If $F(x, y, z) = \sqrt{x^2 + y^2 + z^2}$, then the average value of F over D is the average distance of points in D from the origin. If $F(x, y, z)$ is the density of a solid that occupies a region D in space, then the average value of F over D is the average density of the solid in units of mass per unit volume.

Example 3 Find the average value of $F(x, y, z) = xyz$ over the cube bounded by the coordinate planes and the planes $x = 2$, $y = 2$, and $z = 2$ in the first octant.

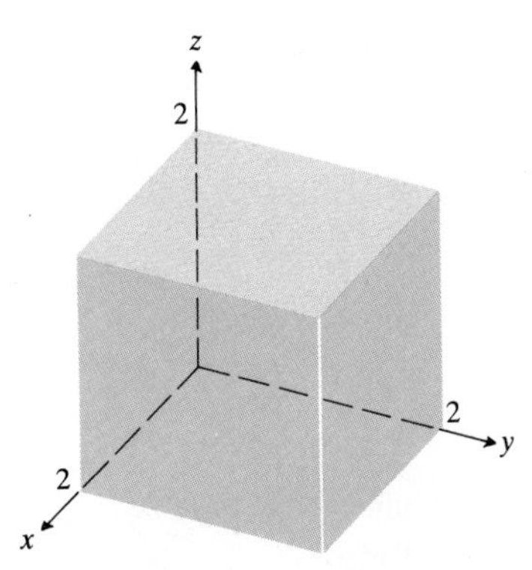

13.36 The solid cube bounded by the planes $x = 2$, $y = 2$, and $z = 2$ is the region of integration in Example 3.

Solution We sketch the cube with enough detail to show the limits of integration (Fig. 13.36). We then use Eq. (6) to calculate the average value of F over the cube.

The volume of the cube is $(2)(2)(2) = 8$.

The value of the integral of F over the cube is

$$\int_0^2\int_0^2\int_0^2 xyz\,dx\,dy\,dz = \int_0^2\int_0^2\left[\frac{x^2}{2}yz\right]_{x=0}^{x=2} dy\,dz = \int_0^2\int_0^2 2yz\,dy\,dz$$

$$= \int_0^2\left[y^2z\right]_{y=0}^{y=2} dz = \int_0^2 4z\,dz = \left[2z^2\right]_0^2 = 8.$$

With these values, Eq. (6) gives

$$\begin{array}{l}\text{Average value of}\\ xyz \text{ over the cube}\end{array} = \frac{1}{\text{volume}}\iiint_{\text{cube}} xyz\,dV = \left(\frac{1}{8}\right)(8) = 1.$$

In evaluating the integral, we chose the order dx, dy, dz, but any of the other five possible orders would have done as well.

EXERCISES 13.4

1. Find the common value of the integrals in Example 2.

2. Write six different iterated triple integrals for the volume of the rectangular solid in the first octant bounded by the coordinate planes and the planes $x = 1$, $y = 2$, and $z = 3$. Evaluate one of the integrals.

3. Write six different iterated triple integrals for the volume of the tetrahedron cut from the first octant by the plane $6x + 3y + 2z = 6$. Evaluate one of the integrals.

4. Write six different iterated triple integrals for the volume of the region in the first octant enclosed by the cylinder $x^2 + z^2 = 4$ and the plane $y = 3$. Evaluate one of them.

Evaluate the integrals in Exercises 5–18.

5. $\displaystyle\int_0^1\int_0^1\int_0^1 (x^2 + y^2 + z^2)\,dz\,dy\,dx$

6. $\displaystyle\int_0^{\sqrt{2}}\int_0^{3y}\int_{x^2+3y^2}^{8-x^2-y^2} dz\,dx\,dy$

7. $\displaystyle\int_1^e\int_1^e\int_1^e \frac{1}{xyz}\,dx\,dy\,dz$

8. $\displaystyle\int_0^1\int_0^{3-3x}\int_0^{3-3x-y} dz\,dy\,dx$

9. $\displaystyle\int_0^1\int_0^\pi\int_0^\pi y\sin z\,dx\,dy\,dz$

10. $\displaystyle\int_{-1}^1\int_{-1}^1\int_{-1}^1 (x + y + z)\,dy\,dx\,dz$

11. $\displaystyle\int_0^3\int_0^{\sqrt{9-x^2}}\int_0^{\sqrt{9-x^2}} dz\,dy\,dx$

12. $\displaystyle\int_0^2\int_{-\sqrt{4-y^2}}^{\sqrt{4-y^2}}\int_0^{2x+y} dz\,dx\,dy$

13. $\displaystyle\int_0^1\int_0^{2-x}\int_0^{2-x-y} dz\,dy\,dx$

14. $\displaystyle\int_0^1\int_0^{1-x^2}\int_3^{4-x^2-y} x\,dz\,dy\,dx$

15. $\displaystyle\int_0^\pi\int_0^\pi\int_0^\pi \cos(u + v + w)\,du\,dv\,dw$ (uvw-space)

16. $\displaystyle\int_1^e\int_1^e\int_1^e \ln r \ln s \ln t\,dt\,dr\,ds$ (rst-space)

17. $\displaystyle\int_0^{\pi/4}\int_0^{\ln\sec v}\int_{-\infty}^{2t} e^x\,dx\,dt\,dv$ (tvx-space)

18. $\displaystyle\int_0^7\int_0^2\int_0^{\sqrt{4-q^2}} \frac{q}{r+1}\,dp\,dq\,dr$ (pqr-space)

19. Figure 13.37 shows the region of integration of the integral

$$\int_{-1}^1\int_{x^2}^1\int_0^{1-y} dz\,dy\,dx.$$

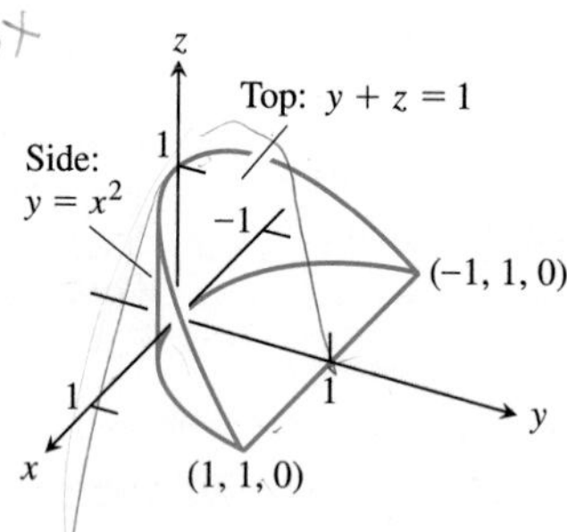

13.37 The region in Exercise 19.

Rewrite the integral as an equivalent iterated integral in the order

a) $dy\,dz\,dx$, b) $dy\,dx\,dz$,
c) $dx\,dy\,dz$, d) $dx\,dz\,dy$,
e) $dz\,dx\,dy$.

20. Figure 13.38 shows the region of integration of the integral

$$\int_0^1\int_{-1}^0\int_0^{y^2} dz\,dy\,dx.$$

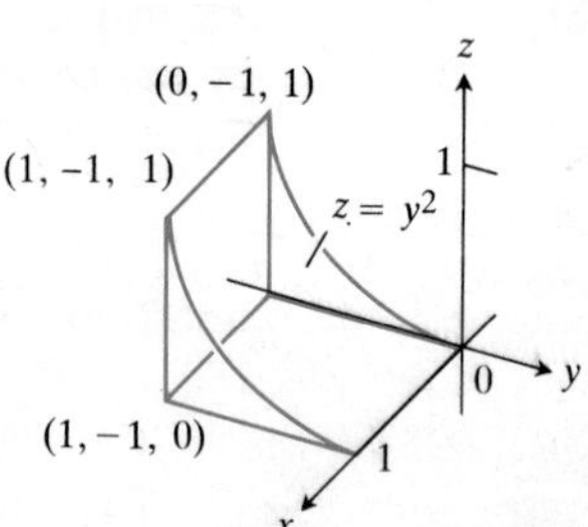

13.38 The region of integration in Exercise 20.

Rewrite the integral as an equivalent iterated integral in the order

a) $dy\,dz\,dx$, b) $dy\,dx\,dz$,
c) $dx\,dy\,dz$, d) $dx\,dz\,dy$,
e) $dz\,dx\,dy$.

Evaluate the integrals in Exercises 21–24 by changing the order of integration in an appropriate way.

21. $\displaystyle\int_0^4\int_0^1\int_{2y}^2 \frac{4\cos(x^2)}{2\sqrt{z}}\,dx\,dy\,dz$

22. $\displaystyle\int_0^1\int_0^1\int_{x^2}^1 12\,xz\,e^{zy^2}\,dy\,dx\,dz$

23. $\displaystyle\int_0^1\int_{\sqrt[3]{z}}^1\int_0^{\ln 3} \frac{\pi e^{2x}\sin \pi y^2}{y^2}\,dx\,dy\,dz$

24. $\displaystyle\int_0^2\int_0^{4-x^2}\int_0^x \frac{\sin 2z}{4-z}\,dy\,dz\,dx$

In Exercises 25–28, find the average value of $F(x, y, z)$ over the given region.

25. $F(x, y, z) = x^2 + 9$ over the cube in the first octant bounded by the coordinate planes and the planes $x = 2$, $y = 2$, and $z = 2$

26. $F(x, y, z) = x + y - z$ over the rectangular solid in the first octant bounded by the coordinate planes and the planes $x = 1$, $y = 1$, and $z = 2$

27. $F(x, y, z) = x^2 + y^2 + z^2$ over the cube in the first octant bounded by the coordinate planes and the planes $x = 1$, $y = 1$, and $z = 1$

28. $F(x, y, z) = xyz$ over the cube in the first octant bounded by the coordinate planes and the planes $x = a$, $y = a$, and $z = a$ $(a > 0)$

Find the volumes of the regions in Exercises 29–42.

29. The region between the cylinder $z = y^2$ and the xy-plane that is bounded by the planes $x = 0$, $x = 1$, $y = -1$, $y = 1$ (Fig. 13.39)

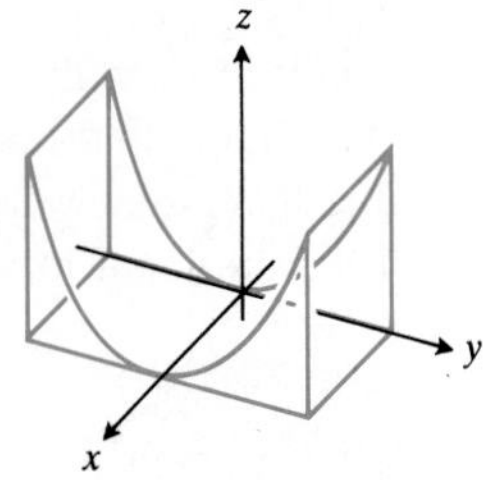

13.39 The region in Exercise 29.

30. The region cut from the first octant by the planes $x + z = 1$ and $y + 2z = 2$ (Fig. 13.40)

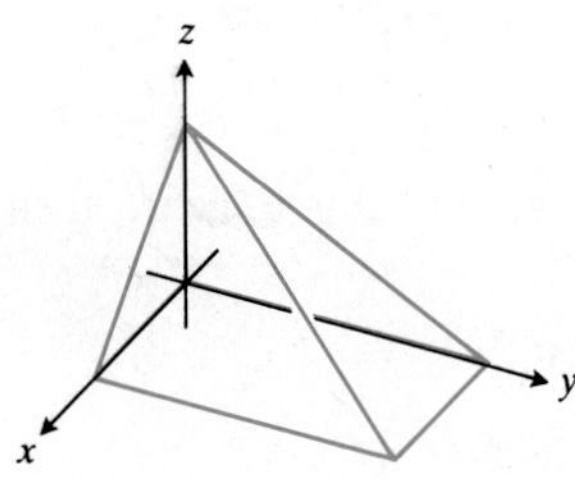

13.40 The region in Exercise 30.

31. The region in the first octant bounded by the coordinate planes, the plane $y + z = 2$, and the cylinder $x = 4 - y^2$ (Fig. 13.41)

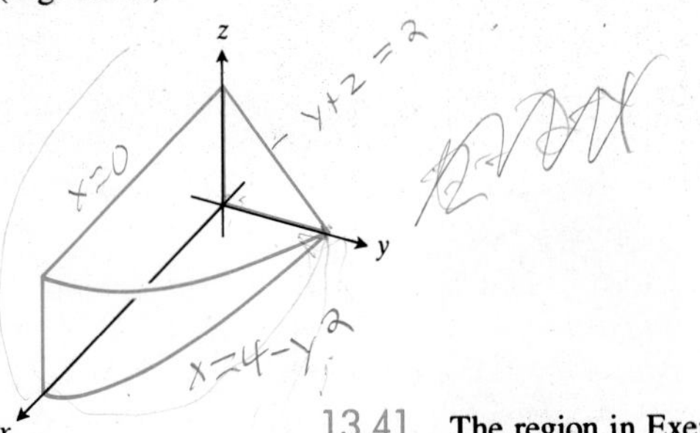

13.41 The region in Exercise 31.

32. The region in the first octant bounded by the coordinate planes, the plane $y = 1 - x$, and the surface $z = \cos(\pi x/2)$, $0 \le x \le 1$ (Fig. 13.42)

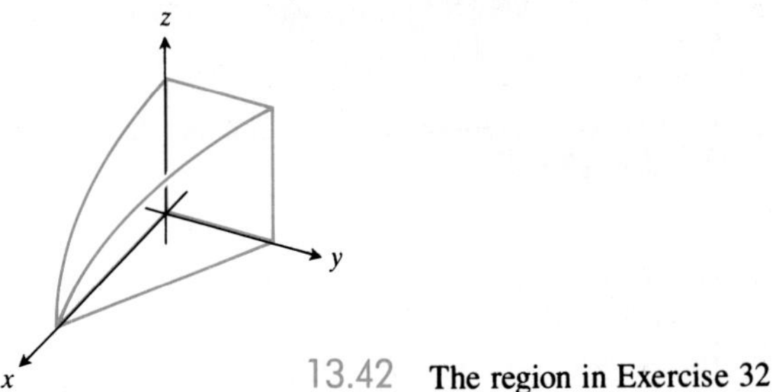

13.42 The region in Exercise 32.

33. The region in the first octant bounded by the coordinate planes and the surface $z = 4 - x^2 - y$ (Fig. 13.43)

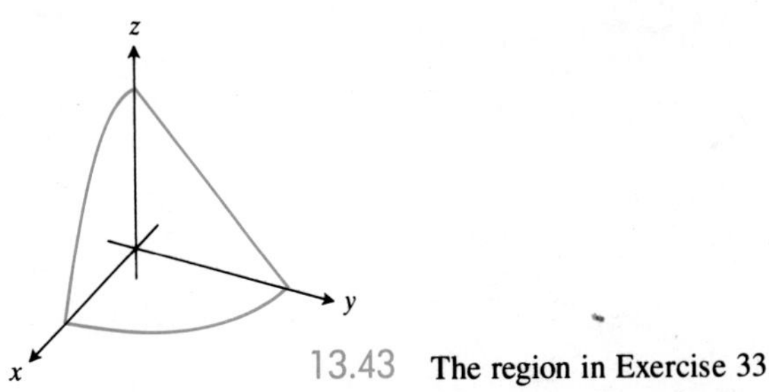

13.43 The region in Exercise 33.

34. The region in the first octant bounded by the coordinate planes, the plane $x + y = 4$, and the cylinder $y^2 + 4z^2 = 16$ (Fig. 13.44)

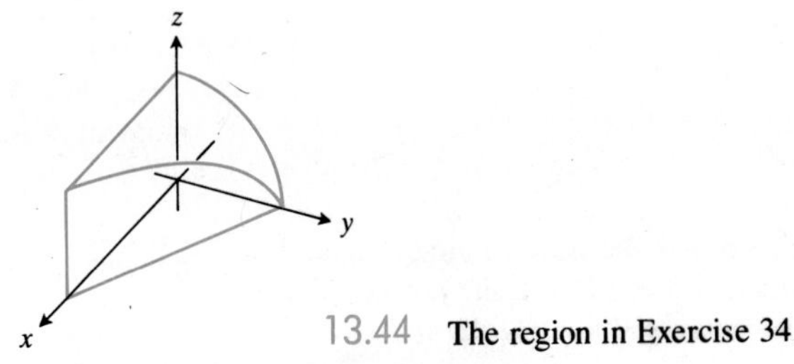

13.44 The region in Exercise 34.

35. The region common to the interiors of the cylinders $x^2 + y^2 = 1$ and $x^2 + z^2 = 1$ (Fig. 13.45)

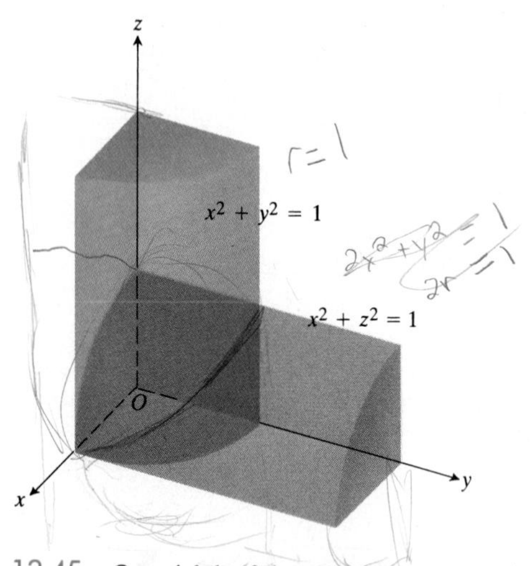

13.45 One-eighth of the region common to the cylinders $x^2 + y^2 = 1$ and $x^2 + z^2 = 1$ in Exercise 35.

36. The tetrahedron cut from the first octant by the plane $x + y/2 + z/3 = 1$

37. The region between the planes $x + y + 2z = 2$ and $2x + 2y + z = 4$ in the first octant

38. The wedge cut from the solid cylinder $x^2 + y^2 \leq 1$ by the half-planes $z = -y,\ y \leq 0$, and $z = 0,\ y \leq 0$

39. The region cut from the solid cylinder $x^2 + y^2 \leq 4$ by the plane $z = 0$ and the plane $x + z = 3$

40. The region cut from the solid elliptical cylinder $x^2 + 4y^2 \leq 4$ by the xy-plane and the plane $z = x + 2$

41. The region bounded in back by the plane $x = 0$, on the front and sides by the parabolic cylinder $x = 1 - y^2$, on the top by the paraboloid $z = x^2 + y^2$, and on the bottom by the xy-plane

42. The wedge-shaped region enclosed on the side by the cylinder $x = -\cos y,\ -\pi/2 \leq y \leq \pi/2$, on the top by the plane $z = -2x$, and on the bottom by the xy-plane

43. Solve for a:

$$\int_0^1 \int_0^{4-a-x^2} \int_a^{4-x^2-y} dz\, dy\, dx = \frac{4}{15}.$$

44. For what value of c is the volume of the ellipsoid $x^2 + (y/2)^2 + (z/c)^2 = 1$ equal to 8π?

45. What domain D in space maximizes the value of the integral

$$\iiint_D (18 - x^2 - y^2 - 2z^2)\, dV?$$

46. What domain D in space minimizes the value of the integral

$$\iiint_D (36 - 9x^2 - 9y^2 - 4z^2)\, dV?$$

13.5 Masses and Moments in Three Dimensions

This section shows how we can calculate the masses and moments of three-dimensional objects in Cartesian coordinates. The formulas we use are similar to the ones we use for two-dimensional objects, the only difference being that we now have three coordinates instead of two. In Section 13.6, we shall discuss moment and mass calculations in spherical and cylindrical coordinates.

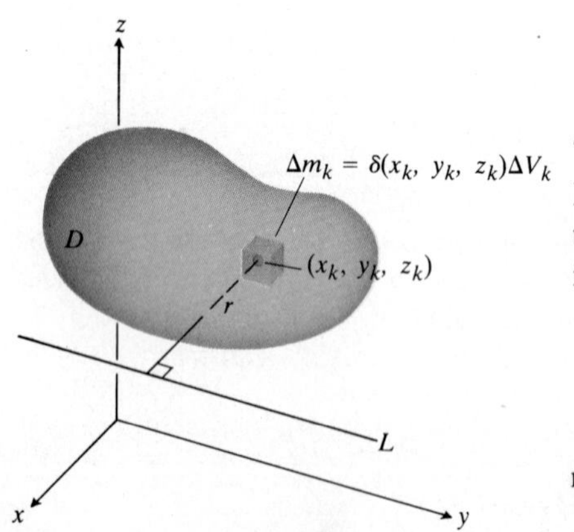

13.46 To define I_L we first imagine D to be subdivided into a finite number of mass elements Δm_k.

Masses and Moments

If $\delta(x, y, z)$ is the density function of an object occupying a region D in space, then the integral of δ over D gives the mass of the object. To see why, imagine subdividing D as in Fig. 13.46. The object's mass is the limit

$$M = \lim \sum_k \Delta m_k = \lim \sum_k \delta(x_k, y_k, z_k)\Delta V_k = \iiint_D \delta(x, y, z)\, dV. \qquad (1)$$

If $r(x, y, z)$ is the distance from the point (x, y, z) in D to a line L, then the moment of inertia of the mass

$$\Delta m_k = \delta(x_k, y_k, z_k)\Delta V_k$$

(shown in Fig. 13.46) about the line L is approximately

$$\Delta I_k = r^2(x_k, y_k, z_k)\Delta m_k.$$

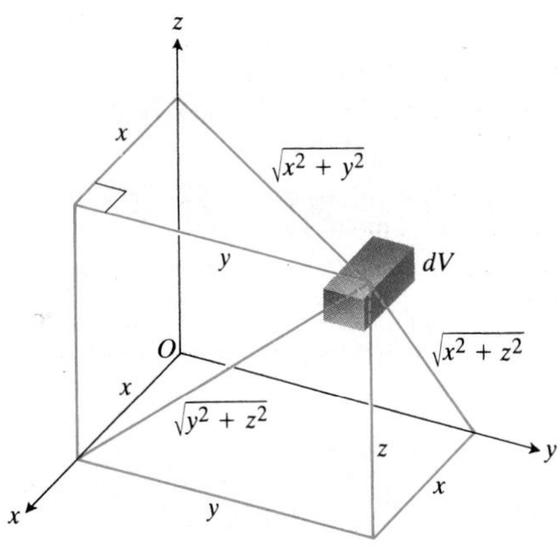

13.47 Distances from dV to the coordinate planes and axes.

The moment of inertia of the entire object about L is

$$I_L = \lim \sum_k \Delta I_k = \lim \sum_k r^2(x_k, y_k, z_k)\delta(x_k, y_k, z_k)\Delta V_k = \iiint_D r^2\delta\, dV.$$

If L is the x-axis, then $r^2 = y^2 + z^2$ (Fig. 13.47) and

$$I_x = \iiint_D (y^2 + z^2)\delta\, dV.$$

Similarly,

$$I_y = \iiint_D (x^2 + z^2)\delta\, dV \quad \text{and} \quad I_z = \iiint_D (x^2 + y^2)\delta\, dV.$$

These and other useful formulas are summarized in Table 13.2.

TABLE 13.2
Mass and moment formulas for objects in space

Mass: $M = \iiint_D \delta\, dV \qquad (\delta(x, y, z) = \text{density})$

First moments about the coordinate planes:

$$M_{yz} = \iiint_D x\,\delta\, dV, \qquad M_{xz} = \iiint_D y\,\delta\, dV, \qquad M_{xy} = \iiint_D z\,\delta\, dV$$

Center of mass:

$$\bar{x} = \frac{\iiint x\,\delta\, dV}{M}, \qquad \bar{y} = \frac{\iiint y\,\delta\, dV}{M}, \qquad \bar{z} = \frac{\iiint z\,\delta\, dV}{M}$$

Moments of inertia (second moments):

$$I_x = \iiint (y^2 + z^2)\delta\, dV, \qquad I_y = \iiint (x^2 + z^2)\delta\, dV,$$
$$I_z = \iiint (x^2 + y^2)\delta\, dV,$$

Moment of inertia about a line L:

$$I_L = \iiint r^2\,\delta\, dV \qquad (r(x, y, z) = \text{distance from point } (x, y, z) \text{ to line } L)$$

Radius of gyration about a line L:

$$R_L = \sqrt{I_L/M}$$

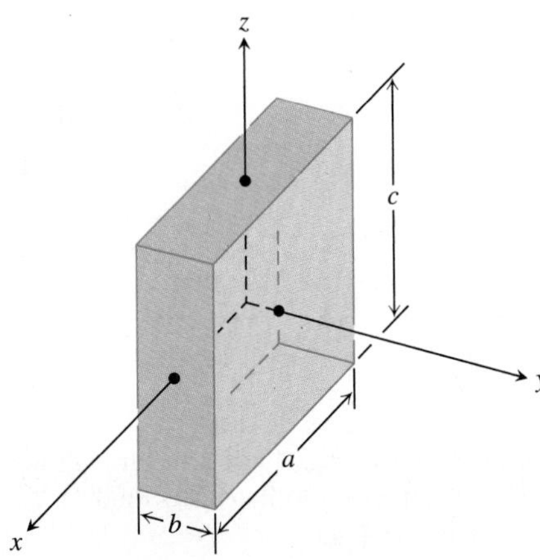

13.48 Example 1 calculates I_x, I_y, and I_z for the block shown here. The origin lies at the center of the block.

Example 1 Find I_x, I_y, I_z for the rectangular solid of constant density δ shown in Fig. 13.48.

Solution The preceding formula for I_x gives

$$I_x = \int_{-c/2}^{c/2}\int_{-b/2}^{b/2}\int_{-a/2}^{a/2} (y^2 + z^2)\delta\, dx\, dy\, dz. \tag{2}$$

We can avoid some of the work of integration by observing that $(y^2 + z^2)\delta$ is an even function of x, y, and z and therefore

$$I_x = 8\int_0^{c/2}\int_0^{b/2}\int_0^{a/2}(y^2+z^2)\delta\,dx\,dy\,dz = 4a\delta\int_0^{c/2}\int_0^{b/2}(y^2+z^2)\,dy\,dz$$

$$= 4a\delta\int_0^{c/2}\left[\frac{y^3}{3}+z^2y\right]_{y=0}^{y=b/2}dz = 4a\delta\int_0^{c/2}\left(\frac{b^3}{24}+\frac{z^2b}{2}\right)dz$$

$$= 4a\delta\left(\frac{b^3c}{48}+\frac{c^3b}{48}\right) = \frac{abc\delta}{12}(b^2+c^2) = \frac{M}{12}(b^2+c^2).$$

Similarly,

$$I_y = \frac{M}{12}(a^2+c^2) \qquad \text{and} \qquad I_z = \frac{M}{12}(a^2+b^2).$$

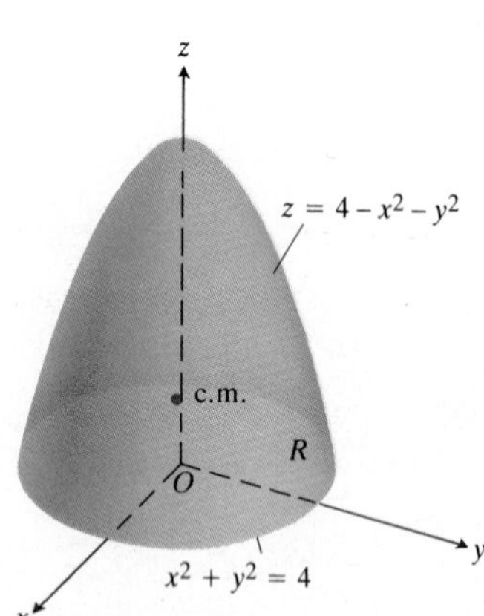

13.49 Example 2 calculates the coordinates of the center of mass of this solid.

Example 2 Find the center of mass of a solid of constant density δ bounded below by the disk $R: x^2+y^2 \le 4$ in the plane $z=0$ and above by the paraboloid $z = 4-x^2-y^2$ (Fig. 13.49).

Solution By symmetry, $\bar{x} = \bar{y} = 0$. To find $\bar{z}$ we first calculate

$$M_{xy} = \iint_R\int_{z=0}^{z=4-x^2-y^2} z\,\delta\,dz\,dy\,dx = \iint_R\left[\frac{z^2}{2}\right]_{z=0}^{z=4-x^2-y^2}\delta\,dy\,dx$$

$$= \frac{\delta}{2}\iint_R(4-x^2-y^2)^2\,dy\,dx$$

$$= \frac{\delta}{2}\int_0^{2\pi}\int_0^2(4-r^2)^2\,r\,dr\,d\theta \qquad \begin{pmatrix}\text{Polar}\\ \text{coordinates}\end{pmatrix}$$

$$= \frac{\delta}{2}\int_0^{2\pi}\left[-\frac{1}{6}(4-r^2)^3\right]_{r=0}^{r=2}d\theta = \frac{16\delta}{3}\int_0^{2\pi}d\theta = \frac{32\pi\delta}{3}.$$

A similar calculation gives

$$M = \iint_R\int_0^{4-x^2-y^2}\delta\,dz\,dy\,dx = 8\pi\delta.$$

Therefore, $\bar{z} = M_{xy}/M = 4/3$, and the center of mass is $(\bar{x}, \bar{y}, \bar{z}) = (0, 0, 4/3)$.

EXERCISES 13.5

Constant Density

The solids in Exercises 1–12 all have constant density $\delta = 1$.

1. Evaluate the integral for I_x in Eq. (2) directly to show that the shortcut in Example 1 gives the same answer. Use the results in Example 1 to find the radius of gyration of the rectangular solid about each coordinate axis.

2. The coordinate axes shown in Fig. 13.50 run through the centroid of a solid wedge parallel to its edges. Find I_x, I_y, and I_z if $a = b = 6$ and $c = 4$.

3. Find the moments of inertia of the rectangular solid shown in Fig. 13.51 with respect to its edges by calculating I_x, I_y, and I_z.

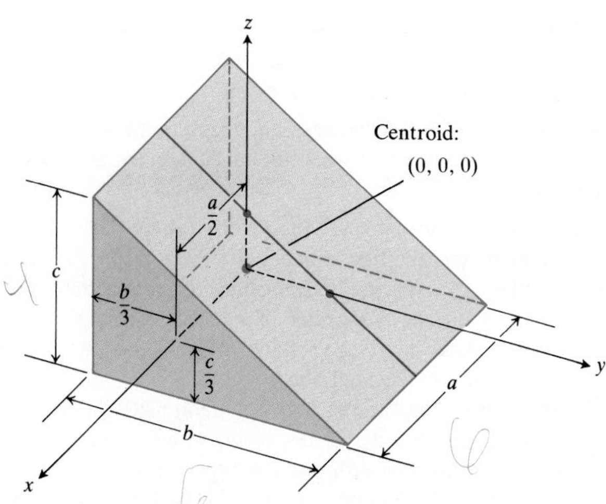

13.50 The figure for Exercise 2.

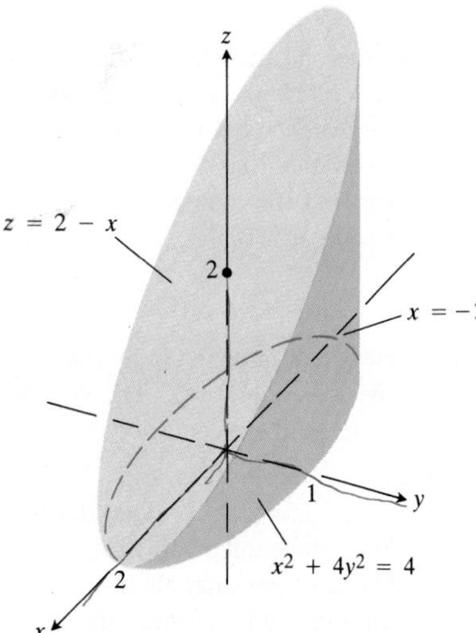

13.52 The solid in Exercise 6.

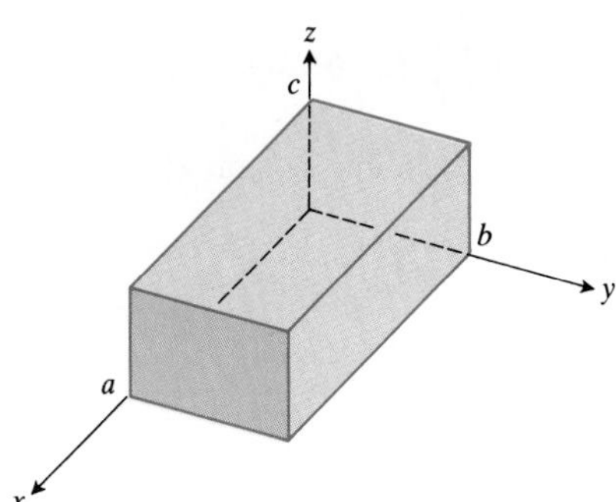

13.51 The rectangular solid in Exercise 3.

4. a) Find the centroid and the moments of inertia I_x, I_y, and I_z of the tetrahedron whose vertices are the points (0, 0, 0), (1, 0, 0), (0, 1, 0) and (0, 0, 1).

b) Find the radius of gyration of the tetrahedron about the x-axis. Compare it with the distance from the centroid to the x-axis.

5. A solid "trough" of constant density is bounded below by the surface $z = 4y^2$, above by the plane $z = 4$, and on the ends by the planes $x = 1$ and $x = -1$. Find the center of mass and the moments of inertia with respect to the three axes.

6. A solid of constant density is bounded below by the plane $z = 0$, on the sides by the elliptical cylinder $x^2 + 4y^2 = 4$, and above by the plane $z = 2 - x$ (Fig. 13.52).

a) Find $\bar{x}$ and $\bar{y}$.

b) Evaluate the integral

$$M_{xy} = \int_{-2}^{2} \int_{-(1/2)\sqrt{4-x^2}}^{(1/2)\sqrt{4-x^2}} \int_{0}^{2-x} z\, dz\, dy\, dx,$$

using integral tables to carry out the final integration with respect to x. Then divide M_{xy} by M to verify that $\bar{z} = 5/4$.

7. a) Find the center of mass of a solid of constant density bounded below by the paraboloid $z = x^2 + y^2$ and above by the plane $z = 4$.

b) Find the plane $z = c$ that divides the solid into two parts of equal volume. This plane does not pass through the center of mass.

8. A solid cube 2 units on a side is bounded by the planes $x = \pm 1$, $z = \pm 1$, $y = 3$, and $y = 5$. Find the center of mass and the moments of inertia and radii of gyration about the coordinate axes. Suppose that the cube's mass is concentrated at the center of mass. How, if at all, does that change the moments of inertia and radii of gyration about the coordinate axes?

9. A wedge shaped like the one in Fig. 13.50 has $a = 4$, $b = 6$, and $c = 3$. Make a quick sketch to check for yourself that the square of the distance from a typical point (x, y, z) of the wedge to the line L: $z = 0$, $y = 6$ is $r^2 = (y - 6)^2 + z^2$. Then calculate the moment of inertia and radius of gyration of the wedge about L.

10. A wedge shaped like the one in Fig. 13.50 has $a = 4$, $b = 6$, and $c = 3$. Make a quick sketch to check for yourself that the square of the distance from a typical point (x, y, z) of the wedge to the line L: $x = 4$, $y = 0$ is $r^2 = (x - 4)^2 + y^2$. Then calculate the moment of inertia and radius of gyration of the wedge about L.

11. A rectangular solid like the one in Fig. 13.51 has $a = 4$, $b = 2$, and $c = 1$. Make a quick sketch to check for yourself that the square of the distance between a typical point (x, y, z) of the solid and the line L: $y = 2$, $z = 0$ is $r^2 = (y - 2)^2 + z^2$. Then find the moment of inertia and radius of gyration of the solid about L.

12. A rectangular solid like the one in Fig. 13.51 has $a = 4$, $b = 2$, and $c = 1$. Make a quick sketch to check for yourself that the square of the distance between a typical point (x, y, z) of the solid and the line $L: x = 4,\ y = 0$ is $r^2 = (x - 4)^2 + y^2$. Then find the moment of inertia and radius of gyration of the solid about L.

Variable Density

13. A solid region in the first octant is bounded by the coordinate planes and the plane $x + y + z = 2$. The density of the solid is $\delta(x, y, z) = 2x$. Find the center of mass.

14. A solid wedge shaped like the one in Fig. 13.50 has dimensions $a = 4$, $b = 6$, and $c = 3$. The density is $\delta(x, y, z) = x + 1$. Find the center of mass and the moments of inertia and radii of gyration about the coordinate axes. Notice that if the density is constant, the center of mass will be $(0, 0, 0)$.

15. A solid cube in the first octant is bounded by the coordinate planes and by the planes $x = 1$, $y = 1$, and $z = 1$. Find its center of mass and the moments of inertia and radii of gyration about the coordinate axes if

$$\delta(x, y, z) = x + y + z + 1.$$

16. A solid in the first octant is bounded by the planes $y = 0$ and $z = 0$ and by the surfaces $z = 4 - x^2$ and $x = y^2$ (Fig. 13.53). Its density function is $\delta(x, y, z) = kxy$.

a) Find the solid's mass.
b) Find $\bar{x}$.
c) Find $\bar{y}$.
d) Find $\bar{z}$.

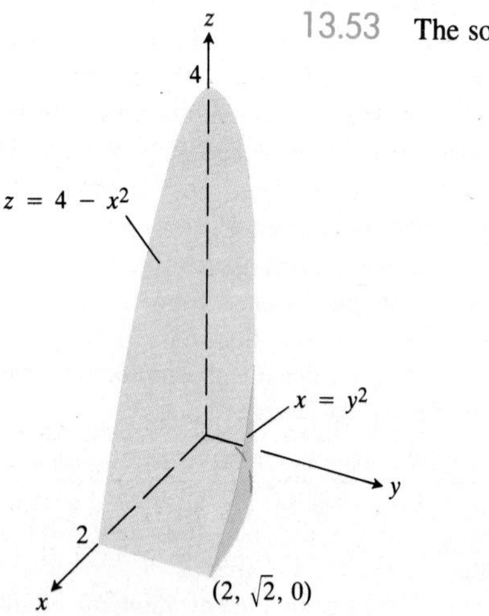

13.53 The solid in Exercise 16.

The Parallel Axis Theorem

The Parallel Axis Theorem (Exercises 13.2) holds in three dimensions as well as in two. Let $L_{\text{c.m.}}$ be a line through the center of mass of a body of mass m and let L be a parallel line h units away from $L_{\text{c.m.}}$. The **Parallel Axis Theorem** says that the moments of inertia $I_{\text{c.m.}}$ and I_L of the body about $L_{\text{c.m.}}$ and L satisfy the equation

$$I_L = I_{\text{c.m.}} + mh^2. \tag{3}$$

As in the two-dimensional case, the theorem gives a quick way to calculate one moment when the other moment and the mass are known.

17. *Proof of the Parallel Axis Theorem*

a) Show that the first moment of a body in space about any plane through the body's center of mass is zero. (*Hint:* Place the body's center of mass at the origin and let the plane be the yz-plane. What does the formula $\bar{x} = M_{yz}/M$ then tell you?)

b) To prove the Parallel Axis Theorem, place the body with its center of mass at the origin, with the line $L_{\text{c.m.}}$ along the z-axis and with the line L perpendicular to the xy-plane at the point $(h, 0, 0)$. Let D be the region of space occupied by the body (Fig. 13.54). Then, in the notation of the figure,

$$I_L = \iiint_D |\mathbf{v} - h\mathbf{i}|^2\, dm. \tag{4}$$

Expand the integrand in this integral and complete the proof.

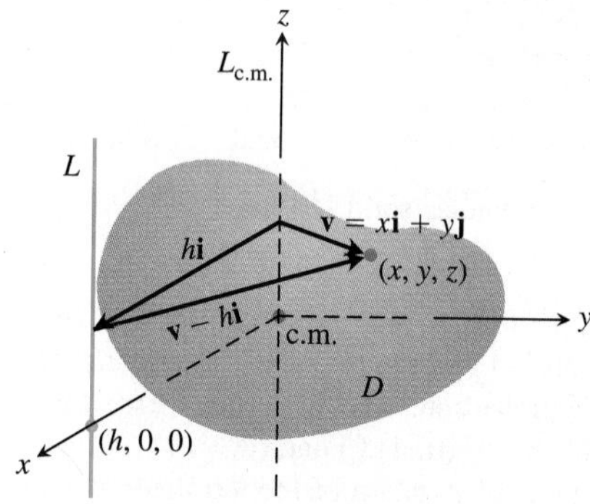

13.54 The diagram for the proof of the Parallel Axis Theorem in Exercise 17.

18. The moment of inertia about a diameter of a solid sphere of constant density and radius a is $(2/5)ma^2$, where m is the mass of the sphere. Find the moment of inertia about a line tangent to the sphere.

19. The moment of inertia of the rectangular solid in Fig. 13.51 about the z-axis is $I_z = abc(a^2 + b^2)/3$.

a) Use Eq. (3) to find the moment of inertia and radius of gyration of the solid about the line parallel to the z-axis through the solid's center of mass.

b) Use Eq. (3) and the result in (a) to find the moment of

inertia and radius of gyration of the solid about the line $x = 0$, $y = 2b$.

20. If $a = b = 6$ and $c = 4$, the moment of inertia of the solid wedge in Fig. 13.50 about the x-axis is $I_x = 208$. Find the moment of inertia of the wedge about the line $y = 4$, $z = -4/3$ (the edge line of the wedge's narrow end).

Pappus's Formula

Pappus's formula (Exercises 13.2) holds in three dimensions as well as in two. Suppose that bodies B_1 and B_2 of mass m_1 and m_2, respectively, occupy nonoverlapping regions in space and that $\mathbf{c}_1$ and $\mathbf{c}_2$ are the vectors from the origin to the bodies' respective centers of mass. Then the center of mass of the union $B_1 \cup B_2$ of the two bodies is determined by the vector

$$\mathbf{c} = \frac{m_1\mathbf{c}_1 + m_2\mathbf{c}_2}{m_1 + m_2}. \qquad (5)$$

As before, this formula is called **Pappus's formula.** As in the two-dimensional case, the formula generalizes to

$$\mathbf{c} = \frac{m_1\mathbf{c}_1 + m_2\mathbf{c}_2 + \cdots + m_n\mathbf{c}_n}{m_1 + m_2 + \cdots + m_n} \qquad (6)$$

for n bodies.

21. Derive Pappus's formula (Eq. 5). (*Hint:* Sketch B_1 and B_2 as nonoverlapping regions in the first octant and label their centers of mass $(\bar{x}_1, \bar{y}_1, \bar{z}_1)$ and $(\bar{x}_2, \bar{y}_2, \bar{z}_2)$. Express the moments of $B_1 \cup B_2$ about the coordinate planes in terms of the masses m_1 and m_2 and the coordinates of these centers.)

22. Figure 13.55 shows a solid made from three rectangular solids of constant density $\delta = 1$. Use Pappus's formula to find the center of mass of

a) $A \cup B$, b) $A \cup C$,
c) $B \cup C$, d) $A \cup B \cup C$.

23. a) Suppose that a solid right circular cone C of base radius a and altitude h is constructed on the circular base of a solid hemisphere S of radius a so that the union of the two solids resembles an ice cream cone. The centroid of a solid cone lies one-fourth of the way from the base toward the vertex. The centroid of a solid hemisphere lies three-eighths of the way from the base to the top. What relation must hold between h and a to place the centroid of $C \cup S$ in the common base of the two solids?

b) If you have not already done so, answer the analogous question about a triangle and a semicircle (Section 13.2, Exercise 53). The answers are not the same.

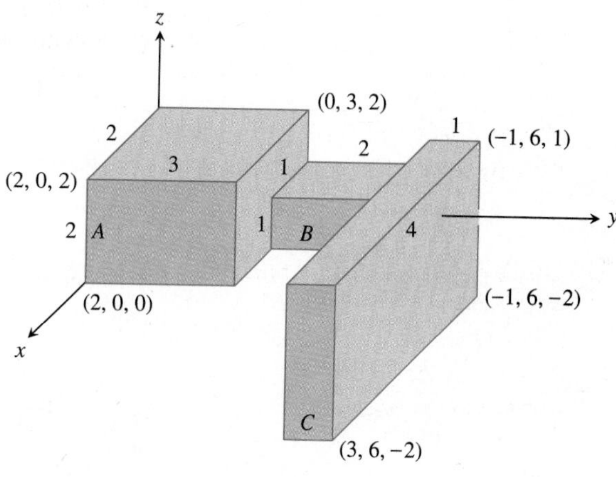

13.55 The solid in Exercise 22.

24. A solid pyramid P with height h and four congruent sides is built with its base as one face of a solid cube C whose edges have length s. The centroid of a solid pyramid lies one-fourth of the way from the base toward the vertex. What relation must hold between h and s to place the centroid of $P \cup C$ in the base of the pyramid? Compare your answer with the answer to Exercise 23. Also compare it to the answer to Exercise 54 in Section 13.2.

13.6 Triple Integrals in Cylindrical and Spherical Coordinates

If we are working with a solid like a cone or a cylindrical shell that has an axis of symmetry, we can often simplify our calculations by taking the axis to be the z-axis in a cylindrical coordinate system. In like manner, if we are working with a shape like a ball or spherical shell that has a center of symmetry, we can usually save time by choosing the center to be the origin in a spherical coordinate system.

This section shows how to calculate masses, moments, and volumes in cylindrical and spherical coordinates.

Cylindrical Coordinates

Cylindrical coordinates (Fig. 13.56) are good for describing cylinders whose axes run along the z-axis and planes that either contain the z-axis or lie perpendicular to

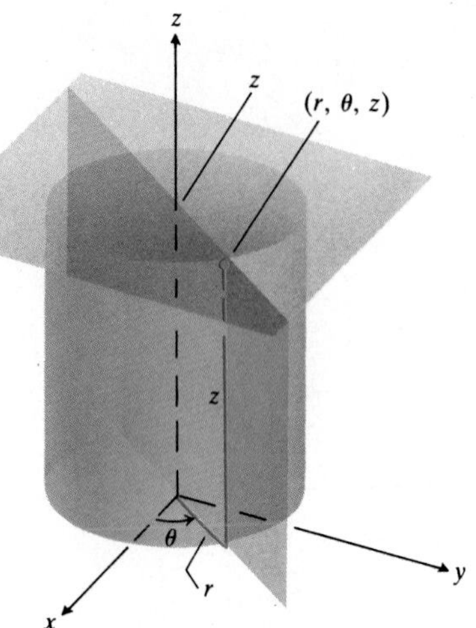

13.56 Cylindrical coordinates and typical surfaces of constant coordinate value.

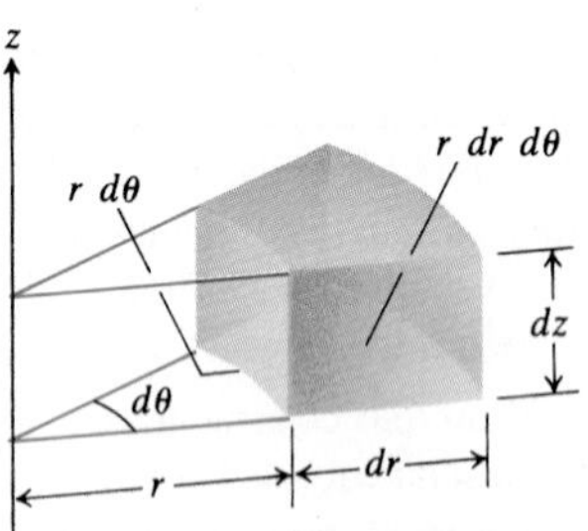

13.57 The volume element in cylindrical coordinates is $dV = dz\, r\, dr\, d\theta$.

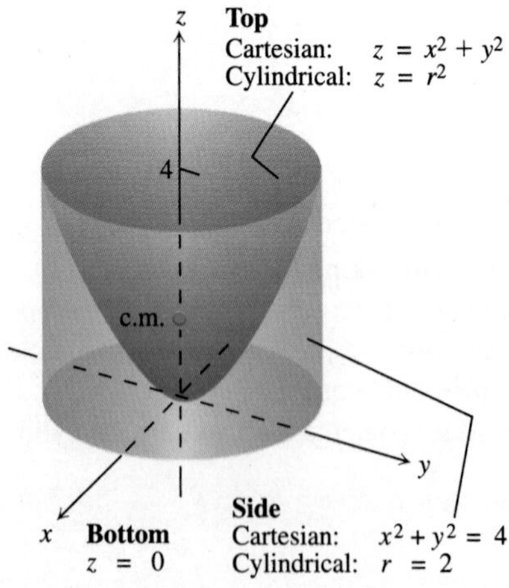

13.58 Example 1 shows how to locate the centroid of this region.

the z-axis. As we saw in Section 10.8, surfaces like these have equations of constant coordinate value:

$$r = 4 \qquad \text{(cylinder, radius 4, axis the } z\text{-axis)},$$

$$\theta = \frac{\pi}{3} \qquad \text{(plane containing the } z\text{-axis)},$$

$$z = 2 \qquad \text{(plane perpendicular to the } z\text{-axis)}.$$

The volume element for subdividing a region in space with cylindrical coordinates is

$$dV = dz\, r\, dr\, d\theta \tag{1}$$

(Fig. 13.57). Triple integrals in cylindrical coordinates are then evaluated as iterated integrals, as in the following example.

How to Integrate in Cylindrical Coordinates

To integrate a continuous function $f(r, \theta, z)$ over a region given by inequalities

$$g_1(r, \theta) \le z \le g_2(r, \theta),$$
$$h_1(\theta) \le r \le h_2(\theta),$$
$$\theta_1 \le \theta \le \theta_2,$$

evaluate the iterated integral

$$\int_{\theta=\theta_1}^{\theta=\theta_2} \int_{r=h_1(\theta)}^{r=h_2(\theta)} \int_{z=g_1(r,\theta)}^{z=g_2(r,\theta)} f(r, \theta, z)\, dz\, r\, dr\, d\theta. \tag{2}$$

Integrate first with respect to z. Multiply by r and integrate with respect to r. Then integrate with respect to θ.

Example 1 Find the centroid of the region enclosed by the cylinder $x^2 + y^2 = 4$, bounded above by the paraboloid $z = x^2 + y^2$, and bounded below by the xy-plane.

Solution We sketch the figure (Fig. 13.58) and find the cylindrical coordinate equations for the bounding surfaces. The coordinate inequalities for the region are

$$g_1(r, \theta) \le z \le g_2(r, \theta): \qquad 0 \le z \le r^2$$
$$h_1(\theta) \le r \le h_2(\theta): \qquad 0 \le r \le 2$$
$$\theta_1 \le \theta \le \theta_2: \qquad 0 \le \theta \le 2\pi.$$

The centroid lies on its axis of symmetry, in this case the z-axis, so $\bar{x} = \bar{y} = 0$.

To find $\bar{z}$, we divide the moment M_{xy} by the mass M. The value of M_{xy} is

$$\begin{aligned} M_{xy} &= \int_0^{2\pi}\int_0^2\int_0^{r^2} z\,dz\,r\,dr\,d\theta = \int_0^{2\pi}\int_0^2 \left[\frac{z^2}{2}\right]_0^{r^2} r\,dr\,d\theta \\ &= \int_0^{2\pi}\int_0^2 \frac{r^5}{2}\,dr\,d\theta = \int_0^{2\pi}\left[\frac{r^6}{12}\right]_0^2 d\theta \\ &= \int_0^{2\pi}\frac{16}{3}\,d\theta = \frac{32\pi}{3}. \end{aligned}$$

The value of M is

$$\begin{aligned} M &= \int_0^{2\pi}\int_0^2\int_0^{r^2} dz\,r\,dr\,d\theta = \int_0^{2\pi}\int_0^2 \Big[z\Big]_0^{r^2} r\,dr\,d\theta \\ &= \int_0^{2\pi}\int_0^2 r^3\,dr\,d\theta = \int_0^{2\pi}\left[\frac{r^4}{4}\right]_0^2 d\theta \\ &= \int_0^{2\pi} 4\,d\theta = 8\pi. \end{aligned}$$

Therefore,

$$\bar{z} = \frac{M_{xy}}{M} = \frac{32\pi}{3}\frac{1}{8\pi} = \frac{4}{3},$$

and the centroid is the point (0, 0, 4/3).

Spherical Coordinates

Spherical coordinates (Fig. 13.59) are good for describing spheres centered at the origin, half-planes hinged along the z-axis, and single-napped cones whose vertices lie at the origin and whose axes lie along the z-axis. Surfaces like these have equations of constant coordinate value:

$\rho = 4$ (sphere, radius 4, center at origin),

$\phi = \dfrac{\pi}{3}$ (cone opening up from the origin, making an angle of $\pi/3$ radians with the positive z-axis),

$\theta = \dfrac{\pi}{3}$ (half-plane, hinged along the z-axis, making an angle of $\pi/3$ radians with the positive x-axis).

The volume element in spherical coordinates is

$$dV = \rho^2 \sin\phi\,d\rho\,d\phi\,d\theta \tag{3}$$

(Fig. 13.60), and triple integrals take the form

$$\iiint f(\rho, \phi, \theta)\,dV = \iiint f(\rho, \phi, \theta)\rho^2 \sin\phi\,d\rho\,d\phi\,d\theta. \tag{4}$$

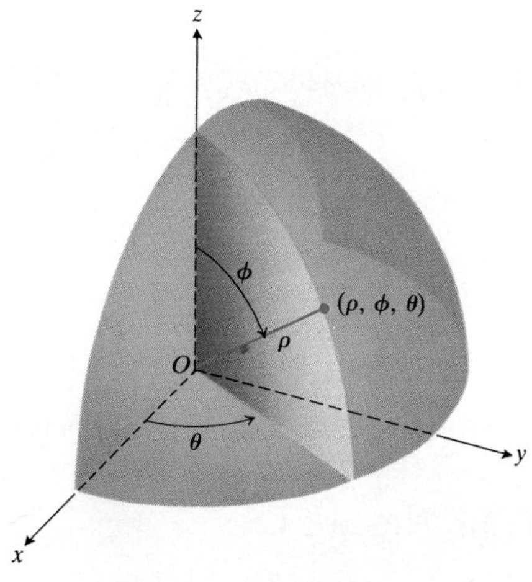

13.59 Spherical coordinates are measured with a distance and two angles.

To evaluate these integrals we usually integrate first with respect to ρ. The procedure for finding the limits of integration for a region D in space is shown in the following box.

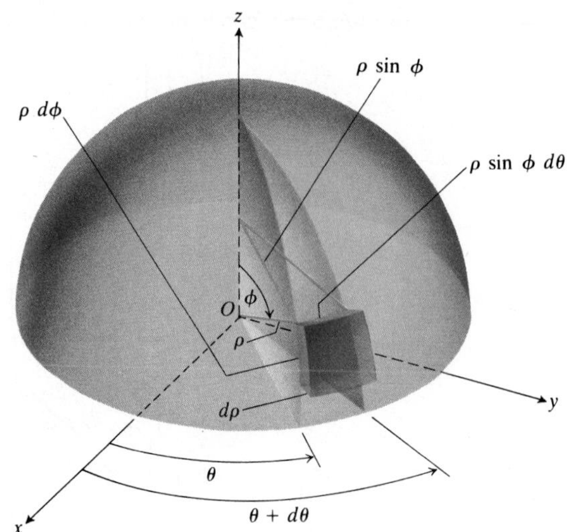

13.60 The volume element in spherical coordinates is

$$dV = d\rho \cdot \rho\, d\phi \cdot \rho \sin\phi\, d\theta$$
$$= \rho^2 \sin\phi\, d\rho\, d\phi\, d\theta.$$

How to Integrate in Spherical Coordinates

To integrate a continuous function $f(\rho, \phi, \theta)$ over a region given by the inequalities

$$g_1(\phi, \theta) \le \rho \le g_2(\phi, \theta),$$
$$h_1(\theta) \le \phi \le h_2(\theta),$$
$$\theta_1 \le \theta \le \theta_2,$$

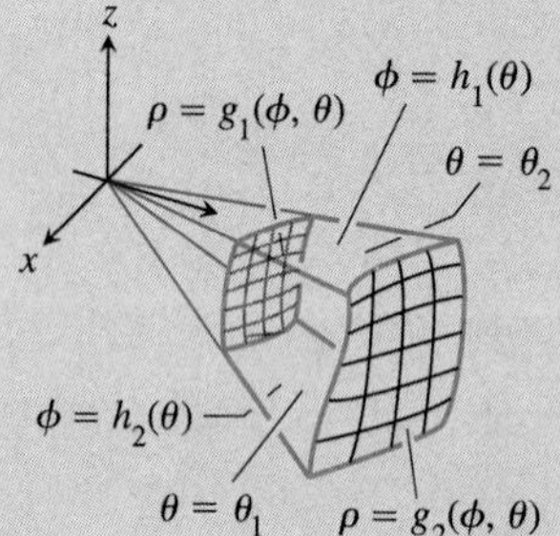

evaluate the iterated integral

$$\int_{\theta=\theta_1}^{\theta=\theta_2} \int_{\phi=h_1(\theta)}^{\phi=h_2(\theta)} \int_{\rho=g_1(\phi,\theta)}^{\rho=g_2(\phi,\theta)} f(\rho, \phi, \theta)\rho^2 \sin\phi\, d\rho\, d\phi\, d\theta. \tag{5}$$

To do so, multiply f by ρ^2 and integrate with respect to ρ. Multiply the result by $\sin\phi$ and integrate with respect to ϕ. Then integrate with respect to θ.

Example 2 Find the volume of the upper region cut from the solid sphere $\rho \le 1$ by the cone $\phi = \pi/3$.

Solution We sketch the cone and sphere (Fig. 13.61). The coordinate inequalities for the region occupied by the solid are

$$g_1(\phi, \theta) \le \rho \le g_2(\phi, \theta): \quad 0 \le \rho \le 1,$$
$$h_1(\theta) \le \phi \le h_2(\theta): \quad 0 \le \phi \le \frac{\pi}{3},$$
$$\theta_1 \le \theta \le \theta_2: \quad 0 \le \theta \le 2\pi.$$

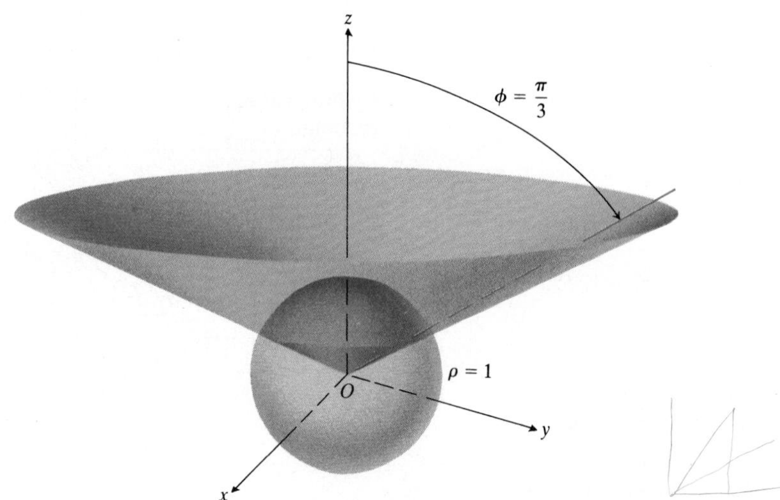

13.61 The region cut from the solid sphere $\rho \le 1$ by the cone $\phi = \pi/3$ (Example 2).

The volume is the integral of $f(\rho, \phi, \theta) = 1$ over this region:

$$\begin{aligned} V &= \int_0^{2\pi}\int_0^{\pi/3}\int_0^1 \rho^2 \sin\phi \, d\rho \, d\phi \, d\theta = \int_0^{2\pi}\int_0^{\pi/3}\left[\frac{\rho^3}{3}\right]_0^1 \sin\phi \, d\phi \, d\theta \\ &= \int_0^{2\pi}\int_0^{\pi/3} \frac{1}{3}\sin\phi \, d\phi \, d\theta = \int_0^{2\pi}\left[-\frac{1}{3}\cos\phi\right]_0^{\pi/3} d\theta \\ &= \int_0^{2\pi}\left(-\frac{1}{6}+\frac{1}{3}\right) d\theta = \frac{1}{6}(2\pi) = \frac{\pi}{3}. \end{aligned}$$

Example 3 Find the moment of inertia about the z-axis of the region in Example 2.

Solution In rectangular coordinates, the moment is

$$I_z = \iiint (x^2 + y^2)\, dV.$$

In spherical coordinates, $x^2 + y^2 = (\rho \sin\phi \cos\theta)^2 + (\rho \sin\phi \sin\theta)^2 = \rho^2 \sin^2\phi$. Hence,

$$I_z = \iiint (\rho^2 \sin^2\phi)\, \rho^2 \sin\phi \, d\rho \, d\phi \, d\theta = \iiint \rho^4 \sin^3\phi \, d\rho \, d\phi \, d\theta.$$

For the region in Example 2, this becomes

$$\begin{aligned} I_z &= \int_0^{2\pi}\int_0^{\pi/3}\int_0^1 \rho^4 \sin^3\phi \, d\rho \, d\phi \, d\theta = \int_0^{2\pi}\int_0^{\pi/3}\left[\frac{\rho^5}{5}\right]_0^1 \sin^3\phi \, d\phi \, d\theta \\ &= \frac{1}{5}\int_0^{2\pi}\int_0^{\pi/3} (1-\cos^2\phi)\sin\phi \, d\phi \, d\theta = \frac{1}{5}\int_0^{2\pi}\left[-\cos\phi + \frac{\cos^3\phi}{3}\right]_0^{\pi/3} d\theta \\ &= \frac{1}{5}\int_0^{2\pi}\left(-\frac{1}{2}+1+\frac{1}{24}-\frac{1}{3}\right) d\theta = \frac{1}{5}\int_0^{2\pi}\frac{5}{24}\, d\theta = \frac{1}{24}(2\pi) = \frac{\pi}{12}. \end{aligned}$$

Coordinate Conversion Formulas (from Section 10.8)

Cylindrical to rectangular	Spherical to cylindrical	Spherical to rectangular
$x = r\cos\theta$	$r = \rho\sin\phi$	$x = \rho\sin\phi\cos\theta$
$y = r\sin\theta$	$z = \rho\cos\phi$	$y = \rho\sin\phi\sin\theta$
$z = z$	$\theta = \theta$	$z = \rho\cos\phi$

Volume: $dV = dx\,dy\,dz$ or $dz\,r\,dr\,d\theta$ or $\rho^2\sin\phi\,d\rho\,d\phi\,d\theta$

EXERCISES 13.6

Evaluate the cylindrical coordinate integrals in Exercises 1–6.

1. $\int_0^{2\pi}\int_0^1\int_r^{\sqrt{2-r^2}} dz\,r\,dr\,d\theta$

2. $\int_0^{2\pi}\int_0^3\int_{r^2/3}^{\sqrt{18-r^2}} dz\,r\,dr\,d\theta$

3. $\int_0^{2\pi}\int_0^{\theta/2\pi}\int_0^{3+24r^2} dz\,r\,dr\,d\theta$

4. $\int_0^{\pi}\int_0^{\theta/\pi}\int_{-\sqrt{4-r^2}}^{3\sqrt{4-r^2}} z\,dz\,r\,dr\,d\theta$

5. $\int_0^{2\pi}\int_0^1\int_r^{1/\sqrt{2-r^2}} 3\,dz\,r\,dr\,d\theta$

6. $\int_0^{2\pi}\int_0^1\int_{-1/2}^{1/2} (r^2\sin^2\theta + z^2)\,dz\,r\,dr\,d\theta$

Evaluate the spherical coordinate integrals in Exercises 7–12.

7. $\int_0^{\pi}\int_0^{\pi}\int_0^{2\sin\phi} \rho^2\sin\phi\,d\rho\,d\phi\,d\theta$

8. $\int_0^{2\pi}\int_0^{\pi/4}\int_0^{2} (\rho\cos\phi)\,\rho^2\sin\phi\,d\rho\,d\phi\,d\theta$

9. $\int_0^{2\pi}\int_0^{\pi}\int_0^{(1-\cos\phi)/2} \rho^2\sin\phi\,d\rho\,d\phi\,d\theta$

10. $\int_0^{3\pi/2}\int_0^{\pi}\int_0^{1} 5\rho^3\sin^3\phi\,d\rho\,d\phi\,d\theta$

11. $\int_0^{2\pi}\int_0^{\pi/3}\int_{\sec\phi}^{2} 3\rho^2\sin\phi\,d\rho\,d\phi\,d\theta$

12. $\int_0^{2\pi}\int_0^{\pi/4}\int_0^{\sec\phi} \rho^3\cos^2\phi\sin\phi\,d\rho\,d\phi\,d\theta$

The integrals we have seen so far suggest that there are preferred orders of integration for cylindrical and spherical coordinates, but this is not really the case. The other orders usually work well and are occasionally better. Evaluate the integrals in Exercises 13–20.

13. $\int_0^{2\pi}\int_0^{3}\int_0^{z/3} r^3\,dr\,dz\,d\theta$

14. $\int_{-1}^{1}\int_0^{2\pi}\int_0^{1+\cos\theta} 4r\,dr\,d\theta\,dz$

15. $\int_0^{1}\int_0^{\sqrt{z}}\int_0^{2\pi} (r^2\cos^2\theta + z^2)r\,d\theta\,dr\,dz$

16. $\int_0^{2}\int_{r-2}^{\sqrt{4-r^2}}\int_0^{2\pi} (r\sin\theta + 1)r\,d\theta\,dz\,dr$

17. $\int_0^{2}\int_{-\pi}^{0}\int_{\pi/4}^{\pi/2} \rho^3\sin 2\phi\,d\phi\,d\theta\,d\rho$

18. $\int_{\pi/6}^{5\pi/6}\int_{\csc\phi}^{2\csc\phi}\int_0^{2\pi} \rho\sin\phi\,d\theta\,d\rho\,d\phi$

19. $\int_0^{1}\int_0^{\pi}\int_0^{\pi/4} 12\rho\sin^3\phi\,d\phi\,d\theta\,d\rho$

20. $\int_{\pi/6}^{\pi/2}\int_{-\pi/2}^{\pi/2}\int_{\csc\phi}^{2} 5\rho^4\sin^3\phi\,d\rho\,d\theta\,d\phi$

21. Set up an iterated triple integral for the volume of the sphere

$x^2 + y^2 + z^2 = 4$ in (a) spherical, (b) cylindrical, and (c) rectangular coordinates.

22. Let D denote the region in the first octant that is bounded below by the cone $z = \sqrt{x^2 + y^2}$ and above by the sphere $x^2 + y^2 + z^2 = 9$. Express the volume of D as an iterated triple integral in (a) cylindrical and (b) spherical coordinates. Then (c) find V.

23. Give the limits of integration for evaluating the integral

$$\iiint f(r, \theta, z)\, dz\, r\, dr\, d\theta$$

as an iterated integral over the region that is bounded below by the plane $z = 0$, on the side by the cylinder $r = \cos\theta$, and on top by the paraboloid $z = 3r^2$.

24. Convert the integral

$$\int_{-1}^{1}\int_{0}^{\sqrt{1-y^2}}\int_{0}^{x} (x^2 + y^2)\, dz\, dx\, dy$$

to an equivalent integral in cylindrical coordinates and evaluate the result.

25. Let D be the smaller cap cut from a solid ball of radius 2 units by a plane 1 unit from the center of the sphere. Express the volume of D as an iterated triple integral in (a) spherical, (b) cylindrical, and (c) rectangular coordinates. Then (d) find the volume by evaluating one of the three triple integrals.

26. Express the moment of inertia I_z of the solid hemisphere $x^2 + y^2 + z^2 \leq 1$, $z \geq 0$, as an iterated integral in (a) cylindrical and (b) spherical coordinates. Then (c) find I_z by evaluating one of these integrals.

Cylindrical Coordinates

In exercises that ask for masses and moments, assume that the density is $\delta = 1$ unless stated otherwise.

27. Find the volume of the region bounded below by the plane $z = 0$, laterally by the cylinder $x^2 + y^2 = 1$, and above by the paraboloid $z = x^2 + y^2$.

28. Find the volume of the region bounded below by the paraboloid $z = x^2 + y^2$, laterally by the cylinder $x^2 + y^2 = 1$, and above by the paraboloid $z = x^2 + y^2 + 1$.

29. Find the volume of the region bounded below by the plane $z = 0$ and above by the paraboloid $z = 4 - x^2 - y^2$.

30. Find the volume of the region enclosed by the cylinder $x^2 + y^2 = 4$ and the planes $z = 0$ and $y + z = 4$. (*Hint:* In cylindrical coordinates, $z = 4 - y$ becomes $z = 4 - r\sin\theta$.)

31. Find the volume of the region bounded above by the paraboloid $z = 5 - x^2 - y^2$ and below by the paraboloid $z = 4x^2 + 4y^2$.

32. Find the volume of the region bounded above by the paraboloid $z = 9 - x^2 - y^2$ and below by the xy-plane and lying *outside* the cylinder $x^2 + y^2 = 1$.

33. Find the volume of the region cut from the solid sphere $x^2 + y^2 + z^2 \leq 4$ by the cylinder $x^2 + y^2 = 1$.

34. Find the volume of the region bounded above by the sphere $x^2 + y^2 + z^2 = 2$ and below by the circular paraboloid $z = x^2 + y^2$.

35. Find the average value of the function $f(r, \theta, z) = r$ over the region bounded by the cylinder $r = 1$ between the planes $z = -1$ and $z = 1$.

36. Find the average value of the function $f(r, \theta, z) = r$ over the solid ball bounded by the sphere $r^2 + z^2 = 1$. (This is the sphere $x^2 + y^2 + z^2 = 1$.)

37. A solid is bounded below by the plane $z = 0$, above by the cone $z = r$, $r \geq 0$, and on the sides by the cylinder $r = 1$. Find the center of mass.

38. Find the centroid of the region that is bounded below by the surface $z = \sqrt{r}$ and above by the plane $z = 2$.

39. Find the moment of inertia and radius of gyration about the z-axis of a thick-walled right circular cylinder bounded on the inside by the cylinder $r = 1$, on the outside by the cylinder $r = 2$, and on the top and bottom by the planes $z = 4$ and $z = 0$.

40. A solid of constant density in the first octant is bounded above by the cone $z = \sqrt{x^2 + y^2}$, below by the plane $z = 0$, and on the sides by the cylinder $x^2 + y^2 = 4$ and the planes $x = 0$ and $y = 0$. Find the center of mass. (*Hint:* $\bar{x} = \bar{y}$.)

41. Find the moment of inertia of a solid circular cylinder of radius 1 and height 2 (a) about the axis of the cylinder, (b) about a line through the centroid perpendicular to the axis of the cylinder.

42. Find the moment of inertia of a right circular cone of base radius 1 and height 1 about an axis through the vertex parallel to the base.

43. Find the moment of inertia of a solid sphere of radius a about a diameter.

44. Find the centroid of the smaller region cut from the solid ball $r^2 + z^2 \leq 1$ by the half-planes $\theta = -\pi/3$, $r \geq 0$, and $\theta = \pi/3$, $r \geq 0$.

45. Show that the centroid of the solid semi-ellipsoid of revolution $(r^2/a^2) + (z^2/h^2) \leq 1$, $z \geq 0$, lies on the z-axis three-eighths of the way from the base to the top. The special case $h = a$ gives a solid hemisphere. Thus the centroid of a solid hemisphere lies on the axis of symmetry three-eighths of the way from the base to the top.

46. Find the moment of inertia of a right circular cone of base radius a and height h about its axis. (*Hint:* Place the cone with its vertex at the origin and its axis along the z-axis.)

47. Show that the centroid of a solid right circular cone is one-fourth of the way from the base to the vertex. (In general, the centroid of a solid cone or pyramid is one-fourth of the way from the centroid of the base to the vertex.)

48. A solid is bounded on the top by the paraboloid $z = r^2$, on the bottom by the plane $z = 0$, and on the sides by the cylinder $r = 1$. Find the center of mass and the moment of inertia and radius of gyration about the z-axis if the density is (a) $\delta(r, \theta, z) = z$, (b) $\delta(r, \theta, z) = r$.

49. A solid is bounded below by the cone $z = \sqrt{x^2 + y^2}$ and above by the plane $z = 1$. Find the center of mass and the moment of inertia and radius of gyration about the z-axis if the density is (a) $\delta(r, \theta, z) = z$, (b) $\delta(r, \theta, z) = z^2$.

50. A solid right circular cylinder is bounded by the cylinder $r = a$ and the planes $z = 0$ and $z = h, h > 0$. Find the center of mass and the moment of inertia and radius of gyration about the z-axis if the density is $\delta(r, \theta, z) = z + 1$.

Spherical Coordinates

In exercises that ask for masses and moments, assume that the density is $\delta = 1$ unless stated otherwise.

51. Find the volume of the portion of the solid sphere $\rho \leq a$ that lies between the cones $\phi = \pi/3$ and $\phi = 2\pi/3$.

52. Find the volume of the region cut from the solid sphere $\rho \leq a$ by the half-planes $\theta = 0$ and $\theta = \pi/6$ in the first octant.

53. Find the volume of the smaller region cut from the solid sphere $\rho \leq 2$ by the plane $z = 1$.

54. Find the volume of the region enclosed by the surface $\rho = 1 - \cos\phi$ (Fig. 13.62).

55. Find the volume of the solid cut from the thick-walled cylinder $1 \leq x^2 + y^2 \leq 2$ by the cones $z = \pm\sqrt{x^2 + y^2}$.

56. Find the volume of the region that lies inside the sphere $x^2 + y^2 + z^2 = 2$ and outside the cylinder $x^2 + y^2 = 1$. $\frac{4\pi}{3}$

57. Find the average value of the function $f(\rho, \phi, \theta) = \rho$ over the solid ball $\rho \leq 1$.

58. Figure 13.63 shows a solid that was made by drilling a conical hole in a solid hemisphere. The equation of the cone is $\phi = \pi/3$. The radius of the hemisphere is $\rho = 2$. Find the center of mass.

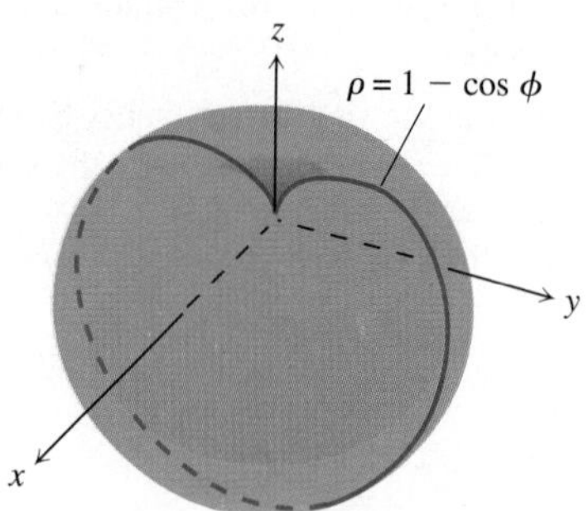

13.62 In spherical coordinates, the surface $\rho = 1 - \cos\phi$ is a cardioid of revolution (Exercise 54).

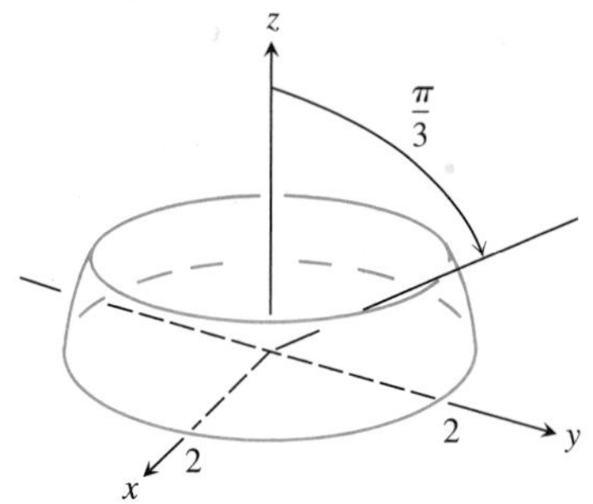

13.63 The solid in Exercise 58.

59. Find the centroid of the solid bounded above by the sphere $\rho = a$ and below by the cone $\phi = \pi/4$.

60. A solid ball is bounded by the sphere $\rho = a$. Find the moment of inertia and radius of gyration about the z-axis if the density is
a) $\delta(\rho, \phi, \theta) = \rho^2$, b) $\delta(\rho, \phi, \theta) = r = \rho\sin\phi$.

13.7 Substitutions in Multiple Integrals

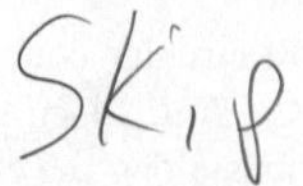

This section shows how to evaluate multiple integrals by substitution. As in single integration, the goal of substitution is to replace complicated integrals by ones that are easier to evaluate. Substitutions can accomplish this by simplifying the integrand, by simplifying the limits of integration, or both.

Substitutions in Double Integrals

The polar coordinate substitution is a special case of a more general substitution method for double integrals, a method that pictures changes in variables as transformations of two-dimensional regions.

Suppose that a region G in the uv-plane is transformed one-to-one into the

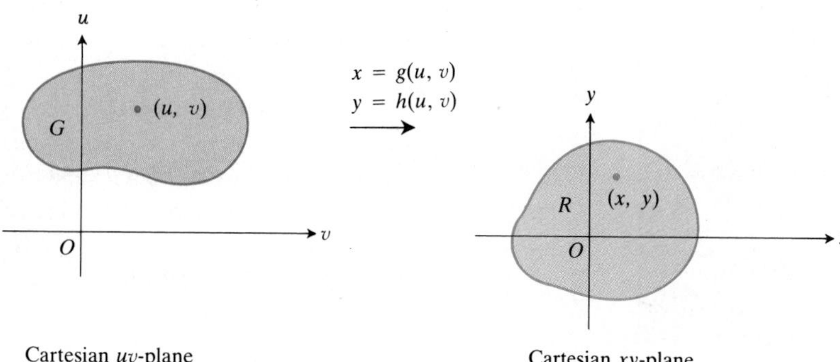

13.64 The equations $x = g(u, v)$ and $y = h(u, v)$ allow us to write an integral over a region R in the xy-plane as an integral over a region G in the uv-plane.

region R in the xy-plane by differentiable equations of the form

$$x = g(u, v), \qquad y = h(u, v),$$

as suggested in Fig. 13.64. Then any function $f(x, y)$ defined on R can be thought of as a function $f(g(u, v), h(u, v))$ defined on G as well. How is the integral of $f(x, y)$ over R related to the integral of $f(g(u, v), h(u, v))$ over G?

Notice the "Reversed" Order

The transforming equations $x = g(u, v)$ and $y = h(u, v)$ go from G to R, but we use them to change an integral over R into an integral over G.

The answer is: If g, h, and f have continuous partial derivatives and $J(u, v)$ (to be discussed in a moment) is zero, if at all, only at isolated points, then

$$\iint_R f(x, y)\,dx\,dy = \iint_G f(g(u, v), h(u, v))\,|J(u, v)|\,du\,dv. \tag{1}$$

The factor $J(u, v)$, whose absolute value appears in this formula, is the determinant

$$J(u, v) = \begin{vmatrix} \dfrac{\partial x}{\partial u} & \dfrac{\partial x}{\partial v} \\[2ex] \dfrac{\partial y}{\partial u} & \dfrac{\partial y}{\partial v} \end{vmatrix} = \frac{\partial(x, y)}{\partial(u, v)}. \tag{2}$$

It is called the **Jacobian determinant** or **Jacobian** of the coordinate transformation $x = g(u, v)$, $y = h(u, v)$, named after the mathematician Carl Jacobi. The alternative notation $\partial(x, y)/\partial(u, v)$ may help you to remember how the determinant is constructed from the partial derivatives of x and y. The derivation of Eq. (1) is intricate and we shall not give it here.

Carl Gustav Jacob Jacobi

Jacobi (1804–1851), one of nineteenth-century Germany's most accomplished scientists, developed the theory of determinants and transformations into a powerful tool for evaluating multiple integrals and solving differential equations. He also applied transformation methods to study nonelementary integrals like the ones that arise in the calculation of arc length. Like Euler, Jacobi was a prolific writer and an even more prolific calculator and worked in a variety of mathematical and applied fields.

For polar coordinates, we have r and θ in place of u and v. With $x = r\cos\theta$ and $y = r\sin\theta$, the Jacobian is

$$J(r, \theta) = \begin{vmatrix} \dfrac{\partial x}{\partial r} & \dfrac{\partial x}{\partial \theta} \\[2ex] \dfrac{\partial y}{\partial r} & \dfrac{\partial y}{\partial \theta} \end{vmatrix} = \begin{vmatrix} \cos\theta & -r\sin\theta \\ \sin\theta & r\cos\theta \end{vmatrix} = r(\cos^2\theta + \sin^2\theta) = r.$$

Hence, Eq. (1) becomes

$$\iint_R f(x, y)\,dx\,dy = \iint_G f(r\cos\theta, r\sin\theta)\,|r|\,dr\,d\theta$$

$$= \iint_G f(r\cos\theta, r\sin\theta)\,r\,dr\,d\theta, \qquad (\text{If } r \geq 0) \tag{3}$$

which is Eq. (7) in Section 13.3.

θ

$\frac{\pi}{2}$

G

O 1 r

Cartesian $r\theta$-plane

$x = r\cos\theta$
$y = r\sin\theta$

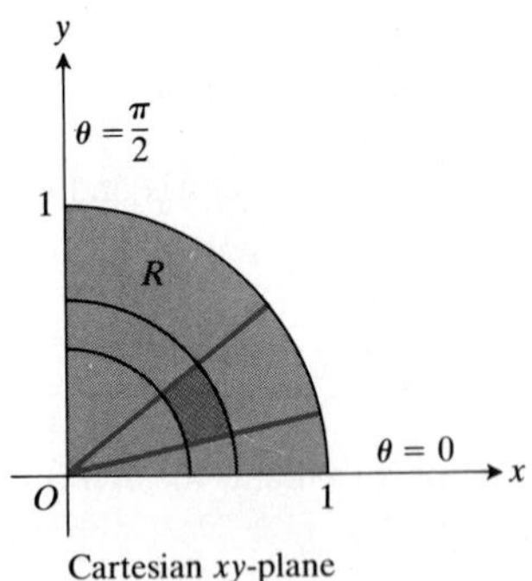

13.65 The equations $x = r\cos\theta$, $y = r\sin\theta$ transform G into R.

Figure 13.65 shows how the equations $x = r\cos\theta$, $y = r\sin\theta$ transform the rectangle $G\colon 0 \le r \le 1,\ 0 \le \theta \le \pi/2$ into the quarter circle R bounded by $x^2 + y^2 = 1$ in the first quadrant of the xy-plane.

Notice that the integral on the right-hand side of Eq. (3) is not the integral of $f(r\cos\theta, r\sin\theta)$ over a region in the polar coordinate plane. It is the integral of the product of $f(r\cos\theta, r\sin\theta)$ and r over a region G in the *Cartesian* $r\theta$-plane.

Here is an example of another substitution.

Example 1 Evaluate

$$\int_0^4 \int_{x=y/2}^{x=(y/2)+1} \frac{2x-y}{2}\,dx\,dy$$

by applying the transformation

$$u = \frac{2x-y}{2}, \qquad v = \frac{y}{2} \tag{4}$$

and integrating over an appropriate region in the uv-plane.

Solution We sketch the region R of integration in the xy-plane and identify its boundaries (Fig. 13.66).

To apply Eq. (1), we need to find the corresponding uv-region G and the Jacobian of the transformation. To find them, we first solve Eqs. (4) for x and y in terms of u and v. Routine algebra gives

$$x = u + v, \qquad y = 2v. \tag{5}$$

We then find the boundaries of G by substituting these expressions into the equations for the boundaries of R (Fig. 13.66):

xy-equations for the boundary of *R*	Corresponding *uv*-equations for the boundary of *G*	Simplified *uv*-equations
$x = y/2$	$u + v = 2v/2 = v$	$u = 0$
$x = (y/2) + 1$	$u + v = (2v/2) + 1 = v + 1$	$u = 1$
$y = 0$	$2v = 0$	$v = 0$
$y = 4$	$2v = 4$	$v = 2$

The Jacobian of the transformation (again from Eqs. 5) is

$$J(u, v) = \begin{vmatrix} \dfrac{\partial x}{\partial u} & \dfrac{\partial x}{\partial v} \\[2mm] \dfrac{\partial y}{\partial u} & \dfrac{\partial y}{\partial v} \end{vmatrix} = \begin{vmatrix} \dfrac{\partial}{\partial u}(u+v) & \dfrac{\partial}{\partial v}(u+v) \\[2mm] \dfrac{\partial}{\partial u}(2v) & \dfrac{\partial}{\partial v}(2v) \end{vmatrix} = \begin{vmatrix} 1 & 1 \\ 0 & 2 \end{vmatrix} = 2.$$

We now have everything we need to apply Eq. (1):

$$\int_0^4 \int_{x=y/2}^{x=(y/2)+1} \frac{2x-y}{2}\,dx\,dy = \int_{v=0}^{v=2} \int_{u=0}^{u=1} u\,|J(u, v)|\,du\,dv$$

$$= \int_0^2 \int_0^1 (u)(2)\,du\,dv = \int_0^2 \Big[u^2\Big]_0^1 dv = \int_0^2 dv = 2.$$

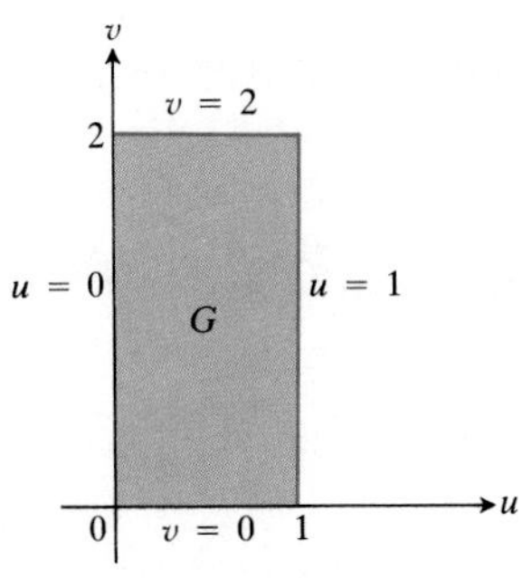

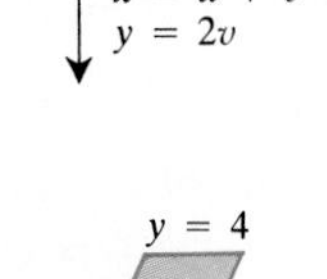

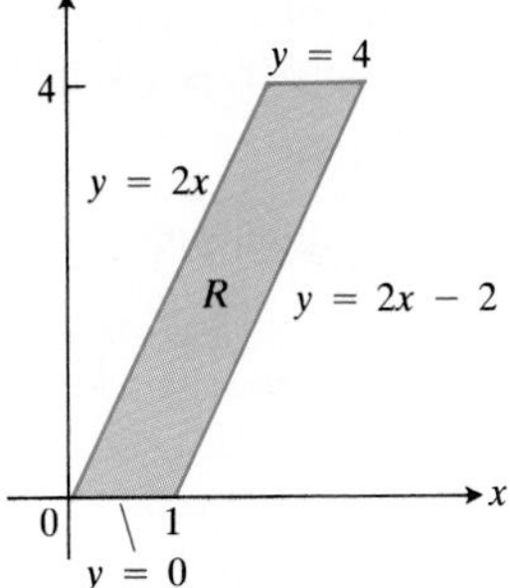

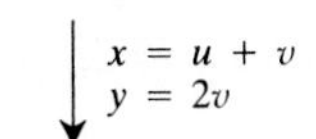

13.66 The equations $x = u + v$ and $y = 2v$ transform G into R. Reversing the transformation by the equations $u = (2x - y)/2$ and $v = y/2$ transforms R into G. See Example 1.

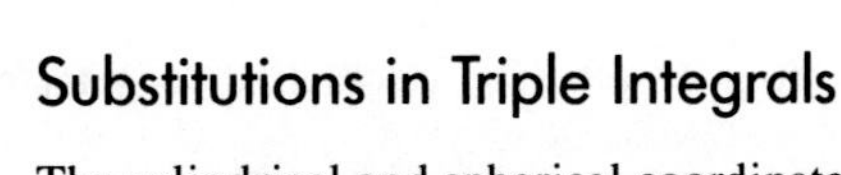

Substitutions in Triple Integrals

The cylindrical and spherical coordinate substitutions are special cases of a substitution method that pictures changes of variables in triple integrals as transformations of three-dimensional regions. The method is just like the method for double integrals, except that now we work in three dimensions instead of two.

Suppose that a region G in uvw-space is transformed one-to-one into the region D in xyz-space by differentiable equations of the form

$$x = g(u, v, w), \qquad y = h(u, v, w), \qquad z = k(u, v, w),$$

as suggested in Fig. 13.67. Then any function $F(x, y, z)$ defined on D can be thought of as a function

$$F(g(u, v, w), h(u, v, w), k(u, v, w)) = H(u, v, w)$$

defined on G. If g, h, and k have continuous first partial derivatives and the Jacobian of the transformation (see below) is never zero or is zero only at isolated points, then the integral of $F(x, y, z)$ over D is related to the integral of $H(u, v, w)$ over G by the equation

$$\iiint_D F(x, y, z)\, dx\, dy\, dz = \iiint_G H(u, v, w)\, |J(u, v, w)|\, du\, dv\, dw. \tag{6}$$

The factor $J(u, v, w)$, whose absolute value appears in this equation, is the **Jacobian determinant**

$$J(u, v, w) = \begin{vmatrix} \dfrac{\partial x}{\partial u} & \dfrac{\partial x}{\partial v} & \dfrac{\partial x}{\partial w} \\[2mm] \dfrac{\partial y}{\partial u} & \dfrac{\partial y}{\partial v} & \dfrac{\partial y}{\partial w} \\[2mm] \dfrac{\partial z}{\partial u} & \dfrac{\partial z}{\partial v} & \dfrac{\partial z}{\partial w} \end{vmatrix} = \frac{\partial(x, y, z)}{\partial(u, v, w)}. \tag{7}$$

As in the two-dimensional case, the derivation of Eq. (6) is complicated and we shall not go into it here.

For cylindrical coordinates, we have r, θ, and z in place of u, v, and w. The transformation from Cartesian $r\theta z$-space to Cartesian xyz-space is given by the equations

$$x = r\cos\theta, \qquad y = r\sin\theta, \qquad z = z$$

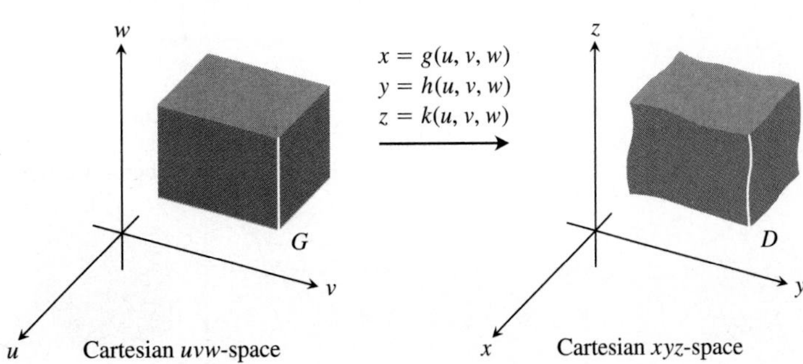

13.67 The equations $x = g(u, v, w)$, $y = h(u, v, w)$, and $z = k(u, v, w)$ allow us to write an integral over a region D in Cartesian xyz-space as an integral over a region G in Cartesian uvw-space.

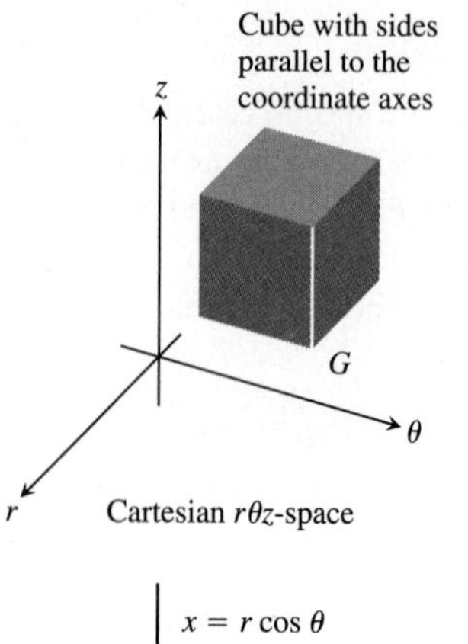

$x = r \cos\theta$
$y = r \sin\theta$
$z = z$

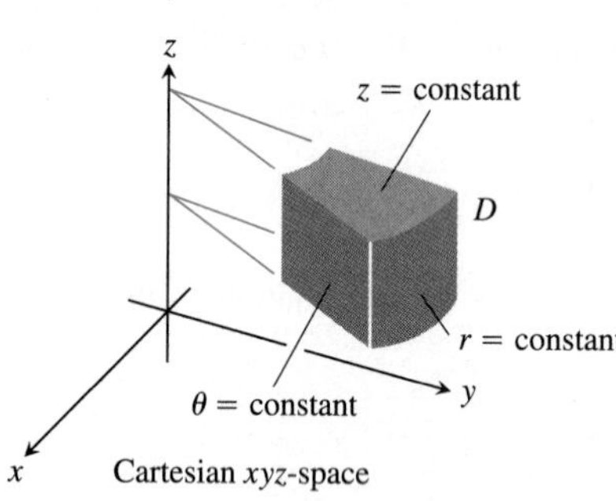

13.68 The equations $x = r\cos\theta$, $y = r\sin\theta$, and $z = z$ transform G into D.

(Fig. 13.68). The Jacobian of the transformation is

$$J(r, \theta, z) = \begin{vmatrix} \frac{\partial x}{\partial r} & \frac{\partial x}{\partial \theta} & \frac{\partial x}{\partial z} \\ \frac{\partial y}{\partial r} & \frac{\partial y}{\partial \theta} & \frac{\partial y}{\partial z} \\ \frac{\partial z}{\partial r} & \frac{\partial z}{\partial \theta} & \frac{\partial z}{\partial z} \end{vmatrix} = \begin{vmatrix} \cos\theta & -r\sin\theta & 0 \\ \sin\theta & r\cos\theta & 0 \\ 0 & 0 & 1 \end{vmatrix} = r\cos^2\theta + r\sin^2\theta = r.$$

The corresponding version of Eq. (6) is

$$\iiint_D F(x, y, z)\, dx\, dy\, dz = \iiint_G H(r, \theta, z)\, |r|\, dr\, d\theta\, dz, \tag{8}$$

and we can drop the absolute value signs whenever $r \geq 0$.

For spherical coordinates, we have ρ, ϕ, and θ in place of u, v, and w. The transformation from Cartesian $\rho\phi\theta$-space to Cartesian xyz-space is given by the equations

$$x = \rho \sin\phi \cos\theta, \qquad y = \rho\sin\phi\sin\theta, \qquad z = \rho\cos\phi$$

(Fig. 13.69). The Jacobian of the transformation is

$$J(\rho, \phi, \theta) = \begin{vmatrix} \frac{\partial x}{\partial \rho} & \frac{\partial x}{\partial \phi} & \frac{\partial x}{\partial \theta} \\ \frac{\partial y}{\partial \rho} & \frac{\partial y}{\partial \phi} & \frac{\partial y}{\partial \theta} \\ \frac{\partial z}{\partial \rho} & \frac{\partial z}{\partial \phi} & \frac{\partial z}{\partial \theta} \end{vmatrix} = \rho^2 \sin\phi, \tag{9}$$

as you will see if you do Exercise 9. The corresponding version of Eq. (6) is

$$\iiint_D F(x, y, z)\, dx\, dy\, dz = \iiint_G H(\rho, \phi, \theta) |\, \rho^2 \sin\phi \,|\, d\rho\, d\phi\, d\theta, \tag{10}$$

and we can drop the absolute value signs because $\sin\phi$ is never negative. Here is an example of another substitution.

Example 2 Evaluate

$$\int_0^3 \int_0^4 \int_{x = y/2}^{x = (y/2)+1} \left(\frac{2x - y}{2} + \frac{z}{3}\right) dx\, dy\, dz$$

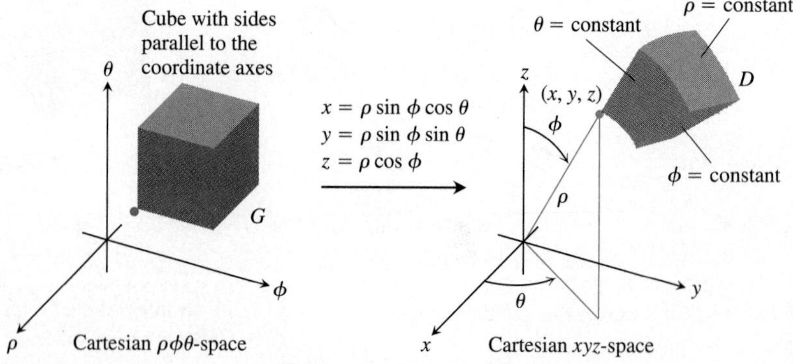

13.69 The equations $x = \rho\sin\phi\cos\theta$, $y = \rho\sin\phi\sin\theta$, and $z = \rho\cos\phi$ transform G into D.

by applying the transformation

$$u = \frac{2x - y}{2}, \qquad v = \frac{y}{2}, \qquad w = \frac{z}{3} \tag{11}$$

and integrating over an appropriate region in uvw-space.

Solution We sketch the region D of integration in xyz-space and identify its boundaries (Fig. 13.70). In this case, the surfaces that bound D are all planes.

To apply Eq. (6), we need to find the corresponding uvw-region G and the Jacobian of the transformation. To find them, we first solve Eqs. (11) for x, y, and z in terms of u, v, and w. Routine algebra gives

$$x = u + v, \qquad y = 2v, \qquad z = 3w. \tag{12}$$

We then find the boundaries of G by substituting these expressions into the equations for the boundaries of D:

xyz-equations for the boundary of D	Corresponding uvw-equations for the boundary of G	Simplified uvw-equations
$x = y/2$	$u + v = 2v/2 = v$	$u = 0$
$x = (y/2) + 1$	$u + v = (2v/2) + 1 = v + 1$	$u = 1$
$y = 0$	$2v = 0$	$v = 0$
$y = 4$	$2v = 4$	$v = 2$
$z = 0$	$3w = 0$	$w = 0$
$z = 3$	$3w = 3$	$w = 1$

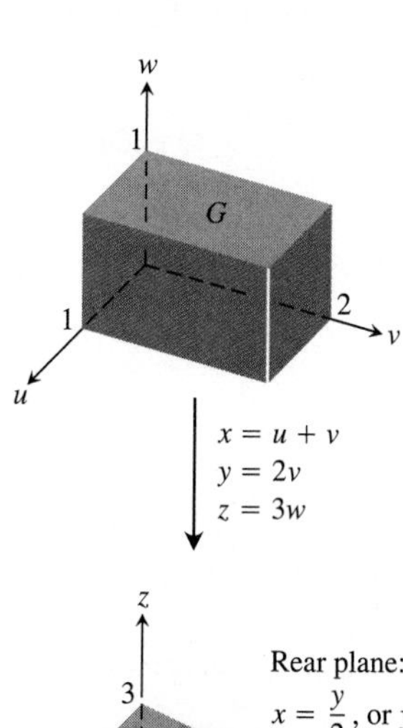

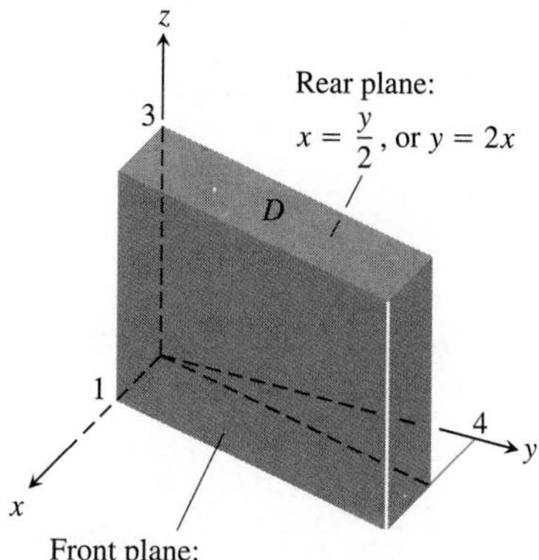

13.70 The equations $x = u + v$, $y = 2v$, and $z = 3w$ transform G into D. Reversing the transformation by the equations $u = (2x - y)/2$, $v = y/2$, and $w = z/3$ transforms D into G. See Example 2.

See Fig. 13.70.

The Jacobian of the transformation, again from Eqs. (12), is

$$J(u, v, w) = \begin{vmatrix} \dfrac{\partial x}{\partial u} & \dfrac{\partial x}{\partial v} & \dfrac{\partial x}{\partial w} \\ \dfrac{\partial y}{\partial u} & \dfrac{\partial y}{\partial v} & \dfrac{\partial y}{\partial w} \\ \dfrac{\partial z}{\partial u} & \dfrac{\partial z}{\partial v} & \dfrac{\partial z}{\partial w} \end{vmatrix} = \begin{vmatrix} 1 & 1 & 0 \\ 0 & 2 & 0 \\ 0 & 0 & 3 \end{vmatrix} = 6.$$

We now have everything we need to apply Eq. (6):

$$\int_0^3 \int_0^4 \int_{x=y/2}^{x=(y/2)+1} \left(\frac{2x - y}{2} + \frac{z}{3}\right) dx\, dy\, dz$$

$$= \int_0^1 \int_0^2 \int_0^1 (u + w)\, |J(u, v, w)|\, du\, dv\, dw$$

$$= \int_0^1 \int_0^2 \int_0^1 (u + w)(6)\, du\, dv\, dw = 6 \int_0^1 \int_0^2 \left[\frac{u^2}{2} + uw\right]_0^1 dv\, dw$$

$$= 6 \int_0^1 \int_0^2 \left(\frac{1}{2} + w\right) dv\, dw = 6 \int_0^1 (1 + 2w)\, dw$$

$$= 6\left[w + w^2\right]_0^1 = 6(2) = 12.$$

Exercise 10 asks you to evaluate the original integral directly to show that its value is also 12.

EXERCISES 13.7

Double Integrals

1. Evaluate the integral

$$\int_0^4 \int_{x=y/2}^{x=(y/2)+1} \frac{2x-y}{2}\, dx\, dy$$

from Example 1 directly by integration with respect to x and y to confirm that its value is 2.

2. Find the Jacobian $\partial(x, y)/\partial(u, v)$ for the transformation

a) $x = u \cos v, \quad y = u \sin v,$

b) $x = u \sin v, \quad y = u \cos v.$

3. a) Solve the system

$$u = x - y, \quad v = 2x + y$$

for x and y in terms of u and v. Then find the value of the Jacobian $\partial(x, y)/\partial(u, v)$.

b) Let R be the region in the first quadrant bounded by the lines $y = -2x + 4$, $y = -2x + 7$, $y = x - 2$, and $y = x + 1$. Evaluate

$$\iint_R (2x^2 - xy - y^2)\, dx\, dy$$

by changing variables with the equations in (a) and integrating over a region G in the uv-plane.

4. a) Find the Jacobian of the transformation $x = u$, $y = uv$, and sketch the region G: $1 \le u \le 2$, $1 \le uv \le 2$ in the uv-plane.

b) Then use Eq. (1) to transform the integral

$$\int_1^2 \int_1^2 \frac{y}{x}\, dy\, dx$$

into an integral over G, and evaluate both integrals.

5. Let R be the region in the first quadrant of the xy-plane bounded by the hyperbolas $xy = 1$, $xy = 9$ and the lines $y = x$, $y = 4x$ (Fig. 13.71). Use the transformation $x = u/v$, $y = uv$ with $u > 0$ and $v > 0$ to rewrite

$$\iint_R \left(\sqrt{\frac{y}{x}} + \sqrt{xy}\right) dx\, dy$$

as an integral over an appropriate region G in the uv-plane. Then evaluate the uv-integral over G.

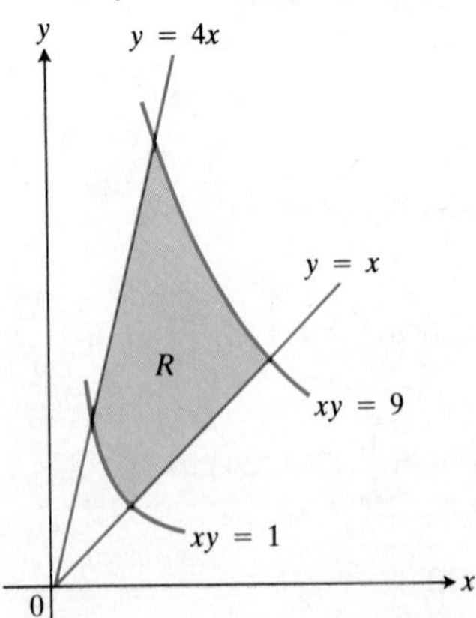

13.71 The xy-region in Exercise 5.

6. The area πab of the ellipse $x^2/a^2 + y^2/b^2 = 1$ can be found by integrating the function $f(x, y) = 1$ over the region bounded by the ellipse in the xy-plane. Evaluating the integral directly requires a trigonometric substitution. An easier way to evaluate the integral is to use the transformation $x = au$, $y = bv$ and evaluate the transformed integral over the disk G: $u^2 + v^2 \le 1$ in the uv-plane. Find the area this way.

7. A thin plate of constant density covers the region bounded by the ellipse $x^2/a^2 + y^2/b^2 = 1$ in the xy-plane. Find the moment of the plate about the origin. (*Hint:* Use the transformation $x = ar \cos \theta$, $y = br \sin \theta$.)

8. Use the transformation $x = u + v/2$, $y = v$, to evaluate the integral

$$\int_0^2 \int_{y/2}^{(y+4)/2} y^3(2x - y)e^{(2x-y)^2}\, dx\, dy$$

by first writing it as an integral over a region G in the uv-plane.

Triple Integrals

9. Evaluate the determinant in Eq. (9) to show that the Jacobian of the transformation from Cartesian $\rho\phi\theta$-space to Cartesian xyz-space is $\rho^2 \sin \phi$.

10. Evaluate the integral in Example 2 by integrating with respect to x, y, and z.

11. Find the volume of the ellipsoid

$$\frac{x^2}{a^2} + \frac{y^2}{b^2} + \frac{z^2}{c^2} = 1.$$

(*Hint:* Let $x = au$, $y = bv$, and $z = cw$. Then find the volume of an appropriate region in uvw-space.)

12. Evaluate

$$\iiint |xyz|\, dx\, dy\, dz$$

over the solid ellipsoid

$$\frac{x^2}{a^2} + \frac{y^2}{b^2} + \frac{z^2}{c^2} \le 1.$$

(*Hint:* Let $x = au$, $y = bv$, and $z = cw$. Then integrate over an appropriate region in uvw-space.)

13. Let D be the region in xyz-space defined by the inequalities

$$1 \le x \le 2, \quad 0 \le xy \le 2, \quad \text{and} \quad 0 \le z \le 1.$$

Evaluate

$$\iiint_D (x^2y + 3xyz)\,dx\,dy\,dz$$

by applying the transformation

$$u = x, \quad v = xy, \quad w = 3z$$

and integrating over an appropriate region G in uvw-space.

14. Assuming the result that the center of mass of a solid hemisphere lies on the axis of symmetry three-eighths of the way from the base toward the top, show, by transforming the appropriate integrals, that the center of mass of a solid semi-ellipsoid $(x^2/a^2) + (y^2/b^2) + (z^2/c^2) \leq 1,\ z \geq 0$ lies on the z-axis three-eighths of the way from the base toward the top. (You can do this without evaluating any of the integrals.)

REVIEW QUESTIONS

1. Define the double integral of a function of two variables over a bounded region in the coordinate plane.
2. How are double integrals evaluated as iterated integrals? Does the order of integration matter? How are the limits of integration determined? Give examples.
3. How are double integrals used to calculate areas, average values, masses, moments, centers of mass, and radii of gyration? Give examples.
4. How can you change a double integral in rectangular coordinates into a double integral in polar coordinates? Why might it be worthwhile to do so? Give an example.
5. Define the triple integral of a function $f(x, y, z)$ over a bounded region in space.
6. How are triple integrals in rectangular coordinates evaluated? How are the limits of integration determined? Give an example.
7. How are triple integrals in rectangular coordinates used to calculate volumes, average values, masses, moments, centers of mass, and radii of gyration? Give examples.
8. How are triple integrals defined in cylindrical and spherical coordinates? Why might one prefer working in one of these coordinate systems to working in rectangular coordinates?
9. How are triple integrals in cylindrical and spherical coordinates evaluated? How are the limits of integration found? Give examples.
10. How are substitutions in double integrals pictured as transformations of two-dimensional regions? Give a sample calculation.
11. How are substitutions in triple integrals pictured as transformations of three-dimensional regions? Give a sample calculation.

MISCELLANEOUS EXERCISES

1. Each of the following integrals represents the area of a region in a Cartesian coordinate plane. Sketch the region. Express the area of the region as a double integral with the order of integration reversed. Then evaluate both integrals.

a) $\displaystyle\int_0^4 \int_{-\sqrt{4-y}}^{(y-4)/2} dx\,dy$ b) $\displaystyle\int_{-1}^1 \int_{u^2}^1 dv\,du$

c) $\displaystyle\int_0^{3/2} \int_{-\sqrt{9-4t^2}}^{\sqrt{9-4t^2}} t\,ds\,dt$ d) $\displaystyle\int_0^2 \int_0^{4-w^2} 2\,w\,dz\,dw$

2. Sketch the region over which the integral

$$\int_0^1 \int_{\sqrt{y}}^{2-\sqrt{y}} x\,y\,dx\,dy$$

is to be evaluated and find its value.

Evaluate the integrals in Exercises 3 and 4.

3. $\displaystyle\int_1^\infty \int_0^{1/(x\sqrt{x^2-1})} 2\,dy\,dx$ **4.** $\displaystyle\int_0^1 \int_0^y e^{-x^2}\,dx\,dy$

5. Evaluate the integral

$$\int_0^\infty \frac{e^{-ax} - e^{-bx}}{x}\,dx.$$

(*Hint:* Use the relation

$$\frac{e^{-ax} - e^{-bx}}{x} = \int_a^b e^{-xy}\,dy$$

to form a double integral and evaluate the integral by changing the order of integration.)

6. Show that

$$\iint \frac{\partial^2 F(x, y)}{\partial x\,\partial y}\,dx\,dy$$

over the rectangle $x_0 \leq x \leq x_1,\ y_0 \leq y \leq y_1$, is

$$F(x_1, y_1) - F(x_0, y_1) - F(x_1, y_0) + F(x_0, y_0).$$

7. Evaluate

$$\int_0^a \int_0^b e^{\max(b^2x^2,\, a^2y^2)}\,dy\,dx,$$

where a and b are positive numbers and

$$\max(b^2x^2, a^2y^2) = \begin{cases} b^2x^2 & \text{if } b^2x^2 \geq a^2y^2 \\ a^2y^2 & \text{if } b^2x^2 < a^2y^2 \end{cases}$$

8. a) Show, by changing to polar coordinates, that

$$\int_0^{a \sin\beta} \int_{y \cot\beta}^{\sqrt{a^2 - y^2}} \ln(x^2 + y^2)\, dx\, dy = a^2\beta\left(\ln a - \frac{1}{2}\right),$$

where $a > 0$ and $0 < \beta < \pi/2$.

b) Rewrite the Cartesian integral with the order of integration reversed.

9. By changing the order of integration, show that the following double integral can be reduced to a single integral:

$$\int_0^x \int_0^u e^{m(x-t)} f(t)\, dt\, du = \int_0^x (x - t) e^{m(x-t)} f(t)\, dt.$$

Similarly, it can be shown that

$$\int_0^x \int_0^v \int_0^u e^{m(x-t)} f(t)\, dt\, du\, dv = \int_0^x \frac{(x-t)^2}{2} e^{m(x-t)} f(t)\, dt.$$

10. Sometimes a multiple integral with variable limits can be changed into one with constant limits. By changing the order of integration, show that

$$\int_0^1 f(x)\left(\int_0^x g(x - y) f(y)\, dy\right) dx$$

$$= \int_0^1 f(y)\left(\int_y^1 g(x - y) f(x)\, dx\right) dy$$

$$= \frac{1}{2}\int_0^1 \int_0^1 g(|x - y|) f(x) f(y)\, dx\, dy.$$

11. The base of a sand pile covers the region in the xy-plane that is bounded by the parabola $x^2 + y = 6$ and the line $y = x$. The height of the sand above the point (x, y) is x^2. Express the volume of sand as (a) a double integral, (b) a triple integral. Then (c) find the volume.

12. Suppose that $f(x, y)$ can be written as a product $f(x, y) = F(x)G(y)$ of a function of x and a function of y. Then the integral of f over the rectangle $R: a \leq x \leq b,\ c \leq y \leq d$ can be evaluated as a product as well, by the formula

$$\iint_R f(x, y)\, dA = \left(\int_a^b F(x)\, dx\right)\left(\int_c^d G(y)\, dy\right). \tag{1}$$

The argument is that

$$\iint_R f(x, y)\, dA = \int_c^d \left(\int_a^b F(x)G(y)\, dx\right) dy \quad \text{(Fubini's theorem)}$$

$$= \int_c^d \left(G(y) \int_a^b F(x)\, dx\right) dy \quad \left(\begin{array}{l}\text{Treating } G(y)\\ \text{as a constant}\end{array}\right)$$

$$= \int_c^d \left(\int_a^b F(x)\, dx\right) G(y)\, dy \quad \left(\begin{array}{l}\text{Algebraic}\\ \text{rearrangement}\end{array}\right)$$

$$= \left(\int_a^b F(x)\, dx\right) \int_c^d G(y)\, dy. \quad \left(\begin{array}{l}\text{The definite}\\ \text{integral}\\ \text{is a constant.}\end{array}\right)$$

When it applies, Eq. (1) can be a time saver. Use it to evaluate the following integrals.

a) $\displaystyle\int_0^{\ln 2} \int_0^{\pi/2} e^x \cos y\, dy\, dx$

b) $\displaystyle\int_1^2 \int_{-1}^1 \frac{x}{y^2}\, dx\, dy$

13. Find the centroid of the triangular region enclosed by the lines $x = 2$, $y = 2$ and the hyperbola $xy = 2$ in the xy-plane.

14. Find the polar moment of inertia about the origin of a thin triangular plate of constant density $\delta = 3$ bounded by the y-axis and the lines $y = 2x$ and $y = 4$ in the xy-plane.

15. Find the centroid of the region between the parabola $x + y^2 - 2y = 0$ and the line $x + 2y = 0$.

16. Find the centroid of the boomerang-shaped region between the parabolas $y^2 = -4(x - 1)$ and $y^2 = -2(x - 2)$.

17. Find the mass and first moments about the coordinate axes of a thin square plate bounded by the lines $x = \pm 1$, $y = \pm 1$ in the xy-plane if the density is $\delta(x, y) = x^2 + y^2 + 1/3$.

18. Find the center of mass and the moments of inertia and radii of gyration about the coordinate axes of a thin plate bounded by the line $y = x$ and the parabola $y = x^2$ in the xy-plane if the density is $\delta(x, y) = x + 1$.

19. Find the moment of inertia and radius of gyration about the x-axis of a thin triangular plate of constant density δ whose base lies along the interval $[0, b]$ on the x-axis and whose vertex lies on the line $y = h$ above the x-axis. As you will see, it does not matter where on the line this vertex lies. All such triangles have the same moment of inertia and radius of gyration.

20. Let $D_{\mathbf{u}}f$ denote the derivative of $f(x, y) = (x^2 + y^2)/2$ in the direction of the unit vector $\mathbf{u} = u_1\mathbf{i} + u_2\mathbf{j}$.

a) Find the average value of $D_{\mathbf{u}}f$ over the triangular region cut from the first quadrant by the line $x + y = 1$.

b) Show in general that the average value of $D_{\mathbf{u}}f$ over a region in the xy-plane is the value of $D_{\mathbf{u}}f$ at the centroid of the region.

21. a) Find the centroid of the region in the polar coordinate plane that lies inside the cardioid $r = 1 + \cos\theta$ and outside the circle $r = 1$.

b) CALCULATOR Sketch the region and show the centroid in your sketch.

22. a) Find the centroid of the plane region defined by the polar coordinate inequalities $0 \leq r \leq a$, $-\alpha \leq \theta \leq \alpha$ $(0 < \alpha \leq \pi)$. How does the centroid move as $\alpha \to \pi^-$?

b) CALCULATOR Sketch the region for $\alpha = 5\pi/6$ and show the centroid in your sketch.

23. Find the centroid of the region in the first quadrant bounded by the rays $\theta = 0$ and $\theta = \pi/2$ and the circles $r = 1$ and $r = 3$.

24. In Section 9.8 we derived the formula

$$A = \int_{\alpha}^{\beta} \frac{1}{2} f^2(\theta)\, d\theta$$

for the area of the fan-shaped region R between the origin and a curve $r = f(\theta)$, $\alpha \le \theta \le \beta$. Show that this formula is also a consequence of the equation

$$\text{Area of } R = \iint_R r\, dr\, d\theta$$

in Section 13.3.

25. The counterweight of a flywheel has the form of the smaller segment cut from a circle of radius a by a chord at a distance b from the center ($b < a$). Find the area of the counterweight and its polar moment of inertia about the center of the wheel.

26. Find the radii of gyration about the x- and y-axes of a thin plate of density $\delta = 1$ enclosed by one loop of the lemniscate $r^2 = 2a^2 \cos 2\theta$.

27. Evaluate $\displaystyle\int_0^{a \sin \beta} \int_{y \cot \beta}^{\sqrt{a^2 - y^2}} \ln (x^2 + y^2)\, dx\, dy$.

28. Integrate the function $f(x, y) = 1/(1 + x^2 + y^2)^2$ over the region enclosed by one loop of the lemniscate $(x^2 + y^2)^2 - (x^2 - y^2) = 0$.

29. The electrical charge distribution on a circular plate of radius R meters is $\sigma(r, \theta) = kr(1 - \sin \theta)$ coulomb/m^2 (k a constant). Integrate σ over the plate to find the total charge Q.

30. *The value of $\Gamma(1/2)$.* As we saw in the Miscellaneous Exercises in Chapter 7, the gamma function,

$$\Gamma(x) = \int_0^{\infty} t^{x-1} e^{-t}\, dt,$$

extends the factorial function from the nonnegative integers to other real values. Of particular interest in the theory of differential equations is the number

$$\Gamma\left(\frac{1}{2}\right) = \int_0^{\infty} t^{(1/2)-1} e^{-t}\, dt = \int_0^{\infty} \frac{e^{-t}}{\sqrt{t}}\, dt. \tag{2}$$

a) If you have not yet done Exercise 16(a) in Section 13.3, do it now to show that

$$I = \int_0^{\infty} e^{-y^2} = \frac{\sqrt{\pi}}{2}.$$

b) Substitute $y = \sqrt{t}$ in Eq. (2) to show that $\Gamma(1/2) = 2I = \sqrt{\pi}$.

31. Find the volume of the solid that is bounded above by the cylinder $z = 4 - x^2$, on the sides by the cylinder $x^2 + y^2 = 4$, and below by the xy-plane (Fig. 13.72).

32. Find the average value of $f(x, y, z) = 30\, xz\sqrt{x^2 + y}$ over the rectangular solid in the first octant bounded by the coordinate planes and the planes $x = 1$, $y = 3$, $z = 1$.

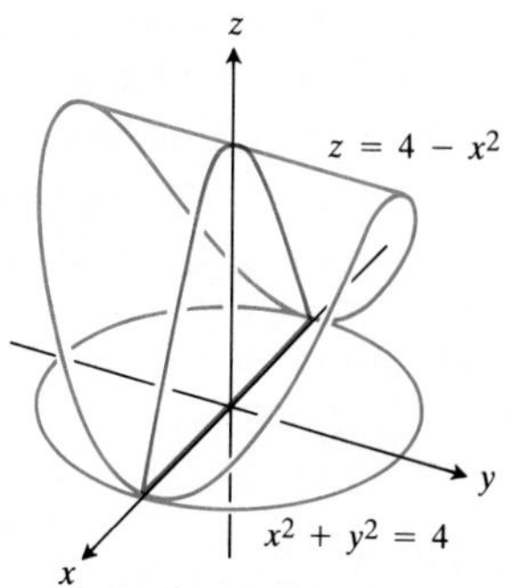

13.72 The solid in Exercise 31.

33. Convert

$$\int_0^{2\pi} \int_0^{1} \int_0^{\sqrt{4 - r^2}} r^2\, dz\, dr\, d\theta$$

to (a) rectangular coordinates, (b) spherical coordinates.

34. Write an iterated triple integral for the integral of $f(x, y, z) = 6 + 4y$ over the region in the first octant bounded by the cone $z = \sqrt{x^2 + y^2}$, the cylinder $x^2 + y^2 = 1$, and the coordinate planes in (a) rectangular coordinates, (b) cylindrical coordinates, (c) spherical coordinates. Then (d) find the integral of f by evaluating one of the triple integrals.

35. Set up an integral in rectangular coordinates equivalent to the integral

$$\int_0^{\pi/2} \int_1^{\sqrt{3}} \int_1^{\sqrt{4 - r^2}} r^3 \sin \theta \cos \theta\, z^2\, dz\, dr\, d\theta.$$

Arrange the order of integration to be z first, then y, then x.

36. The volume of a solid is

$$\int_0^{2} \int_0^{\sqrt{2x - x^2}} \int_{-\sqrt{4 - x^2 - y^2}}^{\sqrt{4 - x^2 - y^2}} dz\, dy\, dx.$$

a) Describe the solid by giving equations for the surfaces that form its boundary.

b) Convert the integral to cylindrical coordinates but do not evaluate the integral.

37. Find the volume of the portion of the solid cylinder $x^2 + y^2 \le 1$ that lies between the planes $z = 0$ and $x + y + z = 2$.

38. A hemispherical bowl of radius 5 cm is filled with water to within 3 cm of the top. Find the volume of water in the bowl.

39. Find the volume of the region in the first octant that lies between the cylinders $r = 1$ and $r = 2$ and that is bounded below by the xy-plane and above by the surface $z = xy$.

40. Find the volume of the region bounded above by the sphere $x^2 + y^2 + z^2 = 2$ and below by the paraboloid $z = x^2 + y^2$.

41. Find the volume of the region bounded above by the paraboloid $z = 3 - x^2 - y^2$ and below by the paraboloid $z = 2x^2 + 2y^2$.

42. Find the volume of the region enclosed by the spherical coordinate surface $\rho = 2 \sin \phi$ (Fig. 13.73).

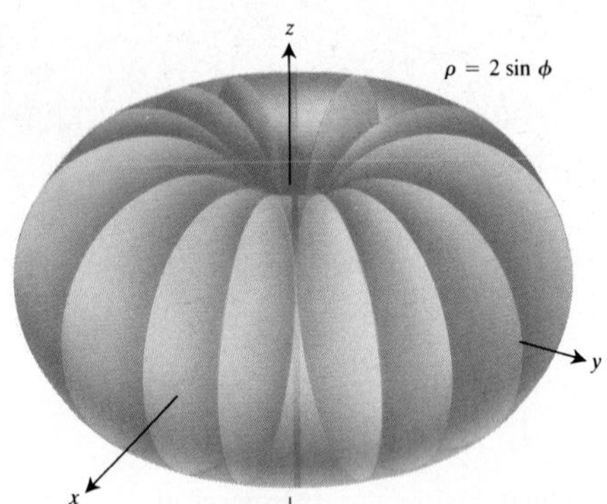

13.73 The spherical coordinate surface $\rho = 2 \sin \phi$ in Exercise 42 is the surface generated by revolving a circle of radius 1 about one of its tangents (in this case, the z-axis).

43. A circular cylindrical hole is bored through a solid sphere, the axis of the hole being a diameter of the sphere. The volume of the remaining solid is

$$V = 2\int_0^{2\pi}\int_0^{\sqrt{3}}\int_1^{\sqrt{4-z^2}} r\, dr\, dz\, d\theta.$$

a) Find the radius of the hole and the radius of the sphere.
b) Evaluate the integral. (See Example 4, Section 5.3.)

44. Find the volume of material cut from the solid sphere $r^2 + z^2 \leq 9$ by the cylinder $r = 3 \sin \theta$.

45. Find the volume of the region enclosed by the surfaces $z = x^2 + y^2$ and $z = (x^2 + y^2 + 1)/2$.

46. *Spherical vs. cylindrical coordinates.* Triple integrals involving spherical shapes do not always require spherical coordinates for convenient evaluation. Some calculations may be accomplished more easily with cylindrical coordinates. As a case in point, find the volume of the region bounded above by the sphere $x^2 + y^2 + z^2 = 8$ and below by the plane $z = 2$ by using (a) cylindrical coordinates, (b) spherical coordinates.

47. Find the moment of inertia about the z-axis of a solid of constant density $\delta = 1$ that is bounded above by the sphere $\rho = 2$ and below by the cone $\phi = \pi/3$ (spherical coordinates).

48. Find the moment of inertia of a solid of constant density δ bounded by two concentric spheres of radii a and b ($a < b$) about a diameter.

49. Find the moment of inertia about the z-axis of a solid of density $\delta = 1$ enclosed by the spherical coordinate surface $\rho = 1 - \cos \phi$ (Fig. 13.62).

50. *A challenge.* Setting up the following integral is straightforward but evaluating it can take hours. Some computer symbolic manipulation programs can evaluate the integral but others cannot. Some can evaluate the integral in one form but not another. One program we know evaluated the integral in less than a minute. Another took 20 minutes.

A square hole of side $2b$ is cut symmetrically through a solid sphere of radius a ($a > b\sqrt{2}$). Find the volume of material removed.

51. Show that if $u = x - y$ and $v = y$, then

$$\int_0^{\infty}\int_0^{x} e^{-sx} f(x-y, y)\, dy\, dx = \int_0^{\infty}\int_0^{\infty} e^{-s(u+v)} f(u, v)\, du\, dv.$$

52. What relationship must hold between the constants a, b, and c to make

$$\int_{-\infty}^{\infty}\int_{-\infty}^{\infty} e^{-(ax^2+2bxy+cy^2)}\, dx\, dy = 1?$$

(*Hint:* Let $s = \alpha x + \beta y$ and $t = \gamma x + \delta y$, where $(\alpha\delta - \beta\gamma)^2 = ac - b^2$. Then $ax^2 + 2bxy + cy^2 = s^2 + t^2$.)

CHAPTER

14

INTEGRATION IN VECTOR FIELDS

Overview This chapter brings together all our previous work with derivatives, integrals, and vector functions to study integration in vector fields. The field concept has proved to be one of the most useful ideas in all of physical science. The mathematics in this chapter is the mathematics we use today to describe the properties of electric charge; explain the behavior of electromagnetic waves, including radio waves, x-rays, and visible light; predict the flow of air around airplanes and rockets; explain the flow of heat in a room or in a star; calculate the amount of work it takes to put a satellite into orbit; and model tropical storms and ocean currents.

14.1 Line Integrals

When a curve $\mathbf{r} = g(t)\,\mathbf{i} + h(t)\,\mathbf{j} + k(t)\,\mathbf{k}$, $a \le t \le b$, passes through the domain of a function $f(x, y, z)$ in space, the values of f along the curve are given by the composite function $f(g(t), h(t), k(t))$. If we integrate this composite with respect to arc length from $t = a$ to $t = b$, we calculate the so-called line integral of f along the curve. Despite the three-dimensional geometry, the line integral is an ordinary integral of a real-valued function over an interval of real numbers. We have evaluated such integrals ever since Chapter 4. The usual techniques apply and nothing new is needed.

The importance of line integrals lies in their application. These are the integrals with which we calculate the moments and masses of springs and curved wires. We also combine them with vectors to calculate the work done by variable forces along paths in space and the rates at which fluids flow along curves and across boundaries. The present section shows how line integrals are defined, evaluated, and used to calculate moments and masses. The next section brings line integrals together with vectors to calculate work and measure fluid flow.

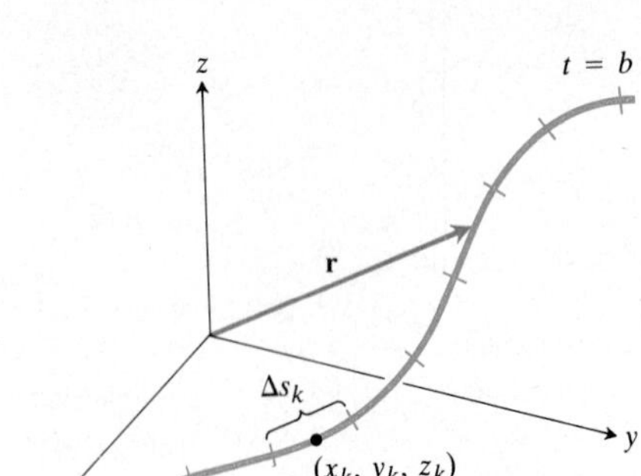

14.1 The curve $\mathbf{r} = g(t)\,\mathbf{i} + h(t)\,\mathbf{j} + k(t)\,\mathbf{k}$, partitioned into small arcs from $t = a$ to $t = b$. The length of a typical subarc is Δs_k.

Definitions and Notation

Line integrals are integrals of functions over curves. We define them in the following way. Suppose that $f(x, y, z)$ is a function whose domain contains the curve

$$\mathbf{r} = g(t)\,\mathbf{i} + h(t)\,\mathbf{j} + k(t)\,\mathbf{k}, \quad a \le t \le b.$$

We partition the curve into a finite number of subarcs (Fig. 14.1). The typical subarc has length Δs_k. In each subarc we choose a point (x_k, y_k, z_k) and form the sum

$$S_n = \sum_{k=1}^{n} f(x_k, y_k, z_k)\,\Delta s_k. \tag{1}$$

If f is continuous and $\mathbf{r}$ is smooth (the functions g, h, and k have continuous first derivatives), then the sums in (1) approach a limit as n increases and the lengths Δs_k approach zero. We call this limit the **line integral of *f* over the curve from *a* to *b*.** If the curve is denoted by a single letter, C for example, the notation for the integral is

$$\int_C f(x, y, z)\,ds \qquad \text{("The integral of } f \text{ over } C\text{")} \tag{2}$$

Evaluation

To evaluate the line integral in Eq. (2), we express ds and the integrand in terms of the curve's parameter t. A remarkable theorem from advanced calculus tells us that we can find the value from any smooth parametrization $x = g(t)$, $y = h(t)$, $z = k(t)$, $a \le t \le b$, for which ds/dt is positive by evaluating the integral

$$\int_{t=a}^{t=b} f(g(t), h(t), k(t))\,\frac{ds}{dt}\,dt = \int_a^b f(g(t), h(t), k(t))\sqrt{\left(\frac{dx}{dt}\right)^2 + \left(\frac{dy}{dt}\right)^2 + \left(\frac{dz}{dt}\right)^2}\,dt. \tag{3}$$

Since the square root in Eq. (3) is the length $|\mathbf{v}(t)|$ of the curve's velocity vector, the integral on the right-hand side can be simplified to give the following result.

THEOREM 1

The Evaluation Theorem for Line Integrals

The integral of a continuous function $f(x, y, z)$ over a space curve C can be calculated from any smooth parametrization $\mathbf{r} = g(t)\,\mathbf{i} + h(t)\,\mathbf{j} + k(t)\,\mathbf{k}$, $a \le t \le b$, of C for which $|\mathbf{v}| \ne 0$ as

$$\int_C f(x, y, z)\,ds = \int_{t=a}^{t=b} f(g(t), h(t), k(t))\,|\mathbf{v}(t)|\,dt. \tag{4}$$

Notice that if f has the constant value 1, then the integral of f over C from a to b gives the length of C from a to b. We calculated lengths in Section 11.3 and will not repeat the calculations here.

In our first example, we integrate a function over two different paths from (0, 0, 0) to (1, 1, 1).

Example 1 Figure 14.2 shows two different paths from the origin to the point (1, 1, 1). One path is the union of the line segments C_1 and C_2. The other path is the

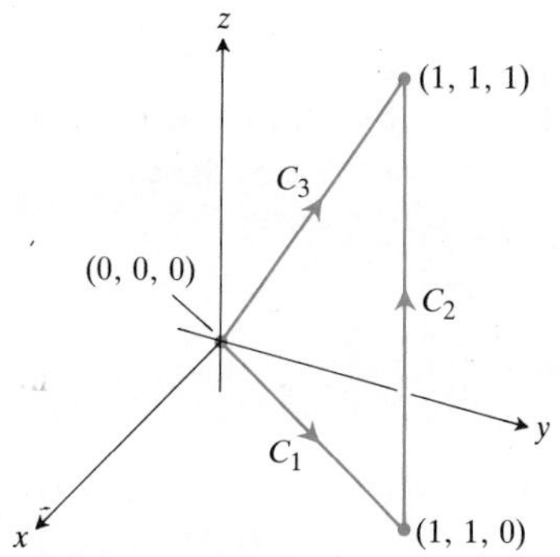

14.2 The paths of integration in Example 1. The arrows show the direction of increasing t in the parametrizations.

line segment C_3. Integrate the function

$$f(x, y, z) = x - 3y^2 + z$$

along each path.

Solution We first parametrize the segments that make up the paths. According to the Evaluation Theorem, we can use any parametrization we want as long as $|\mathbf{v}(t)|$ is never zero and we integrate in the direction of increasing t so that $|\mathbf{v}(t)|\,dt$ is positive. We choose the simplest parametrizations we can think of, checking the lengths of the velocity vectors as we go along:

$$C_1:\quad \mathbf{r} = t\mathbf{i} + t\mathbf{j},\quad 0 \le t \le 1; \qquad |\mathbf{v}| = \sqrt{1^2 + 1^2} = \sqrt{2}$$

$$C_2:\quad \mathbf{r} = \mathbf{i} + \mathbf{j} + t\mathbf{k},\quad 0 \le t \le 1; \qquad |\mathbf{v}| = \sqrt{0^2 + 0^2 + 1^2} = 1$$

$$C_3:\quad \mathbf{r} = t\mathbf{i} + t\mathbf{j} + t\mathbf{k},\quad 0 \le t \le 1; \qquad |\mathbf{v}| = \sqrt{1^2 + 1^2 + 1^2} = \sqrt{3}$$

Having chosen the parametrizations, we use Eq. (4) to integrate $f(x, y, z) = x - 3y^2 + z$ over each path in the direction of increasing t.

$$\text{Path 1:}\quad \int_{C_1 \cup C_2} f(x, y, z)\,ds = \int_{C_1} f(x, y, z)\,ds + \int_{C_2} f(x, y, z)\,ds$$

$$= \int_0^1 f(t, t, 0)\sqrt{2}\,dt + \int_0^1 f(1, 1, t)(1)\,dt$$

$$= \int_0^1 (t - 3t^2 + 0)\sqrt{2}\,dt + \int_0^1 (1 - 3 + t)(1)\,dt$$

$$= \sqrt{2}\int_0^1 (t - 3t^2)\,dt + \int_0^1 (t - 2)\,dt$$

$$= \sqrt{2}\left[\frac{t^2}{2} - t^3\right]_0^1 + \left[\frac{t^2}{2} - 2t\right]_0^1 = -\frac{\sqrt{2}}{2} - \frac{3}{2}.$$

$$\text{Path 2:}\quad \int_{C_3} f(x, y, z)\,ds = \int_0^1 f(t, t, t)(\sqrt{3})\,dt = \int_0^1 (t - 3t^2 + t)(\sqrt{3})\,dt$$

$$= \sqrt{3}\int_0^1 (2t - 3t^2)\,dt = \sqrt{3}\left[t^2 - t^3\right]_0^1 = 0$$

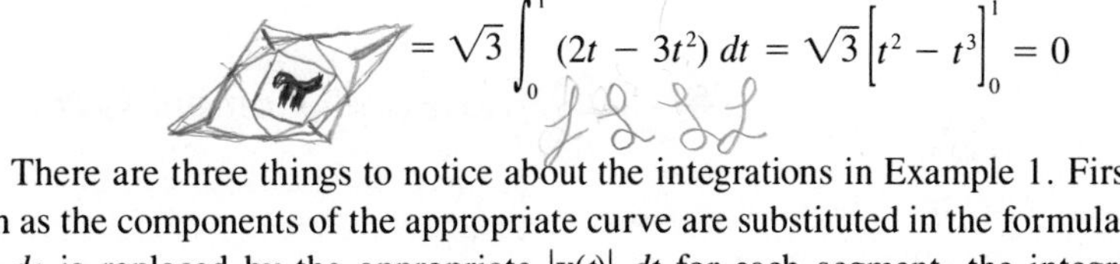

There are three things to notice about the integrations in Example 1. First, as soon as the components of the appropriate curve are substituted in the formula for f and ds is replaced by the appropriate $|\mathbf{v}(t)|\,dt$ for each segment, the integration becomes a straightforward integration with respect to t. Second, the integral of f over path 1 is obtained by integrating f over each section of the path and adding the results. Third, the integrals of f over path 1 and path 2 have different values. For most functions, the value of the line integral along a path joining two points changes when you change the path. For some functions, however, the value of the integral is the same for all paths joining the two points, as we shall see in Section 14.7.

Additivity

Line integrals have all the usual algebraic properties. The line integral of a sum of two functions is the sum of their line integrals, the line integral of a constant times a function is the constant times the line integral of the function, the line integral of

TABLE 14.1
Mass and moment formulas for coil springs, thin rods, and wires

Mass: $M = \int_C \delta(x, y, z)\, ds$

First moments about the coordinate planes:

$$M_{yz} = \int_C x\,\delta\, ds, \qquad M_{xz} = \int_C y\,\delta\, ds, \qquad M_{xy} = \int_C z\,\delta\, ds$$

Coordinates of the center of mass:

$$\bar{x} = M_{yz}/M, \qquad \bar{y} = M_{xz}/M, \qquad \bar{z} = M_{xy}/M$$

Moments of inertia:

$$I_x = \int_C (y^2 + z^2)\,\delta\, ds, \qquad I_y = \int_C (x^2 + z^2)\,\delta\, ds,$$

$$I_z = \int_C (x^2 + y^2)\,\delta\, ds, \qquad I_L = \int_C r^2\,\delta\, ds,$$

$r(x, y, z) =$ distance from point (x, y, z) to line L

Radius of gyration about a line L: $R_L = \sqrt{I_L/M}$.

a function from b to a is the negative of the line integral from a to b, and so on. Included in this list is the property that if a curve C is made by joining together a finite number of curves $C_1, C_2, \ldots, C_n$ end to end, then the line integral of a function over C is the sum of the line integrals over the curves that make it up:

$$\int_C f\, ds = \int_{C_1} f\, ds + \int_{C_2} f\, ds + \cdots + \int_{C_n} f\, ds. \tag{5}$$

We used Eq. (5) in Example 1 to calculate the integral of f over the first path by adding the integrals over C_1 and C_2.

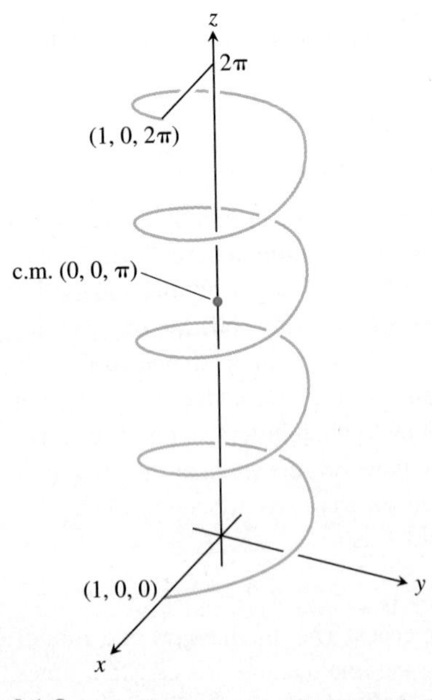

14.3 The helical spring in Example 2.

Mass and Moment Calculations

We can model coil springs and wires as if they were masses distributed along continuously differentiable curves in space. The distribution of mass is modeled by a continuous density function $\delta(x, y, z)$. The spring's or wire's mass, center-of-mass coordinates, and moments are then calculated with the formulas in Table 14.1. The formulas apply to thin rods as well as springs and wires.

Example 2 A coil spring lies along the helix

$$\mathbf{r} = (\cos 4t)\,\mathbf{i} + (\sin 4t)\,\mathbf{j} + t\mathbf{k}, \qquad 0 \le t \le 2\pi.$$

The spring's density is a constant, $\delta = 1$. Find the spring's mass, the coordinates of the spring's center of mass, and the spring's moment of inertia and radius of gyration about the z-axis.

Solution We sketch the spring (Fig. 14.3). Because of the symmetries involved, the center of mass lies at the point $(0, 0, \pi)$ on the z-axis.

For the remaining calculations, we first express ds in terms of t:

$$ds = \sqrt{\left(\frac{dx}{dt}\right)^2 + \left(\frac{dy}{dt}\right)^2 + \left(\frac{dz}{dt}\right)^2}\, dt$$

$$= \sqrt{(-4\sin 4t)^2 + (4\cos 4t)^2 + 1}\, dt = \sqrt{17}\, dt.$$

The formulas in Table 14.1 then give

$$M = \int_{\text{helix}} \delta\, ds = \int_0^{2\pi} (1)\sqrt{17}\, dt = 2\pi\sqrt{17},$$

$$I_z = \int_{\text{helix}} (x^2 + y^2)\delta\, ds = \int_0^{2\pi} (\cos^2 4t + \sin^2 4t)(1)\sqrt{17}\, dt$$

$$= \int_0^{2\pi} \sqrt{17}\, dt = 2\pi\sqrt{17},$$

$$R_z = \sqrt{I_z/M} = \sqrt{2\pi\sqrt{17}/2\pi\sqrt{17}} = 1.$$

Notice that the radius of gyration about the z-axis is the radius of the cylinder around which the helix winds.

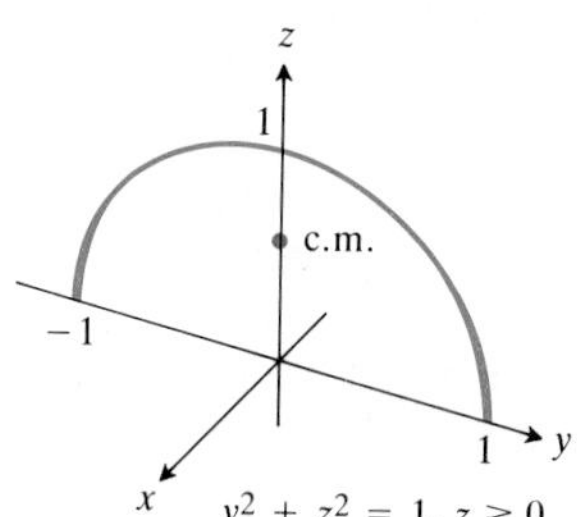

14.4 Example 3 shows how to find the center of mass of a circular arch of variable density.

Example 3 A slender metal arch, thicker at the bottom than at the top, lies along the semicircle $y^2 + z^2 = 1$, $z \geq 0$, in the yz-plane (Fig. 14.4). Find the center of the arch's mass if the density at the point (x, y, z) on the arch is $\delta(x, y, z) = 2 - z$.

Solution We know that $\bar{x} = 0$ and $\bar{y} = 0$ because the arch lies in the yz-plane with its mass distributed symmetrically about the z-axis. To find $\bar{z}$, we parametrize the circle as

$$\mathbf{r} = (\cos t)\,\mathbf{j} + (\sin t)\,\mathbf{k}, \qquad 0 \leq t \leq \pi,$$

and express ds in terms of dt:

$$ds = \sqrt{\left(\frac{dx}{dt}\right)^2 + \left(\frac{dy}{dt}\right)^2 + \left(\frac{dz}{dt}\right)^2}\, dt = \sqrt{(0)^2 + (-\sin t)^2 + (\cos t)^2}\, dt = dt.$$

The formulas in Table 14.1 then give

$$M = \int_C \delta\, ds = \int_C (2 - z)\, ds = \int_0^{\pi} (2 - \sin t)\, dt = 2\pi - 2,$$

$$M_{xy} = \int_C z\,\delta\, ds = \int_C z(2 - z)\, ds = \int_0^{\pi} \sin t(2 - \sin t)\, dt$$

$$= \int_0^{\pi} (2\sin t - \sin^2 t)\, dt = \frac{8 - \pi}{2},$$

$$\bar{z} = \frac{M_{xy}}{M} = \frac{8 - \pi}{2} \cdot \frac{1}{2\pi - 2} = \frac{8 - \pi}{4\pi - 4} = 0.57. \qquad \left(\begin{array}{c}\text{Calculator,}\\ \text{rounded}\end{array}\right)$$

To the nearest hundredth, the center of mass is (0, 0, 0.57).

EXERCISES 14.1

Match the vector equations in Exercises 1–8 with the graphs in Fig. 14.5.

(a)

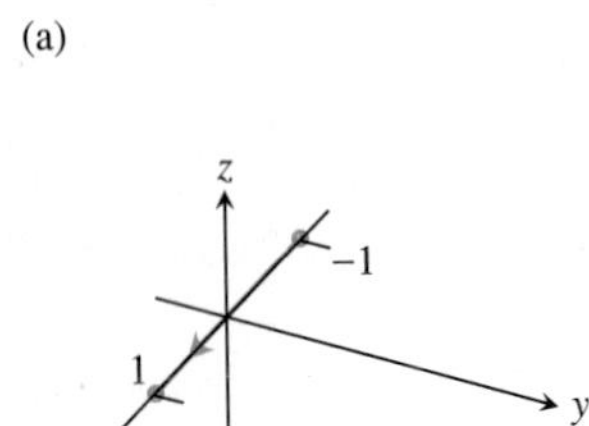

(b)

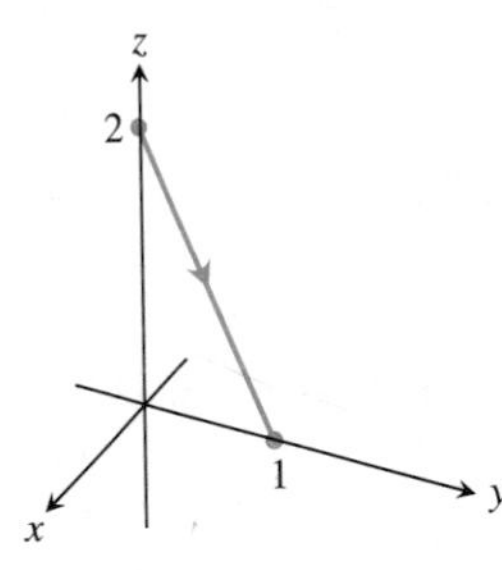

(c)

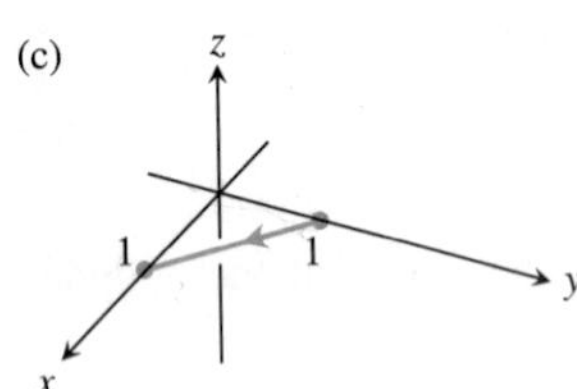

(d)

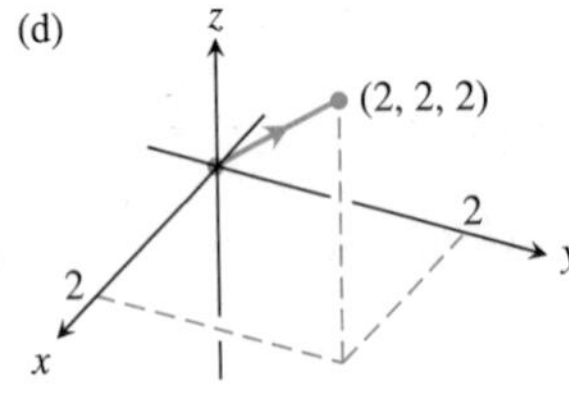

(e)

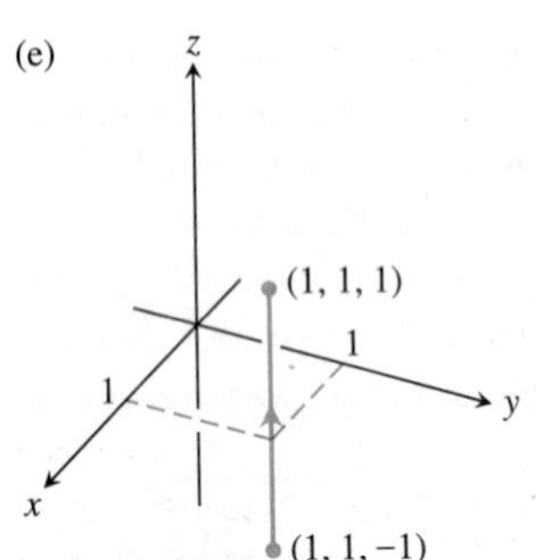

(f)

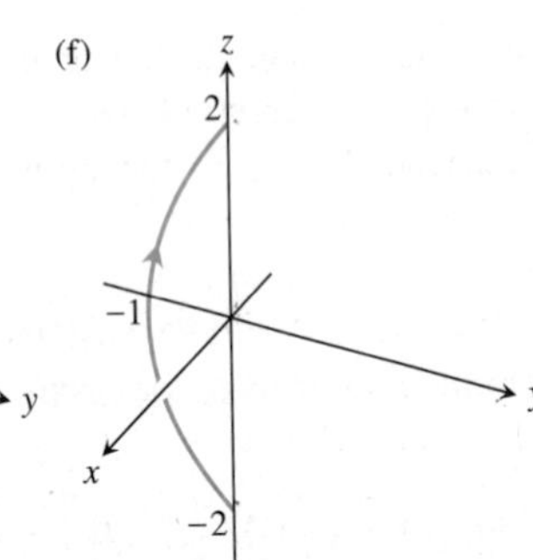

(g)

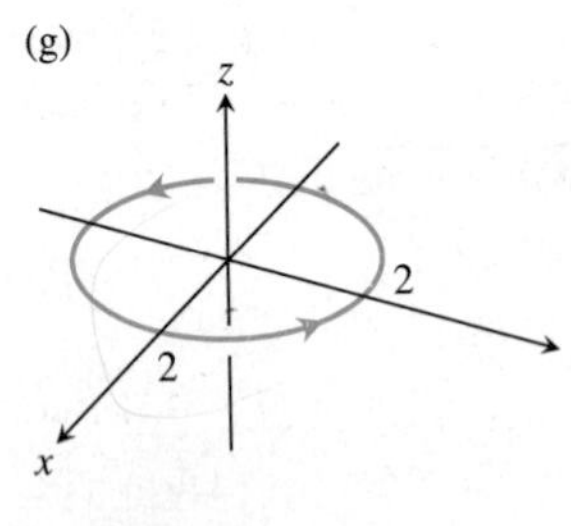

(h)

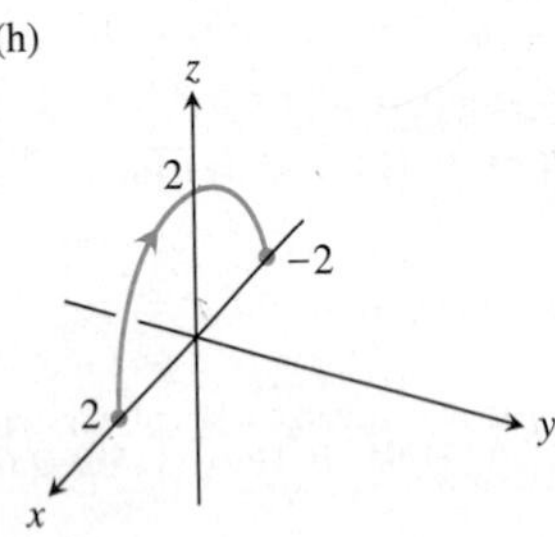

14.5 The graphs for Exercises 1–8.

1. $\mathbf{r} = t\mathbf{i} + (1 - t)\mathbf{j}, \quad 0 \le t \le 1$
2. $\mathbf{r} = \mathbf{i} + \mathbf{j} + t\mathbf{k}, \quad -1 \le t \le 1$
3. $\mathbf{r} = (2\cos t)\,\mathbf{i} + (2\sin t)\,\mathbf{j}, \quad 0 \le t \le 2\pi$
4. $\mathbf{r} = t\mathbf{i}, \quad -1 \le t \le 1$
5. $\mathbf{r} = t\mathbf{i} + t\mathbf{j} + t\mathbf{k}, \quad 0 \le t \le 2$
6. $\mathbf{r} = t\mathbf{j} + (2 - 2t)\,\mathbf{k}, \quad 0 \le t \le 1$
7. $\mathbf{r} = (t^2 - 1)\,\mathbf{j} + 2t\mathbf{k}, \quad -1 \le t \le 1$
8. $\mathbf{r} = (2\cos t)\,\mathbf{i} + (2\sin t)\,\mathbf{k}, \quad 0 \le t \le \pi$

In Exercises 9–14, evaluate the line integral of f along the given curve for the given parameter interval.

9. $f(x, y, z) = x + y$
   $\mathbf{r} = t\mathbf{i} + (1 - t)\,\mathbf{j}, \quad 0 \le t \le 1$
10. $f(x, y, z) = x - y + z - 2$
    $\mathbf{r} = t\mathbf{i} + (1 - t)\,\mathbf{j} + \mathbf{k}, \quad 0 \le t \le 1$
11. $f(x, y, z) = xy + y + z$
    $\mathbf{r} = 2t\mathbf{i} + t\mathbf{j} + (2 - 2t)\,\mathbf{k}, \quad 0 \le t \le 1$
12. $f(x, y, z) = \sqrt{x^2 + y^2}$
    $\mathbf{r} = (4\cos t)\,\mathbf{i} + (4\sin t)\,\mathbf{j} + 3t\mathbf{k}, \quad -2\pi \le t \le 2\pi$
13. $f(x, y, z) = 3z\sqrt{3x^2 + y^2 + z^2}$
    $\mathbf{r} = \mathbf{i} + \mathbf{j} + t\mathbf{k}, \quad -1 \le t \le 1$
14. $f(x, y, z) = \sqrt{3}/(x^2 + y^2 + z^2)$
    $\mathbf{r} = t\mathbf{i} + t\mathbf{j} + t\mathbf{k}, \quad 1 \le t < \infty$
15. Integrate $f(x, y, z) = x + \sqrt{y} - z^2$ over the path from (0, 0, 0) to (1, 1, 1) (Fig. 14.6a) given by
    C_1: $\mathbf{r} = t\mathbf{i} + t^2\mathbf{j}, \quad 0 \le t \le 1$
    C_2: $\mathbf{r} = \mathbf{i} + \mathbf{j} + t\mathbf{k}, \quad 0 \le t \le 1$
16. Integrate $f(x, y, z) = x + \sqrt{y} - z^2$ over the path from (0, 0, 0) to (1, 1, 1) (Fig. 14.6b) given by
    C_1: $\mathbf{r} = t\mathbf{k}, \quad 0 \le t \le 1$
    C_2: $\mathbf{r} = t\mathbf{j} + \mathbf{k}, \quad 0 \le t \le 1$
    C_3: $\mathbf{r} = t\mathbf{i} + \mathbf{j} + \mathbf{k}, \quad 0 \le t \le 1$

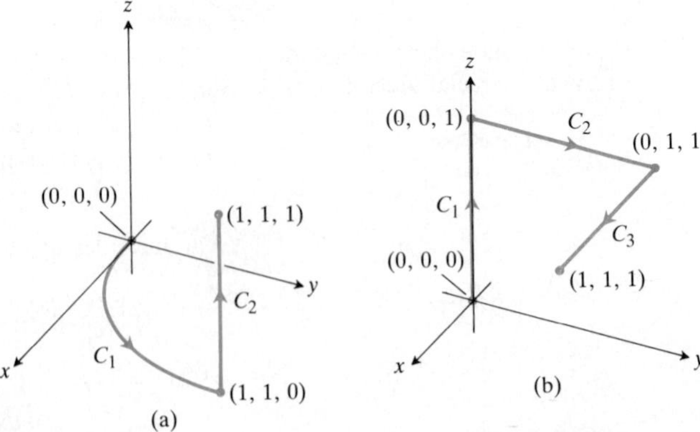

14.6 The paths of integration for Exercises 15 and 16.

17. Integrate $f(x, y, z) = (x + y + z)/(x^2 + y^2 + z^2)$ over the path $\mathbf{r} = t\mathbf{i} + t\mathbf{j} + t\mathbf{k}$, $a \le t \le b$.
18. Integrate $f(x, y, z) = -\sqrt{x^2 + z^2}$ over the circle $\mathbf{r} = (a\cos t)\,\mathbf{j} + (a\sin t)\,\mathbf{k}$, $0 \le t \le 2\pi$, in the direction of increasing t.

19. Find I_x and R_x for the arch in Example 3.

20. Find the mass of a wire that lies along the curve $\mathbf{r} = (t^2 - 1)\mathbf{j} + 2t\mathbf{k}$, $0 \le t \le 1$, if the density is $\delta = (3/2)t$.

21. A circular wire hoop of constant density δ lies along the circle $x^2 + y^2 = a^2$ in the xy-plane. Find the hoop's moment of inertia and radius of gyration about the z-axis.

22. A slender rod of constant density lies along the line segment $\mathbf{r} = t\mathbf{j} + (2 - 2t)\mathbf{k}$, $0 \le t \le 1$ in the yz-plane. Find the moments of inertia and radii of gyration of the rod about the three coordinate axes.

23. A spring of constant density δ lies along the helix $\mathbf{r} = (\cos t)\mathbf{i} + (\sin t)\mathbf{j} + t\mathbf{k}$, $0 \le t \le 2\pi$.
   a) Find I_z and R_z.
   b) Suppose you have another spring of constant density δ that is twice as long as the spring in (a) and lies along the helix for $0 \le t \le 4\pi$. Do you expect I_z and R_z for the longer spring to be the same as those for the shorter one, or should they be different? Check your predictions by calculating I_z and R_z for the longer spring.

24. A wire of density $\delta(x, y, z) = 15\sqrt{y + 2}$ lies along the curve $\mathbf{r} = (t^2 - 1)\mathbf{j} + 2t\mathbf{k}$, $-1 \le t \le 1$. Find its center of mass. Then sketch the curve and center of mass together.

25. Find the mass of a thin wire lying along the curve $\mathbf{r} = \sqrt{2}\,t\mathbf{i} + \sqrt{2}\,t\mathbf{j} + (4 - t^2)\mathbf{k}$, $0 \le t \le 1$, if the density is (a) $\delta = 3t$, (b) $\delta = 1$.

26. Find the center of mass of a thin wire lying along the curve $\mathbf{r} = t\mathbf{i} + 2t\mathbf{j} + (2/3)t^{3/2}\mathbf{k}$, $0 \le t \le 2$, if the density is $\delta = 3\sqrt{5 + t}$.

27. Find the center of mass, and the moments of inertia and radii of gyration about the coordinate axes of a thin wire lying along the curve

$$\mathbf{r} = t\mathbf{i} + \frac{2\sqrt{2}}{3}t^{3/2}\mathbf{j} + \frac{t^2}{2}\mathbf{k}, \quad 0 \le t \le 2,$$

if the density is $\delta = 1/(t + 1)$.

14.2 Vector Fields, Work, Circulation, and Flux

In studies of physical phenomena represented by vectors, we introduce the idea of a vector field on a domain in space, a function that assigns a vector quantity to each point in the domain. For example, we might be studying the velocity field of a moving fluid or the electric field created by a distribution of electric charges. In situations like these, integrals of real-valued functions over closed intervals are replaced by line integrals of functions over paths through the vector field, and the function being integrated is a scalar product of vectors. We can use these line integrals in vector fields to calculate the work done in moving an object along a path against a variable force. (The object might be a vehicle sent into space against Earth's gravitational field.) Or we might want to calculate the work done by a vector field in moving an object along a path through the field (for example, the work done by a particle accelerator in raising the energy of a particle). We can also use line integrals to calculate the rates at which fluids flow along and across curves.

Vector Fields

A **vector field** on a domain in the plane or in space is a function that assigns a vector to each point in the domain. A field of three-dimensional vectors might have a formula like

$$\mathbf{F} = M(x, y, z)\,\mathbf{i} + N(x, y, z)\,\mathbf{j} + P(x, y, z)\,\mathbf{k}. \tag{1}$$

The field is **continuous** if the **component functions** M, N, and P are continuous, **differentiable** if M, N, and P are differentiable, and so on. A field of two-dimensional vectors might have a formula like

$$\mathbf{F} = M(x, y)\,\mathbf{i} + N(x, y)\,\mathbf{j}. \tag{2}$$

If we attach a projectile's velocity vector to each point of the projectile's trajectory in the plane of motion, we have a two-dimensional field defined along the trajectory. If

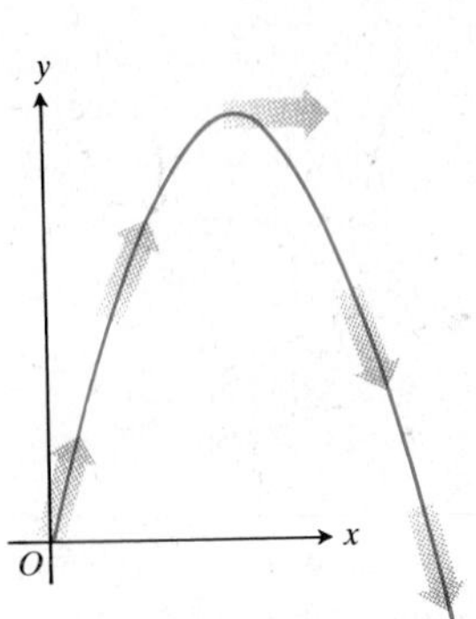

14.7 The velocity vectors $\mathbf{v}(t)$ of a projectile's motion make a vector field along the trajectory.

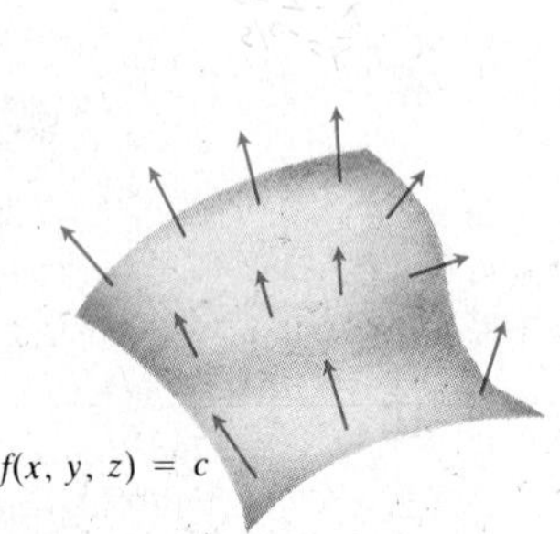

14.8 The field of gradient vectors ∇f on a surface $f(x, y, z) = c$.

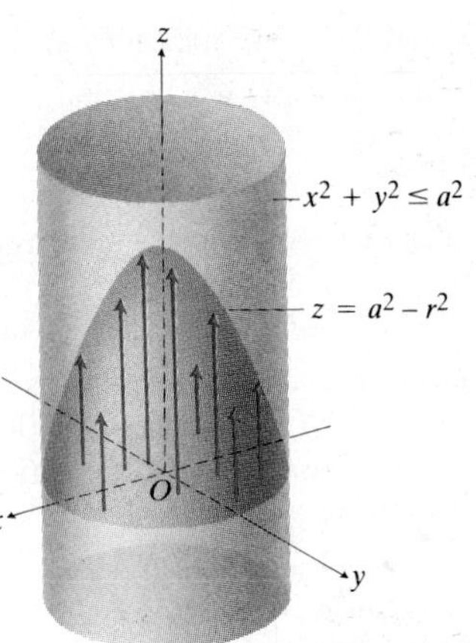

14.9 The flow of fluid in a long cylindrical pipe. The vectors $\mathbf{v} = (a^2 - r^2)\,\mathbf{k}$ inside the cylinder that have their bases in the xy-plane have their tips on the paraboloid $z = a^2 - r^2$.

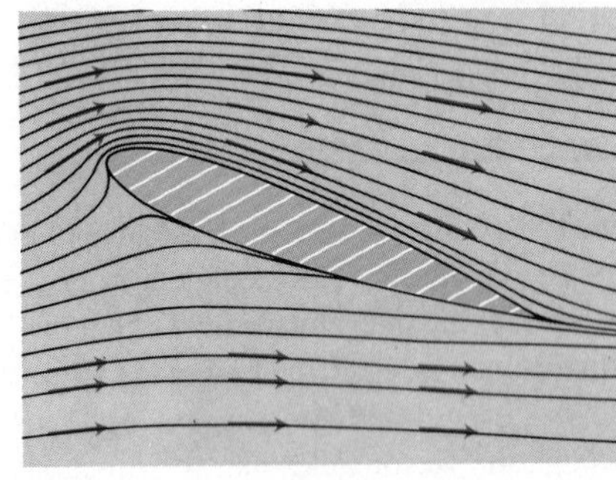

14.10 Velocity vectors of a flow around an airfoil in a wind tunnel. The streamlines were made visible by kerosene smoke. (Adapted from *NCFMF Book of Film Notes,* 1974, the MIT Press with Education Development Center, Inc., Newton, Massachusetts.)

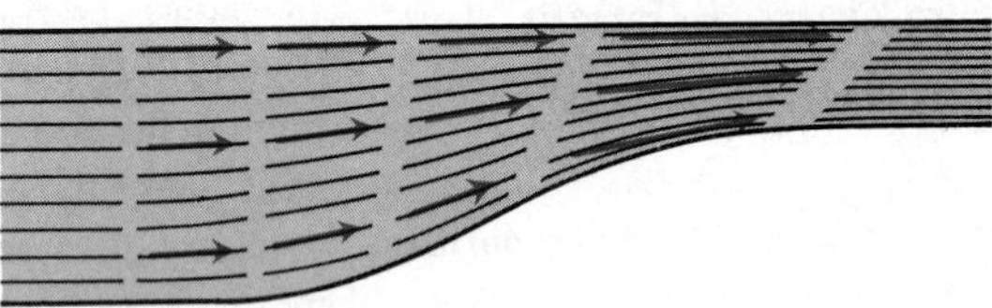

14.11 Streamlines in a contracting channel. The water speeds up as the channel narrows and the velocity vectors increase in length. (Adapted from *NCFMF Book of Film Notes,* 1974, the MIT Press with Education Development Center, Inc., Newton, Massachusetts.)

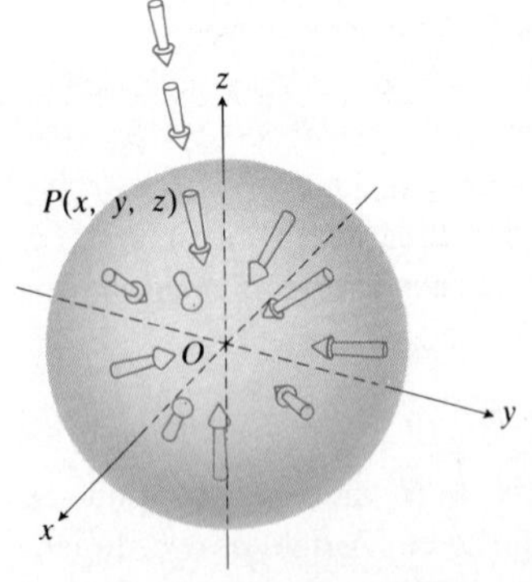

14.12 Vectors in the gravitational field

$$\mathbf{F} = -\frac{GM(x\mathbf{i} + y\mathbf{j} + z\mathbf{k})}{(x^2 + y^2 + z^2)^{3/2}}$$

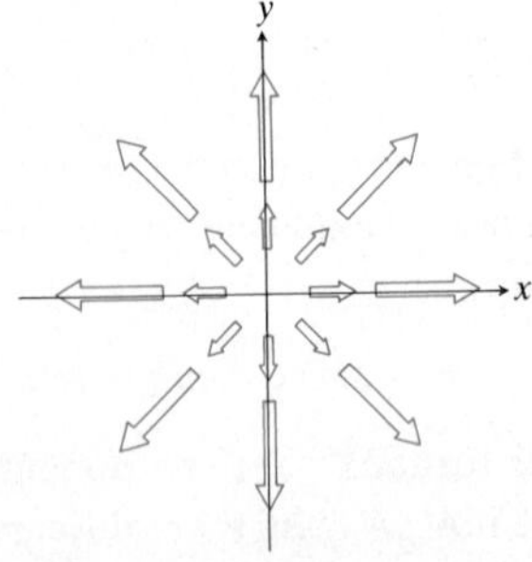

14.13 The radial field $\mathbf{F} = x\mathbf{i} + y\mathbf{j}$ of position vectors of points in the plane. Notice the convention that an arrow is drawn with its tail, not its head, at the point where $\mathbf{F}$ is evaluated.

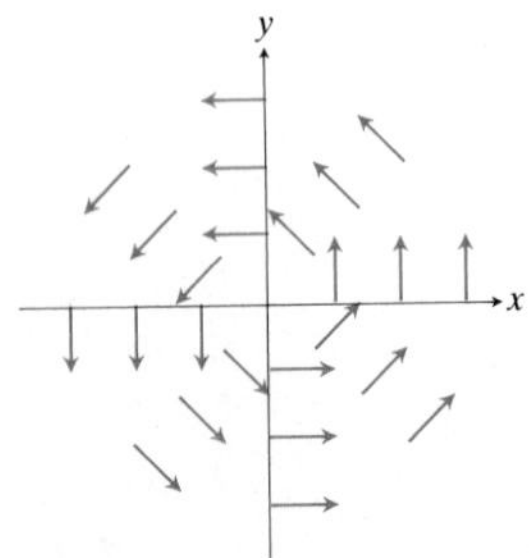

14.14 The circumferential or "spin" field of unit vectors

$$\mathbf{F} = (-y\mathbf{i} + x\mathbf{j})/(x^2 + y^2)^{1/2}$$

in the plane. The field is not defined at the origin.

NASA

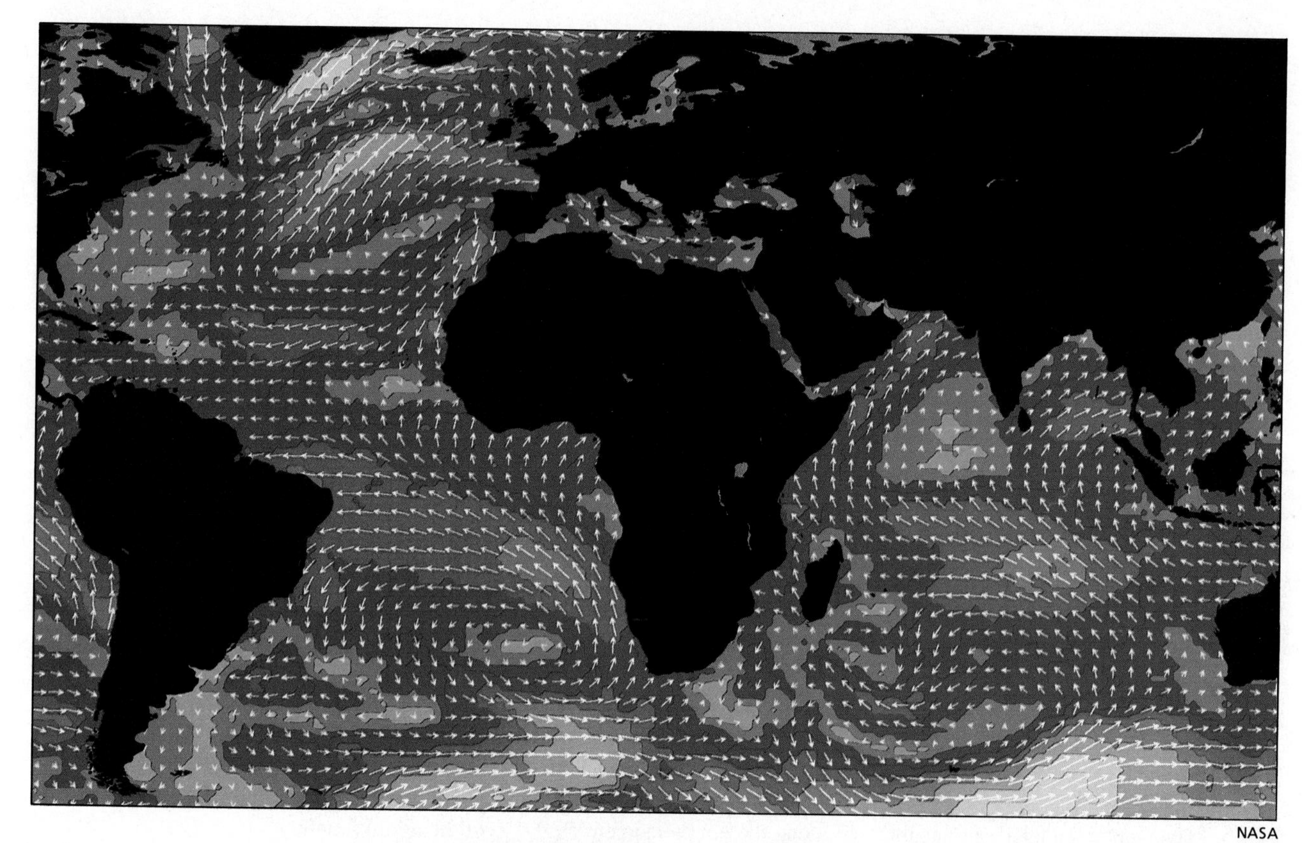

14.15 NASA's Seasat used radar during a 3-day period in September 1978 to take 350,000 wind measurements over the world's oceans. The arrows show wind direction; their length and the color contouring indicate speed. Notice the heavy storm south of Greenland.

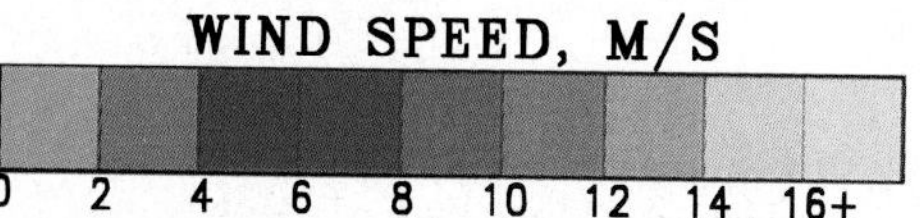

we attach the gradient vector of a scalar function to each point of a level surface of the function, we have a three-dimensional field on the surface. If we attach the velocity vector to each point of a flowing fluid, we have a three-dimensional field defined on a region in space. These and other fields are illustrated in Figures 14.7–14.15. Some of the illustrations give formulas for the fields as well.

To sketch the fields that had formulas, we picked a representative selection of domain points and sketched the vectors attached to them. Notice the convention that the arrows representing the vectors are drawn with their tails, not their heads, at the points where the vector functions are evaluated. This is different from the way we drew the position vectors of the planets and projectiles we studied in Chapter 11, with their tails at the origin and their heads at the planet's and projectile's locations.

The Work Done by a Force over a Curve in Space

Suppose that the vector field $\mathbf{F} = M(x, y, z)\,\mathbf{i} + N(x, y, z)\,\mathbf{j} + P(x, y, z)\,\mathbf{k}$ represents a force throughout a region in space (it might be the force of gravity or an electromagnetic force of some kind) and that

$$\mathbf{r} = g(t)\,\mathbf{i} + h(t)\,\mathbf{j} + k(t)\,\mathbf{k}, \quad a \le t \le b,$$

is a curve in the region. Then the line integral of $\mathbf{F} \cdot \mathbf{T}$, the scalar component of $\mathbf{F}$ in the direction of the curve's unit tangent vector, over the curve is called the work done by **F** over the curve from a to b (Fig. 14.16).

DEFINITION

The **work** done by a continuous force $\mathbf{F} = M(x, y, z)\,\mathbf{i} + N(x, y, z)\,\mathbf{j} + P(x, y, z)\,\mathbf{k}$ over a smooth curve $\mathbf{r} = g(t)\,\mathbf{i} + h(t)\,\mathbf{j} + k(t)\,\mathbf{k}$ from $t = a$ to $t = b$ is the value of the line integral

$$\text{Work} = \int_{t=a}^{t=b} \mathbf{F}\cdot\mathbf{T}\,ds. \tag{3}$$

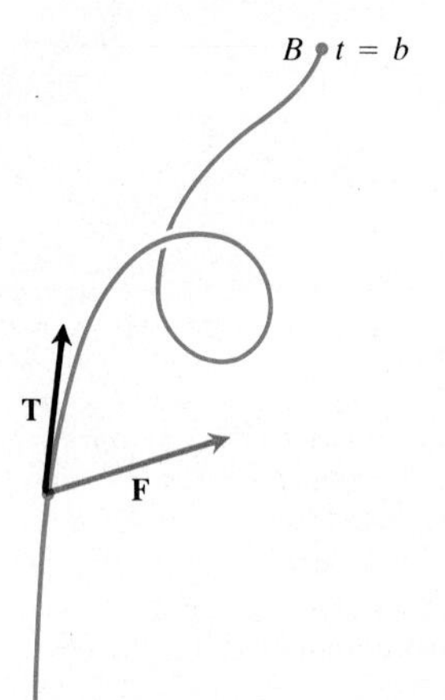

14.16 The work done by a continuous field **F** over a smooth path $\mathbf{r} = g(t)\,\mathbf{i} + h(t)\,\mathbf{j} + k(t)\,\mathbf{k}$ from A to B is the line integral of $\mathbf{F}\cdot\mathbf{T}$ from $t = a$ to $t = b$.

We derive Eq. (3) with the same kind of reasoning we used in Section 5.7 to derive the formula $W = \int_a^b F(x)\,dx$ for the work done by a continuous force of magnitude $F(x)$ directed along an interval of the x-axis. We divide the curve into short segments, apply the constant-force-times-distance formula for work to approximate the work over each curved segment, add the results to approximate the work over the entire curve, and calculate the work as the limit of the approximating sums as the segments become shorter and more numerous. To find exactly what the limiting integral should be, we partition the parameter interval $I = [a, b]$ in the usual way and choose a point c_k in each subinterval $[t_k, t_{k+1}]$. The partition of I determines ("induces," we say) a partition of the curve, with the point P_k being the tip of the position vector **r** at $t = t_k$ and Δs_k being the length of the curve segment P_kP_{k+1} (Fig. 14.17). If $\mathbf{F}_k$ denotes the value of **F** at the point on the curve corresponding to $t = c_k$, and $\mathbf{T}_k$ denotes the curve's tangent vector at this point, then $\mathbf{F}_k\cdot\mathbf{T}_k$ is the scalar component of **F** in the direction of **T** at $t = c_k$ (Fig. 14.18). The work done by **F** along the curve segment P_kP_{k+1} will be approximately

$$\begin{pmatrix}\text{force component in}\\ \text{direction of motion}\end{pmatrix}\times\begin{pmatrix}\text{distance}\\ \text{applied}\end{pmatrix} = \mathbf{F}_k\cdot\mathbf{T}_k\Delta s_k. \tag{4}$$

The work done by **F** along the curve from $t = a$ to $t = b$ will be approximately

$$\sum_{k=1}^{n} \mathbf{F}_k\cdot\mathbf{T}_k\Delta s_k. \tag{5}$$

As the norm of the partition of $[a, b]$ approaches zero, the norm of the induced partition of the curve approaches zero and the sums in (5) approach the line integral

$$\int_{t=a}^{t=b} \mathbf{F}\cdot\mathbf{T}\,ds. \tag{6}$$

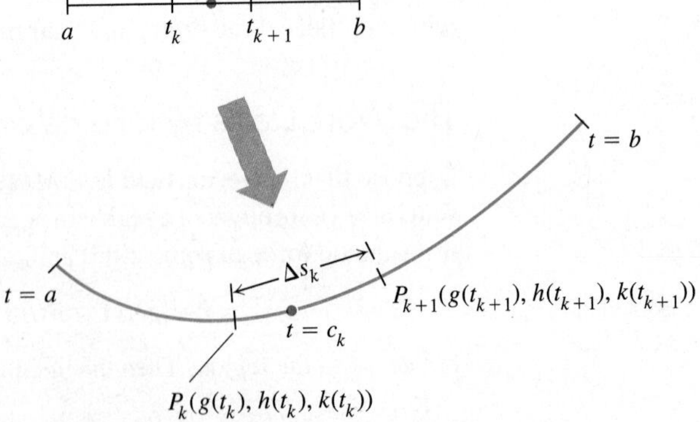

14.17 Each partition of the parameter interval $a \le t \le b$ induces a partition of the curve $\mathbf{r} = g(t)\,\mathbf{i} + h(t)\,\mathbf{j} + k(t)\,\mathbf{k}$.

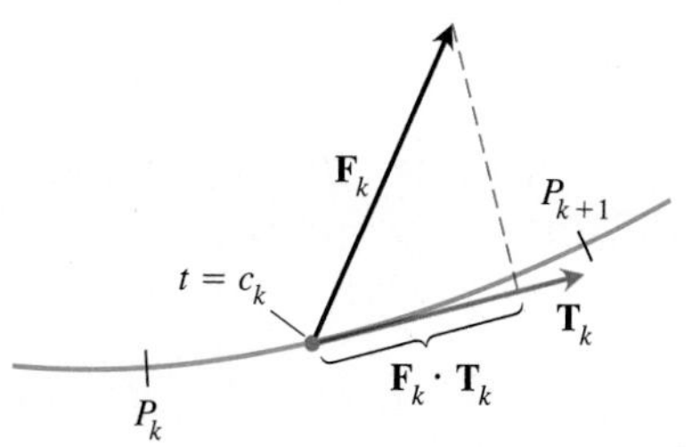

14.18 An enlarged view of the curve segment P_kP_{k+1} in Fig. 14.17, showing the force vector and unit tangent vector at the point on the curve where $t = c_k$.

Notice how the numerical sign of the number we calculate with this integral depends on the direction in which the curve is traversed as t increases. If we reverse the direction of motion, we reverse the direction of $\mathbf{T}$ and change the signs of $\mathbf{F} \cdot \mathbf{T}$ and its integral.

Notation and Evaluation

There are six standard ways to write the work integral in Eq. (3).

Equivalent Ways to Write the Work Integral

$$\text{Work} = \int_{t=a}^{t=b} \mathbf{F} \cdot \mathbf{T}\, ds \qquad \text{(The definition)}$$

$$= \int_{t=a}^{t=b} \mathbf{F} \cdot d\mathbf{r} \qquad \text{(Compact differential form)}$$

$$= \int_{a}^{b} \mathbf{F} \cdot \frac{d\mathbf{r}}{dt}\, dt \qquad \left(\begin{array}{l}\text{Expanded to include } dt\text{;} \\ \text{emphasizes the} \\ \text{velocity vector } d\mathbf{r}/dt\end{array}\right)$$

$$= \int_{a}^{b} \left(M\frac{dg}{dt} + N\frac{dh}{dt} + P\frac{dk}{dt}\right) dt \qquad \left(\begin{array}{l}\text{Emphasizes the} \\ \text{component functions}\end{array}\right)$$

$$= \int_{a}^{b} \left(M\frac{dx}{dt} + N\frac{dy}{dt} + P\frac{dz}{dt}\right) dt \qquad \left(\begin{array}{l}\text{Abbreviates the} \\ \text{components of } \mathbf{r}\end{array}\right)$$

$$= \int_{a}^{b} M\,dx + N\,dy + P\,dz \qquad \left(\begin{array}{l}dt\text{'s canceled; the most} \\ \text{common differential form}\end{array}\right)$$

Despite their variety, these formulas are all evaluated the same way.

Evaluation

To evaluate the work integral, take these steps:

1. Evaluate $\mathbf{F}$ on the curve as a function of t.
2. Find $d\mathbf{r}/dt$.
3. Dot $\mathbf{F}$ with $d\mathbf{r}/dt$.
4. Integrate from $t = a$ to $t = b$.

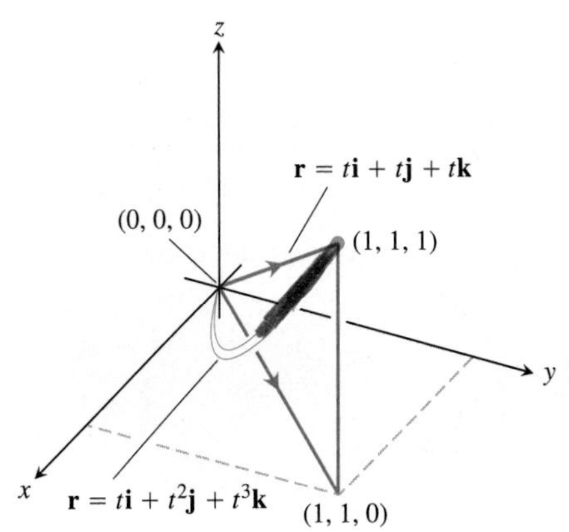

14.19 The curve in Example 1.

Example 1 Find the work done by the vector field

$$\mathbf{F} = (y - x^2)\,\mathbf{i} + (z - y^2)\,\mathbf{j} + (x - z^2)\,\mathbf{k}$$

over the curve

$$\mathbf{r} = t\mathbf{i} + t^2\mathbf{j} + t^3\mathbf{k}, \quad 0 \le t \le 1,$$

from (0, 0, 0) to (1, 1, 1) (Fig. 14.19).

Solution

STEP 1: *Evaluate* **F** *on the curve.*

$$\mathbf{F} = (y - x^2)\,\mathbf{i} + (z - y^2)\,\mathbf{j} + (x - z^2)\,\mathbf{k} = \underbrace{(t^2 - t^2)}_{0}\,\mathbf{i} + (t^3 - t^4)\,\mathbf{j} + (t - t^6)\,\mathbf{k}$$

STEP 2: *Find $d\mathbf{r}/dt$.*

$$\frac{d\mathbf{r}}{dt} = \frac{d}{dt}(t\mathbf{i} + t^2\,\mathbf{j} + t^3\,\mathbf{k}) = \mathbf{i} + 2t\mathbf{j} + 3t^2\,\mathbf{k}$$

STEP 3: *Dot* **F** *with $d\mathbf{r}/dt$.*

$$\begin{aligned}\mathbf{F}\cdot\frac{d\mathbf{r}}{dt} &= ((t^3 - t^4)\,\mathbf{j} + (t - t^6)\,\mathbf{k})\cdot(\mathbf{i} + 2t\mathbf{j} + 3t^2\,\mathbf{k})\\ &= (t^3 - t^4)(2t) + (t - t^6)(3t^2) = 2t^4 - 2t^5 + 3t^3 - 3t^8\end{aligned}$$

STEP 4: *Integrate from $t = 0$ to $t = 1$.*

$$\begin{aligned}\text{Work} &= \int_0^1 (2t^4 - 2t^5 + 3t^3 - 3t^8)\,dt\\ &= \left[\frac{2}{5}t^5 - \frac{2}{6}t^6 + \frac{3}{4}t^4 - \frac{3}{9}t^9\right]_0^1 = \frac{29}{60}.\end{aligned}$$

Flow Integrals and Circulation

If instead of being a force field, the vector field $\mathbf{F} = M\mathbf{i} + N\mathbf{j} + P\mathbf{k}$ represents the velocity field of a fluid flowing through a region in space (a tidal basin, a riverbed, or the turbine chamber of a hydroelectric generator, for example), then the line integral of $\mathbf{F}\cdot\mathbf{T}$ along a curve in the region is called the fluid's flow along the curve.

DEFINITIONS

If $\mathbf{r} = g(t)\,\mathbf{i} + h(t)\,\mathbf{j} + k(t)\,\mathbf{k}$, $a \le t \le b$, is a smooth curve in the domain of a continuous velocity field $\mathbf{F} = M(x, y, z)\,\mathbf{i} + N(x, y, z)\,\mathbf{j} + P(x, y, z)\,\mathbf{k}$, the **flow** along the curve from $t = a$ to $t = b$ is the line integral of $\mathbf{F}\cdot\mathbf{T}$ over the curve from a to b,

$$\text{Flow} = \int_a^b \mathbf{F}\cdot\mathbf{T}\,ds. \qquad (7)$$

The integral in this case is called a **flow integral.** If the curve is a closed loop, the flow is called the **circulation** around the curve.

We evaluate flow integrals the same way we evaluate work integrals.

Example 2 A fluid's velocity field is

$$\mathbf{F} = x\mathbf{i} + z\mathbf{j} + y\mathbf{k}.$$

Find the flow along the helix

$$\mathbf{r} = (\cos t)\,\mathbf{i} + (\sin t)\,\mathbf{j} + t\mathbf{k}, \quad 0 \le t \le \pi/2.$$

Solution

STEP 1: *Evaluate* **F** *on the curve.*

$$\mathbf{F} = x\mathbf{i} + z\mathbf{j} + y\mathbf{k} = (\cos t)\,\mathbf{i} + t\mathbf{j} + (\sin t)\,\mathbf{k}$$

STEP 2: *Find* $d\mathbf{r}/dt$.

$$\frac{d\mathbf{r}}{dt} = (-\sin t)\,\mathbf{i} + (\cos t)\,\mathbf{j} + \mathbf{k}$$

STEP 3: *Find* $\mathbf{F} \cdot (d\mathbf{r}/dt)$.

$$\mathbf{F} \cdot \frac{d\mathbf{r}}{dt} = (\cos t)(-\sin t) + (t)(\cos t) + (\sin t)(1)$$

$$= -\sin t \cos t + t \cos t + \sin t$$

STEP 4: *Integrate from* $t = a$ *to* $t = b$.

$$\text{Flow} = \int_{t=a}^{t=b} \mathbf{F} \cdot \frac{d\mathbf{r}}{dt}\,dt = \int_0^{\pi/2} (-\sin t \cos t + t \cos t + \sin t)\,dt$$

$$= \left[\frac{\cos^2 t}{2} + t \sin t\right]_0^{\pi/2} = \left(0 + \frac{\pi}{2}\right) - \left(\frac{1}{2} + 0\right) = \frac{\pi}{2} - \frac{1}{2}$$

Example 3 Find the circulation of a fluid around the circle

$$\mathbf{r} = (\cos t)\,\mathbf{i} + (\sin t)\,\mathbf{j}, \quad 0 \le t \le 2\pi,$$

if the velocity field is $\mathbf{F} = (x - y)\,\mathbf{i} + x\mathbf{j}$.

Solution

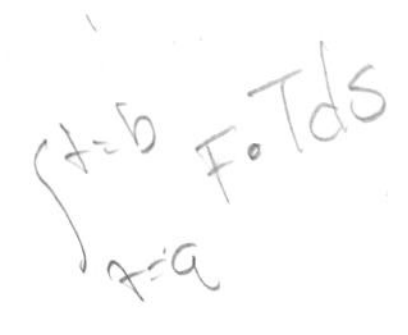

1. On the circle, $\mathbf{F} = (x - y)\,\mathbf{i} + x\mathbf{j} = (\cos t - \sin t)\,\mathbf{i} + (\cos t)\,\mathbf{j}$.

2. $\dfrac{d\mathbf{r}}{dt} = (-\sin t)\,\mathbf{i} + (\cos t)\,\mathbf{j}$

3. $\mathbf{F} \cdot \dfrac{d\mathbf{r}}{dt} = -\sin t \cos t + \underbrace{\sin^2 t + \cos^2 t}_{1}$

4. $\text{Circulation} = \displaystyle\int_0^{2\pi} \mathbf{F} \cdot \frac{d\mathbf{r}}{dt}\,dt = \int_0^{2\pi} (1 - \sin t \cos t)\,dt$

$$= \left[t - \frac{\sin^2 t}{2}\right]_0^{2\pi} = 2\pi$$

Flux Across a Plane Curve

To find the rate at which a fluid is entering or leaving a region enclosed by a curve C in the xy-plane, we calculate the line integral over C of $\mathbf{F} \cdot \mathbf{n}$, the scalar component of the fluid's velocity field in the direction of the curve's outward-pointing normal vector **n**. The value of this integral is called the **flux** of **F** across C. Flux is Latin for *flow,* but many flux calculations involve no motion at all. If **F** were an electric field or a magnetic field, for instance, the integral of $\mathbf{F} \cdot \mathbf{n}$ would still be called the flux of the field across C.

DEFINITION

If C is a smooth closed curve in the domain of a continuous vector field $\mathbf{F} = M(x, y)\,\mathbf{i} + N(x, y)\,\mathbf{j}$ in the plane, and if $\mathbf{n}$ is the outward-pointing unit normal vector on C, the **flux** of $\mathbf{F}$ across C is given by the following line integral:

$$\text{Flux of } \mathbf{F} \text{ across } C = \int_C \mathbf{F} \cdot \mathbf{n}\, ds. \tag{8}$$

Notice the difference between flux and circulation. The flux of $\mathbf{F}$ across C is the line integral with respect to arc length of $\mathbf{F} \cdot \mathbf{n}$, the scalar component of $\mathbf{F}$ in the direction of the outward normal. The circulation of $\mathbf{F}$ around C is the line integral with respect to arc length of $\mathbf{F} \cdot \mathbf{T}$, the scalar component of $\mathbf{F}$ in the direction of the unit tangent vector. Flux is the integral of the normal component of $\mathbf{F}$; circulation is the integral of the tangential component of $\mathbf{F}$.

To evaluate the integral in (8), we begin with a parametrization

$$x = g(t), \qquad y = h(t), \qquad a \le t \le b,$$

that traces the curve C exactly once as t increases from a to b. We can find the outward unit normal vector $\mathbf{n}$ by crossing the curve's unit tangent vector $\mathbf{T}$ with the vector $\mathbf{k}$. But which order do we choose, $\mathbf{T} \times \mathbf{k}$ or $\mathbf{k} \times \mathbf{T}$? Which one points outward? It depends on which way C is traversed as the parameter t increases. If the motion is clockwise, then $\mathbf{k} \times \mathbf{T}$ points outward; if the motion is counterclockwise, then $\mathbf{T} \times \mathbf{k}$ points outward (Fig. 14.20). The usual choice is $\mathbf{n} = \mathbf{T} \times \mathbf{k}$, the choice that assumes counterclockwise motion. Thus, while the value of the arc length integral in the definition of flux in Eq. (8) does not depend on which way C is traversed, the formulas we are about to derive for evaluating the integral in Eq. (8) will assume counterclockwise motion.

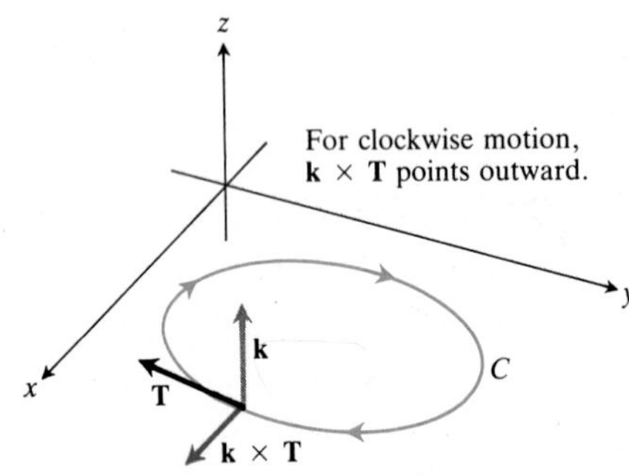

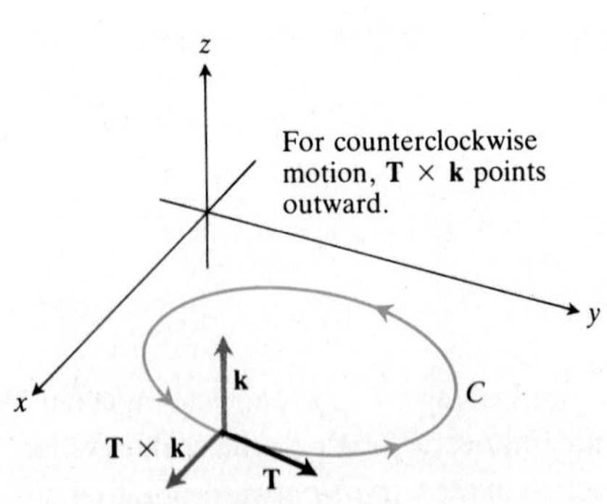

14.20 To find a unit outward normal vector for a curve C in the xy-plane that is traversed counterclockwise as t increases, we take $\mathbf{n} = \mathbf{T} \times \mathbf{k}$.

In terms of components,

$$\mathbf{n} = \mathbf{T} \times \mathbf{k} = \left(\frac{dx}{ds}\,\mathbf{i} + \frac{dy}{ds}\,\mathbf{j}\right) \times \mathbf{k} = \frac{dy}{ds}\,\mathbf{i} - \frac{dx}{ds}\,\mathbf{j}. \tag{9}$$

If $\mathbf{F} = M(x, y)\,\mathbf{i} + N(x, y)\,\mathbf{j}$, then

$$\mathbf{F} \cdot \mathbf{n} = M(x, y)\frac{dy}{ds} - N(x, y)\frac{dx}{ds}.$$

Hence,

$$\begin{aligned} \int_C \mathbf{F} \cdot \mathbf{n}\, ds &= \int_C \left(M\frac{dy}{ds} - N\frac{dx}{ds}\right) ds \\ &= \oint_C M\, dy - N\, dx. \end{aligned} \tag{10}$$

We put a directed circle ⟲ on the last integral as a reminder that the integration around the closed curve C is to be in the counterclockwise direction. To evaluate this integral, we express M, dy, N, and dx in terms of t and integrate from $t = a$ to $t = b$. Strange as it may seem, we do not need to know either $\mathbf{n}$ or ds to find the flux.

The Formula for Calculating Flux Across a Closed Plane Curve

$$\text{Flux of } \mathbf{F} = M\mathbf{i} + N\mathbf{j} \text{ across } C = \oint M\,dy - N\,dx \qquad (11)$$

The integral can be evaluated from any parametrization $x = g(t)$, $y = h(t)$, $a \le t \le b$, that traces C counterclockwise exactly once. To evaluate the integral, express M, dy, N, and dx in terms of t and integrate from $t = a$ to $t = b$.

Example 4 Find the flux of $\mathbf{F} = (x - y)\,\mathbf{i} + x\mathbf{j}$ across the circle $x^2 + y^2 = 1$ in the xy-plane.

Solution The parametrization

$$\mathbf{r} = (\cos t)\,\mathbf{i} + (\sin t)\,\mathbf{j}, \quad 0 \le t \le 2\pi,$$

traces the circle counterclockwise exactly once. We can therefore use this parametrization in Eq. (11). With

$$M = x - y = \cos t - \sin t \qquad dy = d(\sin t) = \cos t\,dt$$

$$N = x = \cos t \qquad dx = d(\cos t) = -\sin t\,dt,$$

we find

$$\text{Flux} = \int_C M\,dy - N\,dx = \int_0^{2\pi} (\cos^2 t - \sin t \cos t + \cos t \sin t)\,dt \quad \text{(Eq. (11))}$$

$$= \int_0^{2\pi} \cos^2 t\,dt = \int_0^{2\pi} \frac{1 + \cos 2t}{2}\,dt = \left[\frac{t}{2} + \frac{\sin 2t}{4}\right]_0^{2\pi} = \pi.$$

The flux of $\mathbf{F}$ across the circle is π. Since the answer is positive, the net flow across the curve is outward. A net inward flow would give a negative flux.

EXERCISES 14.2

1. Give a formula $\mathbf{F} = M(x, y)\,\mathbf{i} + N(x, y)\,\mathbf{j}$ for the vector field in the plane that has the property that $\mathbf{F}$ points toward the origin with magnitude inversely proportional to the square of the distance from (x, y) to the origin. (The field is not defined at $(0, 0)$.)

2. Give a formula $\mathbf{F} = M(x, y)\,\mathbf{i} + N(x, y)\,\mathbf{j}$ for the vector field in the plane that has the properties that $\mathbf{F} = \mathbf{0}$ at $(0, 0)$ and that at any other point (a, b), $\mathbf{F}$ is tangent to the circle $x^2 + y^2 = a^2 + b^2$ and points in the clockwise direction with magnitude $|\mathbf{F}| = \sqrt{a^2 + b^2}$.

3. Sketch the sets of points in the plane where the field

$$\mathbf{F} = \left(\frac{x^2 + 2y^2 - 4}{4}\right)\mathbf{i} + \left(\frac{y - x^2}{4}\right)\mathbf{j}$$

is (a) vertical, (b) horizontal. Then sketch the field on these sets.

4. Draw the (a) vertical and (b) horizontal components of the spin field

$$\mathbf{F} = \frac{(-y\mathbf{i} + x\mathbf{j})}{(x^2 + y^2)^{1/2}}$$

along the circle $x^2 + y^2 = 4$.

In Exercises 5–10, find the work done by the force $\mathbf{F}$ from $(0, 0, 0)$ to $(1, 1, 1)$ over each of the following paths:

a) the line segment $\mathbf{r} = t\mathbf{i} + t\mathbf{j} + t\mathbf{k}$, $0 \le t \le 1$,

b) (Fig. 14.21) the curve $\mathbf{r} = t\mathbf{i} + t^2\mathbf{j} + t^4\mathbf{k}$, $0 \le t \le 1$,

c) the path consisting of the line segment from $(0, 0, 0)$ to $(1, 1, 0)$ followed by the line segment from $(1, 1, 0)$ to $(1, 1, 1)$.

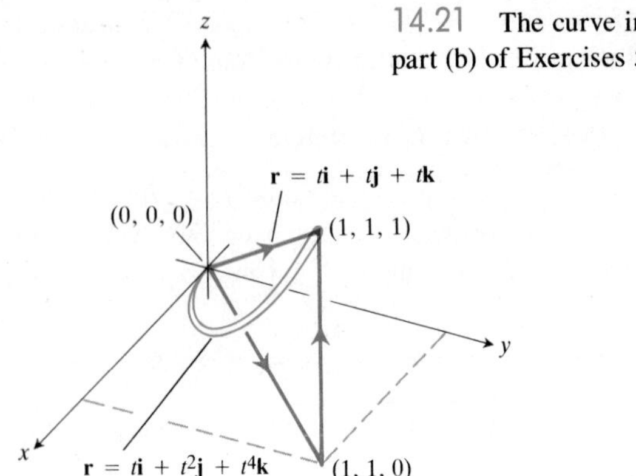

14.21 The curve in part (b) of Exercises 5–10.

5. $\mathbf{F} = 3y\mathbf{i} + 2x\mathbf{j} + 4z\mathbf{k}$

6. $\mathbf{F} = (1/(x^2 + 1))\,\mathbf{j}$

7. $\mathbf{F} = \sqrt{z}\,\mathbf{i} - 2x\mathbf{j} + \sqrt{y}\,\mathbf{k}$

8. $\mathbf{F} = xy\mathbf{i} + yz\mathbf{j} + xz\mathbf{k}$

9. $\mathbf{F} = (3x^2 - 3x)\,\mathbf{i} + 3z\mathbf{j} + \mathbf{k}$

10. $\mathbf{F} = (y + z)\,\mathbf{i} + (z + x)\,\mathbf{j} + (x + y)\,\mathbf{k}$

In Exercises 11–14, find the work done by $\mathbf{F}$ over the curve in the direction of increasing t.

11. $\mathbf{F} = xy\mathbf{i} + y\mathbf{j} - yz\mathbf{k}$
$\mathbf{r} = t\mathbf{i} + t^2\mathbf{j} + t\mathbf{k}, \quad 0 \le t \le 1$

12. $\mathbf{F} = 2y\mathbf{i} + 3x\mathbf{j} + (x + y)\,\mathbf{k}$
$\mathbf{r} = (\cos t)\,\mathbf{i} + (\sin t)\,\mathbf{j} + (t/6)\,\mathbf{k}, \quad 0 \le t \le 2\pi$

13. $\mathbf{F} = z\mathbf{i} + x\mathbf{j} + y\mathbf{k}$
$\mathbf{r} = (\sin t)\,\mathbf{i} + (\cos t)\,\mathbf{j} + t\mathbf{k}, \quad 0 \le t \le 2\pi$

14. $\mathbf{F} = 6z\mathbf{i} + y^2\mathbf{j} + 12x\mathbf{k}$
$\mathbf{r} = (\sin t)\,\mathbf{i} + (\cos t)\,\mathbf{j} + (t/6)\,\mathbf{k}, \quad 0 \le t \le 2\pi$

In Exercises 15–18, $\mathbf{F}$ is the velocity field of a fluid flowing through a region in space. Find the flow along the given curve in the direction of increasing t.

15. $\mathbf{F} = -4xy\mathbf{i} + 8y\mathbf{j} + 2\,\mathbf{k}$
$\mathbf{r} = t\mathbf{i} + t^2\mathbf{j} + \mathbf{k}, \quad 0 \le t \le 2$

16. $\mathbf{F} = x^2\mathbf{i} + yz\mathbf{j} + y^2\mathbf{k}$
$\mathbf{r} = 3t\mathbf{j} + 4t\mathbf{k}, \quad 0 \le t \le 1$

17. $\mathbf{F} = (x - z)\,\mathbf{i} + x\mathbf{k}$
$\mathbf{r} = (\cos t)\,\mathbf{i} + (\sin t)\,\mathbf{k}, \quad 0 \le t \le \pi$

18. $\mathbf{F} = -y\mathbf{i} + x\mathbf{j} + 2\mathbf{k}$
$\mathbf{r} = (-2\cos t)\,\mathbf{i} + (2\sin t)\,\mathbf{j} + 2t\mathbf{k}, \quad 0 \le t \le 2\pi$

19. Find the circulation and flux of the fields

$$\mathbf{F}_1 = x\mathbf{i} + y\mathbf{j} \quad \text{and} \quad \mathbf{F}_2 = -y\mathbf{i} + x\mathbf{j}$$

around and across each of the following curves.
a) The circle $\mathbf{r} = (\cos t)\,\mathbf{i} + (\sin t)\,\mathbf{j}, \quad 0 \le t \le 2\pi$
b) The ellipse $\mathbf{r} = (\cos t)\,\mathbf{i} + (4\sin t)\,\mathbf{j}, \quad 0 \le t \le 2\pi$

20. Find the flux of the fields

$$\mathbf{F}_1 = 2x\mathbf{i} - 3y\mathbf{j} \quad \text{and} \quad \mathbf{F}_2 = 2x\mathbf{i} + (x - y)\,\mathbf{j}$$

across the circle

$$\mathbf{r} = (a\cos t)\,\mathbf{i} + (a\sin t)\,\mathbf{j}, \quad 0 \le t \le 2\pi.$$

In Exercises 21–24, find the circulation and flux of the field $\mathbf{F}$ around and across the closed semicircular path that consists of the semicircular arch $\mathbf{r}_1 = (a\cos t)\,\mathbf{i} + (a\sin t)\,\mathbf{j}$, $0 \le t \le \pi$, followed by the line segment $\mathbf{r}_2 = t\mathbf{i}$, $-a \le t \le a$.

21. $\mathbf{F} = x\mathbf{i} + y\mathbf{j}$

22. $\mathbf{F} = x^2\mathbf{i} + y^2\mathbf{j}$

23. $\mathbf{F} = -y\mathbf{i} + x\mathbf{j}$

24. $\mathbf{F} = -y^2\mathbf{i} + x^2\mathbf{j}$

25. Evaluate the flow integral of the velocity field $\mathbf{F} = (x + y)\,\mathbf{i} - (x^2 + y^2)\,\mathbf{j}$ along each of the following paths from $(1, 0)$ to $(-1, 0)$ in the xy-plane.
a) The upper half of the circle $x^2 + y^2 = 1$
b) The line segment from $(1, 0)$ to $(-1, 0)$
c) The line segment from $(1, 0)$ to $(0, -1)$ followed by the line segment from $(0, -1)$ to $(-1, 0)$

26. The field $\mathbf{F} = xy\mathbf{i} + y\mathbf{j} - yz\mathbf{k}$ is the velocity field of a flow in space. Find the flow from $(0, 0, 0)$ to $(1, 1, 1)$ along the curve of intersection of the cylinder $y = x^2$ and the plane $z = x$.

27. Find the circulation of $\mathbf{F} = 2x\mathbf{i} + 2z\mathbf{j} + 2y\mathbf{k}$ around the closed path that consists of the helix $\mathbf{r}_1 = (\cos t)\,\mathbf{i} + (\sin t)\,\mathbf{j} + t\mathbf{k}$, $0 \le t \le \pi/2$, and the line segments $\mathbf{r}_2 = \mathbf{j} + (\pi/2)(1 - t)\,\mathbf{k}$, $0 \le t \le 1$, and $\mathbf{r}_3 = t\mathbf{i} + (1 - t)\,\mathbf{j}$, $0 \le t \le 1$, traversed in the direction of increasing t.

28. Suppose that a function $f(t)$ is differentiable and positive for $a \le t \le b$, that $\mathbf{r} = t\mathbf{i} + f(t)\,\mathbf{j}$, $a \le t \le b$, and that $\mathbf{F} = y\mathbf{i}$. Show that

$$\int_{t=a}^{t=b} \mathbf{F} \cdot d\mathbf{r}$$

is the area of the region between the graph of f and the t-axis from $t = a$ to $t = b$.

EXPLORER PROGRAM

Vector Fields — Pictures fields of the form $\mathbf{F} = M(x, y)\,\mathbf{i} + N(x, y)\,\mathbf{j}$ and integrates $\mathbf{F} \cdot \mathbf{n}$ over city-block paths in the plane

14.3 Green's Theorem in the Plane

We now come to a theorem that describes the relationship between the way a fluid flows along or across the boundary of a plane region and the way the fluid moves around inside the region. We assume the fluid is incompressible—like water, for example, and not like a gas. The connection between the fluid's boundary behavior and its internal behavior is made possible by the notions of divergence and curl. The divergence of a fluid's velocity field is a measure of the rate at which fluid is being piped into or out of the region at any given point. The curl of the velocity field is a measure of the fluid's rate of rotation at each point.

Green's theorem states that, under conditions usually met in practice, the outward flux of a vector field across the boundary of a plane region equals the double integral of the divergence of the field over the interior of the region. In another form, Green's theorem states that the counterclockwise circulation of a vector field around the boundary of a region equals the double integral of the curl of the vector field over the region.

Green's theorem is one of the great theorems of calculus. It is deep and surprising and has far-reaching consequences. In pure mathematics, it ranks in importance with the Fundamental Theorem of Calculus. In applied mathematics the generalizations of Green's theorem to three dimensions provide the foundation for important theorems about electricity, magnetism, and fluid flow. We shall explore the three-dimensional forms of Green's theorem in Sections 14.5 and 14.6.

Throughout our discussion of Green's theorem, we talk in terms of velocity fields of fluid flows. We do so because fluid flows are easy to picture and the notions of flux and circulation are easy to interpret. We would like you to be aware, however, that Green's theorem applies to any vector field satisfying certain mathematical conditions. It does not depend for its validity on the field's having a particular physical interpretation.

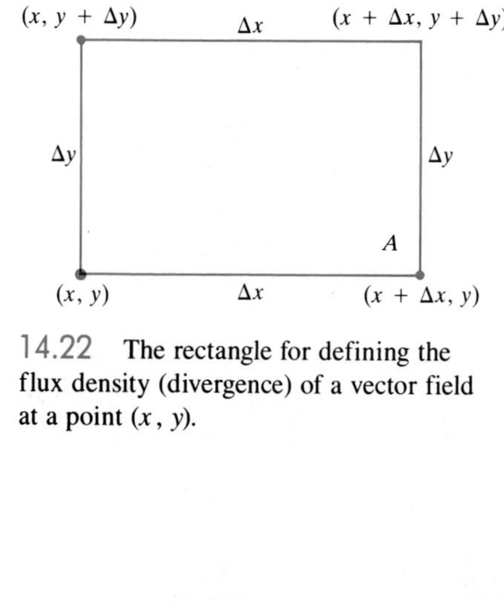

$(x, y + \Delta y)$ Δx $(x + \Delta x, y + \Delta y)$

Δy Δy

A

(x, y) Δx $(x + \Delta x, y)$

14.22 The rectangle for defining the flux density (divergence) of a vector field at a point (x, y).

Flux Density at a Point: Divergence

We need two new ideas for Green's theorem. The first is the idea of the flux density of a vector field at a point, which in mathematics is called the *divergence* of the vector field. We obtain it in the following way.

Suppose that

$$\mathbf{F}(x, y) = M(x, y)\,\mathbf{i} + N(x, y)\,\mathbf{j}$$

is the velocity field of a fluid flow in the plane and that the first partial derivatives of M and N are continuous at each point of a region R. Let (x, y) be a point in R and let A be a small rectangle with one corner at (x, y) that, along with its interior, lies entirely in R (Fig. 14.22). The sides of the rectangle, parallel to the coordinate axes, have lengths Δx and Δy. The rate at which fluid leaves the rectangle across the bottom edge is approximately

$$\mathbf{F}(x, y) \cdot (-\mathbf{j})\,\Delta x = -N(x, y)\,\Delta x. \tag{1}$$

This is the scalar component of the velocity at (x, y) in the direction of the outward normal times the length of the segment. If the velocity is in meters per second, for example, the exit rate will be in meters per second times meters or square meters per second. The rates at which the fluid crosses the other three sides in the directions of their outward normals can be estimated in a similar way. All told, we have

$$\begin{aligned}
&\text{Top:} && \mathbf{F}(x, y + \Delta y) \cdot \mathbf{j}\,\Delta x = N(x, y + \Delta y)\,\Delta x, \\
&\text{Bottom:} && \mathbf{F}(x, y) \cdot (-\mathbf{j})\,\Delta x = -N(x, y)\,\Delta x, \\
&\text{Right:} && \mathbf{F}(x + \Delta x, y) \cdot \mathbf{i}\,\Delta y = M(x + \Delta x, y)\,\Delta y, \\
&\text{Left:} && \mathbf{F}(x, y) \cdot (-\mathbf{i})\,\Delta y = -M(x, y)\,\Delta y.
\end{aligned} \tag{2}$$

Combining opposite pairs gives

$$\text{Top and bottom:} \qquad (N(x, y + \Delta y) - N(x, y))\,\Delta x \approx \left(\frac{\partial N}{\partial y}\Delta y\right)\Delta x, \tag{3}$$

$$\text{Right and left:} \qquad (M(x + \Delta x, y) - M(x, y))\,\Delta y \approx \left(\frac{\partial M}{\partial x}\Delta x\right)\Delta y. \tag{4}$$

Adding (3) and (4) gives

$$\text{Flux across rectangle boundary} \approx \left(\frac{\partial M}{\partial x} + \frac{\partial N}{\partial y}\right)\Delta x\,\Delta y. \tag{5}$$

We now divide by $\Delta x\,\Delta y$ to estimate the total flux per unit area or flux density for the rectangle:

$$\frac{\text{Flux across rectangle boundary}}{\text{Rectangle area}} \approx \left(\frac{\partial M}{\partial x} + \frac{\partial N}{\partial y}\right). \tag{6}$$

Finally, we let Δx and Δy approach zero to define what we call the *flux density* of **F** at the point (x, y).

In mathematics, we call the flux density the *divergence* of **F**. The symbol for it is div **F**, pronounced "divergence of **F**" or "div **F**."

DEFINITION

The **flux density** or **divergence** of a vector field $\mathbf{F} = M\mathbf{i} + N\mathbf{j}$ at the point (x, y) is

$$\text{div}\,\mathbf{F} = \frac{\partial M}{\partial x} + \frac{\partial N}{\partial y}. \tag{7}$$

If water were flowing into a region through a small hole at the point (x_0, y_0), the lines of flow would diverge there (hence the name) and div $\mathbf{F}(x_0, y_0)$ would be positive. If the water were draining out instead of flowing in, the divergence would be negative. See Fig. 14.23.

Example 1 Find the divergence of the vector field

$$\mathbf{F}(x, y) = (x^2 - y)\,\mathbf{i} + (xy - y^2)\,\mathbf{j}.$$

Solution We use the formula in Eq. (7):

$$\text{div}\,\mathbf{F} = \frac{\partial M}{\partial x} + \frac{\partial N}{\partial y} = \frac{\partial}{\partial x}(x^2 - y) + \frac{\partial}{\partial y}(xy - y^2)$$

$$= 2x + x - 2y = 3x - 2y.$$

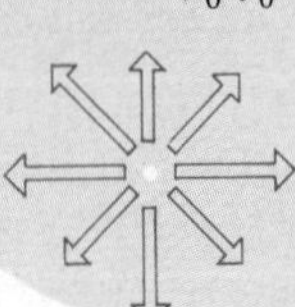

14.23 In the flow of an incompressible fluid across a plane region, the divergence is positive at a "source," a point where fluid enters the system, and negative at a "sink," a point where the fluid leaves the system.

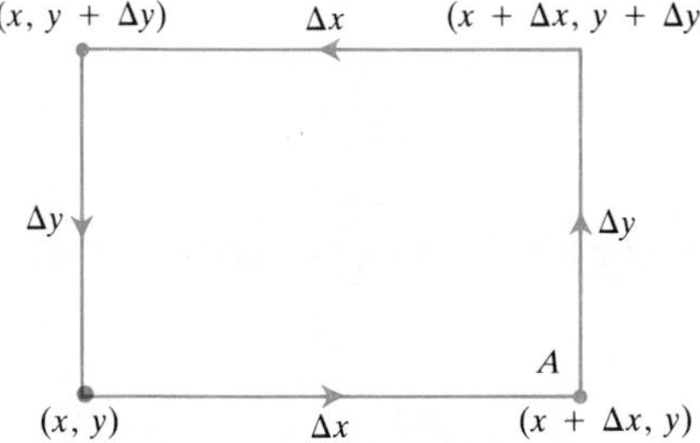

14.24 The rectangle for defining the circulation density (curl) of a vector field at a point (x, y).

Circulation Density at a Point: The Curl

The second of the two new ideas we need for Green's theorem is the idea of circulation density of a vector field $\mathbf{F}$ at a point, which in mathematics is called the *curl* of $\mathbf{F}$. To obtain it, we return to the velocity field

$$\mathbf{F}(x, y) = M(x, y)\,\mathbf{i} + N(x, y)\,\mathbf{j} \tag{8}$$

and the rectangle A. The rectangle is redrawn here as Fig. 14.24.

The counterclockwise circulation of $\mathbf{F}$ around the boundary of A is the sum of flow rates along the sides. For the bottom edge, the flow rate is approximately

$$\mathbf{F}(x, y) \cdot \mathbf{i}\,\Delta x = M(x, y)\,\Delta x. \tag{9}$$

This is the scalar component of the velocity $\mathbf{F}(x, y)$ in the direction of the tangent vector $\mathbf{i}$ times the length of the segment. The rates of flow along the other sides in the counterclockwise direction are expressed in a similar way. In all, we have

$$\begin{aligned} &\text{Top:} && \mathbf{F}(x, y + \Delta y) \cdot (-\mathbf{i})\,\Delta x = -M(x, y + \Delta y)\,\Delta x, \\ &\text{Bottom:} && \mathbf{F}(x, y) \cdot \mathbf{i}\,\Delta x = M(x, y)\,\Delta x, \\ &\text{Right:} && \mathbf{F}(x + \Delta x, y) \cdot \mathbf{j}\,\Delta y = N(x + \Delta x, y)\,\Delta y, \\ &\text{Left:} && \mathbf{F}(x, y) \cdot (-\mathbf{j})\,\Delta y = -N(x, y)\,\Delta y. \end{aligned} \tag{10}$$

We add opposite pairs to get

Top and bottom:

$$-\,(M(x, y + \Delta y) - M(x, y))\,\Delta x \approx -\left(\frac{\partial M}{\partial y}\,\Delta y\right)\Delta x, \tag{11}$$

Right and left:

$$(N(x + \Delta x, y) - N(x, y))\,\Delta y \approx \left(\frac{\partial N}{\partial x}\,\Delta x\right)\Delta y. \tag{12}$$

Adding (11) and (12) and dividing by $\Delta x\,\Delta y$ gives an estimate of the circulation density for the rectangle:

$$\frac{\text{Circulation around rectangle}}{\text{Rectangle area}} \approx \frac{\partial N}{\partial x} - \frac{\partial M}{\partial y}. \tag{13}$$

Finally, we let Δx and Δy approach zero to define what we call the *circulation density* of $\mathbf{F}$ at the point (x, y).

DEFINITION

The **circulation density** or **curl** of a vector field $\mathbf{F} = M\mathbf{i} + N\mathbf{j}$ at the point (x, y) is

$$\operatorname{curl} \mathbf{F} = \frac{\partial N}{\partial x} - \frac{\partial M}{\partial y}. \tag{14}$$

If water is moving about a region in the xy-plane in a thin layer, then the circulation, or curl, at a point (x_0, y_0) gives a way to measure how fast and in what direction a small paddle wheel will spin if it is put into the water at (x_0, y_0) with its axis perpendicular to the plane (Fig. 14.25).

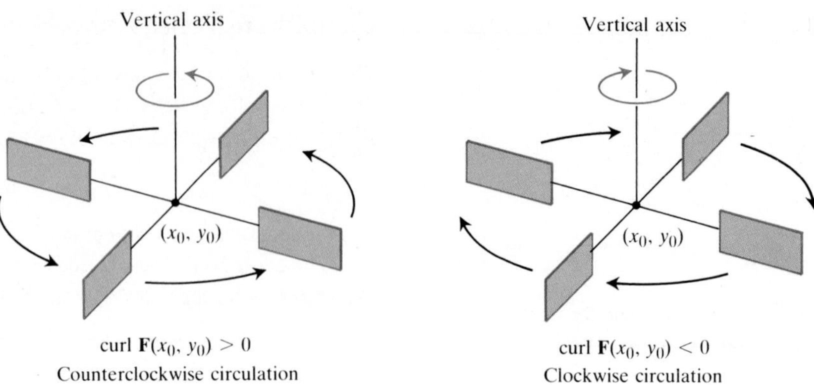

14.25 In the flow of an incompressible fluid over a plane region, the curl measures the rate of the fluid's rotation at a point. The curl is positive at points where the rotation is counterclockwise and negative where the rotation is clockwise.

Example 2 Find the curl of the vector field

$$\mathbf{F}(x, y) = (x^2 - y)\,\mathbf{i} + (xy - y^2)\,\mathbf{j}.$$

Solution We use the formula in Eq. (14):

$$\operatorname{curl} \mathbf{F} = \frac{\partial N}{\partial x} - \frac{\partial M}{\partial y} = \frac{\partial}{\partial x}(xy - y^2) - \frac{\partial}{\partial y}(x^2 - y) = y + 1.$$

Green's Theorem in the Plane

In one form, Green's theorem says that under suitable conditions the outward flux of a vector field across a simple closed curve in the plane (Fig. 14.26) equals the double integral of the divergence of the field over the region enclosed by the curve.

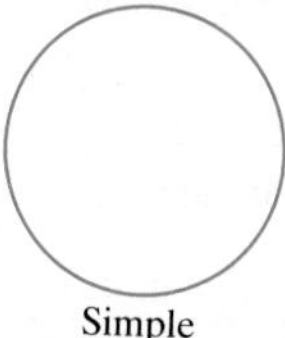

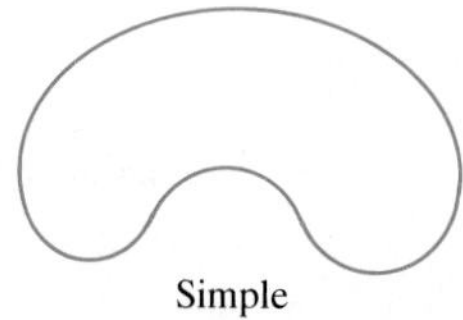

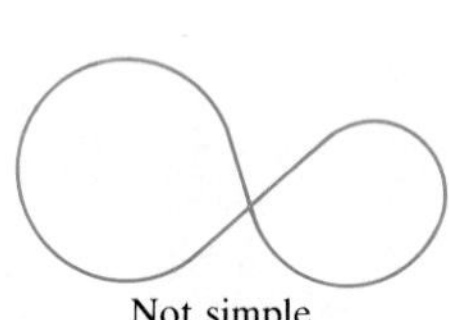

14.26 In proving Green's theorem, we distinguish between two kinds of closed curves, simple and not simple. Simple curves do not cross themselves. A circle is simple but a figure eight is not.

THEOREM 2

Green's Theorem (Flux–Divergence Form)

The outward flux of a vector field $\mathbf{F} = M\mathbf{i} + N\mathbf{j}$ across a simple closed curve C equals the double integral of div $\mathbf{F}$ over the region R enclosed by C:

$$\underbrace{\oint_C M\,dy - N\,dx}_{\text{outward flux}} = \underbrace{\iint_R \left(\frac{\partial M}{\partial x} + \frac{\partial N}{\partial y}\right) dx\,dy}_{\text{divergence integral}}. \tag{15}$$

In another form, Green's theorem says that the counterclockwise circulation of a vector field around a simple closed curve is the double integral of the curl of the field over the region enclosed by the curve.

THEOREM 3

Green's Theorem (Circulation–Curl Form)

The counterclockwise circulation of a vector field $\mathbf{F} = M\mathbf{i} + N\mathbf{j}$ around a simple closed curve C in the plane equals the double integral of curl $\mathbf{F}$ over the region enclosed by C:

$$\oint_C M\,dx + N\,dy = \iint_R \left(\frac{\partial N}{\partial x} - \frac{\partial M}{\partial y}\right) dx\,dy \tag{16}$$

counterclockwise circulation — curl integral

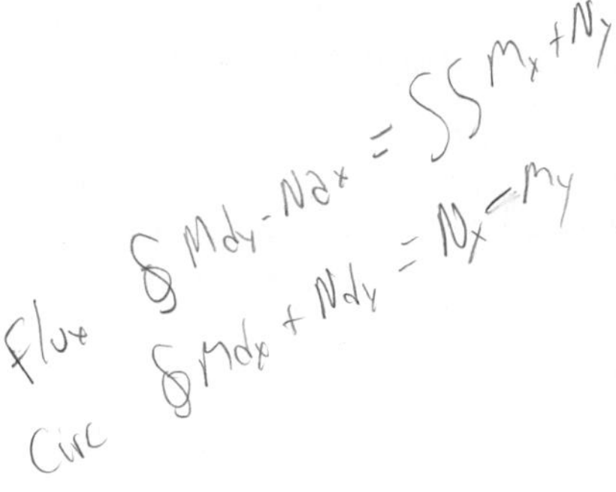

The two forms of Green's theorem are equivalent. Applying Eq. (15) to the field $\mathbf{G}_1 = N\mathbf{i} - M\mathbf{j}$ gives Eq. (16), and applying Eq. (16) to $\mathbf{G}_2 = -N\mathbf{i} + M\mathbf{j}$ gives Eq. (15). We do not need to prove them both. We shall prove the curl form shortly.

We need two kinds of assumptions for Green's theorem to hold. First, we need conditions on M and N to ensure the existence of the integrals. The usual assumptions are that M, N, and their first partial derivatives are continuous at every point of some open region containing C and R. Second, we need geometric conditions on the curve C. It must be simple, closed, and made up of pieces along which we can integrate M and N. The usual assumptions are that C is **piecewise smooth** (consists of a finite number of smooth curves connected end to end). The proof we give for Green's theorem, however, assumes things about the shape of R as well. You can find less restrictive proofs in more advanced texts.

Example 3 Verify both forms of Green's theorem for the field

$$\mathbf{F}(x, y) = (x - y)\,\mathbf{i} + x\mathbf{j}$$

and the region R bounded by the circle

$$C: \quad \mathbf{r} = (\cos t)\,\mathbf{i} + (\sin t)\,\mathbf{j}, \quad 0 \le t \le 2\pi.$$

Solution We first express all functions, derivatives, and differentials in terms of t:

$$\begin{aligned} M &= \cos t - \sin t, & dx &= d(\cos t) = -\sin t\,dt, \\ N &= \cos t, & dy &= d(\sin t) = \cos t\,dt, \end{aligned} \tag{17}$$

$$\frac{\partial M}{\partial x} = 1, \quad \frac{\partial M}{\partial y} = -1, \quad \frac{\partial N}{\partial x} = 1, \quad \frac{\partial N}{\partial y} = 0.$$

The two sides of Eq. (15):

$$\begin{aligned} \oint_C M\,dy - N\,dx &= \int_{t=0}^{t=2\pi} (\cos t - \sin t)(\cos t\,dt) - (\cos t)(-\sin t\,dt) \\ &= \int_0^{2\pi} \cos^2 t\,dt = \pi, \end{aligned}$$

$$\iint_R \left(\frac{\partial M}{\partial x} + \frac{\partial N}{\partial y}\right) dx\,dy = \iint_R (1 + 0)\,dx\,dy = \iint_R dx\,dy = \pi.$$

The Green of Green's Theorem

The Green of Green's theorem was George Green (1793–1841), a self-taught scientist in Nothingham, England. Green's work on the mathematical foundations of gravitation, electricity, and magnetism was published privately in 1828 in a short book entitled *An Essay on the Application of Mathematical Analysis to Electricity and Magnetism*. The book sold all of fifty-two copies (fewer than one hundred were printed), the copies going mostly to Green's patrons and personal friends. A few weeks before Green's death in 1841, William Thompson noticed a reference to Green's book and in 1845 was able to locate a copy. Excited by what he read, Thompson shared Green's ideas with other scientists and had the book republished as a series of journal articles. Green's mathematics provided the foundation on which Thompson, Stokes, Rayleigh, Maxwell, and others built the present-day theory of electromagnetism.

The two sides of Eq. (16):

$$\oint_C M\,dx + N\,dy = \int_{t=0}^{t=2\pi} (\cos t - \sin t)(-\sin t\,dt) + (\cos t)(\cos t\,dt)$$

$$= \int_0^{2\pi} (-\sin t \cos t + 1)\,dt = 2\pi,$$

$$\iint_R \left(\frac{\partial N}{\partial x} - \frac{\partial M}{\partial y}\right) dx\,dy = \iint_R (1 - (-1))\,dx\,dy = 2\iint_R dx\,dy = 2\pi.$$

Using Green's Theorem to Evaluate Line Integrals

If we construct a closed curve C by piecing a number of different curves end to end, the process of evaluating a line integral over C can be lengthy because there are so many different integrals to evaluate. However, if C bounds a region R to which Green's theorem applies, we can use Green's theorem to change the line integral around C into one double integral over R.

Example 4 Evaluate the integral

$$\oint xy\,dy - y^2\,dx$$

around the square C cut from the first quadrant by the lines $x = 1$ and $y = 1$.

Solution We can use either form of Green's theorem to change the line integral into a double integral over the square.

1. *With Eq.* (15): Taking $M = xy$, $N = y^2$, and C and R as the square's boundary and interior gives

$$\oint_C xy\,dy - y^2\,dx = \iint_R (y + 2y)\,dx\,dy = \int_0^1 \int_0^1 3y\,dx\,dy$$

$$= \int_0^1 \left[3xy\right]_{x=0}^{x=1} dy$$

$$= \int_0^1 3y\,dy = \frac{3}{2}y^2\Big]_0^1 = \frac{3}{2}.$$

2. *With Eq.* (16): Taking $M = -y^2$ and $N = xy$ gives the same result:

$$\oint_C -y^2\,dx + xy\,dy = \iint_R (y - (-2y))\,dx\,dy = \frac{3}{2}.$$

Example 5 Calculate the outward flux of the field $\mathbf{F}(x, y) = x\mathbf{i} + y^2\mathbf{j}$ across the square bounded by the lines $x = \pm 1$ and $y = \pm 1$.

Solution Calculating the flux with a line integral would take four integrations, one for each side of the square. With Green's theorem, we can change the line integral to one double integral. With $M = x$, $N = y^2$, C the square, and R the square's interior, we have

$$\text{Flux} = \oint_C M\,dy - N\,dx = \iint_R \left(\frac{\partial M}{\partial x} + \frac{\partial N}{\partial y}\right) dx\,dy \qquad \text{(Green's theorem)}$$

$$= \int_{-1}^{1}\int_{-1}^{1} (1 + 2y)\,dx\,dy = \int_{-1}^{1} \Big[x + 2xy\Big]_{x=-1}^{x=1} dy$$

$$= \int_{-1}^{1} (2 + 4y)\,dy = \Big[2y + 2y^2\Big]_{-1}^{1} = 4.$$

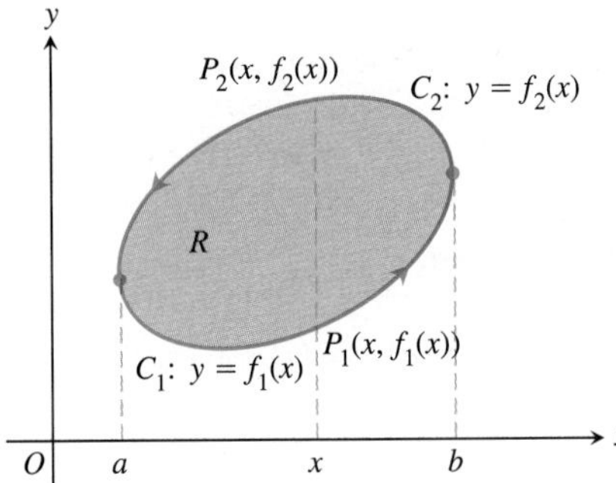

14.27 The boundary curve C is made up of C_1, the graph of $y = f_1(x)$, and C_2, the graph of $y = f_2(x)$.

A Proof of Green's Theorem (Special Regions)

Let C be a smooth simple closed curve in the xy-plane with the property that lines parallel to the axes cut it in no more than two points. Let R be the region enclosed by C and suppose that M, N, and their first partial derivatives are continuous at every point of some open region containing C and R. We want to show that

$$\oint_C M\,dx + N\,dy = \iint_R \left(\frac{\partial N}{\partial x} - \frac{\partial M}{\partial y}\right) dx\,dy. \tag{18}$$

Figure 14.27 shows C made up of two parts:

$$C_1:\quad y = f_1(x), \quad a \le x \le b, \qquad C_2:\quad y = f_2(x), \quad a \le x \le b.$$

For any x between a and b, we can integrate $\partial M/\partial y$ with respect to y from $y = f_1(x)$ to $y = f_2(x)$ and obtain

$$\int_{f_1(x)}^{f_2(x)} \frac{\partial M}{\partial y}\,dy = M(x, y)\Big]_{y=f_1(x)}^{y=f_2(x)} = M(x, f_2(x)) - M(x, f_1(x)). \tag{19}$$

We can then integrate this with respect to x from a to b:

$$\int_a^b \int_{f_1(x)}^{f_2(x)} \frac{\partial M}{\partial y}\,dy\,dx = \int_a^b [M(x, f_2(x)) - M(x, f_1(x))]\,dx$$

$$= -\int_b^a M(x, f_2(x))\,dx - \int_a^b M(x, f_1(x))\,dx$$

$$= -\int_{C_2} M\,dx - \int_{C_1} M\,dx$$

$$= -\oint_C M\,dx.$$

Therefore

$$\oint_C M\,dx = \iint_R \left(-\frac{\partial M}{\partial y}\right) dx\,dy. \tag{20}$$

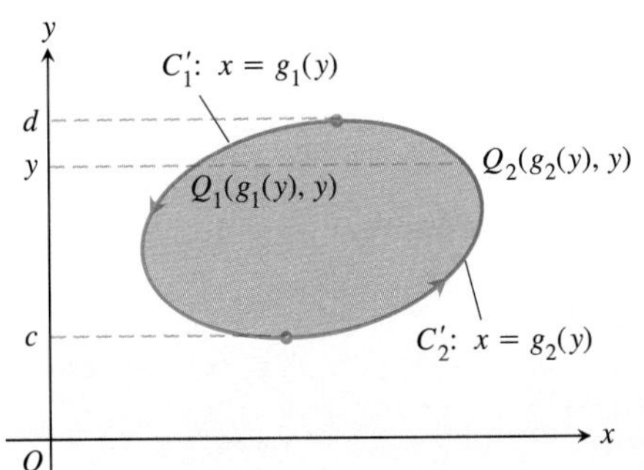

14.28 The boundary curve C is made up of C_1', the graph of $x = g_1(y)$, and C_2', the graph of $x = g_2(y)$.

Equation (20) is half the result we need for Eq. (18). Exercise 29 asks you to derive the other half by integrating $\partial N/\partial x$ first with respect to x and then with respect to y, as suggested by Fig. 14.28. Figure 14.28 shows the curve C of Fig. 14.27 decomposed into the two directed parts

$$C_1':\quad x = g_1(y), \quad d \ge y \ge c, \qquad C_2':\quad x = g_2(y), \quad c \le y \le d.$$

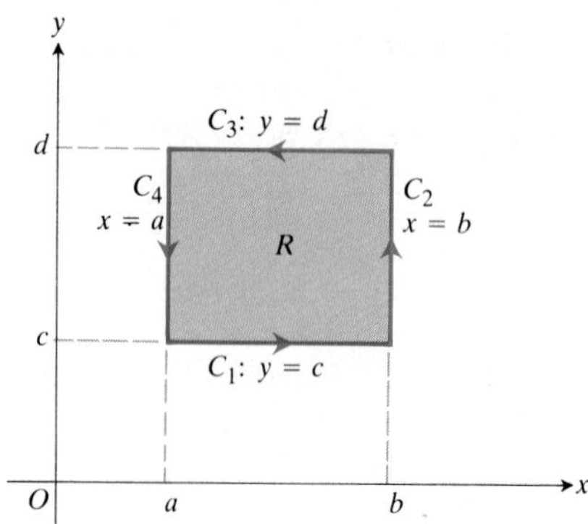

14.29 To prove Green's theorem for the rectangular region shown here, we divide the boundary into four directed line segments.

The result of this double integration is

$$\oint_C N\,dy = \iint_R \frac{\partial N}{\partial x}\,dx\,dy. \tag{21}$$

Combining Eqs. (20) and (21) gives Eq. (18). This concludes the proof.

Extending the Proof to Other Regions

The argument with which we just derived Green's equation does not apply directly to the rectangular region in Fig. 14.29 because the lines $x = a$, $x = b$, $y = c$, and $y = d$ meet the region's boundary in more than two points. However, if we divide the boundary C into four directed line segments,

$$C_1:\quad y = c,\quad a \le x \le b, \qquad C_2:\quad x = b,\quad c \le y \le d,$$

$$C_3:\quad y = d,\quad b \ge x \ge a, \qquad C_4:\quad x = a,\quad d \ge y \ge c,$$

we can modify the argument in the following way.

Proceeding as in the proof of Eq. (21), we have

$$\begin{aligned}\int_c^d \int_a^b \frac{\partial N}{\partial x}\,dx\,dy &= \int_c^d [N(b, y) - N(a, y)]\,dy \\ &= \int_c^d N(b, y)\,dy + \int_d^c N(a, y)\,dy \\ &= \int_{C_2} N\,dy + \int_{C_4} N\,dy.\end{aligned} \tag{22}$$

Because y is constant along C_1 and C_3,

$$\int_{C_1} N\,dy = \int_{C_3} N\,dy = 0,$$

so we can add

$$\int_{C_1} N\,dy + \int_{C_3} N\,dy$$

to the right-hand side of Eq. (22) without changing the equality. Doing so, we have

$$\int_c^d \int_a^b \frac{\partial N}{\partial x}\,dx\,dy = \oint_C N\,dy. \tag{23}$$

Similarly, we can show (Exercise 30) that

$$\int_a^b \int_c^d \frac{\partial M}{\partial y}\,dy\,dx = -\oint_C M\,dx. \tag{24}$$

Subtracting (24) from (23), we again arrive at

$$\oint_C M\,dx + N\,dy = \iint_R \left(\frac{\partial N}{\partial x} - \frac{\partial M}{\partial y}\right) dx\,dy.$$

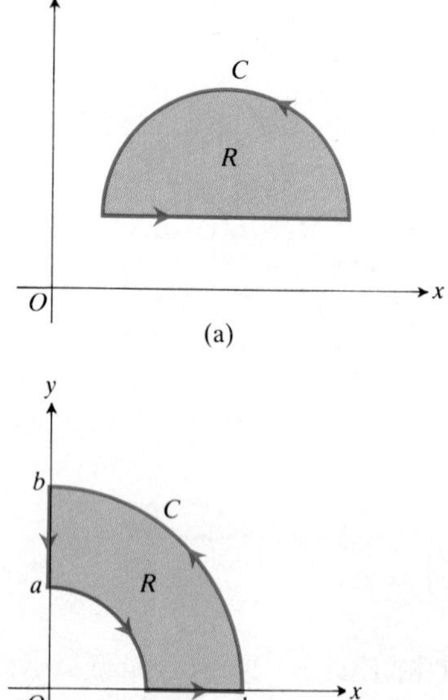

14.30 Other regions to which Green's theorem applies.

Regions like those in Fig. 14.30 can be handled with no greater difficulty. Equation (18) still applies. It also applies to the horseshoe-shaped region R shown in Fig. 14.31, as we see by putting together the regions R_1 and R_2 and their boundaries.

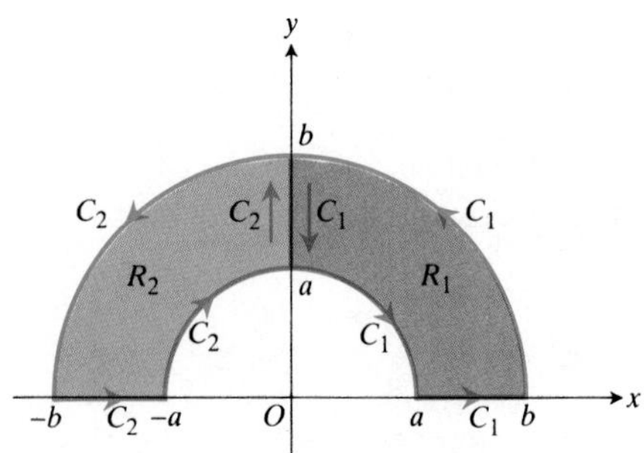

14.31 A region R that combines regions R_1 and R_2.

Green's theorem applies to C_1, R_1, and to C_2, R_2, yielding

$$\int_{C_1} M\,dx + N\,dy = \iint_{R_1} \left(\frac{\partial N}{\partial x} - \frac{\partial M}{\partial y}\right) dx\,dy,$$

$$\int_{C_2} M\,dx + N\,dy = \iint_{R_2} \left(\frac{\partial N}{\partial x} - \frac{\partial M}{\partial y}\right) dx\,dy.$$

When we add these two equations, the line integral along the y-axis from b to a for C_1 cancels the integral over the same segment but in the opposite direction for C_2. Hence

$$\oint_C M\,dx + N\,dy = \iint_R \left(\frac{\partial N}{\partial x} - \frac{\partial M}{\partial y}\right) dx\,dy,$$

where C consists of the two segments of the x-axis from $-b$ to $-a$ and from a to b and of the two semicircles, and where R is the region inside C.

The device of adding line integrals over separate boundaries to build up an integral over a single boundary can be extended to any finite number of subregions. In Fig. 14.32(a), let C_1 be the boundary of the region R_1 in the first quadrant. Similarly for the other three quadrants: C_i is the boundary of the region R_i, $i = 1, 2, 3, 4$. By Green's theorem,

$$\oint_{C_i} M\,dx + N\,dy = \iint_{R_i} \left(\frac{\partial N}{\partial x} - \frac{\partial M}{\partial y}\right) dx\,dy. \tag{25}$$

We add Eqs. (25) for $i = 1, 2, 3, 4$, and get

$$\oint_{r=b} (M\,dx + N\,dy) + \oint_{r=a} (M\,dx + N\,dy) = \iint_{a \le r \le b} \left(\frac{\partial N}{\partial x} - \frac{\partial M}{\partial y}\right) dx\,dy. \tag{26}$$

Equation (26) says that the double integral of $(\partial N/\partial x) - (\partial M/\partial y)$ over the annular ring R equals the line integral of $M\,dx + N\,dy$ over the complete boundary of R in the direction that keeps R on our left as we progress (Fig. 14.32b).

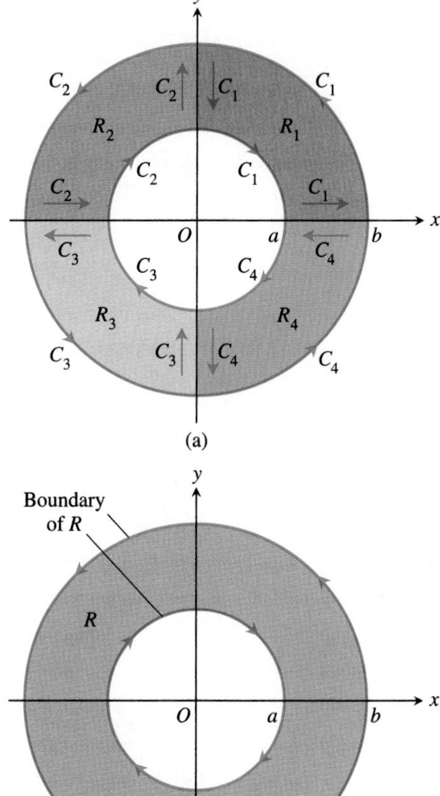

14.32 The annular region R shown here combines four smaller regions. In polar coordinates, we have $r = a$ for the inner circle, $r = b$ for the outer circle, and $a \le r \le b$ for the region itself.

Example 6 Verify the circulation form of Green's theorem (Eq. 16) on the annular ring R: $h^2 \le x^2 + y^2 \le 1$, $0 < h < 1$ (Fig. 14.33), if

$$M = \frac{-y}{x^2 + y^2}, \qquad N = \frac{x}{x^2 + y^2}.$$

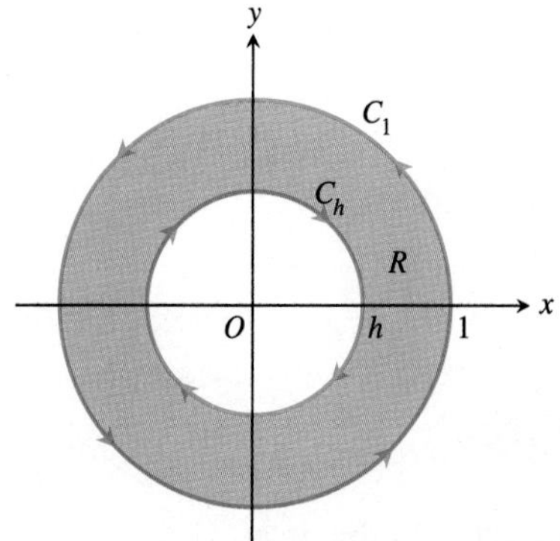

14.33 Green's theorem may be applied to the annular region R by integrating along the boundaries as shown (Example 6).

Solution The boundary of R consists of the circle

$$C_1:\quad x = \cos t,\quad y = \sin t,\quad 0 \le t \le 2\pi,$$

traversed counterclockwise as t increases, and the circle

$$C_h:\quad x = h\cos\theta,\quad y = -h\sin\theta,\quad 0 \le \theta \le 2\pi,$$

traversed clockwise as θ increases. The functions M and N and their partial derivatives are continuous throughout R. Moreover,

$$\frac{\partial M}{\partial y} = \frac{(x^2+y^2)(-1) + y(2y)}{(x^2+y^2)^2} = \frac{y^2 - x^2}{(x^2+y^2)^2} = \frac{\partial N}{\partial x},$$

so

$$\iint_R \left(\frac{\partial N}{\partial x} - \frac{\partial M}{\partial y}\right) dx\,dy = \iint_R 0\,dx\,dy = 0.$$

The integral of $M\,dx + N\,dy$ over the boundary of R is

$$\begin{aligned}\int_C M\,dx + N\,dy &= \oint_{C_1} \frac{x\,dy - y\,dx}{x^2+y^2} + \oint_{C_h} \frac{x\,dy - y\,dx}{x^2+y^2} \\ &= \int_0^{2\pi} (\cos^2 t + \sin^2 t)\,dt - \int_0^{2\pi} \frac{h^2(\cos^2\theta + \sin^2\theta)}{h^2}\,d\theta \\ &= 2\pi - 2\pi = 0.\end{aligned}$$

The functions M and N in Example 6 are discontinuous at (0, 0), so we cannot apply Green's theorem to the circle C_1 and the region inside it. We must exclude the origin. We do so by excluding the points inside C_h.

We could replace the circle C_1 in Example 6 by an ellipse or any other simple closed curve K surrounding C_h (Fig. 14.34). The result would still be

$$\oint_K (M\,dx + N\,dy) + \oint_{C_h} (M\,dx + N\,dy) = \iint_R \left(\frac{\partial N}{\partial x} - \frac{\partial M}{\partial y}\right) dy\,dx = 0,$$

which leads to the surprising conclusion that

$$\oint_K (M\,dx + N\,dy) = 2\pi$$

for any such curve K. We can explain this result by changing to polar coordinates. With

$$x = r\cos\theta,\qquad y = r\sin\theta,$$

$$dx = -r\sin\theta\,d\theta + \cos\theta\,dr,\qquad dy = r\cos\theta\,d\theta + \sin\theta\,dr,$$

we have

$$\frac{x\,dy - y\,dx}{x^2+y^2} = \frac{r^2(\cos^2\theta + \sin^2\theta)\,d\theta}{r^2} = d\theta,$$

and θ increases by 2π as we traverse K once counterclockwise.

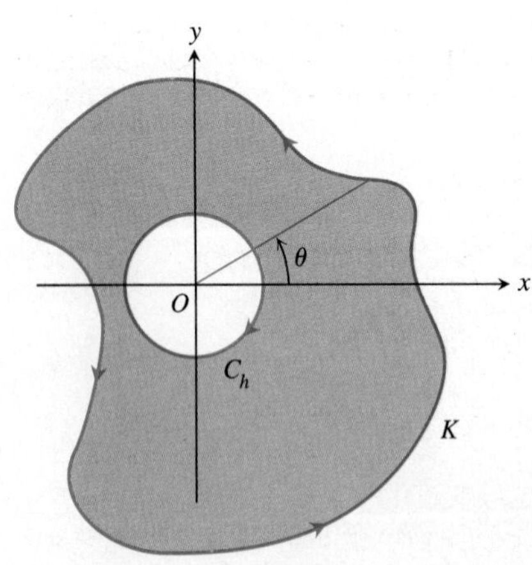

14.34 The region bounded by the circle C_h and the curve K.

The Normal and Tangential Forms of Green's Theorem

As we learned in Section 14.2, the flux of a two-dimensional vector field $\mathbf{F}$ across a closed curve C in the direction of its outer unit normal vector $\mathbf{n}$ is

$$\int_C \mathbf{F} \cdot \mathbf{n}\, ds. \qquad \text{(Flux)}$$

Similarly, the circulation around C in the direction of its unit tangent vector $\mathbf{T}$ is

$$\int_C \mathbf{F} \cdot \mathbf{T}\, ds. \qquad \text{(Circulation)}$$

Hence, the flux form of Green's theorem (Eq. (15)) can be written as

$$\oint_C \mathbf{F} \cdot \mathbf{n}\, ds = \iint_R \left(\frac{\partial M}{\partial x} + \frac{\partial N}{\partial y}\right) dx\, dy \qquad \text{(Normal form)} \qquad (27)$$

and the circulation form of Green's theorem (Eq. (16)) can be written as

$$\oint_C \mathbf{F} \cdot \mathbf{T}\, ds = \iint_R \left(\frac{\partial N}{\partial x} - \frac{\partial M}{\partial y}\right) dx\, dy. \qquad \text{(Tangential form)} \qquad (28)$$

The left-hand sides of these equations are now in vector form. How about the right-hand sides? Can they be expressed in vector form too? The answer is yes, as we shall see in Sections 14.5 and 14.6.

EXERCISES 14.3

In Exercises 1–4, verify Green's theorem by evaluating both sides of Eqs. (15) and (16) for the field $\mathbf{F} = M\mathbf{i} + N\mathbf{j}$. Take the domains of integration in each case to be the disk R: $x^2 + y^2 \leq a^2$ and its bounding circle C: $\mathbf{r} = (a \cos t)\,\mathbf{i} + (a \sin t)\,\mathbf{j}$, $0 \leq t \leq 2\pi$.

1. $\mathbf{F} = -y\mathbf{i} + x\mathbf{j}$

2. $\mathbf{F} = y\mathbf{i}$

3. $\mathbf{F} = 2x\mathbf{i} - 3y\mathbf{j}$

4. $\mathbf{F} = -x^2y\mathbf{i} + xy^2\mathbf{j}$

In Exercises 5–10, use Green's theorem to find the counterclockwise circulation and outward flux for the field $\mathbf{F}$ and curve C.

5. $\mathbf{F} = (x - y)\,\mathbf{i} + (y - x)\,\mathbf{j}$
C: The square bounded by $x = 0$, $x = 1$, $y = 0$, $y = 1$

6. $\mathbf{F} = (x^2 + 4y)\,\mathbf{i} + (x + y^2)\,\mathbf{j}$
C: The square bounded by $x = 0$, $x = 1$, $y = 0$, $y = 1$

7. $\mathbf{F} = (y^2 - x^2)\,\mathbf{i} + (x^2 + y^2)\,\mathbf{j}$
C: The triangle bounded by $y = 0$, $x = 3$, and $y = x$

8. $\mathbf{F} = (x + y)\,\mathbf{i} - (x^2 + y^2)\,\mathbf{j}$
C: The triangle bounded by $y = 0$, $x = 1$, and $y = x$

9. $\mathbf{F} = (x + e^x \sin y)\,\mathbf{i} + (x + e^x \cos y)\,\mathbf{j}$
C: The right-hand loop of the lemniscate $r^2 = \cos 2\theta$

10. $\mathbf{F} = \left(\tan^{-1} \frac{y}{x}\right)\mathbf{i} + \ln(x^2 + y^2)\,\mathbf{j}$
C: The boundary of the region defined by the polar-coordinate inequalities $1 \leq r \leq 2$, $0 \leq \theta \leq \pi$.

11. Find the counterclockwise circulation and outward flux of the field $\mathbf{F} = xy\mathbf{i} + y^2\mathbf{j}$ around and over the boundary of the region enclosed by the parabola $y = x^2$ and the line $y = x$ in the first quadrant.

12. Find the counterclockwise circulation and the outward flux of the field $\mathbf{F} = (-\sin y)\,\mathbf{i} + (x \cos y)\,\mathbf{j}$ around and over the square cut from the first quadrant by the lines $x = \pi/2$ and $y = \pi/2$.

13. Find the outward flux of the field

$$\mathbf{F} = \left(3xy - \frac{x}{1 + y^2}\right)\mathbf{i} + (e^x + \tan^{-1} y)\,\mathbf{j}$$

across the cardioid $r = a(1 + \cos\theta)$, $a > 0$.

14. Find the counterclockwise circulation of the field $\mathbf{F} = (y + e^x \ln y)\,\mathbf{i} + (e^x/y)\,\mathbf{j}$ around the boundary of the region that is bounded above by the curve $y = 2 - x^2$ and below by the curve $y = x^4 + 1$.

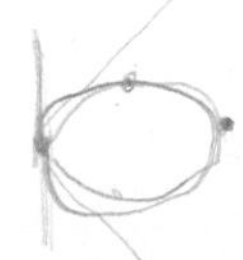

Apply Green's theorem in one of its two forms to evaluate the line integrals in Exercises 15–20.

15. $\oint_C (y^2\,dx + x^2\,dy)$

C: The triangle bounded by $x = 0$, $x + y = 1$, $y = 0$

16. $\oint_C (3y\,dx + 2x\,dy)$

C: The boundary of $0 \le x \le \pi$, $0 \le y \le \sin x$

17. $\oint_C (6y + x)\,dx + (y + 2x)\,dy$

C: The circle $(x - 2)^2 + (y - 3)^2 = 4$

18. $\oint_C (2x + y^2)\,dx + (2xy + 3y)\,dy$

C: Any simple closed curve in the plane for which Green's theorem holds

19. $\oint_C 2xy^3\,dx + 4x^2y^2\,dy$

C: The boundary of the "triangular" region in the first quadrant enclosed by the x-axis, the line $x = 1$, and the curve $y = x^3$

20. $\oint_C (4x - 2y)\,dx + (2x - 4y)\,dy$

C: The circle $(x - 2)^2 + (y - 2)^2 = 4$

21. Let C be the boundary of a region on which Green's theorem holds. Use Green's theorem to calculate

a) $\oint_C f(x)\,dx + g(y)\,dy$,

b) $\oint_C ky\,dx + hx\,dy$ (k and h constants).

22. Show that

$$\oint_C 4x^3y\,dx + x^4\,dy = 0$$

for any closed curve to which Green's theorem applies.

23. Show that

$$\oint_C -y^3\,dx + x^3\,dy$$

is positive for any closed curve C to which Green's theorem applies.

24. Show that the value of

$$\oint_C xy^2\,dx + (x^2y + 2x)\,dy$$

around any square depends only on the area of the square and not on its location in the plane.

25. *Green's theorem and Laplace's equation.* Assuming that all the necessary derivatives exist and are continuous, show that if $f(x, y)$ satisfies the Laplace equation

$$\frac{\partial^2 f}{\partial x^2} + \frac{\partial^2 f}{\partial y^2} = 0,$$

then

$$\oint_C \frac{\partial f}{\partial y}\,dx - \frac{\partial f}{\partial x}\,dy = 0$$

for all closed curves C to which Green's theorem applies. (The converse is also true: If the line integral is always zero, then f satisfies the Laplace equation.)

26. Among all smooth simple closed curves in the plane, oriented counterclockwise, find the one along which the work done by

$$\mathbf{F} = \left(\frac{1}{4}x^2y + \frac{1}{3}y^3\right)\mathbf{i} + x\mathbf{j}$$

is greatest. (*Hint:* Where is curl $\mathbf{F}$ positive?)

27. Green's theorem holds for a region R with any finite number of holes as long as the bounding curves are smooth, simple, and closed and we integrate over each component of the boundary in the direction that keeps R on our immediate left as we go along (Fig. 14.35).

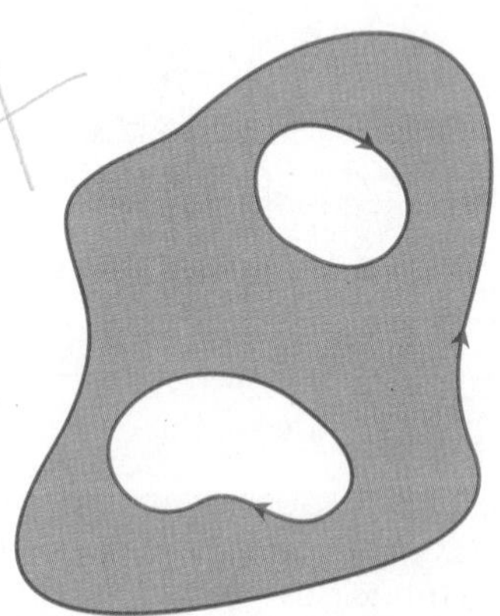

14.35 Green's theorem holds for regions with more than one hole (Exercise 27).

a) Let $f(x, y) = \ln(x^2 + y^2)$ and let C be the circle $x^2 + y^2 = a^2$. Evaluate the flux integral

$$\oint_C \nabla f \cdot \mathbf{n}\,ds.$$

b) Let K be an arbitrary smooth simple closed curve in the plane that does not pass through $(0, 0)$. Use Green's theorem to show that

$$\oint_K \nabla f \cdot \mathbf{n}\,ds$$

has two possible values, depending on whether $(0, 0)$ lies inside K or outside K.

28. *Bendixson's criterion.* The **streamlines** of a planar fluid flow are the smooth curves traced by the fluid's individual particles. The vectors $\mathbf{F} = M(x, y)\mathbf{i} + N(x, y)\mathbf{j}$ of the flow's velocity field are the tangent vectors of the streamlines. Show that if the flow takes place over a *simply connected* region R (no holes or missing points) and that if

$M_x + N_y \neq 0$ throughout R, then none of the streamlines in R is closed. In other words, no particle of fluid ever has a closed trajectory in R. The criterion $M_x + N_y \neq 0$ is called **Bendixson's criterion** for the nonexistence of closed trajectories.

29. Establish Eq. (21) to finish the proof of the special case of Green's theorem.

30. Establish Eq. (24) to complete the argument for the extension of Green's theorem.

Calculating Area with Green's Theorem

If a simple closed curve C in the plane and the region R it encloses satisfy the hypotheses of Green's theorem, then the area of R is given by the following formula.

Green's Theorem Area Formula

$$\text{Area of } R = \frac{1}{2}\oint_C x\,dy - y\,dx. \tag{29}$$

The reason is that by Eq. (15), run backward,

$$\text{Area of } R = \iint_R dy\,dx = \iint_R \left(\frac{1}{2} + \frac{1}{2}\right) dy\,dx$$

$$= \oint_C \frac{1}{2} x\,dy - \frac{1}{2} y\,dx.$$

31. Show that if R is a region in the plane bounded by a piecewise smooth simple closed curve C, then

$$\text{Area of } R = \oint_C x\,dy = -\oint_C y\,dx.$$

32. Suppose that a nonnegative function $y = f(x)$ has a continuous first derivative on $[a, b]$. Let C be the boundary of the region in the xy-plane that is bounded below by the x-axis, above by the graph of f, and on the sides by the lines $x = a$ and $x = b$. Show that

$$\int_a^b f(x)\,dx = -\oint_C y\,dx.$$

33. Let A be the area and $\bar{x}$ the x-coordinate of the centroid of a region R that is bounded by a piecewise smooth simple closed curve C in the xy-plane. Show that

$$\frac{1}{2}\oint_C x^2\,dy = -\oint_C xy\,dx = \frac{1}{3}\oint_C x^2\,dy - xy\,dx = A\bar{x}.$$

34. Let I_y be the moment of inertia about the y-axis of the region in Exercise 33. Show that

$$\frac{1}{3}\oint_C x^3\,dy = -\oint_C x^2y\,dx = \frac{1}{4}\oint_C x^3\,dy - x^2y\,dx = I_y.$$

Use the Green's theorem area formula (Eq. 29) to find the areas of the regions enclosed by the curves in Exercises 35–38.

35. The circle $\mathbf{r} = (a\cos t)\,\mathbf{i} + (a\sin t)\,\mathbf{j}, \quad 0 \le t \le 2\pi$

36. The ellipse $\mathbf{r} = (a\cos t)\,\mathbf{i} + (b\sin t)\,\mathbf{j}, \quad 0 \le t \le 2\pi$

37. The astroid (Fig. 9.46) $\mathbf{r} = (\cos^3 t)\,\mathbf{i} + (\sin^3 t)\,\mathbf{j}, \quad 0 \le t \le 2\pi$

38. The curve (Fig. 9.91) $\mathbf{r} = t^2\,\mathbf{i} + ((t^3/3) - t)\,\mathbf{j}, \quad -\sqrt{3} \le t \le \sqrt{3}$

EXPLORER PROGRAM

Vector Fields	Displays two-dimensional vector fields and integrates the curl over rectangular regions. Also calculates flux across rectangular boundaries

14.4 Surface Area and Surface Integrals

We know how to integrate a function over a flat region in a plane, but what if the function is defined over a curved surface instead? How do we calculate its integral then? The trick to evaluating one of these so-called surface integrals is to rewrite it as a double integral over a region in a coordinate plane beneath the surface (Fig. 14.36). This changes the surface integral into the kind of integral we already know how to evaluate.

Our first step is to find a double integral formula for calculating the area of a curved surface. We will then see how the ideas that arise can be used again to define and evaluate surface integrals. With surface integrals under control, we shall be able to calculate the flux of a three-dimensional vector field through a surface and the masses and moments of thin shells of material. In Sections 14.5 and 14.6 we shall see how surface integrals provide just what we need to generalize the two forms of Green's theorem to three dimensions. One of these generalizations will

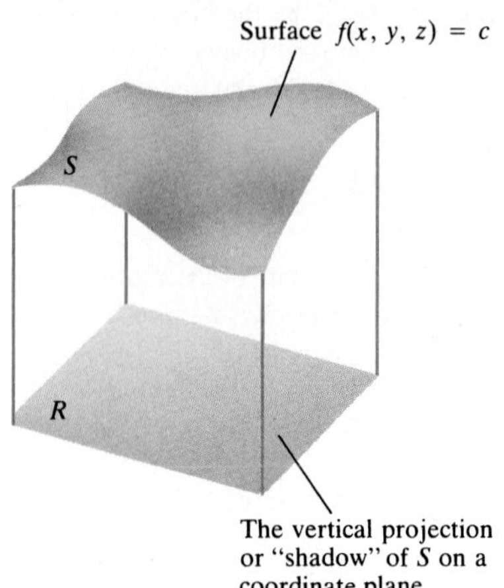

14.36 As we shall soon see, the integral of a function $g(x, y, z)$ over a surface S in space can be calculated by evaluating a related double integral over the vertical projection or "shadow" of S on a coordinate plane.

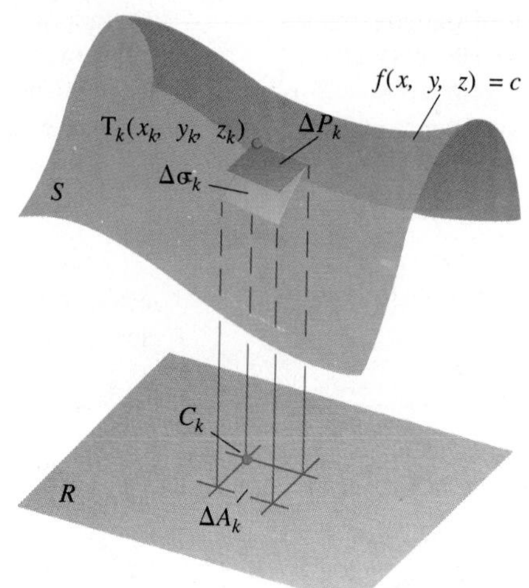

14.37 A surface S and its vertical projection onto a plane beneath it. You can think of R as the shadow of S on the plane. The tangent plate ΔP_k approximates the surface patch $\Delta\sigma_k$ above ΔA_k.

enable us to express the flux of a vector field through a closed surface as a triple integral over the three-dimensional region enclosed by the surface. The other generalization will enable us to express the circulation of a vector field around a closed curve in space as an integral over a surface bounded by the curve. These results have far-reaching consequences in mathematics as well as in the theories of electromagnetism and fluid flow.

The Definition of Surface Area

Figure 14.37 shows a surface S lying above its "shadow" region R in a plane beneath it. The surface is defined by the equation $f(x, y, z) = c$. If the surface is **smooth** (∇f is continuous and never vanishes on S), we can define and calculate its area as a double integral over R. It takes a while to derive the integral, but the integral itself is easy enough to work with.

The first step in defining the area of S is to partition the region R into small rectangles ΔA_k of the kind we would use if we were defining an integral over R. Directly above each ΔA_k lies a patch of surface $\Delta\sigma_k$ that we may approximate with a portion ΔP_k of the tangent plane. To be specific, we suppose that ΔP_k is a portion of the plane that is tangent to the surface at the point $T_k(x_k, y_k, z_k)$ directly above the back corner C_k of ΔA_k. If the tangent plane is parallel to R, then ΔP_k will be congruent to ΔA_k. Otherwise, it will be a parallelogram whose area is somewhat larger than the area of ΔA_k.

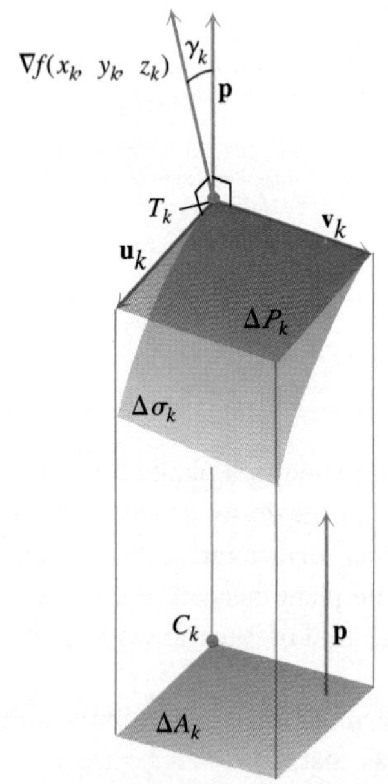

14.38 Magnified view from the preceding figure. The vector $\mathbf{u} \times \mathbf{v}$ (not shown) is parallel to the vector ∇f because both vectors are normal to the plane of ΔP_k.

Figure 14.38 gives a magnified view of $\Delta\sigma_k$ and ΔP_k, showing the gradient vector $\nabla f(x_k, y_k, z_k)$ at T_k and a unit vector $\mathbf{p}$ that is normal to R. The figure also shows the angle γ_k between ∇f and $\mathbf{p}$. The other vectors in the picture, $\mathbf{u}$ and $\mathbf{v}$, lie along the edges of the patch ΔP_k in the tangent plane. Thus, both $\mathbf{u} \times \mathbf{v}$ and ∇f are normal to the tangent plane.

We now need the fact from Section 10.6 that $|(\mathbf{u} \times \mathbf{v}) \cdot \mathbf{p}|$ is the area of the projection of the parallelogram determined by $\mathbf{u}$ and $\mathbf{v}$ onto any plane whose normal is $\mathbf{p}$. In our case, this translates into the statement

$$|(\mathbf{u} \times \mathbf{v}) \cdot \mathbf{p}| = \Delta A_k. \tag{1}$$

Now, $|\mathbf{u} \times \mathbf{v}|$ itself is the area ΔP_k (standard fact about cross products) so Eq. (1)

becomes

$$\underbrace{|\mathbf{u} \times \mathbf{v}|}_{\Delta P_k} \underbrace{|\mathbf{p}|}_{1} \underbrace{|\cos(\text{angle between } \mathbf{u} \times \mathbf{v} \text{ and } \mathbf{p})|}_{\substack{\text{same as } |\cos\gamma_k| \text{ because} \\ \nabla f \text{ and } \mathbf{u}\times\mathbf{v} \text{ are both} \\ \text{normal to the tangent plane}}} = \Delta A_k \tag{2}$$

or

$$\Delta P_k |\cos \gamma_k| = \Delta A_k$$

or

$$\Delta P_k = \frac{\Delta A_k}{|\cos\gamma_k|},$$

provided $\cos \gamma_k \neq 0$. We will have $\cos \gamma_k \neq 0$ as long as ∇f is not parallel to the ground plane and $\nabla f \cdot \mathbf{p} \neq 0$.

Since the patches ΔP_k approximate the surface patches $\Delta \sigma_k$ that fit together to make S, the sum

$$\sum \Delta P_k = \sum \frac{\Delta A_k}{|\cos \gamma_k|} \tag{3}$$

looks like an approximation of what we might like to call the surface area of S. It also looks as if the approximation would improve if we refined the partition of R. In fact, the sums on the right-hand side of Eq. (3) are approximating sums for the double integral

$$\iint_R \frac{1}{|\cos \gamma|}\, dA. \tag{4}$$

We therefore define the **area** of S to be the value of this integral whenever it exists.

A Practical Formula

For any particular surface $f(x, y, z) = c$,

$$|\nabla f \cdot \mathbf{p}| = |\nabla f|\, |\mathbf{p}|\, |\cos \gamma|,$$

so

$$\frac{1}{|\cos \gamma|} = \frac{|\nabla f|}{|\nabla f \cdot \mathbf{p}|}.$$

This combines with Eq. (4) to give a practical formula for area.

The Formula for Surface Area

The area of the surface $f(x, y, z) = c$ over a closed and bounded plane region R is

$$\text{Surface area} = \iint_R \frac{|\nabla f|}{|\nabla f \cdot \mathbf{p}|}\, dA, \tag{5}$$

where $\mathbf{p}$ is a unit vector normal to R and $\nabla f \cdot \mathbf{p} \neq 0$.

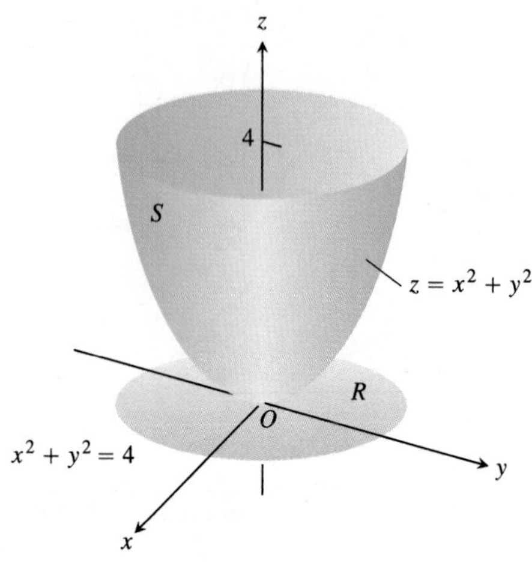

14.39 The area of this parabolic surface is calculated in Example 1.

Thus, the area is the double integral over R of the magnitude of ∇f divided by the magnitude of the scalar component of ∇f normal to R.

We reached Eq. (5) under the assumption that $\nabla f \cdot \mathbf{p} \neq 0$ throughout R and that ∇f is continuous. Whenever the integral exists, however, we may define its value to be the area of the portion of the surface $f(x, y, z) = c$ that lies over R.

Equation (5) agrees with the formulas for surface area in Section 5.5, although we shall not prove this fact.

Example 1 Find the area of the surface cut from the bottom of the paraboloid $x^2 + y^2 - z = 0$ by the plane $z = 4$.

Solution We sketch the surface S and the region R below it in the xy-plane (Fig. 14.39). The surface S is part of the level surface $f(x, y, z) = x^2 + y^2 - z = 0$, and R is the disk $x^2 + y^2 \leq 4$ in the xy-plane. To get a unit vector normal to the plane of R, we can take $\mathbf{p} = \mathbf{k}$.

At any point (x, y, z) on the surface, we have

$$\begin{aligned} f(x, y, z) &= x^2 + y^2 - z, \\ \nabla f &= 2x\mathbf{i} + 2y\mathbf{j} - \mathbf{k}, \\ |\nabla f| &= \sqrt{(2x)^2 + (2y)^2 + (-1)^2} \\ &= \sqrt{4x^2 + 4y^2 + 1}, \\ |\nabla f \cdot \mathbf{p}| &= |\nabla f \cdot \mathbf{k}| = |-1| = 1. \end{aligned}$$

In the region R, $dA = dx\, dy$. Therefore,

$$\begin{aligned} \text{Surface area} &= \iint_R \frac{|\nabla f|}{|\nabla f \cdot \mathbf{p}|}\, dA && \text{(Eq. (5))} \\ &= \iint_{x^2+y^2 \leq 4} \sqrt{4x^2 + 4y^2 + 1}\, dx\, dy \\ &= \int_0^{2\pi} \int_0^2 \sqrt{4r^2 + 1}\, r\, dr\, d\theta && \text{(Polar coordinates)} \\ &= \int_0^{2\pi} \left[\frac{1}{12} (4r^2 + 1)^{3/2} \right]_0^2 d\theta \\ &= \int_0^{2\pi} \frac{1}{12} (17^{3/2} - 1)\, d\theta \\ &= \frac{\pi}{6} (17\sqrt{17} - 1). \end{aligned}$$

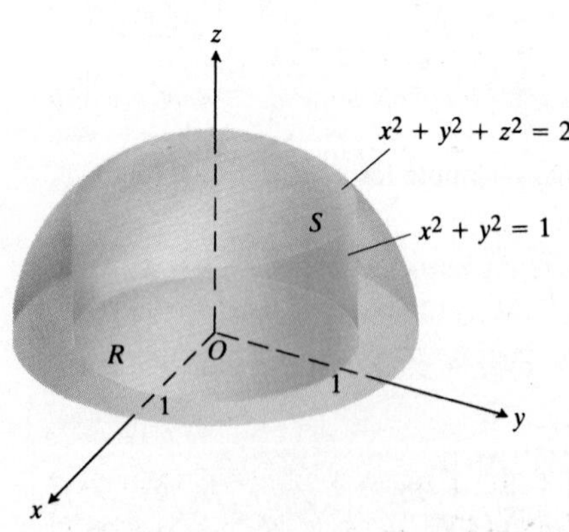

14.40 The cap cut from the hemisphere by the cylinder projects vertically onto the disk $R: x^2 + y^2 \leq 1$ (Example 2).

Example 2 Find the area of the cap cut from the hemisphere $x^2 + y^2 + z^2 = 2$, $z \geq 0$, by the cylinder $x^2 + y^2 = 1$ (Fig. 14.40).

Solution The cap S is part of the level surface $f(x, y, z) = x^2 + y^2 + z^2 = 2$. It projects one-to-one onto the disk $R: x^2 + y^2 \leq 1$ in the xy-plane. The vector $\mathbf{p} = \mathbf{k}$ is normal to the plane of R.

At any point on the surface,

$$f(x, y, z) = x^2 + y^2 + z^2,$$

$$\nabla f = 2x\mathbf{i} + 2y\mathbf{j} + 2z\mathbf{k}$$

$$|\nabla f| = 2\sqrt{x^2 + y^2 + z^2} = 2\sqrt{2}, \quad \left(\begin{array}{l}\text{Because } x^2 + y^2 + z^2 = 2\\ \text{at points of } S\end{array}\right)$$

$$|\nabla f \cdot \mathbf{p}| = |\nabla f \cdot \mathbf{k}| = |2z| = 2z.$$

Therefore,

$$\text{Surface area} = \iint\limits_R \frac{|\nabla f|}{|\nabla f \cdot \mathbf{p}|}\, dA = \iint\limits_R \frac{2\sqrt{2}}{2z}\, dA = \sqrt{2}\iint\limits_R \frac{dA}{z}. \tag{6}$$

What do we do about the z?

Since z is the z-coordinate of a point on the sphere, we can express it in terms of x and y as

$$z = \sqrt{2 - x^2 - y^2}.$$

We continue the work of Eq. (6) with this substitution:

$$\begin{aligned}
\text{Surface area} &= \sqrt{2}\iint\limits_R \frac{dA}{z} = \sqrt{2}\iint\limits_{x^2+y^2\le 1} \frac{dA}{\sqrt{2 - x^2 - y^2}} \\
&= \sqrt{2}\int_0^{2\pi}\int_0^1 \frac{r\,dr\,d\theta}{\sqrt{2 - r^2}} \qquad \text{(Polar coordinates)} \\
&= \sqrt{2}\int_0^{2\pi}\left[-(2 - r^2)^{1/2}\right]_{r=0}^{r=1} d\theta = \sqrt{2}\int_0^{2\pi}\left(\sqrt{2} - 1\right) d\theta \\
&= 2\pi\left(2 - \sqrt{2}\right).
\end{aligned}$$

Surface Integrals

We now show how to integrate a function over a surface, using the ideas we just developed for calculating surface area.

Suppose, for example, that we have an electrical charge distributed over a surface $f(x, y, z) = c$ like the one shown in Fig. 14.41 and that the function $g(x, y, z)$ gives the charge per unit area (charge density) at each point on S. Then we may calculate the total charge on S as an integral in the following way.

We partition the shadow region R on the ground plane beneath the surface into small rectangles of the kind we would use if we were defining the surface area of S. Then directly above each ΔA_k lies a patch of surface $\Delta\sigma_k$ that we approximate with a parallelogram-shaped portion of tangent plane, ΔP_k.

Up to this point the construction proceeds as in the definition of surface area, but now we take one additional step: We evaluate g at (x_k, y_k, z_k) and approximate the total charge on the surface patch $\Delta\sigma_k$ by the product

$$g(x_k, y_k, z_k)\Delta P_k.$$

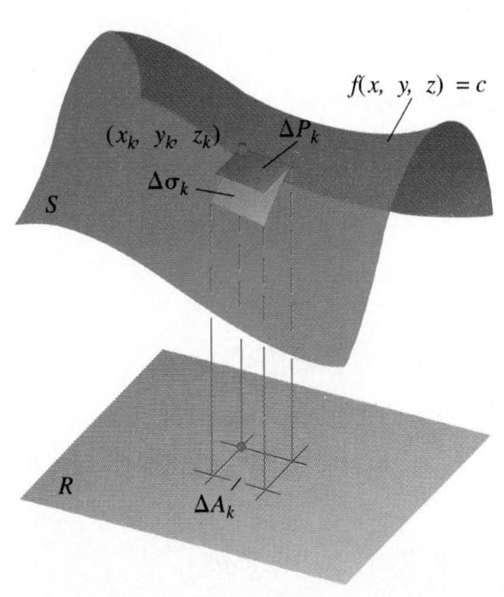

14.41 If we know how an electrical charge is distributed over a surface, we can find the total charge with a suitably modified surface integral.

The rationale is that when the partition of R is sufficiently fine, the value of g throughout $\Delta\sigma_k$ is nearly constant and ΔP_k is nearly the same as $\Delta\sigma_k$. The total charge over S is then approximated by the sum

$$\text{Total charge} \approx \sum g(x_k, y_k, z_k)\,\Delta P_k = \sum g(x_k, y_k, z_k)\,\frac{\Delta A_k}{|\cos\gamma_k|}. \tag{7}$$

If f, the function defining the surface S, and its first partial derivatives are continuous, and if g is continuous over S, then the sums on the right-hand side of Eq. (7) approach the limit

$$\iint_R g(x, y, z)\frac{dA}{|\cos\gamma|} = \iint_R g(x, y, z)\frac{|\nabla f|}{|\nabla f\cdot \mathbf{p}|}\,dA \tag{8}$$

as the rectangular subdivision of R is refined in the usual way. This limit is called the integral of g over the surface S and is calculated as a double integral over R. The value of the integral is the total charge on the surface S.

As you might expect, the formula in Eq. (8) defines the integral of *any* function g over the surface S as long as the integral exists.

DEFINITIONS

If R is the shadow region of a surface S defined by the equation $f(x, y, z) = c$, and g is a continuous function defined at the points of S, then the **integral of g over S** is the integral

$$\iint_R g(x, y, z)\frac{|\nabla f|}{|\nabla f\cdot \mathbf{p}|}\,dA, \tag{9}$$

where $\mathbf{p}$ is a unit vector normal to R and $\nabla f\cdot\mathbf{p}\neq 0$. The integral itself is called a **surface integral.**

The surface integral in (9) takes on different meanings in different applications. If g has the constant value 1, the integral gives the area of S. If g gives the mass density of a thin shell of material modeled by S, the integral gives the mass of the shell.

Algebraic Properties: The Surface Area Differential

We often abbreviate the integral in (9) by writing $d\sigma$ for $(|\nabla f|/|\nabla f\cdot\mathbf{p}|)\,dA$.

The Surface Area Differential and the Differential Form for Surface Integrals

$$d\sigma = \frac{|\nabla f|}{|\nabla f\cdot\mathbf{p}|}\,dA \qquad \iint_S g\,d\sigma \tag{10}$$

surface area differential — differential formula for surface integrals

Surface integrals have all the usual properties of double integrals, the surface integral of the sum of two functions being the sum of their surface integrals and so on. The domain additivity property takes the form

$$\iint_S g\,d\sigma = \iint_{S_1} g\,d\sigma + \iint_{S_2} g\,d\sigma + \cdots + \iint_{S_n} g\,d\sigma.$$

The idea is that if S is partitioned by smooth curves into a finite number of nonoverlapping smooth patches (i.e., if S is **piecewise smooth**), then the integral of a function g over S is the sum of the integrals over the patches. Thus, the integral of a

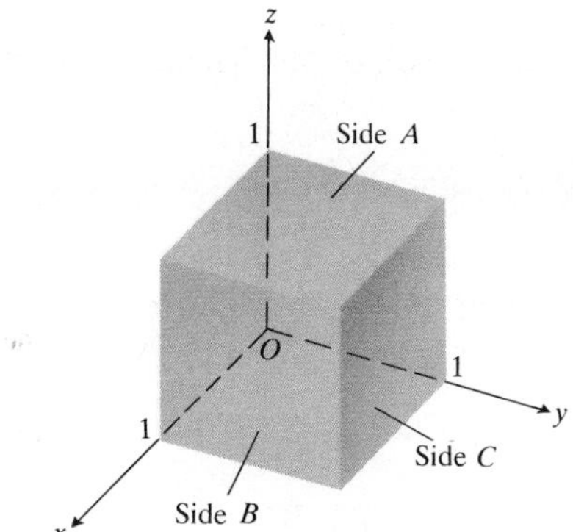

14.42 To integrate a function over the surface of a cube, we integrate over each face and add the results (Example 3).

function over the surface of a cube is the sum of the integrals over the faces of the cube. We integrate over a turtle shell of welded plates by integrating one plate at a time and adding the results.

Example 3 Integrate $g(x, y, z) = xyz$ over the surface of the cube cut from the first octant by the planes $x = 1$, $y = 1$, and $z = 1$ (Fig. 14.42).

Solution We integrate xyz over each of the six sides and add the results. Since $xyz = 0$ on the sides that lie in the coordinate planes, the integral over the surface of the cube reduces to

$$\iint_{\substack{\text{cube}\\ \text{surface}}} xyz\, d\sigma = \iint_{\text{side } A} xyz\, d\sigma + \iint_{\text{side } B} xyz\, d\sigma + \iint_{\text{side } C} xyz\, d\sigma.$$

Side A is the surface $f(x, y, z) = z = 1$ over the square region R_{xy}: $0 \le x \le 1$, $0 \le y \le 1$, in the xy-plane. For this surface and region,

$$\mathbf{p} = \mathbf{k}, \qquad \nabla f = \mathbf{k}, \qquad |\nabla f| = 1, \qquad |\nabla f \cdot \mathbf{p}| = |\mathbf{k} \cdot \mathbf{k}| = 1,$$

$$d\sigma = \frac{|\nabla f|}{|\nabla f \cdot \mathbf{p}|}\, dA = \frac{1}{1}\, dx\, dy = dx\, dy,$$

$$xyz = xy(1) = xy,$$

and

$$\iint_{\text{side } A} xyz\, d\sigma = \iint_{R_{xy}} xy\, dx\, dy = \int_0^1 \int_0^1 xy\, dx\, dy = \int_0^1 \frac{y}{2}\, dy = \frac{1}{4}.$$

Symmetry tells us that the integrals of xyz over sides B and C are also 1/4. Hence,

$$\iint_{\substack{\text{cube}\\ \text{surface}}} xyz\, d\sigma = \frac{1}{4} + \frac{1}{4} + \frac{1}{4} = \frac{3}{4}.$$

Orientation

We call a smooth surface S **orientable** or **two-sided** if it is possible to define a field $\mathbf{n}$ of unit normal vectors on S that varies continuously with position. Any patch or subportion of an orientable surface is orientable. Spheres and other smooth closed surfaces in space (smooth surfaces that enclose solids) are orientable. By convention, we choose $\mathbf{n}$ on a closed surface to point outward.

Once $\mathbf{n}$ has been chosen, we say that we have **oriented** the surface, and we call the surface together with its normal field an **oriented surface.** The vector $\mathbf{n}$ at any point is called the **positive direction** at that point (Fig. 14.43).

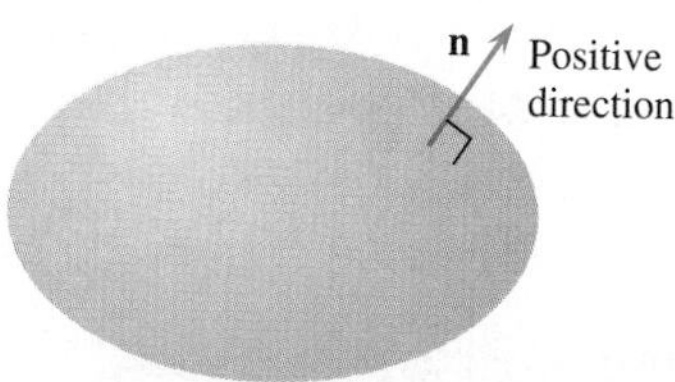

14.43 Smooth closed surfaces in space are orientable. The outward unit normal vector defines the positive direction at each point.

The Möbius band in Fig. 14.44 is not orientable. No matter where we start to construct a continuous unit normal field (shown as the shaft of a thumbtack in the figure), moving the vector continuously around the surface in the manner shown will return it to the starting point with a direction opposite to the one it had when it started out. The vector at that point cannot point both ways and yet it must if the field is to be continuous. We conclude that no such field exists.

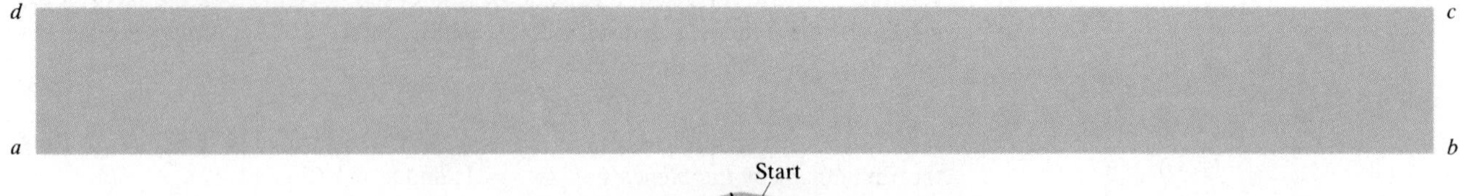

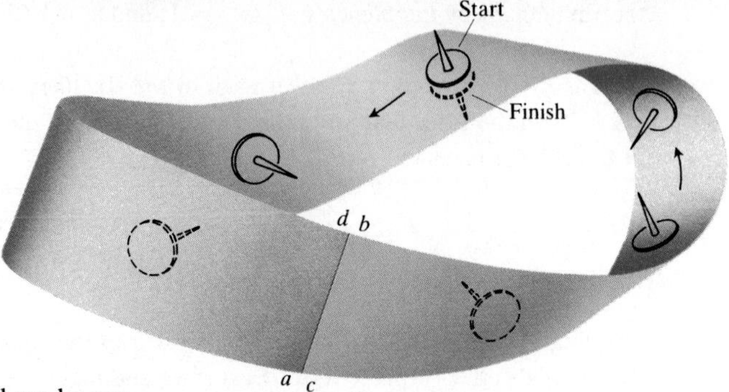

14.44 The Möbius band can be constructed by taking a rectangular strip of paper $abcd$, giving the end bc a single twist to interchange the positions of the vertices b and c, and then pasting the ends of the strip together to match a with c and b with d. The Möbius band is a nonorientable or one-sided surface.

The Surface Integral for Flux

Suppose that $\mathbf{F}$ is a continuous vector field defined over an oriented surface S and that $\mathbf{n}$ is the chosen unit normal field on the surface. We call the integral of $\mathbf{F}\cdot\mathbf{n}$ over S the flux across S in the positive direction. Thus, the flux is the integral over S of the scalar component of $\mathbf{F}$ in the direction of $\mathbf{n}$.

DEFINITION

The **flux** of a three-dimensional vector field $\mathbf{F}$ across an oriented surface S in the direction of $\mathbf{n}$ is given by the formula

$$\text{Flux} = \iint_S \mathbf{F}\cdot\mathbf{n}\,d\sigma. \tag{11}$$

This definition is analogous to the flux of a two-dimensional field $\mathbf{F}$ across a plane curve C. In the plane (Section 14.2), the flux is

$$\int_C \mathbf{F}\cdot\mathbf{n}\,ds,$$

the integral of the scalar component of $\mathbf{F}$ normal to the curve.

If $\mathbf{F}$ is the velocity field of a three-dimensional fluid flow, the flux of $\mathbf{F}$ across S is the net rate at which fluid is crossing S in the chosen positive direction. We shall discuss such flows in more detail in Section 14.5.

If $\mathbf{S}$ is part of a level surface $g(x, y, z) = c$, then $\mathbf{n}$ may be taken to be one of the two fields

$$\mathbf{n} = \pm\frac{\nabla g}{|\nabla g|}, \tag{12}$$

depending on which field gives the preferred direction.

Example 4 Find the flux of $\mathbf{F} = yz\mathbf{j} + z^2\mathbf{k}$ outward through the surface S cut from the semicircular cylinder $y^2 + z^2 = 1$, $z \geq 0$, by the planes $x = 0$ and $x = 1$.

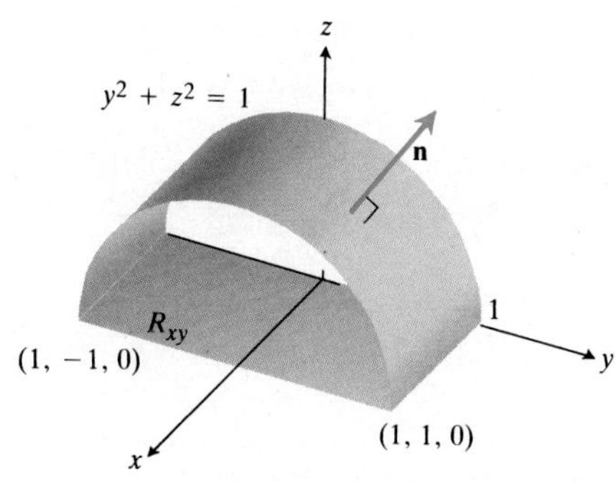

14.45 Example 4 calculates the flux of a vector field outward through this surface. The area of the shadow region R_{xy} is 2.

Solution The outward normal field on S (Fig. 14.45) may be calculated from the gradient of $g(x, y, z) = y^2 + z^2$ to be

$$\mathbf{n} = +\frac{\nabla g}{|\nabla g|} = \frac{2y\mathbf{j} + 2z\mathbf{k}}{\sqrt{4y^2 + 4z^2}} = \frac{2y\mathbf{j} + 2z\mathbf{k}}{2\sqrt{1}} = y\mathbf{j} + z\mathbf{k}.$$

With $\mathbf{p} = \mathbf{k}$, we also have

$$d\sigma = \frac{|\nabla g|}{|\nabla g \cdot \mathbf{k}|}\,dA = \frac{2}{|2z|}\,dA = \frac{1}{z}\,dA.$$

We can drop the absolute value bars because $z \geq 0$ on S.

The value of $\mathbf{F} \cdot \mathbf{n}$ on the surface is given by the formula

$$\begin{aligned} \mathbf{F} \cdot \mathbf{n} &= (yz\mathbf{j} + z^2\mathbf{k}) \cdot (y\mathbf{j} + z\mathbf{k}) \\ &= y^2z + z^3 = z(y^2 + z^2) \\ &= z. && (y^2 + z^2 = 1 \text{ on } S) \end{aligned}$$

Therefore, the flux of $\mathbf{F}$ outward through S is

$$\iint_S \mathbf{F} \cdot \mathbf{n}\,d\sigma = \iint_S (z)\left(\frac{1}{z}\,dA\right) = \iint_{R_{xy}} dA = \text{area}\,(R_{xy}) = 2.$$

Moments and Masses of Thin Shells

In engineering and physics, thin shells of material like bowls, metal drums, and domes are modeled with surfaces. Their moments and masses are calculated with the surface integral formulas in Table 14.2.

TABLE 14.2
Mass and moment formulas for very thin shells

Mass: $M = \iint_S \delta(x, y, z)\,d\sigma \qquad (\delta(x, y, z) = \text{density at } (x, y, z))$

First moments about the coordinate planes:

$$M_{yz} = \iint_S x\,\delta\,d\sigma, \qquad M_{xz} = \iint_S y\,\delta\,d\sigma, \qquad M_{xy} = \iint_S z\,\delta\,d\sigma$$

Coordinates of center of mass:

$$\bar{x} = M_{yz}/M, \qquad \bar{y} = M_{xz}/M, \qquad \bar{z} = M_{xy}/M$$

Moments of inertia:

$$I_x = \iint_S (y^2 + z^2)\,\delta\,d\sigma, \qquad I_y = \iint_S (x^2 + z^2)\,\delta\,d\sigma,$$

$$I_z = \iint_S (x^2 + y^2)\,\delta\,d\sigma, \qquad I_L = \iint_S r^2\,\delta\,d\sigma$$

$r(x, y, z) = \text{distance from point } (x, y, z) \text{ to line } L$

Radius of gyration about a line L: $R_L = \sqrt{I_L/M}$

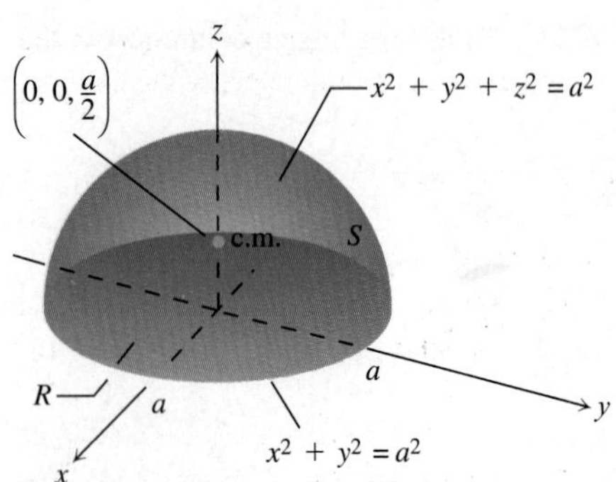

14.46 The center of mass of a thin hemispherical shell of constant density lies on the axis of symmetry halfway from the base to the top (Example 5).

Example 5 Find the center of mass of a thin hemispherical shell of radius a and constant density δ.

Solution We model the shell with the hemisphere

$$f(x, y, z) = x^2 + y^2 + z^2 = a^2, \qquad z \ge 0$$

(Fig. 14.46). The symmetry of the surface about the z-axis tells us that $\bar{x} = \bar{y} = 0$. It remains only to find $\bar{z}$ from the formula $\bar{z} = M_{xy}/M$.

The mass of the shell is

$$M = \iint_S \delta \, d\sigma = \delta \iint_S d\sigma = (\delta)(\text{area of } S) = 2\pi a^2 \delta.$$

To evaluate the integral for M_{xy}, we take $\mathbf{p} = \mathbf{k}$ and calculate

$$|\nabla f| = |2x\mathbf{i} + 2y\mathbf{j} + 2z\mathbf{k}| = 2\sqrt{x^2 + y^2 + z^2} = 2a,$$

$$|\nabla f \cdot \mathbf{p}| = |\nabla f \cdot \mathbf{k}| = |2z| = 2z,$$

$$d\sigma = \frac{|\nabla f|}{|\nabla f \cdot \mathbf{p}|} \, dA = \frac{a}{z} \, dA.$$

Then

$$M_{xy} = \iint_S z \, \delta \, d\sigma = \delta \iint_R z \frac{a}{z} \, dA = \delta a \iint_R dA = \delta a(\pi a^2) = \delta \pi a^3,$$

$$\bar{z} = \frac{M_{xy}}{M} = \frac{\pi a^3 \delta}{2\pi a^2 \delta} = \frac{a}{2}.$$

The shell's center of mass is the point $(0, 0, a/2)$.

EXERCISES 14.4

Surface Area

1. Find the area of the surface cut from the paraboloid $x^2 + y^2 - z = 0$ by the plane $z = 2$.
2. Find the area of the band cut from the paraboloid $x^2 + y^2 - z = 0$ by the planes $z = 2$ and $z = 6$.
3. Find the area of the region cut from the plane $x + 2y + 2z = 5$ by the cylinder whose walls are $x = y^2$ and $x = 2 - y^2$.
4. Find the area of the portion of the surface $x^2 - 2z = 0$ that lies above the triangle bounded by the lines $x = \sqrt{3}$, $y = 0$, and $y = x$ in the xy-plane.
5. Find the area of the surface $x^2 - 2y - 2z = 0$ that lies above the triangle bounded by the lines $x = 2$, $y = 0$, and $y = 3x$ in the xy-plane.
6. Find the area of the surface $x^2 - 2 \ln x + \sqrt{15}y - z = 0$ above the square $R: 0 \le y \le 1, 1 \le x \le 2$, in the xy-plane.
7. Find the area of the cap cut from the sphere $x^2 + y^2 + z^2 = 2$ by the cone $z = \sqrt{x^2 + y^2}$.
8. Find the area of the ellipse cut from the plane $z = cx$ by the cylinder $x^2 + y^2 = 1$.
9. Find the area of the portion of the cylinder $x^2 + z^2 = 1$ that lies between the planes $x = \pm 1/2$ and $y = \pm 1/2$.
10. Find the area of the portion of the surface $x = 9 - y^2 - z^2$ that lies above the ring $1 \le y^2 + z^2 \le 9$ in the yz-plane.
11. Find the area of the surface cut from the paraboloid $x^2 + y + z^2 = 1$ by the plane $y = 0$. (*Hint:* Project the surface on the xz-plane.)
12. Whenever we replace an old definition with a new one, it is a good idea to try out the new one on familiar objects to see that it gives the answers we expect. For instance, is the surface area of a circular cylinder of base radius a and height h still $2\pi ah$? Find out by using Eq. (5) with $\mathbf{p} = \mathbf{k}$ to calcu-

late the area of the surface cut from the cylinder $y^2 + z^2 = a^2$ by the planes $x = 0$ and $x = h > 0$. (Double the area above the xy-plane.)

Surface Integrals

13. Integrate $g(x, y, z) = x + y + z$ over the surface of the cube cut from the first octant by the planes $x = a$, $y = a$, $z = a$.

14. Integrate $g(x, y, z) = y + z$ over the surface of the wedge in Fig. 14.47.

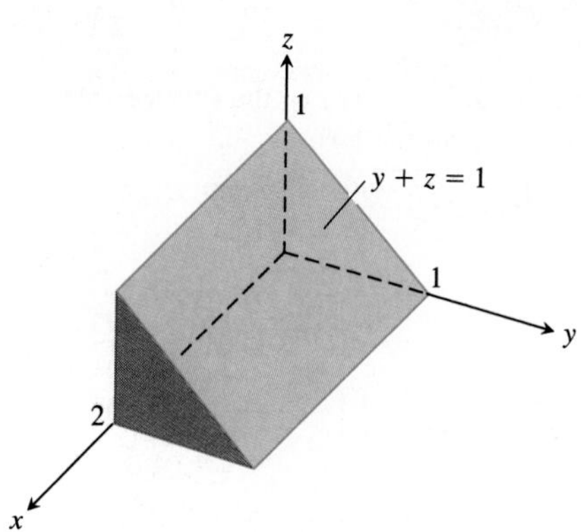

14.47 The wedge in Exercise 14.

15. Integrate $g(x, y, z) = xyz$ over the surface of the rectangular solid cut from the first octant by the planes $x = a$, $y = b$, and $z = c$.

16. Integrate $g(x, y, z) = xyz$ over the surface of the rectangular solid bounded by the planes $x = \pm a$, $y = \pm b$, and $z = \pm c$.

17. Integrate $g(x, y, z) = x + y + z$ over the portion of the plane $2x + 2y + z = 2$ that lies in the first octant.

18. Integrate $g(x, y, z) = x\sqrt{y^2 + 1}$ over the surface cut from the paraboloid $y^2 + 4z = 16$ by the planes $x = 0$, $x = 1$, and $z = 0$.

Flux Across a Surface

In Exercises 19–24, find the flux of the field **F** across the portion of the sphere $x^2 + y^2 + z^2 = a^2$ in the first octant in the direction away from the origin.

19. $\mathbf{F}(x, y, z) = z\mathbf{k}$

20. $\mathbf{F}(x, y, z) = -y\mathbf{i} + x\mathbf{j}$

21. $\mathbf{F}(x, y, z) = y\mathbf{i} - x\mathbf{j} + \mathbf{k}$

22. $\mathbf{F}(x, y, z) = zx\mathbf{i} + zy\mathbf{j} + z^2\mathbf{k}$

23. $\mathbf{F}(x, y, z) = x\mathbf{i} + y\mathbf{j} + z\mathbf{k}$

24. $\mathbf{F}(x, y, z) = \dfrac{x\mathbf{i} + y\mathbf{j} + z\mathbf{k}}{\sqrt{x^2 + y^2 + z^2}}$

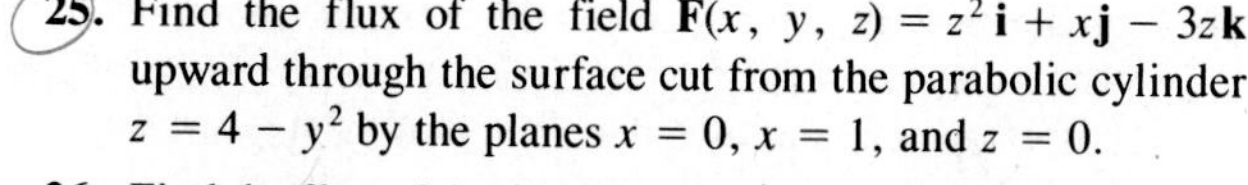

25. Find the flux of the field $\mathbf{F}(x, y, z) = z^2\mathbf{i} + x\mathbf{j} - 3z\mathbf{k}$ upward through the surface cut from the parabolic cylinder $z = 4 - y^2$ by the planes $x = 0$, $x = 1$, and $z = 0$.

26. Find the flux of the field $\mathbf{F}(x, y, z) = 4x\mathbf{i} + 4y\mathbf{j} + 2\mathbf{k}$ outward through the surface cut from the bottom of the paraboloid $z = x^2 + y^2$ by the plane $z = 1$.

27. Let S be the portion of the cylinder $y = e^x$ in the first octant that projects parallel to the x-axis onto the rectangle R_{yz}: $1 \le y \le 2$, $0 \le z \le 1$ in the yz-plane (Fig. 14.48). Let **n** be the unit vector normal to S that points away from the yz-plane. Find the flux of the field $\mathbf{F}(x, y, z) = -2\mathbf{i} + 2y\mathbf{j} + z\mathbf{k}$ across S in the direction of **n**.

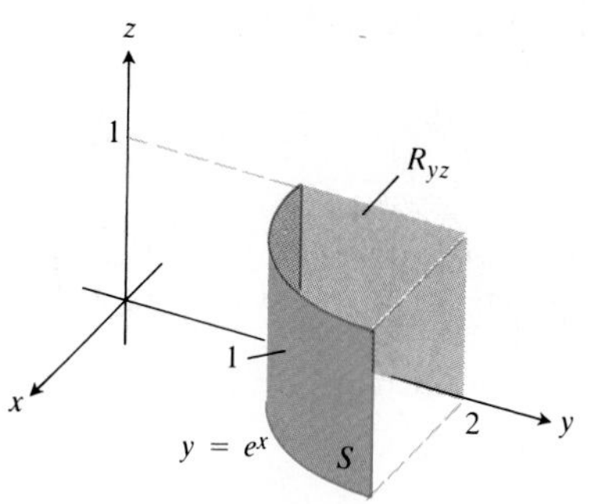

14.48 The surface and region in Exercise 27.

28. Let S be the portion of the cylinder $y = \ln x$ in the first octant whose projection parallel to the y-axis onto the xz-plane is the rectangle R_{xz}: $1 \le x \le e$, $0 \le z \le 1$. Let **n** be the unit vector normal to S that points away from the xz-plane. Find the flux of $\mathbf{F} = 2y\mathbf{j} + z\mathbf{k}$ through S in the direction of **n**.

29. Find the outward flux of the field $\mathbf{F} = 2xy\mathbf{i} + 2yz\mathbf{j} + 2xz\mathbf{k}$ across the surface of the cube cut from the first octant by the planes $x = a$, $y = a$, $z = a$.

30. Find the outward flux of the field $\mathbf{F} = xz\mathbf{i} + yz\mathbf{j} + \mathbf{k}$ across the surface of the upper cap cut from the solid sphere $x^2 + y^2 + z^2 \le 25$ by the plane $z = 3$.

Moments and Masses

31. Find the centroid of the portion of the sphere $x^2 + y^2 + z^2 = a^2$ that lies in the first octant.

32. Find the centroid of the surface cut from the cylinder $y^2 + z^2 = 9$, $z \ge 0$, by the planes $x = 0$ and $x = 3$ (resembles the surface in Example 4).

33. Find the center of mass and the moment of inertia and radius of gyration about the z-axis of a thin shell of constant density δ cut from the cone $x^2 + y^2 - z^2 = 0$ by the planes $z = 1$ and $z = 2$.

34. Find the moment of inertia about the z-axis of a thin shell of constant density δ cut from the cone $4x^2 + 4y^2 - z^2 = 0$, $z \geq 0$, by the circular cylinder $x^2 + y^2 = 2x$ (Fig. 14.49).

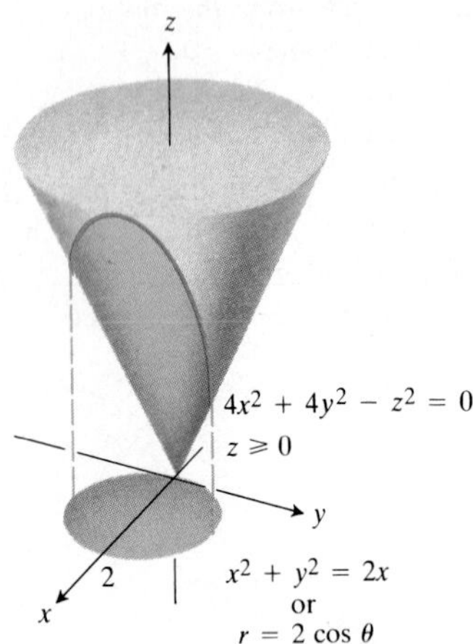

14.49 The surface in Exercise 34.

35. a) Find the moment of inertia about a diameter of a thin spherical shell of radius a and constant density δ. (Work with a hemispherical shell and double the result.)

b) Use the Parallel Axis Theorem (Exercises 13.5) and the result in (a) to find the moment of inertia about a line tangent to the shell.

36. a) Find the centroid of the lateral surface of a solid cone of base radius a and height h (cone surface minus the base).

b) Use Pappus's formula (Exercises 13.5) and the result in (a) to find the centroid of the complete surface of a solid cone (side plus base).

c) A cone of radius a and height h is joined to a hemisphere of radius a to make a surface S that resembles an ice cream cone. Use Pappus's formula and the results in (a) and Example 5 to find the centroid of S. How high does the cone have to be to place the centroid in the plane shared by the bases of the hemisphere and cone?

Special Formulas for Surface Area

If S is the surface defined by a function $z = f(x, y)$ that has continuous first partial derivatives throughout a region R_{xy} in the xy-plane (Fig. 14.50), then S is also the level surface $F(x, y, z) = 0$ of the function $F(x, y, z) = f(x, y) - z$. Taking the unit normal to R_{xy} to be $\mathbf{p} = \mathbf{k}$ then gives

$$|\nabla F| = |f_x\,\mathbf{i} + f_y\,\mathbf{j} - \mathbf{k}| = \sqrt{f_x^{\,2} + f_y^{\,2} + 1},$$
$$|\nabla F \cdot \mathbf{p}| = |(f_x\mathbf{i} + f_y\mathbf{j} - \mathbf{k}) \cdot \mathbf{k}| = |-1| = 1,$$

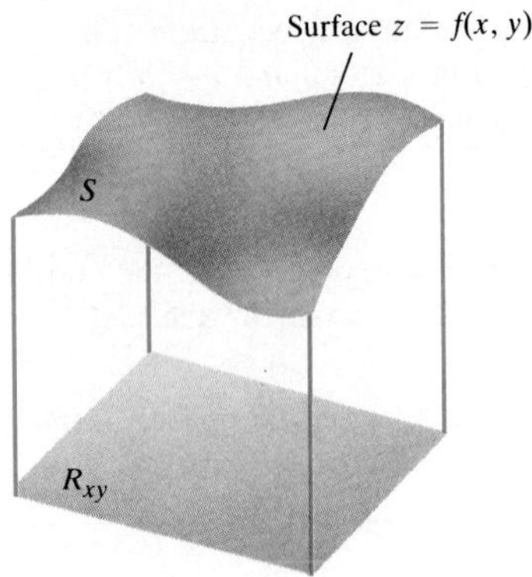

14.50 For a surface $z = f(x, y)$, the surface area formula in Eq. (5) takes the form
$$S = \iint_{R_{xy}} \sqrt{f_x^2 + f_y^2 + 1}\,dx\,dy.$$

and

$$\text{Area of } S = \iint_{R_{xy}} \frac{|\nabla F|}{|\nabla F \cdot \mathbf{p}|}\,dA = \iint_{R_{xy}} \sqrt{f_x^{\,2} + f_y^{\,2} + 1}\,dx\,dy. \tag{13}$$

Similarly, the area of a smooth surface $x = f(y, z)$ over a region R_{yz} in the yz-plane is

$$\text{Area of } S = \iint_{R_{yz}} \sqrt{f_y^{\,2} + f_z^{\,2} + 1}\,dy\,dz, \tag{14}$$

and the area of a smooth surface $y = f(x, z)$ over a region R_{xz} in the xz-plane is

$$\text{Area of } S = \iint_{R_{xz}} \sqrt{f_x^{\,2} + f_z^{\,2} + 1}\,dx\,dz. \tag{15}$$

Use Eqs. (13)–(15) to find the areas of the surfaces in Exercises 37–42.

37. The surface cut from the bottom of the paraboloid $z = x^2 + y^2$ by the plane $z = \sqrt{3}$.

38. The portion of the cone $z = \sqrt{x^2 + y^2}$ that lies over the region between the circle $x^2 + y^2 = 1$ and the ellipse $9x^2 + 4y^2 = 36$ in the xy-plane. (*Hint:* Use formulas from geometry to find the area of the region.)

39. The surface cut from the bottom of the paraboloid $x = 1 - y^2 - z^2$ by the yz-plane.

40. The triangle cut from the plane $2x + 6y + 3z = 6$ by the bounding planes of the first octant. Calculate the area three ways, once with each area formula.

41. The surface in the first octant cut from the cylinder $2y = z^2$ by the planes $x = 1$ and $y = 2$.

42. The portion of the plane $y + z = 4$ that lies above the region cut from the first quadrant of the xz-plane by the parabola $x = 4 - z^2$.

14.5 The Divergence Theorem

The divergence form of Green's theorem in the plane states that the net outward flux of a vector field across a simple closed curve in the plane can be calculated by integrating the divergence of the field over the region enclosed by the curve. The corresponding theorem in three dimensions, called the Divergence Theorem, states that the net outward flux of a vector field across a closed surface in space can be calculated by integrating the divergence of the field over the region enclosed by the surface. In this section, we prove the Divergence Theorem and show how it simplifies the calculation of flux. We also derive Gauss's law for flux in an electric field. The three-dimensional version of the circulation form of Green's theorem will be treated in the next section.

Divergence in Three Dimensions

The **divergence** of a vector field $\mathbf{F} = M(x, y, z)\,\mathbf{i} + N(x, y, z)\,\mathbf{j} + P(x, y, z)\,\mathbf{k}$ is the scalar function

$$\operatorname{div} \mathbf{F} = \frac{\partial M}{\partial x} + \frac{\partial N}{\partial y} + \frac{\partial P}{\partial z}. \tag{1}$$

Div $\mathbf{F}$ has the same physical interpretation as the divergence of a two-dimensional vector field. When $\mathbf{F}$ is the velocity field of a fluid flow, the value of div $\mathbf{F}$ at a point (x, y, z) is the rate at which fluid is being piped in or drained away at (x, y, z). The divergence is the flux per unit volume or flux density at the point. We shall say more about this at the end of the section.

The Divergence Theorem

Mikhail Vassilievich Ostrogradsky (1801–1862) was the first mathematician to publish a proof of the Divergence Theorem. Having been denied his degree at Kharkhov University by the minister for religious affairs and national education (for being an atheist), Ostrogradsky left Russia for Paris in 1822. There he met Laplace, Legendre, Fourier, Poisson, and Cauchy. While working on the theory of heat in the mid-1820s, he formulated the Divergence Theorem as a tool for turning volume integrals into surface integrals.

Carl Friedrich Gauss (1777–1855) had already proved the theorem while working on the theory of gravitation, but his notebooks were not published until many years later. The theorem is sometimes called Gauss's theorem. The list of Gauss's accomplishments in science and mathematics is truly astonishing, ranging from the invention of the electric telegraph (with Wilhelm Weber in 1833) to the development of a wonderfully accurate theory of planetary orbits and to work in non-Euclidean geometry that later became fundamental to Einstein's general theory of relativity.

Del Notation

The divergence of a three-dimensional vector field $\mathbf{F}$ is usually expressed in terms of the symbolic operator ∇ ("del"), defined by the equation

$$\nabla = \mathbf{i}\frac{\partial}{\partial x} + \mathbf{j}\frac{\partial}{\partial y} + \mathbf{k}\frac{\partial}{\partial z}. \tag{2}$$

This operator can be applied to any differentiable vector field $\mathbf{F} = M\mathbf{i} + N\mathbf{j} + P\mathbf{k}$ to give the divergence of $\mathbf{F}$ as a symbolic dot product:

$$\nabla \cdot \mathbf{F} = \frac{\partial M}{\partial x} + \frac{\partial N}{\partial y} + \frac{\partial P}{\partial z} = \operatorname{div} \mathbf{F}. \tag{3}$$

The notation $\nabla \cdot \mathbf{F}$ is read "del dot $\mathbf{F}$."

When applied to a differentiable scalar function $f(x, y, z)$, the del operator gives the gradient of f:

$$\nabla f = \frac{\partial f}{\partial x}\mathbf{i} + \frac{\partial f}{\partial y}\mathbf{j} + \frac{\partial f}{\partial z}\mathbf{k}. \tag{4}$$

This may now be read as "del f" as well as "grad f."

Since

$$\frac{\partial^2 f}{\partial x^2} + \frac{\partial^2 f}{\partial y^2} + \frac{\partial^2 f}{\partial z^2} = \nabla \cdot \nabla f = \nabla^2 f, \tag{5}$$

Laplace's equation,

$$\frac{\partial^2 f}{\partial x^2} + \frac{\partial^2 f}{\partial y^2} + \frac{\partial^2 f}{\partial z^2} = 0,$$

(Exercises 12.3) can now be written as

$$\text{div grad} f = 0 \qquad \text{or} \qquad \nabla^2 f = 0. \tag{6}$$

If we think of a two-dimensional vector field

$$\mathbf{F}(x, y) = M(x, y)\,\mathbf{i} + N(x, y)\,\mathbf{j}$$

as a three-dimensional field whose **k**-component is zero, then

$$\nabla \cdot \mathbf{F} = \frac{\partial M}{\partial x} + \frac{\partial N}{\partial y} = \text{div}\,\mathbf{F}, \tag{7}$$

and the normal form of Green's theorem for two-dimensional fields (about flux and divergence; see Eq. 27 in Section 14.3) becomes

$$\int_C \mathbf{F} \cdot \mathbf{n}\, ds = \iint_R \nabla \cdot \mathbf{F}\, dA. \tag{8}$$

The tangential form of Green's theorem for two-dimensional fields can also be expressed in vector form in del notation, as we shall see in Section 14.6.

As these equations suggest, del notation provides an easy way to express and remember some of the most important relationships in mathematics, engineering, and physics. We shall see more of this as the chapter continues.

The Divergence Theorem

The Divergence Theorem says that under suitable conditions the outward flux of a vector field across a closed surface (oriented outward) equals the triple integral of the divergence of the field over the region enclosed by the surface.

THEOREM 4

The Divergence Theorem

The flux of a vector field $\mathbf{F} = M\mathbf{i} + N\mathbf{j} + P\mathbf{k}$ across a closed oriented surface S in the direction of the surface's outward unit normal field $\mathbf{n}$ equals the integral of $\nabla \cdot \mathbf{F}$ over the region D enclosed by the surface:

$$\underbrace{\iint_S \mathbf{F} \cdot \mathbf{n}\, d\sigma}_{\text{outward flux}} = \underbrace{\iiint_D \nabla \cdot \mathbf{F}\, dV}_{\text{divergence integral}}. \tag{9}$$

As you can see, Eq. (9) has the same form as Eq. (8), which gives the flux–divergence form of Green's theorem for two-dimensional fields. The only difference is that we are now working in three dimensions instead of two.

Proof for Special Regions

To prove the Divergence Theorem, we assume that the components of **F** have continuous partial derivatives. We also assume that D is a convex region with no holes

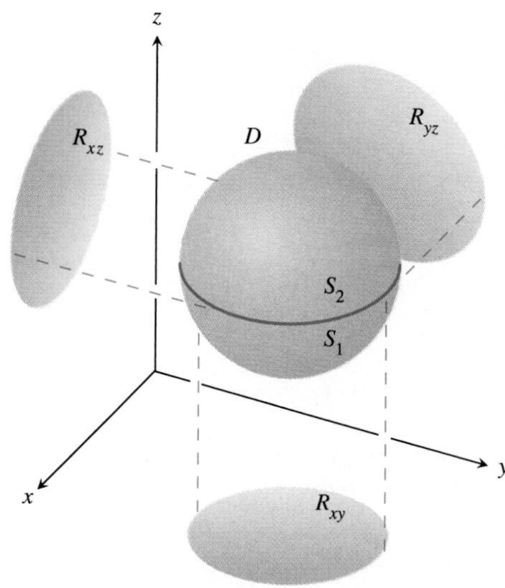

14.51 We first prove the Divergence Theorem for the kind of three-dimensional region shown here. We then extend the theorem to other regions.

or bubbles, such as a solid sphere, cube, or ellipsoid, and that S is a piecewise smooth surface. In addition, we assume that any line perpendicular to the xy-plane at an interior point of the region R_{xy} that is the projection of D on the xy-plane intersects the surface S in exactly two points, producing surfaces S_1 and S_2:

$$S_1: \quad z = f_1(x, y), \qquad (x, y) \text{ in } R_{xy},$$

$$S_2: \quad z = f_2(x, y), \qquad (x, y) \text{ in } R_{xy},$$

with $f_1 \leq f_2$. We make similar assumptions about the projection of D onto the other coordinate planes. See Fig. 14.51.

The components of the unit normal vector $\mathbf{n} = n_1\,\mathbf{i} + n_2\,\mathbf{j} + n_3\,\mathbf{k}$ are the cosines of the angles α, β, and γ that $\mathbf{n}$ makes with $\mathbf{i}$, $\mathbf{j}$, and $\mathbf{k}$ (Fig. 14.52). This is true because all the vectors involved are unit vectors. We have

$$\begin{aligned} n_1 &= \mathbf{n}\cdot\mathbf{i} = |\mathbf{n}|\,|\mathbf{i}| \cos\alpha = \cos\alpha, \\ n_2 &= \mathbf{n}\cdot\mathbf{j} = |\mathbf{n}|\,|\mathbf{j}| \cos\beta = \cos\beta, \\ n_3 &= \mathbf{n}\cdot\mathbf{k} = |\mathbf{n}|\,|\mathbf{k}| \cos\gamma = \cos\gamma. \end{aligned} \tag{10}$$

Thus,

$$\mathbf{n} = (\cos\alpha)\,\mathbf{i} + (\cos\beta)\,\mathbf{j} + (\cos\gamma)\,\mathbf{k}$$

and

$$\mathbf{F}\cdot\mathbf{n} = M\cos\alpha + N\cos\beta + P\cos\gamma.$$

In component form, the Divergence Theorem states that

$$\iint_S (M\cos\alpha + N\cos\beta + P\cos\gamma)\,d\sigma = \iiint_D \left(\frac{\partial M}{\partial x} + \frac{\partial N}{\partial y} + \frac{\partial P}{\partial z}\right) dx\,dy\,dz. \tag{11}$$

We prove the theorem by proving three separate equalities that add up to Eq. (11):

$$\iint M\cos\alpha\,d\sigma = \iiint \frac{\partial M}{\partial x}\,dx\,dy\,dz, \tag{12}$$

$$\iint N\cos\beta\,d\sigma = \iiint \frac{\partial N}{\partial y}\,dx\,dy\,dz, \tag{13}$$

$$\iint P\cos\gamma\,d\sigma = \iiint \frac{\partial P}{\partial z}\,dx\,dy\,dz. \tag{14}$$

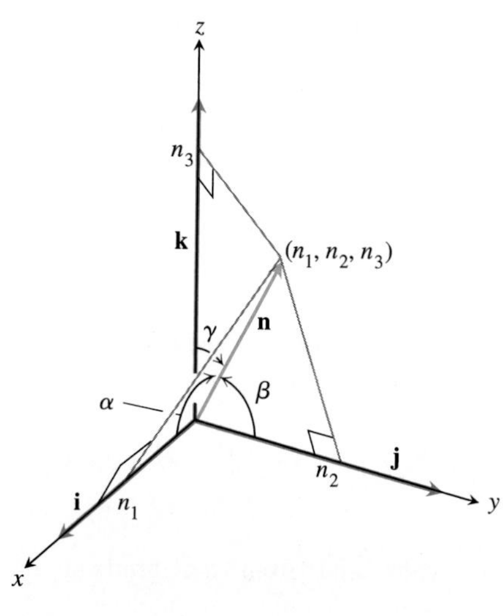

14.52 The scalar components of a unit normal vector **n** are the cosines of the angles α, β, and γ that it makes with **i**, **j**, and **k**.

We prove Eq. (14) by converting the surface integral on the left to a double integral over the projection R_{xy} of D on the xy-plane (Fig. 14.53). The surface S consists of an upper part S_2 whose equation is $z = f_2(x, y)$ and a lower part S_1 whose equation is $z = f_1(x, y)$. On S_2, the outer normal $\mathbf{n}$ has a positive $\mathbf{k}$-component and

$$\cos\gamma\,d\sigma = dx\,dy \quad \text{because} \quad d\sigma = \frac{dA}{|\cos\gamma|} = \frac{dx\,dy}{\cos\gamma}. \tag{15}$$

See Fig. 14.54. On S_1, the outer normal $\mathbf{n}$ has a negative $\mathbf{k}$-component and

$$\cos\gamma\,d\sigma = -dx\,dy. \tag{16}$$

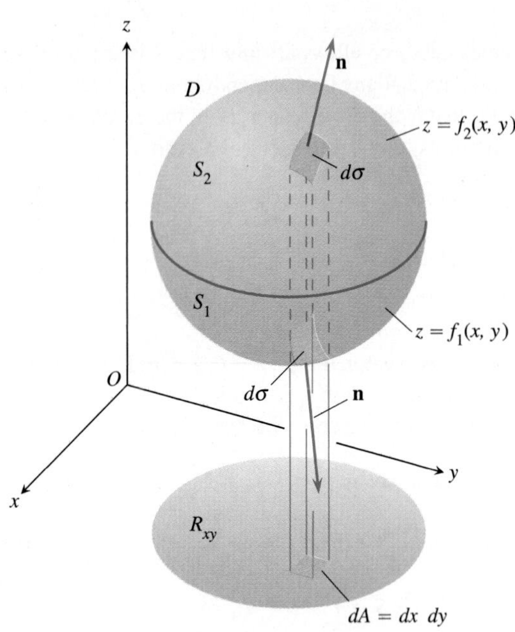

14.53 The three-dimensional region D enclosed by the surfaces S_1 and S_2 shown here projects vertically onto a two-dimensional region R_{xy} in the xy-plane.

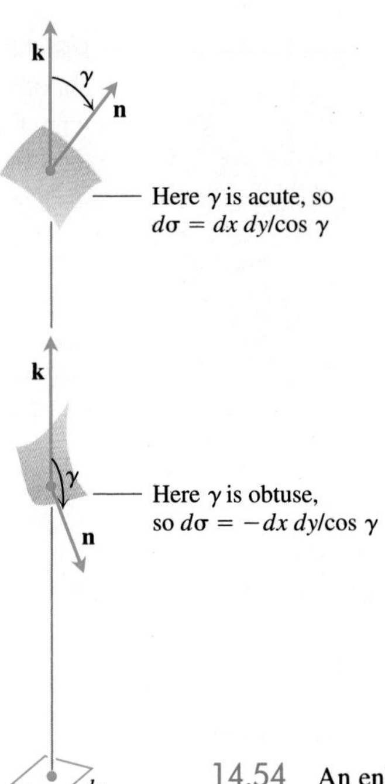

14.54 An enlarged view of the area patches in Fig. 14.53. The relations $d\sigma = \pm dx\, dy/\cos\gamma$ are derived in Section 14.4.

Therefore,

$$\begin{aligned}
\iint_S P\cos\gamma\, d\sigma &= \iint_{S_2} P\cos\gamma\, d\sigma + \iint_{S_1} P\cos\gamma\, d\sigma \\
&= \iint_{R_{xy}} P(x, y, f_2(x, y))\, dx\, dy - \iint_{R_{xy}} P(x, y, f_1(x, y))\, dx\, dy \\
&= \iint_{R_{xy}} [P(x, y, f_2(x, y)) - P(x, y, f_1(x, y))]\, dx\, dy \\
&= \iint_{R_{xy}} \left[\int_{f_1(x, y)}^{f_2(x, y)} \frac{\partial P}{\partial z}\, dz\right] dx\, dy = \iiint_D \frac{\partial P}{\partial z}\, dz\, dx\, dy.
\end{aligned} \tag{17}$$

This proves Eq. (14).

The proofs for Eqs. (12) and (13) follow the same pattern; or just permute x, y, z; M, N, P; α, β, γ, in order, and get those results from Eq. (14).

Example 1 Evaluate both sides of Eq. (9) for the field $\mathbf{F} = x\mathbf{i} + y\mathbf{j} + z\mathbf{k}$ over the sphere $x^2 + y^2 + z^2 = a^2$.

Solution The outer unit normal to S, calculated from the gradient of $f(x, y, z) = x^2 + y^2 + z^2 - a^2$, is

$$\mathbf{n} = \frac{2(x\mathbf{i} + y\mathbf{j} + z\mathbf{k})}{\sqrt{4(x^2 + y^2 + z^2)}} = \frac{x\mathbf{i} + y\mathbf{j} + z\mathbf{k}}{a}.$$

Hence,

$$\mathbf{F} \cdot \mathbf{n}\, d\sigma = \frac{x^2 + y^2 + z^2}{a}\, d\sigma = \frac{a^2}{a}\, d\sigma = a\, d\sigma$$

because $x^2 + y^2 + z^2 = a^2$ on the surface. Therefore

$$\iint_S \mathbf{F} \cdot \mathbf{n}\, d\sigma = \iint_S a\, d\sigma = a \iint_S d\sigma = a(4\pi a^2) = 4\pi a^3.$$

The divergence of $\mathbf{F}$ is

$$\nabla \cdot \mathbf{F} = \frac{\partial}{\partial x}(x) + \frac{\partial}{\partial y}(y) + \frac{\partial}{\partial z}(z) = 3,$$

so

$$\iiint_D \nabla \cdot \mathbf{F}\, dV = \iiint_D 3\, dV = 3\left(\frac{4}{3}\pi a^3\right) = 4\pi a^3.$$

Example 2 Calculate the flux of the field $\mathbf{F} = xy\mathbf{i} + yz\mathbf{j} + xz\mathbf{k}$ outward through the surface of the cube cut from the first octant by the planes $x = 1$, $y = 1$, and $z = 1$.

Solution The Divergence Theorem says that instead of calculating the flux as a sum of six separate integrals, one for each face of the cube, we can calculate the flux by integrating the divergence

$$\nabla \cdot \mathbf{F} = \frac{\partial}{\partial x}(xy) + \frac{\partial}{\partial y}(yz) + \frac{\partial}{\partial z}(xz) = y + z + x$$

over the cube's interior:

$$\begin{aligned}
\text{Flux} &= \iint_{\substack{\text{cube}\\\text{surface}}} \mathbf{F} \cdot \mathbf{n}\, d\sigma \\
&= \iiint_{\substack{\text{cube}\\\text{interior}}} \nabla \cdot \mathbf{F}\, dV && \text{(The Divergence Theorem)} \\
&= \int_0^1 \int_0^1 \int_0^1 (x + y + z)\, dx\, dy\, dz \\
&= \frac{3}{2}. && \text{(Routine integration)}
\end{aligned}$$

The Divergence Theorem for Other Regions

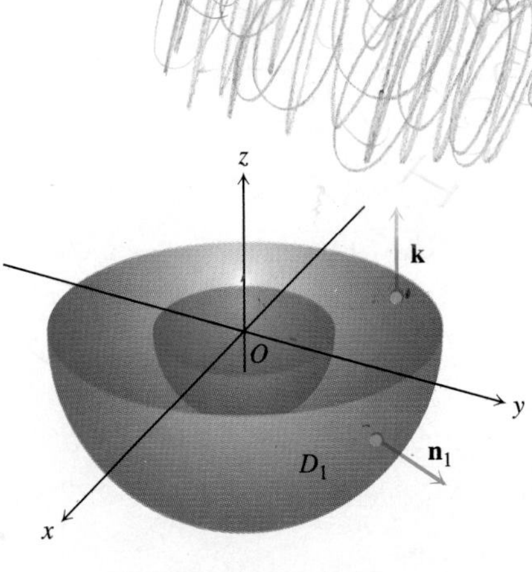

14.55 The lower half of the solid region between two concentric spheres.

The Divergence Theorem can be extended to regions that can be split up into a finite number of simple regions of the type just discussed and to regions that can be defined as limits of simpler regions in certain ways. For example, suppose that D is the region between two concentric spheres and that $\mathbf{F}$ has continuously differentiable components throughout D and on the bounding surfaces. Split D by an equatorial plane and apply the Divergence Theorem to each half separately. The bottom half, D_1, is shown in Fig. 14.55. The surface that bounds D_1 consists of an outer

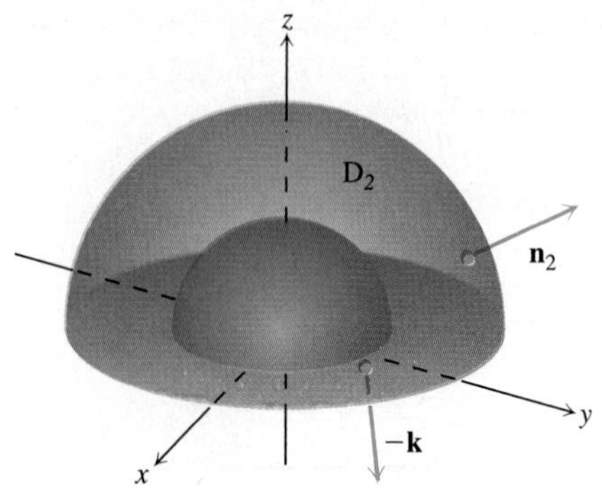

14.56 The upper half of the solid region between two concentric spheres.

hemisphere, a plane washer-shaped base, and an inner hemisphere. The Divergence Theorem says that

$$\iint_{S_1} \mathbf{F} \cdot \mathbf{n}_1 \, d\sigma_1 = \iiint_{D_1} \nabla \cdot \mathbf{F} \, dV_1. \tag{18}$$

The unit normal $\mathbf{n}_1$ that points outward from D_1 points away from the origin along the outer surface, equals $\mathbf{k}$ along the flat base, and points toward the origin along the inner surface. Next apply the Divergence Theorem to D_2, as shown in Fig. 14.56:

$$\iint_{S_2} \mathbf{F} \cdot \mathbf{n}_2 \, d\sigma_2 = \iiint_{D_2} \nabla \cdot \mathbf{F} \, dV_2. \tag{19}$$

As we follow $\mathbf{n}_2$ over S_2, pointing outward from D_2, we see that $\mathbf{n}_2$ equals $-\mathbf{k}$ along the washer-shaped base in the xy-plane, points away from the origin on the outer sphere, and points toward the origin on the inner sphere. When we add Eqs. (18) and (19), the surface integrals over the flat base cancel because of the opposite signs of $\mathbf{n}_1$ and $\mathbf{n}_2$. We thus arrive at the result

$$\iint_{S} \mathbf{F} \cdot \mathbf{n} \, d\sigma = \iiint_{D} \nabla \cdot \mathbf{F} \, dV, \tag{20}$$

with D the region between the spheres, S the boundary of D consisting of two spheres, and $\mathbf{n}$ the unit normal to S directed outward from D.

Example 3 Find the net outward flux of the field

$$\mathbf{F} = \frac{x\mathbf{i} + y\mathbf{j} + z\mathbf{k}}{\rho^3}, \qquad \rho = \sqrt{x^2 + y^2 + z^2}$$

across the boundary of the region $D: 0 < a^2 \le x^2 + y^2 + z^2 \le b^2$.

Solution The flux can be calculated by integrating $\nabla \cdot \mathbf{F}$ over D. We have

$$\frac{\partial \rho}{\partial x} = \frac{1}{2}(x^2 + y^2 + z^2)^{-1/2}(2x) = \frac{x}{\rho}$$

and

$$\frac{\partial M}{\partial x} = \frac{\partial}{\partial x}(x\rho^{-3}) = \rho^{-3} - 3x\rho^{-4}\frac{\partial \rho}{\partial x} = \frac{1}{\rho^3} - \frac{3x^2}{\rho^5}.$$

Similarly,

$$\frac{\partial N}{\partial y} = \frac{1}{\rho^3} - \frac{3y^2}{\rho^5} \quad \text{and} \quad \frac{\partial P}{\partial z} = \frac{1}{\rho^3} - \frac{3z^2}{\rho^5}.$$

Hence,

$$\text{div}\,\mathbf{F} = \frac{3}{\rho^3} - \frac{3}{\rho^5}(x^2 + y^2 + z^2) = \frac{3}{\rho^3} - \frac{3\rho^2}{\rho^5} = 0$$

and

$$\iiint_{D} \nabla \cdot \mathbf{F} \, dV = 0.$$

So the integral of $\nabla \cdot \mathbf{F}$ over D is zero and the net outward flux across the boundary of D is zero. But there is more to learn from this example. The flux leaving D across the inner sphere S_a is the negative of the flux leaving D across the outer sphere S_b (because the sum of these fluxes is zero). This means that the flux of **F** across S_a in the direction away from the origin equals the flux of **F** across S_b in the direction away from the origin. Thus, the flux of **F** across a sphere centered at the origin is independent of the radius of the sphere. What is this flux?

To find it, we evaluate the flux integral directly. The outward unit normal on the sphere of radius a is

$$\mathbf{n} = \frac{x\mathbf{i} + y\mathbf{j} + z\mathbf{k}}{\sqrt{x^2 + y^2 + z^2}} = \frac{x\mathbf{i} + y\mathbf{j} + z\mathbf{k}}{a}.$$

Hence, on the sphere,

$$\mathbf{F} \cdot \mathbf{n} = \frac{x\mathbf{i} + y\mathbf{j} + z\mathbf{k}}{a^3} \cdot \frac{x\mathbf{i} + y\mathbf{j} + z\mathbf{k}}{a} = \frac{x^2 + y^2 + z^2}{a^4} = \frac{a^2}{a^4} = \frac{1}{a^2}$$

and

$$\iint_{S_a} \mathbf{F} \cdot \mathbf{n}\, d\sigma = \frac{1}{a^2} \iint_{S_a} d\sigma = \frac{1}{a^2}(4\pi a^2) = 4\pi.$$

The outward flux of **F** across any sphere centered at the origin is 4π.

Gauss's Law—One of the Four Great Laws of Electromagnetic Theory

There is still more to be learned from Example 3. In electromagnetic theory, the electric field created by a point charge q located at the origin is the inverse square field

$$\mathbf{E}(x, y, z) = q\frac{\mathbf{r}}{|\mathbf{r}|^3} = q\frac{x\mathbf{i} + y\mathbf{j} + z\mathbf{k}}{\rho^3},$$

where **r** is the position vector of the point (x, y, z) and $\rho = |\mathbf{r}| = \sqrt{x^2 + y^2 + z^2}$. In the notation of Example 3,

$$\mathbf{E} = q\mathbf{F}.$$

The calculations in Example 3 show that the outward flux of **E** across any sphere centered at the origin is $4\pi q$. But this result is not confined to spheres. The outward flux of **E** across any surface S that encloses the origin (and to which the Divergence Theorem applies) is also $4\pi q$. To see why, we have only to imagine a large sphere S_a centered at the origin and enclosing the surface S. Since

$$\nabla \cdot \mathbf{E} = \nabla \cdot q\mathbf{F} = q\nabla \cdot \mathbf{F} = 0$$

when $\rho > 0$, the integral of $\nabla \cdot \mathbf{E}$ over the region D between S and S_a is zero. Hence, by the Divergence Theorem,

$$\iint_{\substack{\text{boundary}\\ \text{of } D}} \mathbf{E} \cdot \mathbf{n}\, d\sigma = 0,$$

and the flux of **E** across S in the direction away from the origin must be the same as the flux of **E** across S_a in the direction away from the origin, which is $4\pi q$. This statement, called *Gauss's law,* applies to much more general charge distributions than the one we have assumed here, as you will see in nearly any college physics text.

Gauss's Law

$$\iint_S \mathbf{E} \cdot \mathbf{n}\, d\sigma = 4\pi q \tag{21}$$

The Continuity Equation of Hydrodynamics

If $\mathbf{v}(x, y, z)$ is the velocity field of a fluid flowing smoothly through a region in space, $\delta = \delta(x, y, z, t)$ is the fluid's density at (x, y, z) at time t, and $\mathbf{F} = \delta\mathbf{v}$, then the **continuity equation** of hydrodynamics is the statement that

$$\nabla \cdot \mathbf{F} + \frac{\partial \delta}{\partial t} = 0. \tag{22}$$

The equation evolves naturally from the Divergence Theorem,

$$\iint_S \mathbf{F} \cdot \mathbf{n}\, d\sigma = \iiint_D \nabla \cdot \mathbf{F}\, dV, \tag{23}$$

if the functions involved have continuous first partial derivatives, as we shall now see.

First, the integral

$$\iint_S \mathbf{F} \cdot \mathbf{n}\, d\sigma$$

is the rate at which mass leaves D across S (leaves because $\mathbf{n}$ is the outer normal). To see why, consider a patch of area $\Delta\sigma$ on the surface (Fig. 14.57). In a short time interval Δt, the volume ΔV of fluid that flows across the patch is approximately equal to the volume of a cylinder with base area $\Delta\sigma$ and height $(\mathbf{v}\,\Delta t) \cdot \mathbf{n}$, where $\mathbf{v}$ is a velocity vector rooted at a point of the patch:

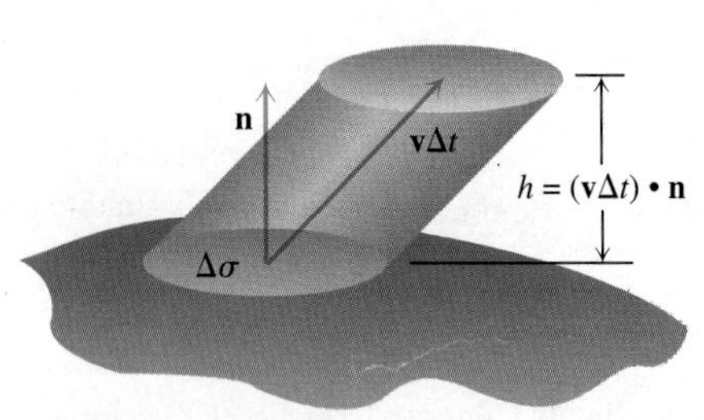

14.57 The fluid that flows upward through the patch $\Delta\sigma$ in a short time Δt fills a "cylinder" whose volume is approximately base × height = $\mathbf{v} \cdot \mathbf{n}\, \Delta\sigma\, \Delta t$.

$$\Delta V \approx \mathbf{v} \cdot \mathbf{n}\, \Delta\sigma\, \Delta t.$$

The mass of this volume of fluid is about

$$\Delta m \approx \delta\mathbf{v} \cdot \mathbf{n}\, \Delta\sigma\Delta t,$$

so the rate at which mass is flowing out of D across the patch is about

$$\frac{\Delta m}{\Delta t} \approx \delta\mathbf{v} \cdot \mathbf{n}\, \Delta\sigma.$$

This leads to the approximation

$$\frac{\Sigma\, \Delta m}{\Delta t} \approx \sum \delta\mathbf{v} \cdot \mathbf{n}\, \Delta\sigma \tag{24}$$

as an estimate of the average rate at which mass flows across S. Finally, letting $\Delta\sigma \to 0$ and $\Delta t \to 0$ gives the instantaneous rate at which mass leaves D across S as

$$\frac{dm}{dt} = \iint_S \delta\mathbf{v} \cdot \mathbf{n}\, d\sigma, \tag{25}$$

which for our particular flow is

$$\frac{dm}{dt} = \iint_S \mathbf{F} \cdot \mathbf{n}\, d\sigma. \tag{26}$$

Now let B be a solid sphere centered at a point Q in the flow. The average value of $\nabla \cdot \mathbf{F}$ over B is

$$\frac{1}{\text{volume of } B} \iiint_B \nabla \cdot \mathbf{F}\, dV. \tag{27}$$

It is a consequence of the continuity of the divergence that $\nabla \cdot \mathbf{F}$ actually takes on this value at some point P in B. Thus,

$$\begin{aligned}(\nabla \cdot \mathbf{F})_P &= \frac{1}{\text{volume of } B} \iiint_B \nabla \cdot \mathbf{F}\, dV = \frac{\iint_S \mathbf{F} \cdot \mathbf{n}\, d\sigma}{\text{volume of } B} \\ &= \frac{\text{rate at which mass leaves } B \text{ across its surface } S}{\text{volume of } B}.\end{aligned} \tag{28}$$

The fraction on the right describes decrease in mass per unit volume.

Now let the radius of B approach zero while the center Q stays fixed. The left-hand side of Eq. (28) converges to $(\nabla \cdot \mathbf{F})_Q$, the right side to $(-\partial\delta/\partial t)_Q$. The equality of these two limits is the continuity equation

$$\nabla \cdot \mathbf{F} = -\frac{\partial \delta}{\partial t}. \tag{29}$$

The continuity equation "explains" $\nabla \cdot \mathbf{F}$: The divergence of $\mathbf{F}$ at a point is the rate at which the density of the fluid is decreasing there.

The Divergence Theorem

$$\iint_S \mathbf{F} \cdot \mathbf{n}\, d\sigma = \iiint_D \nabla \cdot \mathbf{F}\, dV$$

now says that the net decrease in density of the fluid in region D is accounted for by the mass transported across the surface S. In a way, the theorem is a statement about conservation of mass.

EXERCISES 14.5

In Exercises 1–12, use the Divergence Theorem to find the outward flux of $\mathbf{F}$ across the boundary of the region D.

1. $\mathbf{F} = (y - x)\,\mathbf{i} + (z - y)\,\mathbf{j} + (y - x)\,\mathbf{k}$
   D: The cube bounded by the planes $x = \pm 1$, $y = \pm 1$, and $z = \pm 1$

2. $\mathbf{F} = x^2\mathbf{i} + y^2\mathbf{j} + z^2\mathbf{k}$
   a) D: The cube cut from the first octant by the planes $x = 1$, $y = 1$, and $z = 1$
   b) D: The cube bounded by the planes $x = \pm 1$, $y = \pm 1$, and $z = \pm 1$
   c) D: The region cut from the solid cylinder $x^2 + y^2 \le 4$ by the planes $z = 0$ and $z = 1$

3. $\mathbf{F} = y\mathbf{i} + xy\mathbf{j} - z\mathbf{k}$
   D: The region inside the solid cylinder $x^2 + y^2 \le 4$ between the plane $z = 0$ and the paraboloid $z = x^2 + y^2$

4. $\mathbf{F} = x^2\mathbf{i} + xz\mathbf{j} - 3z\mathbf{k}$
   D: The solid sphere $x^2 + y^2 + z^2 \le 4$

5. $\mathbf{F} = x^2\mathbf{i} - 2xy\mathbf{j} + 3xz\mathbf{k}$
   D: The region cut from the first octant by the sphere $x^2 + y^2 + z^2 = 4$

6. $\mathbf{F} = (6x^2 + 2xy)\,\mathbf{i} + (2y + x^2z)\,\mathbf{j} + 4x^2y^3\,\mathbf{k}$
   D: The region cut from the first octant by the cylinder $x^2 + y^2 = 4$ and the plane $z = 3$

7. $\mathbf{F} = 2xz\mathbf{i} - xy\mathbf{j} - z^2\mathbf{k}$
D: The wedge cut from the first octant by the plane $y + z = 4$ and the elliptical cylinder $4x^2 + y^2 = 16$

8. $\mathbf{F} = x^3\mathbf{i} + y^3\mathbf{j} + z^3\mathbf{k}$
D: The solid sphere $x^2 + y^2 + z^2 \leq a^2$

9. $\mathbf{F} = \sqrt{x^2 + y^2 + z^2}\,(x\mathbf{i} + y\mathbf{j} + z\mathbf{k})$
D: The region $1 \leq x^2 + y^2 + z^2 \leq 2$

10. $\mathbf{F} = (x\mathbf{i} + y\mathbf{j} + z\mathbf{k})/\sqrt{x^2 + y^2 + z^2}$
D: The region $1 \leq x^2 + y^2 + z^2 \leq 4$

11. $\mathbf{F} = (5x^3 + 12xy^2)\mathbf{i} + (y^3 + e^y \sin z)\mathbf{j} + (5z^3 + e^y \cos z)\mathbf{k}$
D: The solid region between the spheres $x^2 + y^2 + z^2 = 1$ and $x^2 + y^2 + z^2 = 2$

12. $\mathbf{F} = \ln(x^2 + y^2)\,\mathbf{i} - \left(\frac{2z}{x}\tan^{-1}\frac{y}{x}\right)\mathbf{j} + z\sqrt{x^2 + y^2}\,\mathbf{k}$
D: The thick-walled cylinder $1 \leq x^2 + y^2 \leq 2, -1 \leq z \leq 2$

13. Let S be the closed cubelike surface in Fig. 14.58, with its base the unit square in the xy-plane, its four sides lying in the planes $x = 0$, $x = 1$, $y = 0$, $y = 1$, and its top an arbitrary smooth surface whose identity is unknown. Let $\mathbf{F} = x\mathbf{i} - 2y\mathbf{j} + (z + 3)\,\mathbf{k}$. Suppose the outward flux of $\mathbf{F}$ through side A is 1 and through side B is -3. Find the outward flux of $\mathbf{F}$ through the top.

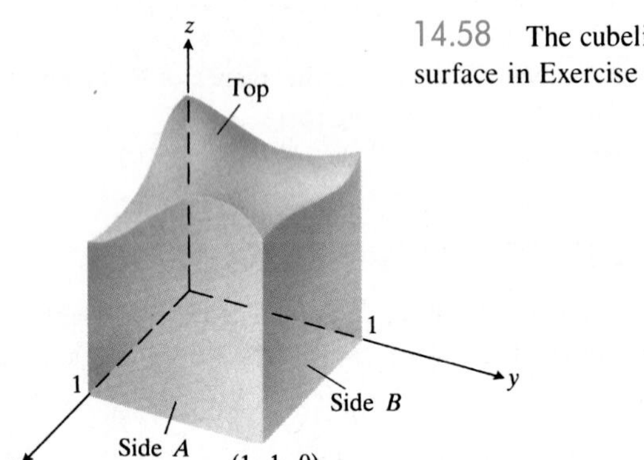

14.58 The cubelike surface in Exercise 13.

14. Let $\mathbf{F}$ be a field whose components have continuous first partial derivatives throughout a portion of space containing a closed and bounded region D and its smooth bounding surface S. Suppose that the length of the vector $\mathbf{F}$ never exceeds 1 on S. What bound can be placed on the numerical size of the integral

$$\iiint_D \nabla \cdot \mathbf{F}\, dV?$$

How do you know?

15. a) Show that the flux of the position vector field $\mathbf{F} = x\mathbf{i} + y\mathbf{j} + z\mathbf{k}$ outward through a smooth closed surface S is three times the volume of the region enclosed by the surface.
b) Let $\mathbf{n}$ be the outward unit normal vector field on S. Show that it is not possible for $\mathbf{F}$ to be orthogonal to $\mathbf{n}$ at every point of S.

16. Among all rectangular solids defined by the inequalities $0 \leq x \leq a$, $0 \leq y \leq b$, $0 \leq z \leq 1$, find the one for which the total flux of $\mathbf{F} = (-x^2 - 4xy)\,\mathbf{i} - 6yz\mathbf{j} + 12z\mathbf{k}$ outward through the six sides is greatest.

17. Let $\mathbf{F} = x\mathbf{i} + y\mathbf{j} + z\mathbf{k}$ and suppose that the surface S and region D satisfy the hypotheses of the Divergence Theorem. Show that the volume of D is given by the formula

$$\text{Volume of } D = \frac{1}{3}\iint_S \mathbf{F} \cdot \mathbf{n}\, d\sigma.$$

18. Show that the outward flux of a constant vector field $\mathbf{F} = \mathbf{C}$ across any closed surface to which the Divergence Theorem applies is zero.

Harmonic Functions and Normal Derivatives

19. A function $f(x, y, z)$ is **harmonic** in a region D in space if throughout D it satisfies the Laplace equation

$$\nabla^2 f = \nabla \cdot \nabla f = \frac{\partial^2 f}{\partial x^2} + \frac{\partial^2 f}{\partial y^2} + \frac{\partial^2 f}{\partial z^2} = 0.$$

Suppose that f is harmonic throughout a bounded region D enclosed by a smooth surface S and that $\mathbf{n}$ is the outward unit normal vector field on S. Let $\partial f/\partial n = \nabla f \cdot \mathbf{n}$ be the derivative of f in the direction of $\mathbf{n}$. This derivative, a real-valued function defined on S, is called the **normal derivative** of f on S.
a) Show that the integral of $\partial f/\partial n$ over S is zero. (*Hint:* Let $\mathbf{F} = \nabla f$.)
b) Show that

$$\iint_S f\frac{\partial f}{\partial n}\, d\sigma = \iiint_D |\nabla f|^2\, dV.$$

(*Hint:* Let $\mathbf{F} = f\,\nabla f$.)

20. Let S be the surface of the portion of the solid sphere $x^2 + y^2 + z^2 \leq a^2$ that lies in the first octant, let $f(x, y, z) = \ln\sqrt{x^2 + y^2 + z^2}$, and let $\partial f/\partial n$ be the normal derivative of f on S defined in Exercise 19. Calculate

$$\iint_S \frac{\partial f}{\partial n}\, d\sigma.$$

21. *Green's first formula.* Suppose that f and g are scalar functions with continuous first- and second-order partial derivatives throughout a closed region D that is bounded by a piecewise smooth surface S. Show that

$$\iint_S f\frac{\partial g}{\partial n}\, d\sigma = \iiint_D (f\,\nabla^2 g + \nabla f \cdot \nabla g)\, dV. \tag{30}$$

Equation (30) is **Green's first formula.** (*Hint:* Apply the Divergence Theorem to the field $\mathbf{F} = f\,\nabla g$.)

22. *Green's second formula* (continuation of Exercise 21). Interchange f and g in Eq. (30) to obtain a similar formula. Then subtract this formula from Eq. (30) to show that

$$\iint_S \left(f\frac{\partial g}{\partial n} - g\frac{\partial f}{\partial n}\right) d\sigma = \iiint_D (f\,\nabla^2 g - g\,\nabla^2 f)\, dV. \tag{31}$$

This equation is **Green's second formula.**

14.6 Stokes's Theorem

As we saw in Section 14.3, the circulation density or curl of a two-dimensional field $\mathbf{F} = M\mathbf{i} + N\mathbf{j}$ at a point (x, y) is described by the scalar quantity $(\partial N/\partial x - \partial M/\partial y)$. In three dimensions, the circulation around a point P in a plane is described not with a scalar but with a vector. This vector is normal to the plane of the circulation (Fig. 14.59) and points in the direction that gives it a right-hand relation to the circulation line. The length of the vector gives the rate of the fluid's rotation, which usually varies as the circulation plane is tilted about P from one position to another. It turns out that the vector of greatest circulation in a flow with velocity field $\mathbf{F} = M\mathbf{i} + N\mathbf{j} + P\mathbf{k}$ is the vector

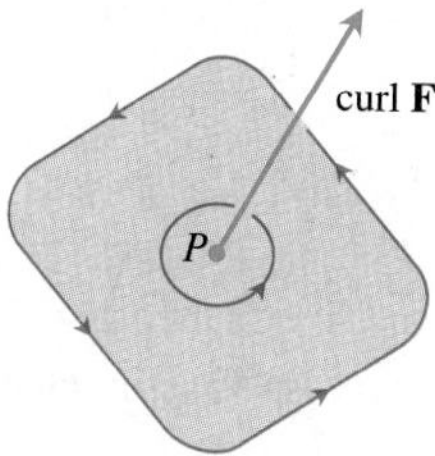

14.59 The circulation vector at a point P in a plane in a three-dimensional fluid flow. Notice its right-hand relation to the circulation line.

$$\text{curl}\,\mathbf{F} = \left(\frac{\partial P}{\partial y} - \frac{\partial N}{\partial z}\right)\mathbf{i} + \left(\frac{\partial M}{\partial z} - \frac{\partial P}{\partial x}\right)\mathbf{j} + \left(\frac{\partial N}{\partial x} - \frac{\partial M}{\partial y}\right)\mathbf{k}. \tag{1}$$

We get this information from Stokes's theorem, the generalization of the circulation–curl form of Green's theorem to space.

In this section we present Stokes's theorem and investigate what it says about circulation. We shall also see how Stokes's theorem sometimes enables us to evaluate a line integral around a loop in space by calculating an easier surface integral. This is similar to the way we used Green's theorem to evaluate a line integral over a complicated curve in the plane by calculating an easier double integral over the region enclosed by the curve. In addition, we shall state the circulation–curl form of Green's theorem in its final vector form. Stokes's theorem has important applications in electromagnetic theory, but we shall not go into them here.

Stokes's Theorem

In del notation, the curl of a three-dimensional vector field $\mathbf{F} = M\mathbf{i} + N\mathbf{j} + P\mathbf{k}$ is written as the symbolic cross product

$$\begin{aligned}\text{curl}\,\mathbf{F} = \nabla \times \mathbf{F} &= \begin{vmatrix} \mathbf{i} & \mathbf{j} & \mathbf{k} \\ \dfrac{\partial}{\partial x} & \dfrac{\partial}{\partial y} & \dfrac{\partial}{\partial z} \\ M & N & P \end{vmatrix} \\ &= \left(\frac{\partial P}{\partial y} - \frac{\partial N}{\partial z}\right)\mathbf{i} + \left(\frac{\partial M}{\partial z} - \frac{\partial P}{\partial x}\right)\mathbf{j} + \left(\frac{\partial N}{\partial x} - \frac{\partial M}{\partial y}\right)\mathbf{k}.\end{aligned} \tag{2}$$

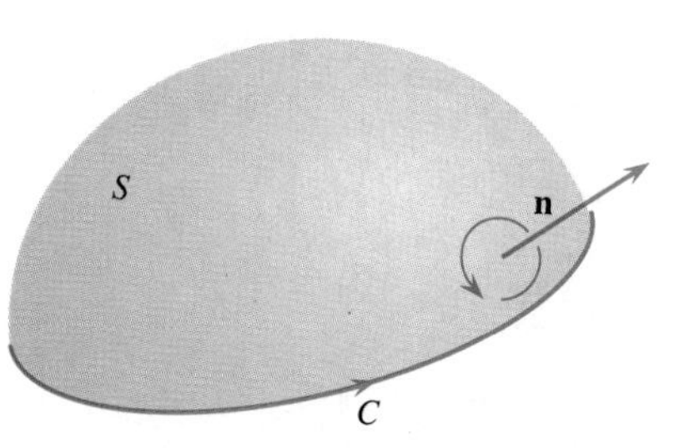

14.60 The orientation of the bounding curve C gives it a right-handed relation to the normal field $\mathbf{n}$.

Stokes's theorem says that, under conditions normally met in practice, the circulation of a vector field around the boundary of an oriented surface in space in the direction counterclockwise with respect to the surface's unit normal vector field $\mathbf{n}$ (Fig. 14.60) is equal to the double integral of the normal component of the curl of the field over the surface.

THEOREM 5

Stokes's Theorem

The circulation of $\mathbf{F} = M\mathbf{i} + N\mathbf{j} + P\mathbf{k}$ around the boundary C of an oriented surface S in the direction counterclockwise with respect to the surface's unit normal vector $\mathbf{n}$ equals the integral of $\nabla \times \mathbf{F} \cdot \mathbf{n}$ over S.

$$\underbrace{\oint_C \mathbf{F} \cdot d\mathbf{r}}_{\text{counterclockwise circulation}} = \underbrace{\iint_S \nabla \times \mathbf{F} \cdot \mathbf{n}\, d\sigma}_{\text{curl integral}} \tag{3}$$

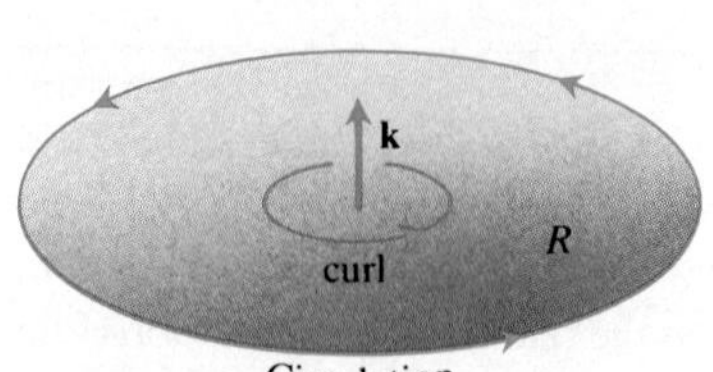

Stokes

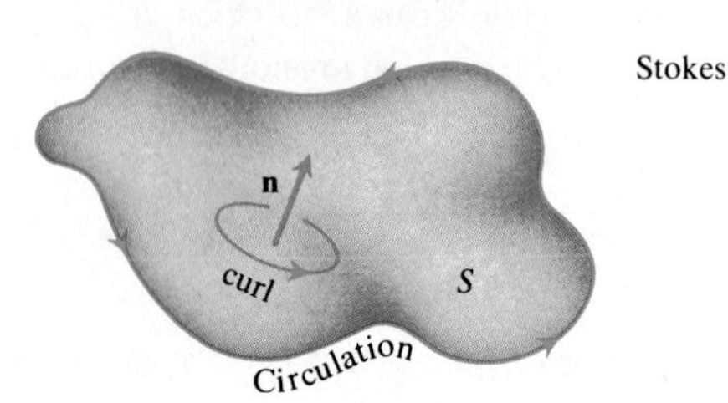

14.61 Green's theorem vs. Stokes's theorem.

Naturally, we need some mathematical restrictions on **F**, C, and S to ensure the existence of the integrals in Stokes's equation. The usual restrictions are that all the functions and derivatives involved be continuous.

If C is a curve in the xy-plane, oriented counterclockwise, and R is the region bounded by C in the xy-plane, then $d\sigma = dx\,dy$ and

$$(\nabla \times \mathbf{F}) \cdot \mathbf{n} = (\nabla \times \mathbf{F}) \cdot \mathbf{k} = \left(\frac{\partial N}{\partial x} - \frac{\partial M}{\partial y}\right). \tag{4}$$

Under these conditions, Stokes's equation becomes

$$\oint_C \mathbf{F} \cdot d\mathbf{r} = \iint_R \left(\frac{\partial N}{\partial x} - \frac{\partial M}{\partial y}\right) dx\,dy,$$

which is the circulation–curl form of the equation in Green's theorem. Conversely, by reversing these steps we can rewrite Green's theorem for two-dimensional fields in del notation as

$$\oint_C \mathbf{F} \cdot d\mathbf{r} = \iint_R \nabla \times \mathbf{F} \cdot \mathbf{k}\, dA. \tag{5}$$

See Fig. 14.61

Example 1 Evaluate the two integrals in Stokes's theorem for the hemisphere

$$x^2 + y^2 + z^2 = 9, \qquad z \geq 0,$$

its bounding circle

$$x^2 + y^2 = 9, \qquad z = 0,$$

and the field

$$\mathbf{F} = y\mathbf{i} - x\mathbf{j}.$$

Solution The integrand of the circulation integral is

$$\mathbf{F} \cdot d\mathbf{r} = (y\mathbf{i} - x\mathbf{j}) \cdot (dx\mathbf{i} + dy\mathbf{j} + dz\mathbf{k}) = y\,dx - x\,dy.$$

By the circulation form of Green's theorem for the plane,

$$\oint_C \mathbf{F} \cdot d\mathbf{r} = \oint_C (y\,dx - x\,dy) = \iint_R \left(\frac{\partial N}{\partial x} - \frac{\partial M}{\partial y}\right) dx\,dy$$

$$= \iint_{x^2+y^2 \leq 9} -2\,dx\,dy = (-2)9\pi = -18\pi.$$

For the curl integral of $\mathbf{F} = y\mathbf{i} - x\mathbf{j}$, we have

$$\nabla \times \mathbf{F} = \left(\frac{\partial P}{\partial y} - \frac{\partial N}{\partial z}\right)\mathbf{i} + \left(\frac{\partial M}{\partial z} - \frac{\partial P}{\partial x}\right)\mathbf{j} + \left(\frac{\partial N}{\partial x} - \frac{\partial M}{\partial y}\right)\mathbf{k}$$

$$= (0 - 0)\,\mathbf{i} + (0 - 0)\,\mathbf{j} + (-1 - 1)\,\mathbf{k} = -2\,\mathbf{k},$$

$$\mathbf{n} = \frac{x\mathbf{i} + y\mathbf{j} + z\mathbf{k}}{\sqrt{x^2 + y^2 + z^2}} = \frac{x\mathbf{i} + y\mathbf{j} + z\mathbf{k}}{3}, \qquad \text{(Outer unit normal)}$$

$$d\sigma = \frac{3}{z}\,dA \qquad \text{(As in Section 14.4, Example 5, with } a = 3\text{)}$$

$$\nabla \times \mathbf{F} \cdot \mathbf{n}\,d\sigma = -\frac{2z}{3}\frac{3}{z}\,dA = -2\,dA,$$

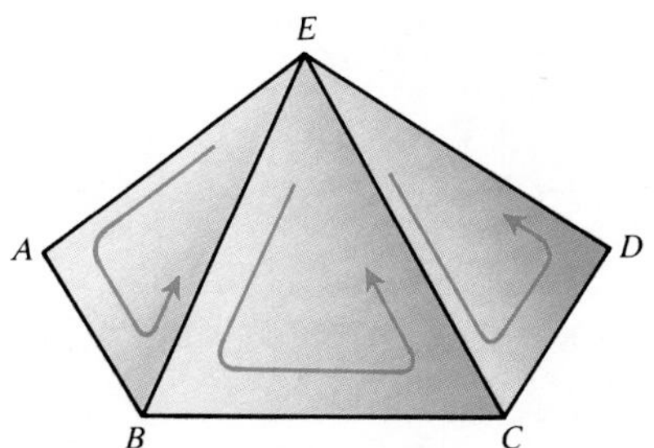

14.62 Part of a polyhedral surface.

and

$$\iint_S \nabla \times \mathbf{F} \cdot \mathbf{n}\, d\sigma = \iint_{x^2+y^2 \le 9} -2\, dA = -18\pi.$$

The circulation around the circle equals the integral of the curl over the hemisphere, as it should.

Proof of Stokes's Theorem for Polyhedral Surfaces

Let the surface S be a polyhedral surface consisting of a finite number of plane regions. (Think of one of Buckminster Fuller's geodesic domes.) We apply Green's theorem to each separate panel of S. There are two types of panels:

1. those that are surrounded on all sides by other panels,
2. those that have one or more edges that are not adjacent to other panels.

The boundary Δ of S consists of edges of type 2 panels that are not adjacent to other panels. In Fig. 14.62, the triangles EAB, BCE, and CDE represent a part of S, with $ABCD$ part of the boundary Δ. Applying Green's theorem to the three triangles in turn and adding the results, we get

$$\left(\oint_{EAB} + \oint_{BCE} + \oint_{CDE}\right) \mathbf{F} \cdot d\mathbf{r} = \left(\iint_{EAB} + \iint_{BCE} + \iint_{CDE}\right) \nabla \times \mathbf{F} \cdot \mathbf{n}\, d\sigma. \quad (6)$$

The three line integrals on the left-hand side of Eq. (6) combine into a single line integral taken around the periphery $ABCDEA$ because the integrals along interior segments cancel in pairs. For example, the integral along the segment BE in triangle ABE is opposite in sign to the integral along the same segment in triangle EBC. Similarly for the segment CE. Hence (6) reduces to

$$\oint_{ABCDEA} \mathbf{F} \cdot d\mathbf{r} = \iint_{ABCDE} \nabla \times \mathbf{F} \cdot \mathbf{n}\, d\sigma.$$

When we apply Green's theorem to all the panels and add the results, we get

$$\oint_{\Delta} \mathbf{F} \cdot d\mathbf{r} = \iint_S \nabla \times \mathbf{F} \cdot \mathbf{n}\, d\sigma. \quad (7)$$

This is Stokes's theorem for a polyhedral surface S.

A rigorous proof of Stokes's theorem for a more general oriented surface S and its bounding curve C is beyond the level of this course. However, the following intuitive argument shows why we would expect Stokes's equation to hold. Imagine a sequence of polyhedral surfaces $S_1, S_2, \ldots, S_n, \ldots$ approximating S. The surface S_n is constructed so that its boundary Δ_n is inscribed in or tangent to C and so that the length of Δ_n approaches the length of C as $n \to \infty$. The faces of S_n are polygonal regions approximating pieces of S, and the area of S_n approaches the area of S as $n \to \infty$. If the first partial derivatives of M, N, and P are continuous throughout an open region D containing S and C, it is plausible to expect that

$$\oint_{\Delta_n} \mathbf{F} \cdot d\mathbf{R} \quad \text{approaches} \quad \oint_C \mathbf{F} \cdot d\mathbf{R}$$

George Gabriel Stokes

Sir George Gabriel Stokes (1819–1903), one of the most influential scientific figures of his century, was Lucasian Professor of Mathematics at Cambridge University from 1849 until his death in 1903. His theoretical and experimental investigations covered hydrodynamics, elasticity, light, gravity, sound, heat, meteorology, and solar physics. He left electricity and magnetism to his friend William Thomson, Baron Kelvin of Largs. It is another one of those delightful quirks of history that the theorem we call Stokes's theorem isn't his theorem at all. He learned of it from Thomson in 1850 and a few years later included it among the questions on an examination he wrote for the Smith Prize. It has been known as Stokes's theorem ever since. As usual, things have balanced out. Stokes was the original discoverer of the principles of spectrum analysis that we now credit to Bunsen and Kirchhoff.

and that

$$\iint_{S_n} \nabla \times \mathbf{F} \cdot \mathbf{n}\, d\sigma_n \quad \text{approaches} \quad \iint_S \nabla \times \mathbf{F} \cdot \mathbf{n}\, d\sigma$$

as $n \to \infty$. But if

$$\oint_{\Delta_n} \mathbf{F} \cdot d\mathbf{R} \to \oint_C \mathbf{F} \cdot d\mathbf{R} \tag{8}$$

and

$$\iint_{S_n} \nabla \times \mathbf{F} \cdot \mathbf{n}\, d\sigma \to \iint_S \nabla \times \mathbf{F} \cdot \mathbf{n}\, d\sigma, \tag{9}$$

and if the left-hand sides of (8) and (9) are equal by Stokes's theorem for polyhedra, then their limits are equal.

Stokes's Theorem for Surfaces with Holes

Stokes's theorem can also be extended to a surface S that has one or more holes in it (like a curved slice of Swiss cheese), in a way analogous to the extension of Green's theorem: The surface integral over S of the normal component of $\nabla \times \mathbf{F}$ is equal to the sum of the line integrals around all the boundaries of S (including boundaries of the holes) of the tangential component of $\mathbf{F}$, where the boundary curves are to be traced in the positive direction induced by the positive orientation of S.

Circulation of a Fluid

Stokes's theorem provides the following vector interpretation for $\nabla \times \mathbf{F}$. As in the discussion of divergence, let $\mathbf{v}$ be the velocity field of a moving fluid, δ the density, and $\mathbf{F} = \delta \mathbf{v}$. Then

$$\oint_C \mathbf{F} \cdot d\mathbf{r}$$

is the circulation of the fluid around the closed curve C. By Stokes's theorem, the circulation is equal to the flux of $\nabla \times \mathbf{F}$ through a surface S spanning C:

$$\oint_C \mathbf{F} \cdot d\mathbf{r} = \iint_S \nabla \times \mathbf{F} \cdot \mathbf{n}\, d\sigma.$$

Suppose we fix a point Q in the domain of $\mathbf{F}$ and a direction $\mathbf{u}$ at Q. Let C be a circle of radius ρ, with center at Q, whose plane is normal to $\mathbf{u}$. If $\nabla \times \mathbf{F}$ is continuous at Q, then the average value of the $\mathbf{u}$-component of $\nabla \times \mathbf{F}$ over the circular disk S bounded by C approaches the $\mathbf{u}$-component of $\nabla \times \mathbf{F}$ at Q as $\rho \to 0$:

$$(\nabla \times \mathbf{F} \cdot \mathbf{u})_Q = \lim_{\rho \to 0} \frac{1}{\pi \rho^2} \iint_S \nabla \times \mathbf{F} \cdot \mathbf{u}\, d\sigma. \tag{10}$$

If we replace the double integral on the right-hand side of Eq. (10) by the circulation, we get

$$(\nabla \times \mathbf{F} \cdot \mathbf{u})_Q = \lim_{\rho \to 0} \frac{1}{\pi \rho^2} \oint_C \mathbf{F} \cdot d\mathbf{r}. \tag{11}$$

The left-hand side of Eq. (11) has its maximum value when $\mathbf{u}$ is the direction of $\nabla \times \mathbf{F}$. When ρ is small, the limit on the right-hand side of Eq. (11) is

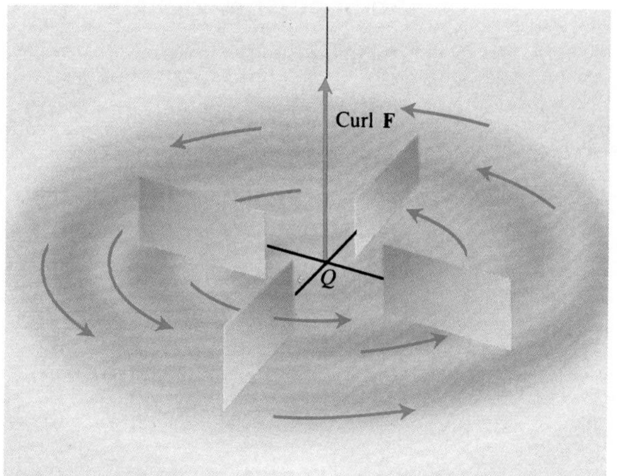

14.63 The paddle wheel interpretation of curl **F**.

approximately

$$\frac{1}{\pi\rho^2}\oint_C \mathbf{F}\cdot d\mathbf{r},$$

which is the circulation around C divided by the area of the disk (circulation density). Suppose that a small paddle wheel of radius ρ is introduced into the fluid at Q, with its axle directed along **u**. The circulation of the fluid around C will affect the rate of spin of the paddle wheel. The wheel will spin fastest when the circulation integral is maximized; therefore it will spin fastest when the axle of the paddle wheel points in the direction of $\nabla\times\mathbf{F}$ (Fig. 14.63).

Example 2 A fluid of constant density δ rotates around the z-axis with velocity $\mathbf{v} = \omega(-y\mathbf{i} + x\mathbf{j})$, where ω is a positive constant called the *angular velocity* of the rotation (Fig. 14.64). If $\mathbf{F} = \delta\mathbf{v}$, find $\nabla\times\mathbf{F}$ and show its relation to the circulation density.

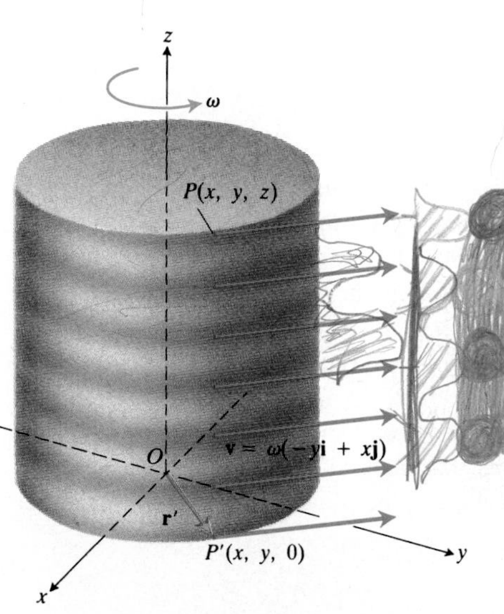

14.64 A steady rotational flow parallel to the xy-plane, with constant angular velocity ω in the positive (counterclockwise) direction.

Solution With $\mathbf{F} = \delta\mathbf{v} = -\delta\omega y\mathbf{i} + \delta\omega x\mathbf{j}$,

$$\begin{aligned}\nabla\times\mathbf{F} &= \left(\frac{\partial P}{\partial y} - \frac{\partial N}{\partial z}\right)\mathbf{i} + \left(\frac{\partial M}{\partial z} - \frac{\partial P}{\partial x}\right)\mathbf{j} + \left(\frac{\partial N}{\partial x} - \frac{\partial M}{\partial y}\right)\mathbf{k}\\ &= (0-0)\,\mathbf{i} + (0-0)\,\mathbf{j} + (\delta\omega - (-\delta\omega))\,\mathbf{k}\\ &= 2\,\delta\omega\mathbf{k}.\end{aligned}$$

By Stokes's theorem, the circulation of **F** around a circle C of radius ρ bounding a disk S in a plane normal to $\nabla\times\mathbf{F}$, say the xy-plane, is

$$\oint_C \mathbf{F}\cdot d\mathbf{r} = \iint_S \nabla\times\mathbf{F}\cdot\mathbf{n}\,d\sigma = \iint_S 2\,\delta\omega\mathbf{k}\cdot\mathbf{k}\,dx\,dy = (2\,\delta\omega)(\pi\rho^2).$$

Thus,

$$(\nabla\times\mathbf{F})\cdot\mathbf{k} = 2\,\delta\omega = \frac{1}{\pi\rho^2}\oint_C \mathbf{F}\cdot d\mathbf{r},$$

in agreement with Eq. (11) with $\mathbf{u} = \mathbf{k}$.

An Important Identity

The identity

$$\text{curl grad} f = \mathbf{0} \qquad \text{or} \qquad \nabla \times \nabla f = \mathbf{0} \tag{12}$$

arises frequently in mathematics and the physical sciences. (We shall use it ourselves in Section 14.7.) It holds for any function $f(x, y, z)$ whose second partial derivatives are continuous. The proof goes like this:

$$\begin{aligned} \nabla \times \nabla f &= \begin{vmatrix} \mathbf{i} & \mathbf{j} & \mathbf{k} \\ \dfrac{\partial}{\partial x} & \dfrac{\partial}{\partial y} & \dfrac{\partial}{\partial z} \\ \dfrac{\partial f}{\partial x} & \dfrac{\partial f}{\partial y} & \dfrac{\partial f}{\partial z} \end{vmatrix} \\ &= (f_{zy} - f_{yz})\,\mathbf{i} - (f_{zx} - f_{xz})\,\mathbf{j} + (f_{yx} - f_{xy})\,\mathbf{k}. \end{aligned} \tag{13}$$

When the second partial derivatives of f are continuous, the mixed second derivatives in parentheses are equal and the vector is zero.

✲ Ampère's Law

In electromagnetic theory, Stokes's theorem gives Ampère's law:

$$\begin{aligned} \oint_C \mathbf{B} \cdot d\mathbf{r} &= \iint_S \nabla \times \mathbf{B} \cdot \mathbf{n}\, d\sigma \\ &= \frac{4\pi}{c} \iint_S \mathbf{J} \cdot \mathbf{n}\, d\sigma \\ &= \frac{4\pi}{c} I. \end{aligned} \tag{14}$$

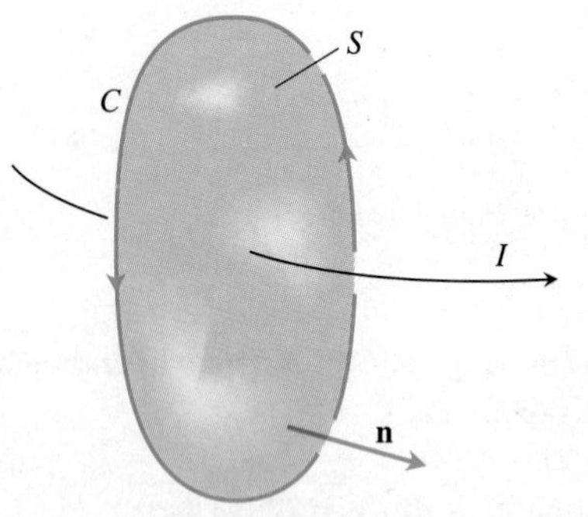

14.65 The directions of C, $\mathbf{n}$, and the current flow in Ampère's law.

In this equation, $\mathbf{B}$ is a magnetic field in space, I is a current passing through a loop C in the domain of $\mathbf{B}$, S is a surface spanning C, $\mathbf{J}$ is the current density on S (current per unit area), and c is the speed of light. The law tells us that the integral of $\mathbf{B}$ around a closed loop is proportional to the current through any surface bounded by the loop. The constant of proportionality is $4\pi/c$. The direction of $\mathbf{n}$ and the direction of I are assumed to have the same right-hand relationship to the direction of integration around C (Fig. 14.65).

Ampère's law is named after its discoverer, André-Marie Ampère (1775–1836), French physicist, mathematician, and natural philosopher known for his contributions to electrodynamics. The ampere, the SI unit of electrical current, is named for him.

Summary of Integral Theorems in del Notation

We close with Table 14.3, which lists the equations from Green's theorem and its generalizations to three dimensions.

TABLE 14.3
Green's theorem and its generalizations

Green's theorem:	$\oint_C \mathbf{F}\cdot\mathbf{n}\,ds = \iint_R \nabla\cdot\mathbf{F}\,dA$
	$\oint_C \mathbf{F}\cdot d\mathbf{r} = \iint_R \nabla\times\mathbf{F}\cdot\mathbf{k}\,dA$
The Divergence Theorem:	$\iint_S \mathbf{F}\cdot\mathbf{n}\,d\sigma = \iiint_D \nabla\cdot\mathbf{F}\,dV$
Stokes's theorem:	$\oint_C \mathbf{F}\cdot d\mathbf{r} = \iint_S \nabla\times\mathbf{F}\cdot\mathbf{n}\,d\sigma$

EXERCISES 14.6

In Exercises 1–6, use the surface integral in Stokes's theorem to calculate the circulation of the field $\mathbf{F}$ around the curve C in the indicated direction.

1. $\mathbf{F} = x^2\mathbf{i} + 2x\mathbf{j} + z^2\mathbf{k}$
C: The ellipse $4x^2 + y^2 = 4$, counterclockwise as viewed from above

2. $\mathbf{F} = 2y\mathbf{i} + 3x\mathbf{j} - z^2\mathbf{k}$
C: The circle $x^2 + y^2 = 9$ in the xy-plane, counterclockwise when viewed from above

3. $\mathbf{F} = y\mathbf{i} + xz\mathbf{j} + x^2\mathbf{k}$
C: The boundary of the triangle cut from the plane $x + y + z = 1$ by the first octant, counterclockwise when viewed from above

4. $\mathbf{F} = (y^2 + z^2)\,\mathbf{i} + (x^2 + z^2)\,\mathbf{j} + (x^2 + y^2)\,\mathbf{k}$
C: The boundary of the triangle cut from the plane $x + y + z = 1$ by the first octant, counterclockwise when viewed from above

5. $\mathbf{F} = (y^2 + z^2)\,\mathbf{i} + (x^2 + y^2)\,\mathbf{j} + (x^2 + y^2)\,\mathbf{k}$
C: The square bounded by the lines $x = \pm 1$ and $y = \pm 1$ in the xy-plane, counterclockwise when viewed from above

6. $\mathbf{F} = x^2y^3\,\mathbf{i} + \mathbf{j} + z\mathbf{k}$
C: The boundary of the circle $x^2 + y^2 = a^2$ in the xy-plane, counterclockwise when viewed from above

7. Let $\mathbf{n}$ be the outer unit normal of the elliptical shell

$$S:\quad 4x^2 + 9y^2 + 36z^2 = 36, \quad z \geq 0,$$

and let

$$\mathbf{F} = y\mathbf{i} + x^2\mathbf{j} + (x^2 + y^4)^{3/2}\sin e^{\sqrt{xyz}}\,\mathbf{k}.$$

Find the value of

$$\iint_S \nabla\times\mathbf{F}\cdot\mathbf{n}\,d\sigma.$$

(*Hint:* One parametrization of the ellipse at the base of the shell is $x = 3\cos t$, $y = 2\sin t$, $0 \leq t \leq 2\pi$.)

8. Let $\mathbf{n}$ be the outer unit normal of the parabolic shell

$$S: 4x^2 + y + z^2 = 4,\ y \geq 0$$

and let

$$\mathbf{F} = \left(-z + \frac{1}{2+x}\right)\mathbf{i} + (\tan^{-1}y)\,\mathbf{j} + \left(x + \frac{1}{4+z}\right)\mathbf{k}.$$

Find the value of

$$\iint_S \nabla\times\mathbf{F}\cdot\mathbf{n}\,d\sigma.$$

9. Let S be the cylinder $x^2 + y^2 = a^2$, $0 \leq z \leq h$, together with its top, $x^2 + y^2 \leq a^2$, $z = h$. Let $\mathbf{F} = -y\mathbf{i} + x\mathbf{j} + x^2\mathbf{k}$. Use Stokes's theorem to calculate the flux of $\nabla\times\mathbf{F}$ outward through S.

10. Evaluate

$$\iint_S \nabla\times(y\mathbf{i})\cdot\mathbf{n}\,d\sigma$$

where S is the hemisphere $x^2 + y^2 + z^2 = 1$, $z \geq 0$.

11. Show that

$$\iint_S \nabla\times\mathbf{F}\cdot\mathbf{n}\,d\sigma$$

has the same value for all oriented surfaces S that span C and that induce the same positive direction on C.

12. Show that if $\mathbf{F} = x\mathbf{i} + y\mathbf{j} + z\mathbf{k}$, then $\nabla\cdot\mathbf{F} = 3$ and $\nabla\times\mathbf{F} = \mathbf{0}$.

13. *div(curl F) = 0.* a) Show that if the necessary partial derivatives of the components of $\mathbf{F} = M\mathbf{i} + N\mathbf{j} + P\mathbf{k}$ are continuous, then $\nabla\cdot\nabla\times\mathbf{F} = 0$.
b) Use the result in (a) to show that

$$\iint_S \nabla\times\mathbf{F}\cdot\mathbf{n}\,d\sigma = 0$$

for any surface to which the Divergence Theorem applies.

14. Use the identity $\nabla \times \nabla f = 0$ and Stokes's theorem to show that the circulations of the following fields around the boundary of any smooth orientable surface in space are zero.

a) $\mathbf{F} = 2x\mathbf{i} + 2y\mathbf{j} + 2z\mathbf{k}$
b) $\mathbf{F} = \nabla(xy^2z^3)$
c) $\mathbf{F} = \nabla \times (x\mathbf{i} + y\mathbf{j} + z\mathbf{k})$
d) $\mathbf{F} = \nabla f$

15. Let $f(x, y, z) = (x^2 + y^2 + z^2)^{-1/2}$. Show that the clockwise circulation of the field $\mathbf{F} = \nabla f$ around the circle $x^2 + y^2 = a^2$ in the xy-plane is zero

a) by taking $\mathbf{r} = (a \cos t)\,\mathbf{i} + (a \sin t)\,\mathbf{j}$, $0 \le t \le 2\pi$, and integrating $\mathbf{F} \cdot d\mathbf{r}$ over the circle, and
b) by applying Stokes's theorem.

16. Let C be a simple closed smooth curve in the plane $2x + 2y + z = 2$, oriented as in Fig. 14.66. Show that

$$\oint_C 2y\,dx + 3z\,dy - x\,dz$$

depends only on the area of the region enclosed by C and not on the position or shape of C.

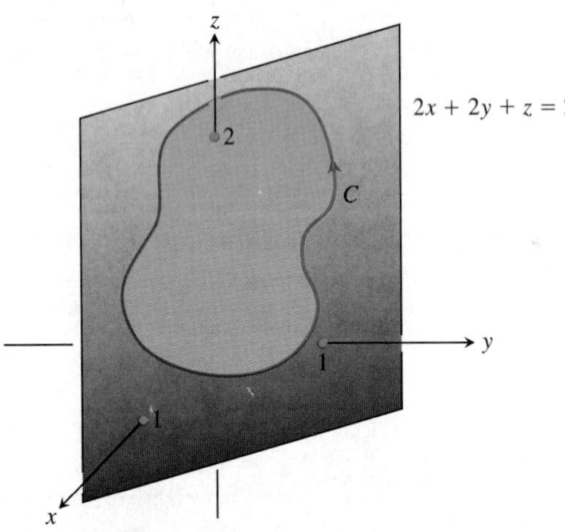

14.66 The curve in Exercise 16.

17. Suppose that a smooth closed oriented surface S in space is the union of two surfaces S_1 and S_2 joined to each other along a simple closed smooth curve C and let $\mathbf{n}$ be the unit normal vector field on S. Show that

$$\iint_S \nabla \times \mathbf{F} \cdot \mathbf{n}\, d\sigma = 0$$

for any differentiable vector field defined on a region containing S and its interior.

18. Find a vector field with twice-differentiable components whose curl is $x\mathbf{i} + y\mathbf{j} + z\mathbf{k}$ or prove that no such field exists.

19. Let R be a region in the xy-plane that is bounded by a piecewise smooth simple closed curve C and suppose that the moments of inertia of R about the x- and y-axes are known to be I_x and I_y. Evaluate the line integral

$$\oint_C \nabla(r^4) \cdot \mathbf{n}\, ds,$$

where $r = \sqrt{x^2 + y^2}$, in terms of I_x and I_y.

20. *Basic properties of curl and divergence.* Let $g(x, y, z)$ be a differentiable scalar function, let $\mathbf{F}$, $\mathbf{F}_1$, and $\mathbf{F}_2$ be differentiable vector fields, and let a and b be arbitrary real constants. Verify the following identities.

a) $\nabla \cdot (a\mathbf{F}_1 + b\mathbf{F}_2) = a\nabla \cdot \mathbf{F}_1 + b\,\nabla \cdot \mathbf{F}_2$
b) $\nabla \times (a\mathbf{F}_1 + b\mathbf{F}_2) = a\nabla \times \mathbf{F}_1 + b\nabla \times \mathbf{F}_2$
c) $\nabla \cdot (g\mathbf{F}) = g\,\nabla \cdot \mathbf{F} + \nabla g \cdot \mathbf{F}$
d) $\nabla \times (g\mathbf{F}) = g\,\nabla \times \mathbf{F} + \nabla g \times \mathbf{F}$
e) $\nabla \cdot (\mathbf{F}_1 \times \mathbf{F}_2) = \mathbf{F}_2 \cdot \nabla \times \mathbf{F}_1 - \mathbf{F}_1 \cdot \nabla \times \mathbf{F}_2$

EXPLORER PROGRAM

Vector Fields	Displays two-dimensional vector fields and integrates the curl over rectangular regions. Also calculates flux across rectangular boundaries

14.7 Path Independence, Potential Functions, and Conservative Fields

This section introduces the notion of path independence of work integrals and discusses the remarkable properties of conservative fields, fields in which all work integrals are path independent. In an electric field, for example, the amount of work it takes to move a charge from one point to another depends only on the locations of the points and not on the path taken to get from one to the other. Electric fields are conservative. Gravitational fields are conservative, too: The amount of work it takes to move a mass from one point to another against gravity depends only on the locations of the points and not on the path taken in between. As we shall see, we can evaluate work integrals in such fields without resorting to parametrizations. We use the fields' so-called potential functions instead.

Path Independence

If A and B are two points in an open region D in space, the work $\int \mathbf{F} \cdot d\mathbf{r}$ done in moving a particle from A to B by a field $\mathbf{F}$ defined on D usually depends on the path taken. For some special fields, however, the integral's value depends only on the points A and B and is the same for all paths from A to B. If this is true for all points A and B in D, we say that the integral $\int \mathbf{F} \cdot d\mathbf{r}$ is path independent in D and that $\mathbf{F}$ is conservative on D.

DEFINITIONS

Let $\mathbf{F}$ be a field defined on an open region D in space, and suppose that for any two points A and B in D the work $\int_A^B \mathbf{F} \cdot d\mathbf{r}$ done in moving from A to B is the same over all paths in D from A to B. Then the integral $\int \mathbf{F} \cdot d\mathbf{r}$ is **path independent in D** and the field $\mathbf{F}$ is **conservative on D.**

The word *conservative* comes from physics, where it refers to fields in which the principle of conservation of energy holds (it does, in conservative fields).

Under conditions normally met in practice, a field $\mathbf{F}$ is conservative if and only if it is the gradient field of a scalar function f; that is, if and only if $\mathbf{F} = \nabla f$ for some f. The function f is then called a potential function for $\mathbf{F}$.

DEFINITIONS

If a scalar function f is differentiable throughout a domain D, the field of vectors ∇f is called the **gradient field** of f on D. If $\mathbf{F}$ is a field defined on D and $\mathbf{F} = \nabla f$ for some scalar function f on D, then f is called a **potential function** for $\mathbf{F}$.

An electric potential is a scalar function whose gradient field is an electric field. A gravitational potential is a scalar function whose gradient field is a gravitational field, and so on. As we shall see, once we have found a potential function f for a field $\mathbf{F}$, we can evaluate all work integrals in the domain of $\mathbf{F}$ by the rule

$$\int_A^B \mathbf{F} \cdot d\mathbf{r} = f(B) - f(A). \tag{1}$$

This is like the Fundamental Theorem of Calculus, and we call it the Fundamental Theorem of Line Integrals. The difference $f(B) - f(A)$ in the values of the potential function in Eq. (1) is called the **potential difference** of B with respect to A or, less precisely, the potential difference *between* B and A.

Conservative fields have other remarkable properties. Saying that a field $\mathbf{F}$ is conservative on D is equivalent to saying that the integral of $\mathbf{F}$ around every closed loop in D is zero. This, in turn, is equivalent in a great many domains to saying that $\nabla \times \mathbf{F} = \mathbf{0}$. We explain these relationships by discussing the implications in the following diagram.

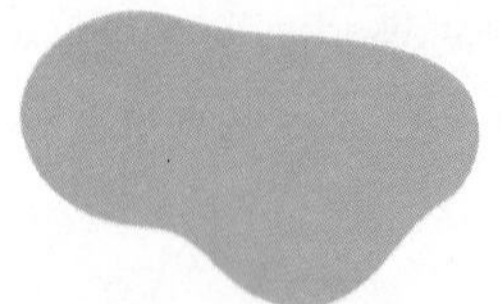

Connected and simply connected.

Connected but not simply connected.

Connected and simply connected.

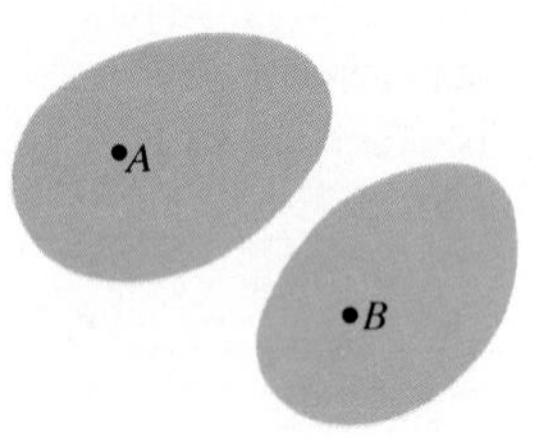

Simply connected but not connected.
No path from A to B lies entirely in the region.

14.67 Connectivity and simple connectivity are not the same thing. In fact, neither implies the other, as these pictures of plane regions illustrate. To make three-dimensional regions with these properties, thicken the plane regions into cylinders.

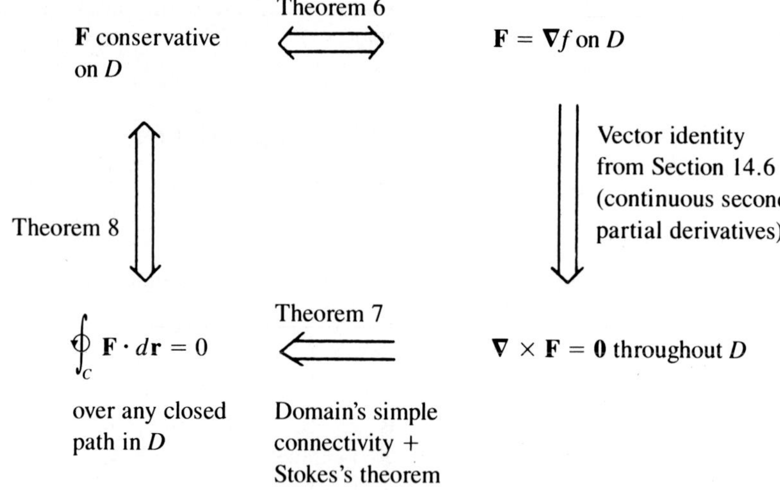

Naturally, we need to impose conditions on the curves, fields, and domains to make these implications hold.

We assume that all of the curves are piecewise smooth.

We also assume that the components of **F** have continuous second partial derivatives. Without this assumption, $\mathbf{F} = \nabla f$ need not imply $\nabla \times \mathbf{F} = \mathbf{0}$. As we saw in the previous section, the implication requires the mixed second derivatives of f to be equal. We assume their continuity to ensure that.

We assume D to be an *open* region in space. This means that every point in D is the center of a solid sphere that lies entirely in D. We need an added restriction on D if $\nabla \times \mathbf{F} = \mathbf{0}$ is to imply that the integral of **F** around every closed path in D is zero. For this, we assume D to be **simply connected.** This means that every closed path in D can be contracted to a point in D without ever leaving D. If D consisted of space with one of the axes removed, for example, D would not be simply connected. There would be no way to contract a loop around the axis to a point without leaving D. (See Exercise 33.) On the other hand, space itself *is* simply connected.

Finally, we assume D to be **connected,** which in an open region means that every point can be connected to every other point by a smooth curve that lies in the region. See Fig. 14.67.

The Implications in the Diagram

We now examine the implications in the preceding diagram.

THEOREM 6

The Fundamental Theorem of Line Integrals

1. Let $\mathbf{F} = M\mathbf{i} + N\mathbf{j} + P\mathbf{k}$ be a vector field whose components are continuous throughout an open connected region D in space. Then there exists a differentiable function f such that

$$\mathbf{F} = \nabla f = \frac{\partial f}{\partial x}\mathbf{i} + \frac{\partial f}{\partial y}\mathbf{j} + \frac{\partial f}{\partial z}\mathbf{k}$$

if and only if for all points A and B in D the value of $\int_A^B \mathbf{F}\cdot d\mathbf{r}$ is independent of the path joining A to B in D.

2. If the integral is independent of the path from A to B, its value is

$$\int_A^B \mathbf{F}\cdot d\mathbf{r} = f(B) - f(A).$$

Proof That $\mathbf{F} = \nabla f$ Implies Path Independence of $\int \mathbf{F}\cdot d\mathbf{r}$ Suppose that A and B are two points in D and that

$$C:\quad \mathbf{r} = g(t)\,\mathbf{i} + h(t)\,\mathbf{j} + k(t)\,\mathbf{k},\quad a \le t \le b,$$

is a smooth curve in D joining A and B. Along the curve, f is a differentiable function of t and

$$\begin{aligned} \frac{df}{dt} &= \frac{\partial f}{\partial x}\frac{dx}{dt} + \frac{\partial f}{\partial y}\frac{dy}{dt} + \frac{\partial f}{\partial z}\frac{dz}{dt} && \text{(Chain Rule)} \\ &= \nabla f\cdot\left[\frac{dx}{dt}\mathbf{i} + \frac{dy}{dt}\mathbf{j} + \frac{dz}{dt}\mathbf{k}\right] = \nabla f\cdot\frac{d\mathbf{r}}{dt} = \mathbf{F}\cdot\frac{d\mathbf{r}}{dt}. && \left(\begin{array}{c}\text{Because}\\ \mathbf{F} = \nabla f\end{array}\right) \end{aligned} \tag{2}$$

Therefore,

$$\int_C \mathbf{F}\cdot d\mathbf{r} = \int_a^b \mathbf{F}\cdot\frac{d\mathbf{r}}{dt}\,dt = \int_a^b \frac{df}{dt}\,dt = f(g(t), h(t), k(t))\Big]_a^b = f(B) - f(A). \quad \text{(Eq. (2))}$$

Thus, the value of the work integral depends only on the values of f at A and B and not on the path in between. This proves Part 2 as well as the forward implication in Part 1.

Proof That Path Independence of $\int \mathbf{F}\cdot d\mathbf{r}$ Implies $\mathbf{F} = \nabla f$

We need to show that if $\mathbf{F} = M\mathbf{i} + N\mathbf{j} + P\mathbf{k}$ is conservative on D then there exists a scalar function f defined on D for which

$$\frac{\partial f}{\partial x} = M, \qquad \frac{\partial f}{\partial y} = N, \qquad \text{and} \qquad \frac{\partial f}{\partial z} = P. \tag{3}$$

To define f, we choose an arbitrary basepoint A in D. We define the value of f at A to be zero and the value of f at every other point B of D to be the value of the integral of $\mathbf{F}\cdot d\mathbf{r}$ along some smooth path in D from A to B. Such paths exist because D is a connected open region, and all such paths assign the same value to $f(B)$ because the integral of $\mathbf{F}\cdot d\mathbf{r}$ is path independent in D. Since the integral of $\mathbf{F}\cdot d\mathbf{r}$ from A to A is zero, the values of f are all defined by the single equation

$$f(B) = \int_A^B \mathbf{F}\cdot d\mathbf{r}. \tag{4}$$

One consequence of this definition is that for any points B and B_0 in D,

$$\begin{aligned} f(B) - f(B_0) &= \int_A^B \mathbf{F}\cdot d\mathbf{r} - \int_A^{B_0} \mathbf{F}\cdot d\mathbf{r} \\ &= \int_{B_0}^A \mathbf{F}\cdot d\mathbf{r} + \int_A^B \mathbf{F}\cdot d\mathbf{r} = \int_{B_0}^B \mathbf{F}\cdot d\mathbf{r}. \end{aligned} \tag{5}$$

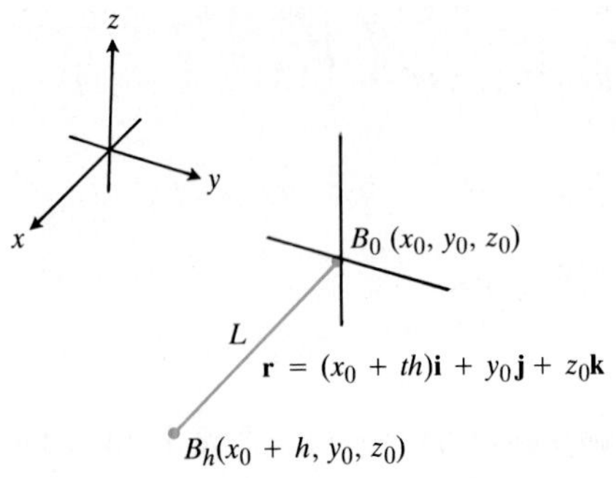

Let us now show that f has partial derivatives that give the components of **F** at every point $B_0(x_0, y_0, z_0)$ in D.

Since D is open, B_0 is the center of a solid sphere that lies entirely in D. We can therefore find a positive number h (it might be small) for which the point $B_h(x_0+h,\ y_0,\ z_0)$ and the line segment L joining B_0 to B_h both lie in D (Fig. 14.68). When we calculate the value of the integral of $\mathbf{F}\cdot d\mathbf{r}$ along L with the parametrization

$$\mathbf{r} = (x_0 + th)\,\mathbf{i} + y_0\,\mathbf{j} + z_0\,\mathbf{k}, \qquad 0 \le t \le 1, \tag{6}$$

we find that

$$\begin{aligned}\int_{B_0}^{B_h} \mathbf{F}\cdot d\mathbf{r} &= \int_0^1 (M\mathbf{i} + N\mathbf{j} + P\mathbf{k})\cdot(h\,dt\,\mathbf{i})\\ &= h\int_0^1 M(x_0 + th,\ y_0,\ z_0)\,dt.\end{aligned} \tag{7}$$

x y z $B_0\ (x_0, y_0, z_0)$ L $\mathbf{r} = (x_0 + th)\mathbf{i} + y_0\mathbf{j} + z_0\mathbf{k}$ $B_h(x_0 + h,\ y_0,\ z_0)$

14.68 The line segment L from B_0 to B_h lies parallel to the x-axis. The partial derivative with respect to x of the function

$$f(x, y, z) = \int_A^{(x,y,z)} \mathbf{F}\cdot d\mathbf{r}$$

at B_0 is the limit of $(1/h)\int_L \mathbf{F}\cdot d\mathbf{r}$ as $h\to 0$.

The difference quotient for defining $\partial f/\partial x$ at B_0 is therefore

$$\begin{aligned}\frac{f(x_0+h,\ y_0,\ z_0) - f(x_0,\ y_0,\ z_0)}{h} &= \frac{f(B_h) - f(B_0)}{h}\\ &= \frac{1}{h}\int_{B_0}^{B_h} \mathbf{F}\cdot d\mathbf{r} && \text{(Eq. (5))}\\ &= \int_0^1 M(x_0 + th,\ y_0,\ z_0)\,dt. && \text{(Eq. (7))}\end{aligned}$$

Since M is continuous, given any $\epsilon > 0$ there exists a $\delta > 0$ such that

$$|th| < \delta \quad \Rightarrow \quad |M(x_0 + th,\ y_0,\ z_0) - M(x_0,\ y_0,\ z_0)| < \epsilon.$$

This implies that whenever $|h| < \delta$

$$\begin{aligned}&\left|\int_0^1 M(x_0 + th,\ y_0,\ z_0)\,dt - M(x_0,\ y_0,\ z_0)\right|\\ &\qquad = \left|\int_0^1 (M(x_0 + th,\ y_0,\ z_0) - M(x_0,\ y_0,\ z_0))\,dt\right|\\ &\qquad \le \int_0^1 \left|M(x_0 + th,\ y_0,\ z_0) - M(x_0,\ y_0,\ z_0)\right| dt < \int_0^1 \epsilon\,dt = \epsilon.\end{aligned}$$

We can find such a positive δ for every positive ϵ, so

$$\begin{aligned}\left.\frac{\partial f}{\partial x}\right|_{(x_0,\,y_0,\,z_0)} &= \lim_{h\to 0}\int_0^1 M(x_0 + th,\ y_0,\ z_0)\,dt\\ &= M(x_0,\ y_0,\ z_0).\end{aligned} \tag{8}$$

In other words, the partial derivative of f with respect to x exists at B_0 and equals the value of M at B_0. Since B_0 was chosen arbitrarily, we can conclude that $\partial f/\partial x$ exists and equals M at every point of D.

The equations $\partial f/\partial y = N$ and $\partial f/\partial z = P$ are derived in a similar way. This concludes the proof. ∎

THEOREM 7

If $\nabla \times \mathbf{F} = \mathbf{0}$ at every point of a simply connected open region D in space, then

$$\oint_C \mathbf{F} \cdot d\mathbf{r} = 0$$

on any piecewise smooth closed path C in D.

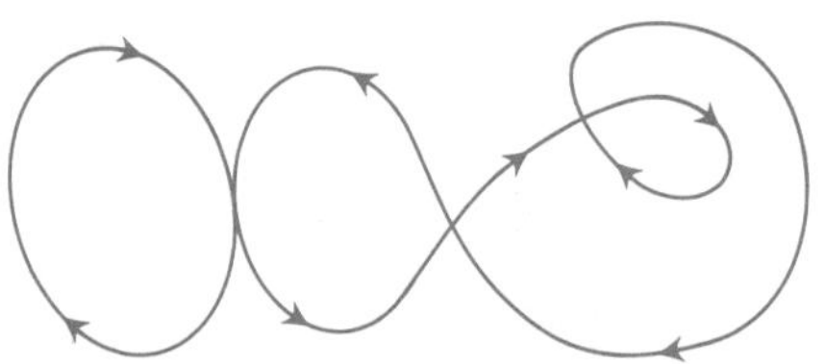

14.69 In a simply connected open region in space, differentiable curves that cross themselves can be divided into loops to which Stokes's theorem applies.

Sketch of a Proof The theorem is usually proved in two steps. The first step is for simple closed curves. A theorem from topology, a branch of advanced mathematics, states that every differentiable simple closed curve C in a simply connected open region D in space is the boundary of a smooth two-sided surface S that also lies in D. Hence, by Stokes's theorem,

$$\oint_C \mathbf{F} \cdot d\mathbf{r} = \iint_S \nabla \times \mathbf{F} \cdot \mathbf{n}\, d\sigma = 0.$$

The second step is for curves that cross themselves, like the one in Fig. 14.69. The idea is to break these into simple loops spanned by orientable surfaces, apply Stokes's theorem one loop at a time, and add the results. ∎

THEOREM 8

The following statements are equivalent:

1. $\int \mathbf{F} \cdot d\mathbf{r} = 0$ around every closed loop in D.
2. The field $\mathbf{F}$ is conservative on D.

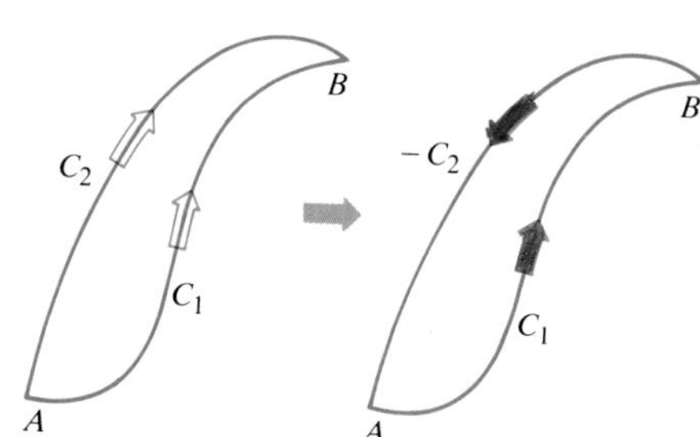

14.70 If we have two paths from A to B, one of them can be reversed to make a loop.

Proof That (1) ⇒ (2) We want to show that for any two points A and B in D the integral of $\mathbf{F} \cdot d\mathbf{r}$ has the same value over any two paths C_1 and C_2 from A to B. We reverse the direction on C_2 to make a path $-C_2$ from B to A (Fig. 14.70). Together, C_1 and $-C_2$ make a closed loop C, and

$$\int_{C_1} \mathbf{F} \cdot d\mathbf{r} - \int_{C_2} \mathbf{F} \cdot d\mathbf{r} = \int_{C_1} \mathbf{F} \cdot d\mathbf{r} + \int_{-C_2} \mathbf{F} \cdot d\mathbf{r} = \int_C \mathbf{F} \cdot d\mathbf{r} = 0.$$

Thus the integrals over C_1 and C_2 give the same value.

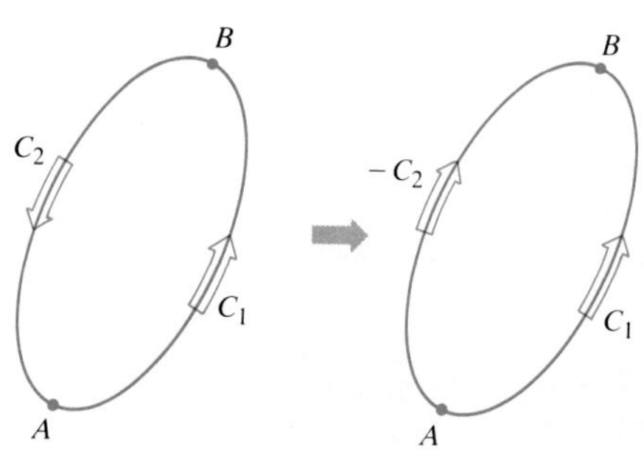

14.71 If A and B lie on a loop, we can reverse part of the loop to make two paths from A to B.

Proof That (2) ⇒ (1) We want to show that the integral of $\mathbf{F} \cdot d\mathbf{r}$ is zero over any closed loop C. We pick two points A and B on C and use them to break C into two pieces: C_1 from A to B followed by C_2 from B back to A (Fig. 14.71). Then

$$\oint_C \mathbf{F} \cdot d\mathbf{r} = \int_{C_1} \mathbf{F} \cdot d\mathbf{r} + \int_{C_2} \mathbf{F} \cdot d\mathbf{r} = \int_A^B \mathbf{F} \cdot d\mathbf{r} - \int_A^B \mathbf{F} \cdot d\mathbf{r} = 0.$$

Finding Potentials for Conservative Fields

As we said earlier, conservative fields get their name from the fact that the principle of conservation of energy holds in them. The test for being conservative, as we have just seen, is that $\nabla \times \mathbf{F} = \mathbf{0}$. In terms of components, this gives us the following test.

Component Test for Conservative Fields

The field $\mathbf{F} = M(x, y, z)\,\mathbf{i} + N(x, y, z)\,\mathbf{j} + P(x, y, z)\,\mathbf{k}$ is conservative if and only if

$$\frac{\partial P}{\partial y} = \frac{\partial N}{\partial z}, \qquad \frac{\partial M}{\partial z} = \frac{\partial P}{\partial x}, \qquad \text{and} \qquad \frac{\partial N}{\partial x} = \frac{\partial M}{\partial y}. \tag{9}$$

When we know that $\mathbf{F}$ is conservative, we usually want to find a potential function for $\mathbf{F}$. This requires solving the equation

$$\nabla f = \frac{\partial f}{\partial x}\mathbf{i} + \frac{\partial f}{\partial y}\mathbf{j} + \frac{\partial f}{\partial z}\mathbf{k} = M\mathbf{i} + N\mathbf{j} + P\mathbf{k} \tag{10}$$

for f. We accomplish this by integrating the three equations

$$\frac{\partial f}{\partial x} = M, \qquad \frac{\partial f}{\partial y} = N, \qquad \frac{\partial f}{\partial z} = P, \tag{11}$$

as in the following example.

Example 1 Show that

$$\mathbf{F} = (e^x \cos y + yz)\,\mathbf{i} + (xz - e^x \sin y)\,\mathbf{j} + (xy + z)\,\mathbf{k}$$

is conservative and find a potential function for it. Then use the potential function to find the value of the integral of $\mathbf{F}$ along paths from $(0, 0, 0)$ to $(-1, \pi/2, 2)$.

Solution To show that $\mathbf{F}$ is conservative, we apply the test in Eq. (9) to

$$M = e^x \cos y + yz, \qquad N = xz - e^x \sin y, \qquad P = xy + z$$

and calculate

$$\frac{\partial M}{\partial z} = y = \frac{\partial P}{\partial x}, \qquad \frac{\partial N}{\partial z} = x = \frac{\partial P}{\partial y}, \qquad \frac{\partial M}{\partial y} = -e^x \sin y + z = \frac{\partial N}{\partial x}.$$

Together, these equalities tell us that $\mathbf{F}$ is conservative and that there is a function f with $\nabla f = \mathbf{F}$.

We find a potential function f by integrating the equations

$$\frac{\partial f}{\partial x} = e^x \cos y + yz, \qquad \frac{\partial f}{\partial y} = xz - e^x \sin y, \qquad \frac{\partial f}{\partial z} = xy + z. \tag{12}$$

We integrate the first equation with respect to x, holding y and z fixed, to get

$$f(x, y, z) = e^x \cos y + xyz + g(y, z).$$

We write the constant of integration as a function of y and z because its value may change if y and z change. We then calculate $\partial f/\partial y$ from this equation and match it with the expression for $\partial f/\partial y$ in Eq. (12). This gives

$$-e^x \sin y + xz + \frac{\partial g}{\partial y} = xz - e^x \sin y,$$

so

$$\frac{\partial g}{\partial y} = 0.$$

Therefore, g is a function of z alone, and

$$f(x, y, z) = e^x \cos y + xyz + h(z).$$

We now calculate $\partial f/\partial z$ from this equation and match it to the formula for $\partial f/\partial z$ in Eq. (12). This gives

$$xy + \frac{dh}{dz} = xy + z, \qquad \text{or} \qquad \frac{dh}{dz} = z,$$

so

$$h(z) = \frac{z^2}{2} + C.$$

Hence,

$$f(x, y, z) = e^x \cos y + xyz + \frac{z^2}{2} + C.$$

We have found infinitely many potential functions for $\mathbf{F}$, one for each value of C.

By the Fundamental Theorem of Line Integrals, the value of the integral of $\mathbf{F}$ over every path from (0, 0, 0) to $(-1, \pi/2, 2)$ is

$$\begin{aligned}\int_{(0,0,0)}^{(-1,\pi/2,2)} \mathbf{F}\cdot d\mathbf{r} &= f(-1, \pi/2, 2) - f(0, 0, 0)\\ &= \left(0 + (-1)\left(\frac{\pi}{2}\right)(2) + \frac{(2)^2}{2} + C\right) - (1 + 0 + 0 + C)\\ &= -\pi + 2 - 1 = -\pi + 1.\end{aligned}$$

Example 2 Show that $\mathbf{F} = (2x - 3)\,\mathbf{i} - z\mathbf{j} + (\cos z)\,\mathbf{k}$ is not conservative.

Solution We apply the component test in Eq. (9) and find right away that

$$\frac{\partial P}{\partial y} = \frac{\partial}{\partial y}(\cos z) = 0, \qquad \frac{\partial N}{\partial z} = \frac{\partial}{\partial z}(-z) = -1.$$

The two are unequal, so $\mathbf{F}$ is not conservative. No further testing is required.

Exact Differential Forms

The forms $M\,dx + N\,dy + P\,dz$ that have been appearing in our line integrals are called differential forms. The most useful differential forms are differentials of scalar functions.

DEFINITIONS

The form

$$M(x, y, z)\,dx + N(x, y, z)\,dy + P(x, y, z)\,dz \tag{13}$$

is called a **differential form.** A differential form is **exact** on a domain D in space if

$$M\,dx + N\,dy + P\,dz = \frac{\partial f}{\partial x}dx + \frac{\partial f}{\partial y}dy + \frac{\partial f}{\partial z}dz = df \tag{14}$$

for some scalar function f throughout D.

Notice that if $M\,dx + N\,dy + P\,dz = df$ on D, then the field $\mathbf{F} = M\mathbf{i} + N\mathbf{j} + P\mathbf{k}$ is the gradient field of f on D. Conversely, if $\mathbf{F} = \nabla f$, then the form $M\,dx + N\,dy + P\,dz$ is exact. The test for the form's being exact is therefore the same as the test for $\mathbf{F}$'s being conservative.

The Test for Exactness of $M\,dx + N\,dy + P\,dz$

The differential form $M\,dx + N\,dy + P\,dz$ is exact if and only if

$$\frac{\partial P}{\partial y} = \frac{\partial N}{\partial z}, \quad \frac{\partial M}{\partial z} = \frac{\partial P}{\partial x}, \quad \text{and} \quad \frac{\partial N}{\partial x} = \frac{\partial M}{\partial y}. \tag{15}$$

This is equivalent to saying that the field $\mathbf{F} = M\mathbf{i} + N\mathbf{j} + P\mathbf{k}$ is conservative or that

$$\nabla \times \mathbf{F} = \mathbf{0}. \tag{16}$$

Example 3 Show that $y\,dx + x\,dy + 4\,dz$ is exact and evaluate the integral

$$\int_{(1,1,1)}^{(2,3,-1)} y\,dx + x\,dy + 4\,dz$$

over the line segment from $(1, 1, 1)$ to $(2, 3, -1)$.

Solution We let

$$M = y, \qquad N = x, \qquad P = 4$$

and apply the test of Eq. (15):

$$\frac{\partial M}{\partial z} = 0 = \frac{\partial P}{\partial x}, \quad \frac{\partial N}{\partial z} = 0 = \frac{\partial P}{\partial y}, \quad \frac{\partial M}{\partial y} = 1 = \frac{\partial N}{\partial x}.$$

These equalities tell us that $y\,dx + x\,dy + 4\,dz$ is exact, so

$$y\,dx + x\,dy + 4\,dz = df$$

for some function f, and the integral's value is $f(2, 3, -1) - f(1, 1, 1)$.

We find f up to a constant by integrating the equations

$$\frac{\partial f}{\partial x} = y, \quad \frac{\partial f}{\partial y} = x, \quad \frac{\partial f}{\partial z} = 4. \tag{17}$$

From the first equation we get

$$f(x, y, z) = xy + g(y, z).$$

The second equation tells us that

$$\frac{\partial f}{\partial y} = x + \frac{\partial g}{\partial y} = x, \qquad \text{or} \qquad \frac{\partial g}{\partial y} = 0.$$

Hence, g is a function of z alone, and

$$f(x, y, z) = xy + h(z).$$

The third of Eqs. (17) tells us that

$$\frac{\partial f}{\partial z} = 0 + \frac{dh}{dz} = 4, \qquad \text{or} \qquad h(z) = 4z + C.$$

Therefore,

$$f(x, y, z) = xy + 4z + C,$$

and the value of the integral is

$$f(2, 3, -1) - f(1, 1, 1) = 2 + C - (5 + C) = -3.$$

In the next chapter, we shall see how exact differential forms sometimes arise in differential equations.

EXERCISES 14.7

Which fields in Exercises 1–6 are conservative and which are not? Explain.

1. $\mathbf{F} = yz\mathbf{i} + xz\mathbf{j} + xy\mathbf{k}$
2. $\mathbf{F} = (y \sin z)\,\mathbf{i} + (x \sin z)\,\mathbf{j} + (xy \cos z)\,\mathbf{k}$
3. $\mathbf{F} = y\mathbf{i} + (x + z)\,\mathbf{j} - y\mathbf{k}$
4. $\mathbf{F} = -y\mathbf{i} + x\mathbf{j}$
5. $\mathbf{F} = (z + y)\,\mathbf{i} + z\mathbf{j} + (y + x)\,\mathbf{k}$
6. $\mathbf{F} = (e^x \cos y)\,\mathbf{i} - (e^x \sin y)\,\mathbf{j} + z\mathbf{k}$

In Exercises 7–12, find a potential function f for the field $\mathbf{F}$.

7. $\mathbf{F} = 2x\mathbf{i} + 3y\mathbf{j} + 4z\mathbf{k}$
8. $\mathbf{F} = (y + z)\,\mathbf{i} + (x + z)\,\mathbf{j} + (x + y)\,\mathbf{k}$
9. $\mathbf{F} = e^{y+2z}(\mathbf{i} + x\mathbf{j} + 2x\mathbf{k})$
10. $\mathbf{F} = (y \sin z)\,\mathbf{i} + (x \sin z)\,\mathbf{j} + (xy \cos z)\,\mathbf{k}$
11. $\mathbf{F} = (\ln x + \sec^2(x + y))\,\mathbf{i} + \left(\sec^2(x + y) + \dfrac{y}{y^2 + z^2}\right)\mathbf{j} + \dfrac{z}{y^2 + z^2}\,\mathbf{k}$
12. $\mathbf{F} = \dfrac{y}{1 + x^2y^2}\,\mathbf{i} + \left(\dfrac{x}{1 + x^2y^2} + \dfrac{z}{\sqrt{1 - y^2z^2}}\right)\mathbf{j} + \left(\dfrac{y}{\sqrt{1 - y^2z^2}} + \dfrac{1}{z}\right)\mathbf{k}$

Evaluate the integrals in Exercises 13–22.

13. $\displaystyle\int_{(0,0,0)}^{(2,3,-6)} 2x\,dx + 2y\,dy + 2z\,dz$
14. $\displaystyle\int_{(1,1,2)}^{(3,5,0)} yz\,dx + xz\,dy + xy\,dz$
15. $\displaystyle\int_{(0,0,0)}^{(1,2,3)} 2xy\,dx + (x^2 - z^2)\,dy - 2yz\,dz$
16. $\displaystyle\int_{(1,1,1)}^{(1,2,3)} 3x^2\,dx + \frac{z^2}{y}\,dy + 2z \ln y\,dz$
17. $\displaystyle\int_{(1,0,0)}^{(0,1,1)} \sin y \cos x\,dx + \cos y \sin x\,dy + dz$
18. $\displaystyle\int_{(0,0,0)}^{(3,3,1)} 2x\,dx - y^2\,dy - \frac{4}{1 + z^2}\,dz$
19. $\displaystyle\int_{(0,2,1)}^{(1,\pi/2,2)} 2\cos y\,dx + \left(\frac{1}{y} - 2x \sin y\right)dy + \frac{1}{z}\,dz$
20. $\displaystyle\int_{(1,2,1)}^{(2,1,1)} (2x \ln y - yz)\,dx + \left(\frac{x^2}{y} - xz\right)dy - xy\,dz$
21. $\displaystyle\int_{(1,1,1)}^{(2,2,2)} \frac{1}{y}\,dx + \left(\frac{1}{z} - \frac{x}{y^2}\right)dy - \frac{y}{z^2}\,dz$
22. $\displaystyle\int_{(-1,-1,-1)}^{(2,2,2)} \frac{2x\,dx + 2y\,dy + 2z\,dz}{x^2 + y^2 + z^2}$
23. Evaluate the integral

$$\int_{(1,1,1)}^{(2,3,-1)} y\,dx + x\,dy + 4\,dz$$

from Example 3 by finding parametric equations for the line segment from $(1, 1, 1)$ to $(2, 3, -1)$ and evaluating the line integral of $\mathbf{F} = y\mathbf{i} + x\mathbf{j} + 4\mathbf{k}$ along the segment. This is permissible in this case because $\mathbf{F}$ is conservative and the integral is path independent.

24. Evaluate

$$\int_C x^2\,dx + yz\,dy + (y^2/2)\,dz$$

along the line segment C joining $(0, 0, 0)$ to $(0, 3, 4)$.

Show that the values of the integrals in Exercises 25 and 26 do not depend on the path taken from A to B.

25. $\displaystyle\int_A^B z^2\,dx + 2y\,dy + 2xz\,dz$

26. $\displaystyle\int_A^B \frac{x\,dx + y\,dy + z\,dz}{\sqrt{x^2 + y^2 + z^2}}$

27. Express $\mathbf{F}$ in the form ∇f if
a) $\mathbf{F} = \dfrac{2x}{y}\,\mathbf{i} + \left(\dfrac{1 - x^2}{y^2}\right)\mathbf{j}$
b) $\mathbf{F} = (e^x \ln y)\,\mathbf{i} + \left(\dfrac{e^x}{y} + \sin z\right)\mathbf{j} + (y \cos z)\,\mathbf{k}$

28. Let $\mathbf{F} = \nabla(x^3y^2)$ and let C be the path in the xy-plane from $(-1, 1)$ to $(1, 1)$ that consists of the line segment from $(-1, 1)$ to $(0, 0)$ followed by the line segment from $(0, 0)$ to $(1, 1)$. Evaluate $\int_C \mathbf{F} \cdot d\mathbf{r}$ in two ways:
a) by finding parametrizations for the segments that make up C and evaluating the integral;
b) by using the fact that $f(x, y) = x^3y^2$ is a potential function for $\mathbf{F}$.

29. Find the work done by $\mathbf{F} = (x^2 + y)\,\mathbf{i} + (y^2 + x)\,\mathbf{j} + ze^z\,\mathbf{k}$ over the following paths from $(1, 0, 0)$ to $(1, 0, 1)$.
a) The line segment $x = 1,\quad y = 0,\quad 0 \le z \le 1$
b) The helix $\mathbf{r} = (\cos t)\,\mathbf{i} + (\sin t)\,\mathbf{j} + (t/2\pi)\,\mathbf{k},\ 0 \le t \le 2\pi$
c) The x-axis from $(1, 0, 0)$ to $(0, 0, 0)$ followed by the parabola $z = x^2,\ y = 0$ from $(0, 0, 0)$ to $(1, 0, 1)$

30. Evaluate $\int_C 2x \cos y\, dx - x^2 \sin y\, dy$ along the following paths C in the xy-plane.
a) The parabola $y = (x - 1)^2$ from $(1, 0)$ to $(0, 1)$
b) The line segment from $(-1, \pi)$ to $(1, 0)$
c) The x-axis from $(-1, 0)$ to $(1, 0)$
d) The astroid $\mathbf{r} = (\cos^3 t)\,\mathbf{i} + (\sin^3 t)\,\mathbf{j},\ 0 \le t \le 2\pi$, counterclockwise from $(1, 0)$ back to $(1, 0)$

31. a) How are the constants a, b, and c related if the differential form
$$(ay^2 + 2czx)\,dx + y(bx + cz)\,dy + (ay^2 + cx^2)\,dz$$
is exact?
b) For what values of b and c will
$$\mathbf{F} = (y^2 + 2czx)\,\mathbf{i} + y(bx + cz)\,\mathbf{j} + (y^2 + cx^2)\,\mathbf{k}$$
be a gradient field?

32. Find a potential function for the gravitational field
$$\mathbf{F} = -GmM\frac{x\mathbf{i} + y\mathbf{j} + z\mathbf{k}}{(x^2 + y^2 + z^2)^{3/2}}$$
(G, m, and M are constants).

33. Show that the curl of
$$\mathbf{F} = \frac{-y}{x^2 + y^2}\,\mathbf{i} + \frac{x}{x^2 + y^2}\,\mathbf{j} + z\mathbf{k}$$
is zero but that
$$\int_C \mathbf{F} \cdot d\mathbf{r}$$
is not zero if C is the circle $x^2 + y^2 = 1$ in the xy-plane. (Theorem 7 does not apply here because the domain of $\mathbf{F}$ is not simply connected. The field $\mathbf{F}$ is not defined along the z-axis so there is no way to contract C to a point without leaving the domain of $\mathbf{F}$.)

EXPLORER PROGRAM

Scalar Fields	Plots two-dimensional scalar fields and their associated gradient fields. Integrates $\nabla f \cdot d\mathbf{r}$ along line segments parallel to the coordinate axes

REVIEW QUESTIONS

1. What is a line integral? What are line integrals used for? Give examples.
2. What is a vector field? Give examples.
3. How do you calculate the work done by a force in moving a particle along a curve?
4. What are the flux density and circulation density of a vector field? How are they related to the divergence and curl of the field?
5. What theorem relates flux to the divergence for a vector field defined in the plane? Give an example.
6. What theorem relates circulation to the curl for a vector field defined in the plane? Give an example.
7. How is surface area calculated? Give an example.
8. What is a surface integral? How are surface integrals used to calculate masses and moments? Give examples.
9. What theorem relates flux to the divergence of a vector field in space? Give an example.
10. What theorem relates circulation to the curl of a vector field in space? Give an example.
11. What does it mean for a vector field to be conservative on a region D in space?
12. What is a potential function for a conservative field? Give an example.
13. How can you evaluate $\int_A^B \mathbf{F} \cdot d\mathbf{r}$ over a path in the domain of a conservative field? Give an example.
14. What special properties do conservative fields have?
15. How do you find a potential function for a conservative field?
16. What is an exact differential form? Give an example.

MISCELLANEOUS EXERCISES

1. Figure 14.72 shows two polygonal paths in space from the origin to the point (1, 1, 1). Integrate $f(x, y, z) = 2x - 3y^2 - 2z + 3$ over each path.

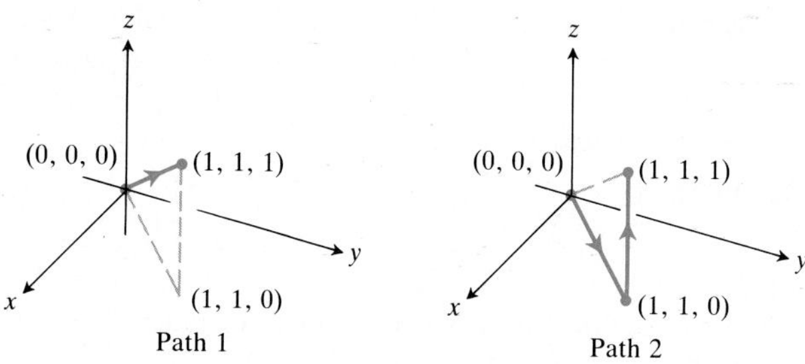

14.72 The paths in Exercise 1.

2. Figure 14.73 shows three polygonal paths from the origin to the point (1, 1, 1). Integrate $f(x, y, z) = x^2 + y - z$ over each path.

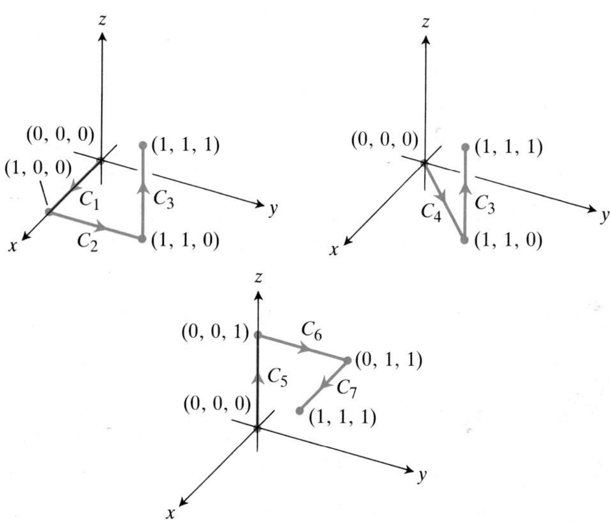

14.73 The paths in Exercise 2.

3. Integrate $f(x, y, z) = \sqrt{x^2 + z^2}$ around the circle $\mathbf{r} = (a \cos t)\mathbf{j} + (a \sin t)\mathbf{k}$, $0 \le t \le 2\pi$, in the direction of increasing t.

4. Integrate $f(x, y, z) = \sqrt{x^2 + y^2}$ over the involute curve $\mathbf{r} = (\cos t + t \sin t)\mathbf{i} + (\sin t - t \cos t)\mathbf{j}$, $0 \le t \le \sqrt{3}$, in the direction of increasing t. (Figure 11.21 shows the curve.)

5. CALCULATOR Find the value of the integral of $f(x, y, z) = y$ over the curve $y = 2\sqrt{x}$, $z = 0$, from (1, 2, 0) to (4, 4, 0) to two decimal places.

6. A slender metal arch lies along the semicircle $y = \sqrt{a^2 - x^2}$ in the xy-plane. The density at the point (x, y) on the arch is $\delta(x, y) = 2a - y$. Find the center of mass.

7. A wire of constant density $\delta = 1$ lies along the curve $\mathbf{r} = (e^t \cos t)\mathbf{i} + (e^t \sin t)\mathbf{j} + e^t\mathbf{k}$, $0 \le t \le \ln 2$. Find $\bar{z}$, I_z, and R_z.

8. A wire of constant density $\delta = 1$ lies along the curve $\mathbf{r} = (t \cos t)\mathbf{i} + (t \sin t)\mathbf{j} + (2\sqrt{2}/3)t^{3/2}\mathbf{k}$, $0 \le t \le 1$. Find $\bar{z}$, I_z, and R_z.

9. Suppose $\mathbf{F}(x, y) = (x + y)\mathbf{i} - (x^2 + y^2)\mathbf{j}$ is the velocity field of a fluid flowing across the xy-plane. Find the flow along each of the following paths from (1, 0) to (−1, 0).
a) The upper half of the circle $x^2 + y^2 = 1$
b) The line segment from (1, 0) to (−1, 0)
c) The line segment from (1, 0) to (0, −1) followed by the line segment from (0, −1) to (−1, 0)

10. Find the circulation of $\mathbf{F} = 2x\mathbf{i} + 2z\mathbf{j} + 2y\mathbf{k}$ along the closed path consisting of the helix $\mathbf{r}_1 = (\cos t)\mathbf{i} + (\sin t)\mathbf{j} + t\mathbf{k}$, $0 \le t \le \pi/2$, followed by the line segments $\mathbf{r}_2 = \mathbf{j} + (\pi/2)(1 - t)\mathbf{k}$, $0 \le t \le 1$, and $\mathbf{r}_3 = t\mathbf{i} + (1 - t)\mathbf{j}$, $0 \le t \le 1$.

Sketch the vector fields in Exercises 11 and 12 in the xy-plane.

11. $\mathbf{F} = (x - y)\mathbf{i} + (x + y)\mathbf{j}$

12. $\mathbf{F} = (x\mathbf{i} + y\mathbf{j})/(x^2 + y^2)$

Use Green's theorem to find the counterclockwise circulation and outward flux for the fields and curves in Exercises 13 and 14.

13. $\mathbf{F} = (2xy + x)\mathbf{i} + (xy - y)\mathbf{j}$
C: The square bounded by $x = 0$, $x = 1$, $y = 0$, $y = 1$

14. $\mathbf{F} = (y - 6x^2)\mathbf{i} + (x + y^2)\mathbf{j}$
C: The triangle made by the lines $y = 0$, $y = x$, and $x = 1$

15. Show that

$$\oint_C \ln x \sin y \, dy - \frac{\cos y}{x}\, dx = 0$$

for any closed curve C to which Green's theorem applies.

16. a) Show that the outward flux of the position vector field $\mathbf{F} = x\mathbf{i} + y\mathbf{j}$ across any closed curve to which Green's theorem applies is twice the area of the region enclosed by the curve.
b) Let $\mathbf{n}$ be the outward unit normal vector to a closed curve to which Green's theorem applies. Show that it is not possible for $\mathbf{F} = x\mathbf{i} + y\mathbf{j}$ to be orthogonal to $\mathbf{n}$ at every point of C.

Apply Green's theorem in one of its two forms to evaluate the line integrals in Exercises 17 and 18.

17. $\displaystyle\int_C 8x \sin y\, dx - 8y \cos x\, dy$

C is the square cut from the first quadrant by the lines $x = \pi/2$ and $y = \pi/2$.

18. $\displaystyle\int_C y^2\, dx + x^2\, dy$

C is the circle $x^2 + y^2 = 4$.

19. a) Find the area enclosed by the curve

$$\mathbf{r} = \left(\frac{2\cos t - \sin t}{2}\right)\mathbf{i} + (\sin t)\,\mathbf{j}, \quad 0 \le t \le 2\pi.$$

b) Find an equation for the curve in Cartesian coordinates and use the equation to identify the curve.

20. Among all rectangular regions $0 \le x \le a$, $0 \le y \le b$, find the one for which the total outward flux of $\mathbf{F} = (x^2 + 4xy)\,\mathbf{i} - 6y\mathbf{j}$ across the four sides is least.

21. Find the area of the elliptical region cut from the plane $x + y + z = 1$ by the cylinder $x^2 + y^2 = 1$.

22. Find the area of the cap cut from the paraboloid $y^2 + z^2 = 3x$ by the plane $x = 1$.

23. Find the area of the cap cut from the top of the sphere $x^2 + y^2 + z^2 = 1$ by the plane $z = \sqrt{2}/2$.

24. a) Find the area of the surface cut from the hemisphere $x^2 + y^2 + z^2 = 4$, $z \ge 0$, by the cylinder $x^2 + y^2 = 2x$ (Fig. 14.74).

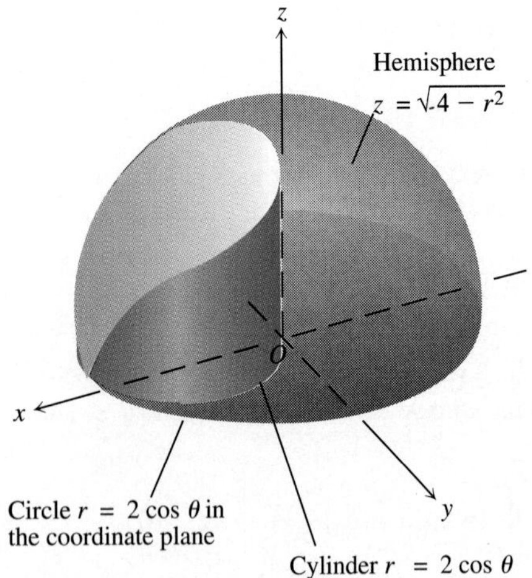

14.74 The surface cut from the hemisphere $x^2 + y^2 + z^2 = 4$, $z \ge 0$, by the cylinder $x^2 + y^2 = 2x$ in Exercise 24. In cylindrical coordinates, the equations of the hemisphere and the cylinder are $z = \sqrt{4 - r^2}$ and $r = 2\cos\theta$.

b) Find the area of the portion of the cylinder that lies inside the hemisphere. (*Hint:* Project onto the xz-plane. Or evaluate the integral $\int h\, ds$, where h is the altitude of the cylinder and ds is the element of arc length on the circle $x^2 + y^2 = 2x$ in the xy-plane.)

25. Find the area of the region in which the plane $(x/a) + (y/b) + (z/c) = 1$ $(a, b, c > 0)$ intersects the first octant. Check your answer with an appropriate vector calculation.

26. Let S be the surface that is bounded on the left by the hemisphere $x^2 + y^2 + z^2 = a^2$, $y \le 0$, in the middle by the cylinder $x^2 + z^2 = a^2$, $0 \le y \le a$, and on the right by the plane $y = a$. Find the flux of the field $\mathbf{F} = y\mathbf{i} + z\mathbf{j} + x\mathbf{k}$ outward across S.

27. Integrate

a) $g(x, y, z) = \dfrac{yz}{\sqrt{4y^2 + 1}}$ b) $g(x, y, z) = \dfrac{z}{\sqrt{4y^2 + 1}}$

over the surface cut from the parabolic cylinder $y^2 - z = 1$ by the planes $x = 0$, $x = 3$, and $z = 0$.

28. Integrate $g(x, y, z) = x^4y(y^2 + z^2)$ over the surface in the first quadrant cut from the cylinder $y^2 + z^2 = 25$ by the coordinate planes and the planes $x = 1$, $y = 4$, and $z = 3$.

29. Find the outward flux of the field $\mathbf{F} = 2xy\mathbf{i} + 2yz\mathbf{j} + 2xz\mathbf{k}$ across the surface of the cube cut from the first octant by the planes $x = 1$, $y = 1$, $z = 1$.

30. Find the outward flux of the field $\mathbf{F} = xz\mathbf{i} + yz\mathbf{j} + \mathbf{k}$ across the entire surface of the upper cap cut from the solid sphere $x^2 + y^2 + z^2 \le 25$ by the plane $z = 3$.

31. Find I_z, R_z, and the center of mass of a thin shell of density $\delta(x, y, z) = z$ cut from the upper portion of the sphere $x^2 + y^2 + z^2 = 25$ by the plane $z = 3$.

32. Find the moment of inertia about the z-axis of the surface of the cube cut from the first octant by the planes $x = 1$, $y = 1$, and $z = 1$ if the density is $\delta = 1$.

In Exercises 33–36, find the outward flux of $\mathbf{F}$ across the boundary of D.

33. $\mathbf{F} = 2xy\mathbf{i} + 2yz\mathbf{j} + 2xz\mathbf{k}$
D: The cube cut from the first octant by the planes $x = 1$, $y = 1$, $z = 1$

34. $\mathbf{F} = xz\mathbf{i} + yz\mathbf{j} + \mathbf{k}$
D: The upper cap cut from the solid sphere $x^2 + y^2 + z^2 \le 25$ by the plane $z = 3$

35. $\mathbf{F} = -2x\mathbf{i} - 3y\mathbf{j} + z\mathbf{k}$
D: The region cut from the solid sphere $x^2 + y^2 + z^2 \le 2$ by the paraboloid $z = x^2 + y^2$

36. $\mathbf{F} = (6x + y)\mathbf{i} - (x + z)\mathbf{j} + 4yz\mathbf{k}$
D: The region in the first octant bounded by the cone $z = \sqrt{x^2 + y^2}$, the cylinder $x^2 + y^2 = 1$, and the coordinate planes

37. Let S be the surface defined by the relations

$$y - \ln x = 0, \quad 1 \le x \le e, \quad 0 \le z \le 1,$$

and let $\mathbf{n}$ be the unit normal field on S that has $\mathbf{n} \cdot \mathbf{j} > 0$. Find the flux of the field $\mathbf{F} = 2y\mathbf{j} + z\mathbf{k}$ across S in the direction of $\mathbf{n}$.

38. CALCULATOR A lead radiation shield 2 in. thick is to be constructed in the shape of a circular cylinder surmounted by a parabolic dome. The equations for the dome and cylinder (measurements in feet) are

Dome: $z = 45 - 0.09x^2 - 0.09y^2, \quad 9 \le z \le 45,$

Cylinder: $x^2 + y^2 = 400, \quad 0 \le z \le 9.$

Lead weighs about 707 lb/ft³. About how many pounds of lead will the shield contain?

39. Find the outward flux of the field $\mathbf{F} = 3xz^2\mathbf{i} + y\mathbf{j} - z^3\mathbf{k}$ across the surface of the solid in the first octant that is bounded by the cylinder $x^2 + 4y^2 = 16$ and the planes $y = 2z$, $x = 0$, and $z = 0$.

40. Find the flux of $\mathbf{F} = (3z + 1)\,\mathbf{k}$ upward across the hemisphere $x^2 + y^2 + z^2 = a^2$, $z \ge 0$ (a) with the Divergence Theorem, (b) by evaluating the flux integral directly.

In Exercises 41 and 42, use the surface integral in Stokes's theorem to find the circulation of the field **F** around the curve C in the indicated direction.

41. $\mathbf{F} = y^2\mathbf{i} - y\mathbf{j} + 3z^2\mathbf{k}$

C: The ellipse in which the plane $2x + 6y - 3z = 6$ meets the cylinder $x^2 + y^2 = 1$, counterclockwise as viewed from above

42. $\mathbf{F} = (x^2 + y)\,\mathbf{i} + (x + y)\,\mathbf{j} + (4y^2 - z)\,\mathbf{k}$

C: The circle in which the plane $z = -y$ meets the sphere $x^2 + y^2 + z^2 = 4$, counterclockwise as viewed from above

Which of the fields in Exercises 43–46 are conservative and which are not?

43. $\mathbf{F} = x\mathbf{i} + y\mathbf{j} + z\mathbf{k}$

44. $\mathbf{F} = (x\mathbf{i} + y\mathbf{j} + z\mathbf{k})/(x^2 + y^2 + z^2)^{3/2}$

45. $\mathbf{F} = x\,e^y\,\mathbf{i} + y\,e^z\,\mathbf{j} + z\,e^x\,\mathbf{k}$

46. $\mathbf{F} = (\mathbf{i} + z\mathbf{j} + y\mathbf{k})/(x + yz)$

Find potential functions for the fields in Exercises 47 and 48.

47. $\mathbf{F} = 2\,\mathbf{i} + (2y + z)\,\mathbf{j} + (y + 1)\,\mathbf{k}$

48. $\mathbf{F} = (z\cos xz)\,\mathbf{i} + e^y\,\mathbf{j} + (x\cos xz)\,\mathbf{k}$

Evaluate the integrals in Exercises 49 and 50.

49. $\displaystyle\int_{(-1,1,1)}^{(4,-3,0)} \frac{dx + dy + dz}{\sqrt{x + y + z}}$

50. $\displaystyle\int_{(1,1,1)}^{(10,3,3)} dx - \sqrt{\frac{z}{y}}\,dy - \sqrt{\frac{y}{z}}\,dz$

In Exercises 51 and 52, find the work done by each field along the paths from (0, 0, 0) to (1, 1, 1) in Fig. 14.72.

51. $\mathbf{F} = 2xy\mathbf{i} + \mathbf{j} + x^2\mathbf{k}$

52. $\mathbf{F} = 2xy\mathbf{i} + x^2\mathbf{j} + \mathbf{k}$

53. Let C be the ellipse in which the plane $2x + 3y - z = 0$ meets the cylinder $x^2 + y^2 = 12$. Show, without evaluating either line integral directly, that the circulation of the field $\mathbf{F} = x\mathbf{i} + y\mathbf{j} + z\mathbf{k}$ around C in either direction is zero.

54. Find the flow of the field $\mathbf{F} = \nabla(xy^2z^3)$

a) once around the curve C in Exercise 53, clockwise as viewed from above,

b) along the line segment from (1, 1, 1) to (2, 1, − 1).

55. Find the work done by

$$\mathbf{F} = \frac{x\mathbf{i} + y\mathbf{j}}{(x^2 + y^2)^{3/2}}$$

over the plane curve $\mathbf{r} = (e^t\cos t)\,\mathbf{i} + (e^t\sin t)\,\mathbf{j}$ from the point (1, 0) to the point $(e^{2\pi}, 0)$ in two ways:

a) by using the parametrization of the curve to evaluate the work integral,

b) by evaluating a potential function for f.

56. Integrate $\mathbf{F} = -(y\sin z)\,\mathbf{i} + (x\sin z)\,\mathbf{j} + (xy\cos z)\,\mathbf{k}$ around the circle cut from the sphere $x^2 + y^2 + z^2 = 5$ by the plane $z = -1$, clockwise as viewed from above.

Finding Areas with Green's Theorem

Use Green's theorem area formula, Eq. (29) in Exercises 14.3, to find the areas of the regions enclosed by the curves in Exercises 57–60.

57. The limaçon $x = 2\cos t - \cos 2t$, $y = 2\sin t - \sin 2t$, $0 \le t \le 2\pi$ (Fig. 14.75)

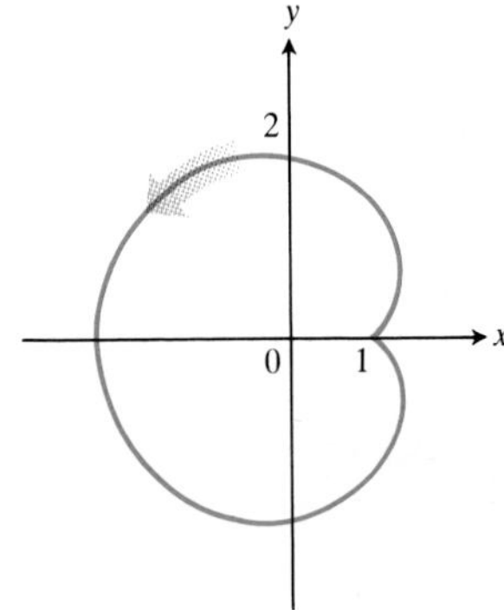

14.75 The limaçon in Exercise 57.

58. The deltoid $x = 2\cos t + \cos 2t$, $y = 2\sin t - \sin 2t$, $0 \le t \le 2\pi$ (Fig. 14.76)

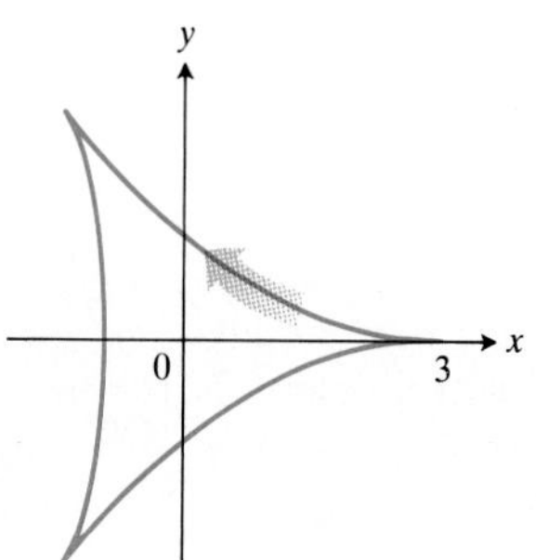

14.76 The deltoid in Exercise 58.

59. The eight curve $x = (1/2)\sin 2t$, $y = \sin t$, $0 \le t \le \pi$ (one loop) (Fig. 14.77)

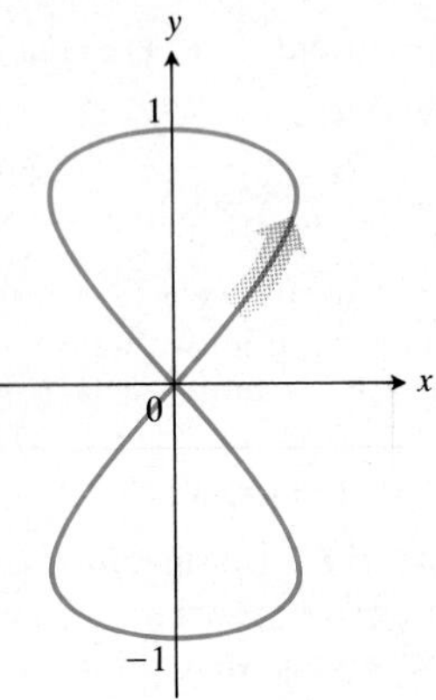

14.77 The eight curve in Exercise 59.

60. The teardrop $x = 2a \cos t - a \sin 2t$, $y = b \sin t$, $0 \le t \le 2\pi$ (Fig. 14.78)

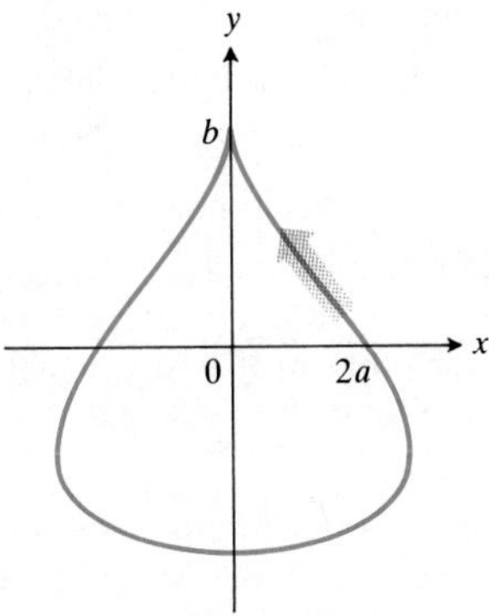

14.78 The teardrop in Exercise 60.

CHAPTER

15

DIFFERENTIAL EQUATIONS

Overview Differential equations arise when we model the effects of change, motion, and growth. They arise when we study moving particles, changing business conditions, oscillating voltages in neural networks, changing concentrations in chemical reactions, and flowing resources in a market economy. Our space program more or less runs on differential equations (implemented by computers), and mathematical models of our environment involve huge systems of differential equations. The basic tool for solving differential equations is calculus. This chapter previews common types of differential equations and the methods we use to solve them.

15.1 Separable First Order Equations

We study differential equations for their mathematical importance and for what they tell us about reality. When we study unchecked bacterial growth, for example, and assume that the rate of increase in the population at any given time t is proportional to the number of bacteria then present, we are assuming that the population size y obeys the differential equation $dy/dt = ky$. If we also know that the size of the population is y_0 when $t = 0$, we can find a formula for the size of the population at any time $t > 0$ by solving the initial value problem

$$\text{Differential equation:} \qquad \frac{dy}{dt} = ky,$$

$$\text{Initial condition:} \qquad y = y_0 \quad \text{when } t = 0.$$

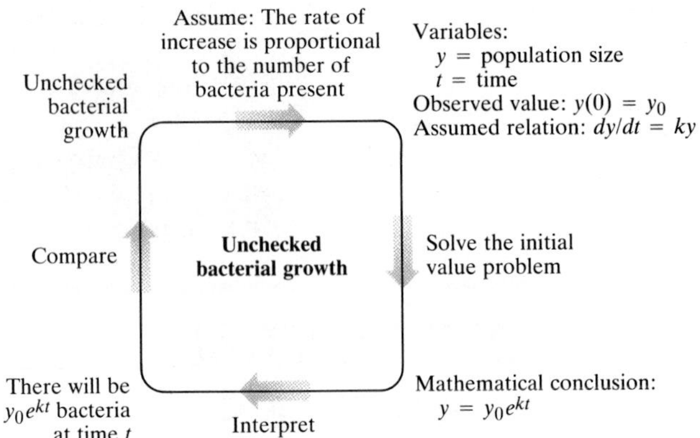

As we saw in Section 6.5, the general solution of the differential equation (the formula that gives all possible solutions) is

$$y = Ae^{kt} \qquad (A \text{ an arbitrary constant})$$

and the particular solution that satisfies the initial condition is $y = y_0 e^{kt}$.

We interpret this as saying that unchecked bacterial growth is exponential. There should be $y_0 e^{kt}$ bacteria present at time t. We know from experience that there are limits to all growth, and we do not expect this prediction to be accurate for large values of t. But for the early stages of growth the formula is a good predictor and gives valuable information about the population's size.

Definitions

A **differential equation** is an equation that contains one or more derivatives of a differentiable function. An equation with partial derivatives is called a **partial differential equation.** The wave equation and the Laplace equations in Exercises 12.3 are partial differential equations. An equation with ordinary derivatives, that is, derivatives of a function of a single variable, is called an **ordinary differential equation.** The equations in this chapter are ordinary differential equations.

The **order** of a differential equation is the order of the equation's highest order derivative. A differential equation is **linear** if it can be put in the form

$$a_n(x)\frac{d^n y}{dx^n} + a_{n-1}(x)\frac{d^{n-1}y}{dx^{n-1}} + \cdots + a_1(x)\frac{dy}{dx} + a_0(x)\,y = F(x), \tag{1}$$

where the a's are functions of x.

Example 1

First order, linear: $\dfrac{dy}{dx} = 5y; \qquad 3\dfrac{dy}{dx} - \sin x = 0$

Third order, nonlinear: $\left(\dfrac{d^3y}{dx^3}\right)^2 + \left(\dfrac{d^2y}{dx^2}\right)^5 - \dfrac{dy}{dx} = e^x$

Almost all the differential equations we have solved so far in this book have been linear equations of first or second order. This will continue to be true, but the equations will now have more variety and model a broader range of applications.

We call a function $y = f(x)$ a **solution** of a differential equation if y and its derivatives satisfy the equation. To test whether a given function solves a particular equation, we substitute the function and its derivatives into the equation. If the equation then reduces to an identity, the function solves the equation; otherwise, it does not.

Example 2 Show that for any values of the arbitrary constants C_1 and C_2 the function $y = C_1 \cos x + C_2 \sin x$ is a solution of the differential equation

$$\frac{d^2y}{dx^2} + y = 0.$$

Solution We differentiate the function twice to find d^2y/dx^2:

$$\begin{aligned} y &= C_1 \cos x + C_2 \sin x \\ \frac{dy}{dx} &= -C_1 \sin x + C_2 \cos x \\ \frac{d^2y}{dx^2} &= -C_1 \cos x - C_2 \sin x. \end{aligned}$$

We then substitute the expressions for y and d^2y/dx^2 into the differential equation to see whether the left-hand side reduces to zero. It does because

$$\frac{d^2y}{dx^2} + y = (-C_1 \cos x - C_2 \sin x) + (C_1 \cos x + C_2 \sin x) = 0.$$

So the function is a solution of the differential equation.

It can be shown that the formula $y = C_1 \cos x + C_2 \sin x$ gives all possible solutions of the equation $d^2y/dx^2 + y = 0$. A formula that gives all the solutions of a differential equation is called the **general solution** of the equation. In this sense, $y = C_1 \cos x + C_2 \sin x$ is the general solution of $d^2y/dx^2 + y = 0$. To **solve** a differential equation means to find its general solution.

Notice that the equation $d^2y/dx^2 + y = 0$ has order two and that its general solution has two arbitrary constants. The general solution of an nth order ordinary differential equation can be expected to contain n arbitrary constants.

Separable Equations

A method that sometimes works for solving a first order differential equation involves treating the derivative dy/dx as a quotient of differentials and rearranging the equation to group the y-terms alone with dy and the x-terms alone with dx. We cannot always accomplish this, but when we can, the solution may then be found by separate integrations with respect to x and y.

DEFINITIONS

A first order differential equation is **separable** if it can be put in the form

$$M(x) + N(y)\frac{dy}{dx} = 0 \tag{2}$$

or in the equivalent differential form

$$M(x)\,dx + N(y)\,dy = 0. \tag{3}$$

When we write the equation this way we say that we have **separated the variables.**

We can solve Eq. (2) by integrating both sides with respect to x to get

$$\int M(x)\,dx + \int N(y)\frac{dy}{dx}\,dx = C. \tag{4}$$

However, the second integral in this equation is equivalent to

$$\int N(y)\,dy$$

and it usually saves time to integrate the equivalent equation

$$\int M(x)\,dx + \int N(y)\,dy = C \tag{5}$$

instead. The result of the integration will express y either explicitly or implicitly as a function of x that solves Eq. (2).

Example 3 Solve the differential equation

$$\frac{dy}{dx} = (1 + y^2)\,e^x.$$

Steps for Solving a Separable First Order Differential Equation

1. Write the equation in the form $M(x)\,dx + N(y)\,dy = 0$.
2. Integrate M with respect to x and N with respect to y to obtain an equation that relates y and x.

Solution We use algebra to separate the variables and write the equation in the form $M(x)\,dx + N(y)\,dy = 0$, obtaining

$$e^x\,dx - \frac{1}{1 + y^2}\,dy = 0.$$

We then integrate to get

$$\int e^x\,dx - \int \frac{1}{1 + y^2}\,dy = C$$

$$e^x - \tan^{-1} y = C \qquad \text{(Constants of integration combined)}$$

$$\tan^{-1} y = e^x - C.$$

In this case, we can solve the resulting equation explicitly for y by taking the tangent of each side:

$$y = \tan(e^x - C).$$

Homogeneous First Order Equations

We can sometimes use a change of variable to transform a differential equation whose variables cannot be separated into one whose variables can be separated. This is the case with homogeneous first order equations.

DEFINITION

A first order differential equation is **homogeneous** if it can be put into the form

$$\frac{dy}{dx} = F\left(\frac{y}{x}\right). \tag{6}$$

We can change Eq. (6) into a separable equation with the substitutions

$$y = vx, \qquad \frac{dy}{dx} = v + x\frac{dv}{dx}. \tag{7}$$

Then Eq. (6) becomes

$$v + x\frac{dv}{dx} = F(v), \tag{8}$$

which can be rearranged algebraically to give

$$\frac{dx}{x} + \frac{dv}{v - F(v)} = 0. \tag{9}$$

With the variables separated, we can now solve the equation by integrating with respect to x and v. We can then return to x and y by substituting $v = y/x$.

Example 4 Show that the equation

$$\frac{dy}{dx} = -\frac{x^2 + y^2}{2xy}$$

is homogeneous and find the solution that satisfies the condition $y(1) = 1$.

Solution Dividing the numerator and denominator of the right-hand side by x^2 gives

$$\frac{dy}{dx} = -\frac{1 + (y/x)^2}{2(y/x)}.$$

This has the form of Eq. (6) with

$$F(v) = -\frac{1 + v^2}{2v}, \qquad \text{where} \quad v = \frac{y}{x},$$

so the original equation is homogeneous. Equation (9) becomes

$$\frac{dx}{x} + \frac{dv}{v + \dfrac{1 + v^2}{2v}} = 0, \qquad \text{or} \qquad \frac{dx}{x} + \frac{2v\,dv}{1 + 3v^2} = 0.$$

The solution of this equation can be written as

$$\ln|x| + \frac{1}{3}\ln(1 + 3v^2) = C \qquad \text{or} \qquad x^3(1 + 3v^2) = \pm e^{3C} = C'.$$

We substitute $v = y/x$ to find the corresponding xy-equation, obtaining

$$x^3\left(1 + 3\frac{y^2}{x^2}\right) = C' \quad \text{or} \quad x^3 + 3xy^2 = C'.$$

The value of C' that gives $y = 1$ when $x = 1$ is

$$(1)^3 + 3(1)(1)^2 = C' \quad \text{or} \quad C' = 4.$$

The corresponding solution (Fig. 15.1) is $x^3 + 3xy^2 = 4$.

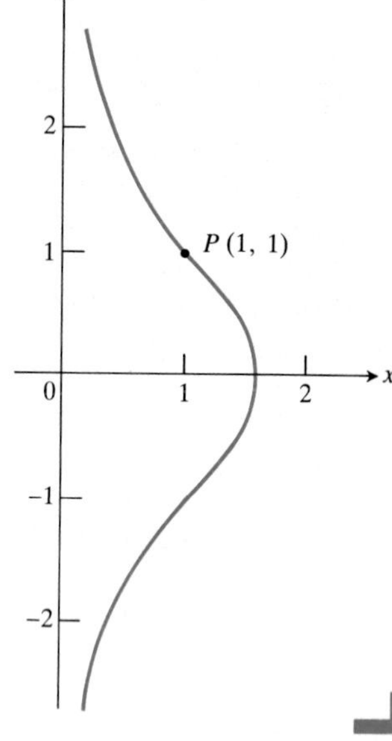

15.1 The solution of $(x^2 + y^2)\,dx + 2xy\,dy = 0$ with $y = 1$ when $x = 1$ is $x^3 + 3xy^2 = 4$ (Example 4).

EXERCISES 15.1

Find the orders of the differential equations in Exercises 1–4. Which equations are linear and which are nonlinear?

1. $y' = x^2 + y^2$

2. $(y''')^2 - 3x^2y'' + y^5 = 10$

3. $y^{(4)} + y = 1$

4. $y'' - \frac{1}{x}y' + \frac{2}{x^2}y = e^x$

In Exercises 5–8, show that each function $y = f(x)$ is a solution of the accompanying differential equation.

5. $xy'' - y' = 0$

a) $y = x^2$ b) $y = 1$ c) $y = C_1x^2 + C_2$

6. $y' + \frac{1}{x}y = 1$

a) $y = \frac{x}{2}$ b) $y = \frac{1}{x} + \frac{x}{2}$ c) $y = \frac{C}{x} + \frac{x}{2}$

7. $2y' + 3y = e^{-x}$

a) $y = e^{-x}$

b) $y = e^{-x} + e^{-(3/2)x}$

c) $y = e^{-x} + Ce^{-(3/2)x}$

8. $yy'' = 2(y')^2 - 2y'$

a) $y = 1$

b) $y = \tan x$

In Exercises 9 and 10, show that the function $y = f(x)$ is a solution of the given differential equation.

9. $y = \frac{1}{x}\int_1^x \frac{e^t}{t}\,dt, \quad x^2y' + xy = e^x$

10. $y = \frac{1}{\sqrt{1+x^4}}\int_1^x \sqrt{1+t^4}\,dt, \quad y' + \frac{2x^3}{1+x^4}y = 1$

In Exercises 11–14, show that each function $y = f(x)$ is a solution of the accompanying initial value problem.

	Differential equation	Initial condition(s)	Solution candidate
11.	$y'' = -32$	$y(5) = 400$, $y'(5) = 0$	$y = 160x - 16x^2$
12.	$2\frac{dy}{dx} + 3y = 6$	$y(0) = 0$	$y = 2\left(1 - e^{-3x/2}\right)$
13.	$y'' + 4y = 0$	$y(0) = 3$, $y'(0) = -2$	$y = 3\cos 2x - \sin 2x$
14.	$y'' - (y')^2 = 1$	$y(1) = 2$, $y'(1) = 0$	$y = 2 - \ln\cos(x - 1)$

Solve the initial value problems in Exercises 15–20.

15. $\frac{dy}{dx} = \frac{2y^2 + 2}{x^2 - 1}, \quad y(0) = 1$

16. $\frac{dy}{dx} = \frac{e^{(y^2 + \sin x)}}{y \sec x}, \quad y(0) = \sqrt{\ln 2}$

17. $\frac{dy}{dx} = \frac{e^{(y + \sqrt{x})}}{\sqrt{x}\,y^3}, \quad y(\ln^2 2) = 0$

18. $\frac{dy}{dx} = (y^2 + 2y + 2)(x + 1)\ln x, \quad y(1) = -1$

19. $\dfrac{dy}{dx} = \sqrt{x^2 - x^2y^2 + 1 - y^2}, \quad y(0) = -1/2$

20. $\dfrac{dy}{dx} = \dfrac{\cot y}{x\sqrt{4x^2 - 1}}, \quad y(1) = 0$

21. CALCULATOR *Newton's law of cooling.* Newton's law of cooling assumes that the temperature T of a small hot object placed in a surrounding cooling medium of constant temperature T_s decreases at a rate proportional to $T - T_s$. An object cooled from 100°C to 40°C in 20 minutes when the surrounding temperature was 20°C. How long did it take the temperature to reach 60°C on the way down?

22. *The snow plow problem.* One morning it began to snow and kept on snowing at a constant rate. A snow plow began to plow at noon. It plowed clean and at a constant rate (volume per unit time). From one o'clock until two o'clock it went only half as far as it had gone between noon and one o'clock. When did the snow begin falling?

Homogeneous Equations

Show that the equations in Exercises 23–26 are homogeneous and find their general solutions.

23. $(x^2 + y^2)\,dx + xy\,dy = 0$

24. $(y^2 - xy)\,dx + x^2\,dy = 0$

25. $(xe^{y/x} + y)\,dx - x\,dy = 0$

26. $(x - y)\,dx + (x + y)\,dy = 0$

Solve the equations in Exercises 27 and 28 subject to the given initial conditions.

27. $\dfrac{dy}{dx} = \dfrac{y}{x} + \cos\dfrac{y - x}{x}, \quad y(2) = 2$

28. $\left(x \sin\dfrac{y}{x} - y\cos\dfrac{y}{x}\right) dx + x\cos\dfrac{y}{x}\,dy = 0, \quad y(2) = \pi$

Orthogonal Trajectories

If every member of one family of curves is a solution of the differential equation

$$M(x, y)\,dx + N(x, y)\,dy = 0$$

and every member of a second family of curves is a solution of the related equation

$$N(x, y)\,dx - M(x, y)\,dy = 0,$$

then each curve in each family is orthogonal to all the curves of the other family. Under these circumstances each family is said to be a family of **orthogonal trajectories** of the other family. If the curves in one family were electric field lines, the curves in the other family would be paths of constant electric potential.

Solve the differential equations in Exercises 29 and 30 to find equations for the families of orthogonal trajectories illustrated in Figs. 15.2 and 15.3.

29. $x\,dx + 2y\,dy = 0$ and $2y\,dx - x\,dy = 0$ (Fig. 15.2)

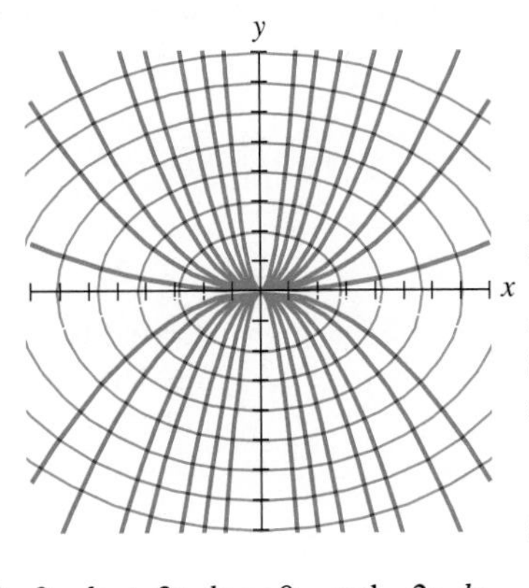

15.2 The curves in Exercise 29.

30. $3x\,dx + 2y\,dy = 0$ and $2y\,dx - 3x\,dy = 0$ (Fig. 15.3)

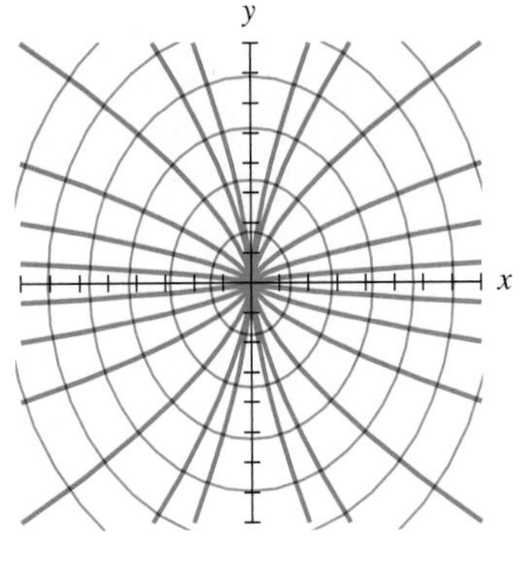

15.3 The curves in Exercise 30.

31. Identify the curves that satisfy the differential equation $y\,dx + x\,dy = 0$ and find equations for their orthogonal trajectories (Fig. 15.4).

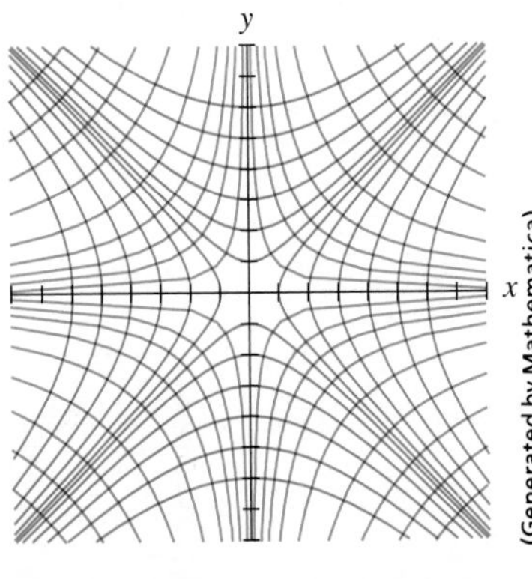

15.4 The curves in Exercise 31.

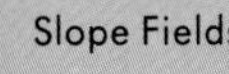

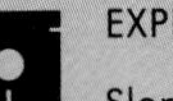

EXPLORER PROGRAM

Slope Fields — Graphs solutions of the initial value problem $y' = f(x, y)$, $y = y_0$ when $x = x_0$, for your choice of f, x_0, and y_0

15.2 Exact Differential Equations

If we write a (not necessarily separable) first order differential equation

$$M(x, y) + N(x, y)\frac{dy}{dx} = 0 \tag{1}$$

in the differential form

$$M(x, y)\, dx + N(x, y)\, dy = 0, \tag{2}$$

the left-hand side has the same form as the differential of a function of x and y. If the left-hand side *is* the differential df of a function $f(x, y)$, then Eq. (2) can be solved in terms of f:

$$\begin{aligned} M(x, y)\, dx + N(x, y)\, dy &= 0 \\ df(x, y) &= 0 \\ f(x, y) &= C. \end{aligned} \tag{3}$$

The equation $f(x, y) = C$ defines y implicitly as one or more differentiable functions of x. Each of these functions, it can be shown, is a solution of the original differential equation. In writing $f(x, y) = C$ we therefore say that we have solved the differential equation.

Example 1 Since

$$d(\sin xy) = y \cos xy\, dx + x \cos xy\, dy = 0,$$

the equation

$$y \cos xy\, dx + x \cos xy\, dy = 0 \tag{4}$$

is equivalent to

$$d(\sin xy) = 0, \qquad \text{or} \qquad \sin xy = C.$$

The equation $\sin xy = C$ defines y implicitly as one or more functions of x that solve Eq. (4).

Not all differential forms are differentials of functions. Before we attempt to solve an equation this way we must know whether such a function f exists and then how to find it when it does. The test for the existence of f is the test for exactness in Section 14.7, repeated here for functions of two variables.

DEFINITION AND TEST

A differential form $M(x, y)\, dx + N(x, y)\, dy$ and the associated equation $M(x, y)\, dx + N(x, y)\, dy = 0$ are both said to be **exact** on a region R in the xy-plane if for some function $f(x, y)$ defined on R,

$$M(x, y)\, dx + N(x, y)\, dy = \frac{\partial f}{\partial x}\, dx + \frac{\partial f}{\partial y}\, dy = df. \tag{5}$$

The form and equation are exact on R if and only if, throughout R,

$$\frac{\partial M}{\partial y} = \frac{\partial N}{\partial x}. \tag{6}$$

Example 2

a) The equation $(x^2 + y^2)\,dx + (2xy + \cos y)\,dy = 0$ is exact because the partial derivatives

$$\frac{\partial M}{\partial y} = \frac{\partial}{\partial y}(x^2 + y^2) = 2y \qquad \text{and} \qquad \frac{\partial N}{\partial x} = \frac{\partial}{\partial x}(2xy + \cos y) = 2y$$

are equal.

b) The equation $(x + 3y)\,dx + (x^2 + \cos y)\,dy = 0$ is not exact because the partial derivatives

$$\frac{\partial M}{\partial y} = \frac{\partial}{\partial y}(x + 3y) = 3 \qquad \text{and} \qquad \frac{\partial N}{\partial x} = \frac{\partial}{\partial x}(x^2 + \cos y) = 2x$$

are not equal.

After we have tested an equation and found it to be exact, we can find the solution $f(x, y) = C$ by taking the steps described in the following example.

Steps for Solving an Equation You Know to Be Exact

1. Match the equation to the form $df = (\partial f/\partial x)dx + (\partial f/\partial y)dy$ to identify $\partial f/dx$ and $\partial f/dy$.
2. Integrate $\partial f/\partial x$ with respect to x, writing the constant of integration as $k(y)$.
3. Differentiate with respect to y and set the result equal to $\partial f/\partial y$ to find $k'(y)$.
4. Integrate to find $k(y)$ and determine f.
5. Write the solution of the exact equation as $f(x, y) = C$.

Example 3

Solve the differential equation

$$(x^2 + y^2)\,dx + (2xy + \cos y)\,dy = 0.$$

Solution We already know that the equation is exact (Example 2), so it is safe to set about finding a function $f(x, y)$ whose differential is the equation's left-hand side. For the equality

$$\frac{\partial f}{\partial x}dx + \frac{\partial f}{\partial y}dy = (x^2 + y^2)\,dx + (2xy + \cos y)\,dy$$

to hold, we must have

$$\frac{\partial f}{\partial x} = x^2 + y^2 \qquad \text{and} \qquad \frac{\partial f}{\partial y} = 2xy + \cos y. \tag{7}$$

The partial derivative $\partial f/\partial x = x^2 + y^2$ was calculated by holding y fixed and differentiating f with respect to x. We may therefore find f by holding y at a constant value and integrating with respect to x. When we do so we get

$$f(x, y) = \int_{y\ \text{const.}} (x^2 + y^2)\,dx = \frac{x^3}{3} + y^2x + k(y). \tag{8}$$

The constant of integration is written as a function of y because its value may change with each new y.

To find $k(y)$, we calculate $\partial f/\partial y$ from Eq. (8) and set the result equal to the known partial derivative $\partial f/\partial y = 2xy + \cos y$:

$$\frac{\partial}{\partial y}\left(\frac{x^3}{3} + y^2x + k(y)\right) = 2xy + \cos y$$

$$2xy + k'(y) = 2xy + \cos y$$

$$k'(y) = \cos y.$$

A single integration with respect to y then shows that $k(y) = \sin y$ plus a constant, so

$$f(x, y) = \frac{x^3}{3} + y^2x + \sin y + \text{a constant}.$$

If we wanted only to find a function whose differential is $(x^2 + y^2)\,dx + (2xy + \cos y)\,dy$, we'd be done now. We have found infinitely many such functions, one for each possible constant. Our goal, however, is to solve the original differential equation, and for that we must take one more step. We must write down the equation $f(x, y) = C$ because it is this equation, and not the formula for f alone, that defines the differential equation's solution functions. The solution of the differential equation is

$$\frac{x^3}{3} + y^2x + \sin y = C. \qquad \text{(Constants combined)}$$

Integrating Factors

It can be shown that every nonexact differential equation $M(x, y)\,dx + N(x, y)\,dy = 0$ can be made exact by multiplying both sides by a suitable **integrating factor** $\rho(x, y)$. In other words, the equation

$$\rho(x, y)\,M(x, y)\,dx + \rho(x, y)\,N(x, y)\,dy = 0$$

is always exact for an appropriate choice of ρ.

Example 4 The equation

$$2y\,dx + x\,dy = 0 \tag{9}$$

is not exact because $\partial(2y)/\partial y = 2$ while $\partial(x)/\partial x = 1$. The equation

$$2xy\,dx + x^2\,dy = 0,$$

obtained by multiplying both sides of Eq. (9) by x, *is* exact because $\partial(2xy)/\partial y = 2x$ and $\partial(x^2)/\partial x = 2x$.

Unfortunately, there is no general technique for finding an integrating factor when you need one, and the search for one can be a frustrating experience. Practice does help, however, and Exercises 19–22 supply integrating factors for you to work with. As you will see there, integrating factors need not be unique.

EXERCISES 15.2

Test the equations in Exercises 1–8 for exactness.

1. $x\,dx - y\,dy = 0$

2. $y\,dx - x\,dy = 0$

3. $\frac{1}{y}\,dx - \frac{x}{y^2}\,dy = 0$

4. $(x + y^2)\,dx + (2xy + 1)\,dy = 0$

5. $(x + e^y)\,dx + (y + xe^y)\,dy = 0$

6. $y\cos xy\,dx + x\cos xy\,dy = 0$

7. $(x + 2y)\,dx - (2x - y)\,dy = 0$

8. $(x + 2y)\,dx - (2x + y)\,dy = 0$

Solve the equations in Exercises 9–18.

9. $(x + y)\,dx + (x + y^2)\,dy = 0$

10. $(2x\, e^y + e^x)\, dx + (x^2 + 1)e^y\, dy = 0$

11. $\dfrac{dy}{dx} = \dfrac{2xy + y^2}{y - 2xy - x^2}$

12. $\dfrac{dy}{dx} = \dfrac{x^3 - y}{x + \sqrt{1 + \sin 2y}}$

13. $\left(e^x + \ln y + \dfrac{y}{x}\right) dx + \left(\dfrac{x}{y} + \ln x + \sin y\right) dy = 0$

14. $\dfrac{dy}{dx} = \dfrac{y \cos xy + 1}{3 - 2y - x \cos xy}$

15. $\left(\cos x \displaystyle\int_0^y \sin^2 t\, dt\right) dx + \sin x \sin^2 y\, dy = 0$

16. $\dfrac{2x - y}{x^2 + y^2}\, dx + \dfrac{x + 2y}{x^2 + y^2}\, dy = 0$

17. $\left(\dfrac{1}{x} + \displaystyle\sum_{n=0}^{\infty} \dfrac{x^{n-1}y^n}{n!}\right) dx + \left(\dfrac{1}{y} + \displaystyle\sum_{n=0}^{\infty} \dfrac{x^n y^{n-1}}{n!}\right) dy = 0$

18. $\left(\displaystyle\sum_{n=0}^{\infty} \dfrac{(-1)^{n+1}}{x^2 y^n}\right) dx + \left(\displaystyle\sum_{n=0}^{\infty} \dfrac{(-1)^{n+1} n}{x y^{n+1}}\right) dy = 0$

Show that the integrating factors ρ in Exercises 19–22 make the differential equations exact. Then solve the equations using those integrating factors.

19. $(xy^2 + y)\, dx - x\, dy = 0; \quad \rho = 1/y^2$

20. $(x + 2y)\, dx - x\, dy = 0; \quad \rho = 1/x^3$

21. $y\, dx + x\, dy = 0,$

a) $\rho = \dfrac{1}{xy}$ b) $\rho = \dfrac{1}{(xy)^2}$

22. $y\, dx - x\, dy = 0,$

a) $\rho = \dfrac{1}{y^2}$ b) $\rho = \dfrac{1}{x^2}$

c) $\rho = \dfrac{1}{xy}$ d) $\rho = \dfrac{1}{x^2 + y^2}$

Solve the initial value problems in Exercises 23–26 for y as a function of x.

23. $\dfrac{dy}{dx} = -\dfrac{x + y^2}{2xy + 1}, \quad y(0) = 2$

24. $\dfrac{dy}{dx} = -\dfrac{x + e^y}{y + xe^y}, \quad y(2) = 0$

25. $\left(\dfrac{1}{x} - y\right) dx + \left(\dfrac{1}{y} - x\right) dy = 0, \quad y(1) = 1$

26. $(2x + y + 1)\, dx + (2y + x + 1)\, dy = 0, \quad y(1) = 5$

Solve the initial value problems in Exercises 27 and 28.

27. $\left(2x + \dfrac{1}{x}\displaystyle\int_1^y \dfrac{\ln t}{t}\, dt\right) dx + \dfrac{\ln x \ln y}{y}\, dy = 0, \quad y(2) = 1$

28. $\left(\sin^{-1} y + \dfrac{y}{\sqrt{1 - x^2}}\right) dx + \left(\sin^{-1} x + \dfrac{x}{\sqrt{1 - y^2}}\right) dy = 0, \quad y(1) = 1$

29. Find the value of a that makes the equation

$$(3x^2 + y^2)\, dx + a\, xy\, dy = 0$$

exact and solve the equation for this value of a.

30. The equation $(x^2 + by^2)\, dx + 6\, xy\, dy = 0$ is not exact but is known to have $\rho = 1/x^2$ as an integrating factor. Use this information to find the value of b and solve the equation.

EXPLORER PROGRAM

Slope Fields — Graphs solutions of the initial value problem $y' = f(x, y)$, $y = y_0$ when $x = x_0$, for your choice of f, x_0, and y_0

15.3 Linear First Order Equations

The equation $dy/dx = ky$, which we use to model bacterial growth, radioactive decay, and temperature change, is a special case of a more general equation called a linear first order equation. This section shows how to solve the more general equation and uses it to predict current flow in an electric circuit.

DEFINITIONS

A differential equation that can be written in the form

$$\frac{dy}{dx} + P(x)y = Q(x), \tag{1}$$

where P and Q are functions of x, is called a **linear first order equation.** Equation (1) is the equation's **standard form.**

Example 1 The standard form of the equation

$$x\frac{dy}{dx} + 3y = x^2$$

is

$$\frac{dy}{dx} + \frac{3}{x}y = x.$$

This is Eq. (1) with $P(x) = 3/x$ and $Q(x) = x$.

The way we solve the equation

$$\frac{dy}{dx} + P(x)\,y = Q(x)$$

is to multiply both sides by an integrating factor $\rho(x)$ that turns the left-hand side of the multiplied equation,

$$\rho\frac{dy}{dx} + P\rho y = \rho Q, \tag{2}$$

into the derivative of the product ρy.

Once we have chosen ρ (we shall do so in a moment) and carried out the multiplication, we can solve Eq. (2) by integrating both sides with respect to x:

$$\rho\frac{dy}{dx} + P\rho y = \rho Q \qquad \text{(Eq. (2))}$$

$$\frac{d}{dx}(\rho y) = \rho Q \qquad \text{(Choice of } \rho\text{)}$$

$$\rho y = \int \rho Q\,dx \qquad \text{(Integration with respect to } x\text{)}$$

$$y = \frac{1}{\rho}\int \rho Q\,dx. \qquad \text{(Solved for } y\text{)}$$

To find ρ, we find the function of x that satisfies the equations

$$\frac{d}{dx}(\rho y) = \rho\frac{dy}{dx} + P\rho y,$$

$$\rho\frac{dy}{dx} + y\frac{d\rho}{dx} = \rho\frac{dy}{dx} + P\rho y, \qquad \text{(Product Rule for derivatives)}$$

$$y\frac{d\rho}{dx} = P\rho y, \qquad \text{(The terms } \rho\tfrac{dy}{dx} \text{ cancel.)}$$

$$\frac{d\rho}{dx} = P\rho, \qquad \text{(The } y\text{'s cancel.)}$$

$$\frac{d\rho}{\rho} = P\,dx, \qquad \text{(Variables separated)}$$

$$\rho = Ce^{\int P\,dx}. \qquad \text{(Equation integrated and solved for } \rho\text{)}$$

Since we do not need the most general function ρ, we may take C to be 1.

We summarize our conclusions with the following theorem.

THEOREM 1

The solution of the equation

$$\frac{dy}{dx} + P(x)y = Q(x) \tag{3}$$

is

$$y = \frac{1}{\rho(x)} \int \rho(x)\, Q(x)\, dx, \tag{4}$$

where

$$\rho(x) = e^{\int P(x)\, dx}. \tag{5}$$

In the formula for ρ we do not need the most general antiderivative of $P(x)$. Any antiderivative will do.

Example 2 Solve the equation $x\dfrac{dy}{dx} - 3y = x^2, \quad x > 0.$

Steps for Solving a Linear First Order Equation

1. Put it in standard form.
2. Find an antiderivative of $P(x)$.
3. Find $\rho = e^{\int P(x)\, dx}$.
4. Use Eq. (4) to find y.

Solution We solve the equation in four steps.

STEP 1: *Put the equation in standard form and identify the functions P and Q.* To do so, we divide both sides of the equation by the coefficient of dy/dx, in this case x, obtaining

$$\frac{dy}{dx} - \frac{3}{x}y = x, \qquad P(x) = -\frac{3}{x}, \qquad Q(x) = x.$$

STEP 2: *Find an antiderivative of* $P(x)$ *(any one will do)*:

$$\int P(x)\, dx = \int -\frac{3}{x}\, dx = -3\int \frac{1}{x}\, dx = -3 \ln |x| = -3 \ln x \qquad (x > 0).$$

STEP 3: *Use Eq. (5) to find* ρ:

$$\rho(x) = e^{\int P(x)\, dx} = e^{-3 \ln x} = e^{\ln x^{-3}} = \frac{1}{x^3}.$$

STEP 4: *Use Eq. (4) to find the solution:*

$$y = \frac{1}{\rho(x)} \int \rho(x)\, Q(x)\, dx = \frac{1}{(1/x^3)} \int \left(\frac{1}{x^3}\right)(x)\, dx \qquad \text{(Values from steps 1–3)}$$

$$= x^3 \int \frac{1}{x^2}\, dx = x^3\left(-\frac{1}{x} + C\right) = Cx^3 - x^2.$$

The solution is the function $y = Cx^3 - x^2$.

RL Circuits

15.5 The *RL* circuit in Example 3.

The diagram in Fig. 15.5 represents an electrical circuit whose total resistance is R ohms and whose self-inductance, shown schematically as a coil, is L henries (hence the name "*RL* circuit"). There is also a switch whose terminals at a and b can be closed to connect a constant electrical source of V volts.

Ohm's law, $V = RI$, has to be modified for such a circuit. The modified form is

$$L\frac{di}{dt} + Ri = V, \tag{6}$$

where i is the intensity of the current in amperes and t is the time in seconds. By solving this equation, we can predict how the current will flow after the switch is closed.

Example 3 The switch in the RL circuit in Fig. 15.5 is closed at time $t = 0$. How will the current flow as a function of time?

Solution Equation (6) is a linear first order differential equation for i as a function of t. When we put it in the standard form

$$\frac{di}{dt} + \frac{R}{L}i = \frac{V}{L}$$

and carry out the solution steps with $P = R/L$ and $Q = V/L$, we find that

$$i = \frac{1}{e^{(R/L)t}}\int e^{(R/L)t}\left(\frac{V}{L}\right) dt = \frac{V}{R} + C\frac{V}{L}e^{-(R/L)t}. \tag{7}$$

Imposing the initial condition that $i(0)$ be 0 determines the value of C to be $-L/R$, so

$$i = \frac{V}{R} + \left(-\frac{L}{R}\right)\frac{V}{L}e^{-(R/L)t} = \frac{V}{R}\left(1 - e^{-(R/L)t}\right). \tag{8}$$

We see from this that the current i is always less than V/R but that it approaches V/R as a **steady-state value:**

$$\lim_{t\to\infty}\frac{V}{R}\left(1 - e^{-Rt/L}\right) = \frac{V}{R}(1-0) = \frac{V}{R}. \tag{9}$$

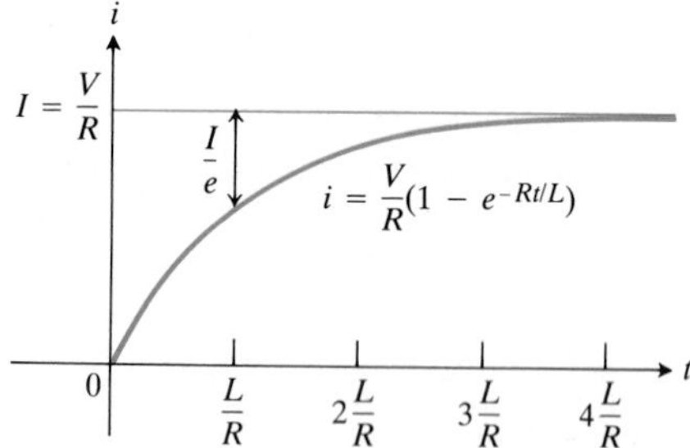

15.6 The growth of the current in the RL circuit in Example 3. I is the current's steady-state value. The number $t = L/R$ is the time constant of the circuit. The current gets to within 5% of its steady-state value in 3 time constants (Exercise 17).

The current $I = V/R$ is the current that will flow in the circuit if either $L = 0$ (no inductance) or $di/dt = 0$ (steady current, $i =$ constant) (Fig. 15.6).

Resistance Proportional to Velocity

In some cases it makes sense to assume that, other forces being absent, the resistance encountered by a moving object, like a car coasting to a stop, is proportional to the object's velocity. The slower the object moves, the less its forward progress is resisted by the air through which it passes. We can describe this in mathematical terms if we picture the object as a mass m moving along a coordinate line with position s and velocity v at time t. The magnitude of the resisting force opposing the motion is then $m\,(dv/dt)$ and we can write

$$m\frac{dv}{dt} = -kv \qquad (k > 0) \tag{10}$$

to say that the force decreases in proportion to velocity. If we rewrite (10) as

$$\frac{dv}{dt} = -\frac{k}{m}v,$$

we can see that the solution is $v = v_0\,e^{-(k/m)t}$.

Resistance Proportional to Velocity

$$v = v_0 e^{-(k/m)t} \tag{11}$$

What can we learn from this equation? For one thing, we can see that if m is something large, like the mass of a 20,000-ton ore boat in Lake Erie, it will take a long time for the velocity to get near zero. For another, we can integrate the equation to find s as a function of t.

Suppose a body is coasting to a stop and the only force acting on it is a resistance proportional to its speed. How far will it coast? To find out, we start with Eq. (11) and solve the initial value problem

Differential equation: $\quad \dfrac{ds}{dt} = v_0 e^{-(k/m)t},$

Initial condition: $\quad s = 0 \quad \text{when} \quad t = 0.$

Integrating with respect to t gives

$$s = -\frac{v_0 m}{k} e^{-(k/m)t} + C. \tag{12}$$

Substituting $s = 0$ when $t = 0$ gives

$$0 = -\frac{v_0 m}{k} + C \qquad \text{and} \qquad C = \frac{v_0 m}{k}.$$

The body's position at time t is therefore

$$s(t) = -\frac{v_0 m}{k} e^{-(k/m)t} + \frac{v_0 m}{k} = \frac{v_0 m}{k}\left(1 - e^{-(k/m)t}\right). \tag{13}$$

To find how far the body will coast, we find the limit of $s(t)$ as $t \to \infty$. Since $\lim_{t\to\infty} e^{-(k/m)t} = 0$, the limit of $s(t)$ is $v_0 m/k$.

$$\text{Distance coasted} = \frac{v_0 m}{k} \tag{14}$$

This is an ideal figure, of course. Only in mathematics can time stretch to infinity. The number $v_0 m/k$ is only an upper bound (albeit a useful one). It is true to life in one respect, at least—if m is large, it will take a lot of energy to stop the body. That is why ocean liners have to be docked by tugboats. Any liner of conventional design entering a slip with enough speed to steer would smash into the pier before it could stop.

Example 4 For a 192-lb ice skater, the k in Eq. (11) is about 1/3 slug/sec and $m = 192/32 = 6$ slugs. How long will it take the skater to coast from 11 ft/sec (7.5 mph) to 1 ft/sec? How far will the skater coast before coming to a complete stop?

Solution We answer the first question by solving Eq. (11) for t:

$$11e^{-t/18} = 1 \qquad \left(\begin{array}{l}\text{Eq. (11) with } k = 1/3,\ m = 6,\\ v_0 = 11,\ v = 1\end{array}\right)$$

$$e^{-t/18} = 1/11$$

$$-t/18 = \ln(1/11) = -\ln 11$$

$$t = 18 \ln 11 = 43 \text{ sec.} \qquad \text{(Calculator, rounded)}$$

We answer the second question with Eq. (14):

$$\text{Distance coasted} = \frac{v_0 m}{k} = \frac{11 \cdot 6}{1/3} = 198 \text{ ft.}$$

EXERCISES 15.3

Solve the differential equations in Exercises 1–8.

1. $e^x \dfrac{dy}{dx} + 2e^x y = 1$

2. $2\dfrac{dy}{dx} - y = e^{x/2}$

3. $x\dfrac{dy}{dx} + 3y = \dfrac{\sin x}{x^2}$

4. $x\dfrac{dy}{dx} + y = x \cos x$

5. $(x-1)^3 \dfrac{dy}{dx} + 4(x-1)^2 y = x + 1$

6. $e^{2x} \dfrac{dy}{dx} + 2e^{2x} y = 2x$

7. $\sin x \dfrac{dy}{dx} + (\cos x)\, y = \tan x$

8. $\cosh x \dfrac{dy}{dx} + (\sinh x)\, y = e^{-x}$

Solve the initial value problems in Exercises 9–12 for y as a function of x.

	Differential equation	Initial condition
9.	$\dfrac{dy}{dx} + 2y = x$	$y(0) = 1$
10.	$x\dfrac{dy}{dx} + 2y = x^3$	$y(2) = 1$
11.	$x\dfrac{dy}{dx} + y = \sin x$	$y(\pi/2) = 1$
12.	$x\dfrac{dy}{dx} - 2y = x^3 \sec x \tan x$	$y(\pi/3) = 2$

13. Solve the initial value problem $dy/dx = ky$, $y(0) = y_0$, with Eq. (4) and see what you get.

14. *Blood sugar.* If glucose is fed intravenously at a constant rate, the change in the overall concentration $c(t)$ of glucose in the blood with respect to time may be described by the differential equation

$$\frac{dc}{dt} = \frac{G}{100V} - kc.$$

In this equation, G, V, and k are positive constants, G being the rate at which glucose is admitted, in milligrams per minute, and V the volume of blood in the body, in liters (around 5 liters for an adult). The concentration $c(t)$ is measured in milligrams per centiliter. The term $-kc$ is included because the glucose is assumed to be changing continually into other molecules at a rate proportional to its concentration.

a) Solve the equation for $c(t)$, using c_0 to denote $c(0)$.

b) Find the steady-state concentration, $\lim_{t\to\infty} c(t)$.

RL Circuits

15. *Current in a closed RL circuit.* How many seconds after the switch in an RL circuit is closed will it take the current i to reach half of its steady-state value? Notice that the time depends only on R and L and not on how much voltage is applied.

16. *Current in an open RL circuit.* If the switch is thrown open after the current in an RL circuit has built up to its steady-state value, the decaying current (Fig. 15.7) obeys the equation

$$L\frac{di}{dt} + Ri = 0, \tag{15}$$

which is Eq. (6) with $V = 0$.

a) Solve Eq. (15) to express i as a function of t.

b) How long after the switch is thrown will it take the current to fall to half its original value?

c) What is the value of the current when $t = L/R$? (The significance of this time is explained in the next exercise.)

17. CALCULATOR *Time constants.* Engineers call the number L/R the *time constant* of the RL circuit in Fig. 15.5. The significance of the time constant is that the current will reach 95% of its final value within 3 time constants of the time the switch is closed (Fig. 15.6). Thus, the time constant gives a

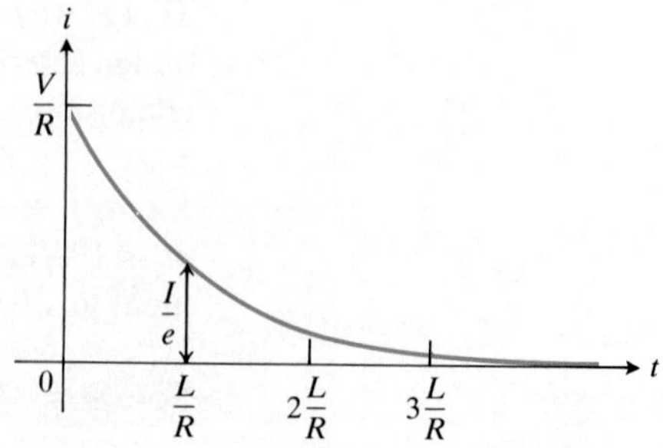

15.7 The current in an RL circuit decays exponentially when the power is turned off (Exercise 16).

built-in measure of how rapidly an individual circuit will reach equilibrium. Find the value of i in Eq. (8) that corresponds to $t = 3L/R$ and show that it is about 95% of the steady-state value $I = V/R$.

Additional Exercises

18. *Continuous compounding.* You have \$1000 with which to open an account and plan to add \$1000 per year. All funds in the account will earn 10% interest per year compounded continuously. If the added deposits are also credited to your account continuously, the number of dollars x in your account at time t (years) will satisfy the initial value problem

Differential equation: $\quad \frac{dx}{dt} = 1000 + 0.10x,$

Initial condition: $\quad x(0) = 1000.$

a) Solve the initial value problem for x as a function of t.

b) CALCULATOR About how many years will it take for the amount in your account to reach \$100,000?

19. For a 145-lb cyclist on a 15-lb bicycle on level ground, the k in Eq. (11) is about 1/5 slug/sec and $m = 160/32 = 5$ slugs. The cyclist starts coasting at 22 ft/sec (15 mph).

a) About how far will the cyclist coast before reaching a complete stop?

b) How long will it take the cyclist's speed to drop to 1 ft/sec?

20. For a 56,000-ton Iowa class battleship, $m = 1{,}750{,}000$ slugs and the k in Eq. (11) might be 3000 slugs/sec. Suppose the battleship loses power when it is moving at a speed of 22 ft/sec (13.2 knots).

a) About how far will the ship coast before it stops?

b) About how long will it take the ship's speed to drop to 1 ft/sec?

EXPLORER PROGRAM

Slope Fields	Graphs solutions of the initial value problem $y' = f(x, y)$, $y = y_0$ when $x = x_0$, for your choice of f, x_0, and y_0

15.4 Linear Homogeneous Second Order Equations

If $F(x) = 0$, the linear equation

$$a_n(x)\frac{d^ny}{dx^n} + a_{n-1}(x)\frac{d^{n-1}y}{dx^{n-1}} + \cdots + a_1(x)\frac{dy}{dx} + a_0(x)y = F(x) \qquad (1)$$

is called **homogeneous;** otherwise it is called **nonhomogeneous.** (This is different from the way we used the term *homogeneous* when we applied it to first order equations in Section 15.1.) This section shows how to solve the homogeneous equation

$$\frac{d^2y}{dx^2} + a\frac{dy}{dx} + by = 0 \qquad (a \text{ and } b \text{ constant}).$$

Section 15.5 shows how to solve the nonhomogeneous equation

$$\frac{d^2y}{dx^2} + a\frac{dy}{dx} + by = F(x).$$

Section 15.6 shows how to use these equations to model oscillation.

Linear Differential Operators

At this point, it is convenient to introduce the symbol D to represent the operation of differentiation with respect to x. That is, we write $Df(x)$ to mean $(d/dx)f(x)$. We define powers of D to mean taking successive derivatives:

$$D^2f(x) = D\{Df(x)\} = \frac{d^2f(x)}{dx^2}, \qquad D^3f(x) = D\{D^2f(x)\} = \frac{d^3f(x)}{dx^3},$$

and so on. A polynomial in D is to be interpreted as an operator that, when applied to $f(x)$, produces a linear combination of f and its successive derivatives. For example,

$$(D^2 + D - 2)f(x) = D^2f(x) + Df(x) - 2f(x) = \frac{d^2f(x)}{dx^2} + \frac{df(x)}{dx} - 2f(x).$$

DEFINITION

A polynomial in D is a **linear differential operator.**

We often denote linear differential operators by single letters. If L_1 and L_2 are two such linear operators, we define their sum and product by the equations

$$(L_1 + L_2)f(x) = L_1f(x) + L_2f(x), \qquad L_1L_2f(x) = L_1(L_2f(x)).$$

Linear differential operators that are polynomials in D with constant coefficients satisfy basic algebraic laws that make it possible for us to treat them like ordinary polynomials so far as addition, multiplication, and factoring are concerned. Thus,

$$(D^2 + D - 2)f(x) = (D + 2)(D - 1)f(x) = (D - 1)(D + 2)f(x).$$

Since this equation holds for any twice-differentiable function f, we can also write the equality between operators:

$$D^2 + D - 2 = (D + 2)(D - 1) = (D - 1)(D + 2).$$

The Characteristic Equation

In the remainder of this section, we consider only linear *second order* equations with constant real-number coefficients. Because the solutions of nonhomogeneous equations depend on the solutions of the corresponding homogeneous equations, we focus on equations of the form

$$\frac{d^2y}{dx^2} + a\frac{dy}{dx} + by = 0 \tag{2}$$

or, in operator notation,

$$(D^2 + aD + b)y = 0. \tag{3}$$

The usual method for solving Eq. (3) is to begin by factoring the operator:

$$D^2 + aD + b = (D - r_1)(D - r_2).$$

We do this by finding the two roots r_1 and r_2 of the equation $r^2 + 2ar + b = 0$.

DEFINITION

The equation

$$r^2 + ar + b = 0 \tag{4}$$

is the **characteristic equation** of

$$\frac{d^2y}{dx^2} + a\frac{dy}{dx} + by = 0. \tag{5}$$

Equation (5) is equivalent to

$$(D - r_1)(D - r_2)y = 0. \tag{6}$$

If we now let

$$(D - r_2)y = u, \tag{7}$$

we get

$$(D - r_1)u = 0. \tag{8}$$

From here, we can solve Eq. (6) in two steps. From Eq. (8), which is first order separable in u and x, we find

$$u = C_1 e^{r_1 x}.$$

We substitute this into (7), which becomes

$$(D - r_2)y = C_1 e^{r_1 x} \quad \text{or} \quad \frac{dy}{dx} - r_2 y = C_1 e^{r_1 x}.$$

This equation is linear in y. Its integrating factor is $\rho = e^{-r_2 x}$ (Section 15.3), and its solution is

$$e^{-r_2 x} y = C_1 \int e^{(r_1 - r_2)x}\, dx + C_2. \tag{9}$$

How we proceed at this point depends on whether r_1 and r_2 are real and unequal, real and equal, or complex.

Real, Unequal Roots

If r_1 and r_2 are real and $r_1 \neq r_2$, the evaluation of the integral in Eq. (9) leads to

$$e^{-r_2 x} y = \frac{C_1}{r_1 - r_2} e^{(r_1 - r_2)x} + C_2 \quad \text{or} \quad y = \frac{C_1}{r_1 - r_2} e^{r_1 x} + C_2 e^{r_2 x}. \tag{10}$$

Since C_1 is an arbitrary constant, so is $C_1/(r_1 - r_2)$, and we can write the solution in Eq. (10) as

$$y = C_1 e^{r_1 x} + C_2 e^{r_2 x}. \tag{11}$$

Example 1 Solve the initial value problem

Differential equation: $\dfrac{d^2y}{dx^2} + \dfrac{dy}{dx} - 2y = 0,$

Initial conditions: $y(0) = 1, \quad y'(0) = -4.$

Solution The characteristic equation $r^2 + r - 2 = 0$ has roots $r_1 = 1$, $r_2 = -2$. Hence, by Eq. (11), the solution of the differential equation is

$$y = C_1 e^{x} + C_2 e^{-2x}.$$

The initial conditions tell us that

$$C_1 + C_2 = 1, \qquad (y(0) = 1)$$

$$C_1 - 2C_2 = -4. \qquad (y'(0) = -4)$$

We solve these equations simultaneously and find that

$$C_1 = -\frac{2}{3}, \qquad C_2 = \frac{5}{3}.$$

The solution of the initial value problem is

$$y = -\frac{2}{3}e^{x} + \frac{5}{3}e^{-2x}.$$

Real, Equal Roots

If $r_1 = r_2$, then $e^{(r_1 - r_2)x} = e^0 = 1$ and Eq. (9) reduces to $e^{-r_2x}y = C_1x + C_2$ or

$$y = (C_1x + C_2)e^{r_2x}. \tag{12}$$

Example 2 Solve the equation $\dfrac{d^2y}{dx^2} + 4\dfrac{dy}{dx} + 4y = 0$.

Solution

$$r^2 + 4r + 4 = (r + 2)^2,$$

$$r_1 = r_2 = -2,$$

$$y = (C_1x + C_2)e^{-2x}.$$

Complex Roots

From the theory of equations, we know that if a and b are real and the roots of the characteristic equation $r^2 + ar + b = 0$ are not real, then the roots must be a pair of complex conjugate numbers

$$r_1 = \alpha + \beta i, \qquad r_2 = \alpha - \beta i. \tag{13}$$

If $\beta \neq 0$ then Eq. (11) applies once again to give

$$y = c_1e^{(\alpha + i\beta)x} + c_2e^{(\alpha - i\beta)x} = e^{\alpha x}(c_1e^{i\beta x} + c_2e^{-i\beta x}), \tag{14}$$

where c_1 and c_2 are complex constants. By Euler's formula (Section 8.7),

$$e^{i\beta x} = \cos \beta x + i \sin \beta x \qquad \text{and} \qquad e^{-i\beta x} = \cos \beta x - i \sin \beta x.$$

Hence, we may replace Eq. (14) by

$$y = e^{\alpha x}((c_1 + c_2) \cos \beta x + i(c_1 - c_2) \sin \beta x). \tag{15}$$

Finally, we introduce new arbitrary constants, $C_1 = c_1 + c_2$ and $C_2 = i(c_1 - c_2)$, to give the solution in Eq. (15) a shorter form:

$$y = e^{\alpha x}(C_1 \cos \beta x + C_2 \sin \beta x). \tag{16}$$

(The constants C_1 and C_2 are real because c_1 and c_2 are complex conjugates.)

Example 3 Solve the initial value problem

Differential equation: $\dfrac{d^2y}{dx^2} + 4\dfrac{dy}{dx} + 6y = 0,$

Initial conditions: $y(0) = -1, \quad y'(0) = 0.$

Solution We solve the characteristic equation $r^2 + 4r + 6 = 0$ with the quadratic formula and find its roots to be $r_1 = -2 + \sqrt{2}i$ and $r_2 = -2 - \sqrt{2}i$. With $\alpha = -2$ and $\beta = \sqrt{2}$, Eq. (16) gives

$$y = e^{-2x}(C_1 \cos \sqrt{2}x + C_2 \sin \sqrt{2}x). \tag{17}$$

A routine differentiation shows that

$$y' = e^{-2x}\left(\left(\sqrt{2}C_2 - 2C_1\right) \cos \sqrt{2}x - \left(\sqrt{2}C_1 + 2C_2\right) \sin \sqrt{2}x\right). \tag{18}$$

We substitute the initial values into Eqs. (17) and (18) and solve the resulting equations simultaneously for C_1 and C_2, finding that

$$C_1 = -1 \qquad \text{and} \qquad C_2 = -\sqrt{2}.$$

The solution of the initial value problem is

$$y = e^{-2x}(-\cos \sqrt{2}x - \sqrt{2} \sin \sqrt{2}x).$$

Example 4 *Imaginary Roots.* Solve the equation

$$\frac{d^2y}{dx^2} + 4y = 0.$$

Solution The roots of the characteristic equation $r^2 + 4 = 0$ are $r = \pm 2i$. With $\alpha = 0$ and $\beta = 2$, Eq. (16) gives

$$y = C_1 \cos 2x + C_2 \sin 2x.$$

Solutions of $\dfrac{d^2y}{dx^2} + a\dfrac{dy}{dx} + by = 0$

Roots of $r^2 + ar + b = 0$	Solution
r_1, r_2 real and unequal	$y = C_1 e^{r_1 x} + C_2 e^{r_2 x}$
r_1, r_2 real and equal	$y = (C_1 x + C_2)e^{r_2 x}$
r_1, r_2 complex conjugates, $\alpha \pm \beta i$	$y = e^{\alpha x}(C_1 \cos \beta x + C_2 \sin \beta x)$

EXERCISES 15.4

Solve the equations in Exercises 1–12.

1. $\dfrac{d^2y}{dx^2} + 2\dfrac{dy}{dx} = 0$
2. $\dfrac{d^2y}{dx^2} + 5\dfrac{dy}{dx} + 6y = 0$
3. $\dfrac{d^2y}{dx^2} + 6\dfrac{dy}{dx} + 5y = 0$
4. $\dfrac{d^2y}{dx^2} - 2\dfrac{dy}{dx} - 3y = 0$
5. $\dfrac{d^2y}{dx^2} - 4\dfrac{dy}{dx} + 4y = 0$
6. $\dfrac{d^2y}{dx^2} + 6\dfrac{dy}{dx} + 9y = 0$
7. $\dfrac{d^2y}{dx^2} - 10\dfrac{dy}{dx} + 25y = 0$
8. $\dfrac{d^2y}{dx^2} - 2\sqrt{2}\dfrac{dy}{dx} + 2y = 0$
9. $\dfrac{d^2y}{dx^2} + \dfrac{dy}{dx} + y = 0$
10. $\dfrac{d^2y}{dx^2} - 6\dfrac{dy}{dx} + 10y = 0$
11. $\dfrac{d^2y}{dx^2} - 2\dfrac{dy}{dx} + 4y = 0$
12. $\dfrac{d^2y}{dx^2} + 8\dfrac{dy}{dx} + 25y = 0$

Solve the initial value problems in Exercises 13–24 ($y'' = d^2y/dx^2$ and $y' = dy/dx$).

13. $y'' - y = 0,\quad y(0) = 1,\quad y'(0) = -2$

14. $2y'' - y' - y = 0,\quad y(0) = -1,\quad y'(0) = 0$

15. $y'' - 4y = 0,\quad y(0) = 0,\quad y'(0) = 3$

16. $y'' - 9y = 0,\quad y(\ln 2) = 1,\quad y'(\ln 2) = -3$

17. $y'' + 2y' + y = 0,\quad y(0) = 0,\quad y'(0) = 1$

18. $4y'' - 2y' + y = 0,\quad y(0) = 4,\quad y'(0) = 2$

19. $4y'' + 12y' + 9y = 0,\quad y(0) = 0,\quad y'(0) = -1$

20. $y'' = 0,\quad y(0) = -3,\quad y'(0) = 5$

21. $y'' + 4y = 0,\quad y(0) = 0,\quad y'(0) = 2$

22. $y'' + 9y = 0,\quad y(0) = 0,\quad y'(0) = \sqrt{3}$

23. $y'' - 2y' + 3y = 0,\quad y(0) = 2,\quad y'(0) = 1$

24. $y'' - 6y' + 10y = 0,\quad y(0) = 7,\quad y'(0) = 1$

25. *Still another way to solve $(d^2y/dx^2) + y = 0$.* Let $y' = dy/dx$ and define the functions c_1 and c_2 by the equations

$$c_1 = y\cos x - y'\sin x, \qquad c_2 = y\sin x + y'\cos x. \tag{19}$$

Show that the condition $y'' + y = 0$ implies that $c_1' = c_2' = 0$ so that c_1 and c_2 are constants. Then eliminate y' from Eqs. (19) to find the general solution for y.

26. *Superposition.* Show that if $y = f_1(x)$ and $y = f_2(x)$ are both solutions of the homogeneous linear differential equation

$$a_n(x)\frac{d^ny}{dx^n} + a_{n-1}(x)\frac{d^{n-1}y}{dx^{n-1}} + \cdots + a_1(x)\frac{dy}{dx} + a_0(x)y = 0 \tag{20}$$

on an interval I then $y = C_1f_1(x) + C_2f_2(x)$ is also a solution of (20) on I. This fact is called the **principle of superposition.** It allows us to construct new solutions of homogeneous equations by superimposing one solution on another. If $y = \cos x$ is one solution of a homogeneous equation and $y = \sin x$ is another, then every linear combination $y = C_1\cos x + C_2\sin x$ is also a solution.

15.5 Second Order Equations; Reduction of Order

This section shows how to solve linear nonhomogeneous equations of the form

$$\frac{d^2y}{dx^2} + a\frac{dy}{dx} + by = F(x) \qquad (a \text{ and } b \text{ real constants}). \tag{1}$$

It also shows how to use an unrelated technique called reduction of order that enables us to solve some nonlinear second order equations by changing them into first order equations to which our earlier techniques may apply.

Linear Nonhomogeneous Second Order Equations

The procedure for solving Eq. (1) has three basic steps. First, we use the techniques of Section 15.4 to find the general solution y_h (h stands for homogeneous) of the **reduced equation**

$$\frac{d^2y}{dx^2} + a\frac{dy}{dx} + by = 0. \tag{2}$$

Then we find a particular solution y_p of the **complete equation,** Eq. (1). Then we add y_p to y_h to form the general solution of the complete equation. Why the procedure works and how we find y_p are the main topics of this section.

The main idea is that we can obtain all solutions of the complete equation (1) by adding a particular solution of Eq. (1) to the general solution of the reduced equation (2). The crux of the matter is that if L is any linear differential operator (not necessarily one of second order or having constant coefficients), then

$$L(k_1y_1 + k_2y_2) = k_1L(y_1) + k_2L(y_2)$$

for any constants k_1 and k_2 and functions y_1 and y_2 to which the operator L can be applied. In the present context, we can take

$$L = D^2 + aD + b$$

and write Eqs. (1) and (2) as $L(y) = F(x)$ and $L(y) = 0$, respectively. The following theorem states the relation between solutions of complete and reduced linear equations.

THEOREM 2 Let $L(y) = F(x)$ be a linear nonhomogeneous differential equation and $L(y) = 0$ be the reduced equation with the same operator L. If y_h is the general solution of the reduced equation and y_p is a particular solution of the complete equation, then the general solution of the complete equation is

$$y = y_h(x) + y_p(x).$$

Proof Let $L(y_h(x)) = 0$ and $L(y_p(x)) = F(x)$. Then

$$L(y_h(x) + y_p(x)) = L(y_h(x)) + L(y_p(x)) = 0 + F(x) = F(x),$$

which establishes that $y = y_h + y_p$ is a solution of the nonhomogeneous equation. To show that all solutions have this form, let y be any solution of the nonhomogeneous equation $L(y) = F(x)$. Then

$$L(y - y_p) = L(y) - L(y_p) = F(x) - F(x) = 0,$$

so that $y - y_p$ satisfies the reduced equation. Therefore, if $y_h = C_1 u_1(x) + C_2\, u_2(x)$ is the general solution of $(D^2 + aD + b)y = 0$, there is some choice of the constants C_1 and C_2 such that

$$y - y_p = C_1 u_1(x) + C_2 u_2(x),$$

which means that

$$y = y_p + C_1 u_1(x) + C_2 u_2(x).$$

We now discuss three ways to find a particular solution of the complete equation (Eq. 1):

1. Inspired guessing (Try this first—it is a real time-saver when it works. Experience will help.)
2. The method of variation of parameters
3. The method of undetermined coefficients

Inspired Guessing

In the following example, we guess that there exists a particular solution of the form $y_p(x) = C$. We find the right value of C by substituting y_p and its derivatives into the differential equation.

Example 1 Find a particular solution of $\dfrac{d^2y}{dx^2} + 2\dfrac{dy}{dx} - 3y = 6.$

Solution We guess that there is a solution of the form $y_p = C$ and substitute y_p and its derivatives $y_p' = 0$, $y_p'' = 0$ into the equation. This leads to

$$\frac{d^2y_p}{dx^2} + 2\frac{dy_p}{dx} - 3y_p = 6, \qquad 0 + 2(0) - 3(C) = 6, \qquad C = -2.$$

The equation has $y_p = -2$ as a solution.

Example 2 Find the general solution of $\dfrac{d^2y}{dx^2} + 2\dfrac{dy}{dx} - 3y = 6.$

We always use these steps:
1. Find y_h.
2. Find y_p.
3. Add y_p to y_h.

Solution We find the solution in three steps.

STEP 1: *Find the general solution y_h of the reduced equation*

$$\frac{d^2y}{dx^2} + 2\frac{dy}{dx} - 3y = 0.$$

The roots of the characteristic equation $r^2 + 2r - 3 = 0$ are $r_1 = 1$ and $r_2 = -3$, so

$$y_h = C_1 e^x + C_2 e^{-3x}.$$

STEP 2: *Find a particular solution y_p of the complete equation:*

$$y_p = -2. \qquad \text{(From Example 1)}$$

STEP 3: *Add y_p to y_h to form the general solution y of the complete equation:*

$$y = y_h + y_p = C_1 e^x + C_2 e^{-3x} - 2.$$

Variation of Parameters

When guessing doesn't work (it usually doesn't), we find a particular solution of the complete equation (Eq. 1) by a method called **variation of parameters.** The method assumes we already know the general solution

$$y_h = C_1 u_1(x) + C_2 u_2(x) \tag{3}$$

of the reduced equation (Eq. 2), which is why we make finding y_h our first solution step.

The method of variation of parameters consists of replacing the constants C_1 and C_2 on the right side of Eq. (3) by functions $v_1(x)$ and $v_2(x)$ and then requiring that the new expression

$$y = v_1u_1 + v_2u_2$$

be a solution of the complete equation (1). There are two functions v_1 and v_2 to be determined, and requiring that Eq. (1) be satisfied is only one condition. It turns out to simplify things if we also require that

$$v_1' u_1 + v_2' u_2 = 0. \tag{4}$$

Then we have

$$y = v_1u_1 + v_2u_2,$$

$$\frac{dy}{dx} = v_1u_1' + v_2u_2' + \underbrace{v_1' u_1 + v_2' u_2}_{0} = v_1u_1' + v_2u_2',$$

$$\frac{d^2y}{dx^2} = v_1u_1'' + v_2u_2'' + v_1' u_1' + v_2' u_2'.$$

If we substitute these expressions into the left-hand side of Eq. (1) and rearrange terms, we obtain

$$v_1\left[\frac{d^2u_1}{dx^2} + a\frac{du_1}{dx} + bu_1\right] + v_2\left[\frac{d^2u_2}{dx^2} + a\frac{du_2}{dx} + bu_2\right] + v_1' u_1' + v_2' u_2' = F(x).$$

The two bracketed terms are zero, since u_1 and u_2 are solutions of the reduced equation (2). Hence Eq. (1) is satisfied if, in addition to Eq. (4), we require that

$$v_1' u_1' + v_2' u_2' = F(x). \tag{5}$$

Equations (4) and (5) may be solved together as a system,

$$v_1' u_1 + v_2' u_2 = 0, \qquad v_1' u_1' + v_2' u_2' = F(x),$$

for the unknown functions v_1' and v_2'. Cramer's rule (Appendix 10) gives

$$v_1' = \frac{\begin{vmatrix} 0 & u_2 \\ F(x) & u_2' \end{vmatrix}}{\begin{vmatrix} u_1 & u_2 \\ u_1' & u_2' \end{vmatrix}} = \frac{-u_2 F(x)}{D}, \qquad v_2' = \frac{\begin{vmatrix} u_1 & 0 \\ u_1' & F(x) \end{vmatrix}}{\begin{vmatrix} u_1 & u_2 \\ u_1' & u_2' \end{vmatrix}} = \frac{u_1 F(x)}{D}, \tag{6}$$

where

$$D = \begin{vmatrix} u_1 & u_2 \\ u_1' & u_2' \end{vmatrix}.$$

Now v_1 and v_2 can be found by integration.

How to Use the Method of Variation of Parameters

In applying the method of variation of parameters to solve the equation

$$\frac{d^2y}{dx^2} + a\frac{dy}{dx} + by = F(x), \tag{1}$$

we can work directly with the equations in Eq. (6). It is not necessary to rederive them. The following are the steps to take.

STEP 1: Solve the reduced equation,

$$\frac{d^2y}{dx^2} + a\frac{dy}{dx} + by = 0,$$

to find the functions u_1 and u_2.

STEP 2: Solve the equations

$$v_1' u_1 + v_2' u_2 = 0$$
$$v_1' u_1' + v_2' u_2' = F(x)$$

together for v_1' and v_2' (as in Eqs. (6)).

STEP 3: Integrate v_1' and v_2' to find v_1 and v_2.

STEP 4: Write down the general solution of Eq. (1) as $y = v_1u_1 + v_2u_2$.

Example 3 Solve the equation $\dfrac{d^2y}{dx^2} + y = \tan x, \quad -\dfrac{\pi}{2} < x < \dfrac{\pi}{2}.$

Solution We first solve the reduced equation

$$\frac{d^2y}{dx^2} + y = 0.$$

The solutions of the characteristic equation $r^2 + 1 = 0$ are $r = \pm i$, so

$$y = C_1 \cos x + C_2 \sin x.$$

With $u_1 = \cos x$ and $u_2 = \sin x$, we have

$$D = \begin{vmatrix} \cos x & \sin x \\ -\sin x & \cos x \end{vmatrix} = \cos^2 x + \sin^2 x = 1.$$

With $F(x) = \tan x$, Eqs. (6) then give

$$\begin{aligned} v_1' &= \frac{-\sin x \tan x}{1} = -\frac{\sin^2 x}{\cos x} = \frac{\cos^2 x - 1}{\cos x} = \cos x - \sec x, \\ v_2' &= \frac{\cos x \tan x}{1} = \sin x. \end{aligned} \tag{7}$$

Hence

$$v_1 = \int (\cos x - \sec x)\, dx = \sin x - \ln|\sec x + \tan x| + C_1,$$

$$v_2 = \int \sin x \, dx = -\cos x + C_2,$$

and

$$\begin{aligned} y &= v_1 u_1 + v_2 u_2 \\ &= (\sin x - \ln|\sec x + \tan x| + C_1) \cos x + (-\cos x + C_2) \sin x \\ &= C_1 \cos x + C_2 \sin x - \cos x \ln|\sec x + \tan x|. \end{aligned}$$

Undetermined Coefficients

The method of variation of parameters is completely general, for it produces a particular solution of Eq. (1) for any continuous function $F(x)$. But the calculations involved can be complicated, and in special cases there may be easier methods to use. For instance, we do not need to use variation of parameters to find a particular solution of

$$\frac{d^2y}{dx^2} - \frac{dy}{dx} + 5y = 3, \tag{8}$$

if we can find the particular solution $y_p = 3/5$ by inspection. And even for an equation like

$$\frac{d^2y}{dx^2} + 3y = e^x, \tag{9}$$

we might guess that there is a solution of the form

$$y_p = Ae^x$$

and substitute $y_p = Ae^x$ and its second derivative into Eq. (9) to find A. If we do so, we find the solution $y_p = (1/4)e^x$.

Again, we might guess that the equation

$$\frac{d^2y}{dx^2} + y = 3x^2 + 4 \tag{10}$$

has a particular solution of the form

$$y_p = Cx^2 + Dx + E.$$

If we substitute this polynomial and its second derivative into Eq. (10) to look for appropriate values for the constants C, D, and E, we get

$$\begin{aligned} 2C + (Cx^2 + Dx + E) &= 3x^2 + 4, \\ Cx^2 + Dx + 2C + E &= 3x^2 + 4. \end{aligned} \tag{11}$$

This latter equation will hold for all values of x if its two sides are identical as polynomials in x; that is, if

$$C = 3, \qquad D = 0, \qquad \text{and} \qquad 2C + E = 4, \tag{12}$$

or

$$C = 3, \qquad D = 0, \qquad E = -2.$$

We conclude that

$$y_p = 3x^2 + 0x - 2 = 3x^2 - 2 \tag{13}$$

is a solution of Eq. (10).

In each of the foregoing examples, the particular solution we found resembled the function $F(x)$ on the right side of the given differential equation. This was no accident, for we guessed the form of the particular solution by looking at $F(x)$ first. The method of first guessing the form of the solution up to certain undetermined constants and then determining the values of these constants by using the differential equation is known as the **method of undetermined coefficients.** It depends for its success on our ability to recognize the form of a particular solution and for this reason, among others, it lacks the generality of the method of variation of parameters. Nevertheless, its simplicity makes it the method of choice in a number of special cases.

We shall limit our discussion of the method of undetermined coefficients to selected equations in which the function $F(x)$ in Eq. (1) is the sum of one or more terms like

$$e^{rx}, \qquad \cos kx, \qquad \sin kx, \qquad ax^2 + bx + c.$$

Even so, we will find that the particular solutions of Eq. (1) do not always resemble $F(x)$ as closely as the ones we have seen.

Example 4 Find a particular solution of

$$\frac{d^2y}{dx^2} - \frac{dy}{dx} = 2 \sin x. \tag{14}$$

Solution If we try to find a particular solution of the form

$$y_p = A \sin x$$

and substitute the derivatives of y_p in the given equation, we find that A must satisfy the equation

$$-A \sin x - A \cos x = 2 \sin x \tag{15}$$

for all values of x. Since this requires A to be equal to -2 and 0 at the same time, we conclude that Eq. (14) has no solution of the form $A \sin x$.

It turns out that the required form is the sum

$$y_p = A \sin x + B \cos x. \tag{16}$$

The result of substituting the derivatives of this new candidate into Eq. (14) is

$$\begin{aligned} -A \sin x - B \cos x - (A \cos x - B \sin x) &= 2 \sin x, \\ (B - A) \sin x - (A + B) \cos x &= 2 \sin x. \end{aligned} \tag{17}$$

Equation (17) will be an identity if

$$B - A = 2 \quad \text{and} \quad A + B = 0, \qquad \text{or} \qquad A = -1, \quad B = 1.$$

Our particular solution is $y_p = \cos x - \sin x$.

Example 5 Find a particular solution of

$$\frac{d^2y}{dx^2} - 3\frac{dy}{dx} + 2y = 5e^x. \tag{18}$$

Solution If we substitute

$$y_p = Ae^x$$

and its derivatives in Eq. (18), we find that

$$\begin{aligned} Ae^x - 3Ae^x + 2Ae^x &= 5e^x, \\ 0 &= 5e^x. \end{aligned}$$

The trouble can be traced to the fact that $y = e^x$ is already a solution of the reduced equation

$$\frac{d^2y}{dx^2} - 3\frac{dy}{dx} + 2y = 0. \tag{19}$$

The characteristic equation of Eq. (19) is

$$r^2 - 3r + 2 = (r - 1)(r - 2) = 0,$$

which has $r = 1$ as a simple root. We may therefore expect Ae^x to "vanish" when substituted into the left-hand side of Eq. (18).

The appropriate way to modify the trial solution in this case is to replace Ae^x by Axe^x. When we substitute

$$y_p = Axe^x$$

and its derivatives into Eq. (18), we get

$$\begin{aligned} (Axe^x + 2Ae^x) - 3(Axe^x + Ae^x) + 2Axe^x &= 5e^x \\ -Ae^x &= 5e^x \\ A &= -5. \end{aligned}$$

Our solution is $y_p = -5xe^x$.

Example 6 Find a particular solution of

$$\frac{d^2y}{dx^2} - 6\frac{dy}{dx} + 9y = e^{3x}. \tag{20}$$

Solution The characteristic equation,

$$r^2 - 6r + 9 = (r - 3)^2 = 0,$$

has $r = 3$ as a *double* root. The appropriate choice for y_p in this case is neither Ae^{3x} nor Axe^{3x}, but Ax^2e^{3x}. When we substitute

$$y_p = Ax^2e^{3x}$$

and its derivatives in the given differential equation, we get

$$(9Ax^2e^{3x} + 12Axe^{3x} + 2Ae^{3x}) - 6(3Ax^2e^{3x} + 2Axe^{3x}) + 9Ax^2e^{3x} = e^{3x}$$

$$2Ae^{3x} = e^{3x}$$

$$A = \frac{1}{2}.$$

Our solution is $y_p = (1/2)x^2e^{3x}$.

When we wish to find a particular solution of Eq. (1), and $F(x)$ has two or more terms, we include a trial function for each term in $F(x)$.

Example 7 Solve the equation

$$\frac{d^2y}{dx^2} - \frac{dy}{dx} = 5e^x - \sin 2x. \tag{21}$$

Solution The roots of the characteristic equation $r^2 - r = 0$ are

$$r_1 = 1, \qquad r_2 = 0,$$

so the general solution of the reduced equation is

$$y_h = C_1e^x + C_2.$$

We now seek a particular solution y_p. That is, we seek a function that will produce $5e^x - \sin 2x$ when substituted into the left side of Eq. (21). One part of y_p is to produce $5e^x$, the other $-\sin 2x$.

Since any function of the form C_1e^x is a solution of the reduced equation, we choose our trial y_p to be the sum

$$y_p = Axe^x + B\cos 2x + C\sin 2x,$$

including xe^x where we might otherwise have included e^x. Substituting the derivatives of y_p in Eq. (21) gives

$$(Axe^x + 2Ae^x - 4B\cos 2x - 4C\sin 2x)$$

$$- (Axe^x + Ae^x - 2B\sin 2x + 2C\cos 2x) = 5e^x - \sin 2x,$$

$$Ae^x - (4B + 2C)\cos 2x + (2B - 4C)\sin 2x = 5e^x - \sin 2x.$$

These equations will hold if

$$A = 5, \qquad (4B + 2C) = 0, \qquad (2B - 4C) = -1,$$

or

$$A = 5, \qquad B = -\frac{1}{10}, \qquad C = \frac{1}{5}.$$

Our particular solution is

$$y_p = 5xe^x - \frac{1}{10}\cos 2x + \frac{1}{5}\sin 2x.$$

The complete solution of Eq. (21) is

$$y = y_h + y_p = C_1 e^x + C_2 + 5xe^x - \frac{1}{10}\cos 2x + \frac{1}{5}\sin 2x.$$

Reduction of Order

We can solve some (not necessarily linear) second order equations by changing them into first order equations with a substitution and solving the resulting first order equations with first order techniques. We did this in Section 6.10 when we solved the nonlinear hanging-cable equation

$$\frac{d^2y}{dx^2} = a\sqrt{1 + \left(\frac{dy}{dx}\right)^2}.$$

The equation is first order in dy/dx and the substitution $p = dy/dx$ changed it into

$$\frac{dp}{dx} = a\sqrt{1 + p^2},$$

which we solved by separation of variables. Having found the solution $p = \sinh ax$ of the associated initial value problem, we replaced p by the original dy/dx and went on to solve the first order equation

$$\frac{dy}{dx} = \sinh ax$$

for y as a function of x. We solved the second order equation by solving two first order equations. Because we reduced the equation's order to solve it, the technique is called **reduction of order.** The general form of a second order differential equation is

$$F\left(x, y, \frac{dy}{dx}, \frac{d^2y}{dx^2}\right) = 0. \tag{22}$$

The equations that can be reduced to first order by substitution normally arise when either the x or the y variable is missing.

TYPE 1: *Equations with dependent variable missing.* When Eq. (22) has the special form

$$F\left(x, \frac{dy}{dx}, \frac{d^2y}{dx^2}\right) = 0, \tag{23}$$

we can reduce it to a first order equation by substituting

$$p = \frac{dy}{dx}, \qquad \frac{d^2y}{dx^2} = \frac{dp}{dx}.$$

Then Eq. (23) takes the form

$$F\left(x, p, \frac{dp}{dx}\right) = 0,$$

which is of the first order in p. If this equation can be solved for p as a function of x, say $p = \phi(x, C_1)$, then we can find y by one additional integration:

$$y = \int \frac{dy}{dx}\, dx = \int p\, dx = \int \phi(x, C_1)\, dx + C_2.$$

Example 8 Solve the equation

$$x\frac{d^2y}{dx^2} + \frac{dy}{dx} = x^2, \qquad x > 0.$$

Solution The dependent variable y does not occur (except in the derivatives) and the substitutions

$$\frac{dy}{dx} = p, \qquad \frac{d^2y}{dx^2} = \frac{dp}{dx}$$

reduce the equation to

$$x\frac{dp}{dx} + p = x^2,$$

a linear first order equation. Dividing by x puts the equation in standard form:

$$\frac{dp}{dx} + \frac{1}{x}p = x. \tag{24}$$

We use Eq. (4) of Section 15.3 with

$$\rho(x) = e^{\int (1/x)dx} = e^{\ln x} = x$$

and $Q(x) = x$ to write the solution of Eq. (24) as

$$p = \frac{1}{x}\int x \cdot x\, dx = \frac{1}{x}\left(\frac{x^3}{3} + C_1\right)$$
$$= \frac{x^2}{3} + \frac{C_1}{x}.$$

We then replace p by dy/dx,

$$\frac{dy}{dx} = \frac{x^2}{3} + \frac{C_1}{x},$$

and integrate again to get

$$y = \frac{x^3}{9} + C_1 \ln x + C_2.$$

(We can omit the usual absolute value bars in the logarithm because $x > 0$.)

TYPE 2: *Equations with independent variable missing.* When Eq. (22) does not contain x explicitly but has the form

$$F\left(y, \frac{dy}{dx}, \frac{d^2y}{dx^2}\right) = 0, \tag{25}$$

the substitutions to use are

$$p = \frac{dy}{dx} \qquad \text{and} \qquad \frac{d^2y}{dx^2} = \frac{dp}{dx} = \frac{dp}{dy}\frac{dy}{dx} = \frac{dp}{dy}p.$$

Then Eq. (25) takes the form

$$F\left(y, p, p\frac{dp}{dy}\right) = 0,$$

which is of the first order in p. Its solution gives p in terms of y, and then a further integration gives the solution of Eq. (25).

Example 9 Solve the initial value problem

$$y\frac{d^2y}{dx^2} + \left(\frac{dy}{dx}\right)^2 = 0, \quad y = 2 \quad \text{and} \quad \frac{dy}{dx} = \frac{1}{2} \quad \text{when } x = 1.$$

Solution We let

$$\frac{dy}{dx} = p, \qquad \frac{d^2y}{dx^2} = \frac{dp}{dy}p.$$

Then we have, in turn,

$$y\left(\frac{dp}{dy}p\right) + p^2 = 0$$

$$y\frac{dp}{dy} + p = 0$$

$$\frac{d}{dy}(yp) = 0$$

$$yp = C_1$$

$$y\frac{dy}{dx} = C_1.$$

The condition that $y = 2$ when $dy/dx = 1/2$ determines C_1:

$$2 \cdot \frac{1}{2} = C_1, \qquad C_1 = 1.$$

Proceeding with $C_1 = 1$, we have

$$y\frac{dy}{dx} = 1,$$

$$\frac{1}{2}y^2 = x + C_2. \qquad \text{(Integrated with respect to } x\text{)}$$

The condition that $y = 2$ when $x = 1$ determines C_2:

$$\frac{1}{2}(2)^2 = 1 + C_2, \qquad C_2 = 1.$$

The solution of the initial value problem is

$$\frac{1}{2}y^2 = x + 1 \qquad \text{or} \qquad y^2 = 2x + 2.$$

Finding the General Solution When One Solution Is Known

When we are fortunate enough to know one solution $u(x)$ of a second order equation (we might have guessed it or identified it from special information), we can sometimes find the equation's general solution by substituting

$$y = uv \tag{26}$$

and solving for the function $v(x)$. The next example shows how this is done.

Example 10 Show that $y = e^x$ is a solution of the equation

$$y'' - 2y' + y = 0,$$

and use the result to find the general solution.

Solution If $y = e^x$, then $y' = y'' = e^x$ and

$$y'' - 2y' + y = e^x - 2e^x + e^x = 0.$$

This shows that $y = e^x$ is one solution. We now substitute $u = e^x$, $y = uv = e^x v$ and calculate the first and second derivatives to get

$$y' = e^x v' + e^x v, \qquad y'' = e^x v'' + 2e^x v' + e^x v.$$

Next we substitute into the equation $y'' - 2y' + y = 0$ and get

$$e^x(v'' + 2v' + v - 2v' - 2v + v) = 0,$$

which simplifies to $v'' = 0$. The solution of this equation is simply $v = C_1 x + C_2$, and this gives

$$y = e^x v = e^x(C_1 x + C_2)$$

as the general solution of $y'' - 2y' + y = 0$.

TABLE 15.1

The method of undetermined coefficients for selected equations of the form $\dfrac{d^2y}{dx^2} + a\dfrac{dy}{dx} + by = F(x)$

If $F(x)$ has a term that is a constant multiple of	And if	Then include this expression in the trial function for y_p
e^{rx}	r is not a root of the characteristic equation	Ae^{rx}
	r is a single root of the characteristic equation	Axe^{rx}
	r is a double root of the characteristic equation	Ax^2e^{rx}
$\sin kx$, $\cos kx$	ki is not a root of the characteristic equation	$B\cos kx + C\sin kx$
	ki is a root of the characteristic equation	$Bx\cos kx + Cx\sin kx$
$ax^2 + bx + c$	0 is not a root of the characteristic equation	$Dx^2 + Ex + F$ (chosen to match the degree of $ax^2 + bx + c$)
	0 is a single root of the characteristic equation	$Dx^3 + Ex^2 + Fx$ (degree one higher than the degree of $ax^2 + bx + c$)
	0 is a double root of the characteristic equation	$Dx^4 + Ex^3 + Fx^2$ (degree two higher than the degree of $ax^2 + bx + c$)

As we saw in the preceding example, $y = e^x$ is a solution of $y'' - 2y' + y = 0$. As you will see if you do Exercise 73, $y = e^x$ is a solution of any linear homogeneous nth order differential equation whose coefficients add to zero. It does not matter whether the coefficients are constant or variable. For instance, $y = e^x$ is a solution of $xy'' + (2 - x)y' - 2y = 0$.

EXERCISES 15.5

Solve the equations in Exercises 1–12 by variation of parameters.

1. $\dfrac{d^2y}{dx^2} + \dfrac{dy}{dx} = x$

2. $\dfrac{d^2y}{dx^2} + y = \tan^2 x, \quad -\dfrac{\pi}{2} < x < \dfrac{\pi}{2}$

3. $\dfrac{d^2y}{dx^2} + y = \sin x$

4. $\dfrac{d^2y}{dx^2} + 2\dfrac{dy}{dx} + y = e^x$

5. $\dfrac{d^2y}{dx^2} + 2\dfrac{dy}{dx} + y = e^{-x}$

6. $\dfrac{d^2y}{dx^2} - y = x$

7. $\dfrac{d^2y}{dx^2} - y = e^x$

8. $\dfrac{d^2y}{dx^2} - y = \sin x$

9. $\dfrac{d^2y}{dx^2} + 4\dfrac{dy}{dx} + 5y = 10$

10. $\dfrac{d^2y}{dx^2} - \dfrac{dy}{dx} = 2^x$

11. $\dfrac{d^2y}{dx^2} + y = \sec x, \quad -\dfrac{\pi}{2} < x < \dfrac{\pi}{2}$

12. $\dfrac{d^2y}{dx^2} - \dfrac{dy}{dx} = e^x \cos x, \quad x > 0$

Solve the equations in Exercises 13–28 by the method of undetermined coefficients.

13. $\dfrac{d^2y}{dx^2} - 3\dfrac{dy}{dx} - 10y = -3$

14. $\dfrac{d^2y}{dx^2} - 3\dfrac{dy}{dx} - 10y = 2x - 3$

15. $\dfrac{d^2y}{dx^2} - \dfrac{dy}{dx} = \sin x$

16. $\dfrac{d^2y}{dx^2} + 2\dfrac{dy}{dx} + y = x^2$

17. $\dfrac{d^2y}{dx^2} + y = \cos 3x$

18. $\dfrac{d^2y}{dx^2} + y = e^{2x}$

19. $\dfrac{d^2y}{dx^2} - \dfrac{dy}{dx} - 2y = 20 \cos x$

20. $\dfrac{d^2y}{dx^2} + y = 2x + 3e^x$

21. $\dfrac{d^2y}{dx^2} - y = e^x + x^2$

22. $\dfrac{d^2y}{dx^2} + 2\dfrac{dy}{dx} + y = 6 \sin 2x$

23. $\dfrac{d^2y}{dx^2} - \dfrac{dy}{dx} - 6y = e^{-x} - 7 \cos x$

24. $\dfrac{d^2y}{dx^2} + 3\dfrac{dy}{dx} + 2y = e^{-x} + e^{-2x} - x$

25. $\dfrac{d^2y}{dx^2} + 5\dfrac{dy}{dx} = 15x^2$

26. $\dfrac{d^2y}{dx^2} - \dfrac{dy}{dx} = -8x + 3$

27. $\dfrac{d^2y}{dx^2} - 3\dfrac{dy}{dx} = e^{3x} - 12x$

28. $\dfrac{d^2y}{dx^2} + 7\dfrac{dy}{dx} = 42x^2 + 5x + 1$

In each of Exercises 29–31, the given differential equation has a particular solution y_p of the form given. Determine the coefficients in y_p. Then solve the differential equation.

29. $\dfrac{d^2y}{dx^2} - 5\dfrac{dy}{dx} = xe^{5x}, \quad y_p = Ax^2e^{5x} + Bxe^{5x}$

30. $\dfrac{d^2y}{dx^2} - \dfrac{dy}{dx} = \cos x + \sin x, \quad y_p = A \cos x + B \sin x$

31. $\dfrac{d^2y}{dx^2} + y = 2 \cos x + \sin x, \quad y_p = Ax \cos x + Bx \sin x$

In Exercises 32–35, solve the given differential equations (a) by variation of parameters and (b) by the method of undetermined coefficients.

32. $\dfrac{d^2y}{dx^2} - 4\dfrac{dy}{dx} + 4y = 2e^{2x}$

33. $\dfrac{d^2y}{dx^2} - \dfrac{dy}{dx} = e^x + e^{-x}$

34. $\dfrac{d^2y}{dx^2} - 9\dfrac{dy}{dx} = 9e^{9x}$

35. $\dfrac{d^2y}{dx^2} - 4\dfrac{dy}{dx} - 5y = e^x + 4$

Solve the differential equations in Exercises 36–45. Some of the equations can be solved by the method of undetermined coefficients, but others cannot.

36. $\dfrac{d^2y}{dx^2} + y = \csc x, \quad 0 < x < \pi$

37. $\dfrac{d^2y}{dx^2} + y = \cot x, \quad 0 < x < \pi$

38. $\dfrac{d^2y}{dx^2} + 4y = \sin x$

39. $\dfrac{d^2y}{dx^2} - 8\dfrac{dy}{dx} = e^{8x}$

40. $\dfrac{d^2y}{dx^2} + 4\dfrac{dy}{dx} + 5y = x + 2$

41. $\dfrac{d^2y}{dx^2} - \dfrac{dy}{dx} = x^3$

42. $\dfrac{d^2y}{dx^2} + 9y = 9x - \cos x$

43. $\dfrac{d^2y}{dx^2} + 2\dfrac{dy}{dx} = x^2 - e^x$

44. $\dfrac{d^2y}{dx^2} - 3\dfrac{dy}{dx} + 2y = e^x - e^{2x}$

45. $\dfrac{d^2y}{dx^2} + y = \sec x \tan x, \quad -\dfrac{\pi}{2} < x < \dfrac{\pi}{2}$

The method of undetermined coefficients can sometimes be used to solve first order ordinary differential equations. Use the method to solve the equations in Exercises 46–49.

46. $\dfrac{dy}{dx} + 4y = x$

47. $\dfrac{dy}{dx} - 3y = e^x$

48. $\dfrac{dy}{dx} + y = \sin x$

49. $\dfrac{dy}{dx} - 3y = 5e^{3x}$

Solve the initial value problems in Exercises 50 and 51.

50. $\dfrac{d^2y}{dx^2} + y = e^{2x}; \quad y(0) = 0, \quad y'(0) = \dfrac{2}{5}$

51. $\dfrac{d^2y}{dx^2} + y = \sec^2 x, \quad -\dfrac{\pi}{2} < x < \dfrac{\pi}{2}; \quad y(0) = y'(0) = 1$

52. *Bernoulli's equation of order 2.* Solve the equation

$$\frac{dy}{dx} + y = (xy)^2$$

by carrying out the following steps: (1) divide both sides of the equation by y^2; (2) make the change of variable $u = y^{-1}$; (3) solve the resulting equation for u in terms of x; (4) let $y = u^{-1}$.

Reduction of Order

Solve the equations in Exercises 53–56.

53. $\dfrac{d^2y}{dx^2} + \dfrac{dy}{dx} = 0$

54. $\dfrac{d^2y}{dx^2} + y\dfrac{dy}{dx} = 0$

55. $y'' = (y')^2$

56. $y'' = 2yy'$

Solve the initial value problems in Exercises 57–61.

57. $x\dfrac{d^2y}{dx^2} + \dfrac{dy}{dx} = 0, \quad y = -3$ and $y' = 2$ when $x = 1$

58. $x\dfrac{d^3y}{dx^3} - 2\dfrac{d^2y}{dx^2} = 0; \quad y = -5,\ y' = 2,$ and $y'' = 3$ when $x = 1$

59. $x\dfrac{d^2y}{dx^2} + 2\dfrac{dy}{dx} = 1, \quad y = 2$ and $y' = 1$ when $x = 2$

60. $2\dfrac{d^2y}{dx^2} + \left(\dfrac{dy}{dx}\right)^2 = -1, \quad y = 2$ and $y' = 0$ when $x = -1$

61. $x\dfrac{d^2y}{dx^2} + \dfrac{dy}{dx} = x^2, \quad y = 0$ and $y' = 1$ when $x = 1$

62. *An integral equation.* Solve the integral equation

$$y(x) + \int_0^x y(t)\,dt = x.$$

(*Hint:* Differentiate first.)

63. *Hooke's law.* A body of mass m hangs at rest from one end of a vertical spring whose other end is attached to a rigid support. The body is pulled down an amount s_0 and then released. Find the body's subsequent position as a function of time t. (*Hint:* Assume Hooke's law and Newton's second law of motion and let s denote the body's displacement from its rest position. Then $m(d^2s/dt^2) = -ks$, where k is the spring constant.)

64. *Air resistance proportional to velocity.* A car suspended from a parachute for a television ad falls through the air under the pull of gravity. If air resistance produces a retarding force proportional to the car's downward velocity and the car starts from rest at time $t = 0$, find the distance fallen as a function of the elapsed time t.

Equations with One Known Solution

In Exercises 65–72, use the given solution $y = u(x)$ to find the equation's general solution.

65. $u = \cos x, \quad y'' + y = 0$

66. $u = x^2, \quad x^2y'' - 2y = 0$

67. $u = e^{-x}, \quad y'' - y' - 2y = 0$

68. $u = \sqrt{x}, \quad 4x^2y'' + y = 0, \quad x > 0$

69. $u = x, \quad x^2y'' + xy' - y = 0$

70. $u = x^3, \quad x^2y'' - 5xy' + 9y = 0$

71. $u = x, \quad (1 - x^2)y'' - 2xy' + 2y = 0$

72. $u = (\sin x)/\sqrt{x}, \quad x^2y'' + xy' + (x^2 - (1/4))y = 0$

73. Show that if $\sum_{k=0}^{n} a_k(x) = 0$, then $y = e^x$ is a solution of $a_n(x)y^{(n)} + a_{n-1}(x)y^{(n-1)} + \cdots + a_1(x)y' + a_0(x)y = 0.$

74. Find the general solution of

$$xy'' - (2x + 1)y' + (x + 1)y = 0.$$

Higher Order Linear Equations with Constant Coefficients

We solve third and higher order linear equations with constant coefficients the same way we solve second order equations. We find the roots of the characteristic equation, use them to construct the general solution y_h of the reduced equation, and add a particular solution y_p of the complete equation to form the general solution $y = y_h + y_p$ of the complete equation.

Find the general solutions of the equations in Exercises 75–78.

75. $y''' - 3y'' + 2y = 3 \sin x + \cos x$

76. $y''' - 7y' + 6y = 6x + 1$

77. $y^{(4)} - 8y'' + 16y = 8x - 16$

78. $y^{(4)} - 4y''' + 6y'' - 4y' + y = 7$

15.6 Oscillation

Linear second order equations with constant coefficients are important because they model oscillation. Oscillation occurs, for example, in atoms and molecules, in machinery and electrical circuits, and in the chemical and electrical systems of our bodies. The classical example of mechanical oscillation is the motion of a weighted spring, and that is the example we shall study in this section. But all oscillations are modeled with the same basic equations, and the mathematics we present here comes up in all the other applications as well.

Undamped Oscillation (Simple Harmonic Motion)

Suppose we have a spring of natural length L and spring constant k, with its upper end fastened to a rigid support (Fig. 15.8). We hang a mass m from the spring. The weight of the mass stretches the spring to a length $L + s$ when allowed to come to rest in a new equilibrium position. By Hooke's law, the tension in the spring is ks. The force of gravity pulling down on the mass is mg. Equilibrium requires

$$ks = mg. \tag{1}$$

How will the mass behave if we pull it down an additional amount x_0 beyond the equilibrium position and release it? To find out, let x, positive direction downward, denote the displacement of the mass from equilibrium t seconds after the motion has started. Then the forces acting on the mass are

$+\,mg$ (weight due to gravity),

$-\,k(s + x)$ (spring tension).

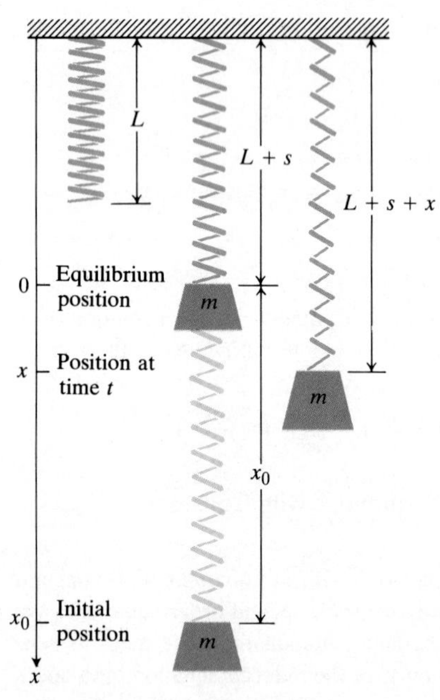

15.8 A spring of natural length L is stretched a distance s by the weight of a mass m. It is then stretched an additional distance x_0 and released. The position of the mass at any subsequent time is described by the solution of an initial value problem.

By Newton's second law, the sum of these forces is $m(d^2x/dt^2)$, so

$$m\frac{d^2x}{dt^2} = mg - ks - kx. \tag{2}$$

Since $mg = ks$ from Eq. (1), Eq. (2) simplifies to

$$m\frac{d^2x}{dt^2} + kx = 0. \tag{3}$$

In addition to satisfying this differential equation, the position of the mass satisfies the initial conditions

$$x = x_0 \quad \text{and} \quad \frac{dx}{dt} = 0 \quad \text{when } t = 0. \tag{4}$$

If we divide both sides of Eq. (3) by m and write ω for $\sqrt{k/m}$, the equation becomes

$$\frac{d^2x}{dt^2} + \omega^2 x = 0. \tag{5}$$

The roots of the characteristic equation $r^2 + \omega^2 = 0$ are $r = \pm\omega i$, so the general solution of Eq. (5) is

$$x = C_1 \cos \omega t + C_2 \sin \omega t. \tag{6}$$

Applying the initial conditions in Eq. (4) determines the constants to be

$$C_1 = x_0 \quad \text{and} \quad C_2 = 0.$$

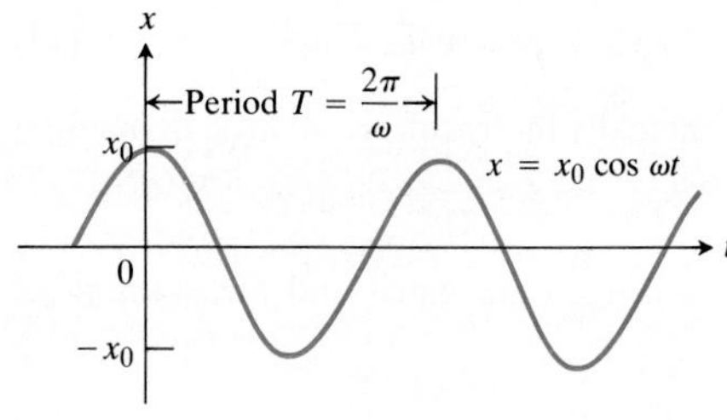

15.9 Simple harmonic motion.

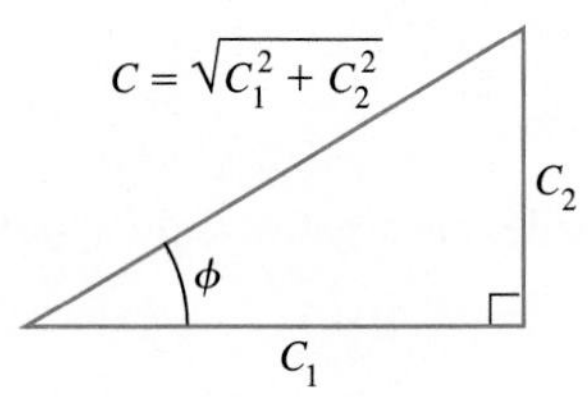

15.10 $C_1 = C\cos\phi$ and $C_2 = C\sin\phi$.

The mass's displacement from equilibrium t seconds into the motion is

$$x = x_0 \cos \omega t. \tag{7}$$

This equation represents a **simple harmonic motion** of amplitude x_0 and period $T = 2\pi/\omega$ (Fig. 15.9).

We normally combine the two terms in the general solution in Eq. (6) into a single term, using the trigonometric identity

$$C \cos(\omega t - \phi) = C \cos\phi \cos \omega t + C \sin\phi \sin \omega t.$$

To apply the identity, we take

$$C_1 = C\cos\phi, \qquad C_2 = C\sin\phi, \tag{8}$$

where

$$C = \sqrt{C_1^2 + C_2^2}, \qquad \tan\phi = \frac{C_2}{C_1}, \tag{9}$$

as in Fig. 15.10. With these substitutions, Eq. (6) becomes

$$x = C\cos(\omega t - \phi). \tag{10}$$

We treat C and ϕ as two new arbitrary constants, replacing the arbitrary constants C_1 and C_2 of Eq. (6). Equation (10) represents a simple harmonic motion of amplitude C and period $T = 2\pi/\omega$ (Fig. 15.11). The angle ϕ is the **phase angle** of the motion.

Damped Oscillation

Suppose that the mass is slowed by a frictional force $c(dx/dt)$ that is proportional to velocity, where c is a positive constant. Then the equation that replaces Eq. (3) is

$$m\frac{d^2x}{dt^2} + c\frac{dx}{dt} + kx = 0 \tag{11}$$

or

$$\frac{d^2x}{dt^2} + 2b\frac{dx}{dt} + \omega^2 x = 0, \tag{12}$$

where

$$2b = \frac{c}{m} \qquad \text{and} \qquad \omega = \sqrt{\frac{k}{m}}.$$

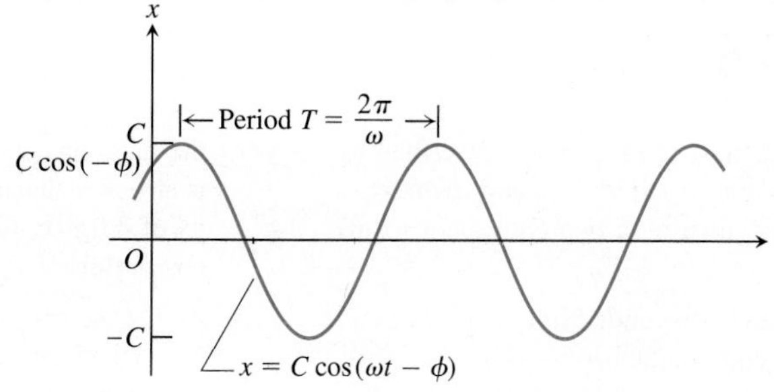

15.11 Graph of displacement vs. time for a simple harmonic motion.

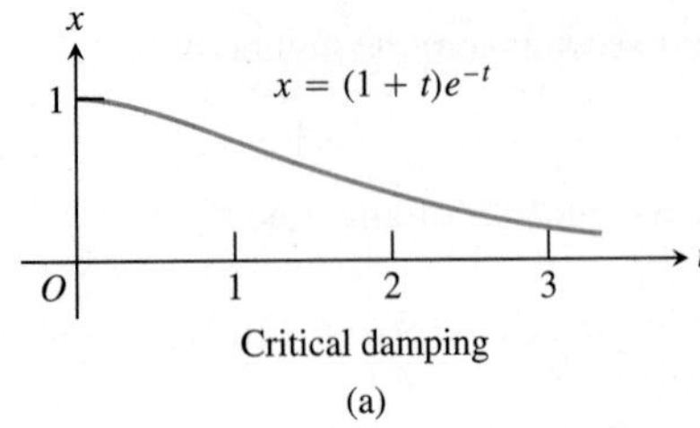

Critical damping
(a)

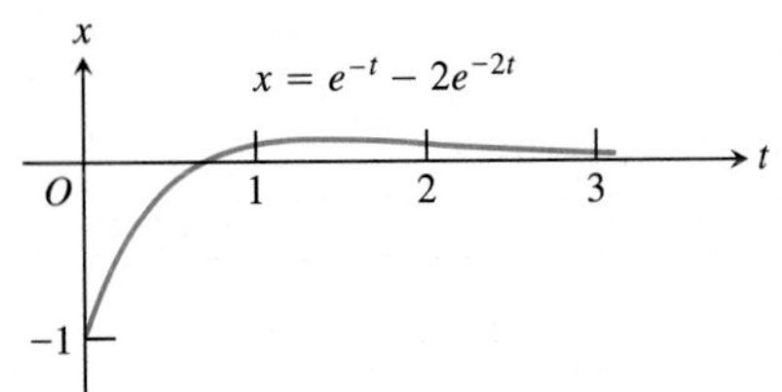

Overcritical damping
(b)

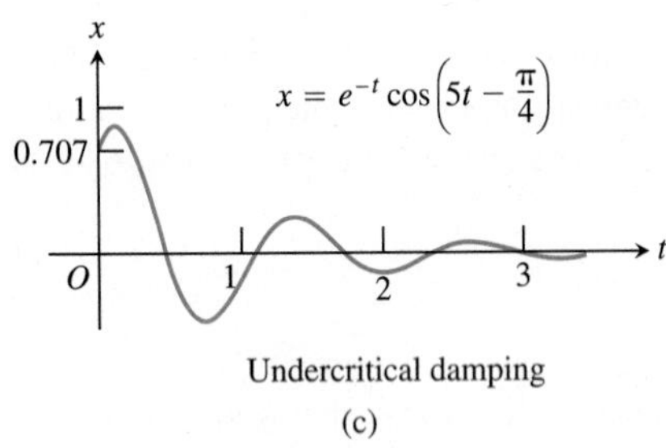

Undercritical damping
(c)

15.12 Examples of damping; x is displacement and t is time.

The roots of the characteristic equation $r^2 + 2br + \omega^2 = 0$ are

$$r_1 = -b + \sqrt{b^2 - \omega^2}, \qquad r_2 = -b - \sqrt{b^2 - \omega^2}. \tag{13}$$

As we shall now see, the mass behaves in three distinctly different ways, depending on the relative sizes of b and ω.

Critical Damping If $b = \omega$, the roots in (13) are equal and the solution of (12) is

$$x = (C_1 t + C_2)e^{-\omega t}. \tag{14}$$

As time passes, x approaches zero. The mass does not oscillate (Fig. 15.12a).

Overcritical Damping If $b > \omega$, the roots in (13) are real and unequal and the solution of (12) is

$$x = C_1 e^{r_1 t} + C_2 e^{r_2 t}. \tag{15}$$

Here again, the mass does not oscillate. Both r_1 and r_2 are negative and x approaches zero as time passes (Fig. 15.12b).

Undercritical Damping If $0 < b < \omega$, let $\omega^2 - b^2 = \alpha^2$. Then

$$r_1 = -b + \alpha i, \qquad r_2 = -b - \alpha i$$

and

$$x = e^{-bt}(C_1 \cos \alpha t + C_2 \sin \alpha t). \tag{16}$$

If we introduce the substitutions (8), we may also write Eq. (16) in the equivalent form

$$x = Ce^{-bt}\cos(\alpha t - \phi). \tag{17}$$

This equation represents damped oscillation. It is analogous to simple harmonic motion, of period $T = 2\pi/\alpha$, except that the amplitude is not constant but is given by Ce^{-bt}. Since this tends to zero as t increases, the oscillations die out as time goes on (Fig. 15.12c). Observe, however, that Eq. (17) reduces to Eq. (10) in the absence of friction. The effect of friction is twofold:

1. $b = c/(2m)$ appears in the exponential **damping factor** e^{-bt}. The larger b is, the more quickly the oscillations tend to become unnoticeable.
2. The period $T = 2\pi/\alpha = 2\pi/\sqrt{\omega^2 - b^2}$ is longer than the period $T_0 = 2\pi/\omega$ in the friction-free system. The motion is slower.

EXERCISES 15.6

1. Suppose the motion of the mass in Fig. 15.8 is described by Eq. (3). Find x as a function of t if $x = x_0$ and $dx/dt = v_0$ when $t = 0$. Express your answer in two equivalent forms (Eqs. 6 and 10).

2. CALCULATOR A 5-lb mass is suspended from the lower end of a spring whose upper end is attached to a rigid support. The weight of the mass extends the spring by 6 in. If, after the mass has come to rest in its new equilibrium position, it is struck a sharp blow that starts it downward with a velocity of 4 ft/sec, find its subsequent motion, assuming there is no friction.

3. *An RLC series circuit.* The electrical series circuit shown in Fig. 15.13 contains a capacitor of capacitance C farads, a coil of inductance L henries, a resistance of R ohms, and a

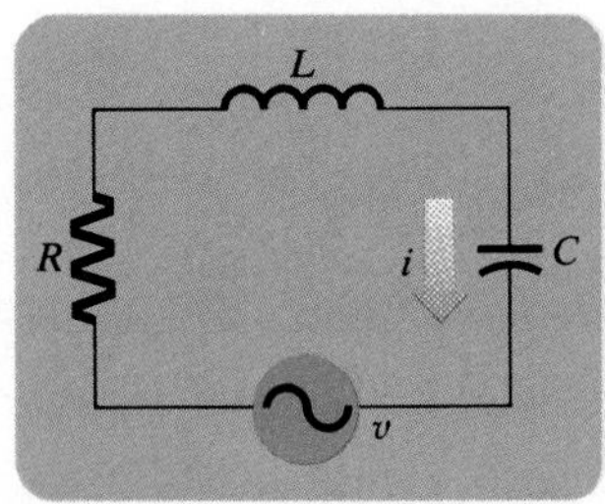

15.13 An *RLC* series circuit (Exercise 3).

generator that produces an electromotive force of v volts. If the current intensity in the circuit at time t is i amperes, the differential equation describing the current i is

$$L\frac{d^2i}{dt^2} + R\frac{di}{dt} + \frac{1}{C}i = \frac{dv}{dt}.$$

Find i as a function of t if

a) $R = 0,\quad 1/(LC) = \omega^2,\quad v = \text{constant};$

b) $R = 0,\quad 1/(LC) = \omega^2,\quad v = V \sin \alpha t,$
$V = \text{constant},\quad \alpha = \text{constant} \neq \omega;$

c) $R = 0,\quad 1/(LC) = \omega^2,\quad v = V \sin \omega t,$
$V = \text{constant};$

d) $R = 50,\quad L = 5,\quad C = 9 \times 10^{-6},\quad v = \text{constant}.$

4. A simple pendulum of length L makes an angle θ with the vertical. As it swings back and forth, its motion, neglecting friction, is described by the differential equation

$$\frac{d^2\theta}{dt^2} = -\frac{g}{L}\sin\theta,$$

where g is the (constant) acceleration of gravity (Fig. 15.14). Solve the differential equation of motion, under the assumption that θ is so small that $\sin\theta$ may be replaced by θ without appreciable error. Assume that $\theta = \theta_0$ and $d\theta/dt = 0$ when $t = 0$.

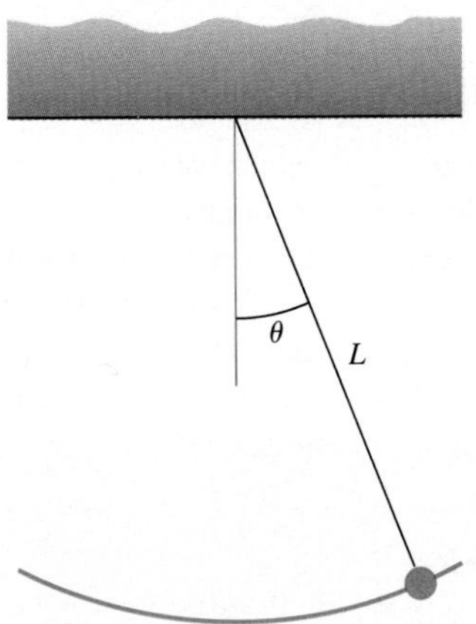

15.14 The pendulum in Exercise 4.

5. A circular disk of mass m and radius r is suspended by a thin wire attached to the center of one of its flat faces (Fig. 15.15). If the disk is twisted through an angle θ, torsion in the wire tends to turn the disk back in the opposite direction. The differential equation for the motion is

$$\frac{1}{2}mr^2\frac{d^2\theta}{dt^2} = -k\theta,$$

where k is the coefficient of torsion of the wire. Find θ as a function of t if $\theta = \theta_0$ and $d\theta/dt = v_0$ when $t = 0$.

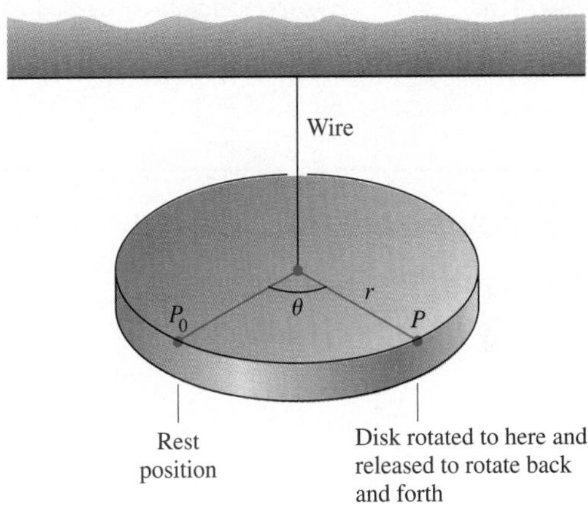

15.15 The suspended disk in Exercise 5.

6. CALCULATOR A cylindrical spar buoy, diameter 1 ft, weight 100 lb, floats partially submerged in an upright position. When it is depressed slightly from its equilibrium position and released, it bobs up and down according to the differential equation

$$\frac{100}{g}\frac{d^2x}{dt^2} = -16\pi x - c\frac{dx}{dt}.$$

Here x (ft) is the vertical displacement of the buoy from its equilibrium position and $c(dx/dt)$ is the frictional resistance of the water. Find c if the period of oscillation is observed to be 1.6 sec. (Take $g = 32$ ft/sec^2.)

7. Suppose the upper end of the spring in Fig. 15.8 is attached not to a rigid support but to a member that itself undergoes up and down motion given by a function of time t, say $y = f(t)$. If the positive direction of y is downward, the differential equation of motion is

$$m\frac{d^2x}{dt^2} + kx = kf(t).$$

Let $x = x_0$ and $dx/dt = 0$ when $t = 0$, and solve for x

a) if $f(t) = A \sin \alpha t$ and $\alpha \neq \sqrt{k/m}$,

b) if $f(t) = A \sin \alpha t$ and $\alpha = \sqrt{k/m}$.

EXPLORER PROGRAMS

Oscillator	Enables you to explore solutions of the equation $x'' + 2bx' + \omega^2 x = 0$. The program can help you to learn more about amplitude, frequency, angular frequency (ω), period, and conservation of energy.
Forced Oscillator	Enables you to explore solutions of the equation $x'' + 2bx' + \omega^2 x = F(t)$, which models the behavior of an oscillator driven by an external force $F(t)$

15.7 Power Series Solutions

When we cannot find a nice compact expression for the solution of a differential equation, we try to get our information in other ways (which may be preferable anyway). One way is to try to find a power series representation for the solution. If we can do so, we immediately have a source of convenient and accurate polynomial approximations of the solution, which may be all we really need. Another way, which works for first order equations, is to graph the solutions directly from the equation. How can we do this without solving the equation first? We use the equation's slope field, as we shall see in Section 15.8. Still another method is to solve the equation numerically to generate tables of values for the solutions of particular initial value problems. The tables are what we usually want anyway and we can get them directly from the equation without the intermediate step of finding the general solution. (Having a computer does help.)

The present section introduces power series solutions. The next section covers graphical solutions, and Section 15.9 deals with numerical methods. Our discussions are too brief to be rigorous but you will get a sense of the main ideas and a chance to practice important techniques on a variety of equations. There is much more known about differential equations than we can describe here—enough to make the field one of the most valuable and far-reaching fields of applied mathematics and current mathematical research.

Equations with Analytic Coefficients

Our first example deals with the equation $y'' + x^2y = 0$, which we cannot solve by any of our previous methods. But we can solve it by assuming that the solution has a power series representation $\Sigma a_n x^n$. The equation $y'' + x^2y = 0$ is a linear second order equation that belongs to a class of equations of the form $y'' + a_1(x)y' + a_0(x)y = 0$, where the coefficient functions $a_1(x)$ and $a_0(x)$ are **analytic** in the neighborhood of $x = 0$ (that is, their Maclaurin series converge to the functions in some interval $|x| < h$ about the origin). In our example, $a_1(x) = 0$ and $a_0(x) = x^2$.

Example 1 Find a power series solution for

$$y'' + x^2y = 0. \tag{1}$$

Solution We assume that there is a solution of the form

$$y = a_0 + a_1x + a_2x^2 + \cdots + a_nx^n + \cdots, \tag{2}$$

and find what the coefficients a_k have to be to make the series and its second derivative

$$y'' = 2a_2 + 3 \cdot 2a_3 x + \cdots + n(n-1)a_n x^{n-2} + \cdots \tag{3}$$

satisfy Eq. (1). The series for x^2y is x^2 times the right-hand side of Eq. (2):

$$x^2y = a_0x^2 + a_1x^3 + a_2x^4 + \cdots + a_n x^{n+2} + \cdots. \tag{4}$$

The series for $y'' + x^2y$ is the sum of the series in Eqs. (3) and (4):

$$\begin{aligned} y'' + x^2y = {} & 2a_2 + 6a_3 x + (12a_4 + a_0)x^2 + (20a_5 + a_1)x^3 \\ & + \cdots + (n(n-1)a_n + a_{n-4})x^{n-2} + \cdots. \end{aligned} \tag{5}$$

Notice that the coefficient of x^{n-2} in Eq. (4) is a_{n-4}. If y and its second derivative y'' are to satisfy Eq. (1), the coefficients of the individual powers of x on the right-hand side of Eq. (5) must all be zero (power series representations are unique, as you saw if you did Exercise 45 in Section 8.6):

$$2a_2 = 0, \qquad 6a_3 = 0, \qquad 12a_4 + a_0 = 0, \qquad 20a_5 + a_1 = 0, \tag{6}$$

and for all $n \geq 4$,

$$n(n-1)a_n + a_{n-4} = 0. \tag{7}$$

We can see from Eq. (2) that

$$a_0 = y(0), \qquad a_1 = y'(0). \tag{8}$$

In other words, the first two coefficients of the series are the values of y and y' at $x = 0$. The equations in (6) and the recursion formula in (7) enable us to evaluate all the other coefficients in terms of a_0 and a_1.

The first two of Eqs. (6) give

$$a_2 = 0, \qquad a_3 = 0.$$

Equation (7) shows that if $a_{n-4} = 0$, then $a_n = 0$; so we conclude that

$$a_6 = 0, \qquad a_7 = 0, \qquad a_{10} = 0, \qquad a_{11} = 0,$$

and whenever $n = 4k + 2$ or $4k + 3$, a_n is zero. For the other coefficients we have

$$a_n = \frac{-a_{n-4}}{n(n-1)}$$

so that

$$a_4 = \frac{-a_0}{4 \cdot 3}, \qquad a_8 = \frac{-a_4}{8 \cdot 7} = \frac{a_0}{3 \cdot 4 \cdot 7 \cdot 8},$$

$$a_{12} = \frac{-a_8}{11 \cdot 12} = \frac{-a_0}{3 \cdot 4 \cdot 7 \cdot 8 \cdot 11 \cdot 12},$$

and

$$a_5 = \frac{-a_1}{5 \cdot 4}, \qquad a_9 = \frac{-a_5}{9 \cdot 8} = \frac{a_1}{4 \cdot 5 \cdot 8 \cdot 9},$$

$$a_{13} = \frac{-a_9}{12 \cdot 13} = \frac{-a_1}{4 \cdot 5 \cdot 8 \cdot 9 \cdot 12 \cdot 13}.$$

The answer is best expressed as the sum of two separate series—one multiplied by a_0, the other by a_1:

$$y = a_0\left(1 - \frac{x^4}{3\cdot 4} + \frac{x^8}{3\cdot 4\cdot 7\cdot 8} - \frac{x^{12}}{3\cdot 4\cdot 7\cdot 8\cdot 11\cdot 12} + \cdots\right)$$
$$+ a_1\left(x - \frac{x^5}{4\cdot 5} + \frac{x^9}{4\cdot 5\cdot 8\cdot 9} - \frac{x^{13}}{4\cdot 5\cdot 8\cdot 9\cdot 12\cdot 13} + \cdots\right). \quad (9)$$

Both series converge absolutely for all values of x, as is readily seen by the ratio test.

Next we show how to find a series solution for an initial value problem of the form

Differential equation: $\quad y' = f(x, y) \quad (10)$

Initial condition: $\quad y = y_0 \quad \text{when} \quad x = x_0. \quad (11)$

If there is a Taylor series in powers of $x - x_0$ that converges to a solution of the problem, it must have the form

$$y = y_0 + y'(x_0)(x - x_0) + \frac{1}{2!}y''(x_0)(x - x_0)^2 + \cdots.$$

The idea is to use the differential equation itself to calculate $y'(x_0) = f(x_0, y_0)$ and higher order derivatives by differentiating both sides of Eq. (10) repeatedly with respect to x and then substituting previously calculated values, as shown in the next example. In the example, $x_0 = 1$, $y_0 = 0$, and $f(x, y) = x + y$.

Example 2 Solve the differential equation

$$y' = x + y \quad (12)$$

subject to the initial condition

$$y = 0 \quad \text{when} \quad x = 1. \quad (13)$$

Solution The Taylor series for $y(x)$, in powers of $x - 1$, is

$$y(x) = y(1) + y'(1)(x - 1) + \frac{1}{2!}y''(1)(x - 1)^2 + \cdots.$$

Substituting $x = 1$ and $y = 0$ in Eq. (12), we get

$$y'(1) = 1 + 0 = 1. \quad (14)$$

If we differentiate both sides of Eq. (12) with respect to x, we get

$$y'' = 1 + y'. \quad (15)$$

We know that $y'(1) = 1$ and now we can calculate $y''(1)$:

$$y''(1) = 1 + y'(1) = 2. \quad (16)$$

By differentiating Eq. (15) with respect to x, we get

$$y''' = 0 + y'' = y''. \quad (17)$$

Since we already know that $y''(1) = 2$, we now find

$$y'''(1) = y''(1) = 2.$$

The pattern continues: all higher order derivatives satisfy

$$y^{(n+1)}(x) = y^{(n)}(x) \qquad \text{for } n \geq 2. \tag{18}$$

At $x = 1$, all of these derivatives have the value 2:

$$y^{(n)}(1) = 2, \qquad \text{for } n \geq 2. \tag{19}$$

When the values of $y(1)$, $y'(1)$, $y''(1)$, and so on are substituted into the Taylor series for $y(x)$, we get

$$\begin{aligned} y(x) &= 0 + 1(x-1) + 2\frac{1}{2!}(x-1)^2 + 2\frac{1}{3!}(x-1)^3 + \cdots \\ &= x - 1 + 2\sum_{n=2}^{\infty} \frac{(x-1)^n}{n!}. \end{aligned} \tag{20}$$

This series brings to mind a comparable series, namely the series for $2e^{x-1}$, which is

$$2e^{x-1} = 2 + 2(x-1) + 2\sum_{n=2}^{\infty} \frac{(x-1)^n}{n!}. \tag{21}$$

Comparing Eqs. (20) and (21), we see that our series representation for $y(x)$, Eq. (20), is the same as

$$y(x) = 2e^{x-1} - 2 - (x-1). \tag{22}$$

Exercise 1 asks you to verify that Eq. (22) does satisfy the differential equation $y' = x + y$ with $y = 0$ when $x = 1$.

The technique used in Example 1 could also have been used in Example 2. To apply the method of Example 1 in solving $y' = x + y$, we could use a series

$$y(x) = b_0 + b_1(x-1) + b_2(x-1)^2 + b_3(x-1)^3 + \cdots$$

together with the series for x as $x = 1 + (x - 1)$. Then

$$y' = b_1 + 2b_2(x-1) + 3b_3(x-1)^2 + \cdots$$

and the equation $y' = x + y$ becomes

$$\begin{aligned} b_1 &+ 2b_2(x-1) + 3b_3(x-1)^2 + \cdots + nb_n(x-1)^{n-1} \\ &+ (n+1)b_{n+1}(x-1)^n + \cdots \\ &= 1 + (x-1) + b_0 + b_1(x-1) + b_2(x-1)^2 + \cdots + b_n(x-1)^n + \cdots \\ &= (1 + b_0) + (1 + b_1)(x-1) + b_2(x-1)^2 + \cdots + b_n(x-1)^n + \cdots. \end{aligned}$$

The coefficients of corresponding powers of $x - 1$ must be the same on both sides of this equation. In other words,

$$\begin{aligned} b_1 &= 1 + b_0, \\ 2b_2 &= 1 + b_1, \\ 3b_3 &= b_2, \quad \ldots, \quad (n+1)b_{n+1} = b_n \quad \text{for } n \geq 2. \end{aligned}$$

These equations can be solved recursively to give all the coefficients in terms of b_0, where $b_0 = y(1)$. In particular, if $y(1) = 0$, then $b_0 = 0$, $b_1 = 1$, $2b_2 = 1 + b_1 = 2$, so $b_2 = 1$; $3b_3 = b_2 = 1$, so $b_3 = 1/3$; and so on.

EXERCISES 15.7

1. Show that $y = 2e^{x-1} - 2 - (x - 1)$ solves the initial value problem
$$y' = x + y, \quad y(1) = 0.$$

2. a) Show that
$$y_1 = \frac{a_0}{x}\left(1 - \frac{x^2}{2!} + \frac{x^4}{4!} - \cdots\right),$$
$$y_2 = \frac{a_1}{x}\left(x - \frac{x^3}{3!} + \frac{x^5}{5!} - \cdots\right)$$
are both solutions of the differential equation
$$xy'' + 2y' + xy = 0.$$
Hence, by the principle of superposition (Section 15.4, Exercise 26), $y = y_1 + y_2$ is a solution of the equation.
b) Express y_1 and y_2 in terms of elementary functions.
c) Solve the *boundary value problem*
Differential equation: $xy'' + 2y' + xy = 0$
Boundary values: $y(\pi/2) = 2$ and $y(\pi) = -1$
on the interval $\pi/2 \le x \le \pi$.

Find series solutions (Taylor or Maclaurin) for the initial value problems in Exercises 3–14.

3. $y' = y, \quad y = 1$ when $x = 0$
4. $y' + y = 0, \quad y = 1$ when $x = 0$
5. $y' = 2y, \quad y = 2$ when $x = 0$
6. $y' + 2y = 0, \quad y = -1$ when $x = 1$
7. $y'' = y, \quad y = 0$ and $y' = 1$ when $x = 0$
8. $y'' + y = 0, \quad y = 1$ and $y' = 0$ when $x = 0$
9. $y'' + y = x, \quad y = 2$ and $y' = 1$ when $x = 0$
10. $y'' - y = x, \quad y = -1$ and $y' = 2$ when $x = 0$
11. $y'' - y = -x, \quad y = 0$ and $y' = -2$ when $x = 2$
12. $y'' - x^2y = 0, \quad y = a$ and $y' = b$ when $x = 0$
13. $y'' + x^2y = x, \quad y = a$ and $y' = b$ when $x = 0$
14. $y'' - 2y' + y = 0, \quad y = 0$ and $y' = 1$ when $x = 0$

15.8 Slope Fields and Picard's Theorem

This section shows how to sketch the solutions of a differential equation $y' = f(x, y)$ without having to solve the equation first. It also describes conditions under which we can be sure that solutions exist and shows how we can sometimes generate polynomial approximations of the solutions by integration.

A **solution curve** is a curve whose equation satisfies the differential equation. For example, any curve whose equation is $y = Ce^x$ is a solution curve for the equation $y' = y$.

A **slope field** assigns to each point $P(x, y)$ in the domain of f the number $f(x, y)$, which is the slope of any solution curve through P.

An **isocline** of a differential equation $y' = f(x, y)$ is a curve with equation $f(x, y) =$ constant. For example, the isoclines of the equation $y' = y$ are the lines $y = C$, where C is a constant.

We could represent a slope field for the equation $y' = f(x, y)$ by the surface $z = f(x, y)$, but here we prefer to use a different approach that helps us visualize what a solution curve might look like without actually solving the differential equation. Figure 15.16 is a graphical representation for a portion of the slope field of the equation

$$y' = x + y. \tag{1}$$

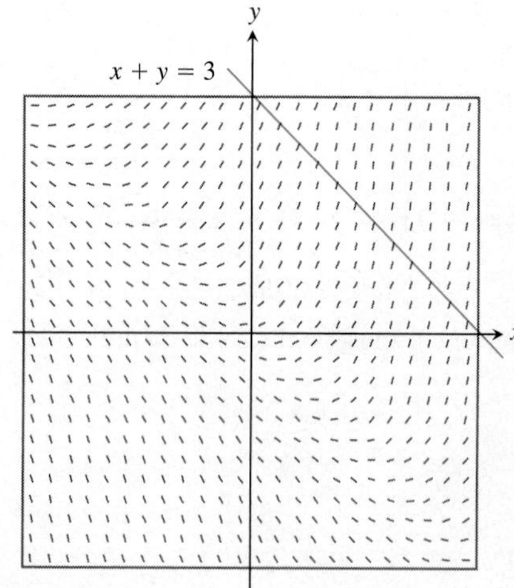

15.16 The slope field for $y' = x + y$, $-3 \le x \le 3$, $-3 \le y \le 3$.

(To simplify the language, we shall say simply that it *is* the slope field.) This figure was done by a computer, but it can be visualized as a pattern of iron filings that have been sprinkled onto a piece of paper and have arranged themselves so that the bit at the point $P(x, y)$ has slope $x + y$. Since iron filings won't actually do that for us, we resort to other approaches.

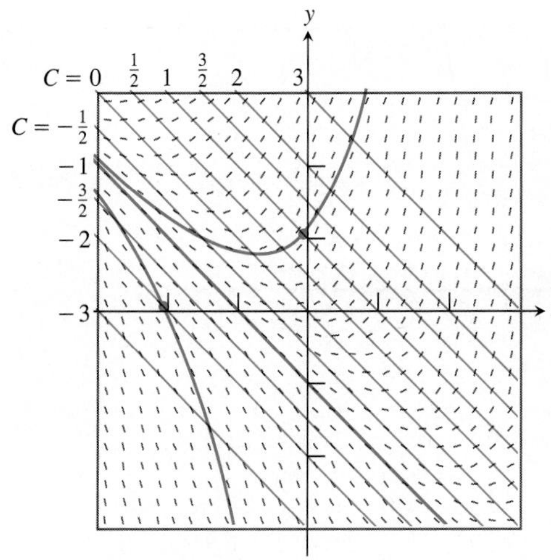

15.17 Slope field, isoclines $x + y = C$, and selected solution curves for the equation $y' = x + y$, $-3 \leq x \leq 3$, $-3 \leq y \leq 3$. The line $x + y = -1$ is both an isocline and a solution curve.

One simple approach is first to construct a set of isoclines. For Eq. (1), each isocline is a straight line satisfying the equation

$$x + y = C,$$

where C is a constant. Through some point on the line $x + y = 3$, for example, we draw a short line segment of slope 3. Having drawn one such segment, we then draw others parallel to it. Figure 15.16 shows 12 such segments on that line. We repeat this process on each of the isoclines, starting with a short segment of slope C at a point on the line $x + y = C$. The resulting pattern is a portion of the slope field for the differential equation $y' = x + y$.

Notice one special line, the line $x + y = -1$, which is both a solution curve and an isocline. Above this line the solution curves are concave up because $y'' = 1 + y' = 1 + x + y$ is positive if $x + y > -1$. Below the line, y'' is negative and the solution curves are concave down.

Figure 15.17 shows selected solution curves and isoclines in the slope field of $y' = x + y$ in the region $|x| \leq 3$, $|y| \leq 3$. The upper curve is for the solution $y = -1 - x + 2e^x$ of the initial value problem $y' = x + y$, $y(0) = 1$. The lower curve is for the solution $y = -1 - x - e^{x+2}$ of the initial value problem $y' = x + y$, $y(-2) = 0$. The curve in between is the special isocline $x + y = -1$. Exercise 21 asks you to show that the solution of the initial value problem $y' = x + y$, $y(x_0) = y_0$ is

$$y = -1 - x + (1 + x_0 + y_0)\, e^{x - x_0}.$$

Picard's Existence and Uniqueness Theorem and Iteration Scheme

We now turn our attention to a theorem of Charles Émile Picard's. The proof hinges on an iterative process that produces a sequence of functions that converge to a solution of the problem $y' = f(x, y)$ with $y(x_0) = y_0$, provided the hypotheses of the theorem are satisfied. The iteration process, known as **Picard's iteration scheme,** is an ingenious method for finding analytic approximations to a solution. We shall illustrate the scheme after we look at the theorem.

Suppose we are given an initial value problem

$$y' = f(x, y), \qquad y(x_0) = y_0, \tag{2}$$

where f is defined and continuous inside a rectangle R in the xy-plane. Does the initial value problem always have at least one solution if (x_0, y_0) is inside R? Might it have more than one solution? The next example shows that the answer to the second question is yes, unless something more is required of f.

Example 1 The initial value problem

$$y' = y^{4/5}, \qquad y(0) = 0, \tag{3}$$

has the obvious solution $y = 0$. Another solution is found by separating the variables and integrating:

$$y = \left(\frac{x}{5}\right)^5.$$

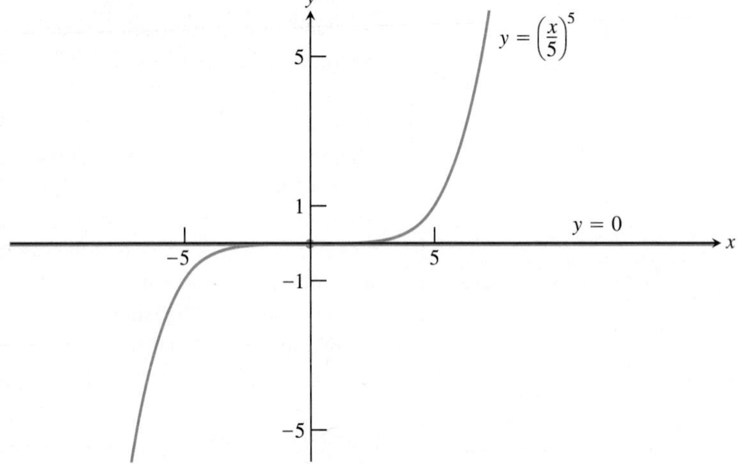

15.18 Part of the graph of $y = (x/5)^5$, one of the solutions of $y' = y^{4/5}$, $y(0) = 0$. Another solution is $y = 0$ (Example 1).

There are many more solutions (Fig. 15.18), among them

$$y = \begin{cases} 0, & x \le 0, \\ \left(\dfrac{x}{5}\right)^5, & x > 0, \end{cases} \quad \text{and} \quad y = \begin{cases} \left(\dfrac{x}{5}\right)^5, & x \le 0, \\ 0, & x > 0. \end{cases}$$

In this example, $f(x, y) = y^{4/5}$ is continuous in the entire xy-plane but its partial derivative $f_y = (4/5)y^{-1/5}$ is not continuous where $y = 0$.

The following theorem gives sufficient conditions for the existence and uniqueness of a solution of the initial value problem in (2).

THEOREM 3

Picard's Existence and Uniqueness Theorem

If (x_0, y_0) is a point in the interior of a rectangle R on which $f(x, y)$ is continuous, then the initial value problem

$$\frac{dy}{dx} = f(x, y), \qquad y(x_0) = y_0 \tag{4}$$

has at least one solution on some open interval of x-values containing x_0. If, in addition, the partial derivative f_y is continuous on R, then the solution is unique on some (perhaps smaller) open interval containing x_0 (Fig. 15.19).

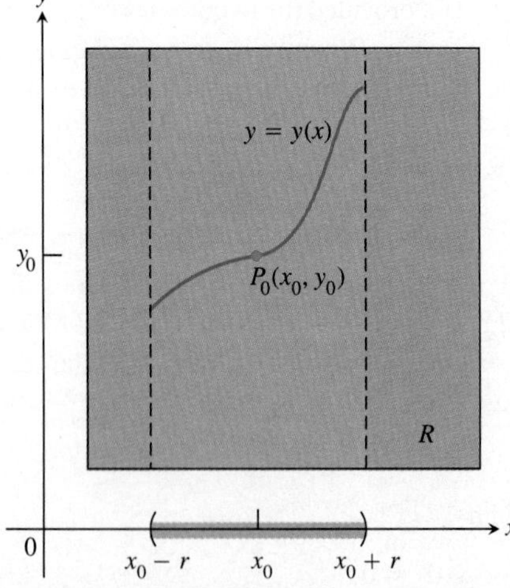

15.19 If f and f_y are continuous in R, and (x_0, y_0) is any point in R, then the initial value problem $y' = f(x, y)$, $y(x_0) = y_0$, has a unique solution on some open interval containing x_0.

We cannot take the time to prove the theorem but we can describe Picard's iteration scheme, which plays a key role in the proof and is important in its own right. We begin by noticing that any solution y of the initial value problem of Eqs. (4) must also satisfy the integral equation

$$y(x) = y_0 + \int_{x_0}^{x} f(t, y(t))\, dt \tag{5}$$

because

$$\int_{x_0}^{x} \frac{dy}{dt}\, dt = y(x) - y(x_0).$$

The converse is also true: If $y(x)$ satisfies Eq. (5), then $y' = f(x, y(x))$ and $y(x_0) = y_0$. So Eqs. (4) may be replaced by Eq. (5). This sets the stage for Picard's iteration method: In the integrand in Eq. (5), replace $y(t)$ by the constant y_0, then integrate and call the resulting right-hand side of Eq. (5) $y_1(x)$:

$$y_1(x) = y_0 + \int_{x_0}^{x} f(t, y_0)\, dt. \tag{6}$$

This starts the process. To keep it going, we use the iterative formulas

$$y_{n+1}(x) = y_0 + \int_{x_0}^{x} f(t, y_n(t))\, dt. \tag{7}$$

The proof of Picard's theorem consists of showing that this process produces a sequence of functions $\{y_n(x)\}$ that converge to a function $y(x)$ that satisfies Eqs. (4) and (5), for values of x sufficiently near x_0. (The proof also shows that the solution is unique: that is, no other method will lead to a different solution.)

The following examples illustrate the Picard iteration scheme, but in most practical cases the computations soon become too burdensome to continue.

Example 2 Illustrate the Picard iteration scheme for the initial value problem

$$y' = x - y, \qquad y(0) = 1. \tag{8}$$

Solution For the problem at hand, Eq. (6) becomes

$$\begin{aligned} y_1(x) &= 1 + \int_0^x (t - 1)\, dt \\ &= 1 + \frac{x^2}{2} - x. \end{aligned} \tag{9a}$$

If we now use Eq. (7) with $n = 1$, we get

$$\begin{aligned} y_2(x) &= 1 + \int_0^x \left(t - 1 - \frac{t^2}{2} + t\right) dt \\ &= 1 - x + x^2 - \frac{x^3}{6}. \end{aligned} \tag{9b}$$

The next iteration, with $n = 2$, gives

$$\begin{aligned} y_3(x) &= 1 + \int_0^x \left(t - 1 + t - t^2 + \frac{t^3}{6}\right) dt \\ &= 1 - x + x^2 - \frac{x^3}{3} + \frac{x^4}{4!}. \end{aligned} \tag{9c}$$

In this example, it is possible to find the exact solution, because

$$\frac{dy}{dx} + y = x$$

is a first order equation that is linear in y. It has an integrating factor e^x and the general solution is

$$y = x - 1 + Ce^{-x}.$$

The solution of the initial value problem is

$$y = x - 1 + 2e^{-x}. \tag{10}$$

If we substitute the Maclaurin series for e^{-x} in Eq. (10), we get

$$\begin{aligned} y &= x - 1 + 2\left(1 - x + \frac{x^2}{2!} - \frac{x^3}{3!} + \frac{x^4}{4!} - \cdots\right) \\ &= 1 - x + x^2 - \frac{x^3}{3} + 2\left(\frac{x^4}{4!} - \frac{x^5}{5!} + \cdots\right), \end{aligned}$$

and we see that the Picard method has given us the first four terms of this expansion in Eq. (9c) in $y_3(x)$.

In the next example, we cannot find a solution in terms of elementary functions. The Picard scheme is one way we could get an idea of how the solution behaves near the initial point.

Example 3 Find $y_n(x)$ for $n = 0, 1, 2,$ and 3 for the initial value problem

$$y' = x^2 + y^2, \qquad y(0) = 0.$$

Solution By definition, $y_0(x) = y(0) = 0$. The other functions $y_n(x)$ are generated by the integral representation

$$\begin{aligned} y_{n+1}(x) &= 0 + \int_0^x [t^2 + (y_n(t))^2]\, dt \\ &= \frac{x^3}{3} + \int_0^x (y_n(t))^2\, dt. \end{aligned}$$

We successively calculate

$$y_1(x) = \frac{x^3}{3},$$

$$y_2(x) = \frac{x^3}{3} + \frac{x^7}{63},$$

$$y_3(x) = \frac{x^3}{3} + \frac{x^7}{63} + \frac{2x^{11}}{2079} + \frac{x^{15}}{59535}.$$

In the next section we introduce numerical methods for solving initial value problems like those in Examples 2 and 3. When we program such a numerical solution for a calculator or computer, it is helpful to have an independent check on the program. For x near zero, we would expect $y_2(x)$ or $y_3(x)$ to provide such a check.

EXERCISES 15.8

In Exercises 1–6, sketch some of the isoclines and part of the slope field. Using the slope field, sketch the solution curve that passes through the point $P(1, -1)$.

1. $y' = x$

2. $y' = y$

3. $y' = 1/x$

4. $y' = 1/y$

5. $y' = xy$

6. $y' = x^2 + y^2$

7. Sketch part of the slope field of the equation $y' = (x + y)^2$. Show isoclines where the slope is 0, 1/4, 1, 4. Sketch (roughly) the solution curves through the points (a) $P(0, 0)$, (b) $Q(-1, 1)$, and (c) $R(1, 0)$. Now make the substitution $u = x + y$ and find the general solution of $y' = (x + y)^2$. Find an equation for the solution curve that passes through the origin. Sketch that solution more accurately.

8. Show that every solution of the equation $y' = (x + y)^2$ has a graph that has a point of inflection but no maximum or minimum. (The graph also has vertical asymptotes, as you may discover by letting $x + y = u$.)

9. Use isoclines to sketch part of the slope field for the equation $y' = x - y$. Include some part of all four quadrants.
   a) Which isocline is also a solution curve?
   b) Use the differential equation to determine where the solution curves are concave up and concave down.

In Exercises 10–13, write an equivalent first order differential equation and initial condition for y.

10. $y = 1 + \int_0^x y(t)\, dt$

11. $y = -1 + \int_1^x (t - y(t))\, dt$

12. $y = \int_1^x \frac{1}{t}\, dt$

13. $y = 2 - \int_0^x (1 + y(t)) \sin t\, dt$

14. What integral equation is equivalent to the initial value problem $y' = f(x)$, $y(x_0) = y_0$?

Use Picard's iteration scheme to find $y_n(x)$ for $n = 0, 1, 2, 3$ in Exercises 15–20.

15. $y' = x, \quad y(1) = 2$

16. $y' = y, \quad y(0) = 1$

17. $y' = xy, \quad y(1) = 1$

18. $y' = x + y, \quad y(0) = 0$

19. $y' = x + y, \quad y(0) = 1$

20. $y' = 2x - y, \quad y(-1) = 1$

21. Show that the solution of the initial value problem

$$y' = x + y, \quad y(x_0) = y_0$$

is

$$y = -1 - x + (1 + x_0 + y_0)\, e^{x - x_0}.$$

22. Verify the formula for $y_3(x)$ in Example 3.

EXPLORER PROGRAM

Slope Fields	Enables you to study solutions of $y' = f(x, y)$ geometrically by constructing their graphs in the slope field defined by the equation

15.9 Numerical Methods

We complete our preview of differential equations by describing three numerical methods for solving the initial value problem $y' = f(x, y)$, $y(a) = y_0$ over an interval $a \leq x \leq b$. These methods do not produce a general solution of $y' = f(x, y)$; they produce tables of values of y for preselected values of x. Instead of being a drawback, however, this is a definite advantage, especially if we want to solve a differential equation like $y' = x^2 + y^2$ whose solution has no closed-form algebraic expression. Also, in solving equations of motion in real situations like launching a rocket or intercepting one in orbit, we find that numerical answers are much more useful than algebraic expressions would be. We shall keep our examples simple but the ideas can be extended to far more complicated equations and systems of equations. Naturally we would not attempt to solve such complicated systems with pencil and paper; instead, we would turn to well-designed computer programs that had been carefully checked for accuracy.

The first method we consider dates back to Euler, the second is an improved Euler method, and the third and most accurate method of the three is a so-called Runge–Kutta method. You will find BASIC computer programs for the three methods following the section exercises.

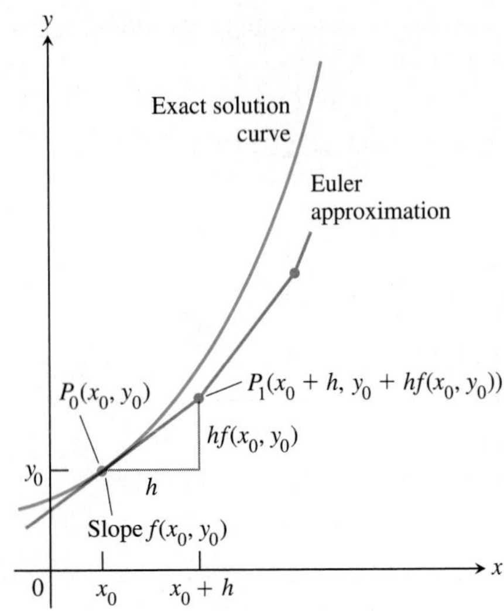

15.20 The Euler approximation to the solution of the initial value problem $y' = f(x, y)$, $y = y_0$ when $x = x_0$. The errors involved usually accumulate as we take more steps.

The Euler Method

The Euler method is a numerical process for generating a table of approximate values of the function that solves the initial value problem

$$y' = f(x, y), \qquad y(x_0) = y_0. \tag{1}$$

The problem furnishes a starting point, $P_0(x_0, y_0)$, and a slope $f(x_0, y_0)$. We know that the graph of the solution must be a curve through P_0 with that slope. If we use the tangent through P_0 to approximate the actual solution curve, the approximation may be fairly good from $x_0 - h$ to $x_0 + h$, for small values of h. Thus, we might choose $h = 0.1$, say, and move along the tangent line from P_0 to P_1 (x_1, y_1), where $x_1 = x_0 + h$ and $y_1 = y_0 + hf(x_0, y_0)$. If we think of P_1 as a new starting point, we can move from P_1 to P_2 (x_2, y_2), where $x_2 = x_1 + h$ and $y_2 = y_1 + hf(x_1, y_1)$. If we replace h by $-h$, we move to the left from P_0 instead of to the right. The process can be continued, but the errors are likely to accumulate as we take more steps (Fig. 15.20).

Example 1 Take $h = 0.1$ and investigate the accuracy of the Euler approximation method for the initial value problem

$$y' = 1 + y, \qquad y(0) = 1, \tag{2}$$

over the interval $0 \le x \le 1$ by letting

$$x_{n+1} = x_n + h, \qquad y_{n+1} = y_n + h(1 + y_n). \tag{3}$$

Solution The exact solution of Eqs. (2) is $y = 2e^x - 1$. Table 15.2 shows the results using Eqs. (3) and the exact results rounded to four decimals for comparison. By the time we get to $x = 1$, the error is about 5.6%.

TABLE 15.2
Euler solution of $y' = 1 + y$, $y(0) = 1$, step size $h = 0.1$

x	y (approx)	y (exact)	Error = y (exact) − y (approx)
0	1	1	0
0.1	1.2	1.2103	0.0103
0.2	1.42	1.4428	0.0228
0.3	1.662	1.6997	0.0377
0.4	1.9282	1.9836	0.0554
0.5	2.2210	2.2974	0.0764
0.6	2.5431	2.6442	0.1011
0.7	2.8974	3.0275	0.1301
0.8	3.2872	3.4511	0.1639
0.9	3.7159	3.9192	0.2033
1.0	4.1875	4.4366	0.2491

Carl Runge

German scientist Carl Runge (1856–1927) was a mathematical physicist of Max Planck's caliber who developed numerical methods for solving the differential equations that arose in his studies of atomic spectra. He used so much mathematics in his research that physicists thought he was a mathematician, and did so much physics that mathematicians thought he was a physicist. Neither group claimed him as their own and it was years before anyone could find him a professorship. In 1904, Felix Klein finally persuaded his Göttingen colleagues to create for Runge Germany's only full professorship in applied mathematics. Runge was the professorship's first and only occupant, there being no one of his accomplishments to assume the post when he died.

The Improved Euler Method

With this method we first get an estimate of y_{n+1}, as in the original Euler method, but call the result z_{n+1}. We then take the average of $f(x_n, y_n)$ and $f(x_{n+1}, z_{n+1})$ in place of $f(x_n, y_n)$ in the next step. Thus

$$z_{n+1} = y_n + hf(x_n, y_n), \tag{4}$$

$$y_{n+1} = y_n + \frac{h}{2}[f(x_n, y_n) + f(x_{n+1}, z_{n+1})]. \tag{5}$$

If we apply this improved method to Example 1, again with $h = 0.1$, we get the following results at $x = 1$:

$$y\text{ (approx)} = 4.4281\ 61693,$$

$$y\text{ (exact)} = 4.4365\ 63656,$$

$$\text{Error} = y\text{ (exact)} - y\text{ (approx)} = 0.0084\ 01963,$$

and the error is less than 2/10 of 1%.

A Runge–Kutta Method*

The Runge–Kutta method we use requires four intermediate calculations, as given in the following equations:

$$\begin{aligned} k_1 &= hf(x_n, y_n) & k_2 &= hf\left(x_n + \frac{h}{2}, y_n + \frac{k_1}{2}\right) \\ k_3 &= hf\left(x_n + \frac{h}{2}, y_n + \frac{k_2}{2}\right) & k_4 &= hf\left(x_n + h, y_n + k_3\right). \end{aligned} \tag{6}$$

We then calculate y_{n+1} from y_n with the formula

$$y_{n+1} = y_n + \frac{1}{6}(k_1 + 2k_2 + 2k_3 + k_4). \tag{7}$$

When we apply this method to the problem of estimating $y(1)$ for the problem $y' = 1 + y$, $y(0) = 1$, still using $h = 0.1$, we get

$$y(1) = 4.4365\ 59490$$

with an error 0.0000 04166, which is less than 1/10,000 of 1%. This is clearly the most accurate of the three methods.

The next example shows that the error in the Runge–Kutta approximation need not continue to increase as the process is continued. In fact, with $h = 0.1$, the difference between the exact solutions and the approximations remain less than 10^{-6} for the two initial value problems:

$$\text{(a) } y' = x - y, \quad y(0) = 1, \qquad \text{(b) } y' = x - y, \quad y(0) = -2.$$

The fact that the differential equation is linear in y is significant in discussing the accuracy of the Runge–Kutta approximation. Such accuracy is not attained for the initial value problem

$$y' = x^2 + y^2, \qquad y(0) = 0.$$

*The method described here is one of many Runge–Kutta methods.

TABLE 15.3

	x	y(Runge–Kutta)	y (true value)	Difference
a) $y' = x - y,\quad y(0) = 1$	0	1	1	0
	0.5	0.7130 61869	0.7130 61319	5.50×10^{-7}
	1.0	0.7357 59549	0.7357 58882	6.67×10^{-7}
	1.5	0.9462 60927	0.9462 60320	6.07×10^{-7}
	2.0	1.2706 71057	1.2706 70566	4.91×10^{-7}
	2.5	1.6641 70370	1.6641 69997	3.73×10^{-7}
	3.0	2.0995 74407	2.0995 74137	2.70×10^{-7}
b) $y' = x - y,\quad y(0) = -2$	0	−2	−2	0
	0.5	−1.1065 30935	−1.1065 30660	-2.75×10^{-7}
	1.0	−0.3678 79775	−0.3678 79441	-3.34×10^{-7}
	1.5	+0.2768 69537	+0.2768 69840	-3.03×10^{-7}
	2.0	0.8646 64472	0.8646 64717	-2.46×10^{-7}
	2.5	1.4179 14816	1.4179 15001	-1.85×10^{-7}
	3.0	1.9502 12796	1.9502 12932	-1.36×10^{-7}

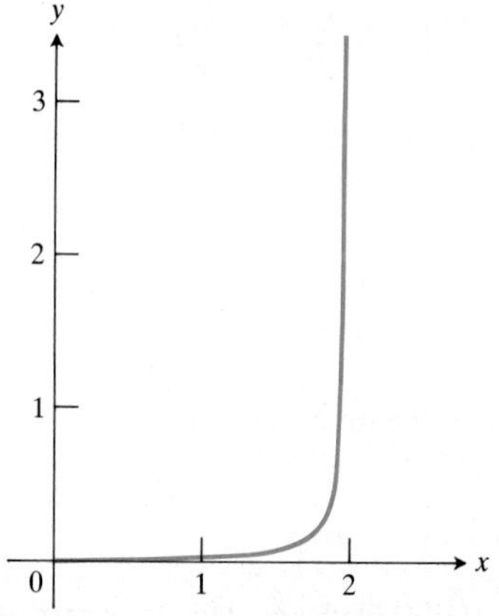

15.21 The graph of a Runge–Kutta solution of the initial value problem $y' = x^2 + y^2$, $y(0) = 0$, for $x > 0$. Data from Table 15.4.

The Runge–Kutta approximation to $y(2.1)$, using $h = 0.1$, is 1.47×10^{11}. The solution curve for this problem has a vertical asymptote at just beyond $x = 2$ (Fig. 15.21). No matter how small we take h, we cannot assert any accuracy for our approximations as the curve approaches this asymptote.

Example 2 Table 15.3 shows the comparison of $y(x)$ as estimated by the Runge–Kutta method with $h = 0.1$ and the true value, for solutions of $y' = x - y$ (a) with $y(0) = 1$ and (b) with $y(0) = -2$.

More points were actually computed and plotted to give the graphs in Fig. 15.22. The upper curve, $y = x - 1 + 2e^{-x}$, is concave up and has a minimum

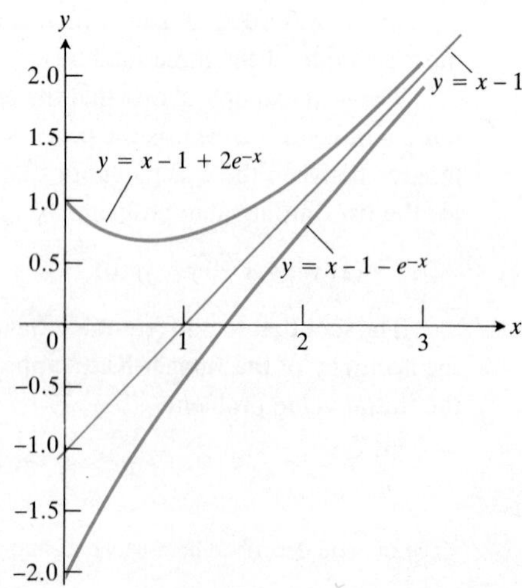

15.22 Two solutions of $y' = x - y$:
a) $y(0) = 1$, $y = x - 1 + 2e^{-x}$,
b) $y(0) = -2$, $y = x - 1 - e^{-x}$

TABLE 15.4
$y' = x^2 + y^2, \quad y(0) = 0$

x	y (Runge–Kutta)	y (actual)
0	0	0
0.5	0.0417 91288	0.0417 91146
1.0	0.3502 33742	0.3502 31844
1.5	1.5174 73414	1.5174 47544
2.0	71.5789 9545	317.2244 00
2.1	1.4700 1E + 11	
2.2	1.6666 7E + **	(meaning, "you broke the bank!")

when $x = y = \ln 2$. The lower curve is concave down, is always rising as x increases, and crosses the x-axis at a value of x near 1.3. Both curves approach the line $y = x - 1$ as $x \to \infty$.

Example 3 Table 15.4 lists Runge–Kutta approximations for the initial value problem

$$y' = x^2 + y^2, \qquad y(0) = 0.$$

We obtained the approximations with a step size of $h = 0.1$. Figure 15.21 shows the graph of y as a function of x.

EXERCISES 15.9

1. Use the Euler method with $h = 1/5$ to estimate $y(1)$ if $y' = y$ and $y(0) = 1$. What is the exact value of $y(1)$?
2. Show that the Euler method leads to the estimate $(1 + (1/n))^n$ for $y(1)$ if $h = 1/n$, $y' = y$, and $y(0) = 1$. What is the limit as $n \to \infty$?
3. Use the improved Euler method with $h = 1/5$ to estimate $y(1)$ if $y' = y$ and $y(0) = 1$.
4. Use the Runge–Kutta method with $h = 1/5$ to estimate $y(1)$ if $y' = y$ and $y(0) = 1$.
5. Show that the solution of the initial value problem $y' = x^2 + y^2$, $y(0) = 1$, increases faster on the interval $0 \le x < 1$ than does the solution of the initial value problem $y' = y^2$, $y(0) = 1$. Solve the latter problem by separation of variables and thus show that the solution of the original problem becomes infinite at a value of x not greater than 1. (The value is about 0.9698 10654.)
6. Solve the initial value problem $y' = 1 + y^2$, $y(0) = 0$, (a) by separation of variables, and (b) by using the substitution $y = -u'/u$ and solving the equivalent problem for u. (Notice the similarity with the initial value problem $y' = x^2 + y^2$, $y(0) = 0$.)

Computer Exercises (or Programmable Calculator)

Find numerical solutions of the initial value problems in Exercises 7–10. Each exercise gives a differential equation in the form $y' = f(x, y)$, a solution interval $a \le x \le b$, the initial value $y(a)$, the step size, and the number of steps. Your answer should be a table showing y vs. x.

7. $y' = x/7, \quad a = 0, \quad b = 4, \quad y(0) = 1, \quad h = 0.1, \quad n = 20$
8. $y' = -y^2/x, \quad a = 2, \quad b = 4, \quad y(2) = 2, \quad h = 0.1, \quad n = 20$
9. $y' = (x^2 + y^2)/2y, \quad a = 0, \quad b = 2, \quad y(0) = 0.1, \quad h = 0.2, \quad n = 10$
10. Repeat Exercise 9, but with $h = 0.1$ and $n = 20$.
11. *Evaluating a nonelementary integral.* The value of the nonelementary integral

$$\int_0^1 \sin(t^2)\, dt$$

is the value of the function

$$y(x) = \int_0^x \sin(t^2)\, dt$$

at $x = 1$. The function, in turn, is the solution of the initial value problem

$$y' = \sin(x^2), \quad y(0) = 0.$$

Thus, by solving the initial value problem numerically on the interval $0 \leq x \leq 1$ we can find the value of the integral as the value of y that corresponds to $x = 1$. Estimate the integral's value by solving the initial value problem with (a) $n = 20$ steps, and (b) $n = 40$ steps.

12. *Continuation of Exercise 11.* Estimate the value of

$$\int_0^1 x^2 e^{-x}\, dx$$

by solving an appropriate initial value problem on the interval [0, 1] with (a) $n = 20$ steps, and (b) $n = 40$ steps.

EXPLORER PROGRAM

First Order Equations — Provides numerical solutions of $y' = f(x, y)$, $y(a) = y_0$ over an interval $a \leq x \leq b$ with a Runge–Kutta method with up to 100 steps. Also prints tables and displays graphs of y and y'.

✲ COMPUTER PROGRAMS: SOLVING FIRST ORDER INITIAL VALUE PROBLEMS

Here are three BASIC computer programs for approximating the solutions of the initial value problem $y' = f(x, y)$, $y = y_0$ when $x = x_0$, over a finite interval starting at x_0.

The Euler Method

PROGRAM	COMMENT
10 INPUT "ENTER INITIAL X-VALUE ", X	Initial value of x
20 INPUT "ENTER INITIAL Y-VALUE ", Y	Initial value of y
30 INPUT "ENTER STEP SIZE ", H	Step size
40 INPUT "ENTER NUMBER OF STEPS ", N	Number of steps
50 DEF FND(X, Y) = 1 + Y	Key in formula for the derivative, $f(x, y)$
60 FOR J = 0 TO N	Each value of J gives one Euler approximation
70 PRINT X;Y	Prints the table
80 Y = Y + H*FND(X, Y)	Next y
90 X = X + H	Next x
100 NEXT J	Returns to line 60 with new value for J
110 END	

An Improved Euler Method

```
10    INPUT "ENTER INITIAL X-VALUE   ", X
20    INPUT "ENTER INITIAL Y-VALUE   ", Y
30    INPUT "ENTER STEP SIZE   ", H
40    INPUT "ENTER NUMBER OF STEPS   ", N
50    DEF FND(X, Y) = 1 + Y
60    FOR J = 0 TO N
70    PRINT X;Y
80    Z = Y + H*FND(X, Y)
90    XN = X + H
100   Y = Y + H*(FND(X, Y) + FND(XN, Z))/2
110   X = XN
120   NEXT J
130   END
```

Runge–Kutta Method

```
10   INPUT "ENTER INITIAL X-VALUE      ", X
20   INPUT "ENTER INITIAL Y-VALUE      ", Y
30   INPUT "ENTER STEP SIZE     ", H
40   INPUT "ENTER NUMBER OF STEPS      ", N
50   DEF FND(X, Y) = X - Y
60   FOR J = 0 TO N
70   PRINT X;Y
80   K1 = H*FND(X, Y)
90   X = X + H/2
100  K2 = H*FND(X, Y + K1/2)
110  K3 = H*FND(X, Y + K2/2)
120  X = X + H/2
130  K4 = H*FND(X, Y + K3)
140  Y = Y + (K1 + 2*K2 + 2*K3 + K4)/6
150  NEXT J
160  END
```

REVIEW QUESTIONS

1. What is a differential equation?
2. What is a solution of a differential equation?
3. Describe methods for solving first order equations that are (a) separable, (b) homogeneous, (c) exact, (d) linear. Give examples.
4. Describe methods for solving linear second order equations with constant coefficients if the equations are (a) homogeneous, (b) nonhomogeneous. Give examples.
5. How do we describe oscillation with second order differential equations? Give an example.
6. How can we sometimes solve differential equations with power series? Give an example.
7. How can we graph the solutions of a differential equation $y' = f(x, y)$ without first solving the equation? Give an example.
8. Describe three numerical methods for solving the initial value problem $y' = f(x, y)$, $y = y_0$ when $x = x_0$.

MISCELLANEOUS EXERCISES

Solve the initial value problems in Exercises 1–20.

1. $e^{y-2}\,dx - e^{x+2y}\,dy = 0, \quad y(0) = -2$
2. $y \ln y\, dx + (1 + x^2)\, dy = 0, \quad y(0) = e$
3. $\dfrac{dy}{dx} = \dfrac{x^2 + y^2}{2xy}, \quad y(5) = 0$
4. $\dfrac{dy}{dx} = \dfrac{y(1 + \ln y - \ln x)}{x(\ln y - \ln x)}, \quad y(1) = 1$
5. $(x^2 + y)\, dx + (e^y + x)\, dy = 0, \quad y(3) = 0$
6. $(e^x + \ln y)\, dx + \left(\dfrac{x + y}{y}\right) dy = 0, \quad y(\ln 2) = 1$
7. $(x + 1)\dfrac{dy}{dx} + 2y = x, \quad y(0) = 1$
8. $x\dfrac{dy}{dx} + 2y = x^2 + 1, \quad y(1) = 1$
9. $\dfrac{d^2y}{dx^2} - \left(\dfrac{dy}{dx}\right)^2 = 1, \quad y(\pi/3) = 0, \quad y'(\pi/3) = \sqrt{3}$
10. $x^2\dfrac{d^2y}{dx^2} + x\dfrac{dy}{dx} = 1, \quad y(1) = 1, \quad y'(1) = 1$
11. $\dfrac{d^2y}{dx^2} - 4\dfrac{dy}{dx} + 3y = 0, \quad y(0) = 2, \quad y'(0) = -2$
12. $\dfrac{d^2y}{dx^2} + 5\dfrac{dy}{dx} + 6y = 0, \quad y(0) = 5/6, \quad y'(0) = -2$
13. $\dfrac{d^2y}{dx^2} + 4\dfrac{dy}{dx} + 4y = 0, \quad y(0) = 0, \quad y'(0) = 7$

14. $\dfrac{d^2y}{dx^2} - 8\dfrac{dy}{dx} + 16y = 0, \quad y(0) = 4, \quad y'(0) = -4$

15. $\dfrac{d^2y}{dx^2} + 2\dfrac{dy}{dx} + 2y = 0, \quad y(0) = 1, \quad y'(0) = 0$

16. $\dfrac{d^2y}{dx^2} - 2\dfrac{dy}{dx} - 4y = 0, \quad y(0) = 1, \quad y'(0) = -1$

17. $\dfrac{d^2y}{dx^2} + 2\dfrac{dy}{dx} = 4x, \quad y(0) = 1, \quad y'(0) = -3$

18. $\dfrac{d^2y}{dx^2} + y = \csc x, \quad y(\pi/2) = 1, \quad y'(\pi/2) = \pi/2$

19. $\dfrac{d^2y}{dx^2} - \dfrac{dy}{dx} - 2y = 3e^{2x}, \quad y(0) = -2, \quad y'(0) = 0$

20. $\dfrac{d^2y}{dx^2} - 2\dfrac{dy}{dx} + 5y = 4e^{-x}, \quad y(0) = 1, \quad y'(0) = -1/2$

21. Find the orthogonal trajectories of the family of curves $x^2 = Cy^3$. (*Caution:* The differential equation for the family should not contain C.)

22. Find the orthogonal trajectories of the circles $(x - C)^2 + y^2 = C^2$.

23. Find the orthogonal trajectories of the parabolas $y^2 = 4C(C - x)$.

24. Which of the following is not an exact first order differential equation?
a) $y^2dx + 2xy\,dy = 0$
b) $(2x \sin y + y^3e^x)\,dx + (x^2 \cos y + 3y^2e^x)\,dy = 0$
c) $(2x \cos y + 3x^2y)\,dx + (x^3 - x^2 \sin y - y)\,dy = 0$
d) $(y \ln y - e^{-xy})\,dx + \left(\dfrac{1}{y} + x \ln y\right) dy = 0$

25. The differential equation $(x^2 - y^3)y' = 2xy$ has an integrating factor of the form y^n. Find n and solve the equation.

26. Solve the equation
$$(x + y + 1)\,dx + (y - x - 3)\,dy = 0$$
by substituting $x = r + a$, $y = b - s$, and choosing values for the constants a and b that allow the resulting equation to assume the form
$$(r + s)\,dr + (r + s)\,ds = 0.$$
Then solve this equation and express its solution in terms of x and y.

27. What values must the constants a, b, and c have to make
$$(ax^2e^y + by^2 + cy)\,dx + (cxy + 2x + x^3e^y)\,dy = 0$$
exact? What is the equation's solution when a, b, and c have these values?

28. The equation
$$(x^2 + y^2)\,dx + cxy\,dy = 0$$
has the integrating factor $1/x^2$. What is the value of the constant c? What is the equation's solution when c has this value?

29. Suppose that P and Q are functions of x, that the function $y = u(x)$ satisfies the equation $y'' + Py' + Qy = 0$, and that u is never zero. Show that the substitution $y = uv$ transforms the equation $y'' + Py' + Qy = F(x)$ into the equation $v'' + v'(P + (2u'/u)) = F(x)/u$. The further substitution $w = v'$ changes this last equation into a first order equation for w as a function of x. Show that this procedure can be applied to the equation $y'' - 2y' + y = e^x$ with $u(x) = e^x$ to lead successively to
$$w' = 1, \quad w = x + C_1, \quad v = \frac{x^2}{2} + C_1x + C_2,$$
and
$$y = uv = e^x\left(\frac{x^2}{2} + C_1x + C_2\right).$$

30. *A rocket with variable mass.* If an external force F acts on a system whose mass varies with time, Newton's law of motion is
$$\frac{d(mv)}{dt} = F + (v + u)\frac{dm}{dt}.$$
In this equation, m is the mass of the system at time t, v is its velocity, and $v + u$ is the velocity of the mass that is entering (or leaving) the system at the rate dm/dt. Suppose that a rocket of initial mass m_0 starts from rest but is driven upward by firing some of its mass directly backward at the constant rate of $dm/dt = -b$ units per second and at constant speed relative to the rocket $u = -c$. The only external force acting on the rocket is $F = -mg$ due to gravity. Under these assumptions, show that the height of the rocket above the ground at the end of t seconds (t small compared with m_0/b) is
$$y = c\left(t + \frac{m_0 - bt}{b} \ln \frac{m_0 - bt}{m_0}\right) - \frac{1}{2}gt^2.$$

31. If y is a solution of $(D^2 + 4)y = e^x$, show that y satisfies the equation $(D - 1)(D^2 + 4)y = 0$. Find the general solution of $(D - 1)(D^2 + 4)y = 0$ and use the result to solve the equation $(D^2 + 4)y = e^x$.

32. The equation $d^2y/dt^2 + 100y = 0$ describes a simple harmonic motion. Find the solution that satisfies the initial conditions that $y = 10$ and $dy/dt = 50$ when $t = 0$. Find the period and amplitude of the motion.

33. Find the first five nonzero terms in the Maclaurin series for the solution of the initial value problem $y' = x^2 + y^2$, $y(0) = 1$. (*Hint:* If
$$y = a_0 + a_1x + a_2x^2 + \cdots + a_nx^n + \cdots$$
and
$$y^2 = c_0 + c_1x + c_2x^2 + \cdots + c_nx^n + \cdots,$$
then
$$c_n = \sum_{k=0}^{n} a_k a_{n-k}$$
and
$$y' = a_1 + 2a_2x + 3a_3x^2 + \cdots + na_nx^{n-1} + \cdots.$$

To satisfy the differential equation, you must have $a_1 = c_0$, $2a_2 = c_1$, $3a_3 = 1 + c_2$, and $na_n = c_{n-1}$ for $n \geq 4$. You can now determine a_0 (from the initial value), c_0, a_1, c_1, a_2, c_2, a_3, and so on.)

34. If you were to solve the initial value problem $y' = y^2$, $y = 1$, when $x = 0$, by finding the Maclaurin series for y as a function of x, for what values of x would you expect the series to converge? (*Hint:* Solve the initial value problem without series and then expand your answer in a series.)

Exercises 35–37 refer to the differential equation $y' = x + \sin y$, whose slope field appears in Figure 15.23.

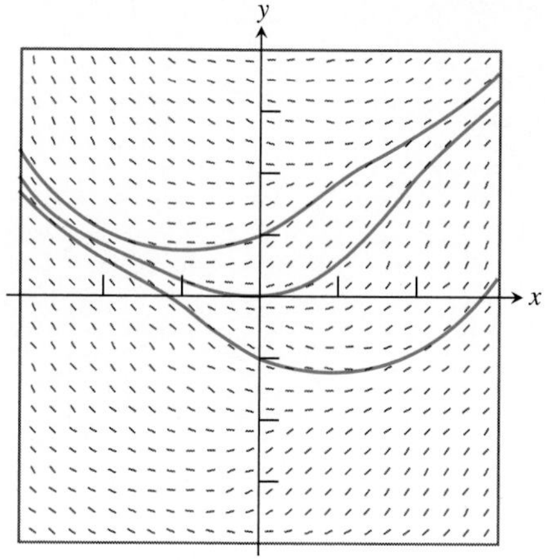

15.23 Solutions to $dy/dx = x + \sin y$ passing through $(0, 0)$, $(0, \pi/2)$, $(0, -\pi/2)$ (Exercises 35–37).

35. For solutions near $y = 0$, we might use the approximation $\sin y \approx y$ and replace the original problem with $dy/dx = x + y$. Solve this equation with $y(0) = 0$. What is the Maclaurin series for your answer?

36. *Series.* Starting with the initial value problem $y' = x + \sin y$, $y(0) = 0$, we can calculate successive derivatives by implicit differentiation. For example, $y'' = 1 + (\cos y)y'$, $y''' = -(\sin y)(y')^2 + (\cos y)y''$. These can in turn be evaluated at $x = 0$ by using the given initial value $y(0) = 0$. Using this procedure, find the terms of the Maclaurin series for $y(x)$ through x^4. (Compare the answer with that of Exercise 35.)

37. Using the method of the preceding exercise, find the terms through x^4 in the Maclaurin series for $y(x)$ if y satisfies $y' = x + \sin y$, and (a) $y(0) = \pi/2$. (b) Repeat for $y(0) = -\pi/2$.

38. Suppose you used a computer to graph solutions of

$$y' = (1 + y^2)\cos x,$$

(a) for $y(0) = 0$ and (b) for $y(0) = 1$. One solution was continuous and very well behaved. The other blew up! By solving the equation for an arbitrary initial value $y(0) = y_0$, find the values of y_0 for which the solution remains bounded. If a solution does not remain bounded, locate the asymptotes of its graph. See Fig. 15.24.

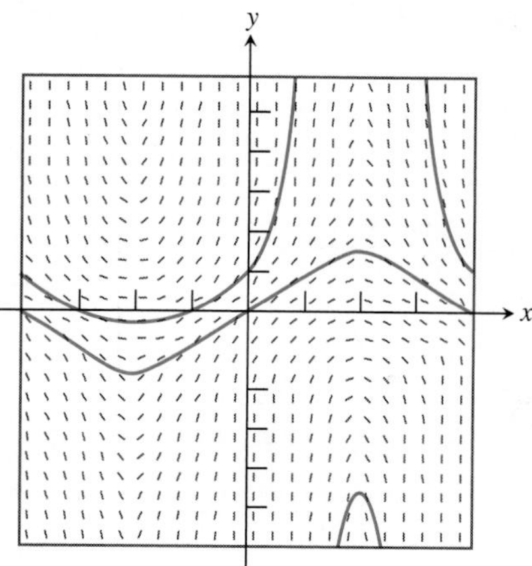

15.24 Solutions: $y = \tan(\sin x + C)$ for $C = 0$ and $\pi/4$ (Exercise 38).

Appendices

A.1 Formulas from Precalculus Mathematics

Algebra

1. Laws of Exponents

$$a^m a^n = a^{m+n}, \qquad (ab)^m = a^m b^m, \qquad (a^m)^n = a^{mn}, \qquad a^{m/n} = \sqrt[n]{a^m}$$

If $a \neq 0$,

$$\frac{a^m}{a^n} = a^{m-n}, \quad a^0 = 1, \quad a^{-m} = \frac{1}{a^m}.$$

2. Zero Division by zero is not defined.

If $a \neq 0$: $\quad \dfrac{0}{a} = 0, \quad a^0 = 1, \quad 0^a = 0$

For any number a: $\quad a \cdot 0 = 0 \cdot a = 0$

3. Fractions

$$\frac{a}{b} + \frac{c}{d} = \frac{ad + bc}{bd}, \qquad \frac{a}{b} \cdot \frac{c}{d} = \frac{ac}{bd}, \qquad \frac{a/b}{c/d} = \frac{a}{b} \cdot \frac{d}{c}, \qquad \frac{-a}{b} = -\frac{a}{b} = \frac{a}{-b},$$

$$\frac{(a/b) + (c/d)}{(e/f) + (g/h)} = \frac{(a/b) + (c/d)}{(e/f) + (g/h)} \cdot \frac{bdfh}{bdfh} = \frac{(ad + bc)fh}{(eh + fg)bd}$$

4. The Binomial Theorem

For any positive integer n,

$$(a + b)^n = a^n + na^{n-1}b + \frac{n(n-1)}{1 \cdot 2} a^{n-2}b^2 + \frac{n(n-1)(n-2)}{1 \cdot 2 \cdot 3} a^{n-3}b^3 + \cdots + nab^{n-1} + b^n.$$

For instance,

$$(a + b)^1 = a + b,$$

$$(a + b)^2 = a^2 + 2ab + b^2,$$

$$(a + b)^3 = a^3 + 3a^2b + 3ab^2 + b^3,$$

$$(a + b)^4 = a^4 + 4a^3b + 6a^2b^2 + 4ab^3 + b^4.$$

5. Difference of Like Integer Powers, $n > 1$

$$a^n - b^n = (a - b)(a^{n-1} + a^{n-2}b + a^{n-3}b^2 + \cdots + ab^{n-2} + b^{n-1})$$

For instance,

$$a^2 - b^2 = (a - b)(a + b),$$
$$a^3 - b^3 = (a - b)(a^2 + ab + b^2),$$
$$a^4 - b^4 = (a - b)(a^3 + a^2b + ab^2 + b^3).$$

6. Completing the Square

If $a \neq 0$, we can rewrite the quadratic $ax^2 + bx + c$ in the form $au^2 + C$ by a process called completing the square:

$$ax^2 + bx + c = a\left(x^2 + \frac{b}{a}x\right) + c \qquad \left(\begin{array}{l}\text{Factor } a \text{ from the}\\ \text{first two terms.}\end{array}\right)$$

$$= a\left(x^2 + \frac{b}{a}x + \frac{b^2}{4a^2} - \frac{b^2}{4a^2}\right) + c \qquad \left(\begin{array}{l}\text{Add and subtract}\\ \text{the square of half}\\ \text{the coefficient of } x.\end{array}\right)$$

$$= a\left(x^2 + \frac{b}{a}x + \frac{b^2}{4a^2}\right) + a\left(-\frac{b^2}{4a^2}\right) + c \qquad \left(\begin{array}{l}\text{Bring out}\\ \text{the } -b^2/4a^2.\end{array}\right)$$

$$= \underbrace{a\left(x^2 + \frac{b}{a}x + \frac{b^2}{4a^2}\right)}_{\text{This is } \left(x + \frac{b}{2a}\right)^2.} + \underbrace{c - \frac{b^2}{4a}}_{\text{Call this part } C.}$$

$$= au^2 + C \qquad (u = x + b/2a)$$

7. The Quadratic Formula

By completing the square on the first two terms of the equation

$$ax^2 + bx + c = 0$$

and solving the resulting equation for x (details omitted), we obtain the formula

$$x = \frac{-b \pm \sqrt{b^2 - 4ac}}{2a}.$$

This equation is called the **quadratic formula.**

The solutions of the equation $2x^2 + 3x - 1 = 0$ are

$$x = \frac{-3 \pm \sqrt{(3)^2 - 4(2)(-1)}}{2(2)} = \frac{-3 \pm \sqrt{9 + 8}}{4},$$

or

$$x = \frac{-3 + \sqrt{17}}{4} \quad \text{and} \quad x = \frac{-3 - \sqrt{17}}{4}.$$

The solutions of the equation $x^2 + 4x + 6 = 0$ are

$$x = \frac{-4 \pm \sqrt{(4)^2 - 4 \cdot 1 \cdot 6}}{2} = \frac{-4 \pm \sqrt{16 - 24}}{2}$$

$$= \frac{-4 \pm \sqrt{-8}}{2} = \frac{-4 \pm 2\sqrt{2}\sqrt{-1}}{2} = -2 \pm \sqrt{2}\,i.$$

The solutions are the complex numbers $-2 + \sqrt{2}\,i$ and $-2 - \sqrt{2}\,i$. Appendix A.6 has more on complex numbers.

Geometry

(A = area, B = area of base, C = circumference, S = lateral area or surface area, V = volume)

1. Triangle

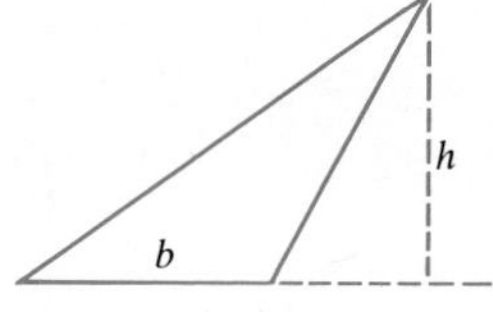

$$A = \frac{1}{2}bh$$

2. Similar Triangles

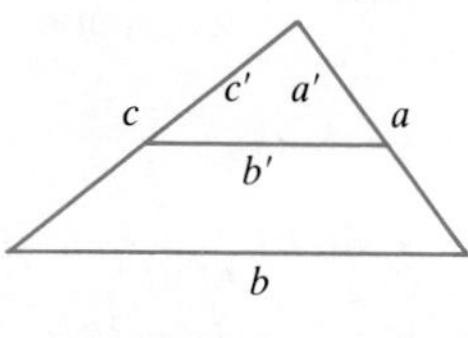

$$\frac{a'}{a} = \frac{b'}{b} = \frac{c'}{c}$$

3. Pythagorean Theorem

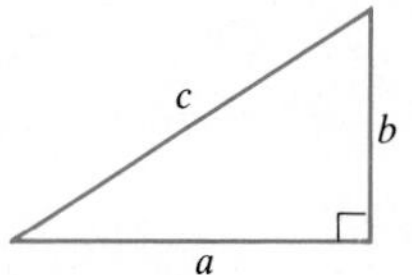

$$a^2 + b^2 = c^2$$

4. Parallelogram

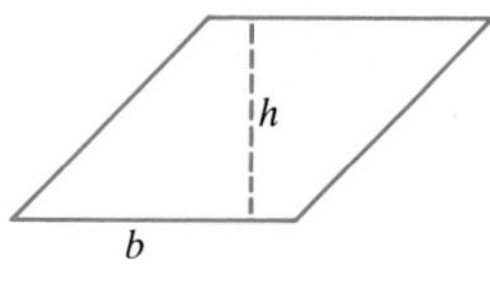

$$A = bh$$

5. Trapezoid

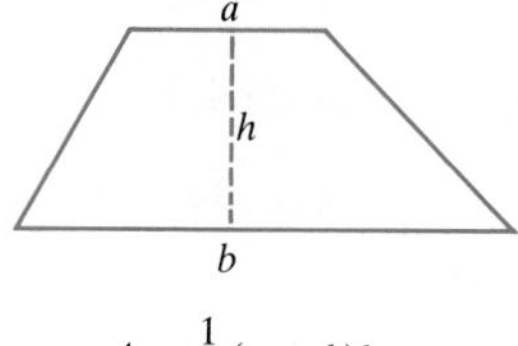

$$A = \frac{1}{2}(a + b)h$$

6. Circle

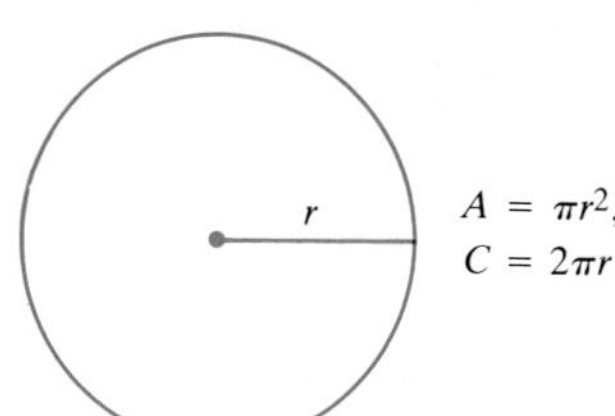

$A = \pi r^2,$
$C = 2\pi r$

7. Any Cylinder or Prism with Parallel Bases

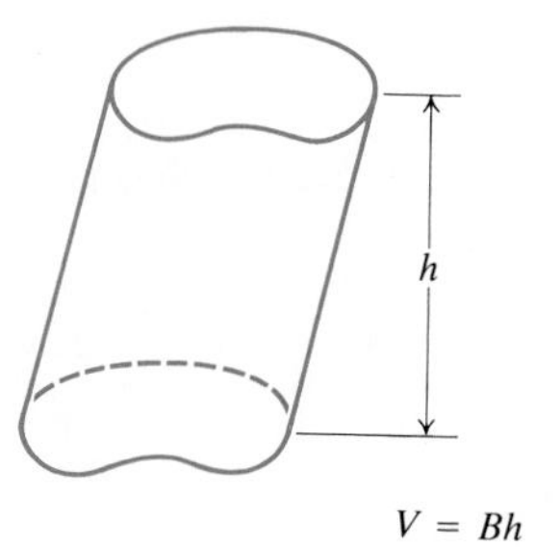

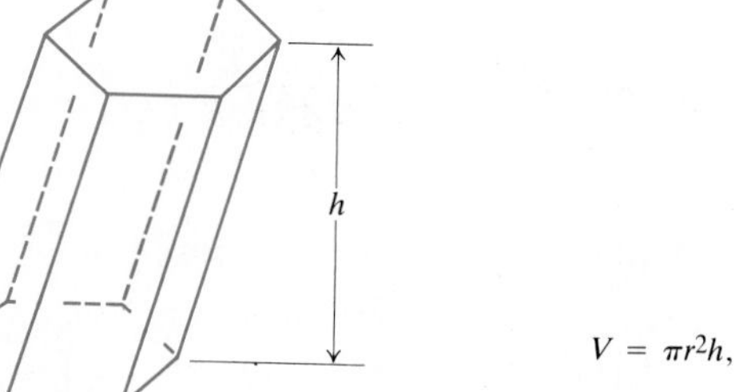

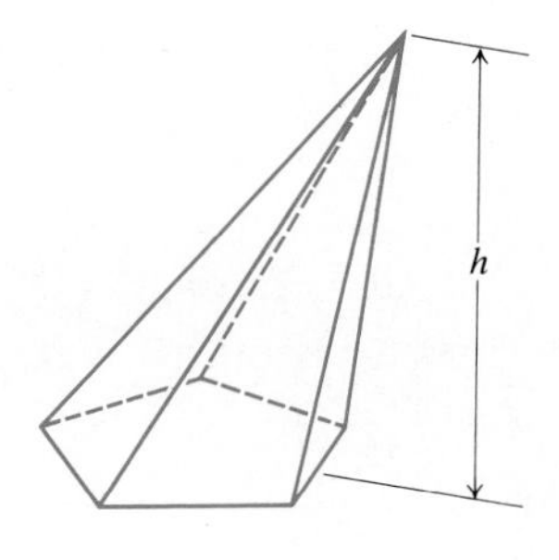

$$V = Bh$$

8. Right Circular Cylinder

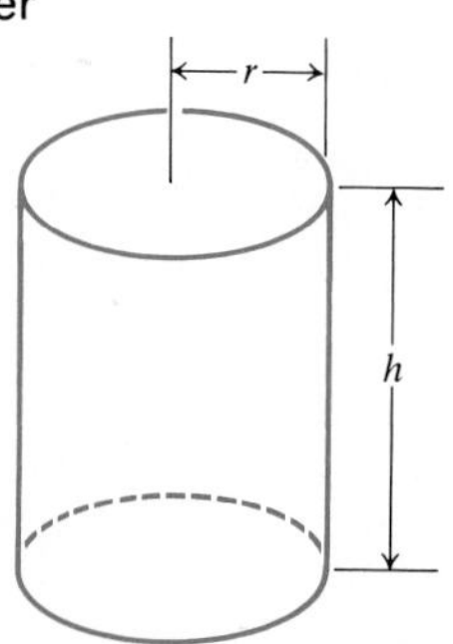

$$V = \pi r^2 h,\ S = 2\pi rh$$

9. Any Cone or Pyramid

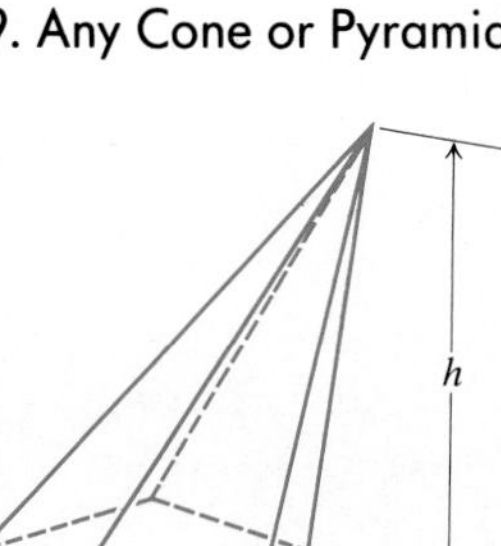

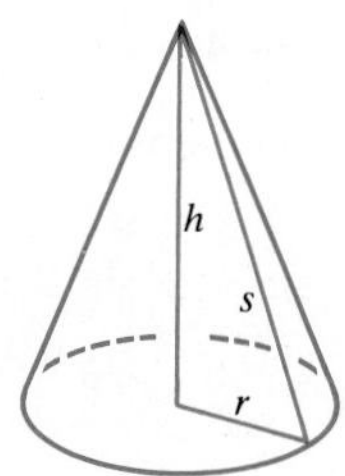

$$V = \frac{1}{3}Bh$$

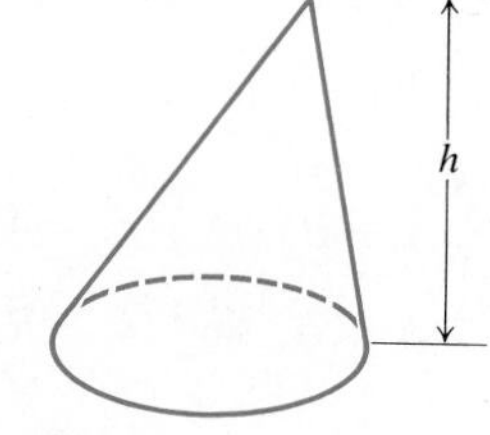

10. Right Circular Cone

$$V = \frac{1}{3}\pi r^2 h,\ S = \pi rs$$

11. Sphere

$$V = \frac{4}{3}\pi r^3,\ S = 4\pi r^2$$

Trigonometry

1. Definitions and Fundamental Identities

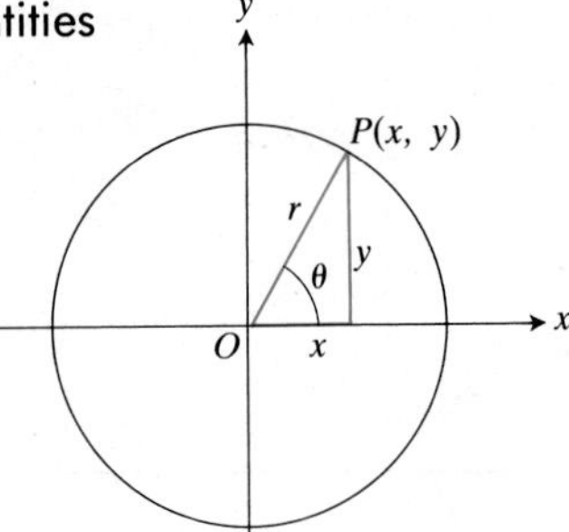

Sine: $\sin\theta = \dfrac{y}{r} = \dfrac{1}{\csc\theta}$

Cosine: $\cos\theta = \dfrac{x}{r} = \dfrac{1}{\sec\theta}$

Tangent: $\tan\theta = \dfrac{y}{x} = \dfrac{1}{\cot\theta}$

2. Identities

$\sin(-\theta) = -\sin\theta, \qquad \cos(-\theta) = \cos\theta$

$\sin^2\theta + \cos^2\theta = 1, \qquad \sec^2\theta = 1 + \tan^2\theta, \qquad \csc^2\theta = 1 + \cot^2\theta$

$\sin 2\theta = 2\sin\theta\cos\theta, \qquad \cos 2\theta = \cos^2\theta - \sin^2\theta$

$\cos^2\theta = \dfrac{1 + \cos 2\theta}{2}, \qquad \sin^2\theta = \dfrac{1 - \cos 2\theta}{2}$

$\sin(A + B) = \sin A\cos B + \cos A\sin B$

$\sin(A - B) = \sin A\cos B - \cos A\sin B$

$\cos(A + B) = \cos A\cos B - \sin A\sin B$

$\cos(A - B) = \cos A\cos B + \sin A\sin B$

$$\tan(A + B) = \frac{\tan A + \tan B}{1 - \tan A\tan B}$$

$$\tan(A - B) = \frac{\tan A - \tan B}{1 + \tan A\tan B}$$

$\sin\left(A - \dfrac{\pi}{2}\right) = -\cos A, \qquad \cos\left(A - \dfrac{\pi}{2}\right) = \sin A$

$\sin\left(A + \dfrac{\pi}{2}\right) = \cos A, \qquad \cos\left(A + \dfrac{\pi}{2}\right) = -\sin A$

$\sin A\sin B = \frac{1}{2}\cos(A - B) - \frac{1}{2}\cos(A + B)$

$\cos A\cos B = \frac{1}{2}\cos(A - B) + \frac{1}{2}\cos(A + B)$

$\sin A\cos B = \frac{1}{2}\sin(A - B) + \frac{1}{2}\sin(A + B)$

$\sin A + \sin B = 2\sin\frac{1}{2}(A + B)\cos\frac{1}{2}(A - B)$

$\sin A - \sin B = 2\cos\frac{1}{2}(A + B)\sin\frac{1}{2}(A - B)$

$\cos A + \cos B = 2\cos\frac{1}{2}(A + B)\cos\frac{1}{2}(A - B)$

$\cos A - \cos B = -2\sin\frac{1}{2}(A + B)\sin\frac{1}{2}(A - B)$

3. Common Reference Triangles

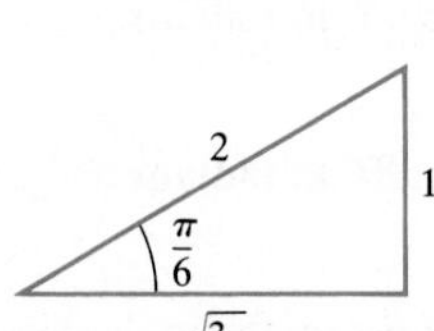

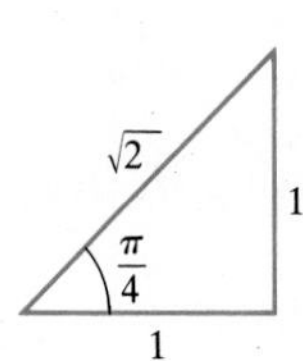

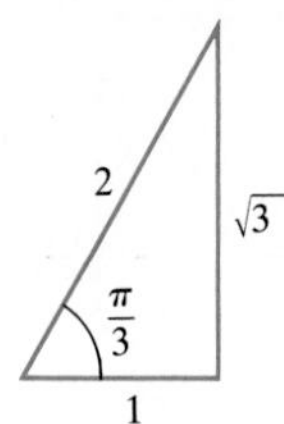

4. Angles and Sides of a Triangle

Law of cosines: $c^2 = a^2 + b^2 - 2ab\cos C$

Law of sines: $\dfrac{\sin A}{a} = \dfrac{\sin B}{b} = \dfrac{\sin C}{c}$

Area $= \dfrac{1}{2}bc\sin A = \dfrac{1}{2}ac\sin B = \dfrac{1}{2}ab\sin C$

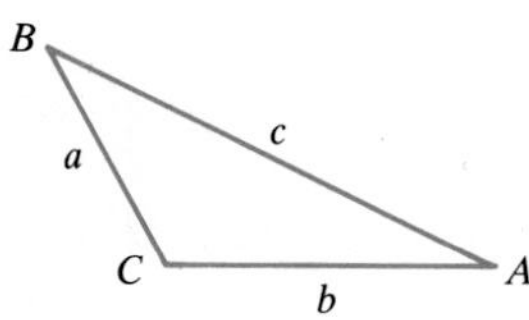

A.2 A Brief Review of Trigonometric Functions

Radian Measure

The **radian measure** of the angle ACB at the center of the unit circle (Fig. A.1) equals the length of the arc that the angle cuts from the unit circle.

If angle ACB cuts an arc $A'B'$ from a second circle centered at C, then circular sector $A'CB'$ will be similar to circular sector ACB. In particular,

$$\frac{\text{Length of arc } A'B'}{\text{Radius of second circle}} = \frac{\text{Length of arc } AB}{\text{Radius of first circle}}. \tag{1}$$

In the notation of Fig. A.1, Eq. (1) says that

$$\frac{s}{r} = \frac{\theta}{1} = \theta \quad \text{or} \quad \theta = \frac{s}{r}. \tag{2}$$

When you know r and s, you can calculate the angle's radian measure θ from this equation. Notice that the units of length for r and s cancel out and that radian measure, like degree measure, is a dimensionless number.

We find the relation between degree measure and radian measure by observing that a semicircle of radius r, which we know has length $s = \pi r$, subtends a central angle of 180°. Therefore,

$$180° = \pi \text{ radians}. \tag{3}$$

We can restate this relation in several useful ways.

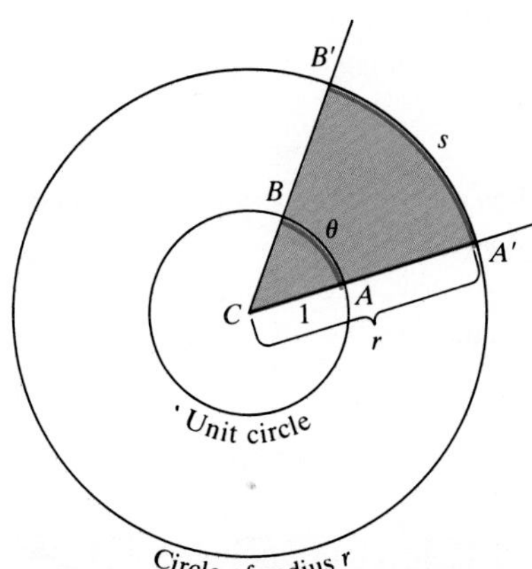

A.1 The radian measure of angle ACB is the length θ of arc AB on the unit circle centered at C. The value of θ can be found from any other circle, however, as the ratio of s to r.

Degrees to radians:

1 degree makes $\dfrac{\pi}{180}$ radians (about 0.02 rad).

To change degrees to radians, multiply degrees by $\dfrac{\pi}{180}$.

Radians to degrees:

1 radian makes $\dfrac{180}{\pi}$ degrees (about 57°).

To change radians to degrees, multiply radians by $\dfrac{180}{\pi}$.

A.2 The angles of two common triangles, in degrees and radians.

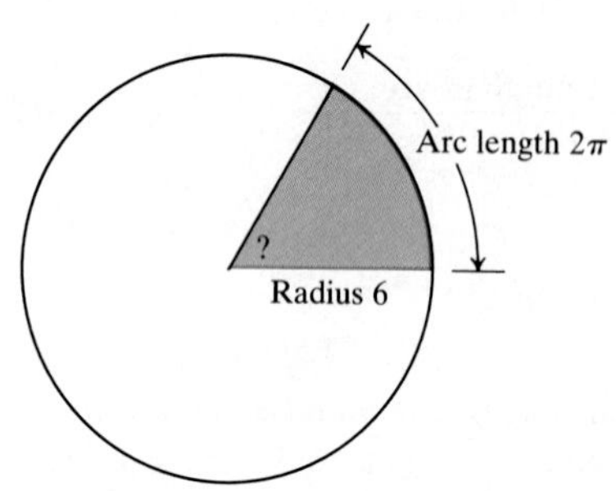

A.3 What is the radian measure of this angle? See Example 2.

Example 1 *Conversions* (Fig. A.2)

Change 45° to radians: $45 \cdot \dfrac{\pi}{180} = \dfrac{\pi}{4}$ rad

Change 90° to radians: $90 \cdot \dfrac{\pi}{180} = \dfrac{\pi}{2}$ rad

Change $\dfrac{\pi}{6}$ radians to degrees: $\dfrac{\pi}{6} \cdot \dfrac{180}{\pi} = 30°$

Change $\dfrac{\pi}{3}$ radians to degrees: $\dfrac{\pi}{3} \cdot \dfrac{180}{\pi} = 60°$

Example 2 An acute angle whose vertex lies at the center of a circle of radius 6 subtends an arc of length 2π (Fig. A.3). The angle's radian measure is

$$\theta = \frac{s}{r} = \frac{2\pi}{6} = \frac{\pi}{3}. \qquad \text{(Eq. (2) with } s = 2\pi,\ r = 6)$$

The equation $\theta = s/r$ is sometimes written

$$s = r\theta. \tag{4}$$

This equation gives a handy way to find s when you know r and θ.

Example 3 An angle of $3\pi/4$ radians at the center of a circle of radius 8 subtends an arc

$$s = r\theta = 8 \cdot \frac{3\pi}{4} = 6\pi \qquad \begin{pmatrix}\text{Eq. (4) with } r = 8 \\ \text{and } \theta = 3\pi/4\end{pmatrix}$$

units long.

Example 4 How long is the arc subtended by a central angle of 120° in a circle of radius 4?

Solution The equation $s = r\theta$ holds only when the angle is measured in radians, so we must find the angle's radian measure before finding s:

$$\theta = 120 \cdot \frac{\pi}{180} = \frac{2\pi}{3} \text{ rad} \qquad \text{(Convert to radians.)}$$

$$s = r\theta = 4 \cdot \frac{2\pi}{3} = \frac{8\pi}{3}. \qquad \text{(Then find } s = r\theta.)$$

The arc is $8\pi/3$ units long.

When angles are used to describe counterclockwise rotations, our measurements can go arbitrarily far beyond 2π radians, or 360°. Similarly, angles that describe clockwise rotations can have negative measures of all sizes (Fig. A.4).

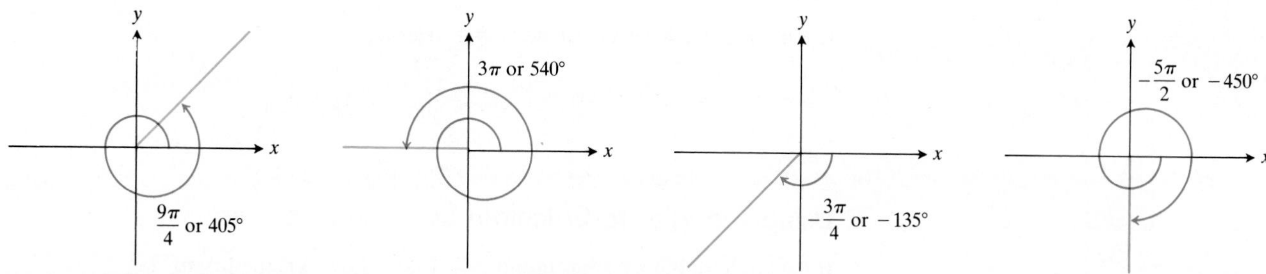

A.4 Angles can have any measure.

The Six Basic Trigonometric Functions

When an angle of measure θ is placed in standard position at the center of a circle of radius r (Fig. A.5), the six basic trigonometric functions of θ are defined in the following way:

$$\text{Sine:} \quad \sin\theta = \frac{y}{r} \qquad \text{Cosecant:} \quad \csc\theta = \frac{r}{y}$$

$$\text{Cosine:} \quad \cos\theta = \frac{x}{r} \qquad \text{Secant:} \quad \sec\theta = \frac{r}{x} \tag{5}$$

$$\text{Tangent:} \quad \tan\theta = \frac{y}{x} \qquad \text{Cotangent:} \quad \cot\theta = \frac{x}{y}$$

As you can see, $\tan\theta$ and $\sec\theta$ are not defined if $x = 0$. In terms of radian measure, this means they are not defined when θ is $\pm\pi/2$, $\pm 3\pi/2$, ... Similarly, $\cot\theta$ and $\csc\theta$ are not defined for values of θ for which $y = 0$, namely $\theta = 0, \pm\pi, \pm 2\pi, \ldots$. Notice also that

$$\tan\theta = \frac{\sin\theta}{\cos\theta}, \qquad \csc\theta = \frac{1}{\sin\theta}$$
$$\sec\theta = \frac{1}{\cos\theta}, \qquad \cot\theta = \frac{1}{\tan\theta} \tag{6}$$

whenever the quotients on the right-hand sides are defined.

Because $x^2 + y^2 = r^2$ (Pythagorean theorem),

$$\cos^2\theta + \sin^2\theta = \frac{x^2}{r^2} + \frac{y^2}{r^2} = \frac{x^2 + y^2}{r^2} = 1. \tag{7}$$

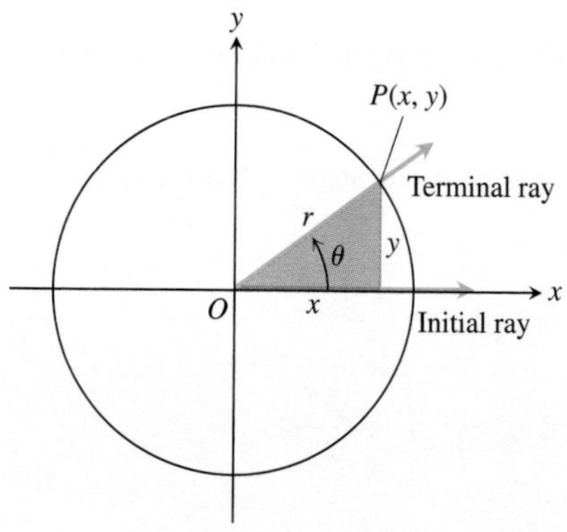

A.5 An angle θ in standard position.

The equation $\cos^2\theta + \sin^2\theta = 1$, true for all values of θ, is probably the most frequently used identity in trigonometry.

The coordinates of the point $P(x, y)$ in Fig. A.5 can be expressed in terms of r and θ as

$$x = r\cos\theta \qquad (\text{Because } x/r = \cos\theta)$$
$$y = r\sin\theta \qquad (\text{Because } y/r = \sin\theta) \tag{8}$$

We use these equations when we study circular motion and when we work with polar coordinates. Notice that if $\theta = 0$ in Fig. A.5, then $x = r$ and $y = 0$, so

$$\cos 0 = 1 \qquad \text{and} \qquad \sin 0 = 0.$$

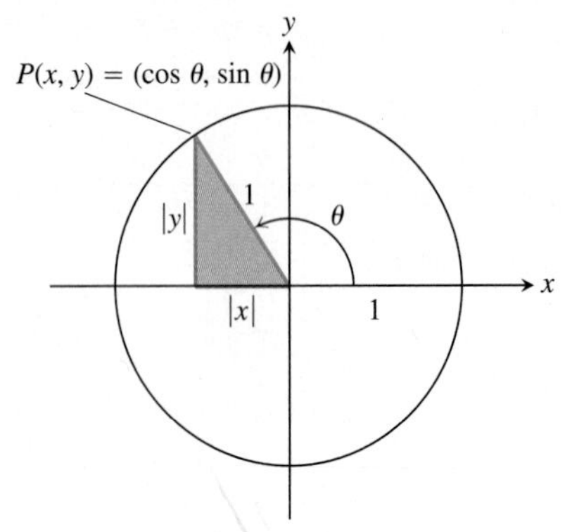

A.6 The acute reference triangle for an angle θ.

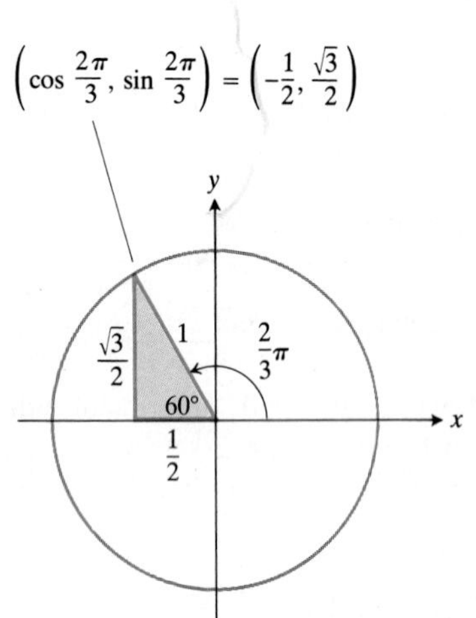

A.7 The triangle for calculating the sine and cosine of $2\pi/3$ radians (Example 5).

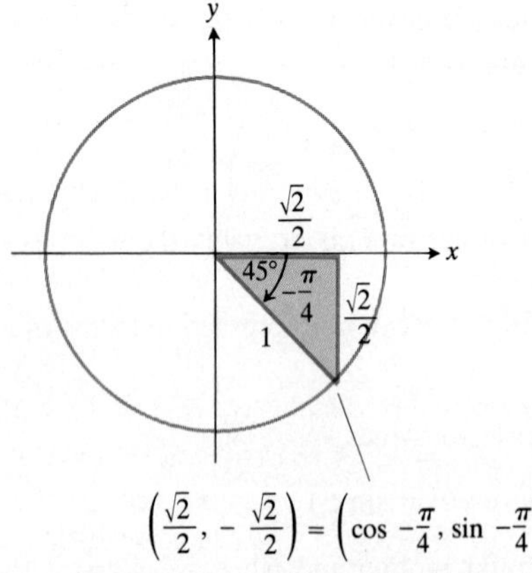

A.8 The triangle for calculating the sine and cosine of $-\pi/4$ radians (Example 6).

If $\theta = \pi/2$, we have $x = 0$ and $y = r$. Hence,

$$\cos\frac{\pi}{2} = 0 \quad \text{and} \quad \sin\frac{\pi}{2} = 1.$$

Using Triangles to Calculate Sines and Cosines

If the circle in Fig. A.5 has radius $r = 1$ unit, Eqs. (8) simplify to

$$x = \cos\theta, \qquad y = \sin\theta.$$

We can therefore calculate the values of the cosine and sine from the acute reference triangle made by dropping a perpendicular from the point $P(x, y)$ to the x-axis (Fig. A.6). The numerical values of x and y are read from the triangle's sides. The signs of x and y are determined by the quadrant in which the triangle lies.

Example 5 Find the sine and cosine of $2\pi/3$ radians.

Solution

STEP 1: Draw the angle in standard position and write in the lengths of the sides of the reference triangle (Fig. A.7).

STEP 2: Find the coordinates of the point P where the angle's terminal ray cuts the circle:

$$\cos\frac{2\pi}{3} = x\text{-coordinate of } P = -\frac{1}{2}$$

$$\sin\frac{2\pi}{3} = y\text{-coordinate of } P = \frac{\sqrt{3}}{2}.$$

Example 6 Find the sine and cosine of $-\pi/4$ radians.

Solution

STEP 1: Draw the angle in standard position and write in the lengths of the sides of the reference triangle (Fig. A.8).

STEP 2: Find the coordinates of the point P where the angle's terminal ray cuts the circle:

$$\cos -\frac{\pi}{4} = x\text{-coordinate of } P = \frac{\sqrt{2}}{2},$$

$$\sin -\frac{\pi}{4} = y\text{-coordinate of } P = -\frac{\sqrt{2}}{2}.$$

Table A.1 gives the values of the sine, cosine, and tangent for selected values of θ.

TABLE A.1
Values of $\sin\theta$, $\cos\theta$, and $\tan\theta$ for selected values of θ

Degrees	-180	-135	-90	-45	0	45	90	135	180
θ (radians)	$-\pi$	$-3\pi/4$	$-\pi/2$	$-\pi/4$	0	$\pi/4$	$\pi/2$	$3\pi/4$	π
$\sin\theta$	0	$-\sqrt{2}/2$	-1	$-\sqrt{2}/2$	0	$\sqrt{2}/2$	1	$\sqrt{2}/2$	0
$\cos\theta$	-1	$-\sqrt{2}/2$	0	$\sqrt{2}/2$	1	$\sqrt{2}/2$	0	$-\sqrt{2}/2$	-1
$\tan\theta$	0	1		-1	0	1		-1	0

Periodicity

When an angle of measure θ and an angle of measure $\theta + 2\pi$ are in standard position, their terminal rays coincide. The two angles therefore have the same trigonometric function values:

$$\begin{aligned}
\cos(\theta + 2\pi) &= \cos\theta \\
\sin(\theta + 2\pi) &= \sin\theta \\
\tan(\theta + 2\pi) &= \tan\theta \\
\cot(\theta + 2\pi) &= \cot\theta \\
\sec(\theta + 2\pi) &= \sec\theta \\
\csc(\theta + 2\pi) &= \csc\theta
\end{aligned} \tag{9}$$

Similarly, $\cos(\theta - 2\pi) = \cos\theta$, $\sin(\theta - 2\pi) = \sin\theta$, and so on.

From another point of view, Eqs. (9) tell us that if we start at any particular value $\theta = \theta_0$ and let θ increase or decrease steadily, we see the values of the trigonometric functions start to repeat after any interval of length 2π. We describe this behavior by saying that the six basic trigonometric functions are **periodic** and that they repeat after a fixed **period** of θ-values.

DEFINITION

A function $f(x)$ is **periodic** with **period** $p > 0$ if $f(x + p) = f(x)$ for every value of x.

Equations (9) tell us that the six basic trigonometric functions are periodic with period 2π. Other periods include 4π, 6π, and so on (positive integer multiples of 2π).

Notice that the tangent and cotangent functions (Fig. A.9) also have period π. Other periods include 2π, 3π, and so on (positive integer multiples of π).

In naming the period of a function it is conventional to name the *smallest* positive value of p for which $f(x + p) = f(x)$ for all x. The longer periods can all be

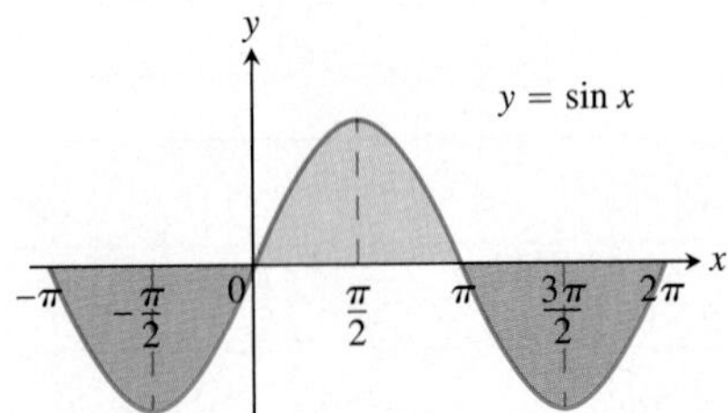

Domain: $-\infty < x < \infty$
Range: $-1 \le y \le 1$

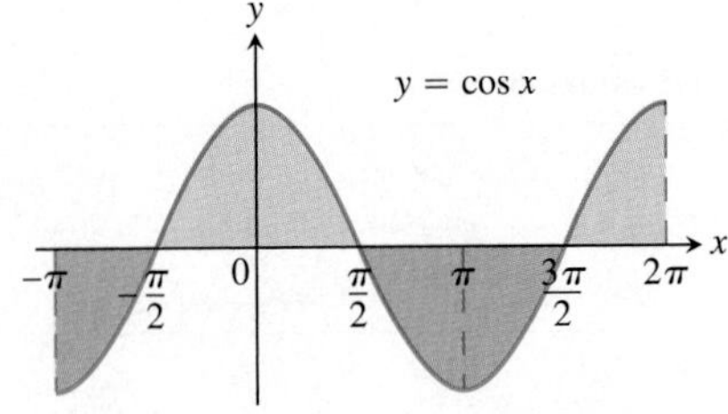

Domain: $-\infty < x < \infty$
Range: $-1 \le y \le 1$

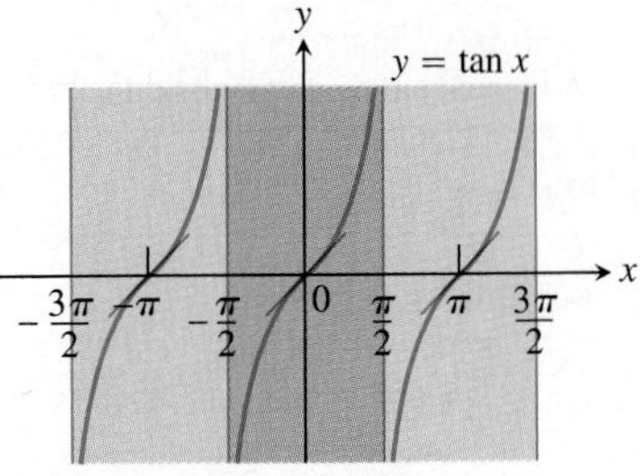

Domain: All real numbers except odd integer multiples of $\pi/2$
Range: $-\infty < y < \infty$

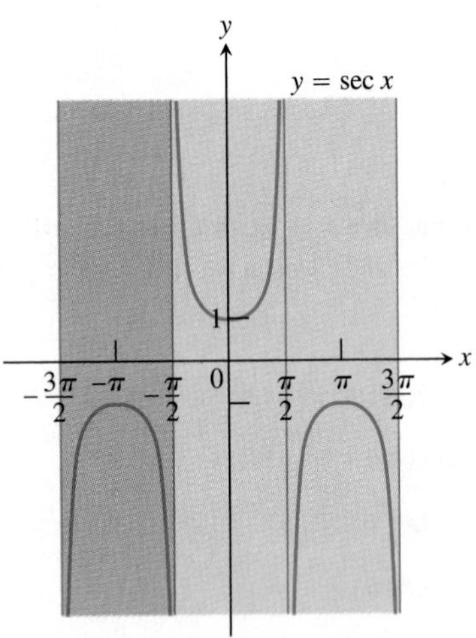

Domain: $x \neq \pm\frac{\pi}{2}, \pm\frac{3\pi}{2}, \ldots$
Range: $y \le -1$ and $y \ge 1$

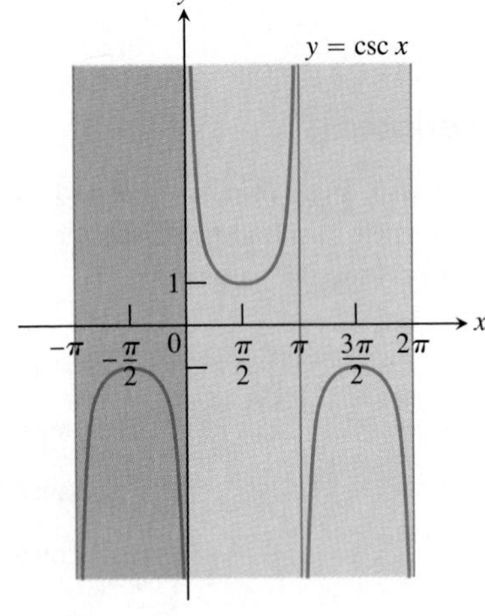

Domain: $x \neq 0, \pm\pi, \pm 2\pi, \ldots$
Range: $y \le -1$ and $y \ge 1$

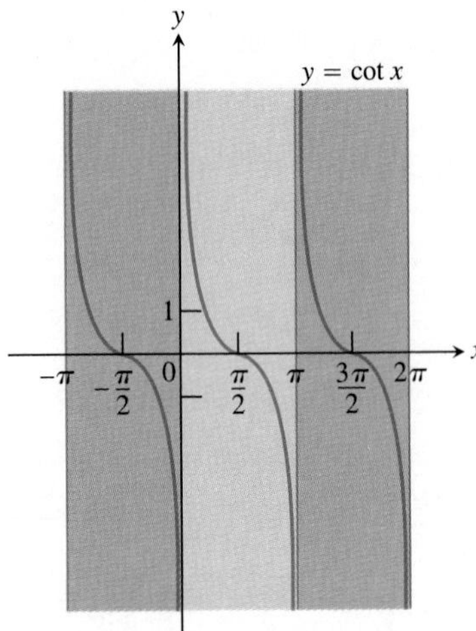

Domain: $x \neq 0, \pm\pi, \pm 2\pi, \ldots$
Range: $-\infty < y < \infty$

A.9 The graphs of the six basic trigonometric functions as functions of radian measure. Each function's periodicity shows clearly in its graph.

constructed from p. Thus, we say that the period of $\tan x$ is π while the period of $\sin x$ is 2π.

The importance of periodic functions stems from the fact that much of the behavior we study in science is periodic. Brain waves and heartbeats are periodic, as are household voltage and electric current. The electromagnetic field that heats food in a microwave oven is periodic, as are cash flows in seasonal businesses and the behavior of rotational machinery. The seasons are periodic—so is the weather. The phases of the moon are periodic, as are the motions of the planets. There is strong evidence that the ice ages are periodic, with a period of 90,000–100,000 years.

If so many things are periodic, why limit our discussion to trigonometric functions? The answer lies in a surprising and beautiful theorem from advanced calculus that says that every periodic function we want to use in mathematical modeling can be written as an algebraic combination of sines and cosines. Thus, once we learn the calculus of sines and cosines, we will know everything we need to know to model the mathematical behavior of periodic phenomena.

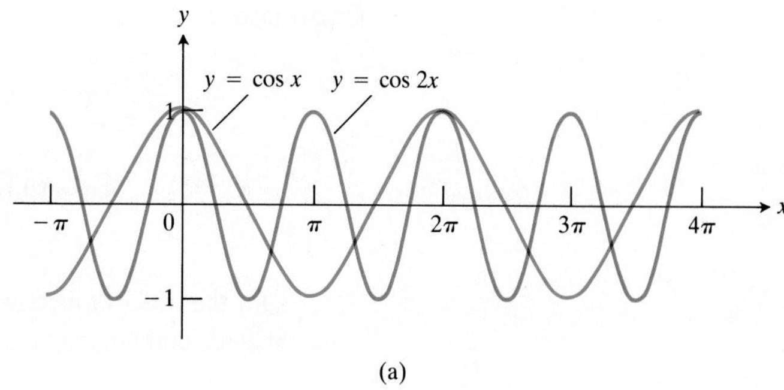

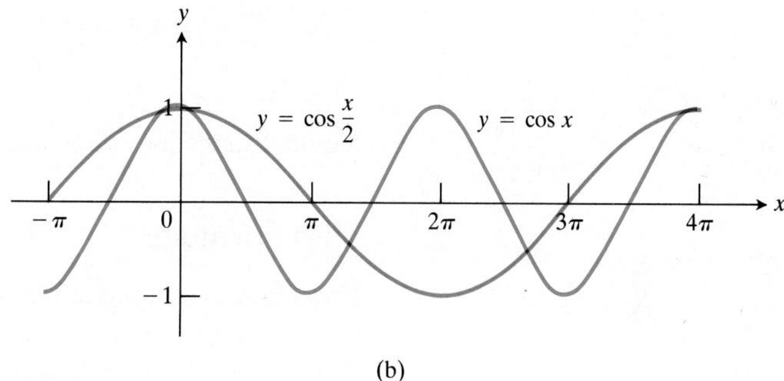

A.10 (a) Multiplying x by a number greater than 1 speeds the cosine up.
(b) Multiplying x by a number less than 1 slows the cosine down.

Graphs

When we graph trigonometric functions in the coordinate plane, we usually denote the independent variable by x instead of θ. Figure A.9 shows how graphs of the six basic trigonometric functions appear as graphs in the xy-plane. Notice that the graph of $\tan x = (\sin x)/(\cos x)$ "blows up" as x approaches odd-integer multiples of $\pi/2$. Notice, too, the similar behavior of $\cot x = (\cos x)/(\sin x)$ as x approaches integer multiples of π.

Figure A.10 shows the graphs of $y = \cos 2x$ and $y = \cos(x/2)$ plotted against the graph of $y = \cos x$. Multiplying x by 2 speeds the cosine up and shortens the period from 2π to π. Multiplying x by 1/2 slows the cosine down and lengthens its period from 2π to 4π.

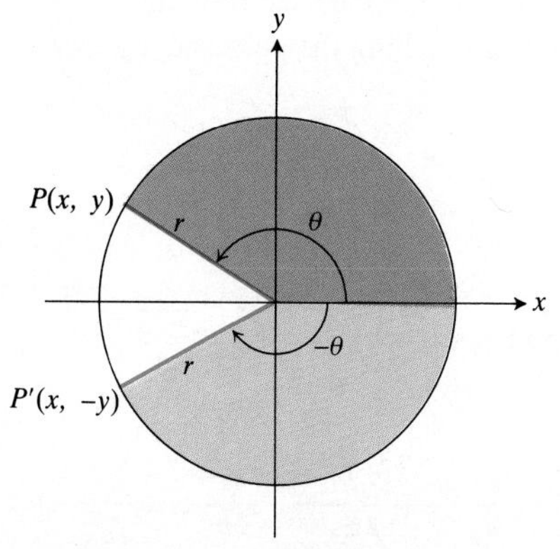

A.11 Angles of opposite sign.

Odd vs. Even

The two angles in Fig. A.11 have the same magnitude but opposite signs. The points where the terminal rays cross the circle have the same x-coordinate, and their y-coordinates differ only in sign. Hence,

$$\cos(-\theta) = \frac{x}{r} = \cos\theta, \qquad \text{(The cosine is an even function.)} \qquad (10a)$$

$$\sin(-\theta) = \frac{-y}{r} = -\sin\theta. \qquad \text{(The sine is an odd function.)} \qquad (10b)$$

Example 7

$$\cos\left(-\frac{\pi}{3}\right) = \cos\frac{\pi}{3} = \frac{1}{2},$$

$$\sin\left(-\frac{\pi}{3}\right) = -\sin\frac{\pi}{3} = -\frac{\sqrt{3}}{2}$$

As for the other basic trigonometric functions, the secant is even and the cosecant, tangent, and cotangent are odd. For the secant and tangent,

$$\sec(-\theta) = \frac{1}{\cos(-\theta)} = \frac{1}{\cos\theta} = \sec\theta, \tag{11}$$

$$\tan(-\theta) = \frac{\sin(-\theta)}{\cos(-\theta)} = \frac{-\sin\theta}{\cos\theta} = -\tan\theta. \tag{12}$$

Similar calculations show that the contangent and cosecant are odd.

Shift Formulas

If you look once again at Fig. A.9, you will see that the cosine curve is the same as the sine curve shifted $\pi/2$ units to the left. Also, the sine curve is the same as the cosine curve shifted $\pi/2$ units to the right. In symbols,

$$\sin\left(x + \frac{\pi}{2}\right) = \cos x, \qquad \cos\left(x - \frac{\pi}{2}\right) = \sin x. \tag{13}$$

Figure A.12(a) shows the cosine shifted to the left $\pi/2$ units to become the reflection of the sine curve across the x-axis. Next to it, Fig. A.12(b) shows the sine curve shifted $\pi/2$ units to the right to become the reflection of the cosine curve across the x-axis. In symbols,

$$\cos\left(x + \frac{\pi}{2}\right) = -\sin x, \qquad \sin\left(x - \frac{\pi}{2}\right) = -\cos x. \tag{14}$$

Example 8 The builders of the Trans-Alaska Pipeline used insulated pads to keep the heat from the hot oil in the pipeline from melting the permanently frozen

A.12 (a) The reflection of the sine as a shifted cosine. (b) The reflection of the cosine as a shifted sine.

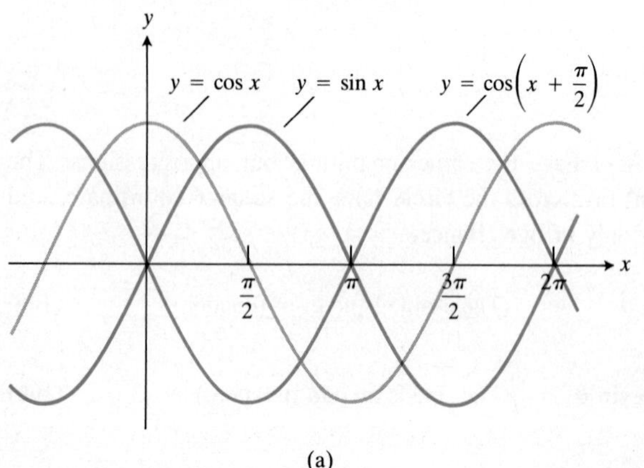

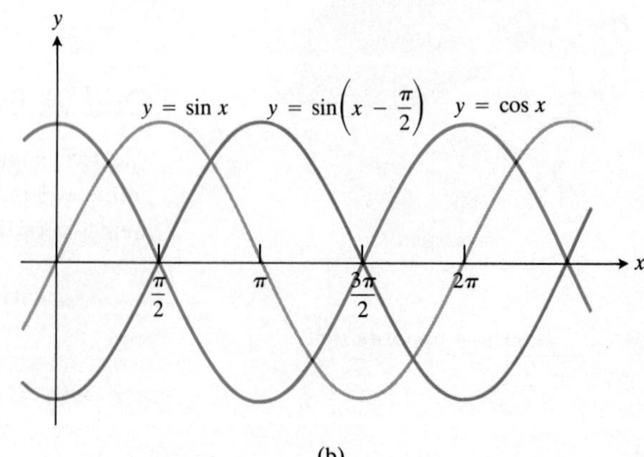

soil beneath. To design the pads, it was necessary to take into account the variation in air temperature throughout the year. The variation was represented in the calculations by a *general sine function*

$$f(x) = A \sin\left[\frac{2\pi}{B}(x - C)\right] + D.$$

In this formula, $|A|$ is the amplitude, $|B|$ is the period, C is the horizontal shift, and D is the vertical shift (Fig. A.13).

Figure A.14 shows how we can use such a function to model temperature data. The data points in the figure are plots of the mean air temperature for Fairbanks, Alaska, based on records of the National Weather Service from 1941 to 1970. The sine function used to fit the data is

$$f(x) = 37 \sin\left[\frac{2\pi}{365}(x - 101)\right] + 25,$$

where f is temperature in degrees Fahrenheit and x is the number of the day counting from the beginning of the year. The fit is remarkably good.

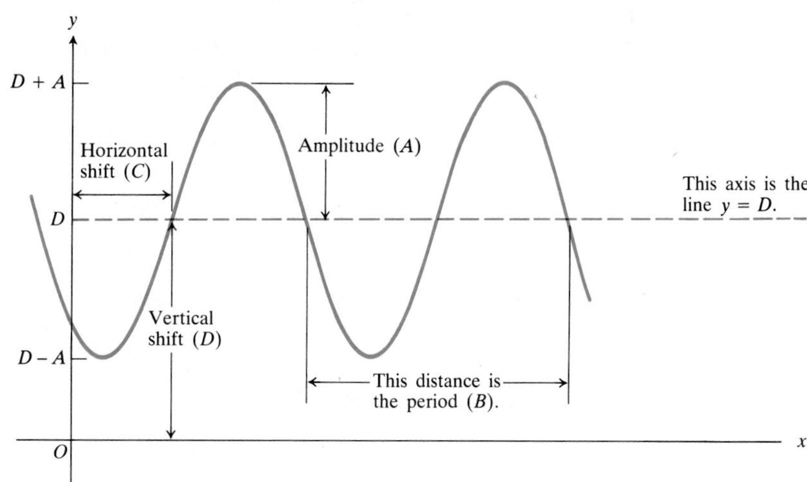

A.13 The general sine curve $y = A \sin[(2\pi/B)(x - C)] + D$, shown for A, B, C, and D positive.

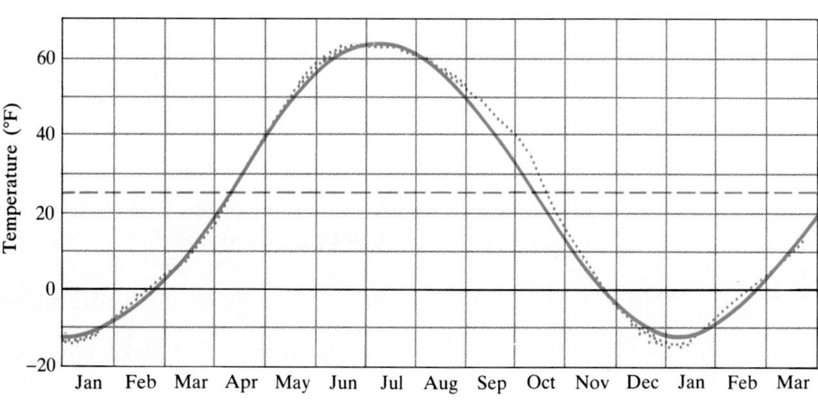

A.14 Normal mean air temperature at Fairbanks, Alaska, plotted as data points. The approximating sine function is

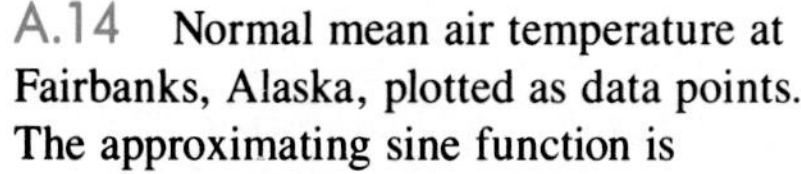

$$f(x) = 37 \sin\left[\frac{2\pi}{365}(x - 101)\right] + 25.$$

(*Source:* "Is the Curve of Temperature Variation a Sine Curve?" by B. M. Lando and C. A. Lando, *The Mathematics Teacher,* 7:6, Fig. 2, p. 535 (September 1977).)

Angle Sum and Difference Formulas

As you may recall from an earlier course,

$$\cos(A+B) = \cos A \cos B - \sin A \sin B, \tag{15}$$

$$\sin(A+B) = \sin A \cos B + \cos A \sin B. \tag{16}$$

These formulas hold for all angles A and B.

If we replace B by $-B$ in Eqs. (15) and (16), we get

$$\begin{aligned} \cos(A-B) &= \cos A \cos(-B) - \sin A \sin(-B) \\ &= \cos A \cos B - \sin A(-\sin B) \\ &= \cos A \cos B + \sin A \sin B, \end{aligned} \tag{17}$$

$$\begin{aligned} \sin(A-B) &= \sin A \cos(-B) + \cos A \sin(-B) \\ &= \sin A \cos B + \cos A(-\sin B) \\ &= \sin A \cos B - \cos A \sin B \end{aligned} \tag{18}$$

Double-angle (Half-angle) Formulas

It is sometimes possible to simplify a calculation by changing trigonometric functions of θ into trigonometric functions of 2θ. There are four basic formulas for doing this, called **double-angle formulas.** The first two come from setting A and B equal to θ in Eqs. (15) and (16):

$$\cos 2\theta = \cos^2\theta - \sin^2\theta, \qquad \text{(Eq. (15) with } A = B = \theta) \tag{19}$$

$$\sin 2\theta = 2\sin\theta\cos\theta. \qquad \text{(Eq. (16) with } A = B = \theta) \tag{20}$$

The other two double-angle formulas come from the equations

$$\cos^2\theta + \sin^2\theta = 1, \qquad \cos^2\theta - \sin^2\theta = \cos 2\theta.$$

We add to get

$$2\cos^2\theta = 1 + \cos 2\theta,$$

subtract to get

$$2\sin^2\theta = 1 - \cos 2\theta,$$

and divide by 2 to get

$$\cos^2\theta = \frac{1 + \cos 2\theta}{2} \tag{21}$$

$$\sin^2\theta = \frac{1 - \cos 2\theta}{2}. \tag{22}$$

When θ is replaced by $\theta/2$ in Eqs. (21) and (22), the resulting formulas are called **half-angle formulas.** Some books refer to Eqs. (21) and (22) by this name as well.

Where to Look for Other Formulas Additional information is available in

1. Appendix A.1 of the present book.
2. *CRC Standard Mathematical Tables* (any recent edition), CRC Press, Inc.

Tips for Graphing sines and cosines: Curve first, scaled axes later

1. The one basic sine and cosine curve:

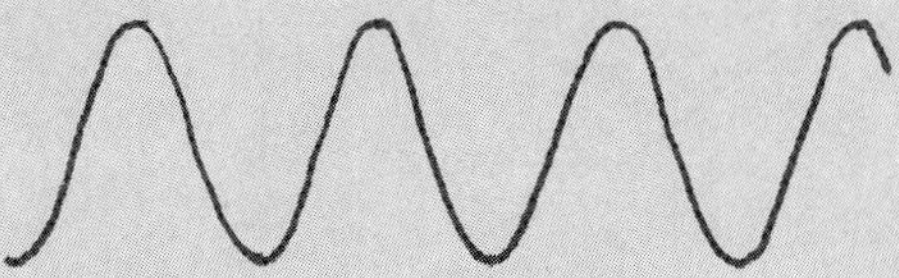

2. The basic curve with approximately scaled axes in different positions:

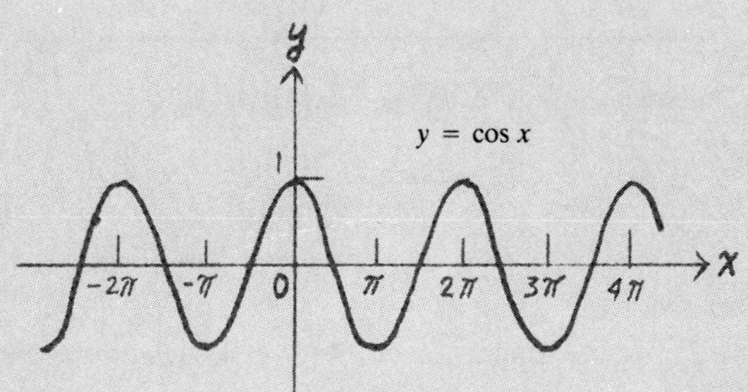

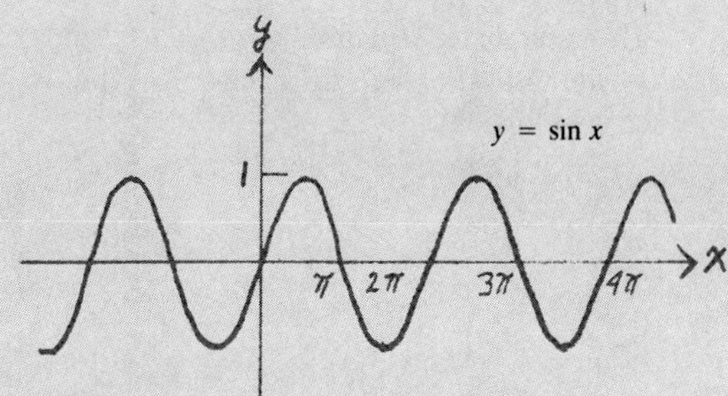

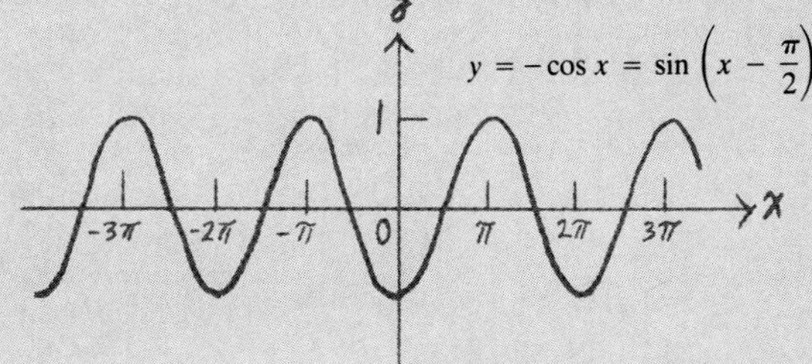

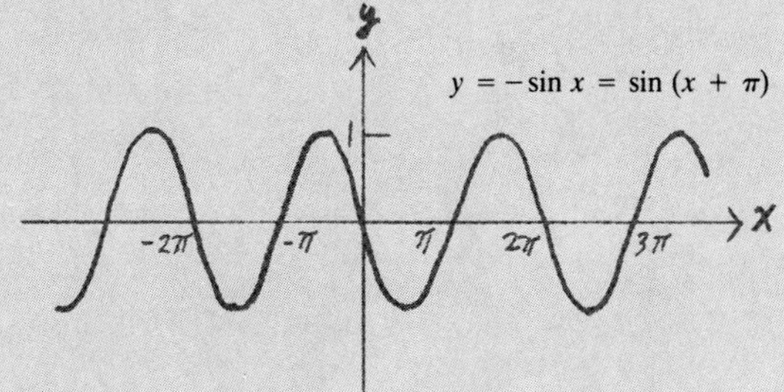

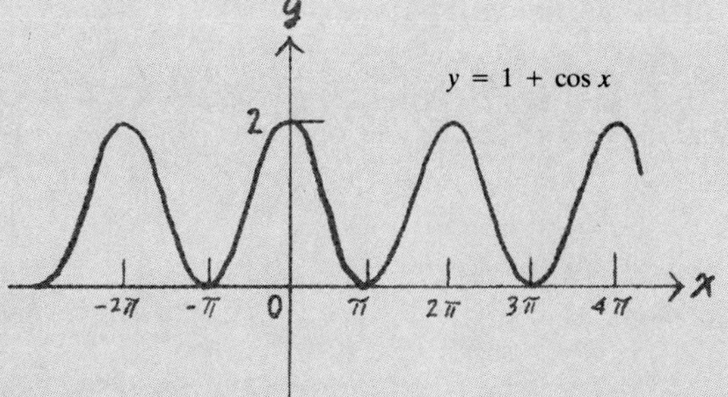

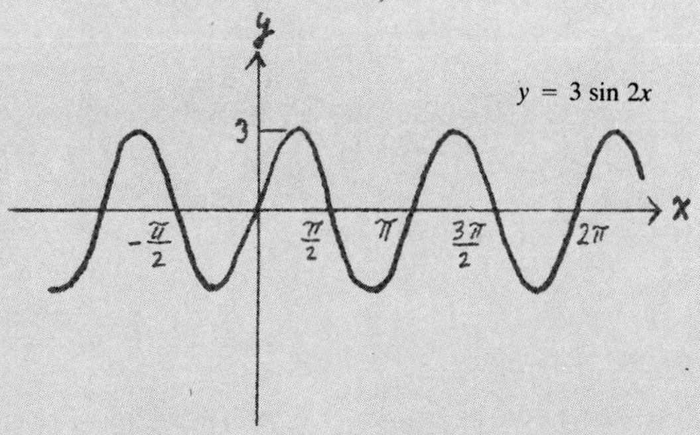

A.3 Symmetry in the xy-Plane

We can use coordinate formulas to describe important symmetries in the coordinate plane. Figure A.15 shows how this is done.

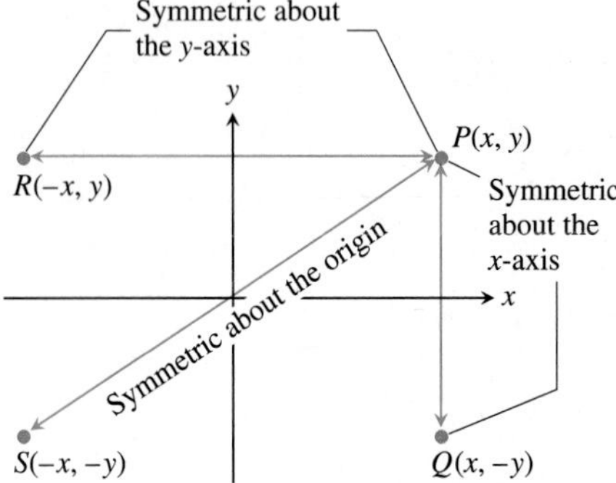

A.15 The coordinate formulas for symmetry with respect to the origin and axes in the coordinate plane.

Example 1 Symmetric points:

$P(5, 2)$ and $Q(5, -2)$	symmetric about the x-axis
$P(5, 2)$ and $R(-5, 2)$	symmetric about the y-axis
$P(5, 2)$ and $S(-5, -2)$	symmetric about the origin

The coordinate relations in Fig. A.15 provide the following symmetry tests for graphs.

Symmetry Tests for Graphs

1. *Symmetry about the x-axis:*
   If the point (x, y) lies on the graph, then the point $(x, -y)$ lies on the graph (Fig. A.16a).
2. *Symmetry about the y-axis:*
   If the point (x, y) lies on the graph, the point $(-x, y)$ lies on the graph (Fig. A.16b).
3. *Symmetry about the origin:*
   If the point (x, y) lies on the graph, the point $(-x, -y)$ lies on the graph (Fig. A.16c).

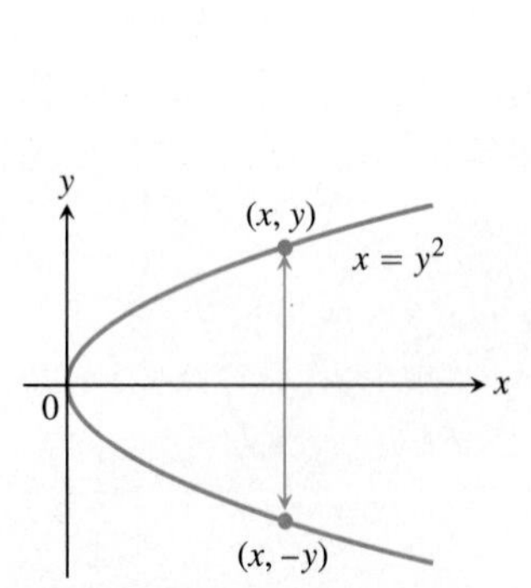

(a) Symmetry about the x-axis

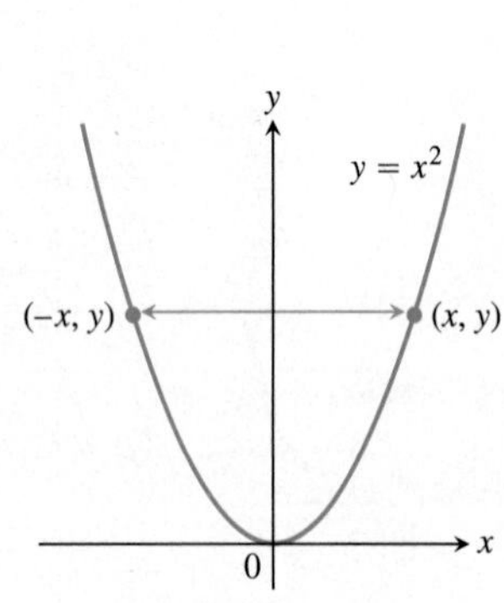

(b) Symmetry about the y-axis

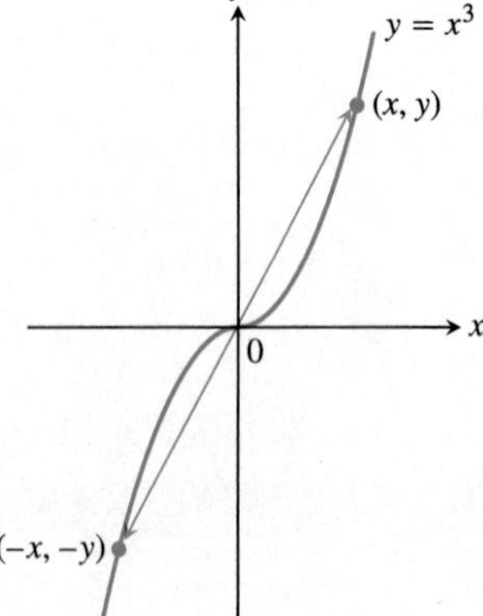

(c) Symmetry about the origin

A.16 Symmetry tests for graphs in the xy-plane.

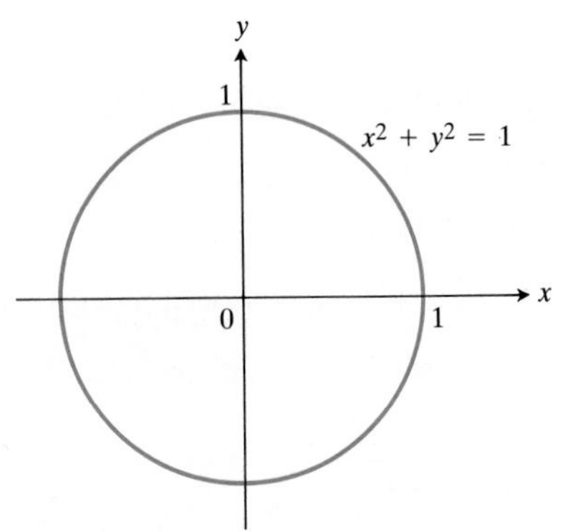

A.17 The graph of the equation $x^2 + y^2 = 1$ is the circle of radius 1, centered at the origin. It is symmetric about both axes and about the origin. Example 2 shows how to predict these symmetries before you graph the equation.

Example 2 The graph of $x^2 + y^2 = 1$ has all three of the symmetries listed above (Fig. A.17).

1. Symmetry about the x-axis:

$$
\begin{aligned}
(x, y) \text{ on the graph} &\Rightarrow x^2 + y^2 = 1 && (\Rightarrow \text{ means "implies."})\\
&\Rightarrow x^2 + (-y)^2 = 1 && (y^2 = (-y)^2)\\
&\Rightarrow (x, -y) \text{ on the graph}
\end{aligned}
$$

2. Symmetry about the y-axis:

$$
\begin{aligned}
(x, y) \text{ on the graph} &\Rightarrow x^2 + y^2 = 1\\
&\Rightarrow (-x)^2 + y^2 = 1 && (x^2 = (-x)^2)\\
&\Rightarrow (-x, y) \text{ on the graph}
\end{aligned}
$$

3. Symmetry about the origin:

$$
\begin{aligned}
(x, y) \text{ on the graph} &\Rightarrow x^2 + y^2 = 1\\
&\Rightarrow (-x)^2 + (-y)^2 = 1\\
&\Rightarrow (-x, -y) \text{ on the graph}
\end{aligned}
$$

A.4 Proofs of the Limit Theorems in Chapter 1

This appendix furnishes the ε-δ proofs for Theorems 1 and 6 in Chapter 1. Theorem 1 tells us that the limit of an algebraic combination of functions is the corresponding combination of the functions' limits, when the latter exists.

THEOREM 1 (SECTION 1.7)

The following rules hold if $\lim_{x\to c} f_1(x) = L_1$ and $\lim_{x\to c} f_2(x) = L_2$.

1. *Sum Rule:*	$\lim\,(f_1(x) + f_2(x)) = L_1 + L_2$
2. *Difference Rule:*	$\lim\,(f_1(x) - f_2(x)) = L_1 - L_2$
3. *Product Rule:*	$\lim f_1(x) \cdot f_2(x) = L_1 \cdot L_2$
4. *Constant Multiple Rule:*	$\lim k \cdot f_2(x) = k \cdot L_2$ (Any number k)
5. *Quotient Rule:*	$\lim \dfrac{f_1(x)}{f_2(x)} = \dfrac{L_1}{L_2}$ if $L_2 \neq 0$

The limits are all taken as $x \to c$, and L_1 and L_2 are real numbers.

We proved the Sum Rule in Section 1.7. We obtain the Difference Rule by replacing $f_2(x)$ by $-f_2(x)$ and L_2 by $-L_2$ in the Sum Rule. The Constant Multiple Rule is the special case $f_1(x) = k$ of the Product Rule. This leaves only the Product and Quotient Rules to prove.

Proof of the Limit Product Rule We need to show that for any $\varepsilon > 0$ there exists a $\delta > 0$ such that for all x in the functions' common domain,

$$0 < |x - c| < \delta \;\Rightarrow\; |f_1(x) f_2(x) - L_1 L_2| < \varepsilon. \tag{1}$$

Suppose then that ε is a positive number, and write $f_1(x)$ and $f_2(x)$ as

$$f_1(x) = L_1 + (f_1(x) - L_1) \qquad \text{and} \qquad f_2(x) = L_2 + (f_2(x) - L_2).$$

Multiply these expressions and subtract L_1L_2:

$$\begin{aligned} f_1(x)\cdot f_2(x) - L_1L_2 &= (L_1 + (f_1(x) - L_1))(L_2 + (f_2(x) - L_2)) - L_1L_2 \\ &= L_1L_2 + L_1(f_2(x) - L_2) + L_2(f_1(x) - L_1) \\ &\quad + (f_1(x) - L_1)(f_2(x) - L_2) - L_1L_2 \\ &= L_1(f_2(x) - L_2) + L_2(f_1(x) - L_1) + (f_1(x) - L_1)(f_2(x) - L_2). \end{aligned} \tag{2}$$

Since f_1 and f_2 have limits L_1 and L_2 as $x \to c$, there exist positive numbers δ_1, δ_2, δ_3, and δ_4 such that for all x,

$$\begin{aligned} 0 < |x - c| < \delta_1 \quad &\Rightarrow \quad |f_1(x) - L_1| < \sqrt{\varepsilon/3}, \\ 0 < |x - c| < \delta_2 \quad &\Rightarrow \quad |f_2(x) - L_2| < \sqrt{\varepsilon/3}, \\ 0 < |x - c| < \delta_3 \quad &\Rightarrow \quad |f_1(x) - L_1| < \varepsilon/(3(1 + |L_2|)), \\ 0 < |x - c| < \delta_4 \quad &\Rightarrow \quad |f_2(x) - L_2| < \varepsilon/(3(1 + |L_1|)). \end{aligned} \tag{3}$$

All four of the inequalities on the right-hand side of (3) will hold for $0 < |x - c| < \delta$ if we take δ to be the smallest of the numbers δ_1 through δ_4. Therefore for all x, $0 < |x - c| < \delta$ implies

$$|f_1(x)\cdot f_2(x) - L_1L_2|$$

$$\le |L_1||f_2(x) - L_2| + |L_2||f_1(x) - L_1| + |f_1(x) - L_1||f_2(x) - L_2|$$

(Triangle inequality applied to Eq. (2))

$$\le (1 + |L_1|)|f_2(x) - L_2| + (1 + |L_2|)|f_1(x) - L_1| + |f_1(x) - L_1||f_2(x) - L_2|$$

$$\le \frac{\varepsilon}{3} + \frac{\varepsilon}{3} + \sqrt{\frac{\varepsilon}{3}}\sqrt{\frac{\varepsilon}{3}} = \varepsilon. \qquad \text{(Values from (3))}$$

This completes the proof of the Limit Product Rule.

Proof of the Limit Quotient Rule We show that

$$\lim_{x\to c} \frac{1}{f_2(x)} = \frac{1}{L_2}.$$

Then we can apply the Limit Product Rule to show that

$$\lim_{x\to c} \frac{f_1(x)}{f_2(x)} = \lim_{x\to c} f_1(x)\cdot\frac{1}{f_2(x)} = \lim_{x\to c} f_1(x)\cdot\lim_{x\to c}\frac{1}{f_2(x)} = L_1\cdot\frac{1}{L_2} = \frac{L_1}{L_2}.$$

To show that $\lim_{x\to c} (1/f_2(x)) = 1/L_2$, we need to show that for any $\varepsilon > 0$ there exists a $\delta > 0$ such that for all x,

$$0 < |x - c| < \delta \quad \Rightarrow \quad \left|\frac{1}{f_2(x)} - \frac{1}{L_2}\right| < \varepsilon.$$

Since $|L_2| > 0$, there exists a positive number δ_1 such that for all x,

$$0 < |x - c| < \delta_1 \quad \Rightarrow \quad |f_2(x) - L_2| < \frac{|L_2|}{2}. \tag{4}$$

For any numbers A and B it can be shown that $|A| - |B| \le |A - B|$ and $|B| - |A| \le |A - B|$, from which it follows that

$$||A| - |B|| \le |A - B|. \tag{5}$$

With $A = f_2(x)$ and $B = L_2$, this gives

$$||f_2(x)| - |L_2|| \le |f_2(x) - L_2|,$$

which we can combine with the right-hand inequality in (4) to get, in turn,

$$||f_2(x)| - |L_2|| < \frac{|L_2|}{2},$$

$$-\frac{|L_2|}{2} < |f_2(x)| - |L_2| < \frac{|L_2|}{2},$$

$$\frac{|L_2|}{2} < |f_2(x)| < \frac{3|L_2|}{2}, \tag{6}$$

$$|L_2| < 2|f_2(x)| < 3|L_2|,$$

$$\frac{1}{|f_2(x)|} < \frac{2}{|L_2|} < \frac{3}{|f_2(x)|}.$$

Therefore $0 < |x - c| < \delta_1$ implies that

$$\left|\frac{1}{f_2(x)} - \frac{1}{L_2}\right| = \left|\frac{L_2 - f_2(x)}{L_2 f_2(x)}\right| \le \frac{1}{|L_2|} \cdot \frac{1}{|f_2(x)|} \cdot |L_2 - f_2(x)|$$

$$< \frac{1}{|L_2|} \cdot \frac{2}{|L_2|} \cdot |L_2 - f_2(x)|. \qquad \text{(Eq. (6))}$$

Suppose now that ε is an arbitrary positive number. Then $\frac{1}{2}|L_2|^2 \varepsilon > 0$, so there exists a number $\delta_2 > 0$ such that for all x,

$$0 < |x - c| < \delta_2 \quad \Rightarrow \quad |L_2 - f_2(x)| < \frac{\varepsilon}{2}|L_2|^2. \tag{7}$$

The conclusions in (6) and (7) both hold for all x such that $0 < |x - c| < \delta$ if we take δ to be the smaller of the positive values δ_1 and δ_2. Combining (6) and (7) then gives

$$0 < |x - c| < \delta \quad \Rightarrow \quad \left|\frac{1}{f_2(x)} - \frac{1}{L_2}\right| < \varepsilon. \tag{8}$$

This concludes the proof of the Limit Quotient Rule. ∎

THEOREM 6 (SECTION 1.9)

The Sandwich Theorem

Suppose that $g(x) \le f(x) \le h(x)$ for all $x \ne c$ in some interval about c and that $\lim_{x\to c} g(x) = \lim_{x\to c} h(x) = L$. Then $\lim_{x\to c} f(x) = L$.

Proof for Right-hand Limits Suppose $\lim_{x\to c^+} g(x) = \lim_{x\to c^+} h(x) = L$. Then for any $\varepsilon > 0$ there exists a $\delta > 0$ such that for all x the inequality $c < x < c + \delta$ implies

$$L - \varepsilon < g(x) < L + \varepsilon \quad \text{and} \quad L - \varepsilon < h(x) < L + \varepsilon. \tag{9}$$

These inequalities combine with the inequality $g(x) \le f(x) \le h(x)$ to give

$$L - \varepsilon < g(x) \le f(x) \le h(x) < L + \varepsilon,$$
$$L - \varepsilon < f(x) < L + \varepsilon, \tag{10}$$
$$-\varepsilon < f(x) - L < \varepsilon.$$

Therefore, for all x, the inequality $c < x < c + \delta$ implies $|f(x) - L| < \varepsilon$. ▄

Proof for Left-hand Limits Suppose $\lim_{x\to c^-} g(x) = \lim_{x\to c^-} h(x) = L$. Then for any $\varepsilon > 0$ there exists a $\delta > 0$ such that for all x the inequality $c - \delta < x < c$ implies

$$L - \varepsilon < g(x) < L + \varepsilon \quad \text{and} \quad L - \varepsilon < h(x) < L + \varepsilon. \tag{11}$$

We conclude as before that for all x the inequality $c - \delta < x < c$ implies $|f(x) - L| < \varepsilon$. ▄

Proof for Two-sided Limits If $\lim_{x\to c} g(x) = \lim_{x\to c} h(x) = L$, then $g(x)$ and $h(x)$ both approach L as $x \to c^+$ and as $x \to c^-$; so $\lim_{x\to c^+} f(x) = L$ and $\lim_{x\to c^-} f(x) = L$. Hence $\lim_{x\to c} f(x)$ exists and equals L. ▄

EXERCISES A.4

1. Suppose that functions $f_1(x)$, $f_2(x)$, and $f_3(x)$ have limits L_1, L_2, and L_3, respectively, as $x \to c$. Show that their sum has limit $L_1 + L_2 + L_3$. Use mathematical induction (Appendix A.5) to generalize this result to the sum of any finite number of functions.

2. Use mathematical induction and the Limit Product Rule in Theorem 1 to show that if functions $f_1(x), f_2(x), \dots, f_n(x)$ have limits $L_1, L_2, \dots, L_n$ as $x \to c$, then $\lim_{x\to c} f_1(x) f_2(x) \cdot \cdots \cdot f_n(x) = L_1 \cdot L_2 \cdot \cdots \cdot L_n$.

3. Use the fact that $\lim_{x\to c} x = c$ and the result of Exercise 2 to show that $\lim_{x\to c} x^n = c^n$ for any integer $n > 1$.

4. *Limits of polynomials.* Use the fact that $\lim_{x\to c}(k) = k$ for any number k together with the results of Exercises 1 and 3 to show that $\lim_{x\to c} f(x) = f(c)$ for any polynomial function
$$f(x) = a_0 x^n + a_1 x^{n-1} + \cdots + a_{n-1} x + a_n.$$

5. *Limits of rational functions.* Use Theorem 1 and the result of Exercise 4 to show that if $f(x)$ and $g(x)$ are polynomial functions and $g(c) \ne 0$, then
$$\lim_{x\to c} \frac{f(x)}{g(x)} = \frac{f(c)}{g(c)}.$$

6. *Composites of continuous functions.* Figure A.18 gives the diagram for a proof that the composite of two continuous functions is continuous. Reconstruct the proof from the diagram. The statement to be proved is this: If g is continuous at $x = c$ and f is continuous at $g(c)$, then $f \circ g$ is continuous at c.

   Assume that c is an interior point of the domain of g and that $g(c)$ is an interior point of the domain of f. This will make the limits involved two-sided. (The arguments for the cases that involve one-sided limits are similar.)

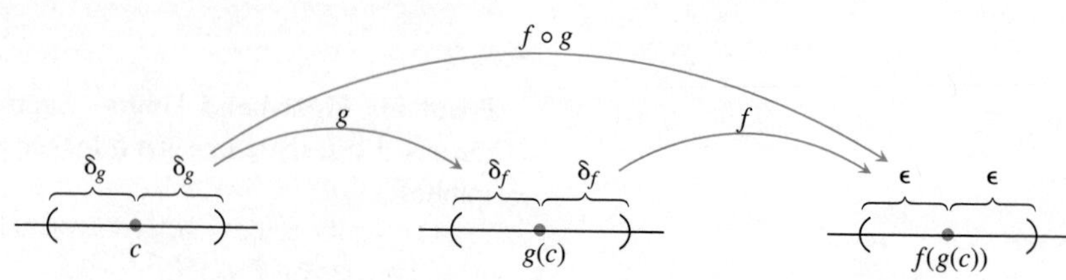

A.18 The diagram for a proof that the composite of two continuous functions is continuous. The continuity of composites holds for any finite number of functions. The only requirement is that each function be continuous where it is applied. In the figure, g is to be continuous at c, and f is to be continuous at $g(c)$.

A.5 Mathematical Induction

Many formulas, like

$$1 + 2 + \cdots + n = \frac{n(n+1)}{2},$$

can be shown to hold for every positive integer n by applying an axiom called the *mathematical induction principle*. A proof that uses this axiom is called a *proof by mathematical induction* or a *proof by induction*.

The steps in proving a formula by induction are

STEP 1: Check that it holds for $n = 1$.

STEP 2: Prove that if it holds for any positive integer $n = k$, then it also holds for $n = k + 1$.

Once these steps are completed (the axiom says), we know that the formula holds for all positive integers n. By step 1 it holds for $n = 1$. By step 2 it holds for $n = 2$, and therefore by step 2 also for $n = 3$, and by step 2 again for $n = 4$, and so on. If the first domino falls, and the kth domino always knocks over the $(k + 1)$st when it falls, all the dominoes fall.

From another point of view, suppose we have a sequence of statements

$$S_1, S_2, \ldots, S_n, \ldots,$$

one for each positive integer. Suppose we can show that assuming any one of the statements to be true implies that the next statement in line is true. Suppose that we can also show that S_1 is true. Then we may conclude that the statements are true from S_1 on.

Example 1 Show that for every positive integer n,

$$1 + 2 + \cdots + n = \frac{n(n+1)}{2}.$$

Solution We accomplish the proof by carrying out the two steps of mathematical induction.

STEP 1: The formula holds for $n = 1$ because

$$1 = \frac{1(1+1)}{2}.$$

STEP 2: If the formula holds for $n = k$, does it also hold for $n = k + 1$? The answer is yes, and here's why: If

$$1 + 2 + \cdots + k = \frac{k(k+1)}{2},$$

then

$$1 + 2 + \cdots + k + (k + 1) = \frac{k(k+1)}{2} + (k + 1) = \frac{k^2 + k + 2k + 2}{2}$$

$$= \frac{(k+1)(k+2)}{2} = \frac{(k+1)((k+1)+1)}{2}.$$

The last expression in this string of equalities is the expression $n(n + 1)/2$ for $n = (k + 1)$.

The mathematical induction principle now guarantees the original formula for all positive integers n.

Notice that all *we* have to do is carry out steps 1 and 2. The mathematical induction principle does the rest.

Example 2 Show that for all positive integers n,

$$\frac{1}{2^1} + \frac{1}{2^2} + \cdots + \frac{1}{2^n} = 1 - \frac{1}{2^n}.$$

Solution We accomplish the proof by carrying out the two steps of mathematical induction.

STEP 1: The formula holds for $n = 1$ because

$$\frac{1}{2^1} = 1 - \frac{1}{2^1}.$$

STEP 2: If

$$\frac{1}{2^1} + \frac{1}{2^2} + \cdots + \frac{1}{2^k} = 1 - \frac{1}{2^k},$$

then

$$\frac{1}{2^1} + \frac{1}{2^2} + \cdots + \frac{1}{2^k} + \frac{1}{2^{k+1}} = 1 - \frac{1}{2^k} + \frac{1}{2^{k+1}} = 1 - \frac{1 \cdot 2}{2^k \cdot 2} + \frac{1}{2^{k+1}}$$

$$= 1 - \frac{2}{2^{k+1}} + \frac{1}{2^{k+1}} = 1 - \frac{1}{2^{k+1}}.$$

Thus, the original formula holds for $n = k + 1$ whenever it holds for $n = k$.

With these two steps verified, the mathematical induction principle now guarantees the formula for every positive integer n.

Other Starting Integers

Instead of starting at $n = 1$, some induction arguments start at another integer. The steps for such an argument are

STEP 1: Check that the formula holds for $n = n_1$ (whatever the appropriate first integer is).

STEP 2: Prove that if the formula holds for any integer $n = k \geq n_1$, then it also holds for $n = k + 1$.

Once these steps are completed, the mathematical induction principle will guarantee the formula for all $n \geq n_1$.

Example 3 Show that $n! > 3^n$ if n is large enough.

Solution How large is large enough? We experiment:

n	1	2	3	4	5	6	7
$n!$	1	2	6	24	120	720	5040
3^n	3	9	27	81	243	729	2187

It looks as if $n! > 3^n$ for $n \geq 7$. To be sure, we apply mathematical induction. We take $n_1 = 7$ in step 1 and try for step 2.

Suppose $k! > 3^k$ for some $k \geq 7$. Then

$$(k+1)! = (k+1)(k!) > (k+1)3^k > 7 \cdot 3^k > 3^{k+1}.$$

Thus, for $k \geq 7$,

$$k! > 3^k \quad \Rightarrow \quad (k+1)! > 3^{k+1}.$$

The mathematical induction principle now guarantees $n! \geq 3^n$ for all $n \geq 7$.

EXERCISES A.5

1. Assuming that the triangle inequality $|a+b| \leq |a| + |b|$ holds for any two numbers a and b, show that

$$|x_1 + x_2 + \cdots + x_n| \leq |x_1| + |x_2| + \cdots + |x_n|$$

for any n numbers.

2. Show that if $r \neq 1$, then

$$1 + r + r^2 + \cdots + r^n = \frac{1 - r^{n+1}}{1 - r}$$

for all positive integers n.

3. Use the Product Rule,

$$\frac{d}{dx}(uv) = u\frac{dv}{dx} + v\frac{du}{dx},$$

and the fact that

$$\frac{d}{dx}(x) = 1$$

to show that

$$\frac{d}{dx}(x^n) = nx^{n-1}$$

for all positive integers n.

4. Suppose that a function $f(x)$ has the property that $f(x_1x_2) = f(x_1) + f(x_2)$ for any two positive numbers x_1 and x_2. Show that

$$f(x_1x_2 \cdots x_n) = f(x_1) + f(x_2) + \cdots + f(x_n)$$

for the product of any n positive numbers $x_1, x_2 \ldots, x_n$.

5. Show that

$$\frac{2}{3^1} + \frac{2}{3^2} + \cdots + \frac{2}{3^n} = 1 - \frac{1}{3^n}$$

for all positive integers n.

6. Show that $n! > n^3$ if n is large enough.

7. Show that $2^n > n^2$ if n is large enough.

8. Show that $2^n \geq 1/8$ for $n \geq -3$.

9. *Sums of squares.* Show that the sum of the squares of the first n positive integers is

$$\frac{n\left(n + \frac{1}{2}\right)(n+1)}{3}.$$

10. *Sums of cubes.* Show that the sum of the cubes of the first n positive integers is $(n(n+1)/2)^2$.

11. Show that the following finite-sum rules hold for every positive integer n.

a) $\displaystyle\sum_{k=1}^{n} (a_k + b_k) = \sum_{k=1}^{n} a_k + \sum_{k=1}^{n} b_k$

b) $\displaystyle\sum_{k=1}^{n} (a_k - b_k) = \sum_{k=1}^{n} a_k - \sum_{k=1}^{n} b_k$

c) $\displaystyle\sum_{k=1}^{n} ca_k = c \cdot \sum_{k=1}^{n} a_k$ (Any number c)

d) $\displaystyle\sum_{k=1}^{n} a_k = n \cdot c$ if a_k has the constant value c

12. *Sums of products of consecutive positive integers.* You are probably already familiar with the sums

$$\sum_{k=1}^{n} k^0 = \sum_{k=1}^{n} 1 = \frac{n}{1} \quad \text{and} \quad \sum_{k=1}^{n} k = \frac{n(n+1)}{2}.$$

Show that

a) $\displaystyle\sum_{k=1}^{n} k(k+1) = \frac{n(n+1)(n+2)}{3}$

b) $\displaystyle\sum_{k=1}^{n} k(k+1)(k+2) = \frac{n(n+1)(n+2)(n+3)}{4}$

c) $\displaystyle\sum_{k=1}^{n} k(k+1)(k+2)(k+3) = \frac{n(n+1)(n+2)(n+3)(n+4)}{5}$

Then show, in general, that if m and n are integers with $m \geq 0$ and $n \geq 1$, then

d) $\displaystyle\sum_{k=1}^{n} \left(\prod_{j=0}^{m} (k+j)\right) = \frac{1}{m+2} \prod_{j=0}^{m+1} (n+j)$

A.6 Complex Numbers

Complex numbers are numbers of the form $a + bi$, where a and b are real numbers and $i = \sqrt{-1}$. The number a is the **real part** of $a + bi$, and b is the **imaginary part.**

Complex numbers $a + bi$ and $c + di$ are equal if and only if $a = c$ and $b = d$. In particular, $a + bi = 0$ if and only if $a = 0$ and $b = 0$ or, equivalently, if and only if $a^2 + b^2 = 0$.

We add and subtract complex numbers by adding and subtracting their real and imaginary parts:

$$(a + bi) + (c + di) = (a + c) + (b + d)i,$$

$$(a + bi) - (c + di) = (a - c) + (b - d)i.$$

We multiply complex numbers the way we multiply other binomials, using the fact that $i^2 = -1$ to simplify the final result:

$$\begin{aligned}(a + bi)(c + di) &= ac + adi + bci + bdi^2 \\ &= ac + adi + bci - bd && (i^2 = -1) \\ &= (ac - bd) + (ad + bc)i. && \left(\begin{matrix}\text{Real and imaginary} \\ \text{parts combined}\end{matrix}\right)\end{aligned}$$

To divide a complex number $c + di$ by a nonzero complex number $a + bi$, multiply the numerator and denominator of the quotient by $a - bi$ (the number $a - bi$ is the **complex conjugate** of $a + bi$):

$$\begin{aligned}\frac{c + di}{a + bi} &= \frac{c + di}{a + bi}\,\frac{a - bi}{a - bi} && \left(\begin{matrix}\text{Multiply numerator and} \\ \text{denominator by the} \\ \text{complex conjugate of} \\ a + bi.\end{matrix}\right) \\ &= \frac{ac - bci + adi - bdi^2}{a^2 - abi + abi - b^2i^2} \\ &= \frac{(ac + bd) + (ad - bc)i}{a^2 + b^2} \\ &= \frac{ac + bd}{a^2 + b^2} + \frac{ad - bc}{a^2 + b^2}\,i.\end{aligned}$$

Example

a) $(2 + 3i) + (6 - 2i) = (2 + 6) + (3 - 2)i = 8 + i$

b) $(2 + 3i) - (6 - 2i) = (2 - 6) + (3 - (-2))i = -4 + 5i$

c) $(2 + 3i)(6 - 2i) = (2)(6) + (2)(-2i) + (3i)(6) + (3i)(-2i)$

$$= 12 - 4i + 18i - 6i^2 = 12 + 14i + 6 = 18 + 14i$$

d) $$\begin{aligned}\frac{2 + 3i}{6 - 2i} &= \frac{2 + 3i}{6 - 2i}\,\frac{6 + 2i}{6 + 2i} \\ &= \frac{12 + 4i + 18i + 6i^2}{36 + 12i - 12i - 4i^2} \\ &= \frac{6 + 22i}{40} = \frac{3}{20} + \frac{11}{20}\,i\end{aligned}$$

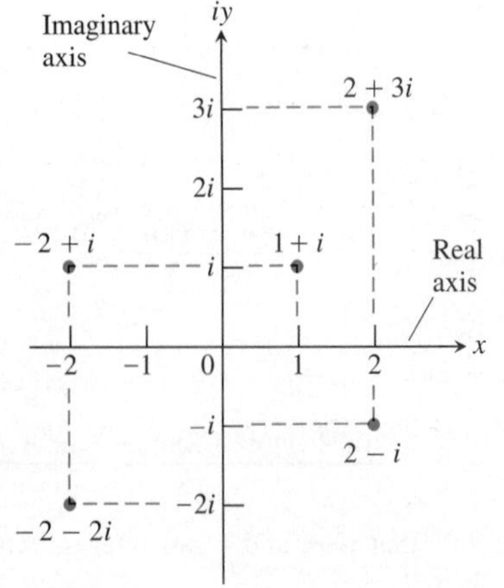

A.19 The complex plane.

We plot complex numbers as points in a plane by identifying the number $x + iy$ with the point (x, y) in the Cartesian plane. We then call the x-axis the **real**

axis, the y-axis the **imaginary axis** (sometimes, ***iy*-axis**), and the relabeled plane the **complex plane** (Fig. A.19).

A.7 Cauchy's Mean Value Theorem and the Stronger Form of L'Hôpital's Rule

In this appendix, we prove the stronger form of l'Hôpital's rule (Theorem 6, Section 3.6). When the limit of a quotient of differentiable functions leads to an indeterminate form, l'Hôpital's rule is the one to apply.

THEOREM 6

L'Hôpital's Rule (Stronger Form)

Suppose that

$$f(x_0) = g(x_0) = 0$$

and that the functions f and g are both differentiable on an open interval (a, b) that contains the point x_0. Suppose also that $g' \neq 0$ at every point in (a, b) except possibly x_0. Then

$$\lim_{x \to x_0} \frac{f(x)}{g(x)} = \lim_{x \to x_0} \frac{f'(x)}{g'(x)}, \tag{1}$$

provided the limit on the right exists.

The proof of the stronger form of l'Hôpital's rule is based on Cauchy's Mean Value Theorem, a mean value theorem that involves two functions instead of one. We prove Cauchy's theorem first and then show how it leads to l'Hôpital's rule.

THEOREM

Cauchy's Mean Value Theorem

Suppose that functions f and g are continuous on $[a, b]$ and differentiable throughout (a, b) and suppose also that $g' \neq 0$ throughout (a, b). Then there exists a number c in (a, b) at which

$$\frac{f'(c)}{g'(c)} = \frac{f(b) - f(a)}{g(b) - g(a)}. \tag{2}$$

Notice that the ordinary Mean Value Theorem (Theorem 3, Section 3.2) is the case $g(x) = x$.

Proof of Cauchy's Mean Value Theorem We apply the Mean Value Theorem of Section 3.2 twice. First we use it to show that $g(a) \neq g(b)$. For if $g(b)$ did equal $g(a)$, then the Mean Value Theorem would give

$$g'(c) = \frac{g(b) - g(a)}{b - a} = 0$$

for some c between a and b. This cannot happen because $g'(x) \neq 0$ in (a, b).

We next apply the Mean Value Theorem to the function

$$F(x) = f(x) - f(a) - \frac{f(b) - f(a)}{g(b) - g(a)} [g(x) - g(a)]. \tag{3}$$

This function is continuous and differentiable where f and g are, and $F(b) = F(a) = 0$. Therefore there is a number c between a and b for which $F'(c) = 0$. In terms of f and g this says

$$F'(c) = f'(c) - \frac{f(b) - f(a)}{g(b) - g(a)} [g'(c)] = 0, \tag{4}$$

or

$$\frac{f'(c)}{g'(c)} = \frac{f(b) - f(a)}{g(b) - g(a)},$$

which is Eq. (2).

Proof of the Stronger Form of L'Hôpital's Rule We first establish Eq. (1) for the case $x \to x_0^+$. The method needs almost no change to apply to $x \to x_0^-$, and the combination of these two cases establishes the result.

Suppose that x lies to the right of x_0. Then $g'(x) \neq 0$ and we can apply Cauchy's Mean Value Theorem to the closed interval from x_0 to x. This produces a number c between x_0 and x such that

$$\frac{f'(c)}{g'(c)} = \frac{f(x) - f(x_0)}{g(x) - g(x_0)}. \tag{5}$$

But $f(x_0) = g(x_0) = 0$, so

$$\frac{f'(c)}{g'(c)} = \frac{f(x)}{g(x)}. \tag{6}$$

As x approaches x_0, c approaches x_0 because it lies between x and x_0. Therefore,

$$\lim_{x \to x_0^+} \frac{f(x)}{g(x)} = \lim_{c \to x_0^+} \frac{f'(c)}{g'(c)} = \lim_{x \to x_0^+} \frac{f'(x)}{g'(x)}.$$

This establishes l'Hôpital's rule for the case where x approaches x_0 from above. The case where x approaches x_0 from below is proved by applying Cauchy's Mean Value Theorem to the closed interval $[x, x_0]$, $x < x_0$.

EXERCISES A.7

Although the importance of Cauchy's Mean Value Theorem lies elsewhere, we can sometimes satisfy our curiosity about the identity of the number c in Eq. (2), as in Exercises 1–4.

In Exercises 1–4, find all values of c that satisfy Eq. (2) in the conclusion of Cauchy's Mean Value Theorem.

1. $f(x) = x$, $g(x) = x^2$, $[a, b] = [-2, 0]$

2. $f(x) = x$, $g(x) = x^2$, $[a, b]$ arbitrary

3. $f(x) = \frac{x^3}{3} - 4x$, $g(x) = x^2$, $[a, b] = [0, 3]$

4. $f(x) = \sin x$, $g(x) = \cos x$, $[a, b] = [0, \pi/2]$

A.8 Limits That Arise Frequently

This appendix verifies the limits in Table 8.1 of Section 8.1.

1. $\lim_{n\to\infty} \frac{\ln n}{n} = 0$
2. $\lim_{n\to\infty} \sqrt[n]{n} = 1$
3. $\lim_{n\to\infty} x^{1/n} = 1 \quad (x > 0)$
4. $\lim_{n\to\infty} x^n = 0 \quad (|x| < 1)$
5. $\lim_{n\to\infty} \left(1 + \frac{x}{n}\right)^n = e^x \quad$ (Any x)
6. $\lim_{n\to\infty} \frac{x^n}{n!} = 0 \quad$ (Any x)

In Formulas 3–6, x remains fixed while $n \to \infty$.

1. $\lim_{n\to\infty} \frac{\ln n}{n} = 0$ We proved this in Section 8.1, Example 9.

2. $\lim_{n\to\infty} \sqrt[n]{n} = 1$ Let $a_n = n^{1/n}$. Then

$$\ln a_n = \ln n^{1/n} = \frac{1}{n} \ln n \to 0. \tag{1}$$

Applying Theorem 3, Section 8.1, with $f(x) = e^x$ gives

$$n^{1/n} = a_n = e^{\ln a_n} = f(\ln a_n) \to f(0) = e^0 = 1. \tag{2}$$

3. If $x > 0$, $\lim_{n\to\infty} x^{1/n} = 1$ Let $a_n = x^{1/n}$. Then

$$\ln a_n = \ln x^{1/n} = \frac{1}{n} \ln x \to 0 \tag{3}$$

because x remains fixed as $n \to \infty$. Applying Theorem 3, Section 8.1, with $f(x) = e^x$ gives

$$x^{1/n} = a_n = e^{\ln a_n} \to e^0 = 1. \tag{4}$$

4. If $|x| < 1$, $\lim_{n\to\infty} x^n = 0$ We need to show that to each $\varepsilon > 0$ there corresponds an integer N so large that $|x^n| < \varepsilon$ for all n greater than N. Since $\varepsilon^{1/n} \to 1$, while $|x| < 1$, there exists an integer N for which

$$\varepsilon^{1/N} > |x|. \tag{5}$$

In other words,

$$|x^N| = |x|^N < \varepsilon. \tag{6}$$

This is the integer we seek because, if $|x| < 1$, then

$$|x^n| < |x^N| \qquad \text{for all } n > N. \tag{7}$$

Combining (6) and (7) produces

$$|x^n| < \varepsilon \qquad \text{for all } n > N, \tag{8}$$

and we're done.

5. For any number x, $\lim_{n\to\infty} \left(1 + \frac{x}{n}\right)^n = e^x$ Let

$$a_n = \left(1 + \frac{x}{n}\right)^n.$$

Then

$$\ln a_n = \ln\left(1 + \frac{x}{n}\right)^n = n \ln\left(1 + \frac{x}{n}\right) \to x,$$

as we can see by the following application of l'Hôpital's rule, in which we differentiate with respect to n:

$$\lim_{n\to\infty} n \ln\left(1 + \frac{x}{n}\right) = \lim_{n\to\infty} \frac{\ln(1 + x/n)}{1/n}$$

$$= \lim_{n\to\infty} \frac{\left(\frac{1}{1 + x/n}\right) \cdot \left(-\frac{x}{n^2}\right)}{-1/n^2} = \lim_{n\to\infty} \frac{x}{1 + x/n} = x.$$

Apply Theorem 3, Section 8.1, with $f(x) = e^x$ to conclude that

$$\left(1 + \frac{x}{n}\right)^n = a_n = e^{\ln a_n} \rightarrow e^x.$$

6. For any number x, $\lim_{n\to\infty} \frac{x^n}{n!} = 0$ Since

$$-\frac{|x|^n}{n!} \leq \frac{x^n}{n!} \leq \frac{|x|^n}{n!},$$

all we need to show is that $|x|^n/n! \rightarrow 0$. We can then apply the Sandwich Theorem for Sequences (Section 8.1, Theorem 2) to conclude that $x^n/n! \rightarrow 0$.

The first step in showing that $|x|^n/n! \rightarrow 0$ is to choose an integer $M > |x|$, so that

$$\frac{|x|}{M} < 1 \quad \text{and} \quad \left(\frac{|x|}{M}\right)^n \rightarrow 0.$$

We then restrict our attention to values of $n > M$. For these values of n, we can write

$$\frac{|x|^n}{n!} = \frac{|x|^n}{1 \cdot 2 \cdot \cdots \cdot M \cdot \underbrace{(M+1)(M+2) \cdot \cdots \cdot n}_{(n-M) \text{ factors}}}$$

$$\leq \frac{|x|^n}{M!M^{n-M}} = \frac{|x|^n M^M}{M!M^n} = \frac{M^M}{M!}\left(\frac{|x|}{M}\right)^n.$$

Thus,

$$0 \leq \frac{|x|^n}{n!} \leq \frac{M^M}{M!}\left(\frac{|x|}{M}\right)^n.$$

Now, the constant $M^M/M!$ does not change as n increases. Thus the Sandwich Theorem tells us that

$$\frac{|x|^n}{n!} \rightarrow 0 \quad \text{because} \quad \left(\frac{|x|}{M}\right)^n \rightarrow 0.$$

A.9 The Distributive Law for Vector Cross Products

In this appendix we prove the distributive law

$$\mathbf{A} \times (\mathbf{B} + \mathbf{C}) = \mathbf{A} \times \mathbf{B} + \mathbf{A} \times \mathbf{C} \tag{1}$$

from Eq. (6) in Section 10.4.

Proof To derive Eq. (1), we construct $\mathbf{A} \times \mathbf{B}$ a new way. We draw $\mathbf{A}$ and $\mathbf{B}$ from the common point O and construct a plane M perpendicular to $\mathbf{A}$ at O (Fig. A.20). We then project $\mathbf{B}$ orthogonally onto M, yielding a vector $\mathbf{B}'$ with length $|\mathbf{B}|\sin\theta$. We rotate $\mathbf{B}'$ 90° about $\mathbf{A}$ in the positive sense to produce a vector $\mathbf{B}''$. Finally, we multiply $\mathbf{B}''$ by the length of $\mathbf{A}$. The resulting vector $|\mathbf{A}|\mathbf{B}''$ is equal to $\mathbf{A} \times \mathbf{B}$ since $\mathbf{B}''$ has the same direction as $\mathbf{A} \times \mathbf{B}$ by its construction (Fig. A.20) and

$$|\mathbf{A}||\mathbf{B}''| = |\mathbf{A}||\mathbf{B}'| = |\mathbf{A}||\mathbf{B}|\sin\theta = |\mathbf{A} \times \mathbf{B}|.$$

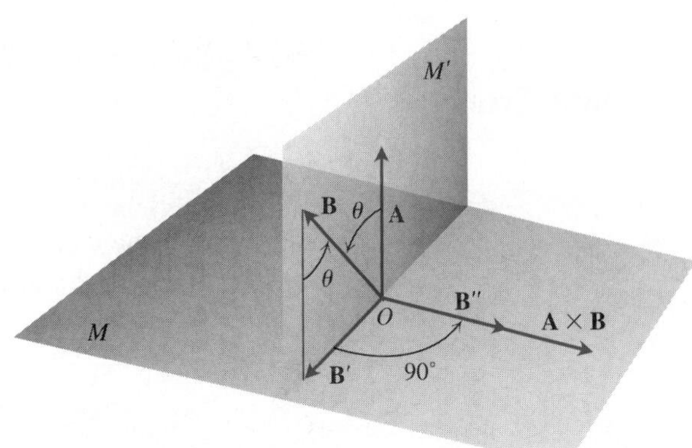

A.20 As explained in the text, $\mathbf{A} \times \mathbf{B} = |\mathbf{A}|\mathbf{B}''$.

Now each of these three operations, namely,

1. projection onto M,
2. rotation about $\mathbf{A}$ through 90°,
3. multiplication by the scalar $|\mathbf{A}|$,

when applied to a triangle whose plane is not parallel to $\mathbf{A}$, will produce another triangle. If we start with the triangle whose sides are $\mathbf{B}$, $\mathbf{C}$, and $\mathbf{B} + \mathbf{C}$ (Fig. A.21) and apply these three steps, we successively obtain

1. a triangle whose sides are $\mathbf{B}'$, $\mathbf{C}'$, and $(\mathbf{B} + \mathbf{C})'$ satisfying the vector equation

$$\mathbf{B}' + \mathbf{C}' = (\mathbf{B} + \mathbf{C})';$$

2. a triangle whose sides are $\mathbf{B}''$, $\mathbf{C}''$, and $(\mathbf{B} + \mathbf{C})''$ satisfying the vector equation

$$\mathbf{B}'' + \mathbf{C}'' = (\mathbf{B} + \mathbf{C})''$$

(the double prime on each vector has the same meaning as in Fig. A.20); and, finally,

3. a triangle whose sides are $|\mathbf{A}|\mathbf{B}''$, $|\mathbf{A}|\mathbf{C}''$, and $|\mathbf{A}|(\mathbf{B} + \mathbf{C})''$ satisfying the vector equation

$$|\mathbf{A}|\mathbf{B}'' + |\mathbf{A}|\mathbf{C}'' = |\mathbf{A}|(\mathbf{B} + \mathbf{C})''. \tag{2}$$

When we use the equations $|\mathbf{A}|\mathbf{B}'' = \mathbf{A} \times \mathbf{B}$, $|\mathbf{A}|\mathbf{C}'' = \mathbf{A} \times \mathbf{C}$, and $|\mathbf{A}|(\mathbf{B} + \mathbf{C})'' = \mathbf{A} \times (\mathbf{B} + \mathbf{C})$, which result from our discussion above, Eq. (2) becomes

$$\mathbf{A} \times \mathbf{B} + \mathbf{A} \times \mathbf{C} = \mathbf{A} \times (\mathbf{B} + \mathbf{C}),$$

which is the law we wanted to establish. ∎

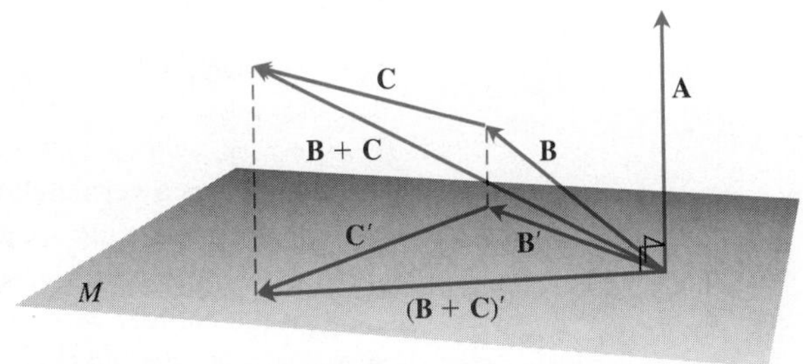

A.21 The vectors, $\mathbf{B}$, $\mathbf{C}$, $\mathbf{B} + \mathbf{C}$, and their projections onto a plane perpendicular to $\mathbf{A}$.

A.10 Determinants and Cramer's Rule

A rectangular array of numbers like

$$A = \begin{bmatrix} 2 & 1 & 3 \\ 1 & 0 & -2 \end{bmatrix}$$

is called a **matrix**. We call A a 2 by 3 matrix because it has two rows and three columns. An m by n matrix has m rows and n columns, and the **entry** or **element** (number) in the ith row and jth column is often denoted by a_{ij}:

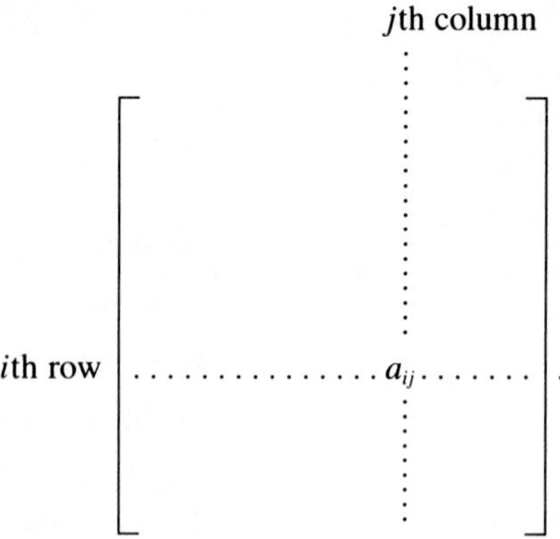

The matrix

$$A = \begin{bmatrix} 2 & 1 & 3 \\ 1 & 0 & -2 \end{bmatrix}$$

has

$$a_{11} = 2, \quad a_{12} = 1, \quad a_{13} = 3,$$
$$a_{21} = 1, \quad a_{22} = 0, \quad a_{23} = -2.$$

A matrix with the same number of rows as columns is a **square matrix.** It is a **matrix of order** $\boldsymbol{n}$ if the number of rows and columns is n.

With each square matrix A we associate a number $\det A$ or $|a_{ij}|$, called the **determinant** of A, calculated from the entries of A in the following way. (The vertical bars in the notation $|a_{ij}|$ do not mean absolute value.) For $n = 1$ and $n = 2$, we define

$$\det [a] = a, \tag{1}$$

$$\det \begin{bmatrix} a_{11} & a_{12} \\ a_{21} & a_{22} \end{bmatrix} = a_{11}a_{22} - a_{21}a_{12}. \tag{2}$$

For a matrix of order 3, we define

$$\det A = \det \begin{bmatrix} a_{11} & a_{12} & a_{13} \\ a_{21} & a_{22} & a_{23} \\ a_{31} & a_{32} & a_{33} \end{bmatrix} = \begin{array}{l} \text{Sum of all signed products} \\ \text{of the form } \pm\, a_{1i}a_{2j}a_{3k}, \end{array} \tag{3}$$

where i, j, k is a permutation of 1, 2, 3 in some order. There are $3! = 6$ such permutations, so there are six terms in the sum. Half of these have plus signs and the other half have minus signs, according to the index of the permutation, where the index is the number we define next. The sign is positive when the index is even and negative when the index is odd.

DEFINITION

Index of a Permutation

Given any permutation of the numbers $1, 2, 3, \ldots, n$, denote the permutation by $i_1, i_2, i_3, \ldots, i_n$. In this arrangement, some of the numbers following i_1 may be less than i_1, and the number of these is called the **number of inversions** in the arrangement pertaining to i_1. Likewise, there is a number of inversions pertaining to each of the other i's; it is the number of indices that come after that particular i in the arrangement and are less than it. The **index** of the permutation is the sum of all of the numbers of inversions pertaining to the separate indices.

Example 1 For $n = 5$, the permutation

$$5 \quad 3 \quad 1 \quad 2 \quad 4$$

has

4 inversions pertaining to the first element, 5,

2 inversions pertaining to the second element, 3,

and no further inversions, so the index is $4 + 2 = 6$.

The following table shows the permutations of 1, 2, 3, the index of each permutation, and the signed product in the determinant of Eq. (3).

Permutation	Index	Signed product
1 2 3	0	$+a_{11}a_{22}a_{33}$
1 3 2	1	$-a_{11}a_{23}a_{32}$
2 1 3	1	$-a_{12}a_{21}a_{33}$
2 3 1	2	$+a_{12}a_{23}a_{31}$
3 1 2	2	$+a_{13}a_{21}a_{32}$
3 2 1	3	$-a_{13}a_{22}a_{31}$

(4)

The sum of the six signed products is

$$a_{11}(a_{22}a_{33} - a_{23}a_{32}) - a_{12}(a_{21}a_{33} - a_{23}a_{31}) + a_{13}(a_{21}a_{32} - a_{22}a_{31})$$

$$= a_{11}\begin{vmatrix} a_{22} & a_{23} \\ a_{32} & a_{33} \end{vmatrix} - a_{12}\begin{vmatrix} a_{21} & a_{23} \\ a_{31} & a_{33} \end{vmatrix} + a_{13}\begin{vmatrix} a_{21} & a_{22} \\ a_{31} & a_{32} \end{vmatrix} = \begin{vmatrix} a_{11} & a_{12} & a_{13} \\ a_{21} & a_{22} & a_{23} \\ a_{31} & a_{32} & a_{33} \end{vmatrix}. \quad (5)$$

The formula

$$\begin{vmatrix} a_{11} & a_{12} & a_{13} \\ a_{21} & a_{22} & a_{23} \\ a_{31} & a_{32} & a_{33} \end{vmatrix} = a_{11}\begin{vmatrix} a_{22} & a_{23} \\ a_{32} & a_{33} \end{vmatrix} - a_{12}\begin{vmatrix} a_{21} & a_{23} \\ a_{31} & a_{33} \end{vmatrix} + a_{13}\begin{vmatrix} a_{21} & a_{22} \\ a_{31} & a_{32} \end{vmatrix} \quad (6)$$

reduces the calculation of a 3 by 3 determinant to the calculation of three 2 by 2 determinants.

Many people prefer to remember the following scheme for calculating the six signed products in the determinant of a 3 by 3 matrix:

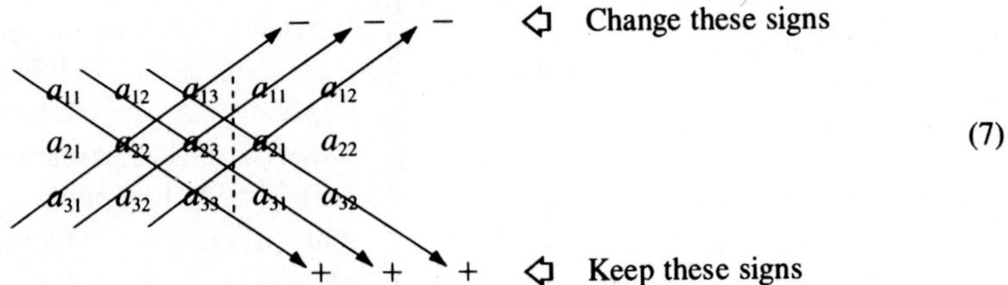

(7)

Minors and Cofactors

The second order determinants on the right-hand side of Eq. (6) are called the **minors** (short for "minor determinants") of the entries they multiply. Thus,

$$\begin{vmatrix} a_{22} & a_{23} \\ a_{32} & a_{33} \end{vmatrix} \text{ is the minor of } a_{11},$$

$$\begin{vmatrix} a_{21} & a_{23} \\ a_{31} & a_{33} \end{vmatrix} \text{ is the minor of } a_{12},$$

and so on. The minor of the element a_{ij} in a matrix A is the determinant of the matrix that remains after we delete the row and column containing a_{ij}:

$$\begin{vmatrix} a_{11} & a_{12} & a_{13} \\ a_{21} & a_{22} & a_{23} \\ a_{31} & a_{32} & a_{33} \end{vmatrix}. \quad \text{The minor of } a_{22} \text{ is } \begin{vmatrix} a_{11} & a_{13} \\ a_{31} & a_{33} \end{vmatrix}.$$

$$\begin{vmatrix} a_{11} & a_{12} & a_{13} \\ a_{21} & a_{22} & a_{23} \\ a_{31} & a_{32} & a_{33} \end{vmatrix}. \quad \text{The minor of } a_{23} \text{ is } \begin{vmatrix} a_{11} & a_{12} \\ a_{31} & a_{32} \end{vmatrix}.$$

The **cofactor** A_{ij} of a_{ij} is $(-1)^{i+j}$ times the minor of a_{ij}. Thus,

$$A_{22} = (-1)^{2+2}\begin{vmatrix} a_{11} & a_{13} \\ a_{31} & a_{33} \end{vmatrix} = \begin{vmatrix} a_{11} & a_{13} \\ a_{31} & a_{33} \end{vmatrix},$$

$$A_{23} = (-1)^{2+3}\begin{vmatrix} a_{11} & a_{12} \\ a_{31} & a_{32} \end{vmatrix} = -\begin{vmatrix} a_{11} & a_{12} \\ a_{31} & a_{32} \end{vmatrix}.$$

The effect of the factor $(-1)^{i+j}$ is to change the sign of the minor when the sum $i + j$ is odd. There is a checkerboard pattern for remembering these sign changes.

$$\begin{matrix} + & - & + \\ - & + & - \\ + & - & + \end{matrix}$$

In the upper left corner, $i = 1, j = 1$ and $(-1)^{1+1} = +1$. In going from any cell to an adjacent cell in the same row or column, we change i by 1 or j by 1, but not both, so we change the exponent from even to odd or from odd to even, which changes the sign from $+$ to $-$ or from $-$ to $+$.

When we rewrite Eq. (6) in terms of cofactors we get

$$\det A = a_{11}A_{11} + a_{12}A_{12} + a_{13}A_{13}. \quad (8)$$

Example 2 Find the determinant of the matrix

$$A = \begin{bmatrix} 2 & 1 & 3 \\ 3 & -1 & -2 \\ 2 & 3 & 1 \end{bmatrix}.$$

Solution 1 The cofactors are

$$A_{11} = (-1)^{1+1}\begin{vmatrix} -1 & -2 \\ 3 & 1 \end{vmatrix}, \qquad A_{12} = (-1)^{1+2}\begin{vmatrix} 3 & -2 \\ 2 & 1 \end{vmatrix},$$

$$A_{13} = (-1)^{1+3}\begin{vmatrix} 3 & -1 \\ 2 & 3 \end{vmatrix}.$$

To find $\det A$, we multiply each element of the first row of A by its cofactor and add:

$$\det A = 2\begin{vmatrix} -1 & -2 \\ 3 & 1 \end{vmatrix} + (-1)\begin{vmatrix} 3 & -2 \\ 2 & 1 \end{vmatrix} + 3\begin{vmatrix} 3 & -1 \\ 2 & 3 \end{vmatrix}$$

$$= 2(-1+6) - 1(3+4) + 3(9+2) = 10 - 7 + 33 = 36.$$

Solution 2 From (7) we find

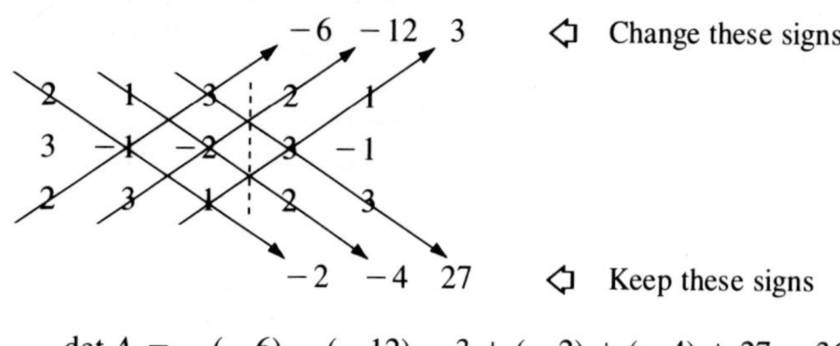

$$\det A = -(-6) - (-12) - 3 + (-2) + (-4) + 27 = 36$$

Expanding by Columns or by Other Rows

The determinant of a square matrix can be calculated from the cofactors of any row or any column.

If we were to expand the determinant in Example 2 by cofactors according to elements of its third column, say, we would get

$$+3\begin{vmatrix} 3 & -1 \\ 2 & 3 \end{vmatrix} - (-2)\begin{vmatrix} 2 & 1 \\ 2 & 3 \end{vmatrix} + 1\begin{vmatrix} 2 & 1 \\ 3 & -1 \end{vmatrix}$$

$$= 3(9+2) + 2(6-2) + 1(-2-3) = 33 + 8 - 5 = 36.$$

Useful Facts About Determinants

FACT 1: If two rows (or columns) of a matrix are identical, the determinant is zero.

FACT 2: Interchanging two rows (or columns) of a matrix changes the sign of its determinant.

FACT 3: The determinant of a matrix is the sum of the products of the elements of the ith row (or column) by their cofactors, for any i.

FACT 4: The determinant of the transpose of a matrix is the same as the determinant of the original matrix. (The "**transpose**" of a matrix is obtained by writing the rows as columns.)

FACT 5: Multiplying each element of some row (or column) of a matrix by a constant c multiplies the determinant by c.

FACT 6: If all elements of a matrix above the main diagonal (or all below it) are zero, the determinant of the matrix is the product of the elements on the main diagonal. (The **main diagonal** is the diagonal from upper left to lower right.)

Example 3

$$\begin{vmatrix} 3 & 4 & 7 \\ 0 & -2 & 5 \\ 0 & 0 & 5 \end{vmatrix} = (3)(-2)(5) = -30.$$

FACT 7: If the elements of any row of a matrix are multiplied by the cofactors of the corresponding elements of a different row and these products are summed, the sum is zero.

Example 4 If A_{11}, A_{12}, A_{13} are the cofactors of the elements of the first row of $A = (a_{ij})$, then the sums

$$a_{21}A_{11} + a_{22}A_{12} + a_{23}A_{13}$$

(elements of second row times cofactors of elements of first row) and

$$a_{31}A_{11} + a_{32}A_{12} + a_{33}A_{13}$$

are both zero.

FACT 8: If the elements of any column of a matrix are multiplied by the cofactors of the corresponding elements of a different column and these products are summed, the sum is zero.

FACT 9: If each element of a row of a matrix is multiplied by a constant c and the results added to a different row, the determinant is not changed. A similar result holds for columns.

Example 5 If we start with

$$A = \begin{bmatrix} 2 & 1 & 3 \\ 3 & -1 & -2 \\ 2 & 3 & 1 \end{bmatrix}$$

and add -2 times row 1 to row 2 (subtract 2 times row 1 from row 2) we get

$$B = \begin{bmatrix} 2 & 1 & 3 \\ -1 & -3 & -8 \\ 2 & 3 & 1 \end{bmatrix}.$$

Since det $A = 36$ (Example 2), we should find that det $B = 36$ as well. Indeed we do, as the following calculation shows:

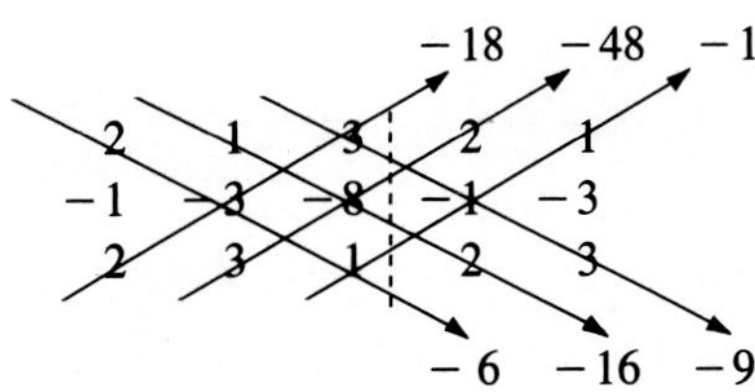

$$\det B = -(-18) - (-48) - (-1) + (-6) + (-16) + (-9)$$
$$= 18 + 48 + 1 - 6 - 16 - 9 = 67 - 31 = 36$$

Example 6 Evaluate the fourth order determinant

$$D = \begin{vmatrix} 1 & -2 & 3 & 1 \\ 2 & 1 & 0 & 2 \\ -1 & 2 & 1 & -2 \\ 0 & 1 & 2 & 1 \end{vmatrix}.$$

Solution We subtract 2 times row 1 from row 2 and add row 1 to row 3 to get

$$D = \begin{vmatrix} 1 & -2 & 3 & 1 \\ 0 & 5 & -6 & 0 \\ 0 & 0 & 4 & -1 \\ 0 & 1 & 2 & 1 \end{vmatrix}.$$

We then multiply the elements of the first column by their cofactors to get

$$D = \begin{vmatrix} 5 & -6 & 0 \\ 0 & 4 & -1 \\ 1 & 2 & 1 \end{vmatrix} = 5(4 + 2) - (-6)(0 + 1) + 0 = 36.$$

Cramer's Rule

If the determinant

$$D = \det A = \begin{vmatrix} a_{11} & a_{12} \\ a_{21} & a_{22} \end{vmatrix}$$

of the coefficient matrix of the system

$$\begin{aligned} a_{11}x + a_{12}y &= b_1, \\ a_{21}x + a_{22}y &= b_2 \end{aligned} \tag{9}$$

of linear equations is 0, the system has either infinitely many solutions or no solution at all. The system

$$\begin{aligned} x + y &= 0, \\ 2x + 2y &= 0 \end{aligned}$$

whose determinant is

$$D = \begin{vmatrix} 1 & 1 \\ 2 & 2 \end{vmatrix} = 2 - 2 = 0$$

has infinitely many solutions. We can find an x to match any given y. The system

$$x + y = 0,$$
$$2x + 2y = 2$$

has no solution. If $x + y = 0$, then $2x + 2y = 2(x + y)$ cannot be 2.

If $D \neq 0$, the system (9) has a unique solution, and Cramer's rule states that it may be found from the formulas

$$x = \frac{\begin{vmatrix} b_1 & a_{12} \\ b_2 & a_{22} \end{vmatrix}}{D}, \qquad y = \frac{\begin{vmatrix} a_{11} & b_1 \\ a_{21} & b_2 \end{vmatrix}}{D}. \tag{10}$$

The numerator in the formula for x comes from replacing the first column in A (the x-column) by the column of constants b_1 and b_2 (the b-column). Replacing the y-column by the b-column gives the numerator of the y-solution.

Example 7 Solve the system

$$3x - y = 9,$$
$$x + 2y = -4.$$

Solution We use Eqs. (10). The determinant of the coefficient matrix is

$$D = \begin{vmatrix} 3 & -1 \\ 1 & 2 \end{vmatrix} = 6 + 1 = 7.$$

Hence,

$$x = \frac{\begin{vmatrix} 9 & -1 \\ -4 & 2 \end{vmatrix}}{D} = \frac{18 - 4}{7} = \frac{14}{7} = 2,$$

$$y = \frac{\begin{vmatrix} 3 & 9 \\ 1 & -4 \end{vmatrix}}{D} = \frac{-12 - 9}{7} = \frac{-21}{7} = -3.$$

Systems of three equations in three unknowns work the same way. If the determinant

$$D = \det A = \begin{vmatrix} a_{11} & a_{12} & a_{13} \\ a_{21} & a_{22} & a_{23} \\ a_{31} & a_{32} & a_{33} \end{vmatrix}$$

of the system

$$\begin{aligned} a_{11}x + a_{12}y + a_{13}z &= b_1, \\ a_{21}x + a_{22}y + a_{23}z &= b_2, \\ a_{31}x + a_{32}y + a_{33}z &= b_3 \end{aligned} \tag{11}$$

is zero, the system has either infinitely many solutions or no solution at all. If $D \neq 0$, the system has a unique solution, given by Cramer's rule:

$$x = \frac{1}{D}\begin{vmatrix} b_1 & a_{12} & a_{13} \\ b_2 & a_{22} & a_{23} \\ b_3 & a_{32} & a_{33} \end{vmatrix}, \qquad y = \frac{1}{D}\begin{vmatrix} a_{11} & b_1 & a_{13} \\ a_{21} & b_2 & a_{23} \\ a_{31} & b_3 & a_{33} \end{vmatrix}, \qquad z = \frac{1}{D}\begin{vmatrix} a_{11} & a_{12} & b_1 \\ a_{21} & a_{22} & b_2 \\ a_{31} & a_{32} & b_3 \end{vmatrix}.$$

The pattern continues in higher dimensions.

EXERCISES A.10

Evaluate the following determinants.

1. $\begin{vmatrix} 2 & 3 & 1 \\ 4 & 5 & 2 \\ 1 & 2 & 3 \end{vmatrix}$

2. $\begin{vmatrix} 2 & -1 & -2 \\ -1 & 2 & 1 \\ 3 & 0 & -3 \end{vmatrix}$

3. $\begin{vmatrix} 1 & 2 & 3 & 4 \\ 0 & 1 & 2 & 3 \\ 0 & 0 & 2 & 1 \\ 0 & 0 & 3 & 2 \end{vmatrix}$

4. $\begin{vmatrix} 1 & -1 & 2 & 3 \\ 2 & 1 & 2 & 6 \\ 1 & 0 & 2 & 3 \\ -2 & 2 & 0 & -5 \end{vmatrix}$

Evaluate the following determinants by expanding according to the cofactors of (a) the third row and (b) the second column.

5. $\begin{vmatrix} 2 & -1 & 2 \\ 1 & 0 & 3 \\ 0 & 2 & 1 \end{vmatrix}$

6. $\begin{vmatrix} 1 & 0 & -1 \\ 0 & 2 & -2 \\ 2 & 0 & 1 \end{vmatrix}$

7. $\begin{vmatrix} 1 & 1 & 0 & 0 \\ 0 & 0 & -2 & 1 \\ 0 & -1 & 0 & 7 \\ 3 & 0 & 2 & 1 \end{vmatrix}$

8. $\begin{vmatrix} 0 & 1 & 0 & 0 \\ 0 & 1 & 1 & 0 \\ 1 & 1 & 1 & 1 \\ 1 & 1 & 0 & 0 \end{vmatrix}$

Solve the following systems of equations by Cramer's rule.

9. $\begin{aligned} x + 8y &= 4 \\ 3x - y &= -13 \end{aligned}$

10. $\begin{aligned} 2x + 3y &= 5 \\ 3x - y &= 2 \end{aligned}$

11. $\begin{aligned} 4x - 3y &= 6 \\ 3x - 2y &= 5 \end{aligned}$

12. $\begin{aligned} x + y + z &= 2 \\ 2x - y + z &= 0 \\ x + 2y - z &= 4 \end{aligned}$

13. $\begin{aligned} 2x + y - z &= 2 \\ x - y + z &= 7 \\ 2x + 2y + z &= 4 \end{aligned}$

14. $\begin{aligned} 2x - 4y \quad &= 6 \\ x + y + z &= 1 \\ 5y + 7z &= 10 \end{aligned}$

15. $\begin{aligned} x \quad - z &= 3 \\ 2y - 2z &= 2 \\ 2x \quad + z &= 3 \end{aligned}$

16. $\begin{aligned} x_1 + x_2 - x_3 + x_4 &= 2 \\ x_1 - x_2 + x_3 + x_4 &= -1 \\ x_1 + x_2 + x_3 - x_4 &= 2 \\ x_1 \quad + x_3 + x_4 &= -1 \end{aligned}$

17. Find values of h and k for which the system

$$2x + hy = 8,$$
$$x + 3y = k$$

has (a) infinitely many solutions, (b) no solution at all.

18. For what value of x will

$$\begin{vmatrix} x & x & 1 \\ 2 & 0 & 5 \\ 6 & 7 & 1 \end{vmatrix} = 0?$$

19. Suppose u, v, and w are twice-differentiable functions of x that satisfy the relation $au + bv + cw = 0$, where a, b, and c are constants, not all zero. Show that

$$\begin{vmatrix} u & v & w \\ u' & v' & w' \\ u'' & v'' & w'' \end{vmatrix} = 0.$$

20. *Partial fractions.* Expanding the quotient

$$\frac{ax + b}{(x - r_1)(x - r_2)}$$

by partial fractions calls for finding the values of C and D that make the equation

$$\frac{ax + b}{(x - r_1)(x - r_2)} = \frac{C}{x - r_1} + \frac{D}{x - r_2}$$

hold for all x.

a) Find a system of linear equations that determines C and D.

b) Under what circumstances does the system of equations in part (a) have a unique solution? That is, when is the determinant of the coefficient matrix of the system different from zero?

A.11 Euler's Theorem and the Increment Theorem

This appendix derives Euler's Mixed Derivative Theorem (Theorem 2, Section 12.3) and the Increment Theorem for Functions of Two Variables (Theorem 3, Section 12.4). Euler first published the Mixed Derivative Theorem in 1734, in one of a series of papers he wrote on hydrodynamics.

THEOREM 2

Euler's Mixed-Derivative Theorem

If $f(x, y)$ and its partial derivatives f_x, f_y, f_{xy}, and f_{yx} are defined throughout an open region containing a point (a, b) and are all continuous at (a, b), then

$$f_{xy}(a, b) = f_{yx}(a, b). \tag{1}$$

Proof The equality of $f_{xy}(a, b)$ and $f_{yx}(a, b)$ can be established by four applications of the Mean Value Theorem (Theorem 3, Section 3.2). By hypothesis, the point (a, b) lies in the interior of a rectangle R in the xy-plane on which f, f_x, f_y, f_{xy}, and f_{yx} are all defined. We let h and k be numbers such that the point $(a + h, b + k)$ also lies in the rectangle R, and we consider the difference

$$\Delta = F(a + h) - F(a), \tag{2}$$

where

$$F(x) = f(x, b + k) - f(x, b). \tag{3}$$

We apply the Mean Value Theorem to F (which is continuous because it is differentiable), and Eq. (2) becomes

$$\Delta = hF'(c_1), \tag{4}$$

where c_1 lies between a and $a + h$. From Eq. (3),

$$F'(x) = f_x(x, b + k) - f_x(x, b),$$

so Eq. (4) becomes

$$\Delta = h[f_x(c_1, b + k) - f_x(c_1, b)]. \tag{5}$$

Now we apply the Mean Value Theorem to the function $g(y) = f_x(c_1, y)$ and have

$$g(b + k) - g(b) = kg'(d_1), \tag{6}$$

or

$$f_x(c_1, b + k) - f_x(c_1, b) = kf_{xy}(c_1, d_1), \tag{7}$$

for some d_1 between b and $b + k$. By substituting this into Eq. (5), we get

$$\Delta = hkf_{xy}(c_1, d_1), \tag{8}$$

for some point (c_1, d_1) in the rectangle R' whose vertices are the four points (a, b), $(a + h, b)$, $(a + h, b + k)$, and $(a, b + k)$. (See Fig. A.22.)

By substituting from Eq. (3) into Eq. (2), we may also write

$$\begin{aligned}\Delta &= f(a + h, b + k) - f(a + h, b) - f(a, b + k) + f(a, b)\\ &= [f(a + h, b + k) - f(a, b + k)] - [f(a + h, b) - f(a, b)]\\ &= \phi(b + k) - \phi(b),\end{aligned} \tag{9}$$

where

$$\phi(y) = f(a + h, y) - f(a, y). \tag{10}$$

The Mean Value Theorem applied to Eq. (9) now gives

$$\Delta = k\phi'(d_2), \tag{11}$$

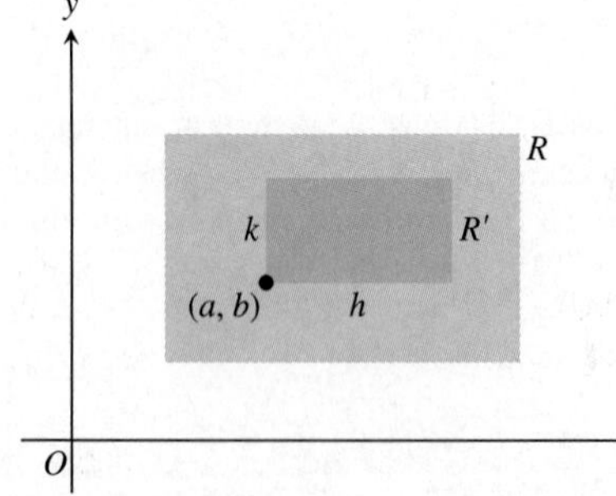

A.22 The key to proving $f_{xy}(a, b) = f_{yx}(a, b)$ is the fact that no matter how small R' is, f_{xy} and f_{yx} take on equal values somewhere inside R' (although not necessarily at the same point).

for some d_2 between b and $b + k$. By Eq. (10),

$$\phi'(y) = f_y(a + h, y) - f_y(a, y). \tag{12}$$

Substituting from Eq. (12) into Eq. (11) gives

$$\Delta = k[f_y(a + h, d_2) - f_y(a, d_2)]. \tag{13}$$

Finally, we apply the Mean Value Theorem to the expression in brackets and get

$$\Delta = khf_{yx}(c_2, d_2), \tag{14}$$

for some c_2 between a and $a + h$.

Together, Eqs. (8) and (14) show that

$$f_{xy}(c_1, d_1) = f_{yx}(c_2, d_2), \tag{15}$$

where (c_1, d_1) and (c_2, d_2) both lie in the rectangle R' (Fig. A.22). Equation (15) is not quite the result we want, since it says only that f_{xy} has the same value at (c_1, d_1) that f_{yx} has at (c_2, d_2). But the numbers h and k in our discussion may be made as small as we wish. The hypothesis that f_{xy} and f_{yx} are both continuous at (a, b) means that $f_{xy}(c_1, d_1) = f_{xy}(a, b) + \varepsilon_1$ and $f_{yx}(c_2, d_2) = f_{yx}(a, b) + \varepsilon_2$, where $\varepsilon_1, \varepsilon_2 \to 0$ as $h, k \to 0$. Hence, if we let h and $k \to 0$, we have $f_{xy}(a, b) = f_{yx}(a, b)$. ∎

The equality of $f_{xy}(a, b)$ and $f_{yx}(a, b)$ can be proved with weaker hypotheses than the ones we assumed. For example, it is enough for f, f_x, and f_y to exist in R and for f_{xy} to be continuous at (a, b). Then f_{yx} will exist at (a, b) and will equal f_{xy} at that point.

THEOREM 3 (SECTION 12.4)

The Increment Theorem for Functions of Two Variables

Suppose that the first partial derivatives of $z = f(x, y)$ are defined throughout an open region R containing the point (x_0, y_0) and that f_x and f_y are continuous at (x_0, y_0). Then the change $\Delta z = f(x_0 + \Delta x, y_0 + \Delta y) - f(x_0, y_0)$ in the value of f that results from moving from (x_0, y_0) to another point $(x_0 + \Delta x, y_0 + \Delta y)$ in R satisfies an equation of the form

$$\Delta z = f_x(x_0, y_0)\Delta x + f_y(x_0, y_0)\Delta y + \varepsilon_1\Delta x + \varepsilon_2\Delta y, \tag{16}$$

in which $\varepsilon_1, \varepsilon_2 \to 0$ as $\Delta x, \Delta y \to 0$.

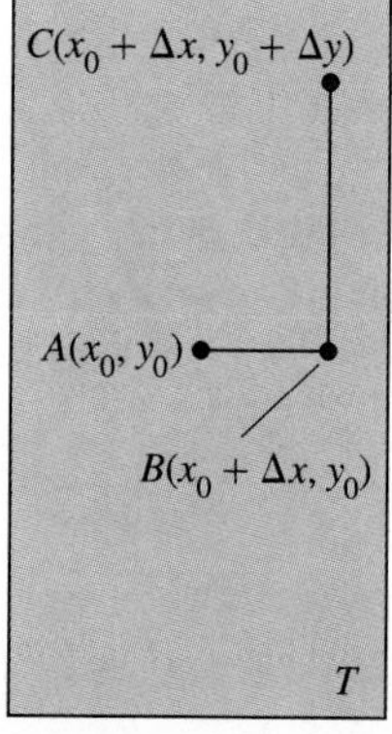

A.23 The rectangular region T in the proof of the Increment Theorem. The figure is drawn for Δx and Δy positive, but either increment might be zero or negative.

Proof We work within a rectangle T centered at $A(x_0, y_0)$ and lying within R, and we assume that Δx and Δy are already so small that the line segment joining A to $B(x_0 + \Delta x, y_0)$ and the line segment joining B to $C(x_0 + \Delta x, y_0 + \Delta y)$ lie in the interior of T (Fig. A.23).

We may think of Δz as the sum $\Delta z = \Delta z_1 + \Delta z_2$ of two increments, where

$$\Delta z_1 = f(x_0 + \Delta x, y_0) - f(x_0, y_0) \tag{17}$$

is the change in the value of f from A to B and

$$\Delta z_2 = f(x_0 + \Delta x, y_0 + \Delta y) - f(x_0 + \Delta x, y_0) \tag{18}$$

is the change in the value of f from B to C (Fig. A.24).

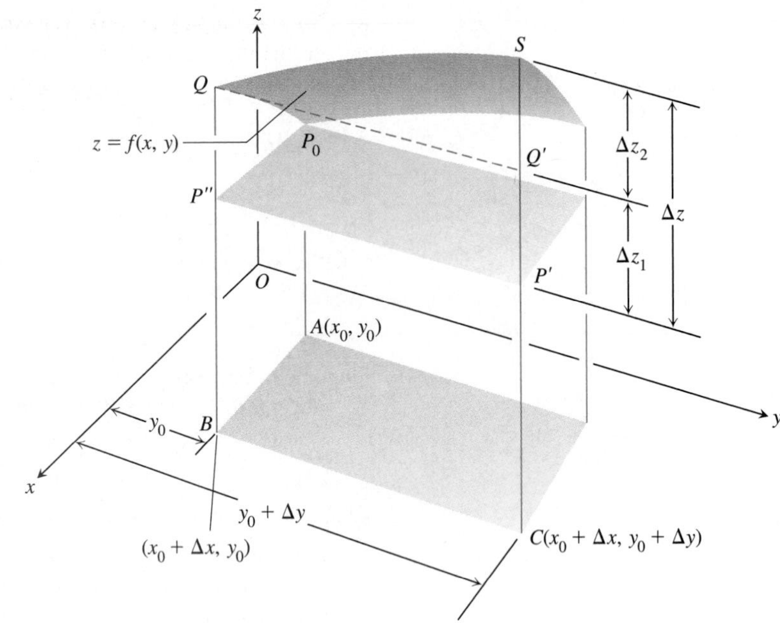

A.24 Part of the surface $z = f(x, y)$ near $P_0(x_0, y_0, f(x_0, y_0))$. The points P_0, P', and P'' have the same height $z_0 = f(x_0, y_0)$ above the xy-plane. The change in z is $\Delta z = P'S$. The change

$$\Delta z_1 = f(x_0 + \Delta x, y_0) - f(x_0, y_0),$$

shown as $P''Q = P'Q'$, is caused by changing x from x_0 to $x_0 + \Delta x$ while holding y equal to y_0. Then, with x held equal to $x_0 + \Delta x$,

$$\Delta z_2 = f(x_0 + \Delta x, y_0 + \Delta y) \\ -f(x_0 + \Delta x, y_0)$$

is the change in z caused by changing y from y_0 to $y_0 + \Delta y$. This is represented by $Q'S$. The total change in z is the sum of Δz_1 and Δz_2.

On the closed interval of x-values joining x_0 to $x_0 + \Delta x$, the function $F(x) = f(x, x_0)$ is a differentiable (and hence continuous) function of x, with derivative

$$F'(x) = f_x(x, y_0).$$

By the Mean Value Theorem (Theorem 3, Section 3.2), there is an x-value c between x_0 and $x_0 + \Delta x$ at which

$$F(x_0 + \Delta x) - F(x_0) = F'(c)\Delta x$$

or

$$f(x_0 + \Delta x, y_0) - f(x_0, y_0) = f_x(c, y_0)\Delta x$$

or

$$\Delta z_1 = f_x(c, y_0)\Delta x. \tag{19}$$

Similarly, the function $G(y) = f(x_0 + \Delta x, y)$ is a differentiable (and hence continuous) function of y on the closed y-interval joining y_0 and $y_0 + \Delta y$, with derivative

$$G'(y) = f_y(x_0 + \Delta x, y).$$

Hence there is a y-value d between y_0 and $y_0 + \Delta y$ at which

$$G(y_0 + \Delta y) - G(y_0) = G'(d)\Delta y$$

or

$$f(x_0 + \Delta x, y_0 + \Delta y) - f(x_0 + \Delta x, y) = f_y(x_0 + \Delta x, d)\Delta y$$

or

$$\Delta z_2 = f_y(x_0 + \Delta x, d)\Delta y. \tag{20}$$

Now, as Δx and $\Delta y \to 0$, we know $c \to x_0$ and $d \to y_0$. Therefore, since f_x and f_y are continuous at (x_0, y_0), the quantities

$$\begin{aligned} \varepsilon_1 &= f_x(c, y_0) - f_x(x_0, y_0), \\ \varepsilon_2 &= f_y(x_0 + \Delta x, d) - f_y(x_0, y_0) \end{aligned} \tag{21}$$

both approach zero as Δx and $\Delta y \to 0$.

Finally,

$$\begin{aligned} \Delta z &= \Delta z_1 + \Delta z_2 \\ &= f_x(c, y_0)\,\Delta x + f_y(x_0 + \Delta x, d)\,\Delta y && \text{(From (19) and (20))} \\ &= [f_x(x_0, y_0) + \varepsilon_1]\,\Delta x + [f_y(x_0, y_0) + \varepsilon_2]\,\Delta y && \text{(From (21))} \\ &= f_x(x_0, y_0)\,\Delta x + f_y(x_0, y_0)\,\Delta y + \varepsilon_1\,\Delta x + \varepsilon_2\,\Delta y, \end{aligned}$$

where ε_1 and $\varepsilon_2 \to 0$ as Δx and $\Delta y \to 0$. This is what we set out to prove. ■

Analogous results hold for functions of any finite number of independent variables. Suppose that the first partial derivatives of

$$w = f(x, y, z)$$

are defined throughout an open region containing the point (x_0, y_0, z_0) and that f_x, f_y, and f_z are continuous at (x_0, y_0). Then

$$\begin{aligned} \Delta w &= f(x_0 + \Delta x, y_0 + \Delta y, z_0 + \Delta z) - f(x_0, y_0, z_0) \\ &= f_x\,\Delta x + f_y\,\Delta y + f_z\,\Delta z + \varepsilon_1\,\Delta x + \varepsilon_2\,\Delta y + \varepsilon_3\,\Delta z, \end{aligned} \tag{22}$$

where

$$\varepsilon_1, \varepsilon_2, \varepsilon_3 \to 0 \qquad \text{when} \qquad \Delta x, \Delta y, \text{ and } \Delta z \to 0.$$

The partial derivatives f_x, f_y, f_z in this formula are to be evaluated at the point (x_0, y_0, z_0).

The result (22) can be proved by treating Δw as the sum of three increments,

$$\Delta w_1 = f(x_0 + \Delta x, y_0, z_0) - f(x_0, y_0, z_0), \tag{23}$$

$$\Delta w_2 = f(x_0 + \Delta x, y_0 + \Delta y, z_0) - f(x_0 + \Delta x, y_0, z_0), \tag{24}$$

$$\Delta w_3 = f(x_0 + \Delta x, y_0 + \Delta y, z_0 + \Delta z) - f(x_0 + \Delta x, y_0 + \Delta y, z_0), \tag{25}$$

and applying the Mean Value Theorem to each of these separately. Two coordinates remain constant and only one varies in each of these partial increments Δw_1, Δw_2, Δw_3. In (24), for example, only y varies, since x is held equal to $x_0 + \Delta x$ and z is held equal to z_0. Since the function $f(x_0 + \Delta x, y, z_0)$ is a continuous function of y with a derivative f_y, it is subject to the Mean Value Theorem, and we have

$$\Delta w_2 = f_y(x_0 + \Delta x, y_1, z_0)\,\Delta y$$

for some y_1 between y_0 and $y_0 + \Delta y$.

A.12
Numerical Tables for sin x, cos x, tan x, e^x, e^{-x}, and ln x

TABLE A.2
Natural trigonometric functions

Angle					Angle				
Degree	**Radian**	**Sine**	**Cosine**	**Tangent**	**Degree**	**Radian**	**Sine**	**Cosine**	**Tangent**
0°	0.000	0.000	1.000	0.000					
1°	0.017	0.017	1.000	0.017	46°	0.803	0.719	0.695	1.036
2°	0.035	0.035	0.999	0.035	47°	0.820	0.731	0.682	1.072
3°	0.052	0.052	0.999	0.052	48°	0.838	0.743	0.669	1.111
4°	0.070	0.070	0.998	0.070	49°	0.855	0.755	0.656	1.150
5°	0.087	0.087	0.996	0.087	50°	0.873	0.766	0.643	1.192
6°	0.105	0.105	0.995	0.105	51°	0.890	0.777	0.629	1.235
7°	0.122	0.122	0.993	0.123	52°	0.908	0.788	0.616	1.280
8°	0.140	0.139	0.990	0.141	53°	0.925	0.799	0.602	1.327
9°	0.157	0.156	0.988	0.158	54°	0.942	0.809	0.588	1.376
10°	0.175	0.174	0.985	0.176	55°	0.960	0.819	0.574	1.428
11°	0.192	0.191	0.982	0.194	56°	0.977	0.829	0.559	1.483
12°	0.209	0.208	0.978	0.213	57°	0.995	0.839	0.545	1.540
13°	0.227	0.225	0.974	0.231	58°	1.012	0.848	0.530	1.600
14°	0.244	0.242	0.970	0.249	59°	1.030	0.857	0.515	1.664
15°	0.262	0.259	0.966	0.268	60°	1.047	0.866	0.500	1.732
16°	0.279	0.276	0.961	0.287	61°	1.065	0.875	0.485	1.804
17°	0.297	0.292	0.956	0.306	62°	1.082	0.883	0.469	1.881
18°	0.314	0.309	0.951	0.325	63°	1.100	0.891	0.454	1.963
19°	0.332	0.326	0.946	0.344	64°	1.117	0.899	0.438	2.050
20°	0.349	0.342	0.940	0.364	65°	1.134	0.906	0.423	2.145
21°	0.367	0.358	0.934	0.384	66°	1.152	0.914	0.407	2.246
22°	0.384	0.375	0.927	0.404	67°	1.169	0.921	0.391	2.356
23°	0.401	0.391	0.921	0.424	68°	1.187	0.927	0.375	2.475
24°	0.419	0.407	0.914	0.445	69°	1.204	0.934	0.358	2.605
25°	0.436	0.423	0.906	0.466	70°	1.222	0.940	0.342	2.748
26°	0.454	0.438	0.899	0.488	71°	1.239	0.946	0.326	2.904
27°	0.471	0.454	0.891	0.510	72°	1.257	0.951	0.309	3.078
28°	0.489	0.469	0.883	0.532	73°	1.274	0.956	0.292	3.271
29°	0.506	0.485	0.875	0.554	74°	1.292	0.961	0.276	3.487
30°	0.524	0.500	0.866	0.577	75°	1.309	0.966	0.259	3.732
31°	0.541	0.515	0.857	0.601	76°	1.326	0.970	0.242	4.011
32°	0.559	0.530	0.848	0.625	77°	1.344	0.974	0.225	4.332
33°	0.576	0.545	0.839	0.649	78°	1.361	0.978	0.208	4.705
34°	0.593	0.559	0.829	0.675	79°	1.379	0.982	0.191	5.145
35°	0.611	0.574	0.819	0.700	80°	1.396	0.985	0.174	5.671
36°	0.628	0.588	0.809	0.727	81°	1.414	0.988	0.156	6.314
37°	0.646	0.602	0.799	0.754	82°	1.431	0.990	0.139	7.115
38°	0.663	0.616	0.788	0.781	83°	1.449	0.993	0.122	8.144
39°	0.681	0.629	0.777	0.810	84°	1.466	0.995	0.105	9.514
40°	0.698	0.643	0.766	0.839	85°	1.484	0.996	0.087	11.43
41°	0.716	0.656	0.755	0.869	86°	1.501	0.998	0.070	14.30
42°	0.733	0.669	0.743	0.900	87°	1.518	0.999	0.052	19.08
43°	0.750	0.682	0.731	0.933	88°	1.536	0.999	0.035	28.64
44°	0.768	0.695	0.719	0.966	89°	1.553	1.000	0.017	57.29
45°	0.785	0.707	0.707	1.000	90°	1.571	1.000	0.000	

TABLE A.3
Exponential functions

x	e^x	e^{-x}	x	e^x	e^{-x}
0.00	1.0000	1.0000	2.5	12.182	0.0821
0.05	1.0513	0.9512	2.6	13.464	0.0743
0.10	1.1052	0.9048	2.7	14.880	0.0672
0.15	1.1618	0.8607	2.8	16.445	0.0608
0.20	1.2214	0.8187	2.9	18.174	0.0550
0.25	1.2840	0.7788	3.0	20.086	0.0498
0.30	1.3499	0.7408	3.1	22.198	0.0450
0.35	1.4191	0.7047	3.2	24.533	0.0408
0.40	1.4918	0.6703	3.3	27.113	0.0369
0.45	1.5683	0.6376	3.4	29.964	0.0334
0.50	1.6487	0.6065	3.5	33.115	0.0302
0.55	1.7333	0.5769	3.6	36.598	0.0273
0.60	1.8221	0.5488	3.7	40.447	0.0247
0.65	1.9155	0.5220	3.8	44.701	0.0224
0.70	2.0138	0.4966	3.9	49.402	0.0202
0.75	2.1170	0.4724	4.0	54.598	0.0183
0.80	2.2255	0.4493	4.1	60.340	0.0166
0.85	2.3396	0.4274	4.2	66.686	0.0150
0.90	2.4596	0.4066	4.3	73.700	0.0136
0.95	2.5857	0.3867	4.4	81.451	0.0123
1.0	2.7183	0.3679	4.5	90.017	0.0111
1.1	3.0042	0.3329	4.6	99.484	0.0101
1.2	3.3201	0.3012	4.7	109.95	0.0091
1.3	3.6693	0.2725	4.8	121.51	0.0082
1.4	4.0552	0.2466	4.9	134.29	0.0074
1.5	4.4817	0.2231	5	148.41	0.0067
1.6	4.9530	0.2019	6	403.43	0.0025
1.7	5.4739	0.1827	7	1096.6	0.0009
1.8	6.0496	0.1653	8	2981.0	0.0003
1.9	6.6859	0.1496	9	8103.1	0.0001
2.0	7.3891	0.1353	10	22026	0.00005
2.1	8.1662	0.1225			
2.2	9.0250	0.1108			
2.3	9.9742	0.1003			
2.4	11.023	0.0907			

TABLE A.4
Natural logarithms

x	$\log_e x$	x	$\log_e x$	x	$\log_e x$
0.0	*	4.5	1.5041	9.0	2.1972
0.1	7.6974	4.6	1.5261	9.1	2.2083
0.2	8.3906	4.7	1.5476	9.2	2.2192
0.3	8.7960	4.8	1.5686	9.3	2.2300
0.4	9.0837	4.9	1.5892	9.4	2.2407
0.5	9.3069	5.0	1.6094	9.5	2.2513
0.6	9.4892	5.1	1.6292	9.6	2.2618
0.7	9.6433	5.2	1.6487	9.7	2.2721
0.8	9.7769	5.3	1.6677	9.8	2.2824
0.9	9.8946	5.4	1.6864	9.9	2.2925
1.0	0.0000	5.5	1.7047	10	2.3026
1.1	0.0953	5.6	1.7228	11	2.3979
1.2	0.1823	5.7	1.7405	12	2.4849
1.3	0.2624	5.8	1.7579	13	2.5649
1.4	0.3365	5.9	1.7750	14	2.6391
1.5	0.4055	6.0	1.7918	15	2.7081
1.6	0.4700	6.1	1.8083	16	2.7726
1.7	0.5306	6.2	1.8245	17	2.8332
1.8	0.5878	6.3	1.8405	18	2.8904
1.9	0.6419	6.4	1.8563	19	2.9444
2.0	0.6931	6.5	1.8718	20	2.9957
2.1	0.7419	6.6	1.8871	25	3.2189
2.2	0.7885	6.7	1.9021	30	3.4012
2.3	0.8329	6.8	1.9169	35	3.5553
2.4	0.8755	6.9	1.9315	40	3.6889
2.5	0.9163	7.0	1.9459	45	3.8067
2.6	0.9555	7.1	1.9601	50	3.9120
2.7	0.9933	7.2	1.9741	55	4.0073
2.8	1.0296	7.3	1.9879	60	4.0943
2.9	1.0647	7.4	2.0015	65	4.1744
3.0	1.0986	7.5	2.0149	70	4.2485
3.1	1.1314	7.6	2.0281	75	4.3175
3.2	1.1632	7.7	2.0412	80	4.3820
3.3	1.1939	7.8	2.0541	85	4.4427
3.4	1.2238	7.9	2.0669	90	4.4998
3.5	1.2528	8.0	2.0794	95	4.5539
3.6	1.2809	8.1	2.0919	100	4.6052
3.7	1.3083	8.2	2.1041		
3.8	1.3350	8.3	2.1163		
3.9	1.3610	8.4	2.1282		
4.0	1.3863	8.5	2.1401		
4.1	1.4110	8.6	2.1518		
4.2	1.4351	8.7	2.1633		
4.3	1.4586	8.8	2.1748		
4.4	1.4816	8.9	2.1861		

*Subtract 10 from $\log_e x$ entries for $x < 1.0$.

A.13
Vector Operator Formulas in Cartesian, Cylindrical, and Spherical Coordinates; Vector Identities

Formulas for Grad, Div, Curl, and the Laplacian

	Cartesian (x, y, z)	Cylindrical (r, θ, z)	Spherical (ρ, ϕ, θ)
	i, **j**, and **k** are unit vectors in the directions of increasing x, y, and z. F_x, F_y, and F_z are the scalar components of $\mathbf{F}(x, y, z)$ in these directions.	$\mathbf{u}_r$, $\mathbf{u}_\theta$, and **k** are unit vectors in the directions of increasing r, θ, and z. F_r, F_θ, and F_z are the scalar components of $\mathbf{F}(r, \theta, z)$ in these directions.	$\mathbf{u}_\rho$, and $\mathbf{u}_\phi$, and $\mathbf{u}_\theta$ are unit vectors in the directions of increasing ρ, ϕ, and θ. F_ρ, F_ϕ, and F_θ are the scalar components of $\mathbf{F}(\rho, \phi, \theta)$ in these directions.
Gradient	$\nabla f = \frac{\partial f}{\partial x}\mathbf{i} + \frac{\partial f}{\partial y}\mathbf{j} + \frac{\partial f}{\partial z}\mathbf{k}$	$\nabla f = \frac{\partial f}{\partial r}\mathbf{u}_r + \frac{1}{r}\frac{\partial f}{\partial \theta}\mathbf{u}_\theta + \frac{\partial f}{\partial z}\mathbf{k}$	$\nabla f = \frac{\partial f}{\partial \rho}\mathbf{u}_\rho + \frac{1}{\rho}\frac{\partial f}{\partial \phi}\mathbf{u}_\phi + \frac{1}{\rho \sin \phi}\frac{\partial f}{\partial \theta}\mathbf{u}_\theta$
Divergence	$\nabla \cdot \mathbf{F} = \frac{\partial F_x}{\partial x} + \frac{\partial F_y}{\partial y} + \frac{\partial F_z}{\partial z}$	$\nabla \cdot \mathbf{F} = \frac{1}{r}\frac{\partial}{\partial r}(rF_r) + \frac{1}{r}\frac{\partial F_\theta}{\partial \theta} + \frac{\partial F_z}{\partial z}$	$\nabla \cdot \mathbf{F} = \frac{1}{\rho^2}\frac{\partial}{\partial \rho}(\rho^2 F_\rho) + \frac{1}{\rho \sin \phi}\frac{\partial}{\partial \phi}(F_\phi \sin \phi) + \frac{1}{\rho \sin \phi}\frac{\partial F_\theta}{\partial \theta}$
Curl	$\nabla \times \mathbf{F} = \begin{vmatrix} \mathbf{i} & \mathbf{j} & \mathbf{k} \\ \frac{\partial}{\partial x} & \frac{\partial}{\partial y} & \frac{\partial}{\partial z} \\ F_x & F_y & F_z \end{vmatrix}$	$\nabla \times \mathbf{F} = \begin{vmatrix} \frac{1}{r}\mathbf{u}_r & \mathbf{u}_\theta & \frac{1}{r}\mathbf{k} \\ \frac{\partial}{\partial r} & \frac{\partial}{\partial \theta} & \frac{\partial}{\partial z} \\ F_r & F_\theta & F_z \end{vmatrix}$	$\nabla \times \mathbf{F} = \begin{vmatrix} \frac{\mathbf{u}_\rho}{\rho^2 \sin \phi} & \frac{\mathbf{u}_\phi}{\rho \sin \phi} & \frac{\mathbf{u}_\theta}{\rho} \\ \frac{\partial}{\partial \rho} & \frac{\partial}{\partial \phi} & \frac{\partial}{\partial \theta} \\ F_\rho & \rho F_\phi & \rho \sin \phi F_\theta \end{vmatrix}$
Laplacian	$\nabla^2 f = \frac{\partial^2 f}{\partial x^2} + \frac{\partial^2 f}{\partial y^2} + \frac{\partial^2 f}{\partial z^2}$	$\nabla^2 f = \frac{1}{r}\frac{\partial}{\partial r}\left(r\frac{\partial f}{\partial r}\right) + \frac{1}{r^2}\frac{\partial^2 f}{\partial \theta^2} + \frac{\partial^2 f}{\partial z^2}$	$\nabla^2 f = \frac{1}{\rho^2}\frac{\partial}{\partial \rho}\left(\rho^2\frac{\partial f}{\partial \rho}\right) + \frac{1}{\rho^2 \sin \phi}\frac{\partial}{\partial \phi}\left(\sin \phi \frac{\partial f}{\partial \phi}\right) + \frac{1}{\rho^2 \sin^2 \phi}\frac{\partial^2 f}{\partial \theta^2}$

Vector Triple Products

$$(\mathbf{A} \times \mathbf{B}) \cdot \mathbf{C} = (\mathbf{B} \times \mathbf{C}) \cdot \mathbf{A} = (\mathbf{C} \times \mathbf{A}) \cdot \mathbf{B}$$

$$\mathbf{A} \times (\mathbf{B} \times \mathbf{C}) = (\mathbf{A} \cdot \mathbf{C})\mathbf{B} - (\mathbf{A} \cdot \mathbf{B})\mathbf{C}$$

Vector Identities for the Cartesian Form of the Operator ∇

In the identities listed here, $f(x, y, z)$ and $g(x, y, z)$ are differentiable scalar functions and $\mathbf{u}(x, y, z)$ and $\mathbf{v}(x, y, z)$ are differentiable vector functions.

$$\nabla \cdot f\mathbf{v} = f\nabla \cdot \mathbf{v} + \mathbf{v} \cdot \nabla f = f\nabla \cdot \mathbf{v} + (\mathbf{v} \cdot \nabla) f$$

$$\nabla \times f\mathbf{v} = f\nabla \times \mathbf{v} + \nabla f \times \mathbf{v}$$

$$\nabla \cdot (\nabla \times \mathbf{v}) = 0$$

$$\nabla \times (\nabla f) = \mathbf{0}$$

$$\nabla(fg) = f\nabla g + g\nabla f$$

$$\nabla(\mathbf{u} \cdot \mathbf{v}) = (\mathbf{u} \cdot \nabla)\mathbf{v} + (\mathbf{v} \cdot \nabla)\mathbf{u} + \mathbf{u} \times (\nabla \times \mathbf{v}) + \mathbf{v} \times (\nabla \times \mathbf{u})$$

$$\nabla \cdot (\mathbf{u} \times \mathbf{v}) = \mathbf{v} \cdot (\nabla \times \mathbf{u}) - \mathbf{u} \cdot (\nabla \times \mathbf{v})$$

$$\nabla \times (\mathbf{u} \times \mathbf{v}) = (\mathbf{v} \cdot \nabla)\mathbf{u} - (\mathbf{u} \cdot \nabla)\mathbf{v} + \mathbf{u}(\nabla \cdot \mathbf{v}) - \mathbf{v}(\nabla \cdot \mathbf{u})$$

$$\nabla \times (\nabla \times \mathbf{v}) = \nabla(\nabla \cdot \mathbf{v}) - (\nabla \cdot \nabla)\mathbf{v} = \nabla(\nabla \cdot \mathbf{v}) - \nabla^2 \mathbf{v}$$

$$(\nabla \times \mathbf{v}) \times \mathbf{v} = (\mathbf{v} \cdot \nabla)\mathbf{v} - \frac{1}{2}\nabla(\mathbf{v} \cdot \mathbf{v})$$

ANSWERS

Chapter 8

Section 8.1, pp. 549–552

1. $a_1 = 0, a_2 = -\frac{1}{4}, a_3 = -\frac{2}{9}, a_4 = -\frac{3}{16}$ **3.** $a_1 = 1, a_2 = -\frac{1}{3}, a_3 = \frac{1}{5}, a_4 = -\frac{1}{7}$ **5.** $1, \frac{3}{2}, \frac{7}{4}, \frac{15}{8}, \frac{31}{16}, \frac{63}{32}, \frac{127}{64}, \frac{255}{128}, \frac{511}{256}, \frac{1023}{512}$ **7.** $2, 1, -\frac{1}{2}, -\frac{1}{4}, \frac{1}{8}, \frac{1}{16}, -\frac{1}{32}, -\frac{1}{64}, \frac{1}{128}, \frac{1}{256}$ **9.** 1, 1, 2, 3, 5, 8, 13, 21, 34, 55 **11.** Converges, 2 **13.** Converges, -1 **15.** Converges, -5 **17.** Diverges **19.** Diverges **21.** Converges, $\frac{1}{2}$ **23.** Converges, 0 **25.** Converges, 0 **27.** Converges, $\sqrt{2}$ **29.** Converges, $\frac{\pi}{2}$ **31.** Converges, 0 **33.** Converges, 0 **35.** Converges, 1 **37.** Converges, e^7 **39.** Diverges **41.** Converges, 1 **43.** Converges, 1 **45.** Diverges **47.** Converges, 0 **49.** Converges, 0 **51.** Converges, e^{-1} **53.** Diverges **55.** Converges, 1 **57.** Converges, 1 **59.** Converges, $e^{2/3}$ **61.** Converges, x **63.** Converges, 0 **65.** Converges, 1 **67.** Converges, $\frac{1}{2}$ **69.** Converges, $\frac{1}{2}$ **71.** 693 **73.** 66 **75.** $x_1 = 1, x_2 \approx 1.540302305, x_3 \approx 1.570791601, x_4 \approx 1.570796327$ **79.** a) $f(x) = x^2 - 2$; $1.414213562 \approx \sqrt{2}$ b) $f(x) = \tan(x) - 1$; $0.7853981635 \approx \frac{\pi}{4}$ c) $f(x) = e^x$; the sequence $1, 0, -1, -2, -3, -4, -5, \ldots$, diverges. **81.** b) 1 **85.** 1 **91.** 0.999998902 **93.** G, 0.73908456 **95.** 0.853748068 **99.** a) -3 b) 0.2

Section 8.2, pp. 561–562

1. $s_n = \frac{2(1-(1/3)^n)}{1-1/3}$, 3 **3.** $s_n = \frac{1-(-1/2)^n}{1-(-1/2)}, \frac{2}{3}$ **5.** $s_n = \frac{1}{2} - \frac{1}{n+2}, \frac{1}{2}$ **7.** $1 - \frac{1}{4} + \frac{1}{16} - \frac{1}{64} + \cdots, \frac{4}{5}$ **9.** $\frac{7}{4} + \frac{7}{16} + \frac{7}{64} + \cdots, \frac{7}{3}$ **11.** $(5+1) + \left(\frac{5}{2} + \frac{1}{3}\right) + \left(\frac{5}{4} + \frac{1}{9}\right) + \left(\frac{5}{8} + \frac{1}{27}\right) + \cdots, \frac{23}{2}$ **13.** $(1+1) + \left(\frac{1}{2} - \frac{1}{5}\right) + \left(\frac{1}{4} + \frac{1}{25}\right) + \left(\frac{1}{8} - \frac{1}{125}\right) + \cdots, \frac{17}{6}$ **15.** 1 **17.** $\frac{1}{9}$ **19.** $2 + \sqrt{2}$ **21.** 1 **23.** Diverges **25.** $\frac{e^2}{e^2 - 1}$ **27.** Diverges **29.** $\frac{3}{2}$ **31.** Diverges **33.** $\frac{\pi}{\pi - e}$ **35.** Diverges **37.** 1 **39.** $a = 1$, $r = -x$, where $|x| < 1$. **41.** $a = 2, r = \frac{x}{3}$, where $|x| < 3$. **43.** $|x| < \frac{1}{2}, \frac{1}{1-2x}$ **45.** $-2 < x < 0, \frac{1}{2+x}$ **47.** $x \neq k(\pi)$, where k is an integer, $\frac{1}{1 - \sin x}$. **49.** $\frac{23}{99}$ **51.** $\frac{7}{9}$ **53.** $\frac{1}{15}$ **55.** $\frac{41333}{33300}$ **57.** 28 m **59.** a) $\sum_{n=-2}^{\infty} \frac{1}{(n+4)(n+5)}$ b) $\sum_{n=0}^{\infty} \frac{1}{(n+2)(n+3)}$ c) $\sum_{n=5}^{\infty} \frac{1}{(n-3)(n-2)}$ **61.** $S_n = \frac{1 + (-1)^{n+1}}{2}$ **63.** 8 m^2 **65.** a) $3\left(\frac{4}{3}\right)^n$ b) $\frac{\sqrt{3}}{4} + \frac{27\sqrt{3}}{64}\left(\frac{4}{9}\right)^2 + \frac{27\sqrt{3}}{64}\left(\frac{4}{9}\right)^3 + \cdots, \frac{2\sqrt{3}}{5}$

Section 8.3, pp. 570–572

1. Converges **3.** Converges **5.** Converges **7.** Diverges **9.** Converges **11.** Diverges **13.** Diverges **15.** Converges **17.** Diverges **19.** Diverges **21.** Converges **23.** Converges **25.** Converges **27.** Converges **29.** Converges **31.** Converges **33.** Diverges **35.** About 41.55

Section 8.4, p. 577

1. Converges **3.** Converges **5.** Diverges **7.** Diverges **9.** Diverges **11.** Converges **13.** Diverges **15.** Converges **17.** Converges **19.** Converges **21.** Converges **23.** Converges **25.** Diverges **27.** Converges **29.** Converges **31.** Converges **33.** Diverges **35.** Converges **37.** Converges **39.** Converges **41.** Diverges

Section 8.5, pp. 584–585

1. Converges **3.** Diverges **5.** Converges **7.** Converges **9.** Diverges **11.** Diverges **13.** Converges absolutely **15.** Converges absolutely **17.** Converges conditionally **19.** Diverges **21.** Converges conditionally **23.** Converges absolutely **25.** Converges absolutely **27.** Diverges **29.** Diverges **31.** Converges absolutely **33.** Converges absolutely **35.** Converges absolutely **37.** Converges conditionally **39.** Diverges **41.** Converges conditionally **43.** Converges absolutely **45.** $|\text{error}| < 0.2$ **47.** $|\text{error}| < 2 \times 10^{-11}$ **49.** 0.54030 **51.** a) $a_n \geq a_{n+1}$ b) $-\frac{1}{2}$

59. $S_1 = -\frac{1}{2}$

$S_2 = -\frac{1}{2} + 1$

$S_3 = -\frac{1}{2} + 1 - \frac{1}{4} - \frac{1}{4} - \frac{1}{6} - \frac{1}{8} - \frac{1}{10} - \frac{1}{12} - \frac{1}{14} - \frac{1}{16} - \frac{1}{18} - \frac{1}{20} - \frac{1}{22} \approx -0.5099$

$S_4 = -\frac{1}{2} + 1 - \frac{1}{4} - \frac{1}{4} - \frac{1}{6} - \frac{1}{8} - \frac{1}{10} - \frac{1}{12} - \frac{1}{14} - \frac{1}{16} - \frac{1}{18} - \frac{1}{20} - \frac{1}{22} + \frac{1}{3} \approx -0.1766$

$S_5 = -\frac{1}{2} + 1 - \frac{1}{4} - \frac{1}{4} - \frac{1}{6} - \frac{1}{8} - \frac{1}{10} - \frac{1}{12} - \frac{1}{14} - \frac{1}{16} - \frac{1}{18} - \frac{1}{20} - \frac{1}{22} + \frac{1}{3} - \frac{1}{24} - \frac{1}{26} - \frac{1}{28} - \frac{1}{30} - \frac{1}{32} - \frac{1}{34} - \frac{1}{36} - \frac{1}{38} - \frac{1}{40} - \frac{1}{42} - \frac{1}{44} \approx -0.512$

$S_6 = -\frac{1}{2} + 1 - \frac{1}{4} - \frac{1}{4} - \frac{1}{6} - \frac{1}{8} - \frac{1}{10} - \frac{1}{12} - \frac{1}{14} - \frac{1}{16} - \frac{1}{18} - \frac{1}{20} - \frac{1}{22} + \frac{1}{3} - \frac{1}{24} - \frac{1}{26} - \frac{1}{28} - \frac{1}{30} - \frac{1}{32} - \frac{1}{34} - \frac{1}{36} - \frac{1}{38} - \frac{1}{40} - \frac{1}{42} - \frac{1}{44} - \frac{1}{46} - \frac{1}{48} - \frac{1}{50} - \frac{1}{52} - \frac{1}{54} - \frac{1}{56} - \frac{1}{58} - \frac{1}{60} - \frac{1}{62} - \frac{1}{64} - \frac{1}{66} \approx -0.511065$

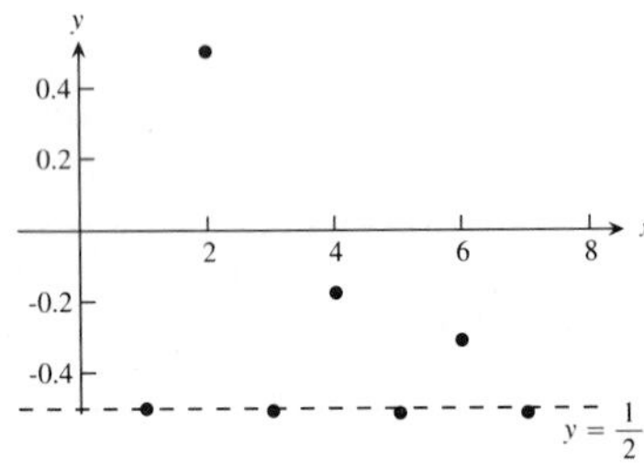

Section 8.6, pp. 592–593

1. a) 1, $-1 < x < 1$ b) $-1 < x < 1$ c) { } **3.** a) 1, $-2 < x < 0$ b) $-2 < x < 0$ c) { } **5.** a) 10, $-8 < x < 12$ b) $-8 < x < 12$ c) { } **7.** a) 1, $-1 < x < 1$ b) $-1 < x < 1$ c) { } **9.** a) 3, $[-3, 3]$ b) $[-3, 3]$ c) { } **11.** a) ∞, for all x b) for all x c) { } **13.** a) ∞, for all x b) for all x c) { } **15.** a) 1, $-1 \leq x < 1$ b) $-1 < x < 1$ c) $\{-1\}$ **17.** a) 4, $[-4, 4)$ b) $(-4, 4)$ c) $\{-4\}$ **19.** a) 3, $-3 < x < 3$ b) $-3 < x < 3$ c) { } **21.** a) 1, $-1 < x < 1$ b) $-1 < x < 1$ c) { } **23.** a) 0, $x = 0$ b) $x = 0$ c) { } **25.** a) 2, $(0, 4]$ b) $0 < x < 4$ c) $\{4\}$ **27.** a) 1, $[-1, 1)$ b) $(-1, 1)$ c) $\{-1\}$ **29.** a) ∞, $(-\infty, -1) \cup (1, \infty)$ b) $x < -1$ or $x > 1$ c) { } **31.** a) $\frac{1}{e}$, $\left(-\frac{1}{e}, \frac{1}{e}\right)$ b) $-\frac{1}{e} < x < \frac{1}{e}$ c) { } **33.** a) 1, $(-1, 1)$ b) $-1 < x < 1$ c) { } **35.** $(1 - \sqrt{2}, 1 + \sqrt{2}), \frac{2}{1 + 2x - x^2}$ **37.** $(-\sqrt{3}, \sqrt{3}), \frac{2}{3 - x^2}$ **39.** $(-\infty, -\sqrt{5}) \cup (\sqrt{5}, \infty)$, $\frac{x^2 - 2}{x^2 + 1}$ **41.** $1 < x < 5, \frac{2}{x - 1}$; $1 < x < 5, \frac{-2}{(x - 1)^2}$ **43.** a) $\frac{x^2}{2} + \frac{x^4}{12} + \frac{x^6}{45} + \frac{17x^8}{2520} + \frac{31x^{10}}{14175}$, when $-\frac{\pi}{2} < x < \frac{\pi}{2}$ b) $1 + x^2 + \frac{2x^4}{3}$, when $-\frac{\pi}{2} < x < \frac{\pi}{2}$

Section 8.7, pp. 606–608

1. $P_0(x) = 0$, $P_1(x) = x - 1$, $P_2(x) = (x - 1) - \frac{1}{2}(x - 1)^2$, $P_3(x) = (x - 1) - \frac{1}{2}(x - 1)^2 + \frac{1}{3}(x - 1)^3$ **3.** $P_0(x) = \frac{1}{2}$, $P_1(x) = \frac{1}{2} - \frac{1}{4}(x - 2)$, $P_2(x) = \frac{1}{2} - \frac{1}{4}(x - 2) + \frac{1}{8}(x - 2)^2$, $P_3(x) = \frac{1}{2} - \frac{1}{4}(x - 2) + \frac{1}{8}(x - 2)^2 - \frac{1}{16}(x - 2)^3$ **5.** $P_1(x) = \frac{\sqrt{2}}{2} + \frac{\sqrt{2}}{2}\left(x - \frac{\pi}{4}\right)$, $P_2(x) = \frac{\sqrt{2}}{2} + \frac{\sqrt{2}}{2}\left(x - \frac{\pi}{4}\right) - \frac{\sqrt{2}}{4}\left(x - \frac{\pi}{4}\right)^2$, $P_3(x) = \frac{\sqrt{2}}{2} + \frac{\sqrt{2}}{2}\left(x - \frac{\pi}{4}\right) - \frac{\sqrt{2}}{4}\left(x - \frac{\pi}{4}\right)^2 - \frac{\sqrt{2}}{12}\left(x - \frac{\pi}{4}\right)^3$ **7.** $P_0(x) = 2$, $P_1(x) = 2 + \frac{1}{4}(x - 4)$, $P_2(x) = 2 + \frac{1}{4}(x - 4) - \frac{1}{64}(x - 4)^2$, $P_3(x) = 2 + \frac{1}{4}(x - 4) - \frac{1}{64}(x - 4)^2 + \frac{1}{512}(x - 4)^3$ **9.** $1 - x + \frac{x^2}{2!} - \frac{x^3}{3!} + \frac{x^4}{4!} - \cdots$ **11.** $3x - \frac{(3x)^3}{3!} + \frac{(3x)^5}{5!} - \cdots$ **13.** $1 - \frac{x^2}{2!} + \frac{x^4}{4!} - \frac{x^6}{6!} + \cdots$ **15.** $1 + \frac{x^2}{2!} + \frac{x^4}{4!} + \frac{x^6}{6!} + \cdots$ **17.** $\frac{x^4}{4!} - \frac{x^6}{6!} + \frac{x^8}{8!} - \frac{x^{10}}{10!} + \cdots$ **19.** $1 + x + \frac{x^2}{2} + \frac{f'''(c)}{3!}x^3$, where c is between 0 and x. **21.** $-x^2 + \frac{f'''(c)}{3!}x^3$, where c is between 0 and x. **23.** $x + \frac{f'''(c)}{3!}x^3$, where c is between 0 and x. **27.** $-0.569679052 < x < 0.569679052$ **29.** $|\text{error}| < 1.67 \times 10^{-10}$, $-10^{-3} < x < 0$ **31.** 0.00018602 **33.** 0.000260416 **35.** $\sin x$, when $x = 0.1$; 0.099833416 **39.** $2x - \frac{8x^3}{3!} + \frac{32x^5}{5!} - \frac{128x^7}{7!} + \frac{512x^9}{9!}$ **45.** a) -1 b) $\left(\frac{1}{\sqrt{2}}\right)(1 + i)$ c) $-i$ **49.** $x + x^2 + \frac{1}{3}x^3 - \frac{1}{30}x^5$, for all x

Section 8.8, p. 614

1. $\cos x = \frac{1}{2} - \frac{\sqrt{3}}{2}(x - \pi/3) - \frac{1}{4}(x - \pi/3)^2 + \frac{\sqrt{3}}{12}(x - \pi/3)^3$ **3.** $1 + x + \frac{x^2}{2!} + \frac{x^3}{3!}$ **5.** $1 - \frac{1}{2}(x - 22\pi)^2 + \frac{1}{4!}(x - 22\pi)^4 - \frac{1}{6!}(x - 22\pi)^6$ **7.** 0.0027; 0.0000003 **9.** 0.00033; 0.000002 **11.** 0.96356; 0.000266 **13.** 0.1; 0.000001 **15.** 0.099944461 **17.** 0.100001 **21.** 500 **23.** 3 **25.** a) $x + \frac{x^3}{6} + \frac{3x^5}{40} + \frac{5x^7}{112}$; 1

b) $\frac{\pi}{2} - x - \frac{x^3}{6} - \frac{3x^5}{40} - \frac{5x^7}{112}$ **27.** $1 - 2x + 3x^2 - 4x^3 + \cdots$
29. $\frac{x^2}{2} + \frac{x^4}{12} + \frac{x^6}{45}$

Section 8 Miscellaneous Exercises, pp. 618–621

1. Converges, 1 **3.** Diverges **5.** Diverges **7.** Converges, $\sqrt{3}$ **9.** Converges, 0 **11.** Converges, ln 2 **13.** Converges, $\frac{1}{e}$ **15.** Diverges **17.** $\frac{e}{e-1}$ **19.** Diverges **21.** Converges conditionally **23.** Converges conditionally **25.** Converges **27.** Converges absolutely **29.** Converges absolutely **31.** Converges **33.** Converges **35.** Diverges **37.** Converges absolutely **39.** Converges **41.** a) 3, $[-5, 1)$ b) $(-5, 1)$ c) $\{-5\}$ **43.** a) ∞, for all x b) for all x c) $\{\}$ **45.** a) 1, $[0, 2]$ b) $[0, 2]$ c) $\{\}$ **47.** a) e, $(-e, e)$ b) $(-e, e)$ c) $\{\}$ **49.** a) $\sqrt{3}$, $(-\sqrt{3}, \sqrt{3})$ b) $(-\sqrt{3}, \sqrt{3})$ c) $\{\}$ **51.** $\frac{2}{3}$ **53.** $\frac{1}{1+x}, \frac{1}{4}, \frac{4}{5}$ **55.** $\sin x, \pi, 0$ **57.** e^x, ln 2, 2 **59.** $2 - \frac{(x+1)}{2 \cdot 1!} + \frac{3(x+1)^2}{2^3 \cdot 2!} + \frac{9(x+1)^3}{2^5 \cdot 3!}$ **61.** Yes, $L^2 = \sin L$ **63.** $n\left(\frac{n+1}{2n}\right)$, $\ln\left(\frac{1}{2}\right)$ **71.** Diverges **75.** b) Yes **77.** a) ∞ **79.** $-\frac{x^2}{2} - \frac{x^4}{12} - \frac{x^6}{45}$ **81.** (1, 3), 1 **83.** ∞, for all x **87.** 0.185330149 **93.** a) $\sum_{n=1}^{\infty} nx^{n-1}$ b) 6 c) $\frac{1}{q}$ **97.** $\frac{1}{2}$ **99.** 1 **101.** 1 **103.** 2 **105.** $\frac{1}{2}$ **107.** 2 **109.** 1 **111.** 0 **113.** ∞

Chapter 9

Section 9.1, pp. 637–640

1. $y^2 = 8x$, $(2, 0)$, $x = -2$ **3.** $x^2 = -6y$, $\left(0, -\frac{3}{2}\right)$, $y = \frac{3}{2}$ **5.** $\frac{x^2}{4} - \frac{y^2}{9} = 1$, $(\pm\sqrt{13}, 0)$, $e = \sqrt{13}/2$; Directrices: $x = \pm 4/\sqrt{13}$; Asymptotes: $y = \pm(3/2)x$ **7.** $\frac{x^2}{2} + y^2 = 1$, $(\pm 1, 0)$, $e = 1/\sqrt{2}$; Directrices: $x = \pm 2$

9. $\frac{x^2}{25} + \frac{y^2}{16} = 1$, $e = \frac{3}{5}$

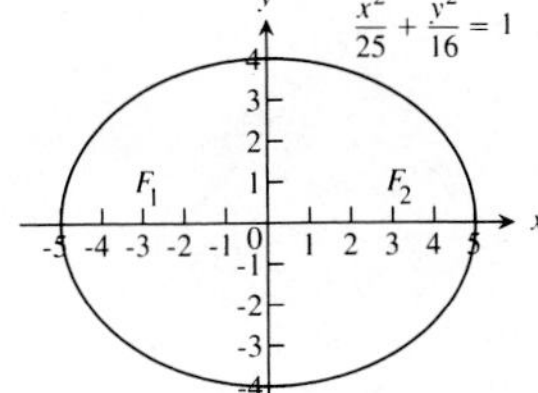

11. $x^2 + \frac{y^2}{2} = 1$, $e = \frac{1}{\sqrt{2}}$

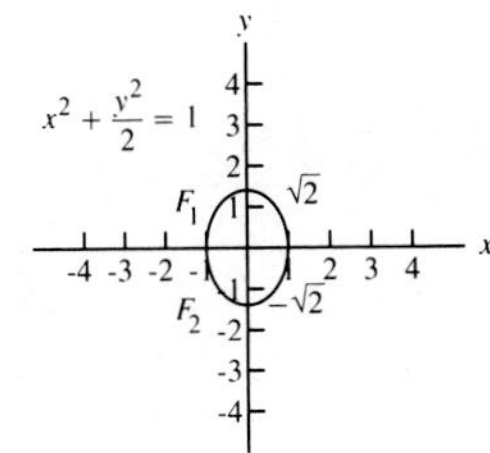

13. $\frac{x^2}{2} + \frac{y^2}{3} = 1$, $e = \frac{1}{\sqrt{3}}$

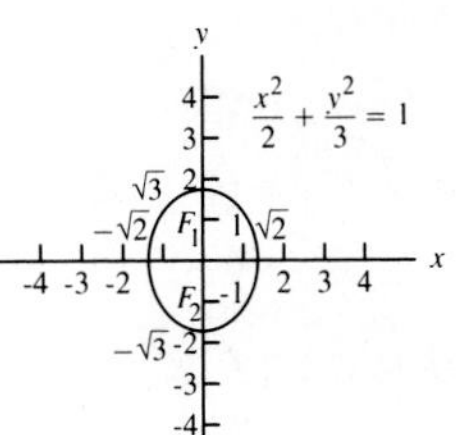

15. $\frac{x^2}{9} + \frac{y^2}{6} = 1$, $e = \frac{\sqrt{3}}{3}$

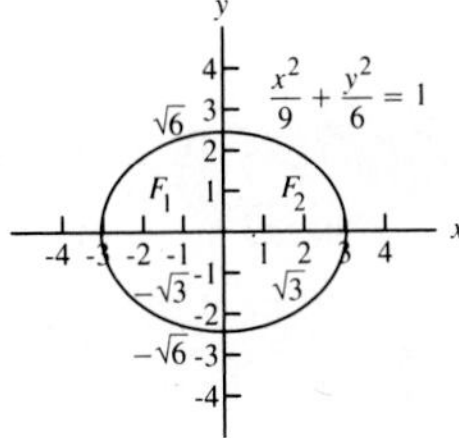

17. $x^2 - y^2 = 1$, $e = \sqrt{2}$, $y = \pm x$

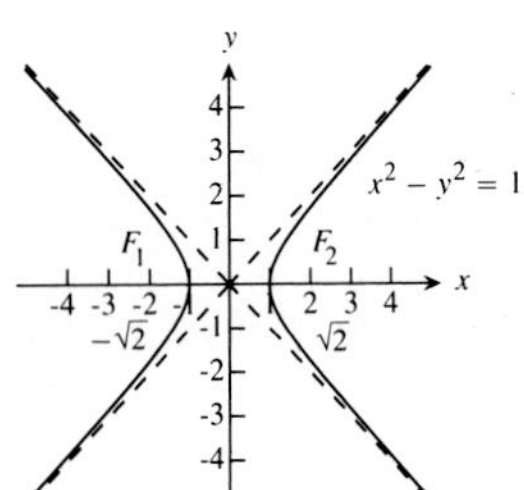

19. $\frac{y^2}{8} - \frac{x^2}{8} = 1$, $e = \sqrt{2}$, $y = \pm x$

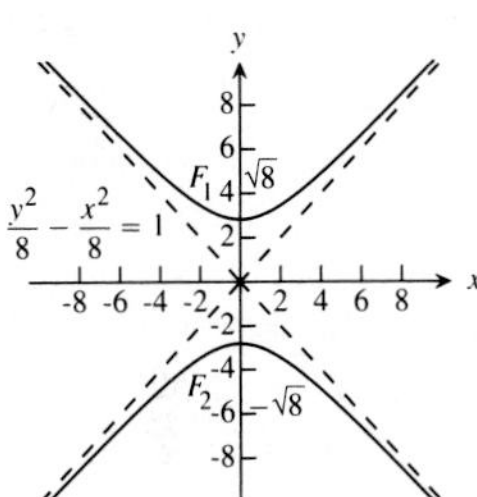

21. $\frac{x^2}{2} - \frac{y^2}{8} = 1$, $e = \sqrt{5}$, $y = \pm 2x$

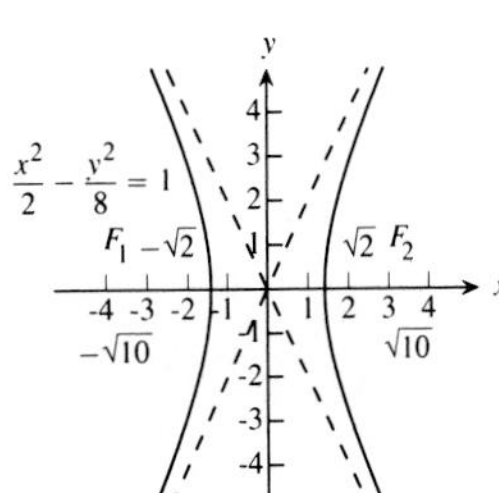

23. $\frac{y^2}{2} - \frac{x^2}{8} = 1$, $e = \sqrt{5}$, $y = \pm\frac{x}{2}$

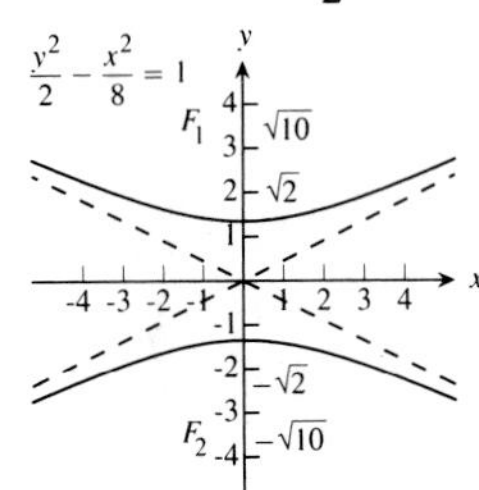

25. $\frac{x^2}{4} + \frac{y^2}{2} = 1$ **27.** $\frac{x^2}{27} + \frac{y^2}{36} = 1$ **29.** $\frac{x^2}{4851} + \frac{y^2}{4900} = 1$ **31.** $e = \frac{\sqrt{5}}{3}, \frac{x^2}{9} + \frac{y^2}{4} = 1$ **33.** $e = \frac{1}{2}, \frac{x^2}{64} + \frac{y^2}{48} = 1$ **35.** $y^2 - x^2 = 1$ **37.** $\frac{x^2}{9} - \frac{y^2}{16} = 1$ **39.** $y^2 - \frac{x^2}{8} = 1$ **41.** $x^2 - \frac{y^2}{8} = 1$ **43.** $e = \sqrt{2}, \frac{x^2}{8} - \frac{y^2}{8} = 1$ **45.** $e = 2, x^2 - \frac{y^2}{3} = 1$ **47.**

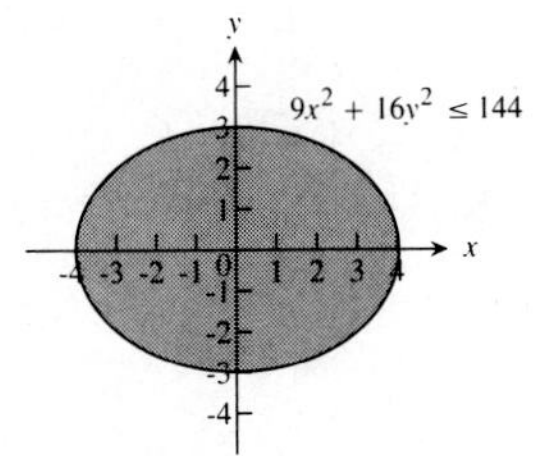

49.

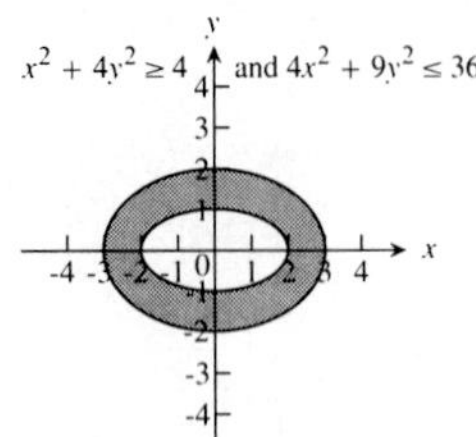

51.

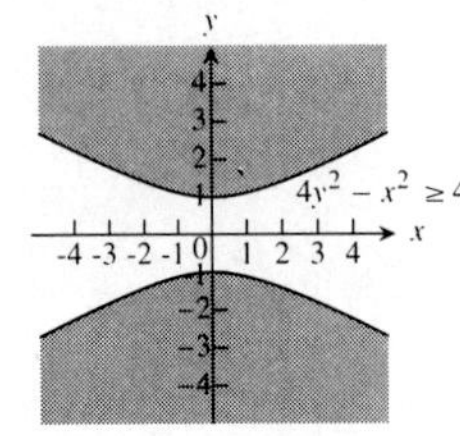

55. At (4, 0), $y = 2x - 8$; at (0, 0), $y = -2x$; at (0, 2), $y = 2x + 2$ **57.** b) 1:1 **59.** Length, $2\sqrt{2}$; height, $\sqrt{2}$ **61.** a) 24π b) 16π **63.** 24π

Section 9.2, pp. 648–649

1. Hyperbola **3.** Parabola **5.** Ellipse **7.** Parabola **9.** Hyperbola **11.** Hyperbola **13.** Ellipse **15.** Ellipse **17.** $x'^2 - y'^2 = 4$, hyperbola **19.** $4x'^2 + 16y' = 0$, parabola **21.** $y'^2 = 1$, parallel horizontal lines **23.** $2\sqrt{2}x'^2 + 8\sqrt{2}y' = 0$, parabola **25.** $4x'^2 + 2y'^2 = 19$, ellipse **27.** $A' = 0.88$, $B' = 0$, $C' = 3.12$, $D' = 0.74$, $E' = -1.20$, $F' = -3$, ellipse **29.** $A' = 0.00$, $B' = 0$, $C' = 5.00$, $D' = 0$, $E' = 0$, $F' = -5$, parallel lines **31.** $A' = 5.05$, $B' = 0$, $C' = -0.05$, $D' = -5.07$, $E' = -6.19$, $F' = -1$, hyperbola **33.** a) $\dfrac{x'^2}{b^2} + \dfrac{y'^2}{a^2} = 1$ b) $\dfrac{y'^2}{a^2} - \dfrac{x'^2}{b^2} = 1$ c) $x'^2 + y'^2 = a^2$ d) $y' = -\dfrac{1}{m}x'$ e) $y' = -\dfrac{1}{m}x' + \dfrac{b}{m}$ **35.** a) $x'^2 - y'^2 = 2$ b) $x'^2 - y'^2 = 2a$ **39.** a) Hyperbola b)

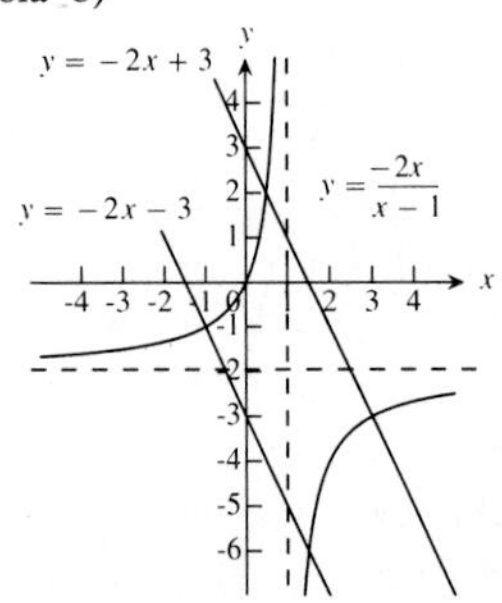

c) $y = -2x - 3$, $y = -2x + 3$ **41.** $x^2 + 4xy + 5y^2 - 1 = 0$, ellipse **49.** There is no unique line L, a directrix.

Section 9.3, pp. 656–657

1. $x^2 + y^2 = 1$

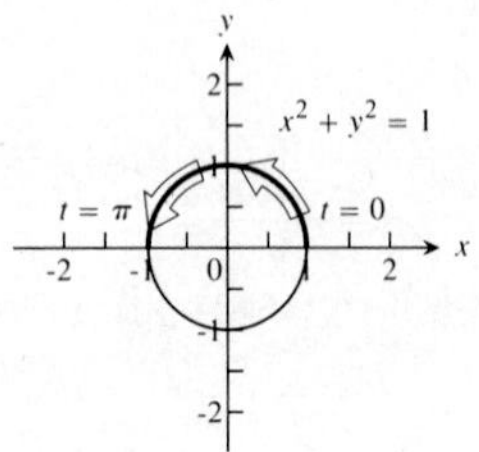

3. $x^2 + y^2 = 1$

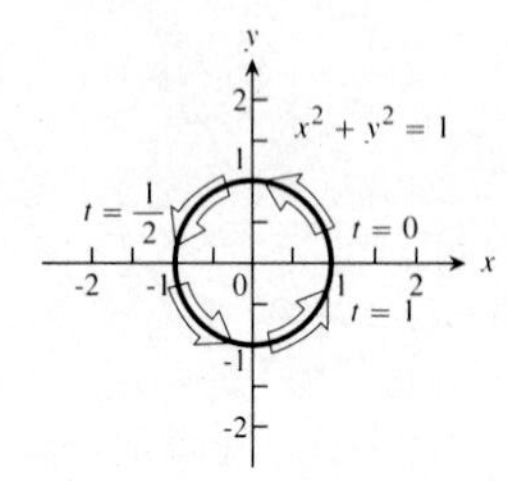

5. $\dfrac{x^2}{16} + \dfrac{y^2}{4} = 1$

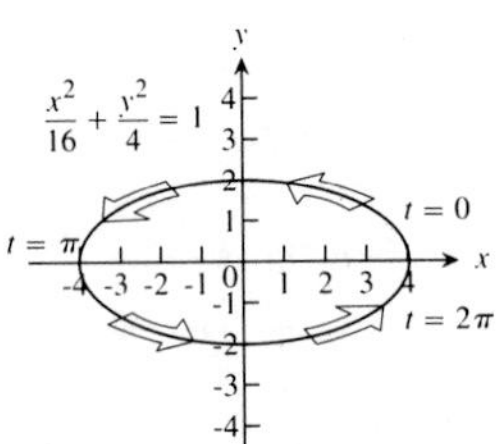

7. $\dfrac{x^2}{16} + \dfrac{y^2}{25} = 1$

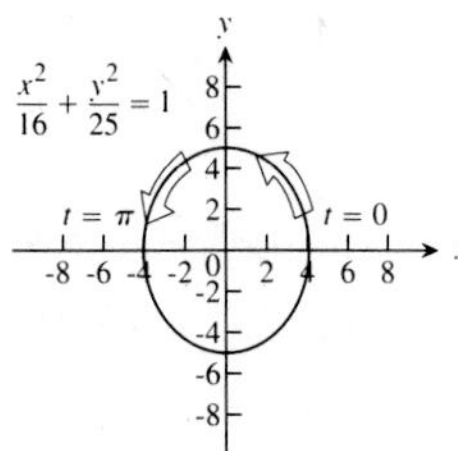

9. $y = x^2$

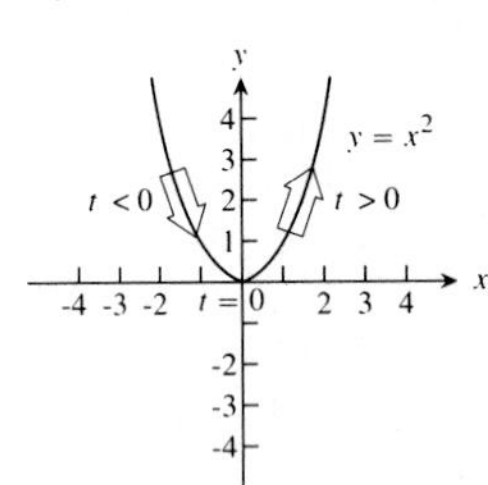

11. $x = y^2$

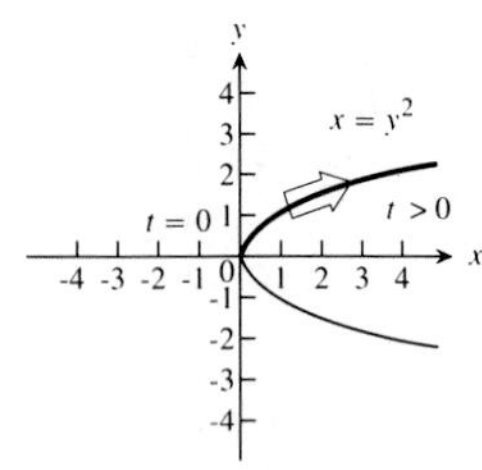

13. $x^2 - y^2 = 1$

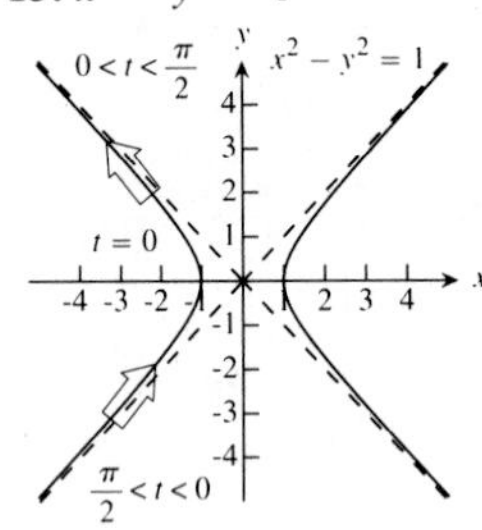

15. $y = 2x + 3$

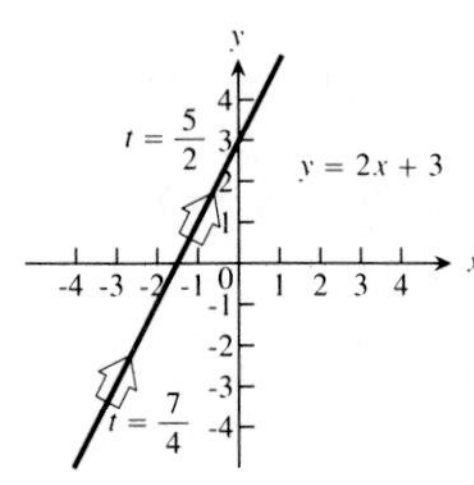

17. $y = 1 - x$

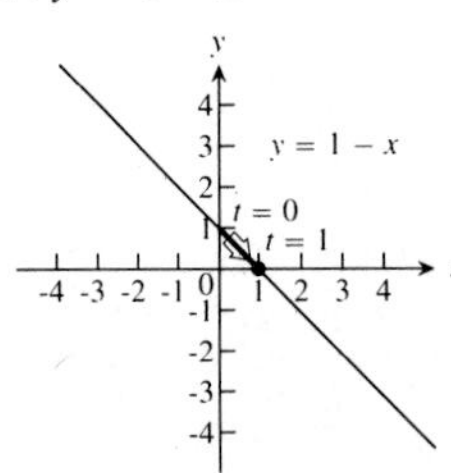

19. $y = \sqrt{1 - x^2}$

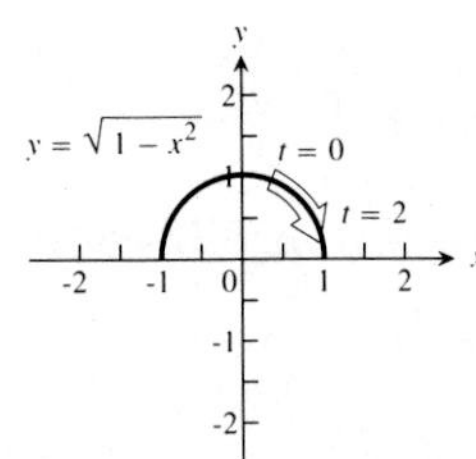

21. $y = \sqrt{x^2 + 1}$, $x \geq 0$

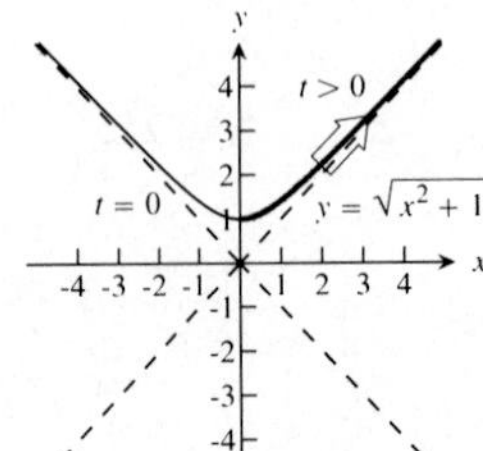

23. $x^2 - y^2 = 1$

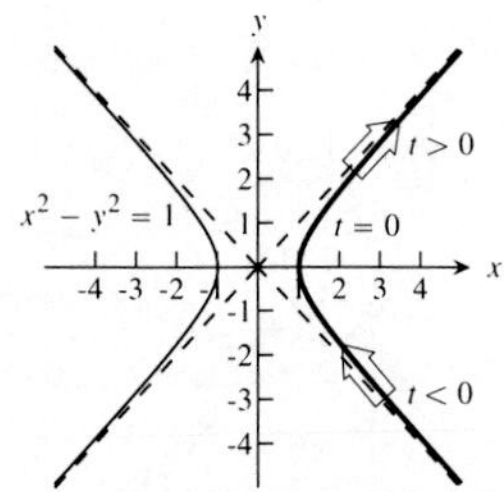

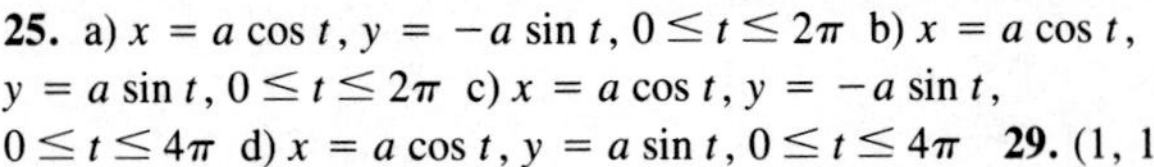

25. a) $x = a\cos t$, $y = -a\sin t$, $0 \leq t \leq 2\pi$ b) $x = a\cos t$, $y = a\sin t$, $0 \leq t \leq 2\pi$ c) $x = a\cos t$, $y = -a\sin t$, $0 \leq t \leq 4\pi$ d) $x = a\cos t$, $y = a\sin t$, $0 \leq t \leq 4\pi$ **29.** (1, 1)

31. b) $x = x_1 t$, $y = y_1 t$ c) $x = -1 + t$, $y = t$, or $x = -t$, $y = 1 - t$

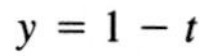

33. a)

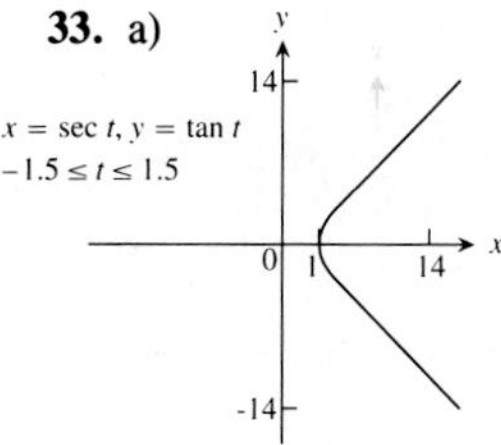

b)

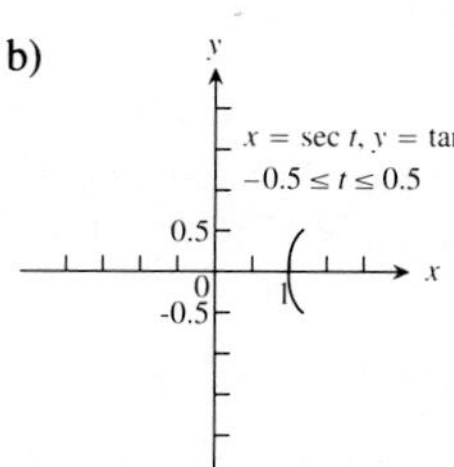

c)

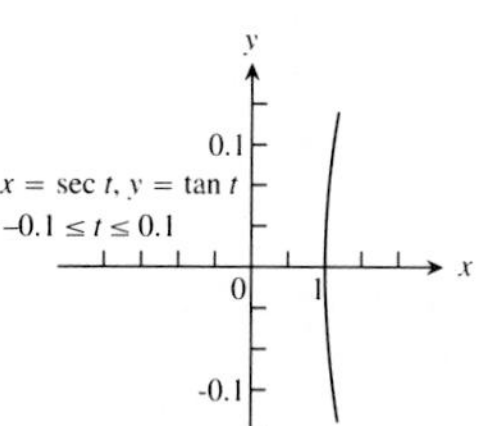

35. a)

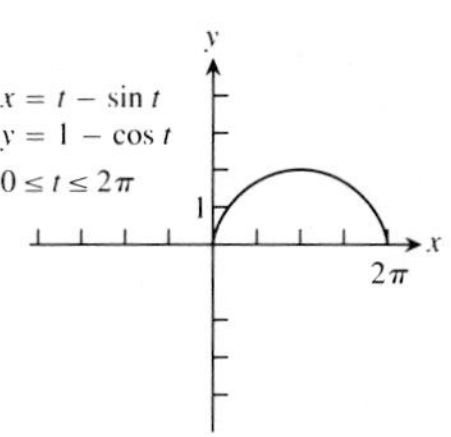

b)

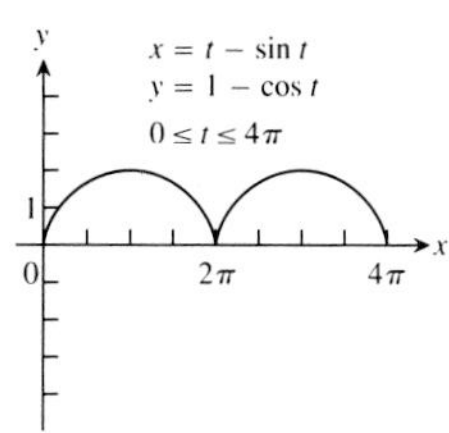

c)

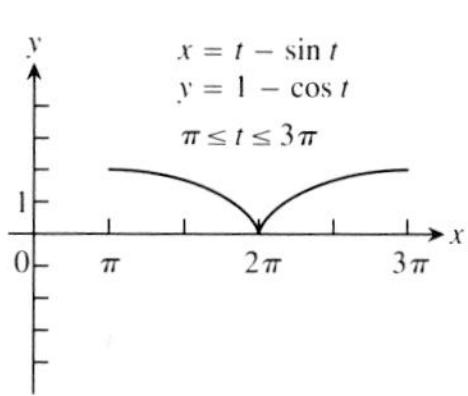

37. a)

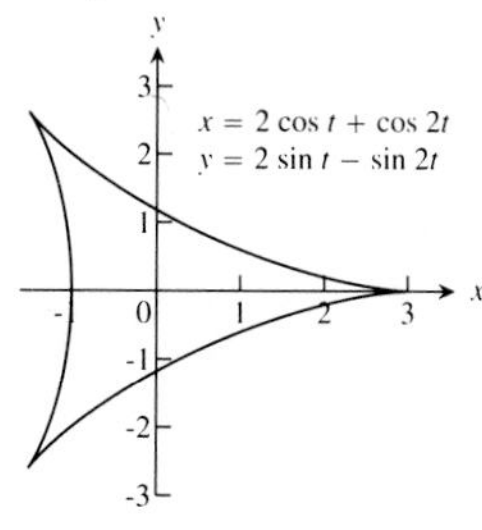

b)

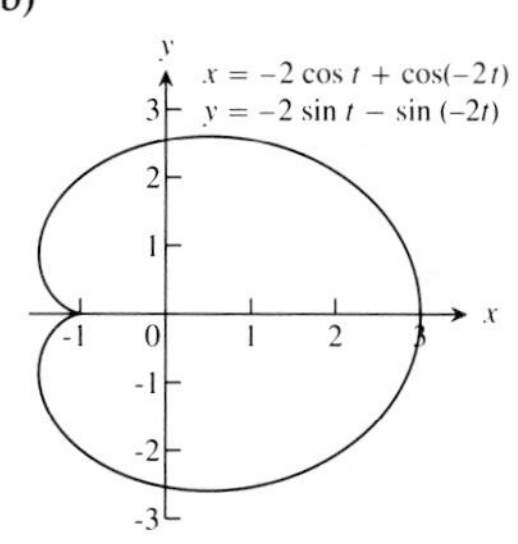

39. a)

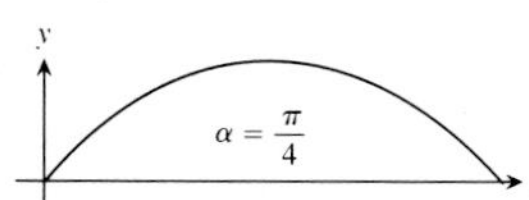

b)

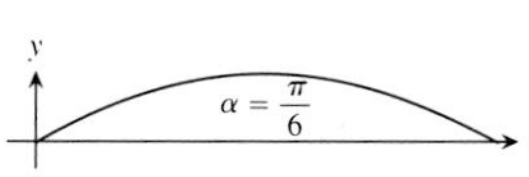

c)

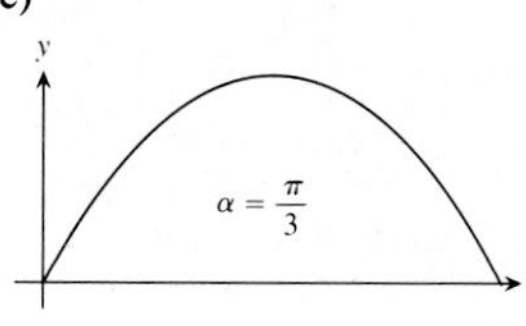

d)

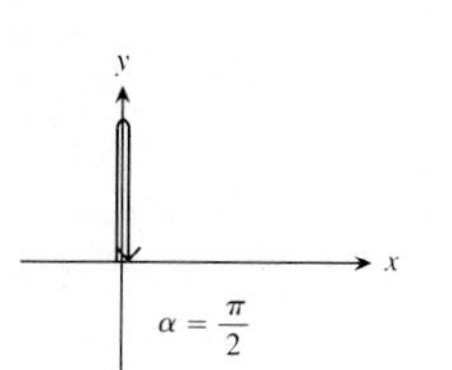

Section 9.4, pp. 662–664

1. $y = -x + 2\sqrt{2}$, $\dfrac{d^2y}{dx^2} = -\sqrt{2}$ **3.** $y = -\dfrac{1}{2}x + 2\sqrt{2}$, $\dfrac{d^2y}{dx^2} = -\dfrac{\sqrt{2}}{4}$ **5.** $y = x + \dfrac{1}{4}$, $\dfrac{d^2y}{dx^2} = -2$ **7.** $y = 2x - \sqrt{3}$, $\dfrac{d^2y}{dx^2} = -3\sqrt{3}$ **9.** $y = x - 4$, $\dfrac{d^2y}{dx^2} = \dfrac{1}{2}$ **11.** $y = \sqrt{3}x - \dfrac{\pi\sqrt{3}}{3} + 2$, $\dfrac{d^2y}{dx^2} = -4$ **13.** 4 **15.** 12 **17.** π^2 **19.** $8\pi^2$ **21.** $\dfrac{52\pi}{3}$ **23.** $3\pi\sqrt{5}$ **25.** a) $(\bar{x}, \bar{y}) = \left(\dfrac{12}{\pi} - \dfrac{24}{\pi^2}, \dfrac{24}{\pi^2} - 2\right)$ b)

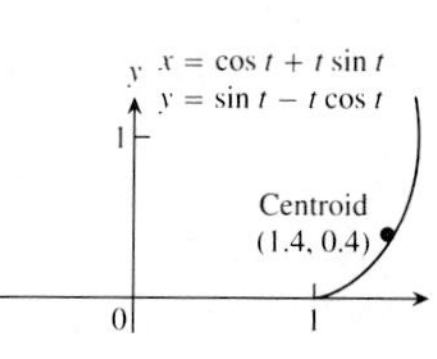

27. a) $(\bar{x}, \bar{y}) = \left(\dfrac{1}{3}, \pi - \dfrac{4}{3}\right)$ b)

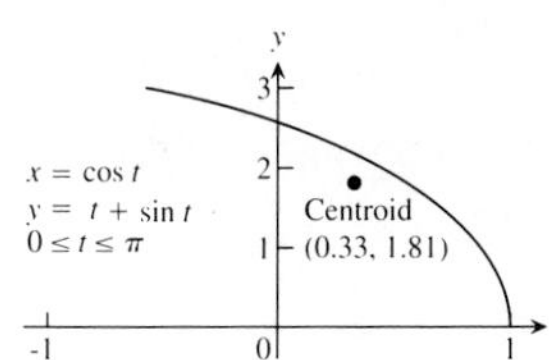

29. a) π b) π **31.** $\left(\dfrac{\sqrt{2}}{2}, 1\right)$, $y = 2x$, $y = -2x$

33.

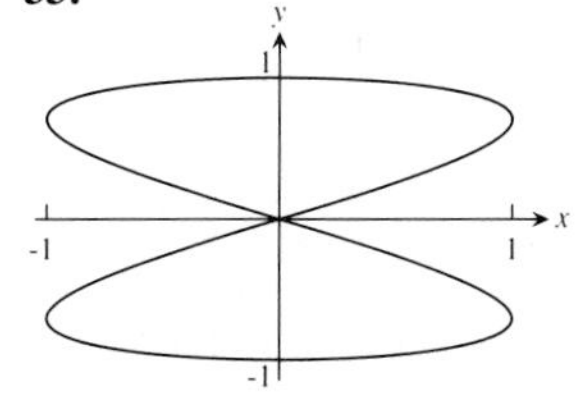

35.

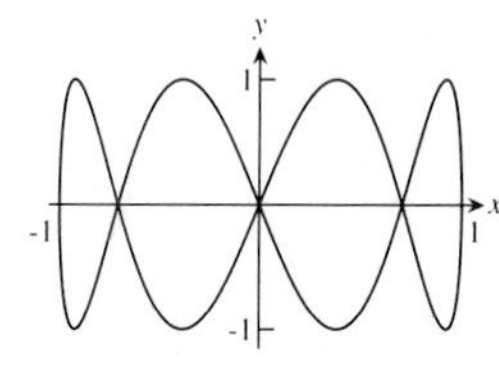

37.

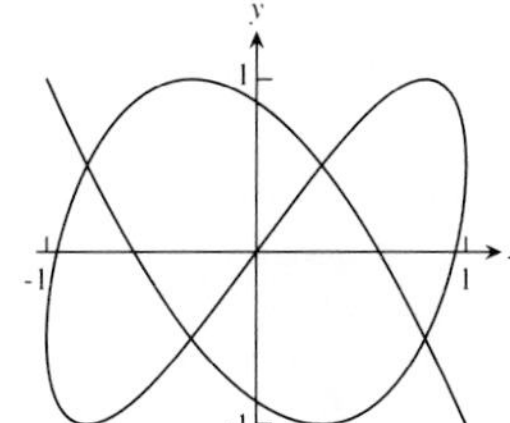

39.

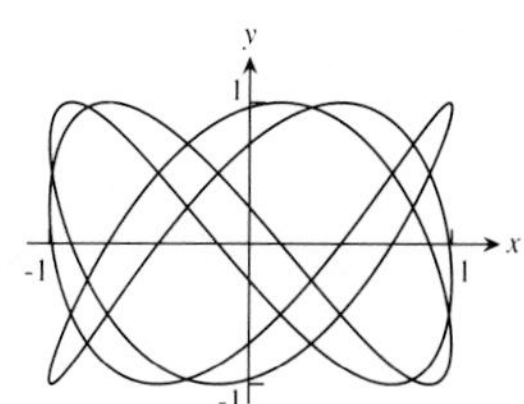

Section 9.5, pp. 669–670

1. a, c; b, d; e, k; g, j; h, f; i, l; m, o; n, p **3.** a) $\left(2, \dfrac{\pi}{2} + 2n\pi\right)$ and $\left(-2, \dfrac{\pi}{2} + (2n + 1)\pi\right)$, n an integer b) $(2, 2n\pi)$ and $(-2, (2n + 1)\pi)$, n an integer c) $\left(2, \dfrac{3\pi}{2} + 2n\pi\right)$ and

$\left(-2, \frac{3\pi}{2} + (2n+1)\pi\right)$, n an integer d) $(2, (2n+1)\pi)$ and $(-2, 2n\pi)$, n an integer

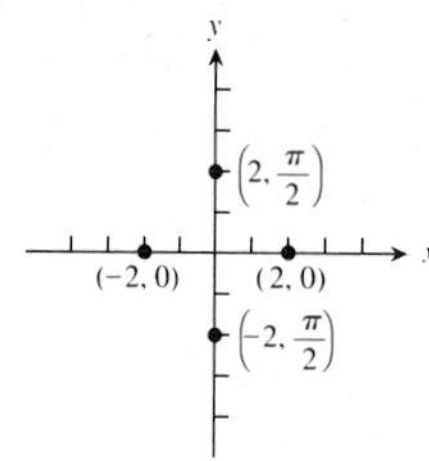

5. a) (1, 1) b) (1, 0) c) (0, 0) d) $(-1, -1)$ e) $\left(\frac{3\sqrt{3}}{2}, -\frac{3}{2}\right)$ f) (3, 4) g) (1, 0) h) $(-\sqrt{3}, 3)$ **7.**

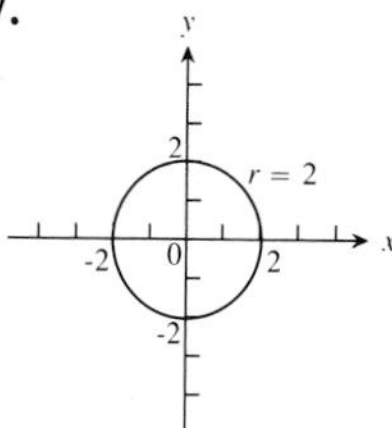

9.

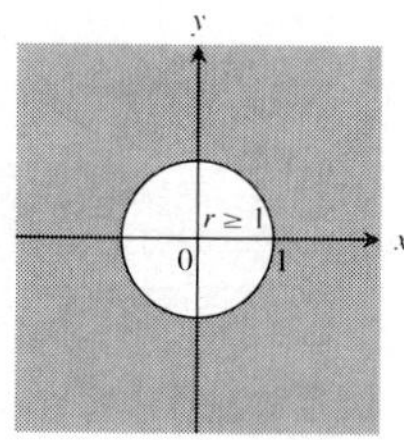

$0 \le \theta \le \frac{\pi}{6}, r \ge 0$

13.

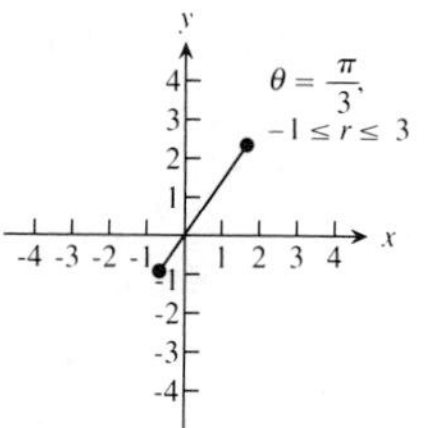

15.

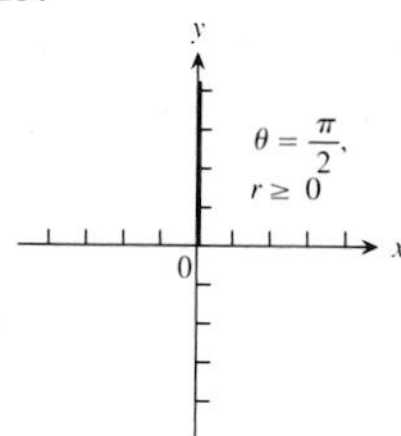

17.

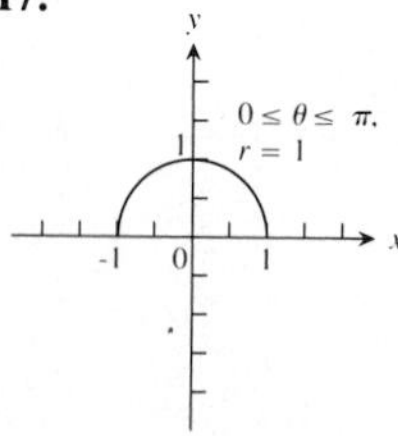

19.

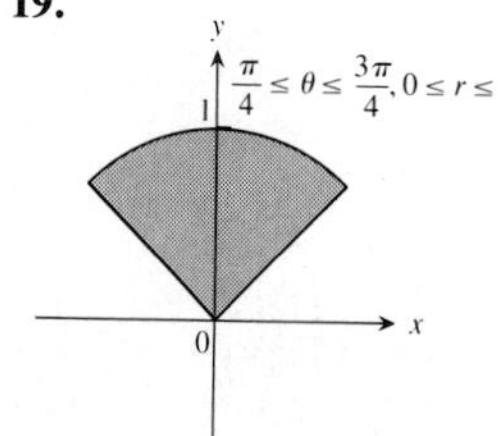

21.

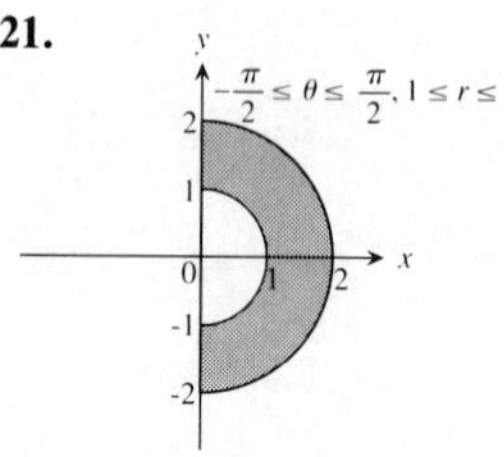

23. $x = 2$, vertical line through (2, 0) **25.** $y = 4$, horizontal line through (0, 4) **27.** $y = 0$, the x-axis **29.** $x + y = 1$, line, $m = -1$, $b = 1$ **31.** $x^2 + y^2 = 1$, circle, $C(0, 0)$, $r = 1$ **33.** $y - 2x = 5$, line, $m = 2$, $b = 5$ **35.** $y^2 = x$, parabola, $V(0, 0)$, opens right **37.** $y = e^x$, natural exponential function **39.** $x^2 + 2xy + y^2 = 1$, parabola, rotated 45° **41.** $(x-1)^2 + (y-1)^2 = 2$, circle, $C(1, 1)$, $r = \sqrt{2}$ **43.** $r \cos\theta = 7$ **45.** $\theta = \frac{\pi}{4}$ **47.** $r = 2$ or $r = -2$ **49.** $4r^2\cos^2\theta + 9r^2\sin^2\theta = 36$ **51.** $r^2\sin^2\theta = 4r\cos\theta$

Section 9.6, pp. 676–677

1.

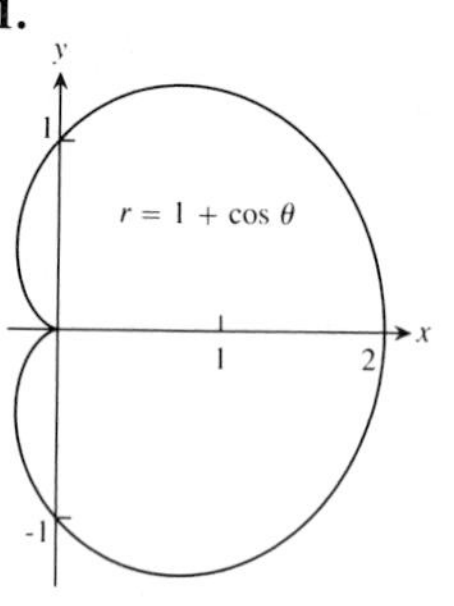

3.

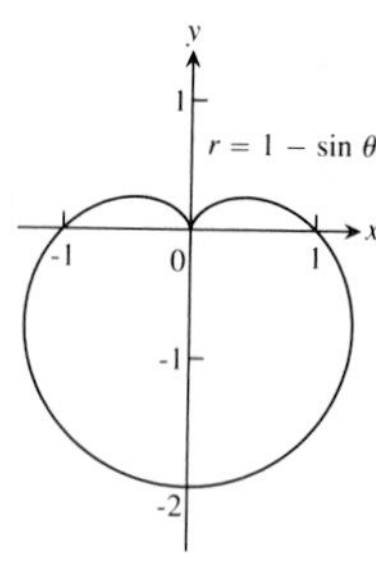

5.

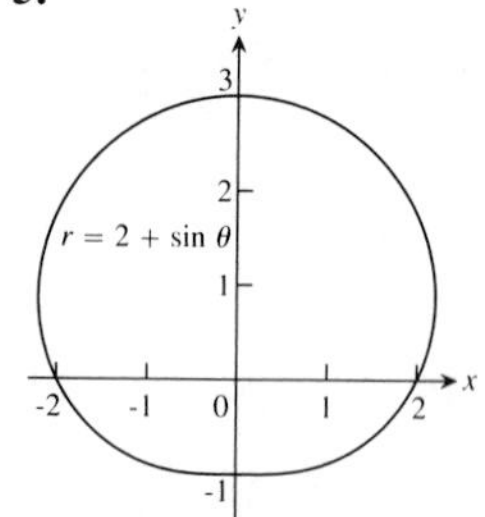

7.

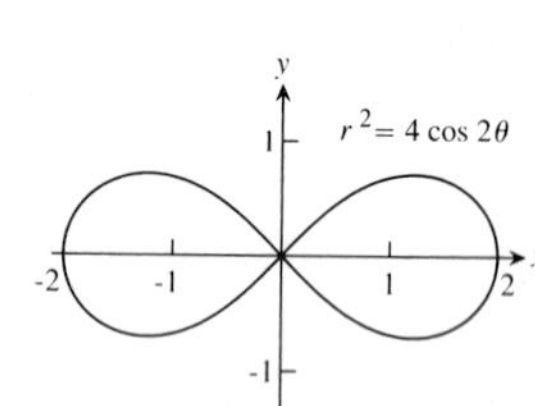

9.

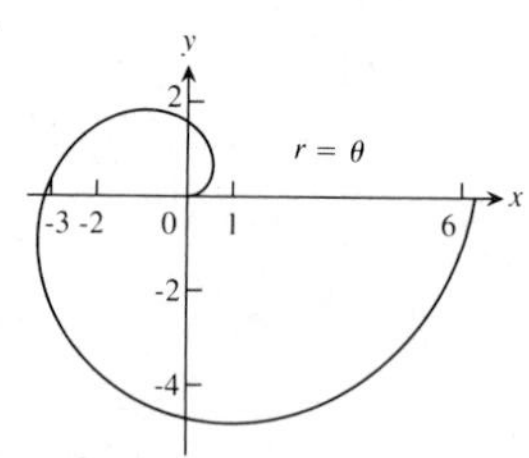

11. Slope at $(-1, \pi/2) = -1$; slope at $(-1, -\pi/2) = 1$

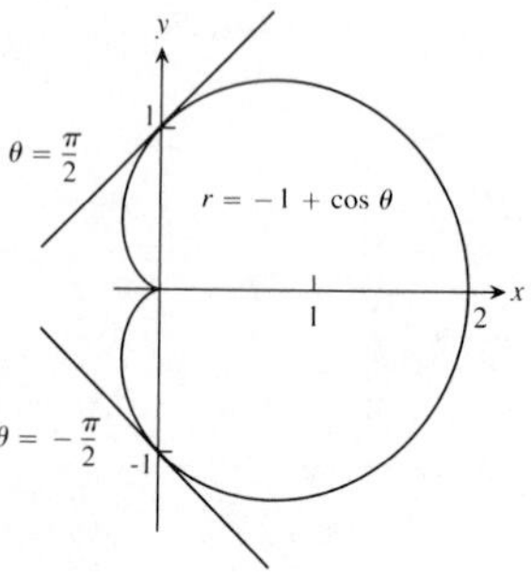

13. Slope at $(1, \pi/4) = -1$; slope at $(-1, -\pi/4) = 1$; slope at $(-1, 3\pi/4) = 1$; slope at $(1, -3\pi/4) = -1$; slope at $(0, 0)$ and $(0, \pi) = 0$; slope at $(0, \pi/2)$ and $(0, 3\pi/2)$ undefined

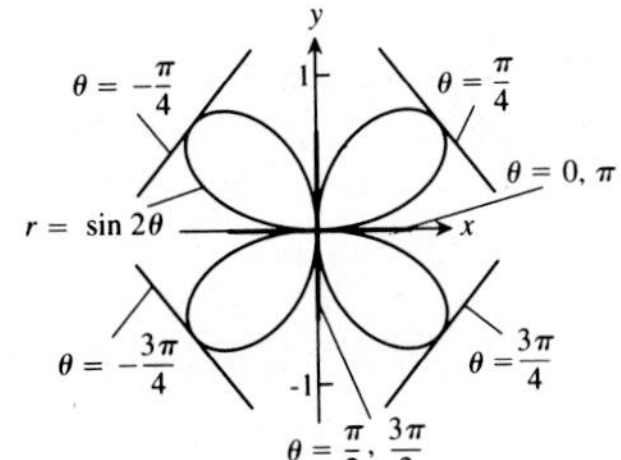

15.

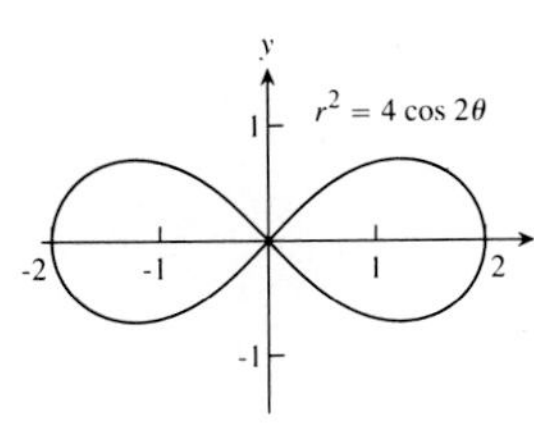

17. a)

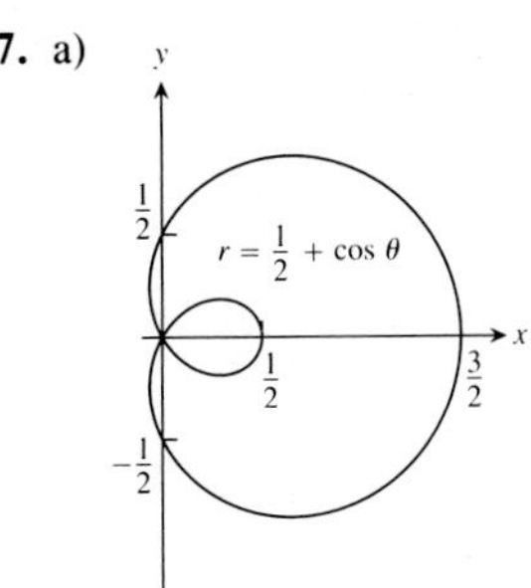

b)

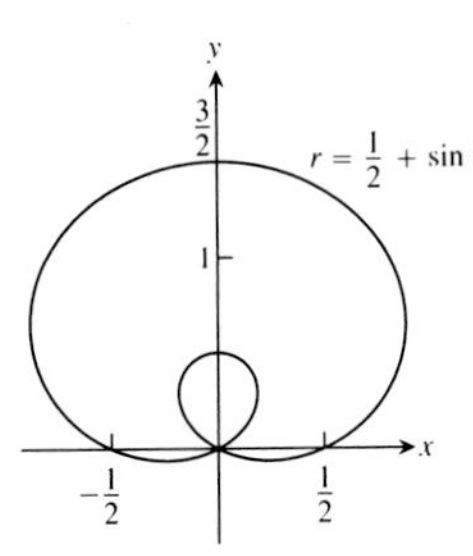

19. a)

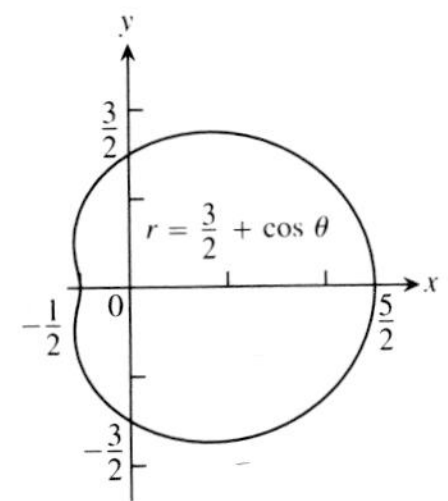

b)

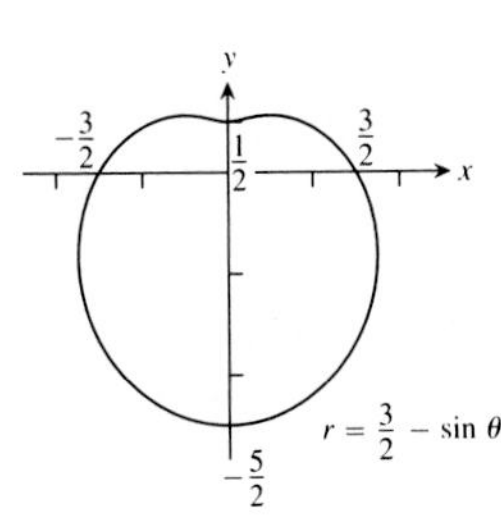

21.

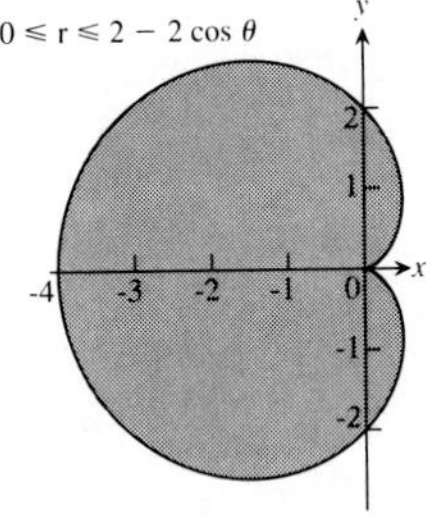

25. $(1, \pi/2)$, $(1, 3\pi/2)$, $(0, 0)$

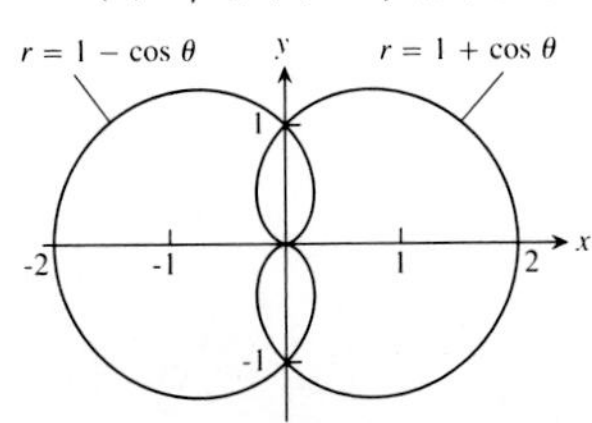

27. $(2\sqrt{2} - 1)$, $\sin^{-1}(3 - 2\sqrt{2}))$, $(0, 0)$, $(2, 3\pi/2)$, $(-2(\sqrt{2} - 1)$, $-\sin^{-1}(3 - 2\sqrt{2}))$

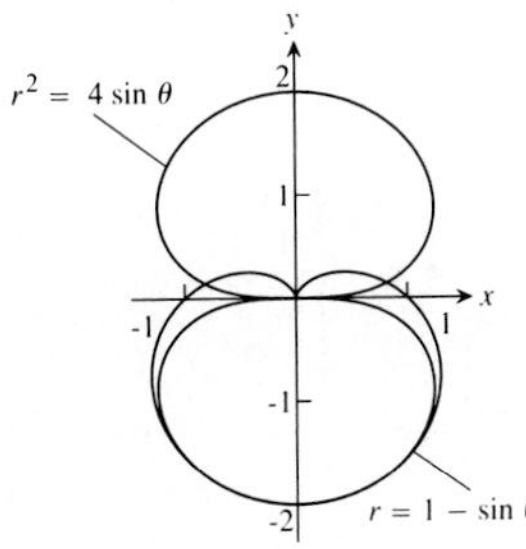

29. $(1, \pi/12)$, $(1, 5\pi/12)$, $(1, 7\pi/12)$, $(1, 11\pi/12)$, $(1, 13\pi/12)$, $(1, 17\pi/12)$, $(1, 19\pi/12)$, $(1, 23\pi/12)$

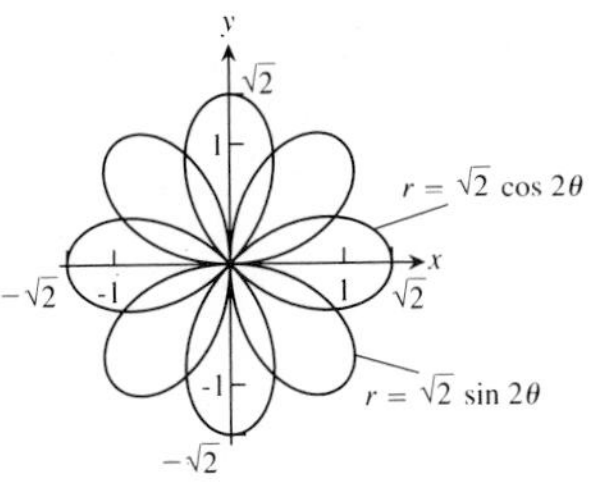

31. $(1, \pi/8)$, $(1, 9\pi/8)$, $(-1, 5\pi/8)$, $(-1, 13\pi/8)$, $(1, 3\pi/8)$, $(1, 7\pi/8)$, $(1, 11\pi/8)$, $(1, 15\pi/8)$

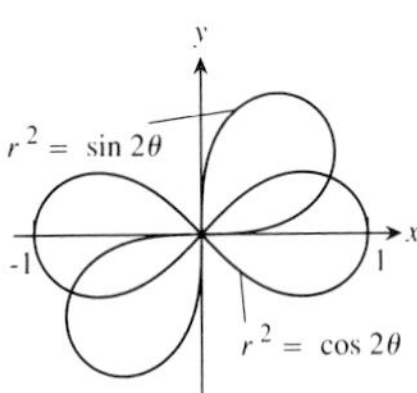

35. $(\pm 1/\sqrt[4]{2}, \pi/8)$, $(0, 0)$

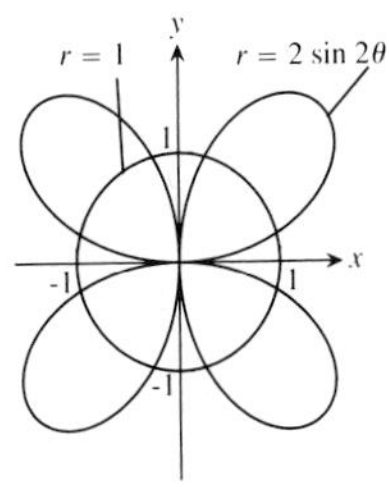

37. $(1, \pi/12)$, $(1, 5\pi/12)$, $(1, 13\pi/12)$, $(1, 17\pi/12)$, $(1, 7\pi/12)$, $(1, 11\pi/12)$, $(1, 19\pi/12)$, $(1, 23\pi/12)$

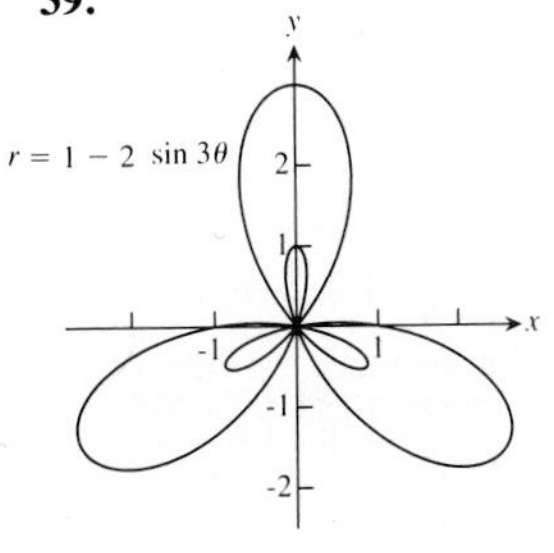

39.

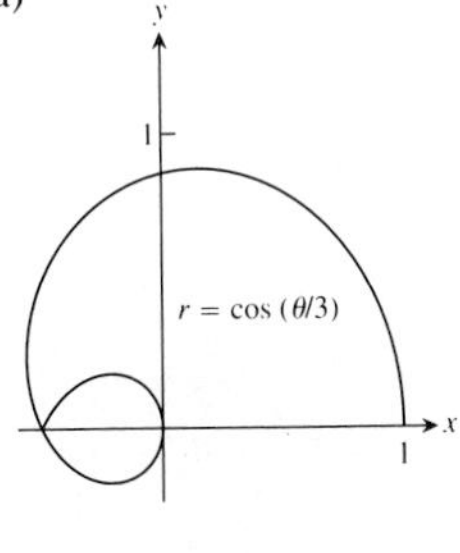

41. a)

b)

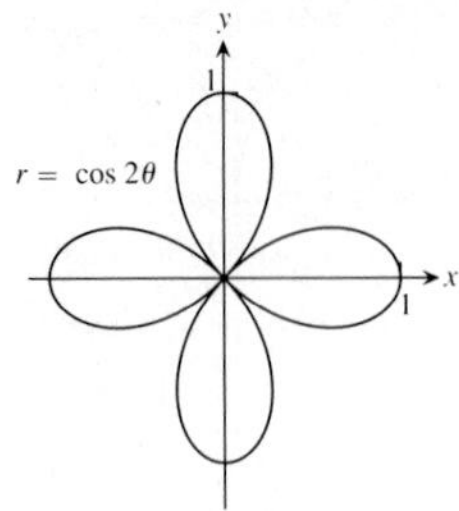

c)

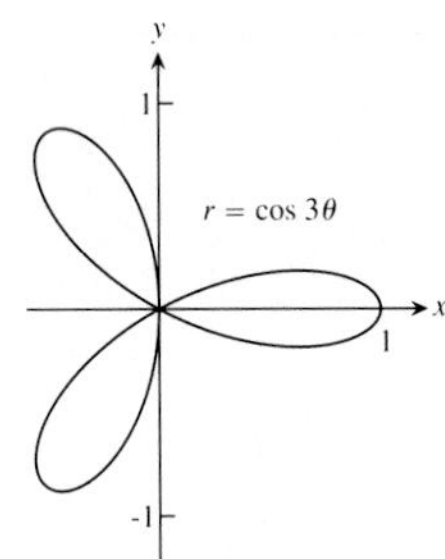

d)

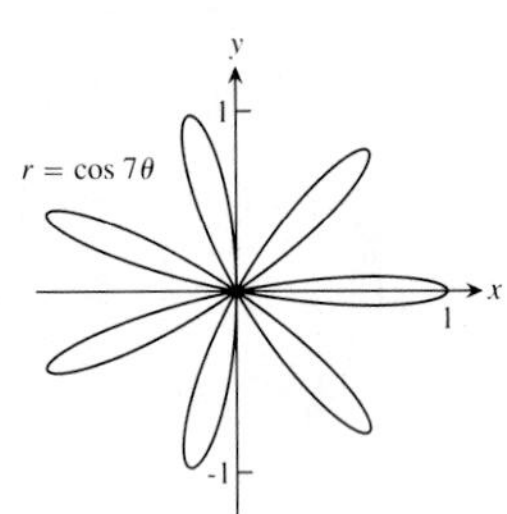

Section 9.7, pp. 682–683

1. $r \cos(\theta - \pi/6) = 5;\ \sqrt{3}x + y = 10$

3. $r \cos(\theta - 4\pi/3) = 3;\ x + \sqrt{3}y = -6$

5. $x + y = 2$ **7.** $y = -1$

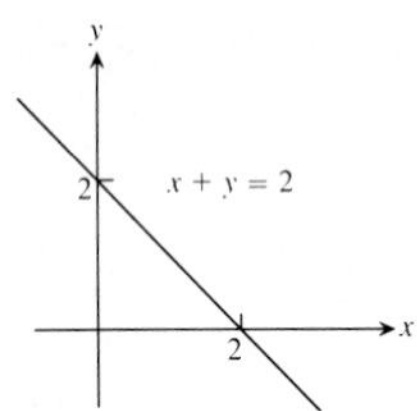

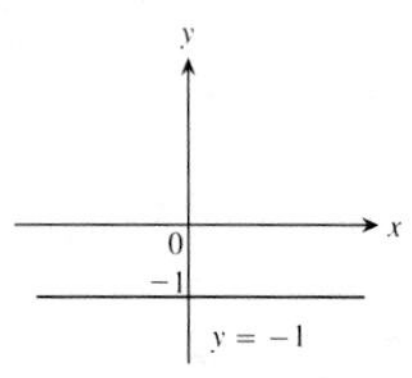

9. $r = 8 \cos \theta$ **11.** $r = 2\sqrt{2} \sin \theta$

13.

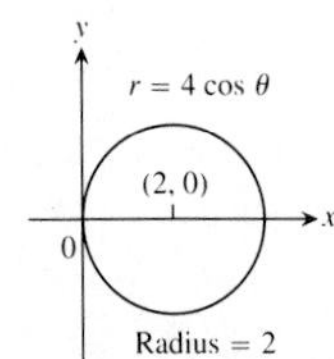

15.

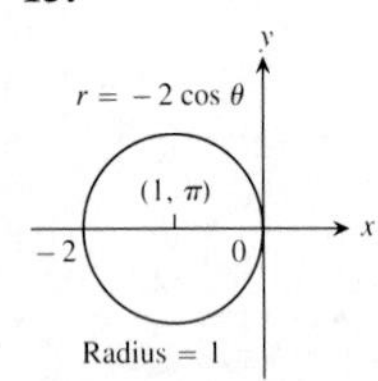

17. $r = \dfrac{2}{1 + \cos \theta}$ **19.** $r = \dfrac{8}{1 + 2\cos \theta}$ **21.** $r = \dfrac{1}{2 + \cos \theta}$

23. $r = \dfrac{10}{5 - \sin \theta}$

25.

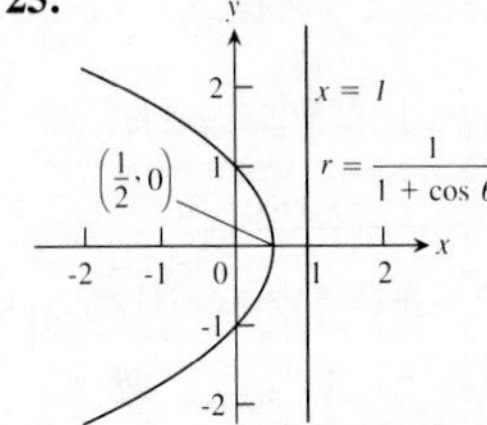

27.

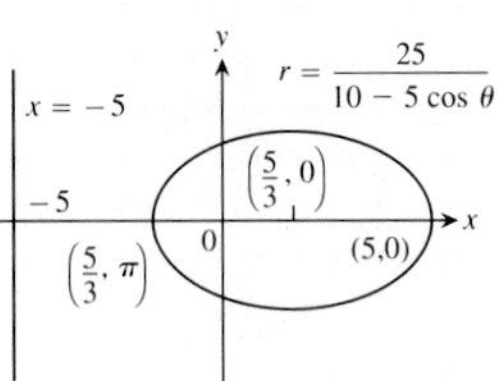

29.

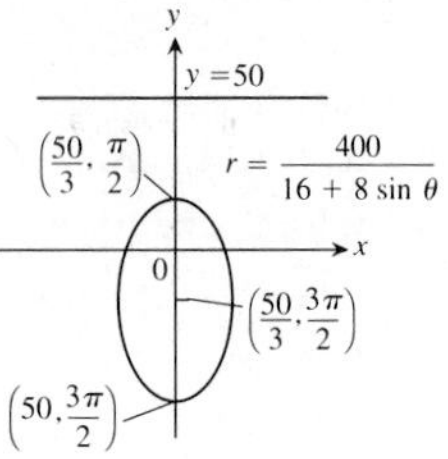

31.

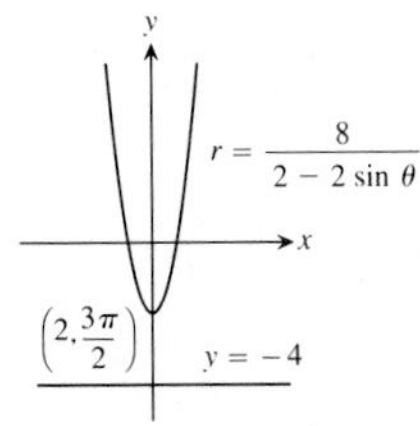

33.

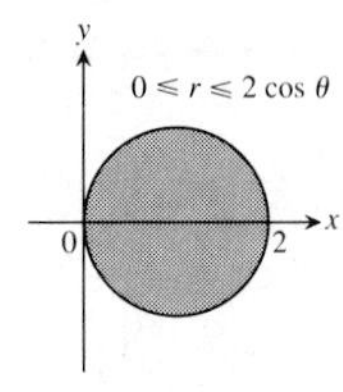

35. b)

Planet	Perihelion	Aphelion
Mercury	0.3075 AU	0.4667 AU
Venus	0.7184 AU	0.7282 AU
Earth	0.9833 AU	1.0167 AU
Mars	1.3817 AU	1.6663 AU
Jupiter	4.9512 AU	5.4548 AU
Saturn	9.0210 AU	10.0570 AU
Uranus	18.2977 AU	20.0623 AU
Neptune	29.8135 AU	30.3065 AU
Pluto	29.6549 AU	49.2251 AU

37. a) $x^2 + (y - 1)^2 = 1;\ y = 1$ b)

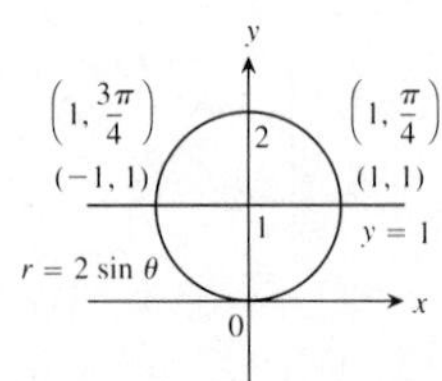

39. $r = \dfrac{4}{1 + \cos \theta}$ **43.** 2 in. apart

45.

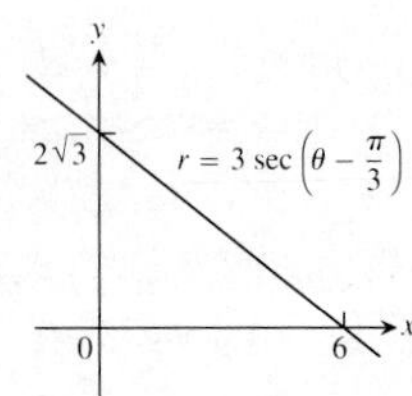

47.

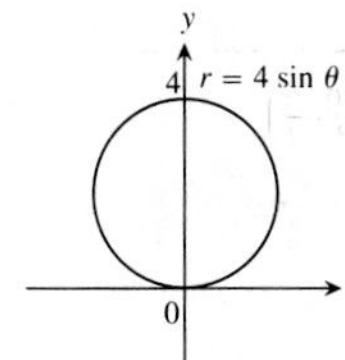

49.

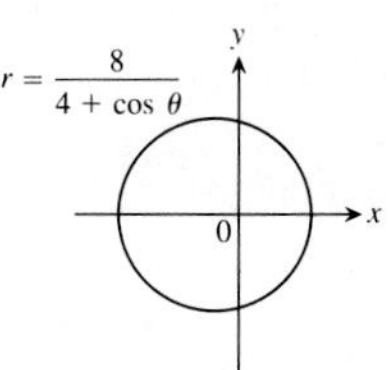

51.

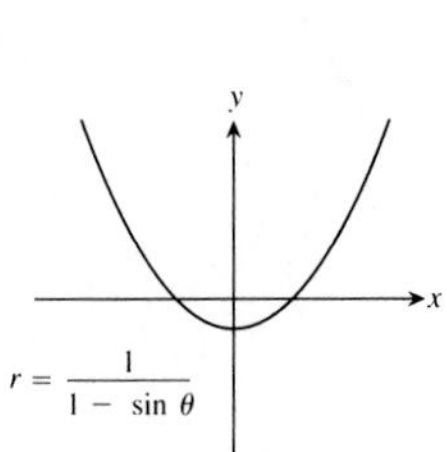

53.

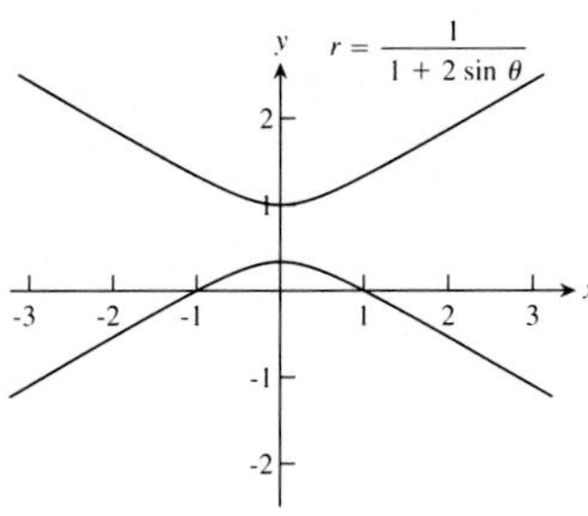

Section 9.8, pp. 689–690

1. 18π **3.** $\pi/8$ **5.** 2 **7.** $\pi/2 - 1$ **9.** $5\pi - 8$ **11.** $3\sqrt{3} - \pi$ **13.** $\pi/3 + \sqrt{3}/2$ **15.** $9\pi - 18$ **17.** a) 1.12 b) Yes **19.** 19/3 **21.** 8 **23.** $\pi/4 + 3/8$ **25.** $1/2 \ln(2 + \sqrt{3})$ **27.** $\pi\sqrt{2}$ **29.** $\pi(2 - \sqrt{2})$ **33.** $\left(\dfrac{5}{6}a, 0\right)$

Section 9 Miscellaneous Exercises, pp. 691–698

1.

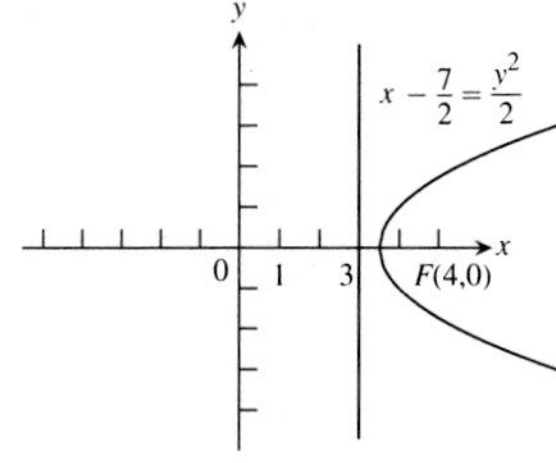

3. $3x^2 + 3y^2 - 16y + 16 = 0$

5. $(0, \pm 1)$ **7.** a) $\dfrac{(y-1)^2}{16} - \dfrac{x^2}{48} = 1$ b) $\dfrac{16\left(y + \frac{3}{4}\right)^2}{25} - \dfrac{2x^2}{75} = 1$

13. 6.5×10^6 mi **15.** a) $\displaystyle\int_0^a \sqrt{a^2 - x^2}\, dx$ b) $\displaystyle\int_0^a b/a\sqrt{a^2 - x^2}\, dx$ c) $b/a(\pi a^2)$

17. $(5, 10\sqrt{10}/3)$, where (0, 0) is midway between the two stations **19.** The listener is on a branch of a hyperbola with foci at the rifle and the target.

23.

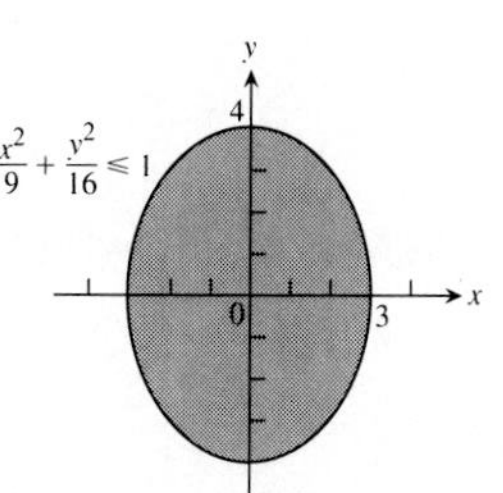

25.

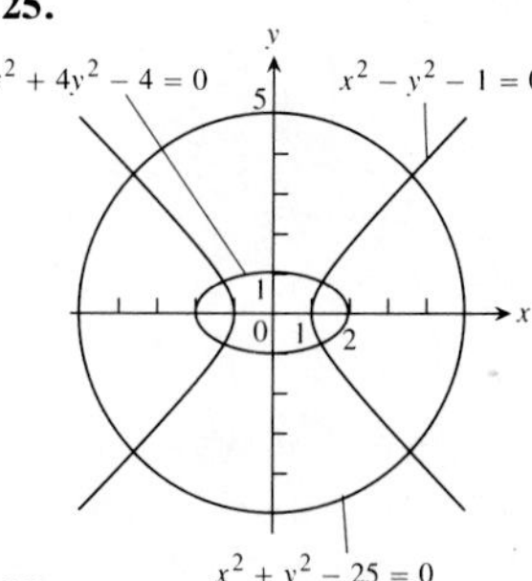

27.

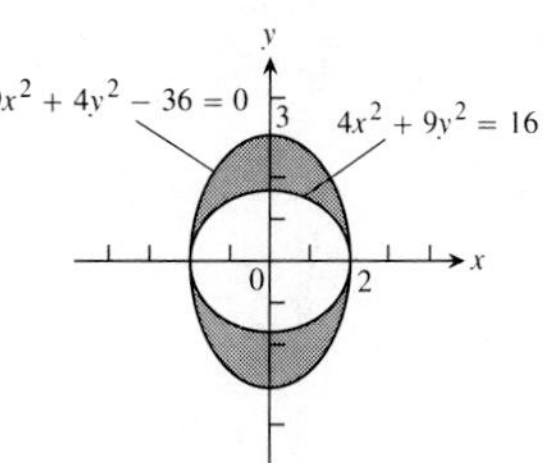

29.

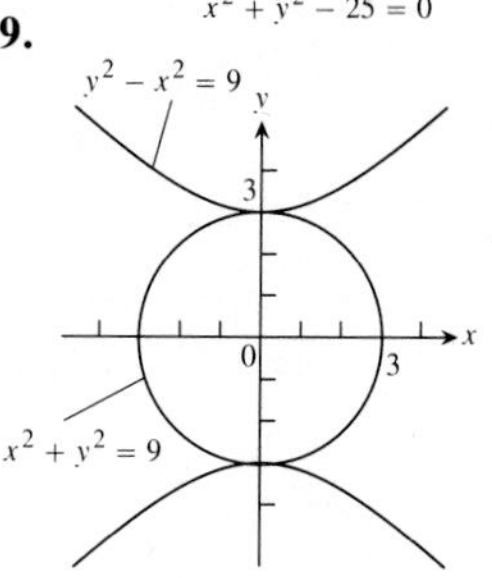

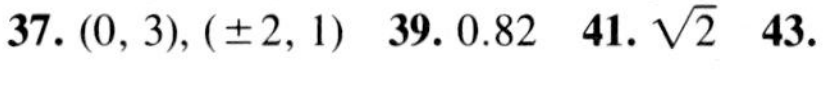

31. Ellipse **33.** Parabola **35.** Ellipse; $5x'^2 + 3y'^2 - 30 = 0$ **37.** (0, 3), $(\pm 2, 1)$ **39.** 0.82 **41.** $\sqrt{2}$ **43.**

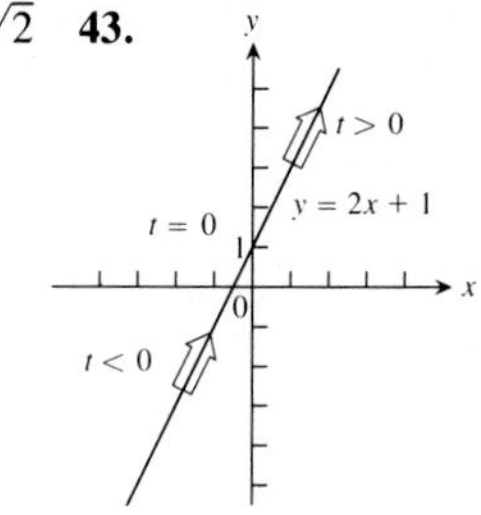

45.

47.

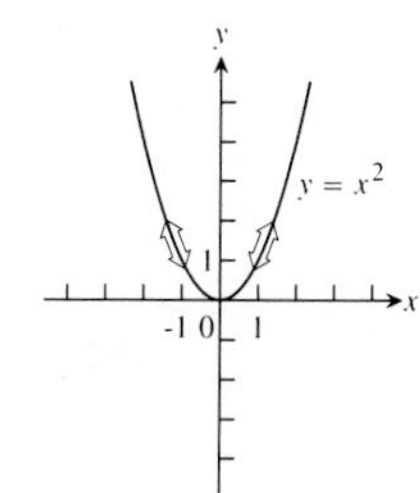

49. $x = 3 \cos t,\ y = 4 \sin t,\ 0 \leq t \leq 2\pi$

51. $x = (a + b)\cos \theta - b \cos\left(\dfrac{a+b}{b}\theta\right)$,

$y = (a + b)\sin \theta - b \sin\left(\dfrac{a+b}{b}\theta\right)$

53. P traces the diameter of the circle. **55.** $y = \dfrac{\sqrt{3}}{2}x + \dfrac{1}{4}$, $\dfrac{d^2y}{dx^2} = \dfrac{1}{4}$ **57.** $3 + \ln 2/8$ **59.** $\dfrac{76\pi}{3}$ **61.** $\left(\pi a, \dfrac{5}{6}a\right)$

63. a) $x = \ln(t + 2),\ y = t^2 + 1$ b) $y = e^{2x} - 4e^x + 5$ c) $x = \ln(\sqrt{y - 1} + 2)$ d) $\dfrac{4}{\ln 2}$ e) 6 **65.** $M_x = \dfrac{36}{5}\sqrt{3}$, $M_y = \dfrac{72}{7}\sqrt{3}$ **67.** d **69.** l **71.** k **73.** i

75.

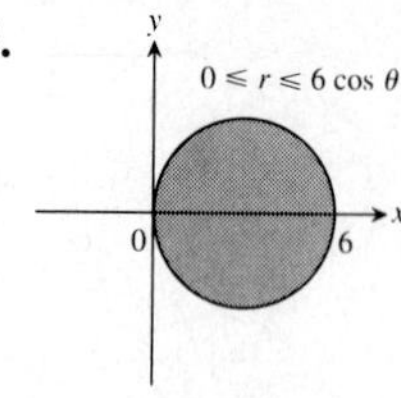

77. (0, 0) **79.** All points of $r = 1 + \sin\theta$
81. $(a, 0), (a, \pi)$ **83.** (0, 0) **85.** $y = x, y = -x$
87. $r\cos(\theta - \pi/4) = 1; r\cos(\theta - 3\pi/4) = 1;$
$r\cos(\theta - 5\pi/4) = 1; r\cos(\theta - 7\pi/4) = 1$

89.

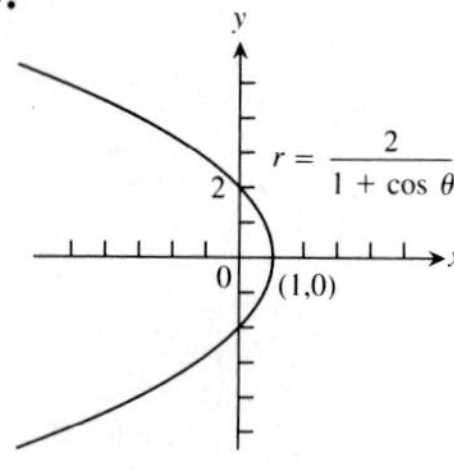

91.

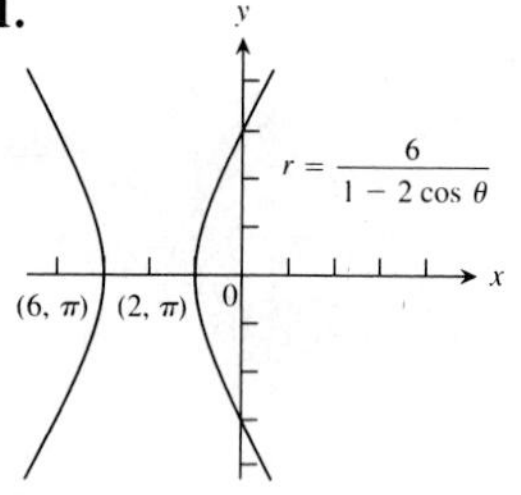

93. $r = \dfrac{4}{1 + 2\cos\theta}$ **95.** $r = \dfrac{2}{2 + \sin\theta}$

97. a) $r = \dfrac{2a}{1 + \cos\left(\theta - \frac{\pi}{4}\right)}$ b) $r = \dfrac{8}{3 - \cos\theta}$

c) $r = \dfrac{3}{1 + 2\sin\theta}$ **99.** $\dfrac{9}{2}\pi$ **101.** $2 + \dfrac{\pi}{4}$ **103.** $2\pi\left(1 - \dfrac{\sqrt{2}}{2}\right)$

105. a) $r = e^{2\theta}$ b)

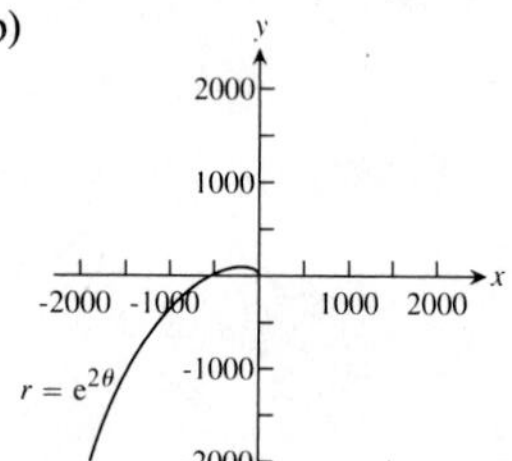

c) $\dfrac{\sqrt{5}e^{4\pi}}{2} - \dfrac{\sqrt{5}}{2}$ **107.** $\dfrac{32\pi - 4\pi\sqrt{2}}{5}$

111. 0 **115.** $\left(0, \pm\dfrac{\pi}{3}\right), \dfrac{\pi}{2}$ **119.** $\dfrac{\pi}{2}$ **121.** $\dfrac{\pi}{4}$

Chapter 10

Section 10.1, pp. 707–708

1. a)

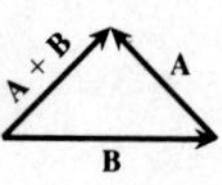

b)

c)

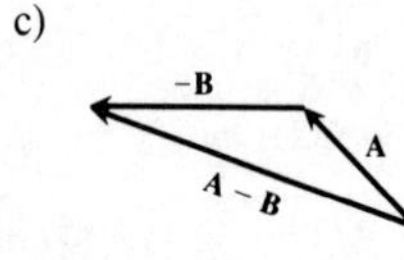

d)

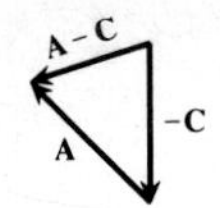

3. $3\,\mathbf{i} - \mathbf{j}$ **5.** $-\mathbf{i} - 9\,\mathbf{j}$

7. $-\ln 2\,\mathbf{i} + (2 - \pi)\,\mathbf{j}$

9. $\mathbf{i} - 4\,\mathbf{j}$

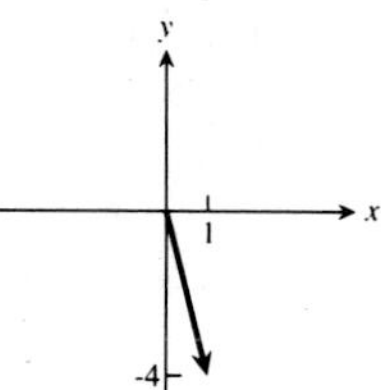

11. $-2\,\mathbf{i} - 3\,\mathbf{j}$

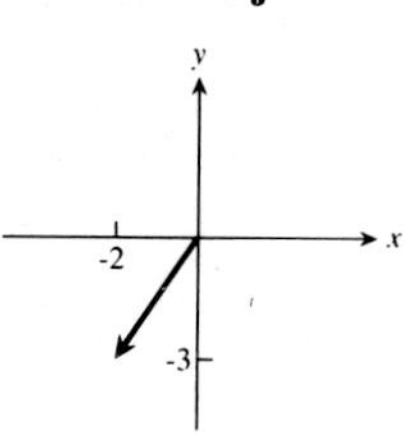

13. $\mathbf{u} = \dfrac{\sqrt{3}}{2}\,\mathbf{i} + \dfrac{1}{2}\,\mathbf{j}$, when $\theta = \dfrac{\pi}{6}$; $\mathbf{u} = -\dfrac{1}{2}\,\mathbf{i} + \dfrac{\sqrt{3}}{2}\,\mathbf{j}$, when $\theta = \dfrac{2\pi}{3}$

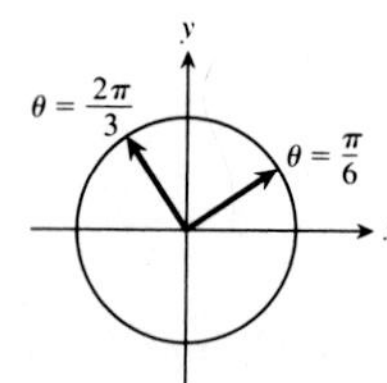

15. $\mathbf{u} = \dfrac{\sqrt{3}}{2}\,\mathbf{i} - \dfrac{1}{2}\,\mathbf{j}$

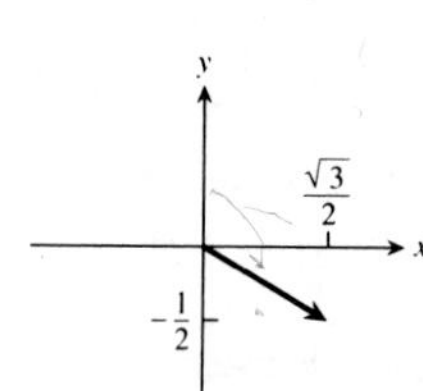

17. $\mathbf{u} = \dfrac{1}{\sqrt{17}}\,\mathbf{i} + \dfrac{4}{\sqrt{17}}\,\mathbf{j}; -\mathbf{u} = -\dfrac{1}{\sqrt{17}}\,\mathbf{i} - \dfrac{4}{\sqrt{17}}\,\mathbf{j}; \mathbf{n} = \dfrac{4}{\sqrt{17}}\,\mathbf{i} - \dfrac{1}{\sqrt{17}}\,\mathbf{j}; -\mathbf{n} = -\dfrac{4}{\sqrt{17}}\,\mathbf{i} + \dfrac{1}{\sqrt{17}}\,\mathbf{j}$

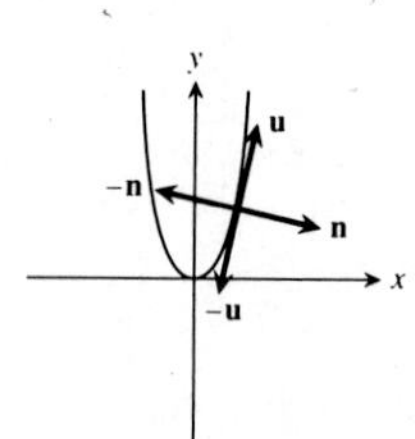

19. $\mathbf{u} = \dfrac{1}{\sqrt{5}}(2\,\mathbf{i} + \mathbf{j}); -\mathbf{u} = \dfrac{1}{\sqrt{5}}(-2\,\mathbf{i} - \mathbf{j}); \mathbf{n} = \dfrac{1}{\sqrt{5}}(-\mathbf{i} + 2\,\mathbf{j});$
$-\mathbf{n} = \dfrac{1}{\sqrt{5}}(\mathbf{i} - 2\,\mathbf{j})$

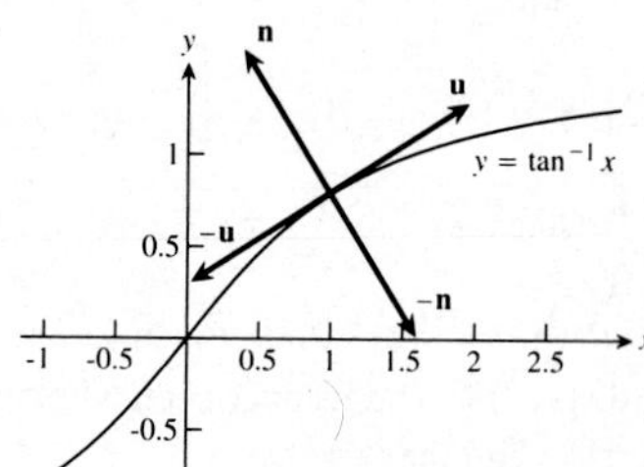

21. $\mathbf{u} = \frac{\pm 1}{5}(-4\mathbf{i} + 3\mathbf{j})$, $\mathbf{v} = \frac{\pm 1}{5}(3\mathbf{i} + 4\mathbf{j})$

23. $\mathbf{u} = \frac{\pm 1}{2}(\mathbf{i} + \sqrt{3}\,\mathbf{j})$, $\mathbf{v} = \frac{\pm 1}{2}(-\sqrt{3}\,\mathbf{i} + \mathbf{j})$

25. $\sqrt{2}\left[\frac{1}{\sqrt{2}}\mathbf{i} + \frac{1}{\sqrt{2}}\mathbf{j}\right]$ **27.** $2\left[\frac{\sqrt{3}}{2}\mathbf{i} + \frac{1}{2}\mathbf{j}\right]$ **29.** $13\left[\frac{5}{13}\mathbf{i} + \frac{12}{13}\mathbf{j}\right]$

31. $|\mathbf{A}| = |3\mathbf{i} + 6\mathbf{j}| = \sqrt{3^2 + 6^2} = 3\sqrt{5} \Rightarrow \mathbf{A} = 3\sqrt{5}\left[\frac{1}{\sqrt{5}}\mathbf{i} + \frac{2}{\sqrt{5}}\mathbf{j}\right]$; $|\mathbf{B}| = |-\mathbf{i} - 2\mathbf{j}| = \sqrt{5} \Rightarrow \mathbf{B} = \sqrt{5}\left[-\frac{1}{\sqrt{5}}\mathbf{i} - \frac{2}{\sqrt{5}}\mathbf{j}\right]$

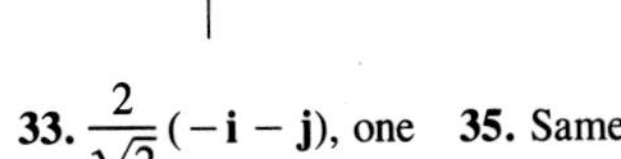

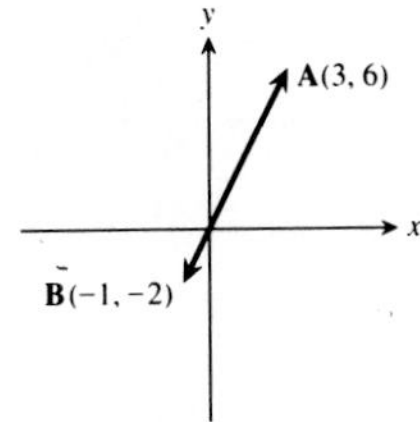

33. $\frac{2}{\sqrt{2}}(-\mathbf{i} - \mathbf{j})$, one **35.** Same

Section 10.2, pp. 716–718

1. A line through the point (2, 3, 0) parallel to the z-axis **3.** The x-axis **5.** The circle, $x^2 + y^2 = 4$ in the xy-plane **7.** The circle, $x^2 + z^2 = 4$ in the xz-plane **9.** The circle, $y^2 + z^2 = 1$ in the yz-plane **11.** The circle, $x^2 + y^2 = 16$ in the xy-plane **13.** a) The first quadrant of the xy-plane b) The fourth quadrant of the xy-plane **15.** a) A solid sphere of radius 1 centered at the origin b) All points that are greater than 1 unit from the origin **17.** a) The upper hemisphere of radius 1 centered at the origin b) The solid upper hemisphere of radius 1 centered at the origin **19.** a) $x = 3$ b) $y = -1$ c) $z = -2$ **21.** a) $z = 1$ b) $x = 3$ c) $y = -1$ **23.** a) $x^2 + (y - 2)^2 = 4$ b) $(y - 2)^2 + z^2 = 4$ c) $x^2 + z^2 = 4$ **25.** a) $y = 3$, $z = -1$ b) $x = 1$, $z = -1$ c) $x = 1$, $y = 3$ **27.** $x^2 + y^2 + z^2 = 25$, $z = 3$ **29.** $0 \le z \le 1$ **31.** $z \le 0$ **33.** a) $(x - 1)^2 + (y - 1)^2 + (z - 1)^2 < 1$ b) $(x - 1)^2 + (y - 1)^2 + (z - 1)^2 > 1$ **35.** $3\left[\frac{2}{3}\mathbf{i} + \frac{1}{3}\mathbf{j} - \frac{2}{3}\mathbf{k}\right]$ **37.** $9\left[\frac{1}{9}\mathbf{i} + \frac{4}{9}\mathbf{j} - \frac{8}{9}\mathbf{k}\right]$ **39.** $5[\mathbf{k}]$ **41.** $4[-\mathbf{j}]$ **43.** $\frac{5}{12}\left[-\frac{4}{5}\mathbf{j} + \frac{3}{5}\mathbf{k}\right]$ **45.** $\sqrt{\frac{1}{2}}\left[\frac{1}{\sqrt{3}}\mathbf{i} - \frac{1}{\sqrt{3}}\mathbf{j} - \frac{1}{\sqrt{3}}\mathbf{k}\right]$ **47.** $3, \frac{2}{3}\mathbf{i} + \frac{2}{3}\mathbf{j} - \frac{1}{3}\mathbf{k}, (2, 2, \frac{1}{2})$ **49.** $7, \frac{3}{7}\mathbf{i} - \frac{6}{7}\mathbf{j} + \frac{2}{7}\mathbf{k}$, (5/2, 1, 6) **51.** $2\sqrt{3}, \frac{1}{\sqrt{3}}\mathbf{i} - \frac{1}{\sqrt{3}}\mathbf{j} - \frac{1}{\sqrt{3}}\mathbf{k}$, (1, −1, −1) **53.** a) $2\mathbf{i}$ b) $4\mathbf{j}$ c) $\sqrt{3}\,\mathbf{k}$ d) $\frac{3}{10}\mathbf{j} + \frac{2}{5}\mathbf{k}$ e) $6\mathbf{i} + 2\mathbf{j} + 3\mathbf{k}$ f) $au_1\mathbf{i} + au_2\mathbf{j} + au_3\mathbf{k}$ **55.** $7\left[\frac{1}{\sqrt{3}}\mathbf{i} + \frac{1}{\sqrt{3}}\mathbf{j} + \frac{1}{\sqrt{3}}\mathbf{k}\right]$ **57.** a) $(-2, 0, 2), 2\sqrt{2}$ b) $(-\frac{1}{2}, -\frac{1}{2}, -\frac{1}{2}), \frac{\sqrt{21}}{2}$ c) $(\sqrt{2}, \sqrt{2}, -\sqrt{2}), \sqrt{2}$ d) $(0, -\frac{1}{3}, \frac{1}{3}), \frac{\sqrt{29}}{3}$ **59.** $(-2, 0, 2), \sqrt{8}$ **61.** $(0, 0, a)$, **a** **63.** a) $\sqrt{y^2 + z^2}$ b) $\sqrt{x^2 + z^2}$ c) $\sqrt{x^2 + y^2}$ **65.** a) $\frac{3}{2}\mathbf{i} + \frac{3}{2}\mathbf{j} - 3\mathbf{k}$ b) $\mathbf{i} + \mathbf{j} - 2\mathbf{k}$ c) (2, 2, 1)

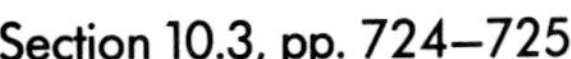

Section 10.3, pp. 724–725

	$\mathbf{A} \cdot \mathbf{B}$	$\lvert\mathbf{A}\rvert$	$\lvert\mathbf{B}\rvert$	$\cos\theta$	$\lvert\mathbf{B}\rvert \cos\theta$	$\text{Proj}_{\mathbf{A}}\mathbf{B}$
1.	10	$\sqrt{13}$	$\sqrt{26}$	$\frac{10}{13\sqrt{2}}$	$\frac{10}{\sqrt{13}}$	$\frac{10}{13}[3\mathbf{i} + 2\mathbf{j}]$
3.	4	$\sqrt{14}$	2	$\frac{2}{\sqrt{14}}$	$\frac{4}{\sqrt{14}}$	$\frac{2}{7}[3\mathbf{i} - 2\mathbf{j} - \mathbf{k}]$
5.	2	$\sqrt{34}$	$\sqrt{3}$	$\frac{2}{\sqrt{3}\sqrt{34}}$	$\frac{2}{\sqrt{34}}$	$\frac{1}{17}[5\mathbf{j} - 3\mathbf{k}]$
7.	$\sqrt{3} - \sqrt{2}$	$\sqrt{2}$	3	$\frac{\sqrt{3} - \sqrt{2}}{3\sqrt{2}}$	$\frac{\sqrt{3} - \sqrt{2}}{\sqrt{2}}$	$\frac{\sqrt{3} - \sqrt{2}}{2}[-\mathbf{i} + \mathbf{j}]$
9.	−25	5	5	−1	−5	$-2\mathbf{i} + 4\mathbf{j} - \sqrt{5}\,\mathbf{k}$
11.	25	15	5	$\frac{1}{3}$	$\frac{5}{3}$	$\frac{1}{9}[10\mathbf{i} + 11\mathbf{j} - 2\mathbf{k}]$

13. $\left[\frac{3}{2}\mathbf{i}+\frac{3}{2}\mathbf{j}\right]+\left[-\frac{3}{2}\mathbf{i}+\frac{3}{2}\mathbf{j}+4\mathbf{k}\right]$

15. $\left[\frac{14}{3}\mathbf{i}+\frac{28}{3}\mathbf{j}-\frac{14}{3}\mathbf{k}\right]+\left[\frac{10}{3}\mathbf{i}-\frac{16}{3}\mathbf{j}-\frac{22}{3}\mathbf{k}\right]$

17. $x + 2y = 4$ **19.** $(-2\mathbf{i}-\mathbf{j})\cdot((x+1)\mathbf{i}+(y-2)\mathbf{j}) = 0 \Rightarrow -2x - y = 0$

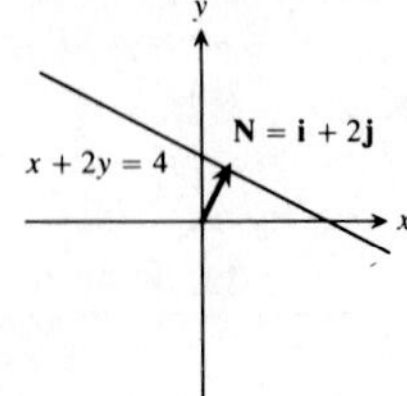

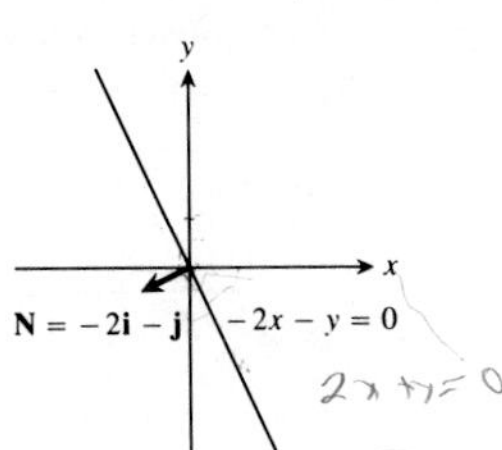

21. $2\sqrt{10}$ **23.** $\sqrt{2}$ **31.** 71.1°, 37.9°, 71.1° **33.** 35.3° **35.** −5 N · m **37.** 3464.10 N · m **39.** 45° or 135° **41.** $\frac{\pi}{3}$ or $\frac{2\pi}{3}$ **43.** 45° or 135°

Section 10.4, p. 730

1. $\mathbf{A}\times\mathbf{B}$, 3, $\frac{2}{3}\mathbf{i}+\frac{1}{3}\mathbf{j}+\frac{2}{3}\mathbf{k}$; $\mathbf{B}\times\mathbf{A}$, 3, $-\frac{2}{3}\mathbf{i}-\frac{1}{3}\mathbf{j}-\frac{2}{3}\mathbf{k}$ **3.** $\mathbf{A}\times\mathbf{B}$, 0, no direction; $\mathbf{B}\times\mathbf{A}$, 0, no direction **5.** $\mathbf{A}\times\mathbf{B}$, 6, $-\mathbf{k}$; $\mathbf{B}\times\mathbf{A}$, 6, $\mathbf{k}$ **7.** $\mathbf{A}\times\mathbf{B}$, $6\sqrt{5}$, $\frac{1}{\sqrt{5}}\mathbf{i}-\frac{2}{\sqrt{5}}\mathbf{k}$; $\mathbf{B}\times\mathbf{A}$, $6\sqrt{5}$, $-\frac{1}{\sqrt{5}}\mathbf{i}+\frac{2}{\sqrt{5}}\mathbf{k}$ **9.** $\mathbf{A}\times\mathbf{B} = \mathbf{k}$

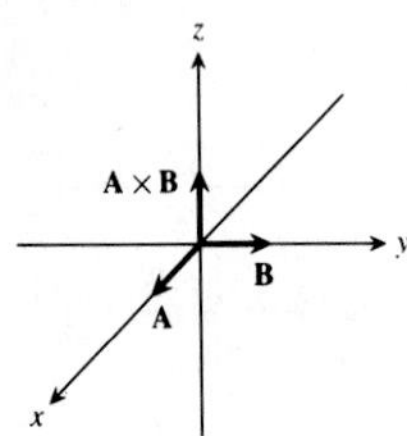

11. $\mathbf{A}\times\mathbf{B} = \mathbf{i}-\mathbf{j}+\mathbf{k}$ **13.** $\mathbf{A}\times\mathbf{B} = 3\mathbf{i}-\mathbf{j}$

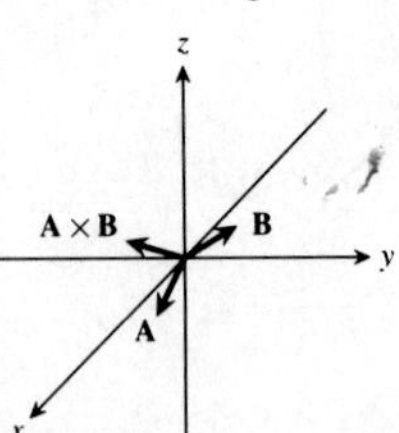

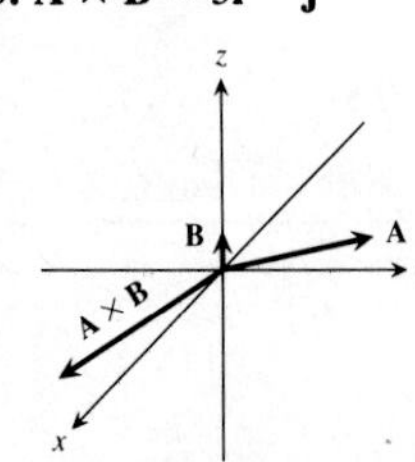

15. a) $\pm(8\mathbf{i}+4\mathbf{j}+4\mathbf{k})$ b) $2\sqrt{6}$ c) $\pm\left[\frac{2}{\sqrt{6}}\mathbf{i}+\frac{1}{\sqrt{6}}\mathbf{j}+\frac{1}{\sqrt{6}}\mathbf{k}\right]$ **17.** a) $\pm(-\mathbf{i}+\mathbf{j})$ b) $\frac{\sqrt{2}}{2}$ c) $\pm\left[\frac{-1}{\sqrt{2}}\mathbf{i}+\frac{1}{\sqrt{2}}\mathbf{j}\right]$ **19.** a) None b) **A** and **C** **21.** $10\sqrt{3}$ ft · lb **23.** 0, 0 **25.** a) $\frac{\mathbf{A}\cdot\mathbf{B}}{\mathbf{B}\cdot\mathbf{B}}\mathbf{B}$ b) $(\pm)(\mathbf{A}\times\mathbf{B})$ c) $\sqrt{\mathbf{A}\cdot\mathbf{A}}\frac{\mathbf{B}}{\sqrt{\mathbf{B}\cdot\mathbf{B}}}$ d) $(\pm)(\mathbf{A}\times\mathbf{B})\times\mathbf{C}$ e) $(\pm)(\mathbf{B}\times\mathbf{C})\times\mathbf{A}$ **27.** 3

Section 10.5, pp. 736–738

1. $x = 3+t, y = -4+t, z = -1+t$ **3.** $x = -2+5t, y = 5t, z = 3-5t$ **5.** $x = 0, y = 2t, z = t$ **7.** $x = 1, y = 1, z = 1+t$ **9.** $x = t, y = -7+2t, z = 2t$ **11.** $x = t, y = 0, z = 0$

13. The direction $\overrightarrow{PQ} = \mathbf{i}+\mathbf{j}+\mathbf{k}$ and $P(0, 0, 0) \Rightarrow x = t, y = t, z = t$, where $0 \le t \le 1$. **15.** The direction $\overrightarrow{PQ} = \mathbf{j}$ and $P(1, 0, 0) \Rightarrow x = 1, y = 1+t, z = 0$, where $-1 \le t \le 0$.

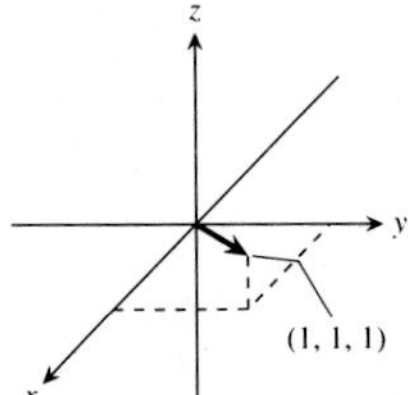

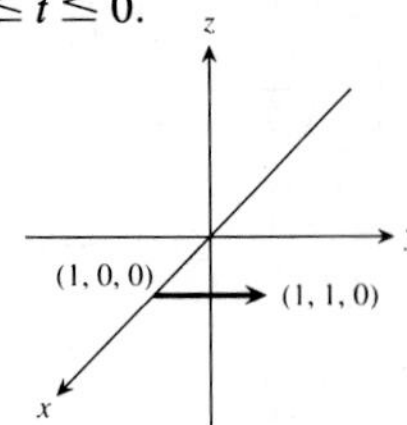

17. The direction $\overrightarrow{PQ} = 2\mathbf{j}$ and P(0, −1, 1) $\Rightarrow x = 0, y = -1+2t, z = 1$, where $0 \le t \le 1$.

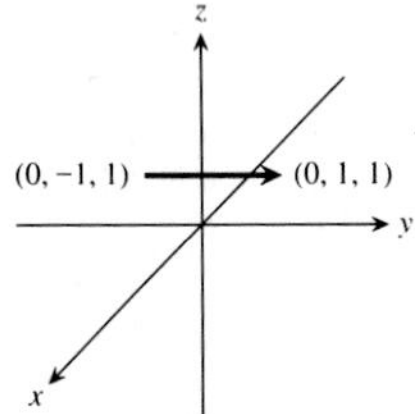

19. The direction $\overrightarrow{PQ} = -\mathbf{i}-2\mathbf{k}$ and $P(2, 2, 0) \Rightarrow x = 2-t, y = 2, z = -2t$, where $0 \le t \le 1$.

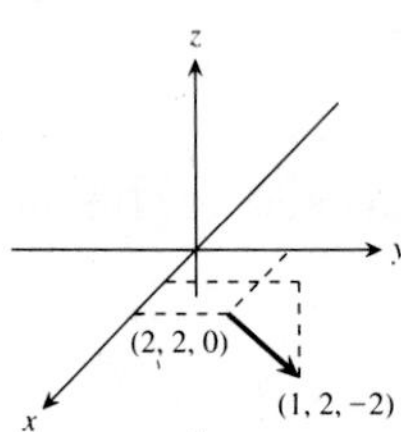

21. $3x - 2y - z = -3$ **23.** $7x - 5y - 4z = 6$ **25.** $x + 3y + 4z = 34$ **27.** $x = 2+2t, y = -4-t, z = 7+3t$; $x = -2-t, y = -2+\frac{1}{2}t, z = 1-\frac{3}{2}t$ **29.** $2\sqrt{30}$ **31.** 0 **33.** $\frac{9\sqrt{42}}{7}$ **35.** 3 **37.** $\frac{19}{5}$ **39.** $\frac{5}{3}$ **41.** (3/2, −3/2, 1/2) **43.** (1, 1, 0) **45.** $\frac{\pi}{4}$ **47.** 101.1° **49.** 47.1° **51.** $x = 1-t, y = 1+t, z = -1$ **53.** $x = 4, y = 3+6t, z = 1+3t$ **55.** (1, −1, 0) **59.** $|\overrightarrow{OA}| = 1.41421356$; $|\overrightarrow{OB}| = 2$; $|\overrightarrow{AB}| = 2.44948974$; the midpoint of AB is (.5, 1, − .5); $\mathbf{A}\cdot\mathbf{B} = 0$; the angle between $\overrightarrow{OA}$ and $\overrightarrow{OB}$ is 1.5707965 radians ≈ 90°; $\mathbf{A}\times\mathbf{B} = 2\mathbf{i}+0\mathbf{j}+2\mathbf{k}$; the line through A and B: $x = 1-t, y = 2t, z = -1+t$; the line through C parallel to AB: $x = -1-t, y = 2t, z = 1+t$; the distance from C to AB is 2.30940108; the equation of the ABC plane is $-4x - 4z = 0$; the distance from S to the ABC plane is

2.12132034. **61.** $|\overrightarrow{OA}| = 4.47213595$; $|\overrightarrow{OB}| = 8.24621126$; $|\overrightarrow{AB}| = 9.79795897$; the midpoint of AB is $(0, 4, 2)$; $\mathbf{A} \cdot \mathbf{B} = -4$; the angle between $\overrightarrow{OA}$ and $\overrightarrow{OB}$ is 1.62032412 radians $\approx 92.8377235°$; $\mathbf{A} \times \mathbf{B} = -32\mathbf{i} + 8\mathbf{j} - 16\mathbf{k}$; the line through A and B: $x = -2 + 4t$, $y = 8t$, $z = 4 - 4t$; the line through C parallel to AB: $x = 2 + 4t$, $y = 4 + 8t$, $z = -2 - 4t$; the distance from C to AB is 3.74165739; the equation of the ABC plane is $32x - 8y + 16z = 0$; the distance from S to the ABC plane is 0.

Section 10.6, p. 746

1. $|(\mathbf{A} \times \mathbf{B}) \cdot \mathbf{C}| = \begin{vmatrix} 2 & 0 & 0 \\ 0 & 2 & 0 \\ 0 & 0 & 2 \end{vmatrix} = 8$, $(\mathbf{A} \times \mathbf{B}) \times \mathbf{C} = 0$, $\mathbf{A} \times (\mathbf{B} \times \mathbf{C}) = 0$
3. $|(\mathbf{A} \times \mathbf{B}) \cdot \mathbf{C}| = \begin{vmatrix} 2 & 1 & 0 \\ 2 & -1 & 1 \\ 1 & 0 & 2 \end{vmatrix} = 7$, $(\mathbf{A} \times \mathbf{B}) \times \mathbf{C} = -4\mathbf{i} - 6\mathbf{j} + 2\mathbf{k}$, $\mathbf{A} \times (\mathbf{B} \times \mathbf{C}) = \mathbf{i} - 2\mathbf{j} - 4\mathbf{k}$
5. a) True b) False c) True d) True e) False f) True g) True h) True i) True **7.** a) 0 b) 2 c) 4 **11.** a) $(0, 9, -3)$ b) $\sqrt{963}$ c) xy, 11; yz, 29; xz, 1

Section 10.7, pp. 756–758

1. $x^2 + y^2 = 4$

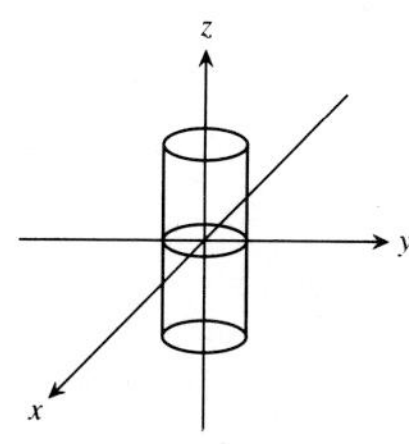

3. $z = y^2 - 1$

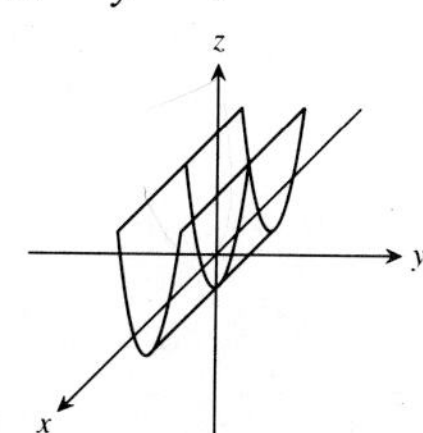

5. $x^2 + 4z^2 = 16$

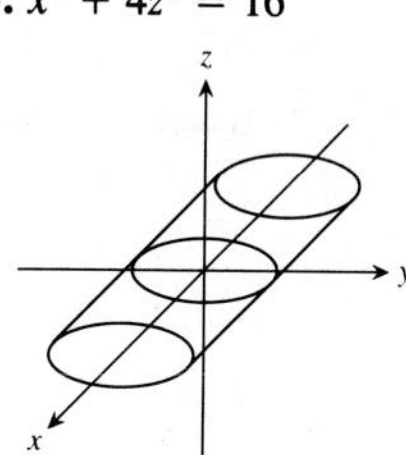

7. $z^2 - y^2 = 1$

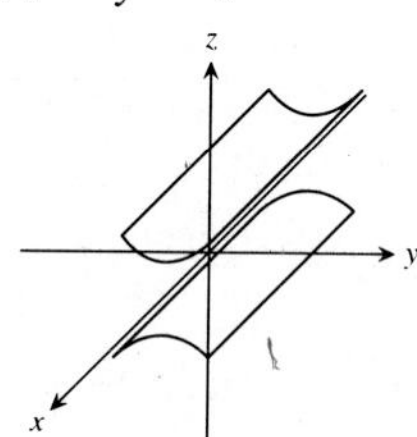

9. $9x^2 + y^2 + z^2 = 9$

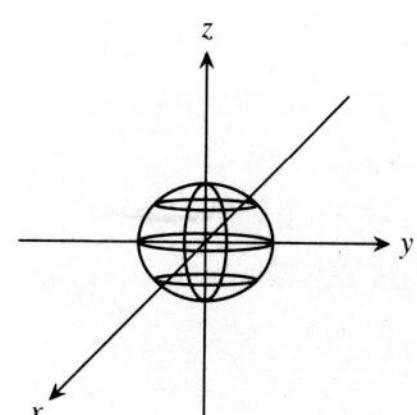

11. $4x^2 + 9y^2 + 4z^2 = 36$

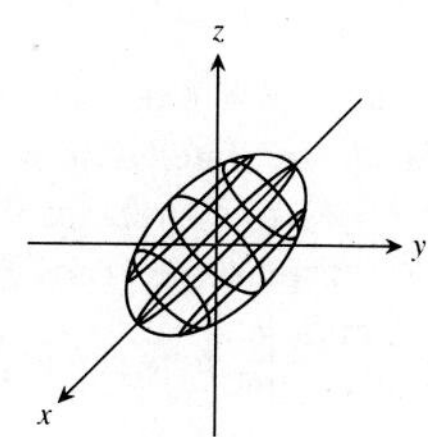

13. $x^2 + 4y^2 = z$

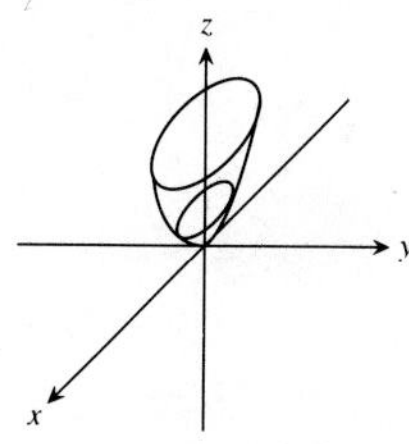

15. $z = 8 - x^2 - y^2$

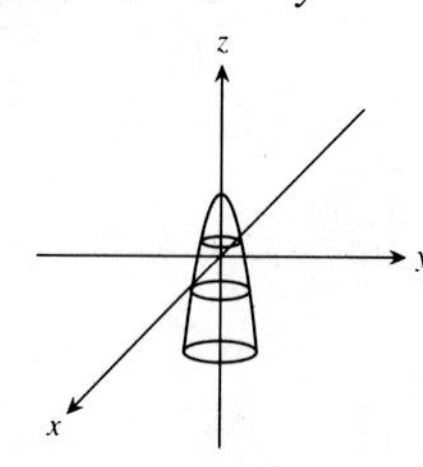

17. $x = 4 - 4y^2 - z^2$

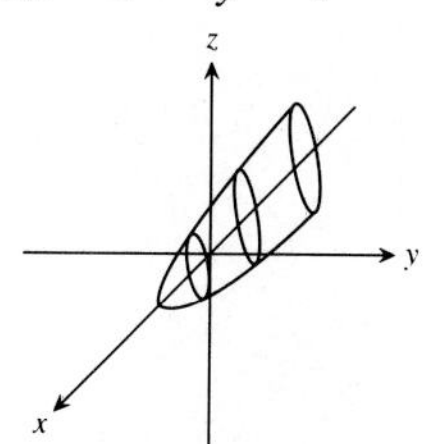

19. $x^2 + y^2 = z^2$

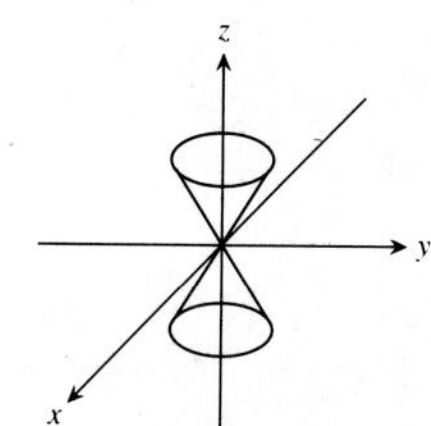

21. $4x^2 + 9z^2 = 9y^2$

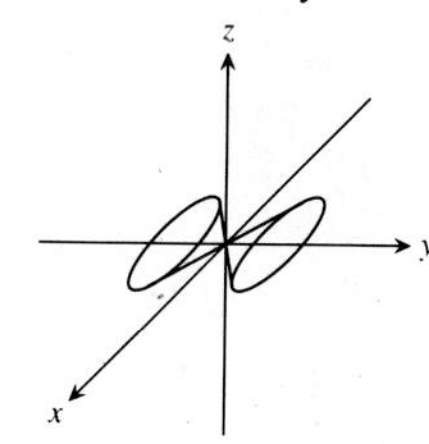

23. $x^2 + y^2 - z^2 = 1$

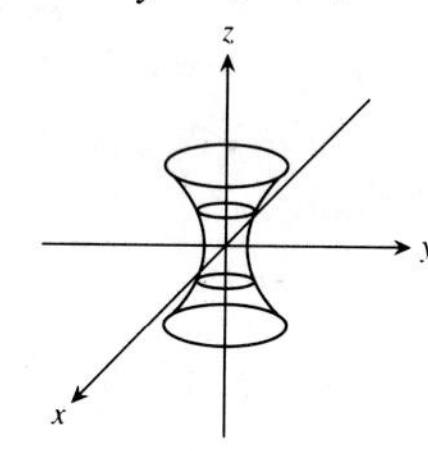

25. $(y^2/4) + (z^2/9) - (x^2/4) = 1$

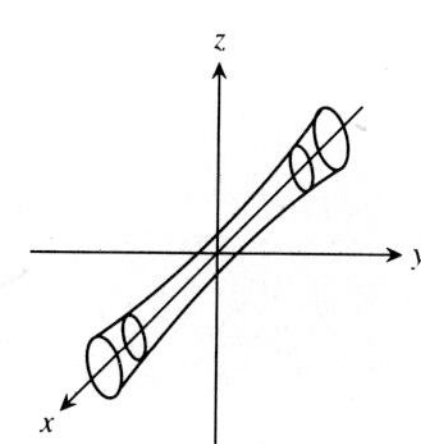

27. $z^2 - x^2 - y^2 = 1$

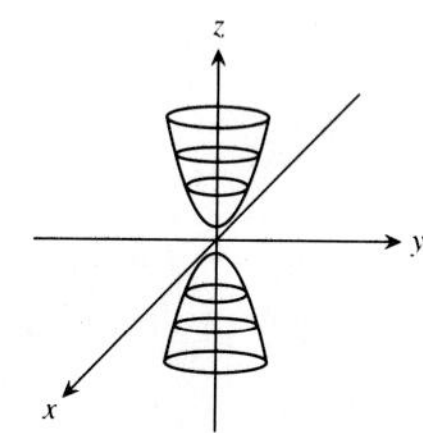

29. $x^2 - y^2 - (z^2/4) = 1$

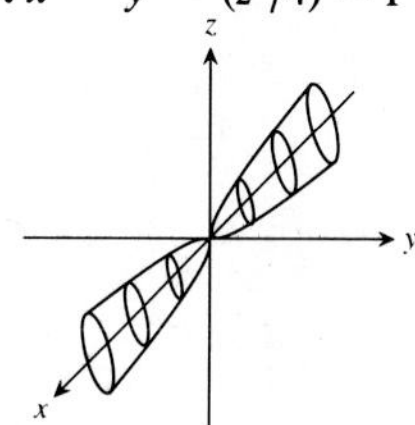

31. $y^2 - x^2 = z$

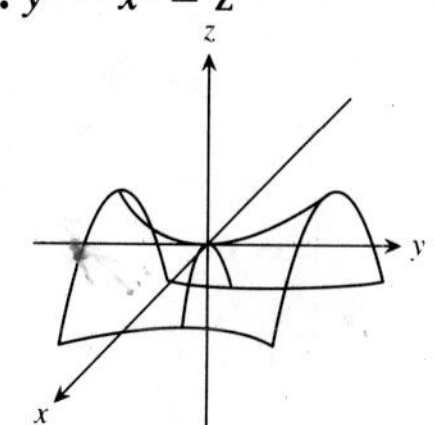

33. $x^2 + y^2 + z^2 = 4$

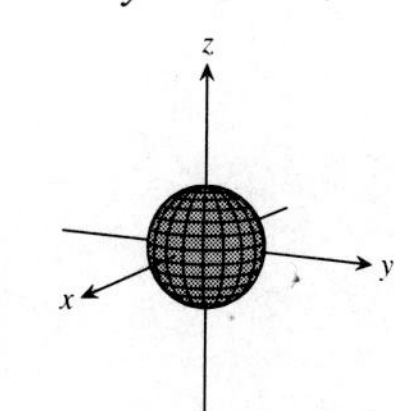

35. $z = 1 + y^2 - x^2$

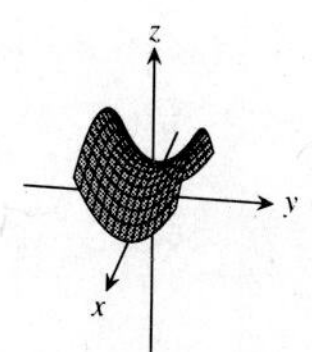

37. $y = -(x^2 + z^2)$

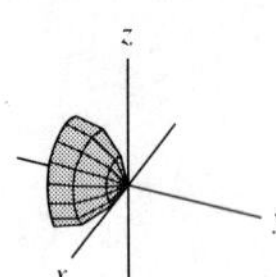

39. $16x^2 + 4y^2 = 1$

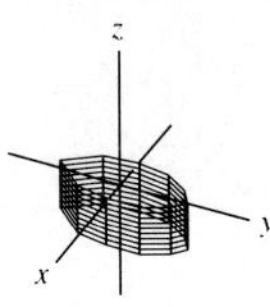

41. $x^2 + y^2 - z^2 = 4$

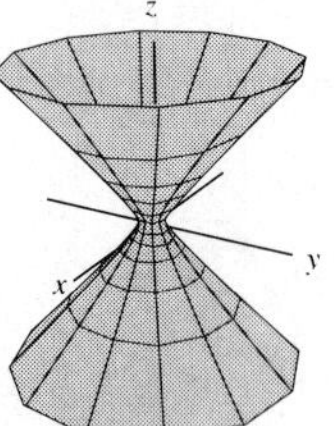

43. $x^2 + z^2 = y$

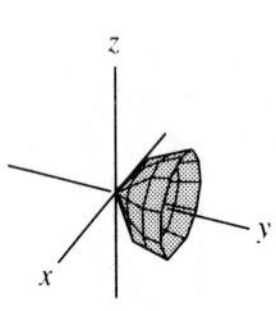

45. $x^2 + z^2 = 1$

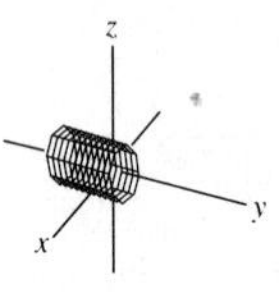

47. $16y^2 + 9z^2 = 4x^2$

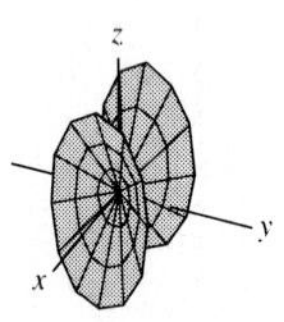

49. $9x^2 + 4y^2 + z^2 = 36$

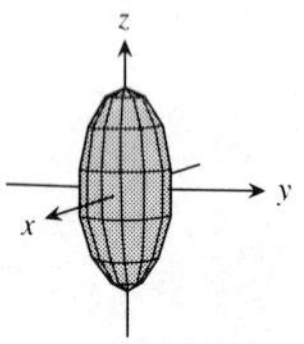

51. $x^2 + y^2 - 16z^2 = 16$

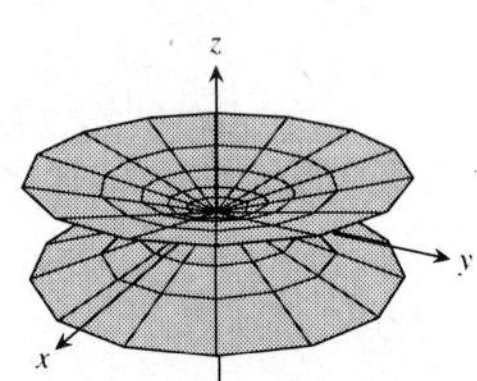

53. $z = -(x^2 + y^2)$

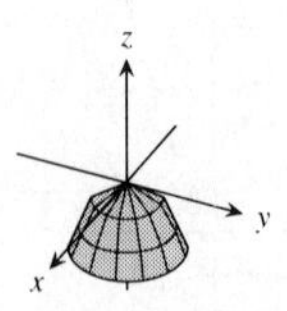

55. $x^2 - 4y^2 = 1$

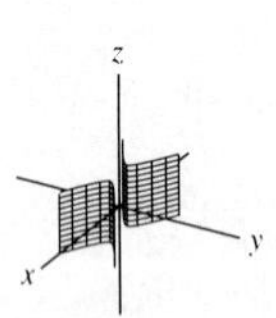

57. $4y^2 + z^2 - 4x^2 = 4$

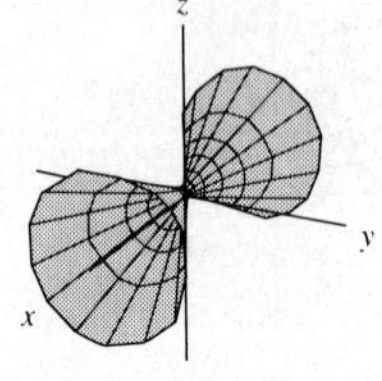

59. $x^2 + y^2 = z$

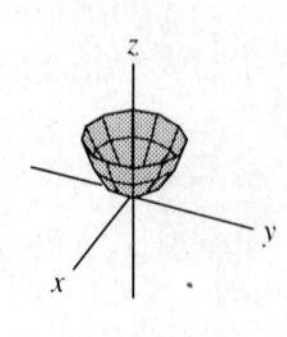

61. $yz = 1$

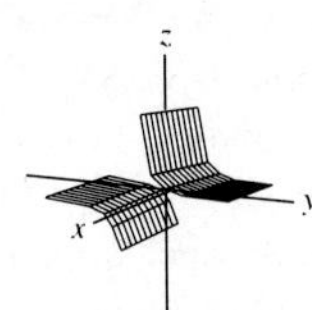

63. $9x^2 + 16y^2 = 4z^2$

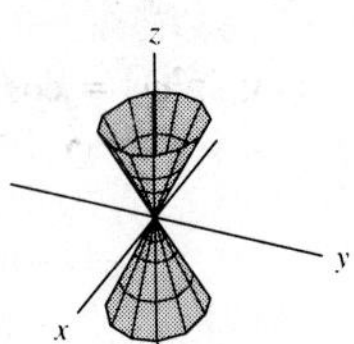

65. a) $\dfrac{2\pi(9 - c^2)}{9}$ b) 8π c) $\dfrac{4\pi\, abc}{3}$, yes

71. $z = y^2$

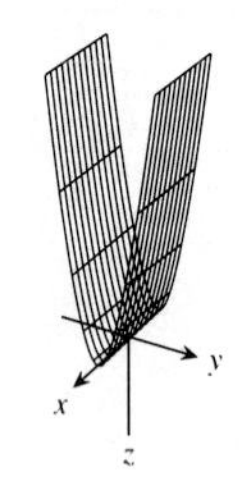

73. $z = x^2 + y^2$

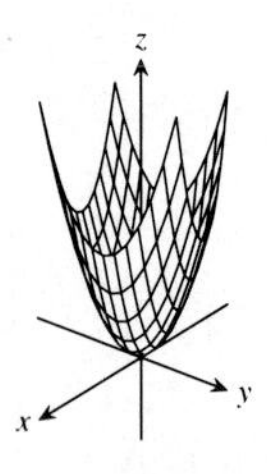

75. $z = \sqrt{1 - x^2}$

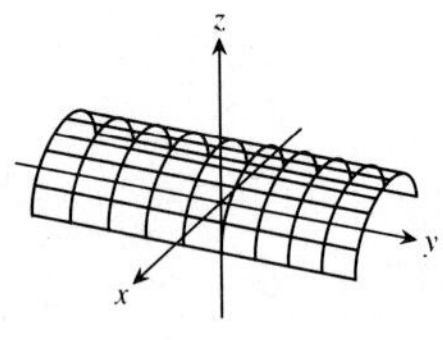

77. $z = \sqrt{x^2 + 2y^2 + 4}$

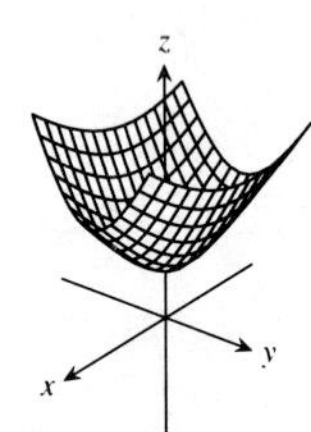

Section 10.8, p. 761

	Rectangular	Cylindrical	Spherical
1.	$(0, 0, 0)$	$(0, 0, 0)$	$(0, 0, 0)$
3.	$(0, 1, 0)$	$(1, \pi/2, 0)$	$(1, \pi/2, \pi/2)$
5.	$(1, 0, 0)$	$(1, 0, 0)$	$(1, \pi/2, 0)$
7.	$(0, 1, 1)$	$(1, \pi/2, 1)$	$(\sqrt{2}, \pi/4, \pi/2)$
9.	$(0, -2\sqrt{2}, 0)$	$(2\sqrt{2}, 3\pi/2, 0)$	$(2\sqrt{2}, \pi/2, 3\pi/2)$

11. $r = 0 \Rightarrow$ rectangular, $x^2 + y^2 = 0$; spherical, $\phi = 0$ and $\phi = \pi$; the z-axis **13.** $z = 0 \Rightarrow$ cylindrical, $z = 0$; spherical, $\phi = \dfrac{\pi}{2}$; the xy-plane **15.** $\rho \cos \phi = 3 \Rightarrow$ rectangular, $z = 3$; cylindrical, $z = 3$; the plane $z = 3$ **17.** $\rho \sin \phi \cos \phi = 0 \Rightarrow$ rectangular, $x = 0$; cylindrical $\theta = \dfrac{\pi}{2}$; the yz-plane

19. $x^2 + y^2 + z^2 = 4 \Rightarrow$ cylindrical, $r^2 + z^2 = 4$; spherical, $\rho = 2$; a sphere centered at the origin with a radius of 2
21. $z = r^2 \cos 2\theta \Rightarrow$ rectangular, $z = r^2(\cos^2\theta - \sin^2\theta) \Rightarrow z = x^2 - y^2$; spherical, $\rho \cos^2\phi = \rho^2 \sin^2\phi \cos 2\theta \Rightarrow \rho = \dfrac{\cos \phi}{\sin^2\phi \cos 2\theta}$; hyperbolic paraboloid **23.** $r = \csc \theta \Rightarrow$ rectangular, $r = \dfrac{r}{y} \Rightarrow y = 1$; spherical, $\rho \sin \phi = \csc \theta \Rightarrow$

$\rho = \dfrac{1}{\sin\phi\sin\theta}$; the plane $y = 1$ **25.** $3\tan^2\phi = 1 \Rightarrow$ rectangular, $3(\sin^2\phi) = \cos^2\phi \Rightarrow 3(\rho^2\sin^2\phi)(\sin^2\theta + \cos^2\theta) = \rho^2\cos^2\phi \Rightarrow 3(\rho\sin\phi\sin\theta)^2 + 3(\rho\sin\phi\cos\theta)^2 = (\rho\cos\phi)^2 \Rightarrow 3x^2 + 3y^2 = z^2$; cylindrical, $3r^2 = z^2$, a cone

27. A right circular cylinder whose generating curve is a circle of radius 4 in the $r\theta$-plane

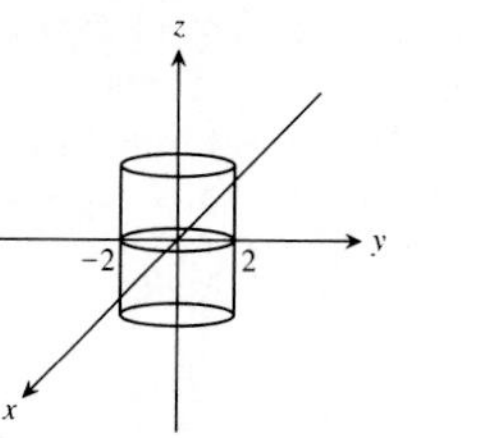

29. A cylinder whose generation curve is a cardioid in the $r\theta$-plane

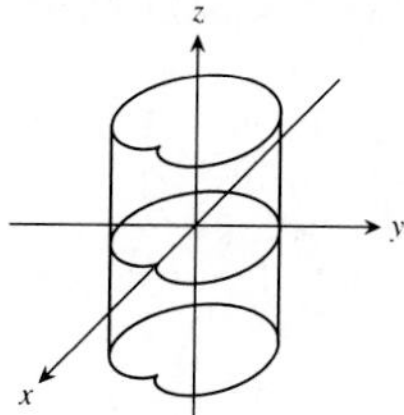

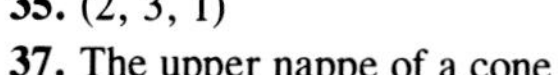

31. A circle contained in the plane $z = 3$ having a radius of 2 and center at (0, 0, 3)

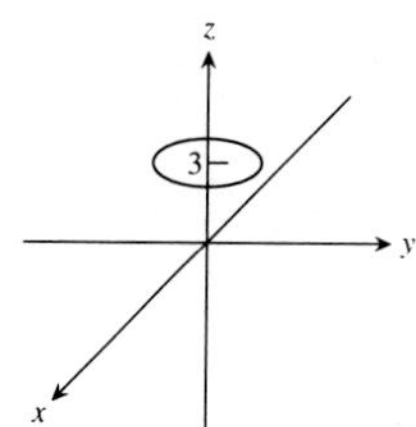

33. A space curve called a helix

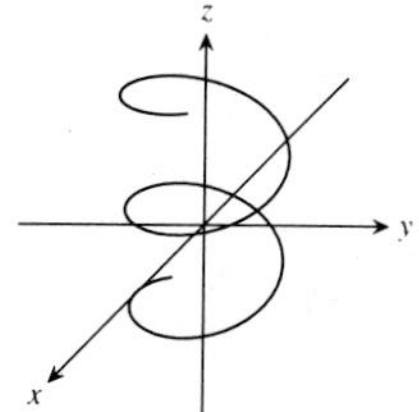

35. (2, 3, 1)

37. The upper nappe of a cone

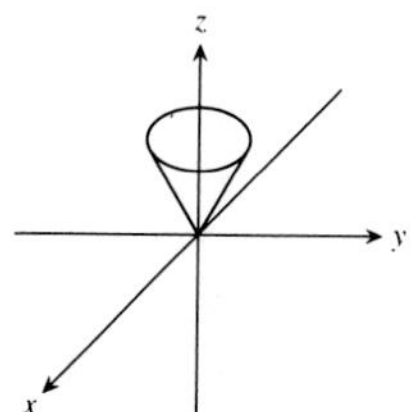

39. A "vertical semicircle" in the $y = x$ plane

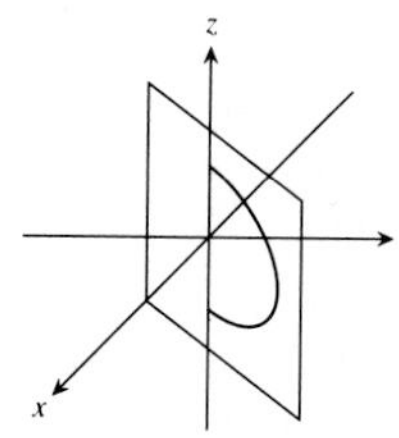

41. A sphere whose center is (0, 0, 1/2) with a radius of 1/2

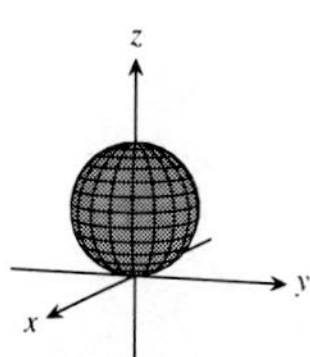

43. A torus-like object centered at the origin

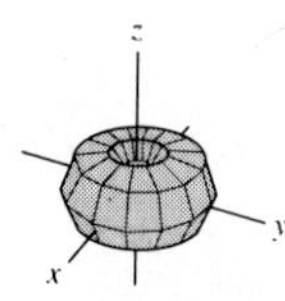

Section 10 Miscellaneous Exercises, pp. 762–766

1. $\theta = 0 \Rightarrow \mathbf{u} = \mathbf{i}$; $\theta = \dfrac{\pi}{2} \Rightarrow \mathbf{u} = \mathbf{j}$; $\theta = \dfrac{2\pi}{3} \Rightarrow$ $\mathbf{u} = -\dfrac{1}{2}\mathbf{i} + \dfrac{\sqrt{3}}{2}\mathbf{j}$; $\theta = \dfrac{5\pi}{4} \Rightarrow \mathbf{u} = -\dfrac{1}{\sqrt{2}}\mathbf{i} - \dfrac{1}{\sqrt{2}}\mathbf{j}$; $\theta = \dfrac{5\pi}{3} \Rightarrow$ $\mathbf{u} = \dfrac{1}{2}\mathbf{i} - \dfrac{\sqrt{3}}{2}\mathbf{j}$

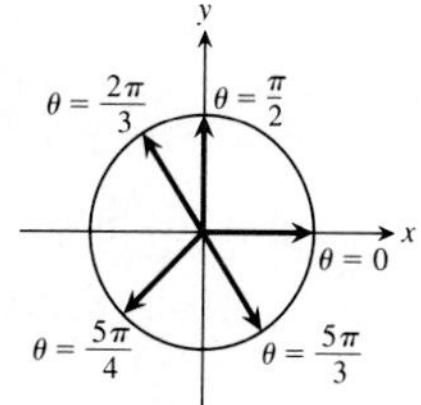

3. Tangents $\pm\left(\dfrac{1}{\sqrt{5}}\mathbf{i} + \dfrac{2}{\sqrt{5}}\mathbf{j}\right)$; normals are $\pm\left(-\dfrac{2}{\sqrt{5}}\mathbf{i} + \dfrac{1}{\sqrt{5}}\mathbf{j}\right)$

5. $2, \dfrac{1}{\sqrt{2}}\mathbf{i} + \dfrac{1}{\sqrt{2}}\mathbf{j}$ **7.** $17, \dfrac{2}{7}\mathbf{i} - \dfrac{3}{7}\mathbf{j} + \dfrac{6}{7}\mathbf{k}$ **11.** 0

13. $\sqrt{2}, 3, 3, 3, -2\mathbf{i} + 2\mathbf{j} - \mathbf{k}, 2\mathbf{i} - 2\mathbf{j} + \mathbf{k}, 3, \dfrac{\pi}{4}, \dfrac{3}{\sqrt{2}}, \dfrac{3}{2}[\mathbf{i} + \mathbf{j}]$ **15.** $\dfrac{4}{3}[2\mathbf{i} + \mathbf{j} - \mathbf{k}] - \dfrac{1}{3}[5\mathbf{i} + \mathbf{j} + 11\mathbf{k}]$

17. $\mathbf{A} \times \mathbf{B} = \mathbf{k}$

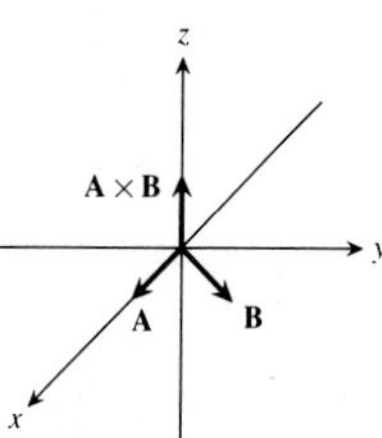

19. $16000\sqrt{3}$ ft · lb **27.** 3 **29.** $\dfrac{\sqrt{78}}{3}$ **31.** $x = 1 - 3t, y = 2, z = 3 + 7t$ **33.** $\sqrt{2}$ **35.** $2x + y - z = 3$ **39.** $\dfrac{\pi}{6}$ **41.** $\dfrac{25}{\sqrt{38}}$

43. $\dfrac{1}{\sqrt{14}}(-2\mathbf{i} - 3\mathbf{j} + \mathbf{k})$ **45.** $\left(\dfrac{4}{3}, -\dfrac{2}{3}, -\dfrac{2}{3}\right)$ **49.** 59.5°

53. $\left(26, 23, -\dfrac{1}{3}\right)$ **55.** $\dfrac{11}{\sqrt{107}}$ **57.** a) (2, 1, 8) b) $\dfrac{3}{\sqrt{15}}$ c) $\dfrac{9}{5}(\mathbf{i} + 2\mathbf{k})$ d) $6\sqrt{6}$ e) $7x + 2y - z = 8$ f) yz, 14; xz, 4; yy, 2

63. b) $\dfrac{6}{\sqrt{14}}$ c) $2x - y + 2x = 8$ d) $x - 2y + z = 3 - 5\sqrt{6}$

67. The y-axis in the xy-plane and the yz-plane in three-dimensional space **69.** A circle centered at (0, 0) with a radius of 2 in the xy-plane; a cylinder parallel with the z-axis in three-dimensional space with the circle as its generating curve **71.** A horizontal parabola opening to the right with its vertex at (0, 0) in the xy-plane; a cylinder parallel with the z-axis in three-dimensional space with the parabola as the generating curve **73.** A horizontal cardioid in the $r\theta$-plane; a cylinder parallel with the z-axis in three-dimensional space with the cardioid as the generating curve **75.** A horizontal lemniscate of length $2\sqrt{2}$ in the $r\theta$-plane; a cylinder parallel with the z-axis in three-dimensional space with the lemniscate as the generating curve. **77.** A sphere with a radius of 2 centered at the origin

79. The upper nappe of a cone whose surface makes a $\pi/6$ angle with the z-axis **81.** The upper hemisphere of a sphere with a radius 1 centered at the origin **83.** $z = 2 \Rightarrow$ cylindrical, $z = 2$; spherical, $\rho \cos \phi = 2$; a plane parallel with the xy-plane **85.** $z = r^2 \cos 2\theta \Rightarrow$ rectangular, $z = r^2(\cos^2\theta - \sin^2\theta) \Rightarrow$ $z = x^2 - y^2$; spherical, $\rho \cos \phi = \rho^2 \sin^2\phi \cos 2\theta \Rightarrow \rho = \dfrac{\cos \phi}{\sin^2\phi \cos 2\theta}$; hyperbolic paraboloid **87.** $\rho = 4 \sec \phi \Rightarrow$ rectangular, $\rho \cos \phi = 4 \Rightarrow z = 4$; cylindrical, $z = 4$, the plane $z = 4$ **89.** 2 **91.** 13 **93.** $\dfrac{11}{2}$ **95.** $\dfrac{25}{2}$

Chapter 11

Section 11.1, pp. 777–779

1. $\mathbf{v} = (-2 \sin t)\,\mathbf{i} + (3 \cos t)\,\mathbf{j} + 4\,\mathbf{k}$, $\mathbf{a} = (-2 \cos t)\,\mathbf{i} - (3 \sin t)\,\mathbf{j}$, speed $= 2\sqrt{5}$, direction $= -\dfrac{1}{\sqrt{5}}\mathbf{i} + \dfrac{2}{\sqrt{5}}\mathbf{k}$, $\mathbf{v}\left(\dfrac{\pi}{2}\right) = 2\sqrt{5}\left[-\dfrac{1}{\sqrt{5}}\mathbf{i} + \dfrac{2}{\sqrt{5}}\mathbf{k}\right]$ **3.** $\mathbf{v} = (-2 \sin 2t)\,\mathbf{j} + (2 \cos t)\,\mathbf{k}$, $\mathbf{a} = (-4 \cos 2t)\,\mathbf{j} - (2 \sin t)\,\mathbf{k}$, speed $= 2$, direction $= \mathbf{k}$, $\mathbf{v}(0) = 2\,\mathbf{k}$ **5.** $\mathbf{v} = (\sec t \tan t)\,\mathbf{i} + (\sec^2 t)\,\mathbf{j} + \dfrac{4}{3}\mathbf{k}$, $\mathbf{a} = (\sec t \tan^2 t + \sec^3 t)\,\mathbf{i} + (2 \sec^2 t \tan t)\,\mathbf{j}$, speed $= 2$, direction $= \dfrac{1}{3}\mathbf{i} + \dfrac{2}{3}\mathbf{j} + \dfrac{2}{3}\mathbf{k}$, $\mathbf{v}\left(\dfrac{\pi}{6}\right) = 2\left[\dfrac{1}{3}\mathbf{i} + \dfrac{2}{3}\mathbf{j} + \dfrac{2}{3}\mathbf{k}\right]$

7. $\theta = \dfrac{\pi}{2}$ **9.** $\theta = \dfrac{\pi}{2}$ **11.** $t = 0, \pi, 2\pi$ **13.** $\dfrac{1}{4}\mathbf{i} + 7\,\mathbf{j} + \dfrac{3}{2}\mathbf{k}$ **15.** $\left(\dfrac{\pi + 2\sqrt{2}}{2}\right)\mathbf{j} + 2\,\mathbf{k}$ **17.** $\mathbf{v}\left(\dfrac{\pi}{4}\right) = \dfrac{\sqrt{2}}{2}\mathbf{i} - \dfrac{\sqrt{2}}{2}\mathbf{j}$, $\mathbf{a}\left(\dfrac{\pi}{4}\right) = -\dfrac{\sqrt{2}}{2}\mathbf{i} - \dfrac{\sqrt{2}}{2}\mathbf{j}$, $\mathbf{v}\left(\dfrac{\pi}{2}\right) = -\mathbf{j}$, $\mathbf{a}\left(\dfrac{\pi}{2}\right) = -\mathbf{i}$

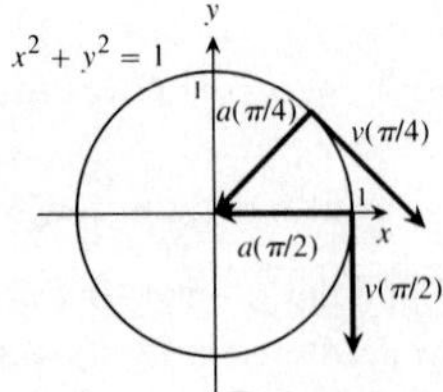

19. $\mathbf{v}(\pi) = 2\,\mathbf{i}$, $\mathbf{a}(\pi) = -\mathbf{j}$, $\mathbf{v}\left(\dfrac{3\pi}{2}\right) = \mathbf{i} - \mathbf{j}$, $\mathbf{a}\left(\dfrac{3\pi}{2}\right) = -\mathbf{i}$

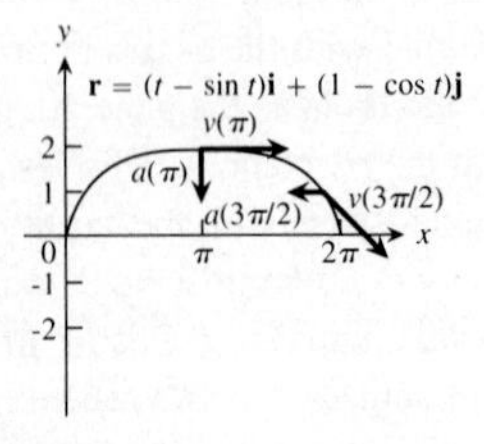

21. $\mathbf{r} = \left(-\dfrac{t^2}{2} + 1\right)\mathbf{i} + \left(-\dfrac{t^2}{2} + 2\right)\mathbf{j} + \left(-\dfrac{t^2}{2} + 3\right)\mathbf{k}$
23. $\mathbf{r} = ((t + 1)^{3/2} - 1)\,\mathbf{i} + (1 - e^{-t})\,\mathbf{j} + (1 + \ln(t + 1))\,\mathbf{k}$
25. $\mathbf{r} = 8t\mathbf{i} + 8t\mathbf{j} + (100 - 16t^2)\,\mathbf{k}$ **27.** $\max|\mathbf{a}| = \min|\mathbf{a}| = 1$

Section 11.2, pp. 784–786

1. 50 sec **3.** a) 72.2 sec, 25,510.2 m b) 4020.3 m c) 6377.6 m **7.** $v_0 = 9.9$ m/s, $\alpha = 18.44°$ or $71.56°$ **9.** a) 189.6 mph b) 120.8 ft · lb **13.** The golf ball will clip the leaves at the top. **15.** 3.72 in. **17.** 2.25 sec, 148.98 ft/sec **21.** $\mathbf{v}(t) = -gt\mathbf{k} + \mathbf{v}_0$, $\mathbf{r}(t) = -\dfrac{1}{2}gt^2\,\mathbf{k} + \mathbf{v}_0 t$ **23.** b) $\mathbf{v}_0$ would bisect $\angle$AOR.

Section 11.3, pp. 790–791

1. $\mathbf{T} = \left(-\dfrac{2}{3}\sin t\right)\mathbf{i} + \left(\dfrac{2}{3}\cos t\right)\mathbf{j} + \dfrac{\sqrt{5}}{3}\mathbf{k}$, 3π **3.** $\mathbf{T} = \dfrac{1}{\sqrt{1 + t}}\mathbf{i} + \dfrac{\sqrt{t}}{\sqrt{1 + t}}\mathbf{k}$, $\dfrac{52}{3}$ **5.** $\mathbf{T} = \dfrac{1}{\sqrt{3}}\mathbf{i} - \dfrac{1}{\sqrt{3}}\mathbf{j} + \dfrac{1}{\sqrt{3}}\mathbf{k}$, $3\sqrt{3}$ **7.** $\mathbf{T} = \left(\dfrac{\cos t - t \sin t}{t + 1}\right)\mathbf{i} + \left(\dfrac{\sin t + t \cos t}{t + 1}\right)\mathbf{j} + \left(\dfrac{\sqrt{2}\,t^{1/2}}{t + 1}\right)\mathbf{k}$, $\dfrac{\pi^2}{2} + \pi$ **9.** $s(t) = 5t$, $L = \dfrac{5\pi}{2}$ **11.** $s(t) = \sqrt{3}\,e^t - \sqrt{3}$, $L = -\dfrac{3\sqrt{3}}{4}$ **13.** $\sqrt{2} + \ln(1 + \sqrt{2})$ **15.** a) Cylinder is $x^2 + y^2 = 1$.
b)

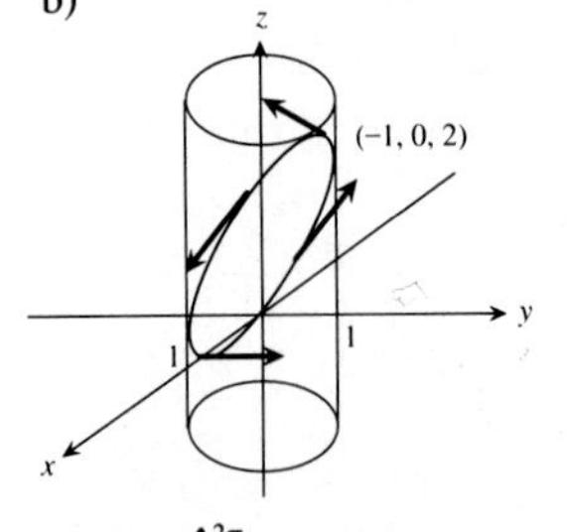

c)

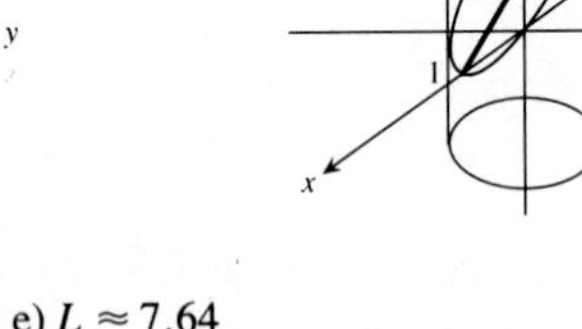

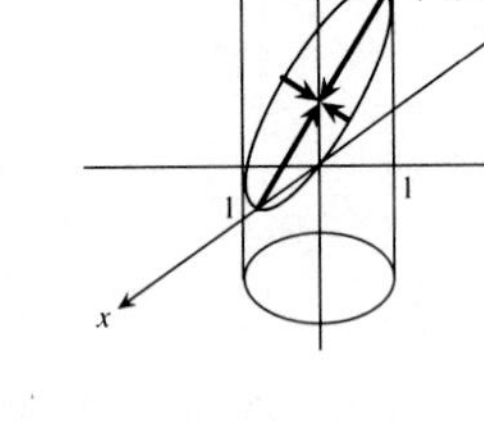

d) $L = \displaystyle\int_0^{2\pi} \sqrt{1 + \sin^2 t}\,dt$ e) $L \approx 7.64$

Section 11.4, pp. 799–800

1. $\mathbf{T} = (\cos t)\,\mathbf{i} - (\sin t)\,\mathbf{j}$, $\mathbf{N} = (-\sin t)\,\mathbf{i} - (\cos t)\,\mathbf{j}$, $\kappa = \cos t$ **3.** $\mathbf{T} = \dfrac{1}{\sqrt{1 + t^2}}\mathbf{i} - \dfrac{1}{\sqrt{1 + t^2}}\mathbf{j}$, $\mathbf{N} = \dfrac{-t}{\sqrt{1 + t^2}}\mathbf{i} - \dfrac{1}{\sqrt{1 + t^2}}\mathbf{j}$, $\kappa = \dfrac{1}{2(\sqrt{1 + t^2})^3}$ **5.** $\mathbf{T} = \dfrac{3 \cos t}{5}\mathbf{i} - \dfrac{3 \sin t}{5}\mathbf{j} + \dfrac{4}{5}\mathbf{k}$, $\mathbf{N} = (-\sin t)\,\mathbf{i} - (\cos t)\,\mathbf{j}$, $\mathbf{B} = \left(\dfrac{4}{5}\cos t\right)\mathbf{i} - \left(\dfrac{4}{5}\sin t\right)\mathbf{j} - \dfrac{3}{5}\mathbf{k}$, $\kappa = \dfrac{3}{25}$, $\tau = -\dfrac{4}{25}$ **7.** $\mathbf{T} = \left(\dfrac{e^t \cos t - e^t \sin t}{e^t\sqrt{2}}\right)\mathbf{i} + \left(\dfrac{e^t \cos t + e^t \sin t}{e^t\sqrt{2}}\right)\mathbf{j}$,

$\mathbf{N} = \left(\frac{-\cos t - \sin t}{\sqrt{2}}\right)\mathbf{i} + \left(\frac{-\sin t + \cos t}{\sqrt{2}}\right)\mathbf{j}$, $\mathbf{B} = \mathbf{k}$, $\kappa = \frac{1}{e^t\sqrt{2}}$, $\tau = 0$ **9.** $\mathbf{T} = \frac{t}{\sqrt{t^2+1}}\mathbf{i} + \frac{1}{\sqrt{t^2+1}}\mathbf{j}$, $\mathbf{N} = \frac{\mathbf{i}}{\sqrt{t^2+1}} - \frac{t\mathbf{j}}{\sqrt{t^2+1}}$, $\mathbf{B} = -\mathbf{k}$, $\kappa = \frac{1}{t(t^2+1)^{3/2}}$, $\tau = 0$

11. $\mathbf{T} = \left(\operatorname{sech}\frac{t}{a}\right)\mathbf{i} + \left(\tanh\frac{t}{a}\right)\mathbf{j}$, $\mathbf{N} = \left(-\tanh\frac{t}{a}\right)\mathbf{i} + \left(\operatorname{sech}\frac{t}{a}\right)\mathbf{j}$, $\mathbf{B} = \mathbf{k}$, $\kappa = \frac{1}{a\cosh^2\frac{t}{a}}$, $\tau = 0$ **13.** $\mathbf{a} = \frac{2t}{\sqrt{1+t^2}}\mathbf{T} + \frac{2}{\sqrt{1+t^2}}\mathbf{N}$

15. $\mathbf{a} = |a|\mathbf{N}$ **17.** $\mathbf{a}(1) = \frac{4}{3}\mathbf{T} + \frac{2\sqrt{5}}{3}\mathbf{N}$ **19.** $\mathbf{a}(0) = 2\mathbf{N}$

21. $\mathbf{r}\left(\frac{\pi}{4}\right) = \frac{\sqrt{2}}{2}\mathbf{i} + \frac{\sqrt{2}}{2}\mathbf{j} - \mathbf{k}$, $\mathbf{T}\left(\frac{\pi}{4}\right) = -\frac{\sqrt{2}}{2}\mathbf{i} + \frac{\sqrt{2}}{2}\mathbf{j}$, $\mathbf{N}\left(\frac{\pi}{4}\right) = -\frac{\sqrt{2}}{2}\mathbf{i} - \frac{\sqrt{2}}{2}\mathbf{j}$, $\mathbf{B}\left(\frac{\pi}{4}\right) = \mathbf{k}$; osculating plane: $z = -1$; rectifying plane: $x + y = \sqrt{2}$; normal plane: $-x + y = 0$

23. Yes, if the car is moving around a circle at a constant speed.

27. $\kappa = \frac{1}{t}$, $\rho = t$ **31.** $\frac{1}{2b}$ **35.** $\left(x - \frac{\pi}{2}\right)^2 + y^2 = 1$ **37.** a) $\kappa = \sin t$ b) $\kappa = \operatorname{sech} t$ **39.** a) $\delta = \frac{1}{2\sqrt{2}}$, $(x+2)^2 + (y-3)^2 = 8$ b) $f'(0) = 1$, $f''(0) = 1$

41.

Components of **v**:	-1.87001408, 0.708992989, 0.999999977
Components of **a**:	-1.69606646, -2.03053933, 0
Speed:	2.23598383
Components of **T**:	-0.836327193, 0.317083237, 0.447230417
Components of **N**:	-0.641065166, -0.76748645, 0
Components of **B**:	0.343243285, -0.28670381, 0.845140807
Curvature:	0.505990677

43.

Components of **v**:	1.99998356, $-796280802 \times 10^{-6}$, -0.162867596
Components of **a**:	0, -1.00000296, $-8.65198672 \times 10^{-3}$
Speed:	2.00660413
Components of **T**:	0.996700614, $-3.96830043 \times 10^{-6}$, -0.0811657837
Components of **N**:	0, -0.999962572, $-8.65163731 \times 10^{-3}$
Components of **B**:	-0.0811627115, $8.62309222 \times 10^{-3}$, -0.99666331
Curvature:	0.248367078

Section 11.5, pp. 809–810

1. 93.17 min **3.** 6765 km **5.** 1655 min **7.** 20,430 km **9.** $1.9966 \times 10^7 r^{-1/2}$ m/s **11.** Circle: $v_0 = \sqrt{\frac{GM}{r_0}}$; Ellipse: $\sqrt{\frac{GM}{r_0}} < v_0 < \sqrt{\frac{2GM}{r_0}}$; Parabola: $v_0 = \sqrt{\frac{2GM}{r_0}}$; Hyperbola: $v_0 > \sqrt{\frac{2GM}{r_0}}$ **15.** $\mathbf{v} = (\cos t - t\sin t)\mathbf{i} + (\sin t + t\cos t)\mathbf{j}$; $\mathbf{a} = (-2\sin t - t\cos t)\mathbf{i} + (2\cos t - t\sin t)\mathbf{j}$; $\kappa = (t^2+1)\sqrt{t^2+1}$

Section 11 Miscellaneous Exercises, pp. 811–814

1.

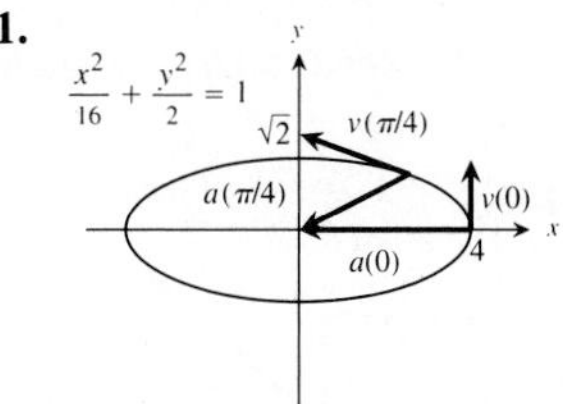

3. $6\mathbf{i} + 8\mathbf{j}$ **5.** $\mathbf{r} = ((\cos t) - 1)\mathbf{i} + ((\sin t) + 1)\mathbf{j} + t\mathbf{k}$

7. $\mathbf{r} = \mathbf{i} + t^2\mathbf{j} + t\mathbf{k}$ **9.** $L = \frac{\pi}{4}\sqrt{1 + \frac{\pi^2}{16}} + \ln\left(\frac{\pi}{4} + \sqrt{1 + \frac{\pi^2}{16}}\right)$

11. $\mathbf{T}(0) = \frac{2}{3}\mathbf{i} - \frac{2}{3}\mathbf{j} + \frac{1}{3}\mathbf{k}$; $\mathbf{N}(0) = \frac{1}{\sqrt{2}}\mathbf{i} + \frac{1}{\sqrt{2}}\mathbf{j}$; $\mathbf{B}(0) = -\frac{1}{3\sqrt{2}}\mathbf{i} + \frac{1}{3\sqrt{2}}\mathbf{j} + \frac{4}{3\sqrt{2}}\mathbf{k}$; $\kappa = \frac{\sqrt{2}}{3}$, $\tau = \frac{1}{6}$; osculating plane: $x - y - 4z = 0$; rectifying plane: $x + y = \frac{8}{9}$; normal plane: $2x - 2y + z = 0$ **13.** $\mathbf{T}(\ln 2) = \frac{1}{\sqrt{17}}\mathbf{i} + \frac{4}{\sqrt{17}}\mathbf{j}$; $\mathbf{N}(\ln 2) = -\frac{4}{\sqrt{17}}\mathbf{i} + \frac{1}{\sqrt{17}}\mathbf{j}$; $\mathbf{B}(\ln 2) = \mathbf{k}$; $\kappa = \frac{8}{17\sqrt{17}}$, $\tau = 0$; osculating plane: $z = 0$; rectifying plane: $4x - y = 4\ln 2 - 2$; normal plane: $x + 4y = \ln 2 + 8$ **15.** $\mathbf{a}(0) = 10\mathbf{T} + 6\mathbf{N}$

17. $\mathbf{T} = \left(\frac{1}{\sqrt{2}}\cos t\right)\mathbf{i} - (\sin t)\mathbf{j} + \left(\frac{1}{\sqrt{2}}\cos t\right)\mathbf{k}$; $\mathbf{N} = \left(-\frac{1}{\sqrt{2}}\sin t\right)\mathbf{i} - (\cos t)\mathbf{j} - \left(\frac{1}{\sqrt{2}}\sin t\right)\mathbf{k}$; $\mathbf{B} = \frac{1}{\sqrt{2}}\mathbf{i} - \frac{1}{\sqrt{2}}\mathbf{k}$; $\kappa = \frac{1}{\sqrt{2}}$, $\tau = 0$ **19.** $\frac{\pi}{2}$ for all t **21.** $\frac{\pi}{3}$ **23.** 26

25. a)

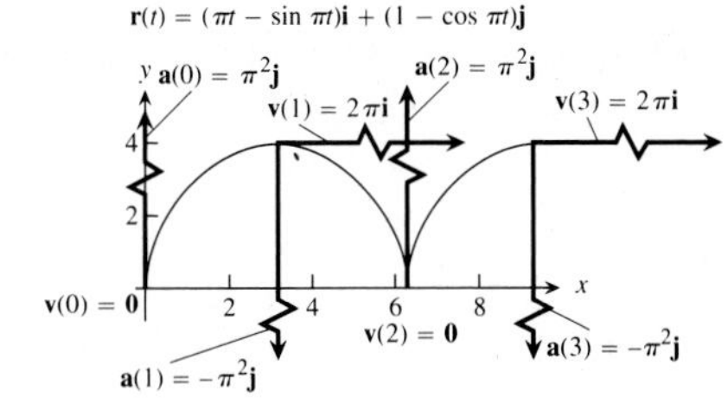

c) 3.14 ft/sec **27.** $\mathbf{r}(t) = \frac{a\cos\theta}{\sqrt{1+\sin^2\theta}}\mathbf{i} + \frac{a\sin\theta}{\sqrt{1+\sin^2\theta}}\mathbf{j} - \frac{a\sin\theta}{\sqrt{1+\sin^2\theta}}\mathbf{k}$; $L = 2\pi a$ **29.** a) 90.91 ft/sec b) 60.30 ft c) 170.46 ft · lb **35.** $\kappa = \pi s$ **37.** $\kappa = \frac{1}{a}$ if $a > 0$

39. a) $\frac{d\theta}{dt} = \frac{\sqrt{2gb\theta}}{\sqrt{a^2 + a^2\theta^2 + b^2}}$ b) $s = \frac{a}{2}\left(\theta\sqrt{c^2+\theta^2} + c^2\ln\left|\theta + \sqrt{c^2+\theta^2}\right| - c^2\ln c\right)$, where $c = \frac{a^2+b^2}{a^2}$ **43.** $\kappa = \frac{f^2 - ff'' + 2(f')^2}{((f')^2 + f^2)^{3/2}}$ **45.** a) $\frac{dx}{dt} = \dot{r}\cos\theta - r\dot{\theta}\sin\theta$, $\frac{dy}{dt} = \dot{r}\sin\theta + r\dot{\theta}\cos\theta$ b) $\frac{dr}{dt} = \dot{x}\cos\theta + \dot{y}\sin\theta$,

$r\frac{d\theta}{dt} = -\dot{x}\sin\theta + \dot{y}\cos\theta$ **47.** a) $\mathbf{u}_\rho = \sin\phi\cos\theta\mathbf{i} + \sin\phi\sin\theta\mathbf{j} + \cos\phi\mathbf{k}$, $\mathbf{u}_\phi = \cos\phi\cos\theta\mathbf{i} + \cos\phi\sin\theta\mathbf{j} - \sin\phi\mathbf{k}$, $\mathbf{u}_\theta = -\sin\theta\,\mathbf{i} + \cos\theta\mathbf{j}$ **49.** b)
c) $7\sqrt{3}$

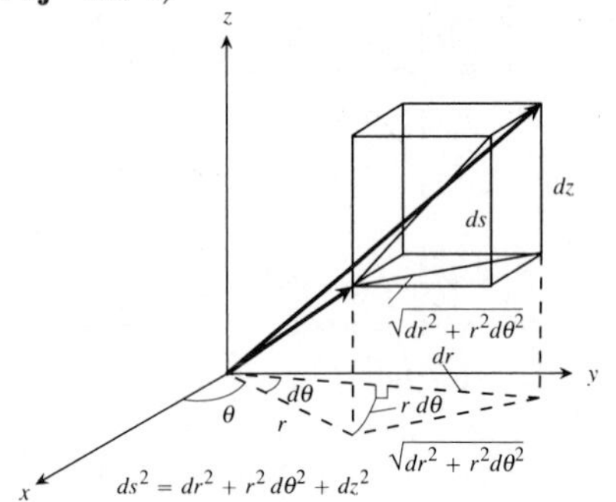

Chapter 12

Section 12.1, pp. 824–825

1. Domain: set of all (x, y) so that $y - x \geq 0 \Rightarrow y \geq x$; range: $z \geq 0$. Level curves are straight lines of the form $y - x = c$, where $c \geq 0$. **3.** Domain: all $(x, y) \neq (0, y)$; range: all real numbers. Level curves are parabolas with vertex (0, 0) and the y-axis as axis. **5.** Domain: all points in the xy-plane; range: all positive real numbers. Level curves are hyperbolas with the x- and y-axes as asymptotes. **7.** Domain: set of all (x, y) so that $-1 \leq y - x \leq 1$; range: $-\frac{\pi}{2} \leq z \leq \frac{\pi}{2}$. Level curves are straight lines of the form $y - x = c$, where $-1 \leq c \leq 1$. **9.** Domain: set of all (x, y) so that $x > 0$ and $y > 0$; range: all real numbers. Level curves are straight lines of the form $y - cx$, where $c > 0$, $x > 0$, and $y > 0$.

11. a)

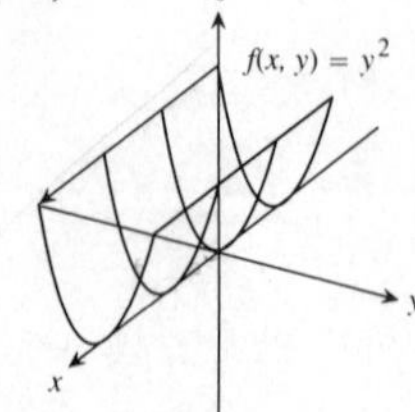

b)

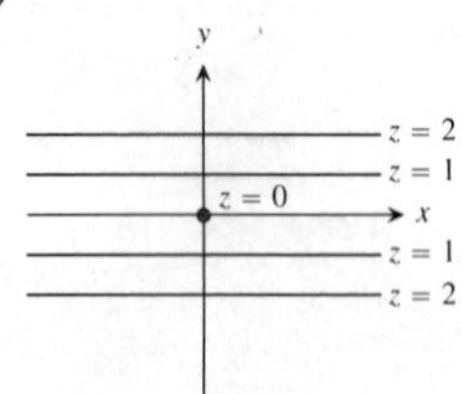

13. a)

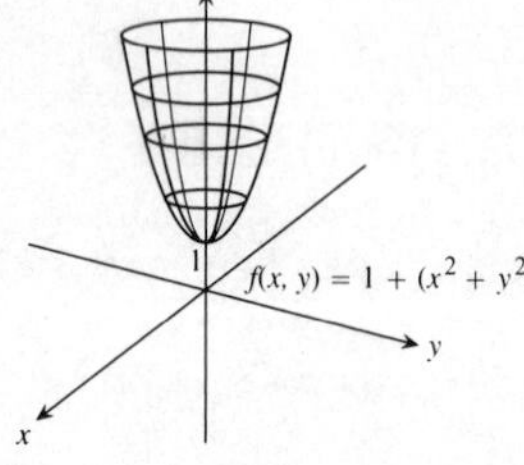

b)

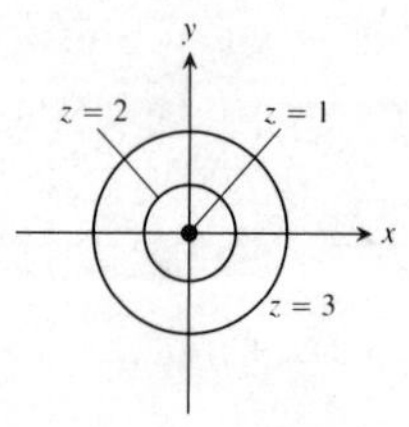

15. a)

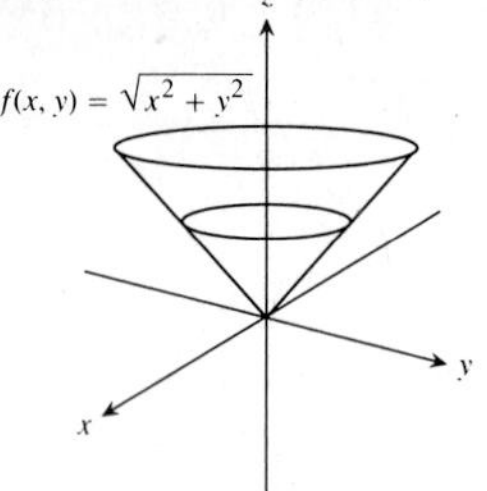

b)

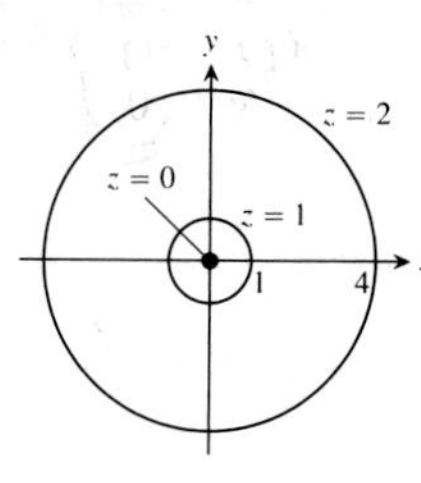

17. a)

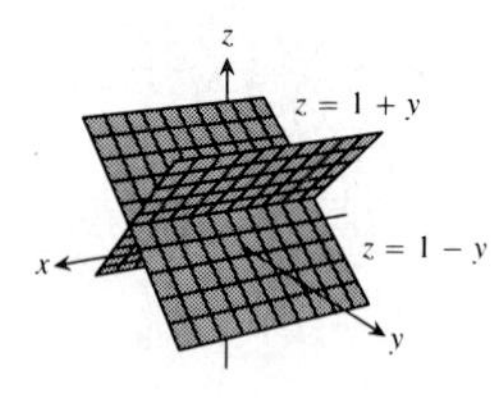

b)

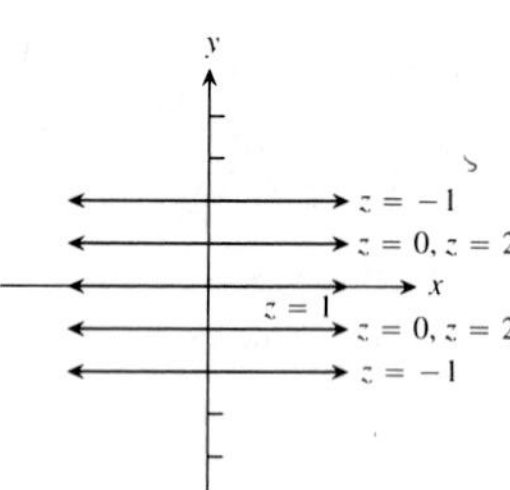

19. a)

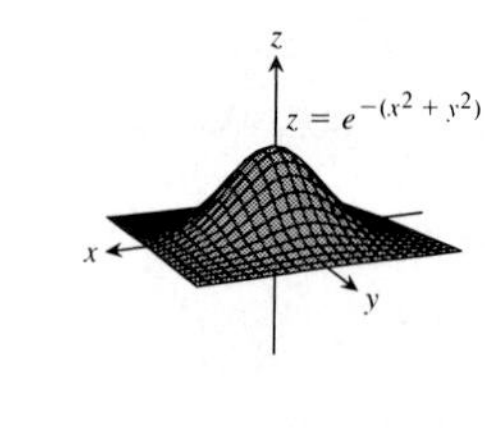

b)

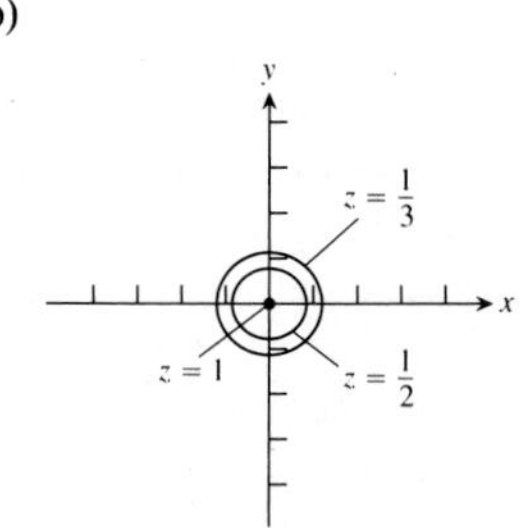

21. f **23.** a **25.** d **27.** Domain: all (x, y, z); range: all real numbers. Level surfaces are spheres with center (0, 0, 0).

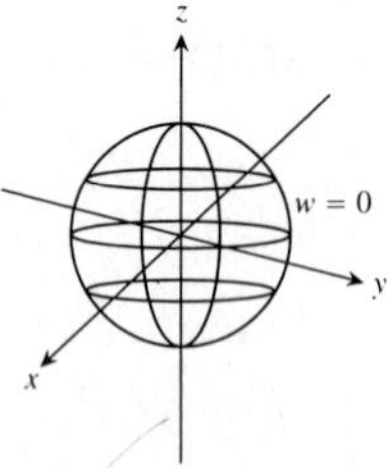

29. Domain: all (x, y, z); range: $-\frac{\pi}{2} < w < \frac{\pi}{2}$. Level surfaces are paraboloids with the z-axis as axis.

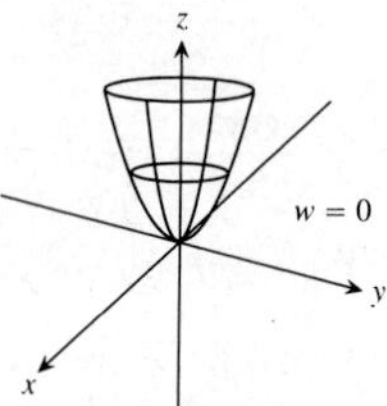

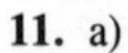

31. Domain: all (x, y, z); range: $0 < w \leq 1$. Level surfaces are pairs of parallel planes perpendicular to the z-axis, equidistant from (0, 0, 0).

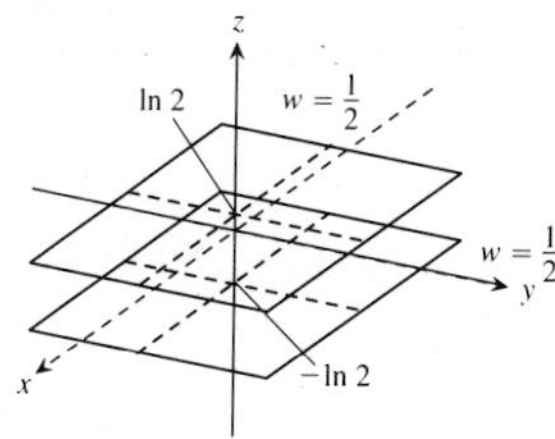

33. Domain: all (x, y, z) so that $z > \sqrt{x^2 + y^2}$; range: all real numbers. Level surfaces are cones with the z-axis as axis.

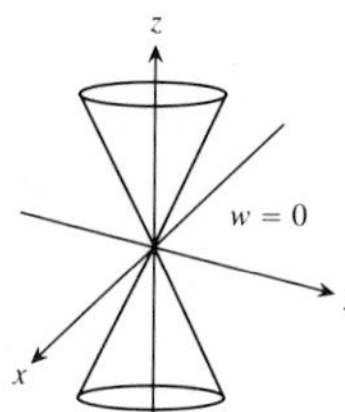

35. $x^2 + y^2 = 10$ **37.** $-2\sqrt{2} - 2\sqrt{2}xy = y - x$ **39.** $z = e^{\sqrt{x-y}-2}$ **41.** $z \ln 2 = x + y$ **43.** 62.43 km south of Nantucket

Section 12.2, pp. 831–833

1. $\frac{5}{2}$ **3.** $\frac{1}{2}$ **5.** 5 **7.** 1 **9.** 1 **11.** 1 **13.** 0 **15.** -1 **17.** 7 **19.** 2 **21.** $\tan^{-1}\left(-\frac{\pi}{4}\right)$ **23.** a) Continuous at all (x, y) b) Continuous at all (x, y) except (0, 0) **25.** a) Continuous at all (x, y) except where $x = 0$ or $y = 0$ b) Continuous at all (x, y) **27.** a) Continuous at all (x, y, z) b) Continuous at all (x, y, z) except the interior of the cylinder $x^2 + y^2 = 1$ **29.** a) Continuous at all (x, y, z) so that $(x, y, z) \neq (x, y, 0)$ b) Continuous at all (x, y, z) except those on the sphere $x^2 + y^2 + z^2 = 1$
31. Consider paths along $y = x$, where $x > 0$ or $x < 0$.
33. Consider paths along $y = kx^2$, k a constant. **35.** Consider paths along $y = kx$, k a constant, $k \neq 1$. **37.** Consider paths along $y = kx^2$, k a constant, $k \neq 0$. **41.** 1 **43.** 0 **45.** Does not exist **47.** $\frac{\pi}{2}$ **49.** $f(0, 0) = \ln 3$ **51.** $\delta = 0.1$ **53.** $\delta = 0.005$ **55.** $\delta = \sqrt{0.015}$ **57.** $\delta = 0.005$

Section 12.3, pp. 840–842

1. $\frac{\partial f}{\partial x} = 4x, \frac{\partial f}{\partial y} = -3$ **3.** $\frac{\partial f}{\partial x} = 2x(y-2), \frac{\partial f}{\partial y} = x^2 - 1$ **5.** $\frac{\partial f}{\partial x} = 2y(xy-1), \frac{\partial f}{\partial y} = 2x(xy-1)$ **7.** $\frac{\partial f}{\partial x} = \frac{x}{\sqrt{x^2+y^2}}, \frac{\partial f}{\partial y} = \frac{y}{\sqrt{x^2+y^2}}$ **9.** $\frac{\partial f}{\partial x} = \frac{-1}{(x+y)^2}, \frac{\partial f}{\partial y} = \frac{-1}{(x+y)^2}$ **11.** $\frac{\partial f}{\partial x} = \frac{-y^2-1}{(xy-1)^2}, \frac{\partial f}{\partial y} = \frac{-x^2-1}{(xy-1)^2}$ **13.** $\frac{\partial f}{\partial x} = \frac{1}{x+y}, \frac{\partial f}{\partial y} = \frac{1}{x+y}$ **15.** $\frac{\partial f}{\partial x} = ye^{xy}\ln y$, $\frac{\partial f}{\partial y} = xe^{xy}\ln y + \frac{e^{xy}}{y}$ **17.** $\frac{\partial f}{\partial x} = -6\cos(3x - y^2)\sin(3x - y^2)$; $\frac{\partial f}{\partial y} = -4y\cos(3x - y^2)\sin(3x - y^2)$ **19.** $\frac{\partial f}{\partial x} = yx^{y-1}, \frac{\partial f}{\partial y} = x^y \ln y$ **21.** $\frac{\partial f}{\partial x} = -f(x), \frac{\partial f}{\partial y} = f(y)$ **23.** $f_x(x, y, z) = y^2$, $f_y(x, y, z) = 2xy, f_z(x, y, z) = -4z$ **25.** $f_x(x, y, z) = 1$, $f_y(x, y, z) = -y(y^2 + z^2)^{-1/2}, f_z(x, y, z) = -z(y^2 + z^2)^{-1/2}$ **27.** $f_x(x, y, z) = \frac{yz}{\sqrt{1 - x^2y^2z^2}}, f_y(x, y, z) = \frac{xz}{\sqrt{1 - x^2y^2z^2}}$, $f_z(x, y, z) = \frac{xy}{\sqrt{1 - x^2y^2z^2}}$ **29.** $f_x(x, y, z) = \frac{1}{x + 2y + 3z}, f_y(x, y, z) = \frac{2}{x + 2y + 3z}$, $f_z(x, y, z) = \frac{3}{x + 2y + 3z}$ **31.** $f_x(x, y, z) = -2xe^{-(x^2+y^2+z^2)}$, $f_y(x, y, z) = -2ye^{-(x^2+y^2+z^2)}, f_z(x, y, z) = -2ze^{-(x^2+y^2+z^2)}$ **33.** $f_x(x, y, z) = \text{sech}^2(x + 2y + 3z), f_y(x, y, z) = 2\,\text{sech}^2(x + 2y + 3z), f_z(x, y, z) = 3\,\text{sech}^2(x + 2y + 3z)$ **35.** $\frac{\partial f}{\partial t} = -2\pi\sin(2\pi t - \alpha), \frac{\partial f}{\partial \alpha} = \sin(2\pi t - \alpha)$ **37.** $\frac{\partial h}{\partial \rho} = \sin\phi\cos\theta, \frac{\partial h}{\partial \phi} = \rho\cos\phi\cos\theta$, $\frac{\partial h}{\partial \theta} = -\rho\sin\phi\sin\theta$ **39.** $W_P(P, V, v, g) = \frac{V}{3}, W_V(P, V, v, g) = \frac{P}{3} + \frac{v^2}{2g}, W_v(P, V, v, g) = \frac{Vv}{g}, W_g(P, V, v, g) = -\frac{Vv^2}{2g^2}$ **41.** $\frac{\partial^2 f}{\partial x^2} = 0, \frac{\partial^2 f}{\partial y^2} = 0, \frac{\partial^2 f}{\partial y \partial x} = \frac{\partial^2 f}{\partial x \partial y} = 1$ **43.** $\frac{\partial^2 g}{\partial x^2} = 2y - y\sin x$, $\frac{\partial^2 g}{\partial y^2} = -\cos y, \frac{\partial^2 g}{\partial y \partial x} = \frac{\partial^2 g}{\partial x \partial y} = 2x + \cos x$ **45.** $\frac{\partial^2 r}{\partial x^2} = \frac{-1}{(x+y)^2}$, $\frac{\partial^2 r}{\partial y^2} = \frac{-1}{(x+y)^2}, \frac{\partial^2 r}{\partial y \partial x} = \frac{\partial^2 r}{\partial x \partial y} = \frac{-1}{(x+y)^2}$ **51.** a) x first b) y first c) x first d) x first e) y first f) y first **53.** -2 **55.** $\frac{\partial A}{\partial b} = \frac{c\cos A - b}{bc\sin A}$ **57.** $v_x = \frac{u^2\ln v - vu}{u^2\ln u\ln v - v^2}$ **65.** $n = 0$ or $n = \frac{1}{2}$

Section 12.4, pp. 852–854

1. a) $L(x, y) = 1$ b) $L(x, y) = 2x + 2y - 1$ **3.** a) $L(x, y) = 1 + x$ b) $L(x, y) = -y + \frac{\pi}{2}$ **5.** a) $L(x, y) = 5 + 3x - 4y$ b) $L(x, y) = 3x - 4y + 5$ **7.** $L(x, y) = 7 + x - 6y$; 0.06 **9.** $L(x, y) = x + y + 1$; 0.08 **11.** $L(x, y) = 1 + x$; 0.0222 **13.** Pay more attention to the width. **15.** 0.31 **17.** $\pm 4.83\%$ **19.** Let $|x - 1| \leq 0.014, |y - 1| \leq 0.014$ **21.** 0.099% **25.** a) $L(x, y, z) = 2x + 2y + 2z - 3$ b) $L(x, y, z) = y + z$ c) $L(x, y, z) = 0$ **27.** a) $L(x, y, z) = x$ b) $L(x, y, z) = \frac{1}{\sqrt{2}}x + \frac{1}{\sqrt{2}}y$ c) $L(x, y, z) = \frac{1}{3}x + \frac{2}{3}y + \frac{2}{3}z$

29. a) $L(x, y, z) = 2 + x$ b) $L(x, y, z) = x - y - z + \frac{\pi}{2} + 1$ c) $L(x, y, z) = x - y - z + \frac{\pi}{2} + 1$ **31.** $L(x, y, z) = 2x - 6y - 2z + 6$; 0.0024 **33.** $L(x, y, z) = x + y - z - 1$; 0.00135 **35.** a) $S_0\left(\frac{1}{100}\,dp + dx - 5\,dw - 30\,dh\right)$ b) More sensitive to a change in height **37.** 6% **39.** ± 319.23 ft^2 **41.** Q is most sensitive to changes in h.

Section 12.5, pp. 860–862

1. $\frac{dw}{dt} = 0, \frac{dw}{dt}(\pi) = 0$ **3.** $\frac{dw}{dt} = 1, \frac{dw}{dt}(3) = 1$ **5.** $\frac{dw}{dt} = 4t\tan^{-1}t + 1, \frac{dw}{dt}(1) = \pi + 1$ **7.** a) $\frac{\partial z}{\partial r} = 4\cos\theta\ln(r\sin\theta) + 4\cos\theta$; $\frac{\partial z}{\partial \theta} = -4r\sin\theta\ln(r\sin\theta) + \frac{4r\cos^2\theta}{\sin\theta}$ b) $\frac{\partial z}{\partial r} = \sqrt{2}(\ln 2 + 2)$; $\frac{\partial z}{\partial \theta} = -2\sqrt{2}\ln 2 + 2\sqrt{2}$ **9.** a) $\frac{\partial w}{\partial u} = 2u + 4uv$; $\frac{\partial w}{\partial v} = -2v + 2u^2$ b) $\frac{\partial w}{\partial u} = 3$; $\frac{\partial w}{\partial v} = -\frac{3}{2}$ **11.** a) $\frac{\partial u}{\partial y} = \frac{z}{(z - y)^2}$; $\frac{\partial u}{\partial z} = \frac{-y}{(z - y)^2}$ b) $\frac{\partial u}{\partial y} = 0$; $\frac{\partial u}{\partial z} = -2$ **13.** $\frac{dz}{dt} = \frac{\partial z}{\partial x}\frac{dx}{dt} + \frac{\partial z}{\partial y}\frac{dy}{dt}$

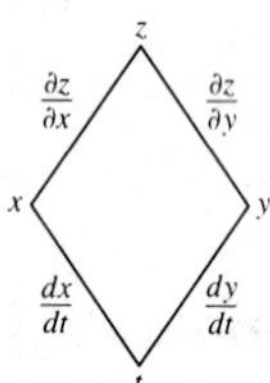

15. $\frac{\partial w}{\partial u} = \frac{\partial w}{\partial x}\frac{\partial x}{\partial u} + \frac{\partial w}{\partial y}\frac{\partial y}{\partial u} + \frac{\partial w}{\partial z}\frac{\partial z}{\partial u}$ $\frac{\partial w}{\partial v} = \frac{\partial w}{\partial x}\frac{\partial x}{\partial v} + \frac{\partial w}{\partial y}\frac{\partial y}{\partial v} + \frac{\partial w}{\partial z}\frac{\partial z}{\partial v}$

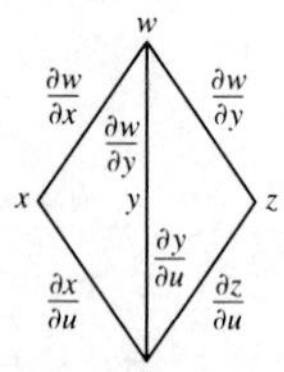

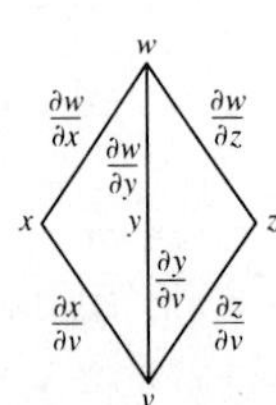

17. $\frac{\partial w}{\partial u} = \frac{\partial w}{\partial x}\frac{\partial x}{\partial u} + \frac{\partial w}{\partial y}\frac{\partial y}{\partial u}$ $\frac{\partial w}{\partial v} = \frac{\partial w}{\partial x}\frac{\partial x}{\partial v} + \frac{\partial w}{\partial y}\frac{\partial y}{\partial v}$

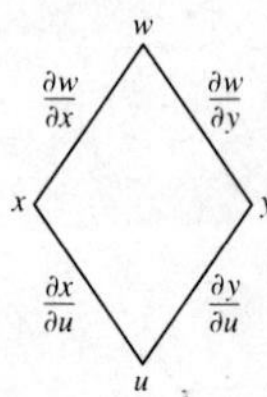

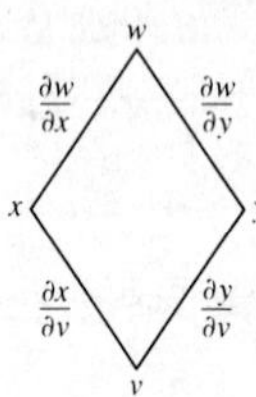

19. $\frac{\partial z}{\partial t} = \frac{\partial z}{\partial x}\frac{\partial x}{\partial t} + \frac{\partial z}{\partial y}\frac{\partial y}{\partial t}$ $\frac{\partial z}{\partial s} = \frac{\partial z}{\partial x}\frac{\partial x}{\partial s} + \frac{\partial z}{\partial y}\frac{\partial y}{\partial s}$

21. $\frac{\partial w}{\partial s} = \frac{dw}{du}\frac{\partial u}{\partial s}$ $\frac{\partial w}{\partial t} = \frac{dw}{du}\frac{\partial u}{\partial t}$

23. $\frac{\partial w}{\partial r} = \frac{\partial w}{\partial x}\frac{dx}{dr} + \frac{\partial w}{\partial y}\frac{dy}{dr} = \frac{\partial w}{\partial x}\frac{dx}{dr}$ since $\frac{dy}{dr} = 0$

$\frac{\partial w}{\partial s} = \frac{\partial w}{\partial x}\frac{dx}{ds} + \frac{\partial w}{\partial y}\frac{dy}{ds} = \frac{\partial w}{\partial y}\frac{dy}{ds}$ since $\frac{dx}{ds} = 0$

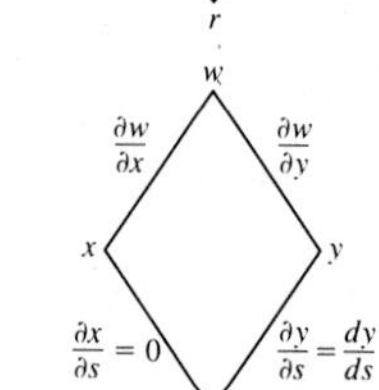

25. $\frac{4}{3}$ **27.** $-\frac{4}{5}$ **29.** $-\frac{3}{4}$ **31.** -1 **33.** 12 **35.** -7 **37.** $\frac{\partial z}{\partial u} = 2$; $\frac{\partial z}{\partial v} = 5$ **39.** 0 **41.** -0.00005 amps/sec **45.** $x = \cos 1$, $y = \sin 1$, $z = 1$ **47.** a) T has a minimum at $\frac{\pi}{4}, \frac{5\pi}{4}$; T has a maximum at $\frac{3\pi}{4}, \frac{7\pi}{4}$ b) $T_{\max} = 6$; $T_{\min} = 2$

Section 12.6, pp. 868–869

1. a) 0 b) $1 + 2z$ c) $1 + 2z$ **3.** a) $\frac{\partial U}{\partial p} + \frac{\partial U}{\partial T}\left(\frac{v}{nR}\right)$ b) $\frac{\partial U}{\partial p}\left(\frac{nR}{v}\right) + \frac{\partial U}{\partial T}$ **5.** a) 5 b) 5 **7.** a) $\cos\theta$ b) $\frac{x}{\sqrt{x^2 + y^2}}$

Section 12.7, pp. 877–879

1. $\nabla f = 3\mathbf{i} + 2\mathbf{j} - 4\mathbf{k}$ **3.** $\nabla f = -\frac{26}{27}\mathbf{i} + \frac{23}{54}\mathbf{j} - \frac{23}{54}\mathbf{k}$

5.

$x^2 + y^2 - z = 0$

$\nabla f = 2\mathbf{i} + 2\mathbf{j} - \mathbf{k}$

7.

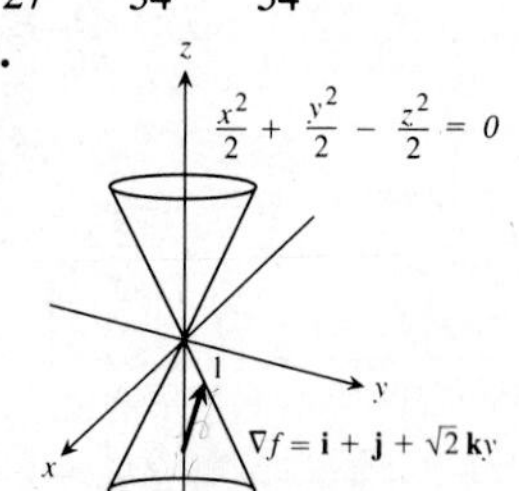

9. **11.**

13. 3 **15.** -2 **17.** $\frac{31}{13}$ **19.** $\mathbf{u} = \mathbf{j}$, $D_{\mathbf{u}}f = 2$ $-\mathbf{u} = -\mathbf{j}$, $D_{-\mathbf{u}}f = -2$ **21.** $\mathbf{u} = \frac{1}{\sqrt{5 + 4\pi^2}}(\mathbf{i} + \mathbf{j} + 2\pi\mathbf{k})$, $D_{\mathbf{u}}f = \sqrt{5 + 4\pi^2}$; $-\mathbf{u} = -\frac{1}{\sqrt{5 + 4\pi^2}}(\mathbf{i} + \mathbf{j} + 2\pi\mathbf{k})$, $D_{-\mathbf{u}}f = -\sqrt{5 + 4\pi^2}$ **23.** $\mathbf{u} = \frac{1}{\sqrt{3}}(\mathbf{i} + \mathbf{j} + \mathbf{k})$, $D_{\mathbf{u}}f = 2\sqrt{3}$; $-\mathbf{u} = -\frac{1}{\sqrt{3}}(\mathbf{i} + \mathbf{j} + \mathbf{k})$, $D_{-\mathbf{u}}f = -2\sqrt{3}$ **25.** Tangent: $-4x + 2z + 4 = 0$; normal line: $x = 2 - 4t$, $y = 0$, $z = t$ **27.** Tangent: $2y + 3z = 7$; normal line: $x = 1$, $y = -1 + 4t$, $z = 3 + 6t$ **29.** Tangent: $2x + 2y + z - 4 = 0$; normal line: $x = 2t$, $y = 1 + 2t$, $z = 2 + t$ **31.** $x = 1$, $y = 1 + 2t$, $z = 1 - 2t$ **33.** $x = 1 + 90t$, $y = 1 - 90t$, $z = 3$

35.

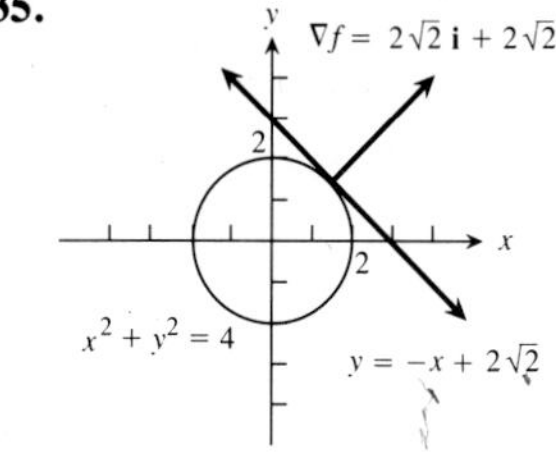

37.

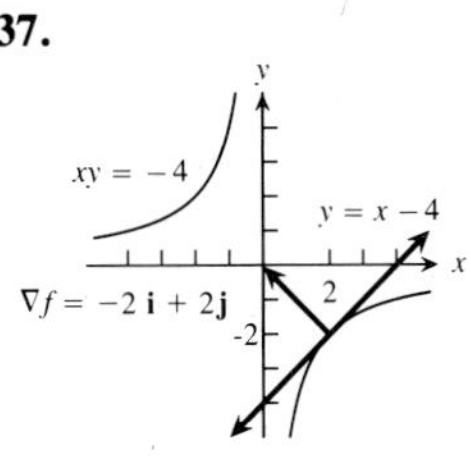

39. $\frac{3\sqrt{2}}{20}$ **41.** $\frac{1}{15}$ **43.** $\mathbf{u} = \frac{7}{\sqrt{53}}\mathbf{i} - \frac{2}{\sqrt{53}}\mathbf{j}$; $-\mathbf{u} = -\frac{7}{\sqrt{53}}\mathbf{i} + \frac{2}{\sqrt{53}}\mathbf{j}$ **45.** At $-\frac{\pi}{4}, \frac{-\pi}{2\sqrt{2}}$; at 0, 0; at $\frac{\pi}{4}, \frac{\pi}{2\sqrt{2}}$ **47.** $-\frac{7}{\sqrt{5}}$ **53.** ∇f: 0, 4, 0; Directional derivative: 2.309401; Plane: $4y = 0$; Normal line: $x = 1$, $y = 4t$, $z = 1$ **55.** ∇f: -1.818396, -1.818396, -0.4161477; Directional derivative: -1.85944; Plane: $-1.818396x - 1.818396y - 0.416477z = -4.05294$; Normal line: $x = 1 - 1.818396t$, $y = 1 - 1.818396t$, $z = 1 - 0.4161477t$

Section 12.8, pp. 887–891

1. $f(-3{,}3) = -5$, absolute minimum **3.** Saddle point at $\left(\frac{6}{5}, \frac{69}{25}\right)$ **5.** Saddle point at $(-2, 1)$ **7.** $f\left(\frac{4}{9}, \frac{2}{9}\right) = -\frac{252}{81}$, absolute maximum **9.** Saddle point at (2, 1) **11.** $f(2, -1) = -6$, absolute minimum **13.** Saddle point at $\left(\frac{1}{6}, \frac{4}{3}\right)$ **15.** $f(1, 0) = 0$, absolute minimum **17.** $f(0, 1) = 4$, absolute maximum **19.** $f\left(-\frac{2}{3}, \frac{2}{3}\right) = \frac{170}{27}$, local maximum **21.** $f(0, 0) = 0$, local minimum; saddle point at $(1, -1)$ **23.** Saddle point at (0, 0); $f(-1, -1) = 1$, local maximum **25.** $f(0, 0) = -1$, local maximum **27.** $f(0, 0) = -4$, absolute minimum **29.** Saddle point at $(n\pi, 0)$, $f(n\pi, 0) = 0$ for every integer n **31.** Absolute maximum is 1 at (0, 0); absolute minimum is -5 at (1, 2) **33.** Absolute maximum is 4 at (0, 2); absolute minimum is 0 at (0, 0). **35.** Absolute maximum is 11 at $(0, -3)$; absolute minimum is -10 at $(4, -2)$. **37.** Absolute maximum is 4 at (2, 0); absolute minimum is $\frac{3\sqrt{2}}{2}$ at $\left(3, -\frac{\pi}{4}\right)$, $\left(3, \frac{\pi}{4}\right)$, $\left(1, -\frac{\pi}{4}\right)$, and $\left(1, \frac{\pi}{4}\right)$. **39.** Hottest is 2.25° at $\left(-\frac{1}{2}, \frac{\sqrt{3}}{2}\right)$ and $\left(-\frac{1}{2}, -\frac{\sqrt{3}}{2}\right)$; coldest is $-0.25°$ at $\left(\frac{1}{2}, 0\right)$. **41.** a) Saddle point at (0, 0) b) Local minimum at (1, 2) c) Local minimum at $(1, -2)$; saddle point at $(-1, -2)$

43. a) i) Absolute minimum is 0 at $t = 0, \frac{\pi}{2}$; absolute maximum is 2 at $t = \frac{\pi}{4}$. ii) Absolute minimum is -2 at $t = \frac{3\pi}{4}$; absolute maximum is 2 at $t = \frac{\pi}{4}$. iii) Absolute minimum is -2 at $t = \frac{3\pi}{4}, \frac{7\pi}{4}$; absolute maximum is 2 at $t = \frac{\pi}{4}, \frac{5\pi}{4}$. b) i) Absolute minimum is 2 at $t = 0, \frac{\pi}{2}$; absolute maximum is $2\sqrt{2}$ at $t = \frac{\pi}{4}$. ii) Absolute minimum is -2 at $t = \pi$; absolute maximum is $2\sqrt{2}$ at $t = \frac{\pi}{4}$. iii) Absolute minimum is $-2\sqrt{2}$ at $t = \frac{5\pi}{4}$; absolute maximum is $2\sqrt{2}$ at $t = \frac{\pi}{4}$. c) i) Absolute minimum is 1 at $t = \frac{\pi}{2}$; absolute maximum is 8 at $t = 0$. ii) Absolute minimum is 1 at $t = \frac{\pi}{2}$; absolute maximum is 8 at $t = 0, \pi$. iii) Absolute minimum is 1 at $t = \frac{\pi}{2}, \frac{3\pi}{2}$; absolute maximum is 8 at $t = 0, \pi, 2\pi$.

45. a) i) Absolute maximum is 12 at $t = \frac{\pi}{2}$; absolute minimum is 9 at $t = 0$. ii) Absolute minimum is 9 at $t = 0, \pi$; absolute maximum is 12 at $t = \frac{\pi}{2}$. iii) Absolute maximum is 12 at $t = \frac{\pi}{2}, \frac{3\pi}{2}$; absolute minimum is 9 at $t = 0, \pi, 2\pi$. b) i) Absolute maximum is $6\sqrt{2}$ at $t = \frac{\pi}{4}$; absolute minimum is 6 at $t = 0, \frac{\pi}{2}$. ii) Absolute maximum is $6\sqrt{2}$; absolute minimum is -6 at $t = \pi$. iii) Absolute maximum is $6\sqrt{2}$ at $t = \frac{\pi}{4}$; absolute minimum is $-6\sqrt{2}$ at $t = \frac{5\pi}{4}$. **47.** $y = 1.5x + 0.7$; $y|_{x=4} = -5.3$ **49.** $y = 1.5x + 0.2$; $y|_{x=4} = 6.2$

51. $y = 0.122x + 3.59$

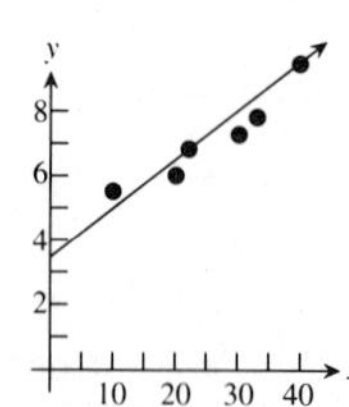

53. a)

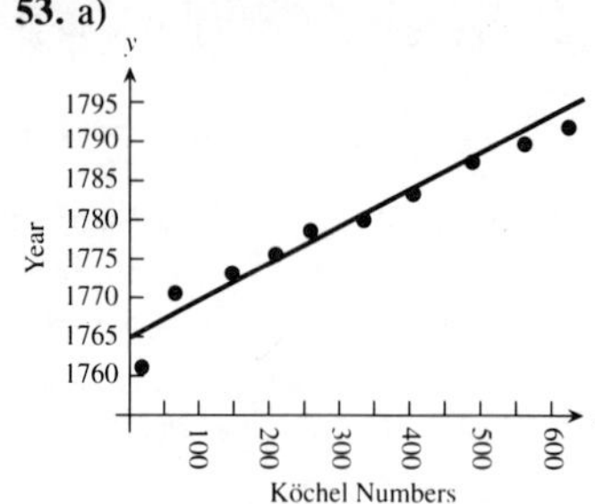

b) $y = 0.0427K + 1764.8$ c) 1780

Section 12.9, pp. 900–902

1. $\left(\pm\sqrt{2}, \frac{1}{2}\right), \left(\pm\sqrt{2}, \frac{1}{2}\right)$ **3.** 39 **5.** $f(0, 3) = 0, f(2, 1) = 4$ **7.** a) 8 b) 64 **9.** $r = 2$ cm, $h = 4$ cm **11.** Minimum $= 0°$, maximum $= 125°$ **13.** $f(0, 0) = 0$ is minimum, $f(2, 4) = 20$ is maximum. **15.** $f(1, -2, 5) = 30$ is maximum, $f(-1, 2, -5) = -30$ is minimum. **17.** $(0, 0, 1)$ **19.** $f\left(\frac{5}{\sqrt{14}}, \frac{10}{\sqrt{14}}, \frac{15}{\sqrt{14}}\right) = 5\sqrt{14}$ is maximum, $f\left(-\frac{5}{\sqrt{14}}, -\frac{10}{\sqrt{14}}, -\frac{15}{\sqrt{14}}\right) = -5\sqrt{14}$ is minimum. **21.** $\frac{4096}{25\sqrt{5}}$ **23.** $U(8, 14) = \$128$ **27.** $(2, 4, 4)$ **29.** Maximum is $1 + 6\sqrt{3}$ at $(\pm\sqrt{6}, \sqrt{3}, 1)$, minimum is $1 - 6\sqrt{3}$ at $(\pm\sqrt{60}, -\sqrt{3}, 1)$. **31.** Maximum is 4 at $(0, 0, \pm 2)$, minimum is 2 at $(\pm\sqrt{2}, \pm\sqrt{2}, 0)$.

Section 12.10, p. 908

1. Quadratic: $1 + x + \frac{1}{2}(x^2 - y^2)$; cubic: $1 + x + \frac{1}{2}(x^2 - y^2) + \frac{1}{6}(x^3 - 3xy^2)$ **3.** Quadratic: $\frac{1}{2}(2x^2) = x^2$; cubic: x^2 **5.** Quadratic: $1 + x + y + x^2 + 2xy + y^2$; cubic: $1 + x + y + x^2 + 2xy + y^2 + x^3 + 3x^2y + 3xy^2 + y^3$ **7.** Quadratic: $1 - \frac{1}{2}x^2 - \frac{1}{2}y^2$; $E(x, y) \le 0.0013$

Section 12 Miscellaneous Exercises, pp. 909–914

1.

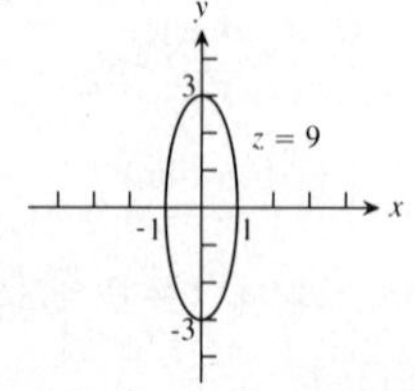

Domain: all points in the xy-plane; range: $f(x, y) \ge 0$. Level curves are ellipses with major axis along the y-axis and minor axis along the x-axis.

3.

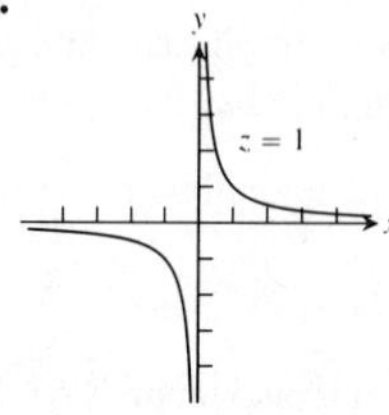

Domain: all (x, y) such that $x \ne 0$ or $y \ne 0$; range: $f(x, y) \ne 0$. Level curves are hyperbolas rotated 45° or 135°.

5.

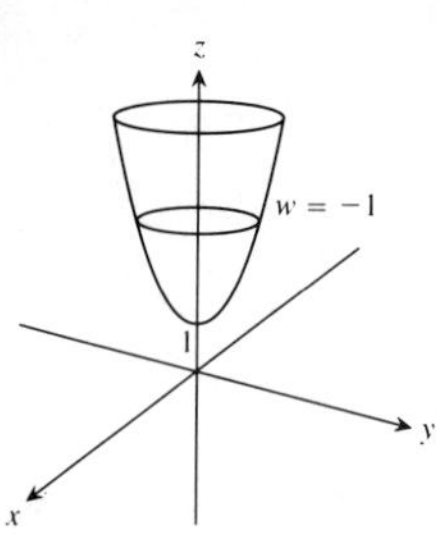

Domain: all (x, y, z) such that $(x, y, z) \ne (0, 0, 0)$; range: all real numbers. Level surfaces are paraboloids of revolution with the z-axis as axis.

7.

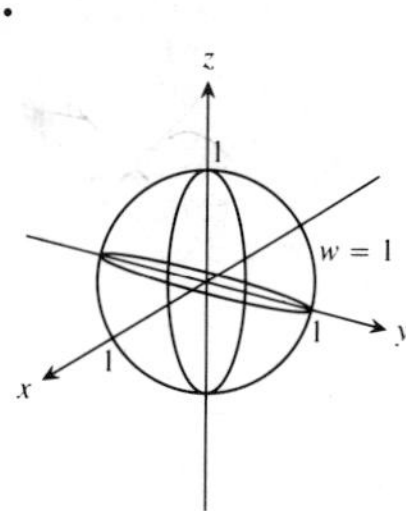

Domain: all (x, y, z) such that $(x, y, z) \ne (0, 0, 0)$; range: $f(x, y, z) > 0$. Level surfaces are spheres with center $(0, 0, 0)$ and radius $r > 0$.

9. -2 **11.** 2 **13.** 0 **17.** a) Does not exist b) Not continuous at $(0, 0)$ **19.** $\frac{\partial g}{\partial r} = \cos\theta + \sin\theta, \frac{\partial g}{\partial\theta} = -r\sin\theta + r\cos\theta$ **21.** $\frac{\partial f}{\partial R_1} = -\frac{1}{R_3^2}$ **23.** $\frac{\partial P}{\partial n} = \frac{RT}{V}, \frac{\partial P}{\partial R} = \frac{nT}{V}, \frac{\partial P}{\partial T} = \frac{nR}{V}, \frac{\partial P}{\partial V} = -\frac{nRT}{V^2}$ **25.** $\frac{\partial^2 f}{\partial x^2} = 0, \frac{\partial^2 f}{\partial y^2} = \frac{2x}{y^3}, \frac{\partial^2 f}{\partial y\partial x} = \frac{\partial^2 f}{\partial x\partial y} = -\frac{1}{y^2}$ **27.** $(\partial^2 f, \partial x^2) = -30x + \frac{2 - 2x^2}{(x^2 + 1)^2}, \frac{\partial^2 f}{\partial y^2} = 0, \frac{\partial^2 f}{\partial y\partial x} = \frac{\partial^2 f}{\partial x\partial y} = 1$ **29.** $f_{xy}(0, 0) = -1, f_{yx}(0, 0) = 1$ **31.** $L(x, y) = \frac{1}{2} + \frac{1}{2}x - \frac{1}{2}y$, $|E(x, y)| \le 0.02$ **33.** a) $L(x, y, z) = y - 3z$ b) $L(x, y, z) = x + y + 2z - 1$ **35.** Be more careful with the diameter. **37.** $dl = 0.038$, % change in $V = -4.17\%$, % change in $R = -20\%$, % change in l $= 15.83\%$ **39.** a) 5% **41.** -1 **43.** $\frac{\partial w}{\partial r} = 2, \frac{\partial w}{\partial s} = 2 - \pi$ **45.** -1 **47.** $-(\sin 1 + \cos 2)\sin 1 + (\cos 1 + \cos 2)\cos 1 - 2(\sin 1 + \cos 1)\sin 2$ **49.** $\frac{\partial w}{\partial x} = \cos\theta\frac{\partial w}{\partial r} - \frac{\sin\theta}{r}\frac{\partial w}{\partial\theta}; \frac{\partial w}{\partial y} = \sin\theta\frac{\partial w}{\partial r} + \frac{\cos\theta}{r}\frac{\partial w}{\partial\theta}$ **59.** $\mathbf{u} = -\frac{\sqrt{2}}{2}\mathbf{i} - \frac{\sqrt{2}}{2}\mathbf{j}, -\mathbf{u} = \frac{\sqrt{2}}{2}\mathbf{i} + \frac{\sqrt{2}}{2}\mathbf{j}, (D_{\mathbf{u}}f)P_0 = \frac{\sqrt{2}}{2}, (D_{-\mathbf{u}}f)P_0 = -\frac{\sqrt{2}}{2}, (D_{\mathbf{u}}f)P_0 = -\frac{7}{10}$ **61.** $\mathbf{u} = \frac{2}{7}\mathbf{i} + \frac{3}{7}\mathbf{j} + \frac{6}{7}\mathbf{k}, -\mathbf{u} = -\frac{2}{7}\mathbf{i} - \frac{3}{7}\mathbf{j} - \frac{6}{7}\mathbf{k}, (D_{\mathbf{u}}f)P_0 = 7, (D_{-\mathbf{u}}f)P_0 = -7, (D_{\mathbf{u}_1}f)P_0 = 7$ **63.**

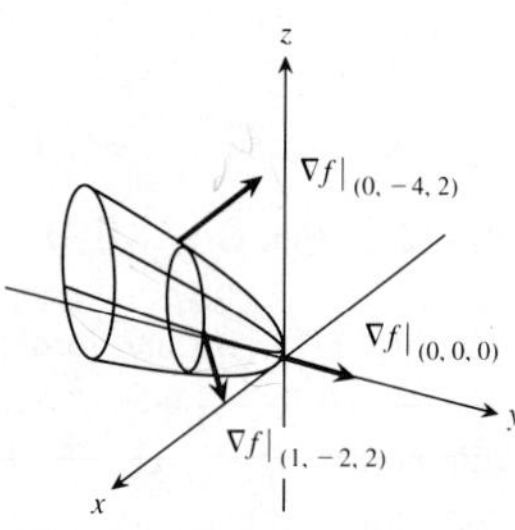

65. Tangent: $4x - y - 5z = 4$; normal line: $x = 2 + 4t$, $y = -1 - t, z = 1 - 5t$

67.

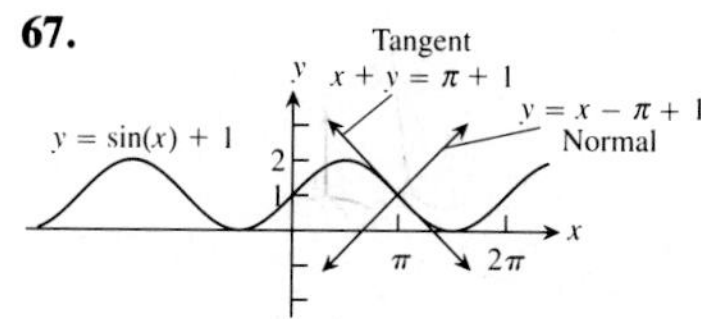

69. $x = 1 - 2t$, $y = 1$, $z = \frac{1}{2} + 2t$ **71.** $(t, -t \pm 4, t)$, t a real number **73.** $\frac{\pi}{\sqrt{2}}$ **75.** $\frac{14}{5}$ **87.** $f(-2, -2) = -8$, absolute minimum **89.** $f(0, 0) = 0$, saddle point; $f(6, 18) = 108$, local maximum **91.** $f(0, 0) = 15$, saddle point; $f(1, 1) = 14$, local minimum **93.** Absolute maximum is 28 at (0, 4); absolute minimum is $-\frac{9}{4}$ at (3/2, 0). **95.** Absolute maximum is 18 at (2, − 2); absolute minimum is $-\frac{17}{4}$ at $\left(-2, \frac{1}{2}\right)$. **97.** Absolute maximum is 8 at (−2, 0); absolute minimum is −1 at (1, 0). **99.** Absolute maximum is 4 at (1, 0); absolute minimum is −4 at (0, −1). **101.** Maximum **103.** $(3, \pm 3\sqrt{2})$ **105.** 50 is maximum; − 50 is minimum. **107.** (1, 1, 1), (1, −1, −1), (−1, −1, −1), (−1, 1, −1) **109.** $V = \frac{\sqrt{3}\,abc}{2}$ **111.** $\mathbf{w} = e^{-c^2\pi^2 t} \sin \pi x$ **113.** 0.213%

Chapter 13

Section 13.1, pp. 924–926

1. 16

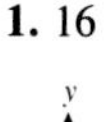

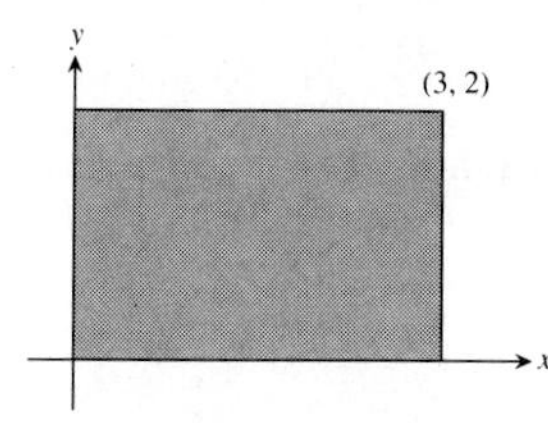

3. 1

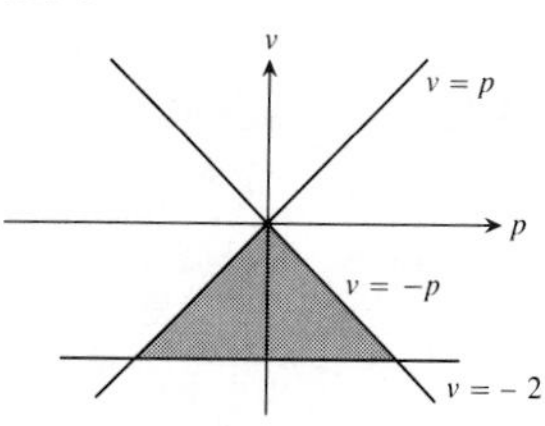

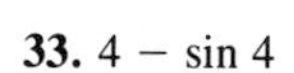

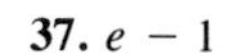

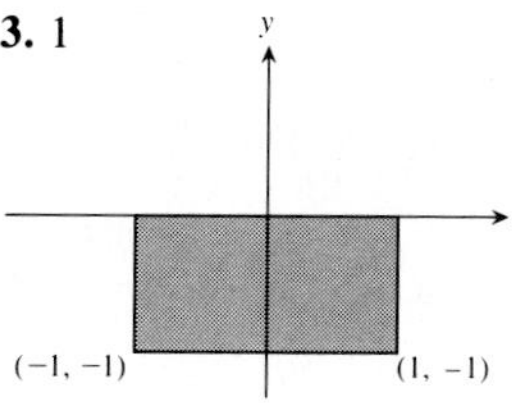

5. $\frac{\pi^2}{2} + 2$

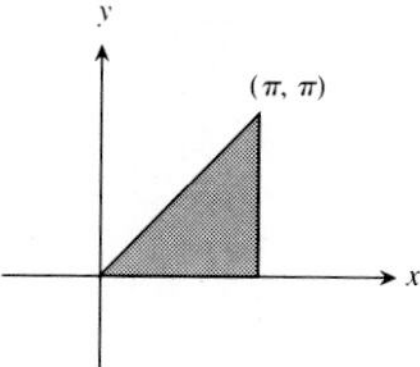

7. $8 \ln 8 - 16 + e$

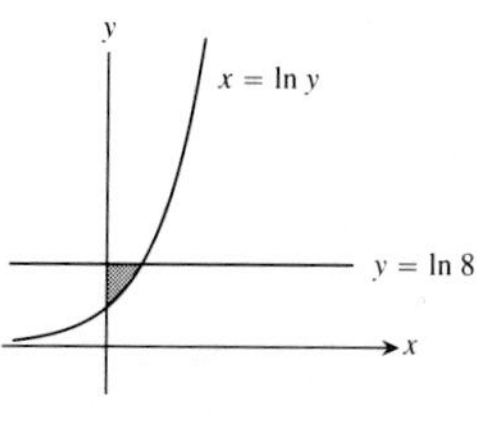

9. $9 - 9e$

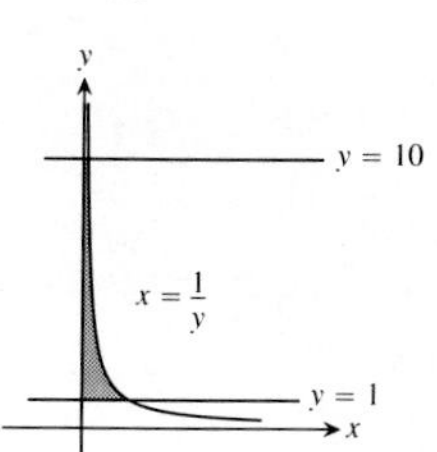

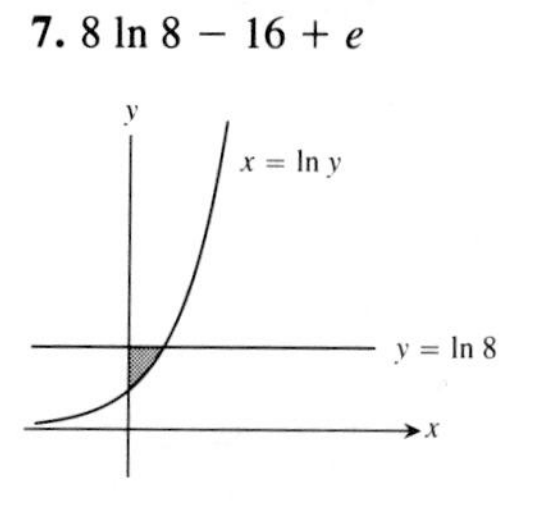

11. $\frac{3}{2} \ln 2$ **13.** $(\ln 2)^2$ **15.** $-\frac{1}{10}$

17. 8

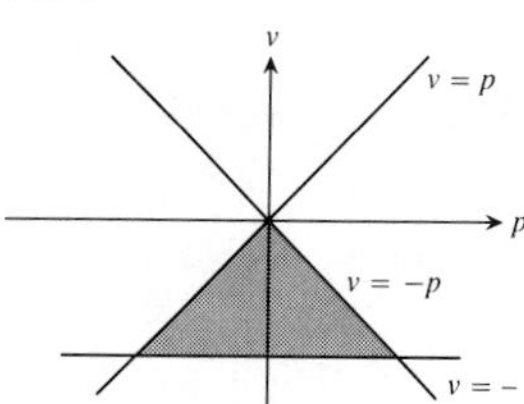

19. 2π

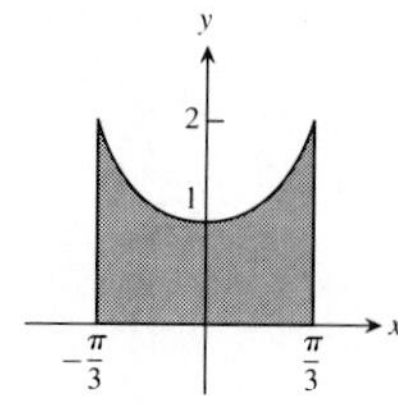

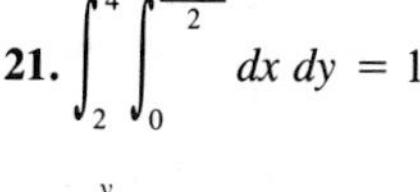

21. $\int_2^4 \int_0^{\frac{4-y}{2}} dx\, dy = 1$

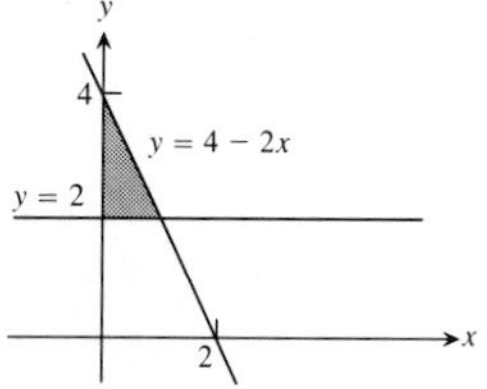

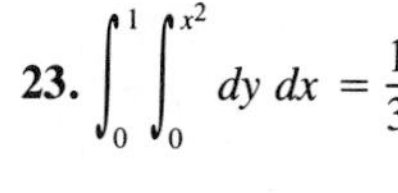

23. $\int_0^1 \int_0^{x^2} dy\, dx = \frac{1}{3}$

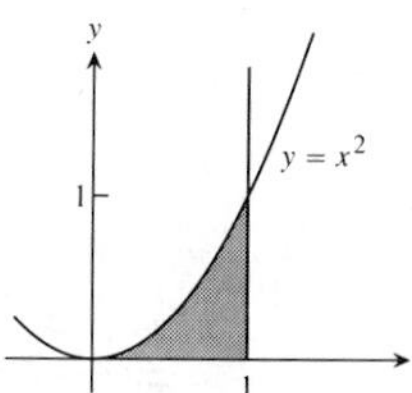

25. $\int_0^9 \int_0^{\frac{1}{2}\sqrt{9-y}} 16x\, dx\, dy = 81$ **27.** 1 **29.** π^2

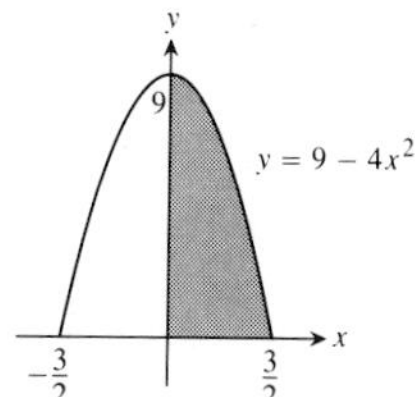

31. 2

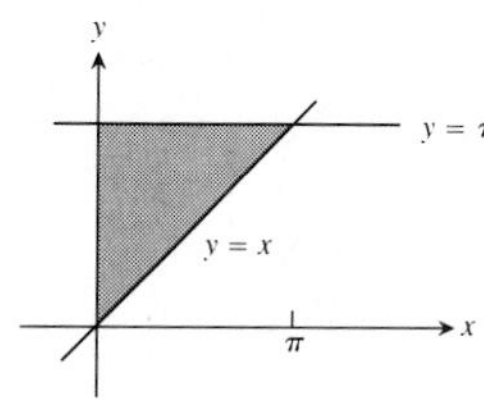

33. $4 - \sin 4$

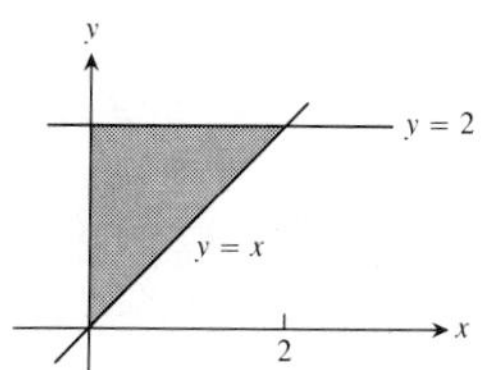

35. 2

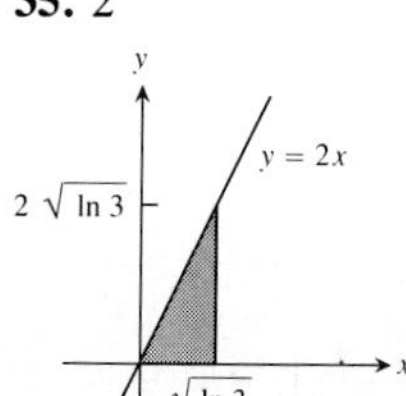

37. $e - 1$

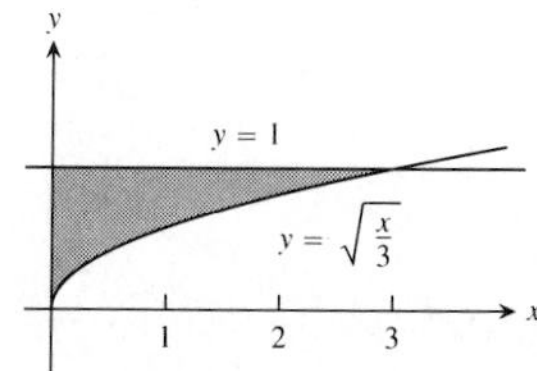

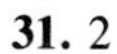

39. $\frac{4}{3}$ **41.** $\frac{625}{12}$ **43.** 16 **45.** 20 **47.** $2(1 + \ln 2)$

49. $\frac{20\sqrt{3}}{9}$

51. $\int_0^1 \int_x^{2-x} (x^2 + y^2)\, dy\, dx = \frac{4}{3}$

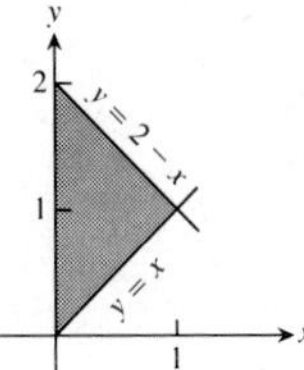

53. Interior of the ellipse $x^2 + 2y^2 = 4$ **55.** 0.603 **57.** 0.233

13. $\frac{3}{2}$

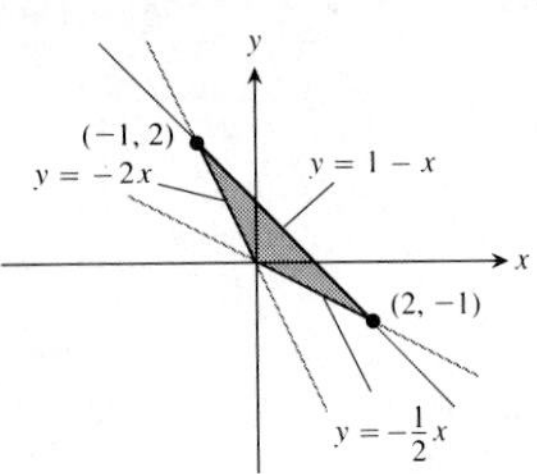

15. $(\bar{x}, \bar{y}) = \left(\frac{5}{14}, \frac{38}{35}\right)$ **17.** $(\bar{x}, \bar{y}) = \left(\frac{64}{35}, \frac{5}{7}\right)$ **19.** $(\bar{x}, \bar{y}) = \left(0, \frac{4}{3\pi}\right)$ **21.** $(\bar{x}, \bar{y}) = \left(\frac{4a}{3\pi}, \frac{4a}{3\pi}\right)$ **23.** $(\bar{x}, \bar{y}) = \left(\frac{\pi}{2}, \frac{\pi}{8}\right)$ **25.** $\frac{4}{3}ab(a^2 + b^2)$ **27.** $(\bar{x}, \bar{y}) = \left(-1, \frac{1}{4}\right)$ **29.** $I_x = \frac{64}{105}$, $R_x = \sqrt{\frac{8}{7}}$ **31.** $(\bar{x}, \bar{y}) = \left(\frac{3}{8}, \frac{17}{16}\right)$ **33.** $\bar{x} = \frac{11}{3}, \bar{y} = \frac{14}{27}$, $I_y = 432, R_y = 4$ **35.** $\bar{x} = 0, \bar{y} = \frac{13}{31}, I_y = \frac{7}{5}, R_y = \sqrt{\frac{21}{31}}$ **37.** $\bar{x} = 0, \bar{y} = \frac{7}{10}, I_x = \frac{9}{10}, I_y = \frac{3}{10}, I_0 = \frac{6}{5}, R_x = \frac{3\sqrt{6}}{10}, R_y = \frac{3\sqrt{2}}{10}, R_0 = \frac{3\sqrt{2}}{5}$ **39.** a) 0 b) $\frac{4}{\pi^2}$ **41.** $\frac{8}{3}$ **43.** If $0 < a \le \frac{5}{2}$, then the appliance will have to be tipped more than 45° to fall over. **45.** $(\bar{x}, \bar{y}) = \left(\frac{2}{\pi}, 0\right)$ **47.** a) By the hint, $M_{\text{c.m.}} = M_y = 0$. b) $I_L = I_y - 2M_y h + mh^2$, but $M_y = 0$ and $I_{\text{c.m.}} = I_y$. **49.** Use the domain additivity property for M, M_x, and M_y. **51.** a) $\left(\frac{7}{5}, \frac{31}{10}\right)$ b) $\left(\frac{19}{7}, \frac{18}{7}\right)$ c) $\left(\frac{9}{2}, \frac{19}{8}\right)$ d) $\left(\frac{11}{4}, \frac{43}{16}\right)$ **53.** In order for c.m. to be on the common boundary, $h = a\sqrt{2}$. In order for c.m. to be inside T, $h > a\sqrt{2}$.

Section 13.2, pp. 934–937

1. 2

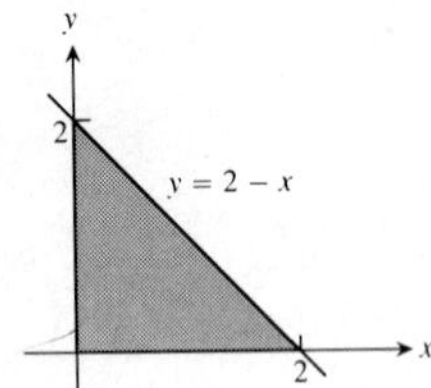

3. 4

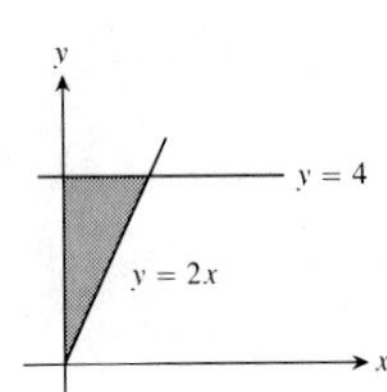

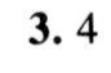

5. $\frac{1}{3}$

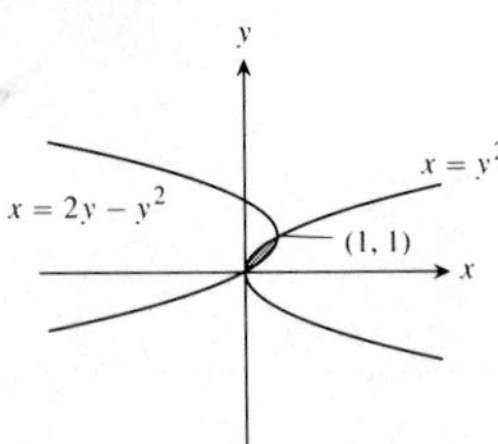

7. $\pi + 2$

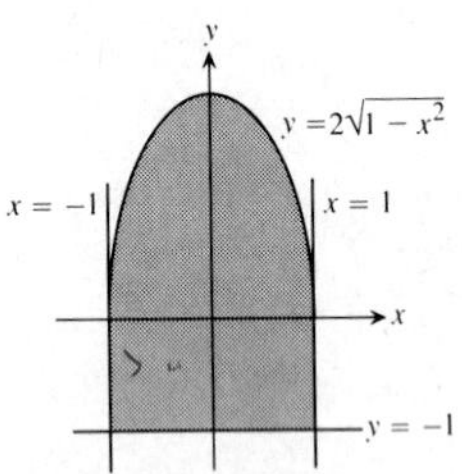

9. 12

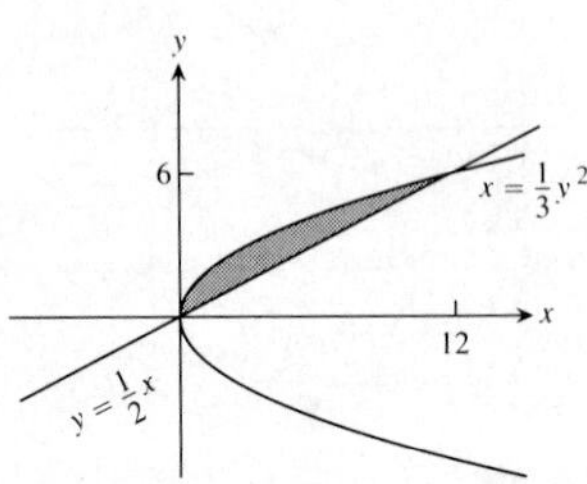

11. $\sqrt{2} - 1$

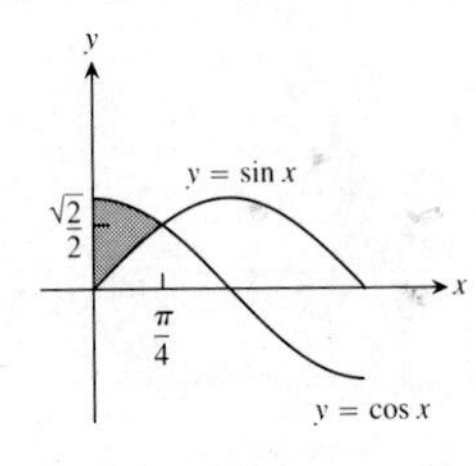

Section 13.3, pp. 942–943

1. $\frac{\pi}{2}$ **3.** $\frac{\pi}{8}$ **5.** πa^2 **7.** $\frac{2}{3}$ **9.** $3 \ln (2 + \sqrt{3})$ **11.** $\frac{5\pi}{6}$ **13.** $\frac{\pi}{2} + 1$ **15.** π **17.** $2(\pi - 1)$ **19.** 12π **21.** $\frac{3\pi}{8} + 1$ **23.** $\frac{8}{3}$ **25.** $\pi \ln 4$ **27.** 4 **29.** $\bar{x} = \frac{5}{6}, \bar{y} = 0$ **31.** $\frac{2}{3}a(3\sqrt{3} - \pi)$, assuming $\delta(r) = \frac{1}{r}$ **33.** $\frac{4}{3} + \frac{5\pi}{8}$ **35.** $\frac{2}{3}$ **37.** $\frac{1}{2}(a^2 + 2h^2)$

Section 13.4, pp. 948–950

1. 1 **3.** $\int_0^1 \int_0^{2-2x} \int_0^{\frac{6-6x-3y}{2}} dz\, dy\, dx$; $\int_0^2 \int_0^{1-\frac{y}{2}} \int_0^{\frac{6-6x-3y}{2}} dz\, dx\, dy$;

$\int_0^1\int_0^{3-3x}\int_0^{\frac{6-6x-2z}{3}} dy\,dz\,dx$; $\int_0^3\int_0^{1-\frac{z}{3}}\int_0^{\frac{6-6x-2z}{3}} dy\,dx\,dz$;
$\int_0^2\int_0^{3-\frac{3y}{2}}\int_0^{\frac{6-3y-2z}{6}} dx\,dz\,dy$; $\int_0^3\int_0^{2-\frac{2z}{3}}\int_0^{\frac{6-3y-2z}{6}} dx\,dy\,dz$; the value of all six integrals is 1. **5.** 1 **7.** 1 **9.** $\frac{\pi^3}{2}(1-\cos 1)$

11. 8 **13.** $\frac{7}{6}$ **15.** 0 **17.** $\frac{1}{2}-\frac{\pi}{8}$

19. a) $\int_{-1}^1\int_0^{1-x^2}\int_{x^2}^{1-z} dy\,dz\,dx$ b) $\int_0^1\int_{-\sqrt{1-z}}^{\sqrt{1-z}}\int_{x^2}^{1-z} dy\,dx\,dz$
c) $\int_0^1\int_0^{1-z}\int_{-\sqrt{y}}^{\sqrt{y}} dx\,dy\,dz$ d) $\int_0^1\int_0^{1-y}\int_{-\sqrt{y}}^{\sqrt{y}} dx\,dz\,dy$
e) $\int_0^1\int_{-\sqrt{y}}^{\sqrt{y}}\int_0^{1-y} dz\,dx\,dy$ **21.** 2 sin 4 **23.** 4 **25.** $\frac{31}{3}$ **27.** 1

29. $\frac{2}{3}$ **31.** $\frac{20}{3}$ **33.** $\frac{32}{15}$ **35.** $\frac{16}{3}$ **37.** 2 **39.** 12π **41.** $\frac{2}{7}$

43. $a = 3$ or $a = \frac{13}{3}$ **45.** Interior of the ellipsoid $x^2+y^2+2z^2 = 18$

Section 13.5, pp. 952–955

1. $R_x = \sqrt{\frac{b^2+c^2}{12}}$, $R_y = \sqrt{\frac{a^2+c^2}{12}}$, $R_z = \sqrt{\frac{a^2+b^2}{12}}$

3. $I_x = \frac{M}{3}(b^2+c^2)$, $I_y = \frac{M}{3}(a^2+c^2)$, $I_z = \frac{M}{3}(a^2+b^2)$

5. $\bar{x} = \bar{y} = 0$, $\bar{z} = \frac{12}{5}$, $I_x = \frac{7904}{105}$, $I_y = \frac{4832}{63}$, $I_z = \frac{256}{65}$

7. a) $(\bar{x}, \bar{y}, \bar{z}) = \left(0, 0, \frac{8}{3}\right)$ b) $c = 2\sqrt{2}$ **9.** $I_L = 1386$, $R_L = \sqrt{\frac{77}{2}}$

11. $I_L = \frac{40}{3}$, $R_L = \sqrt{\frac{5}{3}}$ **13.** $(\bar{x}, \bar{y}, \bar{z}) = \left(\frac{4}{5}, \frac{2}{5}, \frac{2}{5}\right)$ **15.** $(\bar{x}, \bar{y}, \bar{z}) = \left(\frac{8}{15}, \frac{8}{15}, \frac{8}{15}\right)$, $I_x = I_y = I_z = \frac{11}{6}$, $R_x = R_y = R_z = \sqrt{\frac{11}{15}}$

17. a) By the hint, $M_{\text{c.m.}} = M_{yz} = 0$. b) $I_L = I_{yz} - 2M_{yz}h + mh^2$, but $M_{yz} = 0$ and $I_{\text{c.m.}} = I_{yz}$ **19.** a) $I_{\text{c.m.}} = \frac{abc(a^2+b^2)}{12}$, $R_{\text{c.m.}} = \sqrt{\frac{a^2+b^2}{12}}$ b) $I_L = \frac{abc(a^2+7b^2)}{3}$, $R_L = \sqrt{\frac{a^2+7b^2}{3}}$

21. Use the domain additivity property for M, M_{yz}, M_{xz}, and M_{xy}. **23.** a) $h = a\sqrt{3}$ b) $h = a\sqrt{2}$

Section 13.6, pp. 960–962

1. $\frac{4\pi(\sqrt{2}-1)}{3}$ **3.** $\frac{17\pi}{5}$ **5.** $\pi(6\sqrt{2}-8)$ **7.** π^2 **9.** $\frac{\pi}{3}$

11. 5π **13.** $\frac{3\pi}{10}$ **15.** $\frac{\pi}{3}$ **17.** 2π **19.** $\left(\frac{8-5\sqrt{2}}{2}\right)\pi$

21. a) $8\int_0^{\frac{\pi}{2}}\int_0^{\frac{\pi}{2}}\int_0^2 \rho^2 \sin\phi\, d\rho\, d\phi\, d\theta$ b) $8\int_0^{\frac{\pi}{2}}\int_0^2\int_0^{\sqrt{4-r^2}} r\, dz\, dr\, d\theta$
c) $8\int_0^2\int_0^{\sqrt{4-x^2}}\int_0^{\sqrt{4-x^2-y^2}} dz\, dy\, dx$

23. $\int_{-\frac{\pi}{2}}^{\frac{\pi}{2}}\int_0^{\cos\theta}\int_0^{3r^2} f(r, \theta, z)\, r\, dz\, dr\, d\theta$

25. a) $\int_0^{2\pi}\int_0^{\frac{\pi}{3}}\int_{\sec\phi}^2 \rho^2 \sin\phi\, d\rho\, d\phi\, d\theta$ b) $\int_0^{2\pi}\int_0^{\sqrt{3}}\int_1^{\sqrt{4-r^2}} r\, dz\, dr\, d\theta$
c) $\int_{-\sqrt{3}}^{\sqrt{3}}\int_{-\sqrt{3-x^2}}^{\sqrt{3-x^2}}\int_1^{\sqrt{4-x^2-y^2}} dz\, dy\, dx$ d) $\frac{5\pi}{3}$ **27.** $\frac{\pi}{2}$ **29.** 8π

31. $\frac{5\pi}{2}$ **33.** $\frac{4\pi(8-3\sqrt{3})}{3}$ **35.** $\frac{2}{3}$ **37.** $(\bar{x}, \bar{y}, \bar{z}) = \left(0, 0, \frac{3}{8}\right)$

39. $I_z = 30\pi$, $R_z = \sqrt{\frac{5}{2}}$ **41.** a) π b) $\frac{7\pi}{6}$ **43.** $\frac{8\pi a^5}{15}$

45. $M = \int_0^{2\pi}\int_0^a\int_0^{h\sqrt{1-\frac{r^2}{a^2}}} r\, dz\, dr\, d\theta = \frac{2\pi a^2 h}{3}$ and
$M_{xy} = \int_0^{2\pi}\int_0^a\int_0^{h\sqrt{1-\frac{r^2}{a^2}}} rz\, dz\, dr\, d\theta = \frac{\pi a^2 h^2}{4}$

47. $M = \int_0^{2\pi}\int_0^a\int_0^{h\sec\phi} \rho^2 \sin\phi\, d\rho\, d\phi\, d\theta = \frac{\pi h^3}{6}(\sec^4 a - 1)$ and
$M_{xy} = \int_0^{2\pi}\int_0^a\int_0^{h\sec\phi} \rho^3 \sin\phi \cos\phi\, d\rho\, d\phi\, d\theta = \frac{3h}{4}M$

49. a) $(\bar{x}, \bar{y}, \bar{z}) = \left(0, 0, \frac{4}{5}\right)$, $I_z = \frac{\pi}{12}$, $R_z = \sqrt{\frac{1}{3}}$
b) $(\bar{x}, \bar{y}, \bar{z}) = \left(0, 0, \frac{5}{6}\right)$, $I_z = \frac{\pi}{14}$, $R_z = \sqrt{\frac{5}{14}}$

51. $\frac{2\pi a^3}{3}$ **53.** $\frac{5\pi}{3}$ **55.** $\frac{4(2\sqrt{2}-1)\pi}{3}$ **57.** $\frac{3}{4}$ **59.** $\bar{x} = \bar{y} = 0$, $\bar{z} = \frac{3(2+\sqrt{2})a}{16}$

Section 13.7, pp. 968–969

1. 2 **3.** a) $x = \frac{u+v}{3}$, $y = \frac{v-2u}{3}$, $J = \frac{1}{3}$ b) $\frac{33}{4}$ **5.** $8 + \frac{52\ln 2}{3}$

7. $\frac{\pi ab(a^2+b^2)}{4}$ **9.** $\rho^2 \sin\phi$ **11.** $\frac{4\pi abc}{3}$ **13.** $2 + 3\ln 2$

Section 13 Miscellaneous Exercises, pp. 969–972

1. a) $\displaystyle\int_{-2}^{0}\int_{2x+4}^{4-x^2} dy\,dx = \frac{4}{3}$ b) $\displaystyle\int_{0}^{1}\int_{-\sqrt{v}}^{\sqrt{v}} du\,dv = \frac{4}{3}$

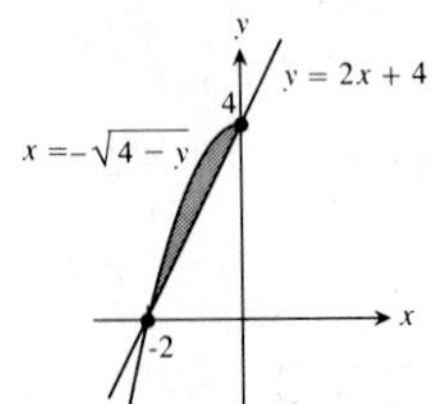

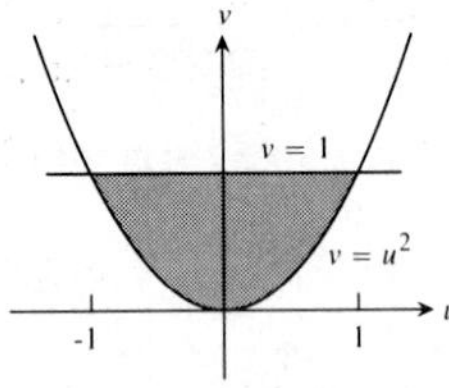

c) $\displaystyle\int_{-3}^{3}\int_{0}^{\frac{\sqrt{9-s^2}}{2}} t\,dt\,ds = \frac{9}{2}$ d) $\displaystyle\int_{0}^{4}\int_{0}^{\sqrt{4-z}} 2w\,dw\,dz = 8$

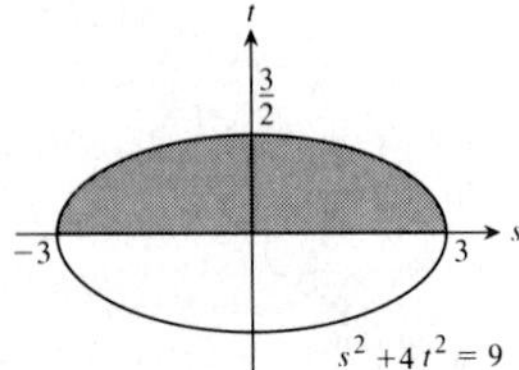

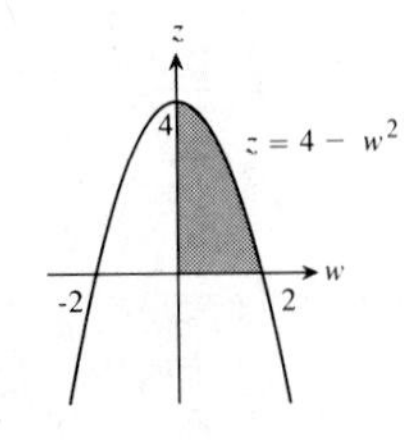

3. π **5.** $\ln\left(\frac{b}{a}\right)$ **7.** $\dfrac{e^{a^2b} + e^{ab^2} - 2}{2ab}$ **11.** a) $\displaystyle\int_{-3}^{2}\int_{x}^{6-x^2} x^2\,dy\,dx$

b) $\displaystyle\int_{-3}^{2}\int_{x}^{6-x^2}\int_{0}^{x^2} dz\,dy\,dx$ c) $\dfrac{125}{4}$ **13.** $\bar{x} = \bar{y} = \dfrac{1}{2 - \ln 4}$

15. $\left(-\frac{12}{5}, 2\right)$ **17.** $M = 4$, $M_x = M_y = 0$ **19.** $I_x = \dfrac{\delta}{12}bh^3$,

$R_x = \dfrac{h}{\sqrt{6}}$ **21.** $\bar{x} = \dfrac{15\pi + 32}{6\pi + 48}$, $\bar{y} = 0$

23. $\bar{x} = \dfrac{13}{3\pi}$, $\bar{y} = \dfrac{13}{3\pi}$ **25.** Area $= a^2\cos^{-1}\left(\frac{b}{a}\right) - b\sqrt{a^2 - b^2}$,

$I_0 = \dfrac{a^4}{2}\cos^{-1}\left(\frac{b}{a}\right) - \dfrac{b^3}{2}\sqrt{a^2 - b^2} - \dfrac{b^3}{6}(a^2 - b^2)^{\frac{3}{2}}$

27. $a^2\beta\left(\ln a - \frac{1}{2}\right)$ **29.** $\dfrac{2\pi kR^3}{3}$ coulomb **31.** 4π

33. a) $4\displaystyle\int_{0}^{1}\int_{0}^{\sqrt{1-x^2}}\int_{0}^{\sqrt{4-x^2-y^2}} \sqrt{x^2 + y^2}\,dz\,dy\,dx$

b) $4\displaystyle\int_{0}^{\frac{\pi}{2}}\int_{0}^{\frac{\pi}{6}}\int_{0}^{2} \rho^3\sin^2\phi\,d\rho\,d\phi\,d\theta + 4\int_{0}^{\frac{\pi}{2}}\int_{\frac{\pi}{6}}^{\frac{\pi}{2}}\int_{0}^{\csc\phi} \rho^3\sin^2\phi\,d\rho\,d\phi\,d\theta$

35. $\displaystyle\int_{0}^{1}\int_{\sqrt{1-x^2}}^{\sqrt{3-x^2}}\int_{1}^{\sqrt{4-x^2-y^2}} z^2xy\,dz\,dy\,dx + \int_{1}^{\sqrt{3}}\int_{0}^{\sqrt{3-x^2}}\int_{1}^{\sqrt{4-x^2-y^2}}$

$z^2xy\,dz\,dy\,dx$ **37.** 2π **39.** $\dfrac{15}{8}$ **41.** $\dfrac{3\pi}{2}$ **43.** a) Sphere radius = 2; hole radius = 1. b) $4\sqrt{3}\pi$ **45.** $\dfrac{\pi}{4}$ **47.** $\dfrac{8\pi}{3}$ **49.** $\dfrac{64\pi}{35}$

Chapter 14

Section 14.1, pp. 978–979

1. c **3.** g **5.** d **7.** f **9.** $\sqrt{2}$ **11.** $\dfrac{13}{2}$ **13.** 0

15. $\dfrac{1}{6}(5\sqrt{5} + 9)$ **17.** $+\infty$ **19.** $I_x = 2\pi - 2$, $R_x = 1$

21. $I_z = 2\pi\delta a^3$, $R_z = a$ **23.** a) $I_z = 2\pi\sqrt{2}\delta$, $R_z = 1$ b) $I_z = 4\pi\sqrt{2}\delta$, $R_z = 1$ **25.** a) $4\sqrt{2} - 2$ b) $\sqrt{2} + \dfrac{1}{2}\ln(1 + \sqrt{2})$ **27.** $(\bar{x}, \bar{y}, \bar{z}) = \left(1, \frac{16}{15}, \frac{2}{3}\right)$, $I_x = \dfrac{29}{5}$, $I_y = \dfrac{64}{15}$, $I_z = \dfrac{56}{9}$, $R_x = \sqrt{\dfrac{29}{10}}$, $R_y = \sqrt{\dfrac{32}{15}}$, $R_z = \sqrt{\dfrac{28}{9}}$

Section 14.2, pp. 987–988

1. $F(x, y) = -\dfrac{kx}{x^2 + y^2}i - \dfrac{ky}{x^2 + y^2}j$, any $k > 0$

3. a) 

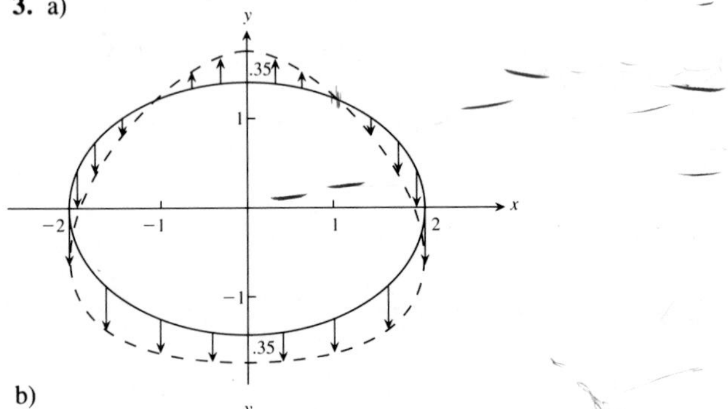

b)

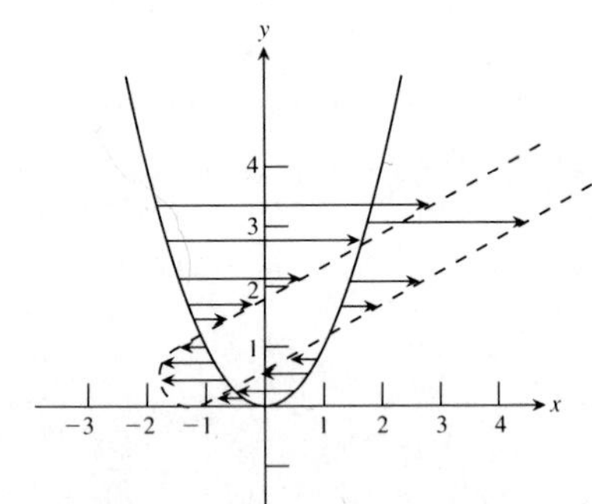

5. a) $\dfrac{9}{2}$ b) $\dfrac{13}{3}$ c) $\dfrac{9}{2}$ **7.** a) $\dfrac{1}{3}$ b) $-\dfrac{1}{5}$ c) -1 **9.** a) 2 b) $\dfrac{3}{2}$ c) $-\dfrac{1}{2}$ **11.** $\dfrac{1}{2}$ **13.** $-\pi$ **15.** 48 **17.** π **19.** a) F_1: circ. = 0, flux = 2π; F_2: circ. = 2π, flux = 0 b) F_1: circ. = 0, flux = 8π; F_2; circ. = 8π, flux = 0 **21.** Circ. = 0, flux = πa^2

23. Circ. = πa^2, flux = 0 **25.** a) $-\dfrac{\pi}{2}$ b) 0 c) 1 **27.** $\dfrac{9}{2}$

Section 14.3, pp. 999–1001

1. Flux = 0, circ. = $2\pi a^2$ **3.** Flux = $-\pi a^2$, circ. = 0

5. Flux = 2, circ. = 0 **7.** -9 **9.** Flux = $\dfrac{1}{2}$, circ. = $\dfrac{1}{2}$

11. Flux $= \frac{1}{5}$, circ. $= -\frac{1}{12}$ **13.** $\frac{3}{2}\pi a^2$ **15.** 0 **17.** -16π **19.** $\frac{2}{33}$ **35.** πa^2 **37.** $\frac{3}{8}\pi$

Section 14.4, pp. 1010–1012

1. $\frac{13}{3}\pi$ **3.** 4 **5.** $6\sqrt{6} - 2\sqrt{2}$ **7.** $2\pi(2 - \sqrt{2})$ **9.** $\frac{2}{3}\pi$ **11.** $\frac{\pi}{6}(5\sqrt{5} - 1)$ **13.** $9a^3$ **15.** $\frac{abc}{4}(ab + ac + bc)$ **17.** 2 **19.** $\frac{\pi a^3}{6}$ **21.** $\frac{\pi a^2}{4}$ **23.** $\frac{\pi a^3}{2}$ **25.** -32 **27.** -4 **29.** $3a^4$ **31.** $\left(\frac{a}{2}, \frac{a}{2}, \frac{a}{2}\right)$ **33.** $(\bar{x}, \bar{y}, \bar{z}) = \left(0, 0, \frac{14}{9}\right)$, $I_z = \frac{15\pi\sqrt{2}\delta}{2}$, $R_z = \sqrt{\frac{5}{2}}$ **35.** a) $2\pi a^4\delta$ b) $2\pi a^4\delta + 4\pi a^2\delta$ **37.** $\frac{\pi}{6}\left((4\sqrt{3} + 1)^{\frac{3}{2}} - 1\right)$ **39.** $\frac{\pi}{6}(5\sqrt{5} - 1)$ **41.** $2\sqrt{5}$

Section 14.5, pp. 1021–1022

1. -16 **3.** -8π **5.** 3π **7.** $\frac{76}{3}$ **9.** 12π **11.** $12\pi(4\sqrt{2} - 1)$

Section 14.6, pp. 1029–1030

1. 4π **3.** $-\frac{5}{6}$ **5.** 0 **7.** -6π **9.** $2\pi a^2$

Section 14.7, pp. 1039–1040

1. Conservative **3.** Not conservative **5.** Not conservative **7.** Conservative **9.** Conservative **11.** Not conservative **13.** 49 **15.** -16 **17.** 1 **19.** $\ln\left(\frac{\pi}{2}\right)$ **21.** 0 **27.** a) $F = \nabla\left(\frac{x^2 - 1}{y}\right)$ b) $F = \nabla(e^x \ln y + y \sin z)$ **29.** a) 0 b) 0 c) 0 **31.** a) $c = b = 2a$ b) $c = b = 2$

Section 14 Miscellaneous Exercises, pp. 1041–1044

1. Path 1: $2\sqrt{3}$; path 2: $1 + 3\sqrt{2}$ **3.** $4a$ **5.** 11.08 **7.** $\bar{z} = \frac{2}{3}$, $I_z = \frac{14}{3}$, $R_z = \sqrt{\frac{7}{3}}$ **9.** a) $-\frac{a^2\pi}{2}$ b) 0 c) 1 **13.** $\frac{3}{2}$ **17.** 0 **19.** a) π b) An ellipse **21.** $\pi\sqrt{3}$ **23.** $2\pi\left(1 - \frac{1}{\sqrt{2}}\right)$ **25.** $\frac{abc}{2}\sqrt{\frac{1}{a^2} + \frac{1}{b^2} + \frac{1}{c^2}}$ **27.** a) $\frac{3}{4}$ b) 2 **29.** 3 **31.** $(\bar{x}, \bar{y}, \bar{z}) = \left(0, 0, \frac{98}{24}\right)$, $I_z = 640\pi$, $R_z = 2\sqrt{2}$ **33.** 3 **35.** $\frac{4\pi}{3}(2 - 2\sqrt{2})$ **37.** 3 **39.** $\frac{8}{3}$ **41.** 0 **43.** Conservative **45.** Not conservative **47.** $f(x, y, z) = y^2 + yz + 2x + z$ **49.** 0 **51.** Path 1: $\frac{3}{2}$; path 2: $\frac{5}{3}$ **57.** 6π **59.** $\frac{2}{3}$

Chapter 15

Section 15.1, pp. 1050–1051

1. First order, nonlinear **3.** Fourth order, nonlinear **5.** a) $xy'' - y' = x(2) - 2x = 0$ b) $xy'' - y' = x(0) - 0 = 0$ c) $xy'' - y' = x(2C_1) - 2C_1x = 0$ **7.** a) $2y' + 3y = 2(-e^{-x}) + 3(e^{-x}) = e^{-x}$ b) $2y' + 3y = 2\left(-e^{-x} - \frac{3}{2}e^{-\left(\frac{3}{2}\right)^x}\right) + 3\left(e^{-x} + e^{-\left(\frac{3}{2}\right)^x}\right) = e^{-x}$ c) $2y' + 3y = 2\left(-e^{-x} - \frac{3}{2}Ce^{-\left(\frac{3}{2}\right)^x}\right) + 3\left(e^{-x} + Ce^{-\left(\frac{3}{2}\right)^x}\right) = e^{-x}$ **9.** $x^2y' + xy = -\int_1^x \frac{e^t}{t}\,dt + e^x + \int_1^x \frac{e^t}{t}\,dt = e^x$ **11.** $y' = 160 - 32x$, $y'' = -32$, $y(5) = 400$, $y'(5) = 0$ **13.** $y' = -6\sin 2t - 2\cos 2t$, $y'' = -12\cos 2t + 4\sin 2t$, $y(0) = 3$, $y'(0) = -2$, $y'' + 4y = 0$ **15.** $y = \tan\left(\ln\left|\frac{x - 1}{x + 1}\right| + \frac{\pi}{4}\right)$ **17.** $e^{-y}(y^3 + 3y^2 + 6y + 6) = 2e^{\sqrt{x}} + 38e^{-2} - 2$ **19.** $y = \sin\left[\frac{x}{2}\sqrt{x^2 + 1} + \frac{1}{2}\sinh^{-1}x - \frac{\pi}{6}\right]$ **21.** 10 min **23.** $\ln|x| = -\frac{1}{4}\ln\left(2\left(\frac{y}{x}\right)^2 + 1\right) + C$ **25.** $\ln|x| + e^{-\frac{y}{3}} = C$ **27.** $\ln|x| - \ln\left|\sec\left(\frac{y}{x} - 1\right) + \tan\left(\frac{y}{x} - 1\right)\right| = \ln 2$ **29.** $y = Cx^2$, a family of parabolas **31.** $\frac{x^2}{2} - \frac{y^2}{2} = C$, a family of hyperbolas

Section 15.2, pp. 1054–1055

1. Exact **3.** Exact **5.** Exact **7.** Not exact **9.** $\frac{1}{2}x^2 + xy + \frac{1}{3}y^3 + C = 0$ **11.** $x^2y + xy^2 - \frac{1}{2}y^2 + C = 0$ **13.** $e^x + x\ln y + y\ln x - \cos y + C = 0$ **15.** $\left(\frac{y}{2} - \frac{1}{4}\sin 2y\right)\sin x + C = 0$ **17.** $\ln x^2y^2 + \sum_{n=1}^{\infty}\frac{x^ny^n}{n \cdot n!} + C = 0$ **19.** $\frac{1}{2}x^2 + \frac{x}{y} + C = 0$ **21.** a) $\ln|xy| + C = 0$ b) $y = \frac{C}{x}$ **23.** $\frac{1}{2}x^2 + xy^2 + y - 2 = 0$ **25.** $\ln|xy| - xy + 1 = 0$ **27.** $x^2 + \frac{1}{2}(\ln x)(\ln y)^2 - 4 = 0$ **29.** $x^3 + xy^2 + C = 0$

Section 15.3, pp. 1060–1061

1. $y = c^{-x} + Ce^{-2x}$ **3.** $y = \frac{-\cos x + C}{x^3}$ **5.** $y = \frac{\frac{1}{3}x^3 - x + C}{(x - 1)^4}$ **7.** $y = (\ln|\sec x| + C)\csc x$ **9.** $y = \frac{x}{2} - \frac{1}{4} + \frac{3}{4}e^{-2x}$ **11.** $y = \frac{-\cos x + \frac{\pi}{2}}{x}$ **13.** $y = y_0e^{kx}$ **15.** $t = \frac{L}{R}\ln 2$ **17.** $.95\left(\frac{V}{R}\right)$ (calculator, rounded) **19.** a) 550 ft b) 77.28 sec (calculator, rounded)

Section 15.4, pp. 1065–1066

1. $y = C_1 + C_2e^{-2x}$ **3.** $y = C_1e^{-x} + C_2e^{-5x}$ **5.** $y = (C_1x + C_2)e^{2x}$ **7.** $y = (C_1x + C_2)e^{5x}$ **9.** $y = e^{-\frac{1}{2}x}\left(C_1\cos\left(\frac{\sqrt{3}}{2}x\right) + C_2\sin\left(\frac{\sqrt{3}}{2}x\right)\right)$ **11.** $y = c^x(C_1\cos(\sqrt{3}x) + C_2\sin(\sqrt{3}x))$ **13.** $y = -\frac{1}{2}e^x + \frac{3}{2}e^{-x}$ **15.** $y = \frac{3}{4}e^{2x} - \frac{3}{4}c^{-2x}$ **17.** $y = xe^{-x}$ **19.** $y = 2xe^{-\frac{3}{2}x}$ **21.** $y = \sin 2x$ **23.** $y = e^x\left(2\cos\sqrt{2}x - \frac{1}{\sqrt{2}}\sin\sqrt{2}x\right)$ **25.** Multiply $C_1 = y\cos x - y'\sin x$ by $\cos x$ and multiply $C_2 = y\sin x + y'\cos x$ by $\sin x$. Add the equations together to obtain $C_1\cos x + C_2\sin x = y(\cos^2 x + \sin^2 x) = y$.

55. $y = -\ln|x + C_1| + C_2$ **57.** $y = \ln x^2 - 3$ **59.** $y = \frac{1}{2}x - \frac{2}{x} + 2$ **61.** $y = \frac{1}{9}x^3 + \frac{2}{3}\ln|x| - \frac{1}{9}$ **63.** $s = s_0\cos\left(\sqrt{\frac{k}{m}}t\right)$ **65.** $y = C_1\sin x + C_2\cos x$ **67.** $y = Ae^{2x} + Be^{-x}$ **69.** $y = C_1x^{-1} + C_2x$ **71.** $y = C_1\left(\frac{x}{2}\ln\left|\frac{x+1}{x-1}\right| - 1\right) + C_2x$ **73.** For all k, $y^{(k)} = y = e^x$, so $\sum_{k=0}^{n} a_n(x)y^{(k)} = \sum_{k=0}^{n} a_n(x)e^x = e^x\left(\sum_{k=0}^{n} a_n(x)\right) = 0.$ **75.** $y = C_1e^x + C_2e^{(1+\sqrt{3})x} + C_3e^{(1-\sqrt{3})x} + \frac{7}{13}\sin x + \frac{4}{13}\cos x$ **77.** $y = C_1e^{-2x} + C_2xe^{-2x} + C_3e^{2x} + C_4xe^{2x} + \frac{1}{2}x - 1$

Section 15.5, pp. 1078–1079

1. $y = 1 - x + \frac{1}{2}x^2 + C_2e^{-x} + C_1$ **3.** $y = C_1\cos x + C_3\sin x - \frac{1}{2}x\cos x$ **5.** $y = C_1e^{-x} + C_2xe^{-x} + \frac{1}{2}x^2e^{-x}$ **7.** $y = C_1e^x + C_2e^{-x} + \frac{1}{2}xe^x$ **9.** $y = e^{-2x}(C_1\cos x + C_2\sin x) + 2$ **11.** $y = C_1\cos x + C_2\sin x + \cos x\ln|\cos x| + x\sin x$ **13.** $y = C_1e^{5x} + C_2e^{-2x} + \frac{3}{10}$ **15.** $y = C_1 + C_2e^x + \frac{1}{2}(\cos x - \sin x)$ **17.** $y = C_1\cos x + C_2\sin x - \frac{1}{8}\cos 3x$ **19.** $y = C_1e^{2x} + C_2e^{-x} - 6\cos x - 2\sin x$ **21.** $y = C_1e^x + C_2e^{-x} + \frac{1}{2}xe^x - x^2 - 2$ **23.** $y = C_1e^{-2x} + C_2e^{3x} - \frac{1}{4}e^{-x} + \frac{49}{50}\cos x + \frac{7}{50}\sin x$ **25.** $y = C_1 + C_2e^{-5x} + x^3 - \frac{3}{5}x^2 + \frac{6}{25}x$ **27.** $y = C_1 + C_2e^{3x} + \frac{1}{3}xe^{3x} + 2x^2 + \frac{4}{3}x$ **29.** $y = C_1 + C_2e^{5x} + \left(\frac{1}{10}x^2 - \frac{1}{25}x\right)e^{5x}$ **31.** $y = C_1\cos x + C_2\sin x + x\left(\sin x - \frac{1}{2}\cos x\right)$ **33. a)** $y_p = (x - 1)e^x + \frac{1}{2}e^{-x}$ **b)** $y = (C_1 + C_2e^x) + xe^x + \frac{1}{2}e^{-x}$ **35. a)** $y_p = -\frac{1}{8}e^x - \frac{4}{5}$ **b)** $y = C_1e^{5x} + C_2e^{-x} - \frac{1}{8}e^x - \frac{4}{5}$ **37.** $y = C_1\cos x + C_2\sin x - \sin x\ln|\csc x + \cot x|$ **39.** $y = C_1 + C_2e^{8x} + \frac{1}{8}xe^{8x}$ **41.** $y = C_1 + C_2e^x - \frac{1}{4}x^4 - x^3 - 3x^2 - 6x$ **43.** $y = C_1 + C_2e^{-2x} + \frac{1}{6}x^3 - \frac{1}{4}x^2 + \frac{1}{4}x - \frac{1}{3}e^x$ **45.** $y = C_1\cos x + C_2\sin x + x\cos x - \sin x + \sin x\ln|\sec x|$ **47.** $y = Ce^{3x} - \frac{1}{2}e^x$ **49.** $y = Ce^{3x} + 5xe^{3x}$ **51.** $y = 2\cos x + \sin x - 1 + \sin x\ln|\sec x + \tan x|$ **53.** $y = -C_1e^{-x} + C_2$

Section 15.6, pp. 1082–1083

1. $x = C\sin(\omega t + \phi)$, where $C = \sqrt{x_0^2 + \left(\frac{v_0}{\omega}\right)^2}$ and $\phi = \tan^{-1}\left(\frac{v_0}{\omega x_0}\right)$ **3. a)** $i = C_1\cos\omega t + C_2\sin\omega t$ **b)** $i = C_1\cos\omega t + C_2\sin\omega t + \frac{\alpha V}{L(\omega^2 - \alpha^2)}\cos\alpha t$ **c)** $i = C_1\cos\omega t + C_2\sin\omega t + \frac{V}{2L}t\cos\alpha t$ **d)** $i = e^{-5t}(C_1\cos 149t + C_2\sin 149t)$ (rounded) **5.** $0 = \theta_0\cos\omega t + \frac{v_0}{\omega}\sin\omega t$, $\omega = \sqrt{\frac{2k}{mr^2}}$ **7. a)** $x = C_1\cos\omega t + C_2\sin\omega t + \frac{A\omega^2}{\omega^2 - \alpha^2}\sin\alpha t$, where $C_1 = x_0$ and $C_2 = -\frac{\alpha\omega A}{\omega^2 - \alpha^2}$ **b)** $x = x_0\cos\omega t + \frac{A}{2}\sin\omega t - \frac{\omega A}{2}t\cos\omega t$

Section 15.7, p. 1088

1. $y' = 2e^{x-1} - 1$, and $x + y = x + 2e^{x-1} - 2 - (x - 1) = 2e^{x-1} - 1$. Also, $y(1) = 0$. **3.** $y = 1 + x + \frac{1}{2}x^2 + \cdots = \sum_{n=0}^{\infty}\frac{x^n}{n!}$ **5.** $y = 2 + 2^2x + \frac{2^3}{2!}x^2 + \frac{2^4}{3!}x^3 + \cdots = 2\sum_{n=0}^{\infty}\frac{(2x)^n}{n!}$ **7.** $y = x + \frac{x^3}{3!} + \frac{x^5}{5!} + \cdots = \sum_{n=0}^{\infty}\frac{x^{2n+1}}{(2n+1)!}$ **9.** $y = 2 + x - \frac{2x^2}{2!} + \frac{2x^4}{4!} + \cdots = 2 + x + 2\sum_{n=1}^{\infty}\frac{(-1)^n x^{2n}}{n!}$ **11.** $y = -(x - 2) - 2\sum_{n=1}^{\infty}\frac{(x-2)^{2n}}{(2n)!} - 3\sum_{n=1}^{\infty}\frac{(x-2)^{2n+1}}{(2n+1)!}$ **13.** $y = \sum_{n=0}^{\infty} a_n\frac{x^n}{n!}$, where $a_0 = a$, $a_1 = b$, $a_2 = 0$, $a_3 = 1$, $a_4 = -2a$, $a_5 = -6b$, and $a_n = (n-2)(n-3)a_{n-4}$ for $n \geq 6$.

Section 15.8, pp. 1092–1093

1. The isoclines are the vertical lines x = constant.

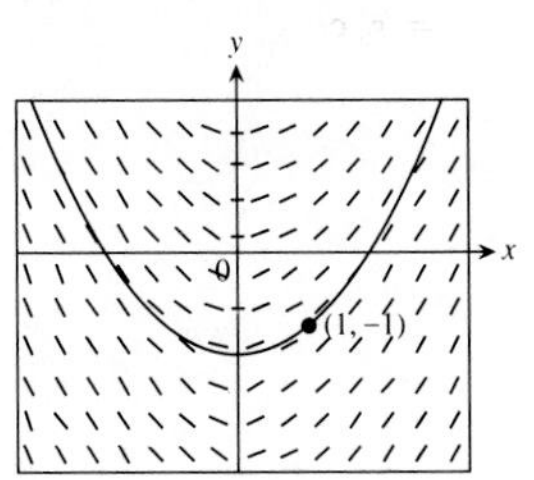

3. The isoclines are the vertical lines x = constant.

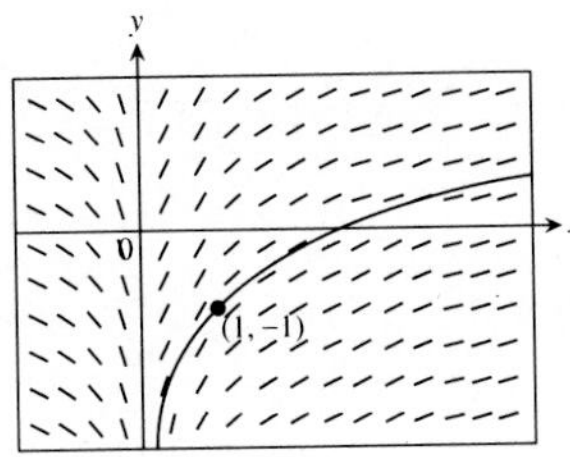

5. The isoclines are the hyperbolas xy − constant.

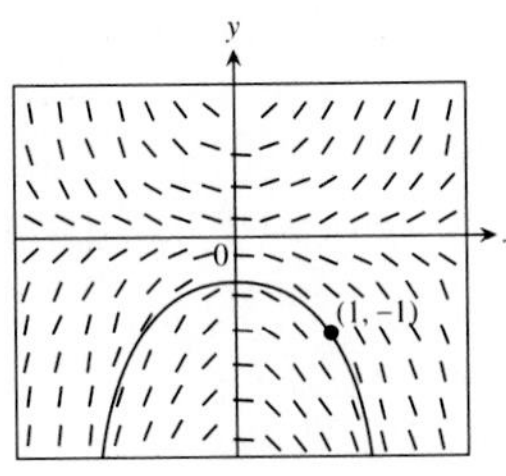

7. The isoclines are the diagonal lines $x + y$ = constant; $y = \tan(x + C) - x$; passes through (0, 0) when $C = 0$.
9. a) $y = x - 1$ b) Concave up for $y > x - 1$; concave down for $y < x - 1$

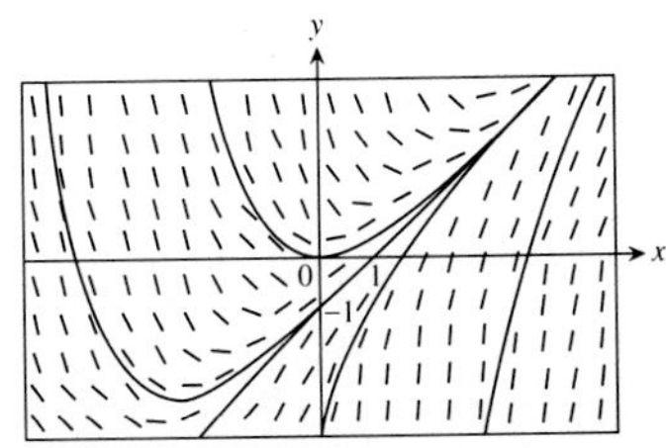

11. $y' = x - y; y(1) = -1$ **13.** $y' = -(1 + y)\sin x; y(0) = 2$
15. $y_0 = 2; y_1 = (1/2)x^2 + 3/2; y_2 = (1/2)x^2 + 3/2; y_3 = (1/2)x^2 + 3/2$
17. $y_0 = 1; y_1 = (1/2)x^2 + 1/2; y_2 = 5/8 + (1/4)x^2 + (1/8)x^4; y_3 = 29/48 + (5/16)x^2 + (1/16)x^4 + (1/48)x^6$ **19.** $y_0 = 1; y_1 - 1 + x + (1/2)x^2; y_2 = 1 + x + x^2 + (1/6)x^3; y_3 = 1 + x + x^2 + (1/3)x^3 + (1/24)x^4$ **21.** $y = (x_0 + y_0 + 1)e^{x - x_0} - (x + 1)$

Section 15.9, pp. 1097–1098

1. $y \approx 2.48832$. The exact value is e. **3.** $y \approx 2.702708163$
5. $y = \dfrac{1}{x - 1}$. Since $x^2 + y^2 \geq y^2$, the solution to $y' = x^2 + y^2$ grows faster than that of $y' = y^2$, which grows infinite at $x = 1$.

7.

Method	Approximation
Euler	1.271428571
Improved Euler	1.285714286
Runge–Kutta	1.285714286

9.

Method	Approximation
Euler	2.073343406
Improved Euler	2.211571264
Runge–Kutta	2.202834625

11.

Method	(a)	(b)
Euler	0.289456778	0.299806204
Improved Euler	0.310493553	0.310324591
Runge–Kutta	0.310268270	0.310268300

Section 15 Miscellaneous Exercises, pp. 1099–1101

1. $y = \ln|2 - e^{-x}| - 2$ **3.** $\ln\left|x^2\left(\dfrac{x^2 - y^2}{y^2}\right)\right| + C = 0$
5. $\dfrac{1}{3}x^3 + xy + e^y - 9 = 0$ **7.** $y = (x + 1)^{-2}\left(\dfrac{1}{3}x^3 + \dfrac{1}{2}x^2 + 1\right)$
9. $y = \ln\left|\dfrac{\sec x}{2}\right|$ **11.** $y = 2e^{-x}$ **13.** $y = 7xe^{-2x}$
15. $y = e^{-x}(\cos x + \sin x)$ **17.** $y = -\dfrac{1}{2} + \dfrac{3}{2}e^{-2x} + x^2 - x$
19. $y = -\dfrac{7}{6}e^{2x} - \dfrac{5}{6}e^{-x} + \dfrac{3}{2}xe^{2x}$ **21.** $y^2 = -\dfrac{3}{2}x^2 + C$
23. $y^2 = 4B(B - x)$ **25.** $-x^2y^{-1} - \dfrac{1}{2}y^2 + C = 0$
27. $x^3e^y + xy^2 + 2xy + C = 0$ **29.** $y = e^x\left(\dfrac{1}{2}x^2 + C_1 + xC_2\right)$
31. $y = \dfrac{1}{5}e^x + C_2\cos 2x + C_3\sin 2x$ **33.** $a_0 = 1, a_1 = 2, a_3 = \dfrac{13}{3}, a_4 = \dfrac{19}{4}$ **35.** $y = \displaystyle\sum_{k=2}^{\infty}\frac{x^n}{n!}$ **37.** $a_0 = 0, a_1 = 0, a_2 = \dfrac{1}{2}, a_3 = \dfrac{1}{12}, a_4 = \dfrac{1}{288}$

Appendix

Section A.4, pp. A-20

1. If $F_1(t) \to L_1, F_2(t) \to L_2, \ldots, F_n(t) \to L_n$ as $t \to c$, then $F_1(t) + F_2(t) + \cdots + F_n(t) \to L_1 + L_2 + \cdots + L_n$ as $t \to c$, for any positive integer n.

Section A.5, pp. A-23

7. Statement true for $n \geq 5$.

Section A.7, pp. A-26

1. $c = -1$ **3.** $c = \dfrac{-1 + \sqrt{37}}{3}$

Section A.10, pp. A-37

1. -5 **3.** 1 **5.** a) -7 b) -7 **7.** a) 38 b) 38 **9.** $x = -4, y = 1$ **11.** $x = 3, y = 2$ **13.** $x = 3, y = -2, z = 2$
15. $x = 2, y = 0, z = -1$ **17.** a) $h = 6, k = 4$ b) $h = 6, k \neq 4$

Index

Note: Numbers in parentheses refer to problems on the pages indicated.

A Brief Table of Integrals

1. $\int u\,dv = uv - \int v\,du$

2. $\int a^u\,du = \frac{a^u}{\ln a} + C, \quad a \neq 1, \quad a > 0$

3. $\int \cos u\,du = \sin u + C$

4. $\int \sin u\,du = -\cos u + C$

5. $\int (ax+b)^n\,dx = \frac{(ax+b)^{n+1}}{a(n+1)} + C, \quad n \neq -1$

6. $\int (ax+b)^{-1}\,dx = \frac{1}{a}\ln|ax+b| + C$

7. $\int x(ax+b)^n\,dx = \frac{(ax+b)^{n+1}}{a^2}\left[\frac{ax+b}{n+2} - \frac{b}{n+1}\right] + C, \quad n \neq -1, -2$

8. $\int x(ax+b)^{-1}\,dx = \frac{x}{a} - \frac{b}{a^2}\ln|ax+b| + C$

9. $\int x(ax+b)^{-2}\,dx = \frac{1}{a^2}\left[\ln|ax+b| + \frac{b}{ax+b}\right] + C$

10. $\int \frac{dx}{x(ax+b)} = \frac{1}{b}\ln\left|\frac{x}{ax+b}\right| + C$

11. $\int (\sqrt{ax+b})^n\,dx = \frac{2}{a}\frac{(\sqrt{ax+b})^{n+2}}{n+2} + C, \quad n \neq -2$

12. $\int \frac{\sqrt{ax+b}}{x}\,dx = 2\sqrt{ax+b} + b\int \frac{dx}{x\sqrt{ax+b}}$

13. (a) $\int \frac{dx}{x\sqrt{ax+b}} = \frac{2}{\sqrt{-b}}\tan^{-1}\sqrt{\frac{ax+b}{-b}} + C, \quad \text{if } b < 0$

(b) $\int \frac{dx}{x\sqrt{ax+b}} = \frac{1}{\sqrt{b}}\ln\left|\frac{\sqrt{ax+b}-\sqrt{b}}{\sqrt{ax+b}+\sqrt{b}}\right| + C, \quad \text{if } b > 0$

14. $\int \frac{\sqrt{ax+b}}{x^2}\,dx = -\frac{\sqrt{ax+b}}{x} + \frac{a}{2}\int \frac{dx}{x\sqrt{ax+b}} + C$

15. $\int \frac{dx}{x^2\sqrt{ax+b}} = -\frac{\sqrt{ax+b}}{bx} - \frac{a}{2b}\int \frac{dx}{x\sqrt{ax+b}} + C$

16. $\int \frac{dx}{a^2+x^2} = \frac{1}{a}\tan^{-1}\frac{x}{a} + C$

17. $\int \frac{dx}{(a^2+x^2)^2} = \frac{x}{2a^2(a^2+x^2)} + \frac{1}{2a^3}\tan^{-1}\frac{x}{a} + C$

18. $\int \frac{dx}{a^2-x^2} = \frac{1}{2a}\ln\left|\frac{x+a}{x-a}\right| + C$

19. $\int \frac{dx}{(a^2-x^2)^2} = \frac{x}{2a^2(a^2-x^2)} + \frac{1}{2a^2}\int \frac{dx}{a^2-x^2}$

20. $\int \frac{dx}{\sqrt{a^2+x^2}} = \sinh^{-1}\frac{x}{a} + C = \ln|x+\sqrt{a^2+x^2}| + C$

Continued

21. $\displaystyle\int \sqrt{a^2+x^2}\,dx = \frac{x}{2}\sqrt{a^2+x^2} + \frac{a^2}{2}\sinh^{-1}\frac{x}{a} + C$

22. $\displaystyle\int x^2\sqrt{a^2+x^2}\,dx = \frac{x(a^2+2x^2)\sqrt{a^2+x^2}}{8} - \frac{a^4}{8}\sinh^{-1}\frac{x}{a} + C$

23. $\displaystyle\int \frac{\sqrt{a^2+x^2}}{x}\,dx = \sqrt{a^2+x^2} - a\sinh^{-1}\left|\frac{a}{x}\right| + C$

24. $\displaystyle\int \frac{\sqrt{a^2+x^2}}{x^2}\,dx = \sinh^{-1}\frac{x}{a} - \frac{\sqrt{a^2+x^2}}{x} + C$

25. $\displaystyle\int \frac{x^2}{\sqrt{a^2+x^2}}\,dx = -\frac{a^2}{2}\sinh^{-1}\frac{x}{a} + \frac{x\sqrt{a^2+x^2}}{2} + C$

26. $\displaystyle\int \frac{dx}{x\sqrt{a^2+x^2}} = -\frac{1}{a}\ln\left|\frac{a+\sqrt{a^2+x^2}}{x}\right| + C$

27. $\displaystyle\int \frac{dx}{x^2\sqrt{a^2+x^2}} = -\frac{\sqrt{a^2+x^2}}{a^2x} + C$

28. $\displaystyle\int \frac{dx}{\sqrt{a^2-x^2}} = \sin^{-1}\frac{x}{a} + C$

29. $\displaystyle\int \sqrt{a^2-x^2}\,dx = \frac{x}{2}\sqrt{a^2-x^2} + \frac{a^2}{2}\sin^{-1}\frac{x}{a} + C$

30. $\displaystyle\int x^2\sqrt{a^2-x^2}\,dx = \frac{a^4}{8}\sin^{-1}\frac{x}{a} - \frac{1}{8}x\sqrt{a^2-x^2}\,(a^2-2x^2) + C$

31. $\displaystyle\int \frac{\sqrt{a^2-x^2}}{x}\,dx = \sqrt{a^2-x^2} - a\ln\left|\frac{a+\sqrt{a^2-x^2}}{x}\right| + C$

32. $\displaystyle\int \frac{\sqrt{a^2-x^2}}{x^2}\,dx = -\sin^{-1}\frac{x}{a} - \frac{\sqrt{a^2-x^2}}{x} + C$

33. $\displaystyle\int \frac{x^2}{\sqrt{a^2-x^2}}\,dx = \frac{a^2}{2}\sin^{-1}\frac{x}{a} - \frac{1}{2}x\sqrt{a^2-x^2} + C$

34. $\displaystyle\int \frac{dx}{x\sqrt{a^2-x^2}} = -\frac{1}{a}\ln\left|\frac{a+\sqrt{a^2-x^2}}{x}\right| + C$

35. $\displaystyle\int \frac{dx}{x^2\sqrt{a^2-x^2}} = -\frac{\sqrt{a^2-x^2}}{a^2x} + C$

36. $\displaystyle\int \frac{dx}{\sqrt{x^2-a^2}} = \cosh^{-1}\frac{x}{a} + C = \ln\left|x+\sqrt{x^2-a^2}\right| + C$

37. $\displaystyle\int \sqrt{x^2-a^2}\,dx = \frac{x}{2}\sqrt{x^2-a^2} - \frac{a^2}{2}\cosh^{-1}\frac{x}{a} + C$

38. $\displaystyle\int (\sqrt{x^2-a^2})^n\,dx = \frac{x(\sqrt{x^2-a^2})^n}{n+1} - \frac{na^2}{n+1}\int (\sqrt{x^2-a^2})^{n-2}\,dx, \quad n \neq -1$

39. $\displaystyle\int \frac{dx}{(\sqrt{x^2-a^2})^n} = \frac{x(\sqrt{x^2-a^2})^{2-n}}{(2-n)a^2} - \frac{n-3}{(n-2)a^2}\int \frac{dx}{(\sqrt{x^2-a^2})^{n-2}}, \quad n \neq 2$

40. $\displaystyle\int x(\sqrt{x^2-a^2})^n\,dx = \frac{(\sqrt{x^2-a^2})^{n+2}}{n+2} + C, \quad n \neq -2$

41. $\displaystyle\int x^2\sqrt{x^2-a^2}\,dx = \frac{x}{8}(2x^2-a^2)\sqrt{x^2-a^2} - \frac{a^4}{8}\cosh^{-1}\frac{x}{a} + C$

42. $\displaystyle\int \frac{\sqrt{x^2-a^2}}{x}\,dx = \sqrt{x^2-a^2} - a\sec^{-1}\left|\frac{x}{a}\right| + C$

43. $\displaystyle\int \frac{\sqrt{x^2-a^2}}{x^2}\,dx = \cosh^{-1}\frac{x}{a} - \frac{\sqrt{x^2-a^2}}{x} + C$

44. $\displaystyle\int \frac{x^2}{\sqrt{x^2-a^2}}\,dx = \frac{a^2}{2}\cosh^{-1}\frac{x}{a} + \frac{x}{2}\sqrt{x^2-a^2} + C$

45. $\displaystyle\int \frac{dx}{x\sqrt{x^2-a^2}} = \frac{1}{a}\sec^{-1}\left|\frac{x}{a}\right| + C = \frac{1}{a}\cos^{-1}\left|\frac{a}{x}\right| + C$

46. $\displaystyle\int \frac{dx}{x^2\sqrt{x^2-a^2}} = \frac{\sqrt{x^2-a^2}}{a^2x} + C$

47. $\displaystyle\int \frac{dx}{\sqrt{2ax-x^2}} = \sin^{-1}\left(\frac{x-a}{a}\right) + C$

48. $\displaystyle\int \sqrt{2ax-x^2}\,dx = \frac{x-a}{2}\sqrt{2ax-x^2} + \frac{a^2}{2}\sin^{-1}\left(\frac{x-a}{a}\right) + C$

49. $\displaystyle\int (\sqrt{2ax-x^2})^n\,dx = \frac{(x-a)(\sqrt{2ax-x^2})^n}{n+1} + \frac{na^2}{n+1}\int (\sqrt{2ax-x^2})^{n-2}\,dx,$

50. $\displaystyle\int \frac{dx}{(\sqrt{2ax-x^2})^n} = \frac{(x-a)(\sqrt{2ax-x^2})^{2-n}}{(n-2)a^2} + \frac{(n-3)}{(n-2)a^2}\int \frac{dx}{(\sqrt{2ax-x^2})^{n-2}}$

51. $\displaystyle\int x\sqrt{2ax-x^2}\,dx = \frac{(x+a)(2x-3a)\sqrt{2ax-x^2}}{6} + \frac{a^3}{2}\sin^{-1}\frac{x-a}{a} + C$

52. $\displaystyle\int \frac{\sqrt{2ax-x^2}}{x}\,dx = \sqrt{2ax-x^2} + a\sin^{-1}\frac{x-a}{a} + C$

53. $\displaystyle\int \frac{\sqrt{2ax-x^2}}{x^2}\,dx = -2\sqrt{\frac{2a-x}{x}} - \sin^{-1}\left(\frac{x-a}{a}\right) + C$

54. $\displaystyle\int \frac{x\,dx}{\sqrt{2ax-x^2}} = a\sin^{-1}\frac{x-a}{a} - \sqrt{2ax-x^2} + C$

55. $\displaystyle\int \frac{dx}{x\sqrt{2ax-x^2}} = -\frac{1}{a}\sqrt{\frac{2a-x}{x}} + C$

56. $\displaystyle\int \sin ax\,dx = -\frac{1}{a}\cos ax + C$

57. $\displaystyle\int \cos ax\,dx = \frac{1}{a}\sin ax + C$

58. $\displaystyle\int \sin^2 ax\,dx = \frac{x}{2} - \frac{\sin 2ax}{4a} + C$

59. $\displaystyle\int \cos^2 ax\,dx = \frac{x}{2} + \frac{\sin 2ax}{4a} + C$

60. $\displaystyle\int \sin^n ax\,dx = \frac{-\sin^{n-1} ax\cos ax}{na} + \frac{n-1}{n}\int \sin^{n-2} ax\,dx$

61. $\displaystyle\int \cos^n ax\,dx = \frac{\cos^{n-1} ax\sin ax}{na} + \frac{n-1}{n}\int \cos^{n-2} ax\,dx$

62. (a) $\displaystyle\int \sin ax\cos bx\,dx = -\frac{\cos(a+b)x}{2(a+b)} - \frac{\cos(a-b)x}{2(a-b)} + C, \quad a^2 \neq b^2$

(b) $\displaystyle\int \sin ax\sin bx\,dx = \frac{\sin(a-b)x}{2(a-b)} - \frac{\sin(a+b)x}{2(a+b)}, \quad a^2 \neq b^2$

(c) $\displaystyle\int \cos ax\cos bx\,dx = \frac{\sin(a-b)x}{2(a-b)} + \frac{\sin(a+b)x}{2(a+b)}, \quad a^2 \neq b^2$

Continued

63. $\displaystyle\int \sin ax \cos ax\, dx = -\frac{\cos 2ax}{4a} + C$

64. $\displaystyle\int \sin^n ax \cos ax\, dx = \frac{\sin^{n+1} ax}{(n+1)a} + C, \quad n \neq -1$

65. $\displaystyle\int \frac{\cos ax}{\sin ax}\, dx = \frac{1}{a}\ln|\sin ax| + C$

66. $\displaystyle\int \cos^n ax \sin ax\, dx = -\frac{\cos^{n+1} ax}{(n+1)a} + C, \quad n \neq -1$

67. $\displaystyle\int \frac{\sin ax}{\cos ax}\, dx = -\frac{1}{a}\ln|\cos ax| + C$

68. $\displaystyle\int \sin^n ax \cos^m ax\, dx = -\frac{\sin^{n-1} ax \cos^{m+1} ax}{a(m+n)} + \frac{n-1}{m+n}\int \sin^{n-2} ax \cos^m ax\, dx,$

$n \neq -m \quad$ (If $n = -m$, use No. 86.)

69. $\displaystyle\int \sin^n ax \cos^m ax\, dx = \frac{\sin^{n+1} ax \cos^{m-1} ax}{a(m+n)} + \frac{m-1}{m+n}\int \sin^n ax \cos^{m-2} ax\, dx,$

$m \neq -n \quad$ (If $m = -n$, use No. 87.)

70. $\displaystyle\int \frac{dx}{b + c\sin ax} = \frac{-2}{a\sqrt{b^2 - c^2}}\tan^{-1}\left[\sqrt{\frac{b-c}{b+c}}\tan\left(\frac{\pi}{4} - \frac{ax}{2}\right)\right] + C, \quad b^2 > c^2$

71. $\displaystyle\int \frac{dx}{b + c\sin ax} = \frac{-1}{a\sqrt{c^2 - b^2}}\ln\left|\frac{c + b\sin ax + \sqrt{c^2 - b^2}\cos ax}{b + c\sin ax}\right| + C, \quad b^2 < c^2$

72. $\displaystyle\int \frac{dx}{1 + \sin ax} = -\frac{1}{a}\tan\left(\frac{\pi}{4} - \frac{ax}{2}\right) + C$

73. $\displaystyle\int \frac{dx}{1 - \sin ax} = \frac{1}{a}\tan\left(\frac{\pi}{4} + \frac{ax}{2}\right) + C$

74. $\displaystyle\int \frac{dx}{b + c\cos ax} = \frac{2}{a\sqrt{b^2 - c^2}}\tan^{-1}\left[\sqrt{\frac{b-c}{b+c}}\tan\frac{ax}{2}\right] + C, \quad b^2 > c^2$

75. $\displaystyle\int \frac{dx}{b + c\cos ax} = \frac{1}{a\sqrt{c^2 - b^2}}\ln\left|\frac{c + b\cos ax + \sqrt{c^2 - b^2}\sin ax}{b + c\cos ax}\right| + C, \quad b^2 < c^2$

76. $\displaystyle\int \frac{dx}{1 + \cos ax} = \frac{1}{a}\tan\frac{ax}{2} + C$

77. $\displaystyle\int \frac{dx}{1 - \cos ax} = -\frac{1}{a}\cot\frac{ax}{2} + C$

78. $\displaystyle\int x\sin ax\, dx = \frac{1}{a^2}\sin ax - \frac{x}{a}\cos ax + C$

79. $\displaystyle\int x\cos ax\, dx = \frac{1}{a^2}\cos ax + \frac{x}{a}\sin ax + C$

80. $\displaystyle\int x^n \sin ax\, dx = -\frac{x^n}{a}\cos ax + \frac{n}{a}\int x^{n-1}\cos ax\, dx$

81. $\displaystyle\int x^n \cos ax\, dx = \frac{x^n}{a}\sin ax - \frac{n}{a}\int x^{n-1}\sin ax\, dx$

82. $\displaystyle\int \tan ax\, dx = \frac{1}{a}\ln|\sec ax| + C$

83. $\displaystyle\int \cot ax\, dx = \frac{1}{a}\ln|\sin ax| + C$

84. $\displaystyle\int \tan^2 ax\, dx = \frac{1}{a}\tan ax - x + C$

85. $\displaystyle\int \cot^2 ax\, dx = -\frac{1}{a}\cot ax - x + C$

86. $\displaystyle\int \tan^n ax\, dx = \frac{\tan^{n-1} ax}{a(n-1)} - \int \tan^{n-2} ax\, dx, \quad n \neq 1$

87. $\displaystyle\int \cot^n ax\, dx = -\frac{\cot^{n-1} ax}{a(n-1)} - \int \cot^{n-2} ax\, dx, \quad n \neq 1$

88. $\displaystyle\int \sec ax\, dx = \frac{1}{a}\ln|\sec ax + \tan ax| + C$

89. $\displaystyle\int \csc ax\, dx = -\frac{1}{a}\ln|\csc ax + \cot ax| + C$

90. $\int \sec^2 ax\,dx = \frac{1}{a}\tan ax + C$

91. $\int \csc^2 ax\,dx = -\frac{1}{a}\cot ax + C$

92. $\int \sec^n ax\,dx = \frac{\sec^{n-2} ax \tan ax}{a(n-1)} + \frac{n-2}{n-1}\int \sec^{n-2} ax\,dx, \quad n \neq 1$

93. $\int \csc^n ax\,dx = -\frac{\csc^{n-2} ax \cot ax}{a(n-1)} + \frac{n-2}{n-1}\int \csc^{n-2} ax\,dx, \quad n \neq 1$

94. $\int \sec^n ax \tan ax\,dx = \frac{\sec^n ax}{na} + C, \quad n \neq 0$

95. $\int \csc^n ax \cot ax\,dx = -\frac{\csc^n ax}{na} + C, \quad n \neq 0$

96. $\int \sin^{-1} ax\,dx = x\sin^{-1} ax + \frac{1}{a}\sqrt{1-a^2x^2} + C$

97. $\int \cos^{-1} ax\,dx = x\cos^{-1} ax - \frac{1}{a}\sqrt{1-a^2x^2} + C$

98. $\int \tan^{-1} ax\,dx = x\tan^{-1} ax - \frac{1}{2a}\ln(1+a^2x^2) + C$

99. $\int x^n \sin^{-1} ax\,dx = \frac{x^{n+1}}{n+1}\sin^{-1} ax - \frac{a}{n+1}\int \frac{x^{n+1}\,dx}{\sqrt{1-a^2x^2}}, \quad n \neq -1$

100. $\int x^n \cos^{-1} ax\,dx = \frac{x^{n+1}}{n+1}\cos^{-1} ax + \frac{a}{n+1}\int \frac{x^{n+1}\,dx}{\sqrt{1-a^2x^2}}, \quad n \neq -1$

101. $\int x^n \tan^{-1} ax\,dx = \frac{x^{n+1}}{n+1}\tan^{-1} ax - \frac{a}{n+1}\int \frac{x^{n+1}\,dx}{1+a^2x^2}, \quad n \neq -1$

102. $\int e^{ax}\,dx = \frac{1}{a}e^{ax} + C$

103. $\int b^{ax}\,dx = \frac{1}{a}\frac{b^{ax}}{\ln b} + C, \quad b > 0, \; b \neq 1$

104. $\int xe^{ax}\,dx = \frac{e^{ax}}{a^2}(ax-1) + C$

105. $\int x^n e^{ax}\,dx = \frac{1}{a}x^n e^{ax} - \frac{n}{a}\int x^{n-1}e^{ax}\,dx$

106. $\int x^n b^{ax}\,dx = \frac{x^n b^{ax}}{a\ln b} - \frac{n}{a\ln b}\int x^{n-1}b^{ax}\,dx, \quad b > 0, \; b \neq 1$

107. $\int e^{ax}\sin bx\,dx = \frac{e^{ax}}{a^2+b^2}(a\sin bx - b\cos bx) + C$

108. $\int e^{ax}\cos bx\,dx = \frac{e^{ax}}{a^2+b^2}(a\cos bx + b\sin bx) + C$

109. $\int \ln ax\,dx = x\ln ax - x + C$

110. $\int x^n(\ln ax)^m\,dx = \frac{x^{n+1}(\ln ax)^m}{n+1} - \frac{m}{n+1}\int x^n(\ln ax)^{m-1}\,dx, \quad n \neq -1$

111. $\int x^{-1}(\ln ax)^m = \frac{(\ln ax)^{m+1}}{m+1} + C, \quad m \neq -1$

112. $\int \frac{dx}{x\ln ax} = \ln|\ln ax| + C$

113. $\int \sinh ax\,dx = \frac{1}{a}\cosh ax + C$

114. $\int \cosh ax\,dx = \frac{1}{a}\sinh ax + C$

115. $\int \sinh^2 ax\,dx = \frac{\sinh 2ax}{4a} - \frac{x}{2} + C$

116. $\int \cosh^2 ax\,dx = \frac{\sinh 2ax}{4a} + \frac{x}{2} + C$

117. $\int \sinh^n ax\,dx = \frac{\sinh^{n-1} ax \cosh ax}{na} - \frac{n-1}{n}\int \sinh^{n-2} ax\,dx, \quad n \neq 0$

Continued

118. $\displaystyle\int \cosh^n ax\,dx = \frac{\cosh^{n-1} ax \sinh ax}{na} + \frac{n-1}{n}\int \cosh^{n-2} ax\,dx, \qquad n \neq 0$

119. $\displaystyle\int x \sinh ax\,dx = \frac{x}{a}\cosh ax - \frac{1}{a^2}\sinh ax + C$

120. $\displaystyle\int x \cosh ax\,dx = \frac{x}{a}\sinh ax - \frac{1}{a^2}\cosh ax + C$

121. $\displaystyle\int x^n \sinh ax\,dx = \frac{x^n}{a}\cosh ax - \frac{n}{a}\int x^{n-1}\cosh ax\,dx$

122. $\displaystyle\int x^n \cosh ax\,dx = \frac{x^n}{a}\sinh ax - \frac{n}{a}\int x^{n-1}\sinh ax\,dx$

123. $\displaystyle\int \tanh ax\,dx = \frac{1}{a}\ln(\cosh ax) + C$

124. $\displaystyle\int \coth ax\,dx = \frac{1}{a}\ln|\sinh ax| + C$

125. $\displaystyle\int \tanh^2 ax\,dx = x - \frac{1}{a}\tanh ax + C$

126. $\displaystyle\int \coth^2 ax\,dx = x - \frac{1}{a}\coth ax + C$

127. $\displaystyle\int \tanh^n ax\,dx = -\frac{\tanh^{n-1} ax}{(n-1)a} + \int \tanh^{n-2} ax\,dx, \qquad n \neq 1$

128. $\displaystyle\int \coth^n ax\,dx = -\frac{\coth^{n-1} ax}{(n-1)a} + \int \coth^{n-2} ax\,dx, \qquad n \neq 1$

129. $\displaystyle\int \operatorname{sech} ax\,dx = \frac{1}{a}\sin^{-1}(\tanh ax) + C$

130. $\displaystyle\int \operatorname{csch} ax\,dx = \frac{1}{a}\ln\left|\tanh\frac{ax}{2}\right| + C$

131. $\displaystyle\int \operatorname{sech}^2 ax\,dx = \frac{1}{a}\tanh ax + C$

132. $\displaystyle\int \operatorname{csch}^2 ax\,dx = -\frac{1}{a}\coth ax + C$

133. $\displaystyle\int \operatorname{sech}^n ax\,dx = \frac{\operatorname{sech}^{n-2} ax \tanh ax}{(n-1)a} + \frac{n-2}{n-1}\int \operatorname{sech}^{n-2} ax\,dx, \qquad n \neq 1$

134. $\displaystyle\int \operatorname{csch}^n ax\,dx = -\frac{\operatorname{csch}^{n-2} ax \coth ax}{(n-1)a} - \frac{n-2}{n-1}\int \operatorname{csch}^{n-2} ax\,dx, \qquad n \neq 1$

135. $\displaystyle\int \operatorname{sech}^n ax \tanh ax\,dx = -\frac{\operatorname{sech}^n ax}{na} + C, \qquad n \neq 0$

136. $\displaystyle\int \operatorname{csch}^n ax \coth ax\,dx = -\frac{\operatorname{csch}^n ax}{na} + C, \qquad n \neq 0$

137. $\displaystyle\int e^{ax}\sinh bx\,dx = \frac{e^{ax}}{2}\left[\frac{e^{bx}}{a+b} - \frac{e^{-bx}}{a-b}\right] + C, \qquad a^2 \neq b^2$

138. $\displaystyle\int e^{ax}\cosh bx\,dx = \frac{e^{ax}}{2}\left[\frac{e^{bx}}{a+b} + \frac{e^{-bx}}{a-b}\right] + C, \qquad a^2 \neq b^2$

139. $\displaystyle\int_0^\infty x^{n-1}e^{-x}\,dx = \Gamma(n) = (n-1)!, \qquad n > 0.$

140. $\displaystyle\int_0^\infty e^{-ax^2}\,dx = \frac{1}{2}\sqrt{\frac{\pi}{a}}, \qquad a > 0$

141. $\displaystyle\int_0^{\pi/2} \sin^n x\,dx = \int_0^{\pi/2} \cos^n x\,dx = \begin{cases} \dfrac{1\cdot 3\cdot 5\cdots(n-1)}{2\cdot 4\cdot 6\cdots n}\cdot\dfrac{\pi}{2}, & \text{if } n \text{ is an even integer} \geq 2, \\ \dfrac{2\cdot 4\cdot 6\cdots(n-1)}{3\cdot 5\cdot 7\cdots n}, & \text{if } n \text{ is an odd integer} \geq 3 \end{cases}$